3rd International Conference on Trends in Material Science and Inventive Materials (ICTMIM 2021)

IOP Conference Series: Materials Science and Engineering
Volume 1126

Coimbatore, India
12 - 13 March 2021

Editors:

K Geetha
Morris Abraham G. Ezra
Ivan Izoni

ISBN: 978-1-7138-4610-9
ISSN: 1757-8981

Printed from e-media with permission by:

Curran Associates, Inc.
57 Morehouse Lane
Red Hook, NY 12571

proceedings.com

Some format issues inherent in the e-media version may also appear in this print version.

Institute of Physics
Dirac House, Temple Back
Bristol BS1 6BE UK

Phone: 44 1 17 929 7481
Fax: 44 1 17 920 0979

techtracking@iop.org

Additional copies of this publication are available from:

Curran Associates, Inc.
57 Morehouse Lane
Red Hook, NY 12571 USA
Phone: 845-758-0400
Fax: 845-758-2633
Email: curran@proceedings.com
Web: www.proceedings.com

TABLE OF CONTENTS

IOP Conf. Series: Materials Science and Engineering 1126 (2021) 011001 doi:10.1088/1757-899X/1126/1/011001

PREFACE

With an extended gratification, we happily welcome you to the proceedings of 2021 3rd International Conference on Trends in Material Science and Inventive Materials (ICTMIM 2020) at Coimbatore, India. The main objective and feature of the conference is to bring together academicians, researchers and industrialist to share and exchange their research experiences and results in different aspects of material science, computational mechanics and innovative materials.

The editorial team of 3rd ICTMIM 2021 has mainly preferred the state-of-the-art discussions on the practical challenges encountered in the emerging innovative material science paradigm and solutions adopted to it. The 3rd ICTMIM 2021 proceedings has been committed to remain both informative and research stimulating to cope up with the current issues present in innovative materials and general material science domain. Delegates from different countries have made the conference to be truly international in scope. Totally, 3rd ICTMIM 2021 has received 292 papers out of which 86 papers were selected for the conference. It has been strictly followed that, each contributed paper has refereed by at least 2-3 international reviewers before being accepted for publication.

3rd ICTMIM 2021 conference program consists of technical sessions and discussions on most of the recent material science research topics with eminent session chairs, program chairs, plenary lectures, and keynote speakers by covering a wide range of topics. The conference totally had 86 oral presentations, which are delivered by the conference participants by leveraging a great opportunity for the academicians, scholars ad industrialists to gain a wider knowledge on the state-of-the-art progress in material science research. Also, we are pleased to extend our gratitude to the review committee members, who have delivered their professional and technical expertise to the fullest to improve the publication quality of the research papers submitted to 3rd ICTMIM 2021. Furthermore, we would like to thank the conference organizers, local organizing committee members, faculty and non-faculty members, and volunteers for making this technical gathering to be a more enjoyable and informative event.

At last, we would like to extend our gratitude to the publication support delivered by the IOP team in this entire journey from conference to publication and our institution JCT College of Engineering and Technology for their continual support in successfully organizing this conference event.

Virtual Conference Disclaimer

We are happy to introduce you the 2021 3rd International Conference on Trends in Material Science and Inventive Materials (ICTMIM 2021), which was successfully held on March 12-13, 2021. By introducing a vast difference from the previous times of ICTMIM conference, 3rd ICTMIM 2021 was carried out in the form of virtual conference due to the adverse impact of COVID-19. Due to the travel restriction across the globe, the conference has been hosted online in very flexible way to gather the research scholars and experts all around the world with an aim to continue disseminating the recent and advanced research in the field of material science and also develops academic and research exchange among material science researchers even during the unprecedented times. Prior to the conference day, authors are provided with the detailed instructions to join their corresponding meeting link according to the conference presentation schedule. Furthermore, in order to avoid the technical delay from the conference participants during presentation, participants are also instructed to submit their pre-recorded presentation. Amidst the existing unprecedented pandemic situation, 3rd ICTMIM 2021 has provided the conference participants to share and exchange their innovative knowledge with the other attendees and make this event more successful and informative as like the previous conferences.

Guest Editors

Dr. K.Geetha Dean Academics and Research, JCT College of Engineering and Technology, Coimbatore, India

Dr. Morris Abraham G Ezra, Associate Professor, Department of Electrical and Electronic Engineering, Lee Kong Chian Faculty of Engineering and Science, University Tunku Abdul Rahman, ungai Long Campus, Jalan Sungai Long, Bandar Sungai Long, Kajang, Selangor, Malaysia.

Dr. Ivan Izoni, Publishing Information Technology Department, Lviv Polytechnic National, University, Ukraine.

Chief Patron

Dr. V. J. Arulkarthick, Principal, JCT college of Engineering and Technology, Coimbatore,India.

Conference Chair

Dr. K. Geetha,Dean Academics and Research, JCT College of Engineering and Technology, Coimbatore, India.

Technical Program Committee

Dr. Hui-Ming Wee Distinguished Professor, Department of Industrial and Systems Engineering, Chaplain, Chung Yuan Christian University (CYCU), Taiwan.
Dr. Vinod Singh Yadav, Coordinator, Associate Dean- Students Welfare Warden- Satellite Campus, National Institute of Technology, Uttarakhand.
Dr. Arindam Indra, Assistant Professor, Department of Chemistry, IIT BHU, Varanasi, Uttar Pradesh-221005, India.
Dr. Vijay Kumar Dalla, Assistant Professor, Mechanical Engineering Department, and Professor-in Charge, Institute Sports and Athletics National Institute of Technology Jamshedpur, Jamshedpur-831014, Jharkhand.
Dr. Omkar Singh Kushwaha, Senior Pr. Scientist, Chemical Engineering Department Indian Institute of Technology, Madras, Chennai.
Dr. Pitamber Mahanandia, Associate Professor, Department of Physics & Astronomy, National Institute of Technology, Rourkela, Sundargarh, Odisha, INDIA.
Dr. A. Ravi Kant (Assoc. Prof.), Dept. of Electrical & Electronics Engg., Sasi Institute of Technology & Engg., Tadepalligudem, West Godavari, AP.

IOP Conf. Series: Materials Science and Engineering 1126 (2021) 011001 doi:10.1088/1757-899X/1126/1/011001

Dr. Masoud Rahiminezhad Galankashi, Department of Materials, Manufacturing and Industrial Engineering, Universiti Teknologi, Malaysia.

IOP Conf. Series: Materials Science and Engineering 1126 (2021) 011001 doi:10.1088/1757-899X/1126/1/011001

IOP Conf. Series: Materials Science and Engineering 1126 (2021) 011001 doi:10.1088/1757-899X/1126/1/011001

Dr. Anna Mavroudi, Norwegian University of Science and Technology, Trondheim, Norway.
Dr. Md Masud Rana, Assistant Prof, Dept of ECE, KUET, Khulna, Bangladesh. MdMasud.
Dr. Arun Das, Computer Software, Staff Software Engineer at Visa, San Francisco Bay Area, California,
Dr. Prajoy Podder, Assistant Professor, Department of Electronics and Communication Engineering. Khulna University of Engineering & Technology (KUET). Khulna-9203, Bangladesh.

Local Organizing Committee
Dr. A. Kumar, HOD/Civil, JCT College of Engineering and Technology, Coimbatore, India.
Dr. G. Rajiv Suresh Kumar, HOD/CSE, JCT College of Engineering and Technology, Coimbatore, India.
Dr. S. Venkatesh Babu, HOD/PE, JCT College of Engineering and Technology, Coimbatore, India.
Dr. V. Vijaya Gopal, HOD/PCE, JCT College of Engineering and Technology, Coimbatore, India.
Dr. D. Vedhavinodha, HOD/ECE, JCT College of Engineering and Technology, Coimbatore, India.
Dr. P. Suresh Kumar, HOD/Automobile, JCT College of Engineering and Technology, Coimbatore, India.
Dr. G. Gnanavel, HOD/Food Tech, JCT College of Engineering and Technology, Coimbatore, India.
Dr. G. Mahesh, HOD/Mech, JCT College of Engineering and Technology, Coimbatore, India.

Peer review declaration

All papers published in this volume of *IOP Conference Series: Materials Science and Engineering* have been peer reviewed through processes administered by the Editors. Reviews were conducted by expert referees to the professional and scientific standards expected of a proceedings journal published by IOP Publishing.

- **Type of peer review: Double-blind**
- **Conference submission management system:**
  Easy Chair and our own online Submission system
- **Number of submissions received:**
  292
- **Number of submissions sent for review:**
  280
- **Number of submissions accepted:**
  86
- **Acceptance Rate (Number of Submissions Accepted / Number of Submissions Received X 100):**
  30%
- **Average number of reviews per paper:**
  6
- **Total number of reviewers involved:**
  3
- **Any additional info on review process:**

The pre-screening process is led by the Review Committee Head and is focused on the following metrics: Conference Scope, Technical English, Paper Quality, and Similarity Index. Reviewers will be chosen based on the consistency of their analysis and their previous review experience. All of the papers obtained will be reviewed for plagiarism using the crosscheck iThenticate method. The papers will be allocated to reviewers by the technical committee, with each reviewer receiving a maximum of six papers. Acceptance will be based on the recommendations of two reviewers.

- **Contact person for queries:**
  Dr. K. Geetha
  kgeetha.jctcet@gmail.com

A Review on Modelling, Design and Optimization of Forging Process

Dhiraj Kumar*, and Sharifuddin Mondal
Department of Mechanical Engineering, National Institute of Technology Patna, Bihar (800005), India.
* dhiraj.me15@nitp.ac.in

Abstract. Forging is a core manufacturing process. It gives a product of superior mechanical properties and produces minimum wastage of material. This makes it a better manufacturing process as compared to other processes like casting etc. An aligned grain flow defines a good forging. The effectiveness of a manufacturing process is characterized by low cost, high productivity and high quality of product. This necessitates the use of three dimensional computer graphics and powerful Computer Aided Engineering (CAE) software to analyze the forging process. The present paper discusses on the various research works carried out in different fields of forging. The scopes of further researches on different problems of forging are also elaborated.
Keywords: FEM; FEA; DEFORM; Forging; Optimization technique.

1. Introduction

Forging is one of the most useful metal working process by which one can easily give a metal work to a useful shape through dies & tools. The process consists of hammering or pressing the metal object. The steps of forging are shown in flow chart form in fig.1. It is one of the oldest and useful metal work methods whose origin can be found on about some thousands of years back. Forging process produces high degree of structurally reliable and strength to weight ratio products, which make it one of the most suitable manufacturing processes to be considered for production of product with high strength and long-term usages. To reduce the cost of forging process, the method should be designed in such a way that best parameter set can be used for producing better quality (high strength & high reliability) products as compared to other production methods. But establishing a sequence for new product design is not as simple as machining process. It requires a lot of trials & errors to produce a satisfactory product set up. Empirical trail & error method is widely used for forging process, but it is very time consuming and costly method. With the availabilities of highly efficient numerical techniques like FEM, FEA etc., the computer simulations are used to test the performance of the manufacturing processes. With the advancement of analysis techniques like the finite element method (or analysis) (FEM (or A)), computer simulation was introduced in early 1980's as a possible alternative. Many successful attempts have be made to optimize the die design and to produce better quality forged part. Various techniques are adopted in literature like Finite Element Analysis (FEA) [2, 3, 4, 7, 9, 11, 13, 15, 16, 17, 19, 20, 23, 24, 27, 32, 33, 35, 38, 49, 52, 53, 54, 65, 66, 67, 68], Finite Element Method (FEM) [1, 6, 8, 10, 14, 21, 30, 31, 43, 50, 64, 69], Finite Volume Method (FVM) [25], Finite Element Modelling [60, 63], Taguchi Method [23, 61], Grey Taguchi Method [20], Rigid Plastic (Rigid Visco-plastic) Method [4, 14, 27], ANOVA [22, 39], Dynamic Recrystallisation (DRX) Model [6, 31, 56], Serve Plastic Deformation (SPD) [25,30,32], Response Surface Method (RSM) [29, 39, 47, 63], Equal Channel Angular Pressing (ECAP) [32, 38], Equal Channel Angular Extrusion (ECAE) [30, 34], Latham and Cockcroft Formula [12, 57], Monte Carlo Method [5, 19], Pseudo Inverse Approach [18], Archard Wear Theory [2], Johanson-Cook Type Material Model [8], Slab Method [69] & soft computing tools like ANN (Artificial Neural Network) [5, 62], Genetic Algorithm (GA) [29, 62], etc. Among these methods, FEA is widely used in metal forming analysis due to its capabilities to model the complicated shapes of die and parts in forging process. FEA has been adopted by many researchers for optimization of die design and die design process. Lee & Jou [2] utilized the FEA combined with Archard's Wear Model to explain the wear mechanism in warm forging. Wear amount is predicted in outer race of automotive transmission in design stage of warm forging die. Lv et al. [4] utilized FEA to explore the complete information of forging process like material flow in die, forging load, strain in forging, die temperature etc. Lu et al. [9] utilized the FEA combined with Dynamic Material Models (DMM) to work on the hot forging process (HFP). In this work, author predicted the distribution and variation of deformation zones in external HFP and also determined & optimized the parameters of external HFP. Hino et al. [15] used FEA for developing a design methodology, which is used to reduce the number of stages in multistage forging process. Equbal et al. [20] used FEA combined with Grey Taguchi Method and ANOVA Method to optimize the forging load and billet temperature loss individually and also find the significant parameters. Christiansen et al. [23] applied FEA combined with Porous Plasticity Models to forecast and optimize V-shaped angled lower die. Salcedo et al. [25] used FVM to predict different die design configurations for the isothermal forging of cam. Different aspects of forging techniques like modelling; design and optimization of forging processes are discussed in different subsections.

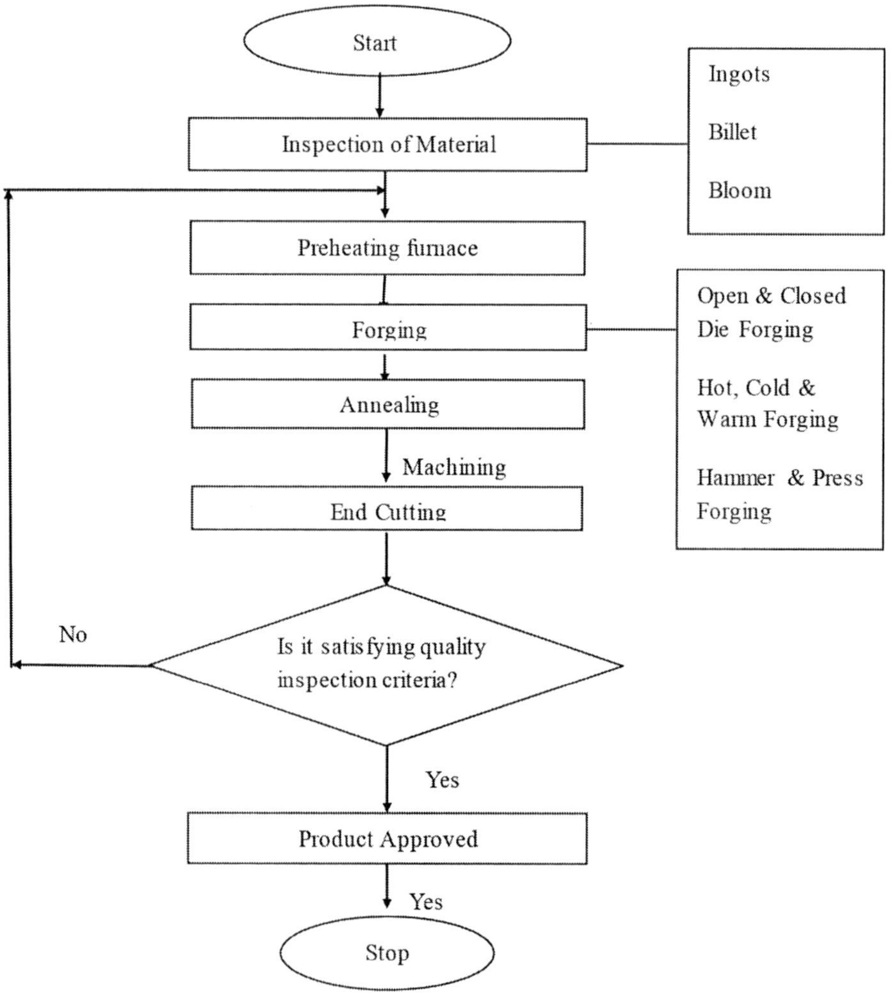

Fig. 1 Flow Diagram of Forging Process

Rests of the paper are organized as follows. In section 2, the main works on modelling, optimization and design are discussed. In section 3, a brief discussion is made on the reported works to outline the future directions of researches on forging processes. Concluding remarks are made in section 4.

2. Literature review

The research works carried out on forging can be classified into different categories based on types of researches like modeling and simulation of process, optimization of process, materials and design of process, etc. The different techniques used, the modification done, case studies & results observed on these areas are being outlined in different subsections.

2.1. Modelling and simulation

Viability of modelling and simulation is very important to the industrialists. Many researchers have used DEFORM software (finite element based) for performance analysis of forging processes through simulation. The steps of simulation are shown in flow chart form in fig. 2. Li et al. [1] utilized the FEM, Lagrangian method, PM (Penalty Method) & SIM (Selective Integration Method) to material flow, heat transfer, forming equipment response to the work piece, die stress and die deflection. Lee & Jou [2] calculated the wear velocity and also predicted the die life in warm forging using Archard's wear model and FEA. Then FEA combined with modified wear model and experimental data is used to calculate wear amount of automotive transmission design stage. Lv et al. [4] utilized the FEA combined with 3D rigid-Viscoplastic FEM to improve the quality of gas turbine compressor blade. Sun et al. [5] utilized the Artificial Neural Network, Monte Carlo Method, Cellular Automata (CA) Method and Empirical Models to analyze the deformation behavior for optimization of deformation process of TA15 titanium alloy by micro-structure evolution model. Ding et al. [6] utilized the constitutive equation and dynamic recrystallisation model combined with FEM to work on the manufacturing of AM50 alloy sheet and obtained a uniform micro-structure with moderate recrystallized grain size. Multi-pass hot rolling (rapid roller speed) is recommended for this process at a high temperature. Zamani et al. [8] utilized the FEM, Johanson-Cook Type Material Model and Lagrange Mesh Formulation to reduce tool wear and increase tool life by the assistance of the laser. Also, by the application of optimization of both laser and machining parameters, forces are reduced significantly. Lu et al. [9] utilized the FEA and Dynamic Materials Model (DMM) to analyze and optimize the unstable deformation in hot forging process parameters and finally resolved the unstable deformation.

Equbal et al. [10] utilized the FEM and Thermo Mechanical Simulator (GLEEBLE-3500) to analyze the deformation behavior of die/work piece interface. Hot compression experiment and FE simulation are found to be more effective means to study deformation behavior. Yukawa et al. [11] utilized the FEA, Compression Test and Oxide scale thickness Method to analyze and understand the heat transfer coefficient and different contact pressure in a function of thickness of three oxide layers. Gisbert et al. [13] utilized the FEA, slab method, ring compression test, shear friction model and coulomb friction model to develop an analytical model in bimetallic ring cylinder assemblies for evaluating the forging load. Also, coulomb friction model combined with above approach, two new analytical models developed for a better simulation and nearer to the actual problems. Ou et al. [3] utilized the ABAQUS software combined with finite element analysis to study the aerofoil cross-section of forging. In this study, authors found that aerofoil thickness error was mainly due to die elasticity, aerofoil bow, twist error and press deflection. Afazov et al. [7] utilized the FEDES combined with FEA, NPM (Nearest Point Method), Distance method using the elements, ESFM (Element Shape Function Method),

Distance method using field of point to explore and understand the FEDES (Finite Element Data Exchange System). It is found that FEDES can be very useful for the simulation of manufacturing chains. Kakimoto & Arikawa [12] utilized the Forge-2D & Forge-3D combined with Numerical simulation (By Latham & Cockcroft formula) and High Temperature Tensile Test to analyze the surface cracking. It is found out that surface cracking prediction is by the numerical simulation based on temperature, strain rate and fracture parameters. Also, damage parameter critical value is calculated by numerical simulation (based on the result of reduction in area).

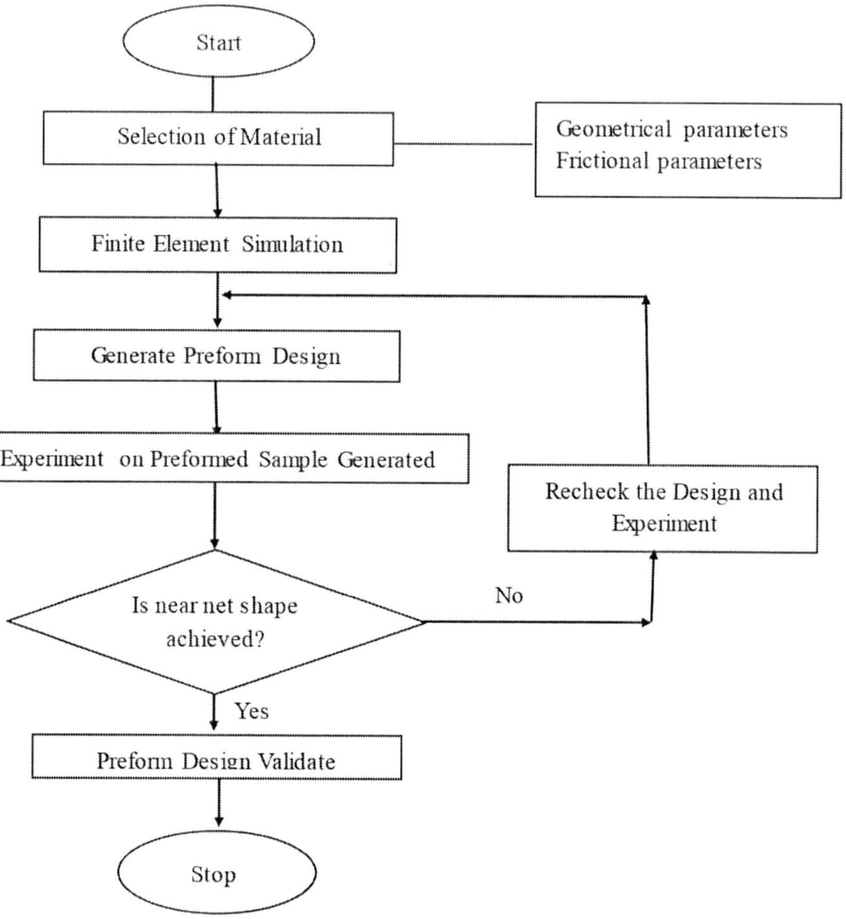

Fig. 2 Flow Diagram of Simulation

2.2. Optimization

Optimization is the process of making the best or most effective use of a situation or resource. Optimization is very important for forging processes to make the process efficient. Many researchers worked on the optimization of forging techniques in recent pasts. The steps of Taguchi method are shown in flow chart form in fig. 3. Choi et al. [14] utilized the DEFORM software combined with FEM (3D-Rigid plastic) to work on the feed rate and roundness in bar. It is found that roundness in bar depends on rotational angle. Also, feed rate balances the quality of product and productivity in forging. Hino et al. [15] utilized the FEA to develop a new design method. This design method is important for multistage forging process to reduce the numbers of stages. This method is also applied to the other metal forming processes like multistage deep drawing. Recker et al. [16] applied the FEA to analyze the microstructural properties. The study gives a rough, quick approximation of properties and quality of work piece in the process. Also, surface temperature is measured at the respective location. Cheng-Liang et al. [17] utilized the FEA and Orthogonal experiment method combined with DEFORM software to improve the process design and design variable like width of rib, thickness of rib, length of rib and inclination of rib. Halouani et al. [18] utilized the Pseudo inverse approach (PIA) and B-spline curve to explore the preform design. It is found that, to optimize preform design and respective parameters, simulation approach is much faster than the incremental approach. At the optimum preform shape, modified parameters improve the objective function. KroiB et al. [19] applied the FEA and Monte Carlo method to work on the interaction between process design, tool and machine in cold forging process. To achieve high work piece accuracy, optimized value of influencing parameters like work piece dimensions and forging load should be used. Equbal et al. [20] used the FEA, Grey Taguchi method and ANOVA method combined with DEFORM software to analyze the process parameters e.g. billet temperature and forging load. It is found that low billet temperature & low forging load are significant. Zhang et al. [21] applied FEM combined with Taguchi method to explore the precision forming process. It is found that strength and toughness of forging (rib-web forging) increased and utilization of material is improved by this precision forming process. Rib-Web forgings are obtained by multiobjective design process. Bai & Yang [22] used the vibration-assisted micro forging combined with ABAQUS software to optimize better metal foil surface finish due to large vibration amplitude and low initial static stress. Christiansen et al. [23] utilized the FEA and Porous plasticity model to optimize the V-shaped angled lower dies (Lower die angle in the range 90-120 degree ensures best result) in ingot forging.

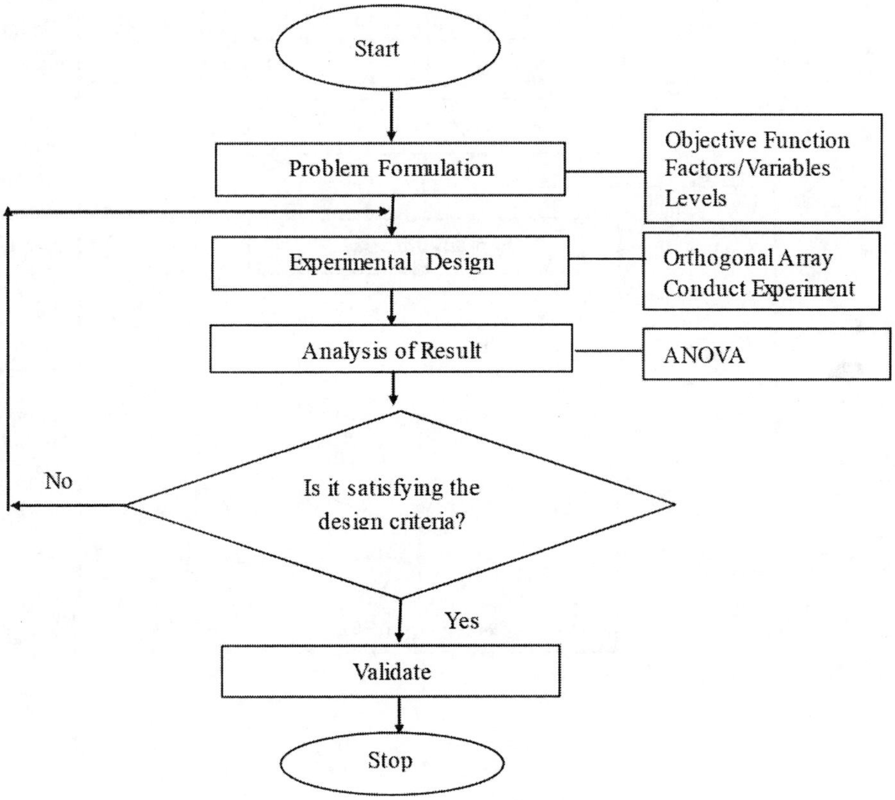

Fig. 3 Flow Diagram of Taguchi Method

Chval & Cechura [24] applied FEA to improve the technical parameters of mechanical forging process. In this study, authors found that stiffness increased by 7.5% and weight reduced by 10% by the application of this new approach. Stiffness and weight both are essential technical parameters for increasing accuracy and reducing cost of production of the process in the forging process. Salcedo et al. [25] utilized the FVM (Finite volume method), severe plastic deformation (SPD) process and Cockcroft Latham's method to complete die filling, minimum imparted damage, minimum required forging force, maximum accumulated plastic strain & found that complete die filling is better in two stroke than one stroke forging configuration. Wang et al. [26] used the High temperature low-cycle fatigue (HTLCF) method to show that innermost layer is the best fatigue resistance performances, while the shortest HTLCF life was observed at the outermost layer for the same condition. The distribution of grains was more homogeneous and the content of δ-phase along the grain boundary was lower at the innermost layer.

2.3. Material and design

Material and Design is another important area of forging researches. Many researchers worked on the material and design of forging processes for better quality products. The steps of general forging design process are shown as flow chart form in fig. 4.

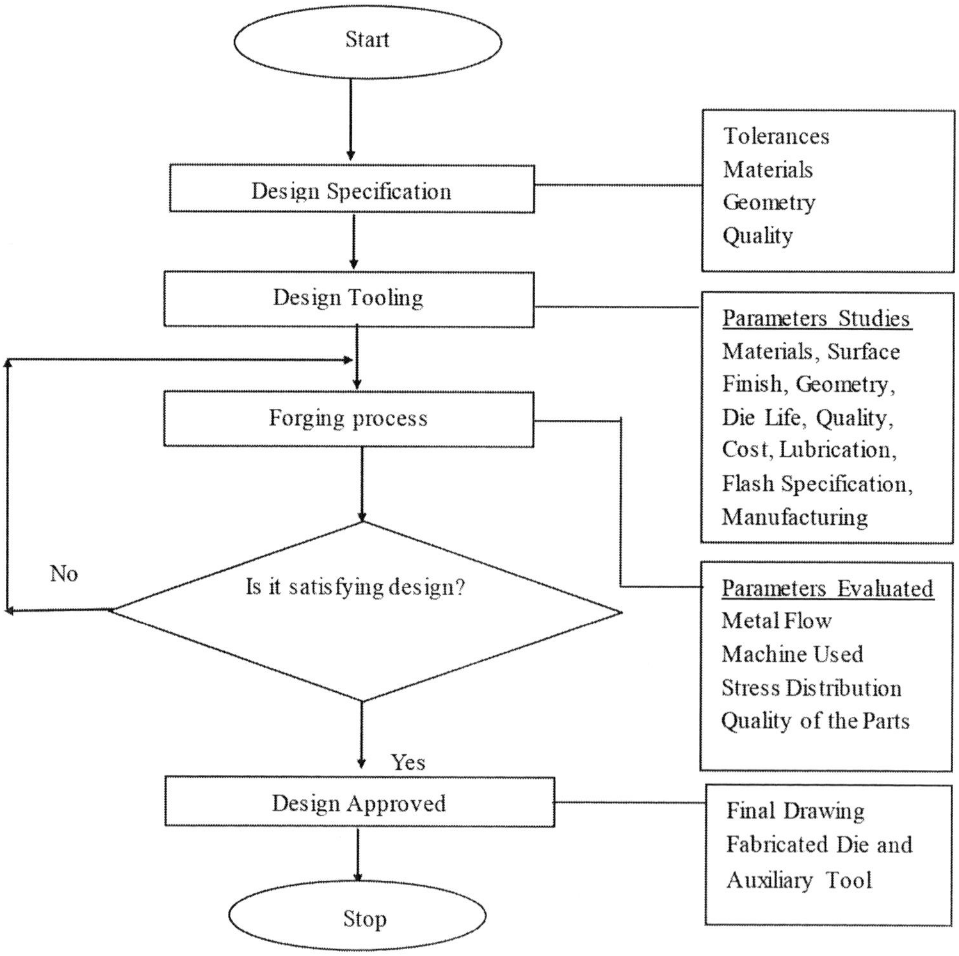

Fig. 4 Flow Diagram of General Forging Design Process

Deng et al. [27] utilized the FEA (3D-FE model) and ring compression test combined with DEFORM software to optimize the spur bevel gear geometry. Lower forming load and better filling of gear shape are analyzed by numerical simulation. Finite element analysis helps to determine the following detailed information like forming load, effective strain, forming panorama, and metal flow combined with process design. Kaynar et al. [28] used the OM (Optical Microscopy) & SEM for determination of final microstructure and mechanical properties (like hardness and tensile test). Authors also analyzed the following microstructural compounds such as coarse precipitates, ferrite and pearlite. Lee et al. [29] utilized the Equivalent static load (ESL_S), Design of experiment (DOE), Response surface method (RSM), Backward tracing method, Sensitive analysis method and Genetic algorithm (GA) to optimize the work piece shape. Post-process in the forging is not required for the flash removal and found that forging produces fine quality products. Puertas et al. [30] used FEM combined with SPD and ECAE to explore and understand the Francis turbine blade design. In this study, Francis turbine blades are manufactured by AA1050 (micrometric microstructure). It is found that mechanical properties are better in comparison to conventional forging. Also, moderate forging temperatures are required in comparison to conventional forging. Suresh et al. [31] applied FEM, Dynamic recrystallization (DRX) Model and Inter-crystalline cracking combined with DEFORM software to study the development of microstructure. For estimating the local & average strain rate in forging envelop, FE based simulation is used. At high speed & low temperature, material flow is localized and hence changed microstructure in forging. Perez et al. [32] used FEA combined with SPD and ECAP to improvement in forging ability and micro-hardness and decrease in material grain size. For achieving mechanical component with no cracks & high hardness, pre-processing of ECAP is not carried out due to combination of low temperature of forging & ECAP. Bagherpour et al. [33] used the FEA combined with DEFORM software to optimize maximum angle of inclination that is higher in the fixed inlet channel die. For achieving maximum inclination angle, minimizing the power (for given friction factor) & maximizing the angle of distortion are applied. Jiang et al. [34] used the ECAE, SEM, OM & EBSD (Electron Back Scatter Diffraction) combined with DEFORM, TEXEVAI software to analyze and explore the microstructure of β-Mg17Al12 and found that eutectic network completely disappears in fine Mg-alloy particle and non-based texture stay stable. It shows that good balance of strength and ductility. To optimize the grain size, texture and morphology of eutectic network, MDIF combined with heat treatment is applied. Ozturk et al. [35] used FEA combined with ANSYS software to work on the design and manufacturing aspect of product in cold forging. For minimizing the total cost of cold forging parts, authors developed a concurrent design optimization method, which is related to the design and manufacturing parameters. Li et al. [36] utilized the OlympusGXSI optical microscope, ZEISS Evo 5.0 SEM & ED X-ray spectroscopy to explore the magnesium alloys from AZ91 alloy (weight reduction). For increasing the ignition point (above 800° C) and its strength & elongation, 2% calcium is added in the AZ91 alloy after liquid forging and extrusion. Ghosh et al. [37] utilized the EBSD (Electron Back Scatter Diffraction) & TEM (Transmission Electron Microscopy) to improve the mechanical property combined with high ductility and micro-structure of Nb–Ti stabilized IF steel by multi-axial forging process. Fuertes et al. [38] applied FEA and ECAP to analyze the mechanical property of aluminium alloy such as AA1050 (low value of strain hardening). To increase in hardness of a connecting rod, it is manufactured from ultra-fine-grained material and found that hardness is increased by 20%. Balta et al. [39] utilized the CDFW (Continuous Drive Friction Welding), RSM and ANOVA to explore the mechanical properties like tensile strength, elongation & petal crack length of weld. In this study, error (predicted value – measured value) are found 1.06% for tensile strength, 13.37% for elongation and 2.44% for petal crack length and hence weld quality predominantly depends on tensile strength. Also, weld quality is identified by petal test.

3. Discussion & Future Directions

The main perspectives of manufacturing goods are quality of product, minimizing the cost of product, the lead time, and the manufacturing defect. For defect free manufacturing with minimum load, preform design, optimum die design and process parameters should also be included.

The most of the researchers have utilized finite element analysis for the examination of forging operation. It is more advantageous over other methods of analysis like archard wear theory, dynamic recrystallization method, nearest point method, Johanson-Cook type material model, slab method, pseudo inverse approach, compression test method, etc. due to the fact that FEA gives near net shape products. The advantages of FEA is that it gives detailed information about the different properties in computer screen. So one can decide about the system easily without testing on real products. So it saves a lot of effort and time. It simulates many things such as work piece & tool temperature, material property dependent on strain rate, deformation heat transfer, capabilities & characteristics of strain hardening for analysis of microstructure.

Many researchers worked on different aspects of forging process. Li et al. [1] focused on the verification of material parameters like flow stress, strain and temperature to analyze the transfer of heat, equipment of forging & flow of material. But other aspects like robustness, computation effects are not studied. Further research on forging should be focused on the improvement in robustness, computational efficiency, user-friendliness, graphics and post-processing techniques. Also focus should be on microstructural analysis such as size of grain, transformation of phase, fracture & precipitation. Ou et al. [3] worked on process and design parameters to minimize the forging error, defects, total cost and material loss. Further investigation should be done on improving optimization system using multi-weighing factor scheme to reduce forging error. Sun et al. [5] worked on rate of dislocation density, the size & recrystallization of primary α- grain process to analyze the distortion behavior and optimize deformation process of TA15 titanium alloy by micro-structure evolution model. The work was conducted only on two phase regions. Further research should be focused on multi-phase region to improve the result. Hino et al. [15] worked on process and design parameters of forging process to develop effective tool for the reduction of the number of stages in multistage forging process. The developed technique can be applied for envisaging the other metal forming processes such as multistage deep drawings.

With the advancement of technologies, forging industry is becoming more energy efficient and environment friendly. This makes it a very valued and responsible manufacturing process as compared to other similar processes. In next decades, forging will be zero liability and source of high paying jobs for employees in surrounding areas. So, virtual environment of forging is an important area of research, which can be examined.

Other specific areas where upgradation of technology is needed are process controls & sensors, software, process modeling & optimization, die design & die modeling, lubrication, maintenance in real time, and equipment of primary & secondary processing etc.

4. Conclusion

Many research works have been done to improve quality of manufacturing goods, minimize the cost and lead time, defect free forging etc. mainly through the use of FEA. This method is very beneficial for optimization process, predicting defect, analysis of die, forging load etc. Many researchers confirmed the result of FEA. In addition to FEA, so many other methods like Archard wear theory, Dynamic recrystallisation method, nearest point method, Johanson-Cook type material model, Taguchi method, Compression test method, Slab method, ANOVA method, Design of experiment method etc. are used in literature. The results are used to minimize the cost, increase the product quality, reduce cycle time & shop floor trail and remove defects. A few researchers have also worked upon Tribological properties of forged part – Wear, Tensile strength, Toughness and Hardness. FEA is a very useful tool for forging as it requires least effort for flawed die design or erroneous process plan. The researches have also been done on the control of the virtual environment of forging like damage model. This is a very important area of researches, which can be examined. Further improvement in efficiency of computation, graphics & post processing methods, robustness, and user & environmental friendliness are needed. Different aspects of current and future research scopes in the field of forging are also discussed.

References

[1] Li J, Jinn JT, Wu WT, Oh SI (2001) recent development and application of three-dimensional finite element modeling in bulk forming process. Journal of Materials Processing Technology 113: 40-45

[2] Lee RS, Jou JL (2003) Application of numerical simulation for wear analysis of warm forging dies. Journal of Materials Processing Technology 140:43–48

[3] Ou H, Lan J, Armstrong CG, Price MA (2004) An FE simulation and optimization approach for the forging of aeroengine components. Journal of Materials Processing Technology 151:208–216

[4] Lv C, Zhang L, Mu Z, Tai Q, Zheng Q (2008) 3D FEM simulation of the multi-stage forging process of a gas turbine compressor blade. Journal of materials processing technology 198:463–470

[5] Sun ZC, Yang H, Han GJ, Fan XG (2010) A numerical model based on internal-state variable method for the microstructure evolution during hot-working process of TA15 titanium alloy. Materials Science and Engineering A 527:3464–3471

[6] Ding H, Hirai K, Homma T, Kamado S (2010) Numerical simulation for microstructure evolution in AM50 Mg alloy during hot rolling. Computational Materials Science 47:919–925

[7] Afazov SM, Becker AA, Hyde TH (2012) Development of a finite element data exchange system for chain simulation of manufacturing processes. Advances in Engineering Software 47:104–113

[8] Zamani H, Hermani JP, Sonderegger B, Sommitsch C (2013) 3D Simulation and Process Optimization of Laser Assisted Milling of Ti6Al4V. Procedia CIRP 8:75 – 80

[9] Lu SQ, Li X, Wang KL, Liu SB, Fu MW (2013) A method for prediction of unstable deformation in hot forging process by simulation. Trans. Nonferrous Met. Soc. China 23:3739-3747

[10] Equbal MI, Talukdar P, Kumar V, Ohdar RK (2014) Deformation behavior of microalloyed steel by using thermo mechanical simulator and finite element method. Procedia Materials Science 6:674 – 681

[11] Yukawa N, Nakashima Y, Ishiguro T, Abe E, Ishikawa T, Choda T (2014) Modeling of heat transfer coefficient of oxide scale in hot forging. Procedia Engineering 81:492 – 497

[12] Kakimoto H, Arikawa T (2014) Prediction of surface cracks in hot forging by numerical simulation. Procedia Engineering 81:474 – 479

[13] Gisbert C, Bernal C, Camacho AM (2015) Improved analytical model for the calculation of forging forces during compression of bimetallic axial assemblies. Procedia Engineering 132:298 – 305

[14] Choi SK, Chun MS, Tyne CJV, Moon YH (2006) Optimization of open die forging of round shapes using FEM analysis. Journal of Materials Processing Technology 172:88–95

[15] Hino R, Sasaki A, Yoshida F, Toropov VV (2008) A new algorithm for reduction of number of press-forming stages in forging processes using numerical optimization and FE simulation", International Journal of Mechanical Sciences 50:974–983
[16] Recker D, Franzke M, Hirt G (2011) Fast models for online-optimization during open die forging. CIRP Annals - Manufacturing Technology 60:295–298
[17] Cheng-Liang HU, Li-fen M, Zhen Z, Bing GU, Bing C (2012) Optimization for extrusion process of aluminum controller housing. Trans. Nonferrous Met. Soc. China 22: S48-S53
[18] Halouani A, Li YM, Boussad A, Guo YQ (2012) Optimization of cold forging performs tools using Pseudo Inverse Approach", Transactions of Nonferrous Metals Society of China 22: S207–S213
[19] KroiB T, Engel U, Merklein M (2013) Comprehensive approach for process modeling and optimization in cold forging considering interactions between process, tool and press. Journal of Materials Processing Technology 213:1118– 1127
[20] Equbal MI, Kumar R, Shamim M, Ohdar RK (2014) A grey-based Taguchi method to optimize hot forging process. Procedia Materials Science 6:1495 – 1504
[21] Zhang J, Wu D, Zhou J, Wang J (2014) Multi-objective optimization of process parameters for 7050 aluminum alloy rib-web forgings' precise forming based on Taguchi method. Procedia Engineering 81:558 – 563
[22] Bai Y, Yang M (2014) Optimization of metal foils surface finishing using vibration assisted micro-forging. Journal of Materials Processing Technology 214:21– 28
[23] Christiansena P, Martins PAF, Bay N, Hattel JH (2014) Multi-objective optimization of die geometry in ingot forging. Procedia Engineering 81:2457 – 2462
[24] Chval Z, Cechura M (2014) Optimization of power transmission on mechanical forging presses. Procedia Engineering 69:890 – 896
[25] Salcedo D, Luis CJ, Luri R, León J, Puertas I, Fuertes JP (2015) Design and optimization of the dies for the isothermal forging of a cam. Procedia Engineering 132:1069 – 1076
[26] Wang RZ, Zhang XC, Liu F, Yao LL, Tu ST (2015) Fatigue and fracture behaviors of nickel-based forging GH4169 super alloy and optimization of model parameters. Procedia engineering 130:1088-1096
[27] Deng X, Hua L, Han X, Song Y (2011) Numerical and experimental investigation of cold rotary forging of a 20CrMnTi alloy spur bevel gear. Materials and Design 32:1376–1389
[28] Kaynar A, Gündüz S, Turkmen M (2013) Investigation on the behavior of medium carbon and vanadium micro alloyed steels by hot forging test. Materials and Design 51:819–825
[29] Lee JJ, Jung UJ, Park GJ (2013) Shape optimization of the Work piece in the forging process using equivalent static loads. Finite Elements in Analysis and Design 69:1–18
[30] Puertas I, Luis Perez CJ, Salcedo D, Leon J, Fuertes JP, Luri R (2013) Design and mechanical property analysis of AA1050 turbine blades manufactured by equal channel angular extrusion and isothermal forging. Materials and Design 52:774–784
[31] Suresh K, Rao KP, Prasad YVRK, Hort N, Kainer KU (2014) Study of hot forging behavior of as-cast Mg–3Al–1Zn–2Ca alloy towards optimization of its hot workability. Materials and Design 57:697–704
[32] Perez CJL, Perez DS, Arbizu IP (2014) Design and mechanical property analysis of ultrafine grained gears from AA5083 previously processed by equal channel angular pressing and isothermal forging. Materials and Design 63:126–135
[33] Bagherpour E, Ebrahimi R, Qods F (2015) An analytical approach for simple shear extrusion process with a linear die profile. Materials & Design 83:368–376
[34] Jiang MG, Yana H, Chen RS (2015) Microstructure, texture and mechanical properties in an as-cast AZ61 Mg alloy during multi-directional impact forging and subsequent heat treatment. Materials and Design 87:891–900
[35] Ozturk M, Kocaoglan S, Sonmez FO (2016) Concurrent design and process optimization of forging. Computers and Structures 167:24–36
[36] Li F, Peh WY, Nagarajan V, Ho MK, Danno A, Chua BW, Tan MJ (2016) Development of non-flammable high strength AZ91 + Ca alloys via liquid forging and extrusion. Materials and Design 99:37–43

[37] Ghosh S, Singh AK, Mula S (2016) Effect of critical temperatures on microstructures and mechanical properties of Nb–Ti stabilized IF steel processed by multiaxial forging. Materials and Design 100:47–57
[38] Fuertes JP, Luis CJ, Luri R, Salcedo D, Leon J, Puertas I (2016) Design, simulation and manufacturing of a connecting rod from ultra-fine grained material and isothermal forging. Journal of Manufacturing Processes 21:56–68
[39] Balta B, Arici AA, Yilmaz M (2016) Optimization of process parameters for friction weld steel tube to forging joints. Materials and Design 103:209–222
[40] Kaur J, Pabla BS, Dhami SS (2016) A review on field areas of research in forging process using FEA. International Journal of Engineering Research & Technology, ISSN: 2278-0181
[41] Patel MBV, Thakkar DHR, Mehta PSB (2016) Review of Analysis on Forging Defects for Quality Improvement in Forging Industries. Journal of Emerging Technologies and Innovative Research (ISSN-2349-5162), Dec 2014 (Volume 1 Issue 7)
[42] Equbal MI, Talukdar P, Odhar RK (2013) Application of optimization techniques in metal forging- A review and reflection. International Journal of Scientific & Engineering Research, ISSN 2229-5518, Volume 4
[43] Hawryluk M, Jakubik J (2016) Analysis of forging defects for selected industrial die forging processes. Engineering Failure Analysis 59:396–409
[44] Koc M, Vazquez V, Witulski T, Altan T (1996) Application of the Finite Element Method to Predict Material Flow and Defects in the Semi-Solid Forging of A356 Aluminum Alloys. Journal of Materials Processing Technology 59:106-112
[45] Joshi TC, Prakash U, Dabhade VV (2015) Micro structural development during hot forging of Al 7075 powder. Journal of Alloys and Compounds 639:123–130
[46] Paschke H, Yilkiran T, Lippold L, Brunotte K, Weber M, Braeuer G, Behrens BA (2015) Adapted surface properties of hot forging tools using plasma technology for an effective wear reduction. Wear 330-331:429–438
[47] Wang W, Zhao J, Zhai R (2016) A forming technology of spur gear by warm extrusion and the defects control. Journal of Manufacturing Processes 21:30–38
[48] Jia Z, Liu Y, Liu W, Zhang C, Yang J, Wang L, Zhao K (2015) A spectrum selection method based on SNR for the machine vision measurement of large hot forgings. Optik 126:5527–5533
[49] Langner J, Stonis M, Behrens BA (2016) Investigation of a moveable flash gap in hot forging. Journal of Materials Processing Technology 231:199–208
[50] Gronostajski Z, Kaszuba M, Polak S, Zwierzchowski M, Niechajowicz A, Hawryluk M (2016) The failure mechanisms of hot forging dies. Materials Science & Engineering A657:147–160
[51] Neag A, Favier V, Bigot R, Atkinson HV (2016) Comparison between numerical simulation of semisolid flow into a die using FORGE and in situ visualization using a transparent sided die. Journal of Materials Processing Technology 229:338–348
[52] Fan L, Wang Z, Wang H (2014) 3D finite element modeling and analysis of radial forging processes. Journal of Manufacturing Processes 16:329–334
[53] Bai Q, Lin J, Jiang J, Dean TA, Zou J, Tian G (2015) A study of direct forging process for powder superalloys. Materials Science & Engineering A621:68–75
[54] Matsumoto R, Osumi Y, Utsunomiya H (2014) Reduction of Friction of Steel Covered with Oxide Scale in Hot Forging. Journal of Materials Processing Technology 214:651– 659
[55] Zhang YC, Han JX, Fu XB, Zhang FL (2014) Measurement and control technology of the size for large hot forgings. Measurement 49:52–59
[56] Li J, Liu Y, Liu B, Wang Y, Liang X, He Y (2014) Microstructure characterization and mechanical behaviors of a hot forged high Nb containing PM-TiAl alloy. Materials Characterization 95:148-156
[57] Wang J, Langlois L, Rafiq M, Bigot R, Lu H (2014) Study of the hot forging of weld cladded work pieces using upsetting tests. Journal of Materials Processing Technology 214:365– 379
[58] Maderbacher H, Oberwinkler B, Ganser HP, Tan W, Rollett M, Stoschka M (2013) The influence of microstructure and operating temperature on the fatigue endurance of hot forged Inconel 718 components. Materials Science & Engineering A 585 (2013) 123–131

[59] Battabyal M, Schaublin R, Spatig P, Walter M, Reith M, Baluc N (2013) Microstructure and mechanical properties of a W–2wt. %Y2O3 composite produced by sintering and hot forging. Journal of Nuclear Materials 442: S225–S228
[60] Wang S, Fang S, Shi Z, Jiang J, Zhou X, Lin J (2017) Direct power forging of PM nickel-based superalloy: desification and recrystallisation. International Journal of Advanced Manufacturing Technology 88:2661-2670
[61] Pinarbasi M, Sel C, Alagas HM, Yuzukirmizi M (2013) integrated definition modeling and Taguchi analysis of flexible manufacturing system: aircraft industry application. International Journal of Advanced Manufacturing Technology 68:2169- 2183
[62] Noorossana R, Tajbakhsh SD, Saghaei A (2009) An artificial neural network approaches to multiple-response optimization. International Journal of Advanced Manufacturing Technology 40:1227-1238
[63] Abosaf M, Essa K, Alghawail A, Tolipov A, Su S, Pham D (2017) Optimisation of multi-point forming process parameters. International Journal of Advanced Manufacturing Technology doi: 10.1007/s00170-017-0155-y
[64] Stellin T, Tijum RV, Engel U (2016) Modelling and experimental study of a microforging process from metal strip for the reduction of defects in mass production. Production engineering: research and development 10:103–112
[65] Behrens BA, Denkena P, Park HS, Henjes J, Kroning S, Yilkiran T, Klassen A (2014) Application of methods for ecological optimization of crank shaft forging process. Production engineering: research and development 8:253–261
[66] Santangelo A, Blanke P, Hadifi T, Wolter FE, Behrens BA (2013) Fast 3D inverse simulation of hot forging process via Medial Axis Transformation: an approach for preform estimation in hot die forging. Production engineering: research and development 7:409–416
[67] Phanitwong W, Thipprakmas S (2017) Finite element analysis of a taper U-shaped in the U-bending process. International Journal of Manufacturing Technology and Management. Vol. 31, No. 4, pp. 275-287
[68] Patel C, Kore SD (2017) Finite element analysis of crowbarring effect during electromagnetic forming. International Journal of Manufacturing Technology and Management. Vol. 31, No. 4, pp. 288-299
[69] Sheth D, Das S, Chatterjee A, Bhattacharya A (2017) Modeling of Closed-Die Forging for Estimating Forging Load. Journal of the Institution of Engineers (India): Series C. 98(1):53–61

IOP Conf. Series: Materials Science and Engineering 1126 (2021) 012002 doi:10.1088/1757-899X/1126/1/012002

Comparative analysis of the compressive strength of concrete under different curing methods

M Murillo[1*], D Abudinen[1], M del Río[1], N Serrato[1], L Patrón[1] and J Ramírez[1]
[1] Civil and Environmental Department, Universidad de la Costa, Barranquilla, 080002, Colombia
dabudine@cuc.edu.co

Abstract. Concrete is the most widely used material in construction, which possesses different characteristics and possible manufacturing methods. Concrete is, also, characterized by its high resistance to compression and durability, factors that could have been directly influenced by the curing method in early ages. The objective of this research is to demonstrate the importance of the concrete curing process, as well as to analyze the influence of the chosen curing method on the compression resistance. 48 concrete test samples were manufactured, by the standard requirements. Subsequently, the samples were divided into eight batches, subjected to different types of curing: immersion curing, twice a day; outdoor curing; total immersion curing in different waters, application of commercial curing agent; with polyethylene foil coating and without curing. Performing compression testing at 7 and 28 days. Among the results obtained, the batch that underwent the method of curing with polyethylene coating presented a more efficient effect in terms of resistance to compression; followed by the techniques of immersion in water.

1. Introduction

Concrete is the most widely used agglomerate worldwide. It is known as a basic material and construction tool in civil works, mainly for its excellent performance and low cost. Besides, concrete is ductile enough to adapt to any type of project in the construction area, regardless of the size of the project [1]. In particular, the binder itself consists of mixing the cement with an exact dose of water, to keep the concrete in a saturated state. So that the spaces originally filled with water are filled with moisturizing products. If water is not present, the hydration stops, leaving excessive porosity that reduces the mechanical properties and durability of the concrete [2-3], and a variable proportion of mineral additives such as aggregates (sand and gravel) [4]. The hydration provides the compound to be applied easily and resistance to extreme conditions [5]. Globally, the ready-mixed concrete industry is projected to exceed $600 billion in revenues by 2025 [6].

On the other hand, it is essential to know that fly ash (FA) is a subproduct of coal-fired power plants whose use is important to reduce the environmental impact of the cement industry. Therefore, FA has been widely used in the construction sector as a mineral additive for the manufacturing of concrete, mineral filling for asphalt concrete, sub-base for pavement layers, among others. It should be noted that, in the field of construction, this material is known as a useful supplementary cementing material [7].

Considering how valuable the load-bearing of any material is, it is crucial to think about the strength conditions. Whose ownership is responsible for the quality control of the material and the concern of designers [8]. Concrete has a great success in its level of compressive load capacity. However, like many other materials, concrete has porosity qualities [9]. Porosity affects the strength of concrete structures during their performance, As well as allowing fluid infiltration into the material, this infiltration can contribute to the future cracking of the compound. In this sense, it is then known that cracks in concrete are inevitable, difficult to detect, and almost impossible to repair. Cracking results in a latent threat to concrete structures since in the long term they can suffer from dangerous faults [10].

Curing is a process in which hydraulic cement concrete develops hardening properties through the hydration of cement in the presence of water and heat [11]. Without effective curing, significant defects related to the long-term durability of the concrete can occur. These defects usually manifest in the form of visible cracks, micro-cracks, and a weak surface. A good state of curing favors the development of the strength and durability of the concrete, as well as can effectively improve the microstructure of the concrete and its ability to resist erosion from external environment. The curing, inherent to the concrete curing process, is directly proportional to the mechanical performance, in addition to its resistance to environmental agents [8].

The curing process can be carried out through different methods or techniques used, being chosen according to the specific conditions of the environment, economy, ease of application, and other factors of the project [12]. Several studies have now been conducted to demonstrate the viability of several curing agents under different conditions. According to studies conducted by engineers at Landmark University in Nigeria [5], curing depends on the type of agent used in the sample. The samples, in this case, are cylindrical and with conventional properties previously dosed, curing time, and environmental conditions. Other studies have also provided evidence of the variety and permutability of benefits in different properties of the compound [5]. This means that different methods provide different benefits separately, but at the same time show ineffectiveness in other factors. According to the researched literature, two steam curing methods were found [13-16]; burning ash method [17]; microwave curing [18] which consist of techniques commonly used in the practice of curing concrete with heating and atmospheric vapor pressure which according to research has a great advantage in a higher compressive strength at an early age and a lower compressive strength at a later age.

This current research aims to evaluate the strength of treated concrete under eight different curing methods: standard projection, which consists of hydrating the surface of the sample set [19-20]; outdoor curing, involving the non-intervention of external agents [5, 21]; spraying of two types of liquid compounds forming curing membranes; immersion in seawater and immersion in freshwater [22-23], two separate methods in which only the source of the water where the samples are intended to be immersed varies; plastic or vinyl wrappers [14, 24-25]; and immersion in freshwater with lime [1, 26]. The practice of these methods will be further developed in the subparagraphs that require the development of these techniques. The most effective method will be determined by compressive strength test performed on each cylindrical sample, as explained in the methodology.

2. Experimental program

The compressive strength, at 7 and 28 days of hydraulic concrete cylinders that were induced to different curing methods, was evaluated according to the requirements and procedure guidelines established by the ASTM standard [27]. In addition to estimating the results in a laboratory, these results were compared to check if there is an influence between the chosen curing method on the compression resistance values of the concrete. Each curing method had three cylinders for its application, so forty-eight samples were produced (24 to be tested at 7 days and 24 at 28 days).

2.1 Materials

For the experimental program of this research, samples of concrete were made in cylindrical form (20 cm in height and 10 cm in diameter), dosed with two cementitious materials (Portland and FA cement). The samples are composed of the following materials: structural cement type grey, fly ash (from Mount Líbano), coarse sand, fine sand, gravel TM 3-8", water, a plasticizing additive, and a retardant additive.

The physical and chemical characterization tests of the materials were performed before design under ambient conditions of 24.2 °C and 62.1% humidity. The physical characterization procedures were carried out by ASTM C33 [28] and ASTM C131-03 [29], and the procedures related to chemical characterization were carried out under ASTM 1218 [30]. Table 1 shows the properties of the aggregates used for this concrete mixture.

Table 1. Physico-chemical characterization of the concrete component aggregates.

Chemical Properties	**Fine Aggregate**	**Coarse Aggregate**
Sulphate-SO4 [%]	0,088	0,712
Chlorides-Cl [%]	0,0002	0,00117
Physical Properties		
Equivalent percentage of baked sand [%]	68	-
Wear Resistance [%]	-	22,9
Light Particles [%]	0	0
Density [kg/m3]	2.616	2.539

The fly ash was taken as a supplementary cementing of the mixture to mitigate the environmental impact, aiming at LED certification and to improve the durability factor of said concrete. The characterization of the ashes used in this research is presented in Table 2.

Table 2. Physico-chemical properties of fly ash used.

Chemical Properties	**(%)**	**Chemical Properties**	**(%)**
Silicon Oxide – Sio2	44,5	Sulfur Oxide – SO3	3,2
Aluminum Oxide – Al2O3	19,3	Sodium Oxide – Na2O	1,1
Ferric Oxide – Fe2O3	14,5	Potassium Oxide – K2O	1,3
%Sio2 + %Al2O3 + %Fe2O3	78,2	Equivalent Bases – Na2Oeq	2
Calcium Oxide – Cao	6,7	Fire Losses	4,6
Magnesium Oxide – MgO	3,1	Humidity	1
Physical Properties			
Sieve fineness N°325 (45 um)	24		
Density (g/ml)	2,5		

The dosing of the raw materials for the designed concrete mixture was performed for a net capacity of 100L according to the ACI method [31] and is presented in Table 3.

Table 3. Designed mixing dosage.

Raw Materials	**Dry weights**	
	Quantity (kg)	**Tolerance (%)**
Portland Cement	246	±4
Fly Ashes	43	±4
Coarse Sand	468	±3
Fine Sand	468	±3
Gravel TM 3-8”	950	±3
Water	165	±1

Additive 1 (Plasticizer)	1,44	±3
Additive 2 (Super Plasticizer)	2,17	±3

The plasticizing and super plasticizing additives were added as aqueous solutions that allowed to regulate, balance, and maintain the design of the mixture. Note that the calculation did not alter the water-cement conditions. Thus, the first additive (plasticizer) achieved the mixture to obtain greater malleability, reduced segregation, and maintained its viscosity; and the second additive (superplasticizer) acted as reducer and retardant of setting water. The mixture was considered considering the characteristic temperature conditions of the coast zone (Barranquilla-Atlántico).

2.2 Methodology

2.2.1. Mixing. The previously dosed materials were placed in a concrete mixer for a period of 15 minutes, ensuring a uniform mixing process.

2.2.2. Trials in a fresh state. The tests were carried out to verify the quality and efficiency of the mixture, to guarantee the correct mixing process of the concrete. The settlement test according to ASTM C143-90 [32] and the density and air content tests according to ASTM C231-09 [33] were performed. The procedure of these trials is illustrated in Figures 1 and 2. The results showed a settlement of 6 ± 1 in, a density of 2.116 kg/m3 and air content of 2%.

Figure 1. Settlement test illustration.

Figure 2. Air content test illustration.

2.2.3. Manufacture of concrete cylinders. We proceeded to manufacture and form concrete cylinders in plastic molds that had dimensions of 10 cm in diameter and 20 cm in length (Figure 3). After 24 hours, the cylinders were uncoupled and prepared for the application of the different curing techniques already mentioned as an objective of this investigation (Figure 4).

Figure 3. Concrete sample formwork.

Figure 4. Uncoated concrete cylinders.

2.2.4. Application of curing methods. The curing methods that were applied in this study correspond to conventional on-site and unconventional methods. Eight curing methods were chosen: immersion in drinking water, immersion in seawater, immersion in drinking water + lime + temperature control, application of brand name curing agent #1. Application of brand name curing agent #2, coating with polyethylene plastic foil, standard manual spraying of fresh water and non-application of curing (6 specimens for each of the curing techniques). Each of these procedures was performed on each trio of samples during the 7 and 28 days after they were unwrapped.

Table 4 specifies the nomenclature and description of the eight curing techniques chosen for the investigation.

Table 4. Nomenclature for selected and applied curing techniques.

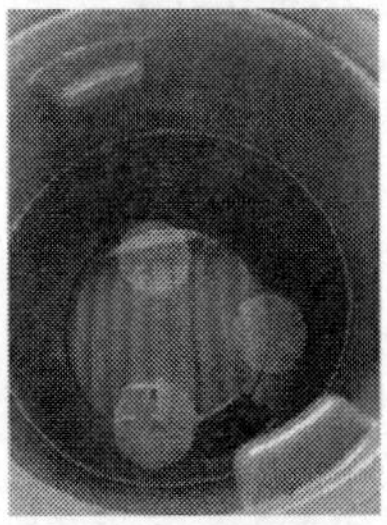	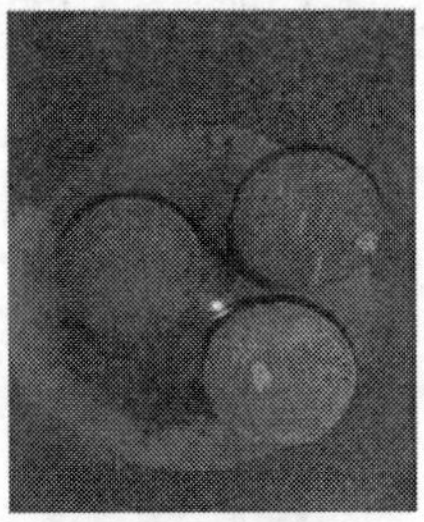	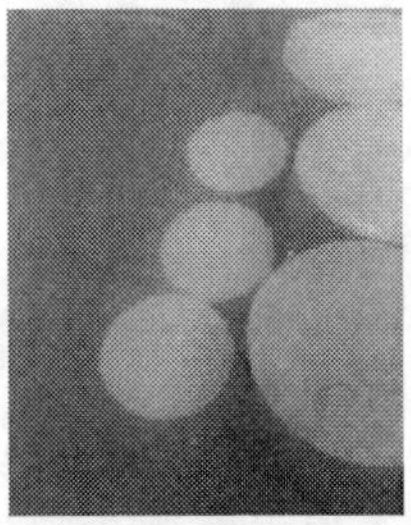	
T1 - Immersion in tap water	T2 - Immersion in sea water	T3 - Immersion in tap water + lime + temperature control	T4 - Curing agent 1

T5 - Curing agent 2	T6 - Polyethylene plastic sheets	T7 - Manual water sprinkling	T8 - Without curing

2.2.5 Hardened State Tests. The compressive strength tests were implemented on each of the samples. Two machines were used to carry out these tests. The first one is a cylinder compression frame, with a capacity of 1500 kN, of the reference CT-1500 of the Dirimpex brand. The second machine is a super-automatic console Automax reference 50-C10F04 also of the brand Dirimpex. These machines worked together to carry out the compression tests, which were carried out according to the ASTM C39 standard. Three specimens (of the six specimens of concrete cured under each curing method) failed within 7 days of stripping, and the other three specimens corresponding to each method failed within 21 days (28 days after stripping).

3. Analysis of Results

3.1 Compressive strength (f'c) at 7 days.

Based on the results obtained from the compressive strength tests of the concrete samples subjected to the curing process during the first 7 days, showed that the batch of samples submitted under the curing technique corresponding to polyethylene plastic sheets (T6) presented a higher resistance value, reaching an average value of 26.7 MPa, followed by the curing agent of brand #1 (T4), which had to undergo an average compression of 26.1 MPa to fail. Moreover, the method of flooding with drinking water with lime + temperature control (T3), presented the weakest yield, with a value of f'c of 22,8 MPa. These resistance values are slightly lower than those achieved by the samples without receiving any curing method (25,5 MPa).

3.2 Compressive strength (f'c) at 28 days

According to the values given by the compressive strength tests carried out on the sample groups at 28 days, the samples cured with polyethylene foil (T6) repeat as the most resistant. This time with a value of 32.7 MPa. Followed by the T6, are the drinking water and seawater flooding techniques (T1-T2), with values of 32.2 and 32 MPa, respectively. These two techniques did not present the same efficiency in the seven-day tests. It is pertinent to note that all the samples increased their performance considerably concerning the previous test, so it is easy to deduce that these methods must incorporate into the concrete through a longer rest time.

This was the case for the samples to which the drinking waterflood technique with a solution and temperature control (T3) was applied, which showed the most unfavorable results in the first test, while in the latter they delivered a more resistant product than the samples to which the method related to the spraying with drinking water (T7) and the technique with curing agent #1 (T4) were applied, which provided a final compressive strength of only 27.4 MPa, being further outweighed by the uncured samples. In other words, it could be said that the method with the curing agent of the trademark #1 has counterproductive effects on the concrete in the long term, since in the previous trial (at 7 days) presented an immediate efficiency, which, with the passage of days, was diminished.

Figure 5 shows an illustration of the characteristic failures of the cylinders tested at both 7 and 28 days, their appearance being very similar for each of the groups of samples subjected to the different curing techniques studied.

Figure 5. Typical cylinder failures tested at 7 and 28 days.

Figure 6 summarises the average compressive strength values of groups of samples by curing technique used at 7 and 28 days, respectively:

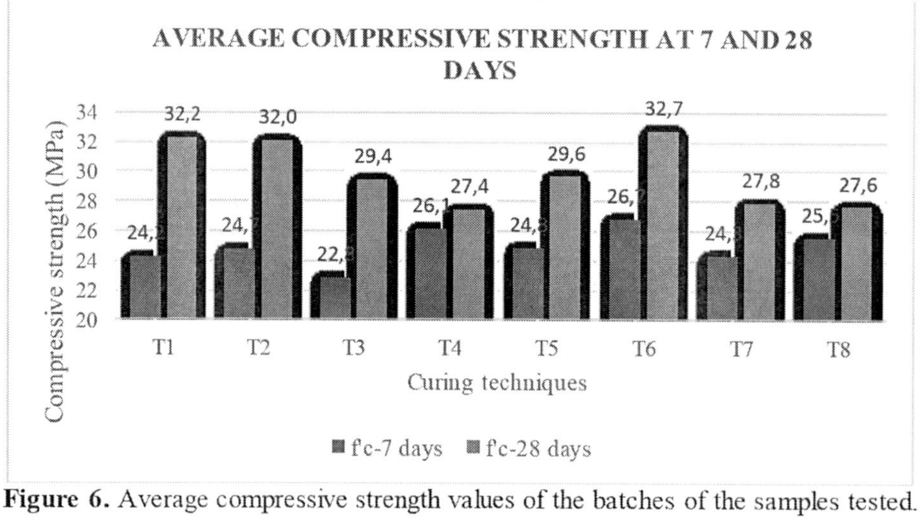

Figure 6. Average compressive strength values of the batches of the samples tested.

Figure 7 illustrates the efficiency of each of the curing techniques selected for the analysis, contrasted with the samples that were not subjected to any curing process (T8). It is worth highlighting the similarity of the technique corresponding to the manual spraying of water (T7), which is a method commonly used in works, with the samples uncured. That fact could be supported as it is a manual process with a lot of care to consider, it is difficult to guarantee its correct and constant procedure.

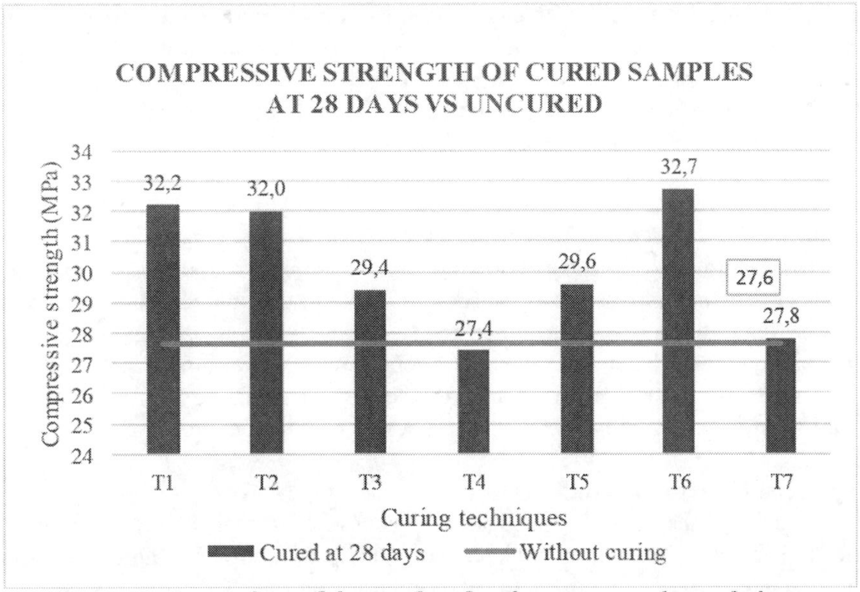

Figure 7. Comparison of the results of curing versus curing techniques.

Table 5 presents a simple statistical analysis of the results obtained, given from the value of the standard deviation, rating these results according to the criterion of ACI 214 [34], as: excellent, very good, good and, only in a poor case (T7), indicating for the latter a considerable dispersion in the values released. It should be noted that resistance was worked with at 28 days, as the results of the ACI 214 report are valid only for this age.

Table 5. Statistical analysis of the results obtained.

Technique	*f'c* av. (MPa)	Standard Deviation σ (MPa)	ACI 214 Classification
T1	32,2	0,428	Excellent
T2	32	1,112	Excellent
T3	29,4	1,483	Very Good
T4	27,4	1,926	Good
T5	29,6	1,687	Very Good
T6	32,7	0,877	Excellent
T7	27,8	2,672	Poor
T8	27,6	1,69	Very Good

Based on the results of the investigation, the peak of efficacy was evident in the technique using polyethylene plastic sheets (T6), which showed favorable results from its compression trial applied at 7 days (26.7 MPa) ratifying with an average resistance of 32.7 MPa after 28 days.

At the same time, when evaluating the experimental results, it was possible to observe the negative influence on the compression resistance of the commercial curing agent #1 (T4), which was evidenced at 28 days with an average resistance of 27,7 MPa, differing from samples that were not cured (T8) by only 0.5 MPa (26.7 MPa), with the latter being the least effective curing technique of all those studied here. For this reason, perhaps a better control or greater application of the product is suggested.

About the manual spraying technique for drinking water (T7), it can be concluded that although it is the most frequently used standard technique in the local construction industry and from which better results were expected, yielded an average compressive strength at 28 days also very similar to not applying any type of curing (27.8 MPa versus 26.7 MPa). With the result, the recommendation to change the curing methods in the usual constructions is generated, putting in alert the quality of the concrete in a matter of resistance to the compression.

Finally, the technique of immersion in seawater (T2) was found to increase the resistance to 28 days of the samples by a value of 32 MPa, demonstrating the benefits of compression resistance in concrete. However, it is worth noting that these analyses were carried out to a specific standard without reinforcement, so that the response of reinforced concrete to the application of the technique of curing by immersion in seawater is opened to other investigations, which is presumed to have corrosion and carbonation consequences due to the concentration of salts, lowering the pH of the concrete. It is necessary to emphasize that the application of this technique would be appropriate to the demand for environmental awareness and the inevitable use of water in construction.

4. Conclusions and Recommendations

- From the results of the present investigation, it is concluded that the techniques where immersion in water (T1, T2 and T3) and the technique of coating with polyethylene plastic around specimens (T6) are the most effective methods to perform curing on site.

- The cylinders with curing agent 1 (T4), presented an adequate behavior to the resistance to the compression at 7 days, but its lack of efficacy is evident in more advanced days and to be able to maintain the humidity in the concrete because resistance to compression was obtained very similar to the resistance that was obtained without carrying out the curing process.

- The compression resistance of the cylinders with the seawater curing technique yielded positive results. However, the seawater is not recommended for use with concrete as the reinforcing steel is usually in conjunction with the concrete and could end up with problems of accelerated corrosion and carbonation.

- Positive results according to ACI 214 on the standard deviation of the compressive strength of concrete cylinders are evident, which supports an adequate experimental procedure. However, the T7 technique gave "poor" confidence because when simulating spraying conditions on-site, it probably was not performed uniformly, as it happens in construction sites.

- The need for the curing process in the concrete is evident independently of the technique used to achieve greater resistance to compression.

- Finally, despite the results found in the research, it is recommended to continue analyzing the influence of concrete curing methods on other parameters such as durability, water absorption, carbonate penetration, among others; and it is also recommended to carry out a microstructural analysis of the samples tested to further deepen the interpretation of the results obtained, as well as taking a greater number of samples per batch.

Acknowledgments
The authors thanks to company ULTRACEM, for providing the inputs, facilities and equipment properly calibrated for the conduct of the tests and experimental development of the investigation.

References

[1] W. Zhang, Q. Zheng, A. Ashour and H. Baoguo 2019 Self-healing cement concrete composites for resilient infrastructures: A review *Composites Part B: Engineering* **189** 107892

[2] M. López, L. Kahn and K. Kirtis 2005 Internal curing in high performance concretes - a new paradigm *Ingeniería de Construcción* **20** 2 pp 117-126

[3] B. Liu, G. Luo and Y. Xie 2018 Effect of curing conditions on the permeability of concrete with high ume mineral admixtures *Construction and Building Materials* **167** pp 359-371

[4] S. Rodríguez and N. Torres 2019 Evaluation of internal curing effects on concrete *Ingeniería e investigación* **39** 2

[5] O. D. Atoyebi, P. P. Ikubanni, A. Adesina, O. V. Araoye and I. E. E. Davies 2020 Effect of curing methods on the strength of interlocking paving blocks *Cogent Engineering* **13** pp 2-6

[6] C. R. Gagg 2014 Cement and concrete as an engineering material: An historic appraisal and case study analysis *Engineering Failure Analysis* **40** pp 114-140

[7] YukoOgawa, P. T. Bui, K. Kawai and R. Sato 2020 Effects of porous ceramic roof tile waste aggregate on strength development and carbonation resistance of steam-cured fly ash concrete *Construction and Building Materials* **236**

[8] M. López, L. Kahn and K. Kirtis 2005 Internal curing in high performance concretes - a new paradigm *Ingeniería de Construcción* **20** 2 pp 117-126

[9] M.-F. Ba, C.-x. Qian, X.-j. Guo and X.-y. Han 2011 Effects of steam curing on strength and porous structure of concrete with low water/binder ratio *Construction and Building Materials* **25** 1 pp 123-128

[10] Y. Xie, X. Wang, G. Long and C. Ma 2019 Quantitative analysis of the influence of subfreezing temperature on the mechanical properties of steam-cured concrete *Construction and Building Materials* **206** pp 504-511

[11] R. K 2008 Effect of curing method on characteristics of cold bonded fly ash aggregates *Cement and Concrete Composites* **30** 9 pp 848-853

[12] B. Liua, J. Jianga, S. Shenb, F. Zhoua, J. Shia and Z. He 2020 Effects of curing methods of concrete after steam curing on mechanical strength and permeability *Construction and Builidng Materials* **256**

[13] J. Shia, B. Liua, S. Shenb, J. Tana, J. Daia and R. Ji 2020 Effect of curing regime on long-term mechanical strength and transport properties of steam-cured concrete *Construction and Building Materials* **255** 119407

[14] L. GuangCheng, W. Meng, X. YouJun and M. KunLin 2014 Experimental investigation on dynamic mechanical characteristics and microstructure of steam-cured concrete *Science China Technological Sciences* **57** 10 pp 1902-1908

[15] C. Zou, G. Long, C. Ma and Y. Xie 2018 Effect of subsequent curing on surface permeability and compressive strength of steam-cured concrete *Construction and Building Materials* **188** pp 424-432

[16] Y. Ogawaa, P. T. Buib, K. Kawaia and R. Sato 2020 Effects of porous ceramic roof tile waste aggregate on strength development and carbonation resistance of steam-cured fly ash concrete *Construction and Building Materials* **236** 117462

[17] Y. Lv, G. Ye y G. D. Schutter 2019 Utilization of miscanthus combustion ash as internal curing agent in cement-based materials: Effect on autogenous shrinkage *Construction and Building*

Materials **207** pp 585-591

[18] M. El-Feky, M. Kohail, A. El-Tair and I. Serag 2020 Effect of microwave curing as compared with conventional regimes on the performance of alkali activated slag pastes *Construction and Building Materials* **253**

[19] M. Nematollahzade, A. Tajadini, I. Afshoon and F. Aslani 2020 Influence of different curing conditions and water to cement ratio on properties of self-compacting concretes *Construction and Building Materials* **237** 117570

[20] Zhi-min, L. Guang-Cheng and X. you-jun 2012 Influence of subsequent curing on water sorptivity and pore structure of steam-cured concrete *Journal of Central South University* **19** pp 1155-1162

[21] M. Wang, Y. Xie, G. Long and C. Ma 2019 Microhardness characteristics of high-strength cement paste and interfacial transition zone at different curing regimes *Construction and Building* Materials **221** pp 151-162

[22] J.-G. Teng, Y. Xiang, T. Yu and Z. Fang 2019 Development and mechanical behaviour of ultra-high-performance seawater sea-sand concrete *SAGE Journals* pp 2-8 22(14):3100-3120.

[23] Dasara, D. Patah, H. Hamada, Y. Sagawa and D. Yamamoto 2020 Applicability of seawater as a mixing and curing agent in 4-year-old concrete *Building and Construction Materials* **259**

[24] Debska and L. Licholai 2016 The effect of the type of curing agent on selected properties of epoxy mortar modified with PET glycolisate *Construction and Building Materials* **124** pp 11-19

[25] Falliano, D. D. Domenico, G. Ricciardia and E. Gugliandolo 2018 Experimental investigation on the compressive strength of foamed concrete: Effect of curing conditions, cement type, foaming agent and dry density *Construction and Building Materials* **165** pp 735-749

[26] B. Liu, G. Luo and Y. Xie Effect of curing conditions on the permeability of concrete with high ume mineral admixtures *Construction and Building Materials* **167** pp. 359-371

[27] ASTM International 2018 Standard Test Method for Compressive Strength of Cylindrical Concrete Specimens *ASTM Concrete and Aggregates* 04.02

[28] ASTM International 2018 Standard Test Method for Resistance to Degradation of Small-Size Coarse Aggregate by Abrasion and Impact in the Los Angeles Machine *ASTM Concrete and Aggregates* **04.02**

[29] ASTM International 2018 Standard Specification for Concrete Aggregates *Concrete and Aggregates* **04.02**

[30] ASTM Subcommittee: C09.60 2018 Standard Test Method for Air Content of Freshly Mixed Concrete by the Pressure Method *Concrete and Aggregates* **04.02**

[31] ACI Committee 212 2016 ACI PRC-212.3-16 Report on Chemical Admixtures for Concrete *American Concrete Institute*

[32] ASTM International 2016 Standard Test Method for Refractive Index and Refractive Dispersion of Hydrocarbon *Liquids Petroleum Products, Liquid Fuels, and Lubricants (II):* C1234 – D4176 **05.01**

[33] ASTM Subcommittee: C09.60 2018 Standard Test Method for Slump of Hydraulic-Cement Concrete *Concrete and Aggregates* **04.02**

[34] C. A. 214 2017 Guía para la evaluación de resultados de ensayos de resistencia del concreto American Concrete Institute ACI 214

Study on mechanical properties of mortars containing steel shot and sea sand as fine aggregate replacement

H Cabrera[1], C Daza[2], C Pacheco-Bustos[2], M Murillo[1]
[1] Civil and Environmental Department, Universidad de la Costa, Barranquilla, Colombia
[2] Civil and Environmental Engineering Department, Universidad del Norte, Barranquilla, Colombia

Abstract. In this study, the use of solid waste known in the metallurgical market as steel shot that is generated in the production of metallic elements has been analyzed. Normally the destination of this solid waste is the sanitary landfills of the cities or the security cells since it is classified as hazardous waste. In addition, the use of materials with high availability and that were not previously used in the construction industry, such as sea sand was used in the manufacture of mortar cubes as a fine aggregate. A granulometric analysis of each of the filling materials followed by the casting of 63 mortar cubes with four different fine aggregate replacement ratios (0%, 30%, 70% and 100%) were carried out. After the curing period, the compressive strength and the weight of the resulting mortars have been determined. The results from this study indicated that by completely replacing the fine aggregate with the steel shot waste the density and the compressive strength of the resulting mortars were increased, which indicates that this new material can be used in the construction industry.

1. Introduction

The increase in the number of inhabitants worldwide, as well as the development of industrialization processes, mainly in the developed countries, which generate a high demand for raw material to maintain its production and commercialization processes [1], brings with it the increase of waste production and exploitation of natural resources to meet the basic needs generated by the high demand for new infrastructure [2]. The current consumer mentality encouraged the generation of larger volumes of waste, which, for the most part, are destined to final disposal areas to be stored. Among the main activities that have been influenced by the increase in population, and which in turn requires a greater amount of resources to be able to be carried out, is construction. The construction of new structures to meet the basic needs of the growing population, such as housing, brings with it an increase in the exploitation of natural resources, which generates an accelerated reduction of their availability in the environment. Current constructions are made of different types of materials, with concrete and mortar being the most widely used mixtures worldwide, becoming the second most consumed element on the planet, after water [3].

It is easy to show this increase or growth in the demand for new materials in developing countries. In this study, Colombia will be used as an example, whose figures in the first quarter of 2017 indicate that the construction sector increased by 3.3 % compared to the same quarter of 2016 (according to the Civil Works Investment Indicator of the National Administrative Department of Statistics (DANE)

[4]. This growth is evidenced by the increase in construction policies implemented by the last governments and by the need to overcome existing oversupply, mainly in corporate and hotel construction [5]. Taking into account this construction boom, the production of cement (the main component of the mortar) and the exploitation of the raw material required for its elaboration increased. This assumption is based on the upward trend in cement production in Colombia, which increased from 852,100 tons in 2012 to 976,900 tons of production in 2017 [6], directly increasing the exploitation of quarries for the extraction of stony materials (fine and coarse aggregate).

Nowadays, research has been carried out to find alternatives for concrete production, implementing new resources, mainly waste generated by the activity of the human being.

Analysis as the use of e-plastic waste as a replacement for the coarse aggregate [7], or some studies about the sand replacement for different waste materials like quarry waste [8] or iron ore tailing [9] represent the need to reduce polluting elements that affect the environment. On the other hand, some studies have discovered the possible replacement for cement with other kinds of waste in the concrete, like ceramic waste [10] or marble and granite waste [11], among others.

In order to reduce the negative impact caused by the uncontrolled generation of solid waste in industrial steel cleaning processes, these materials have been used in the preparation of mortar mixtures [12]. Also, the use of raw material with wide availability, such as sea sand, encouraged the realization of studies that allow its application in construction [13].

Regarding industrial waste, several studies have been carried out and then some of the most relevant and interesting for this research will be shown: materials such as residual sand foundry have been studied to analyze their effects on concrete mixtures as a replacement for the fine aggregate. This is the case of the replacements of 10%, 20%, 30%, 40% and 50% of the fine aggregate by residual sand from aluminum industries, being a viable option to reduce the exploitation of fine aggregate required for the manufacture of concrete [14]. On the other hand, it has been found that by including residual sand from cast iron in concrete mixtures, an increase in water absorption is evidenced, which generates a decrease in the compressive strength in the specimens tested every time it increases the amount of the residual sand found in the mixture [15], for this reason, is not recommending exceeding a 20% replacement [16]. In the same way, in others studies that have used residual sand from cast iron and that have used replacements from 15% of fine aggregate, the compressive strength improves [17].

Another waste with conditions very similar to the one used in this project is steel slag, which is a waste generated in the industries dedicated to the manufacture of iron and steel [18]. The application of the replacement of this waste by the fine aggregate above 50% in the concrete reduces the workability of the mixture; however, by implementing replacements between 15% and 30%, improvements in the compressive strength of concrete are evidenced [19]. The steel slag replacements above 30% have shown a reduction in the compressive strengths obtained compared to the control mixture [20].

Based on the above, the present study aims to find an alternative to reuse a metallurgical by-product (blast residue) and to implement sea sand in the manufacture of resistant mortars according to ASTM C109M [21] evaluating the mortar resistance that includes steel-shot as a replacement of the fine aggregate, in order to reduce the obvious environmental impact generated by the erroneous disposal of this waste in landfills.

2. Methodology

The present research is of experimental quantitative type. The filling materials (sea sand, rock sand and metallic grit residue) were obtained in different ways. The sea sand was obtained in the excavations carried out in the construction of the Monticello condominium, in the vicinity of the township of Salgar (Atlántico-Colombia). Regular sand was acquired in the city of Barranquilla (Atlántico-Colombia), and finally the residue of metallic grit originated from the Polyuprotec plant located in the municipality of Malambo (Atlántico-Colombia), which is generated in the process of cleaning oxides and paintings of metallic elements. The cement used was a general use type from the cement company Argos. The chemical parameters of the cement used indicates that it has a maximum percentage [%] of magnesium oxide (MgO) of 6 and a maximum percentage [%] of sulfur trioxide (SO3) of 3.5, thus fulfilling the requirements of ASTM C-1157 Type UG. The procedures stipulated in the international standard ASTM C33 [22] were taken to obtain the gradation of fine aggregates (sea sand and conventional sand or AM and AP for its acronym in Spanish) and alternative material (granular residue (RG for its acronym in Spanish)), as well as the values of the fineness modulus of each filling material.

Subsequently, two different mortar mixtures were made, one mixture with AM and one with AP (defined as "control" (without replacement)), as stipulated in ASTM C109M [21]. With the purpose of mix with AM, taking into account the granulometric behavior found and the fineness modulus calculated, it was necessary to sieve approximately 10000 g of the material to comply with the gradation established in the standard and to make comparisons with the cubes elaborated with AP. With the mentioned mixtures, mortar cubes were made (Figure 1) by replacing the fine filler material with RG in the following percentages: 30%, 70%, and 100%. The replacement of the fine aggregate by RG in the defined percentages was carried out by weight [g], taking the retained material in each sieve, established in ASTM C109M [21] and performing the respective AM or AP replacement by RG. Following the established in the same standard, the relation water/cement used is 0.48. Subsequently, the cubes were tested at 7, 14 and 21 days, in order to make comparisons of the values of the obtained resistances.

Figure 1. Ilustration of the a) cubes molds and the b) casted mortars cubes.

The amount of chloride ion was determined for sea sand (AM) by the Standard Methods of the American Public Health Association (APHA). The result obtained was 14.7 mg / L, for a 2: 1 solution.

After the 63 cubes with the different replacements were tested, we proceeded to perform the analysis of the results obtained, plotting the compressive strengths in MPa to compare which one is the optimal mixture based on the weight versus compressive strength.

3. Results and Discussions

3.1 Granulometric analysis

The graphs related to the granulometry of the fine materials and the residue used for the elaboration of the mortar cubes according to ASTM C109M [21] are presented in Figure 2, to calculate the fineness modulus of each material and corroborate that the values obtained are within the range established by ASTM C33 [22].

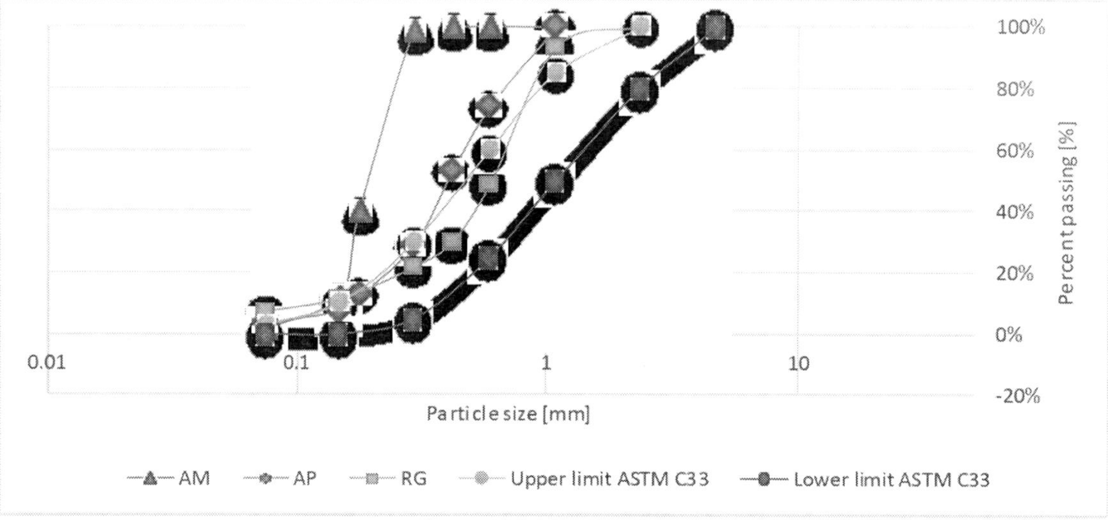

Figure 2. Fine aggregate granulometric curves.

Analyzing Figure 2, the verticality of the graph representing the AM indicates that the granulometry or distribution of the materials is mostly concentrated between the No. 80 and 100 sieves (292.2 g and 140.7 g respectively) showing homogeneity of the particles of the sample. In contrast, a more varied behavior was identified in the samples of AP and RG, these granulometries being the most acceptable for the formation of mortars and concretes, according to ASTM C33 [22], evidencing that 51% of the granular residue was retained between the Sieves No. 16 and No. 30.

Taking into account the granulometric behavior and the amount of material retained per sieve, the fineness modulus of each of the aggregate samples (AM, AP and RG) (Table 1) was calculated. The fineness module (MF) is a value that allows having an approximation of the level of fineness (thickness) that owns the material used, important data to analyze subjects of water demand in the mixtures.

Table 1. Fineness Modulus obtained for each of the aggregates.

Fine aggregate	Fineness Module [FM]
Sea Sand	1.5
Conventional sand	3.2
Granular residue	2.9

With the fineness modulus calculated for each material, a comparison was made with the ASTM C33 [22] acceptable values, which indicates that, for use of fine aggregate in mortars and concretes, the values of the fineness module must be within the range of 2.4 and 3.0. On the other hand, the RG, with a modulus of fineness of 2.9, is in the range acceptable by the norm and literature found, where it is established that one of the characteristics that allows the high resistance of concrete and mortars is that its modules of fineness are close to 3.00, avoiding homogeneities in the granulometric classification [23], affirmation that favors the use of the RG that presents a modulus of fineness of 2.9. However, AM with a modulus of low fineness, generates the need to include larger amounts of water in the mixtures so that the molten cubes handle the same consistency [24].

3.2 Compressive strength

In total 63 mortar cubes were molded, distributed as follows: 9 cubes using only AM as filling material, 9 cubes using AP only, 9 cubes using RG only and 36 cubes, replacing the implemented fine aggregate (AM or AP) with RG. In Table 2, the quantities of material used for each of the constituted mixtures are presented, corresponding to both AP-processed and AM-processed cubes. Taking into account the granulometry required by the standards [25-26], a sieving was performed to comply with the gradation of the fine aggregate to be used in the elaboration of the mortar cubes, for both the AM and AP controls, as well as the cubes that have thin material replacement. The water/cement ratio was 0.48 for all the mixtures following the amounts of material established by ASTM C109M [21].

Table 2. Fine material quantity for control cubes and cubes with replacement.

	0% replace [Control]	30% replacement		70% replacement		100% replacement
	Sand [AP-AM]	Sand [AP-AM]	Granular residue	Sand [AP-AM]	Granular residue	Granular residue
Sieve [No.]	Amount Withheld [g]					
16	0	0	0	0	0	0
30	41	29	12	12	29	41
40	603	422	181	181	422	603
50	1056	739	317	317	739	1056
100	362	253	109	109	253	362
Total	2062	1443	619	619	1443	2062

Subsequently, the measurements of the dimensions and weights of each of the cubes (high, width, length, weight) were taken employing a king's foot and a digital balance. The cubes were tested in the compression test machine, as shown in Figure 3. The results of the characterization and test results under compression are presented in Table 3.

Table 3. Weights and resistances of mortar cubes with different replacement percentages.

	Days of curing	0% (Control)		30%		70%		100%	
		Weight [g]	Resistance [MPa]	Weight [g]	Resistance [MPa]	Weight [g]	Resistance [MPa]	Weight [g]	Resistance [MPa]
Sea-sand	7	273.8	8.3	300.5	12	330.0	17.8	415	18
	14	294.8	10.9	293.6	10.9	349.2	12.4	409	23.2
	21	290.7	13.2	320.9	15.6	340.9	14.6	383	26
Regular sand	7	304.7	11.9	345.5	15.3	408.0	17.5	415	18
	14	316.0	12.1	349.7	12.9	423.7	29.1	409	23.2
	21	304.0	12.38	341.5	21.7	439.5	26.9	383	26

Figure 3. Cubes tested with rock-sand and 70% of replacement by waste of shot.

Having the results of the compression tests of the cubes, a comparison was made between the resistances obtained between AM mortar cubes with 0% replacement and mortar cubes made with AP with 0% replacement, as shown in Figure 4.

IOP Conf. Series: Materials Science and Engineering 1126 (2021) 012003 doi:10.1088/1757-899X/1126/1/012003

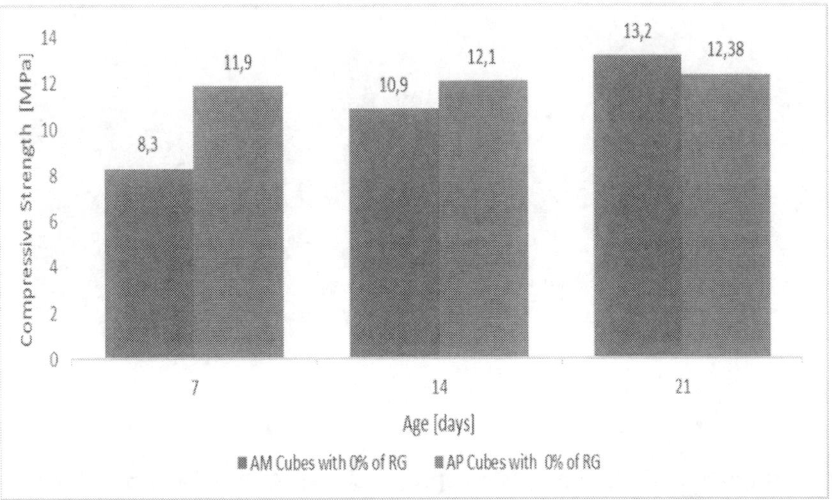

Figure 4. Comparison of compressive strength obtained in mortar cubes made with AM and AP.

As can be seen in Figure 4, it is evident that, after 21 days of curing, the mortar cubes constituted by AM have higher resistances (13.2 MPa). Nevertheless, it is identified that the cubes with this material, in the first days of curing, have inferior resistances to the cubes elaborated with AP, requiring more time to reach the expected resistances.

Then, the behavior of the compressive strengths obtained in each mix design (AM or AP) with their respective replacements by RG, as shown in Figures 5 and 6, was identified.

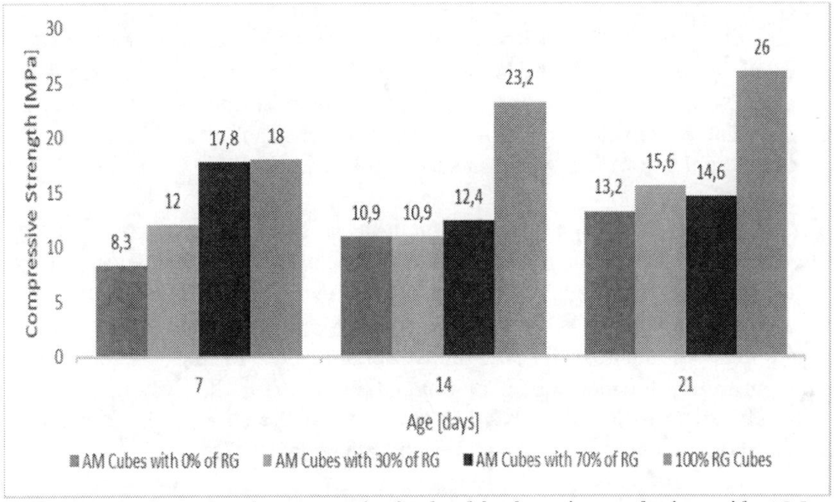

Figure 5. Compressive strength obtained in the mixture design with AM.

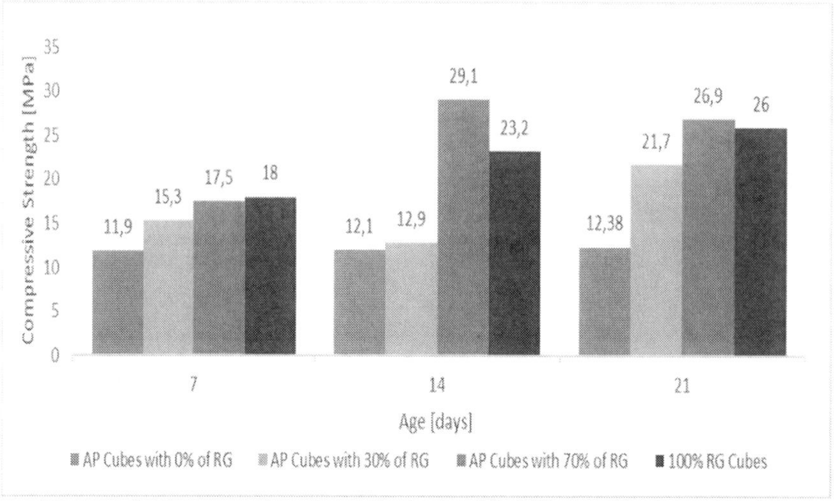

Figure 6. Compressive strength obtained in Mixture Design with AP.

As can be observed in the results of Table 3 and represented in Figures 4-6, the higher resistances were obtained in the cubes with 70% replacement of the fine aggregate by RG, in the mixture with AP (26.9 MPa). In contrast, the cubes with less resistance were those obtained with the mixture of AP without replacement by RG (12.38 MPa).

Analyzing the resistance obtained in the different days of curing, there are variations in the values obtained in the mixtures that present 30% and 70% replacement of the filling material by RG, presenting reductions of the resistances at 14 days of curing and increments again at 21 days. However, it is identified that in the two mixtures used in the present investigation (AM and AP) the obtained resistances, applying any percentage of replacement by RG, generate resistances superior to the control cubes. On the other hand, the results obtained are inverse to the resistances found when using mixing designs with other waste, like steel slag [18] or residual sand foundry [16], where better results are obtained with low replacement percentages. The aforementioned facilitates the reduction of the quantities of waste of existing shot when it is required in greater volumes for the elaboration of mortars, reducing the amount of this waste in safety cells.

On the other hand, as the percentage of replacement of the fine aggregate by RG increases, the weight also increases, due to the difference of densities that appear between the sands and the GR. However, it is important to note that the weights decrease as they increase the days of curing the mortar, which generates a new research topic to find the cause of this weight loss. Due to this, these mortars become very dense and are not suitable for tall constructions, due to the great weight that would be contributing for purposes of dimensioning of foundations and for the system of seismic resistance to lateral load (lateral loading is directly related to the mass of the structure). Therefore, we are looking for lighter constructions to avoid oversizing by high self-weight [27].

4. Conclusions and Recommendations

In the present investigation, it can be concluded that the granular residue (RG) complies with the Colombian normative requirements of granulometry and compressive strength for its use as mortar filling material. Concerning the compressive strength obtained in the cubes with different fine aggregates, it is recommended, in future investigations, to analyze the reasons for differences in the obtained values of resistances, because, granulometrically, both mixes are standardized according to the established in the norm ASTM C778 [28], deducing that this difference is due by different parameters than the mentioned one.

Finally, the use of RG could be proposed for prefabricated products on an industrial scale, with a high compressive strength and at a lower cost to the market, such as; Separator cases on tracks, containment structures such as gabions, foundations for tanks or silos, including walls and/or slabs for protection against gamma rays or x-rays. However, due to the density of the elements with 100% replacement percentages, their use is not very feasible for elements such as beams or columns because they would increase the dead load of the buildings. For foundation topics, the relationship between strength and size of foundations could be analyzed to find possible benefits of using materials, such as GR, in the concrete to be used. Likewise, it is recommended to analyze the effects that the foundations can generate with RG of steel on the ground and the possible effects that this residue can generate on the steel inside the structure, to validate in environmental and structural issues the application of this new material in construction.

References

[1] United Nations 1992 United Nations Conference on Environment and Development Río de Janeiro, Brazil, 3 to 14 June 1992

[2] Silva, D., Gameiro, F., and de Brito, J. 2013 Mechanical Properties of Structural Concrete Containing Fine Aggregates from Waste Generated by the Marble Quarrying Industry *In Journal of Materials in Civil Engineering* **26** https://doi.org/10.1061/(asce)mt.1943-5533.0000948

[3] Meyer, C. 2009 The greening of the concrete industry *Cement and Concrete Composites* **31** 601–605. https://doi.org/10.1016/j.cemconcomp.2008.12.010

[4] DANE 2017 Indicador de inversion en Obras Civiles. Retrieved June 23, 2017, from https://www.dane.gov.co/index.php/estadisticas-por-tema/construccion/indicador-de-inversion-en-obras-civiles

[5] Flórez, G. 2016 Así está la construcción en Colombia | *Economía* | *Portafolio*. Retrieved June 15, 2017, from http://www.portafolio.co/economia/asi-esta-la-construccion-en-colombia-501002

[6] Bogotá DC 2017 Comunicado de Prensa Estadísticas de Cemento Gris (ECG)

[7] Sabău, M., and Vargas, J. R. 2018 Use of e-plastic waste in concrete as a partial replacement of coarse mineral aggregate *Computers and Concrete* **21** (4) 377–384. https://doi.org/10.12989/cac.2018.21.4.377

[8] Gómez-Balbuena, D. N., López-Lara, T., Hernandez-Zaragoza, J. B., Ortiz-Mena, R. G., Navarro-Rojero, M. G., Horta-Rangel, J. and Rojas-Gonzalez, E. 2018 Polymer-Cement Mortar with Quarry Waste as Sand Replacement *Advances in Materials Science and Engineering* 1–10. https://doi.org/10.1155/2018/3984835

[9] Chavan, R. R., and Kulkarni, D 2013 Performance of Copper Slag on Strength Properties As Partial Replace of Fine Aggregate in Concrete Mix Design *Int. J. Adv. Engg. Res. Studies* **II**-IV 95–8

[10] Karthikeyan, B. and Dhinakaran, G. 2017 Strength and durability studies on high strength concrete using ceramic waste powder *Structural Engineering and Mechanics* **61** (2) 171–181

[11] Ghorbani, S., Taji, I., Tavakkolizadeh, M., Davodi, A., and Brito, J. 2018 Improving corrosion resistance of steel rebars in concrete with marble and granite waste dust as partial cement replacement. *Construction and Building Materials* **185** 110–119. https://doi.org/10.1016/j.conbuildmat.2018.07.066

[12] Thomson Gale, M. F. S., and ruiz, D. D. P. 2011 Revista educación en ingeniería *Revista Educación en Ingeniería* **6**

[13] Huiguang, Y., Yan, L., Henglin, L., and Quan, G. 2011 Durability of sea-sand containing concrete: Effects of chloride ion penetration *Mining Science and Technology* **21** 123–127. https://doi.org/10.1016/j.mstc.2010.07.003

[14] Ganesh Prabhu, G., Hyun, J. H., and Kim, Y. 2014 Effects of foundry sand as a fine aggregate

in concrete production *Construction and Building Materials* **70** 514–521. https://doi.org/10.1016/j.conbuildmat.2014.07.070
[15] Khatib, J. M., Herki, B. A., and Kenai, S. 2013 Capillarity of concrete incorporating waste foundry sand *Construction and Building Materials* **47** 867–871. https://doi.org/10.1016/j.conbuildmat.2013.05.013
[16] Basar, H. M. and Deveci Aksoy 2012 The effect of waste foundry sand (WFS) as partial replacement of sand on the mechanical, leaching and microstructural characteristics of ready-mixed concrete *Construction and Building Materials* **35** 508–515. https://doi.org/10.1016/j.conbuildmat.2012.04.078
[17] Singh, G., and Siddique, R. 2012 Effect of waste foundry sand (WFS) as partial replacement of sand on the strength, ultrasonic pulse velocity and permeability of concrete *Construction and Building Materials* **26** (1) 416–422. https://doi.org/10.1016/j.conbuildmat.2011.06.041
[18] Dash, M. K., Patro, S. K., and Rath, A. K. 2016 Sustainable use of industrial-waste as partial replacement of fine aggregate for preparation of concrete – A review *International Journal of Sustainable Built Environment* **5** (2) 484–516. https://doi.org/10.1016/j.ijsbe.2016.04.006
[19] Qasrawi, H., Shalabi, F., and Asi, I. 2009 Use of low CaO unprocessed steel slag in concrete as fine aggregate *Construction and Building Materials* **23** (2) 1118–1125 https://doi.org/10.1016/j.conbuildmat.2008.06.003
[20] John, A. and John, E. 2013 Study on the partial replacement of fine aggregate using induction furnace slag *American Journal of Engineering Research* **4** 1–5
[21] ASTM 2016 ASTM Standard C109M *Standard Test Method for Compressive Strength of Hydraulic Cement Mortars*
[22] ASTM International 2018 ASTM C33M-18 *Standard Specification for Concrete Aggregates* https://doi.org/10.1520/C0033_C0033M-18
[23] Carpio, C. 2007 Concretos de alta resistencia dentro de los no convencionales *Administración para el diseño* **16**
[24] Sanchez de Guzman, D. 2001 Tecnología del Concreto y del Mortero *Bhandar Editores ltda, Ed.* Bogotá
[25] ASTM 2003 Standard Specification for Concrete Aggregates
[26] ASTM 2006 Standard Specification for Standard Sand
[27] Joel Novas Cabrera, I. A. (n.d.). 2010 Sistemas constructivos prefabricados aplicables a la construcción de edificaciones en países en desarrollo *Universidad Politécnica de Madrid*
[28] ASTM International 2017 ASTM C778-17 *Standard Specification for Standard Sand* **6** https://doi.org/10.1520/C0778-17

ABC recombination model for quantum dot laser

Ivan B. Karomi[1*] *Ammar T. Zakar*[1] *and Mohammed S. Al-Ghamdi*[2]
[1]*University of Mosul, College of Education for Pure Science Mosul, Iraq 41002*
[2] *Department of Physics, Faculty of Science, King Abdulaziz University, P.O. Box 80203, Jeddah 21589, Saudi Arabia*

*e-mail: ivanbahnam@uomosul.edu.iq

Abstract. In this paper, we applied the ABC model in quantum dot (QD) semiconductor laser for the first time. We used a 1000μm cavity length InAsP/GaAs quantum dot laser emitting at 761nm, which was improved at Cardiff University. The ABC model is used to estimate the carrier losses that are caused by spontaneous emission and Auger recombination in semiconductor materials. It is shown that the ABC model is applicable in such lasers. The results show that the Shockley-Read-Hall (A) is 2.03 x10^9 sec^{-1}. The radiative coefficient (B) is 2.28 x10^{-14} cm^{-3} .sec^{-1} and the Auger recombination (C) is around 8 x10^{-37} cm^{-6} .sec^{-1}. The results are very close to the actual findings as measured by several different methods. Moreover, the measurement method is feasible, which can pave the path for the use of this procedure to determine the losses mechanism in semiconductor lasers.

1. Introduction

Quantum dot (QD) lasers have shown superior properties in both operation and applications in optoelectronic devices, due to their reduced threshold current [1] and high efficiency [2]. Therefore, understanding the losses coefficients in these lasers is essential because they control the efficiency of such devices; hence, they offer the possibility of finding an effective method to enhance device performance. The ABC model has attracted significant attention because it offers an excellent fitting model of external quantum efficiency (EQE) [3, 4]. Traditionally, the ABC model has been applied for light emitting diodes, such as GaInN, where there is a clear drop of the light emission efficiency with the current pumping the device (i.e., efficiency droop) [5, 6, 7]. The main reason for this drop is the self-heating effect [8, 9]. The ABC model is used to evaluate the losses coefficients, namely Shockley-Read-Hall, radiative coefficients and Auger recombination, only from the optical power–current relation and the radiative recombination time of the carriers [10] instead of using complicated experiments and calculations to find these coefficients. In this paper, we have applied the ABC model for a 1000μm cavity length uncoated facets InAsP QD laser to calculate the basic losses coefficients (i.e., Shockley-Read-Hall, Radiative and Auger recombination coefficients). The InAsP QD laser showed a high quality optical device and it offers many promising applications in optoelectronic devices and photonic integrated circuits [11, 12].

2. Theoretical part

The ABC-model is often employed to investigate the effeciency of a light emitting diode by using the power-current relationship where a drop of EQE happens in a high driven current. In this model, the EQE can be written as[13]:

$$EQE = \eta \frac{Bn}{A + Bn + C} \quad .1$$

Here, n is the carrier density, η is the light efficiency, A Auger recombination, B radiative coefficient, and C is SRH coefficient. The number of variables can be minimized provided equation 1, which is written in normalized optical power $(p = p_{out}/p_{max})$ P_{max} corresponding to the $EQE_{max,}$ and thus EQE_{max} can be written as [10]:

$$EQE_{max} = \eta \frac{Q}{Q + p^{1/2} + p^{-1/2}} \quad .2$$

Here $Q = B/(A\ C)^{1/2}$ represents a unitless combination of (A), (B), and (C) factors, and is called the quality factor. For the purpose of evaluating of quality factor Q, it is possible to refurbish equation 2 in terms of (EQE_{max}/EQE), while taking into account normalized power P:

$$\frac{EQE_{max}}{EQE} = \frac{Q + p^{1/2} + p^{-1/2}}{Q + 2} \quad .3$$

Plotting the (EQE_{max} / EQE) ratio against the normalized optical power allows us to calculate the Q-factor. In addition, the differential recombination lifetime (t) of the carriers can be expressed by [13]:

$$t = \frac{A^{-1}}{1 + 2Qp^{1/2} + 3p} \quad .4$$

Equation 4 allows us to evaluate the (SRH) recombination coefficient A by plotting the differential recombination lifetime (t) against normalized power. Furthermore, after evaluating Q and A, B and C can be calculated provided that the active recombination volume of the device (V_r) is recognized by means of the following equations [13]:

$$B = A^2 Q(2 + Q)\left(\frac{eV_r}{I_{max}}\right) \quad .5$$

$$C = A^3 (2 + Q)^2 \left(\frac{eV_r}{I_{max}}\right)^2 \quad .6$$

Here, I_{max} is the value of the driven current caused by EQE_{max}.

3. Experimental part

The samples were grown by the metal-organic chemical vaper deposition method for arsenides and phosphides to form self-organized QDs. The active area of the device has five layers of InAsP QD in a GaInP quantum well, separated by a matrix layer of AlGaInP. For more details about the growth conditions and sample structure, see [14]. Figure 1a. shows the transmission electron microscopy of the active region of the structure. A laser of 1000μm cavity length with 50μm oxide isolated striped width is used to measured the optical power- current curve of the sample, see Figure 1b. To determine the EQE from the (P-I) curves, the following equation was used above the threshold current (I_{th}) [15]:

$$EQE = \frac{\Delta p}{\Delta I}\left(\frac{e}{hf}\right) \quad .7$$

Here, e is the electron charge, h Planck constant, f frequency of light, and $\Delta p/\Delta I$ is the gradient of the (P- I) above the threshold. A multi-section sample was used to measure the radiative recombination time in the InAsP QD laser. We used the segmented contact method discribed in [16]. The spontaneous emission spectrum I_{Sp}^{meas} can be calculated by means of:

$$I_{Sp}^{meas} = \left[\frac{I_{meas}^2 L1}{I_{meas}\, L(1+2) - 2I_{meas}\, L1}\right] \frac{1}{l} ln\left[\frac{I_{meas}\, L(1+2)}{I_{meas}\, L1} - 1\right] \quad .8$$

Here, l is the length of the section (300μm), $I_{meas}\, L(1+2)$ is the amplified spontaneous emission (ASE) when both section 1 and section 2 are pumped, $I_{meas}L1$ is the ASE of section 1 (as discribed in Figure 1c.). It is then possible to find the radiative current density (J) from the area under the curve of the spactra. The radiative recombination time (t_{spon}) can then be estimated from the following expression [17]:

$$J = \frac{2Ne}{t_{spon}} \qquad .9$$

Here, J stands for the radiative current density, which can be determined from the spontaneous emission spectrum in equation (9); and N is the carrier density, which in QD is typically around 3 x10^{10} cm^{-2}.

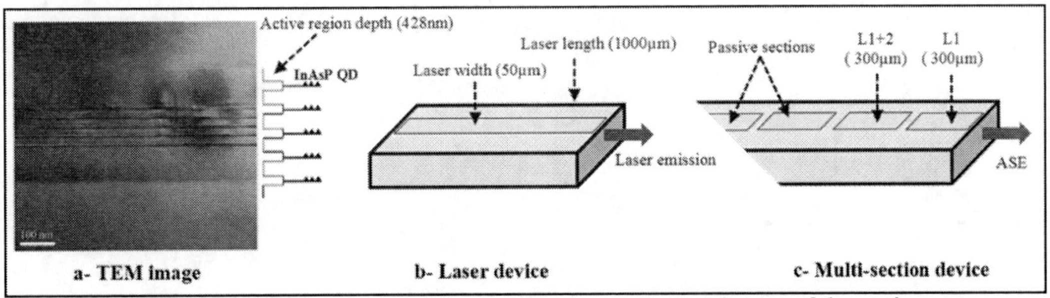

Figure 1. **(a)** Transmission electron microscopy (TEM) image of the active area, **(b)** laser device and **(c)** a multi-section device.

4. Results and discussion

Figure 2 shows the (current- power) curves for 1000μm cavity length of an InAsP QD laser at room temperature under pulse operation. The output optical power is measured in real units (mW) using an integrated sphere. The laser threshold current is around 220 mA and the slope efficiency (Δp/ΔI) can be determined at any point at the curve; hence, we can determine EQE by means of equation (7). It can be seen that at a high pump level of above around1100 mA, the slope efficiency starts to drop. This could be due to increasing the self head, even though the device is under pulse operation, or due to gain saturation at high pumped level or increasing homogenous broadening. We start from this point to apply the ABC model in the InAsP QD laser.

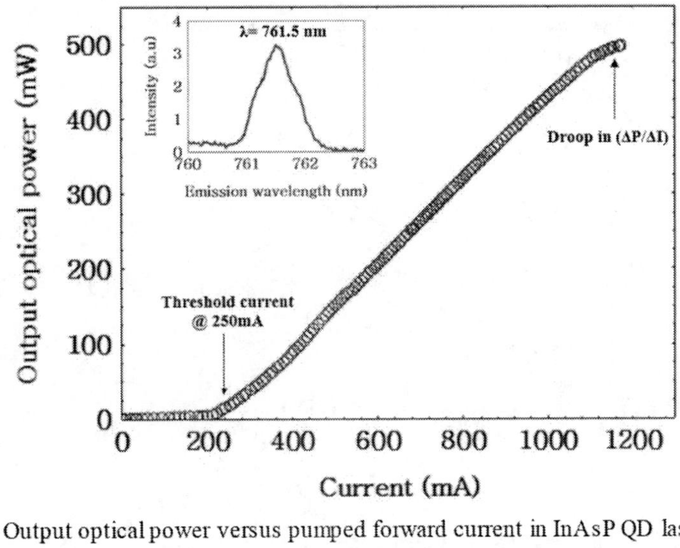

Figure 2. Output optical power versus pumped forward current in InAsP QD laser at room temperature under pulse mode operation, inset is the optical emission of the laser.

It is now possible to find Δp and ΔI above the threshold current (220 mA) of the laser, to measure the (Δp/ΔI) and EQE at each level of pumped current. Hence, it became possible to find the EQEmax. To find the normalized optical power (P=Pout/Pmax), the EQE is plotted against output optical power in Figure 3. Pmax is found to be near 373 mW and EQEmax is approximately 34%. To find Q-factor in equation (3), we plot EQEmax/ EQE ratio against $p^{1/2} + p^{-1/2}$ after normalizing the optical power; as shown in Figure 4. Q was calculated from the linear fitting line in Figure 4 and is found to be 0.7 ±0.03.

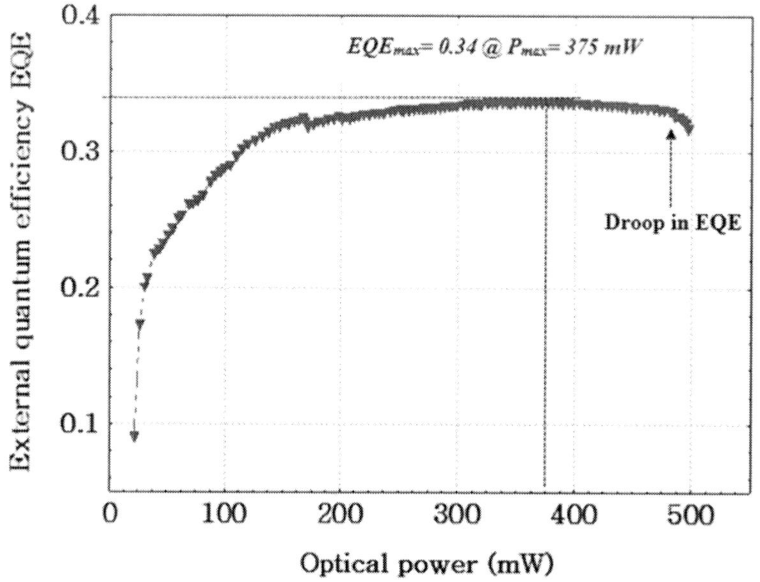

Figure 3. EQE against output optical power in InAsP QD laser at room temperature

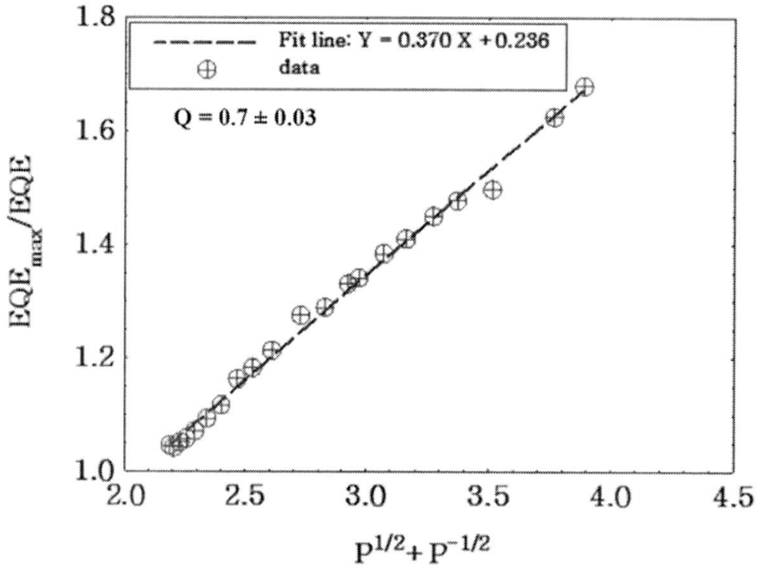

Figure 4. (EQE_{max} / EQE) ratio against the normalized optical power $P^{1/2} + P^{-1/2}$ in the QD laser

Furthermore, to calculate the radiative recombination time, the calibrated spontaneous emission spectra were found at different pumped currents for the multi-section device by means of equation 8. Figure 5 represents the spontaneous emission spectra in real units for InAsP QD. The area under the curve provides the radiative recombination rate. By multiplying this rate by the electron charge yielding the radiative current density (J), t_{spon} can be calculated via equation 9. Additionally, to find SHR recombination coefficient A, we plot the differential carrier lifetime (i.e. radiative recombination time) against the corresponding normalized power in the form shown in Figure 6. By fitting experimental data in Figure 6 in equation 4, it is found that A is approximately 2.03 x10^9 sec^{-1}

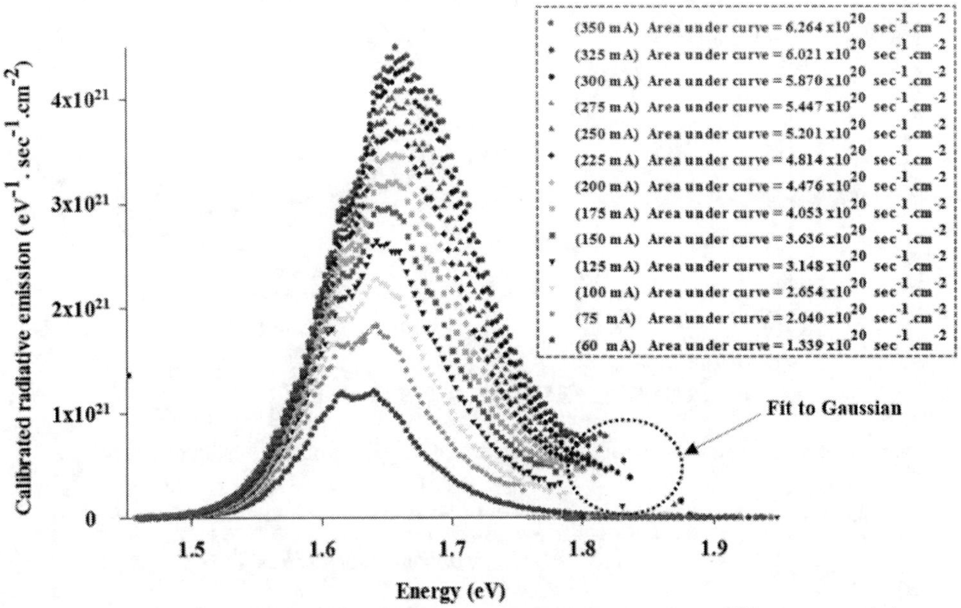

Figure 5. Spontaneous emission spectra in real unit for InAsP QD at different pumped currents

We used equation (5) and equation (6) to calculate Radiative and Auger recombination coefficients (B) and (C). To find the I_{max}, which is the value of the laser current caused by the maximum EQE, we plot in Figure 7 EQE versus driven current. It is found that the I_{max} is roughly 950 mA.

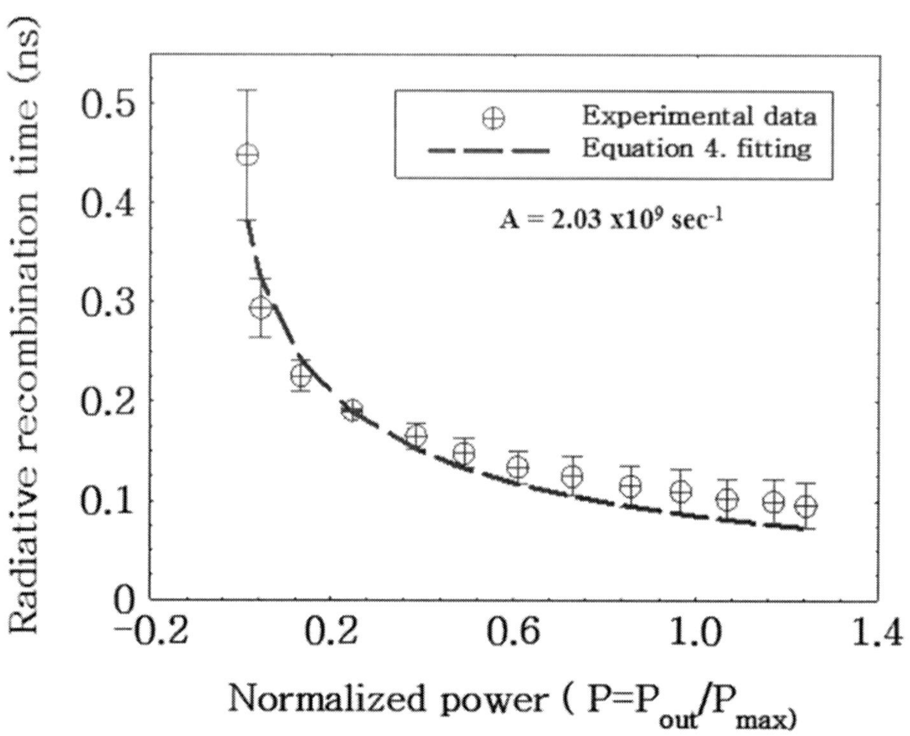

Figure 6. The differential recombination lifetime against normalized optical power

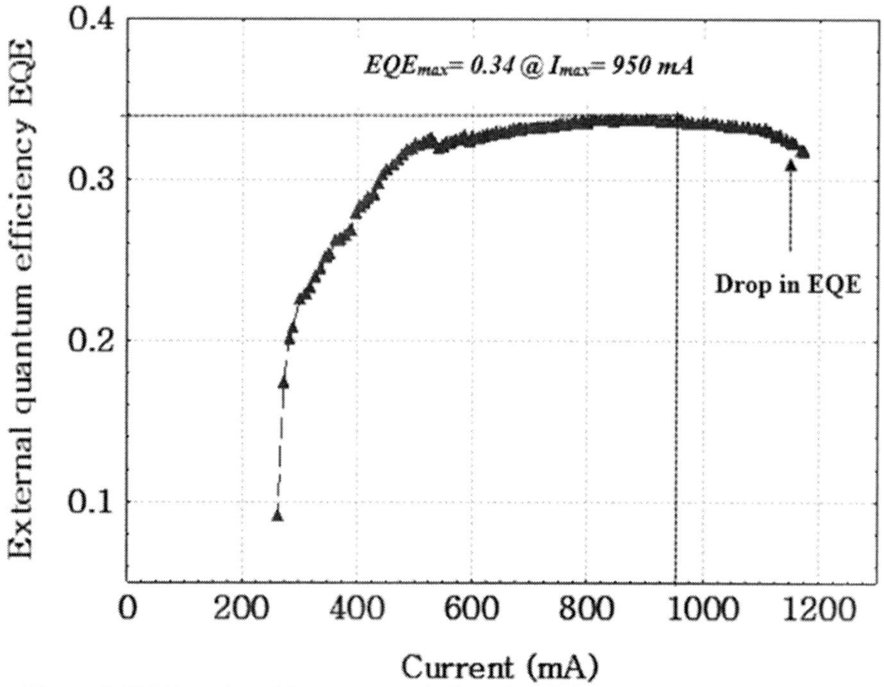

Figure 7. EQE against driven current in InAsP QD laser at room temperature

It is now possible to calculate the radiative coefficient (B) and Auger recombination coefficient (C) because all of the operators in equation 5 and equation 6 are now known. We yield the radiative coefficient B at around 2.89 $x10^{-14}$ cm^{-3}.sec^{-1} and Auger recombination coefficient C at approximately 8 $x10^{37}$ cm^{-6}.sec^{-1}. Now, if we invoke the relation $Q = B/(A\ C)^{1/2}$ to calculate Q, we get Q = 0.7, which is consistent with the result from Figure 4. Finally, the model has been applied successfully to the InAsP QD laser.

5. Conclusion

The ABC recombination model was applied for the first time to the InAsP quantum dot laser. Shockley-Read-Hall, Radiative and Auger recombination coefficients were determined. The results show that the Shockley-Read-Hall (A) is 2.03 xl09 sec-1. Radiative coefficient (B) is 2.28 xl0-14 cm-3 .sec-1 and the Auger recombination (C) is around 8 xl0-37 cm-6 .sec-1. Moreover, the measurement method is feasible which can be used to determine the losses mechanism. We believe that the ABC model with further adjustment could be an appropriate model to calculate important loss coefficients (Shockley-Read-Hall, Radiative and Auger recombination coefficients) that currently require complicated experimental setups.

Acknowledgments

We thank prof. Peter M. Smowton from Cardiff University for designing the sample, Dr Andrey B. Krysa form University of Sheffield for growing the sample and Dr Richard Beanland from University of Warwick for the TEM image.

References

[1] S.G.Lia, Q.Gongb, C.F.Caob, X.Z.Wanga, R.C.Wanga, J.P.Fana, C.X.Zhanga, L.Z.Xiaa, Temperature dependent characteristics of InAs/GaAs quantum dot laser grown by gas source molecular beam epitaxy, Infrared Physics & Technology, 54(5) (2011) 445-448. https://doi.org/10.1016/j.infrared.2011.07.002

[2] Xiaodong Huang, A. Stintz, C.P. Hains, G.T. Liu, J. Cheng, K.J. Malloy, Very low threshold current density room temperature continuous-wave lasing from a single-layer InAs quantum-dot laser, IEEE Photonics Technology Letters 12(3) (2000) 227 – 229. DOI: 10.1109/68.826896

[3] Y. C. Shena, G. O. Mueller, S. Watanabe, N. F. Gardner, A. Munkholm, M. R. Krames, Auger recombination in InGaN measured by photo-luminescence. Appl. Phys. Lett. 91 (2007) 141101. https://doi.org/10.1063/1.2785135

[4] Titkov, I.E., Karpov, S.Yu., Yadav, A., Zerova, V.L., Zulonas, M., Galler, B., Strassburg, M., Pietzonka, I., Lugauer, H.-J., Rafailov, E.U., Temperature-dependent internal quantum efficiency of blue high-brightness light-emitting diodes, IEEE JQE 50 (2014) 911–920. DOI:10.1109/JQE.2014.2359958

[5] P.Prajoon, D. Nirmal, M.Anuja Menokey, J. Charles Pravin, A modified ABC model in InGaN MQW LED using compositionally step graded Alternating Barrier for efficiency improvement, Superlattices and Microstructures 96 (2016) 55-163. https://doi.org/10.1016/j.spmi.2016.05.013

[6] M. A. Hopkins D. W. E. Allsopp, M. J. Kappers, R. A. Oliver, C. J. Humphreys, The ABC model of recombination reinterpreted: Impact on understanding carrier transport and efficiency droop in InGaN/GaN light emitting diodes, Journal of Applied Physics 122 (2017) 234505. https://doi.org/10.1063/1.4986434

[7] Thi Huong Ngo, Bernard Gil, Benjamin Damilano, Kaddour Lekhal, Philippe De Mierry, Internal quantum efficiency and Auger recombination in green, yellow and red InGaN-based light emitters grown along the polar direction, Superlattices and Microstructures 103 (2017)

245-251. https://doi.org/10.1016/j.spmi.2017.01.026

[8] Mukai, T., M. Yamada and S. Nakamura. "Characteristics of InGaN-Based UV/Blue/Green/Amber/Red Light Emitting Diodes", Jpn. J. Appl. Phys. 38 (1999) 3976–3981. https://iopscience.iop.org/article/10.1143/JJAP.38.3976

[9] Krames, M. R. et. al., High-brightness AlGaInN light-emitting diodes, Proc. SPIE 3938 (2000) 2–12. https://spie.org/Publications/Proceedings/Paper/10.1117/12.382822?SSO=1

[10] Karpov, S., ABC-model for interpretation of internal quantum efficiency and its droop in III-nitride LEDs: a review, Opt Quant Electron 47(2015) 1293–1303. https://doi.org/10.1007/s11082-014-0042-9

[11] Karomi, I., Smowton, P.M., Shutts, S., Krysa, A.B., Beanland, R.. "InAsP quantum dot lasers grown by MOVPE" Optics Express 23(21) (2015) 27282-27291. https://doi.org/10.1364/OE.23.027282

[12] Krysa, A.B., Roberts, J.S., Devenson, J., Beanland, R., Karomi, I., Shutts, S. Smowton , P.M., InAsP/AlGaInP/GaAs QD laser operating at ~770 nm, Journal of Physics: Conference Series 740(1) (2016) 012008. https://iopscience.iop.org/article/10.1088/1742-6596/740/1/012008

[13] Felix Nippert, Sergey Karpov, Ines Pietzonka, Bastian Galler, Alexander Wilm, Thomas Kure1, Christian Nenstiel, Gordon Callsen1, Martin Stra burg, Hans-Jürgen Lugauer, Axel Hoffmann, Determination of recombination coefficients in InGaN quantum-well light-emitting diodes by small-signal time-resolved photoluminescence, Japanese Journal of Applied Physics 55 (2016) 05FJ01-05 https://iopscience.iop.org/article/10.7567/JJAP.55.05FJ01

[14] A. B. Krysa et al., Growth and characterisation of InAsP/AlGaInP QD laser structures, 2016 Compound Semiconductor Week (CSW) [Includes 28th International Conference on Indium Phosphide & Related Materials (IPRM) & 43rd International Symposium on Compound Semiconductors (ISCS), Toyama, (2016) 1-2, DOI: 10.1109/ICIPRM.2016.7528566.

[15] P. M. Smowton and P. Blood, The differential efficiency of quantum-well lasers, IEEE J. Sel. Top. Quantum Electron. 3(2) (1997) 491–498. DOI: 10.1109/2944.605699

[16] Blood, P., Lewis, G., Smowton, P., Summers, H., Thomson, J., and Lutti, J., Characterization of Semiconductor Laser Gain Media by the Segmented Contact Method, IEEE Journal of selected topics in quantum electronics, 9(5) (2003) 1275–1282. DOI: 10.1109/JSTQE.2003.819472

[17] Blood, P., Gain and recombination in quantum dot lasers, IEEE Journal of selected topics in quantum electronics, 15(3) (2009) 808–818. DOI: 10.1109/JSTQE.2008.2011998

IOP Conf. Series: Materials Science and Engineering 1126 (2021) 012005 doi:10.1088/1757-899X/1126/1/012005

Study on mechanical and wear properties of hypereutectic Al-18Si-3.6Cu-0.34Ce alloy

Krishnaraddi Gangal[1], **Dr. Devendra K**[2] **and Chandrashakar Kaggalagowdra**[3]

[1]Department of Mechanical Engineering. Rural Engineering Collage, Hulkoti-582205, India

E-mail: krishnargangal@gmail.com (Krishnaraddi Gangal)

[2]DepartmentofMechanicalEngineering,SKSVMACET,Laxmeshwar,India

[3]DepartmentofMechanicalEngineering,SKSVMACET,Laxmeshwar,India

Abstract. In the present study an effort was made to find the mechanical and wear properties of a heat treatable hyper eutectic aluminium-silicon alloy. The alloy was prepared by conventional casting method in the form of ingots, and then chemical compositions of the alloy analysis are made by spectro spark emission spectrometer. The as cast alloy test samples are T6 heat treated by solutionizing at 500°C for 1h followed by water quenching and isothermal ageing treatment at 210°C for different aging time was achieved. The tensile test was carried out for as cast and cast aged conditions of different aging time of 1h, 3h, 5h and 7h. It was observed from tensile test the ultimate tensile strength of 1h (193.83Mpa) and 3h (168.68Mpa) aged (T6 treated) samples improved the UTS by 22% and 7% respectively as compared to as cast sample (158Mpa). Dry sliding Wear test was conducted at constant sliding speed of 1m/s, 2m/s, 3m/s and 4m/s. with varying load 10-40N. From the dry sliding wear study confirmed that wear rate and coefficient of friction are load dominant, In spite at higher sliding speed and load lower friction was noticed.

Keywords: Alloy, Hypereutectic, T6 treatment, Wear Rate.

1. Introduction

Aluminum and its alloys are the second most structural material used after the steel due to its versatile engineering properties and have low density of 2.7gm/cm^3 Its almost one third of steel (7.83 gm/cm^3) weight. Commercially the Al-Si base alloys are most widely used because of their special traits such as high strength to light weight ratios, good castability, better wear resistance, corrosion resistant and machinability [1]. Hyper eutectic Aluminum-Silicon alloys are the most prominent metal composites to manufacture the wear resistant spares used in automobile industry such as engine piston cylinder body, engine head and aerospace industry [1–5] along with this; hyper eutectic aluminum silicon alloys are potentially being used at high temperature environments [6].

The presence of Rare earth metals such as Ce in the hypereutectic Aluminum-Silicon base alloy, the primary silicon crystal were significantly refined, eutectic Silicon phase of large α-Al to smaller α-Al grain size phase were modified to fine fibrous structure [7].

The rare earth metal Ce can remarkably improve the primary Silicon crystals size and morphology of coarse star like and polygonal shape to finer Si blocky shape with Ce concentration. Ce was an effective modifier for eutectic silicon [7,8]. Hence improves the mechanical and wear properties of the alloy.

Hypereutectic Aluminium-Silicon Alloy with high percentage of silicon (18 wt %) leads the alloy to excellent casting characteristics. Inclusions of Silicon in a pure aluminum greatly enhance fluidity, with hot tear resistance and better feeding characteristics. To the large extent Aluminium-Silicon compositions are used in most of the casting exercise [8]. And copper (3.6 wt%) was used to produce the heat treatable superior metal of light weight to high strength for automotive applications, Copper is one of the effective precipitation strengthening factor in Aluminium-silicon alloy. Addition of Copper about 5% assists the alloy with attractive strength and toughness after forced or natural aging. By forced aging (T6 treatment) will improve the inert metallic compound dissolution and nucleation rate, hardness improved with copper addition, because of precipitation hardening [9, 10]. Earlier studies confirmed that varying the copper proportion up to 5% will improve the hardness and also investigated the T6 heat treatment effects on Al-Si-Cu alloy and concluded that T6 treatment play the key role on its mechanical and macrostructure properties [11].

2. Experimentation

2.1. Chemical composition. In the present study, the hypereutectic ranged alloy, with light weight heat treatable 3xx.x grade Al-18Si-3.6Cu-0.34Ce was prepared by conventional casting method. The castings are in the ingot form of 30mm diameter and 190mm length. The chemical constituent of the prepared alloy was examined with the help of spark emission spectrometer (Make - PAN Alytical XRD and XRF Instrumentation, Model QSN 750-II single or multi matrix system.) at Karnataka material testing and research, summary of chemical composition are listed in Table1.

Table 1. Summary of chemical composition (wt %) of Al-18Si-3.6Cu-0.34Ce alloy.

Alloy	Si	Cu	Ce	Al
Al-18Si-3.6Cu-0.34Ce	18	3.6	0.34	Balance

2.2. Tensile test. The tensile test specimens are prepared from the as cast ingots as well as from cast aged ingot bars according to ASTM 8E standards. These specimens are tested for as cast and T6 heat treated conditions; precipitation hardening was achieved by the guidelines of ASM hand book. Heat treatment was attained by using muffle furnace having temperature accuracy of ±3°C. This process involves solutionizing followed by homogenization. Solutionizing was done at 510°C for a soaking time of 1hr Followed by sub zero temperature water quenching and homogenizing at 210°C for different aging time of 1, 3, 5 and 7 hours followed by air cooling. For all conditions of the alloy samples tensile test was carried out on UTM machine (model TUF-C-20000). The effect of aged condition results will be compared with the as cast condition result and the same is plotted as shown in Figure 2. For a 1hour T6 heat treated sample the tensile stress verses strain is shown in Figure 1.

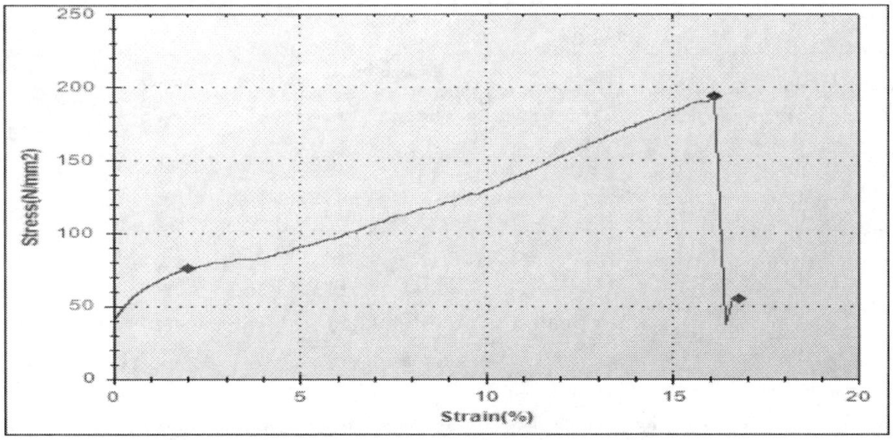

Figure 1. Stress Vs Strain for 1 hour T6 treated tensile test sample result.

2.3. Wear test. The dry sliding wear test experiments was conducted using pin on disk wear experiment setup [Make - DUCOM instruments] as per the ASTMG99-05 measures. The wear testing pin samples dimension of 30mm length × 10mm diameter rounded bars with flat surface are prepared by machining the ingot bars on lathe machine. The wear test was conducted at constant speed of 1m/s, 2m/s, 3m/s and 4m/s with varying normal load from 10 to 40N for a predetermined sliding distance of 2000m. During the wear test, temperature and the frictional force developed on the test specimens are noted continually from the display on controlling unit of pin on disk wear testing machine setup. These sample pins weight are noted before and after the wear test using an electron balance instrument with accuracy of 10^{-7}kg. Weight loss due wear was recorded and Samples are washed with the acetone to avid oxidation of worn surfaces. Weight loss method was applied to find the volumetric loss and then volumetric wear rate was determined. Using these data volumetric wear rate verses load and speed. Coefficient of friction verses load and speeds were drawn.

3. Results and Discussions

3.1. Mechanical Test. Tensile test was conducted for as cast and cast aged hyper eutectic Al-18Si-3.6Cu-0.34Ce alloy. Figure 2. Show the comparative ultimate tensile strength of an alloy at different test condition.

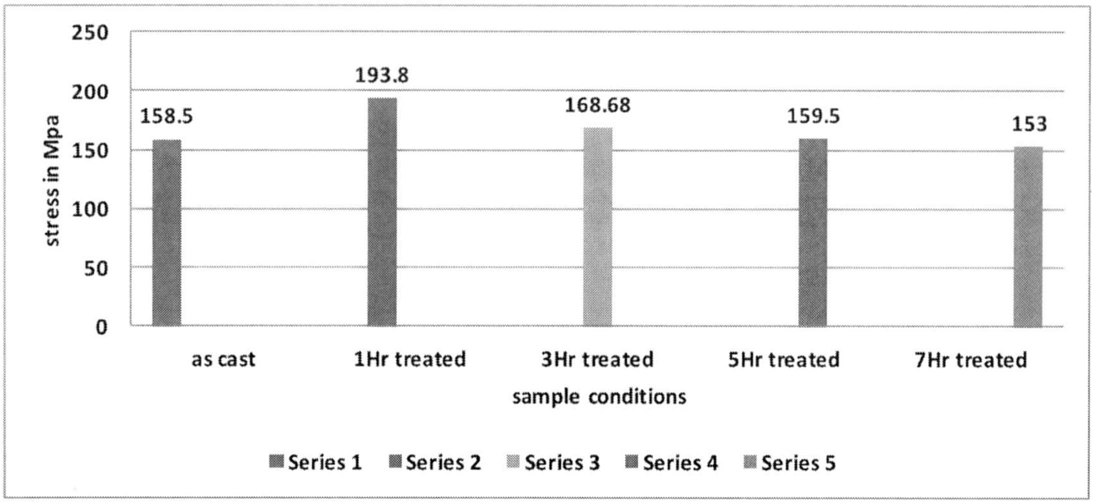

Figure 2. The tensile test result of an alloy.

From the UTS test results it was observed that an improvement in the Ultimate tensile Strength of 1hr (193.8Mpa) and 3 hr (168.7Mpa) aged sample by 22% and 7% as compared to that of as cast condition(158.5Mpa) respectively due the re-precipitation with redistribution of $CuAl_2$ phase improve the hardness and tensile strength by T6 treatment of alloy [12].

3.2. Wear study of as cast alloy. Dry sliding wear study was conducted on a polished disk of hard counter surface of EN-32 with HRC 62-65, room temperature of 30°C and humidity around 60%. At constant speed with varying load test was carried out for as cast condition; from the wear test at different level of speeds of 1m/s, 2m/s,3m/s and 4m/s. with varying load 10-40N.

The trend shows in Figure 2(a, b) that with the normal load the volumetric wear rate was also increased for every speed level, and other side of observation at higher sliding speeds of 2-3m/s lower volumetric wear rate was recorded.

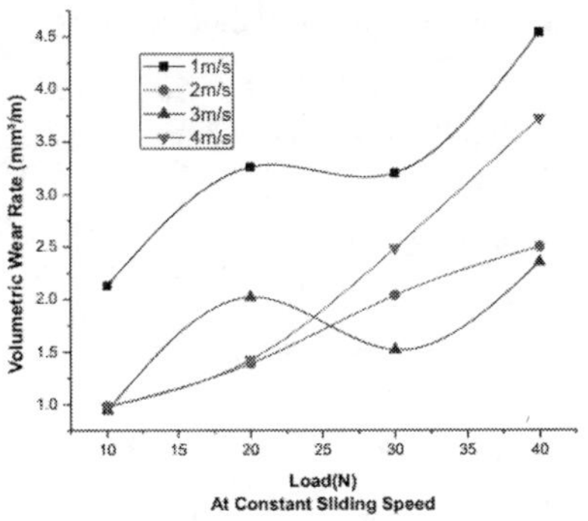

Figure 2.a.Volumetric wear rate
At constant sliding speed.

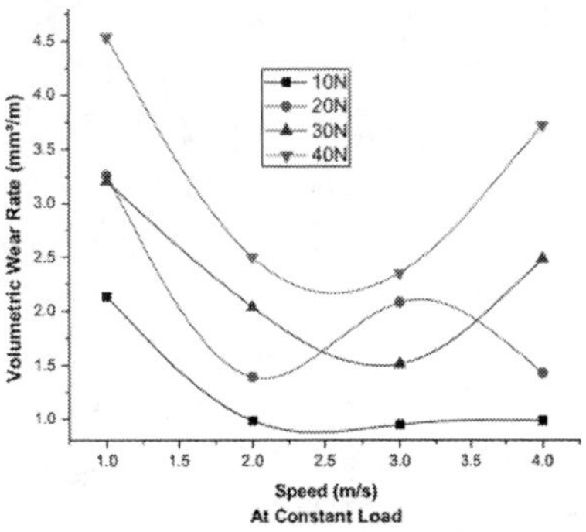

Figure 2.b. Volumetric wear rate
At constant load.

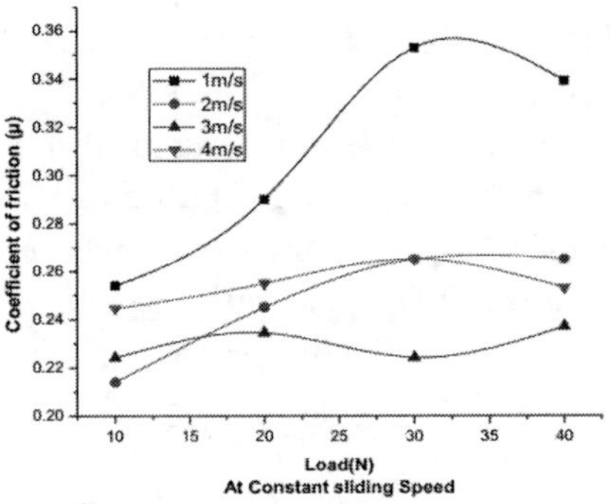

Figure 3.a. Change in Coefficient of friction
At constant speed with varying load.

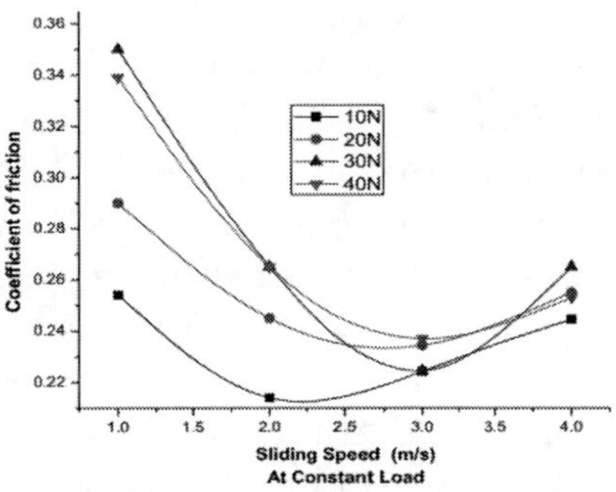

Figure 3.b. Change in Coefficient of friction
At constant load with varying speed.

From the Figure 3(a, b). The Coefficient of friction during wear was increased with load at lower constant sliding speed, and moderate change was notice at higher sliding speeds of 3m/s and 4m/s. hence wear decreases with speed during constant load due to reduced contact time.

4. Conclusions:

1. Refinement and re-precipitation of primary and eutectic Si crystals of alloy by T6 heat treatment; The UTS of 1hr and 3hr aged are notably enhanced by 22% and 7% respectively as that of As cast condition.
2. Volumetric wear rate of as cast alloy is a load dominant, at constant sliding speed of 2-3m/s, moderate change in wear rate was noticed for varying load of 10-40N.
3. Coefficient of friction at constant sliding speed despite increase with load. At higher sliding speed alloy shows low friction.

5. References

[1] Naglaa Fathy1. Microstructural Evolution of Hyper-Eutectic Al-18% Si Alloy during Semi-Solid Isothermal Heat Treatment. Int'l Journal of Research in Chemical, Metallurgical and Civil Engg. ISSN 2349-1442 EISSN 2349-1450.

[2] Qiao J G, Gao G H, Gao Z Y. The dual modification on hypereutectic Al-Si alloy. J. Internal Combustion Engine & Parts, 2004, 4: 18.

[3] Meng F Q. Effects of rare earth and phosphorus on micro-structures and properties of hypereutectic Al-Si alloy. J. Rare Earths, 1998, 19(2): 61.

[4] Wang Z H, Mao X M, Zhang J L. Effects of modifier Sr2- PM on hypereutectic Al-Si alloy. J. Special Casting & Nonferrous Alloys, 2005, 25(4): 241.

[5] Kapranos P, Kirkwood D H, Atkinson H V. Thixo formingof an automotive part in A390 hypereutectic Al-Si alloy. J. Mater. Process. Technol., 2003, 135: 271.

[6] Joon-Young Heo1,etl. Effects of Heat Treatment on the Microstructures and highTemperature Mechanical Properties of Hypereutectic Al–14Si–Cu–Mg Alloy Manufactured by Liquid Phase Sintering Process.©The Korean Institute of Metals and Materials 2018, https://doi.org/10.1007/s12540-018-0068-9.

[7] Qinglin Li a,b, Tiandong Xia a,b,⇑, Yefeng Lan a,b, Wenjun Zhao a,b, Lu Fan a, Pengfei Li a Effect of rare earth cerium addition on the microstructure and tensile properties of hypereutectic Al–20%Si alloy.Journal of Alloys and Compounds 562 (2013) 25–32.

[8] Hesham Elzanaty. Effect of Composition on the Microstructure, Tensile and Hardness Properties of Al–xSi Alloys. Journal of Materials Science & Surface Engineering Vol. 2 (2), 2015, pp 126-129.

[9] Xiao-hui Ao1,etl. Effect of Ce addition on microstructures and mechanical properties of A380 aluminum alloy prepared by squeeze-casting.https://doi.org/10.1007/s12613-018-1602-y.

[10] Mousa Javidani and Daniel Larouche*. Application of Cast Al-Si Alloys in Internal Combustion Engine Components.https://www.researchgate.net/publication/261798413.DOI: 10.1179/1743280413Y.0000000027.

[11] MUZAFFER ZEREN and ERDEM KARAKULAK* Study on hardness and microstructural characteristics of sand cast Al–Si–Cu alloys. Bull. Mater. Sci., Vol. December 2009, pp. 617–620. © Indian Academy of Sciences.

[12] Joon-Young Heo1,etl. Efects of Heat Treatment on the Microstructures and High Temperature Mechanical Properties of Hypereutectic Al–14Si–Cu–Mg Alloy Manufactured by Liquid Phase Sintering Process.https://doi.org/10.1007/s12540-018-0068-9.

IOP Conf. Series: Materials Science and Engineering 1126 (2021) 012006 doi:10.1088/1757-899X/1126/1/012006

Characterization of Silver Nanoparticles on Porous silicon by Laser ablation in Liquid.

[a] **Mohammed Ibrahim Ismael,** [a] **Ghazwan Ghazi Ali,** [b] **Uday Muhsin Nayef**

[a] Physics Department, Education College for Pure Science, University of Mosul, Iraq
[b] Applied science Department, University of Technology, Baghdad, Iraq.
dr.ghazwan39@uomosul.edu.iq

Abstract. In this article, Porous silicon was prepared using a photo-electrochemical etching method of n-type silicon at etching times (5 and 15 min.). The silver nanoparticles were coated on porous silicon by laser ablation in liquid with energy 400, 600, and 800 mJ. Atomic force microscope images showed an increase of pores size with etching time and it becomes smaller in diameter after laser ablation. Moreover, Scanning electron microscope displayed an aggregates formation when increasing laser energy. The strong peak of absorption of AgNPs found around 415-420 nm. The absorption amplitude increased when laser energy increases. Therefore, the nanostructures on the porous silicon properties play a crucial role in the formation of the homogeneity and this improves the structural stability of PSi layer.

1.Introduction

The silver nanoparticle has attracted great attention recently due to its excellent properties such as chemical [1], gas concentrators [2], biosensors, and high carrier mobility [3]. It is one of the noble metals and resistant to corrosion and oxidation even at increased temperature. The installation of AgNPs using the laser ablation method requires minimal reagents, and sometimes Ag pure metal can be performed either in the liquid medium or in aerosols [4, 5]. Silver target has been immersed in an aqueous solution in a container [6]. Laser ablation technique is one of the physical techniques for nanoscale (top-down) by a high-energy laser and short pulses affecting the metal. This method is very easy and less expensive, as the generated nanoparticles are homogeneous. Laser ablation is a very important technique for improving optical devices due to its high yield, high quality, and ease of used [7, 8]. Additionally, this technique can be created from homogenous layers along the substrate [9]. Recently, several articles reported have studied silver nanoparticles on the porous silicon material due to the suitable material for use in photodetectors, molecular sensors waveguides, light- emitting diodes, photovoltaic devices and high photoluminescence efficiency in the visible region [10]. The combinations of silver nanostructures and porous silicon layer are more significant due to the high quality of silver nanoparticles on a PSi layer can introduce a large matching of lattice constants [11]. This might lead to a new type of active material due to the luminescence properties of the PSi layer deposited with silver nanostructure could be suitable for the optical and electrical and properties [12,13]. This work provides more insight and a better understanding of the growth dynamic of silver nanostructures within the pores. The obtained results were investigated using an Atomic force microscope (AFM), Scanning electron microscope (SEM), and UV- spectrophotometer.

2.Experimental part

Figure 1 illustrates a schematic diagram of the Photo-electrochemical etching technique of porous silicon (n-type) with an orientation (100) and resistivity of (0.01-0.02 Ω cm). The sample of porous silicon was dipped into an aqueous solution of HF (20% wt) and ethanol (99%) (1:1). The applied current was fixed 15 mA/cm2 and etching time was used (5 and 20 min). Additionally, ethanol solution is a key role to increase the wet strength of the porous silicon substrate which is important to enhance the lateral homogeneity of the PSi surface. Furthermore, the etching cell was made from

Teflon connected with a rubber O-ring of 1.5 cm2 diameter to permit it to touch the silicon wafer. The current was fixed by two electrodes: The first electrode is a stainless-steel that was placed below the

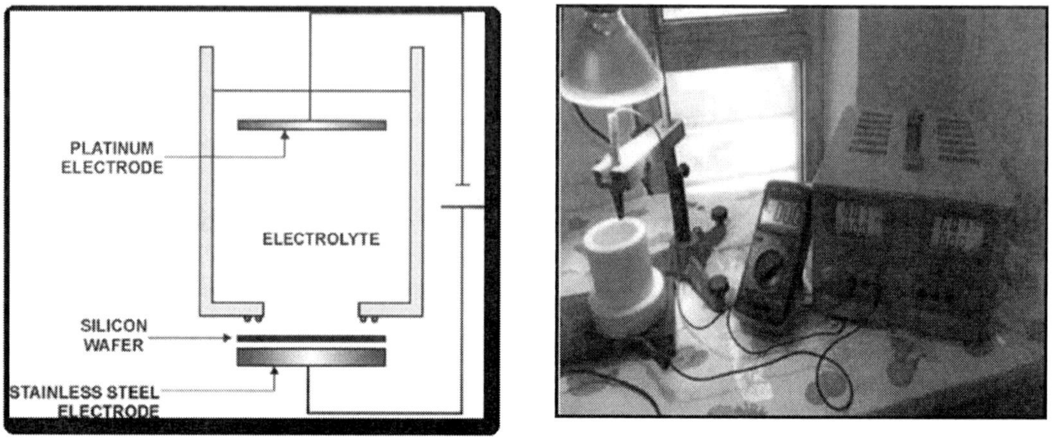

Figure 1. Schematic diagram of Porous silicon cell .

silicon layer and the second electrode has made of gold and is connected to the Hydrofluoric acid solution.

Afterwards, silver nanoparticles coated on the surface of PSi by drop - casting. Silver nanoparticles (Ag NPs) can be produced by the high-power laser, it is immersed (with high purity) in a glass container of 3ml of ionic water. The target is radiated vertically by the pulsed laser Nd: YAG has a wavelength of 1064nm and different energies of 400,600,800m J with the number of pulses of 200 pulses. To achieve the maximum intensity of the laser, a lens with a distance of 12cm is used as Figure 2, it shows the Setup of laser ablation. Besides, this method is done as a result of the chemical reaction between (atoms - ions) as well as in the form of silver clusters that are bonded with oxygen atoms in the water. During the cooling process of the plasma, during the union process, the silver atom is formed with the oxygen atom and the silver oxide particles.

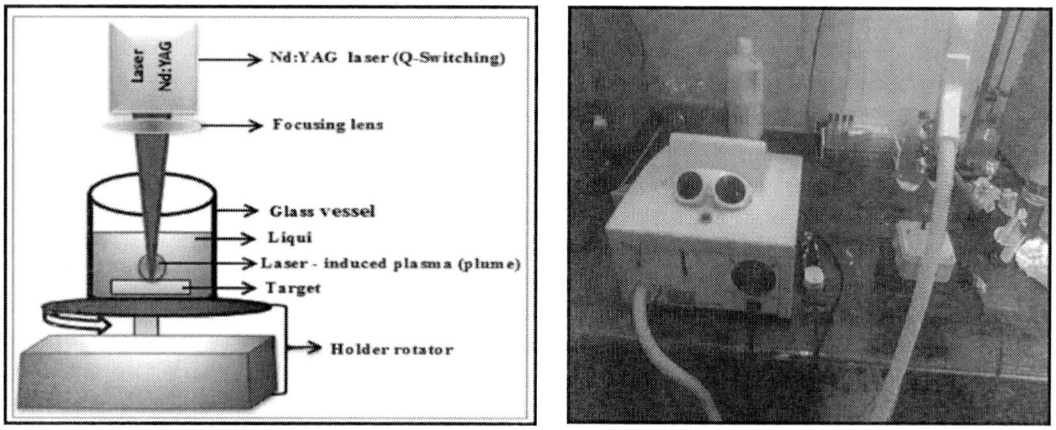

Figure 2. The setup of the laser ablation.

3.Results and discussion part

3.1.Morphological Properties of Silver Nanoparticles on porous silicon

Figure 3 shows AFM images at etching time 5 and 15 min with the concentration of hydrofluoric acid 15%HF and applied current density 20 mA/cm^2. It reveals that the porous silicon has a sponge-like structure 2D and 3D. It can be seen that the shape of the pore is spherical and change to a cylindrical shape with increasing etching time [14]. Additionally, the pore size increases with etching time due to the chemical dissolution of walls between the pores leads to an excess number of holes. Subsequently, the surface area increases with increasing etching time [15, 16]. Figures 4 and 5 show AFM images of silver nanoparticles (AgNPs) coated on porous silicon substrate by drop-casting for energy 400 and 800 mJ with 200 pules at 5min and 15 min. It is clear, the amount of seeds of silver nanoparticles makes partially fills or covers the entire pores [17]. The average size of the pore decreases with laser energy (20.52 to 15.10 nm) at etching time 5 min and (46.12 to 23.3 nm) at etching time 15 min. respectively due to the growth of seeds on the edge of the other sides and the particle size of the silver is larger than pore size, this is reduced the pore diameter. Tables 1 and 2 illustrate parameters of the AgNPs of porous silicon before and after laser ablation.

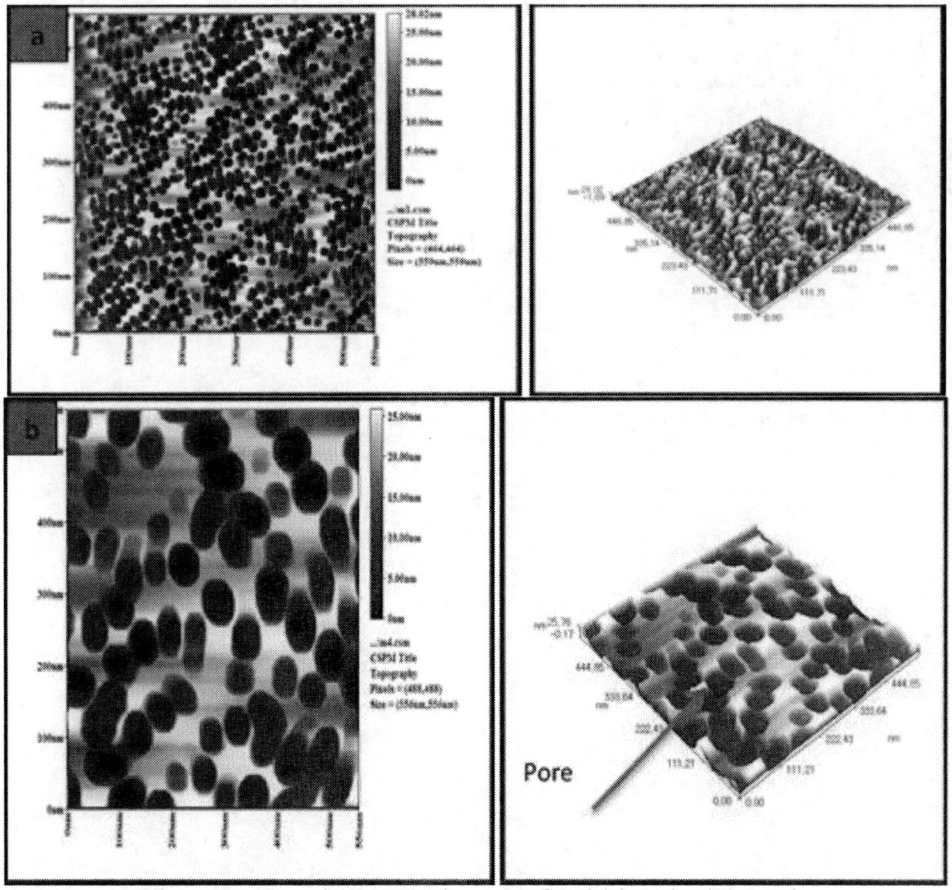

Figure 3. 2D and 3D AFM images of PSi (a) 5min. (B) 15min.

IOP Conf. Series: Materials Science and Engineering 1126 (2021) 012006 doi:10.1088/1757-899X/1126/1/012006

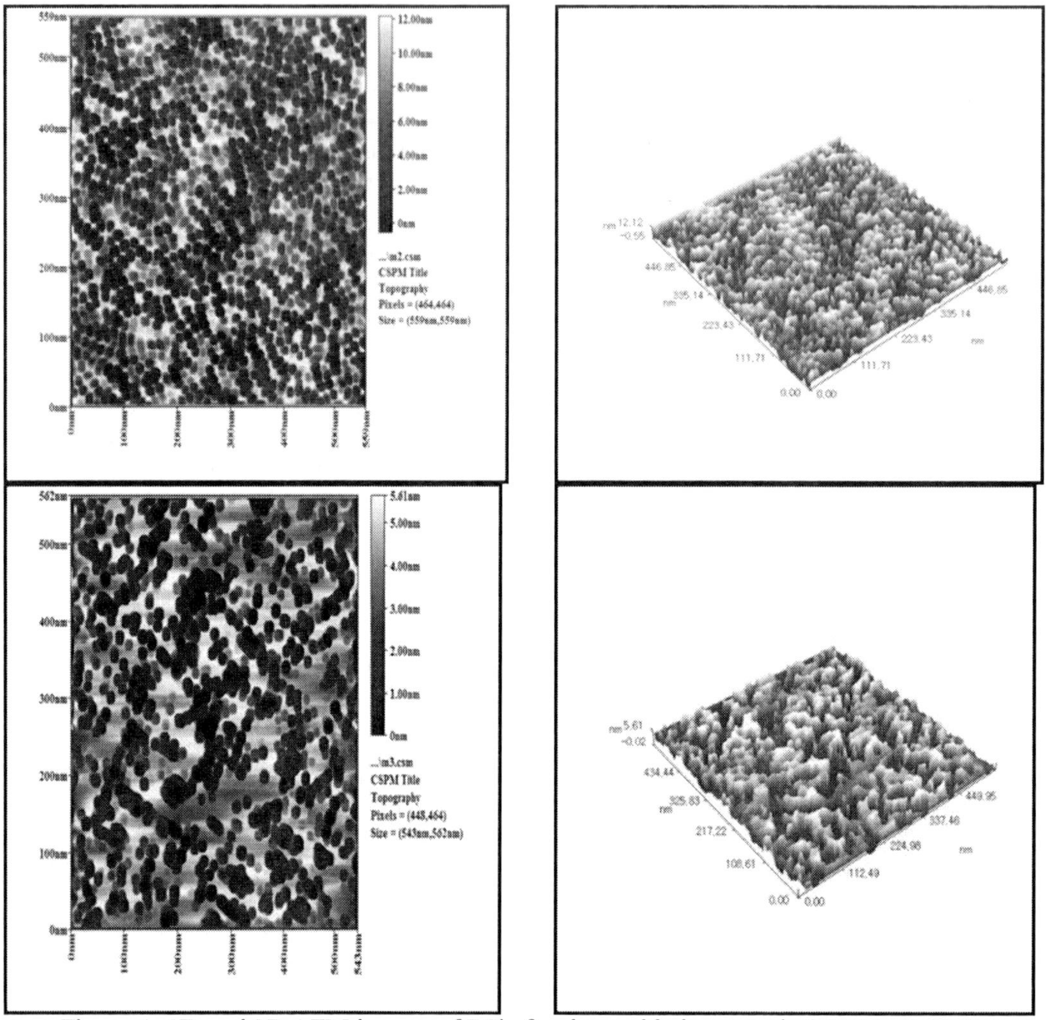

Figure 4. 2D and 3D AFM images of PSi after laser ablation at 5min (a) 400mJ (b) 800mJ

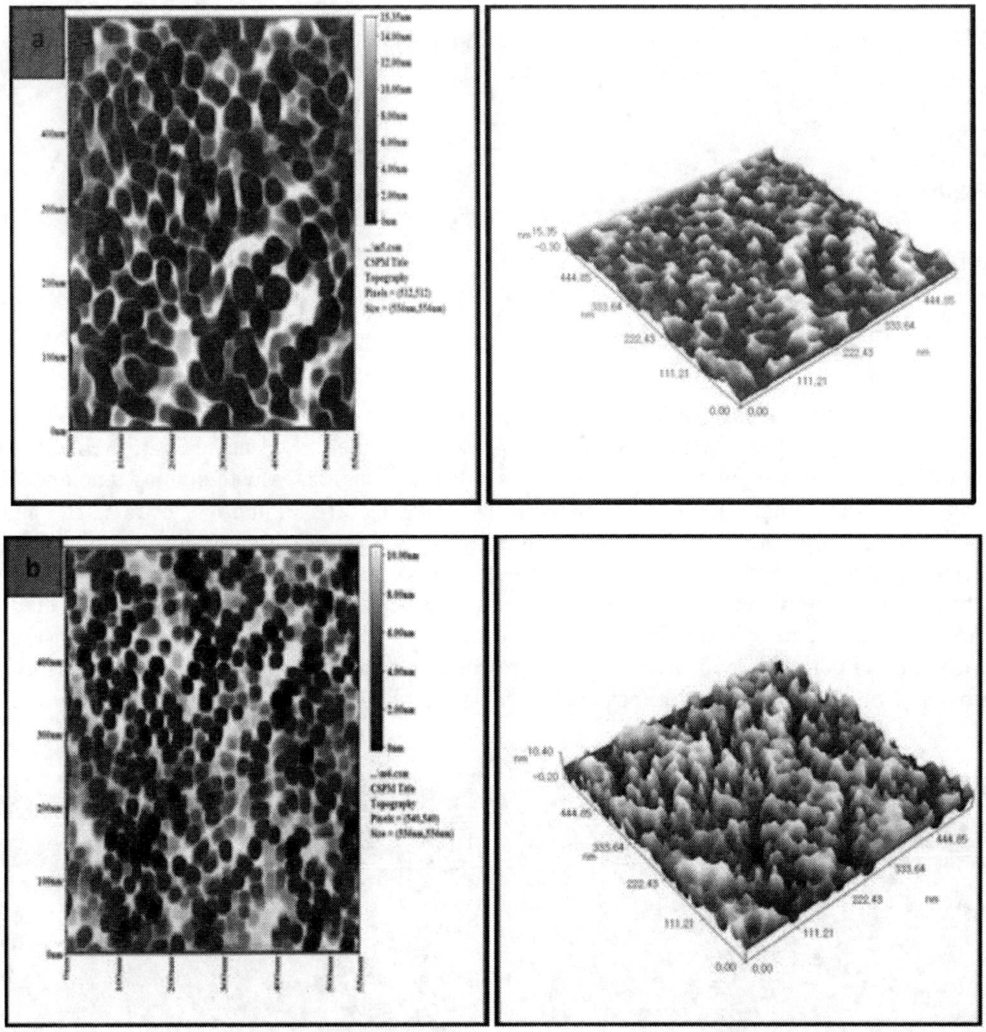

Figure 5. 2D and 3D AFM images of PSi after laser ablation at 15min (a) 400mJ (b) 800mJ

Table 1. The parameters of morphology characteristics of PSi samples before laser ablation

Etching time(min.)	Average Diameter (nm)	Average Roughness (nm)
5	20.52	6.48
15	46.12	7.42

IOP Conf. Series: Materials Science and Engineering 1126 (2021) 012006 doi:10.1088/1757-899X/1126/1/012006

Table 2. The parameters of morphology characteristics of PSi samples after laser ablation

Laser energy (mJ)	**Etching time(min.)**	**Average Diameter (nm)**	**Average Roughness (nm)**
400	5	19.34	3.17
800		15.10	1.41
400	15	31.32	3.51
800		23.30	2.65

Scanning electron microscope (SEM) is one of the most important characteristics of the morphological study. Figure 6a exhibits SEM images of porous silicon samples at 5 min with 15%HF and applied current density 20 mA/cm^2, it indicates that the pore diameter increases with etching time and the layer of porous silicon has a uniform pore-like network [18]. Figure 6 (b, c, and d) illustrates the images of the AgNPs on porous silicon. It is clear that the pore has a pore-like structure and spherical shape. Furthermore, figure(c) shows AgNPs has aggregates formed on the surface of porous silicon and the pore shape convert from spherical shape to flower and star shape with increasing the laser energy(400 600, and 800 mJ) due to surface Plasmon resonance (SPR) and the interface between silver nanoparticles and porous silicon. nanoparticles provide good coverage of the crystalline surface on the PSi substrate, which can improve the structural stability of the PSi substrate [19].

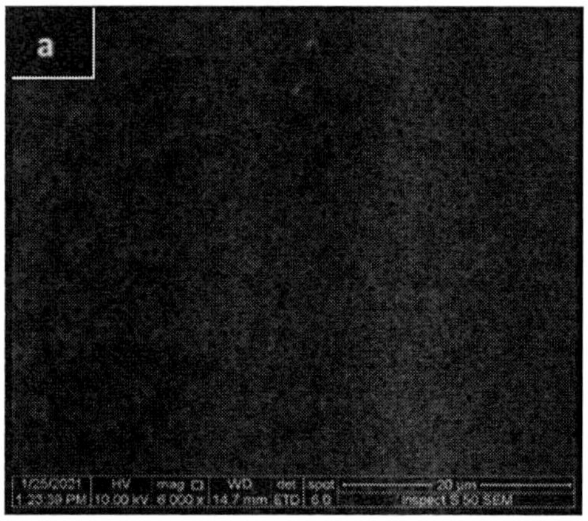

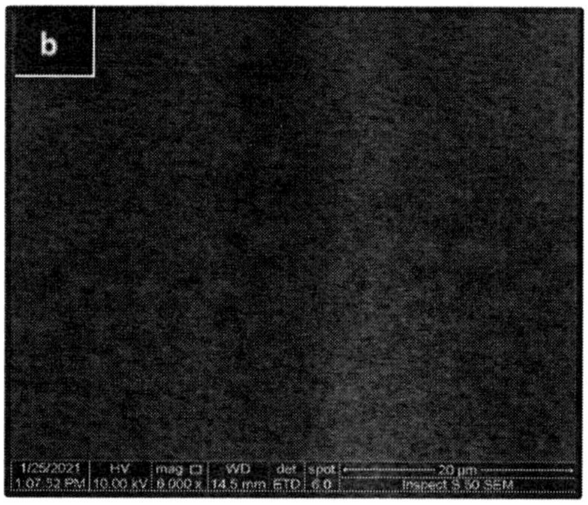

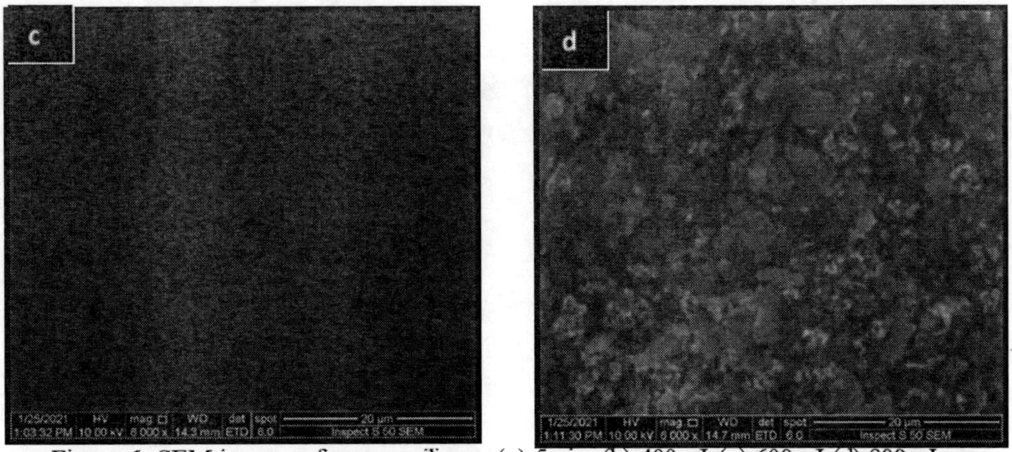

Figure 6. SEM images of porous silicon (a) 5min. (b) 400mJ (c) 600mJ (d) 800mJ

3.2. Optical Properties of Silver Nanoparticles (AgNPs)

UV spectrophotometer is an important technique used to study optical properties, it is easy, simple, and sensitive for different kinds of nanoparticles. Figure 7a shows the absorption spectra of silver nanoparticles as a function of the wavelength. The results indicate that the absorbance was increased with the laser energy (400,600 and 800 mJ) due to the Surface Plasmon Resonance (SPR) of the free electrons. It is clear the strong peak of absorption at 407-412 nm in the ultraviolet region [20]. Figure 7b exhibits the absorption coefficient of AgNPs as functions of wavelength. The absorption coefficient increases linearly as the laser energy for all the samples. This could be due to the absorbance depend on the concentration of suspension of the nanostructure. Furthermore, Figure 7c shows the $(\alpha\, h\nu)^2$ and photon energy (hv) of AgNPs with different energy(400,600 and 800mJ). The energy gap (E_g) can be determined via the Tauc formula[21]:

$$\alpha h\upsilon = C(h\upsilon - Eg)^r \qquad \text{...........(1)}$$

where C is constant, α is the absorption coefficient, hυ is photon energy, and r is the value of 1/2 of direct transition and 2 of indirect transition. The results show that the energy gap decreases with laser energy due to the increase of particle size of AgNPs from 2.68eV at 400mJ, 2.61eV at 600mJ, and 2.56eV.

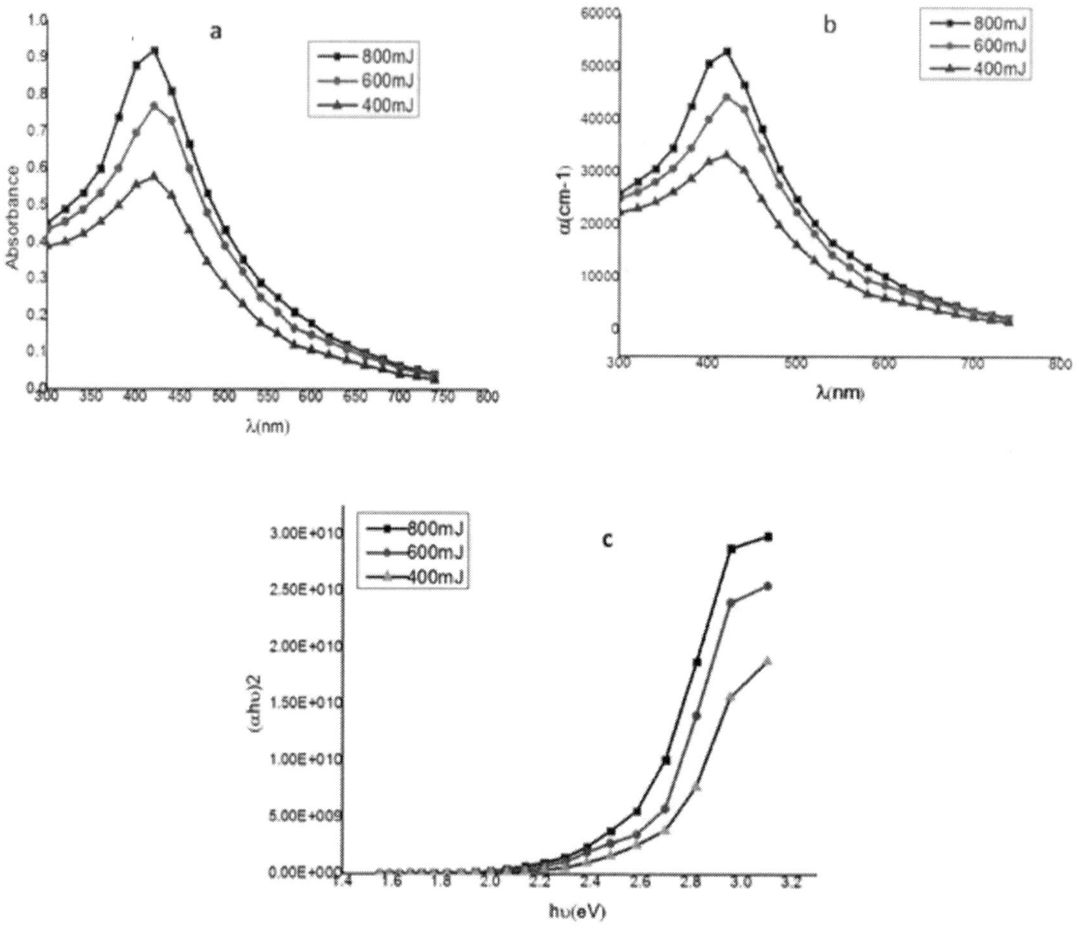

Figure 7. Optical properties of silver nanoparticles after laser ablation (a) Absorbance a against wavelength (b) absorption coefficient of AgNPs as functions of wavelength ,(c) $(\alpha h\upsilon)^2$ as a function of photon energy with laser energy 400 , 600 and 800mJ.

4. Conclusions

We have presented a detailed study on silver nanoparticles coated on porous silicon prepared by drop-casting in liquid with high power laser. The AFM images revealed that pore width increases linearly with etching time and has a pore-like network. Moreover, SEM images illustrate an aggregates formation with increasing laser energy. Our main conclusion is that silver nanostructure on porous silicon enhances the surface properties of the porous silicon and, thus, offer the possibility of enhancing the performance of porous silicon.

References

[1] Prabhu S and Poulose E 2012 Silver nanoparticles: mechanism of antimicrobial action, synthesis, medical applications and toxicity effects Int. Nano Lett. 2 1–10A

[2] Junejo Y, Baykal A, Safdar M and Balouch A 2014 A novel green synthesis and characterization of AgNPs with its ultra-rapid catalytic reduction of methyl green dye *Appl. Surf. Sci.* **290** 499–503.

[3] Zheng X, Xu W, Corredor C, Xu S, An J and Zhao B 2007 Laser-induced growth of monodisperse silver nanoparticles with tunable surface plasmon resonance properties and a

wavelength self-limiting effect *J. Phys. Chem. C* **111** 14962–14967.

[4] Pyatenko A , Shimokawa K , Yamaguchi M , Nishimura O and Suzuki M 2004 Synthesis of silver nanoparticles by laser ablation in pure water. *Appl. Phys. A* **79** 803–806.

[5] Nguyen T , Lee K. and Lee B 2010 Fabrication of Ag nanoparticles dispersed in PVA nanowire mats by microwave irradiation and electro-spinning *Mater. Sci. Eng. C* **30** 944–950.

[6]] Huang X H , Jain P K and El-Sayed I H 2007 Gold nanoparticles: Interesting optical properties and recent applications in cancer diagnostic and therapy *Nanomedicine* **2** 681–693.

[7] Yasir Y K, Ghazwan Gh A and Marwan H Y 2021 Irradiation Effects on The Sensitivity of ZnO Thin Films Synthesized on Glass Substrate by Sol-gel Method *Iraqi J. of Sci.* **62** 130-137.

[8] Krysa A B , , Roberts J S, Devenson J , Beanland R , Karomi I , Shutts S and P M Smowton 2016 InAsP/AlGaInP/GaAs QD laser operating at ~770 nm *J. of Phys. IOP Conf. Series* **740** 012008.

[9] Simakin A V, Voronov V V, Kirichenko N A and Shafeev G A 2004 Nanoparticles produced by laser ablation of solids in liquid environment *Appl. Phys. A* **79** 1127-1132.

[10] Ghazwan Gh , Abdul Khaliq A and Marwan H 2019 Study of Porous Silicon Behavior via Neutron Irradiation: Fabrication and Characterization *Results in Physics* **14** 1-7.

[11] Pyatenko A , Shimokawa K , Yamaguchi M, Nishimura O and Suzuki M 2004 Synthesis of silver nanoparticles by laser ablation in pure water *Appl. Phys. A* **79** 803-806.

[12] Pyatenko A, Yamaguchi M, Suzuki M 2007 Synthesis of spherical Silver nanoparticles with controllable sizes in aqueous solutions *J. Phys. Chem. C* **111** 7910–7917.

[13] Ghazwan Gh A , Ivan B K and Abdul Khaliq A 2020 Properties of P-Type Porous Silicon Bombarded by Neutrons Nucl. Inst. and Methods in *Phys. Research B* **468** 23-27.

[14] Das R , Nath S , Chakdar D, Gope G and Bhattacharjee R 2009 Preparation of silver nanoparticles and their characterization *J. of nanotechnology* **5** 1-6.

[15] Naderi N and Hashim M 2012 Effect of Surface Morphology on Electrical Properties of Electrochemically- Etching Porous Silicon Photodetectors *Int .J. Electrochem. Sci*, **7** 11512-11518.

[16] Ghazwan Gh A , Abdul Khaliq A and Ivan B K 2018 The Effect of the ZnO Thickness Layer on the Porous Silicon Properties Deposited by Chemical Vapor Deposition" *AIP Conf. Proceedings* **2034** 45-53.

[17] Petr S , Nikola S 2019 Methods of Gold and Silver Nanoparticles Preparation *Materials* **15** 2333-41.

[18] Williams N K 2007 Design Fabrication and Characterization of Porous Silicon Multilayer Optical Devices *University of Rovira. Virgili, Spain PhD. Thesis*.

[19] Garcia M 2012 Surface plasmons in metallic nanoparticles: fundamentals and applications *J. Phys. D Appl. Phys*. **45** 389501

[20] Taleb A , Petit C and Pileni M P 1998 Optical properties of self-assembled 2D and 3D superlattices of silver nanoparticles *J. of Phys. Chem. B* **102** 2214-2220.

[21] Link S A and El-Sayed M A 2003 Optical properties and ultrafast dynamics of metallic nanocrystals *Annual review of Phys. Chem.* **54** 331-366.

IOP Conf. Series: Materials Science and Engineering 1126 (2021) 012007 doi:10.1088/1757-899X/1126/1/012007

Comparison of the most harmful defects present in a steel pipe that can cause fatigue failure

Zakaria MIGHOUAR[1,2], Hamza KHATIB[1], Laidi ZAHIRI[1] and Khalifa MANSOURI[1]

[1]SSDIA Laboratory, Hassan II University of Casablanca, ENSET of Mohammedia, Post Box No. 159, 28830, Mohammedia, Morocco
[2]Corresponding author's e-mail address: z.mighouar@gmail.com

Abstract. Steel pipes used for conveying pressurized fluids are often defective due to manufacturing error, improper handling or construction materials. These defects may be in the form of cracks which propagate until the mechanical structure is fractured, or in the form of dent which produce a local concentration of stress and thus weaken the structure. In this paper, we will present a numerical model for estimating the accumulation of damage that takes into account the effect of the loading history. This model will be used to evaluate the harmfulness of a dent on the one hand, of a semi-elliptical crack on the other hand. These defects wil be considered to be present in a metal tube subjected to internal pressures caused by the water hammer phenomenon. A parametric study will make it possible to conclude the most harmful defects among the defects studied.

1. Introduction

Over the last 50 years, demand for energy has steadily increased and oil and gas transmission pipelines have become important networks for transporting large amounts of pressurized fluid over long distances [1]. In order to improve the profitability of this mode of transport, manufacturers are increasing the operating pressure and the pipe diameter [2]. From 1910 to 2000, the diameter of the largest pipelines increased fourfold and the transport pressure increased sixty times.

As with any other structure, failure can occur in in-service oil transport steel pipelines. However, due to good design guidelines, materials and operating practices, transmission pipelines may have a comprehensive safety device.

The safe operation of the pipeline and its high reliability depend on various factors, such as: mechanical damage or external interference, fatigue cracking, material defects, welding cracks, inadequate welding, internal or external corrosion, and the aging of the physical condition of the pipeline metal and welded joints. Damage from human error or vandalism is also not unusual [3, 4].

In the oil industry, damage resulting from contact with foreign objects is often referred to as mechanical damage. Defects in the pipeline may occur in a number of forms. A combination of two or more defects is also common [5]. To ensure proper monitoring of the safe operation of these structures, tools are used to assess the harmfulness of defects [6]. These tools depend on the types of defects and failure modes. Limit analysis is generally used for the assessment of non-severe defects such as corrosion defects or dents [3]. In the case of serious defects such as cracks in welds, sharp notches or a combination of a dent and a notch, a mechanical fracture approach is usually used.

Previous work has been carried out by the authors on the study of the main defects that are likely to exist in the pipes, in particular those due to impact with a foreign body (dent), welding (crack in the welded area) and corrosion [7-9]. The results of these analyzes made it possible to identify the most dangerous defects which could lead to the rupture of metal pipes under pressure. Several numerical models have also been developed which allow the determination of the harmfulness of the defect most commonly encountered in metal pipes under internal pressure.

These works have shown that the most harmful dent defect is the one which has a rectangular shape oriented longitudinally with respect to the pipe with unconstrained configuration. Unconstrained dent is a dent caused by a foreign body that has been removed after impact. For the harmful crack defect, it is a longitudinal semi-elliptical crack present in the internal wall of the pipe. The parameters of the dent defect are presented in figure 1 and those of the crack are presented in figure 2.

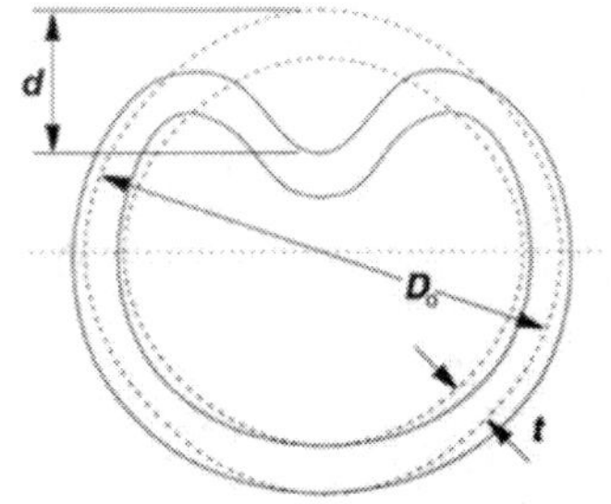

Figure 1. Dent parameters.

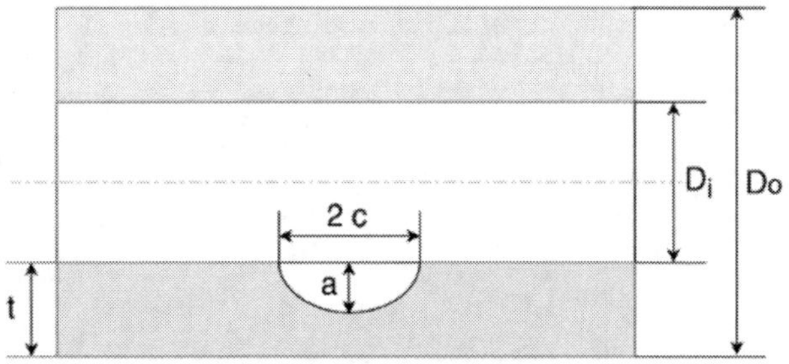

Figure 2. Crack parameters.

In the previous studies, cracking and indentation defects are investigated separately. In this paper, a model estimation of the damage caused by the defect in the pipe is applied to both of these defects in order to conclude on the most harmful defect. This numerical damage assessment model, which takes into account the loading history, is presented.

The pipes considered in the parametric study are subjected to internal pressures caused by the water hammer phenomenon (figure 3). High levels of dynamic loading, such as water hammer pressure waves, make the defect even more damaging [10]. The range of defect dimensions used in this numerical model is that encountered in practice.

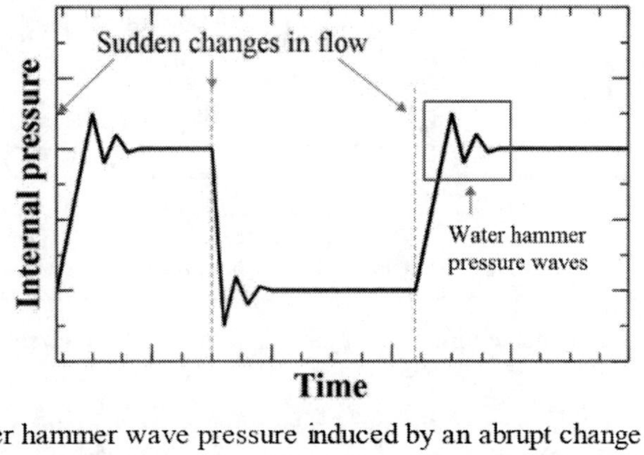

Figure 3. Water hammer wave pressure induced by an abrupt change in fluid velocity.

2. Material and methods

Fatigue loading of pipe structures varies by nature. During fatigue loading, the material in the pipeline undergoes alterations resulting in damage [11-12]. It is therefore essential for models to take explicit account of the variability of fatigue loading [13].

To this end, authors have proposed a nonlinear model for fatigue damage accumulation [9]. It's based on the work of Thun et al. [14]. The effect of the load interaction is considered in this numerical model. The algorithm for this numerical model is shown in Figure 4. Cumulative damage under variable amplitude loading levels may then be estimated. The leakage or rupture of the structure is to be achieved when the cumulative damage is equal to the unit.

It should be noted that this algorithm is to be adapted to the type of defect being studied (dent or crack). The fatigue life estimation model for the dent case uses as additional parameters the dent depth (d/D_0) and the endurance limit of the material (σ_D). If the studied defect is a crack, then the model for estimating the fatigue life takes as an additional parameter the dimension of the crack (a/c), the constants of the law of Paris (C and m) and the critical stress intensity factor (K_{IC}). Details on these two models are provided on [7] and [8].

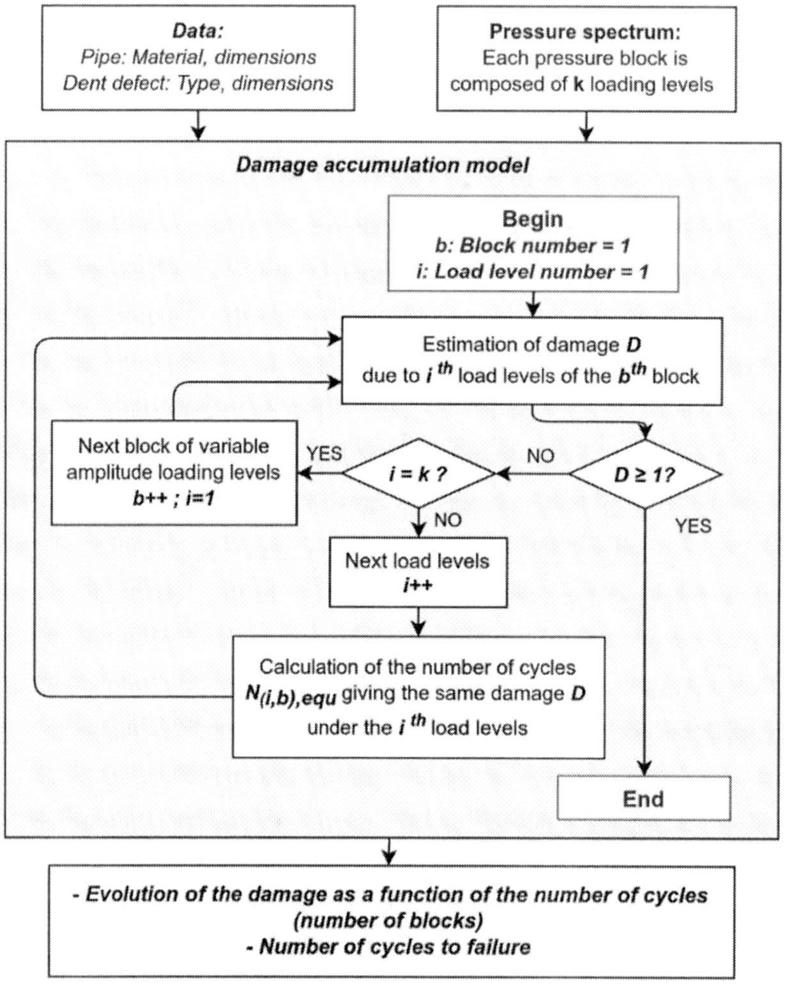

Figure 4. Damage accumulation model that takes into account load history.

This model takes the parameters of the pipe, the defect as well as the internal load as inputs. The material used is steel API X52 and its mechanical properties are shown in Table 1. The dimensions of the pipe can be seen in Figure 5.

Table 1. Mechanical properties of API X52.

Parameter	Young modulus **E**	Poisson's ratio **υ**	Yield strength σ_E	Ultimate tensile strength **UTS**
Value	**200** GPa	**0.3**	**410** MPa	**498** MPa

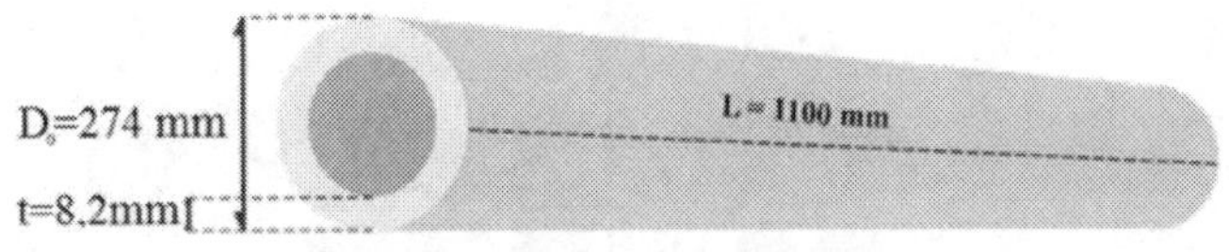

Figure 5. Pipe dimensions.

The pipe is subjected to an internal pressure of variable amplitude as shown in Figure 6. Pressure peaks are caused by a water hammer phenomenon following a sudden change in the valve state (opening/closing) controlling the transport of water in the pipe. This curve is produced using a characteristic method. Details of the calculation are available in [8].

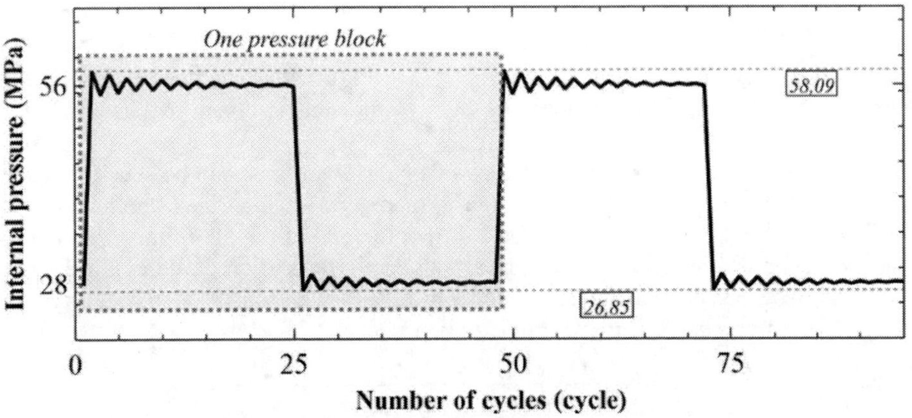

Figure 6. Internal loading of the pipe.

The range of defect dimensions used in this numerical model is that encountered in practice. Namely, dent defect are considered dangerous from a depth of d/D_0=2.5%. From a value of d/D_0=20%, the dent generally leads to an immediate rupture after impact and therefore the determination of the number of rupture cycles is no longer meaningful. The most common semi-elliptical crack defects in the field vary between a/c=0.4 and a/c=0.9.

A parametric study involving 14 cases has been carried out. The d/D_0 dimensions considered in this study are: 2.5, 5, 7.5, 10, 12.5, 15, 17.5 and 20% for the dent defect. The a/c dimensions considered are 0.4, 0.5, 0.6, 0.7, 0.8, and 0.9 for the crack defects. The initial crack depth is set to 2.5 mm to allow a comparison of these two types of defect in a similar order of magnitude.

3. Results and discussion

The damage accumulation model is validated in a previous study led by authors [9]. The results provided by this model are presented in figure 7 and figure 8.

For the calculation of the deviations, let us take as a reference the value of the number of failure cycles obtained for the dimension of the defect immediately below (example: the value of the result obtained for the penetration of the depth d/D of 20 per cent is compared to that of 17.5 per cent).

Figure 7 shows a deviation of -40.2, -29.1, -22.9, -18.6, -13, -12.3 and -11% for d/D_0 dent depth of 5, 7.5, 10, 12.5, 15, 17.5 and 20%, respectively. This shows that the deeper the defect, the more harmful it is. One can also notice that the d/D_0 parameter varies exponentially with the number of cycles.

From figure 8, one notice a deviation of 167.2, 73.2, 54.4, 37 and 34.5% for cracks of dimension a / c of 0.5, 0.6, 0.7, 0.8 and 0.9, respectively. This shows that the higher the ratio, the less harmful the defect. It is also noticeable that the a/c parameter varies exponentially with the number of cycles.

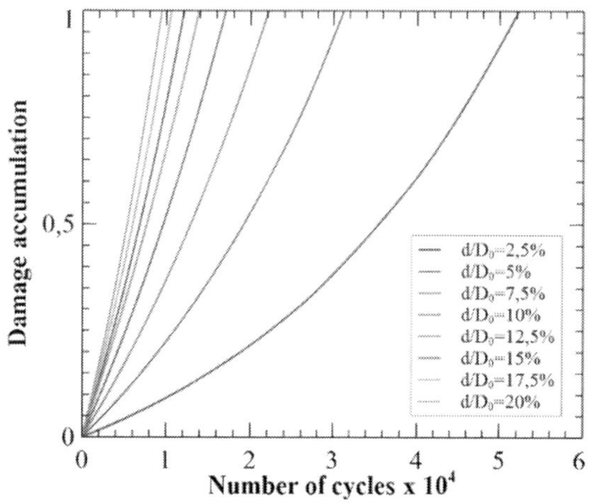

Figure 7. Damage accumulation for pipe presenting an unconstrained longitudinal dent.

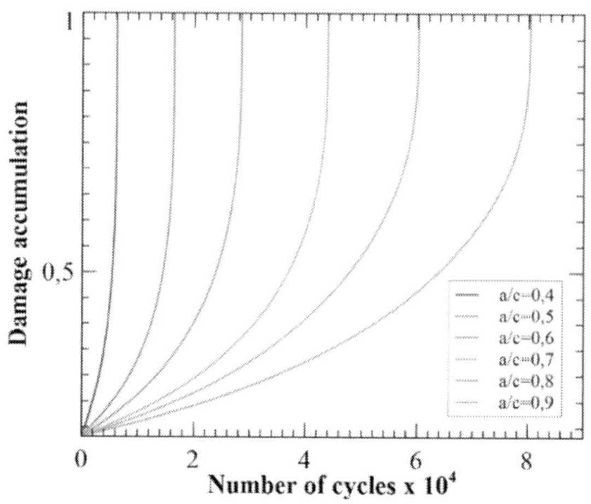

Figure 8. Damage accumulation for pipe presenting an internal longitudinal crack.

By simple reading and comparing the curves of figures 7 and 8, it can be seen that the shape of the evolution curve of the accumulation of damage is more curved for crack defects. A low number of cycles can cause a large variation in damage towards the end of the life of the pipe. This phenomenon is much more accentuated in the case of pipes with semi-elliptical crack defects located on the inner surface of the pipe in the longitudinal direction than in the case of pipes with a rectangular longitudinal dent with an unconstrained configuration.

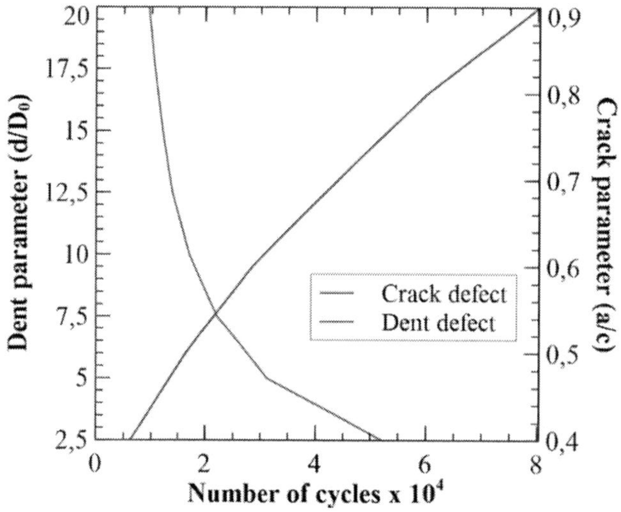

Figure 9. Number of cycles according to the dimensional parameters of the defects studied.

To better compare the two types of defects, one will retain the values of the number of cycles to fatigue and put them on the same scale. The result of this process is shown in Figure 9. This figure can make it possible to approach the equivalence in terms of harmfulness due to two different types of defects. For example, for the defects studied and from a reading of this figure, one can say that a dent defect of d/D_0=7.5% is equivalent in terms of harmfulness to a crack of a/c=0.55. It can also be concluded from this figure that, among the cases studied, the crack defect is more harmful than the dent defect.

4. Conclusion and perspectives

A nonlinear numerical model for estimating damage accumulation, taking into account the effect of the loading history, was presented. This model has been previously developed and validated by the authors and is used in this paper to study the effect of the presence of a defect in a steel pipe. Two defects are studied: an unconstrained longitudinal dent and a semi-elliptical internal crack. Each of these two defects represents the most harmful defect of its type. A parametric study concluded that:

- The increase in the number of pressure cycles causes more fatigue in a structure with a crack defect than with the same structure with a dent defect;
- In the cases studied, it has been found that the crack defect is more harmful than the dent defect.

The developed numerical model makes it possible to find equivalence between two different types of defects in terms of harmfulness. In future work, it will be used to classify the most important defects that may be present in a steel pipe subjected to water hammer pressure waves.

References

[1] Zhang C, Sun X, Li Y, Zhang X, Yang X and Li F 2018 Hydraulic characteristics of transporting a piped carriage in a horizontal pipe based on the bidirectional fluid-structure interaction *Mathematical Problems in Engineering* pp 1–27

[2] Cristoffanini C, Karkare M and Aceituno M 2014 Transient simulation of long distance tailings and concentrate pipelines for operation training *Proc. SME Annual Meeting/Exhibit (Salt Lake City, UT)* pp 210–14

[3] MMI Engineering 2008 Oil and gas pipelines *United States Geological Survey Open File (Report* 2008-1150, California Geological Survey Preliminary Report 25 version 1.0).

[4] Irfan O M, and Omar H M 2017 Experimental study and prediction of erosion-corrosion of aa6066 aluminum using artificial neural network *International Journal of Mechanical & Mechatronics Engineering* **17** pp 17–31

[5] Vilkys T, Rudzinskas V, Prentkovskis O, Tretjakovas J, Visniakov N and Maruschak P 2018 Evaluation of failure pressure for gas pipelines with combined defects *Metals* **8** 346

[6] Zhuravska I, Lernatovych D and Burenko O 2019 Detection the places of the heat energy leak on the underground thermal pipelines using the computer system *Advances in Science, Technology and Engineering Systems Journal* **4** pp 01–09

[7] Zahiri L, Mighouar Z, Khatib H, Mansouri K and Salhi B 2017 Fatigue life analysis of dented pipes subjected to internal pressure *Int. Review of Mechanical Engineering* **11** pp 587–96

[8] Mighouar Z, Zahiri L, Khatib H, Mansouri K and El Majid Z 2018 Effect of water hammer on pipes containing a crack defect *International Journal of Mechanical & Mechatronics Engineering* **18** pp 25–31

[9] Mighouar Z, Zahiri L, Khatib H and Mansouri K 2020 Damage accumulation model for cracked pipes subjected to water hammer *Advances in Science, Technology and Engineering Systems Journal* **5** pp 523–30

[10] Twyman J 2016 Wave speed calculation for water hammer analysis *Obras y Proyectos* **20** pp 86–92

[11] Benin A, Nazarova S and Uzdin A 2019 Designing scenarios of damage accumulation *Proc. International Scientific Conference Energy Management of Municipal Facilities and Sustainable Energy Technologies*, pp 600–10

[12] Belegundu A, Nayak S, Loverich J and Grissom M 2016 Vibration-based damage accumulation modeling *Proc. 28th Conf. on Mechanical Vibration and Noise (North Carolina, USA)*

[13] Gao H, Huang H Z, Zhu S P, Li Y F and Yuan R 2014 A modified nonlinear damage accumulation model for fatigue life prediction considering load interaction effects *The Scientific World Journal*

[14] Thun H, Ohlsson U and Elfgren L 2011 A deformation criterion for fatigue of concrete in tension *Structural Concrete* **12** pp 187–97

Recycled PVC with chicken feathers as bio-load

Abdelghani Lakhdar*, Aziz Moumen and Khalifa Mansouri
SSDIA Laboratory, Hassan II University of Casablanca, ENSET of Mohammedia, 28830, Morocco.
lakhdarabdelghani11@gmail.com, Moumenaziz1@gmail.com, khmansouri@hotmail.com

Abstract. The poultry industry is responsible for a significant amount of waste. This bio-load is generally transformed into food with low nutritional value, or else incinerated. Not only are these options unsustainable, but it is also a missed opportunity to contribute to the circular economy. In this article, we try to contribute to the development of methods and strategies necessary to transform poultry feathers from waste for disposal into valuable raw materials. The chicken feathers are collected and washed several times, after drying in an oven, and finally crushed to obtain a flour which will be a bio-filler for the PVC used in this flexible article. Different percentages are used to find the one that most improves the mechanical characteristics of recycled PVC. The tensile test is carried out on test specimens made of recycled PVC without additives and recycled PVC reinforced with 5, 10 and 15% chicken feathers. It is deduced from the test results that the value of 10% was the most adequate to improve the mechanical characteristics of recycled PVC.

1. Introduction

Millions of tons of waste are produced around the world by poultry processing factories, this large amount of feathers produced with their accumulation creates a disposal problem leading to environmental pollution. In most cases, its waste is incinerated. However, this method has significant ecological drawbacks. [1-7]

In addition, a common use of its feathers is their transformation into feather meal for animal feed, or mixed with cardboard to obtain a new material. [2]

In this article we have used chicken feathers (figures 1 and 2) as the bio-load for flexible PVC, which are composed of 98% keratin. and we added it with three different percentages to see both the effect of chicken feathers on PVC, as well as the percentage to improve the recyclability of this material.

Figure 1. Chicken feathers.

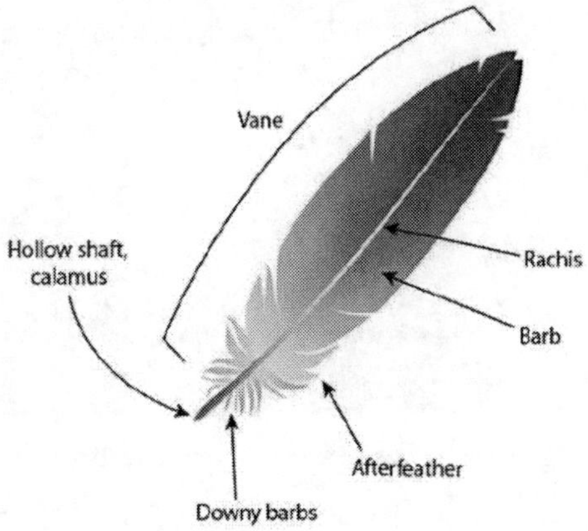

Figure 2. The different parts of the feather. [3]

The study is carried out experimentally on flexible PVC samples already aged according to an aging model validated in a previous study [8-9].

Recycling PVC means trying to improve its mechanical characteristics by adding bio-fillers [10-17] Experimentally, the aged PVC was crushed and sieved, then mixed with three different percentages of 5, 10 and 15% chicken feather powder, already prepared by collecting the feathers, washing several times, drying in the oven, by grinding and sifting.

The mixture was reground together to obtain a homogeneous powder, so that plates 2mm thick could be extruded from which tensile samples were taken. the experiments are then carried out on a traction machine on the PVC without additive and on the samples of PVC at three percentages to see the best of them, which makes it possible to improve the mechanical characteristics of this material [2]. The percentage considered most suitable is 10%, it has greatly improved the flexibility and reduced the stiffness compared to the other two percentages.

2. Experimental Study

2.1. Sample preparation

The feathers are collected on an altar of chickens. this waste is found everywhere. They are washed several times as shown in figure 3, and dried in an oven for 5 minutes at 150 ° C then they are crushed and sieved to finally obtain a powder of chicken feathers Figure 4.

Figure 3. Cleaning process. [2]

Figure 4. Chicken feather waste powder.

The prepared powder is added to the already aged flexible PVC with three percentages of 5.10, and 15%, the tensile test pieces are then prepared by the mixtures of flexible PVC with each percentage of the bio-filler as shown in figure 5 below.

Figure 5. Flexible PVC with different percentages of chicken feather.

IOP Conf. Series: Materials Science and Engineering 1126 (2021) 012008 doi:10.1088/1757-899X/1126/1/012008

2.2 Tensile test

The test is carried out on the traction machine of figure 6, on the different samples. These tests follow a well-defined operating mode among the parameters to be adjusted, it is the speed which is for a flexible PVC is set at 25mm / min and the applied preload was 5.6N

Figure 6. Tensile test machine.

3. Results and discussion

The results of the tensile test on the flexible PVC recycled without additives in figure 7, show the effect of this recycling operation on the mechanical properties of this material, in particular the strong degradation of the elongation at break and the rigidity which has strongly increases.

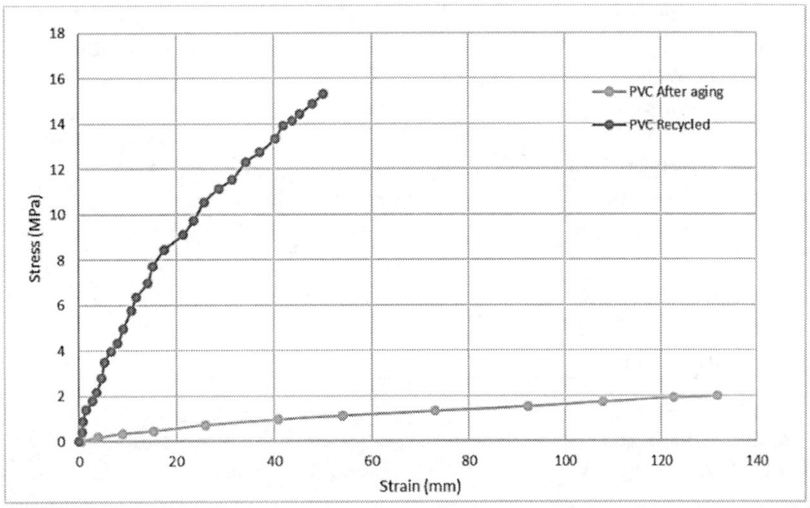

Figure 7. Tensile test curves of the flexible PVC aged and recycled without additives.

The tests applied to the other bio-filled PVC test pieces show different results which vary between the improvement and the degradation of the mechanical properties of the flexible PVC according to the added percentage. What we are looking for is that the recycled material retains or increases its flexibility with a slight variation in stiffness.

In figure 8, and table 1 we see the different tensile curves of the different specimens as well as the different values of these tests

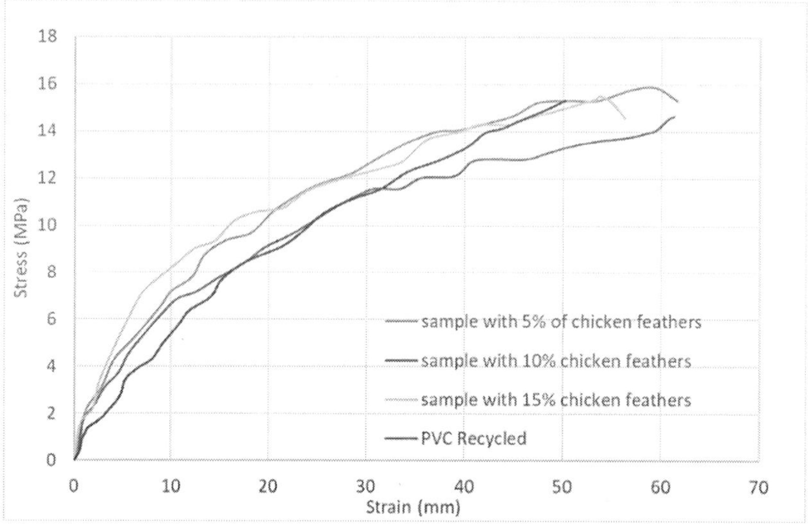

Figure 8. Tensile test curves of the flexible PVC with & without bio-load.

Table 1. Stress and strain at break of flexible PVC with and without bio-loads.

	Stress at break (MPa)	Strain at break (mm)
Flexible PVC recycled without bio-loads	15,32	50,23
Flexible PVC bio-loaded with 5% of chicken feathers	15,31	61,56
Flexible PVC bio-loaded with 10% of chicken feathers	14,65	61,25
Flexible PVC bio-loaded with 15% of chicken feathers	14,56	56,25

From the results found, it is seen that the addition of 5% chicken feather powder improves the elongation at break with a rigidity which remains constant compared to that of flexible PVC recycled without additives.

The addition of 10% of the bio-load improved the elongation at break and also improved the stiffness by reducing the stress at break.

For the latter, which is reinforced by 15% of the bio-load, it also increased the elongation at break but with a value lower than that found by the addition of 10%, and it reduced the stress at the rupture which makes the material less rigid

Finally, we can say that from these results the addition of 10% chicken feather powder is the most appropriate value to improve the mechanical characteristics of recycled flexible PVC.

4. Conclusion

In this study, we tried to validate the effect of adding a bio-filler in the form of chicken feathers on the mechanical properties of aged flexible PVC, and also to find the most suitable percentage that will be added in as a bio-filler.

All this to validate the effect of this bio-filler on the recyclability of this material, to improve recycling.

The percentage of 10% is the most suitable, which has improved the mechanical characteristics of flexible PVC the most by increasing the elongation at break and by reducing the rigidity by reducing the stress at break.

The results of adding this percentage will then be compared with the other results obtained by adding the other bio-fillers already analyzed (cow horns and coconut) on the PVC, then will have a comparative study of the mechanical characteristics of the new compound material. 90% already aged PVC and 10% bio-fillers that will be recycled for the first time.

A future recycling study will be carried out several times with the addition of the most suitable bio-load for upgrading PVC.

References

[1] W.F. Schmidt, ''Innovative Feather Utilization Strategies,''in Poultry Waste Management Conference, Springdale, Arkansas (1998).

[2] Ansarullah, Ramli Rahim, Asniawaty Kusno, Baharuddin and Nurul Jamala "Utilization of waste of chicken feathers and waste of cardboard as the material of acoustic panel maker" Earth and Environmental Science 126 (2018) 012036, 1-8, doi:10.1088/1755-1315/126/1/012036

[3] Saha, Subhasish, " Exploration of keratinolytic actinobacteria for the bioconversion of poultry feather waste into poultry feed supplement", doctoral thesis in microbiology, Department of Microbiology, Bharathidasan University, India, 2009, 142p.

[4] Mingjiang Zhan, Richard P. Wool, " Mechanical Properties of Chicken Feather Fibers", Department of Chemical Engineering and Center for Composite Materials, University of Delaware, Newark, Delaware 19716, p8, DOI 10.1002/pc.21112

[5] Aranberri, I.; Montes, S.; Azcune, I.; Rekondo, A.; Grande, H.-J. Fully Biodegradable Biocomposites with High Chicken Feather Content. *Polymers* 2017, *9*, 593. https://doi.org/10.3390/polym9110593

[6] Reddy, N., Yang, Y. Structure and Properties of Chicken Feather Barbs as Natural Protein Fibers. J Polym Environ 15, 81–87 (2007). https://doi.org/10.1007/s10924-007-0054-7

[7] Menandro N. Acda, "Waste Chicken Feather as Reinforcement in Cement-Bonded Composites", Philippine Journal of Science, 139 (2): 161-166, December 2010

[8] A. Lakhdar, M. Jammoukh, L. Zahiri, K. Mansouri, A. Moumen, B. Salhi, "Numerical and Experimental Study of the Behavior of PVC Material Subjected to Aging," in 2020 1st International Conference on Innovative Research in Applied Science, Engineering and Technology (IRASET), IEEE: 1–6, 2020.

[9] Abdelghani Lakhdar, Aziz Moumen, Laidi Zahiri, Mustapha Jammoukh, Khalifa Mansouri, "Experimental and Numerical Study of the Mechanical Behavior of Bio-Loaded PVC Subjected to Aging", Adv. Sci. Technol. Eng. Syst. J. **5**(5), 607-612 (2020); DOI: 10.25046/aj050574

[10] A. Moumen, M. Jammoukh, L. Zahiri, K. Mansouri, "Numerical modeling of the thermo mechanical behavior of a polymer reinforced by horn fibers," International Journal of Advanced Trends in Computer Science and Engineering, 9(4), 6541–6548, 2020, doi:10.30534/ijatcse/2020/342942020.

[11] Md Sah, Muhammad Hanafi. (2017). Mechanical Properties of Coconut Shell Powder Reinforced PVC Composites in Automotive Applications. Journal of Mechanical

Engineering. 14. 49-61

[12] A. Moumen, M. Jammoukh, L. Zahiri, K. Mansouri, "Numerical modeling of the thermo mechanical behavior of a polymer reinforced by horn fibers," International Journal of Advanced Trends in Computer Science and Engineering,9(4), 6541–6548, 2020, doi:10.30534/ijatcse/2020/342942020.

[13] M. Jammoukh, K. Mansouri, B. Salhi, Industrial and ecological effect of a bio-load on polymers, 10p (2018).

[14] C.D.E.S. Activités, L. De Vinylplus, "PROGRESS REPORT 2020 REPORTING ON 2019 ACTIVITIES" 2017", 40:7-12, 2020.

[15] I. Janajreh M. Alshrah Samih Zamzam, Mechanical Recycling of PVC Plastic Waste Streams from Cable Industry: A Case Study, Sustainable Cities and Society, 15 p (2015).

[16] PVC W.V. Titow, PVC technology, Springer, 31 december 1984.

[17] PVC W. V. Titow, PVC technology. Springer. pp.6 ,2011

IOP Conf. Series: Materials Science and Engineering 1126 (2021) 012009 doi:10.1088/1757-899X/1126/1/012009

Prediction of the mechanical properties of Polypropylene reinforced with Snail Shell Powder with a Deep Neural Network Model and the Finite Element Method

Aziz Moumen*, Zineb Laabid, Abdelghani Lakhdar and Khalifa Mansouri

SSDIA Laboratory, Hassan II University of Casablanca, ENSET of Mohammedia, 28830, Morocco.
Moumenaziz1@gmail.com,Laabid.zineb@gmail.com,lakhdarabdelghanil1@gmail.com and khmansouri@hotmail.com.

Abstract. Neural networks have led to the evolution of the processing methodology of computational sciences. The problems like bio composites modeling and prediction are difficult to model with classical mathematical and statistical tools because of the data inherent noise. NN's processing capability in the forecasting, recognition, modeling, system analysis and control can give fast characterization, modeling and prediction of bio composites properties, provided as long as datasets are available. Using Matlab®, a neural network model was evaluated to characterize the optimal properties of the ANS reinforced the Polypropylene. The feed forward multilayer model provided best results in comparison with the finite element method and the experimental tensile tests. The trained neural network is able to provide a best prediction of such bio composite based on natural particles having more advantages to the environment, economy and the sustainable development.

Key words: Bio composites,Natural Particles, Neural network, Feed forward multilayer, Finite Element Method.

1. Introduction

Neural network approach can be trained to give intelligent decisions by using its knowledge base[1]–[4], Researchers confirmed that particles reinforced polymers properties are non-homogeneous , inconsistent and varying [5]–[10]. Based on learning and adaptation, the neural structure and the problem definition can serve in prototype development of novel bio-composites and generate several decisions reserved to human expert.

NN have shown good results towards modeling composites property. Research results illustrate that with an appropriate NN aid and best base research knowledge we will be able to the fast characterization, modeling and prediction of novel bio composites materials. Neural networks will help the appropriate characterization of eco composites by using real data to develop an efficient system in the correct domain [11].

In this study, we will predict using the neural network the Young modulus of snail shell powder reinforced Polypropylene in comparison with the finite element method which is known for its efficiency in predicting complex mechanical behaviors [12]–[15].

2. Material and methods

2.1 Sample preparation

The collected snails are from rural areas of Casablanca (Morocco). We carefully removed the shells. Firstly, we washed shells with the tap water in order to remove, sand, mud and the other impurities. Secondly, we cleaned the shells and dried them at 80C° in an oven during 24 h. Finally, the bio load was sieved to take only particles under 10 µm. The figures 1,2 and 3 present respectively the structure of Polypropylene, Snails collected and Snail shell powder.

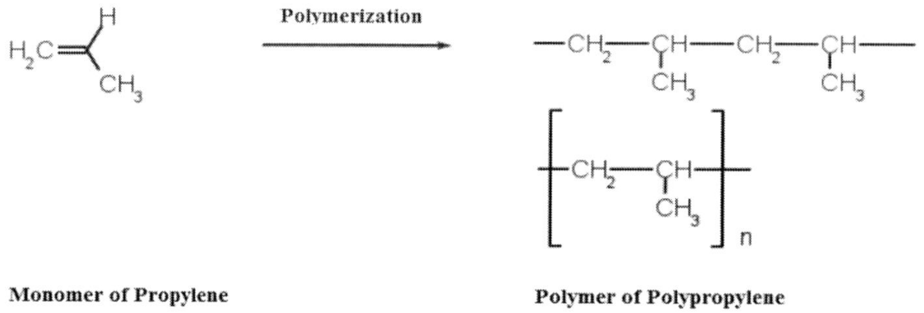

Figure 1. Polypropylene structure

Figure 2. Snail shell collected

Figure 3. Snail shell powder

2.2 Composites processing

To study the Snail shell effect on the properties of the polypropylene matrix, compounding was performed in the Leistritz GmbH extruder. Mixing was done using 125 rpm in the speed of screw. Polypropylene was introduced at the principal hopper and the snail shell along a side-feeder at 40 rpm of the screw speed. The water bath is used to cool the compounds. Then, the samples were made on a molding injection machine (e-Victory Engel). The injection barrel temperature is fixed at 200 C°, and the nozzle is set at 180 C° with 45°C at the mold. The figure 4 shows the polypropylene in pure state and reinforced with snail shell powder.

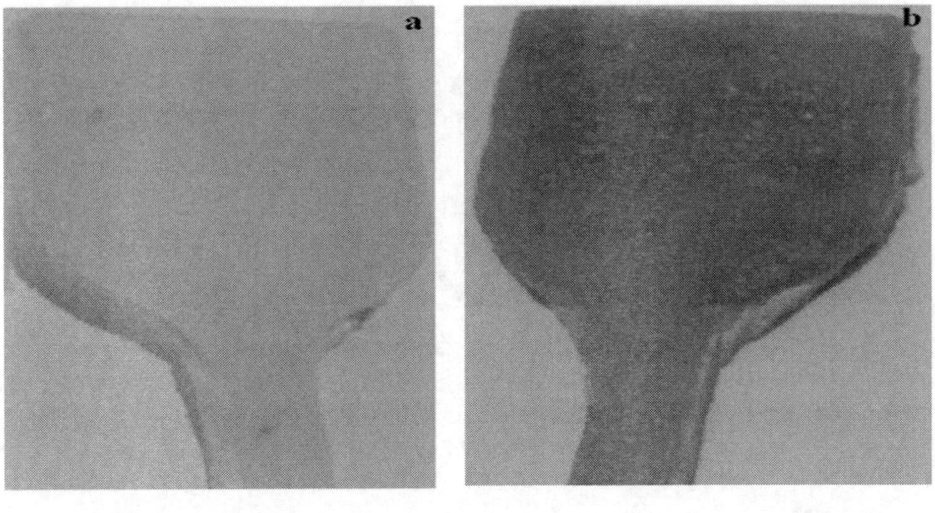

Figure 4. Polypropylene samples: (a) Pure PP, (b) Snail shell reinforced PP.

2.3 Tensile test

The test is performed in the tensile test machine presented in figure 5.The test follows a well-defined operating mode among the parameters to be adjusted with a speed set at 25mm / min and an applied preload of 5.6N.

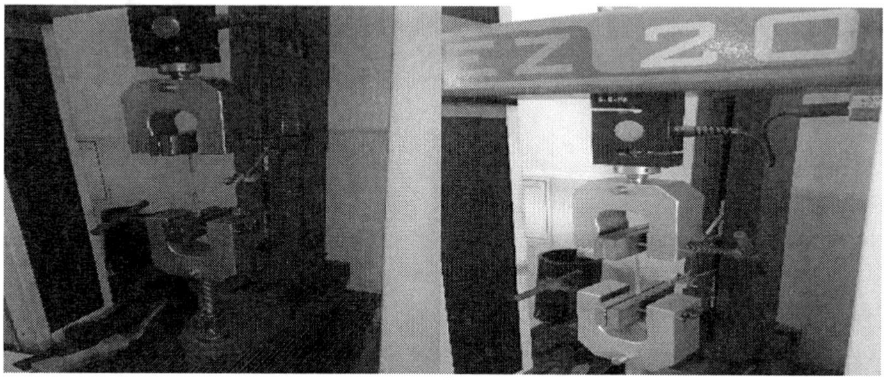

Figure 5. Tensile test machine

2.4. Finite element method

The finite element method is one of the tools of applied mathematics. In numerical analysis, the finite element method is used to numerically solve partial differential equations. It can for example represent analytically the dynamic behavior of certain physical systems (mechanical, thermodynamics, acoustics, etc.). In mathematics, this involves replacing a complicated problem for which a priori we do not know a solution, by a more simple that we know how to solve.

The geometry and the mesh generated for the snail shell reinforced polypropylene by tetra quadratic elements are presented in figures 6 and 7.

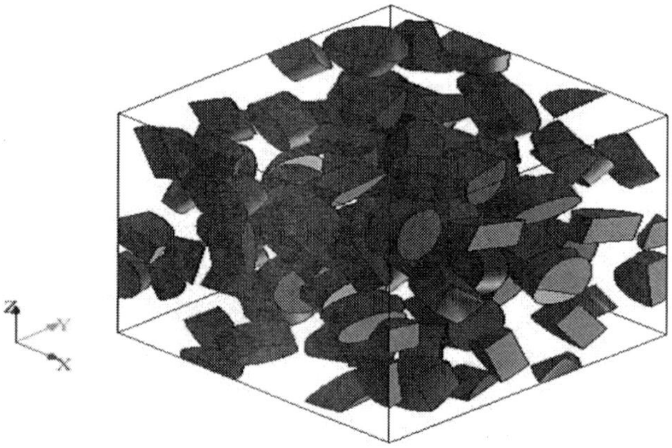

Figure 6. Geometry of Snail Shell reinforced Polypropylene

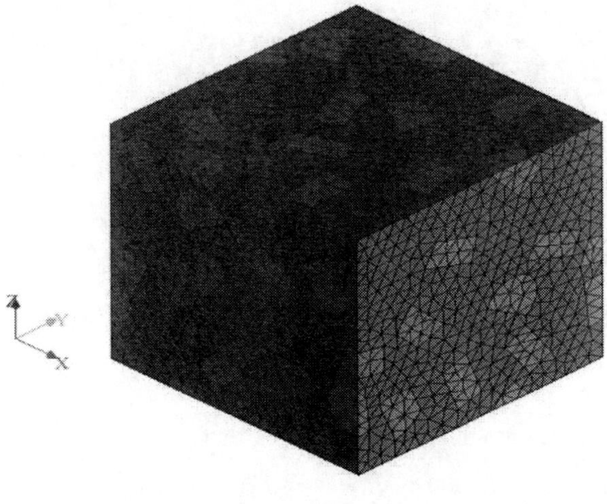

Figure 7. Mesh of Snail Shell reinforced Polypropylene

2.5 Neural network model

The principal concept of the neural network is taken from the neuroscience. The architecture of the multilayer is shown in figure 8. The input is the first network layer and the output is the last one. The hidden layers present the layers between input and output.

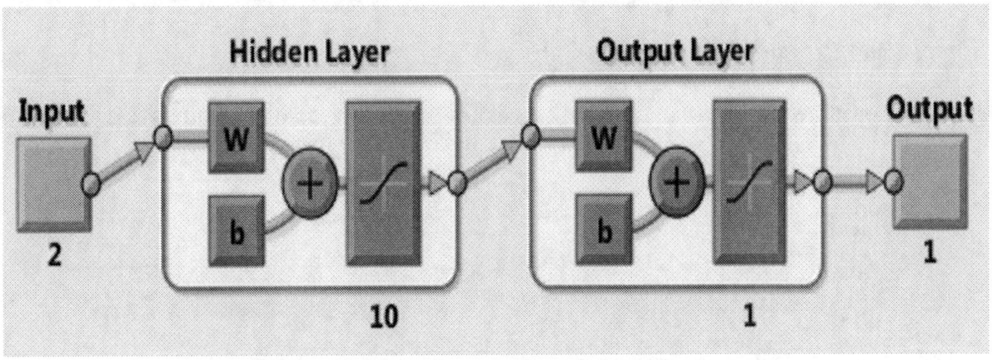

Figure 8. Neural network model architecture

3. Results and discussion

3.1. Experimental results

The figure 9 illustrates the tensile curves of pure PP and reinforced by Snail Shell Powder.

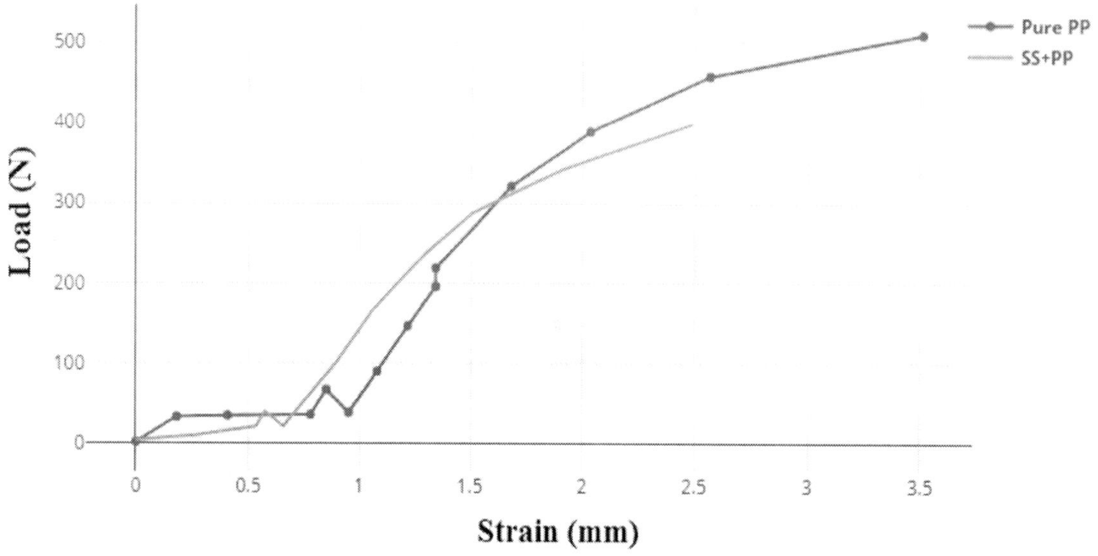

Figure 9. Tensile curves of pure PP and reinforced by Snail Shell Powder

The results of the tensile test on pure PP and reinforced with Snail Shell Powder illustrated in the figure 9 clearly show that the reinforced PP is more rigid with a Young modulus of 1981 MPa at 30% of particles compared to the pure one that presents 1034 MPa.

3.2. Finite element results

Figure 10 shows the finite element tensile curves for the two types of Polypropylene

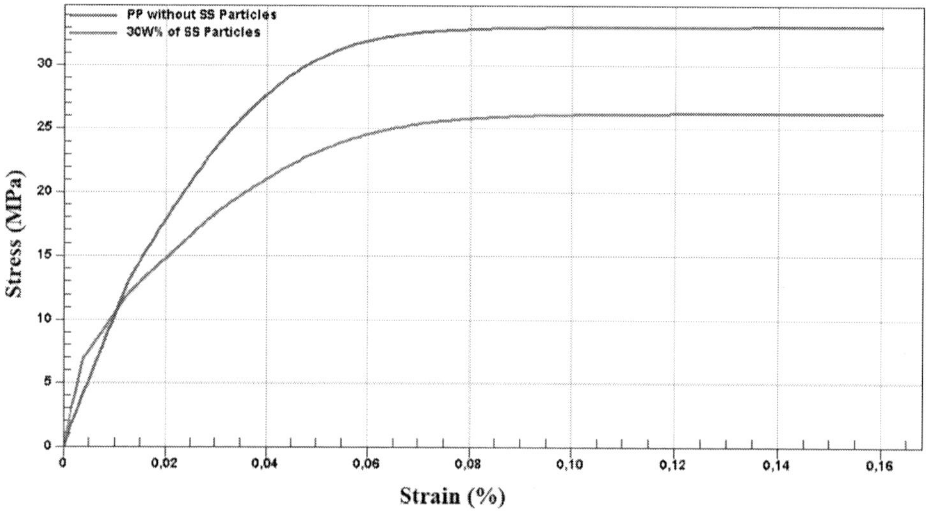

Figure 10. Finite element curves of SS reinforced PP

It is observed in the figure 10 that the results of the FEM confirm the experimental tests. The Young modulus increase to 1945 MPa in comparison with the pure state of PP (1034 MPa) leading to an increase of the rigidity of the reinforced PP at 30% of particles.

IOP Conf. Series: Materials Science and Engineering 1126 (2021) 012009 doi:10.1088/1757-899X/1126/1/012009

3.3. Neural network model

The evaluation of the ANN performance can be done by two statistical indicators provided by Matlab.They are the correlation coefficient (R) and mean square error (MSE) .Figures 11, 12 and 13 present respectively the MSE, training state and R coefficient at 6 epochs.

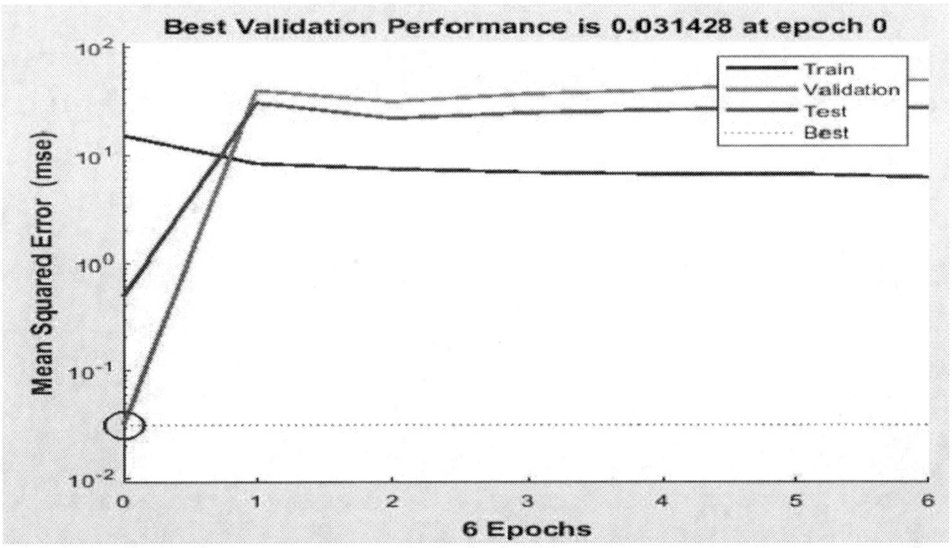

Figure 11. Neural network training performance

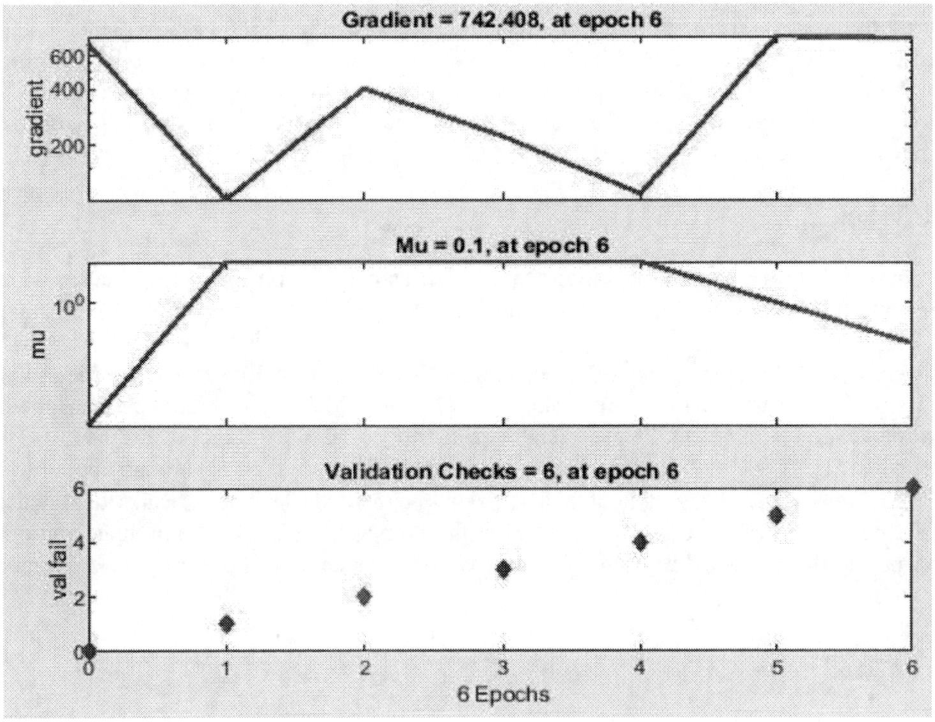

Figure 12. Neural network training state

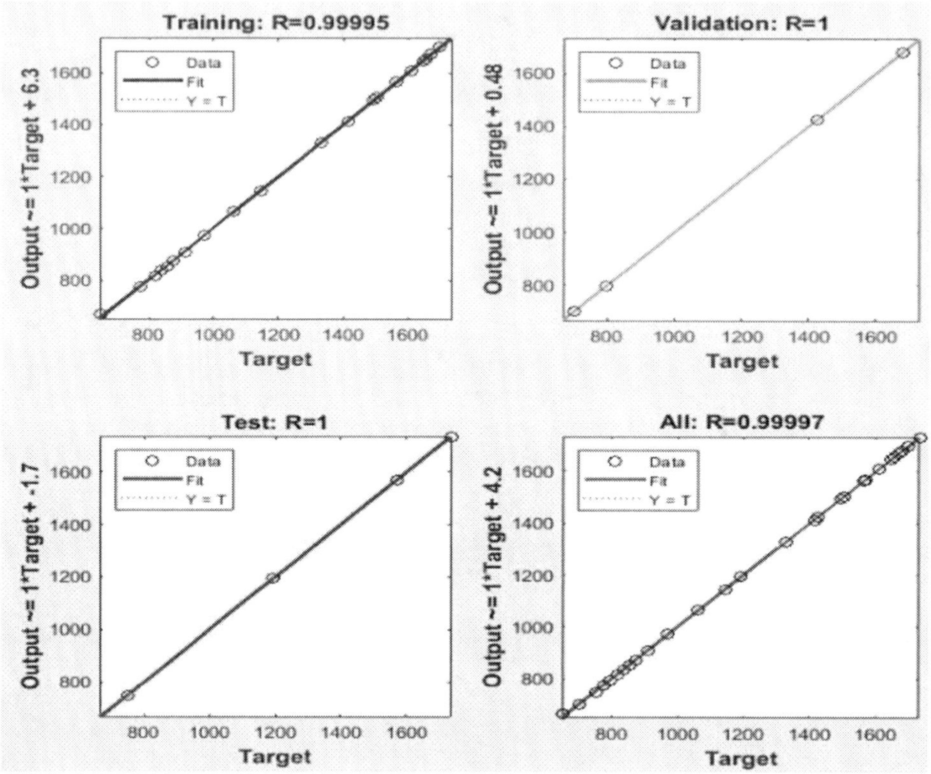

Figure 13.Neural network correlation coefficient

The more the R is closest to 1 and the more the value of the MSE is lower, the better the ANN is performant. Lowest value of the MSE (0.031428) is observed for the Young modulus prediction (1963 MPa). The best accuracy is also determined in terms of the coefficient of correlation, with an R equal to 0.99995 for the training and 1 for the validation and test which confirm the performance of the neural network model chosen.

4. Conclusion

In this study, we tried to confirm the effect of bio loading the polypropylene matrix with snail shell powder with neural network model and the finite element method in comparison with the experimental tensile test.

It was observed that the rigidity of the bio composite obtained at 30% of particles led to an increase of Young modulus from 1034 MPa to 1945, 1963 and 1981 MPa by the finite element method, neural network model and experimental tensile test respectively.

The neural network model proved a good efficiency to predict the Young modulus of the bio composite compared to the well-known finite element method. The best results obtained in the test phase illustrate that NNs are able to generalize the complicated link between input and output set of data learned in the phase of training which allow us to make a satisfactory predictive model for the Young modulus.

References

[1] H. Su, W. Qi, C. Yang, J. Sandoval, G. Ferrigno, and E. De Momi, "Deep neural network approach in robot tool dynamics identification for bilateral teleoperation," *IEEE Robot. Autom. Lett.*, vol. 5, no. 2, pp. 2943–2949, 2020.

[2] H.-T. Bang, S. Park, and H. Jeon, "Defect identification in composite materials via thermography and deep learning techniques," *Compos. Struct.*, vol. 246, p. 112405, 2020.

[3] Y.-C. Hsu, C.-H. Yu, and M. J. Buehler, "Using deep learning to predict fracture patterns in crystalline solids," *Matter*, vol. 3, no. 1, pp. 197–211, 2020.

[4] Y. Gao, H. Yao, H. Wei, and Y. Liu, "Physics-based Deep Learning for Probabilistic Fracture Analysis of Composite Materials," in *AIAA Scitech 2020 Forum*, 2020, p. 1860.

[5] A. Moumen, M. Jammoukh, L. Zahiri, and K. Mansouri, "Study Of The Optimal Micromechanical Behavior Of A Polymer Reinforced By Snail Shell Particles Using The Mori-Tanaka Numerical Model," in *2020 IEEE International conference of Moroccan Geomatics (Morgeo)*, 2020, pp. 1–6, doi: 10.1109/Morgeo49228.2020.9121908.

[6] A. MOUMEN, A. LAKHDAR, M. JAMMOUKH, L. ZAHERI, and K. MANSOURI, "Optimization of the Mechanical and Morphological Properties of Polypropylene Bio-Loaded by Argan Nut Shell Particles with Different Theoretical and Numerical Models," in *2020 IEEE 2nd International Conference on Electronics, Control, Optimization and Computer Science (ICECOCS)*, 2020, pp. 1–6.

[7] A. Belaadi, M. Boumaaza, S. Amroune, and M. Bourchak, "Mechanical characterization and optimization of delamination factor in drilling bidirectional jute fibre-reinforced polymer biocomposites," *Int. J. Adv. Manuf. Technol.*, vol. 111, no. 7, pp. 2073–2094, 2020.

[8] A. Sharma, S. A. Kumar, and V. Kushvaha, "Effect of aspect ratio on dynamic fracture toughness of particulate polymer composite using artificial neural network," *Eng. Fract. Mech.*, vol. 228, p. 106907, 2020.

[9] Q. Sun and T. Ertekin, "Screening and optimization of polymer flooding projects using artificial-neural-network (ANN) based proxies," *J. Pet. Sci. Eng.*, vol. 185, p. 106617, 2020.

[10] J. S. Chohan *et al.*, "Mechanical Strength Enhancement of 3D Printed Acrylonitrile Butadiene Styrene Polymer Components Using Neural Network Optimization Algorithm," *Polymers (Basel).*, vol. 12, no. 10, p. 2250, 2020.

[11] G. Balokas, S. Czichon, and R. Rolfes, "Neural network assisted multiscale analysis for the elastic properties prediction of 3D braided composites under uncertainty," *Compos. Struct.*, vol. 183, pp. 550–562, 2018.

[12] A. Moumen, M. Jammoukh, L. Zahiri, and K. Mansouri, "Numerical modeling of the thermo mechanical behavior of a polymer reinforced by horn fibers," *Int. J. Adv. Trends Comput. Sci. Eng.*, vol. 9, no. 4, pp. 6541–6548, 2020, doi: 10.30534/ijatcse/2020/342942020.

[13] A. Lakhdar, M. Jammoukh, L. Zahiri, K. Mansouri, A. Moumen, and B. Salhi, "Numerical and Experimental Study of the Behavior of PVC Material Subjected to Aging," in *2020 1st International Conference on Innovative Research in Applied Science, Engineering and Technology (IRASET)*, 2020, pp. 1–6.

[14] A. Moumen, A. Lakhdar, and K. Mansouri, "Numerical study of the mechanical behavior of polyamide 66 reinforced by argan nut shell particles with the finite element method and the mori-tanaka model," *Int. J. Adv. Trends Comput. Sci. Eng.*, vol. 9, no. 5, pp. 7723–7730, 2020, doi: 10.30534/ijatcse/2020/115952020.

[15] K. M. Abdelghani Lakhdar, Aziz Moumen, Laidi Zahiri, Mustapha Jammoukh, "Experimental and Numerical Study of the Mechanical Behavior of Bio-Loaded PVC Subjected to Aging," *Adv. Sci. Technol. Eng. Syst. J.*, vol. 5, no. 5, pp. 607–612, 2020.

IOP Conf. Series: Materials Science and Engineering 1126 (2021) 012010 doi:10.1088/1757-899X/1126/1/012010

Towards the Numerical Implementation of Neural Network to Predict the Mechanical Characteristics of Bio Composites

Zineb Laabid*, Aziz Moumen, Abdelghani Lakhdar and Khalifa Mansouri

SSDIA Laboratory, Hassan II University of Casablanca, ENSET of Mohammedia, 28830, Morocco.

laabid.zineb@gmail.com,Moumenaziz1@gmail.com,lakhdarabdelghani11@gmail.com, khmansouri@hotmail.com.

Abstract. Bio composites are a new category of materials using natural based components in their constituents. The study and simulation of the behavior of these innovative materials occupies an important place in the field of scientific research. Discovering and using new methods has always been the goal of researchers. In recent years, artificial intelligence has been very successful and is used in several fields. it represents a big part of today's industrial revolution. Smart solutions are more and more favored over conventional solutions as they give more precise results in a short time. We can find them in different sectors, such as banking, commerce, transport and industry, especially in materials science.The intersection of the artificial intelligence with materials engineering, gives extraordinary results. This smart method was able to boost the discovery of new materials, and to solve the most complex problems encountered when determining the mechanical properties of bio composites. What characterizes theEco-composites is their light in weight, their sustainable development, and that they are environmentally friendly. However, the determination of their mechanical properties is not obvious. Certainly, solutions based on homogenization methods or even on the finite element method have given good results, but the complexity of the microstructure of these materials limits the determination of their characteristics. In our paper, we hilight the use of Deep Learning that is an artificial intelligence machine learning method that relies on neural networks to predict the mechanical behavior of a polypropylen bioloaded by the natural fibers.

Key Words : Deep learning, artificial intelligent, bio composites, innovant materials,mecanical properties.

1. Introduction

Discovered in the mid-19th century, composite materials have attracted the attention of researchers and have been the subject of much research. Between the determination of the properties of these and the way in which they are manufactured, the interest of researchers was also to arrive at results showing that these new materials can be more advantageous by exploiting them in such a way as to meet the demands of the green world. These new materials have been very successful given the attractive advantages they represent. Aside from their sustainable development and light weight, these materials have proven to be environmentally friendly.

For example, in the automotive sector, by using these materials in the construction of car parts, this will result in vehicles that weigh even less than vehicles today, which will save in terms of energy, and therefore, we will have less CO2 released.

They are also used in the aircraft engines, in sports equipment for major competitions and others. The areas where we can find them are numerous and this is due to the quality and usefulness of the benefits they represent, and even more, the presence of the latter has grown lately and this since we started to talk about bio composite materials.

In this paper, we will study the effect of using deep neural network in predicting the final mechanical properties of these bio composites.

2. Discussion

2.1. BioComposites

Bio composites or eco composites are biological and ecological materials, which combine natural fibres with biodegradable or non-biodegradable polymers. Several current studies focus on the use of natural fillers for the bio filling of polymers such as Alpha fibres , Doom , snail shell and horn fibres[1]–[6] etc. These bio fibres have become an alternative to conventional fibres affecting the environment and more advantageous in terms of recycling, sustainable development and green chemistry.

Table 1. Mechanical properties of different bio loads [3], [7]–[11]

Material	Density (g/cm3)	Tensile strengh (Mpa)	Young's Modulious (Gpa)	Elongation at break (%)
Spiker Silc	1.3	1300-2000	30	28-30
Flax	1.45	500-900	50-70	1.5-4.0
Hemp	1.48	350-800	30-60	1.6-4.0
Kenaf	1.3	400-700	25-50	1.7-2.1
Jute	1.3	300-700	20-50	1.2-3.0
Bamboo	1.4	500-740	30-50	2
Sizal	1.5	300-500	10-30	2-5
Coconut	1.2	150-180	4-6	20-40
Glass fiber	2.5	1200-1800	72	2.5
Carbon Fiber	1.4	4000	235	2
Kevlar 49	1.44	3600-4100	130	2.8

Polypropylene studied as matrix is the most widely used polymer in industry, because it is cheaper, available, has interesting thermo mechanical properties [3] and also it is an easily recyclable form of plastic. So the homogenization of polyropylene with a natural charge will give birth to new materials which will have a significant effect on the environment because they will disadvantage the use of fossil fuels and also release greenhouse gas [8]. in addition to the environmental benefits, these materials are very light, very stiff and have a very high strength to weight ratio. Considering the large number of advantages represented by these new materials, a great deal of research has been carried out to study them and follow their serial life. Their manufacture attracted the curiosity of some researchers and others were drawn to the determination of their properties [12].

What characterizes one material from another is its mechanical property, and knowing these properties will help us to choose the best meets of our needs and meeting our specifications. However, several parameters and instructions define and differentiate one bio composite from another, so the mechanical properties of bio composite are primarily dependent on these parameters; There are the properties of the materials that constitute the new, the process followed in the manufacture of the latter and also its microstructure.

2.2. Determination of mechanical characteristics

2.2.1. Homogenization methods

To determine the mechanical properties of a bio composite, there are several leading methods to use. Apart from the experimental study which gives us the desired result, there is theoretical and numerical homogenization methods [2], [13]–[15] .Of course, experience allows us to have a safer and more correct results, but itis still very expensive and takes a lot of time.

Hence the need to plan ahead and use theoretical and numerical methods which will give approximate results and which are close to experimental ones but with a certain degree of error. Among these methods, we can cite the self-coherent method which is a means for modelling the behaviour of polycrystalline materials, the MoriTanaka mean field method and the micromechanical modulus of Chamis. These methods have been able to give acceptable results, on the other hand, in terms of precision and response time, these methods are very limited. As we can use the finite element method which is a safer and quite often used in the determination of the mechanical characteristics of materials and in particular of bio composites and this by defining a volume element large enough to remove the effect of RVE sizes.

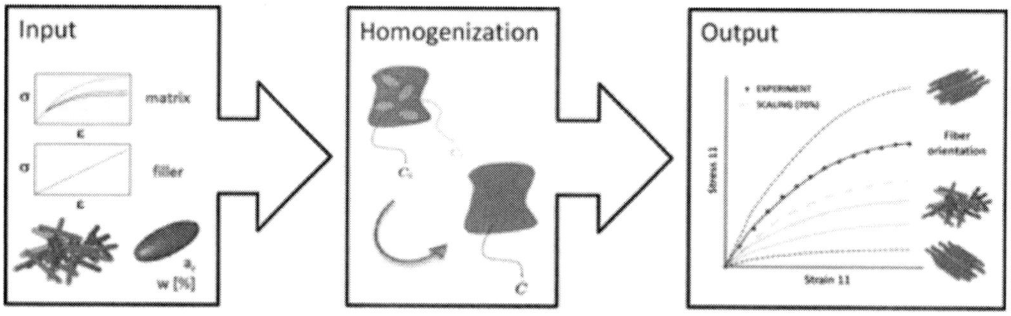

Figure 1. Mean field homogenization method

2.2.2. Artificial intelligence

Bio composites are characterized by very complex microstructures and determining their mechanical characteristics is not so simple, because it will put us in front of a nonlinear problems very specific to the material itself, which generates an enormous time during the simulation and determining the properties thereof. On top of that, and until fairly recently, the birth of a new material took a long time, even years, until artificial intelligence came to replace the existing methods.

Artificial intelligence has boosted not only the discovery of new materials but also the determination of their mechanical characteristics. This purely mathematical method is based on algorithms that solve the most complex problems by relying on results pre-established by the finite element method and by experiment. For example, the creation of metallic glass which is lighter and stronger than steel, and which can be used in smartphones. So thanks to artificial intelligence, we multiply the number of tests on combinations in virtual ways in a short time. We were then be able to study the stability and classification of the crystalline structures of materials using deep learning [16]–[19] and also to optimize the processes of shaping by NNAs of]. This smart method was chosen by other researchers, who tested it and approved its contribution to excellent results.[20]

2.3. Analyses and approach

In his paper "Prediction of Mechanical Properties of Materials with Deep Neural Network", the author chose to predict the mechanical properties of materials by using a ConvNet-type neural network. The ConvNet is an artificial neural network that uses an image processing system.

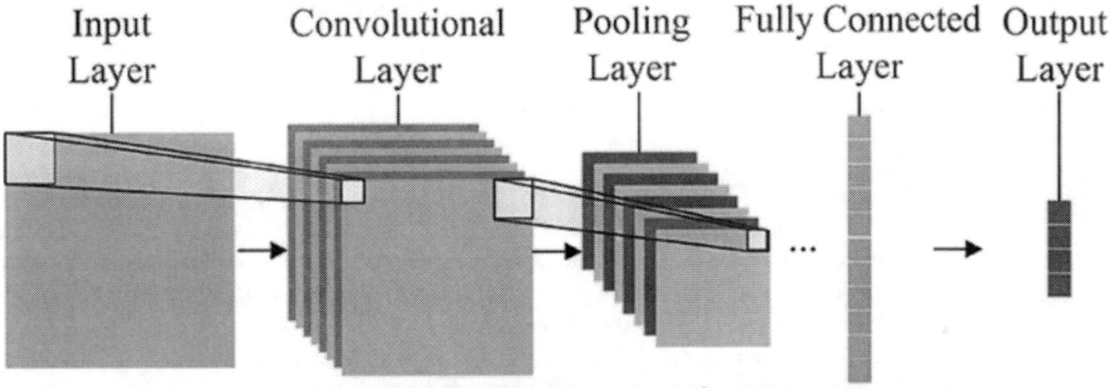

Figure 2. General structure of a CNN

The layers of this network consist of an input ,hidden and output output layers that include multiple convolutional layers, aggregation,normalization and fully connected layers. Removing limitations and increasing efficiency for image processing results in a much more efficient, easier to train, and specialized system for image processing and natural language processing. And to predict the mechanical properties using ConvNet, first they generated a huge dataset that contains images describing the microstructure of the material, its properties and those of the constituents. This database results from the representation of the voluminal elements by using the finite element method. From this database the chosen convnet was able to capture the mapping of the microstructural characteristics of composites and predict their properties in a robust and efficient manner[21].

By adopting the same method, the properties of heterogeneous materials which are characterized by complex microstructures were predicted, based on their microstructural images. With the help of DNN the explicit relationship between the mechanical property on a macroscopic scale and the microstructure of the material is easily determined. So this made it possible to obtain very precise results in a minimum of time.

By exploiting artificial intelligence, we can predict the mechanical characteristics of eco-composites. The ConvNet is just one architecture of the Deep Neural Network among dozens of structures. Each structure has its own algorithm and responds to a very specific need. To predict the behavior or owners of the materials, there is also the Feed Forward Backpropagation. It is a neural model of supervised learning that relies on error correction to converge towards the safest outcome.

3. Conclusion

In this paper, we were able to first highlight the bio-composite materials and their advantages , in particular their contributions to a green world with environmentally friendly characteristics. We studied the mechanical characteristics of materials that define them.

The classical methods which are based on homogenization and the artificial intelligence method which use learning as a solution to determine the properties of bio composites, can be combinatorial methods, and this by relying on the results of simulations of the finite element method to create databases in order to give the neural network the possibility of doing a supervised or unsupervised learning according to our choice and predicting the behavior of bio composites in minimum time and with less error is one of the dilemmas that researchers are studying.

Future scope

Polypropylenes bio loaded with natural fibers are taking the interest of current researchs, and finding significant results by highlighting an architecture often neglected by researchers, will be the objective of future research and will surely lead to desirable solutions by combining the powerful efficiency of the Deep learning neural network with finite element method.

References

[1] N. J. Arockiam, M. Jawaid, and N. Saba, *Sustainable bio composites for aircraft components*. Elsevier Ltd, 2018.
[2] A. Moumen, M. Jammoukh, L. Zahiri, and K. Mansouri, "Study Of The Optimal Micromechanical Behavior Of A Polymer Reinforced By Snail Shell Particles Using The Mori-Tanaka Numerical Model," in *2020 IEEE International conference of Moroccan Geomatics (Morgeo)*, 2020, pp. 1–6, doi: 10.1109/Morgeo49228.2020.9121908.
[3] A. Moumen, M. Jammoukh, L. Zahiri, and K. Mansouri, "Numerical modeling of the thermo mechanical behavior of a polymer reinforced by horn fibers," *Int. J. Adv. Trends Comput. Sci. Eng.*, vol. 9, no. 4, pp. 6541–6548, 2020, doi: 10.30534/ijatcse/2020/342942020.
[4] B. W. Li, H. P. Zhao, X. Q. Feng, W. W. Guo, and S. C. Shan, "Experimental study on the mechanical properties of the horn sheaths from cattle," *J. Exp. Biol.*, vol. 213, no. 3, pp. 479–486, 2010.
[5] A. Moumen, A. Lakhdar, and K. Mansouri, "Numerical study of the mechanical behavior of polyamide 66 reinforced by argan nut shell particles with the finite element method and the mori-tanaka model," *Int. J. Adv. Trends Comput. Sci. Eng.*, vol. 9, no. 5, pp. 7723–7730, 2020, doi: 10.30534/ijatcse/2020/115952020.
[6] K. M. Abdelghani Lakhdar, Aziz Moumen, Laidi Zahiri, Mustapha Jammoukh, "Experimental and Numerical Study of the Mechanical Behavior of Bio-Loaded PVC Subjected to Aging," *Adv. Sci. Technol. Eng. Syst. J.*, vol. 5, no. 5, pp. 607–612, 2020.
[7] X. Improvement on the interfacial compatibility of jute fiber-reinforced polypropylene composites by different surface treatmentsLiu, S.-J. Hao, Y.-H. Cui, and H. Chen, "Improvement on the interfacial compatibility of jute fiber-reinforced polypropylene composites by different surface treatments," *J. Ind. Text.*, vol. 49, no. 7, pp. 906–922, 2020.
[8] A. Lakhdar, A. Moumen, L. Zahiri, M. Jammoukh, and K. Mansouri, "Experimental and numerical study of the mechanical behavior of bio-loaded PVC subjected to aging," *Adv. Sci. Technol. Eng. Syst.*, vol. 5, no. 5, pp. 607–612, 2020, doi: 10.25046/AJ050574.
[9] S. Rabe, G. Sanchez-Olivares, R. Pérez-Chávez, and B. Schartel, "Natural keratin and coconut fibres from industrial wastes in flame retarded thermoplastic starch biocomposites," *Materials (Basel).*, vol. 12, no. 3, p. 344, 2019.
[10] Martijanti, Sutarno, and A. L. Juwono, "Bamboo fibers, fabrication of bamboo fiber reinforced composites, and their mechanical properties-a review," *Int. J. Emerg. Trends Eng. Res.*, vol. 8, no. 6, pp. 2346–2359, 2020, doi: 10.30534/ijeter/2020/24862020.
[11] A. MOUMEN, A. LAKHDAR, M. JAMMOUKH, L. ZAHERI, and K. MANSOURI, "Optimization of the Mechanical and Morphological Properties of Polypropylene Bio-Loaded by Argan Nut Shell Particles with Different Theoretical and Numerical Models," in *2020 IEEE 2nd International Conference on Electronics, Control, Optimization and Computer Science (ICECOCS)*, 2020, pp. 1–6.

[12] B. Wang, B. Zhou, X. Zhang, and B. Wang, "Microstructure and mechanical properties of an alpha keratin bovine hoof wall," *J. Mech. Behav. Biomed. Mater.*, vol. 104, p. 103689, 2020.

[13] I. Park, J. Moon, S. Bae, J. E. Oh, and S. Yoon, "Application of micro-CT to Mori-Tanaka method for non-randomly oriented pores in air-entrained cement pastes," *Constr. Build. Mater.*, vol. 255, p. 119342, 2020.

[14] Lee, Doojin. "Local anisotropy analysis based on the Mori-Tanaka model for multiphase composites with fiber length and orientation distributions." *Composites Part B: Engineering* 148 (2018): 227-234.

[15] P. Samadian, C. Butcher, and M. J. Worswick, "New mean-field homogenization schemes for the constitutive modelling of the elastic and elastoplastic deformation behavior of multi-phase materials," *Mater. Today Commun.*, vol. 24, p. 100707, 2020.

[16] S. Ye, B. Li, Q. Li, H. P. Zhao, and X. Q. Feng, "Deep neural network method for predicting the mechanical properties of composites," *Appl. Phys. Lett.*, vol. 115, no. 16, 2019, doi: 10.1063/1.5124529.

[17] X. Li, Z. Liu, S. Cui, C. Luo, C. Li, and Z. Zhuang, "Predicting the effective mechanical property of heterogeneous materials by image based modeling and deep learning," *Comput. Methods Appl. Mech. Eng.*, vol. 347, pp. 735–753, 2019, doi: 10.1016/j.cma.2019.01.005.

[18] Y.-C. Hsu, C.-H. Yu, and M. J. Buehler, "Using deep learning to predict fracture patterns in crystalline solids," *Matter*, vol. 3, no. 1, pp. 197–211, 2020.

[19] H. Su, W. Qi, C. Yang, J. Sandoval, G. Ferrigno, and E. De Momi, "Deep neural network approach in robot tool dynamics identification for bilateral teleoperation," *IEEE Robot. Autom. Lett.*, vol. 5, no. 2, pp. 2943–2949, 2020.

[20] A. K. F. Hassan, R. J. Jassim, and M. M. Jafaar, "Neural Network and Genetic Algorithm Based Finite Element Model for Optimal Die Shape Design in Al-1100 Cold Forward Extrusion," *Univers. J. Mech. Eng.*, vol. 4, no. 2, pp. 25–33, 2016, doi: 10.13189/ujme.2016.040202.

[21] D. W. Abueidda, M. Almasri, R. Ammourah, U. Ravaioli, I. M. Jasiuk, and N. A. Sobh, "Prediction and optimization of mechanical properties of composites using convolutional neural networks," *Compos. Struct.*, vol. 227, no. August, p. 111264, 2019, doi: 10.1016/j.compstruct.2019.111264.

Design and Simulation of Vibration Isolator by Hexahedron Finite Element

K . Viswanath Allamraju & Kosana Sai Adi Narayana

Department of Mechanical Engineering, Institute of Aeronautical Engineering, Dundigal, Hyderabad-500040

Akvn87@gmail.com

Abstract. Composite material and Vibration are the main research areas in which today's industries are growing. Vibration Isolator is a material which is used in mechanical industries to reduce vibration of machine by placing them at the bottom of small and heavy machineries under dynamic conditions. In designers and engineering life, composite materials help to increase material efficiency resulting in the reduction of cost and for utilization of resources as best as possible. Generally, composite materials are wide in Automobile sectors, manufacturing industries. In this experiment, study on isolator involves different types of composite materials. The design model was designed using CATIA and analyses have been performed using ANSYS Workbench. Anti-vibration mounts, also known as AV mounts, are key components in reducing noise and vibration in heavy machinery and equipment. AV mounts are a low cost, effective component, with a range of different uses. Machinery used in a variety of industries ranging from agriculture to the defence sector can be subject to excessive noise and vibrations which affects the longevity, efficiency and safety of the equipment used. Anti-vibration mounts possess effective shock absorption properties and act as a vibration dampener to reduce noise and improve the lifespan of machine applications.

1. Introduction

In this competitive world major problem for the manufacturing industries are based on the vibrations. The market noise and vibration ranks high. Vibration reduction is a method of avoiding unwanted vibration and noise. Anti-vibration mounts are key components in reducing noise and vibration in heavy machinery and equipment. Vibration mounts can be found in automobiles and power plants, specifically in engine and gearbox. Vibration reduction can be achieved through the use of isolation mounts.

Anti-vibration mounts possess effective shock absorption properties and act as a vibration dampener to reduce noise and improve the lifespan of machine applications. To overcome the problem of vibration and noise various methods are used such as vibration damping, isolation, and vibration absorption. To overcome the above problems composites materials are widely used in different industry. Range of Isolation Mounts is available in different sizes and rubber hardness' to suit each application, many of which are held in stock. Standard parts are manufactured from colored natural rubber.

2. Literature Survey

Here Comparative study of the two conditions using materials and also without mount and with mount is analyzed. Analytical calculations are done to the mount as single degree of freedom mass spring damper system and the frequency response[1]. In this experiment a study of vibration isolation is carried out analytically and numerically is calculated using FEM. Elevators is the aim at isolating the traction machine, cabin and structure with an AVM. Here ANSYS is used to calculate the dynamic and static parameter of the mount and analytically justify the structure as stable [2]. The stiffness coefficient K and damping coefficient C for the rubber pieces empirically calculated for the generator mounts. The study is to measure the vibration and the stresses developed in the Anti-vibration rubber mounts of a generator. The stresses were calculated numerically by ANSYS (FEA) [3]. The study of AVM to measure the stresses developed in the AVM mounts of a generator which are result from vibration of a generator at duty cycle. [4]. Static and dynamic analysis was done on the existing model and newly designed models. In dynamic transient condition, the vibration was transferred through the three models with a static load and dynamic load condition [5].This research works in reduction in vibration of engine with the help of composite material and to show the elastic properties and to reduce the vibration compared to other materials and rubber pad[6].Vibration is inevitable in various mechanical system because of dynamic motion to get advantage of system, it may prevent by mounting system or methods. It is also provide the accommodation to the system. The material properties of anti-vibration mount play vital role to arrest and isolate the vibration [7]. Vibration isolation involves the insertion of a resilient member or isolator in between the vibrating mass and source of vibration so that reduction in dynamic response of the system is achieved under specific conditions of vibration excitation. Many factors such as maximum moving dimension (MMD) of vehicle, overall weight of the cargo, dynamic/shock load, and natural frequency govern the selection of vibration isolators and design of superstructure [8]. And also researchers had done work on composite materials both experimentally and simulations studies of disc brakes which are made up of various materials[9-12]. In this paper simulation of spring isolator is proposed by using ANSYS WORKBENCH.

2.1. Simulation

A. Static Structural Model

Below image Fig.A represents the model of the project – Spring Isolator.

Fig.A - Spring Isolator

1. Aluminum Alloy

The Vibration Isolator made of Aluminum Alloy material with the volume of 1.5482e-005 m³, mass 4.2885e-002 kg and number of bodies are 3. Hex Dominant is the method used for mesh control. The force applied on the body is -2000N towards Y Axis and 0N towards both the X and Z Axis.

Common Material Properties:

It has density of 2770 kg/m³. The Young's Modulus, Bulk Modulus and Shear Modulus is 7.1e+10 Pa, 6.9608e+010 Pa and 2.6692e+010 Pa respectively. The Poisson's Ratio is 0.33. The Tensile Yield Strength and Compressive Yield Strength is 2.8e+008 Pa. The Tensile Ultimate Strength and Compressive Ultimate Strength of the material is 3.1e+008 Pa and 0 Pa.

1.1 Total Deformation:

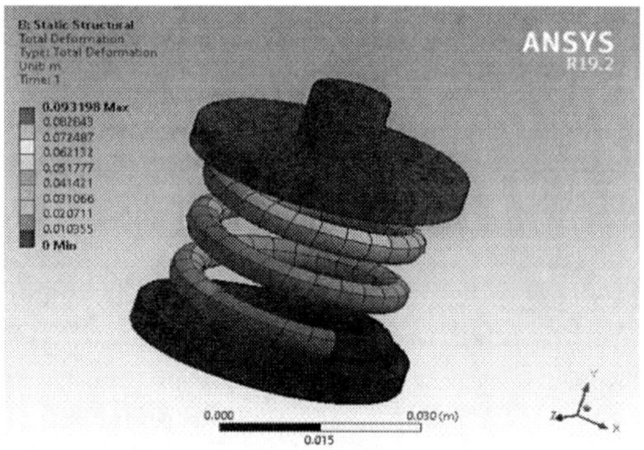

Fig.1.1 – Total Deformation

Fig.1.1 represents the Total Deformation of Isolator Model observed on applying the force of -2000N on Aluminum Alloy towards Y Axis at minimum is 0 [m] and maximum is 9.3198e-002 [m]. The average total deformation is 4.9386e-002 [m].

1.2 Directional Deformation:

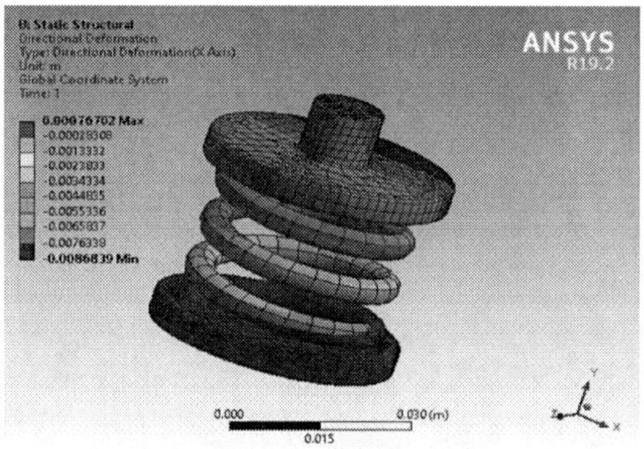

Fig.1.2 – Directional Deformation

Fig.1.2 represents the directional deformation of IsolatorModel observed on applying the force of -2000N on the Aluminum Alloy towards Y Axis at minimum is -8.6839e-003 [m] and at Maximum is 7.6702e-004 [m]. The average Directional Deformation is -3.5163e-003 [m].

IOP Conf. Series: Materials Science and Engineering 1126 (2021) 012011 doi:10.1088/1757-899X/1126/1/012011

1.3 Equivalent Elastic Strain:

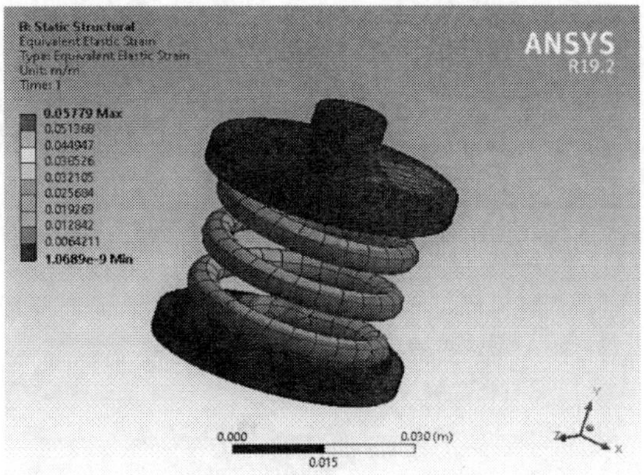

Fig.1.3 – Equivalent Elastic Strain

Fig.1.3 represents the Equivalent Elastic Strain of IsolatorModel observed on applying the force of -2000N on Aluminum Alloy towards Y Axis at minimum is 1.0689e-009 [m/m] and maximum is 5.779e-002 [m/m]. The average Equivalent Elastic Strain is 2.8527e-003 [m/m].

1.4 Maximum Principal Elastic Strain:

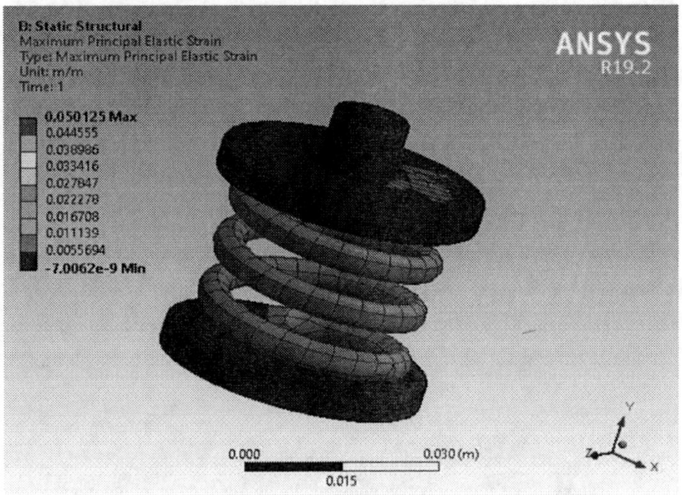

Fig.1.4 – Maximum Principal Elastic Strain

Fig.1.4represents the Maximum Principal Elastic Strain of Isolator Model observed on applying the force of-2000N on Aluminum Alloy towards Y Axis at minimum is -7.0062e-009 [m/m] and maximum is 5.0125e-002 [m/m]. The average strain is 1.8815e-003 [m/m].

IOP Conf. Series: Materials Science and Engineering 1126 (2021) 012011 doi:10.1088/1757-899X/1126/1/012011

1.5 Minimum Principal Elastic Strain:

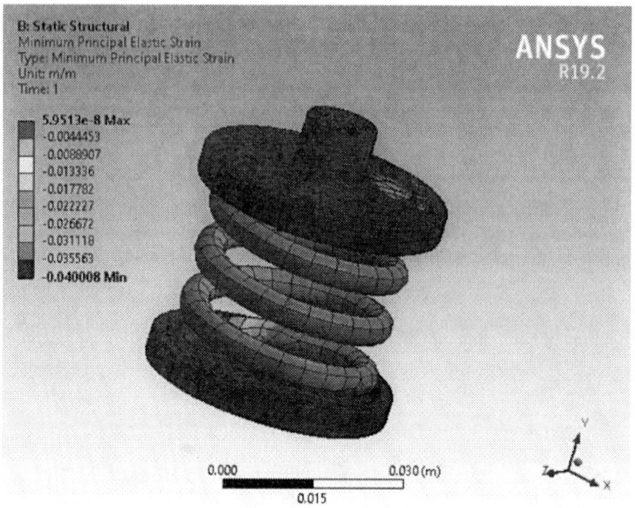

Fig.1.5Minimum Principal Elastic Strain of Spring Model

Fig.1.5 represents the Minimum principal elastic strain of the spring observed on applying the forceof–2000N on the Aluminum Alloy towards Y Axis at minimum is -4.0008e-002 [m/m] and maximum is 5.9513e-008 [m/m]. The Average Elastic Strain is -1.9804e-003 [m/m].

2. Gray Cast Iron

The Vibration Isolator made of Gray Cast Iron with the volume of 1.5482e-005 m³, mass 0.11147 kg and number of bodies are 3. Hex Dominant is the method used for mesh control. The force applied on the body is -2000N towards Y Axis and 0N towards both the X and Z Axis.

Common Material Properties:

It has density of 7200 kg/m³. The Young's Modulus, Bulk Modulus and Shear Modulus is 1.1e+011 Pa, 8.3333e+010 Pa and 4.2969e+010 Pa respectively. The Poisson's Ratio is 0.28. The Tensile Yield Strength and Compressive Yield Strength is 0 Pa. The Tensile Ultimate Strength and Compressive Ultimate Strength of the material is 2.4e+008 Pa and 8.2e+008 Pa.

2.1 Total Deformation:

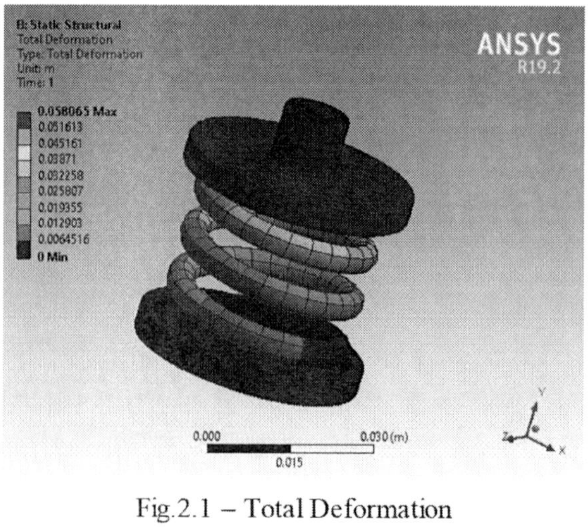

Fig.2.1 – Total Deformation

Fig.2.1 shows the Total Deformation of Isolator Model observed on applying the force of -2000N on Gray Cast Iron towards Y Axis at minimum is 0 [m] and maximum is 5.8065e-002 [m]. The average total deformation is 3.0732e-002 [m].

2.2 Directional Deformation:

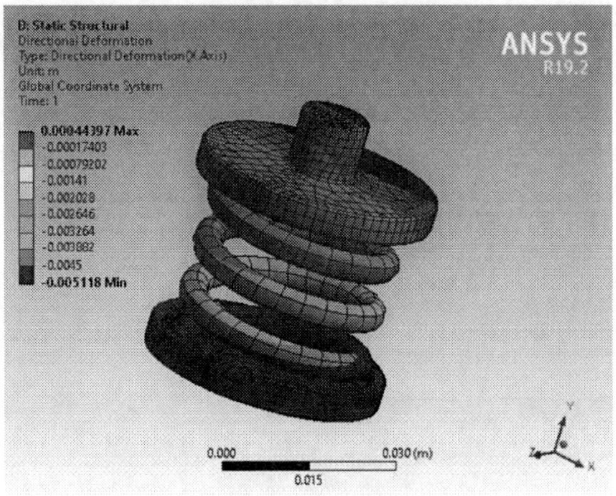

Fig.2.2 – Directional Deformation

Fig.2.2 represents the directional deformation of Isolator Modelobserved on applying the force of -2000N on the Gray Cast Iron towards Y Axis at minimum is -5.118e-003 [m] and at Maximum is 4.4397e-004 [m]. The average Directional Deformation is -2.1765e-003 [m].

2.3 Equivalent Elastic Strain:

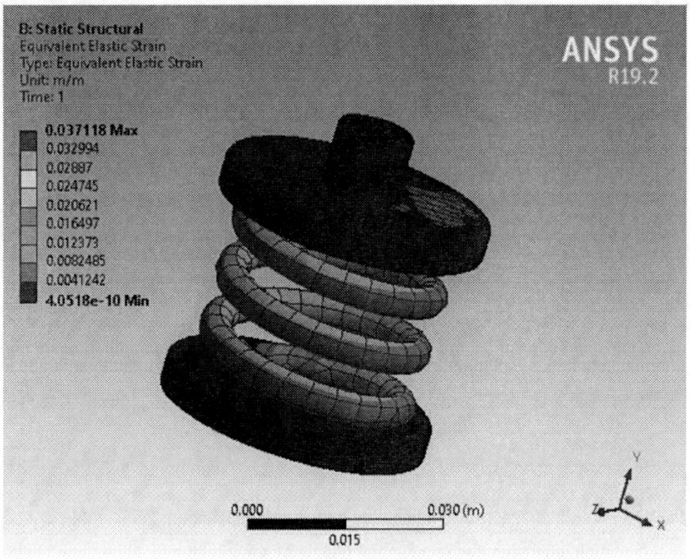

Fig.2.3 – Equivalent Elastic Strain

Fig.2.3 represents the Equivalent Elastic Strain of Isolator Model observed on applying the force of -2000N on Gray Cast Iron towards Y Axis at minimum is 4.0518e-010 [m/m] and maximum is 3.7118e-002 [m/m]. The average Equivalent Elastic Strain is 1.8492e-003 [m/m].

IOP Conf. Series: Materials Science and Engineering 1126 (2021) 012011 doi:10.1088/1757-899X/1126/1/012011

2.4 Maximum Principal Elastic Strain:

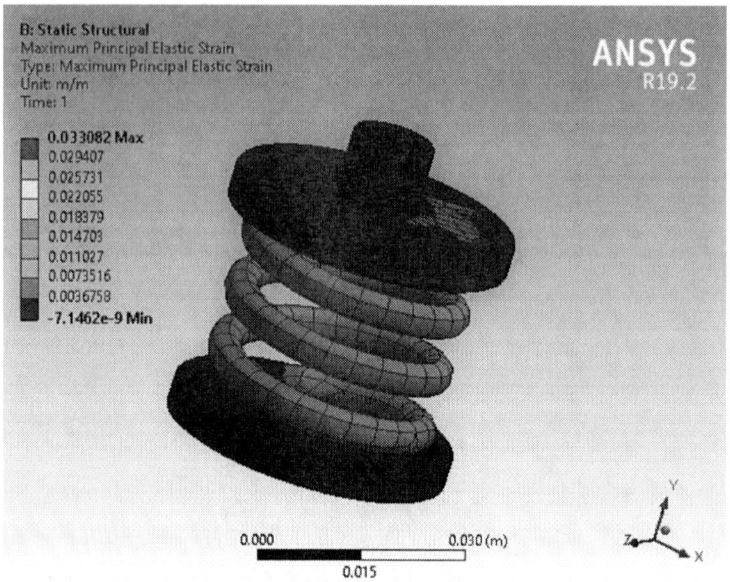

Fig.2.4 – Maximum Principal Elastic Strain

Fig.2.4 represents the Maximum Principal Elastic Strain of Isolator Model observed on applying the force of -2000N on Gray Cast Iron towards Y Axis at minimum is -7.1462e-009 [m/m] and maximum is 3.3082e-002 [m/m]. The average strain is 1.1697e-003 [m/m].

2.5 Minimum Principal Elastic Strain:

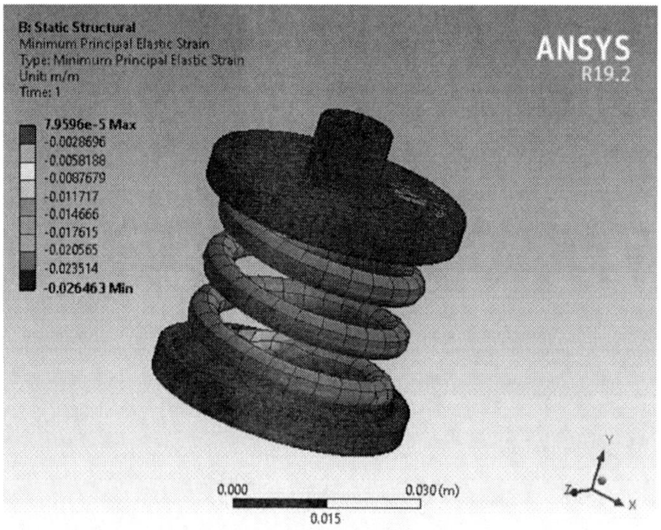

Fig.2.5 – Minimum Principal Elastic Strain

Fig.2.5 represents the Minimum Principal Elastic Strain of Spring Model observed on applying the force of–2000N on the Gray Cast Iron towards Y Axis at minimum is -2.6463e-002 [m/m] and maximum is 7.9596e-005 [m/m]. The Average Elastic Strain is -1.2361e-003 [m/m].

References

[1] V.S.Chavan, R.Askhedkar, S.B.Sanap - Analysis Of Anti Vibration Mounts For Vibration Isolation In Diesel Engine Generator Set - International Journal of Engineering Research and Applications (IJERA) ISSN: 2248-9622 - Vol. 3, Issue 3, May-Jun 2013 - pp.1423-1428

[2] Patel Sanket Bharat, Shivprakash B Barve, Sandeep G Thorat and Samata S Mujumdar - Design and Analysis of Anti-vibration mount for G+3 Elevator - International Journal of Current Engineering and Technology (IJCET) ISSN 2277 - 4106 (Special Issue-3) - April 2014

[3] Hazim Nasir Ghafil - Vibration Analysis of a Generator Anti-Vibration Rubber Mounts - Journal of Computer Science & Computational Mathematics - Volume 6, Issue 4 - December 2016

[4] Hazim Nasir GHAFIL - Vibration Analysis of a Generator Anti-Vibration Rubber Mounts - Vol.9 (2017) No.1, ISSN 1821-1259 - Dec 2016

[5] A. Krishnamoorthy and M. Jayavel - Design and Analysis of Vibration Isolation Pad of a Heavy Load Machine and To Perform the Progressive Rate Frequency Analysis - ARPN Journal of Engineering and Applied Sciences - VOL. 12, NO. 16 - AUGUST 2017

[6] G. C. Mekalke , S. R. Basavaraddi - Vibration Mitigation of Engine by Using Composite Materials - International Journal of Engineering and Advanced Technology (IJEAT) ISSN: 2249 – 8958 - Volume-8 Issue-3 - February 2019

[7] Sangeetha N, Mohanraj K - Research on Design and Method to Predict Fatigue Life of an Anti-Vibration Mount - International Journal of Engineering and Advanced Technology (IJEAT) ISSN: 2249 – 8958 - Volume-8 Issue-6S - August 2019

[8] Shaikh Hannan Ahmed, Manoj V. Amale - Nonlinear Transient Vibration Analysis of Mounting Superstructure for Road Generated Shock Load Response - International Research Journal of Engineering and Technology (IRJET) - e-ISSN: 2395-0056, p-ISSN: 2395-0072 - Volume: 07 Issue: 01 - Jan 2020.

[9] K Viswanath Allamraju 'Contact stress Analysis of Tyre and Mild Steel Plate' (IJEAT)', ISSN: 2249- 8958 (Online), Volume-9, Issue-2, December 2019, Page No 5059-5061.

[10] K Viswanath Allamraju , Study of Mechanical behaviour of Hybrid Jute Nano Fiber Composite, Materials Today: Proceedings, Volume 5, Issue 9, Part 3, 2019, Pages 20750-20753.

[11] K Viswanath Allamraju , Mechanical behaviour of nano jute filler glass fiber self healing and epoxy resin composite, Materials Today: Proceedings, Volume 5, Issue 9, Part 3, 2018, Pages 20278-20284.

[12] K Viswanath Allamraju , Structural and thermal analysis of brake with drilled holes of various materials, IJITEE, ISSN: 2278-3075, Volume-8 Issue-10, August 2019, DOI: 10.35940/ijitee.J9439.0881019.

IOP Conf. Series: Materials Science and Engineering 1126 (2021) 012012 doi:10.1088/1757-899X/1126/1/012012

Influence of manufacturing parameters on the mechanical properties of nano-reinforced CFRP by carbon nanofibers

P. Santos[1,a)], **Alberto Maceiras**[1], **P. N. B. Reis**[1]

[1] C-MAST, Depart. of Electromechanical Engineering, University of Beira Interior, 6201-100 Covilhã, Portugal

[a)]Corresponding author: paulo.sergio.santos@ubi.pt

Abstract. One of the most relevant problems of nano-reinforced resins is the uniform dispersion of nanoparticles. This problem is even more relevant when they are involved in the production of laminated composites. In this case, in addition to the agglomeration of nanoparticles, air bubbles are also very frequent during the manufacturing process of laminated composites with nano-reinforced resins. These air bubbles are even more frequent in the manual lay-up process. In this context, this study intends to study some manufacture parameters in order to maximize the mechanical properties. For this purpose, the mixer rotation speed, the dispersion time of the nanoparticles and the vacuum time applied to the system were analyzed in detail. It was found that the mixing time is very dependent on the rotation speed used and there is an ideal vacuum time. A short vacuum time allows air bubbles to exist in the resin, while a long time promotes a decrease in mechanical properties due to the removal of resin in the laminates and, consequently, the fibers are not fully wet.

1. Introduction

Nowadays, polymer matrix composites (PMC) are replacing, more and more, the traditional metallic materials and this trend extends to the most diverse industrial applications. This is consequence of their high specific strength and stiffness, good fatigue performance, corrosion resistance, and low processing costs [1].

The manual lay-up process, also called the wet lay-up method, is the simplest and oldest method and it is particularly suitable for large components/structures. In this process, each layer is handled manually, layer by layer, up to the desired thickness. Although brushes are often used to distribute the resin evenly onto the fibers, rollers are used to ensure complete wetting of the fibers and to remove air bubbles. However, the large amount of air bubbles that a laminate can present is the main drawback of this process. In this case, the vacuum bagging process proves to be a good alternative, because, as it is a closed mold process, it is able to manufacture components/structures with higher mechanical performance and low cost [2,3].

On the other hand, the open literature reports that the mechanical properties of nanocomposites are dependent on the content of nanoparticles, type of resin and manufacturing process. In this case, for example, in order to maximize mechanical properties, good dispersion is essential, which often involves great technical difficulties and high costs to achieve them. For example, Reis et al. [4] compared two dispersion processes, direct method (DM) and indirect method (IDM), in terms of mechanical properties, and they concluded that the nanocomposites obtained by the indirect method had lower mechanical properties than the neat resin due to the residual acetone.

Therefore, the present study intends to analyze the effect of some manufacturing parameters on the mechanical properties of nano-reinforced CFRP by CNFs. The influence of the rotation speed, mixing time and vacuum application time will be analyzed in detail for an epoxy resin. In fact, epoxy resins are thermosetting polymers used as matrices due to their chemical resistance, high specific strength, dimensional stability, and excellent adhesion to fillers, among others. Epoxies are two-component materials: resin and hardener, where resins are normally diglycidyl ethers of bisphenol-A and/or bisphenol-F; and hardeners are the curing agents. There are different types of hardeners, but the most commons are based on amines, and these amines when reacting with the epoxy rings are able to start the polymerization and subsequent crosslinking [5]. On the other hand, CNFs present superior mechanical and chemical properties over commonly used carbon fibers owing to their unique small-size characteristics. In addition, they have excellent electrical and high thermal conductivity. Therefore, CNFs can be considered for extending the physical and mechanical properties of conventional carbon fiber composites for lightweight structural applications.

2. Materials and Experimental Procedure

An SR 8100 epoxy resin and a SD 8822 hardener, both supplied by Sicomim, were used to produce nanocomposites and carbon/epoxy laminates reinforced with CNFs. Their average diameter is about 130 nm, length between 20 and 200 μm, and average specific surface area around 54 m^2/g. The SR 8100 epoxy is composed of a mixture of bisphenol A (DGEBA), bisphenol F (DGEBF) epoxy resins, and 1,6-bis(2,3-epoxypropoxy)hexane, as diluent. The hardener SD 8822 is a mixture of amines, 2-methylpentane-1,5-diamine, m-phenylenebis(methylamine), and trimethylolpropane tris[poly(propylene glycol), amine terminated] ether [5].

In terms of nanocomposites three different manufacturing processes were adopted, which are shown in Table 1

Table 1. Manufacturing process of nanocomposites.

Manuf. Process	Detail used during the manufacturing process
1	Resin and hardener mixed manually.
2	Resin and hardener mixed manually, followed by vacuum for 10 minutes.
3	The resin mixture with 0.5 wt.% CNFs was conducted at 1000 rpm for 3 h and, at same time, submitted to an ultrasonic bath (frequency of 40 kHz). The hardener was subsequently added at 150 rpm for 10 minutes and, finally, the mixture was degassed for 10 minutes in a vacuum chamber.

After the manufacture process reported in Table 1, the mixture is carefully poured into a cardboard mold with dimensions of $100\times130\times3\pm0.09$ mm^3 and cured at room temperature for 24 hours. Finally, a post-cure at 40ºC was applied for 24 hours.

In terms of composite laminates, eight ply laminates of woven bi-directional carbon 195-1000P (195 g/m^2), all in the same direction, were prepared by hand lay-up and the overall dimensions of the plates were $150\times150\times1.5\pm0.18$ mm^3. The resin used was prepared by process 2. The system was placed inside a vacuum bag and a load of 2.5 kN was applied during 24 h in order to maintain a constant fiber volume fraction and uniform laminate thickness. The bag remained attached to a vacuum pump to eliminate any air bubbles existing in the composite, and the vacuum time of 5 s, 1 min, 2 min, 3 min, 5 min and 10 min was analyzed. Finally, the post-cure was followed according to manufacturer datasheet (epoxy resin) in an oven at 40 ºC during 24 h.

Finally, similar composites laminates were produced, but the epoxy resin was, in this case, enhanced by 0.5 wt.% of CNFs. Nine different manufacturing processes were studied, which are summarized in Table 2. In all conditions, the manufacturing process is completed with a post-cure in an oven at 40 °C for 24 h.

Three-point bending (3PB) static tests were performed at room temperature and using specimens cut nominally to 60×10×3 mm^3 and 60×10×2 mm^3 from those plates, respectively, for resin and composite laminates as shown in Figure 1. The specimens were tested with a span of 50 mm and 40 mm, respectively, for resin and composite laminates, according to the European Standard EN ISO 178:2003. An Autograph AGS-X universal testing machine, from Shimadzu, with a 10 kN load cell and a displacement rate of 2 mm/min was used to test six different samples for each configuration.

Table 2. Manufacturing process of composite laminates.

Manuf. Process	Detail used during the manufacturing process
1	Control samples obtained by: Resin and hardener mixed manually, followed by vacuum for 10 minutes. The bag remained attached to a vacuum pump for 2 min.
2	Composite laminates were manufactured based on process 3, reported in Table 1, but conducted at <u>400 rpm for 3 h</u>. Thereafter, the system was placed in a vacuum bag, a 2.5 kN load was applied for 24 h and vacuum for 2 min.
3	Composite laminates were manufactured based on process 3, reported in Table 1, but conducted at <u>700 rpm for 3 h</u>. Thereafter, the system was placed in a vacuum bag, a 2.5 kN load was applied for 24 h and vacuum for 2 min.
4	Composite laminates were manufactured based on process 3, reported in Table 1, but conducted at <u>700 rpm for 4 h</u>. Thereafter, the system was placed in a vacuum bag, a 2.5 kN load was applied for 24 h and vacuum for 2 min.
5	Composite laminates were manufactured based on process 3, reported in Table 1, but conducted at <u>700 rpm for 5 h</u>. Thereafter, the system was placed in a vacuum bag, a 2.5 kN load was applied for 24 h and vacuum for 2 min.
6	Composite laminates were manufactured based on process 3, reported in Table 1, but conducted at <u>1000 rpm for 3 h</u>. Thereafter, the system was placed in a vacuum bag, a 2.5 kN load was applied for 24 h and vacuum for 2 min.
7	Composite laminates were manufactured based on process 3, reported in Table 1, but conducted at <u>1000 rpm for 4 h</u>. Thereafter, the system was placed in a vacuum bag, a 2.5 kN load was applied for 24 h and vacuum for 2 min.
8	Composite laminates were manufactured based on process 3, reported in Table 1, but conducted at <u>1000 rpm for 5 h</u>. Thereafter, the system was placed in a vacuum bag, a 2.5 kN load was applied for 24 h and vacuum for 2 min.
9	Composite laminates were manufactured based on process 3, reported in Table 1, but conducted at <u>1000 rpm for 3 h</u>. The system, after 2 min in a vacuum chamber, was subsequently placed in a vacuum bag and a 2.5 kN load was applied for 24 h. Finally, the bag remained attached to a vacuum pump for another 2 minutes.

The flexural strength was calculated as the nominal stress in the middle span section obtained using maximum value of the load (equation (1)), while the bending stiffness modulus was obtained by linear regression of the load-displacement curves considering the interval in the linear segment with a correlation factor greater than 95% according with equation (2) [6]. Finally, the bending strain is obtained according with equation (3):

$$\sigma = \frac{3\,P\,L}{2\,b\,h^2} \tag{1}$$

$$E = \frac{\Delta P\,L^3}{48\,\Delta u\,I} \tag{2}$$

$$\varepsilon_f = \frac{6\,S\,h}{L^2} \tag{3}$$

where P is the load, L the span length, b the width, h the thickness of the specimen, I the moment of inertia of the cross-section, ΔP and Δu are, respectively, the load range and flexural displacement range in the middle span for an interval in the linear region of the load versus displacement plot and S is the deflexion.

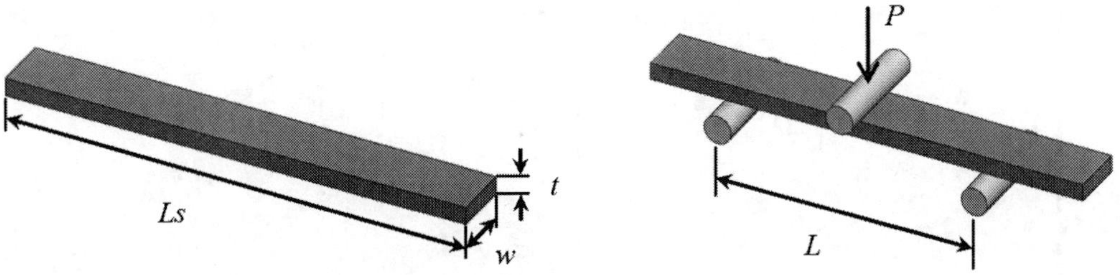

Figure 1. a) Specimens geometry; b) Schematic view of the three-point bending apparatus.

3. Results and Discussion

Flexural static tests were performed to obtain the effect of some manufacture parameters on the mechanical performance. In this context, Figure 2 shows the stress-strain curves and the results obtained for the procedures described in Table 1.

From Fig. 2a), it is possible to observe a linear increase of the bending stress with the bending strain, followed by a non-linear behaviour in which the maximum load is reached. It is also noticed that, when the resin is enhanced by CNFs, higher bending stress and modulus are achieved, while the bending strain decreases. After the peak load is reached, the bending stress decreases, arising the impending failure of the specimen. Figures 2b) to 2d) show the main bending properties in terms of average values (marks), and the bands represent, respectively, the maximum and minimum values for each condition. In terms of bending stress, Fig. 2b, it is noticed that the highest value was obtained with the resin enhanced by CNFs (114 MPa), reaching values around 12.6% higher than with the neat resin obtained without vacuum (process 1). However, when compared to the neat resin that involved vacuum, this improvement was only around 7.9%. Similar behaviour was observed for the bending modulus with values around 20.9% and 5.6%, respectively. Finally, the effect on the bending strain is not so evident due to the dispersion of the values observed. However, a small downward trend is observed for nano-enhanced resin. These results are in line with the open literature, because nanomaterials have a large surface area per volume unit compared to macro level materials and, in addition to their unique surface effects, increased chemical activity and particular physical properties, they promote synergy between constituents (polymer and CNFs) [6–8]. Simultaneously, the benefit of the vacuum is evident, where improvements in bending stress around 4.3% are obtained compared to the procedure that did not involve vacuum (process 1). Regardless of the resin/hardener mixture being made manually, and with particular care to avoid air bubbles, this process is not free from defects that need to be removed. In this context, the vacuum proved to be a viable technique for this purpose [2,3].

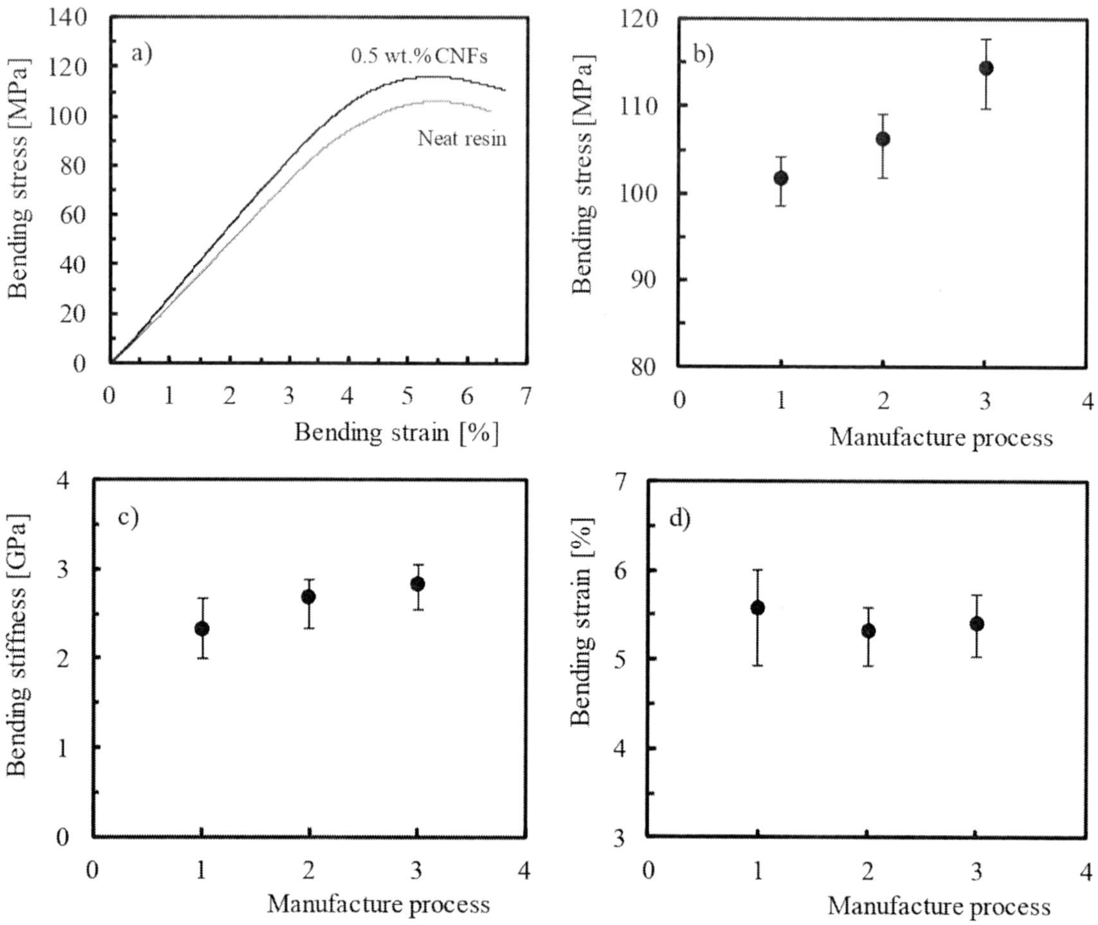

Figure 2. a) Representative bending stress-strain curves; b) Bending stress vs manufacture process; c) Bending modulus vs manufacture process; d) Bending strain vs manufacture process.

Figure 3 shows the fracture surface, showing a typical behaviour of brittle polymers with low resistance to crack propagation. On the other hand, the absence of air bubbles and a good dispersion of CNFs are also evident, confirming the benefits previously reported with the nanoparticles used. In fact, according to the open literature, agglomerations/aggregations are responsible for stress concentrations in nanocomposites and reduce the interfacial area between the polymeric matrix and nanoparticles [8–11]. On the other hand, only a few polymer molecules are able to penetrate between the nanoparticles, which promotes an increase in viscosity [12].

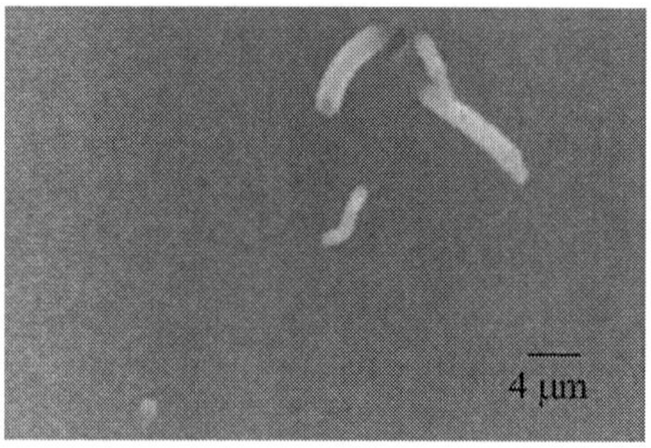

Figure 3. SEM image of the fracture surface.

In terms of composite laminates, Figure 4 shows the effect of the vacuum time on the bending properties. This study involved only laminates produced with neat resin obtained according to process 2 in Table 1. Therefore, the vacuum time is studied after the lamination of the composites. In fact, one of the most common defects of epoxy resins is the formation of air bubbles, so it is crucial to analyse the time for removing air by applying vacuum.

Fig. 4a) shows that all curves present a nearly fragile behaviour with a linear region up to the maximum load, followed by a significant drop after this value. Fibres breakage in compression side is the main damage mechanism observed, which is in line with the open literature [13,14]. According to Giancaspro et al. [14], carbon fibre composites fail mainly on the compression side, because this damage mechanism is consequence of the high compressive stress concentration in the pin load contact region associated to the low compressive strength of the carbon fibres [13,14].

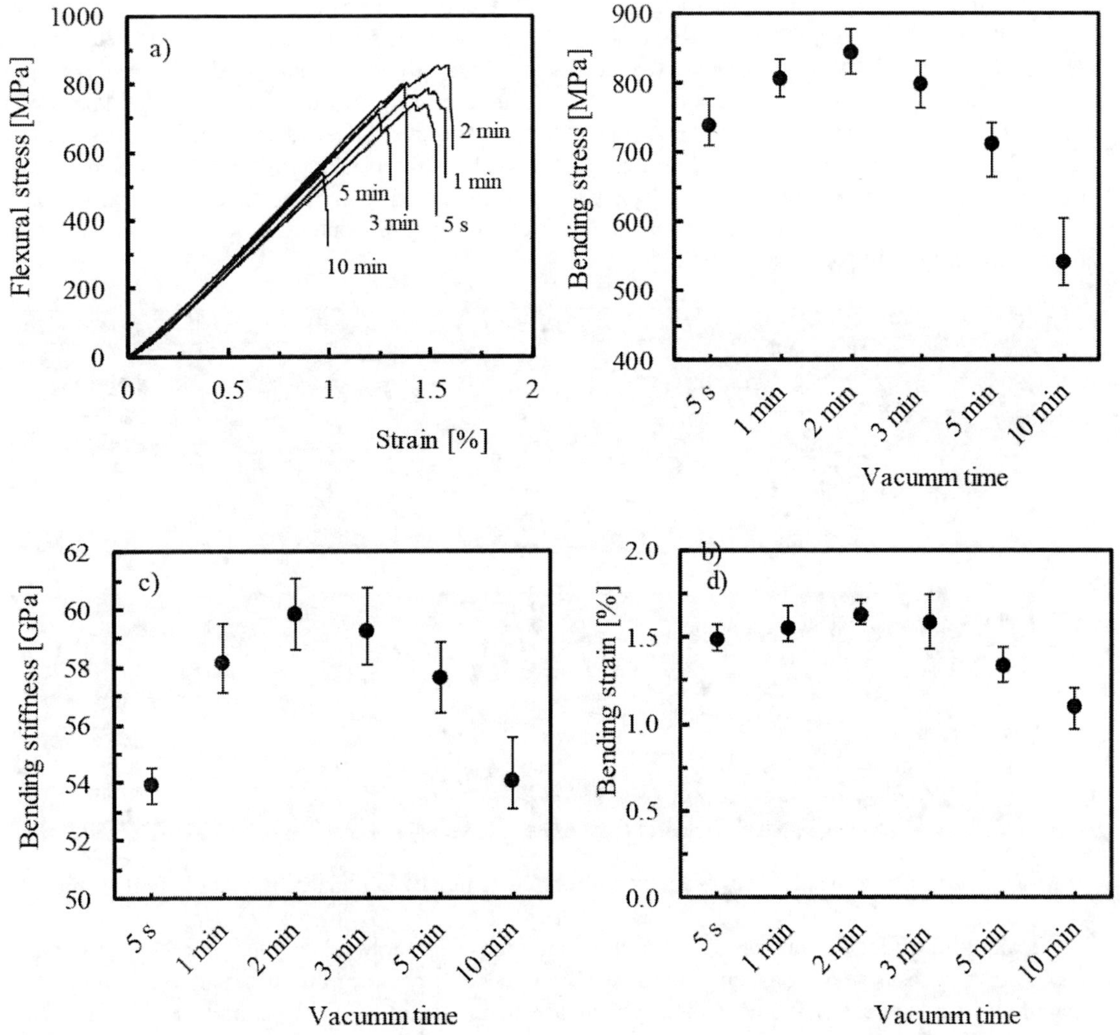

Figure 4. Effect of vacuum time on the: a) Bending stress-strain curves; b) Bending stress vs vacuum time; c) Bending modulus vs vacuum time; d) Bending strain vs vacuum time.

From the other figures, Fig 4b) to 4d), it is possible to observe that the bending properties increase up to a vacuum time of 2 min, reaching their maximum values, after which they start to decrease. For

IOP Conf. Series: Materials Science and Engineering 1126 (2021) 012012 doi:10.1088/1757-899X/1126/1/012012

example, compared to the values obtained for a vacuum time of 5 s, the bending stress increases around 14.1%, the bending modulus about 11% and the bending strain around 10.1%. On the other hand, the same comparison for the vacuum time of 10 min promotes decreases of 26.8% and 25.7%, respectively, for bending stress and bending strain, while the bending modulus remains very similar (comparisons in relation to the vacuum time of 5 s).

In order to explain this phenomenon, Fig. 5 illustrates typical cross sections of specimens produced with vacuum times of 1min and 2 min. As shown in Fig. 5a), it is possible to observe some voids, corresponding to air bubbles, in laminates that were produced with 1 min of vacuum, while for laminates subject to 2 min of vacuum, the absence of these voids is notorious (Fig. 5b)). Therefore, the increase in the mechanical properties previously reported up to 2 min of vacuum is a consequence of the decrease in voids. On the other hand, for higher vacuum times, the mechanical properties no longer decrease due to voids, but as a consequence of the lower amount of resin in the laminate. Vacuum suction was carried out at the end of the bag and, in this case, the resin was sucked into the absorbent fabric that delimits the laminates. In this case, with the increase in the vacuum time, more resin is removed from the laminate and, consequently, the remaining resin is not enough to completely wet the fibers. Because the load transfer between fibers is not effective, due to the lack of resin, the mechanical properties of the laminates decrease. This can be justified by the lower thickness of the laminates that is observed with the increase in the vacuum time, because they were all subject to the same 2.5 kN load during the manufacturing process. For example, average thicknesses around 1.47 mm, 1.45 and 1.36 mm were measured for laminates subjected to vacuum times of 3 min, 5 min and 10 min, respectively, which is a clear evidence of the lower resin content. In this context, the vacuum time of 2 min proved to be the ideal.

a)
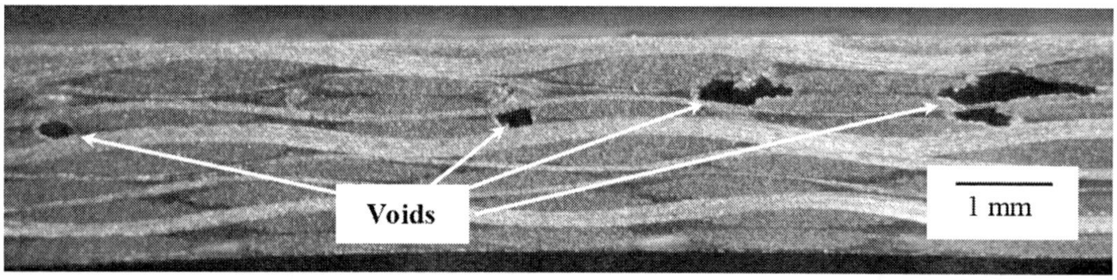

b)

Figure 5. a) Laminates with defects: air bubbles 1 min; b) Laminates without defects 2 min.

Finally, the effect of rotation speed and mixing time was analyzed to obtain the best dispersion of CNFs in the resin and, consequently, in the composite laminates. For this purpose, nine different manufacturing processes were studied, which are summarized in Table 2, and the results obtained shown in Fig. 6.

From Fig. 6a) it is possible to observe a typical profile to those obtained in Fig. 4a), where all curves present a nearly fragile behaviour with a linear region up to the maximum load, followed by a significant drop after this value. Only three curves are illustrated, but they are representative of all the others. As reported previously, fibres breakage in compression side is also the main damage mechanism observed.

In terms of bending stresses, it is noticed that composite laminates produced according to process 5, reported in Table 2, show improvements about 11.3% higher than the value obtained for the control samples (resin and hardener mixed manually, followed by vacuum for 10 minutes, the bag remained connected to a vacuum pump for 2 min and a load of 2.5 kN was applied). This is clear evidence of the benefits obtained with CNFs, which were described in detail earlier when only the resin was analysed. In this context, CNFs have a large surface area per unit of volume, good physical and mechanical properties and promote synergy with the matrix [6–8]. However, the bending strength can be further improved by degassing the system (resin and fibers) after mixing with CNFs (beyond 2 min in the vacuum bag). In this case, for process 9, the bending stress is around 16.5% higher when compared to the control laminates (808.5 MPa). This benefit is related to the effect of the vacuum on the mechanical properties, as previously discussed [2,3]. It is worth mentioning that another vacuum step performed after mixing the resin with CNFs, in addition to the vacuum in curing, increased the flexural strength by about 4.7%.

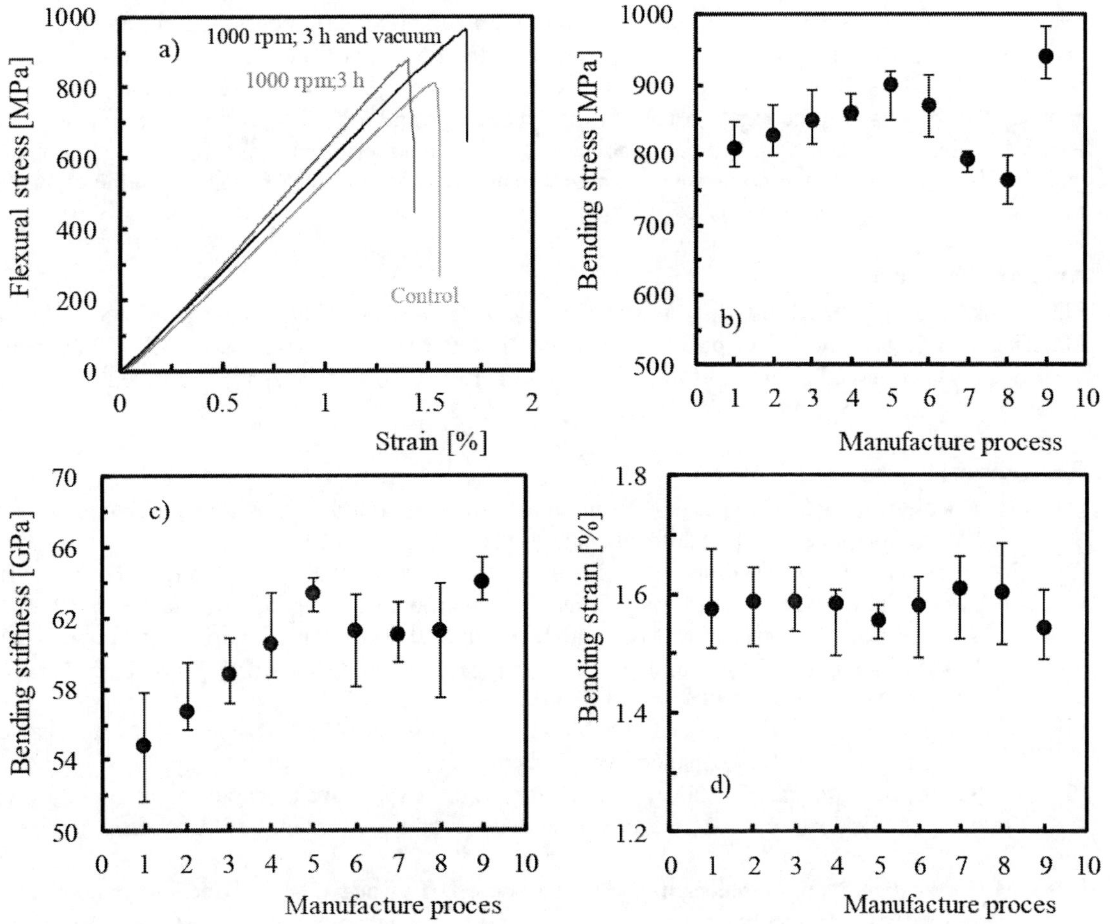

Figure 6. Effect of the manufacture parameters on: a) Typical bending stress-strain curves; b) Bending stress; c) Bending modulus; d) Bending strain.

In terms of bending modulus, the same tendency was observed. In this case, the benefits obtained with CNFs, compared to the composites with neat resin, were around 15.6% higher, and the extra step of vacuum promoted improvements of 16.6%. In this example, comparing only the extra step of vacuum, it promoted an improvement around 6.4%. On the other hand, from Fig. 6d), it is possible to observe that the bending strain is not as affected as the bending stress and bending modulus, because the values are much more uniform. However, the processes that promoted higher bending stress and bending modulus were those that, as expected, presented lower values of bending strain. For example, compared to the control samples (1.57 %), the bending strain decreased between 1.3% and 1.9%.

Conclusions

CNFs were used to improve the mechanical properties of a commercial epoxy resin and a laminate composed by eight layers of bi-directional carbon woven prepared by hand lay-up process.

It was possible to conclude that the rotation speed of the mixer, dispersion time of the CNFs and vacuum time applied, both in the resin after its preparation and in the laminate after its manufacture, influence the mechanical properties of these materials.

In terms of resin, the best mechanical properties were obtained when 0.5 wt.% of CNFs were mixed at 1000 rpm for 3 h, simultaneously using an ultrasound bath, followed by degassing for 10 minutes in a vacuum chamber. For example, the resin enhanced by CNFs reached values around 12.6% higher than the bending stress obtained for neat resin produced without vacuum. Finally, in relation to laminated composites, it was found that the manufacturing process that promoted the best mechanical properties involved two vacuum stages. In this case, resin and CNFs were mixed at 1000 rpm for 3 h followed by 2 min in a vacuum chamber. The system composed of fibres and nano-enhanced resin was, subsequently, placed in a vacuum bag and a 2.5 kN load was applied for 24 h. Finally, the bag remained attached to a vacuum pump for another 2 minutes. The benefits obtained with CNFs, compared to the composites with neat resin, were around 15.6% higher, and the extra step of vacuum promoted improvements of 16.6%.

Acknowledgements

This work was supported by the project Centro-01-0145-FEDER-000017 - EMaDeS - Energy, Materials and Sustainable Development, co-financed by the Portugal 2020 Program (PT 2020), within the Regional Operational Program of the Center (CENTRO 2020) and the European Union through the European Regional Development Fund (ERDF).

References

[1] B. Middleton (2016) Composites: Manufacture and Application. Design and Manufacture of Plastic Components for Multifunctionality, Elsevier, pp. 53–101.

[2] M. Sussmann, M. Amirkhosravi, M. Pishvar, M.C. Altan (2018) Fabrication of High Quality, Large Wet Lay-Up/Vacuum Bag Laminates by Sliding a Magnetic Tool, Polymers, 10, 992.

[3] K. Abdurohman, T. Satrio, N.L. Muzayadah, Teten (2018) A comparison process between hand lay-up, vacuum infusion and vacuum bagging method toward e-glass EW 185/lycal composites, J. Phys.: Conf. Ser., 1130, 012018.

[4] J.A.M. Ferreira, P.N.B. Reis, J.D.M. Costa, C. Capela (2014) Assessment of the mechanical properties of nanoclays enhanced low T_g epoxy resins, Fiber Polym., 15, pp. 1677-1684.

[5] P. Santos, A. Maceiras, S. Valvez, P.N.B. Reis (2021) Mechanical characterization of different epoxy resins enhanced with carbon nanofibers, Frattura ed Integrità Strutturale, 55, pp. 198-212.

[6] M.H. Woldemariam, G. Belingardi, E.G. Koricho, D.T. Reda (2019) Effects of nanomaterials and particles on mechanical properties and fracture toughness of composite materials: a short review, AIMS Materials Science, 6(6), pp. 1191-1212.

[7] P.N.B. Reis, J.A.M. Ferreira, Z.Y. Zhang, T. Benameur, M.O.W. Richardson (2013) Impact response of Kevlar composites with nanoclay enhanced epoxy matrix, Composites: Part B 46, pp. 7-14.

[8] Y. Zare (2016) The roles of nanoparticles accumulation and interphase properties in properties of polymer particulate nanocomposites by a multi-step methodology, Compos. Part A Appl. Sci. Manuf., 91, pp. 127-32.
[9] J. Oberdisse (2006) Aggregation of colloidal nanoparticles in polymer matrices, Soft Matter, 2(1), pp. 29-36.
[10] V. Padmanabhan, A.L. Frischknecht, M.E. MacKay (2012) Effect of chain stiffness on nanoparticle segregation in polymer/nanoparticle blends near a substrate, Macromol. Theory Simulations, 21(2), pp. 98-105.
[11] X. Ma, Y. Zare, K.Y. Rhee (2017) A Two-Step Methodology to Study the Influence of Aggregation/Agglomeration of Nanoparticles on Young's Modulus of Polymer Nanocomposites, Nanoscale Res. Lett., 12, pp. 0-6.
[12] M.S.P. Shaffer, X. Fan, A.H. Windle (1998) Dispersion and packing of carbon nanotubes, Carbon N. Y., 36(11), pp. 1603-12.
[13] P.N.B. Reis, J.A.M. Ferreira, F.V. Antunes, and J.D.M. Costa (2007) Flexural behaviour of hybrid laminated composites, Compos. Part A-Appl. S., 38, 1612-1620.
[14] J.W. Giancaspro, C.G. Papakonstantinou, P.N. Balaguru (2010) Flexural response of inorganic hybrid composites with E-Glass and Carbon fibers, J. Eng. Mater. Technol., 132, 021005-1.

Experimental Study of Tensile, Flexural and Impact Strength of Basalt/Glass FRP Hybrid Composite with Carbon Black

MITHILESH K M [a] and Dr. K MAHESH DUTT [b]

a Department of Mechanical Engineering, PES university- Electronics City Campus, Bengaluru, India.
b Department of Mechanical Engineering, Dayananda Sagar Academy of Technology and Management Bengaluru, India and affiliated to Visvesvaraya Technological University, Belagavi, Karnataka, India.
Email: mithileshkm@pes.edu

Abstract. Metal components have been replaced by composite ones in the modern engineering aspects due to various reasons such as strength to weight ratio, ease of manufacturing as well as aesthetic value. Currently majority of composites are made by using synthetic fibres which again is a concern for the environment because of non-biodegradability issues with synthetic fibres. Hence natural fibres appear to be an alternative option to synthetic fibres. But the major drawback would be the strength of the fibres and its behaviour with its surrounding environment hence hybrid appears to be the possible solution in order to balance both the constraints of strength and environment. This study describes the mechanical properties of a hybrid composite comprising of Basalt fibre and Glass fibre developed by using hand lay-up techniques. Woven Basalt fibres and Glass fibres were used for fabricating the hybrid composites, carbon black as nano fillers, vinyl ester resin as matrix with different fibre stacking sequences. The result shows that the reinforcement of woven Basalt and Glass fibres hybrid composites with carbon black has a significant influence on the mechanical properties of the composites. The stacking sequence also has an effect on the strength of the specimen. The tensile strength of hybrid composite having stacking sequence B/3G/B/3G/B increase by 15.49 % with the addition of 9% carbon black whereas the stacking sequence G/3B/G/3B/G has a decrease in tensile strength by 21.77%. The flexural strength of the hybrid B/3G/B/3G/B decrease by 72.5% with the addition of carbon black while the hybrid G/3B/G/3B/G shows an increase by 32.38%. The highest impact strength is of basalt with 6% composite.

Keywords— Basalt, Hybrid composite, Stacking sequence

1. Introduction

Hybrid composites are newly developed composites when contrasted with regular poylmer composites. Hybrids can have more than one reinforcement stage and unimatrix stage or single reinforcement stage with multiple matrix stages or numerous reinforcement and different matrix stages. They have better adaptability with relatively other fiber reinforced composites. Regularly it contains a high modulus fiber with low modulus fibre. The high-modulus fiber gives the rigidness and stress bearing characteristics, though the low-modulus fiber makes the composite more tear tolerant and keeps the material cost low. The mechanical properties of a hybrid composite can be differed by varying volume ratio and stacking sequence of different plies [1,2]. The enthusiasm for natural fiber-strengthened polymer composite materials is quickly becoming a go for in terms of industrial applications and research. They are sustainable, cost effective and totally or somewhat recyclable. Plants, for example, cotton, hemp, jute, bamboo, just as wood, utilized from time immemorial as a wellspring of lignocellulose filaments, are increasingly more frequently connected as the support of composites. Their accessibility, sustainability, low thickness, and cost just as an agreeable mechanical properties make them an alluring environmental choice to glass, carbon and man-made strands utilized for the assembling of composites [3]. The natural fibre-containing composites are used in transportation, military applications, building and construction industries (ceiling panelling, partition boards), packaging, consumer products, etc. [4,5].

2. Experimental procedure

2.1 Materials

2.1.1 Glass fibre

Glass fibre is a material containing various incredibly fine strands of glass. Glass fibre has generally practically identical mechanical properties to different strands, for example, polymers and carbon fibre. In spite of the fact that not as solid or as inflexible as carbon fibre, it is a lot less expensive and altogether less fragile when utilized in composites. Properties are shown in table 1.

Table 1: Glass fibre properties

Property	Value
Density	2.60x103 kg/m^3
Tensile Strength	3.45 GPa
Specific strength	1.33 kg/m^3
Elastic Modulus	85 N/m^2

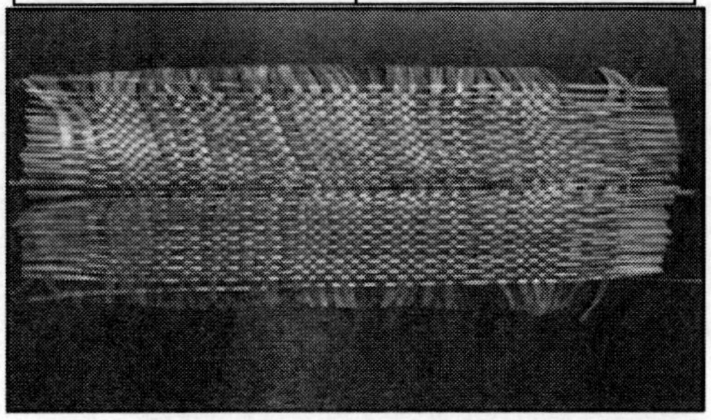

Figure 1: E Glass fibre 360 GSM

2.1.2 Basalt Fibre

Basalt fibre is a material produced using incredibly fine filaments of basalt, which is made out of the minerals plagioclase, pyroxene, and olivine. It is like carbon fibre and fiberglass, having preferable physical mechanical properties over fiberglass, yet being fundamentally less expensive than carbon fibre. It is utilized as a flame resistant material in the aviation and car enterprises and can likewise be utilized as a composite to create items, for example, camera tripods. Properties are shown in table 2.

Table 2: Basalt fibre properties

Property	Value
Density	2.50x103 kg/m^3
Tensile Strength	4.15-4.80GPa
Specific strength	1.57 kg/m^3
Elastic Modulus	100-110 N/m^2

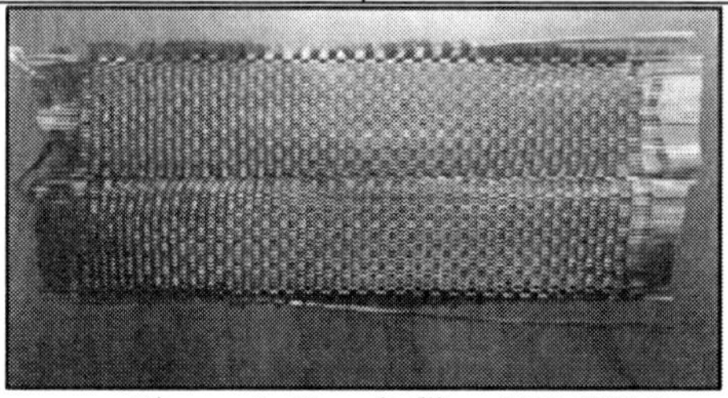

Figure 2: Basalt fiber 360 GSM

2.1.3 Carbon Black

Carbon black is a material manufactured by the inadequate burning of substantial oil based goods, for example, FCC tar, coal tar, ethylene breaking tar, and a little sum from vegetable oil. Carbon black is a type of Para crystalline carbon that has a high surface-region to-volume proportion, though lower than that of initiated carbon. It is not at all like ash in its a lot larger surface-area to-volume proportion and fundamentally lower (insignificant and non-bioavailable) PAH (polycyclic sweet-smelling hydrocarbon) content. Be that as it may, carbon black is generally utilized as a model compound for diesel residue for diesel oxidation tests. Carbon black is principally utilized as fortifying filler in tires and different rubber products. In plastics, paints, and inks carbon black is utilized as a shading color.

Figure 3: Carbon Black

2.1.4 Vinyl Ester Resin

Vinyl Ester is a resin produced by way of the esterification of an epoxy resin with an unsaturated monocarboxylic acid. The reaction product is then dissolved in a reactive solvent, including styrene, to 35–45 percent content by using weight percentage. It can be used as an substitute to polyester and epoxy materials in matrix or composite substances, in which its traits, strengths, and bulk cost intermediate between polyester and epoxy. Vinyl ester has decrease resin viscosity (about 200 cps) than polyester (approximately 500cps) and epoxy (approximately 900cps). The promoter, catalyst and accelerator used in this research work are Di-methyl acetamyde, methyl ethyl ketone peroxide and cobalt naphthalate respectively.

2.2 Preparation of FRP Hybrid Composites

The materials required for fabrication were E- glass fibre of 360 gsm, Basalt fibre of 360 gsm, Carbon black, Vinyl Ester resin, catalyst, promoter, accelerator and Mylar sheet. Fabrication of glass and basalt with vinyl ester resin composite with varying composition of carbon black was done by hand lay-up. Glass and basalt fibre hybrid composites of size 250x250x3 mm were fabricated by varying the percentage of carbon black in the vinyl ester resin and by varying the stacking sequence. The fibres were arranged in the sequence B/3G/B/3G/B and G/3B/G/3B/G. Carbon black was varied in the percentage of 0, 3, 6 and 9 by weight of the resin. The fibre and resin amount was calculated for the fibre resin ratio of 60:40 by weight as shown in Figure 1.The prepared laminates were then cut into different shapes and dimension by a wire hacksaw based on the ASTM standards. The specimen was cut into dog bone shape having dimension 33x6x3mm for tensile test, flat bar shape of dimensions 100x10x3mm and 64x8x3mm for flexural and impact test respectively. The specimens were then tested based on ASTM standards D638 and D790 for tensile and flexural strength respectively in a computer assisted universal testing machine. The impact test was performed on a digital Izod test machine according to ASTM D256.

Cutting of fibers

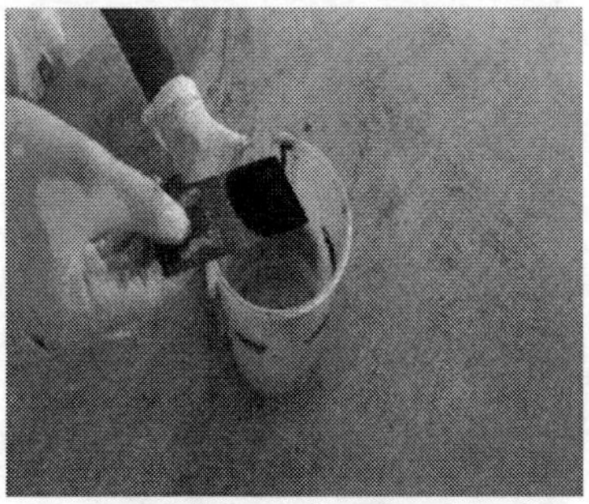
Addition of carbon black

Stirring of resin

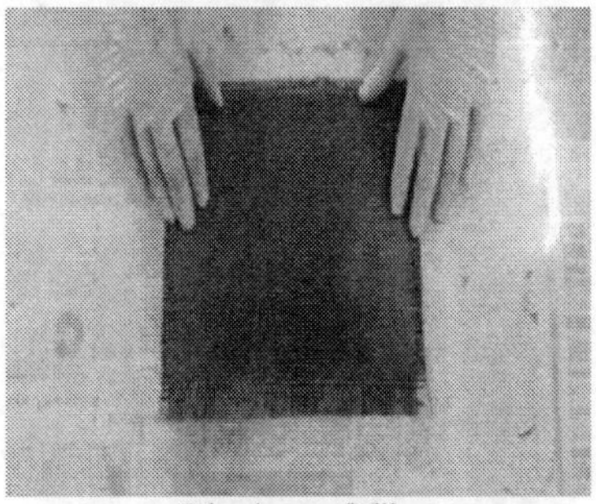
Placing of fibers

Application of resin

Spreading of resin

Figure 4: The Handlay Up process employed.

2.3 *Testing*

2.3.1 *Tensile Test*

The tensile test for the specimens was done according to ASTM D638. The specimens of size 216 mm x 19 mm x 3mm were tested with a guage length of 105mm at a cross head speed of 1 mm/ min. The fixtures consideredifor the tensile testing is shown in the Figure 5.

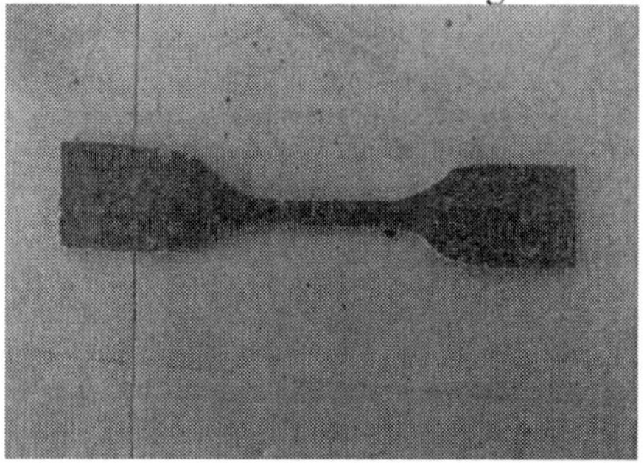

a) Tensile Specimen

b) Specimen fixed UTM for tensile test

Figure 5: Tensile Test specimen and loading

2.3.2 *Flexural Test*

The mode of flexural test to know the mechanical properties of polymeric composites is widely accepted because of the relative ease of the test method, instrumentation and testing equipment needed. The flexural strengths of the specimens were determined using the three-point bending test as per ASTM-D790. The specimens of dimensions 80mm x 8mm x 3mm were tested with a span length of 50 mm. Specimen and the method of fixing are shown in figure 6

a) Flexural Specimen

b) Specimen fixed UTM for flexural test

Figure 6: Flexural Test specimen and loading

2.3.3 *Impact Test*

Impact test was carried out using the Izod Impact Testing Machine supplied by M/s International Equipments, Mumbai, India. The test was conducted as per ASTM D256.The specimens of dimensions 64 mm x 8 mm x 3 mm were tested. The impact strength was determined using the equation. Figure 7 shows the Impact testing machine.

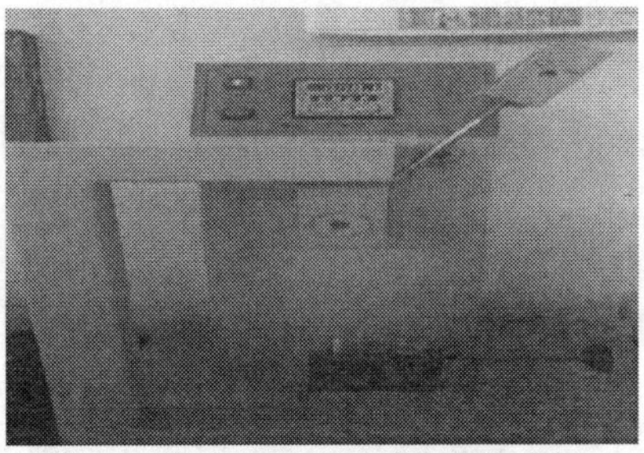
Figure 7: Impact testing Machine

Table 3: ASTM Standard adopted for testing

Sl.No	ASTM standards	Tests
1	D638	Tensile test
2	D790	Flexural test
3	D256	Impact test

3. Results and Discussion

3.1 Tensile test

It was observed that the combination of glass and basalt having G/3B/G3B/G with carbon black 0% had the highest tensile strength with a value of 274.28 Mpa. Basalt composite with 0% and 3 % carbon black have the same tensile strength of 228 MPa. Glass composite having the 3% carbon black had the least tensile strength of 138 Mpa. The tensile strength of hybrid composite having stacking sequence B/3G/B/3G/B was increased by 15.49 % with the addition of 9% carbon black whereas the stacking sequence G/3B/G/3B/G had a decrease in tensile strength by 21.77%. The tensile strength decrease for glass composite by 73.72% with the addition of 9% carbon black while for basalt there is an increase of 10.52 % for the addition of carbon black. It was observed that the addition of 3% carbon black to basalt has no significant effect on the tensile strength since the values obtained were in close proximity to each other. It was also observed that the tensile strength drops for 9% carbon black for individual fibre composites. However, there was a steady increase in the tensile strength for hybrid composites with 9% carbon black. This can be attributed to the increase in strength of two different fibres stacked together and the addition of carbon black. Variations is shown in the Figure 8.

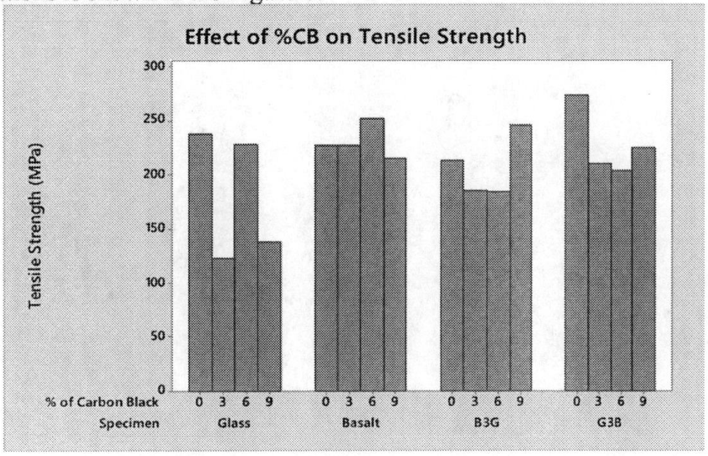

Figure 8: Variation of Tensile Strength.

3.2 Flexural test

It is observed that glass composite with 6% carbon black has a flexural strength of 304.08 MPa. This is 86.88% higher than glass with 0% carbon black. It is also observed that the flexural strength of composites is very similar for 0% and 3% carbon black. Basalt with 6% carbon black is the best in its category with a flexural strength of 242.79 MPa. The flexural strength of the hybrid B/3G/B/3G/B decrease by 72.5% with the addition of carbon black while the hybrid G/3B/G/3B/G shows an increase by 32.38%. The glass composite shows an increase by 86.88% flexural strength with the addition of carbon black while basalt shows an increase of 40.69%. Variations is shown in the Figure 9.

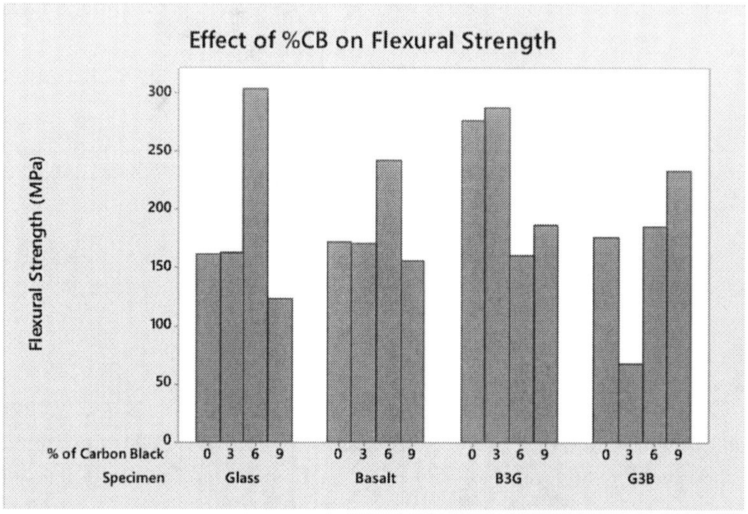

Figure 9: Variation of Flexural Strength.

3.3 Impact Test

It is observed that the impact strength of basalt fibre composite with 6% carbon black is the highest with a value of 0.4083 J/mm2. Hybrid composites with stacking sequence of B/3G/B/3G/B have similar impact strength of 0.3 J/mm2.the different values are shown in the Figure 10.

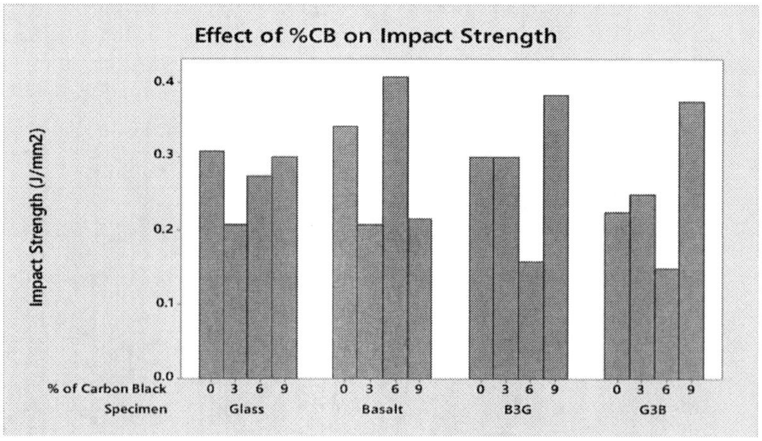

Figure 10: Variation of Impact Strength.

3.4 Main Effects Plots

The main effects plot for the composite with different carbon back percentages are show in the Figure 11and Figure 12 This was done using MINITAB® software and the analysis results showed that the carbon black with 6 % had major effect on the composite behaviour and it was close to the values without addition of carbon back to the composite and the other two percentages 3 and 9 filled FRP Composite did not feature favourable results. Also the Hybrid composite with stacking sequence B3G had the most effect when we add the carbon black as indicate by the plot.

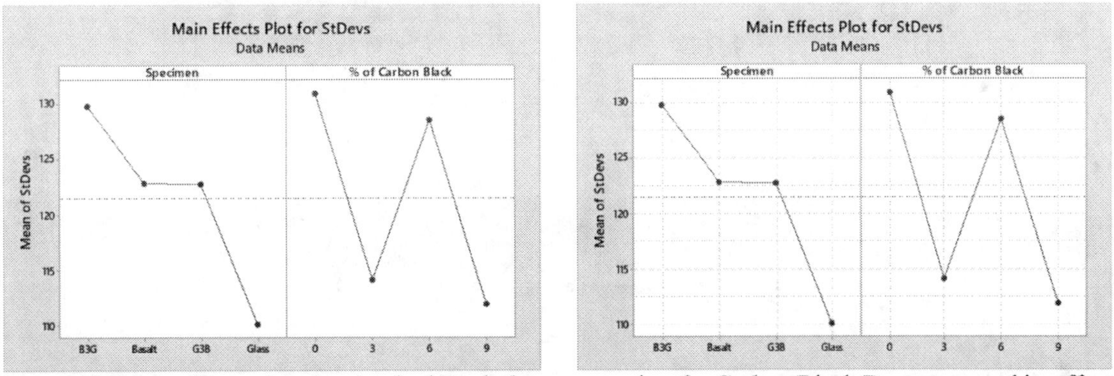

Figure 11: Main effects plots for standard Deviations comparing the Carbon Black Percentage and its effect on different stacking sequence.

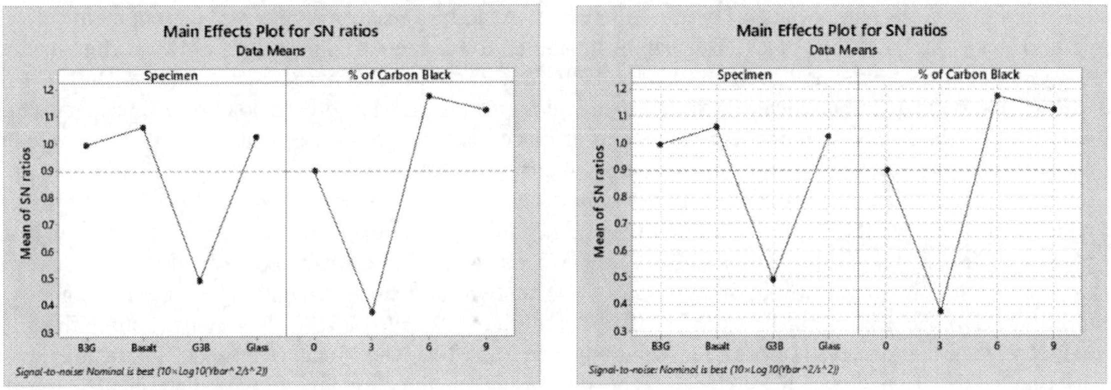

Figure 12 Main Effect plots for SN ratio comparing the Carbon Black Percentage and its effect on different stacking sequence.

3.5 Residual Plot

From the Figure 13 the residuals are random scattered and it follows linear assumption meaning the carbon black has an effect on the mechanical properties of the hybrid composite. Also the normal probability plot indicates the data are in line with the predicted model and the experimental results are good and acceptable

IOP Conf. Series: Materials Science and Engineering 1126 (2021) 012013 doi:10.1088/1757-899X/1126/1/012013

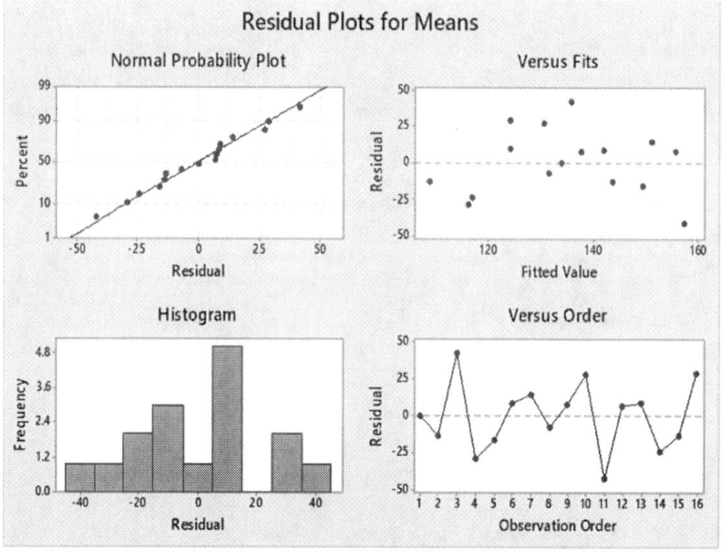

Figure 13 Residual Plots of the data obtained.

4. Conclusion

Stacking sequence has an effect on the strength of the specimen. The properties are anisotropic. This is due to the difference in orientation of plies. It is observed that the tensile strength drops for 9% carbon black for individual fibre composites. However, there is a steady increase in the tensile strength for hybrid composites with 9% carbon black. This can be attributed to the increase in strength of two different fibres stacked together. The flexural strength of B/3G/B/3G/B (276.43 MPa) is higher than G/3B/G/3B/G (176.51 MPa). This is due to the dominance exerted by the glass fibre which has greater load bearing capacity than basalt fibre. The steady drop and rise in the load displacement and stress strain curves can be explained as a result of incomplete breakage of fibres. A drop in curve indicates the breakage of fibres. The sudden rise indicates that there is a layer of fibre still yet to break and hence the load is increased further until there is no more rise in curve. The impact strength of B/3G/B/3G/B composites for all compositions of carbon black is greater than the impact strength of G/3B/G/3B/G. This can again be attributed to the dominance exerted by the glass fibre in the respective composites. The highest tensile strength is of hybrid composite G/3B/G/3B/G with 0% carbon black (274 MPa). Aluminium P/M alloy 201AB has a tensile strength of 303 MPa and Stainless steel SS316 has a tensile strength of 262 MPa. The Model main plots and residual plots also agrees the same. Hence this hybrid composite can replace the above mentioned materials for specific applications. The highest flexural strength is of glass with 6% carbon black (304.08 MPa). LCF40-PEEK Polyether ketone, 40% Long Carbon Fibre Reinforced has a flexural strength of 304 MPa. Hence this 6% carbon black glass composite can replace this material depending upon the need of application. The highest impact energy is of basalt fibre with 6% carbon black (9.8 J). Sintered Ultimet 04,304 has impact energy of 10.8 J which is very close to 6% carbon black. The basalt fibre with 6% carbon black can replace the above mentioned material.

References

[1] Kuruvilla Joseph.et.al "A Review on Sisal Fibre Reinforced Polymer Composites" Revista Brasileira De Engenharia Agrícola E Ambiental,V.3, N.3, P.367-379, 1999
[2] Gururaja M N and A N Hari Rao "A Review on Recent Applications and Future Prospectus of Hybrid Composites" International Journal of Soft Computing and Engineering (IJSCE) ISSN: 2231-2307, Volume-1, Issue-6, January 2012, pp.352-355.
[3] Girisha.C.et.al "Mechanical Performance of Natural Fibre-Reinforced Epoxy-Hybrid Composites" Engineering Research and Applications (IJERA) ISSN: 2248-9622, Vol. 2, Issue 5, September- October 2012, pp.615-619.
[4] Mohd Yussni Hashim.et.al "Plant Fibre Reinforced Polymer Matrix Composite: A Discussion on Composite Fabrication and Characterization Technique" Tun Hussein Onn University of Malaysia (UTHM).
[5] Girisha.C.et.al "Effect of Alkali Treatment, Fibre Loading and Hybridization On Tensile Properties of Sisal Fibre, Banana Empty Fruit Bunch Fibre and Bamboo Fibre Reinforced Thermoset Composites" International Journal of Engineering Science & Advanced Technology ISSN: 2250– 3676, Volume-2, Issue-3, pp.706 – 711.
[6] Mohd Yussni Hashim.et.al "Mercerization Treatment Parameter Effect on Natural Fibre Reinforced Polymer Matrix Composite: A Brief Review", World Academy of Science, Engineering and Technology 68 2012
[7] R.A. Braga, "Analysis of the mechanical and thermal properties of jute and glass fibre as reinforcement epoxy hybrid composites" Materials Science and Engineering C 56 (2015) 269–273.
[8] Dr. Richard Parnas, Dr. Montgomery Shaw, "Basalt Fibre Reinforced Polymer Composites", the New England Transportation Consortium August, 2007, NETCR63
[9] I.D.G. Ary Subagia, Yonjig Kim, Leonard D. Tijing, Cheol Sang Kim, Ho KyongShon, "Effect of stacking sequence on the flexural properties of hybrid composites reinforced with carbon and basalt fibres" Composites: Part B 58 (2014) 251–258.
[10] Rohan Muni Bajracharya, Allan C. Manalo, WarnaKarunasena, Kin-tak Lau, "Experimental and theoretical studies on the properties of injection moulded glass fibre reinforced mixed plastics composites", Composites: Part A 84 (2016) 393– 405.
[11] PatilDeogonda, Vijaykumar N Chalwa, "Mechanical Property of Glass Fibre Reinforcement Epoxy Composites", International Journal of Scientific Engineering and Research (IJSER), 2347-3878.
[12] Huaian Zhang, Yiming Yao, Deju Zhu, BarzinMobasher, Liang Huang, "Tensile mechanical properties of basalt fibre reinforced polymer composite under varying strain rates and temperatures", Polymer Testing 51 (2016) pp. 29-39
[13] M.D. Samper, R. Petrucci, L. Sanchez-Nacher, R. Balart, J.M. Kenny, "Properties of composite laminates based on basalt fibres with epoxidized vegetable oils", Materials and Design 72 (2015) pp.9–15.
[14] B. Soares, R. Preto, L. Sousa, L. Reis, "Mechanical behaviour of basalt fibres in a basalt-UP composite", XV Portuguese Conference on Fracture, PCF 2016,10-12 February 2016, pp.82-89.
[15] Qinglin Wu, Kai Chi, Yiqiang Wu, Sunyoung Lee," Mechanical, thermal expansion, and flammability properties of co-extruded wood polymer composites with basalt fibre reinforced shells", Materials and Design 60 (2014) pp.334–342.
[16] L. M. P. Durao, D. J. S. Goncalves, J. M. R. S. Tavares, V. H. C. de Albuquerque and A. Torres Marques - "Drilling process of composite laminates, a tool based analysis" Presented at the 14TH European conference on composite materials, 7-10 June 2010, Budapest, Hungary, pp. 219-227.
[17] R.M. Kulkarni, H. N. Narasimha Murthy, G.B.Rudrakshi, Sushilendra. "Effect of drilling parameters in drilling of glass fibre reinforced with vinyl ester and carbon black Nano composite". International journal of scientific & technology research, volume3, issue7, July 2014.
[18] P. Amutakkaanan, V. Manikandan, M. Uthayakumar."Analysis of delamination in drilling of basalt fibre reinforced polymer composites" Materials Physics and Mechanics 24 (2015) pp. 1-8

[19] M. K. A MohdAriffin, M. I. Mohd Ali, S. M. Sapuan and N. Ismail- "An optimize drilling process for an aircraft composite structure using design of experiments". Scientific Research and Essay Vol.4, October 2009; pp. 1109-1116.
[20] A. Langella and M. Durante – "Comparison of Tensile Strength of Composite Material Elements with Drilled and Moulded-in Holes". Appl Compos Mater, 2008. vol15; pp 227–239
[21] Sadasiva Rao T., Rajesh V. and Venu Gopal A- "Taguchi based Grey Relational Analysis to Optimize Face Milling Process with Multiple Performance Characteristics". International Conference on Trends in Industrial and Mechanical Engineering March 24-25, 2012, Dubai.
[22] A. Al-Refaie, L. Al-Durgham, and N. Bata- "Optimal Parameter Design by Regression Technique and Grey Relational Analysis". Proceedings of the World congress on Engineering 2010 vol III WCE 2010, London, U.K.9.
23] H. Hocheng and C.C. Tsao – "The path towards delamination-free drilling of composite materials". Journal of Materials Processing Technology vol 167, 2005; pp. 251–264.
[24] Murthy B.R.N, Lewlyn L.R. Rodrigues and Anjaiah Devineni- "Process Parameters Optimization in GFRP Drilling through Integration of Taguchi and Response Surface Methodology". Research Journal of Recent Sciences, Vol. 1(6), June 2012; pp. 7 -15.
[25] U.A. Khashaba- "Delamination in drilling GFR-thermoset composites" Composite Structures 6, 2004; pp. 313–327.
[26] Sathishrao.v. And Dr. Lewlyn.L.R. Rodrigues. "Influence of Machining Parameters on Tool Wear in Drilling of GFRP Composites –Taguchi Analysis and ANOVA Methodology". Proc. of the Intl. Conf. on Advances in Mechanical and Robotics Engineering -- MRE 2014
[27] M.C. Murugesh and K. Sadashivappa. "Study on influence of filler material on glass fibre/epoxy composite laminating during drilling". International Journal of Advances in Engineering & Technology, March 2012, Vol. 3, Issue 1, pp. 233-239
[28] Sakthivel1, S. Vijayakumar, Channankaiah, D. Athikesavan "A review on thrust force and torque in drilling of glass reinforced polymer composites". Journal of Chemical and Pharmaceutical Sciences January-March 2015,Volume 8 Issue 1

An environment friendly synthesis and characterization of Zinc Oxide nanoparticles using *Mentha viridis* leaf extract

Rashedul Alam Chowdhury[a, 1*], Md. Mahmudul Hassan[a], Shaon Das[a, c], Sajib Aninda Dhar[b], Md. Moniruzzaman[a]

[a]*Department of Materials and Metallurgical Engineering, Bangladesh University of Engineering and Technology, Zahir Raihan Road, Dhaka, 1000, Bangladesh*

[b]*Pilot Plant and Process Development Centre (PP & PDC), Bangladesh Council of Scientific and Industrial Research, Dr. Qudrat I Khuda Road, Dhaka, 1205, Bangladesh*

[c]*Nuclear Safety and Security and safeguard Division, Bangladesh Atomic Energy Regulatory Authority (BAERA), E-12/A, Shahid Shahabuddin Shorok, Dhaka,1207, Bangladesh*

[1]* Corresponding author.

E-mail address: rashedulachowdhury@gmail.com

Abstract. Zinc Oxide Nanoparticles (ZnO Nps) are one of the most predominant inorganic nanoparticles, which have diverse applications in the field of cosmetic, coating, medicine and electronic sensors. Morphology and particle size of nanoparticles mainly control its advantages and drawbacks in applications. In this study, an environment friendly and economic synthesis of ZnO Nps was carried out using *Mentha viridis* leaf extract as reducing and stabilizing agent to investigate the morphology, particle size and other properties of the synthesized ZnO nanoparticles. A series of characterization methods: ultraviolet-visible (UV-Vis) spectroscopy, x-ray diffraction analysis, field emission scanning electron microscopy (FESEM) and energy dispersive x-ray analysis (EDX) were used to confirm the characteristics of the synthesized nanoparticles. It was evident from the XRD analysis that ZnO Nps possess a polycrystalline hexagonal structure (as per the International Centre for Diffraction Data (ICDD) number #98-002-9272) with an average crystallite size of 32.3 nm. The spectra showing peaks in the EDX data indicated the presence of Zn (50.35 %) and O (49.65 %) in higher percentage and FE-SEM analysis revealed the morphology and Average Length and diameter of the nanorods are 156 nm and 48 nm respectively of the synthesized ZnO Nps. The analysis of the UV visible absorbance spectrum revealed the absorbance band peak at 372nm and confirmed the presence of ZnO Nps. From all these analyses, it is evident that *Mentha viridis* leaf extract can act as a potential reducing agent in the green synthesize ZnO Nps.

1. Introduction

Nanotechnology has framed as a vibrant research field in the recent past. Various science and engineering disciplines pave their way to incorporate nanomaterial with their research topics due to the outstanding properties of nanoparticles. Nanoparticles gain significant attention for their unique surface to volume ratio due to their diminutive size and concerning the bulk form, they exhibit most extensive variations in applicable properties which lead to a diverse application in the field of research such as photonics, semiconductor, medical and packaging [1, 2, 3]. In recent years, the ecofriendly and nontoxic processes have gained great attention in the research arena for producing various metal and metal oxide Nanoparticles. Green synthesis has adopted various bio-friendly materials such as plant extracts, fungi, algae to use as reducing and capping agent to synthesize nanoparticles [4,5,6]. The use of plant extracts offers a green route to control the variance of the shape, size and morphology of the particles [7]. Here plant biodiversity has been a great source for phytochemicals. Flavonoids and so many diverse types of aliphatic or aromatic hydrocarbons are such kind of phytochemicals [8]. Among the various metal and metal oxide nanoparticles, Zinc Oxide Nanoparticle is one of the most intriguing materials which can be produced by Green synthesis process. There are two distinct reasons for this interest: one is due to previous physical and chemical processes were environmentally hazardous and the other reason is, it has extended wide application in Agriculture, Biotechnology, Electronic sensor, Packaging Industry [9,10,11].

In this study, leaves of *Mentha viridis* used as a plant extract to synthesize ZnO Nanoparticles. *Mentha viridis* belongs to Mentha genus and widely available as an essential plant in the Indian subcontinent area. This plant is widely recognized as a medicinal plant because of its use in digestive disorders, headache and tiredness, anti-inflammatory and antimicrobial [12,13,14]. Abdelhakim Bouyahya et el. have shown the contribution of different phytochemicals present in *Mentha viridis* as antidiabetic, dermatoprotective, antioxidant and antibacterial properties [15]. So, the main purposes of choosing leaves of *Mentha viridis* as plant extract are: cost-effective, widely available, non-hazardous and most importantly rich in effective phytochemicals who will contribute to the synthesis as reducing agent. For the first time, this study synthesizes the ZnO Nps in which *Mentha viridis* propagate the system.

2. Materials and Methods

2.1. Materials

Locally produced *Mentha viridis* were collected from Keranigonj which situated at Dhaka, Bangladesh. The leaves were in fresh condition. Precursor of the synthesis was Zinc acetate powder, producer 'Sigma-Aldrich'. HPLC grade Methanol was used which is produced by Loba Chemi, India (purity-99.8%). Water which was used for the solution and cleaning was also HPLC grade. (Producer PubChem)

2.2. Preparation of Mentha viridis leaves extract

The leaves of *Mentha viridis* were cut to 8-10 mm and washed with running tap water at first, then washed again with deionized water thoroughly. 25 gm of leaves was added to 100 ml 70% HPLC gradient grade Methanol solution. The solution heated to 1 hour with continuous stirring on the hotplate at 65°C. Subsequently, the solution was kept to cool. Whatman No.1 filter paper was to attain the clear extract of Mentha viridis leaves. It was kept at 4°C for the later use.

2.3. Synthesis of Zinc Oxide Nano particles using Mentha viridis leaves

For the synthesis process, 10 ml *Mentha viridis* leaves extract solution was added to 90 ml of 1M zinc acetate aqueous solution. The solution heated to 60 °C and stirred continuously for 1 hour. During this process pH of the mixture was maintained in between 8 to 8.5 by instantly prepared 1M NaOH solution to achieve smaller size nanoparticles. Subsequently, the solution dried at 90°C for 1 day. Afterwards dried powder transferred to a porcelain crucible for calcination and heated to 500°C for 2 hours. Finally synthesized dried powder of ZnO Nps was preserved for further studies.

2.4. *Characterization methods*

2.4.1. *X-ray diffraction*

Synthesized Zinc oxide Nanoparticles was investigated by X-ray diffractometer. Thus we got the crystal orientation and phase. Rietveld refinement method was used to determine the size of the crystal, parameter of lattice and analyze the phase of the particles.

Table 1: Operating parameters of XRD

Operating parameters	Values
Target	Cu
λ	1.5406A°
Scanning Speed	0.1°/min
Step Differences	0.02°
Voltage Source	40 kv
Tube current	40 mA
Range of 2θ	20°-80°
Mirror	Gobel mirror

2.4.2. *FESEM and EDX*

A field emission scanning Microscope was used to analyze the morphology of producing Zinc oxide nanoparticles. Size, shape, distribution system, thus identified by the image of SEM. Linear intercept method was used to determine the average grain size. To attain the elemental composition EDX analysis was procured by the same machine. Operating parameter: Voltage 15 kV, Probe Current 1.00 nA and acquisition range of time 50-100s.

2.4.3. *UV-Vis*

To obtain the absorption spectrum UV-Visible absorption spectrophotometer (Shimadzu UV-1800, Japan) was used. The characteristic absorption spectrum confirms the stable production of the ZnO. Scanning parameter was the 600-200 nm with the speed of 200nm/min.

3. Result and Analysis

3.1. *X-ray diffraction analysis*

Figure 1 exhibits the XRD spectrum of ZnO nanoparticles synthesized from the leaf extracts of *Mentha viridis*. The sharp diffraction peaks were identified at the 2θ values of 31.73°, 34.42°, 36.25°, 47.53°, 56.56°, 62.78°, 66.33°, 67.82° and 69.1°. These peaks were corresponded to the lattice planes of (100), (002), (101), (102), (110), (103), (200), (112) and (201) respectively. These distinct peaks conformed to the International Centre for Diffraction Data (ICDD) file no: 36-1451.

The formation of narrow and sharp characteristic diffraction peaks ascertains the crystallinity of biosynthesized ZnO. Furthermore, no characteristic diffraction peak other than ZnO was identified. That indicates the formation of pure ZnO. The average crystallite size of the synthesized ZnO NPs was 32.3 nm.

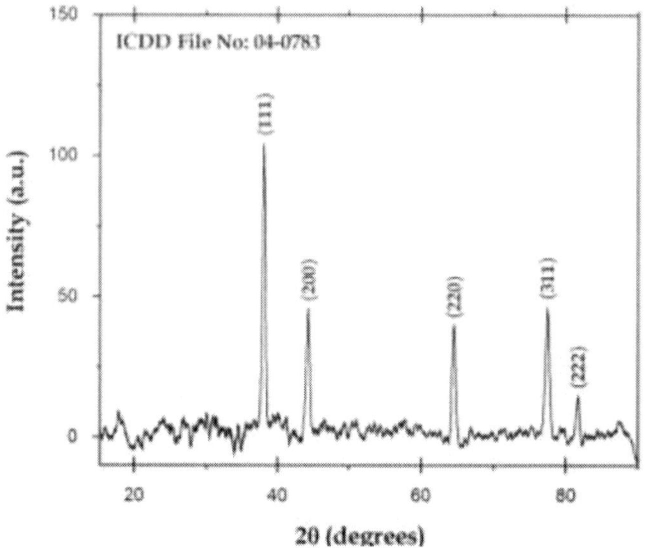

Figure 1: X-ray Dirrfractometer pattern of Zinc oxide nanoparticles prepared by using *Mentha viridis*

Table 2: The rietveld refinement parameters of crystallographic data of Zinc oxide and goodness of fit parameters

Parameters	Value
Crystal orientation	Hexagonal
Space group	P63mc (186)
Quantitative analysis	100 %
Lattice Parameter, a (Å)	3.25145
c (Å)	5.20834
Cell volume, V (Å3)	47.685
Crystallite size (Lvol-IB) K:1	32.3 nm
(LVol-FWHM) K:0.89	38.4
Goodness of fit parameters (R indices)	
Rexp	9.31
Rwp	16.84
Rp	13.79
Goodness of fit	1.809

3.2. FE-SEM and EDX analysis

In the Figure 2 FESEM image of Synthesized ZnO nanoparticles, morphology of the nanoparticles clearly depicts the nanorod like structure. Some of the nanorods are clustered and some of them are free. Average Length and diameter of the nanorods are 156 nm and 48 nm respectively. Nanorod of ZnO represents 1D array and used vastly in the field of electronics [16]. Formation and growth of the ZnO nanorod mainly supported by the lowest surface energy of the facet of the nanorod [17]
Energy-dispersive x-ray spectroscopy (EDX) analysis one of the effective techniques to determine the elemental composition of particles. Actually the surface plasma resonance causes an absorption peak.

EDX confirms the presence of 80.12 % Zn and 19.88 % O in mass percentage scale in Figure 3, represents the purity of the Zinc oxide.

Figure 2: FESEM micrographs of Zinc Oxide prepared by *Mentha viridis* leaf extracts

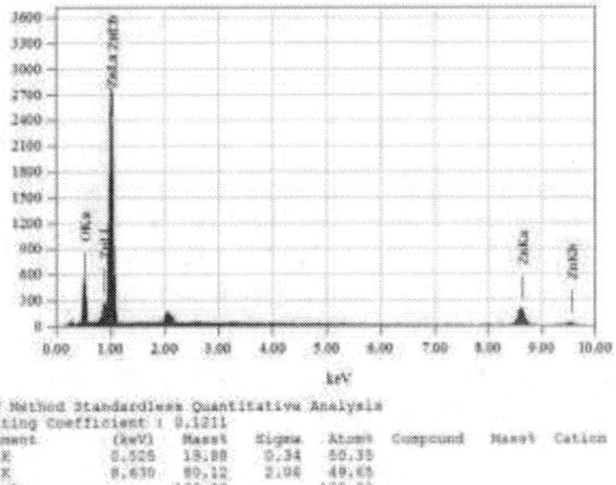

Figure 3: EDX of ZnO nanoparticles by *Mentha viridis*

3.3. *UV-vis spectrum analysis*

In this study, optical property of green synthesized ZnO Nps investigated by absorbance spectrum from UV-vis characterization. Previous studies indicated that an absorption spectrum around 370nm confirms the successful synthesis of ZnO Nps From Fig. 4, absorption peak obtained at 372nm confirmed the presence of ZnO Nps. Absorbance peak by ZnO [18,19,20] Nps mainly depends on particle size and shape which depends on several factors: calcination temperature, pH value, concentration of the precursor etc. [21,22,23]

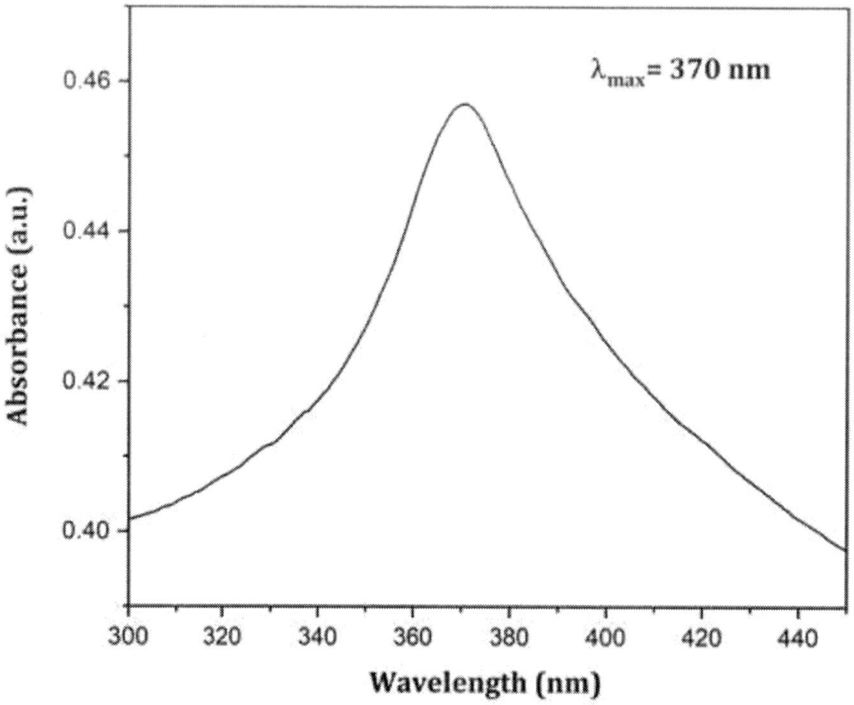

Figure4: Spectrum for Zinc oxide nanoparticles mediated by *Mentha viridis* leaf extract

4. Conclusion

This paper attempted an environment friendly synthesis of ZnO Nps mediated by the extract of leaves of Mentha Viridis. The characterization results show the shape of the Zinc Oxide particle like Rod, which is varies in the length within range of 80-135 nm and within diameter of 20-35 nm. XRD results show the stable phase is hexagonal wurtzite. Elemental Zinc and Oxygen are found in composition of particles which has been confirmed by the EDX result. The study can be further extended to the antimicrobial impact of ZnO Nps.

Acknowledgements

We have greatly acknowledged the support of Department of MME, BUET and BCSIR (Bangladesh Council of Scientific and Industrial Research), an organisation of the Ministry of Science and Technology, Bangladesh.

References

[1] C. Mahendra, M. Murali, G. Manasa, P. Ponnamma, M.R. Abhilash, T.R. Lakshmeesha, A. Satish, K.N. Amruthesh, M.S. Sudarshana, Antibacterial and antimitotic potential of bio-fabricated zinc oxide nanoparticles of Cochlospermum religiosum (L.), Microb. Pathog. 110 (2017) 620–629.
[2] A.K. Singh, P. Pal, V. Gupta, T.P. Yadav, V. Gupta, S.P. Singh, Green synthesis, characterization and antimicrobial activity of zinc oxide quantum dots using Eclipta alba, Mater. Chem. Phys. 203 (2018) 40–48.
[3] S. Gunalan, R. Sivaraj, V. Rajendran, Green synthesized ZnO nanoparticles against bacterial and fungal pathogens, Prog. Nat. Sci. Mater. Int. 22 (2012) 693–700.
[4] M. Gholami-Shabani, M. Shams-Ghahfarokhi, Z. Gholami-Shabani, A. Akbarzadeh, G. Riazi, S. Ajdari, A. Amani, M. Razzaghi-Abyaneh, Enzymatic synthesis of gold nanoparticles using sulfite reductase purified from Escherichia coli: a green ecofriendly approach, Process Biochem. 50 (2015) 1076–1085.
[5] S. Vijayakumara, C. Krishnakumar, P. Arulmozhi, S. Mahadevan, N. Parameswari, Biosynthesis, characterization and antimicrobial activities of zinc oxide nanoparticles from leaf extract of Glycosmis pentaphylla (Retz.) DC, Microb. Pathog. 116 (2018) 44–48.
[6] A. Raja, S. Ashokkumar, R. Pavithra Marthandam, J. Jayachandiran, C.P. Khatiwada, K. Kaviyarasu, R. Ganapathi Raman, M. Swaminathan, Eco-friendly preparation of zinc oxide nanoparticles using Tabernaemontana divaricata and its photocatalytic and antimicrobial activity, J. Photochem. Photobiol. B Biol. 181 (2018) 53–58.
[7] P. Rajiv, S. Rajeshwari, R. Venckatesh, Spectrochim. Acta A 112 (2013) 384–387.
[8] Doble M, Kruthiventi AK. Green chemistry and engineering. Cambridge: Academic Press; 2007.
[9] Dagdeviren, C., Hwang, S.W., Su, Y., Kim, S., Cheng, H., Gur, O., Haney, R., Omenetto, F. G., Huang, Y. and Rogers, J.A. (2013) Transient, Biocompatible Electronics and Energy Harvesters Based on ZnO. Small, 9, 3398-3404.
[10] Wang, L., Kang, Y., Liu, X., Zhang, S., Huang, W. and Wang, S. (2012) ZnO Nanorod Gas Sensor for Ethanol Detection. Sensors & Actuators, B: Chemical, 162, 237-243.
[11] Cross, S.E., Innes, B., Roberts, M.S., Tsuzuki, T., Robertson, T.A. and McCormick, P. (2007) Human Skin Penetration of Sunscreen Nanoparticles: In-Vitro Assessment of a Novel Micronized Zinc Oxide Formulation. Skin Pharmacology and Physiology, 20, 148-154
[12] Bouyahya, A., Abrini, J., Et-Touys, A., Bakri, Y. and Dakka, N (2017) Indigenous knowledge of the use of medicinal plants in the North-West of Morocco and their biological activities. European Journal of Integrative Medicine 13, 9-25.
[13] Mkaddem, M., Bouajila, J., Ennajar, M., Lebrihi, A., Mathieu, F., Romdhane, M., (2009) Chemical composition and antimicrobial and antioxidant activities of Mentha (Longifolia L. and viridis) essential oils. J. Food Sci. 74 (7), M358–M363
[14] Shaikh, S., Yaacob, H. B. and Rahim, Z. H. A. (2014). Prospective role in treatment of major illnesses and potential benefits as a safe insecticide and natural food presertive of mint (Mentha spp.): a review. Asian J. Biomed. Pharm. Sci., 4 (35): 1-12
[15] Abdelhakim Bouyahyaa, Fatima Lagrouha, Nasreddine El Omarib, Ilhame Bouraisa, Meryem El Jemlic, Ilias Marmouzic, Najoua Salhic, My El Abbes Faouzic, Omar Belmehdid, Nadia Dakkaa, Youssef Bakri. (2014) Essential oils of Mentha viridis rich phenolic compounds show important antioxidant, antidiabetic, dermatoprotective, antidermatophyte and antibacterial properties. Biocatalysis and Agricultural Biotechnology Volume 23, January 2020, 101471
[16] V.Desai, M.A.Haque (2006) Mechanical properties of ZnO nanowires. Sensors and Actuators A 134,169–176

[17] Sini Kuriakose, Biswarup Satpatib and Satyabrata Mohapatra (2014) Enhanced photo catalytic activity of Co doped ZnO Nano disks and nanorods prepared by a facile wet chemical method. Physical Chemistry Chemical Physics issue 25
[18] Samira Shahriyari Rada Ali, Mohamadi Sanib, Sharareh Mohsenic (2019) Biosynthesis, characterization and antimicrobial activities of zinc oxide nanoparticles from leaf extract of Mentha pulegium (L.) Microbial Pathogenesis Volume 131, June 2019, Pages 239-245
[19] Surabhi Siva Kumar, Putcha Venkateswarlu, Vanka Ranga Rao & Gollapalli Nageswara Rao (2013) Synthesis, characterization and optical properties of zinc oxide nanoparticles.International Nano Letters volume 3, Article number: 30
[20] Solabomi Olaitan Ogunyemi,Yasmine Abdallah,Muchen Zhang,Hatem Fouad,Xianxian Hong,Ezzeldin Ibrahim (2019) Green synthesis of zinc oxide nanoparticles using different plant extracts and their antibacterial activity against Xanthomonas oryzae. Artificial Cells, Nanomedicine, and Biotechnology Volume 47,Issue 1 Pages 341-352
[21] Goh, E. G., Xu, X., & McCormick, P. G. (2014). Effect of particle size on the UV absorbance of zinc oxide nanoparticles. Scripta Materialia, 78-79, 49-52.
[22] Surabhi Siva Kumar, Putcha Venkateswarlu, Vanka Ranga Rao & Gollapalli Nageswara Rao (2013) Synthesis, characterization and optical properties of zinc oxide nanoparticles.International Nano Letters volume 3, Article number: 30
[23] Faezeh Malek Mohammadi & Nahid Ghasemi (2018) Influence of temperature and concentration on biosynthesis and characterization of zinc oxide nanoparticles using cherry extract. Journal of Nanostructure in Chemistry volume 8, 93–102

Effect of Welding Parameters on Angular Distortion of Submerged Arc Welded Low Carbon Steel Plates.

Kashish[a], Charu Singh Chaudhary[b], Pradeep Khanna[c]
[a, b] Student, [c] Associate Professor, Department of Mechanical Engineering, NSUT, New Delhi, India.

kashishbadwal@gmail.com

Abstract: Submerged Arc Welding is an important industrial joining process used extensively for welding of thick plates. Hence, finds utility in heavy fabrication industries like ship building and pressure vessel industry. The process is characterized as a high heat input process which sometimes leads to certain weld abnormalities of which angular distortion is the one. The present work is an attempt to analyse the effect of input welding parameters like Wire Feed Rate, Voltage and Welding Speed on angular distortion which is a response parameter. Angular Distortion is caused because of the non uniform thermal stresses experienced by the top and bottom surfaces of the weld with top surface getting the maximum and the bottom surface getting the minimum thermal effects. This causes top surface to experience more contraction as compared to bottom resulting in distortion of weld across the line of weld known as angular distortion. This distortion may result in rejection, scrap or misfit of the weldment with parent structure. It is therefore important to keep this defect to a minimum and the present study is an attempt to work in this direction. The results are expected to be of practical importance from the industrial viewpoint.

Keywords: Submerged Arc Welding, Angular Distortion, Input Parameters, Response Parameters. Mathematical Modelling.

1. Introduction:

Submerged arc welding is one of the extensively used arc welding process which was developed in early 1940s to meet the demands of heavy fabrication industries like ship welding. The process since then has gained popularity owing to its ability to get automated and making long continuous joints on these sections. The process can join plates up to a thickness of 16 mm without the need of any specific edge preparation because of deep penetration capability and high deposition rates[1]. It is also suited for circumferential butt-welds on pipes, pressure vessels and rail wagons in the horizontal position and finds a wide industrial applications[2]. An arc is created between the work piece and a bare solid wire, which is consumed during the welding operation. The electrode is in the form of a spool and the shielding is provided by a blanket of granular, fusible flux, which covers both the arc and molten metal by forming a slag blanket to prevent oxidation[3]. It acts as a thermal insulator, permitting deeper heat penetration[4]. The process involves minimum heat loss, provides flexibility of automated and semi automated welding with no weld spatters, ensure increased protection of weld from atmospheric gas contamination, etc[5]. All the above listed features make SAW a prevalent and popular metal joining process. However, a disadvantage of SAW is that; because of large heat inputs, it often results in unacceptable levels of geometric imperfections such as shrinkage and angular distortion[6]. Angular distortion as shown in fig. 1, is caused when the stresses generated from non uniform thermal expansion/contraction exceed the yield strength of the parent metal, plastic deformation of the metal occurs and this plastic deformation causes the plate edges to rise about the weld line in across the weld[7]. Such welding distortions can lead to rejection of the weldment with high rectification costs[8]. Therefore, distortion needs to be controlled at all stages of the fabrication. The welding distortions are strongly influenced by welding process parameters. Hence, estimating the magnitude and distribution of welding distortions and characterizing the effects of the welding conditions can be of significant importance[9]. The major input factors influencing welding distortion are; weld current, weld voltage, and travel speed. One of the methods to avoid the angular distortion during the fabrication process is to provide an initial angle in the negative direction[10]. This could be an effective way to reduce the angular distortion produced in the workpiece up to a certain level. However, it is difficult to predict the value of negative angle to be given to counter the final angular distortion. This can however be

done by conducting structured preliminary experiments by using some statistical tool to establish relationship between angular distortion and the input parameters.
In the present work, an attempt has been made to develop a mathematical model relating the various input welding parameters with the angular distortion produced. The statistical method of Design of Experiments was found useful for the purpose, where, full factorial technique with three factors, each at two levels was adopted. The developed model can be used to predict the value of angular distortion produced for a given set of input conditions selected from the working range.

Fig. 1 Angular Distortion

2. Experimental setup and Experimentation:

The experimental setup used to carry out the present investigative work is shown in fig. 2. It consists of a weld power source with flat V-I characteristics with ampere rating of 800 amps, 100 % duty cycle, 440 volts, three phase.

Fig. 2 The welding setup

The weld carriage is mounted on the guide rail with the facility of having different welding speeds. A total of eight butt welds were made with mild steel plates of thickness 10 mm. The range of input parameters with respective codes is shown in table 1. Lower limit of the parameter is coded as (-1) and upper limit is coded as (+1).

IOP Conf. Series: Materials Science and Engineering 1126 (2021) 012015 doi:10.1088/1757-899X/1126/1/012015

Table-1 input parameters and their values.

S no.	Input Parameters	Symbols	Unit	Levels	
				(-1)	(+1)
1.	Welding speed	A	cm/min	10	20
2.	Arc voltage	B	V	30	40
3.	Wire feed rate	C	m/min	5	7

The experimentation was carried out as per the combinations shown in table 2. All the welds were cleaned especially on the back side to ensure flatness while measuring angular distortion. The values of angular distortion measured on each weldment are also shown in table-2.

Table-2 the values of angular distortion measured as per combinations.

S. No.	Welding speed	Voltage	Wire feed rate	Angular distortion
1.	-1	-1	-1	1.25
2.	+1	+1	+1	2.15
3.	+1	+1	-1	2.96
4.	-1	+1	+1	4.2
5.	-1	+1	-1	4.0
6.	-1	-1	+1	4.88
7.	+1	-1	+1	3.01
8.	+1	-1	-1	3.85

2.1 Measurement of Angular Distortion:

The angular distortion of the weldments was measured by using the setup shown in fig 3. The setup consists of a properly levelled surface plate, a vernier height gauge and some suitable weights. The weld specimen was placed on the surface plate with one of the edges pressed against it. In this position initial reading R_1 is taken with the help of vernier height gauge as shown. The same edge is now raised by an amount of angular distortion when a suitable weight is placed on the other plate of the joint. In this position reading R_2 of the raised edge is again taken. The angular distortion can now be easily determined by using the relation, $\sin\theta = (R_2\text{-}R_1)/l$, where l is the width of one plate which is 100 mm in this case. By using same procedure, angular distortion of all the weldments was measured.

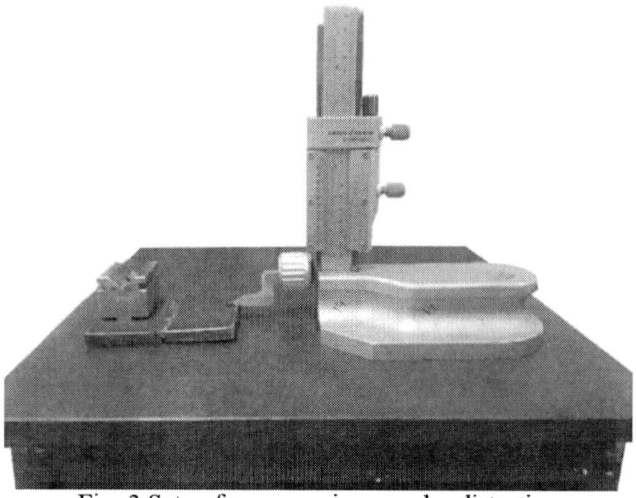

Fig. 3 Setup for measuring angular distortion

3 Results:

The values of angular distortion measured were used in design of expert software. The mathematical model developed is given below-

Angular Distortion = 3.3 – 0.28A + 0.02V + 0.28C – 0.48AB – 0.67AC – 0.43BC

This equation can be used to predict values of angular distortion at different values of input parameters A, B and C within their operating range. The equation also shows the interaction effect of variables. The higher order terms, insignificant terms and triple interaction terms have been ignored for the sake of calculation convenience.

The graphical results produced by the software are given Fig. 4-8 in the next section.

4 Analysis of Results:

The graphical results produced by the software are discussed under two categories-

1. Direct effect plots.
2. Interaction plots (surface plots).

4.1 Direct effects of input parameters on angular distortion:

Fig. 4 shows the direct effects of input parameters on angular distortion. From the figure it is evident that wire feed rate has a positive effect on angular distortion. The probable reason could be that with the increase in wire feed rate, the welding current also increases which increases the depth of penetration of heat into the work piece. More amount of heat causes more thermally generated stresses causing more angular distortion.

Voltage is also found to have a positive effect on angular distortion but in a very less magnitude than the wire feed rate. The probable reason could be that with the increase in voltage the arc length increases which results in the redistribution of the heat in the arc giving more of a spreading effect. As the heat now spreads in a wider area, the angular distortion increases but very less as compared to its value with the increase in wire feed rate where the heat could penetrate much deeper into the weld.

Welding speed has been found to have negative effect on the angular distortion. The probable reason could be that with the increase in speed, the heat input per unit length of the weld decreases sharply. With less amount of heat, lesser thermally generated stresses are produced, resulting in sharply reducing angular distortion with the increase in welding speed.

IOP Conf. Series: Materials Science and Engineering 1126 (2021) 012015 doi:10.1088/1757-899X/1126/1/012015

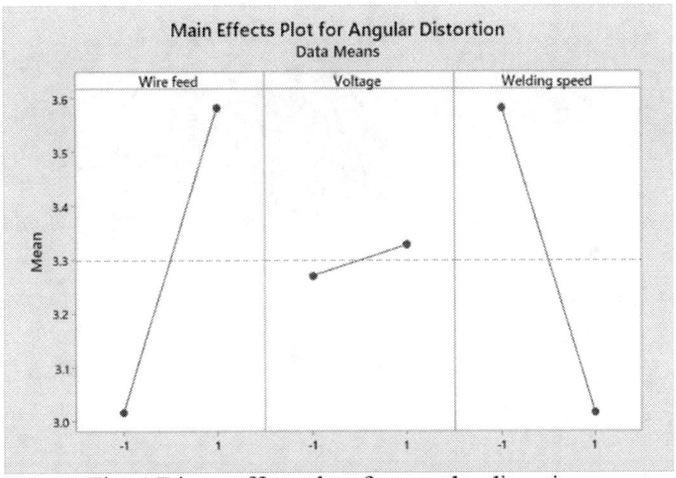

Fig. 4 Direct effect plots for angular distortion.

4.2 Interaction effects of input parameters on angular distortion:

It can be observed from fig. 5 that the input parameters are non-parallel and intersecting each other in all the given graphs. Thus, these parameters interact with one another.

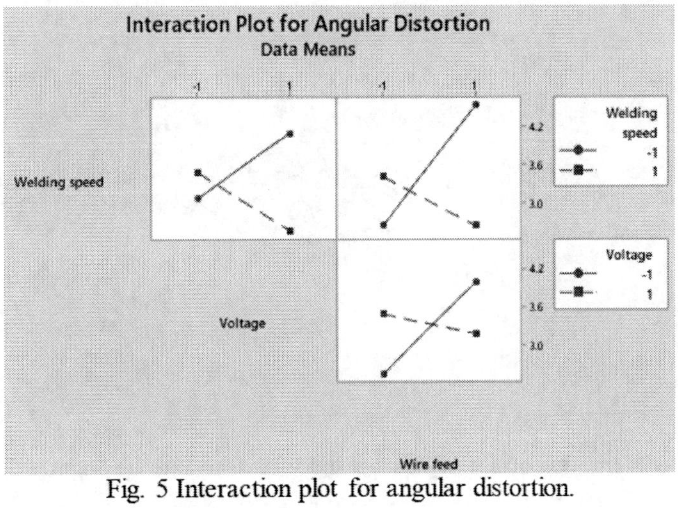

Fig. 5 Interaction plot for angular distortion.

4.2.1 Interaction effect of welding speed and voltage:
From fig. 6 it is observed that at all values of welding speed, the angular distortion increased with the increase in voltage except at maximum speed where the trend appears to be opposite. The probable reason could be that at maximum speed the arc spreading effect doesn't have any effect on the spread of heat.

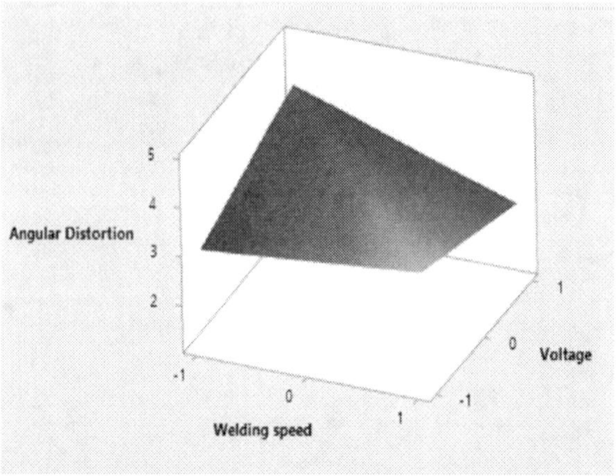

Fig. 6 Interaction effect of voltage and welding speed on angular distortion.

4.2.2 Interaction effect of wire feed rate and voltage:

From fig. 7 it can be observed that with an increase in wire feed rate and voltage, angular distortion increases in general. It is already inferred that wire feed rate has a pronouncing positive effect on angular distortion, but from the figure it is also observed that at maximum voltage, the angular distortion decreased with the increase in wire feed rate. The probable reason could be that the spreading effect of voltage might have outweighed the penetrating effect of current.

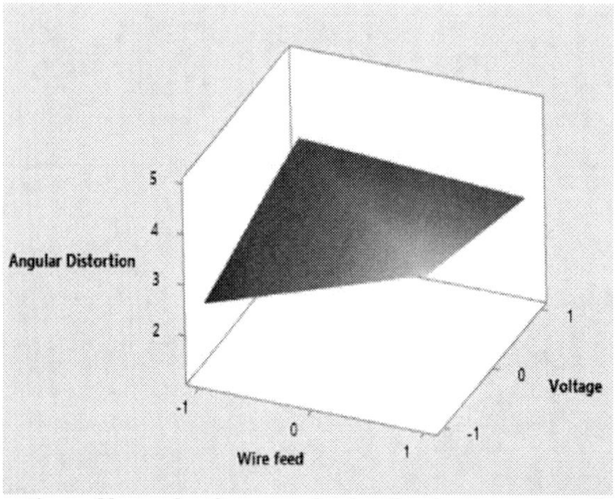

Fig. 7 Interaction effect of voltage and wire feed rate on angular distortion.

4.2.3 Interaction effect of welding speed and wire feed rate:

From fig. 8 it can be concluded that wire feed rate has a positive effect on angular distortion whereas, welding speed has a negative effect. The Wire feed rate controls the amperage as well as the amount of weld penetration. Therefore, with an increase in wire feed rate, current increases as discussed above. The maximum angular distortion can be at minimum welding speed and maximum wire feed rate.

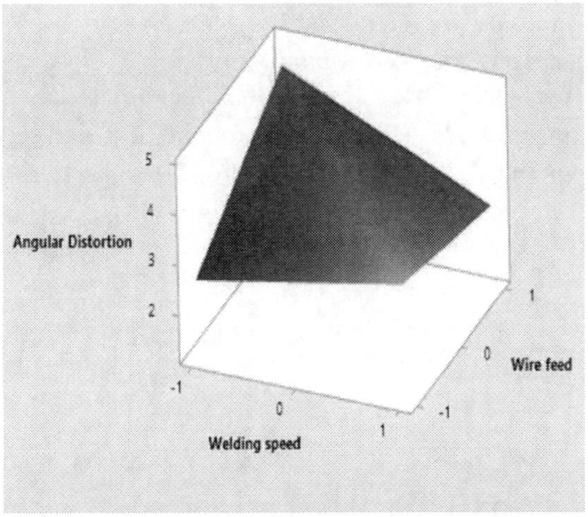

Fig. 8 Interaction effect of wire feed rate and welding speed on angular distortion.

5. Conclusions:

The following conclusions can be drawn from the above investigation-

1. Welding speed has a negative effect on angular distortion whereas, wire feed rate and voltage have a positive effect on angular distortion.
2. The maximum angular distortion attained is 4.85° at welding speed of 10 cm/min and wire feed rate of 7m/min.
3. The minimum angular distortion attained is 2.50° at wire feed rate 7m/min of and voltage of 40 V.

References:

1. https://www.millerwelds.com/resources/article-library/submerged-arc-welding-can-greatly-improve-deposition-rates#:~:text=Higher%20deposition%20rates%20offer%20productivity%20gains&text=Deposition%20rate%20refers%20to%20the,wire%20size%2C%20type%20and%20polarity.
2. https://www.yourarticlelibrary.com/welding/submerged-arc-welding/submerged-arc-welding-saw-equipment-and-applications/96695
3. Peter Smith 2007 *The Fundamentals of Piping Design*, Fabrication pp 171-189 (Gulf Publishing Company)
4. Singh Gagandeep and Singh Parminder 2019 A Review on Submerged Arc Welding and its Various Parameters (Journal of Emerging Technologies and Innovative Research: India)
5. https://www.oerlikon-welding.com/processes/submerged-arc-welding-process
6. Ahmet Irfan Yukler, Memduh Kurtulmus and Ezgi Dogan 2018 *Angular Distortion in Butt Arc Welding* (International Journal of Engineering Science and Application: Turkey)
7. https://www.twi-global.com/technical-knowledge/job-knowledge/distortion-types-and-causes-033#:~:text=If%20the%20stresses%20generated%20from,dimensions%20and%20distorts%20the%20structure.

8. C.M. Davies, J. Ahn, M. Tsunori, D Dye and K.M. Nikbin 2014 *The Influence of Pre existing Deformation on GMA Welding Distortion in Thin Steel Plates* (Journal of Materials Engineering and Performance: Europe)
9. Tso-Liang Tenga, Chih-Cheng Ling 1998 *Effect of Welding Conditions on Residual Stresses due to Butt Welds* (International Journal of Pressure Vessels and Piping: China)
10. V.Vel Murugan and V. Gunraj 2005 *Effects of Process Parameters on Angular Distortion of Gas Metal Arc Welded Structural Steel Plates* (Material Science Welding journal)

Removal of CTAB by synthesized ZnO nanoparticles

Mata Subhashita[1],Namballa Lakshmi Mounika[2],Thanusha Punugoti[3],Divya Kurimilla[4],Akkinenni Hitesh[5],Meena Vangalapati*[6]

[1,2,4,5]M.Tech student,Department of Chemical Engineering,Andhra University
[3]B.Tech student,Department of Mechanical Engineering,Andhra University
[*6]Professor,Department of Chemical Engineering,Andhra University
Corresponding Author E-Mail: meenasekhar2002@yahoo.com

Abstract. In this study, the catalytic mechanism of cetrimonium bromide (CTAB), a cationic surfactant,by pure ZnO nanoparticles was investigated.Further observed ZnO nanoparticles synthesization by precipitation method in aqueous solution with sulfates of metallic precursors.Adding,XRD(X-Ray Diffraction) and SEM(Scanning Electron Microscope) were observed for the crystalline and structural analysis of synthesized ZnO nanoparticles.SEM shows obtained ZnO nanoparticles are of size 180.4 nm.For phase identification of a crystalline material XRD was used and to observe chemical properties of ZnO nanoparticles,studied FTIR (Fourier Transmission of Infra-Red).The maximum percentage removal of CTAB was found to be 95% by varying different conditions by time,dosage,pH and temperature parameters.

1. Introduction

Cetrimonium bromide ([(C16H33)N(CH3)3]Br;CTAB) is a quaternary ammonium surfactant which is a cation used as bacteria and fungi antiseptic agent. CTAB is main component in some buffers used in extraction of DNA.These are mostly found in shampoos,cosmetics,few household products like topical antiseptics.These are dangerous omnipresent water contaminants and pollutants in environment due to their high usage in laundry and cleaning products.These get deposited easily on land and into water systems,as household waste or as a industrial waste[1].These surfactants are toxic to animals, our ecosystems,and to us humans, and can increase the diffusion of other environmental contaminants.Therefore there is a need to degrade or remove using many processes like advanced oxidation process, Fenton oxidation, electro coagulation, adsorption, and photo catalysis[2][3][4][5]. In the present work catalytic mechanism is used and the catalyst used in the work is pure ZnO nano particles (ZnO np).These nano particles has wide range of potential applications.Various techniques to prepare these nanoparticles like precipitation method,thermal decomposition method, pulse laser deposition and hydrothermal method,chemical co-precipitation method, sol-gel method.[6][7]

2. Materials and Methods

2.1. Materials

Coomassic Brilliant Blue R-250($C_{47}H_{50}N_3NaO_7S_2$) anionic dye was analytical reagent grade and 0.579 g of the dye was used without further purification of 10^{-3}M solution by diluting in distilled water.A stock solution of 10ppm was prepared from standard CTAB of 98% purity,storing it in fridge. Standard solutions were prepared by accurate dilution and with an acetate buffer (CH3COOH/ CH3COONa) a buffer solution was prepared with 4-4.5 pH,this was adjusted with a pH meter.

Choloroform used as an solvent and 0.5 M of sodium sulfate solution was also used to accelerate separation of phase in extraction.[8]

2.2. Equipment

UV–Visible spectrophotometer with quartz cells of path length 10 mm for all absorption spectra and spectrometric measurements,pH-meter equipped with a calomel electrode for required pH measurements,Magnetic stirrer.

2.3. Analytical Procedure

100ml CTAB of 10ppm was transferred to a conical flask, and adding 5 ml of sodium sulfate solution,5 ml of acetate buffer,2.5 ml of blue dye solution and 15ml of chloroform to the flask and then shaken vigorously with a magnetic stirrer for 15 min covering the flask.The two phases gets separated in a separating funnel. The absorbance of the extract recovered was measured taking chloroform as reference at 627nm wavelength.[9]

2.4. Synthesis of ZnO np

By precipitation method,ZnO np were prepared. 0.2M of ZnSO4.7H2O was dissolved in 100ml of distilled water under continuous heating stirrer.Then 0.1M of Potassium Hydroxide(KOH) was dissolved in 10ml of distilled water.Then addition of this solution drop wise to zinc sulfate solution until pH reaches to 4.6 and was stirred for 4hrs under 600rpm and the temperature at 80oC is maintained.After this,to remove impurities the solution was washed with distilled water up to 5-6 times,filtered and centrifuged for 15 min at 4000rpm.Then the precipitate obtained when this solution was filtered was dried in hot air oven at 60oC for 6hrs.[10][11]

2.5. Characterization of ZnO np

2.5.1. SEM and XRD Analysis of ZnO np. SEM producing images of a ZnO np by scanning the surface with a focused beam of electrons interacting with atoms in the ZnO np, producing various signals that contain information about composition of the ZnO np and the surface topography.SEM achieves better resolution for more than 1nanometer.SEM (Scanning Electron Microscope) image of ZnO np shown in 'Figure 1' clearly indicating the small structures and different sizes of pure ZnO np..The range of 180.4 to 284.1 nm of ZnO np providing higher resolution and very high stable probe currents for optimum imaging.[12]

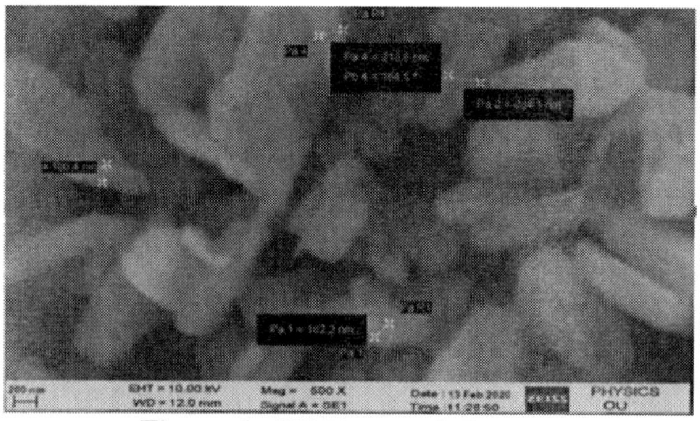

Figure 1. SEM image of ZnO np.

X-ray diffraction (XRD) relies on the dual wave particle nature of X-rays,gives information about the structure of crystalline materials.A primary use of this technique is to identify and characterize compounds by diffraction pattern basis. ZnO np has the hexagonal wurtzite structure and the peak positions in the 'Figure 2' below indicate the formation of the same.The peaks at 2θ values of

IOP Conf. Series: Materials Science and Engineering 1126 (2021) 012016 doi:10.1088/1757-899X/1126/1/012016

31.5o,3o,36o,47.5o,56.5o,62.5o and 68o and the corresponding lattice points with preferred orientations are (1 0 0), (0 0 2),(1 0 1),(1 0 2),(1 1 0),(1 0 3) and (1 1 2).Moreover,we can observe peak positions are not being exactly same in all four samples and there is some difference in the position of the peaks.[13]

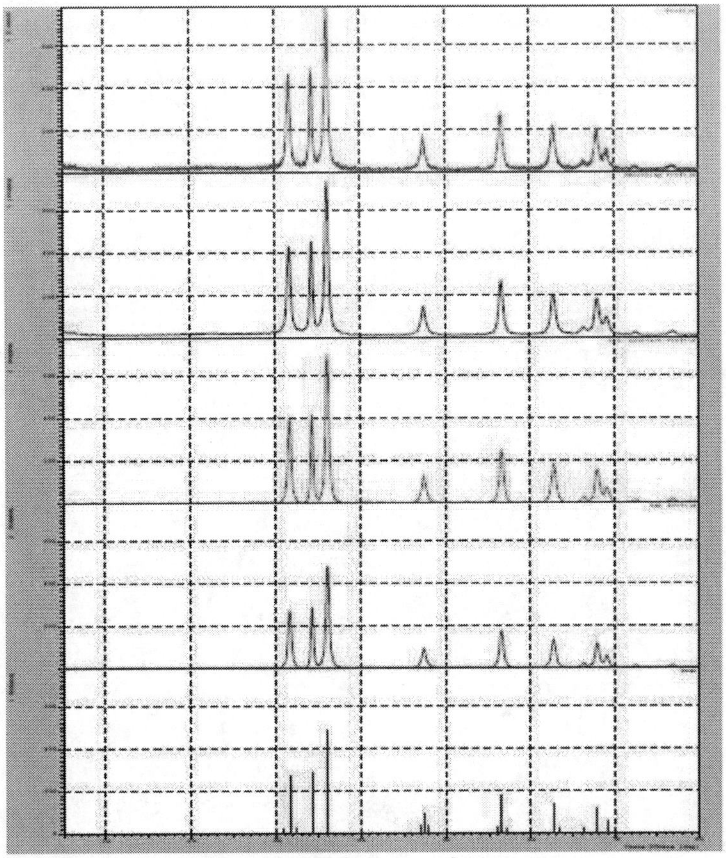

Figure 2. XRD image of ZnO np.

3. Results and Discussions

3.1. FTIR analysis for ZnO np

'Figure 3' presents FTIR spectrum of ZnO np .The broad peak at 3740 - 3000 cm^{-1} in higher energy region and peak in the lower range at 1524 - 1691 cm^{-1} is due to O-H stretching and O-H bending respectively.Zn-O stretching of nano particles is in the range of 552 - 417 cm^{-1}.All other peaks are characterising prepared ZnO np. The bands near 1900 - 2354 cm^{-1} may be due to CO on the surface of oxide the .Lastly,might be the peroxide formation (M-O-O-M) at the bands at 780 - 980 cm^{-1}.

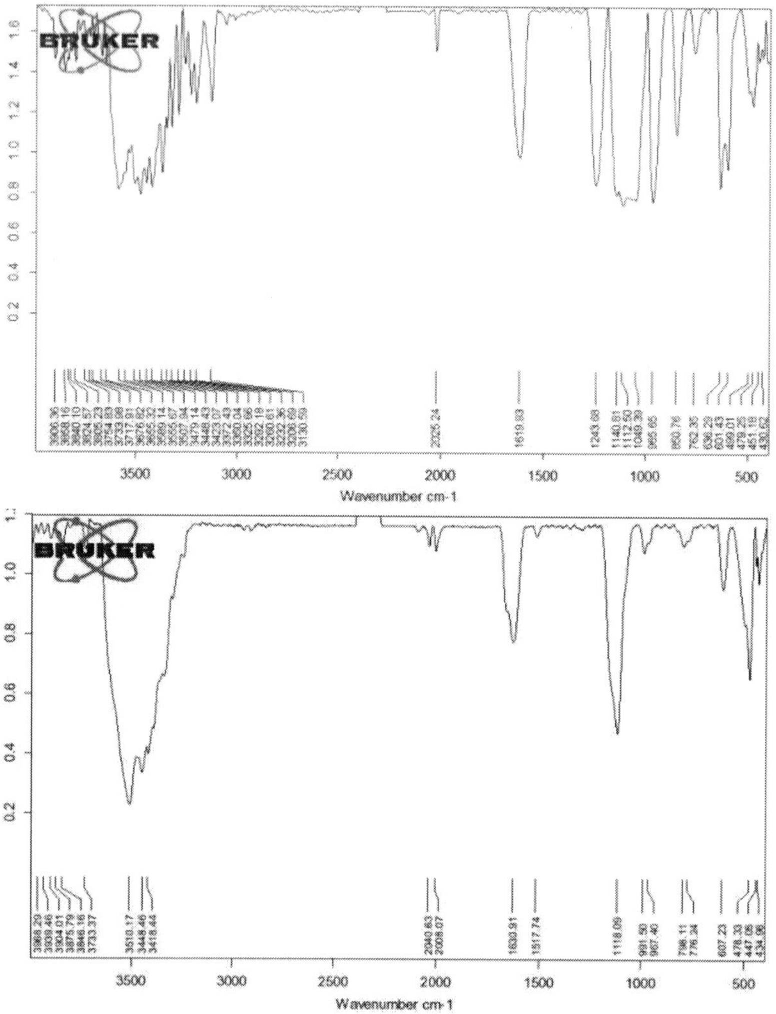

Figure 3. FTIR pattern for ZnO np before catalysis &FTIR pattern for ZnO np after catalysis

3.2. *Experimental Parameters*

Batch equilibrium experiments were carried out by varying contact time, pH, initial surfactant concentration, adsorbent dosage and temperature. In each experiment, accurately weighed nanoparticle was added to one litre of stock solution in a beaker and the mixture was agitated on a magnetic stirrer. [14][15]

3.2.1. Effect of Contact time 0.25 gm/L of ZnO np was taken in a beaker. The stock solution of CTAB with 10 ppm concentration was prepared and one litre of this solution was added to the beaker. The beaker was then kept on magnetic stirrer at room 313K and was kept for shaking for 20min.After every 2 min, the sample is analyzed for determination of the amount of CTAB present in the solution.The data showed that to achieve an equilibrium a contact time of 14 min was required and with further increase in contact time there was no significant change in concentration of CTAB.The percentage removal of CTAB was 95.07%.

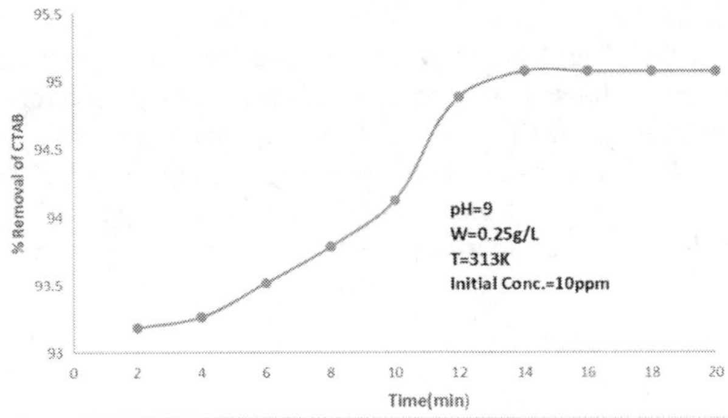

Figure 4. Effect of Contact time on % Removal of CTAB

3.2.2. Effect of catalytic dosage One litre of stock solution of CTAB of 10ppm concentration was taken in a beaker for seven batches. The above procedure was repeated for different catalytic dosages of 0.1,0.2,0.3,0.4,0.5 ,0.6 and 0.7 g/L at the equilibrium contact time of 14 min.With an increase of catalytic dosage from 0.1 to and 0.4g/L then remaining constant the % removal increased to 95.69% for ZnO np for 10 ppm CTAB solution. This can be due to increased surface area and large number of available catalytic sites.

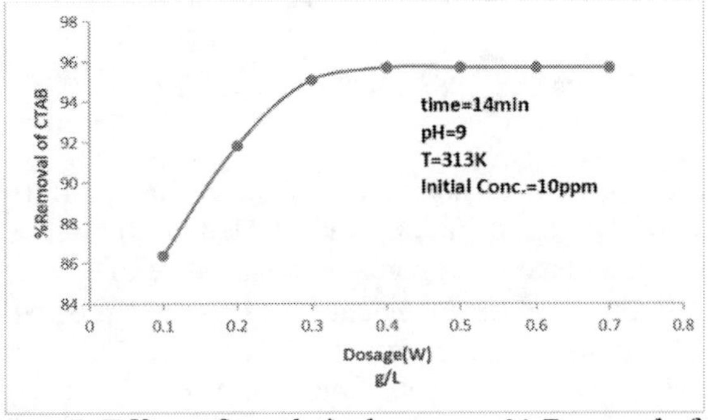

Figure 5. Effect of catalytic dosage on % Removal of CTAB

3.2.3. Effect of pH One litre of stock solution was taken in each of the five beakers. The pH of the aqueous solution was adjusted by adding required amounts of 0.1 N HCL and 0.1N NaOH. The pH of the aqueous solution was set to 8, 9,10,11and 12 in each of the beakers. Then 0.4 g/L of ZnO np was added to each of the beakers. The beakers were then agitated on magnetic stirrer for the optimum contact time of 14 min at 313K temperature. The samples thus obtained is separated and analyzed for remaining amount of CTAB present in the solution.Optimum pH was taken as 10 for ZnO np.

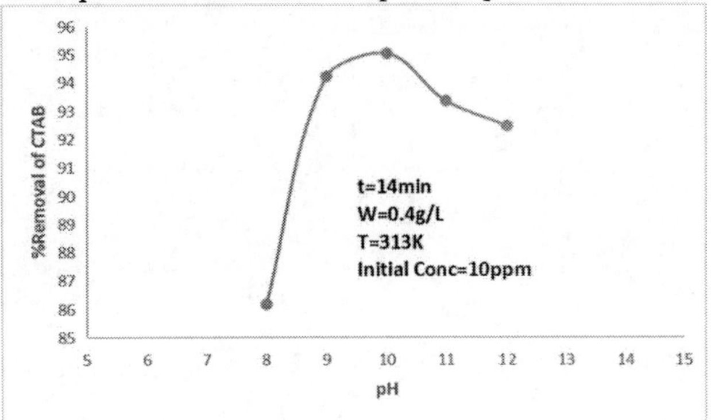

Figure 6. Effect of pH on % Removal of CTAB

IOP Conf. Series: Materials Science and Engineering 1126 (2021) 012016 doi:10.1088/1757-899X/1126/1/012016

3.2.4. Effect of Initial Concentration One litre of stock solution was taken in each of the five beakers with different concentrations ranging from 6ppm,7ppm,8ppm,9ppm,and 10 ppm. 0.4 g/L of ZnO np was added to each of the .The beakers were then kept for stirring for equilibrium contact time of 14 min at 313K temperature. The samples thus collected are, separated and was analyzed for remaining amount of surfactant present in the solution.The percentage removal of CTAB decreased with an increase in initial CTAB concentration.The percentage removal of CTAB for ZnO np the decrease from 96.36% to 95.07 %.At lower concentrations, the percentage removal was higher at higher initial surfactant concentrations as all surfactant present in solution can interact with the binding sites.

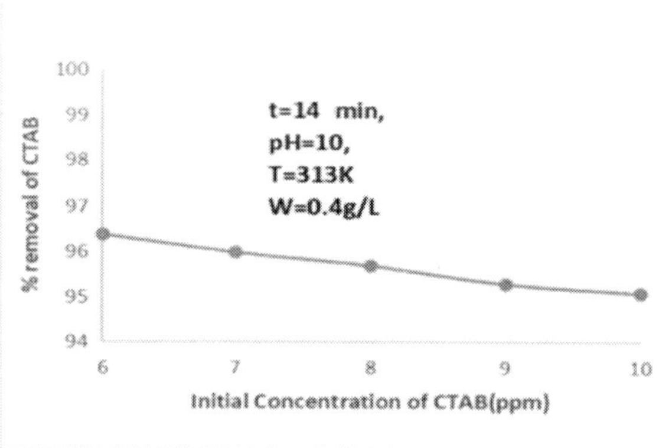

Figure 7. Effect of intial concentration on % Removal of CTAB

3.2.5. Effect of Temperature One litre of stock solution , each of different concentrations from 6-10 ppm, were taken in beakers. 0.4 g/L of ZnO np was added to each of the beakers.The beakers were then kept for at optimum contact time of 14 min at a temperature of 303 K. The samples thus collected is separated and analyzed for remaining amount of surfactant present in the solution. The same procedure was repeated for different temperatures 308K, 313K, 318K and 323K.With an increase in temperature, the % removal was increased from 93.69% to 95.57% for ZnO np at the initial concentration of 10ppm.

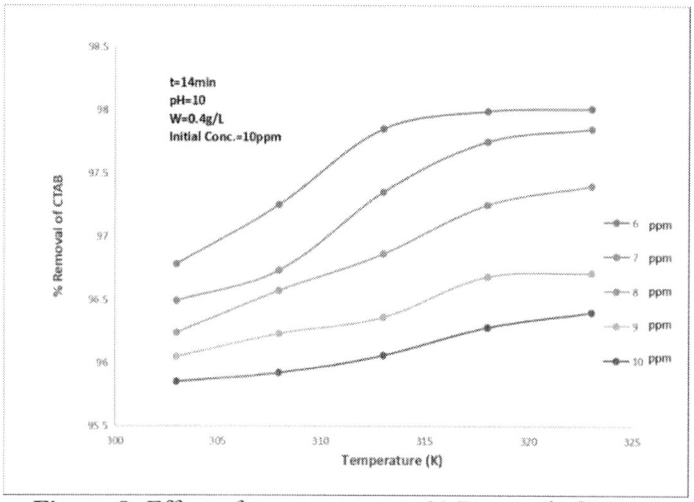

Figure 8. Effect of temperature on % Removal of CTAB

4. Conclusion

The current work focused on synthesized ZnO np,were characterized using SEM, XRD analysis and characterized before catalysis and after catalysis through FTIR analysis. Using SEM,the particles size of ZnO np was found to be 180.4nm.These were used as catalyst to remove CTAB,cationic surfactant which is mostly found in industrial or household waste water. The optimum conditions for various parameters such as time(14min), dosage(0.4g/L), pH(10).Initial concentration and Temperature at five different temperatures at five different ppm were observed that at the lower concentrations with increase in temperature's there is increase in removal percentage.By using these optimum parameters we can effectively remove the CTAB with removal percentage of 95%.

References

[1] Mehmath yalcin. (2004) The Adsorption Kinetics Of CTAB into Powdered Active Carbon kluwer Academic Publishers,pp 339-348.
[2] Matthew J. Scott, Malcolm N. Jones. (2000) Review The biodegradation of surfactants in the environment ,elsevier,1508,pp 235-251.
[3] Abdelhafidh Dhouib,Naïma Hamad,Ilem Hassaïri,Sami Sayadi.(2002) Degradation of anionic surfactants by Citrobacter braakii,pp 1245-1250.
[4] Amir Hossein mahvi, A.Maleki and B.Roshani.(2004) Removal of anionic surfactants in detergent waste water by chemical coagulation,, Pakistan journal of biological sciences,pp 2222-2226.
[5] M. A. Aboulhassan, Souabi, Yaacoubi and M.Baudu. (2006) Removal of surfactant from industrial wastewaters by coagulation flocculation process, , Int.J. Environ. Sci.Tech.,pp 327-332.
[6] Thutiyaporn Thiwawong,Korakot Onlaor, Natpasit Chaithanatkun and Benchapol Tunhoo. (2018) Preparation of Copper Doped Zinc Oxide Nanoparticles by Precipitation Process for Humidity Sensing Device, AIP Conference Proceedings.
[7] Muhammad Sajjad, Inam Ullah, Khan, Jamshid Khan, Yaqoob Khan and Muhammad Tauseef Qureshi (2018) Structural and optical properties of pure and copper doped zinc oxide nanoparticles,Results in Physics,pp 1301–1309.
[8] Yakout S. M and Nayl A. A. (2009) Removal of cationic surfactant (CTAB) from aqueous solution on to activated carbon obtained from corncob , in Applied Science Innovations Pvt. Ltd., pp107-116.
[9] GarapatiVamsiKrishna,ShaikIbrahim,BejawadaSurendra,Dr.MeenaVangalapati.(2017) Optimization of parameters for cationic surfactant removal from domestic waste water by activated corn cobpowder"International Journal of Engineering Sciences & Research Technology.
[10] Jyotsna Chauhan, Neelmani Shrivastav, Ashish Dugaya, Devendra Pandey. (2017) Synthesis and characterization of Ni and Cu doped Zno, MOJ Polymer Science.
[11] Shaveta Thakur, Neha Sharma, Anamika Varkiaand Jitender Kumar. (2014) Structural and optical properties of Cu/ZnO-Np and thin films",Advances in Applied Science Research,pp 18-24.
[12] Muhammad Sajjad, Inam Ullah, Khan, Jamshid Khan, Yaqoob Khan and Muhammad Tauseef Qureshi (2018) Structural and optical properties of pure and copper doped zinc oxide nanoparticles,Results in Physics,pp 1301-1309.
[13] Khan, Noreen, Kanwal and Hussain (2017) Comparative synthesis, characterization of cu doped ZnO nanoparticles and their antioxidant, antibacterial, antifungal and photoadsorption surfactant degradation activities", Digest Journal of Nanomaterials and Biostructures,pp 877 - 889 .
[14] Ahmad Said , Supwatul Hakim M, and Yuli Rohyami. (2014) The effect of contact time and pH on Methylene Blue Removal By Volcanic Ash ,international conference on Chemical,Biological and Environmental Sciences, kanpur, pp12-13.
[15] TesfayeTeka and Semegn Enyew. (2014) Study on effect of different parameters on adsorption efficiency of low cost activated orange peels for the removal of methylene blue dye, International Journal of Innovation and Scientific Research,pp 106-111.

IOP Conf. Series: Materials Science and Engineering 1126 (2021) 012017 doi:10.1088/1757-899X/1126/1/012017

Removal of Sodium Dodycl Sulfate by using copper doped ZnONanoparticles

[1]Dudekula Parveen, [2]Basheera Hussain Khatoon, [3]Vundavalli Venkata Sravanth, [4]A.Ajay Raj, [5]Venkata Rao Poiba, [6]Meena Vangalapati

[1,2,3,4]M.Tech student, Department of Chemical Engineering, Andhra University
[5]Assistant Professor, Department of Chemical Engineering, Andhra University
[*6]Professor, Department of Chemical Engineering, Andhra University
Corresponding Author E-Mail: meenasekhar2002@yahoo.com

Abstract. In present study Sodium Dodycl Sulfate(SDS), an anionic surfactant by using copper doped ZnO nanoparticles(Cu/ZnO-Np) are investigated. Furthur (Cu/ZnO-Np) were synthesized by using precipitation method in aqeous solution with sulfates of metallic precursors. By using XRD(X-ray diffraction) the crystalline analysis of nanoparticles were observed and by using SEM(scanning electron microscopy) the structure of nanoparticles were observed. For Cu/ZnO-Np the SEM size is 107.7nm. For phase identification and crystalline material XRD was used and to observe chemical properties of Cu/ZnO-Np, studied FTIR(Fourier Transmission of Infra-Red). The maximum percentage removal of Cu/ZnO-Np was found to be 98.5% by varying different conditions by time, dosage, pH, initial concentration and temperature parameters.

1.Introduction

Surfactants are reduce the surface tension between liquid and solid or two liquids such as emulsions, detergents, wetting agents etc. [1]. Commonly surfactants are deposited in so many ways on land and also into water systems, whether as part of an intended process or as industrial and household waste [2]. alkali and chelating agents that can be damaging to plants and should not be applied to soils[3][4]. For a considerable period the continuous work will be done by commercial soil wetting agents, by soil micro-organisms they will eventually degraded. To protect the aquatic organisms care should be taken that reduction of pumping these materials into streams. The application of sludge, wastewater irrigation, and remediation processes the Anionic Surfactants can be found in soil. Relatively high concentrations of surfactants together with multi metals can represent an environmental risk. On trace metal mobility the surfactant application is unlikely to have a significant effect at low concentrations [5]. Sodium Dodycl Sulfate can be removed using many processes like catalysis method, adsorption method, electro cogulation method[6]. Copper doped ZnO nanoparticles(Cu/ZnO-Np) used as a catalyst in this present research.

2.Material and Methods

2.1Materials: Sodium Carbonate, Zinc Sulfate, Copper Sulfate, Sodium Hydroxide, Hydrochloric Acid, Toulene, Glacial Acidic Acid, Acridine Orange.

2.2Equipments: Magnetic stirrer, Separating funnel, Glass beakers, Measuring Jars, Ultra-Violet Spectrometer and pH meter.

2.3Synthesis of Cu/ZnO-Np

By using Precipitation method Cu/ZnO-Np was prepared. 0.1M of Sodium Carbonate was dissolved under 100ml distilled water under continuous stirring and heating. Then in 100ml of distilled water 0.2M of Zinc Sulfate was dissolved. Then in Zinc Sulfate solution 0.2M of Copper Sulfate is added. This solution is slowly added to Zinc Sulfate solution upto PH reaches to 4. This mixing solutions stirred for 4 hrs under 600rpm and maintained Temperature is 80°C. For the removal of impurities the precipitate was washed with distilled water upto 4 to 5 times. And then filtered and centrifuged for 20 mins at 500rpm, that precipitate was dried at hot air oven at 60°C for 6 hrs[7],[8],[9].

2.4Determinsation of Sodium Dodycl Sulfate

By using spectrophotometric method Sodium Dodycl Sulfate is determined by acridine orange. A 100μL of glacial acetic acid and acridine orange each is added to 10ml sample of synthetic water and added 5ml of toluene and these solution is kept in separation funnel and it is shaken for one minute and allowed to settle for 5mins then separation takes place. Take 2.5 ml of toluene layer presented in the top and the absorbance is measured directly by using UV-Spectrophotometer at the wavelength of 467nm [10],[11],[12].

2.5Experimental procedure for removal of Sodium Dodycl Sulfate

Sodium Dodycl Sulfate was prepared by taking 1g Sodium Dodycl Sulfate sample into 1 liter of distilled water and it is taken as sample of experimentation. Consider 300ml of Sodium Dodycl Sulfate solution from stock solution and add 0.25 gm of Cu/ZnO-Np in a beaker. Then kept this solution on continuous stirring for 20minss at 600 rpm. Collect the sample for every mins and observed optical density of Sodium Dodycl Sulfate solution. And also observed the removal percentage of Sodium Dodycl Sulfate increases up to certain time then it is constant for time increases. Then go for other parameter to get maximum removal percentage of Sodium Dodycl Sulfate. Then 50ml of sample was taken and add different dosages of nanoparticles. When dosage of Cu/ZnO-Np increases removal percentage also increases. And we got maximum percentage removal of Sodium Dodycl Sulfate from water was observed. And we had experimental procedures with different PH values. The Anionic Surfactant removal was increases with the increase of temperature. Temperature values are 313K, 323K, 333K, 343K. Obtained maximum percentage removal of Sodium Dodycl Sulfate from water by varying different parameters at different conditions.

2.6Characterization of Cu/ZnO- Np

2.6.1 Scanning Electron Microscope(SEM) and X-Ray Diffraction(XRD) Analysis of Cu/ZnO-Np

SEM producing images of Cu/ZnO-Np by scanning the surface with a focused beam of electrons interacting with atoms in the Cu/ZnO-Np, producing various signals and contain information about comparision of the Cu/ZnO-Np and the surface topology[13],[14],[15]. SEM achieved better resolutions for more than one nanometer. SEM image of Cu/ZnO-Np shown in figure 1 clearly indicating the small structures and different sizes of Cu/ZnO-Np The produced nanoparticles were in the range of 107.7 nm to 150nm providing higher resolutions.
X-Ray diffraction (XRD) relies on the dual wave particle nature of X-Ray, gives information about the structure of the crystalline material. The main use of this technique is to identify and characterize the compounds by diffraction pattern basis [16]. Cu/ZnO-Np also has the hexagonal wurtzite structure and Peak positions of all four samples are not exactly same there

IOP Conf. Series: Materials Science and Engineering 1126 (2021) 012017 doi:10.1088/1757-899X/1126/1/012017

is some difference in the position of the peaks represented in figure 2. The peaks at 2θ values of 14°,16°,22°,28°,30°,33° 36° and 53°.Corresponding lattice points with preferred orientations which are belongs to four strongest peaks (1 0 0),(0 0 2),(1 0 1) and (1 1 0).

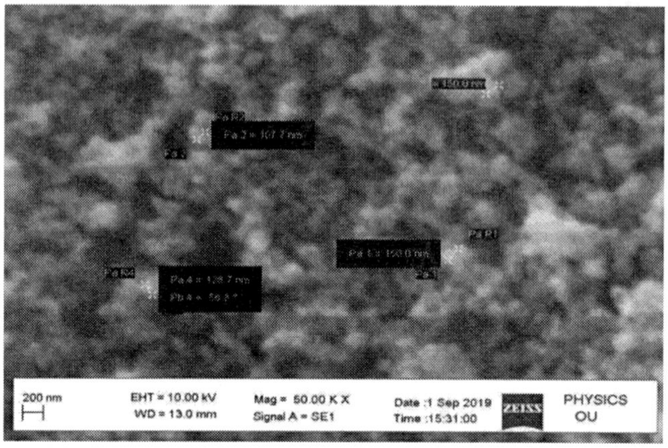

FIGURE 1: Scanning electron microscope image of Cu/ZnO- np

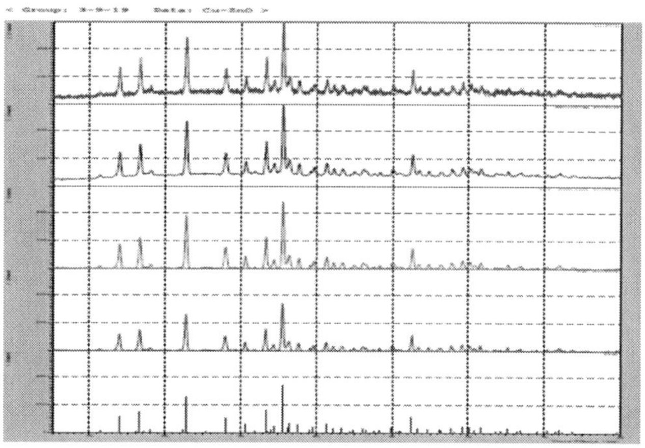

FIGURE 2: X-ray diffraction image of Cu/ZnO-Np

3. Results and Discussions

3.1 FTIR analysis for Cu/ZnO- np

Figure 3 and 4 presents FTIR spectrum of Cu/ZnO-Np. The broad peak at 3740 – 3000 cm^{-1} in higher energy region and due to O – H stretching and peak in the range at 1640 – 1680 cm^{-1} sulfate (O-S-O). All other peaks are attributed to the characteristics of the prepared Cu/ZnO nanoparticles. The bands appear near at 1020 – 1250 cm^{-1} indicates the N-H deformation. The bands appear near at 610 – 700 cm^{-1} indicates the O – H out of plane bending.

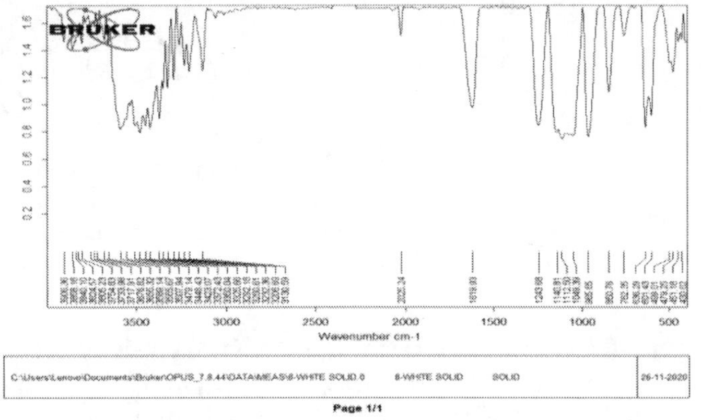

FIGURE 3: FTIR pattern for Cu/ZnO-Np before catalysis

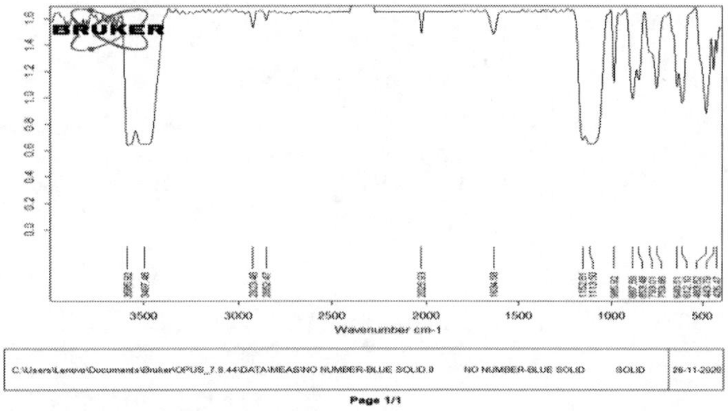

FIGURE 4: FTIR pattern for Cu/ZnO-Np after catalysis

3.2 Effect of various parameters

3.2.1 Effect of time for the removal of SDS: Time course profiles for the Catalysis of SDS solutions of 2,4,6,8,10 and 12mins are shown in Fig.5. The data obtained from the catalysis of SDS onto used Cu/ZnO-Np showed that a contact time 4 mins was required to achieve an equilibrium catalysis, %removal is 65.77 and there was no significant change in concentration of the SDS with further increase in contact time. For further studies of catalysis with other parameters with this Cu/ZnO-Np as a catalyst, the equilibrium time of 4 mins has been chosen for contact period.

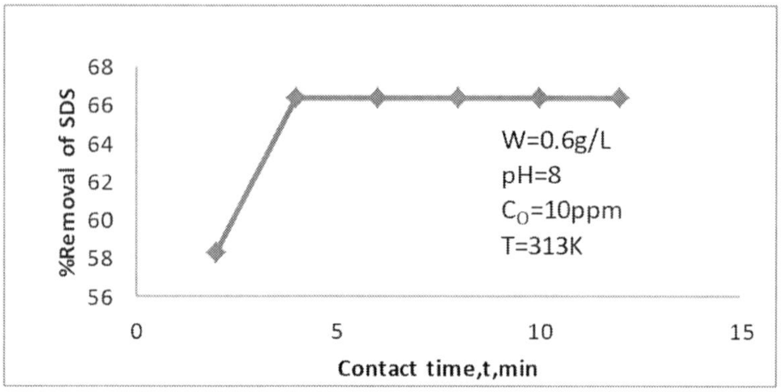

FIGURE 5: Effect of contact time on % removal of SDS by using Cu/ZnO-Np

3.2.2 Effect of dosage for the removal of SDS: For studying the effect of catalytic dosage on removal of SDS the catalytic dosage is varied from 0.1 to 0.7 g/L, fixing other parameters constant. The contact time was 4 mins for Cu/ZnO-Np. The Fig.6 shows an increase in percent SDS removal with an increase in catalytic dosage. The % removal increased from Cu/ZnO-Np 65.77% to 70.57% for 10 ppm SDS solution. With an increase of catalytic dosage from 0.1 to 0.7 g/L, the increase in Anionic Surfactant removal with catalytic dosage will increased surface area and large number of available catalytic sites.

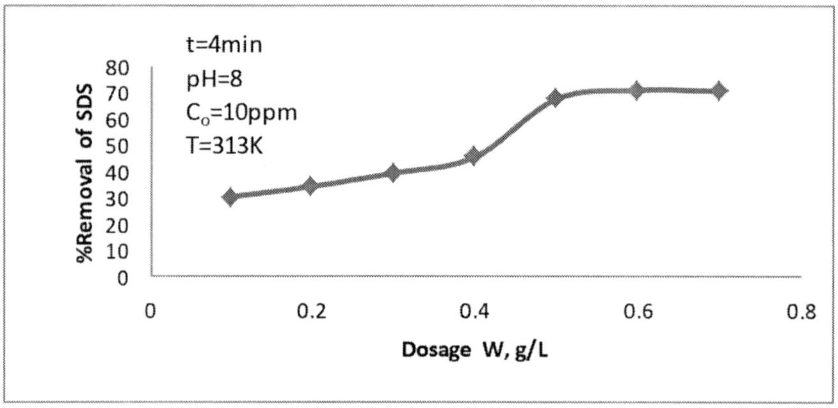

FIGURE 6: Effect of Dosage on % removal of SDS by using Cu/ZnO-Np

3.2.3 Effect of pH for the removal of SDS: To study the effect pH on the removal of SDS by changing the pH from 4 to 8 and fixing the other parameters are time 4mins, Catalytic dosage 0.6g/L. By increase in pH the percentage removal also increases and highest %removal is takes place at pH 7 shown in fig 7 and the %removal is 74.77. Further increases in the pH the %removal is decreases.

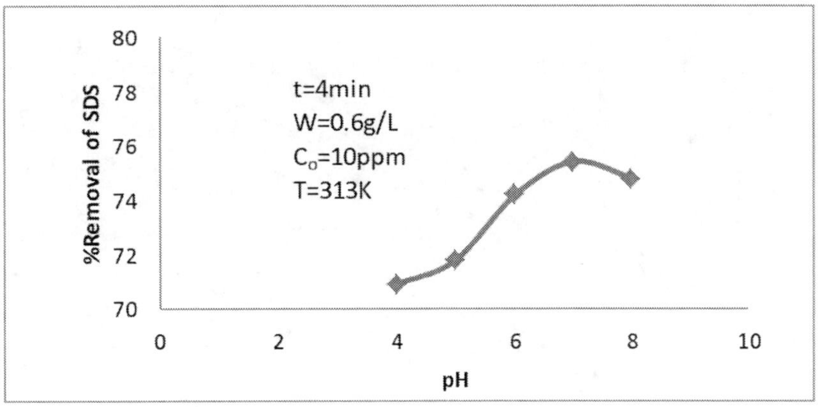

FIGURE 7: Effect of pH on % removal of SDS by using Cu/ZnO-Np

3.2.4 Effect of Initial Concentration for the removal of SDS: To study the effect of Initial Concentration on the removal of SDS by changing the concentration from 2ppm to 10ppm and fixing the other parameters are time 4mins, Catalytic Dosage 0.6 g/L, pH 7. By increase in the Initial Concentration of SDS the % removal of SDS is decreases. The highest %removal takes place at 2ppm as shown in fig 8 i. e, 93.10.

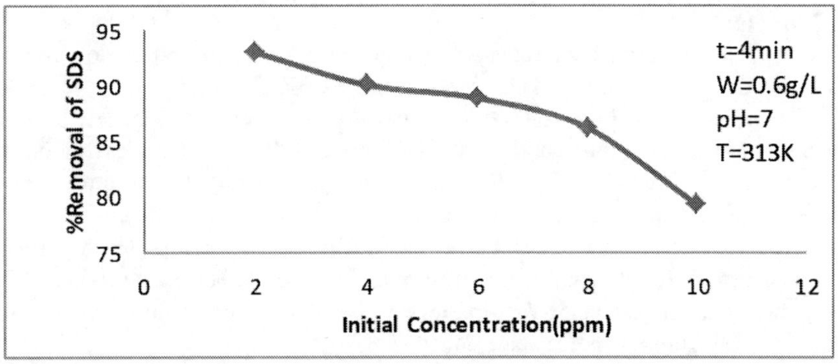

FIGURE 8: Effect of Initial Concentration on % removal of SDS by using Cu/ZnO-Np

3.2.5 Effect of Temperature for the removal of SDS: To study the effect of Temperature on the removal SDS by changing the Temperature from 313K to 353K and fixing the other parameters are time 4 mins, Catalytic Dosage 0.6g/L, pH 7 and also Initial concentrations are varying from 2ppm to 10ppm. The highest %removal is takes place at 353K as shown in fig 9 and the %removal is 98.5. By increase in the Temperature % removal also increases, because catalysis is takes place at highest temperatures.

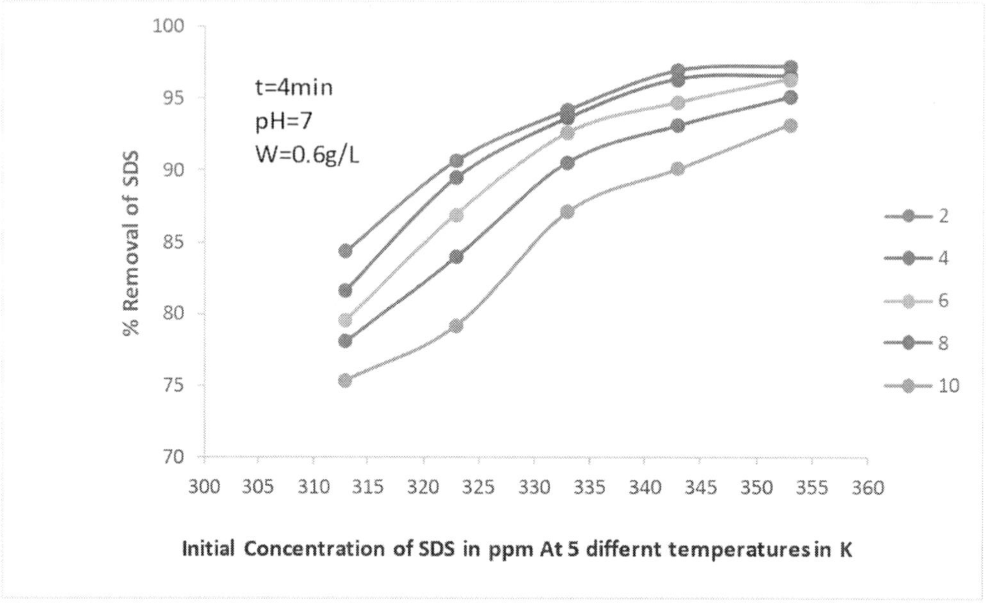

FIGURE 9: Effect of Temperature on % removal of SDS by using Cu/ZnO-Np

4. Conclusion

The current research was focused to find the optimized condition of the removal of SDS by Cu/ZnO-Np as catalyst. The nano sized particles of Cu/ZnO were synthesized using precipitation method with sulfates of metallic precursors. The particle size was determined by using SEM analysis and found to be 107.7nm to 150nm. XRD clearly confirms the crystalline structure of synthesized Cu/ZnO-Np as hexagonal wurtzite structure and FTIR shows the different functional groups present in them. The result present in this research was showed how the nanoparticle will effect on the different parameters like optimum time as 4 min, Dosage as 0.6g/L, Ph as 7, Temperature as 353K, Initial concentration AS 2ppm of SDS from aqueous solution. Optimum parameters we can effectively remove the SDS from aqueous solution with removal percentage of 98.50%.

5. References

[1] Matthew J. Scott, Malcolm N. Jones 2000 The biodegradation of surfactants in the environment, Elsevier Science, 235-251.

[2] Bilaspur (Chhattisgarh), A Study of Spectrophotometric Determinsation of Ion Association Complex, Formed by Anionic Surfactant (SDS) Sodium Dodecyl by Using Crystal Violet as A Cationic Dye in Region oriental journal of chemistry, ISSN: 0970-020 X, CODEN: OJCHEG,Vol. 30, No. (3):Pg. 1335-1341, (2014).

[3] J. Rivera-Utrillaa,, J. Me´ndez-Dı´aza, M. Sa´ nchez- Poloa,b, M.A. Ferro-Garcı´aa, I. Bautista- Toledoa 2006 Removal of the surfactant sodium dodecylbenzenesulphonate from water by simultaneous use of ozone and powdered activated carbon: Comparison with systems based on O3 and O3/H2O2 Elsevier, Vol. 40, pp. 1717- 1725.

[4] Kanchi S., Niranjan T., K. Babu Naidu, *Naidu Venkatasubba N., Monitoring the Status of Anionic Surfactant (SDS)s in Various Water Systems in Urban and Rural Areas of Tirupati, Andhra Pradesh, South India, International Journal of Research in Chemistry and Environment Vol. 2, Issue 3, (144-156) , ISSN 2248-9649, (2012).

[5] Abdelhafidh Dhouib , Naı¨ma Hamad, Ilem Hassaı¨ri, Sami Sayadi 2002 Degradation of

Anionic Surfactant (SDS)s by Citrobacter braakii 1245-1250.
[6] Paritosh das, Anjali Pal, Manas Bandyopadhyay 2002 Catalysis of Anionic Surfactant (SDS) by a low-cost adsorbent Journal of Environmental Science and Health, vol 37(5):925-38.
[7] Thutiyaporn Thiwawong , Korakot Onlaor , Natpasit Chaithanatkun and Benchapol Tunhoo 2018 Preparation of Copper Doped Zinc Oxide Nanoparticles by Precipitation Process for Humidity Sensing Device, AIP Conference Proceedings.
[8] Ruby Chauhan, Ashavani Kumar and Ram Pal Chaudhary 2010 Synthesis and Characterization of Cu/ZnO-Np", J. Chem. Pharm. Res., vol.2 178-183
[9] Chauhan, Shrivastav , Dugaya and Pandey 2017 Synthesis and Characterization of Ni and Cu Doped ZnO Journal of Nanomedicine & Nanotechnology, vol.8.
[10] Shaveta Thakur, Neha Sharma, Anamika Varkiaand Jitender Kumar 2014 Structural and optical properties of Cu/ZnO-Np and thin films",Advances in Applied Science Research, vol.5 18-24.
[11] Dilip B.Patil*, A. Kshirsagar and Ajay P. Ganorkar 2005 Estimation Of Surfactants At Ppm Level From Synthetically Polluted Water, Jr. of Industrial Pollution Control 21 (2), pp 293- 298.
[12] Asok Adak, Anjali pal & Manas Bandyopadhyay 2005 Spectrophotometric determinsation of Anionic Surfactant (SDS) in waste water using acridine orange Indian Journal of Chemical Technology, Vol. 12, pp. 145-148.
[13] Varunkumar, Rafikul Hussain, Gurumurthy Hegde, Anita Ethiraj 2017 Effect of calcination temperature on Cu doped NiO nanoparticles prepared via wet-chemical method: Structural, optical and morphological studies Materials Science in SemiconductorProcessing,vol.66149–15.
[14] Ruby Chauhan, Ashavani Kumar, Ram Pal Chaudhary 2011 structure and optical properties of Zn1-x NixO nanoparticles by coprecipitation method, Journal of Optoelectronics and Biomedical Materials, Vol. 3 17-23.
[15] Muhammad Sajjad, Inam Ullah, Khan, Jamshid Khan, Yaqoob Khan and Muhammad Tauseef Qureshi 2018 Structural and optical properties of pure and copper doped zinc oxide nanoparticles, Results in Physics , vol.9 1301–1309.
[16] Jyotsna Chauhan, Neelmani Shrivastav, Ashish Dugaya, Devendra Pandey 2017 Synthesis and characterization of Ni and Cu doped Zno, MOJ Polymer Science, vol.1.

Tribological and Corrosion analysis of Co-20Al-GNSA composites produced through powder metallurgy process

G R Raghav[1*], U Arunachalam[2], R Sujith[3]

[1]Associate Professor, Department of Mechanical Engineering, SCMS School of Engineering and Technology, Vidyanagar, Karukutty, Ernakulam- 683576, Kerala, India
[2] Assistant Professor, Department of Mechanical Engineering, University College of Engineering, Nagercoil, 629004,Tamilnadu, India

[3] Assistant Professor, Department of Mechanical Engineering, SCMS School of Engineering and Technology, Vidyanagar, Karukutty, Ernakulam- 683576, Kerala, India

[*] E-mail: raghavmechklnce@gmail.com

Abstract. In this study,Co-20Al-GNSA metal matrix composites were produced using mechanical alloying process. The Co-20Al-GNSA composites were mixed using a high-energy ball mill at a constant speed of 350 rpm for 2hours. The composite powders were then characterized for their morphological study using Scanning Electron Microscope. The composite powders are then compressed and sintered at 500MPa and 700°C respectively. The density and compressive strength of the composite materials shows decrement values whereas the wear resistance of the composite materials has increased considerably. The mechanism of wear was identified as abrasive wear. The electrochemical corrosion test also reveals that the Co-20Al-10GNSA composites have better corrosion resistance. The weight-loss corrosion test also shows that the composites with 10GNSA content have better corrosion resistance.

Keywords: Wear, Corrosion, Powder Metallurgy, Microhardness.

1. Introduction

The exploitation of hybrid composites for the potential replacement of conventional metals has been drastically increased in severalapplications such asautomobile industries, commercial industrial applications, and also in aerospace industries where enhanced mechanical, wear and corrosion resistance properties are expected. Hence it was the main objective of the researchers to develop materials with lesser density with better tribological and corrosion performance[1–3].

In the process of developing a composite material, it is very important to select the matrix materials and reinforcements with the good wet ability to improve the bonding of the composite materials. Another important factor is the selection of the fabrication method and its working parameters as per the matrix and reinforcement materials.Now a day the production economy is one of the important factors due to the economic thoughtfulnessof the industries[4–7]. They prefer low-cost composites, to reduce the production cost due to the raw materials. The Co is a material with very good mechanical and corrosion resistance property whereas it is very costly. So it is necessary to tailor the mechanical properties using low-cost reinforcements such as Al and GNSA. The most generally used reinforcements are SiC, Al_2O_3, and TiO_2.The ceramic particles Al_2O_3 and TiO_2 donot have good wetting characteristics[8–10]. Thereby to improve the wettability the Ground Nut Shell Ash (GNSA) is utilized as secondary reinforcement.

The uniform dispersion of reinforcement was another area thathas to be addressed while selecting a fabrication process. There are many methods such as stir casting to produce composite materials whereas achieving uniform dispersion is not possible due to cluster formation. But it is possible to produce composite materials with a uniform dispersion of reinforcements using the powder metallurgy process. Another reason for choosing powder metallurgy was their cost-effectiveness and their reliability for the production of high melting point materials[5,8,11,12].

In this study,a range of combinations of Co-20Al-GNSA hybridcomposites was formed via the powder metallurgy process. The hybrid composites were then made-up into 10mm cylindrical pellets using a die setup made up of high-speed steel die. The muffle furnace was utilized to harden the compacted green pellets usinga sintering operation. The microhardness, density, and compressive strength of the composite materials were studied and reported. Thepin on disc apparatus and electrochemical workstation were utilized to study the wear and corrosion resistance propertiesrespectively. Thus the main purpose of this effort is to develop hybrid composite materials with better mechanical, improved wear, and corrosion resistance properties that can be employed in automobile, industrial,and aerospaceapplications.

2. Materials and Methods

The chemicals used in this work are of research-grade (99.5% purity).The composite powders are synthesized using a high-energy ball mill that comprises tungsten carbide balls. The ball milling process was carried out for a duration of 2hours at a speed of 350 rpm in the existence of toluene as a process control agent to acquire homogenous hybrid composites . The homogenously unified composite powders are then packed down using a uniaxial hydraulic press at 500 Mpa and sintered at 700°C to produce a 10 mm cylindrical pellet. The composite materials are characterized using SEM to find out the morphology of the composite materials.The microhardness of the Co-20Al-GNSA hybrid composites was carried out at 0.5 kg load using Vickers hardness equipment and standard deviation values are considered and reported. The density of the Co-20Al-GNSA hybrid composites was calculated based on the Archimedes principle.The 10 mm diameter composite pellets are compressed using the universal testing machine (UTM) at a uniform and gradual speed rate of 3mm min^{-1}[8,13,14]. The wear analysis was carried out at constant load, constant speed, and sliding distance of 10N, 1.5 m/s, and 1000m respectively. The electrochemical corrosion analysis was carried out using a bio-logic electrochemical workstation. The workstain consists of three electrodes, the platinum counter electrode, Al/AgCl reference electrode and composite pellets as working electrode. The scan was carried out at 5mV/s. The composite pellets are immersed in 3% NaCl solution for 1 hour so as to stabilize the open circuit potential.The weight-loss corrosion analysis was carried out with various corrosive media such as 0.1N HCl, 0.1N H_2SO_4 and 3% NaCl solution. The composite pellets are immersed in the corrosive media for 24 hours. The composite specimens are weighed before and after the test and the weight loss is calculated.[13,15].

3. Results and Discussion

3.1 Density and Microhardness.

The SEM image of The Co-20Al-5GNSA hybrid composites is shown in Figure.1. The SEM image is taken in secondary electron mode. It is observed that there is uniform dispersion of Co, Al, and GNSA particles in hybrid composite material. Figure.2 exhibits the microhardness and density graphs of Co-20Al-GNSA The density of the Co-20Al-GNSA hybrid composites increases slightly with the addition of GNSA particles. The density of Co-20Al-2.5GNSA composite was 5.2 g/cm^3 which increase slightly to 5 g/cm^3 for Co-20Al-5GNSA hybrid composites. The further addition of GNSA reinforcements resulted in a decrease in the density of the Co-20Al-10GNSA hybrid composites (5 g/cm^3). The microhardness of the Co-20Al-GNSA hybrid composite increases linearly up to 5% GNSA reinforcement further addition of GNSA particles has resulted in a slight reduction in the microhardness of the composite material.

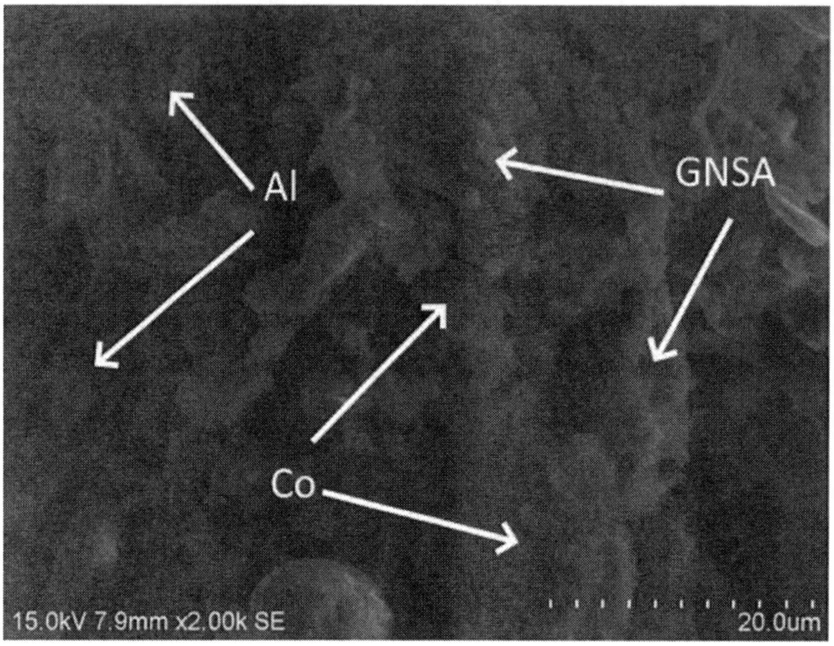

Figure.1 SEM image of Co-20Al-5GNSA hybrid composite

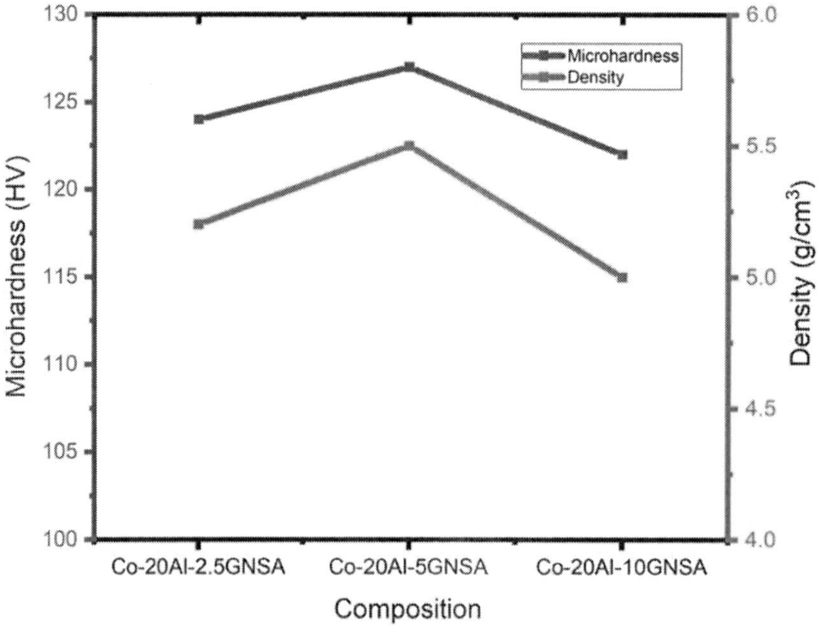

Figure.2 Microhardness and Density of Co-20Al-GNSA hybrid composites

3.2 Compressive Strength

The Compressive strength of Co-20Al-5GNSA hybrid composites is 112 MPa which is better compared to Co-20Al-2.5GNSA (110 MPa) and Co-20Al-10GNSA (108 MPa) hybrid composites. The compressive strength increased gradually with GNSA addition till it reaches 5% GNSA, whereas further addition of GNSA does not influence the compressive strength of the hybrid composites. Figure.3 represents the compressive strength of the Co-20Al-GNSA hybrid composites.

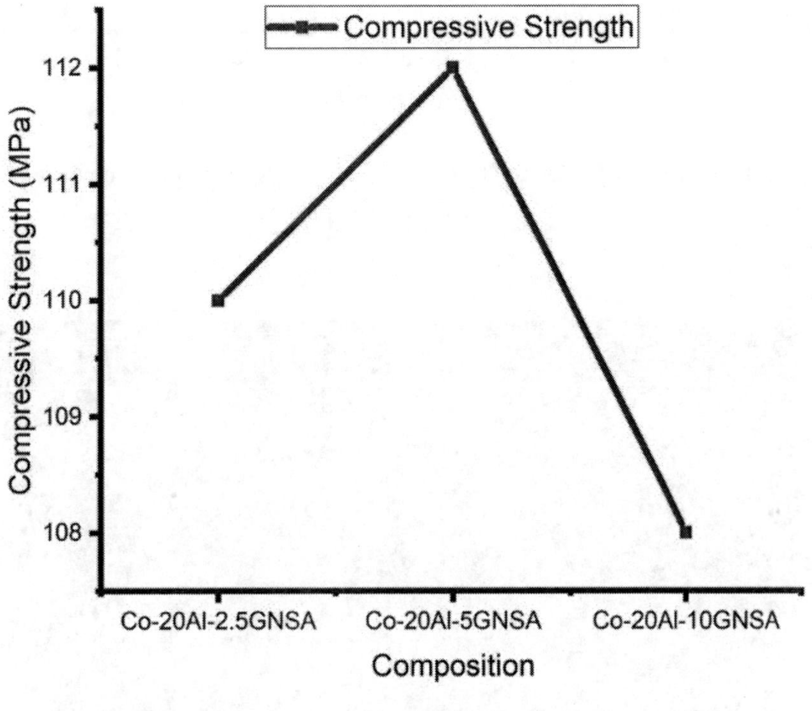

Figure.3 Compressive Strength of Co-20Al-GNSA hybrid composites

3.3 Wear and COF Analysis

The wear analysis results of Co-20Al-GNSA hybrid composites are represented in Figure.4. The wear test was carried out at a constant load of 10N, constant sliding speed of 1.5 m/s, and constant sliding distance of 1000m. The wear loss of the Co-20Al-10GNSA hybrid composites exhibited better wear resistance and COF values compared to that of other specimens. The Co-20Al-2.5GNSA has produced a COF value of 0.9 whereas the COF value of Co-20Al-10GNSA is 0.6. Hence it can be confirmed that the addition of GNSA particles has a good influence in increasing the wear resistance of the hybrid composite materials. Figure.5 represents the SEM image of Co-20Al-10GNSA hybrid composites after the wear test. From the pattern of wear track, it is evident that the major wear mechanism is abrasive wear.

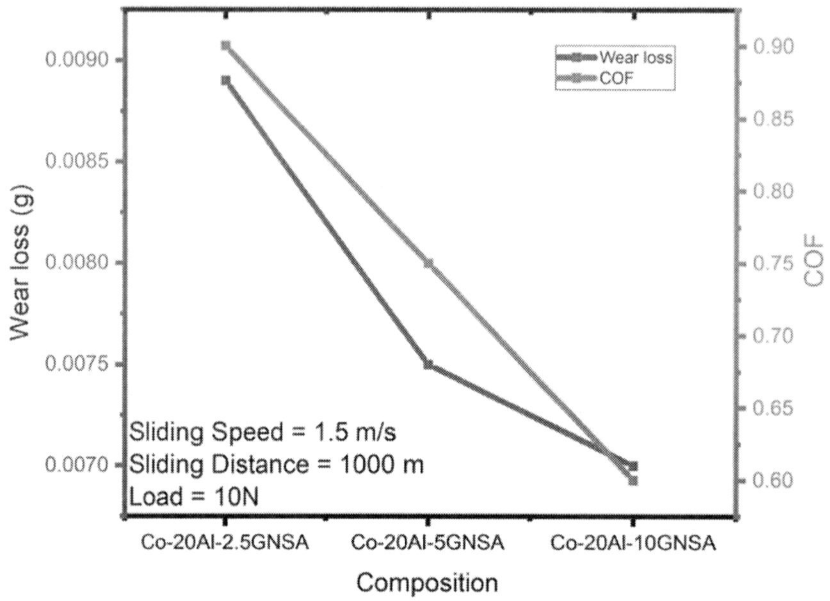

Figure. 4 Wear loss and COF of Co-20Al-GNSA hybrid composites

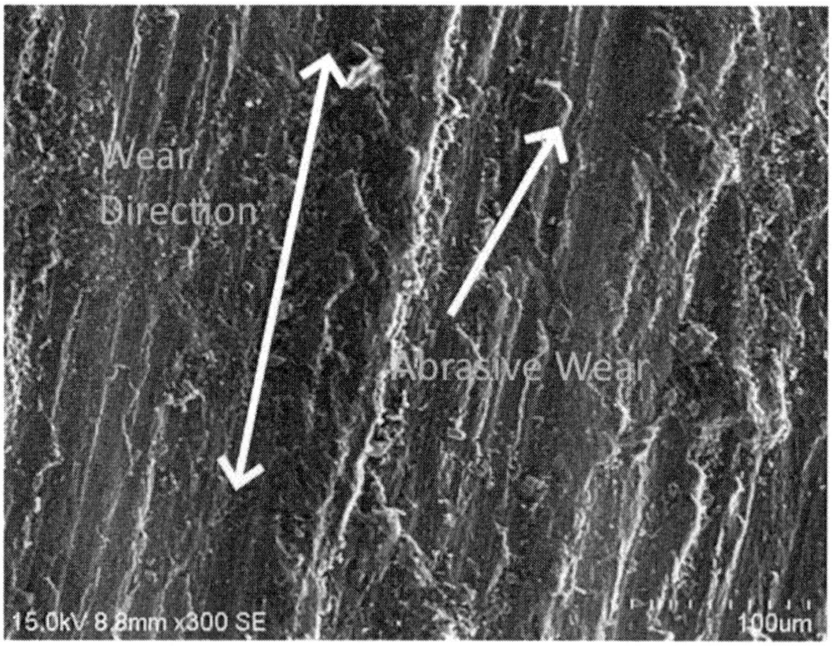

Figure .5 SEM image of Co-20Al-10GNSA hybrid composite after wear test

3.4 Corrosion Analysis

Figure .6 shows the weight-loss corrosion graphs of Co-20Al-GNSA hybrid composites at various Corrosive media such as 3% NaCl, 0.1N HCl, and 0.1N H_2SO_4. From the graph, it can be concluded that the weight loss of the composites decreases with the increase in GNSA content. The weight loss of the Co-20Al-10GNSA hybrid composite is better compared to other combinations in all kinds of corrosive media. The weight loss was maximum for 0.1 N H_2SO_4 for all samples compared to other corrosive media. The weight loss was minimum in 3% NaCl solution.The electrochemical corrosion analysis was carried out using three-electrode systems using the composite pellets as the working electrode. The potentiodynamic polarization results shows that the Co-20Al-10GNSA hybrid composites have exhibited better E_{corr} (-0.442 V) and i_{corr} values (1.5 mA/cm^2) compared to that of Co-20Al-5GNSA (-0.448V& 1.7 mA/cm^2) and Co-20Al-2.5 GNSA (-0.453V & 1.9 mA/cm^2). It is evident that the E_{corr} values are shifted to more positive side and the i_{corr} values decreases with the increase in GNSA content which confirms the increase in corrosion resistance.

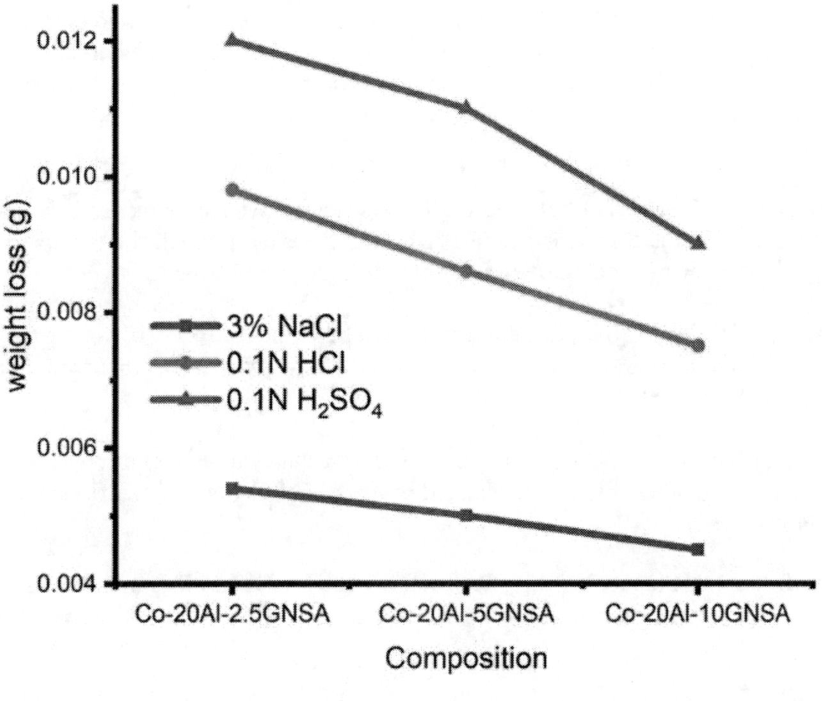

Figure.6 Weight loss corrosion results of Co-20Al-GNSA hybrid composites

4. Conclusions

The Co-20Al-GNSA hybrid composites were amalgamated by employing a high-energy ball mill.

- The Microhardness and density of the Co-20Al-5GNSA hybrid composites have improved compared to other samples.
- The compressive strength of the Co-20Al-5GNSA hybridcomposite was 112 MPa and superior compared to other combinations.
- The wear analysis authenticates that Co-20Al-10GNSA hybrid composites exhibited better wear resistance and coefficient of friction.
- The potentiometric polarization analysis shows that the Co-20Al-10GNSA hybrid composites have enhanced corrosion resistance due to the existence of GNSA particles.
- The weight-loss corrosion analysis also proves that the Co-20Al-10GNSA hybrid composites have better corrosion resistance.

From the conclusion of this study, it can be accomplished that the Co-20Al-10GNSA hybrid composites have experienced a slight decrease in density, microhardness,and compressive strength but exhibits better wear and corrosion resistance. Hence it can be concluded that Co-20Al-10GNSA hybrid composites have better tribological and corrosion resistance with decent mechanical properties which may be considered for potential industrial applications.

References:

[1] Gopinath S, Prince M and Raghav G R 2020 Enhancing the mechanical, wear and corrosion behaviour of stir casted aluminium 6061 hybrid composites through the incorporation of boron nitride and aluminium oxide particles *Mater. Res. Express* **7** 016582

[2] Raghav G R, Balaji A N, Muthukrishnan D, Sruthi V and Sajith E 2018 An experimental investigation on wear and corrosion characteristics of Mg-Co nanocomposites *Mater. Res. Express* **5** 066523

[3] Kumar N, Bharti A and Saxena K K 2021 A re-investigation: Effect of powder metallurgy parameters on the physical and mechanical properties of aluminium matrix composites *Mater. Today Proc.*

[4] Albert T, Sunil J, Christopher A S, Jegan R, Prabhu P A and Selvaganesan M 2020 Preparation and characterization of aluminium-titanium carbide (Al-TiC) composite using powder metallurgy *Mater. Today Proc.*

[5] Karthikeyan N, Krishnan B R, VembathuRajesh A and Vijayan V 2020 Experimental analysis of Al-Cu-Si metal matrix composite by powder-metallurgy process *Mater. Today Proc.* S2214785320365251

[6] Li K C, Prior D J, Waddell J N and Swain M V 2015 Comparison of the microstructure and phase stability of as-cast, CAD/CAM and powder metallurgy manufactured Co–Cr dental alloys *Dent. Mater.* **31** e306–15

[7] Zhou Z, Liu B, Guo W, Fu A, Duan H and Li W 2021 Corrosion behavior and mechanism of FeCrNi medium entropy alloy prepared by powder metallurgy *J. Alloys Compd.* **867** 159094

[8] Tan L, He G, Li Y, Liu F, Nie Y and Jiang L 2018 Flow behaviors and microstructural evolutions of a novel high-Co powder metallurgy superalloy during hot working *J. Mater. Process. Technol.* **262** 221–31

[9] Rodrigues W C, Broilo L R, Schaeffer L, Knörnschild G and Espinoza F R M 2011 Powder metallurgical processing of Co–28%Cr–6%Mo for dental implants: Physical, mechanical and electrochemical properties *Powder Technol.* **206** 233–8

[10] Mihalcea E, Vergara-Hernández H J, Jimenez O, Olmos L, Chávez J and Arteaga D 2021 Design and characterization of Ti6Al4V/20CoCrMo–highly porous Ti6Al4V biomedical bilayer processed by powder metallurgy *Trans. Nonferrous Met. Soc. China* **31** 178–92

[11] Venkatesh V S S and Deoghare A B 2020 Fabrication and mechanical behaviour of Al-Kaoline metal matrix composite fabricated through powder metallurgy technique *Mater. Today Proc.*

[12] Raghav G R, Balaji A N, Selvakumar N, Muthukrishnan D and Sajith E 2019 Effect of tungsten reinforcement on mechanical, tribological and corrosion behaviour of mechanically alloyed Co-25C Cermet nanocomposites *Mater. Res. Express* **6** 1165e4

[13] Raghav G R, kumar R A, Muthukrishnan D, Nagarajan K J, E S and V S 2020 Synthesis and mechanical characterization of Fe–BN–TiC nanocomposites *Eng. Res. Express* **2** 025036

[14] Sudha G T, Stalin B, Ravichandran M and Balasubramanian M 2020 Mechanical Properties, Characterization and Wear Behavior of Powder Metallurgy Composites - A Review *Mater. Today Proc.* **22** 2582–96

[15] Raghav G R, Janardhanan S, Sajith E, Chandran V and Sruthi V 2021 Mechanical and tribological performance of Al-Fe-SiC-Zr hybrid composites produced through powder metallurgy process *Mater. Res. Express* **8** 016533

IOP Conf. Series: Materials Science and Engineering 1126 (2021) 012019 doi:10.1088/1757-899X/1126/1/012019

Third Order Band Pass Filter With Double-Gate MOSFET: A Simulation Perspective

Danurshan Naidoo and Viranjay M. Srivastava

Department of Electronic Engineering,
Howard College, University of KwaZulu-Natal,
Durban - 4041, South Africa.

viranjay@ieee.org

Abstract. It is an extended version of the mathematical analysis of the proposed 3rd order Band Pass Filter (BPF). This device design uses the Double-Gate (DG) MOSFET's which provides a better performance and reliability of the band pass filter. This 3rd order BPF design improves the device performance characteristics, such as enhanced power efficiency and switching capabilities, due to the properties of DG MOSFET technology. Simulation analysis of the proposed filter model has been performed. The BPF model has been designed for lower and upper cut-off frequencies of *100 kHz* and *1.7 MHz*, respectively. The 3rd order BPF filter has been modeled to have Butterworth characteristics, thus having a maximally flat pass-band response.

1. Introduction

A system is required to pass signals that vary within the desired frequency range while rejecting all undesired frequency components of an electrical signal. Band-pass filters are frequency selective devices commonly used in applications such as wireless communication systems and audio amplifier circuits [1, 2]. The proposed system should have high performance and low power consumption features to support modern electronic devices while maintaining a compact device size. Considering the increasing demand for the downscaling of components to produce more compact analog electronic devices, a Band Pass Filter (BPF) using Double-Gate (DG) MOSFET has been proposed by *Naidoo and Srivastava* [3]. That circuit perspective has been extended to the simulation analysis in this present work. Traditional components such as operational amplifiers and single-gate MOSFETs have an increased power consumption and a larger electronic package size. The use of the DG MOSFET leads to a reduction in the filter circuit's size, therefore reducing the overall weight of the circuit [4]. The performance and reliability of the filter also improved. Since the DG MOSFET has a comparatively smaller width and offers better gate control over the channel than single-gate MOSFET, DG MOSFET provides improved power efficiency and switching capabilities.

Sawigun et. al. [5] have analyzed the steps involved in designing a compact nano-power fourth-order band-pass filter. This filter has an adjustable centre frequency range of *125 Hz* to *16 kHz* and is operated from *0.5 V* supply. The filter was cascaded with a second-order circuit network and fabricated in *0.18-μm* CMOS technology. The band-pass filter design achieved a center frequency of 1 kHz, dynamic range of *55 dB*, and *2 nW* power consumption. *Yang et. al.* [6] have proposed a *14.4 nW* fourth-order band-pass filter for biomedical applications. The pass-band of the designed filter was *523-1024 Hz*. The filter achieved a power consumption of *14.4 nW* and a dynamic range of *55 dB*, this provides the best figure of merit in comparison to circuitry for similar applications at the time of this work. *D'Amico et. al.* [7, 8] have designed 3rd order *132 MHz* low-pass filter at a *28-nm* CMOS technology, which uses an op-amp two-stage topology, which operates from a 0.9 V supply voltage while consuming *340 μW* and performs high linearity ($IIP_3 = 11.5$ *dBm* at *21* and *22 MHz* input tones) and large Signal-to-Noise ratio (*58 dB*).

Bao et. al. [9] have designed a 3rd order inductor-free memristive chaotic circuit, which was derived from a 2nd order active BPF by replacing a resistor with an improved memristor and it consists of three op-amps, two multipliers, three capacitors, and six resistors. *Ahmadi et. al.* [10] have designed a wide-bandwidth 2nd order voltage-mode all-pass filter derived from a single transistor BPF, and it has one transistor, two resistors, and two energy storage elements. This was implemented in *0.13-μm* CMOS technology to have *55 ps* group delay across *6 GHz* bandwidth while consuming *18.5 mW* from a *1.5 V* supply. *Yesil and Minaei* [11] have presented two MOSFET only 2nd order voltage and transadmittance-mode BPF utilizing five transistors without any passive elements. As a result, both circuits possess low-power consumption and occupy a small chip area. In this present design, the authors used a four double-gate transistor with a *12 V* power supply and function generator.

Modern devices require more components to be integrated on a single chip (VLSI, ULSI, etc.) to meet the low power and high-performance requirements. Single-Gate (SG) MOSFETs have a large physical size, leading to issues like an increased cost due to the large system construction. Issues such as short channel effects and increased leakage current need to be considered as they limit the device's performance [12-14]. The MOS devices used in the electronic development industry needs to offer an improvement in their performance and performance characteristics to counter the issues experienced due to the short channel effects, this leads to the use of DG MOSFETs. The scaling of MOSFETs offers advantages such as smaller device dimensions and improved performance. The DG MOSFET provides an improvement in the device scaling due to its *25 nm* gate lengths compared to the *65 nm* characteristic of MOSFET devices [4, 13, 15].

Here the 3rd order BPF has been realized by cascading a first-order high pass filter stage with a second-order low pass filter stage. The third order band-pass filter was initially designed using operational amplifiers. The mathematical analysis of the designed filter was performed to determine the filter's overall transfer function in ref. [3]. The transfer function of the individual filter stages was derived to calculate the desired cut-off frequencies' component values. The band-pass filter was designed using a differential structure to implement the DG MOSFET technology, the operational amplifiers were replaced with the BF998 IC to achieve this circuit [16-18]. The parameters of the designed filter were compared to existing filters to determine its feasibility. The proposed model offers improved device performance characteristics, such as enhanced power efficiency and switching capabilities, due to the DG MOSFET technology. The system has been designed for a broader range of RF applications, it can be utilized for applications in the Low Frequency (LF) and Medium Frequency (MF) range due to its larger pass-band [19-21]. The use of the DG MOSFET allows the system to maintain a small-scale design construction due to the reduction in the circuit's size and overall weight.

The work presented in the paper has been organized as follows: Section 3 describes the system design for the model. Section 3 has the simulation analysis and its discussion. Section 4 has its further extension as hardware implementation. Finally, Section 5 concludes the work and recommends the future aspect of the work.

2. System Description

The active components which can be used in filter designs include operational amplifiers (LM353) and SG MOSFETS (2N7000 N-Channel enhancement mode). However, some of these components have large power requirements and insufficient performance capabilities due to modern electronic devices' demands. Therefore, DG MOSFETs are suitable replacements since they improve the device's performance characteristics while having a small-scale design construction. The process to design the 3rd order BPF with DG MOSFET is shown in figure 1. It has been designed using a cascaded high pass filter stage and low pass filter stage. A first-order high pass filter stage will be connected in series with a second-order low pass filter stage to realize the third-order band-pass filter. The band-pass filter will

IOP Conf. Series: Materials Science and Engineering 1126 (2021) 012019 doi:10.1088/1757-899X/1126/1/012019

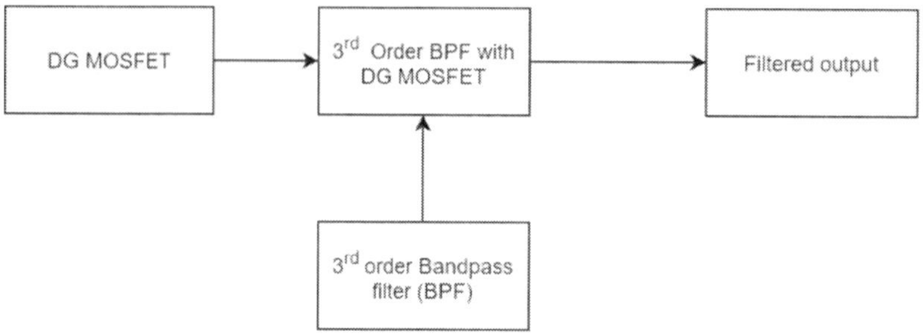

Figure 1. The design process for the third-order band-pass filter with Double Gate (DG) MOSFET.

first be designed with the use of operational amplifiers [3]. Thereafter, the operational amplifiers in the circuit design will be replaced with the MOSFET technology (BF998) to realize the third-order band-pass filter.

The high-level design block diagram for the 3rd order BPF with DG MOSFET is shown in figure 2. The filter stages will be tested individually to ensure they generate the required frequency response before cascading them. The first stage filter will only pass frequencies above *100 kHz* while attenuating any low-frequency signals. The second stage filter will allow frequencies below *1.7 MHz* to pass while attenuating any high-frequency signals. The first stage of the filter requires the input signal from the function generator. The second stage will be cascaded with the first stage to realize the third-order band-pass filter. The filtered output can then be seen on the oscilloscope. The cascaded band-pass filter will only allow frequencies in the pass-band range of *100 kHz - 1.7 MHz* to pass. Any other frequency signals will be attenuated, these signals will be significantly attenuated the further away they are from the pass-band.

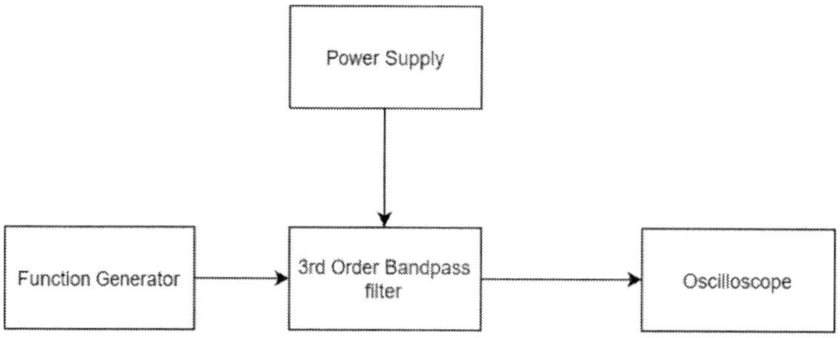

Figure 2. High-level design block diagram.

The pass-band ripple is the amount of variation in the amplitude within the designated pass-band of the band-pass filter [1, 2]. The filter was modeled to have a Butterworth filter response. The filter model's magnitude response will be analyzed to verify if it has a maximally flat pass-band response. Figure 3 shows the low-level design block diagram for the 3rd order BPF with DG MOSFET. Both of the filter stages will be powered by a *12 V* power supply. The function generator will supply the input signal to the high pass filter stage. The second stage will be cascaded with the first stage to realize the third order band-pass filter. Figure 4 show the basic structures of double-gate MOSFET. This device consists of four terminals, this includes the source, drain, and two gate terminals. The key improvement of the DG MOSFET over the single gate MOSFET is that this device has a very small Si channel width, and the two gate terminals provide the user with two platforms to control the device.

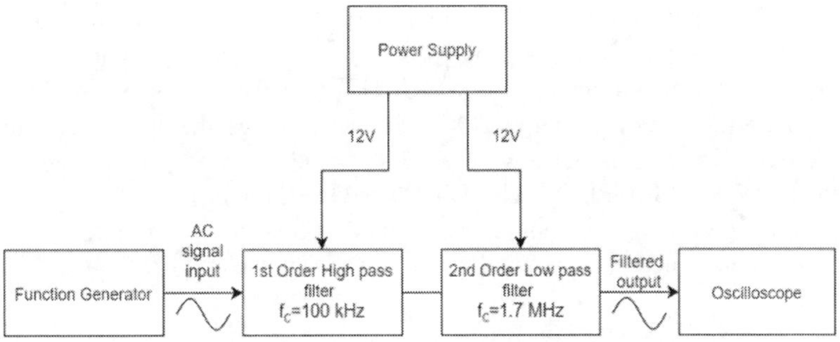

Figure 3. Low-level design block diagram.

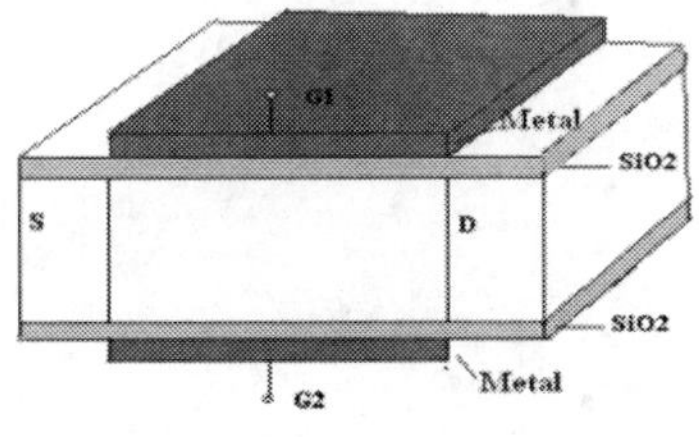

(a)

(b)

Figure 4. (a) Structure of Double-Gate MOSFET [3], and (b) BF998 DG MOSFET [16, 17].

3. Simulation Analysis and Its Discussion

The small-signal model for the first-order high-pass filter section and second-order low-pass filter section, respectively, are shown in figure 5. Ref. [18] has been used as a guide to formulate the small-signal models for the 3rd order BPF. In the small-signal models, transistors Q_1 and Q_2 represent the first-order high-pass filter section, and transistors Q_3 and Q_4 represent the second-order low-pass filter section. Kirchhoff's voltage law was used to derive the voltage gain equations for these models, the capacitors were short-circuited.

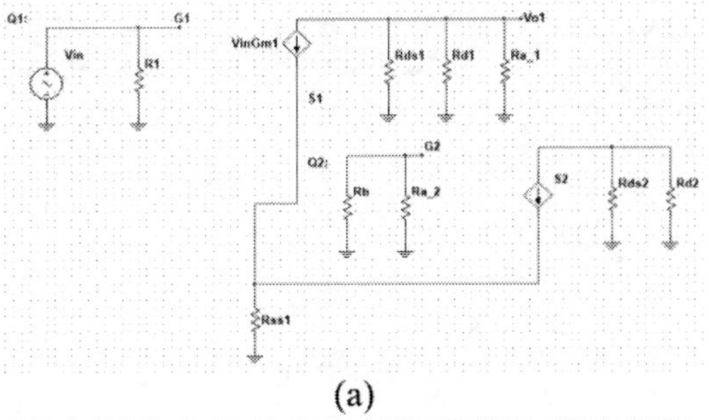

(a)

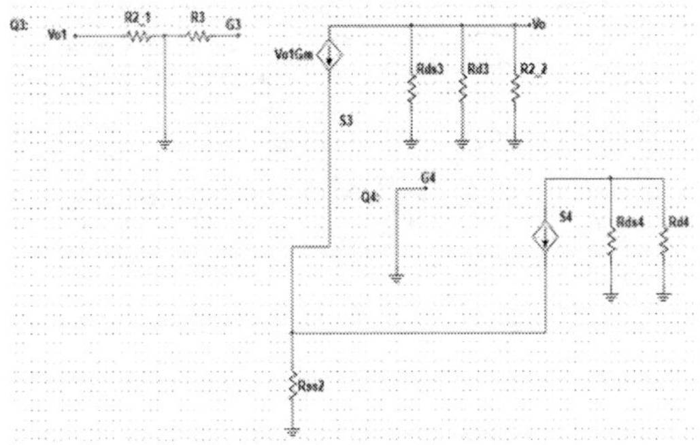

(b)

Figure 5. Small signal model of (a) 1st order HPF section and (b) 2nd order LPF section [18].

The proposed solution, which is to utilize the BF998 Silicon N-channel dual-gate MOSFET, is shown in figure 6. This MOSFET leads to a reduction in the filter circuit's size, and therefore reduces the overall weight of the circuit. The band-pass filter's performance and reliability will also be improved if the BF998 is used instead of SG MOSFET. The system will be suitable for high-frequency RF applications since utilizing a DG device reduces the input capacitance. The gate to channel coupling is doubled in the BF998 MOSFET, which is an improvement on the short channel effects experienced by SG MOSFETS. The BF998 MOSFET can be utilized for applications in the VHF and UHF band, including television tuners and professional communications equipment [22]. This BF909 MOSFET has been used as an alternative for simulation purposes in this work [23].

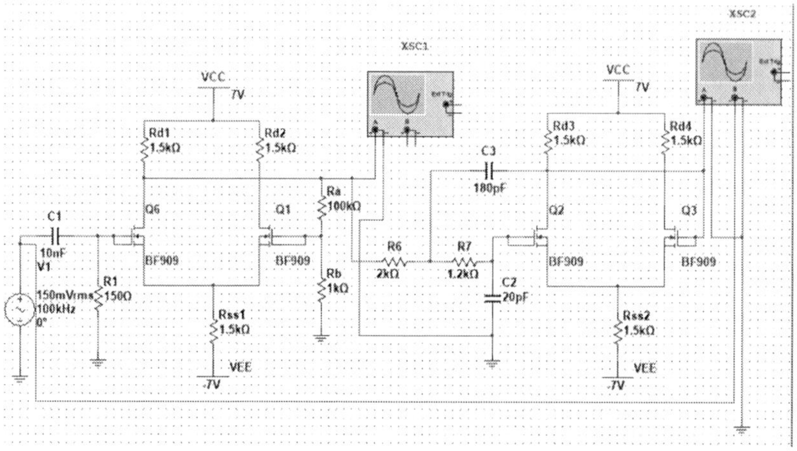

Figure 6. Circuit diagram of implemented the band-pass filter.

In ref. [3], figure 7 shows the frequency response generated for the 3rd order BPF. It correlates closely with the theoretical frequency response of the BPF. The magnitude response was determined to be a combination of the first order high pass and second-order low pass magnitude responses in that work. Here, in this work, figure 7(a) shows the magnitude response, indicating the cut-off frequencies and bandwidth that the system achieved. The lower and upper cut-off frequencies are indicated using cursors 1 and 2, respectively. The cursor menu shows the cut-off frequency and *-3 dB* frequency point in variables x_1 and y_1 for the lower cut-off frequency and x_2 and y_2 for the lower cut-off frequency. The bandwidth of the band-pass filter is the difference between the upper and lower cut-off frequencies. Therefore, the dx variable in the cursor menu can analyze the difference between the cut-off frequencies x_1 and x_2. The bandwidth was found to be *1.65 MHz*, which correlates closely to the value from the ref's mathematical analysis [3].

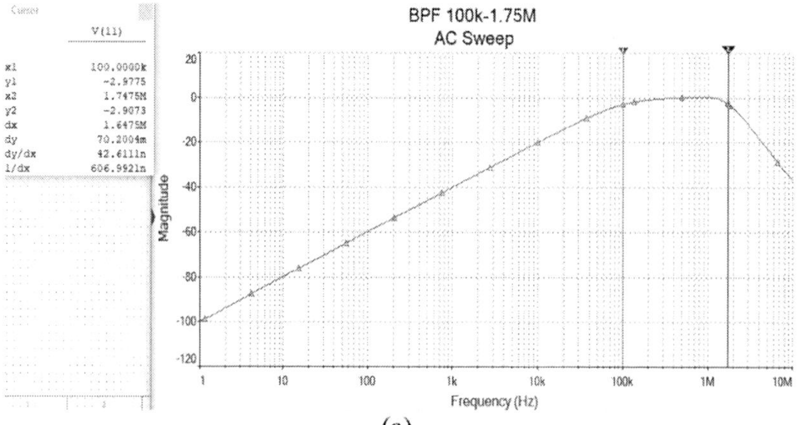

(a)

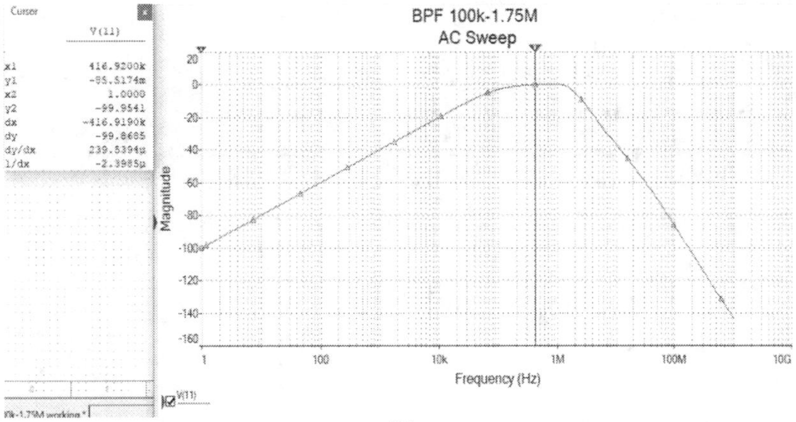

(b)

Figure 7. (a) Cut-off frequencies and bandwidth, and (b) Centre frequency of proposed 3rd order BPF with DG MOSFET.

The magnitude response, as shown in figure 7(b), indicates the center frequency of approximately *416 kHz* for the filter. The center frequency value from the mathematical analysis performed in ref. [3] was *412 kHz*, which correlates closely to the center frequency of the filter's magnitude response. Figure 8 shows the bode plot response of the band-pass filter with DG MOSFET. The bode plot was used to analyze the pass-band ripple. The pass-band ripple was found to be approximately *0.2 dB*, thus the filter demonstrates ideal Butterworth characteristics.

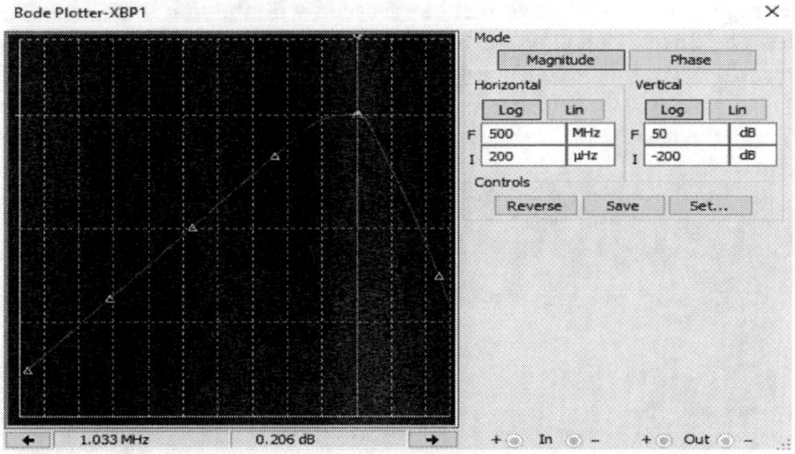

Figure 8. Pass-band ripple of BPF filter with DG MOSFET.

The comparison of band-pass filter implementations for the different technologies has been detailed in Table I. The existing works as described in ref. [24] and ref. [25] have been designed for Very Low Frequency (VLF) applications, these works offer a limited filter pass-band of *20-26 kHz* and *10 kHz - 20 kHz*, respectively. Both of these works have a 4th order filter response and were designed using operational amplifiers. The developed system can be utilized for a broader range of RF applications, namely the Low Frequency (LF) and Medium Frequency (MF) range due to its larger pass-band. The frequency band used for Amplitude Modulation (AM) radio is *550 kHz - 1720 kHz* (*0.55 MHz - 1.72 MHz*). Therefore, the proposed filter model has the required bandwidth for this frequency band. This filter is also suitable for Low Frequency (LF) aviation band applications (*200 kHz* to *415 kHz*). The 3rd order BPF model improves device performance characteristics, such as enhanced power efficiency and switching capabilities, due to the DG MOSFET technology.

Table 1. Comparison of band-pass filter implementations for the different technologies

Parameters	[24]	[25]	This work
Technology used	CMOS	CMOS	DG MOSFET
Filter type	Butterworth	Butterworth	Butterworth
Filter order	4	4	3
Lower cut-off freq.	20 kHz	10 kHz	100 kHz
Higher cut-off freq.	26 kHz	20 kHz	1.7 MHz
Bandwidth	6 kHz	10 kHz	1.6 MHz
Quality factor	3.8	1.4142	0.258
Pass-band Ripple	0.1 dB	0.1 dB	0.2 dB

4. Further Extension as Hardware Implementation

The implemented BPF circuit on a breadboard is shown in figure 9. The BF998 MOSFETs were soldered onto adapters to use on a breadboard. The filter circuit requires the *12 V* power supply module to power the system. The system also requires the function generator to supply the input signal and an oscilloscope to analyze the filtered output signal. The two filter stages would be tested individually to ensure they generate the required frequency response before cascading them.

The filter in the first stage will only pass above *100 kHz* while attenuating any low-frequency signals. The second stage filter will allow frequencies below *1.7 MHz* to pass while attenuating any high-frequency signals. The function generator will supply the input signal to the input capacitor of the high-pass filter stage. The second stage will be cascaded with the first stage to realize the third-order band-pass filter. The filtered output will then be able to be analyzed on the oscilloscope. The cascaded band-pass filter will only allow frequencies in the pass-band range of *100 kHz-1.7 MHz* to pass. Any other frequency signals will be attenuated, these signals will be significantly attenuated the further away they are from the pass-band. The hardware circuit will be communicated in a future article.

Figure 9. Hardware circuit of the filter system.

5. Conclusions and Future Recommendations

The design of the DG MOSFET based 3rd order BPF has been presented using simulation analysis. The designed BPF achieves a pass-band range of *100 kHz* to *1.7 MHz*, which is suitable for RF applications. The 3rd order BPF model offers an improvement in the device performance characteristics, such as enhanced power efficiency and switching capabilities, due to the DG MOSFET technology.

In the future, the hardware circuit will be fully tested using an oscilloscope to analyze the parameters of the 3rd order BPF. The measured results from the hardware circuit will then be compared to the simulation results to determine if the design was successfully implemented.

6. References

[1] Mohan P V A 2013 *VLSI Analog Filters*, (New York: Springer Science & Business Media, 1st Ed.).
[2] Winder S 2002 *Analog and Digital Filter Design*, (Elsevier, 2nd Ed.).
[3] Naidoo D and Srivastava V M 2021 Third order band pass filter with double-gate MOSFET: A circuit perspective *Proc. 11th IEEE Int. Conf. on Computer Communication and Informatics (India)* pp 1-6.
[4] Srivastava V M and Singh G 2013 *MOSFET Technologies For Double-Pole Four Throw Radio Frequency Switch*, (Switzerland: Springer International Publishing, Switzerland).
[5] Sawigun C, Ngamkham W, and Serdijn W 2014 A 2014 "A 0.5-V, 2- nW,55-dB DR, fourth-order bandpass filter using single branch biquads: An efficient design for FoM enhancement *Microelectronics Journal* **45** pp 367–374.
[6] Yang M, Liu J, Xiao Y, and Liao H 2010 14.4 nW fourth-order bandpass filter for biomedical applications *Electronics Letters* **46(14)** pp 973-974.
[7] Amico S D, Matteis M D, Donno A, and Baschirotto A 2019 A 0.9 V 3rd -order single-opamp analog filter in 28 nm bulk-CMOS *Analog Integrated Circuits and Signal Processing* **98** pp 155–167.
[8] Matteis M D, Donno A, Marinaci S, Amico S D, and Baschirotto A 2017 A 0.9V 3rd-order single-opamp analog filter in 28 nm CMOS-bulk *7th IEEE International Workshop on Advances in Sensors and Interfaces (IWASI), Vieste, Italy* pp 155-158.
[9] Bao B, Wang N, Xu Q, Wu H, and Hu Y 2017 A simple third-order memristive band pass filter chaotic circuit *IEEE Transactions on Circuits and Systems II: Express Briefs* **64(8)** pp 977 – 981.
[10] Ahmadi P, Maundy B, Elwakil A S, Belostotski L, and Madanayake A 2016 A new second-order all-pass filter in 130-nm CMOS *IEEE Transactions on Circuits and Systems II: Express Briefs* **63(3)** pp 249-253.
[11] Yesil A and Minaei S 2018 High-order realisation of MOSFET-only band-pass filters for RF applications *IET Circuits, Devices & Systems* **12(4)** pp 467-477.
[12] Nel B J and Perinpanayagam S 2017 A brief overview of SiC MOSFET failure modes and design reliability *Procedia CIRP* **59** pp 280-285.
[13] Sedra A S and Smith K C 2014 *Microelectronic circuits: Theory and applications* (Oxford University Press, 7th Ed.).
[14] Passi V and Raskin J P 2017 Review on analog/radio frequency performance of advanced silicon MOSFETs *Semiconductor Science and Technology* **32(12)** art 123004.
[15] Srivastava V M, Yadav K S, and Singh 2013 Explicit model of cylindrical surrounding double-gate MOSFETs *WSEAS Trans. on Circuits and Systems* **12(3)** pp 81-90.
[16] Discrete Semiconductors, 1996 BF998; BF998R Silicon N-channel dual-gate MOS-FETs, Datasheet, *Philips Semiconductor*.
[17] Discrete Semiconductors, 1999 BF992 Silicon N-channel dual gate, MOS-FETs, Datasheet, *Philips Semiconductor*.
[18] Ramdhani S and Srivastava V M 2020 Realization with Fabrication of double-gate MOSFET based third order high pass filter *International Journal of Electrical and Electronic Engineering & Telecommunications* **9(4)** pp 213-222.
[19] Cameron R J, Kudsia C M, and Mansour R R 2018 *Microwave Filters for Communication Systems: Fundamentals, Design, and Applications* (John Wiley & Sons, Inc. 2nd Ed.).
[20] Srivastava V M, Yadav K S, and Singh G 2010 Performance of double-pole four-throw double-gate RF CMOS switch in 45 nm technology *Int. J. of Wireless Engineering and Technology* **1(2)** pp 47-54.

[21] Kumar R, Kumar M, and Srivastava V M 2012 Design and noise optimization of RF low noise amplifier for IEEE standard 802.11a WLAN *Int. J. of VLSI Design and Communication Systems* **3(2)** pp 165-173.

[22] Ramdhani S and Srivastava V M, 2020 Third order high pass filter with double-gate MOSFET: a circuit perspective *International Conference on Computer Communication and Informatics (India)* pp 1-6.

[23] Product data sheet, 2007 BF909, BF909R, N-channel dual gate MOS-FETs, Datasheet, *NXP Semiconductors*, Rev. 02.

[24] Sadiq A A, Othman N, Jamil M, Youseffi M, Denyer M, Zakaria W, and Tomari M R 2018 Fourth-order butterworth active bandpass filter design for single-sided magnetic particle imaging scanner *Journal of Telecommunication, Electronic and Computer Engineering* **10** pp 17-21.

[25] Myo Z, Aung Z, and Naing Z 2009 Design and implementation of active band-pass filter for low frequency RFID (radio frequency identification) system *Lecture Notes in Engineering and Computer Science*, **2174**.

Preparation of superfine cinnamon bark nanocrystalline powder using high energy ball mill and estimation of structural and antioxidant properties.

Archana[1], Aman Kr. Abhay [1], Singh Kr. Rakesh [1], Kr. Nishant [1], Prasad Birendra [2]

1. Aryabhatta centre for Nanoscience and Nanotechnology, School of Engineering and Technology, Aryabhatta Knowledge University, Patna, India, Pin-800001
2. Dept. of Biotechnology, Patna University, Patna, India, Pin-800005
Corresponding author-rakeshsinghpu@gmail.com (Dr. Rakesh Kumar Singh)

Abstract: Application of Food material for medicinal use is become a common and safe approach to treat various diseases. Although Nanoscience is seeming to most promising area to be explore at every aspect of existing science including medicine and pharmaceuticals. Medicinal properties and application of various spices are well explored science but their nanopowder synthesis and effect is not very much known. In this present work, we have used a commonly known spice from Indian kitchen know as Cinnamon, for synthesis of nano powder high energy ball mill instrument was used. The crystallographic study, functional group analysis, were done using modern characterization equipment such as XRD (X-ray diffraction), FTIR (Fourier transform infrared spectroscopy), SEM (Scanning Electron Microscope), and UV-Visible spectroscopy. XRD measurement confirms that crystal structure of powder milled for 5 hours and 10 hours were different. Similarly, morphology of differently milled sample found to be different from general Cinnamon powder. This might be due the formation of different fractions of particles were formed as a result of deterioration of cohesion bond due to high energy milling. The present study suggested that cinnamon superfine powder could be a potential source of natural antioxidant and thus could be useful as therapeutic agents and also open new window for the progress of surface science of food materials, which are beneficial for biomedical engineering, pharmaceutical, health, and medicine industries.

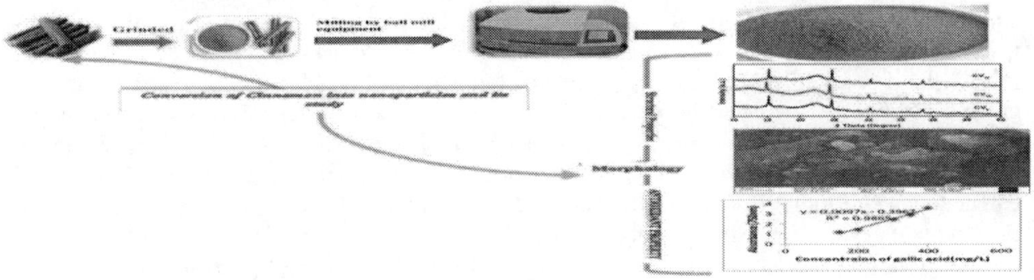

Graphical Abstract

Keywords- Cinnamon, Fine powder, Nanoparticle, Crystal structure, Antioxidant Properties.

Introduction

Herbal products are compounds produced by plants that have numerous nutritional and medicinal properties for the development of drugs and healthy food. The food powders are being produced by different pharmaceutical industries and are consumed by mass people across the globe. The knowledge of active functional ingredient at the nanoscale and their antioxidant behavior was not reported at large to the best of my knowledge. The bark of cinnamon is widely used as a powerful spice all over the world from the very start of human civilization as a crude powder [1-3]. Cinnamon bark powder is used not only for cooking and flavoring agent but also for conventional and contemporary medicines. This is also used in fragrances and essence industries. Due to its pungent smell, it is used for different varieties of foodstuffs products. The active component of cinnamon is cinnamaldehyde, which is responsible for its aroma and flavor cinnamon. While cinnamic acid, eugenol, and coumaric acid cinnamyl alcohol are active components, responsible for the antioxidant, antifungals, anti-inflammatory, antipyretic, properties [4-5]. Some researchers also reported that it act as anti-diabetic, anticancer, and memory enhancer [5]. It enhances glucose uptake by activating the insulin receptor kinase activity, auto-phosphorylation of the insulin receptors. In this present study, we focused on the different structures(morphological properties) of Cinnamon food powder and their antioxidant behaviour. Such Food source obtained from the bark as medicine is widely used in recent days as it fewer side effects as compared to artificially synthesize drugs. Nanomaterials, Nanoscience, and nanotechnology are the newly emerging field of nanomedicine. Antioxidant activities, such as total phenolic content, OH scavenging rate, superoxide scavenging ability, are also measured. Nanoparticles usually have improved or different properties than the bulk/crude size of material of the same substance and have different properties and an antioxidant behaviour .For centuries, Cinnamon has also been used as a health-promoting agent for the treatment of diseases. Very Small amounts of cinnamon bark nanoparticles can contribute to being a potent antioxidant [7]. we have tried to address the possible scientific aspect for the industrial production of superfine cinnamon bark powder for its numerous applications. The main objective of this present research is to achieve an understanding about change in structural parameters and morphology of cinnamon bark powder during whole process of grinding by high energy ball milling and their potential applications in food science, pharmaceutical, and Biomedical sectors.[8-10].

Material and Methods

Materials required

In this present research for antioxidant property evalution, we used, Ethanol,Tris-HCl buffer, DMSO, Gallic Acid,Folin–Ciocalteu reagent, Hydrogen peroxide, pyrogallol (1, 2, 3-trihydroxybenzene)

Preparation of Cinnamon bark Powder using Ball Mill

The high energy Ball milling is the equipment in which the powdered sample of cinnamon bark sample was placed in the stainless-steel jar of 50ml with steel ball 5 mm sizes. The sample of cinnamon bark was powder was prepared using a kitchen based mixture grinder, for size reduction and this powder is called as general/crude powder of cinnamon (say, C0). The ratio of weight of balls to powder is at 20:1 in the 50 ml stainless-steel jar was filled to about 1/3 of its capacity by volume. During the process of milling, the balls were rotated horizontally with constant speed of 500 rpm for up to 5 hours (say, C_5) and 10 h (say, C_{10}), respectively. The rotational direction of ball milling was changed after every 30 min interval and a constant temperature at 25°C was maintained by an air-cooling system to balance the overheating of ball milling equipment. Similar method was also used earlier to prepare turmeric, bitter guard, Ginger etc. [11,12,13]. Physical properties of Cinnamon bark fine particles were measured using X-ray diffraction, Scanning electron microscope, Fourier transform infrared spectroscopy, and UV-Visible-NIR spectrophotometer. Biomedical investigations such as Total phenolic content and antioxidant etc. were carried out to determine its potential application. The Structural, morphological, the functional group determination and other properties like antioxidant are discussed in detail in the result and discussion in later part of the journal.

Total phenolic content and antioxidant activity
Total Phenolic Content finding

The total phenolic content of Cinnamon crude powder and nanocrystalline powder was obtained using Ultraviolet-Visible spectrophotometer technique. The Folin-Ciocalteu reagent method has been used [16]. The concentration used for working of prepared food powder was taken as1mg/ml and 100μg/ml and dissolved in Dimethyl sulfoxide. The absorbance of the solution was obtained at the radiation of a wavelength of 750 nm [13]. The total phenolic content (TPC) was found, using the formula

$$\mathbf{TPC} = \frac{(X*V*DF)}{M}$$

Where symbol X represents the concentration in ppm, V is the volume of a solution of extract in-unit ml, DF = Dilution Factor of the sample solution, and M is the mass of the sample in gram.

Hydrogen Peroxide scavenging Activity assay

The fundamental scavenging process of each extract solution was obtained using the H_2O_2 process [14]. About 2 mL of extract liquid solution (10–100 μg/mL) in methanol was mixed to 4.0 mL of H_2O_2 (20 mM) solution in phosphate buffer. The pH of the buffer solution was maintained at a numerical value of 7.4. Then at an interval of 10 min, the optical absorbance was found of radiation of wavelength (λmax) 230 nm. This radiation absorbance was noted not in favour of the phosphate buffer blank solution. The scavenging percentage of H_2O_2 was determined using the equation:

$$\textbf{\% scavenging of } H_2O_2 = \frac{[(A_0 - A_1)]}{A_1} * 100$$

In this equation symbol, A_0 represents the absorbance of the control. The control was taken as phosphate buffer with Hydrogen peroxide). The symbol A1 represents the absorbance of the extracts.

Superoxide radicals scavenging activity

In this present experiment, the original pyrogallol (1, 2, 3-trihydroxy benzene) technique was used for superoxide dismutase. This is generally used for measuring superoxide-scavenging of other antioxidants also [15-16]. The procedure of these experimental activities is explained further. A pyrogallol solution (in 1 M HCl) is thoroughly mixed with pH 7.4 Tris-HCl buffer. Again concentration of pyrogallol (25mM) is added in the reaction solution, in which •O2- were scavenged. Further optical absorbance at wavelength 325 nm is measured every 30 seconds for 5 min at 37 °C. The advanced pyrogallol method is a reliable and economic superoxide-scavenging assay suitable for all types of antioxidants.

RESULTS AND DISCUSSIONS

X-ray diffraction (XRD) Measurement

Structural analysis for synthesized Cinnamon bark food powder were considered using X-ray diffractometer (D8 Advance, Bruker Germany) work at 40kV, 40mA and at room temperature. The diffraction pattern were analysed in range from 10 degrees to 50 degrees at a rate of 2° /min scanning rate, which is been presented in figure 1.

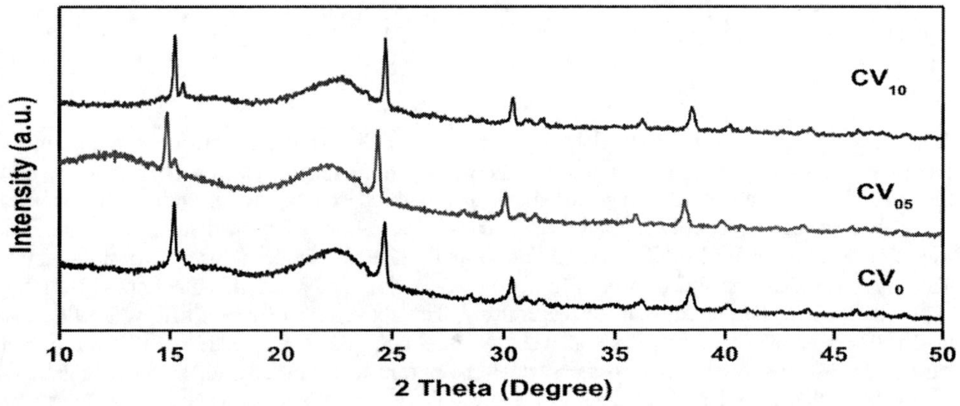

Figure.1 Figure.1. XRD pattern of Cinnamon powder

Mean crystalline size corresponding to different peaks were determined using Scherrer's equation [17].

$$D = \frac{0.94\lambda}{\beta Cos\theta} \quad (1)$$

Where D =crystalline size, K (form factor=0.9) β (FWHM) which is equal to twice the diffraction angle θ, and λ= wavelength for Cu-Kα radiation having value 1.54056Å).

Table 1. XRD data of general Cinnamon powder prepared by mixture grinder (C_0)

2θ of Peak (Degree)	FWHM	d-spacing (Å)	Size of crystals (nm)
15.19	0.12	5.82	121.56
15.55	0.12	5.69	121.51
24.65	0.19	3.60	49.04
28.43	0.28	3.13	28.55
30.33	0.14	2.94	79.99
31.63	0.38	2.82	20.55
34.94	0.67	2.56	10.99

Table 2. XRD data of Cinnamon powder milled for 5hr using Ball milling (C_{05})

2θ of Peak (Degree)	FWHM	d-spacing (Å)	Size of crystals (nm)
15.21	0.4800	5.81	79.56
24.34	0.5760	3.65	79.46
28.18	0.0480	3.16	36.07
30.73	1.1520	2.90	28.69
31.38	1.1520	2.84	29.40
39.15	1.1520	2.35	27.78

Table 3. XRD data of Cinnamon powder prepared by 10hr using Ball milling (C_{10})

2θ of Peak (Degree)	FWHM	d-spacing (Å)	Size of crystals (nm)
15.59	0.14	5.81	79.54
22.62	0.96	3.92	7.31
24.69	0.14	3.60	79.48
31.06	0.28	2.87	28.45
40.15	0.28	2.24	29.43
43.83	0.88	2.06	29.81

The structural data obtained from XRD for general/crude Cinnamon powder, for zero hour milled let it denote by CV_0 (conventional powder obtained by general grinding) which is been shown in table-1. Similarly (CV05) and (CV10) when milled by using high energy ball at 5hr and 10hr respectively in Table 2-3. Further structural parameter was evaluated for above mentioned samples were evaluated like interplanar distance, FWHM, crystallite size and diffraction angle by using Origin 8.0 software. The crystallite size was found to be in range of 29-79 nm ,27-79 nm and 10.99 nm -121.56 nm for 10hr, 5hr and zero hour milled sample respectively. The above result reflects that milling time plays a vital role in crystal structure of nanopowder. Table (1-3) shows the values of different structural parameters like d-spacing, peak positions etc for different hours milled samples respectively, which may indicate that considerable changes in the crystal structure of Cinnamon powder, and distortion of the cell wall structure of powder. Some of the diffraction peaks disappear in 5hr and 10hr nanopowder. XRD analysis also shows that at high milling rate say 5 hrs and 10 hrs some peaks disappears showing amorphous nature of cinnamon (Figure 3 and Table 1-3).. Peak intensity did not

change significantly, and this shows the crystallinity of cellulose of materials was not destroyed at large. Thus, angular peak position with the different crystalline size of ground powder for 0hr, 5hr, and 10 hr does not significantly change. This result reveals that some of the crystalline regions were destroyed during grinding for different time duration. Such XRD details of some food powders are also reported [18]. The intense peak at 15° and 24°, which might be due to the presence of crystalline cellulose having assigned planes of (110) and (200). Crystalline size of few crystal of general Food powder (Fig-1 and Table-1) is above 100nm. But when the same powder was milled for 5hr and 10 hr respectively, sizes reduced to less than 100nm (table 2-3). This shows the more grinding produces superfine nanoscale powder.
XRD analysis also shows that at high milling rate say 5 hrs and 10 hrs some peaks disappears showing amorphous nature of cinnamon .

FTIR analysis

Functional group of different sizes of Cinnamon bark powder were measured using Fourier Transform Infrared Spectroscope, and pellets were prepared using potassium bromide (KBr) as a binder, Spectra were observed with (PerkinElmer UK) at room temperature.

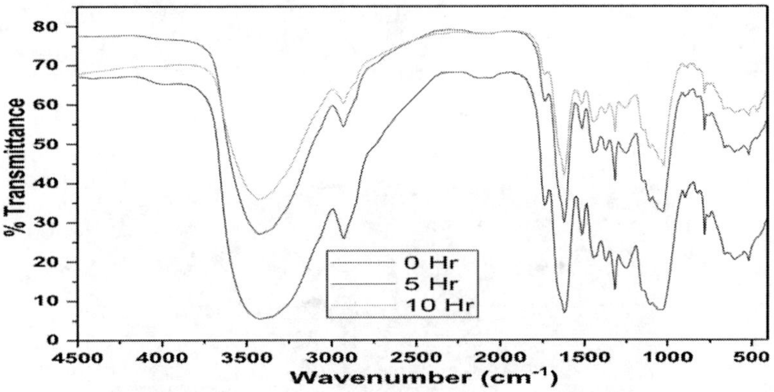

Figure 2: FTIR spectrum for Cinnamon powder samples

Table-4: Details of wave number and Functional group of different Food powder

FTIR spectra of C_0		**FTIR spectra of C_{05}**		**FTIR spectra of C_{10}**	
Wavenumber (Cm^{-1})	**Functional Group**	**Wavenumber (Cm^{-1})**	**Functional Group**	**Wavenumber (Cm^{-1})**	**Functional Group**
3299 Cm^{-1}	O-H	3400	O-H	3400	O-H
2930 Cm^{-1}	N-H	2936 Cm^{-1}	N-H	2936 Cm^{-1}	N-H
2058 Cm^{-1}	N=C=N	2065 Cm^{-1}	N=C=N	2059 Cm^{-1}	N=C=N
1619Cm^{-1}	C=N	1620 Cm^{-1}	C=N	1620 Cm^{-1}	C=N
1515 Cm^{-1}	N-O	1516 Cm^{-1}	N-O	1516 Cm^{-1}	N-O
1450 Cm^{-1}	CH_3	1449 Cm^{-1}	CH_3	1451 Cm^{-1}	CH_3
1375 Cm^{-1}	C-H	1378 Cm^{-1}	C-H	1378 Cm^{-1}	C-H
1317 Cm^{-1}	NO_2	1318 Cm^{-1}	NO_2	1318 Cm^{-1}	NO_2
1248Cm^{-1}	C-O	1246 Cm^{-1}	C-O	1246 Cm^{-1}	C-O
1157 Cm^{-1}	C-O-C	1157 Cm^{-1}	C-O-C	1154 Cm^{-1}	C-O-C
1020 Cm^{-1}	C-OH	1017 Cm^{-1}	C-OH	1017 Cm^{-1}	C-OH
894 Cm^{-1}	C-Cl	896 Cm^{-1}	C-Cl	896 Cm^{-1}	C-Cl
782 Cm^{-1}	C-H	782 Cm^{-1}	C-H	780 Cm^{-1}	C-H
517 Cm^{-1}	C-Br	515 Cm-1	C-Br	515 Cm^{-1}	C-Br

FTIR spectra of ball-milled cinnamon powder is been shown in fig.2 and data are interpreted in table 4 at different durations of (a) 0 h (b) 5 h (c) 10 h. The FTIR spectra were recorded in range of 400 to

IOP Conf. Series: Materials Science and Engineering 1126 (2021) 012020 doi:10.1088/1757-899X/1126/1/012020

4000 cm-1 wave number. Table 4 depicts the materials present in the prepared material. The results shows that there is no new chemical bond was found. Though, very few of characteristics bonds were found to change in transmittance or absorbance range. Further result shows that crystallite size is not the function of functional group. Grinding couldn't effect the intramolecular structure of the prepared material significantly. This can only destroys the cohesive amorphous region of surface of crystallites. The chemical bond present show the presence of cellulose in cinnamon powder [3, 7].Such research were also found by other , shows change number of bond, surface area, density, porosity changes, as reported by some research group [19]. SEM morphology supports these findings, which are shown in figure-3.

Surface morphology of Cinnamon powder

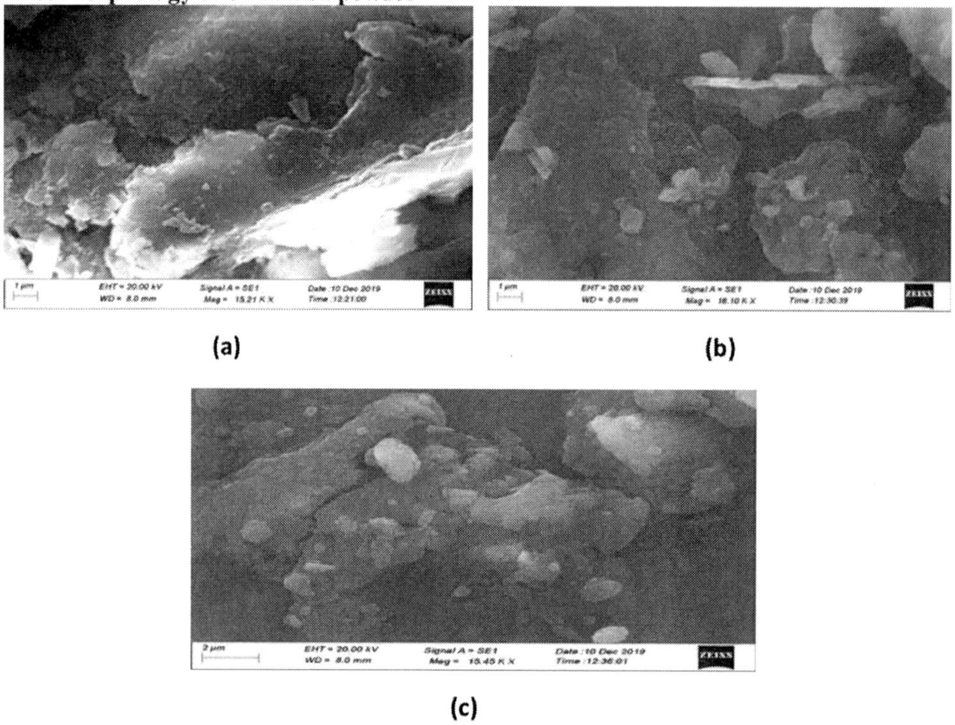

Figure 3. (a-c) - SEM images for prepared Cinnamon Nanopowders

Scanning electron microscope (EVO 18 Zeis UK) was used to study the morphology of synthesized material. The standard parameters were maintained i.e sample were analysed at 20 KV. The sem images of different milling our sample is been shown in Fig 3(a-c).External surface morphologies and large agglomerations of particles show that molecular structure is distortion shape of different micron sizes Further the result shows large agglomerations and distorted shape of micron size . Milling hours also effect the morphology of the prepared material. This observation shows that griding is also effective in changing the original structure of cinnamon powder. Surface morphology after milling at different hr. changed significantly, which can have a considerable impact on Physical-Chemical behaviour for various industrial applications. High energy Milling weakened the cohesive force and broke the particles into different size fractions, and they combined aggregation results of various shapes of particles, as shown in fig.3 (a-c). Different surface structures were also reported by some of the research groups by pressure grinding/ball milling [20].

UV-VIS NIR Measurement

The UV-visible (Lambda 950, PerkinElmer, UK) spectrum were analysed in the wavelengths 200 to 550 nm and results are shown in figure 4. Proteins biomaterials are present in animal tissue and plants that strongly absorb radiation in the range of about 280 nm [21-22]. It may be due to the amino acids, which made up the proteins and absorb the radiation in UV range. Such type of UV absorption can be used for the identification of protein bands in the food materials. In the present research, we have taken food nanoparticles (Cinnamon), and it has found maximum absorbance in the range of 280 nm radiation. The spectra also reveal that as the milling hours are increased, it also shows more absorption. This study reveals that absorption depends on the degree of superfine powder.

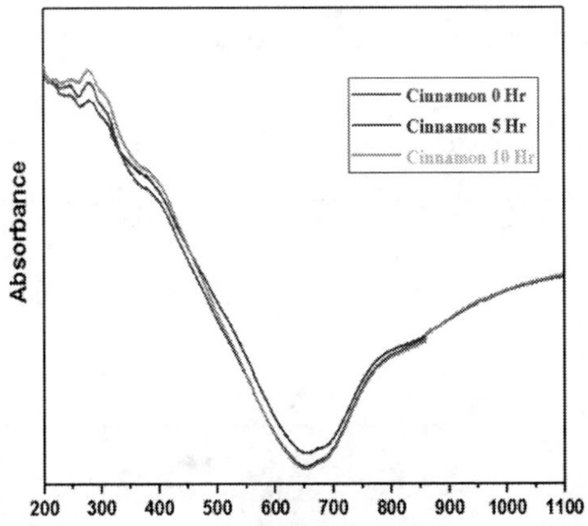

(a)

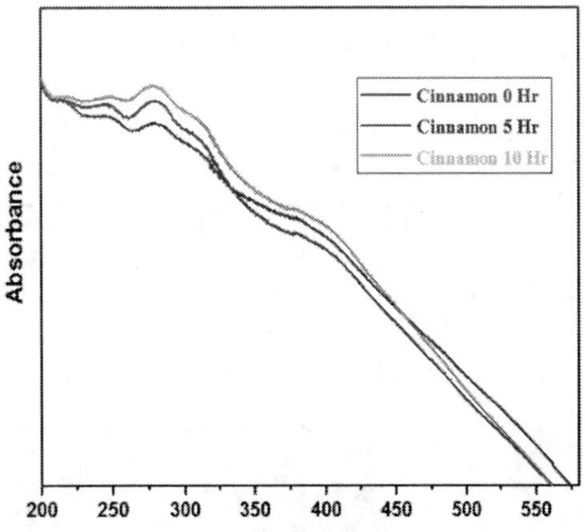

(b)

Figure 4: UV-VIS-NIR spectrum of Ginger Cinnamon powder.

ANTIOXIDANT PROPERTIES

Total phenolic content

In antioxidant activities, Phenolic compounds exhibit redox behaviour. The free radical activities are due to the hydroxyl group, which are present in the materials. As mentioned in the materials method section that phenolic content was calculated by the Folin–Ciocalteu reagent method [13]. The experimental findings were obtained from a calibration curve of equations

$$y = 0.0097\ X - 0.3967,\ R^2 = 0.9865$$

of gallic acid. The range of concentration is taken between 0 – 450 μg/mL and articulated in gallic acid equivalents (GAE) per gram dry extract weight (Table 5 and fig-5). In Which Y is the absorbance, X is the total phenolic content, and R square is regression coefficient. R_2 is statistically measured that reflects the proportion of the variance for a dependent to relative variable. R squared shows to what extent the variance of one variable explains the variance of a second.

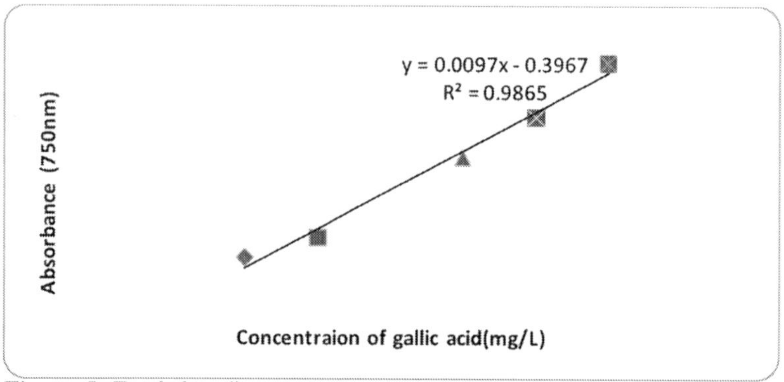

Figure 5: Total phenolic content standard curve with reference to gallic acid.

Table 5- In vitro antioxidant activity assay and Total Phenolic Contents (TPC) of superfine cinnamon particles (Value of mean ±SD of up to 3 replications)

Details(group)	Total phenolic content in mg, **GAE/g(±SD)**	Hydroxyl radicals %scavenging activity**±SD)**	Superoxide radicals % scavenging ability assay**(±SD)**
General Cinnamon powder(1mg/ml) in DMSO	49.58 ± 0.077	21.69 ± 1.72	33.9 ± 0.22
Superfine Cinnamon powder of 5hr milling (1 mg/ml) in DMSO	49.69 ± 0.070	37.6 ± 1.28	45.52 ± 1.66
Superfine Cinnamon powder of 10hr milling(1mg/ml) in DMSO	60.28 ± 0.18	61.47 ± 0.64	56.44 ± 1.66
General Cinnamon powder(100µg/ml) in DMSO	22.68 ± 1.03	16.52 ± 1.51	21.22 ± 0.82
Superfine Cinnamon powder of 5hr milling(100µg/ml) in DMSO	37.68 ± 0.52	26.57 ± 1.25	40.76 ± 0.67
Superfine Cinnamon powder of 10hr milling(100µg/ml) in DMSO	44.94 ± 0.37	49.15 ± 2.27	47.7 ± 0.20

Superfine Cinnamon bark powder has total phenolic content ranges amid 49.58 mg to 60 .28 mg G/g for crude /general Cinnamon powder (C_0), (C_5), (C_{10}) prepared food powder materials for the concentration of the sample 1mg/ml. whereas Total Phenolic Content value ranges are 22.68 to 44.94 GAE/g for concentration 100 µg/ml of cinnamon superfine powder. The outcome shows that total phenolic contents concentration depend in present materials and surface structure of food Nanopowders [table-5]. This measurement shows that surface science and related crystal structure changes the TPC activities significantly. In this present study, Surface reactivity was found to increase for three prepared materials.

Free radical scavenging

a) Hydroxyl radicals scavenging activity

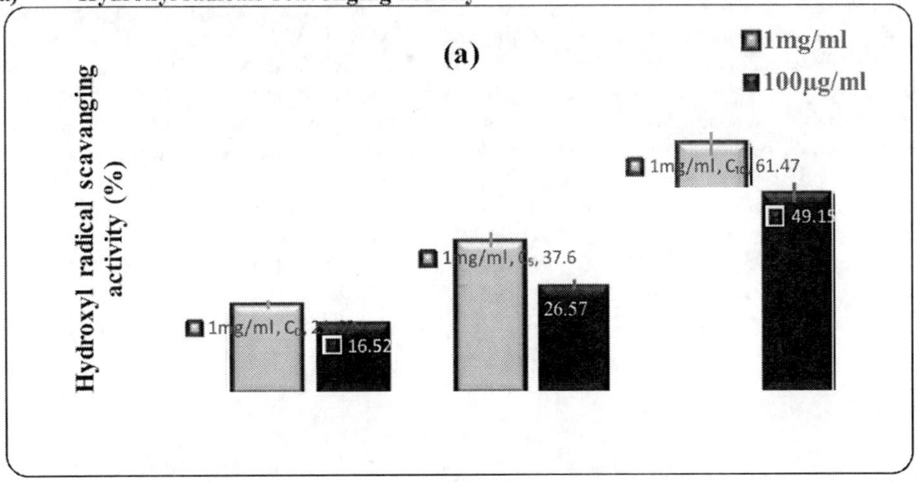

Fig. 6- Effect of cinnamon powder on the Hydroxyl radial scavenging effect (Value of mean ±SD of 3 replications)

The in vitro analysis of antioxidant capability of general/crude Cinnamon powder (C_0) and Cinnamon superfine powder after five hours of milling (C_5) and 10 hours of milling powder (C_{10}) was enhancing as the size of the particle and surface area of the material changes. The scavenging activity was highest i.e. 61.47 ± 0.64 % found in the sample which is milled for 10 hours with the concentration of working solution was 1mg/ml DMSO. Whereas the scavenging capability of crude/general Cinnamon powder at this concentration is attain as 21.59 ± 1.72%, which is quite low as compared to superfine cinnamon powder, as given in figure 6 and table 5. While for the concentration of 100µg/ml, largest scavenging was obtained 49.15± 2.27 % and minimum scavenging was found 16.52 ± 1.51 % for 10hr milled and 0 hr milled Cinnamon food powder.

Pyrogallol superoxide radicals scavenging activity

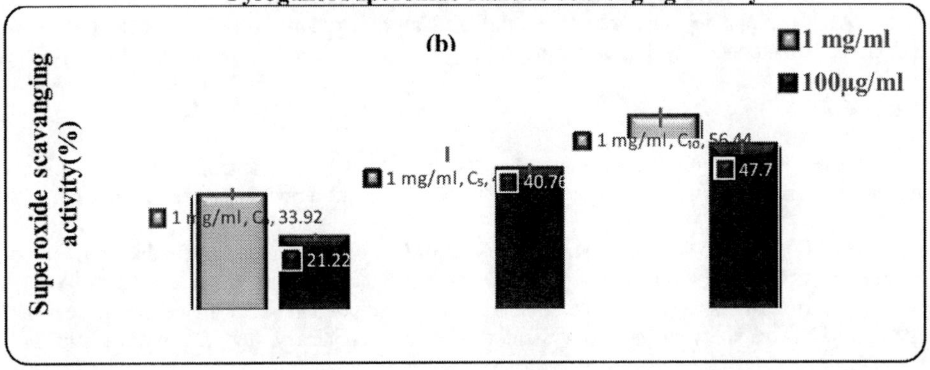

Figure7. In vitro Superoxide scavenging activity of different sizes of Cinnamon powder of different size and concentrations in DMSO.

The superoxide emitted, by some Biological & Photochemical reactions and attacking important biological organs is a toxic species. The elemental oxygen and OH radical is the constituent of superoxide radical which may increases oxidative stress, pathological incident, destructions of cell, etc. These radicals may also produced by auto-oxidation. In the current research study the scavenging activity of superfine Cinnamon powder was recorded, through spectrophotometer and data obtained in this process shows that the binding capacity to the O_2 radicals generated by the metabolic reaction and is highest in 10 hours of milled superfine cinnamon bark powder when 1mg of powder dissolved in 1ml DMSO is 53.9 ± 0.0012 % as shown in figure 7 and table 5. Whereas the scavenging capability at 0 hours milled fine cinnamon bark powder (1mg/ml) I DMSO is 39.2 ± 0.0037 %. (Value of mean ±SD of 3 replications. Thus, from table-5, fig.6-7, we found that Hydroxyl radicals and superoxide scavenging ability increases with increase in concentration and decrease in size of superfine cinnamon powder. Therefore, present studies suggest that medicinal plant products like spices condiments, fruit seed powder could be engineered for commercially processing and production of food Nanopowders which is seemed to be a promising good effect on health and could be used as new functional materials by pharmaceutical sector and Biomedical Engineering. Such antioxidant capability help improve the nutritional value of prepared food materials. Based on reported work, In a nutshell, it may be concluded that the major effect in the present studies is that the grinding produces a small crystalline size of cinnamon food particles. Reduction in size increases surface reactivity and agglomerations and increases the efficiency of antioxidant properties. Similar results are also reported for green tea powder and other food powder [23-25]

CONCLUSIONS

Superfine Cinnamon food Nanocrystalline powder of different morphology and crystal structures were successfully prepared using high energy ball milled for prominent industrial and scientific significance. The crystal structure, functional group, were evaluated using modern scientific tools such as X-Ray Diffraction (XRD), Fourier Transforms Infrared Spectroscopy (FTIR), Scanning Electron Microscope(SEM) and found considerable change in position of wave number, inter-planar distance and reactivity's at the surfaces. The present study reveals grinding produces a new surface structure, which is beneficial for physicochemical behaviour. Optical measurement shows that UV absorbance was found at about 280nm.The phenolic content, Hydroxyl radicals and superoxide radicals scavenging activity were found to increase as the milling hour, and superfine behaviour increases. Antioxidant properties also depend on the concentration of food powder. The current research finding opens a new aspect for the development of surface science of food nanopowders and its potential biomedical applications, drug discovery, health sciences, and medicine industries.

ACKNOWLEDGEMENT

Authors are grateful to the Department of Education, Govt. of Bihar and Aryabhatta Knowledge University (AKU), Patna for support in the establishment and functioning of Aryabhatta Centre for Nanoscience& Nanotechnology, Patna, Bihar, India.

REFERENCES

1. Määttä-Riihinen, K. R.; Kähkönen, M. P.; Törrönen, A. R.; Heinonen, I. M. Catechins and procyanidins in berries of Vaccinium species and their antioxidant activity. J Agric Food Chem 2005; 8485– 8491.
2. Lu J, Zhang K, Nam S, Anderson RA, Jove R, Wen W. Novel angiogenesis inhibitory activity in cinnamon extract blocks VEGFR2 kinase and downstream signalling, Carcinogenesis 2010; 481-488.
3. Molania T, Moghadamnia A, Aghel S, et al. The effect of cinnamaldehyde on mucositis and salivary antioxidant capacity in gamma-irradiated rats (a pre-liminary study). DARU J Pharm Sci, 2012: 20-89.
4. Hong J W, Yang GE, Kim YB, Eom SH, Lew JH, Kang H. Anti-inflammatory activity of cinnamon water extract in vivo and in vitro LPS-induced models. BMC Complement Altern Med 2012; 237.
5. Roussel A, Hininger I, Benaraba R, Ziegenfuss TN, Anderson RA. Antioxidant effects of a cinnamon extract in people with impaired fasting glucose that are overweight or obese. J Am Coll Nutr 2009; 16-21.

6. Ziegenfuss TN, Hofheins JE, Mendel RW, Landis J, Anderson RA. Effect of a water- soluble cinnamon extract on body composition and features of the metabolic syndrome in pre-diabetic men and women. J Int Soc Sports Nutr 2006; 45-53.
7. Anderson RA, Broadhurst CL, Polansky MM, Schmidt WF, Khan, et al. Isolation and characterization of polyphenol type-A polymers from cinnamon with insulin-like biological activity. J Agric Food Chem; 2004; 65-70.
8. Wu GC, Zhang M, Wang YQ, Mothibe KJ, Chen X, et al. Production of silver carp bone powder using superfine grinding technology: suitable production parameters and its properties. J Food Eng, 2012;730-735.
9. Huang ZQ, Xie XL, Chen Y, Lu JP, Tong ZF. Ball-milling treatment effect on physicochemical properties and features for cassava and maize starches. CR Chim, 2008;73–79.
10. Li G, Guo W, Gao X, Wang Y, Sun S. Effect of superfine grinding on physicochemical and antioxidant properties of soybean residue powder. Food Sci Nutr 2020; 1208-1214.
11. Aman AK ,Singh RK, M, Kirti M. Effect of high energy ball milling grinding on physico-chemical, structural and morphological studies of bitter melon (momordicacharantia) nanopowder. Int J Recent Sci Res 2017;19258-19263.
12. Aman AK, Singh RK, Kumar R, Ghosh AK. Effect of high energy ball milling grinding on physico-chemical, morphological and optical properties of Curcuma longa nanoparticles powders. Int J Pharm Sci Res, 2018; 672-77.
13. Archana, Abhay Kr. Aman, Rakesh Kr. Singh , Nishant Kr, Amber Jabeen , Effect of superfine grinding on structural, morphological and antioxidant properties of ginger (Zingiberofficinale) nano crystalline food powder, doi: https://doi.org/10.1016/j.matpr.2020.09.028
14. Prabhakar K R, Veeresh V P, Vipan K, Sudheer M, Priyadarsini KI, Satish RBSS, Unnikrishnan MK. Bioactivity-guided fractionation of Coronopusdidymus: A free radical scavenging perspective. Phyto medicine, 2006; 591-595.
15. Moreno CS. Methods used to evaluate the free radical scavenging activity in foods and biological systems, Review. Food Sci. Technol Int, 2002;121-137.
16. Li X. Improved Pyrogallol Autoxidation Method: A Reliable and Cheap Superoxide-Scavenging Assay Suitable for All Antioxidants. J. Agric. Food Chem 2012; 60:6418−6424.
17. Theivasanthi T, Alagar M. An Insight Analysis of Nano sized powder of Jackfruit Seed. Nano Biomed Eng 2011;163-168.
18. Muruganantham S, Anbalagan G, Ramamurthy N. FTIR and SEM- EDS Comparative Analysis of Medicinal Plants, Eclipta alba hassk and Ecliptaprostratalinn. Romanian J. Biophys, 2009;285-294.
19. Yang Y, Wei XZ, Sun P, Wan JM. Preparation, Characterization and Adsorption Performance of a Novel Anionic Starch Microsphere. Molecules, 2010; 2872-2885.
20. Ehsan NZ, Peyman NM, Elham A, Iman S. Conductive and Biodegradable Polyaniline/Starch Blends and their Composites With Polystyrene. Iranian Polymer J, 2011; 319-328.
21. Shenghua H, Yibing Q, Elfalleh W, Lin L, JieC,Ying M. Effect of ball milling on the physicochemical properties of maize starch, Biotechnology Reports, 2014; 54-59.
22. Ming J, Chen L, Hong H, Li J. Effect of superfine grinding on the physic chemical, morphological and thermo gravimetric properties of Lentinusedodes mushroom powder,. J. Sci Food Agri, 2015; 2431-2437.
23. Hu J, Chem Y, Ni D. Effect of superfine grinding on quality and antioxidant properties of fine green tea powder LWT-Food Science and Technologies 2012 ;8-12.
24. Chen H, Weiss J, Shahidi F. Nanotechnology in nutraceuticals and Functional. foods Food Tech, 2006; 30-36.
25. Zhao X, Zhu H, Zhang G, Tang W. Effect of superfine grinding on the physicochemical properties and antioxidant activity of red grape pomace powders. Powder Technology, 2015; 838-844.

IOP Conf. Series: Materials Science and Engineering 1126 (2021) 012021 doi:10.1088/1757-899X/1126/1/012021

Salt additive extraction of cinnamon oil using microwave radiation

Piyush I Modi[1,2,3], Jigisha K Parikh[2], Meghal A Desai[2]

[1]Chemical Engineering Department, Pacific School of Engineering, Surat-394305, Gujarat, India.
[2]Chemical Engineering Department, SardarVallabhbhai National Institute of Technology, Surat-395007, Gujarat, India.
desai_ma@yahoo.co.in

Abstract
Inorganic salts have the ability to disturb the plant cell structure which may improve the process for isolation of the essential oil. In the present work, potassium chloride (KCl) was utilized as an additive in extracting cinnamon oil from *Cinnamomum zeylanicum* using microwave irradiation. The parametric study was performed for the four parameters namely; concentration of KCl salt, extraction time, solid loading, and microwave power. The parameters were then optimized using the Taguchi method and significance of each parameter was determined by analysis of variance. The highest recovery of cinnamon oil was found to be 5.43 (%, w/w) at 10 g KCl salt concentration, 40 min extraction time and 800 W microwave power. This approach has reduced the extraction time, utility requirements and energy consumption thereby showing the greener and cleaner aspects. Moreover, this process would be easily adopted at an industrial scale due to minor changes in existing process.

1. Introduction

The spices are principally engaged in cooking as a major flavoring ingredient globally. Flavors of these spices are the resultant of the essential oil present in it. Since long back, Asian countries have very good market due to major cultivation of spices [1]. The essential oils have wide footstep in various industries like; fragrances, pharmaceutical, pesticides and preservative products (Jeyaratnam et al., 2016). Among all essential oils, cinnamon (*Cinnamomum zeylanicum*) is of specific interest. Cinnamon belongs to the Lauranceae family and small brown scrolls with classic look are made available in tropical countries [3]. Due to its sweet and pungent taste with presence of fine aroma, cinnamaldehyde (CA), one of the major components, is valued more in the oil[4].

Cinnamaldehyde (CA) was first isolated in 1834 from cinnamon oil by Andre Dumas and Peligot [5]. It has broad spectrum activities of antifungal formulation. It has also been reported to have antidiabetic, antioxidant, antimicrobial, anti-inflammatory and anticancer properties [6]; [7]; [8]; [9]; [10]. The essential oil from the bark of cinnamon can be extracted by convectional method like hydrodistillation[11], steam distillation [12], solvent extraction [13] and soxhlet extraction [14]. Over the time, extraction methods have been bettered to improve the processing conditions, maximize the recovery of the oil, lowering the operating cost and obtain the superior quality of the product.

In research articles, cinnamon oil extraction has been carried out with organic solvent instead of water [13], superheated water [15] and sonication [16]. However, a large quantity of organic solvents, longer extraction time and higher energy consumption are the major disadvantages of these methods which can lead to higher environmental impact and a higher pressure pauses an issue of safety. Though sonication is considered as a novel technique, the exposure of the sonicator probe to the higher thermal energy could lead to rapid erosion and finally inefficient operation.

In this era, microwave assisted extraction (MAE) has acquired the diverse advantages of reduction in time and energy consumption [17]. Further, it has the potential to be scaled up because of a low break-even point [18]. In MAE, electromagnetic waves are penetrating in the biological tissues and then extract the phytochemical or essential oil from plant material at a higher rate [19].

For extraction of cinnamon oil, addition of salt was incorporated with water and cinnamon. The idea was to further improve the oil yield, reduce the operation time, energy requirement and utility requirement compared to the conventional microwave distillation. The various parameters were studied for additive assisted MAE of cinnamon oil. The parameters have been optimized by the Taguchi method which was used for the three levels and three factors i.e.; extraction time, KCl salt loading and microwave power. An analysis of variance (ANOVA) was applied to identify the most significant parameter for the extraction process. The various factors were compared for MAE method performed with and without KCl salt.

2. Materials and methods

2.1. Materials

Cinnamon sticks have been procured from a nearby area of Surat, India and stored in an air tight container. An experiment was performed with a freshly grounded powdered sample (1000 μm) of cinnamon. Potassium chloride (98% pure) was supplied by M/s Finar Chemicals Pvt. Ltd., Ahmedabad, India to carry out the extraction method. The experiments were performed using tap water.

2.2 Methods

2.2.1 Microwave assisted extraction (MAE)

An experiment was conducted in a modified microwave oven (Samsung ME83HD-B, 1250 W, 2450 MHz, India) equipped with Dean and Stark apparatus. The appropriate volume of salt solution was kept in a round bottom flask (1 L) with a desired solid loading to avoid charring and overheating of cinnamon powder. Microwave irradiation has caused water molecules and salt to achieve the excited state and generated heat due to dipole moment and ionic conduction. Because of rapid heat generation leading to instant vaporization in cells, the plant cells have been severely damaged so that, the volatile components came out along with water vapors and have been directed towards the condensing section. The condensed mixture was then collected and separated. For further studies extracted cinnamon oil was measured and stored in vial. The experiment was performed twice to check the accuracy and reproducibility.

2.2.2 Parametric study

The extraction parameters are assuming an important role in the extraction process. Here, four parameters, namely KCl salt loading (5-25 g), extraction time (20-60 min), solid loading (15-35 g) and microwave power (300-800 W) have been chosen to assess their impact on the recovery of the oil from the bark of cinnamon. KCl salt can impact the heating phenomena thereby the recovery of the oil hence it was varied in the range of 5-25 g. Extraction time to complete the extraction process was varied from 20-60 min. Solid loading was varied from 15 to 35 g to obtained the upmost capacity and avoid charring of the system. To supply the heat for the system, microwave power was varied from 300 to 800 W.

2.2.3 Optimization of process parameters using the Taguchi method

The Taguchi method was used to plan and perform well defined set of experiments. Three important parameters, namely KCl salt loading (5-25 g), extraction time (20-60 min) and microwave power (300-800 W) were studied at three levels. To obtain the yield of essential oil, the signal to noise (S/N) ratio for the "larger-the-better" case was selected [20]. ANOVA study was performed to identify the most significant process parameters and contribution of each parameter.

3. Result and discussion

3.1 Parametric study

3.1.1 Effect of KCl salt loading

The structure of water molecule is influenced when salts are dissolved in it. At some concentrations of salt, the microwaves have the maximum impact to the pure water because dielectric component increases dipole movements [21]. In conventional extraction method salt additive process was used for cinnamon and it achieved remarkable result [4]. In the present experiment, KCl salt was used as an additive salt amount (5-25 g) in water while keeping remaining factors constant viz. extraction time: 50 min, solid loading: 25 g and microwave power: 800 W. KCl salt loading from 5 g to 20 g has increased the yield of cinnamon oil as shown in Figure 1. It might be that the chloride salt has differed the plant cell wall structure to improve the extraction [22]. At 20 g KCl salt loading, the maximum yield of oil obtained so further experiments have been conducted with 20 g KCl salt.

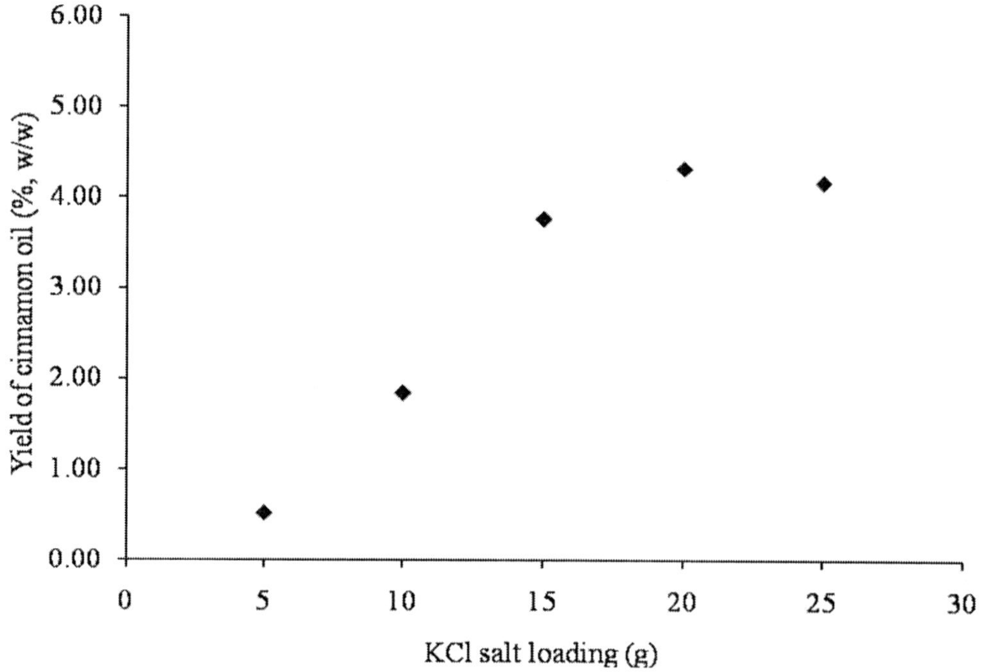

Figure 1: Effect of KCl salt loading on yield of cinnamon oil

3.1.2 Extraction time

Sufficient extraction time is mandatory to complete the extraction and to minimize the need of unnecessary energy consumption. To identify an extent of impact of the extraction time on the recovery of the oil, the experiments were performed with parameters like; KCl salt loading 20 g, solid loading 25 g and microwave power 800 W. As shown in Figure 2, the extraction of the oil was increased with an increase in extraction time. However, it was decreased after 30 min of extraction time due to the dilution effect which requires more time to get vaporized [19]. In comparison with other methods, MAE has a benefit of a short time period of extraction without affecting the oil composition [2].

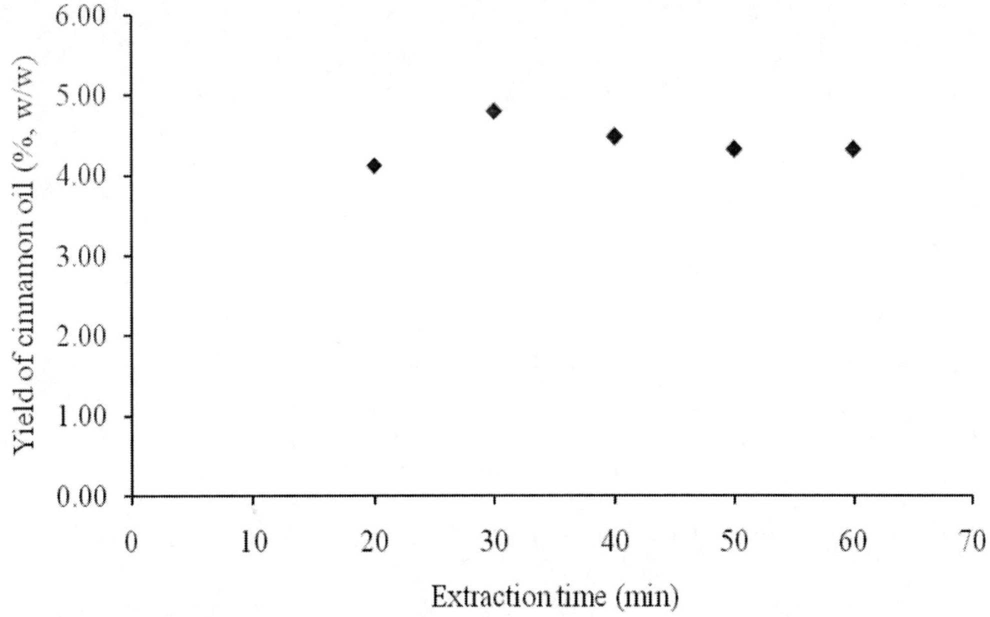

Figure 2: Effect of extraction time on yield of cinnamon oil

3.1.3 Solid loading

Solid loading is an effective parameter on the extraction process to run the system at its highest possible capacity. It was varied from 15 to 35 g and other parameters were fixed i.e.; KCl salt loading 20 g, extraction time 30 min and microwave power 800 W. Results showed that with solid loading up to 20 g the extraction yield was improved and subsequently the recovery was reduced (Figure 3). A higher amount of solid loading does not attain a desired magnitude of radiation per particle. It could lead to the non-uniform heating in the flask. Further, a higher amount of solid loading requires more time to achieve the complete extraction. This type of observation has been made while extracting piperine from *Piper nigrum* [23]. Microwave power absorbed per unit volume can be estimated by the Lambert's law equation [24] as shown in equation 1.

$$P = 2\beta P_0 \exp(-2\beta z) \quad (1)$$

Where,

P: Power absorbed per unit volume,

P_0: Transmitted power flux,

β: Attenuation constant,

z: Distance from the sample surface

As per the equation, with increase in the depth of the sample power would reduce in an exponential manner leading to uneven heating in the sample mixture. Therefore, more time is required with an increase in the amount of the plant material.

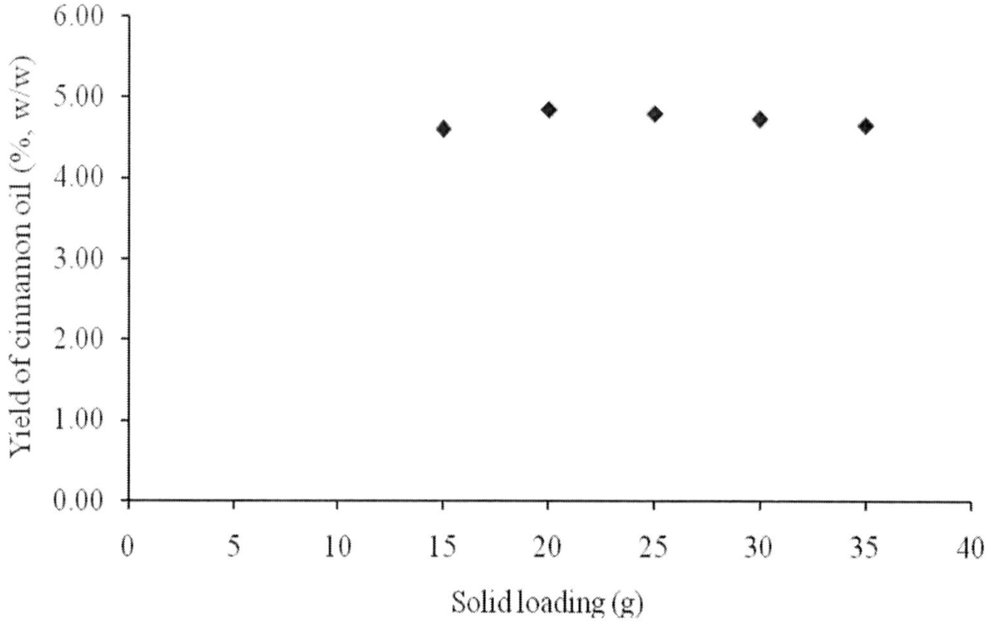

Figure 3: Effect of solid loading on yield of cinnamon oil

3.1.4 Microwave power

Microwave power provides the energy to the flask solution where this energy is converted into a thermal energy in the dielectric material like aqueous solution and accordingly increases the temperature of the solution. Due to absorption of microwaves by the solution which contains KCl salt and water molecules [25], the heating processes is escalated and a faster extraction is achieved. With an increase in the intensity of the microwave power, heating process is further enhanced. Microwave irradiation power can be represented by equation 2.

$$Q_{abs} = mc_p(\Delta T/\Delta t) \quad (2)$$

Where,

Q_{abs}: absorbed power per unit weight of mixture,
m: mass of mixture,
c_p: specific heat of mixture,
ΔT: temperature increase of mixture,
Δt: time period to increase the temperature

In the present study, microwave power was studied from 180 to 800 W and other parameters were maintained at a constant level like; KCl salt loading 20 g, extraction time 30 min and solid loading 20 g. Microwave irradiation power is used to increase the specific absorbed power to convert the energy and destroy the cell wall. Because of this, water molecules into plant matrices have broken and thereby accelerated the extraction rate [26]. Figure 4 shows that the recovery of the cinnamon oil was increased with an increase in the microwave power.

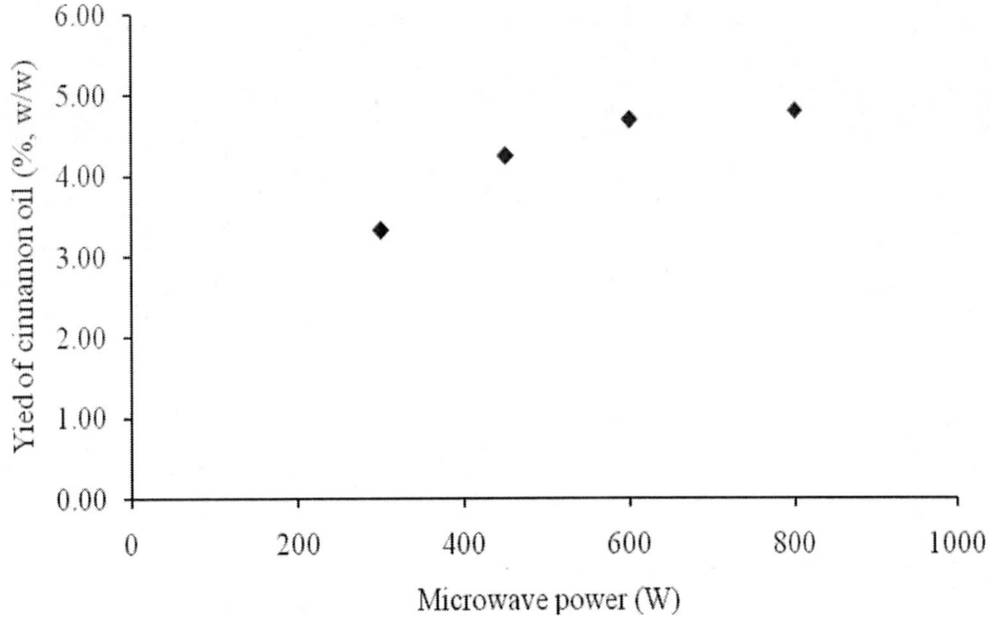

Figure 4: Effect of microwave power on yield of cinnamon oil

3.2 Optimization of process parameters using Taguchi method

Experiments were conducted on the basis of the L_9 array of the Taguchi method for three variables at three levels as shown in Table 1. The array is orthogonal in nature thereby showing unbiasedness towards the parameters (Table 2). Based on the array, the experiments have been conducted and the yield of cinnamon was calculated followed by S/N ratio determination as per equation 3. The experimental planning and statistical analysis of the result was carried out using Minitab 17.0.

$$\frac{S}{N} = -10\log\left(\frac{1}{n}\sum_{i=1}^{n}\frac{1}{y_i^2}\right) \quad (3)$$

Where,

n: number of repetitions,

y_i: yield of the ith experiment.

Total S/N ratio was calculated (Table 2) to find out the optimum conditions. To observe the effect of each parameter the plot was constructed using the total S/N ratio as shown in Figure 5. The optimum parameters for the present study have been obtained as KCl salt loading 10 g (level 1), extraction time 40 min (level 3) and microwave power 800 W (level 3). As the variation in response due to change in the KCl salt loading was minimal, its loading was kept minimum (10 g) to reduce its consumption.

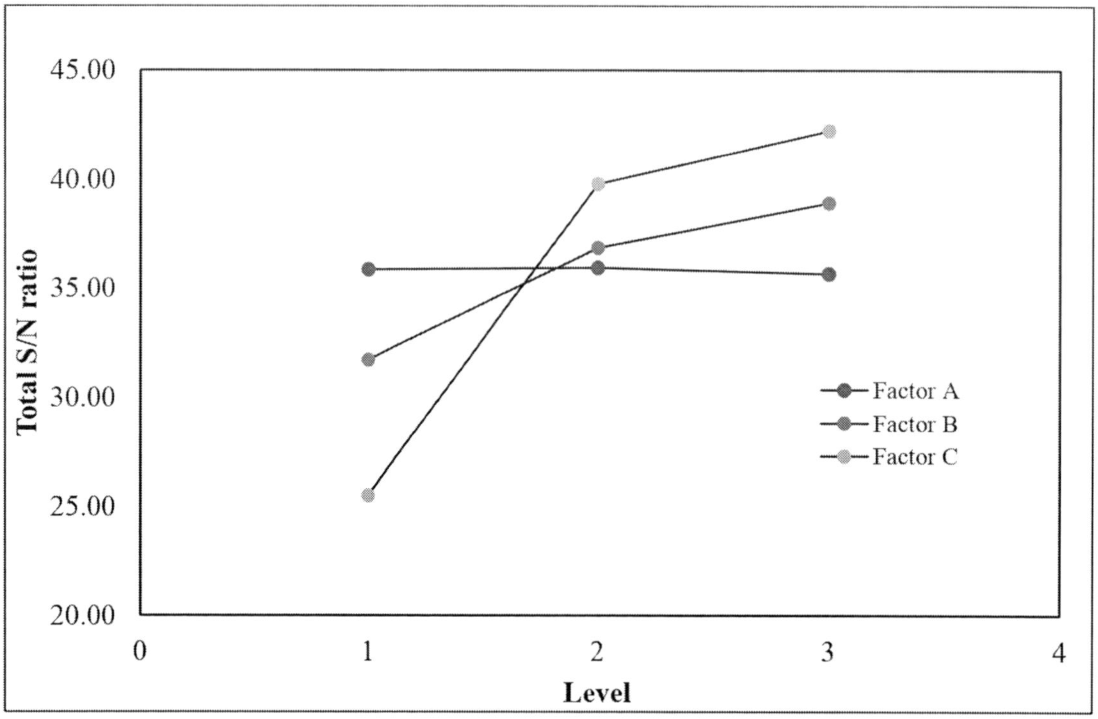

Figure 5: Main effect plot of S/N ratio

Table 1: Factors and their levels

Factors	Levels		
	1	2	3
A: KCl salt loading (g)	10	15	20
B: Extraction time (min)	20	30	40
C: Microwave power (W)	300	600	800

Table 2: Experimental layout using L_9 orthogonal array and the response

Experiment	Factors			Yield (%, w/w)		S/N Ratio
	KCl loading (g)	Extraction time (min)	Microwave power (W)	1	2	
1	1 (10)	1 (20)	1 (300)	2.30	2.10	6.82
2	1	2 (30)	2 (600)	5.55	5.00	14.41
3	1	3 (40)	3 (800)	5.38	5.43	14.66
4	2 (15)	1	3	4.50	5.05	13.54
5	2	2	1	2.50	2.80	8.42
6	2	3	2	4.85	5.20	14.01
7	3 (20)	1	2	3.95	3.50	11.37
8	3	2	3	5.25	4.85	14.05

9	3	3	1	3.45	3.10	10.27
Total S/N ratio (level 1)	35.89	31.73	25.51	-	-	-
Total S/N ratio (level 2)	35.97	36.88	39.79	-	-	-
Total S/N ratio (level 3)	35.69	38.93	42.24	-	-	-
10 (Confirmation experiment)	1 (10)	3 (40)	3 (800)	5.45	5.40	14.69

3.3 Analysis of variance (ANOVA)

ANOVA study has been conducted for evaluating the impact of the factors involved in the study on the extraction yield of cinnamon oil (Table 3). For factor C (Microwave power), *F* value was considerably higher than other two factors and statistically insignificant for factor A (KCl salt loading). The percentage contribution of factor C (Microwave power) was 81.92% (Table 3). The parameters can be arranged based on their impact on yield of cinnamon oil as follows; microwave power > extraction time >KCl salt loading. A little variation in the microwave power could impact the yield of cinnamon oil in the extraction process at a greater extent.

Table 3: Analysis of variance (ANOVA)

Factors	**S**	**DF**	**V**	**F**	**% CF**
KCl salt loading (A)	0.01	2.00	0.01	0.00	0.02
Extraction time (B)	9.16	2.00	4.58	3.24	13.80
Microwave power (C)	54.40	2.00	27.20	19.24	81.92
Error	2.83	2.00	1.41	-	4.26
Total	66.41	8.00			100.00

To confirm the optimization conditions, the experiment was conducted under optimized conditions; 10 g KCl salt loading, 40 min extraction time and 800 W microwave power. The maximum yield of cinnamon oil (5.43%, w/w) was obtained using these conditions and it was the highest among the outcomes obtained using L_9 array. The results of this study has substantiated the application of the Taguchi method to optimize the processing conditions for cinnamon oil recovery using MAE.

3.4 Comparison of energy and environmental impact

In the present study, the results have been compared for MAE techniques with and without KCl salt as shown in Table 4. Extraction in presence of KCl salt required 40 min for 20 g solid loading with an energy consumption of 26.67 W·h while without KCl salt, extraction was complete in 50 min using 33.33 W·h energy by treating the same amount (20 g) of plant material. The recovery of the cinnamon oil was 5.43% and 4.83% for with and without KCl, respectively. It was observed that reduction in time as well as energy was possible by incorporating KCl salt in MAE of the oil from cinnamon bark with improved yield.

In addition, it is important to consider the CO_2 emission during the extraction to justify the green aspect of the process. If the power generation is based on non-renewable sources like coal or fossil fuel, 800 g of CO_2 emits in the environment during generation of 1 kW power. The CO_2 emission was calculated using the equation (4) [27]. In case of MAE without salt, ~20% more CO_2 emission was observed.

$$E_{CO_2} = E_{Comsumed} \times 800 \tag{4}$$

Where, E_{CO_2} represent the amount of carbon dioxide emission in kg.

Table 4: Comparison of microwave assistance extraction with and without KCl

Factors	Without KCl	KCl
Yield of essential oil (%, w/w)	4.83	5.43
Extraction time (min)	50	40
Power consumption (W·h per g of plant material)	33.33	26.67
CO_2 emission (g per g of plant material treated)	26.67	21.34

4. Conclusion

The extraction of cinnamon oil from cinnamon bark by microwave assisted extraction with KCl salt was effectively carried out and statistical treatment was successful using the Taguchi method. The maximum yield (5.43%, w/w) of cinnamon oil was obtained under the optimum conditions (10 g KCl salt loading, 40 min extraction time and 800 W microwave power) in case of single response study. The most important parameter contributing towards the extraction yield was found to be the microwave power and its contribution was 85.57%. A reduction in carbon emission was around 20 % with KCl salt addition during MAE. A comparison of both the methods i.e.; with and without KCl salt in MAE has shown promising result for salt assisted MAE like minimum extraction time, energy and environment impact. Thus, KCl salt addition for microwave assisted extraction can be employed for extraction of oil from several natural plant materials and for considering opportunities at an industrial scale.

References

[1] R.O.B. Wijesekera, C. O. Chichester 1978 *C R C Crit. Rev. Food Sci. Nutr.* **10** 1–30.
[2] N. Jeyaratnam, A.H. Nour, R. Kanthasamy, A.H. Nour, A.R. Yuvaraj, J.O. Akindoyo 2016 *Ind. Crops Prod.* **92** 57–66.
[3] L. Malsawmtluangi, B.P. Nautiyal, T. Hazarika, R.S. Chauhan, A. Tava 2016 *J. Essent. Oil Res.* **28** 551–56.
[4] L. Pistelli, S. Giovanelli, P. Margari, C. Chiappe 2016 *RSC Adv.* **6** 52421–26.
[5] S. Shreaz, W.A. Wani, J.M. Behbehani, V. Raja, M. Irshad, M. Karched, I. Ali, W.A. Siddiqi, L.T. Hun 2016 *Fitoterapia.* **112** 116–131.
[6] F. Mustaffa, J. Indurkar, M. Shah, S. Ismail, S.M. Mansor 2013 *Nat. Prod. Res.* **27** 888–95.
[7] J. Asghari, B. Ondruschka, M. Mazaheritehrani 2011 *J. Med. Plants Res.* **5** 495–506.
[8] H.S. Youn, J.K. Lee, Y.J. Choi, S.I. Saitoh, K. Miyake, D.H. Hwang, J.Y. Lee 2008 *Biochem. Pharmacol.* **75** 494–502.
[9] J. Cocchiara, C.S. Letizia, J. Lalko, A. Lapczynski, A.M. Api 2005 *Food Chem. Toxicol.* **43** 867–923.
[10] S. T. Chang, P. F. Chen, S. C. Chang 2001 *J. Ethnopharmacol.* **77** 123–27.
[11] R. Wang, R. Wang, B. Yang 2011 *J. Food Process Eng.* **34** 175–85.
[12] P. Hili, C.S. Evans, R.G. Veness 1997 Lett. Appl. Microbiol. **24** 269–75.
[13] A. Jayatilaka, S.K. Poole, C.F. Poole, T.M.P. Chichila 1995 *Anal. Chim. Acta.* **302** 147–62.
[14] G. Singh, S. Maurya, M.P. deLampasona, C.A.N. Catalan 2007 *Food Chem. Toxicol.* **45** 1650–61.
[15] M.H. Eikani, F. Golmohammad, Z.B. Sadr, H.S. Amoli, M. Mirza 2013 *J. Essent. Oil Bear. Plants* **16** 740–48.
[16] P.I.Modi, J. K. Parikh, M.A. Desai 2019 *Ind. Crop. Prod* **142** 111838.
[17] H.G. Lee, Y. Jo, K. Ameer, J.H. Kwon 2018 *Food Sci. Biotechnol.* **27** 1607–17.
[18] M.A. Desai, J. K. Parikh 2015 *ACS Sustain. Chem. Eng.* **3** 421–31.
[19] M.A. Desai, J. K. Parikh 2012 *Sep. Sci. Technol.* **47** 1963–70.
[20] S.M. Kapadiya, J. K. Parikh, M.A. Desai 2018 *Ind. Crop. Prod.* **112** 626–32.

[21] C. Gabriel, S. Gabriel, E.H. Grant, B.S.J. Halstead, D.M.P. Mingos 1998 Chem. Soc. Rev. **29** 213–23.
[22] G. Flamini, B. Melai, L. Pistelli, C. Chiappe 2015 *RSC Adv.* **5** 69894–98.
[23] G. Raman, V.G. Gaikar 2002 *Ind. Eng. Chem. Res.* **41** 2521–28.
[24] K.G. Ayappa, H.T. Davis, G. Crapiste, E.A. Davis, J. Gordon 1991 *Chem. Eng. Sci.* **46** 1005–1016.
[25] P. Pongmalai, S. Devahastin, N. Chiewchan, S. Soponronnarit 2015 *Sep. Purif. Technol.* **144** 37–45
[26] G.J. Swamy, K. Muthukumarappan 2017 *Food Chem.* **220** 108–14.
[27] M.A. Ferhat, B.Y. Meklati, J. Smadja, F. Chemat 2006 *J. Chromatogr. A.* **1112** 121–26.

Effect of Welding Parameters on the Weld Bead Profile of Submerged Arc Welded Low Carbon Steel Plates.

Charu Singh Chaudhary[a] , Kashish[b] , Pradeep Khanna[c]

[a, b] Students, [c] Associate Professor, Department of Mechanical Engineering, NSUT, New Delhi, India.
kashishbadwal@gmail.com

Abstract: Welding of thick and long steel sections with minimal requirements of etch preparation has posed a challenge to the fabricators to have a reliable and good quality joint. Such requirements are successfully catered to by the submerged arc welding process which was developed in early 40s to weld thick hull sections of naval ships. The process has since been used extensively in heavy fabrication industries like ship building, pressure vessel and rail industries. The strength and quality of weld produced is decided by its bead parameters like depth of penetration, bead width and reinforcement height. It has been found that weld parameters like wire feed rate, voltage welding speed nozzle to plate distance, torch angle and flux composition have influence on bead parameters. Present work is an attempt to develop a mathematical relation between bead parameters and input parameters so that an empirical relation can be developed between them. Full factorial approach with 2 levels and 3 factors is used and graphical depictions are used to analyse the result

Keywords: Submerged arc welding, Bead parameters, Input parameters, mathematical modeling, Factorial approach

1. Introduction

Submerged arc welding is a widely used welding operation that finds its utility in heavy fabrication industries [1]. The process can join plates up to a thickness of 16 mm without the need of any specific edge preparation because of deep penetration capability and high deposition rates [2]. It also facilitates higher utilization of the electrode wire and has gained popularity due to its ability to get automated and its potential to generate a uniform spatter-free finished weld. The process involves the emergence of an electrical arc between a constantly fed electrode and metal pieces to be joined. This arc is covered by a layer of granular flux in order to ensure shielding of the molten metal from atmospheric contamination [3]. The inherent advantages associated with the process has made it one of the major fabrication processes in industry. All these features have made SAW a chief metal joining process employed in the industry to manufacture a reliable, acceptable and defect-free weld component [4]. The quality, acceptability and reliability of the weldment rely upon the geometric design of the weld bead produced which in turn depends upon the different process parameters that are involved in the metal joining process [5].

Depth of penetration, height of reinforcement and bead width (fig.1) of a welded joint are important weld bead parameters. in determining the quality, operation efficiency, and strength of welded structure. These parameters should be selected according to the shape of the base materials to provide necessary strength [6]. Welding parameters like arc voltage, arc welding speed and wire feed rate are the major variables that influence the characteristics of the bead that are mentioned above. The mechanical and metallurgical properties of the welded joints are influenced because of the above listed parameters. Since the solidification of liquid metal results in weld bead, the surface tension is an important contributing factor in deciding the final bead geometry [7]. The total shrinkage is affected by the bead cross-sectional area,its height and width which to a great extent determines the residual stresses and distortion [7]. Therefore, the input parameters should be precisely chosen so as to obtain the desired geometrical design of the bead [8].

These input parameters, if not carefully controlled, might result in the formation of an inappropriate bead design/outline leading to failure of welded component [9]. Therefore, the study of weld bead geometry is of significant importance as it decides the stress carrying capacity of a weld, controls and predicts the weld quality, maintains the overall efficiency of a weldment and helps establish a relation of the weld bead geometry with the input process variables[10].

The aim of this paper is to relate the process parameters of welding and the bead characteristics using a mathematical model obtained from experimental results to anticipate the bead geometry parameters accurately. Graphical depictions have been used to analyze the results obtained for drawing constructive and logical conclusions.

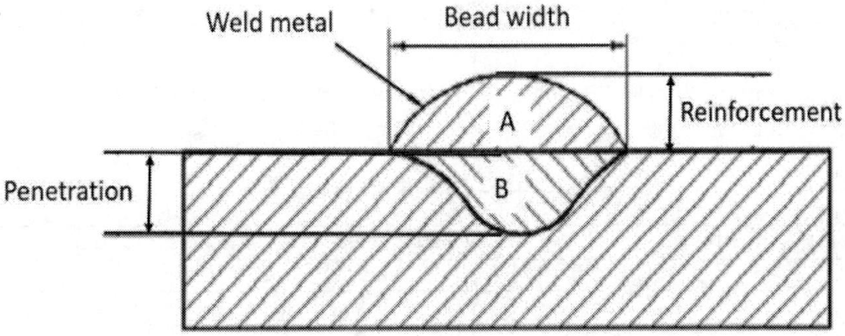

Fig. 1 Weld bead characteristics

2. Experimental Setup and Experimentation

The setup used to carry out the present investigative work is shown in fig. 2. It consists of a weld power source with flat V-I characteristics with ampere rating of 800 amps, 100 % duty cycle, 440 volts, three phase.

Fig. 2 The welding set up

The weld carriage is mounted on the guide rail with the facility of having different welding speeds. The system ensures a constant welding speed in all the runs to ensure the repeatability. A total of eight butt welds were made with mild steel plates of thickness 10 mm. The range of input parameters with respective codes is shown in table 1. The lower and upper limits of the input variables are represented by (-1) and (+1) respectively.

Table-1 Input parameters and their values

S no.	Input parameter	Symbol	Unit	Levels	
				(-1)	(+1)
1.	Welding speed	A	cm/min	10	20
2.	Voltage	B	V	30	40
3.	Wire feed rate	C	m/min	5	7

A total of eight experiments were conducted with two replicates. The combination of input parameters used for each run according to the design matrix is shown in table-2.

Table-2 Design matrix

S. No.	Welding speed	Voltage	Wire feed rate
1.	-1	-1	-1
2.	+1	+1	+1
3.	+1	+1	-1
4.	-1	+1	+1
5.	-1	+1	-1
6.	-1	-1	+1
7.	+1	-1	+1
8.	+1	-1	-1

Weld specimens were cut from each of the weldments with the help of a band saw cutting machine. These specimens were then polished by using rough to fine grade emery papers followed by fine polishing on a disc polishing machine with diamond paste. Nital was used as an etchant to observe the bead profile which can be seen in Fig 3. The bead profiles were projected on the screen of a profile projector and were measured by using a DRO System with a resolution of 0.01 mm. The results so obtained in respect of depth of penetration, bead width and reinforcement are recorded in table 3.

IOP Conf. Series: Materials Science and Engineering 1126 (2021) 012022 doi:10.1088/1757-899X/1126/1/012022

Fig. 3 Bead profile

Table 3 The values of bead width, reinforcement height and depth of penetration as per design matrix.

S. No.	Wek Spe	Vol				
1.	-1	-1	-1	9.38	2.47	3.50
2.	-1	+1	+1	19.87	3.19	6.18
3.	+1	+1	+1	14.61	1.28	5.89
4.	+1	-1	-1	9.14	1.82	1.76
5.	+1	+1	-1	11.09	0.61	4.76
6.	-1	-1	+1	12.56	3.56	3.69
7.	-1	+1	-1	14.49	2.46	2.32
8.	+1	-1	+1	8.66	1.86	3.37

3. Development of mathematical models

Minitab 19 software was used to develop mathematical models for the response parameters depth of penetration, width and reinforcement respectively as given in eq 1-3.

Results-

The following section explains the result obtained with respect to bead width, reinforcement height and depth of penetration.

Depth of penetration = 3.93+0.01A+0.85B+0.84C+0.52AB-0.16AC+0.4BC (1)

Reinforcement height = 2.16 - 0.76A - 0.27B + 0.31C - 0.17AB – 0.13AC + 0.03 BC (2)

Bead width = 12.48-1.6A+02.54B+1.45C-0.56AB-0.69AC+0.77BC (3)

The above equations can be used to estimate the values of different response parameters for different values of input variables A, B and C. The above mathematical models include the main as well as the interaction effects of input parameters.

4. Results and discussions

The results obtained after performing the analysis through Minitab software are presented in the graphical form under two categories-

1. Direct effects
2. Two factor interaction effects

4.1 Direct effects of input parameters

Direct or main effects refer to the influence of a particular parameter on the response when taken alone. Though such trends may not be realistic because of the possibility of interaction among the input parameters but still such trends provide broader idea of individual effects of the parameters.

4.1.1 Direct effects of input parameters on depth of penetration: Fig. 4 clearly depicts that with an increase in wire feed rate and voltage, the depth of penetration strongly increases. The probable cause could be that with an increase in wire feed rate the heat input into the base metal increases and the digging force of arc also increases resulting in increase in depth of penetration. With the increase in voltage, the overall heat input into the weld increases which results in increase in depth of penetration. Welding speed however was found to have negligible effect on the depth of penetration for the present selected range of speed.

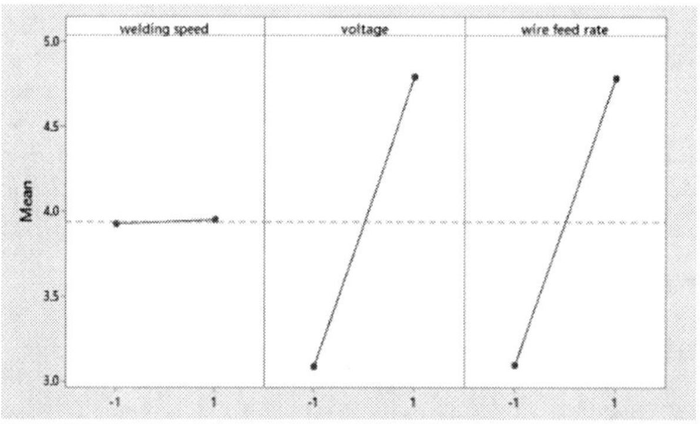

Fig. 4 Direct effects of parameters on depth of penetration

4.1.2 Direct effects of input parameters on bead width: It can be observed from fig. 5 that voltage and wire feed rate have a positive effect on bead width while welding speed has a strong negative effect. An increase in voltage is associated with increase in arc length which results in more spread of the arc spreading the molten metal over a wider area thereby increasing the bead width. An increase in wire feed rate results in the increase in welding current which increases the bead width because of high heat input. When welding speed increases, the heat input per unit length decreases sharply which results in reduced bead width.

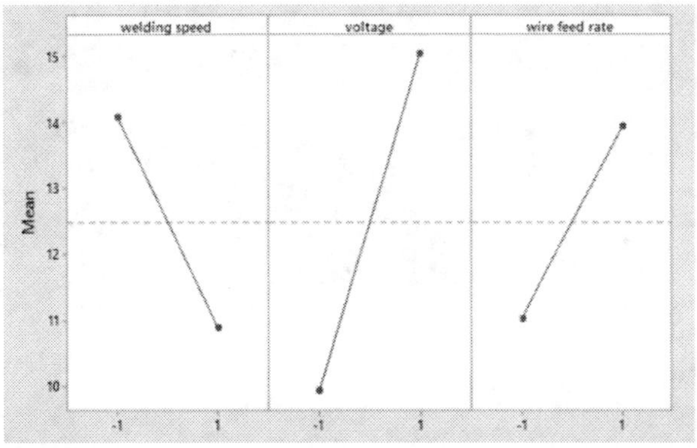

Fig. 5 Direct effects of input parameters on bead width

4.1.3 Direct effects of input parameters on reinforcement height: Fig. 6 depicts that reinforcement height strongly decreases with an increase in welding speed because less molten filler metal gets an opportunity to deposit into the surface of the workpiece. Height of reinforcement decreases with an increase in voltage due to associated increase in arc length which distributes the molten metal over a wider area thereby flattening the bead. There is an increase on reinforcement height with an increased wire feed rate because current increases which results in deposition of more melting of filler metal thereby increasing the reinforcement

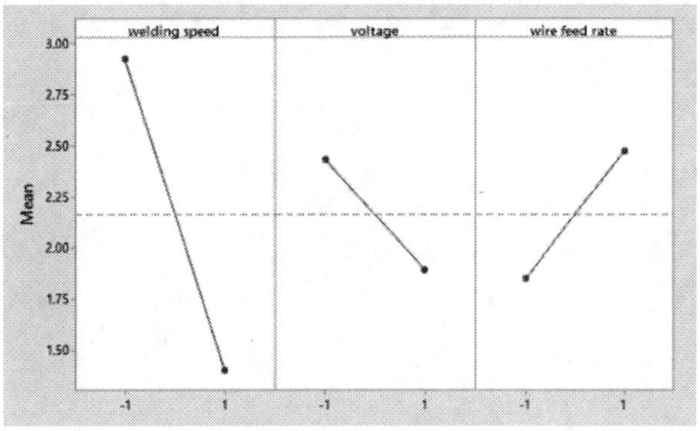

Fig. 6 Direct effects of input parameters on reinforcement height

4.2 Interaction effects of input parameters

Interaction effects refer to more realistic situations during experimentations as there is always a strong possibility of interaction among the input parameters which may lead to unpredictable effects on the response parameter. Therefore, consideration of interaction effects is of practical importance in any statistical investigation of experiments.

4.2.1 Interaction effects of input parameters on depth of penetration: It is evident from fig. 7 that both voltage and wire feed rate have a positive effect on depth of penetration. Therefore, minimum depth of penetration is obtained at a point where voltage and wire feed rate are minimum and maximum penetration are obtained where wire feed rate and voltage are maximum.

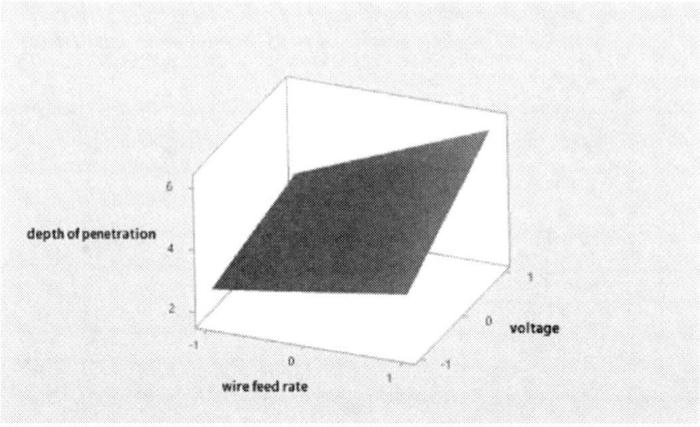

Fig. 7 Interaction effects of wire feed rate and voltage on depth of penetration.

Fig. 8 shows that voltage is found to have a positive effect on penetration while welding speed is having a negative effect on penetration. Also, depth of penetration is minimum for maximum welding speed and minimum voltage while depth of penetration reaches its maximum value at a point where both voltage and welding speed attain a maximum value. No significant change in depth of penetration when voltage increases for least magnitude of welding speed.

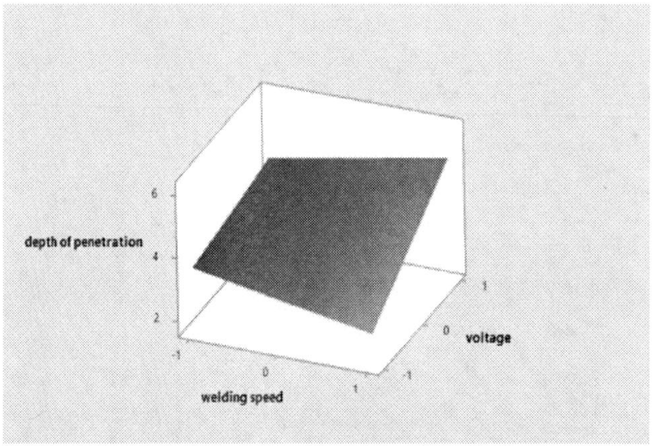

Fig. 8. Interaction effect of voltage and welding speed on depth of penetration

Fig. 9 shows that when welding speed is minimum and wire feed rate is maximum, depth of penetration is maximum. Also, minor deviation is observed in the magnitude of depth of penetration when welding speed increases for maximum value of wire feed rate.

IOP Conf. Series: Materials Science and Engineering 1126 (2021) 012022 doi:10.1088/1757-899X/1126/1/012022

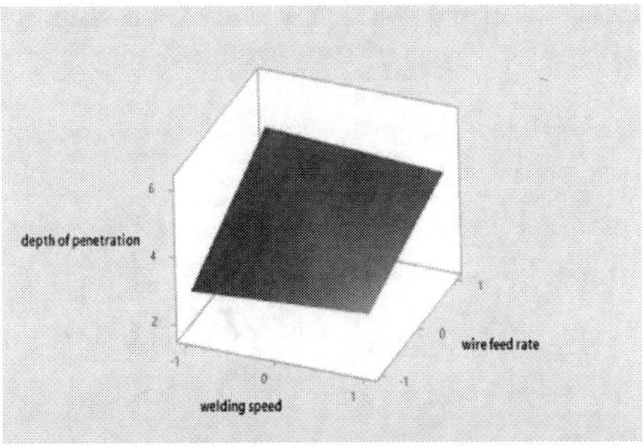

Fig.9 Interaction effect of wire feed rate and welding speed on depth of penetration

4.2.2 Interaction effects of input parameters on bead width: It can be seen from Fig. 10 that both voltage and wire feed rate have a positive effect on bead width. It can be inferred that bead width is minimum for minimum values of both parameters. Whereas, bead width is maximum for maximum values of both input parameters.

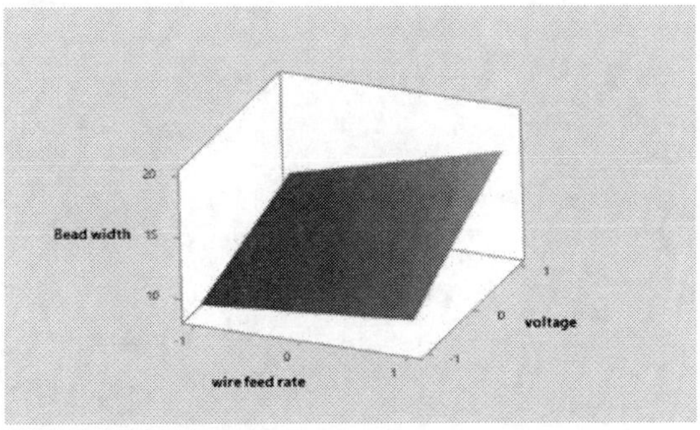

Fig. 10 Interaction effect of voltage and wire feed rate on bead width

It can be concluded from fig.11 that at low welding speed and high value of wire feed rate, bead width increases. While, at maximum welding speed and minimum wire feed rate, bead width is minimum. This concludes that wire feed rate has positive and welding speed has negative effect on bead width.

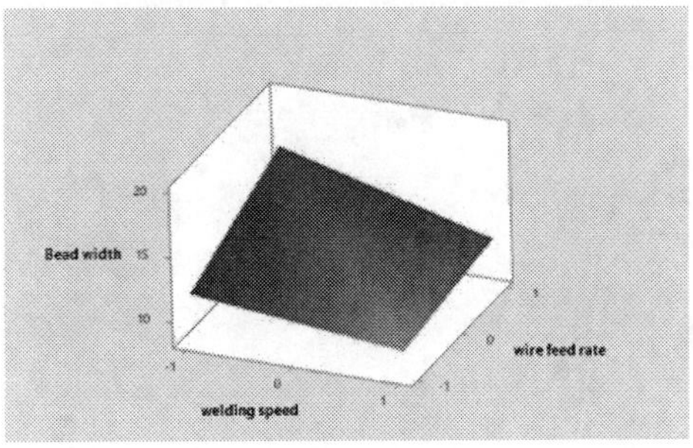

Fig. 11 Interaction effect of welding speed and wire feed rate on bead width

It can be concluded from fig. 12 that at low welding speed and high voltage, bead width increases. Whereas, at maximum welding speed and minimum voltage, bead width is minimum. This concludes that voltage has positive and welding speed has negative effect on bead width.

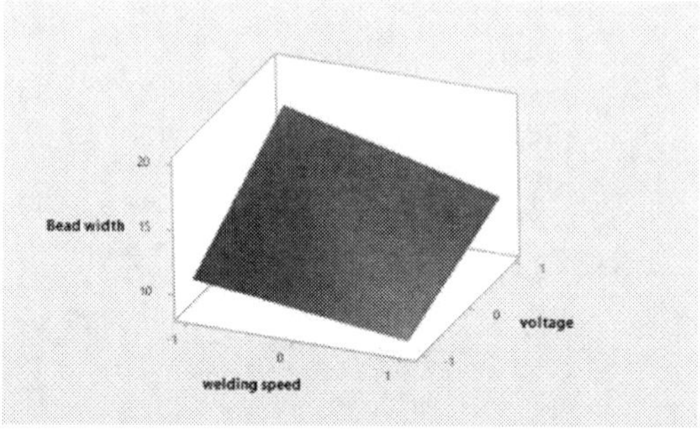

Fig. 12 Interaction effect of voltage and welding speed on bead width

4.2.3 Interaction effects of input parameters on Reinforcement height: Fig. 13 depicts that for the given ranges of voltage and wire feed rate, the reinforcement height does not show significant changes. Although, the reinforcement is found to increase moderately with an increase in wire feed rate and is seen to have slight decrease with an increase in voltage.

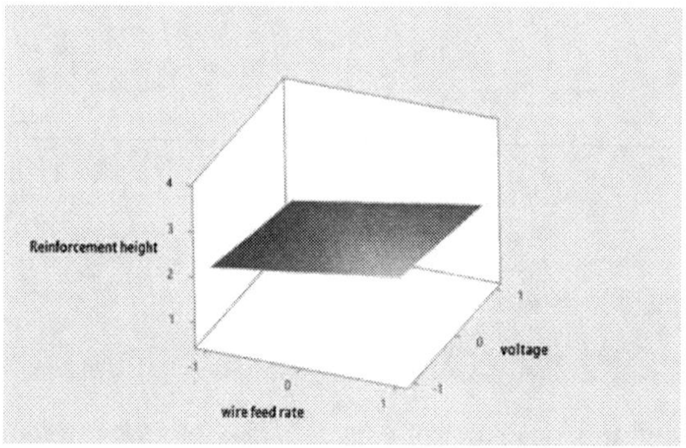

Fig. 13 Interaction effect of voltage and wire feed rate on reinforcement height

It can be seen from fig. 14 that welding speed has a strong negative effect on reinforcement height while voltage has a weak negative effect on reinforcement height.

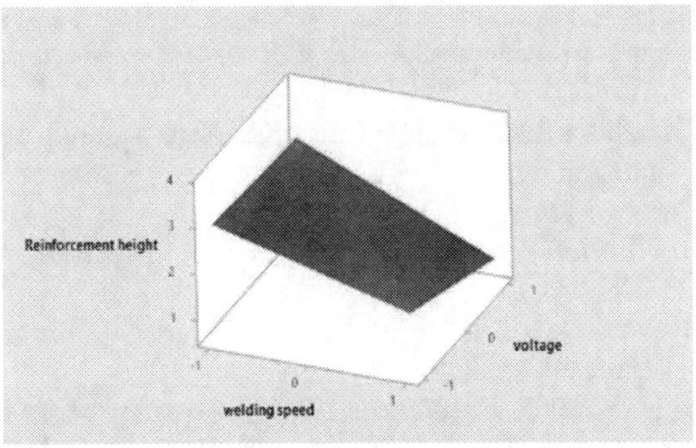

Fig. 14 Interaction effect of voltage and welding speed on reinforcement height

Fig. 15 shows that welding speed has a strong negative effect on reinforcement height whereas wire feed rate has a strong positive effect on reinforcement height.

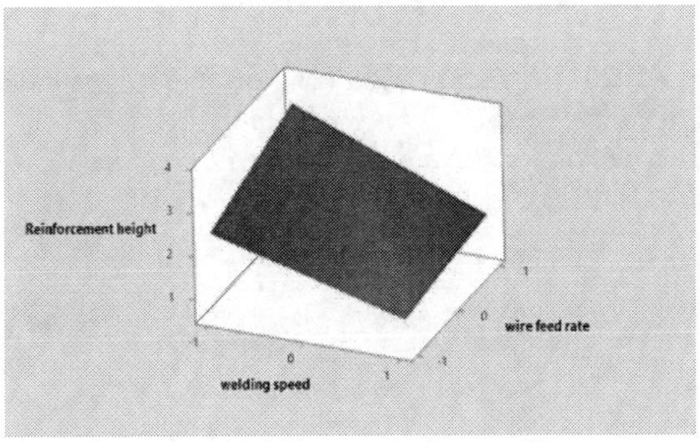

Fig. 15 Interaction effect of welding speed and wire feed rate on reinforcement height

Conclusions

Following are the conclusions on the basis of investigative work carried out so far-

1. Full factorial technique was found to be satisfactory in bringing out the relationships between input and output parameters.
2. The graphical results produced by the software are convenient to interpret and assist in making logical predictions.
3. The mathematical models in respect of depth of penetration, bead width and reinforcement height have been developed incorporating the direct and interactive effects of the input parameters.
4. Depth of penetration is found to increase with wire feed rate and voltage and decrease with wire feed rate.'
5. Bead width is found to have increased strongly with voltage and wire feed rate whereas decrease with increase in welding speed.
6. Reinforcement height is found to reduce strongly with welding speed and moderately with voltage whereas wire feed rate has a strong positive effect on reinforcement.

References-

[1] Singh Gurpinder and Kumar Pradeep 2017 *Effect of Input Parameters on Bead Geometry in Submerged Arc Welding of SS-316L* (India: IJLTET)

[2] Dinovitzer, Malcolm, Xiaohu Chen, Jeremy Laliberte, Xiao Huang, and Hanspeter Frei. "Effect of wire and arc additive manufacturing (WAAM) process parameters on bead geometry and microstructure." *Additive Manufacturing* 26 (2019): 138-146.

[3] Peter Smith 2007 *The Fundamentals of Piping Design*, Fabrication Assembly and Erection Gulf Publishing Company pp 171-189

[4] Ghosh A, Chattopadhyaya S. and Sarkar P.K. 2007 *Effects of Input Parameters on Weld Bead Geometry of Saw Process* (Bangladesh: ICME)

[5] Murugan N and Gunaraj V 2005 *Prediction and Control of Weld Bead Geometry and Shape Relationships in Submerged Arc Welding of Pipes* (India: ResearchGate)

[6] https://www.keyence.com/ss/products/measure/welding/trouble/quality.jsp

[7] Nagesh DS and Datta GL 2002 *Prediction of Weld Bead Geometry and Penetration in Shielded Metal Arc Welding using Artificial Neural Networks* (India: Elsevier)

[8] Akkas N, Karayel D, Ozkan SS, Ogur A and Topal B 2013 *Modeling and Analysis of the Weld Bead Geometry in Submerged Arc Welding by using Adaptive Neurofuzzy Inference System* (Turkey: Hindawi Publishing Corporation)

[9] Sarolkar AD and Kolhe KP 2017 *Effect of Process Parameters on Weld Bead Geometry and Micro Hardness of Welding AA 6082 using GTAW Process* (India: JETIR)

[10] Mistry PKJ 2016 *Effect of Process Parameters on Bead Geometry and Shape Relationship of Gas Metal Arc Weldments* (India: IJARMET)

Mathematical Analysis of Process Parameters in Drilling of Various Aluminium Matrix Composites Using TOPSIS

S Senthil Babu[1]
Research Scholar, VELS Institute of Science, Technology & Advanced Studies, Chennai. India.
rssenthilbabu@gmail.com
C Dhanasekaran[2]
Professor & Head, Department of Mechanical Engineering VELS Institute of Science, Technology & Advanced Studies, Chennai. India.
dhans.se@velsuniv.ac.in

Abstract. Drilling is one of the major material removal processes to produce holes and the key parameters involved in the process are speed of the drilling tool and feed rate of the tool into the work piece to be machined. A conventional study would include the diameter of the drill bit as a key geometric feature that affects the output parameters: thrust, torque, roundness error and the surface roughness. Drilling experiments are performed on three different hybrid composite specimen of Al6061 alloy. Keeping the diameter constant and by varying the point angle of the tool, drilling is performed and the responses of the output parameters are recorded and tabulated. Using the data from the experiment, analysis of the data is done by one of the multi-criteria decision analysis method, TOPSIS. This method is used to determine the optimal condition to drill a particular specimen of the composite. In TOPSIS method, the data is ranked based on performance of the drill, to find the parameters which are ideal to obtain a hole with minimum roughness, circularity error and thrust force. But it is enough to keep the goal as minimum roughness because the other parameters will invariably get minimized as they are dependent

1. Introduction

Metal matrix Composites (MMCs) usually consist of a low density metal such as Aluminium or magnesium reinforced with particulate or fibres of a ceramic such as Silicon Carbide and graphite. Compared with unreinforced metals, MMCs offer high specific strength and stiffness, higher operating temperature, and greater wear resistance, as well as the opportunity to tailor these properties for a particular application. Current markets for MMCs are primarily in the military and aerospace applications. Experimental MMC components have been developed for the use in aircraft, satellites, jet engines, missiles and space shuttles. The development is very important because it demonstrates that the MMCs are at least not prohibitively expensive for a very cost sensitive application [1,2]. Other commercial applications include cutting tools and circuit breaker contacts. Metal matrix composites with high specific stiffness and strength could be used in applications in which saving weight is an important factor [3]. Included in this category are robots, high speed machinery, and high-speed rotating shafts for ships and land vehicles. Good wear resistance, along with high specific strength, also favours MMC use in automotive engine and braking parts. Based on information now in public domain, the following military applications appear attractive: high temperature fighter aircraft engines and structures; high-temperature missile structures. Research and Development priorities include cheaper processes, cheaper materials, and coatings.

Dewangan et al studied the parameters of EDM and optimized the data set using TOPSIS and sensitivity analysis, to determine the best fit answer for the given constraints of weights of the data that is considered. They optimized the surface integrity for a given set of input parameters like voltage, material used, dielectric used etc [4]. They have also extended their concept to fuzzy TOPSIS, which is out of the scope of this paper. We shall be using the same concept as theirs and also as indicated in the forthcoming section so as to complete our analysis.
Tripathy and Tripathy used TOPSIS to optimize the material removal rate on using EDM in Aluminium matrix composites. They have used the concept illustrated beneath this section to optimize the material removal rate for a given composition of the material.[5].

2. Experimental Setup and Procedure

The three different specimen materials used in our study are aluminium alloy Al6061 with 5% weight fraction of boron carbide powders as first reinforcement along with 5% weight fraction of silicon carbide, 5% of Graphite powder and 5% of Mica as second reinforcement by stir casting technique and the specimen were prepared as a square slab of dimensions 100mm x 100mm x 10mm.
For the aim of finding the degree of influence of the drilling parameters, three drilling factors (spindle speed, feed rate and point angle) each at three levels are taken into consideration as shown in Table 1. Taguchi [6] design of experiments has been selected to optimize the multiple performance characteristics. Accordingly, Taguchi based L27 orthogonal array is selected. The drilling experiments are conducted on the prepared hybrid composites specimen by vertical CNC machining centre and TiAlN coated carbide drill bits are used for drilling [7]. The response characteristics considered to study the parameters on drilling of Al Hybrid MMC Composite are thrust force, surface roughness [8] and circularity error. Three different levels of the drilling parameters are shown in Table 1.

Table 1. Process Parameters and their levels

Factor	Level 1	Level 2	Level 3
Point Angle A [°]	108	118	128
Speed B [rpm]	1000	2000	3000
Feed C [mm/rev]	0.05	0.10	0.15

3. Measurement of Responses

The dynamometer with dynaware software is employed to record the response parameter, thrust force [9,10], throughout the drilling of the hybrid composite specimen. The surface quality of the drilled hole was characterized by its surface roughness value [11]. This response parameter, drilled hole surface roughness is measured with the help of surface roughness tester at Kosaka labs, Chennai. It was measured by moving the probe parallel to every hole axis from three totally different points and also the mean values are taken for analysis.
The next response parameter, circularity error was measured using three dimensional coordinate measuring machine (CMM) as given in [12] by V.N.Gaintonde at al. The experiments were conducted on three different specimen and the results of the three responses were tabulated as shown in Table 2.

Table 2. Responses from three different specimen

Point angle (°)	Feed (mm/rev)	Speed (rpm)	Al6061/Sic/B4C			Al6061/Graphite/B4C			Al6061/Mica/B4C		
			circularity error(mm)	Surface Roughness (μm)	Thrust force (N)	circularity error(mm)	Surface Roughness (μm)	Thrust force (N)	circularity error(mm)	Surface Roughness (μm)	Thrust force (N)
108	0.05	1000	0.064	0.7665	98.8848	0.0562	1.4355	274.8	0.0784	1.1422	241.2
108	0.05	2000	0.0618	0.7514	76.518	0.066	1.0562	264.3	0.0445	1.5644	227.5
108	0.05	3000	0.0439	0.274	58.86	0.04	1.1148	241.2	0.0231	1.1262	209.5
108	0.1	1000	0.0882	0.9334	168.352	0.0353	1.8246	340.8	0.0536	1.8752	354.2
108	0.1	2000	0.0849	0.891	104.808	0.042	1.7827	303.3	0.0470	1.9397	313.6
108	0.1	3000	0.0671	0.7723	95.378	0.0509	1.4819	282.4	0.0196	1.5550	292.1
108	0.15	1000	0.1201	3.4664	294.3	0.0348	3.0583	407.9	0.0595	1.9912	450.4
108	0.15	2000	0.1185	1.4565	235.44	0.042	1.9669	387.2	0.0299	1.5971	450.3
108	0.15	3000	0.1063	1.169	211.896	0.0424	1.8305	365.9	0.0153	1.5012	448.2
118	0.05	1000	0.0669	1.0198	111.834	0.0346	1.6672	310.2	0.0579	1.7992	271.2
118	0.05	2000	0.0573	0.8379	91.8216	0.046	1.5822	298.2	0.0410	1.5240	256.4
118	0.05	3000	0.0448	0.5254	76.518	0.0227	1.3661	276.9	0.0416	1.2170	254.9
118	0.1	1000	0.1081	1.5102	235.44	0.0558	2.189	413.4	0.1253	2.3888	371.3
118	0.1	2000	0.0975	1.5061	176.58	0.059	2.0217	404.1	0.0685	1.8169	368.9
118	0.1	3000	0.0703	1.321	117.72	0.0594	1.9978	401.7	0.0418	2.2370	360.7
118	0.15	1000	0.1281	2.498	412.02	0.0255	2.6211	464.4	0.0293	2.8296	500.0
118	0.15	2000	0.1165	2.0726	353.16	0.0267	2.6184	456.9	0.0193	2.6780	481.3
118	0.15	3000	0.1106	1.8948	235.44	0.0559	2.2641	447.9	0.0182	2.8229	466.9
128	0.05	1000	0.0518	0.8175	164.808	0.0528	1.1464	326.1	0.1308	1.5418	290.7
128	0.05	2000	0.0513	0.589	141.264	0.0573	1.0929	316.2	0.0460	1.6354	275.3
128	0.05	3000	0.0597	0.7084	117.72	0.0339	1.3138	315.3	0.0355	1.3708	273.5
128	0.1	1000	0.0627	1.5839	276.642	0.0615	1.5096	434.4	0.0751	1.4081	418.6
128	0.1	2000	0.0623	1.2628	211.896	0.0656	1.4988	431.7	0.0508	1.7911	398.8
128	0.1	3000	0.0564	0.9847	188.352	0.1481	1.2129	427.2	0.0656	1.4644	377.9
128	0.15	1000	0.1122	2.1193	423.792	0.062	2.0189	478.8	0.1674	1.6271	773.0
128	0.15	2000	0.0747	1.9929	347.274	0.0626	1.8341	470.7	0.0791	1.7091	769.8
128	0.15	3000	0.0734	1.743	300.186	0.0677	1.7182	467.1	0.0536	1.7611	536.5

4. Concept of TOPSIS

Technique for Order Preference by Similarity to Ideal Solution is what TOPSIS stands for. This technique is a popular way to analyze the variance in data and rank them as per the criterion specified by the method [13]. The basic technique involves this concept- the chosen alternative should have the shortest distance from the ideal solution and farthest distance from the negative ideal solution [14]. Decision matrix D consists of m alternatives and n attributes or criteria. It is shown below.

$$D = \begin{matrix} \\ A_1 \\ \vdots \\ A_m \end{matrix} \begin{matrix} \begin{matrix} x_1 & \dots & x_n \end{matrix} \\ \begin{bmatrix} x_{11} & \cdots & x_{1n} \\ \vdots & \ddots & \vdots \\ x_{m1} & \cdots & x_{mn} \end{bmatrix} \end{matrix}$$

Hypothesis 1: Each Attribute in the Decision Matrix takes either monotonically increasing or monotonically decreasing utility.

Hypothesis 2: A set of weights for the attributes is required. The manipulation of the data set to compute and the order the data set from best to worst by ranking involves a series of 6 steps as elucidated below.

STEP 1: Construct the normalized decision matrix to transform the various dimensional attributes to the non-dimensional attributes which allow comparison between the attributes

$$r_{ij} = \frac{x_{ij}}{\sqrt{\sum_{i=1}^{m} x_{ij}^2}}$$

STEP 2: Construct the weighted normalized decision matrix

$$V = \begin{bmatrix} v_{11} & \cdots & v_{1n} \\ \vdots & \ddots & \vdots \\ v_{m1} & \cdots & v_{mn} \end{bmatrix} \begin{bmatrix} w_1 r_{11} & \cdots & w_n r_{1n} \\ \vdots & \ddots & \vdots \\ w_1 r_{m1} & \cdots & w_n r_{mn} \end{bmatrix}$$

STEP 3: Determine the ideal and Negative-ideal solutions

$$A^+ = \{(max_i v_{ij} | j \in J), (min_i v_{ij} | j \in J') | i = 1,2, \ldots m\} = \{v_1^+, v_2^+, \ldots\ldots v_n^+\}$$

$$A^- = \{(min_i v_{ij} | j \in J), (max_i v_{ij} | j \in J') | i = 1,2, \ldots m\} = \{v_1^-, v_2^-, \ldots\ldots v_n^-\}$$

Where $J = \{j = 1,2, \ldots . n | \ j \ associated \ with \ benefit \ criteria$

and $J' = \{j = 1,2, \ldots . n | \ j \ associated \ with \ cost \ criteria$

STEP 4: Calculate the separation measure

$$Ideal \ separation \ S_i^+ = \sqrt{\sum_{j=1}^{n} (v_{ij} - v_j^+)^2} \quad , \ i= 1,2,3,\ldots.m$$

$$Negative - Ideal \ separation \ S_i^- = \sqrt{\sum_{j=1}^{n} (v_{ij} - v_j^-)^2} \ , \ i= 1,2,3,\ldots.m$$

STEP 5: Calculate the relative closeness to the ideal solution

$$C_i^* = \frac{S_i^-}{(S_i^+ + S_i^-)}, \ 0 < C_i^+ < 1 \ , \ i = \ 1,2,3, \ldots . m$$

$$C_i^* = 1 \ if \ A_i = A^+$$

$$C_i^* = 0 \ if \ A_i = A^-$$

STEP 6: Rank the preference order. A set of alternatives can now be preference ranked according to the descending order of C_i^*.

Table 3. Normalized Vectors of TOPSIS table for Specimen 1

Normalised vectors			Weighted Normalised vectors			Dij+	Dij-	C_i	Rank
			0.25	0.5	0.25				
Circularity error(mm)	Surface roughness (µm)	Thrust force (N)	Circularity error(mm)	Surface Roughness (µm)	Thrust force (N)				
0.1466	0.0994	0.0854	0.0366	0.0497	0.0214	0.0350	0.1921	0.8458	6
0.1415	0.0974	0.0661	0.0354	0.0487	0.0165	0.0328	0.1950	0.8560	5
0.1005	0.0355	0.0509	0.0251	0.0178	0.0127	0.0000	0.2266	1.0000	1
0.2020	0.1210	0.1454	0.0505	0.0605	0.0364	0.0550	0.1747	0.7604	13
0.1944	0.1155	0.0905	0.0486	0.0577	0.0226	0.0474	0.1823	0.7935	10
0.1537	0.1001	0.0824	0.0384	0.0501	0.0206	0.0358	0.1917	0.8426	7
0.2751	0.4493	0.2543	0.0688	0.2247	0.0636	0.2175	0.0283	0.1153	27
0.2714	0.1888	0.2034	0.0679	0.0944	0.0509	0.0957	0.1366	0.5881	18
0.2435	0.1515	0.1831	0.0609	0.0758	0.0458	0.0757	0.1563	0.6736	16
0.1532	0.1322	0.0966	0.0383	0.0661	0.0242	0.0514	0.1758	0.7738	11
0.1312	0.1086	0.0793	0.0328	0.0543	0.0198	0.0380	0.1892	0.8327	8
0.1026	0.0681	0.0661	0.0257	0.0341	0.0165	0.0167	0.2103	0.9263	2
0.2476	0.1957	0.2034	0.0619	0.0979	0.0509	0.0960	0.1336	0.5818	19
0.2233	0.1952	0.1526	0.0558	0.0976	0.0381	0.0892	0.1389	0.6089	17
0.1610	0.1712	0.1017	0.0403	0.0856	0.0254	0.0707	0.1575	0.6902	14
0.2934	0.3238	0.3560	0.0733	0.1619	0.0890	0.1700	0.0628	0.2697	26
0.2668	0.2686	0.3051	0.0667	0.1343	0.0763	0.1391	0.0918	0.3977	24
0.2533	0.2456	0.2034	0.0633	0.1228	0.0509	0.1181	0.1101	0.4826	22
0.1186	0.1060	0.1424	0.0297	0.0530	0.0356	0.0422	0.1858	0.8147	9
0.1175	0.0763	0.1220	0.0294	0.0382	0.0305	0.0274	0.2011	0.8800	3
0.1367	0.0918	0.1017	0.0342	0.0459	0.0254	0.0322	0.1946	0.8580	4
0.1436	0.2053	0.2390	0.0359	0.1027	0.0597	0.0976	0.1315	0.5739	20
0.1427	0.1637	0.1831	0.0357	0.0818	0.0458	0.0729	0.1546	0.6797	15
0.1292	0.1276	0.1627	0.0323	0.0638	0.0407	0.0544	0.1736	0.7615	12
0.2570	0.2747	0.3661	0.0642	0.1373	0.0915	0.1485	0.0878	0.3715	25
0.1711	0.2583	0.3000	0.0428	0.1292	0.0750	0.1288	0.1016	0.4409	23
0.1681	0.2259	0.2593	0.0420	0.1130	0.0648	0.1098	0.1190	0.5201	21

The experimental results on specimen 1 are normalized by the TOPSIS algorithm and the normalized values are tabulated in Table 3. Among the three responses, a higher weightage of 50% is given to surface roughness of the drilled hole and 25% weightages are given to both thrust force and roundness error. Based on this weightage, weighted normalized vectors are found and tabulated in the same table. Finally, by TOPSIS algorithm, the calculations are made to rank the 27 experiments and that also included in the table.

From the results we can see that the experiment 3 is ranked 1, which is having the parameters of 108°-point angle, 0.05 mm/rev feed rate and 3000 rpm speed.

Table 4. Normalized Vectors of TOPSIS table for Specimen 2

Normalised vectors			Weighted Normalised vectors			**Dij+**	**Dij-**	**C$_i$**	**Rank**
			0.25	0.5	0.25				
Circularity error(mm)	Surface Roughness (μm)	Thrust force (N)	circularity error(mm)	Surface Roughness (μm)	Thrust force (N)				
0.1900	0.1520	0.1399	0.0475	0.0760	0.0350	0.0350	0.1187	0.7724	10
0.2231	0.1118	0.1346	0.0558	0.0559	0.0336	0.0367	0.1296	0.7793	7
0.1352	0.1180	0.1228	0.0338	0.0590	0.0307	0.0149	0.1409	0.9041	1
0.1193	0.1932	0.1735	0.0298	0.0966	0.0434	0.0439	0.1169	0.7269	12
0.1420	0.1888	0.1544	0.0355	0.0944	0.0386	0.0425	0.1145	0.7291	11
0.1721	0.1569	0.1438	0.0430	0.0785	0.0360	0.0332	0.1197	0.7828	6
0.1176	0.3238	0.2077	0.0294	0.1619	0.0519	0.1086	0.0962	0.4697	27
0.1420	0.2083	0.1972	0.0355	0.1041	0.0493	0.0542	0.1073	0.6645	16
0.1433	0.1938	0.1863	0.0358	0.0969	0.0466	0.0470	0.1114	0.7033	13
0.1170	0.1765	0.1580	0.0292	0.0883	0.0395	0.0350	0.1228	0.7783	8
0.1555	0.1675	0.1519	0.0389	0.0838	0.0380	0.0349	0.1187	0.7729	9
0.0767	0.1447	0.1410	0.0192	0.0723	0.0353	0.0170	0.1411	0.8924	2
0.1886	0.2318	0.2105	0.0472	0.1159	0.0526	0.0697	0.0909	0.5661	22
0.1994	0.2141	0.2058	0.0499	0.1070	0.0514	0.0631	0.0937	0.5974	20
0.2008	0.2115	0.2046	0.0502	0.1058	0.0511	0.0622	0.0942	0.6024	19
0.0862	0.2775	0.2365	0.0215	0.1388	0.0591	0.0876	0.1062	0.5479	23
0.0903	0.2773	0.2327	0.0226	0.1386	0.0582	0.0872	0.1052	0.5468	24
0.1890	0.2397	0.2281	0.0472	0.1199	0.0570	0.0746	0.0886	0.5429	25
0.1785	0.1214	0.1661	0.0446	0.0607	0.0415	0.0280	0.1308	0.8234	4
0.1937	0.1157	0.1610	0.0484	0.0579	0.0403	0.0308	0.1309	0.8095	5
0.1146	0.1391	0.1606	0.0286	0.0696	0.0401	0.0191	0.1352	0.8762	3
0.2079	0.1598	0.2212	0.0520	0.0799	0.0553	0.0475	0.1100	0.6985	14
0.2218	0.1587	0.2198	0.0554	0.0794	0.0550	0.0495	0.1082	0.6861	15
0.5006	0.1284	0.2175	0.1252	0.0642	0.0544	0.1089	0.0979	0.4735	26
0.2096	0.2138	0.2438	0.0524	0.1069	0.0610	0.0679	0.0912	0.5732	21
0.2116	0.1942	0.2397	0.0529	0.0971	0.0599	0.0607	0.0971	0.6152	18
0.2289	0.1819	0.2379	0.0572	0.0910	0.0595	0.0592	0.0982	0.6241	17

The experimental results of specimen 2 are normalized similarly and the values are tabulated in Table 4. Again 50% weightage is given to drilled hole roughness 25% weightages are given to thrust force and roundness error each. In the same manner the weighted normalized vectors are found and tabulated. Finally, the ranking of all the experimental trials is done by TOPSIS algorithm and is tabulated in the last column of Table 4.

For the specimen 2, the optimum result (rank 1) is same as that of specimen 1. From the results we can see that the experiment 3 is ranked 1, which is having the parameters of 108°-point angle, 0.05 mm/rev feed rate and 3000 rpm speed.

Table 5. Normalized Vectors of TOPSIS table for Specimen 3

Normalised vectors			Weighted Normalised vectors			Dij+	Dij-	C_i	Rank
			0.25	0.5	0.25				
Circularity error(mm)	Surface Roughness (µm)	Thrust force (N)	circularity error(mm)	Surface Roughness (µm)	Thrust force (N)				
0.2269	0.1200	0.1129	0.0567	0.0600	0.0282	0.0458	0.1260	0.7331	10
0.1287	0.1643	0.1065	0.0322	0.0822	0.0266	0.0313	0.1280	0.8035	6
0.0669	0.1183	0.0980	0.0167	0.0592	0.0245	0.0057	0.1524	0.9640	1
0.1552	0.1970	0.1657	0.0388	0.0985	0.0414	0.0510	0.1081	0.6793	15
0.1361	0.2038	0.1467	0.0340	0.1019	0.0367	0.0500	0.1125	0.6922	14
0.0568	0.1633	0.1367	0.0142	0.0817	0.0342	0.0247	0.1381	0.8483	4
0.1722	0.2092	0.2107	0.0430	0.1046	0.0527	0.0623	0.0972	0.6095	20
0.0865	0.1678	0.2107	0.0216	0.0839	0.0527	0.0389	0.1245	0.7618	9
0.0442	0.1577	0.2097	0.0110	0.0788	0.0524	0.0342	0.1357	0.7989	7
0.1675	0.1890	0.1269	0.0419	0.0945	0.0317	0.0475	0.1125	0.7033	12
0.1188	0.1601	0.1200	0.0297	0.0800	0.0300	0.0285	0.1293	0.8192	5
0.1203	0.1278	0.1193	0.0301	0.0639	0.0298	0.0203	0.1383	0.8719	2
0.3626	0.2509	0.1738	0.0906	0.1255	0.0434	0.1053	0.0606	0.3652	26
0.1981	0.1909	0.1726	0.0495	0.0954	0.0432	0.0561	0.1009	0.6428	17
0.1209	0.2350	0.1688	0.0302	0.1175	0.0422	0.0639	0.1075	0.6271	19
0.0849	0.2972	0.2339	0.0212	0.1486	0.0585	0.0962	0.1048	0.5214	23
0.0559	0.2813	0.2252	0.0140	0.1407	0.0563	0.0875	0.1127	0.5628	21
0.0527	0.2965	0.2185	0.0132	0.1483	0.0546	0.0941	0.1137	0.5472	22
0.3786	0.1620	0.1360	0.0946	0.0810	0.0340	0.0869	0.0920	0.5141	24
0.1330	0.1718	0.1288	0.0333	0.0859	0.0322	0.0356	0.1226	0.7750	8
0.1028	0.1440	0.1280	0.0257	0.0720	0.0320	0.0209	0.1356	0.8666	3
0.2172	0.1479	0.1959	0.0543	0.0740	0.0490	0.0519	0.1084	0.6764	16
0.1470	0.1881	0.1866	0.0368	0.0941	0.0467	0.0487	0.1096	0.6923	13
0.1899	0.1538	0.1768	0.0475	0.0769	0.0442	0.0451	0.1127	0.7143	11
0.4844	0.1709	0.3617	0.1211	0.0855	0.0904	0.1309	0.0632	0.3254	27
0.2288	0.1795	0.3602	0.0572	0.0898	0.0901	0.0858	0.0869	0.5031	25
0.1550	0.1850	0.2510	0.0388	0.0925	0.0628	0.0578	0.1034	0.6414	18

Similarly, The above Table 5 shows the results for the specimen 3. The maximum weight assigned is for the roughness values because that is the main response to be minimized, followed by thrust force and roundness error so as to utilize minimal effort to obtain the best result, and also as a concern to keep the tool from having more wear. After the data is manipulated, the result is obtained.
From the above tables, it is clear that the optimal result is the one with least point angle, lowest feed rate and highest spindle speed, that is 108°-point angle, 0.05 mm/rev feed and 3000 rpm speed. After manipulation is done by the steps indicated in the concept of TOPSIS, the above result is obtained. It can also be gauged that the minimum roughness value is the major criteria to be treated as optimal result.

The C_i values represent the overall quality measure of all the responses, because the C_i values indicate the relative closeness to the ideal solution.

Higher values of C_i show that the particular set of parameters in the corresponding trial would produce better responses. The combination of drilling parameters for the trial number 3 was found to have highest C_i value in our experiments on specimen1.

The main effect plots for relative closeness C_i of specimen 1 are shown in Fig 1. Higher values give better results. It shows that the minimum point angle, minimum feed and maximum speed are desirable parameters for optimum results of drilling. This fig also shows that the influence of point angle in the performance of drilling is less when compared with speed and feed rate

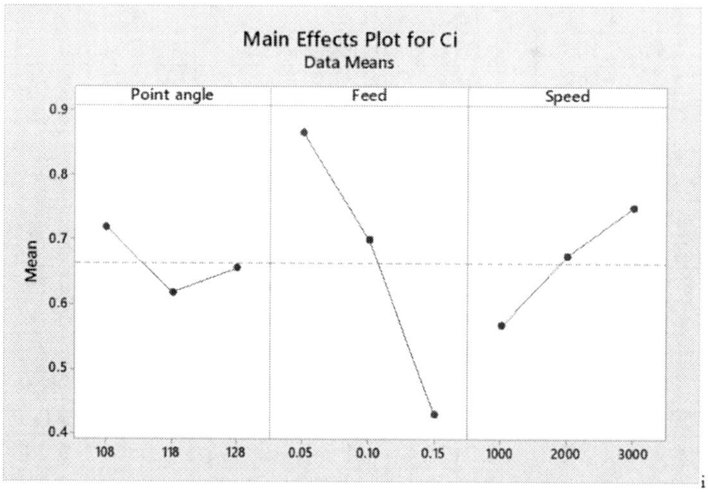

Fig 1. Main effects plot for relative closeness C_i for specimen 1

Main effect plots for relative closeness C_i of specimen 2 is shown in Fig 2. This figure shows that the influence of point angle, feed rate and spindle speed for specimen 2 is almost similar to specimen1. Fig 3 shows the main effects plot for specimen3.

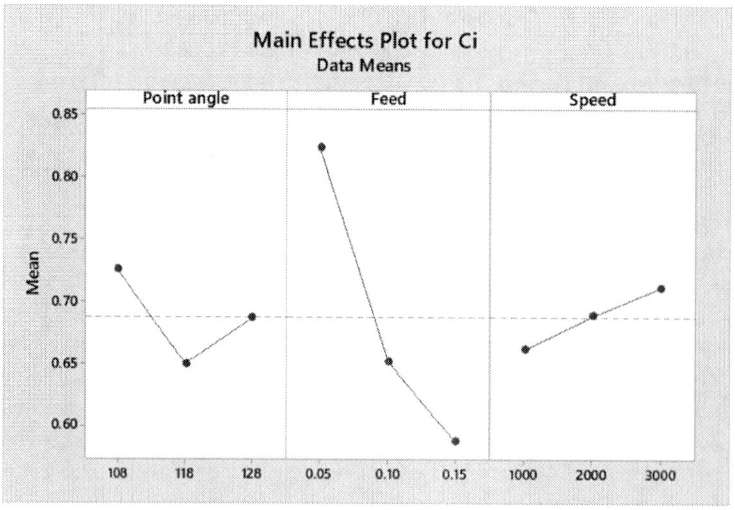

Fig 2. Main effects plot for relative closeness C_i for specimen 2

Influence of Point angle and Speed for specimen 3 is slightly differ from specimen 1 and 2 as shown in Fig.3, but the influence of feed rate remain same.

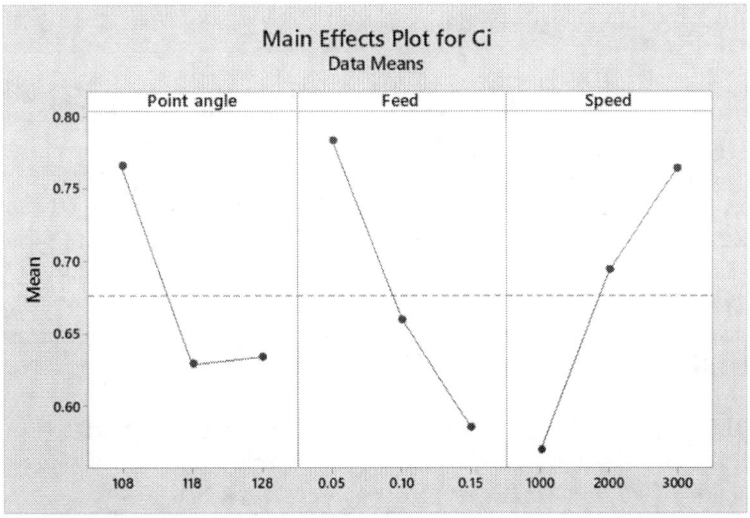

Fig 3. Main effects plot for relative closeness C_i for specimen 3

5. Conclusion

An attempt has been made to optimize the drilling parameters in drilling experiments of three different hybrid composite specimen of Al6061 based on L27 orthogonal array in vertical CNC Machining centre. The conclusions drawn from the experimental analysis are summarized below:

Optimal process parameters are obtained by the concept of TOPSIS which makes the multiple performance characteristics as a beneficial feature using simplified optimization procedure, and the optimal parameters are observed as point angle of 108°, feed rate of 0.05mm/rev and drilling speed of 3000 rpm.

The optimum drilling parameters produce the minimal roughness inside the drilled surfaces over the entire data set considered. Since we are considering roughness as the main criterion for deciding which is the best condition for drilling, maximum weightage is assigned to it for bringing the weighted normalized vectors.

Since the results are consistent for all the three specimens, we can conclude that the influence of reinforcements on the quality of drilling is minimal.

References

[1] Muhammad Aamir et al, 2020 A review: Drilling performance and hole quality of aluminium alloys for aerospace applications, *Journal of Materials Research and Technology* **9**.6, 12484-12500.

[2] Saravanan S et al, 2020 Tribological behaviour of aluminum alloy (AA7075) based hybrid composites, *IOP Conference Series: Materials Science and Engineering* Vol. **923**. No. 1.

[3] Mavhungu S T et al, 2017 Aluminum matrix composites for industrial use: advances and trends, *Procedia Manufacturing* **7** 178-182.

[4] Dewangan S, Gangopadhyay S and Biswas C K, 2015 Study of surface integrity and dimensional accuracy in EDM using Fuzzy TOPSIS and sensitivity analysis, *Measurement* **63** : 364-376.

[5] Tripathy S, and Tripathy D K, 2016 Multi-attribute optimization of machining process parameters in powder mixed electro-discharge machining using TOPSIS and grey relational analysis, *Engineering Science and Technology, an International Journal* **19**.1 : 62-70.
[6] Vankanti, Vinod Kumar, and Venkateswarlu Ganta, 2014 Optimization of process parameters in drilling of GFRP composite using Taguchi method, *Journal of Materials Research and Technology* **3**.1 : 35-41.
[7] Narale M N et al, 2018 Experimental investigation and optimization of multiple performance characteristics of Al-Mica-B4C hybrid reinforced composite in drilling operation, *Materials Today: Proceedings* **5**.9 : 19763-19772.
[8] Samy G S, and Thirumalai Kumaran S, 2017 Measurement and analysis of temperature, thrust force and surface roughness in drilling of AA (6351)-B4C composite, *Measurement* **103**: 1-9.
[9] Rubi C, Sarala J, Udaya Prakash and Rajkumar C, 2020 Optimization of Process Parameters using Taguchi Technique for Drilling Aluminium Matrix Composites (LM6/B4C), *IOP Conference Series: Materials Science and Engineering*. Vol. **912**. No. 3.
[10] Juliyana S, Jebarose and Udaya Prakash J, 2020 Drilling parameter optimization of metal matrix composites (LM5/ZrO2) using Taguchi Technique, *Materials Today: Proceedings* **33** :3046-3050.
[11] Senthil babu S, Vinayagam B K, 2015 Modeling and Analysis of Surface Roughness and Thrust Force in Drilling of Al Based Metal Matrix / Hybrid Composites, *International Review on Modelling and Simulations (I.RE.MO.S.)*, Vol. **8**, N. 4 ISSN 1974-9821.
[12] Khaled Giasin and Sabino Ayvar-Soberanis, 2017 An investigation of burrs, chip formation, hole size, circularity and delamination during drilling operation of GLARE using ANOVA, *Composite Structures* **159** : 745-760.
[13] Arun Prasad K R et al, 2020 Optimization of turning parameters for Magnesium Silicon Carbide using TOPSIS method, *IOP Conference Series: Materials Science and Engineering*. Vol. **912**. No.3.
[14] Ficko M et al, 2020 Multi-Response Optimisation Of Turning Process Parameters with GRA And TOPSIS Methods, *International Journal of Simulation Modelling* **19**.4: 547-558.

Research and design of VR-based mechanical simulation interaction and network experiment platform

Zhenghong Liu[1], Huiliang Zhao[2], Jie Liu[1], Yuliang Xu[1*]

1. School of Mechanical Engineering, Guiyang University, Guiyang 550005, China
2. Guizhou Minzu University, Guiyang 550025, China
Corresponding author's e-mail:415803168@qq.com

Abstract. This article takes mechanical basic experimental courses as the research object, and makes full use of virtual simulation, computer graphics, Internet and other technologies to innovate experimental teaching content and methods. First, construct a mechanical basic experiment teaching system based on four levels: guided cognition, basic experiment, comprehensive design experiment and research and innovation experiment. Then, combined with the existing virtual simulation related hardware equipment of the college, a software and hardware virtual experiment platform integrating virtual 3D display and interaction is built. Finally, with the help of Unreal Engine 4 (UE4) and network technology, a Web-based 3D virtual experiment platform is designed. Through the research of this article, a basic mechanical experiment system can be constructed, the experimental teaching training program can be improved, and the engineering practice education level can be further improved. The virtual reality (VR) experiment platform proposed in this paper can improve existing teaching methods, expand experiment content, solve problems such as insufficient educational resources, and improve students' interest and initiative in learning. The research of the network virtual platform enables students to break through the limitations of time and space, and realize different levels of mechanical experimental courses, such as online learning and hands-on operation of guided cognition, basic experiments, comprehensive design experiments, etc.

1. Introduction

Virtual Reality (VR) is a virtual simulation environment with three-dimensional effects, which has the characteristics of immersive interaction and multi-channel perception. The emergence and maturity of VR technology provides users with an interactive method closer to the real world. This advanced interactive method is widely used in various fields. In terms of teaching, it is mainly used in VR laboratories, VR libraries, and VR classroom, etc. With the popularization of the Internet and the continuous maturity of its technology, the combination of networking and VR technology has emerged, mainly in VR online classrooms, virtual museums, etc.

Since VR technology was first proposed by the United States, the United States is still in a leading position in related research[1]. Medicine School of Stanford University designed a 3D online learning platform on frog biology for middle school students; MIT built a VR-based media teaching and research room; Umich used VR modeling language to build an "art museum"; The University of Wisconsin System is a VR-based COVIRDS system that interacts with virtual modeling scenes through gesture operation and voice recognition to achieve rapid virtual conceptual design, thereby improving design efficiency[2]. The University of Calgary in Canada has developed the VLAB (Virtual Laboratory) platform, which can quickly build biological simulation models for users[3]; The network virtual course system developed by Kim and others in South Korea can simulate physical phenomena such as water waves and electromagnetic induction; The University of Leeds in the United Kingdom Khalili and others developed the WebSet (Web-based Standard Education Tools) platform, which is based on VR technology to simulate the operation process and realize remote medical related teaching[4]; Georgiou and others in Greece developed a network virtual chemistry laboratory to enable students to familiarize themselves with chemistry in advance Experimental instruments and equipment to better carry out chemical experiments[5].

Compared with foreign countries where the research on VR technology and application is more advanced, my country's VR technology research started late, and the research time is shorter. There are still some gaps in basic equipment and technical theory. However, the development of VR technology in my country has been stronger in recent years. Some excellent works have emerged in the field of combining with network information. In the process of constructing national quality course resources, colleges and universities across the country have built a lot of better online course resources, but these course resources are basically stored and shared in the form of office documents, Flash animations, pictures, videos and other multimedia forms, so that teaching effect has been improved[6-11]. Only a small part of the course resources building an online learning platform by VR, and this part of the resource is very popular. Beihang University is one of the earliest universities in China to conduct VR technology research, and has established the State Key Laboratory of VR Virtual Reality Technology and Systems, which is mainly engaged in scientific research and application technology development in VR technology, visualization technology, image information processing, distributed systems, artificial intelligence, etc. Zhejiang University scholars use multiple client-side scripting languages such as JavaScript to achieve user interaction, download and run Java Applet, Active X and other programs from a web browser, and use Java Applet to call JDBC and specified data resources on the Internet establishing a connection to complete the interaction between the server and the user[12]; Based on Culd3D technology, Dalian University of Technology has built a network resource for high-quality mechanical courses, and users can browse all mechanical parts on the Internet; Tongji University has developed VR-based construction related course network laboratory, which can simulate the appearance of building structure[13-15].

Most of the existing experimental projects in universities at home and abroad are confirmatory experiments to verify basic theoretical knowledge. Students just complete the experiments mechanically and cannot understand the principles well. Unable to complete innovative experimental projects, lack of training of engineering technology and engineering experience. The main problems are as follows: i. The teaching method lacks novelty. The traditional teaching method is that the teacher explains the principle of the experiment, and then demonstrates an experiment. The students complete the experiment mechanically according to the teacher's handouts. Without the process of conception, design, participation, and interaction of the experiment, students' creative thinking cannot be cultivated. ii. The experiment content lacks systematic. Traditional experiments are mainly dependent on theoretical teaching. Experimental projects are mainly to verify unilateral theoretical knowledge points. The lack of integration among various knowledge points and the lack of systematic are not conducive to students' overall understanding of the curriculum theoretical knowledge system.

With the continuous increase of network resources for high-quality courses and the continuous improvement of the application of VR technology, the cost of virtual network teaching will become lower and lower, so it will become more and more popular. As a teacher, the focus is how to make high-quality online courses[16]. The network virtual teaching platform based on Internet technology and VR technology will develop into a comprehensive learning platform integrating learning, discussion, feedback and evaluation, and will be established in large numbers worldwide.

This article takes mechanical basic experimental courses as the research object, and makes full use of virtual simulation, computer graphics, Internet and other technologies to innovate experimental teaching content and methods. First, build a systematic experimental curriculum content system; then, combine the existing virtual simulation related hardware equipment of the college, to build a software and hardware virtual experiment platform integrating virtual 3D display and interaction; finally, integrate network technology to design a Web-based 3D virtual experiment platform.

The main contributions of this article are as follows:

(1) Construct an experimental teaching system and standardize experimental teaching content. Construct a mechanical basic experiment teaching system framed by four levels: guided cognition, basic experiment, comprehensive design experiment and research and innovation experiment. The system covers all basic mechanical experiments from simple to deep, simple to comprehensive, and focuses on holistic and systematic experiments, which can meet the training requirements of experimental courses for students of different mechanical majors at different learning stages.

(2) Combine with VR technology to improve experimental teaching methods. Traditional professional course teaching is taught in the form of physical models, blackboards, or slides. It cannot or is difficult to visualize the conversion process between the three-dimensional form of the model and the two-dimensional plane, which makes classroom teaching dull. i. In the teaching of "Mechanical Drawing", because students are just beginning to get involved with 3D, it is difficult to generally accept it. With the help of virtual reality experimental platform, students can get started quickly and stimulate learning initiative and interest. ii. In the course teaching of "Mechanical Principles" and "Mechanical Design", virtual reality technology and network technology are used to simulate the experimental process of mechanism cognition, mechanism diagram mapping, dynamic balance experiment, reducer disassembly and assembly, with the help of relevant external equipment , So that the experimenter is completely immersed in the virtual environment for interactive experimental operations, not only can solve the problem of insufficient educational resources, but also enable students to perform experimental operations without time and space constraints, which is conducive to giving full play to the experimenters' innovative ability. iii. In the course teaching of "Modern Cutting and Processing Technology", "Jig Design", "Machinery Manufacturing Technology", the operation control of large and precision machinery is involved, especially the disadvantages of danger and high cost in certain machinery manufacturing processes. Carrying out the research of virtual mechanical manufacturing processing simulation experiment has certain practical value for the learning of mechanical manufacturing processing knowledge and operation training.

(3) Research the network virtual experiment platform and enrich the experimental teaching methods. Research the context and interactive methods of related experiments and publish them in the form of web pages to build a web-based virtual experiment platform. Enable students to break through the limitations of time and space, and complete experimental preparatory work, experimental homework and other learning outside the laboratory. At the same time, the interactive operation and learning materials are integrated, such as adding multimedia courseware, processing videos, operation instructions, equipment photos, linking learning websites and other auxiliary content of the corresponding machine tool working principle and operation in the mechanical processing virtual experiment system.

2. Construction of experimental teaching system of mechanical basics

Construct a mechanical basic experiment teaching system framed by four levels: guided cognition, basic experiment, comprehensive design experiment and research and innovation experiment. Take mechanical cognition, mechanical design and mechanical research as the main line, broaden the foundation, and focus on cultivating students' comprehensive innovation and practical ability. The guided cognitive experiments mainly include mechanism, intersecting line model mapping, mechanism cognition, mechanical structure and parts cognition, mechanism diagrams; Basic experiments mainly include combined body model surveying and mapping, mechanism sketch mapping, dynamic balance experiment, Sliding bearing pressure experiment; Comprehensive design experiment mainly includes computer drawing, speed fluctuation adjustment, mechanical transmission design and performance analysis, reducer disassembly experiment; Research innovative experiment mainly includes mechanism innovation design, mechanical transmission scheme design.

By consulting materials, investigating experts and other methods, combined with the project team members' years of front-line teaching experience, construct the mechanical basic experimental teaching system as shown in the figure below. The system has a clear level and reasonable system, covering all mechanical basic experiments, and can meet the requirements of mechanical engineering. The training requirements of experimental courses for students of different majors at different learning stages.

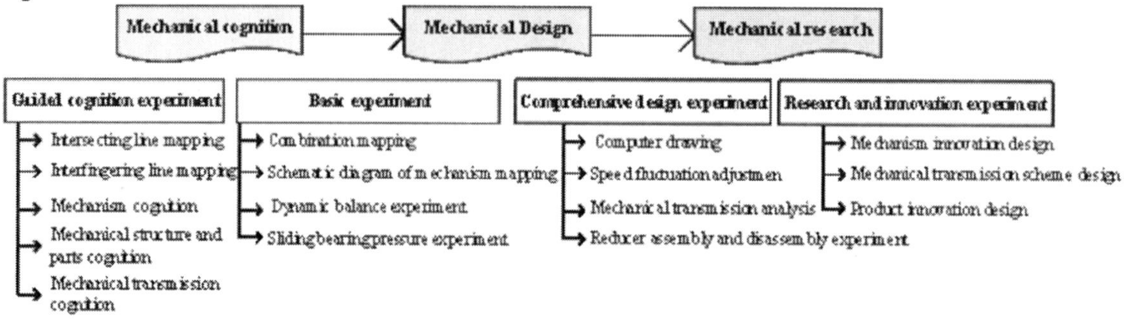

Figure 1. Experimental teaching system of mechanical basics

3. Virtual simulation and interactive experiment construction based on VR

Based on VR technology, build a virtual simulation and interactive experiment platform that combines hardware and software. On the basis of constructing the experimental teaching system, firstly use 3D modeling, 3D scanning and other modeling techniques to construct the 3D model of the experimental equipment and parts; then combine the constructed model with the existing ring screen and 3D projection of the college virtual simulation laboratory Fusion of hardware devices such as instrument, 3D glasses, etc.; finally, through interactive process development, a virtual, realistic, interactive virtual experiment platform combining software and hardware is constructed.

Through modeling, the equipment, related parts, scenes, etc. included in the experiment are digitized, and the existing virtual reality hardware equipment of the college is integrated to analyze and develop the interactive process and complete the research of interactive virtual experiment based on VR. In addition to satisfying the experimental content of "Mechanical Drawing", "Mechanical Principles", "Mechanical Design", "Mechanical Manufacturing Technology" and other courses. The realistic virtual experiment system can continuously add new content with the development of technology, which can involve the models of various parts in the machinery industry, the processing process of parts from rough to finished products, various processing methods of parts, and dynamic simulation of the principle and processes of parts processing, etc.

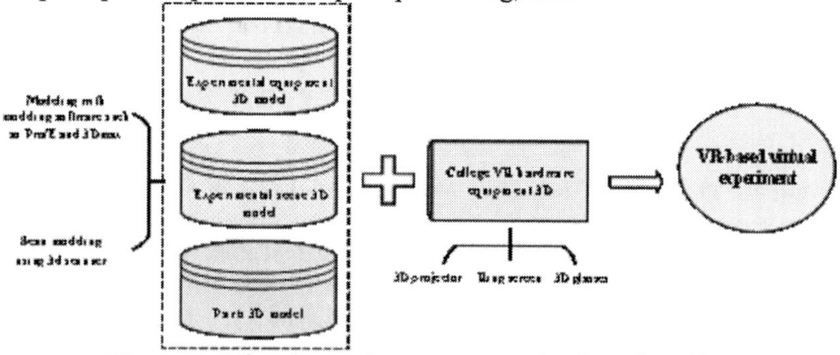

Figure 2. Virtual experiment construction based on VR

4. Research on the network platform corresponding to VR experiment

Based on the UE4 game technology engine in virtual reality technology, using C++ scripting language for interactive process development, integrating network technology, designing a Web-based 3D virtual experiment platform through business process and demand analysis, architecture design, function design and other steps.

(1) Function design of network experiment platform: The function design of the network experiment platform is carried out through demand analysis. The function modules are shown in the

figure below. The experiment platform has three main functions: online virtual experiment, experiment teaching management and platform management.

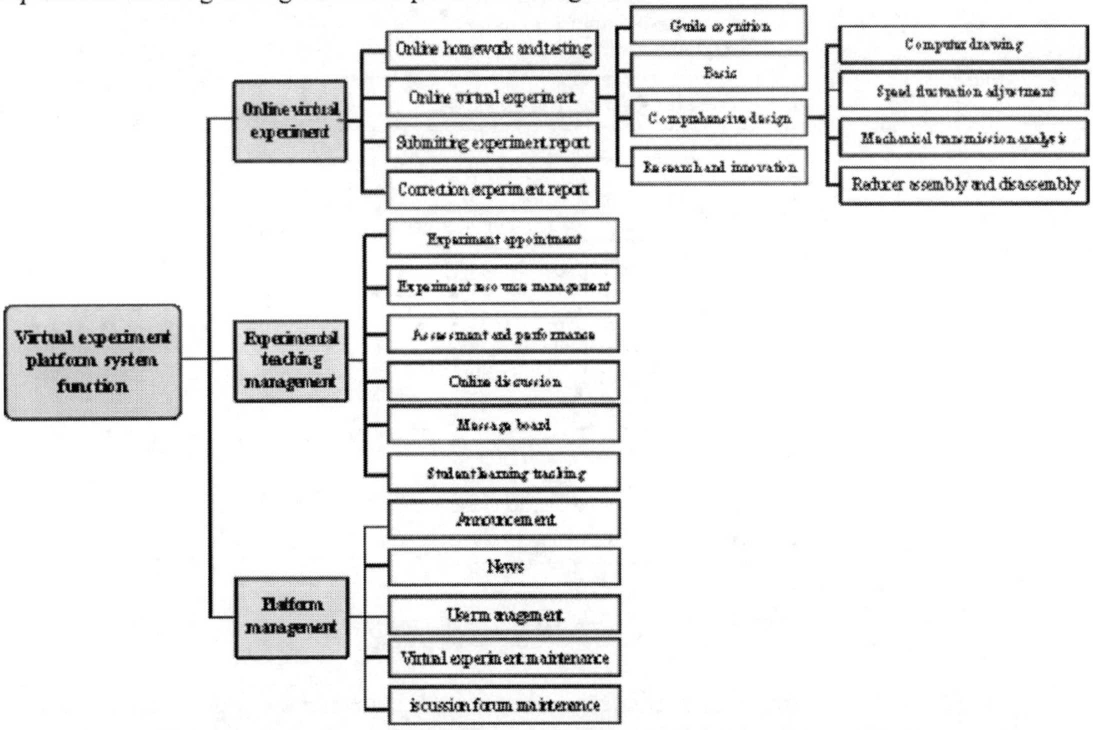

Figure 3. Function module diagram of network virtual experiment platform

Students enter the experiment scene under the guidance of the teacher. Take the reducer assembly and disassembly experiment as an example. The platform has the following main functions: i. By dragging and dropping the mouse, the viewing angle of each component can be changed, and the scroll wheel can be used to achieve the zoom function of each component. ii. Use the mouse to click the relevant buttons in the scene to realize the assembly and disassembly demonstration of the reducer as a whole and its parts and components. iii. Use the mouse to drag and drop the corresponding parts for interactive assembly and give feedback on the operation sequence. If it is correct, continue the assembly. If it is wrong, a text and sound error prompt appears (the prompt can be hidden by mouse click), and the disassembly operation is similar. iv. Click the relevant button to measure the corresponding shaft diameter, gear tooth number, gear modulus and other parameters.

In order to improve the experimental effect, the virtual platform also has auxiliary experimental functions. For example, online assignments and tests: set the time and questions, conduct online tests for students, and students can submit real-time feedback on grades and correct answers and analysis after submission. Experiment report submission and correction: After the experiment is completed, students first submit the pdf version of the experiment report, and then the teacher corrects the experiment report on the experiment platform. After the correction is completed, the teacher informs the students to check their own experiment report on the experiment platform, thus realizing paperless experiment of report submission, correction, and review.

(2) Function realization of network experiment platform: Taking the disassembly and assembly experiment of the reducer of the virtual experiment platform as an example, the design and implementation of the virtual experiment are described in detail. The experiment can be divided into three steps: the construction of the 3D component model, the design and realization of the interactive function of the model, and the publishing network. The specific process is shown in Figure 4.

Build a three-dimensional model of parts: The construction of the three-dimensional model of parts is the basis for the entire virtual experiment display and interaction, and to run other functions. The quality of the model directly affects the fineness and realism of the virtual experiment display and determines the smoothness and speed of the system operation to a large extent.

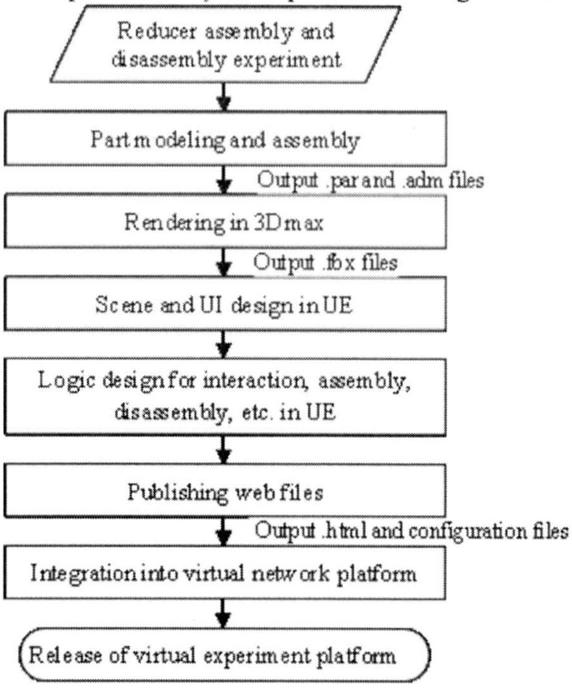

Figure 4. Flow chart of the realization of the network experiment design of the reducer disassembly and assembly

Making materials, textures and scenes: First, import the built 3D model of the reducer (completed through related parametric modeling software or 3D Max) into 3D Max, and assign shaders and textures to it. You can also edit the material directly in UE4. Then build scenes such as lighting and environment in UE4. The correct material information can make the model closer to the real object, and the lighting and environment settings can also obtain more realistic effects.

Virtual interaction design based on UE4 (The interactive control function in the virtual experiment is mainly developed in UE4, which is also the most critical and complex part of the entire virtual experiment): First, export the reducer 3D model file of .fbx containing the texture map from 3D Max. Then import the file from UE4. Finally, use C++ programming or blueprint technology for interactive design in UE4. UE4 supports the C++ scripting language, which can realize the interactive logic control between model objects or other buttons through scripts and can also control the input and output by calling the API interface. Mount the script to the corresponding Game Object and set the corresponding trigger mode to realize the interactive function of assembly, disassembly and related parameter measurement of the reducer model. Some simple interactive functions can also be designed directly through blueprints.

The project webpage is released (UE4 supports project releases on multiple platforms): First add corresponding scenarios. Then set the relevant parameters of the publishing webpage, and then the fully functional reducer disassembly and assembly experiment project can be published as a webpage.

Combine existing hardware equipment to build a network-based virtual experimental software and hardware platform to realize virtual 3D display and interaction of reducer disassembly and assembly projects.

5. Conclusion

This article proposes firstly construct a mechanical basic experimental teaching system to prepare for follow-up research. Then build a VR virtual experiment platform that integrates software and hardware to realize virtual teaching. Finally, on the basis of the constructed digital model, the corresponding network virtual platform is studied with the help of UE4 technology. The key to the design of this paper is: (1) The establishment of a virtual experimental environment based on VR is the key to improving teaching methods such as physical models, blackboards, or slides. The realistic and interactive virtual system can greatly stimulate students' learning Interest, improve students' initiative and creativity. However, the establishment of a virtual environment with good interaction requires the integration of software and hardware, which has many technical difficulties and needs to be addressed. (2) The research on the network experiment platform allows students to break through the limitations of time and space, complete experimental preparatory training, experimental homework and other learning outside the laboratory, while integrating interactive operations and learning materials. In order to release the virtual experimental network platform, UE4 engine technology is needed to use C++ programming or blueprint technology for interactive design in UE4, which is also the difficulty of this project.

Based on advanced technologies such as virtual reality, interactive design, digitization, and the Internet, conduct innovative research on the teaching content, teaching methods, and teaching methods of mechanical basic experimental courses. Research on the network virtual platform based on UE4 engine technology can realize the integration of interactive operation and learning materials. At the same time, it enables students to break through the limitations of time and space, and further enhance their interest and enthusiasm in learning. It has a certain degree of innovation in terms of technical use and functional design.

Acknowledgements

This work was financially supported by Guiyang University Teaching Research Project (Grant No.: JT2019520206). Project supported by the Natural Science Foundation of the Guizhou Higher Education Institutions of China (Grant No. [2018]152, No. [2017]239). Project supported by the Humanity and Social Science Foundation of the Guizhou Higher Education Institutions of China (Grant No. 2018qn46).

References

[1] Chang-ming Tang, Tao Jin, Research on product display based on VR-platform with virtual reality technology[J], Modern Electronics Technique 2021, 44(02): 51-55.
[2] Tuikka T, Salmela M. Facilitating designer-customer communication in the worldwide web[J]. Journal of Internet Research, 1998, 5(12): 56-58.
[3] Barceló J A. Virtual reality and scientific visualization. Working with Models and Hypothese[J]. International Journal of Modern Physics C, 2001, 1(12): 569-580.
[4] Kim J H, Park S T, Lee H, et al. Virtual reality simulations in physics education[J]. Interactive Multimedia Electronic Journal of Computer-Enhanced Learning, 2001, 3(2): 1-7.
[5] Georgiou J. A virtual reality laboratory for distance education in chemistry[J]. International Journal of Social Sciences, 2000(2):69-75.
[6] Xuan Zhou, Rui-fang Sui, Current status of virtual reality and augmented reality-based training tools in ophthalmology teaching[J]. Basic & Clinical Medicine, 2020, 40(12), 1744-1748.
[7] Pan-pan Tang, Research on precision parts design of laser VR virtual reality technology[J]. Automation &. Instrumentation. 2020, (04), 163-166.
[8] Bao Cai, Kun-jv Shi, Wen-hua Zhu. Lathe Simulation System Based on Virtual Reality Technology[J]. Computer Systems & Applications, 2018, 27(5): 86–90.
[9] Min-jie Chen, Rong-jin Yang, Meng-feng Shen. Research on application of virtual reality technology in experimental teaching of hydraulic transmission[J], Experimental Technology and Management2018, 35(04): 136-139.

[10] Wen Lei, Qing-kui Chen, Xiao-long Zhu, et al. Simulative Teaching System of "Agricultural. Mechanics" Based on Virtual Reality[J], Computer Systems & Applications, 2018, 27(04): 76-81.

[11] Yang Cao. Application of Virtual Reality in Chemical Engineering Simulation[J], Chemical. Engineering Design Communications, 2018, 44(03): 59.

[12] Hong-fei Zhan, Xin-jian Gu, Jian-rong Tan. Selection and application of tools for developing enterprise-modeling system based on internet/intranet[J]. Computer Integrated Manufacturing Systems, 2000(01): 61-65.

[13] Bo-hua Ma. Design of VR technology based 3D integration simulation system for architectural landscape features[J], Modern Electronics Technique, 2020, 43(20): 153-156

[14] Yao-dong Chen, Zhe Cui, Luo-xi Hao. Research on the Methodology of Evidence-based Design Based on VR Technology[J], 2019, 30(02): 123-129.

[15] Core M R, Stube J W. Computers and information systems[J]. McGraw-Hill Book Company, 2000(2): 36-39.

[16] Simon M, John M. E-Learning initiatives to support prescribing[J]. British Journal of Clinical Pharmacology, 2012, 74(4): 621-631

IOP Conf. Series: Materials Science and Engineering 1126 (2021) 012025 doi:10.1088/1757-899X/1126/1/012025

Study on Design and Analysis of Industrial Y Type Strainer

Mohammed Ather Patel[1] Shaik Himam Saheb[2]

[1,2] Department of Mechanical Engineering, Faculty of Science and Technology,
The ICFAI Foundation for Higher Education, Hyderabad
himam.mech@gmail.com

Abstract: Industrial Strainers are widely being used in Various Heavy Mechanical Industries such as Petro Chemical, Oil and Gas, Steel industry, Water Treatment and Process Industries. The Main purpose being Removal of unwanted Materials in pipelines is a never ending one. Whether the Flowing Material as Sea Water, Oil, Paint or variety of Food or Chemical products. The cause of damage might be dirt, foreign matter or even Clamps of the Product. The Main widely used Industrial Strainers are Basket, Magnetic, Duplex, Y-Type, Tee type and conical strainer. The material used in overall Manufacturing of the Different Strainers is MS (Mild Steel), SS (Stainless Steel). The most efficient of all the strainers Duplex Strainer, because of its large design ratio and larger Material Handling Capacity. The Main Purpose of this Research is to Study Design and Materials involved in the manufacturing of Y-Type are being mainly used in the fluid pipelines in order to Filter Out the impurities, Foreign Particles and protect the downstream.

Key Words: Industrial Strainers, Mechanical Industries, Basket, Magnetic, Duplex, Y-Type, Tee Type, Conical strainer, Fluid pipelines, Foreign Particles.

1. Introduction:

In the Oil and Gas Industry, a efficient filtration or separation system is basic to the dependability and execution of the actual resources that interaction oil based goods. Streamlining the dependability of these filtration frameworks is fundamental in amplifying your operational presentation. Y-Type Strainers are utilized in numerous mechanical applications to shield plant apparatus to harm from foreign substances like residue, scales, sand, strong particles and different prospects. It is utilized to eliminate solids from streaming gases or liquids utilizing stressing component.

Anurag Gupta et al.,[1] examined the common factors of new product planning, design & development of Y- type strainer. It additionally states about the process required for concept choice manufacturing & establishes criteria for new product achievement, the market research tools available for coordinating the Client/customer needs into the innovative process. Yashwanth et.al., [3][4] But they have not worked on the improvement of pressure drop valve of the strainer. If we consider the general length of process pipe which give process fluid to number of equipment and to ensure these equipment Y strainers are introduced before the same. Pavan et al.,[2][5] have concentrated deeply the internal chain of importance operations of concurrent engineering in the new product development process. The reason of this paper is to develop a dynamic planning strategy that is innovative, efficient and flexible for new product improvement by using the concurrent design idea.

Saeed Ovaysi et al.,[6] have studied a dynamic particle-based model for direct pore-level displaying of incompressible thick fluid flow in disordered media. The model is equipped of simulating flow directly in three-dimensional high resolution micro-CT pictures of rock tests. It is depend on moving particle semi-implicit (MPS) strategy. They modify this procedure to improve its stability for flow in porous media issues. Alexander Grahn et al.,[7] have studied Pressure Drop Model for the CFD Simulation of Clogged

Containment Sump Strainers. The present study aims at modeling the pressure drop of flows through growing cakes of compressible fibrous materials, which may shape on the upstream side of regulation sump strainers after a loss-of-coolant incident.

A.Grahna et al.,[8] have implemented a strainer model for calculating the pressure drop across beds of compressible, fibrous materials. The presented study aims at modeling the pressure drop of flows across growing cakes of compressible, fibrous materials and at the implementation of the model into a general-purpose three-dimensional (3D) computational fluid dynamics (CFD) code. Computed pressure drops are compared with experimentally found values. The ability of the CFD implementation to simulate 3D flows with a non consistently conveyed molecule stage is exemplified utilizing a stage-like channel geometry with a horizontally strainer plate.J. S. Andrade, et al.,[9] has considered Inertial Effects on Fluid Flow through Disordered Porous Media. They had explored the cause of the deviations from the traditional Darcy law by numerical simulation of the Navier-Stokes conditions in two dimensional disordered permeable media.

K. Ann-Sofi Jonson et al.,[10] have studied Fluid Flow in Compressible Porous Media Steady-State Conditions. In this article a model describing liquid stream and pressure factor prompted varieties in porosity under fixed conditions is created. In forthcoming article the powerful conduct during filtration and wet squeezing of compressible permeable media are introduced. Liquid course through inflexible permeable media is for the most part described by Darcy's law. The corresponding expression for compressible materials is determined in this article. Chwan P. Kyan et al.,[11] have considered Flow of Single-stage Fluids through Fibrous Beds. A pore model for the flow of a single stage fluid through a bed of random fibers is proposed. A effective pore number, Ne, represents the impact of dead space on flow; deflection number, N6, the impact of fiber avoidance on pressing factor drop. Test information were acquired with glass, nylon, and Dacron fibers of 8-to 28-micron width and with fluids of thickness ranging from 1 to 22 cp. A generalized up grating factor. Reynolds number equation is presented.

Stephen Whitaker et al.,[12] have considered Fluid Motion in Porous Media. In tackling the issue of incompressible flow in porous media, one is defied with the way that the outcome is truly settled i.e., Darcy's law gives a precise depiction of the flow. Because of this, it is not difficult to continue along an assortment of approaches, some of which likely could be mistaken or entirely instinctive, to the correct final result. We will attempt to maintain a strategic distance from this entanglement in the current investigation and build up as cautiously as possible a legitimate, correct course to the final result.

B. F. Ruth et al.,[13][14] have featured the nature of fluid flow through filter sept and its significance in the filtration condition. The consequences of examinations upon fluid flow through an variety of septa are summarized. It is indicated that Poiseuille's law governs fluid flow through channel septa under the states of pressure and rate of flow conventionally experienced during filtrations.

Chandrashekhar, et al., [15-16] The verification that flow is thick all through the whole filtration cycle comprises a critical commitment to the hypothesis and arithmetic of filtration, and sets up a firm test and hypothetical reason for the developed in a previous paper. Strainers are fundamentally used in various enterprises, for example, Pharmaceutical industry, Metals and Mining industry, Water and waste management, Process Industries, Fire fighting industry, Power industry, Chemical industry, Oil and Gas Industry, Pulp and Paper industry, Refineries and Petrochemical Plants. Strainers are used to eliminate

IOP Conf. Series: Materials Science and Engineering 1126 (2021) 012025 doi:10.1088/1757-899X/1126/1/012025

dangerous parts that may cause inadequate or complete breakdown of tasks in the event that they get into the System.

Ernst et. al., [17] Strainers are intended to catch solid particles and other solid unfamiliar substances inside a fluid and prevent them from proceeding through the system. When working appropriately, strainers help prevent potential harm to different pieces of the system. With the better performing in strainer, cost of ownership decreases while life-cycle of the equipment being used is extended, profiting makers over the long run. Stainless steel is the preferred body, basket, and screen material for the drug, food processing, and Chemical industry due to its protection from consumption and simplicity of cleaning. stainless steel costs around multiple times the amount of as Cast Iron.

David et. al., [18] Y Type Strainer is Critical wherever clean fluids are required. While Clean Fluids can help Maximize the Strength, Reliability and Life of any Mechanical Systems, they're particularly huge with Solenoid Valves. This is because Solenoid Valves are exceptionally touchy to Dirt and will just limit appropriately with clean Liquids or Air. Any possibility that any Solids enter the flow; it can agitate and even damage the entire System. Y strainer is a remarkable Component. Despite guaranteeing the introduction of solenoid valves, they are furthermore helping ensure various kinds of mechanical equipment.

1.1 Y Type Strainers: Y Type strainers have a bolt Cover alongside blow off Drain connections. Flanged closes/butt-weld closes, Socket Weld Screws end makes it minimized in size. Despite the fact that it easy to install its sizes doesn't permit it to hold a high amount of Dirt. The manufactured Y type strainers are easy to carry and introduce due to its low weight and minimized size. It makes bigger size and pressing factor of manufactured strainers as well. (Figure 1)

Y Strainers is shape like a "Y" and is utilized to Filter, strain, out particulates from Liquid, Gas and Steam. This mechanical stressing measure is made possible by means of a Filter Element involved cross section, metal, or a wedge wire stressing element. The most widely recognized sort of stressing element is a wire mesh. Some likewise "blow-off valves" that make the cleaning process easy in applications with more Dirt and soil. The strainer itself has a compact, Y shape Structure. The Y shape has better stream qualities then for instance a T Strainer, in light of the fact that the liquid flow through the channel with less alter of direction.

A straight forward Y Type Strainer can keep these segments, which are probably the most significant and costly pieces of the pipeline, shielded from the existences of pipe scale, rust, drugs or some other sort of debris. Y Strainer is accessible in a bunch of plans that can oblige any industries or application.

Figure 1

End

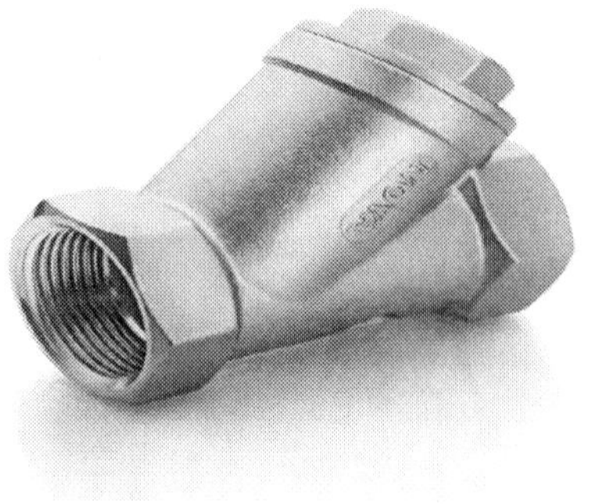

2. Raw Materials

Y Type Strainer, which are Utilized eliminating unfamiliar Particles from Fluids, Gas or Steam Lines. Known for their superior and Longer Assistance life, Y Type Strainers are better solution to Different Industry Businesses.

Table 1: Specification of Y Strainer

Type	Y Type Strainer
Size Range	25mm to 400mm
Pressure Rating	150 to 1500
Body Materials	Alloy 20, Cast Iron, WCB, CF 8.
Material	M.S., MSRL , SS 304/316, Brass, Monal
End Connection	Flanged as Per ASNI B 16.5 Butt Welded

This product is designed to eliminate unfamiliar material from pipe lines and protect the pumps, meters, valves, and other mechanical equipment, which is called as Strainer. "Y" Type Strainers are named after their shape and typically for filtration. Large filtration area can be utilized with fine mesh. The Strainer improves the medium, and draws out the life of valves; secure costly pumps, meters and different types of equipment. It is reasonable for Gas, water, Air, oil, steam and other liquids. Y-strainer can be introduced in horizontal or vertical position (Downward stream) with the screen component pointing descending. This permits the strainer screen to gather material in the strainer at the absolute bottom of the screen.

3. Parts of Strainer

3.1 Body:

SS (Stainless steel) is an austenitic Chromium-Nickel Stainless Steel with superior Corrosion resistance. The low Carbon contents less to Carbide precipitation during Welding. This use is serious destructive conditions. (Figure 3)

3.2 Filter Element:

The filter element is the central component of industrial filters. The channel component is the focal segment of mechanical channels. The filtration take in process, for example, maintenance limit, dirt holding limit and pressure loss are dictated by the filter element components and the filter media utilized in them. (Figure 4)

3.3 Flange

A Flange is an outside or interior edge for strength, as the Flange of an iron pillar, for example, an I-shaft or a T-bar; or for connection to another object, as the flange on the finish of a line, steam cylinder and Flange of a rail vehicle or cable car wheel (Figure 5). The word "Flange" is additionally utilized for a sort of hardware used to frame Flange. Flange with Pipes can be can be disassembled and assembled easily. Elliptical plate Fit inside the Strainer. This is Well Welded and Place in the Y Strainer. Plate is joined along with Mesh. (Figure 6)

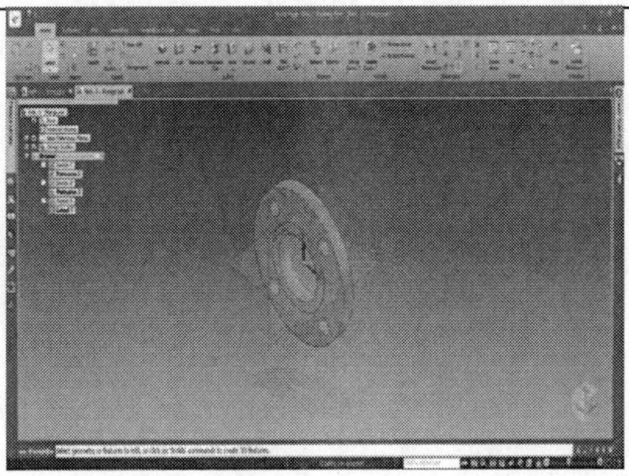

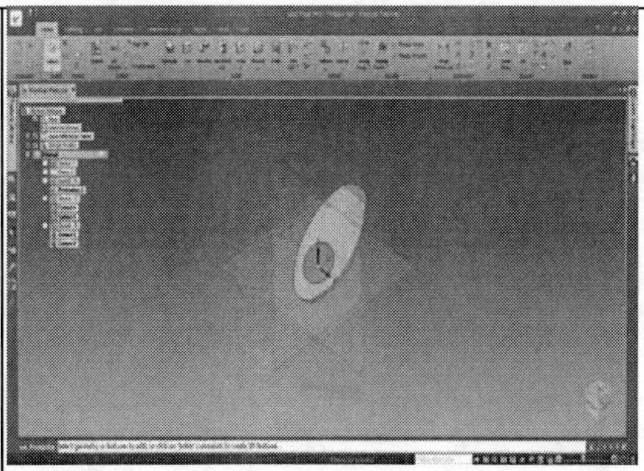

3.4 Bolts & Nuts:

A bolt is a form of threaded fastener with an external male thread (Figure 7) and A nut is a type of fastener with a Female threaded hole internal (Figure 8). Nut and Bolt held together while Vibration or rotation the object.

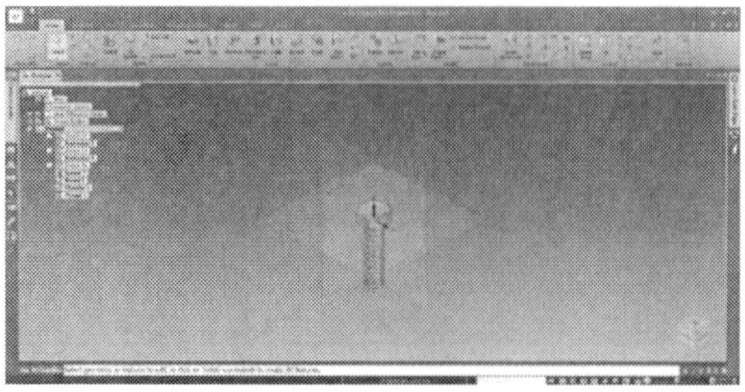

Figure 5: Bolt Design for Y Strainer

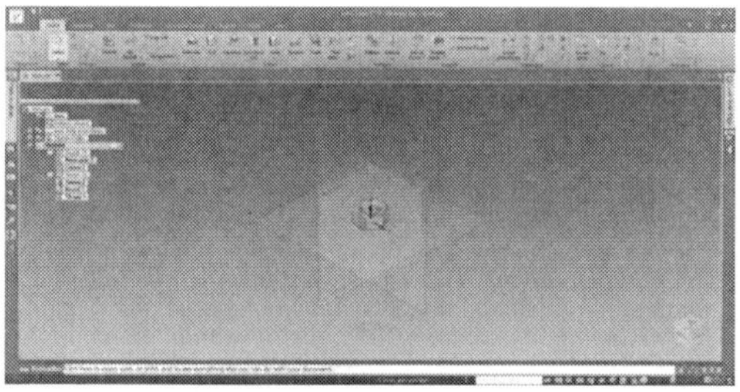

Figure 6: Nut Design for Y Strainer

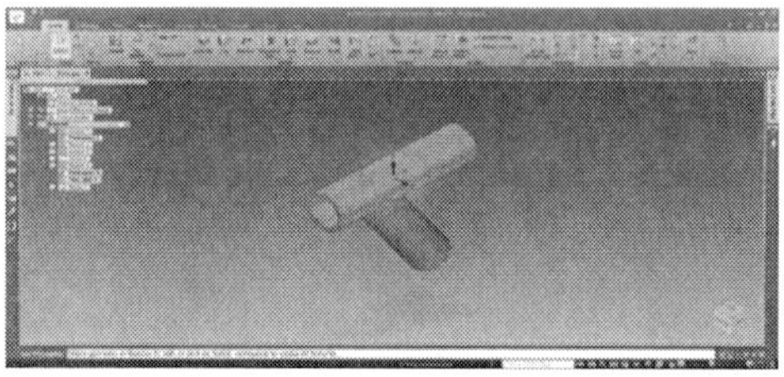

Figure 7: Body of the Y Strainer

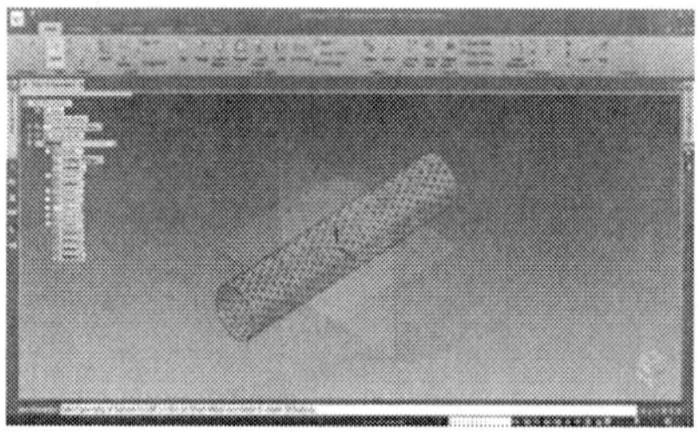

Figure 8: Filter Element of the Y Strainer

3.5 Gasket

A gasket is a mechanical seal which occupies the space between at least two mating surfaces, for the most part to stop leakage from or into the joined items while under compression. (Figure 9)

3.6 Shell Flange

A Flange is a plate for covering or shutting the finish of a Pipe. A Flange joint is an association of Pipes, where the interfacing pieces have spines by which the parts are joined together. (Figure 10)

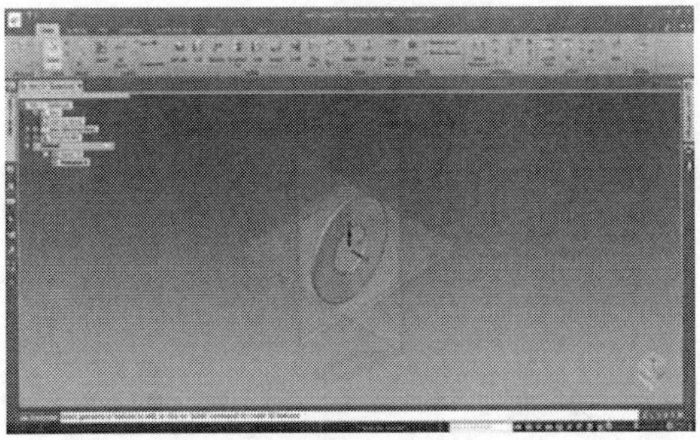

Figure 9: Gasket for the Y Strainer

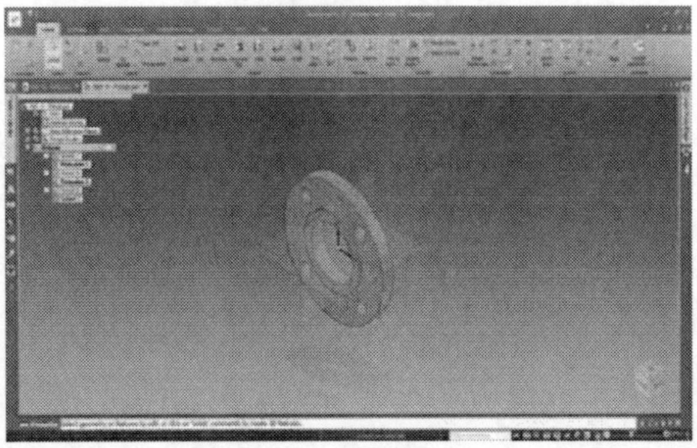

3.7 Shell Cover

Flange is a plate for covering or shutting the finish of a pipe. It has the Drain Plug, where it assists with flushing the Liquid. (Figure 11)

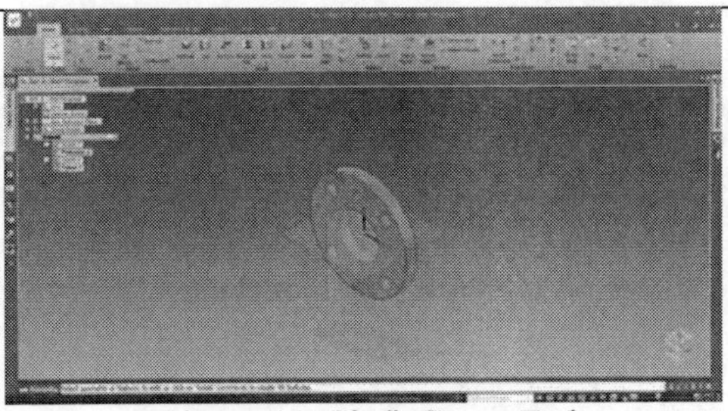

Figure 11: Shell Cover End

Figure 12: Drain with plug

3.8 Drain with plug

The insertion of a plug into a drainage outlet allows the container to be filled with water or other fluids to Drain out. (Figure 12)

IOP Conf. Series: Materials Science and Engineering 1126 (2021) 012025 doi:10.1088/1757-899X/1126/1/012025

4. Assembly of Y Strainer:

Figure 13: Assembly of Y Strainer 1

Figure 14: Assembly of Strainer 2

5. Calculation:

We will calculate Pressure Loss Using $\boldsymbol{Cv}$ Factor:

ΔP = $\boldsymbol{[Q/Cv]^2\ (133.66)}$ **- (1)**

Cv = Flow Coefficient

Q = Flow in M^2/Hr

ΔP = Pressure Drop

The Pressure Loss in the Y strainer can be Calculated with Flow rate and $\boldsymbol{Cv}$ Factor for the Y Strainer. We will consider 2, 4 & 8 psi pressure to find the Pressure loss in Y Strainer with Equation (1).

Case (1):

We will obtain the Pressure loss for 2 Psi.

ΔP = $[Q/Cv]^2$ (133.66)

2Psi = 13.7 Kpa

ΔP = $[13.7Kpa/30]^2$ (133.66)

ΔP = 27.87 Kpa

Case (2):

We will obtain the Pressure loss for 4 Psi.

$\Delta P = [Q/Cv]^2\ (133.66)$

4Psi = 27.5 Kpa

$\Delta P = [27.5Kpa/30]^2\ (133.66)$

ΔP = 112.38 Kpa

Case (3):

We will obtain the Pressure loss for 8 Psi.

$\Delta P = [Q/Cv]^2\ (133.66)$

8 Psi = 55.15 Kpa

$\Delta P = [55.15Kpa/30]^2\ (133.66)$

ΔP = 455.69 Kpa

- The Pressure loss in the Y Strainer from above result as 27.8, 112.3 & 455.69 Kpa is obtained. Pressure loss increases or decreases, Difference occurs with the difference in Size of the mesh .Hence Maximum Pressure loss can be up to 500Kpa to run for Long Period of time.
- The Higher the Mesh size the More stability is attained in the Outflow of the water.
- If it is smaller at the inlet Pressure will be more.
  P>Outlet
  P< Outlet
- Depending upon the Foreign Material entering along the Liquid (Microns) the life of the Strainer is obtained.

Calculate Pressure Drop in Strainers:

- The purpose of the Guideline is to Lay down the Procedure for estimation of Pressure Drop across Y Strainers in a Piping system.
- For Pressure drop estimation Graphs are referred.
- This Graph gives Pressure Drop based on Specific gravity of Water and Viscosity of Water.
- Screen openings chart indicates the Percentage open Area of Perforated plate and Meshed screens.
- For Fluids of High viscosity and finer weave Meshes viscosity and Density correction Factor chart will be referred.

Pressure Drop: For maximum efficiency, a differential pressure gauge installed across the inlet and outlet will indicate pressure loss due to clogging and may be used as a guide to determined when cleaning is required. Normally, when differential pressure reached 2-8 Psi, screen must be cleaned. (Figure 15)

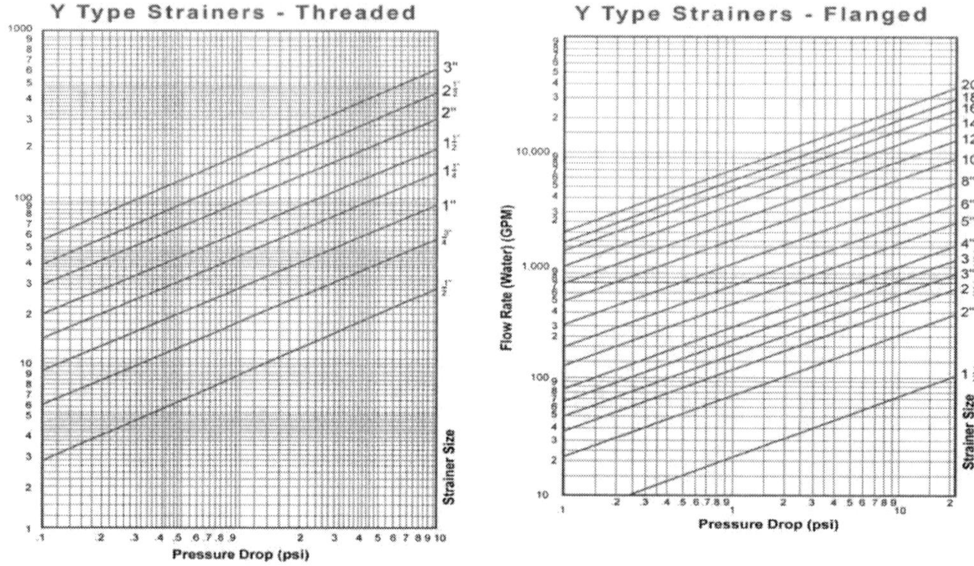

Figure 15: Pressure Drop Chart

Chemical Analysis and Material Properties:

The Test Report of the Y Strainer shows the result of the Chemical Analysis and Material Properties. This report shows the Ultimate Tension Strength, Yield Strength, Elongation and Percentage of Chemical Component in the Material. If the Pressure is more then, Liquid will not allow to Flow properly and strainer gasket will fails and the mesh will fail its purpose to sustain the foreign particle.

Manufacturing Process: Steps Involved in the manufacturing Y Strainer are material selection (Figure 16), plasma cutting (Pipe & Sheet) , MIG & Arc Welding, Buffing and Grinding , Perforated Sheet, Mesh (On Customer Enquiry), Assemble and Fitting, Hydro Test, Quality Control Check, Final Inspection, Paint, Packing, Dispatch.

The applications of industrial strainers are Heavy Duty Construction, Thermal Plant, Cement Industries, Process Industry, Pharma-Industries, Power Industry, Oil and Gas Industry, Food Industries, Water and Waste Management, Hydro-Carbon and Steel Industry, Refineries and Petrochemical Plants.

Figure 16: Parts of Y Strainer

6. Conclusion

Y Strainer is designed and studied in the above research paper. The material used is SS Stainless strainer (316 L) with negligible losses. The future studies can be done with a different material for the further development. Strainers are not long, at this point kept to a Simple cast body with a Wire Mesh screen, yet are a Technical, Highly Refined, Designed bit Carefully of Equipment. In some cases they work at 1,500 degrees F and 10,000 psig or at cryogenic Temperatures. They are modified with Steam coats, Cover Lifting davits, Magnets, Motorized Cleaning gadgets and Automatic vent Valves. They are Supplied with Screwed, Flanged, Socket weld, Butt weld, Ring joint and Silver brazing end Connections. Appropriately, the execution of a Strainer should be thoroughly examined and engineered. While it is acceptable practice to utilize a Strainer to secure Down-stream Equipment, it is imperative to carefully consider the choices Available. Picking the right Strainer can Save Money by Protecting Equipment, yet in addition by keeping Operations and Maintenance Costs at the very least.

References

1. Anurag Gupta Pushpdant Jain, Prabhsah Jain [2] (Dec. 2013) inthe paper titled "Product Planning & Development of Y-TypeStrainer Used in Thermal Power Plant and Process Plant" reported inthe International Journal of Engineering Inventions, e-ISSN: 2278-7461, p-ISSN: 2319-6491 Volume 3, Issue 5 (December 2013) PP:42-50.
2. Pavan Kishore Mamaduri , Chandra Sekhar Akula , H S Shaik, "Numerical investigation for influence of pre-twist on stress behavioral characteristics of curved blade" Vibro engineering PROCEDIA, Vol. 22, 2019, p. 153-157.
3. M. Yashwanth Kumar, Shaik H S & M. Venkata Ramana Reddy, Transient Thermal Analysis of the Turbine Blade, Global Journal of Researches in Engineering: A Mechanical and Mechanics Engineering Volume 20 Issue 3 Version 1.0 Year 2020 Online ISSN: 2249-4596 & Print ISSN: 0975-5861
4. Kona, Ravisandeep & Saheb, Sk & Kumar, Yashwanth. (2016). Study Flow Analysis on Hull of a Maya-AUV. International Journal of Scientific Research and Development, Volume 3, Issue10.
5. A. Chandrashekhar,H.s, S. Ranjan MishraTushar Srivastava M. L. Pavan Kishore, Static Structural Analysis of Hybrid Honeycomb Structures Using FEA, Recent Trends in Mechanical Engineering, Springer Nature Switzerland AG, 363-375, 31 October 2020, 978-981-15-7556-3
6. Saeed Ovaysil, Mohammad Piri2 June 2010) in the paper titled "Direct pore-level modeling of incompressible fluid flow in porous media" reported in the Journal of Computational Physics, 229 (2010) 7456–7476.

7. Alexander Grahn1, Eckhard Krepper2, Frank-Peter Weiß3 (August 2010) in the paper titled "Implementation of a Pressure Drop Model for the CFD Simulation of Clogged Containment Sump Strainers" reported in the Journal of Engineering for Gas Turbines and Power by ASME, Vol. 132 / 082902-1
8. A. Grahna1, E. Krepper1, S. Alt2, W. Kastner2 (April-2008)in the paper titled "Implementation of a strainer model for calculating the pressure drop across beds of compressible, fibrous materials" reported in the Nuclear Engineering and Design by ELSEVIER, 238(2008) 2546–2553.
9. J. S. Andrade, Jr., U. M. S. Costa, M. P. Almeida, H. A. Makseand H. E. Stanley [6] (June 1999) in the paper title "Inertial Effects on Fluid Flow through Disordered Porous Media" reported in the journal of Physics Review Letters by American Physical Society, Volume 82, 0031-9007/99/82(26)/5249(4).
10. K. Ann-Sofi Jonsson and Bengt T. L. Jonsson (Sept. 1992) inthe paper title "Fluid Flow in Compressible Porous Media: I: SteadyState Conditions" reported in the AIChE Journal Vol. 38, No. 9.
11. Chwan P. Kyan, Darshanlal T. Wasan, and Robert C. Kintner (1970) in the paper title "Flow of Single-phase Fluids through Fibrous Beds" reported in the journal Ind. Eng. Chem. Fundamen.,9 (4), pp 596–603.
12. Stephen Whitaker (1969) in the paper title "Fluid Motion in Porous Media" reported in the journal of Industrial and Engineering Chemistry Vol. 61 No. 12.
13. B. F. Ruth (July 1935) in the paper title "Studies in Filtration- IV. Nature of Fluid Flow Through Filter Sept and Its Importance in the Filtration Equation" reported in the journal of Industrial and Engineering Chemistry Vol. 27 No. 07.
14. B. P. Ruth with G. H. Montillon and R. E. Montonna (Feb1993) in the paper title "Studies in Filtration- II. Fundamental Axiomof Constant - Pressure Filtration" reported in the journal of Industrial and Engineering Chemistry Vol. 25 No. 02.
15. A.Chandra Shekhar, S.Shahab, S.HimamSaheb, "Analysis of Geneva mechanism using Dwell Symmetrical Coupler Curve Mechanism" Materials Today: Proceedings, 2020,.
16. Avinash Malladi, Chandrashekhar A, "WIFI for Vehicular communication systems", Journal of Emerging Technologies and Innovative Research International Open Access Journal. Vol. 2, No, 6, (June 2015), pp 1939 -1943.
17. Ernst H (2002) "Success factors of new product development: a review of the empirical literature" International Journal of management Reviews, Volume 4, Issue 4, pp 1 – 40.
18. David Atkinson BEng (2007) "Global New Product Introduction and Development in the Automotive Sector", University of Wolver hampton.

IOP Conf. Series: Materials Science and Engineering 1126 (2021) 012026 doi:10.1088/1757-899X/1126/1/012026

Re-Design and cost reduction in fabrication of a novel 3 DoF active ankle joint.

Ranjeet Ranjan, Richa Pandey, Ojas Suryawanshi, Purushottam Kumar, Sushil Kumar

irla Institute of Technology
richapandey@bitmesra.ac.in

Abstract— The mechanism of active ankle is a novel 3 DoF parallel mechanism that operates in an almost spherical manner. It is a project by DFKI Germany and currently it is being used in medical rehabilitation. The project was re-engineered with techniques like CAD and Rapid-Prototyping to reduce the cost of fabrication. In the present paper, a comprehensive study of the design, and fabrication of the active ankle is provided. Various motions of human ankle joint such as adduction-abduction, inversion-eversion, and plantarflexion-dorsiflexion, was achieved by the prototype. This project was modeled in CAD software like SolidWorks, accordingly mounting were prepared for the available motors. Various parts of Active Ankle prototype was printed using Olivetti 3D printer by feeding the .gcode file created by slicing software Cura. The paper also discuss about the integration of Dynamixel motors with MATLAB for control. Finally, the paper talks about various application of parallel manipulator in the area of medical rehabilitation.

1.INTRODUCTION

Our project is a 3 DoF parallel mechanism. A parallel manipulator (PM) can be defined as a closed-loop mechanism in which the end effector (mobile platform) is connected to the base by at least two kinematic chains. Its geometry provides various advantages like better control and robust construction. Closed chain in each leg offers more stiffness and orientation maneuverability and this is the reason it has many applications. We here have tried to replicate the work by 3D printing the structure and making it more economical in the process.

Fig 1.1: Assembly

There are many other spherical manipulators such as AGILE EYE and AGILE WRIST. Joint axis of these type of manipulator are required to intersect at a single point[2]. However, due to manufacturing and assembling errors, it is difficult to achieve an accurate intersection of all joint axes.

Our prototype version is a 3D printed one and it is firm belief of authors that to really boost exoskeleton technology to a common man it needs to get comfortable, effective and easily replaceable.

The exoskeleton has to have many customizable parts, as inter-subject variability still remains a challenging issue [5]. So, 3D printing could serve a good option. Although it has some limitation discussed in the further section.

Human ankle has 3 DoF. That is rotation in 3 mutual perpendicular axes are possible, namely-inversion, eversion; dorsiflexion, plantarflexion; abduction adduction. An ankle joint in an exoskeleton is very important as it acts as a final interface between ground and body so the complete load transfer occurs through this joint. Also it should be easy to control and rigid enough to let the load get transfer. Hence, we were interested in ankle joint.

Further the section 2 discuses upon set-up of our project, section 3 talks about 3D printing aspects. Section 4 gives an overview of control, and section 5 and 6 informs about the limitations and future works to be done respectively.

2.SET-UP

2.1 Actuators used

Actuators used becomes a major part of your designing process. That is it drives the direction in which the designing is led. We here used Dynamixel AX-12A servo actuator. With fully integrated DC Motor, Reduction Gearhead, Controller, Driver. The Robotics Dynamixel AX-12A handles the sensor management and position control. To strengthen the assembly, horn and nuts are fixed inside. For easier cable management extra holes are given. The new design of this actuator increases the durability and reduces the damage rate. Operational voltage is 9V – 12V. Gear ratio 254:1. Weight is 54.6g.

According to this actuator we had alter the design from the original [1],[2] keeping the basic architecture same. We had design an different structure on the first try and the printer was just not able to print the 3 mm cantilever platform for motor mounting. So we had to design a more solid design as can be seen now.

The current design is made by keeping in mind that the axis of all three motors must be mutually perpendicular, and also by considering the rod and rod end bearing and the distance between the two links attached to same motor.

Fig 2.1: AX-12A

IOP Conf. Series: Materials Science and Engineering 1126 (2021) 012026 doi:10.1088/1757-899X/1126/1/012026

1. Dimensions of actuator:

- 32 X 40 X 50 (mm)

2. Various features of motors:

- Easy to track its shaft position, speed, temperature, voltage and load.
- It also let us control the speed and strength of the motor's response based on the control algorithm used to maintain the shaft position.
- It can also be used as a wheel so that you can create a robot with a combination of wheel type and joint type in a single module

2.2 Actuators used

There were total five, 3D printed parts, one base, one end effector and three couplers. In future we also want to increase the count of 3D printed parts used. More on 3D printing is given in further section.

2.3 SMPS

Fig 2.2: SMPS

The figure of SMPS is shown in Figure 2.2 .A switched-mode power supply, is an electronic power supply that incorporates a switching regulator to convert electrical power efficiently [3]. It transfers power from a AC source mainly mains to a DC supply in our case it was 12 V.
This was the voltage rating of actuator used. It provided a stable power source for our experiment.

2.4 U2D2

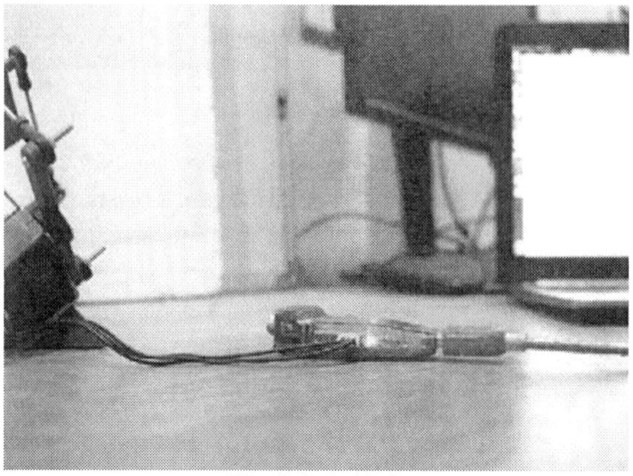

Fig 2.3: U2D2

U2D2 is a small size USB communication converter that enables to control and operate DYNAMIXEL with PC, this is the interface between laptop to communicate with the motors. It basically sends and reads position of motors as shown in Figure 2.3.Figure 2.4 shows the setup of the system

Laptop (For control)
Intel - Core i7 3.16GHz 8Gb Ram Linux 16

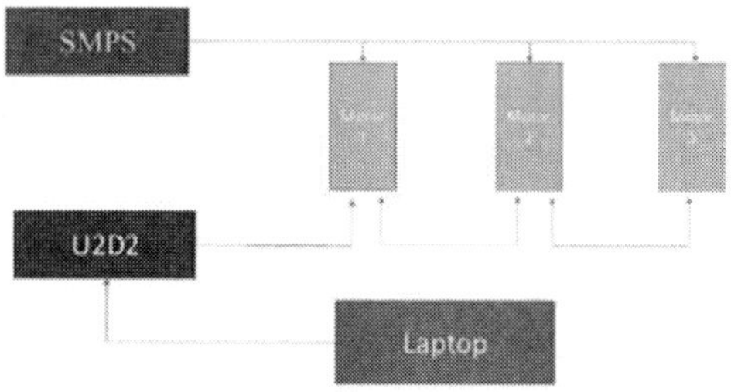

Fig 2.4: Setup of the system

3. 3D PRINTING

3D printing is a manufacturing process where a 3D printer creates a three-dimensional object by depositing material by depositing material layer by layer in accordance to the object's 3D digital model.[3] In recent years this technology has gained momentum and future is definitely looking bright for it. This significantly reduced cost of manufacturing compared to when machining metals. But the question of strength perpendicular to layers still remains.

3.1 Fused Deposition Modelling (FDM)

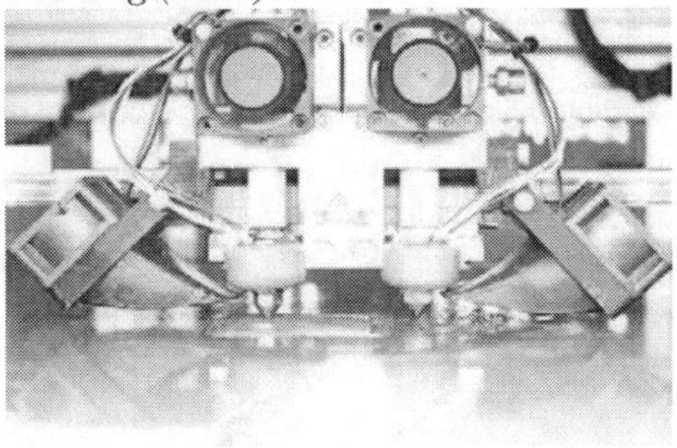

Fig 3.1: FDM

In FDM, an object is built by selectively depositing melted material in a predetermined path layer-by-layer. The materials used are thermoplastic polymers and come in a filament form of melted plastic. FDM Technology works with specialized 3D printers and production-grade thermoplastics to build strong, durable and dimensionally stable parts with the best accuracy and repeatability of any 3D printing technology. FDM is the most widely used 3D Printing technology it represents the largest installed base of 3D printers globally and is often the first technology people are exposed to. FDM have many advantages few of them are the technology is clean and simple-to-use. Supported production-grade thermoplastics are mechanically and environmentally stable. Complex geometries and cavities that would otherwise be problematic become practical with FDM technology. Our printer used this 3D printing technique and was quite effective, as final assembly was done by using that parts itself.

3.2Slicer

Cura slices 3D models. It translates the 3D STL, OBJ or 3MF file into a format that the printer can understand. Fused filament fabrication (FFF) 3D printers print one layer upon another to build up the 3D object. Cura takes the 3D model and works out how those layers are placed on the print bed and creates a set of instructions for the printer to follow layer by layer.

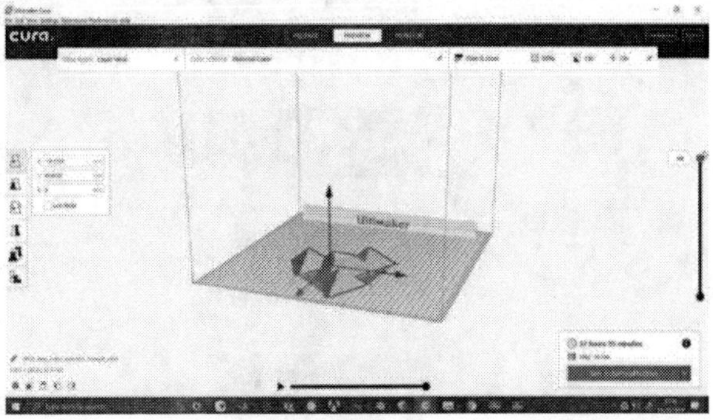

Fig 3.2: Cura

Cura generates instructions for your 3D printer. They are called G-Code, a text document that ends with the file extension. gcode. By opening the file and we can actually read through quite a bit of the code and understand what it's telling the printer to do.

<u>Infill density</u>: It's the amount of plastic that is being used on the inside of the print. More infill density means that there is higher amount of plastic on the inside of the print, making it stronger. And for the end use parts, higher densities can be used. We used 60% infill density

<u>Infill pattern:</u> Ultimaker cura also have settings to change the pattern of the infill structure of printing parts. These patterns are selected accordingly it's uses. Here are some pattern options available in Cura:

- Grid: strong 2D infill
- Lines: quick 2D infill

- Triangles: Strong 2D infill
- Zig-zag: a grid shaped infill, printing continuously in one diagonal direction.
- Cross: Flexible 3D print.

The settings which worked for us was infill pattern zig-zag, wall thickness 0.8mm nozzle temperature 210 degrees Celsius, with printing speed to be 60mm/s and build plate adhesion type to be Raft. It took around 14 hrs. to print the base part. [4]

3. *Filament*

PLA and ABS are two commonly used filament/material in 3D printing. Not having much difference but certain factors that separate these materials in terms of performance, quality, process etc.

Like PLA is used for higher strength [6], higher rigidity and strong layer bond while ABS is considered for higher impact resistance, higher flexibility and higher temperature resistance (Tg) in terms of performance. PLA filaments gives sharper details (features, corners, surfaces). In terms of process, PLA is taken for lower wrapping, better odor, and less particle emissions while ABS is taken for lower risk of jamming.

4. CONTROL

4.1 Kinematics

Kinematics of this architecture is done earlier in the papers [1],[2]. The basic idea being that the distance between the end effector points and the coupler point remains same as the length of the rod. Writing position vector of the 12 points mentioned above and subtracting it gives us the required equitation. Then eliminating sin(q_x)and cos(q_x), q_x being the angle made by motor whose axis is in x direction half angel substitution and latter solving a quadratic equitation to get the angles.

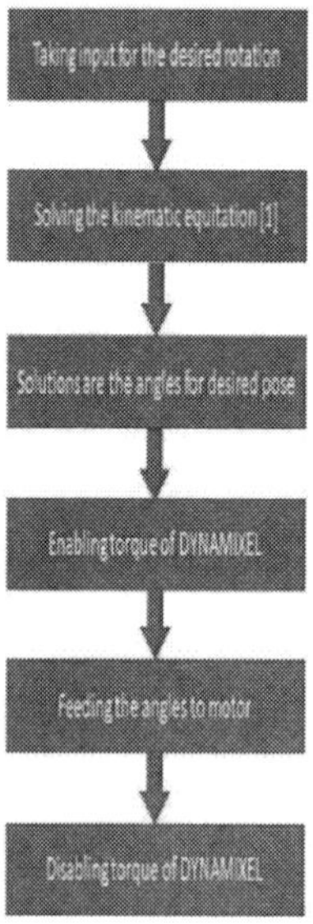

Fig 4.1: Schematic of control

4.2 Dynamixel SDK

We are currently controlling our prototype by MATLAB. We have developed code to interface the kinematics and the motor control. We are using SDK DYNAMIXEL controller. Our control model has overshooting problems and we are planning to tune it via a PID controller. Currently we are able to get motor data from the kinematic solution and feed directly to the motor. So that we can control the orientation of the end effector. The baud rate used was 2000000. The protocol 1.0 was used for communication. DYNAMIXEL does the Asynchronous Serial Communication with 8-bit, 1 Stop bit, and None Parity.

Fig 5.1: Dynamixel setup code

5. Limitations

Humans have others joints too with 3DoF such as hip joint and shoulder joint. In case of shoulder it has 3 rotational degree of freedom but can also has elevation, depression and retraction, protraction motion, this is because of the architectural property of human anatomy. Whereas in case of hip joint it has 3 DoF and this type spherical parallel mechanism can be used for its actuation. However, it does not take into account the compensation of misalignment. Misalignment is an issue of high concern as axis misalignment between human anatomical axis and exoskeleton axis can cause detrimental effect on user [5]. Alos to install our present prototype, we need a big volume. Work could be done to minimize the space required.

6. Future work

The future of using parallel manipulator in exoskeleton is very vast, and that too fabrication with 3D printing cannot be ignored. The future of our project can be integration with actual human subject and checking the effectiveness of model. Also, it can be printed by carbon fiber to increase model's strength.

7. Conclusion

In conclusion we come to a perspective that by Rapid Prototyping the structure it makes the parts easily replaceable and highly cost efficient. We also look forward to make the assembly more compact and finally test on a human subject this paper servers more as an overview of our work until now and journey of our first working prototype.

8. References

[1] Kumar, Shivesh, Abhilash Nayak, Bertold Bongardt, Andreas Mueller, and Frank Kirchner. "Kinematic analysis of active ankle using computational algebraic geometry." In *Computational Kinematics*, pp. 117-125. Springer, Cham, 2018.
[2] Kumar, Shivesh, Bertold Bongardt, Marc Simnofske, and Frank Kirchner. "Design and kinematic analysis of the novel almost spherical parallel mechanism active ankle." *Journal of Intelligent & Robotic Systems* 94, no. 2 (2019): 303-325.
[3] Basso, Christophe. *Switch-mode power supplies spice simulations and practical designs*. McGraw-Hill, Inc., 2008.
[4] Ultimaker Cura User guide.
[5] Näf, Matthias B., Karen Junius, Marco Rossini, Carlos Rodriguez-Guerrero, Bram Vanderborght, and Dirk Lefeber. "Misalignment compensation for full human-exoskeleton kinematic compatibility: State of the art and evaluation." *Applied Mechanics Reviews* 70, no. 5 (2018).
[6] Rodríguez-Panes, Adrián, Juan Claver, and Ana María Camacho. "The influence of manufacturing parameters on the mechanical behaviour of PLA and ABS pieces manufactured by FDM: A comparative analysis." *Materials* 11, no. 8 (2018): 1333.

IOP Conf. Series: Materials Science and Engineering 1126 (2021) 012027 doi:10.1088/1757-899X/1126/1/012027

Effect of concentration, size, granularity, shear time and temperature on rhological properties of coal water slurries

Rishi Jain, Shreyash Yadav, Sachin Bisht, Jagadeeshwar Kodavaty, *Suresh Kumar Yatirajula

Department of Chemical Engineering, University of Petroleum and Energy Studies, Bidholi, Via Premnagar, Dehradun, India 248007

*Department of Chemical Engineering, IIT Dhanbad, Jharkand, India 826004

E-mail: j.kodavaty@ddn.upes.ac.in

Abstract. Coal water slurries (CWS) were proposed as a substitute for the fuel oils century ago. As there is a huge demand for energy, it is necessary for us to use such concepts with good understanding of the alternative fuels. As the use of CWS depends on physio-chemical properties of coal and rheological properties of the slurry, coal obtained from Dugda coal washery, Jharkand, India were subjected to examination. These coals are ground and segregated according to the sizes and the samples of CWS were prepared. The prepared samples were tested for their physio-chemical properties first. various samples of coal water slurries were prepared and examined for rheology of the samples by varying their concentration, size and granularity of the particles, shearing time and temperature. A detailed analysis on stress strain relation was analyzed during this study. It is found that the stresses would increase with concentration of CWS when the strain rates are varied. It is found that the size of the particles would play active role in deciding the shear forces when the samples were subjected to varying shear rates with their self assembly. The granularity of the particles in the sample has significant effect in rheological properties of the CWS samples. The temperature has no effect.

1. Introduction

Fine coal particles suspended in water is termed as coal water slurry (CWS). Presence of water in CWS has advantages like reduction of harmful emissions into atmosphere, making explosion resistant and can be used as a substitute for liquid fuel. In any size of heating and power stations the CWS can be used as fuel. CWS is proven to be compatible fuel which is suitable for existing gas, oil and coal boilers with little modifications. Use of CWS as a substitute for oil in boilers was an ancient idea which was implemented a century ago. These came into light when people started using coal tailings in the boiler without de-watering them and usually as a coal oil mixture [1].

It was proposed that the fuel oil can be replaced with CWS based on the calorific values. As the cost of CWS is cheaper than the oil, it will have its own impact on technoeconomical aspects [2]. Usually the CWS fuel that is used in the boilers and gasifiers has an apparent viscosity of 800 to 1200 m Pas with a loading of 70% of dry solids [3]. It is suggested that the typical concentration of CWS consists of 60-75 % coal, 25-40 % water, and about 1 % chemical additives [4].

The important factors effecting the CWS being the coal properties, size of the coal particles and their distribution, solid loading of coal. Apart from this, the other factors being the method of preparation and the additives present in CWS and importantly the rheological behavior of the slurry [5]. The three important factors in combustion of coal burning are the size of the particles, concentration and viscosity of the slurry. However, the characterization of coal water mixture becomes necessary before it is sent for burning in industrial boiler. Characteristics of CWS is more complex than that of coal characteristics itself since CWS requires the physio-chemical properties along with the rheological understanding [4, 6].

The transportation of CWS plays crucial role in many industrial applications ranging from their combustion to gasification. Hence a correct prediction of the rheological properties of coal water slurries minimizes the power consumption [7]. It is known that for several years, the Atlantic Research Corporation has been developing coal-water slurries as a low-cost liquid boiler fuel.

The CWS are highly-loaded and stable that need to be transported, stored, pumped, and burned like residual fuel oil. To achieve the required rheological properties in highly-loaded suspensions requires a very carefully selected distribution of particle sizes. The effect of particle sizes and their distribution, the method of preparation of the slurry and the effect of its rheological properties are very important factors to be considered in the analysis of behavior of CWS [8].

The CWS usually behave as non-Newtonian nature and its rheological properties are functions of water content, coal particle size distribution, etc. As the viscosity and stress strain relation being important factors in the transportation of CWS, a particular coal water slurry should have high loading with low viscosity value. Particle size distribution exceeding a limit determines the apparent viscosity of CWS that increases with the increase of solid concentration [9, 10].

Coal type is not important parameter when preparing CWS but its water absorption capacity plays an important role in the rheological properties of CWS hence the effect of coal properties on the viscosity of CWS should be determined [11]. In the wide particle size distribution of coal, the particle interactions does have its effect on the rheological properties as they modify the volume fractions of coal when small particles fills the gaps between the course particles. The rank of coals also has the probability of altering the volume fractions.

2. Materials and methods

Coal from Dugda coal washery, Jharkhand, India were taken for analysis. Coal collected as samples were graded to requisite size for preliminary test, such as, proximate analysis, calorific value, ultimate analysis and in the preparation of coal water slurry (CWS). These samples are ground in a crusher and classified according to Taylors standard meshes after grinding. The stack of meshes that were used have mesh sizes 48, 65 ,100 ,200 that has openings of 295, 147, 104, 74 μm, respectively.

2.1. Proximate analysis of coal

2.1.1. Moisture content The inherent moisture content of the coal varies depending on the surrounding conditions of temperature and humidity.To determine the moisture content, one gram of coal that pass through 72 mesh sieve was weighed and taken in a pair of watch glasses. The coal is spread evenly in the watch glass and then kept in an air oven with temperature maintained at 110$\pm$ 2 °C for one hour with the lid partially open. Before the removal of the sample from the oven, lid is covered and samples are taken out and placed in a desiccator. These samples are cooled for 15 minutes and weighed. The difference in weight is reported as the moisture content.

IOP Conf. Series: Materials Science and Engineering 1126 (2021) 012027 doi:10.1088/1757-899X/1126/1/012027

2.1.2. Ash content To determine the ash content, one gram of the coal that pass through 72 mesh sieve was weighed and placed in a crucible. This crucible is placed in a muffle furnace for one hour at 750 °C. The residue left over is weighed and is reported as ash content of coal.

2.1.3. Volatile matter To determine the ash content, approximately one gram of coal that pass through 72 mesh sieve was weighed and taken in a watch glass. The coal is spread evenly in the watch glass and then kept in an air oven with temperature maintained at 110± 2 °C for one hour. This sample is weighed for loss in moisture content and then immediately placed in a crucible. This crucible is fitted with a cover and placed in the furnace at 750 °C for 7 minutes. The loss in weight after moisture removal and furnace heating is reported as the volatile matter.

2.1.4. Fixed carbon The percentage of fixed carbon is obtained by subtracting the sum of moisture percent, ash percent and volatile matter percent from 100.

2.2. Ultimate analysis of coal
Vario EL III CHNS Analyzer from Elemental Germany was used for determination of nitrogen, carbon, hydrogen, sulfur and Oxygen.

2.3. Swelling index of coal
To determine the swelling index 1 gm of air dried coal that passed through 72 mesh size is heated in a crucible with a lid until volatile matter ceases to evolve. Rate of heating is maintained such that the temperature of the inner surface of the base of the crucible reaches 800 within 1.5 minutes and 820 within 2.5 minutes of lighting the burner. After 2.5 minutes the crucible is cooled and the coke button is removed .The shape and size of the coke button is compared with the standard coke profile number from 1 to 9 at the interval of 0.5. This is called swelling number or free swelling index (FSI).

2.4. Hard grove grind ability index (HGI) of coal
50 grams of coal that pass through 30 mesh is collected. This sample is ground in standard mill for 60 revolutions. After grinding, the coal is screened through a 200 mesh sieve. Hard grove grind ability index (HGI) is then calculated as

$$G = 16 + 6.93W \tag{1}$$

where G is hard grove grind ability index and W is weight of the sample passing through the 200 mesh sieve.

2.5. Specific gravity of coal
Weight of water (W2) and kerosene(W3) that were filled in a 25 ml specific gravity bottle were noted down. In another 25 ml specific gravity bottle, the coal power that pass through 200 mesh screen is filled partially (approximately 50% of bottle) and its weight is noted down(W4). This bottle containing coal is slowly filled with kerosene until the coal inside the bottle is saturated with kerosene oil. After makeup of kerosene, the bottle is weighed and the weight is noted down(W5). If W1 is the weight of the empty specific gravity bottle, the specific gravity of the sample is calculated as

$$Specificgravity = \frac{(W4 - W1)\,(W3 - W1)}{[(W3 - W1) - (W5 - W4)]\,(W2 - W1)} \tag{2}$$

2.6. Preparation of coal water slurry (CWS)

Coal water slurry samples were prepared by blending coal of different particle sizes with water on weight basis. The Rosin-Rammler relationship was used for representing the size distribution of powdered coal, as given in equation 3

$$R = 100e^{\left(-\frac{X}{K}\right)^n} \tag{3}$$

where R is cumulative weight percent of particular fraction of total mass, X is mean particle size in a fraction in micron, n is dispersion constant and K is absolute size constant [12].

In the present study, the values of dispersion constant were chosen as 1.5, 1.6, 1.7, 1.8 and 1.9 and the values of absolute constant were chosen as 50, 60, 70, 80, 90 μm and the cumulative weight fraction is calculated for different mean particle size using equation 3.

2.7. Rheological experiments

Anton Paar (Rheolab QC) Quality Control Rheometer was used to determine the rheological properties of the CWS at low and high shear rates. The present investigation is mainly aimed to study the effect of coal concentration, coal particle size, size distribution, shear rate, shear time at room temperature and also at varying temperature on the Coal water slurry. The investigations were carried out with coal sample mainly weakly caking type having 33.63%ash and 18.96% volatile matter. The coal is chosen from weakly caking type because the CWS have to undergo through combustion in the boiler or in the furnaces.

In the present work, initially the physio-chemical properties of the coal were estimated. Then the coal water slurries were prepared with different concentrations of the powdered coal according to the absolute size constant values. The rheological analysis is performed by varying the absolute size constant values and the concentrations of the samples. Selective sizes of the coal fractions were chosen based on their sizes and the CWS were prepared. The rheological analysis on these samples were performed to test the granularity effect. The effect of shearing times at two different concentrations of the CWS were estimated. The effect of temperature on CWS were determined at varying concentrations and at various temperatures.

3. Results and discussion

3.1. Proximate analysis of coal

Proximate analysis is one of the significant test that determines the quality of the coal. This test is important for both supplier and user to determine the ratio of combustible and incombustible matter in the given coal. Table 1 gives the various constituents present in the given coal analyzed for proximate analysis with the procedure as described in section 2.1.

Table 1: Proximate analysis of the coal sample

S.No	Content	% Composition
1	Moisture	0.97
2	Volatile matter 5	18.96
3	Ash 8	33.63
4	Fixed carbon 15	46.43

3.2. Ultimate analysis of coal

It is an elemental analysis that determines the quantities of carbon, nitrogen, hydrogen, sulfur and oxygen content that were present in a solid fuel. It is also one of the measure that signifies the quality of the coal. Table 2 shows the ultimate analysis of the coal sample as described in section 2.2.

Table 2: Ultimate analysis of the coal sample

S.No	Composition	value
1	% C	54.55
2	% H	3.37
3	% N	1.36
4	% S	0.44
5	% O	5.68
6	Gross calorific value	4258 kcal/kg

3.3. Coal indices

The free swelling index is an indication of its use as a coking coal. It is very significant in determining the plastic properties of the coal. Coking index is one of the significant indices that classifies bituminous coals in the classification of coals. It signifies the binding strength among coal particles and also the binding strength between the coal particles and inert particles present. Grindability index is a measure of grindability nature of the coal. The minimum grindability index implies the hardness of the coal and hence is less grindable. Specific gravity of the coal is one of the measure that is used to classify and quantify the coals available in the market. Table 3 shows the various indices that are measured using the methods described through Sections 2.3, 2.4 and 2.5.

Table 3: Coal Indices values

S.No	Indexing	value
1	Free swelling Index	1
2	Caking index	9
3	Hard grove Grindability Index	81
4	Specific gravity	1.5

3.4. Packing density of Coal sample with varying size parameter

Packing density of particles is a measure of size and shape of the particles. Irregular particles often occupy packing densities less compared to the spherical particles. The packing densities of various sized particles were measured and presented in the Table 4 given below to approximate their similarities towards regular shapes.

Table 4: Proximate analysis of the coal sample

Absolute size constant(K)	50	60	70	80	90
Packing density (g/ml)	0.64	0.67	0.72	0.75	0.80

3.5. Rheological analysis of coal water slurries

After the preliminary analysis of coal, for coal water slurries, it is important to analyze the flow behavior. The flow behavior analysis of CWS was carried out at various concentrations of coal, various size constants various shearing times and various temperatures. This analysis include the dependency of shear rate with viscosity or shear stress.

3.5.1. Rheological properties of CWS at a value of absolute size constant and at different concentrations At a given absolute size constant and varying concentrations of coal, the rheological analysis is done by varying shear rate of 0 to 1000 s^{-1}.

It is evident from Figure 1 that the CWS solutions at all values of absolute constant K exhibit shear thickening nature. At low shears these solutions have less tendency to obstruct the flow field. However, at high shears, these particles become reluctant to adopt to the flow field there by exhibiting the resistance. Due to this fact the stresses will be developed in the flow resulting the shear thickening nature of the fluid.

At any value of K, an increasing trend of shear is evident with an increase in concentration of the CWS. At higher concentrations it is observed that the sher stresses exhibits non trendy behavior with the shear rates. This might be due to the excessive concentration of the slurries that actually effect the flow behavior abnormally. Even with the increase in the value of absolute size constant K, the viscosities have almost remain unchanged with little change in their shear stress values, not to the order of magnitude. However, at all values of K and higher strains, the transformation from likely Newtonian to non Newtonian behavior might be due to the spacial rearrangement of particles in the mixture when the voids are filled by the small particles present.

3.5.2. Rheological properties of CWS at a concentration with varied values of absolute size constant To investigate the effect of K on rheological properties, at a given concentrations and varying absolute size constants of coal, the rheological analysis is done by varying shear rate of 0 to 1000 s^{-1}.

It is again very clear from the Figure 2 that all the samples shows shear thickening behavior at varied value of K and at different concentrations. It can be also seen that there is shift of stresses almost two orders of magnitude with an increase in concentration from 55 % to 70% of CWS. It is known the presence of solid particles effect the flow field by increasing the viscosity and shear, and hence the increase is proportional to the number density.

It is also evident from the Figure 2 that though the K values has no significant effect on the stresses, the variations of the K values however does not follow any any trend. It could be observed that at any concentration of the CWS, the variation of K never followed any trend as shown via Figures 2a to 2d. This nature of the CWS might be due to spacial arrangement of the particles with in the voids that determines the development of stresses when the fluid is under the influence of the shear strains. It is now important to know weather the size, shape and density of the particles have effect in the determination of flow properties of the CWS by looking at the above results.

IOP Conf. Series: Materials Science and Engineering 1126 (2021) 012027 doi:10.1088/1757-899X/1126/1/012027

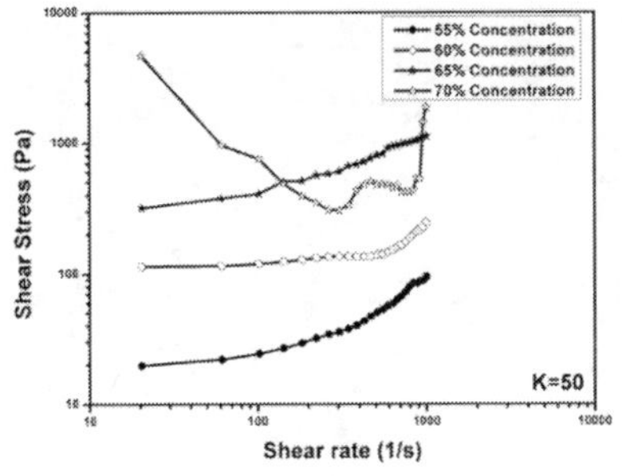

(a)

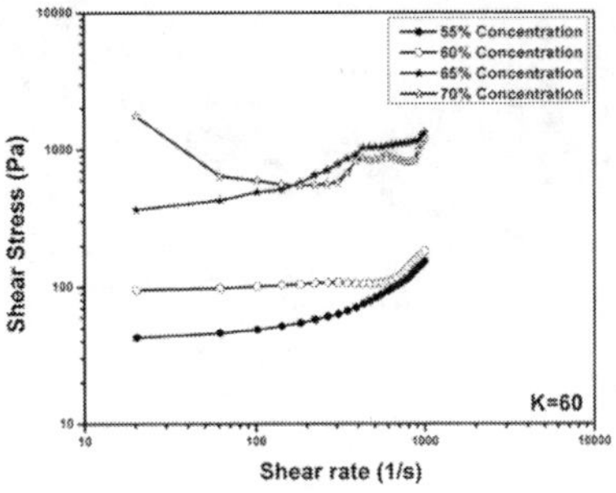

(b)

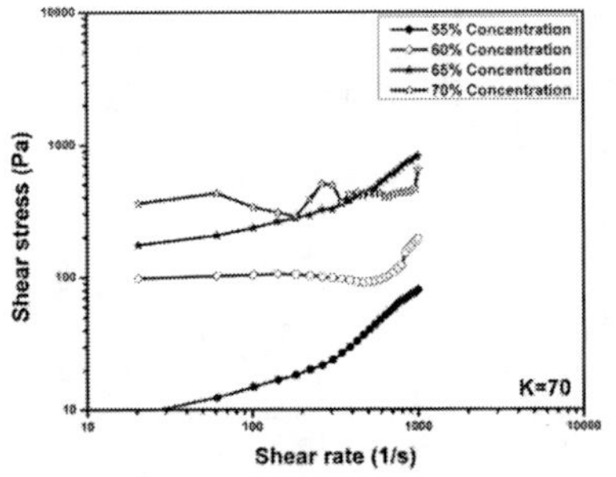

(c)

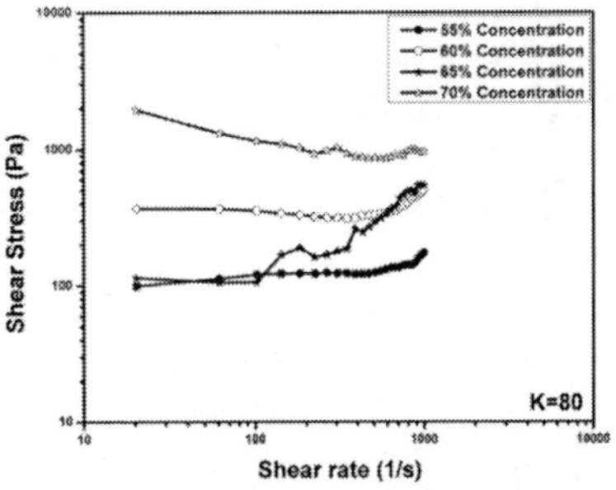

(d)

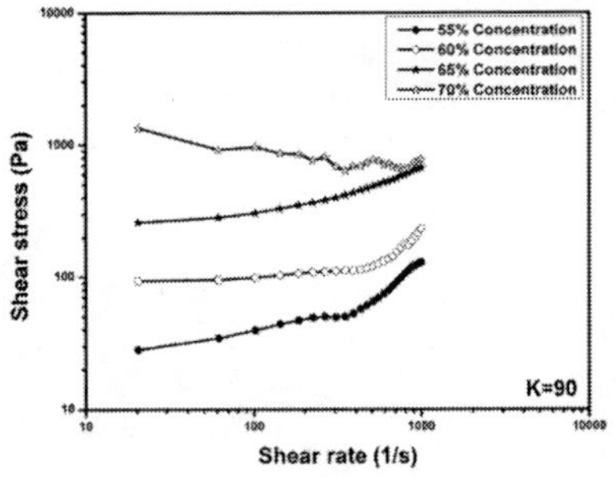

(e)

Figure 1: Variation of shear stress with shear rate for various concentrations of CWS at (a) K=50 (b) K=60 (c) K=70 (d) K=80 (e) K=90

3.5.3. Effect of granularity on rheological properties of CWS To test the effect of the granularity of the particles on the rhological properties of the coal water slurries, two different sizes of the particles were chosen. The one being the particles that pass through the 100 mesh screen in the Taylor's standard screens and the other being the particles that pass through that of 200 mesh screen. The former sample contains a mixture of sizes that vary from a very large size of 0.149 mm to a smaller sizes up to microns. It is a mixture of different sized particles having different shapes. The later sample contains particles that has a maximum size of 0.074 mm to particles of micron sizes. The later size particles are more or less have similarity in shapes as their size is very less compared to the former sample.

Figure 3 shows the variations of shear stresses with the varying shear rates. It is evident

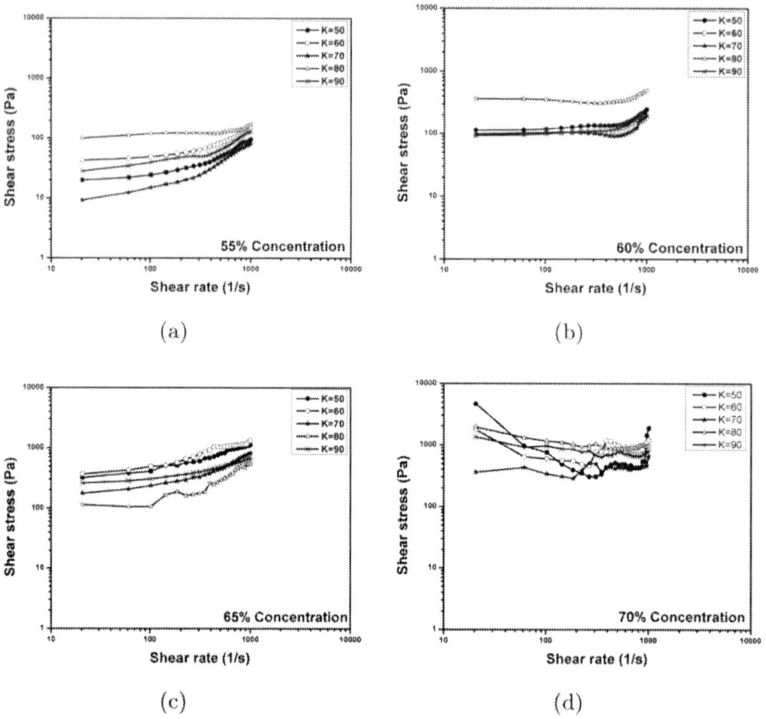

Figure 2: Variation of shear stress with shear rate for various K values at (a) 55% concentration (b) 60% concentration (c) 65% concentration (d) 70% concentration

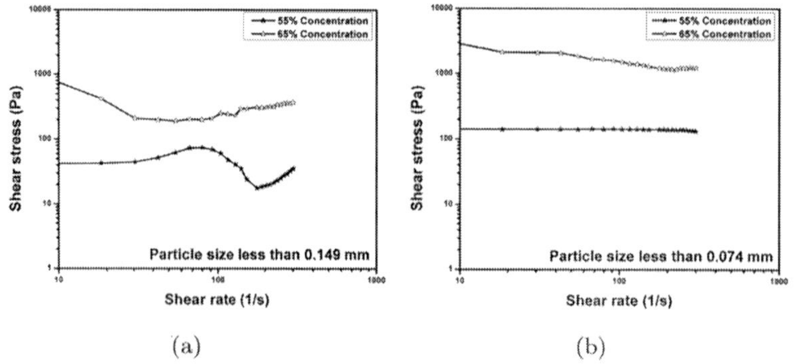

Figure 3: Variation of shear stress with shear rate for granularity of the particles whose sizes are (a) less than 0.149 mm (b) less than 0.074 mm

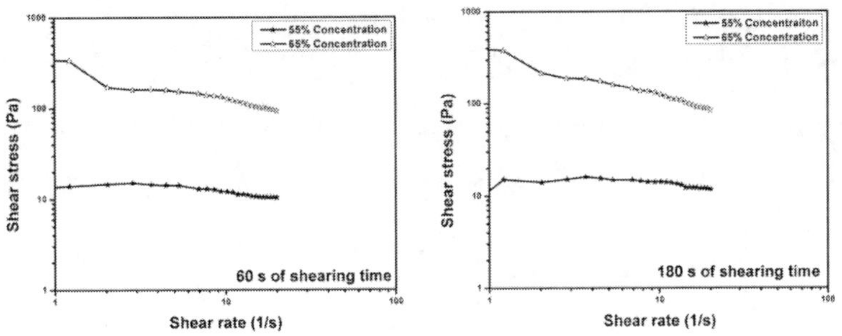

(a) (b)

Figure 4: Variation of shear stress with shear rate at shearing times of (a) 60 seconds (b) 180 seconds

from Figure 3a that the stresses varies abnormally with the variation in shear rates. This might be due to the inter particle interactions and self assembly during the straining of CWS as these mixtures have different sized and shaped particles varying from a larger size of 0.149 mm to micron size. This nature of trend is observed for both concentrations of 55 % and 60% of CWS. However, from Figure 3b shows that the variations of stresses are nullified. This nature could happen when there is more or less uniform sized and shaped particles, that particularly does not effect the self assembly of particles in the flow field. From the above results it can be concluded that the size and shape of the particles have its effect on the rheological properties of CWS.

3.5.4. Effect of shearing time on CWS To investigate the effect of shearing time on coal water slurries, the CWS samples of 55% and 65% concentrations were chosen. These samples were kept under the influence of shear for a time period of 60 seconds and 180 seconds. Then they are subjected to the rheological analysis by varying the shear strain and measuring the corresponding shear stress. It is to be noted that the variations of viscosity with shear rate is analyzed at lower shear rates up to 300 s^{-1}.

It can be seen from Figure 4a and Figure 4b that the shear stresses are almost similar for both the concentrations of the CWS at 55% and at 65%. There is no significant interplay among the particles where the shearing effect is observed. Hence it can be said that the shearing has no effect on the rheological properties of CWS at almost any concentrations.

3.5.5. Effect of temperature on CWS To investigate the effect of temperature on CWS, two different concentrations of CWS, 55% and 65% were chosen and the viscosity of these samples at temperature of 30^0C, 50^0C and 70^0C were measured.

It can be seen from Figure 5a and Figure 5b that the viscosity is decreasing with an increase in the shear rate. The decrease is evident in both the cases with the two orders of magnitude. However, as the concentration of the CWS has its effect on viscosity, the 65 % concentration slurry have higher values of viscosity when compared to that of the 55 % concentrations of CWS. It is very clear from Figure 5a and Figure 5b that the viscosity is not a function of temperature rather it is function of concentration of CWS.

4. Conclusions

The proximate analysis of the coal showed a moisture content of 0.97%, volatile matter of 18.96%, ash content of 33.63% and fixed carbon of 46.43%. The ultimate analysis of the coal

IOP Conf. Series: Materials Science and Engineering 1126 (2021) 012027 doi:10.1088/1757-899X/1126/1/012027

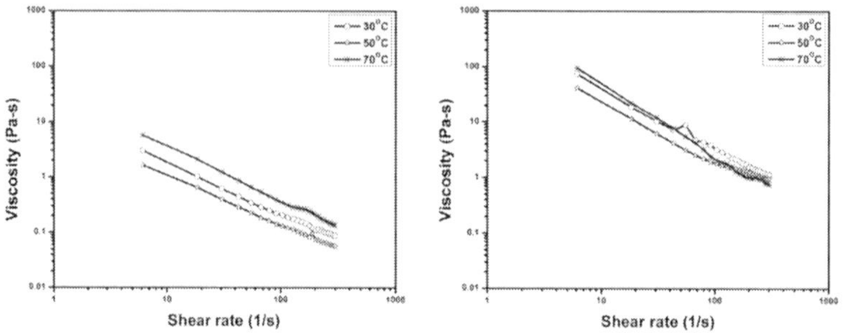

(a) (b)

Figure 5: Variation of viscosity with shear rate at temperatures of 30^0C, 50^0C and 70^0C for (a) 55 % CWS concentration (b) 65 % CWS concentration

sample showed 54.55% C, 3.37%H, 1.36%N, 0.44%S, 5.68 %O with a gross calorific value of 4958 kcal/kg. The analyzed coal sample showed a free swelling index of 1, caking index of 9, hard groove grindability index of 81 with a specific gravity of 1.5. The rheological analysis of CWS at a value of absolute size constant and at varied concentrations showed unchanged values of viscosity. However at higher strains near to 1000 s^{-1}, shift was observed from Newtonian nature to non-Newtonian nature which might be due to the self assembly of the particles of different sizes that were present in the sample. At a given concentration with varied absolute size constants, no significant effect on stresses variations were observed. However, no trend was observed with an increase in the absolute size constant which might be due to the self assembly of small particles within the voids of the bigger particles. The granularity of the particles has shown significant effect on rheological properties. Smaller particles due to their similarity in size and shape showed a complete Newtonian behavior. The shearing time of the CWS has no effect on the stress strain relation. The temperature has no significant effect on the rheology of the CWS. However, there is no trend of stress strain relation with temperature changes.

References

[1] Pommier L, Frankiewicz T and Weissberger W 1983 Coal water slurry fuels: an overview Tech. rep. Occidental Research Corp., Irvine, CA (USA)
[2] Xu Z, Chong L, Wang W, Chen Y, Zhang R *et al.* 2009 Coal water mixture preparation technology and application in replacing oil to generate electricity *2009 Asia-Pacific Power and Energy Engineering Conference* (IEEE) pp 1–5
[3] Cheng J, Zhou J, Li Y, Liu J and Cen K 2008 *Fuel* **87** 2620–2627
[4] Boylu F, Dincer H and Ateşok G 2004 *Fuel Processing Technology* **85** 241–250
[5] Roh N S, Shin D H, Kim D C and Kim J D 1995 *Fuel* **74** 1220–1225
[6] Tiwari K K, Basu S K, Bit K C, Banerjee S and Mishra K K 2004 *Fuel Processing Technology* **85** 31–42
[7] Hai-long Y, Chao Z, Jian-zhong L and Ke-fa C 2008 *Energy & fuels* **22** 1170–1173
[8] Laskowski J 1999 *Coal Perparation* **21** 105–123
[9] Natoli J, Mahar R C and Bobsein B R 1985 Polycrylate thickeners for coal water slurries: Slurry formation, stability and rheology *International Chemical Engineering Symposium Series* vol 95 pp 17–36
[10] KUROSE N 1988 New energy and industrial technology development organization *Proceedings of the VIII. International Symposium on Alcohol Fuels* (Sanbi Insatsu Co., Ltd., Tokyo)
[11] Tang C, Liu W, Xu Y and Liu Q 1998 *Power Eng* **18** 70–4
[12] Delagrammatikas G and Tsimas S 2004 *Chemical Engineering Communications* **191** 1362–1378

IOP Conf. Series: Materials Science and Engineering 1126 (2021) 012028 doi:10.1088/1757-899X/1126/1/012028

Comparative analysis on strains in asphalt pavement design using linear elastic and viscoelastic theories.

O Mora[1], F Palomino[1], M Figueroa[1], D Borrero[1] A Castro[2] C Cordoba[1], J Heilbron[1]

[1] Civil and Ambiental Department, Universidad de la Costa, Calle 58 # 55 - 66. Barranquilla, Colombia
[2] Civil and Ambiental Department, Universidad del Norte, km 5 vía puerto Colombia. Barranquilla, Colombia
omora@cuc.edu.co

Abstract. In Colombia it is common to design pavements using the AASHTO 93 method and to complement it with an elastic analysis of the deformations that cause fatigue and rutting; this has repercussions on the behavior of the structure since it does not take into account the viscoelastic behavior of the asphalt mixtures, In this research, a comparison of three structures at different velocity ranges is made to compare the variation in fatigue and rutting concerning the traditional method of analysis in Colombia and to analyze the differences that may occur in linear elastic analysis and viscoelastic analysis of rutting and fatigue.

1. Introduction

In Colombia, over the years, empirical methodologies have been used for the design of flexible pavement structures, such as the AASHTO 93 methodology [13], however, these designs have an empirical nature due to the design methodology was born from a large-scale test in the United States where through statistical analysis they determined pavement thicknesses from the recorded data and to an extent do not reflect the reality in the Colombian context where wheatear and traffic operating conditions may be different from those considered in the AASHTO Road Test.

In Colombia, the empirical mechanistic methodology is used for the verification of AASHTO pavement designs, which is known as the rational design methodology for flexible pavements. This methodology is based on the calculation of stress and strain to determine the fatigue and rutting life of asphalt pavement structures. This procedure is widely used in Colombia to design flexible pavements, compute layer thicknesses, and analyze strains related to fatigue and rutting in the pavement structure.

One of the oldest failure modes in flexible pavements is fatigue [10] [14]. Fatigue in flexible pavements is cracked in the bottom of the asphalt layer that over time are reflected in the surface, either longitudinally, transversely, or can turn into a system of interconnected cracks. Fatigue life in pavements is governed by the tensile strains generated at the bottom of the asphalt layer, because if the strains due to vehicular traffic exceed the maximum strain allowed by the asphalt concrete, the pavement fatigues, causing top-down cracks or bottom-up cracks [4] [6]. It is relevant to determine the impact of performing a viscoelastic analysis of the fatigue in asphalt caused by different load application speeds or temperature variations to know in what proportion it affects the analysis of these strains only as an elastic material.

On the other hand, rutting is a manifestation of both permanent strain and a surface impairment in flexible pavements, pavement design should aim to eliminate or reduce rutting in the pavement surface over a given period. The rutting criterion is based on limiting the vertical strain produced at the top of the subgrade because if the produced strain is greater than the allowed by the subgrade, rutting will occur [11]. Rutting can cause problems in the operating conditions of the road and it is not known what the impact of analyzing it only from an elastic perspective without considering the viscoelastic behavior of the asphalt material, variations in temperature or the frequency of load application can vary the dynamic modulus of the asphalt concrete affecting rutting behavior.

Most analytical models used for the design of flexible pavements are based on a linear elastic theory which assumes that each layer is homogeneous, isotropic, linearly elastic with a constant modulus of elasticity and the surface layer is free of shear and normal stresses. [12] However, it is interesting to know the response of the asphalt pavement in terms of stresses and strains, considering viscoelastic theories to compare with traditional elastic linear methods.

2. Input Data and Analysis Methodology

For making a comparison of fatigue and rutting in flexible pavements, a series of predefined variables were selected to calculate the thicknesses of the structure through the AASHTO 93 method [1], which is one of the most widely used in Colombia. The analysis methodology to consider the elastic and viscoelastic behavior of the material consists of predefining the input variables for the AASHTO 93 method, among the main ones are the traffic load, the elastic modulus of the granular layers, pavement type, and the average temperature, in addition to the remaining parameters such as serviceability and others specific to the AASHTO design method, these are shown in the following Table 1 in which some recommendations of INVIAS an AASHTO 93are specified. [1] [2] [3] [5] INVIAS is the governmental body responsible for roads in Colombia.

Table 1. Parameters and Values in AASHTO Design

Parameters		**Values**	**Recommendations**
Reliability	R%	95%	Main roads
Standard deviation	So	0.49	Considering transit variance
Initial serviceability	PSI Initial	4.2	Flexible pavements
final serviceability	PSI Final	2.5	Main roads
Subgrade resilient modulus	MR	10.500	Correlation CBR 7%
Asphalt structural coefficient	a1	0.3	Recommended by INVIAS for temperatures between 20-30 Celsius. [2]
Base structural coefficient	a2	0.14	Recommended by AASHTO for 28.000 psi granular base [1]
Subbase structural coefficient	a3	0.10	Recommended by AASHTO for 14.000 psi granular subbase [1]

Asphalt drainage coefficient	m1	1	Recommended by AASHTO for asphalt mix [1]
Base drainage coefficient	m2	1	Selected Value
Subbase drainage coefficient	m3	1	Selected Value

The procedure for the comparative analysis of strains taking into account the elastic and viscoelastic behavior of the asphalt pavement begins with the definition of the input data among which it was chosen to vary the traffic levels in three groups to obtain three pavement designs of different thicknesses using the AASHTO 93 method to compare fatigue and rutting in three structures of different thicknesses, Also, it has been chosen to vary the load application frequency and not the temperature, which will be kept constant, this is possible because the 3D Move Analysis software has the master curve of the selected asphalt, this allows us to know the asphalt modulus by varying the temperature or the load application frequency, for this analysis five levels of traffic speed were taken into account from 20 kph to 100 kph, which are the operating range of the roads in the region. The procedure for the comparative analysis of fatigue and rutting deformations is summarized in the following figure 1.

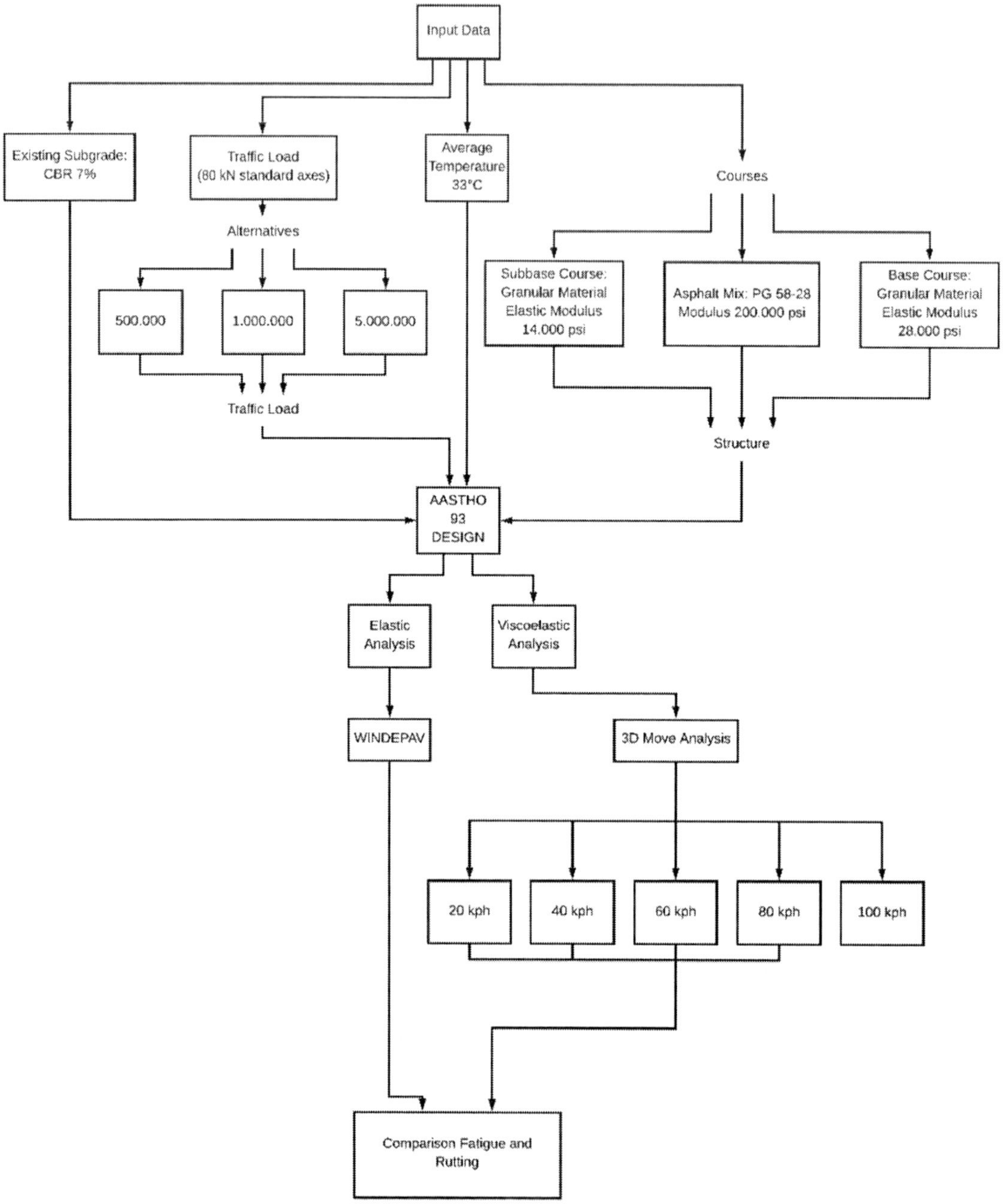

Figure 1. Comparative analysis procedures schema

The three pavement structures have a surface course with PG 58-28 asphalt, a base course in granular material with a modulus of 28,000 psi, a subbase in granular material with a modulus of 14,000 psi over a subgrade with a CBR percentage of 7% or a resilient modulus of 10,500 psi using the Heukelomp and Klomp correlation [8]. The asphalt modulus for the elastic analysis consists of the value at the average temperature in the Caribbean region of Colombia and an average speed of 60 kph as an intermediate value between urban and extra-urban roads.

IOP Conf. Series: Materials Science and Engineering 1126 (2021) 012028 doi:10.1088/1757-899X/1126/1/012028

The temperature chosen for the thickness designs is an average of 33°C [7] (because it is the average temperature in the Caribbean region where Barranquilla is) and varying the speed from 20 kph to 100 kph for the evaluation considering the viscoelastic behavior of the asphalt.

The software used for the pavement design is "Método AASHTO para el diseño de pavimentos" by engineer Luis R. Vasquez which allows the calculation of the thicknesses of the structures, an image of the configuration of the software is shown in Figure 2 the software is only available in Spanish. For the fatigue and rutting analysis considering only the elastic behavior of the material, the WINDEPAV software, and for the viscoelastic analysis the 3D MOVE ANALYSIS software [9].

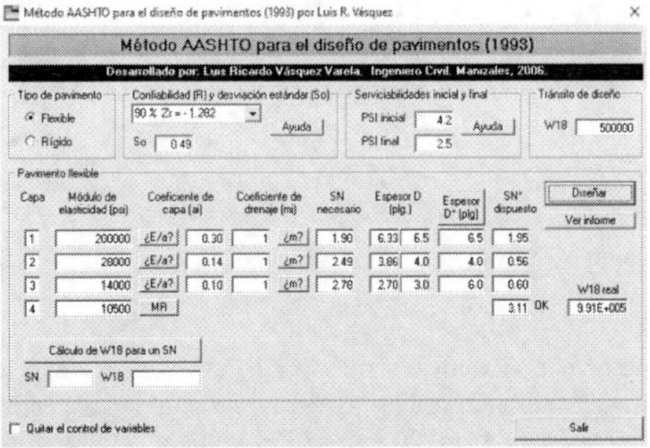

Figure 2. Software configuration for AASTHO design

The master curve of the selected asphalt is the following shown in Figure 3 which is entered into the 3D MOVE ANALYSIS software used for the deformation analysis considering the viscoelastic behavior of the asphalt.

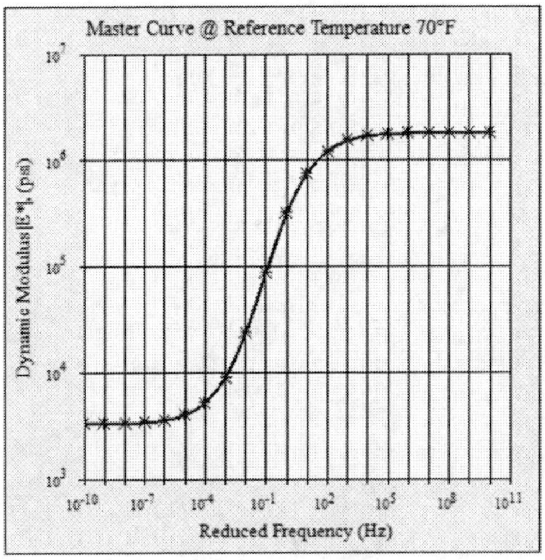

Master Curve equation

$$\log E^* = \delta + \frac{\alpha}{1 + e^{\beta + \gamma[\log(t) - c(\log(\eta) - \log(\eta_{T_r}))]}}$$

Shifting equation

$$log|a(T)| = aT^2 + bT + c$$

Fitting Parameters

Parameter	Value
δ	2.7403
α	-0.9535
β	0.5203
γ	0.8715
c	1.5542

Shifting Parameters

Parameter	Value
a	0.0004
b	-0.1424
c	8.2668

Figure 3. Master Curve for PG 58-22 asphalt.

3. Results

The three different structures obtained from the software for the AASHTO 93 method are as follows, Figure 4. The characteristics of the layers are the same for the structures, the only thing that changes are the value of the number of standard axes that the structure must support.

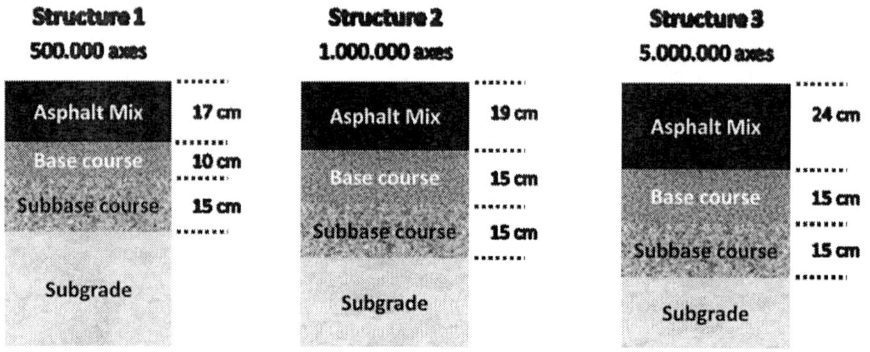

Figure 4. Final design structures at different traffic levels using AASHTO 93 method design.

The results of the deformation analyses in WINDEPAV and 3D MOVE ANALYSIS software are summarized in the following table according to the type of analysis performed in each of the software and for the different speed values from 20 kph to 100 kph, the values of the deformations of the tensile strains at the bottom of the asphalt layer (fatigue) and the compressive strains above the subgrade (rutting) are shown in Table 2 the values are expressed in microstrain.

Table 2. Fatigue and Rutting strains in the three structures in different analysis

	Strain	Structure 1	Structure 2	Structure 3
Elastic Analysis	Fatigue	-287	-246	-193
	Rutting	486	388	305
Viscoelastic Analysis 20 KPH	Fatigue	-350	-291	-235
	Rutting	627	500	425
Viscoelastic Analysis 40 KPH	Fatigue	-311	-256	-206
	Rutting	580	461	388
Viscoelastic Analysis 60 KPH	Fatigue	-283	-236	-189
	Rutting	555	440	365
Viscoelastic Analysis 80 KPH	Fatigue	-266	-222	-176
	Rutting	535	420	350
Viscoelastic Analysis 100 KPH	Fatigue	-254	-210	-168
	Rutting	521	414	339

4. Analysis

First, we compare the values for fatigue in the shown Figure 5.

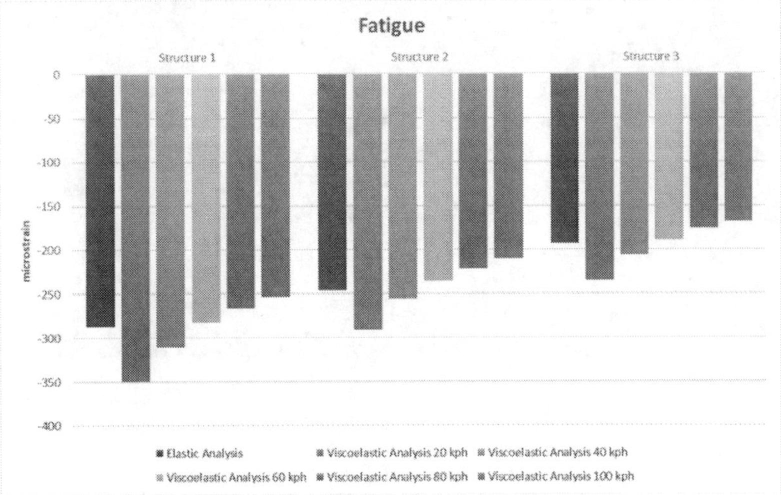

Figure 5. Fatigue strain comparison.

For fatigue, it can be seen that the design of pavements with the AASHTO 93 method and only an elastic analysis of the asphalt layers can consider fatigue strains much lower than the real behavior of the asphalt for speeds below 40 kph, being very similar for the value of 60 kph because the modulus considered in the elastic analysis and the calculation of layer thicknesses employing the AASHTO 93 method is that of an asphalt layer for a modulus of 200,000 psi equivalent to the value of 60 kph and a temperature of 33°C according to its master curve. However, if the variations in the viscoelastic behavior of the asphalt are not taken into account, fatigue values may be much lower than the real values for the operating conditions of the road at lower speeds.

It is also important to clarify that when the analysis is carried out at low speeds, a greater deformation is obtained in the viscoelastic analysis compared to the elastic one, given that at lower speeds, the asphalt performs with a lower dynamic modulus [15] and therefore more strains are generated.

While for rutting we can observe that not considering the viscoelastic behavior of the asphalt considering a fully elastic behavior will outcome in rutting results always lower than the results obtained under viscoelastic behavior, even if the comparison is made at the same load application speed and the same average environment temperature. Note in Figure 6 that the value of the viscoelastic analysis rutting at 60 kph is much higher than the elastic analysis in all three structures regardless of the volume of traffic that the structure must bear.

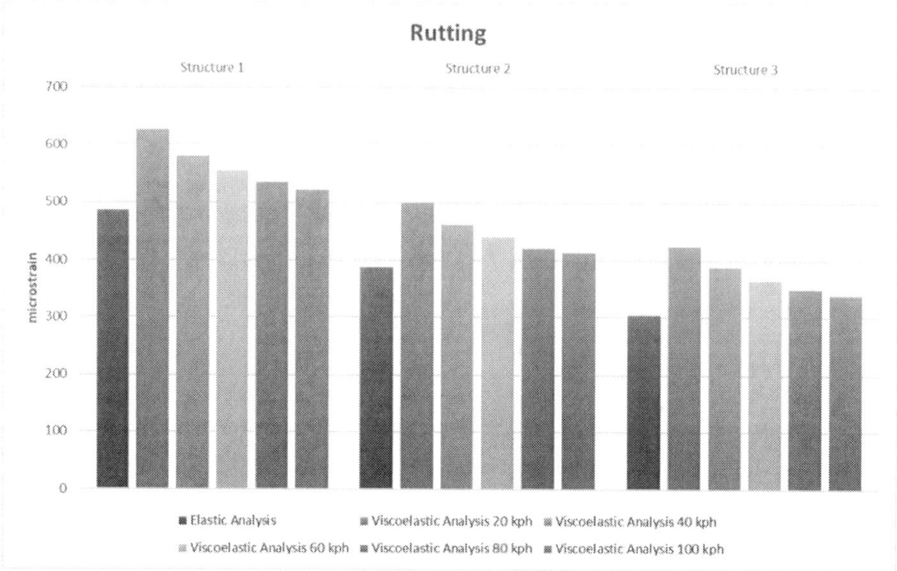

Figure 6. Rutting strain comparison.

5. Conclusion

In Colombia, most flexible pavement designs are carried out using the AASHTO 93 method to determine the thicknesses of the structure, and then a check is made by the rational methodology in which stresses and strains are calculated at the bottom of the asphalt layer and above the subgrade to estimate the service life of the pavement due to fatigue and rutting, respectively, but only considering the linear elastic behavior of the asphalt. This represents a serious problem especially for rutting analysis in flexible pavements used in warm regions, where it is more prone to suffer from this degradation due to the lower modulus value of the asphalt mix at high temperatures [15] However, fatigue can represent a problem for speeds lower than the design speed of the flexible pavement because of the cases where values well above those estimated in the elastic analysis are found and an advantage for values higher than the design speed of the pavement where fatigue values are estimated well above the values considering the viscoelastic behavior.

Due to the variation in results and the differences between viscoelastic and elastic analysis for different speed and temperature values, it is necessary to review pavement designs that are currently performed only under elastic analysis and that do not consider operating conditions below and above the preset speed and temperature values.

Therefore, it demonstrates the importance of performing designs under viscoelastic analysis and avoiding underestimation of rutting and fatigue values in flexible pavements that may represent problems in the short or medium-term due to variations in the operating conditions of the roads. Besides, it should be noted that one of the main limitations of this research is that the analysis is being carried out with theoretical assumptions that may vary from reality, among these variations are the characteristics of the materials and the characteristics of the traffic, It is therefore recommended for future research to analyze the results with field measurements of a pavement structure to analyze the behavior of the asphalt in real conditions and compare them with the proposed design for the structure, and that the analysis of the stresses and deformations only considers the elastic behavior of the asphalt, likewise, for future research to carry out this analysis using different types of asphalt; such as modified asphalt, warm mixes, recycled mixes, cold mixes, among others.

References

[1] AASHTO. (1993). Guide for Design of Pavement Structures. Washington D.C: American Association of State Highway and Transportation Officials.
[2] INVIAS (2014) Manual de diseño de pavimentos asfálticos para vías con medios y altos volúmenes de tránsito. Instituto nacional de Vías - INVIAS
[3] INVIAS. (2008). Guía metodológica para el diseño de obras de rehabilitacion de pavimentos asfálticos en carreteras. Bogotá: Instituto nacional de Vías - INVIAS
[4] Huang, Y. H. (1993). Pavement analysis and design.
[5] INVIAS. (2013). Especificaciones Generales de Construcción de Carreteras en Colombia. Bogotá: Instituto nacional de Vías – INVIAS
[6] Behiry, A. E. A. E. M. (2012). Fatigue and rutting lives in flexible pavement. Ain Shams Engineering Journal, 3(4), 367-374
[7] CIOH - Centro de Investigaciones Oceanográficas e Hidrográficas
[8] Heukelom, W., & Klomp, A.J.G. (1962). Dynamic testing as a means of controlling pavements during and after construction. Proc., of 1st Int. Conf. on Structural Design of Asphalt Pavements, USA.
[9] Nitharsan, R. (2011). Development of Windows-Based Version of the 3D-Move Analysis Software for Pavement Response Analysis (Doctoral dissertation).
[10] Moghaddam, T. B., Karim, M. R., & Abdelaziz, M. (2011). A review on fatigue and rutting performance of asphalt mixes. Scientific Research and Essays, 6(4), 670-682.
[11] Ekwulo, E. O., & Eme, D. B. (2009). Fatigue and rutting strain analysis of flexible pavements designed using CBR methods. African Journal of Environmental Science and Technology, 3(12).
[12] Nagakumar. N (2013). Applications of layered theory for the analysis of flexible pavement. International Journal of Research in Engineering and Technology
[13] Mora, O., Murillo, M., Rosanía, T., Castañeda Amashta, A. G., Pinto, R., & Padilla-Muñoz, A. (2020). Analysis of CBR design value selection methods on flexible pavement design: Colombia case study.
[14] Usman, R. S., Setyawan, A., & Suprapto, M. (2018, March). Prediction of pavement remaining service life based on repetition of load and permanent deformation. In IOP Conference Series: Materials Science and Engineering (Vol. 333, No. 1, p. 012089). IOP Publishing.
[15] Mora (2016). Desarrollar de una metodología para diseño de pavimentos que permita incorporar el efecto de la rugosidad en la carga dinámica transmitida al pavimento. Tesis de maestria. Universidad del Nortefd

IOP Conf. Series: Materials Science and Engineering 1126 (2021) 012029 doi:10.1088/1757-899X/1126/1/012029

Solar Energy Based LED Powered Light and Inverter Circuit with Different Types of Sensors

E. Siva Senthil[a], R. Yogeshwaran[b], B.S. Yeshwanth[b], S. Saira Banu[c], Ashwin Sudhakaran[d], Allwin Sudhakaran[d]

Corresponding Author: E. Siva Senthil, Email ID: sivasenthil.e@kahedu.edu.in
[a] Associate Professor & Head ,[c] Associate Professor, [b,c,d]Research Scholar, Department of Physics, Karpagam Academy of Higher Education, Coimbatore, TamilNadu, India

Abstract. In this article we have designed and constructed a simple yet effective Solar Energy Based LED Powered Light and Inverter Circuit with Different Types of Sensors. The Solar Energy Based LED Powered Light and Inverter Circuit with Different Types of Sensors are constructed using simple electronic components and the voltage vs time graph during the sunny, cloudy, and rainy days are analyzed. The observed values are plotted in graph and the constructed device is found to be working perfectly which shows that the device can be used for various domestic and commercial solar applications.

1. Introduction

Every single living thing on earth relies on lot of power paced in our sun. plants utilize this energy for photosynthesis, giving new oxygen and cleaning air. This energy additionally gives light and warmth giving fundamental portion of nutrient D for well evolved creatures. The sun additionally impacts water cycle and earth's climate. People convert this energy for many other applications such as cooking, heating and controlling regular machines.

Solar power systems have greater efficiency and thus affordable for residential and commercial customers particularly taking into account late oil spill in Gulf, more people look for alternative source of energy that will save both money and environment. By harnessing energy from natural renewable resources like sun, we are ecologically ensured by diminishing the use of other hurtful assets, for example, oil and coal.

A portion of the significant uses of sun powered vitality incorporates, solar powered buildings, distillations, water heater, power generation, pumps, cooker, solar furnaces, sun based animal products drier, thermal power production and green houses.

A material which discharges light energy when power or electric flow is gone through it is known as electroluminescence. In 21st century, the greatest discoveries include luminescence, incandescence, chemiluminescence, cathodoluminescence, triboluminescence, and photoluminescence.

Most commonly, semiconducting materials were made up of organic, inorganic or with metal dopants. This is the most common principles of all Light Emitting Diodes (LED). These light emitting diodes are p-n junction diodes which emits light in forward bias condition.

2. System Design and Description

Requirements

Solar panel

A 30 watts solar panel is used which is made up of polycrystalline material. The output obtained from this 30-watt solar panel is 15 volts, but in general we get 10-15 Volts. In case of peak hours i.e. from morning 10 o'clock to evening 4 o'clock it would-be equal to the efficiency level, where as it would be low during the cloudy days.

Solar energy based powered LED circuit

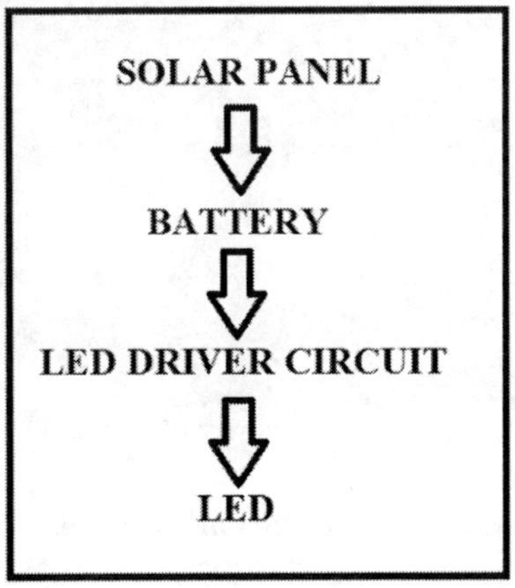

Batteries

Ordinary lead acid batteries are used to overcome the disadvantages such as low life time, high toxicity, low discharge, low efficiency and non-recyclability. In our case we used 12-volt Li-ion battery fixed under the solar panel to store D.C output.

Charge controller

Since the solar panel doesn't give a constant output, it can be achieved using Charge controller. The power cannot be directly sent to the battery as it is stabilized by means of charge controller. It is also used to prevent the overload of the battery and to extend the battery life time.

LED

The LED works behind the principle of electroluminescence. It states that in radioactive recombination of electron and pores produces photons. In this circuit a 12V LED is used.

Construction and Working

A 30 watts solar panel is constructed by using 30 individual poly crystalline silicon solar cells these individual solar cells are associated arrangement to get the voltage output in case of the current output we connect solar cells in parallel the solar cells produce a output of 10 to 15V this output is connected to a 12V lead acid battery which acts as a storage device. And also, a charge controller is fixed intermediately to control the output DC power and the battery is connected with variable resistor-controlled LED circuit. By using the regulator, we adjust the resistance value and the LED input is controlled here after the brightness level of the LED light increases or reduces which is depending upon the adjustable value of the resistance.

Fig 1. Constructed Solar energy based powered LED circuit

Solar energy-based inverter circuit with different types of sensor

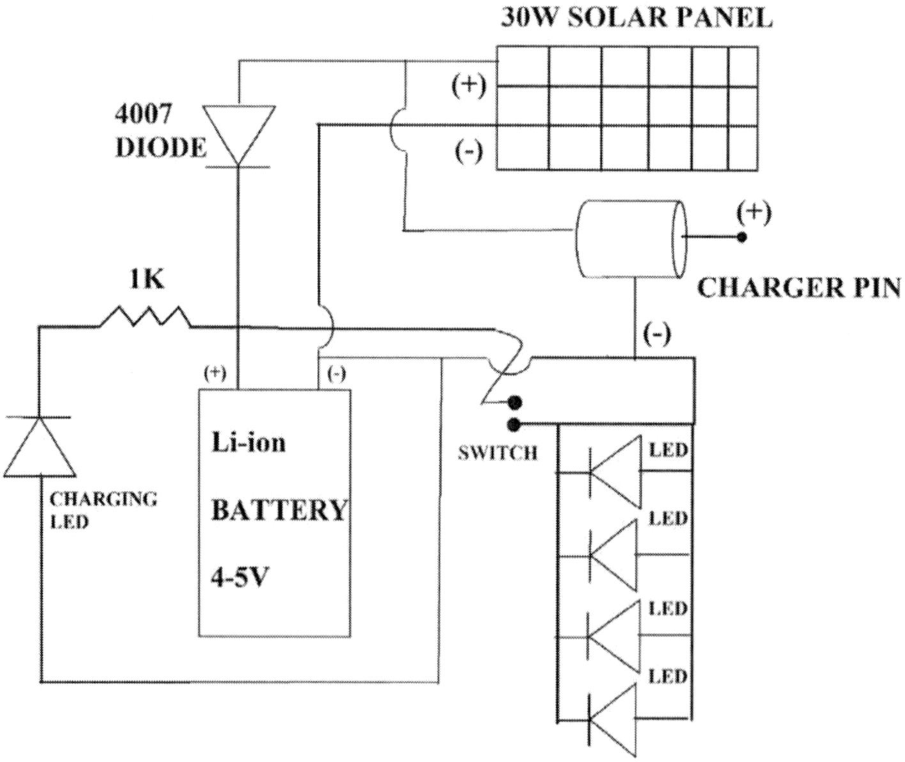

Fig 2. Circuit Diagram of Solar energy-based inverter circuit with different types of sensors

IOP Conf. Series: Materials Science and Engineering 1126 (2021) 012029 doi:10.1088/1757-899X/1126/1/012029

Block diagram

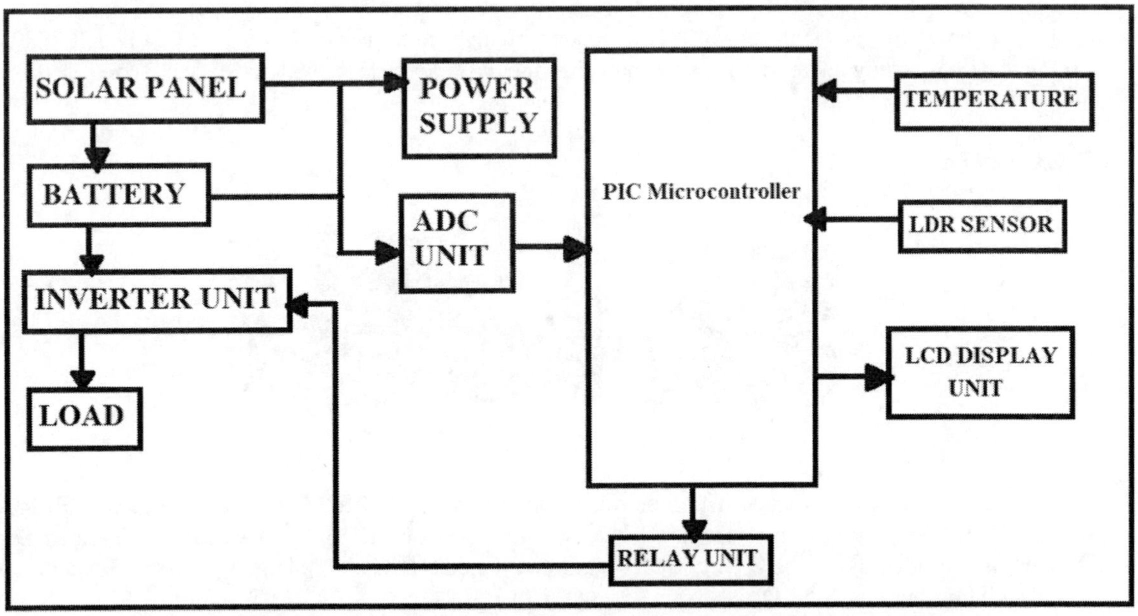

Fig 3. Square Diagram of Solar energy-based inverter circuit with different types of sensor

Fig 4. Circuit constructed Solar energy-based inverter circuit with different types of sensor

IOP Conf. Series: Materials Science and Engineering 1126 (2021) 012029 doi:10.1088/1757-899X/1126/1/012029

Components required

LM7085

In order to give uniform voltage, a popular voltage controller 7805 IC is used. The IC 7805 is placed in 78xx arrangement where xx represents the voltage yield providing +5 volts-controlled force supply with heat sink arrangement.

PIC16F877A

Fig 5. Chip diagram of PIC16F877A

PIC16F877A is a 8 piece microcontroller with compactable PIC16CS5X containing an EEPROM memory with self-programming capacity. The sequential port of PIC16F877A can be modified as Serial Peripheral Interface (SPI™), 2-wire Inter-Integrated Circuit (I^2C™) and as Universal Asynchronous Receiver Transmitter (USART).

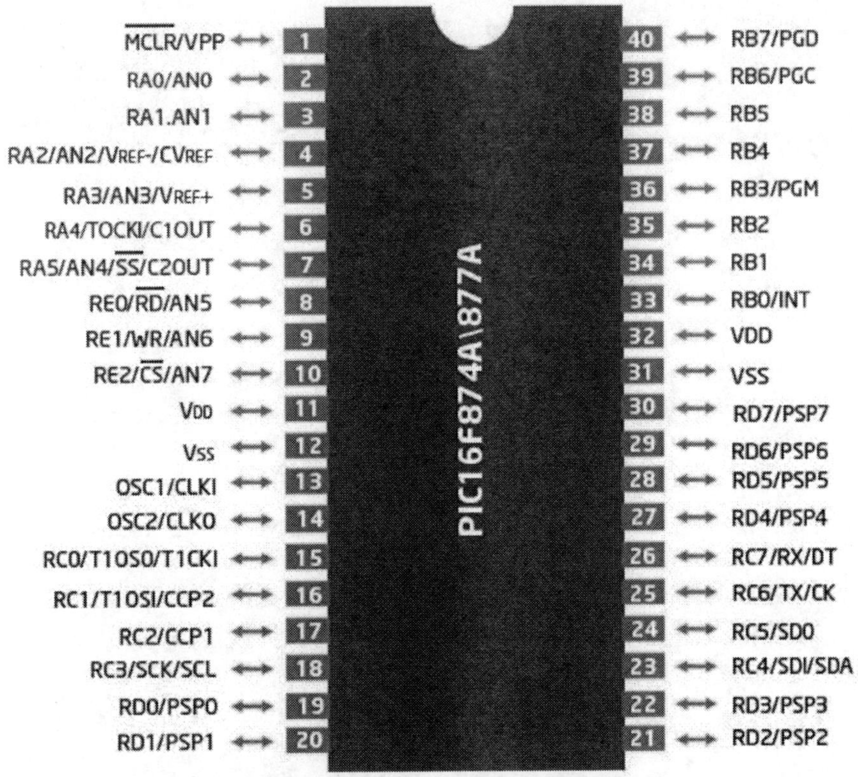

Fig 6. Pin Diagram of PIC16F877A

IC LM35

The IC LM35 is a temperature detector which converts the voltage to its relative centigrade temperature. By using this gadget, the user can also sense the temperature in kelvin instead of oC which has a huge benefit. This device doesn't involve any exterior calibration or trimming with workable temperature range of -55^{0}C to 150^{0}C.

BC547 Transistor

The NPN transistor BC547 is forward bias when the signal is connected to base and reverse biased on grounded base. It has an increase estimation of enhancement limits ranging from 110 to 800.

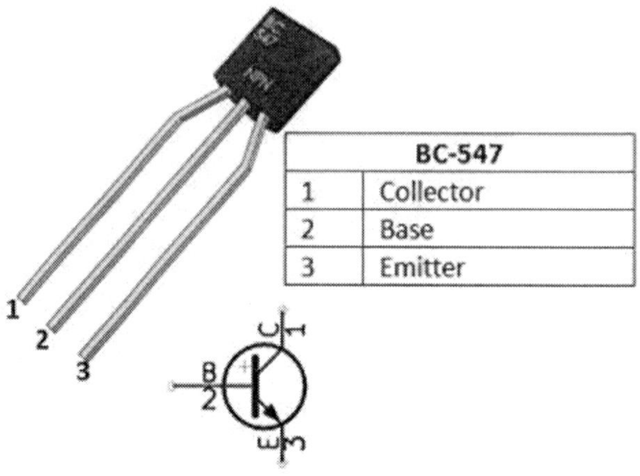

Fig 7. Chip Diagram of BC-547

A base current of 5mA has to be supplied since 100mA is the most extreme measure of collector current that can be utilized in this transistor.

Fig 8. Leaded LDR

Photo resistor

A photo resistor is a segment which are sensitive to light, changing its resistance when exposed to sunlight. These values for resistance may change over numerous sets based on the wavelength and intensity of the light.

In order to increase their light sensitive properties, photo resistors are made using semiconductors like cadmium.

IOP Conf. Series: Materials Science and Engineering 1126 (2021) 012029 doi:10.1088/1757-899X/1126/1/012029

CT COIL

In order to quantify the exchanging current, a Current transformer coil (CT) is used which creates current in the secondary coil relative to the current in the primary coil.

Construction and working

Solar panel is the primary energy source. The 3 watts solar panel is constructed using poly-crystalline silicon solar cells and the output is connected to the DC battery. A DC battery of 12V is placed to store the electricity energy by the solar panel and the electrical energy is stored in the DC battery now the DC current is converted in AC by using the inverter circuit here the sample inverter circuit is constructed using the n channel MOSFET, diode, capacitor and inductor coil. The relay unit is fixed with the circuit with the circuit to control the progression of the DC current, after the inverter changes over the DC to AC current (the 12V of DC is changed over into 220V of AC by utilizing the inverter) it directly given to the lead, where the AC light is associated.

The other part of the circuit comprises of LDR sensor which used to recognize the presence of the sun light once the light does not fall on the sensor the circuit gets initiated and the light associated with the circuit will glow.

The current transformer coil is connected to the AC current to find the amount of the current that used by the load connects to the circuit, which acts as the current sensor. LM35 integrated with the temperature sensor and it is used to find the amount of temperature in the circuit. In addition to that, the pulse with modulation is added in the pica microcontroller to save power and its used other for some applications.

These sensors are associated with the control unit where the programmed pica micro controller is connected, the PIC16F877A is programmed with the pulse modulation to increase the battery time and the temperature readings, voltage readings and the amount of energy that utilized by the AC lights are displayed in the LCD display and the thereafter the graph is plotted between

3. Result and Discussion

i. The solar power LED lights circuit with regulator. In this project we constructed the solar LED light with help of solar panel, charge controller, LED driver and regulator circuit. The 30W solar panel is constructed for the power source with a charge regulator to control the output power connected to the batteries. The LED driver is used to provide the constant current source for the LED's and the regulator circuit for adjustment of the brightness of the LED with the help of six adjustable resistance connected to control the brightness of the LED's its similar to the motion sensor. The six ways adjustable brightness will be useful for the time we need low brightness and we need high brightness we can adjust the regulator for the maximum brightness.

ii. The observed different voltage readings from the solar panel is plotted as graph with respect to the time, the voltage reading are noted for the several days which includes the sunny days, cloudy days and some days without proper sun light, there are three different graphs are plotted for sunny, cloudy and rainy days and we observed the different types of output values with respect to the time the output is high sunny days and moderate in cloudy and very low in extreme cloudy days.

Voltage and Time

The graph plotted between the voltage and time, as the voltage differs based on the temperature obtained on various intervals on time.

Graph plotted for Sunny day

1st Day

This graph shows the output of the solar panel during the sunny days and we get the different values which includes peak values and maximum output voltages, the peak value obtained during 12 PM - 1 PM where the sun is exactly at the center and the peak value reaches 14V and after the output reduces gradually when the sun starts to set in evening.

Voltage Vs Time graph

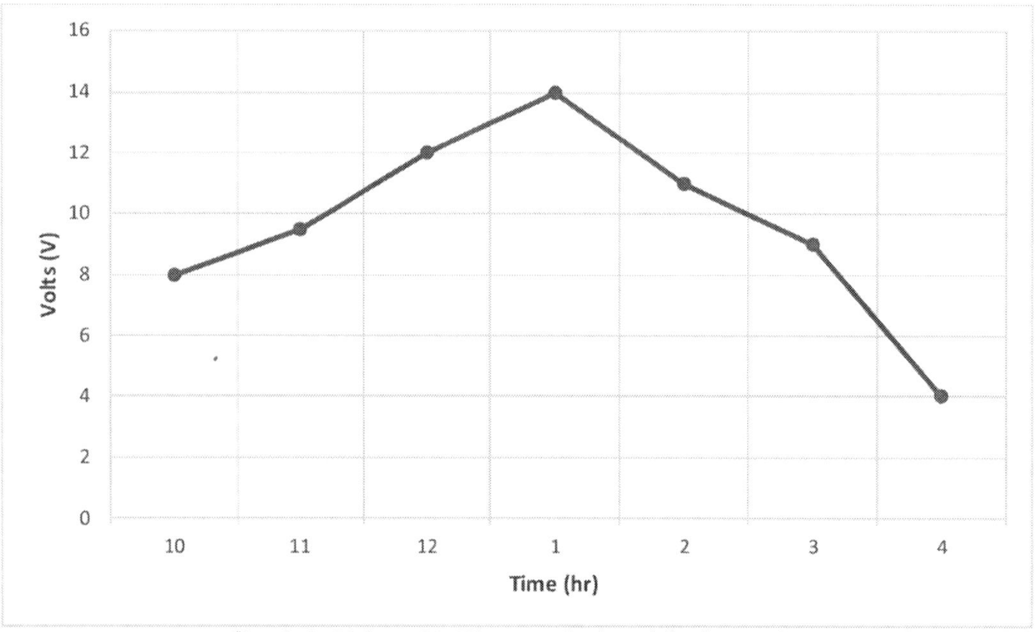

Graph 1. Voltage Vs Time graph plotted for Sunny day

Graph plotted for Cloudy day

2nd Day

The readings are noted for the cloudy day where the direct rays from the sun fall on the clouds and it gets scattered, thus the scattered rays fall on the solar panel and thus the voltage is not high as the energy we get in sunny day but it is moderate as the peak value reaches 5V during the cloudy days.

Voltage Vs Time graph

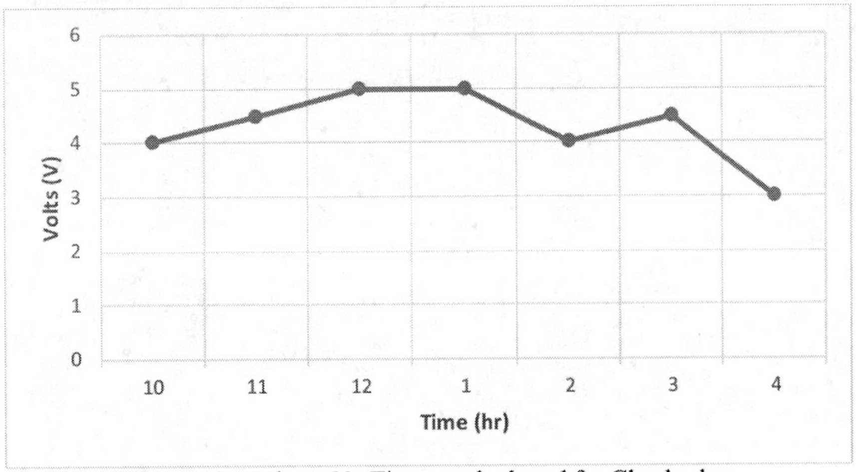

Graph 2. Voltage Vs Time graph plotted for Cloudy day

Graph plotted for the Cloudiest day or Rainy day

3rd Day

During the cloudy day or rainy day, the temperature will be very low is the solar panel produced electricity but it very low as compare to the sunny days.

Voltage Vs Time Period

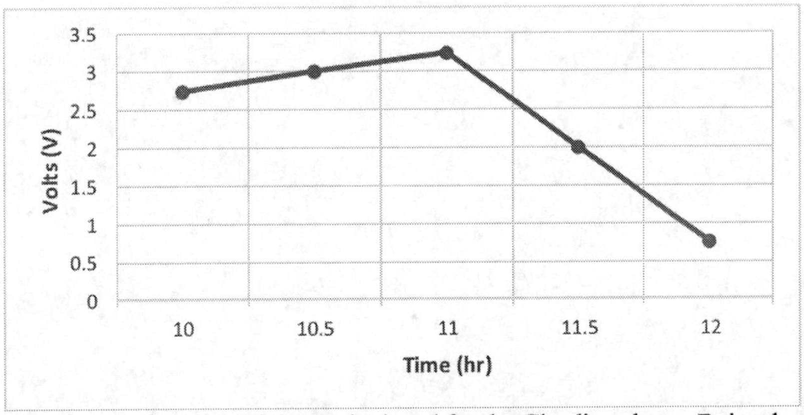

Graph 3. Voltage Vs Time graph plotted for the Cloudiest day or Rainy day

Fig 9. Off condition of Solar energy-based inverter circuit with different types of sensor

Fig 10. Display showing the readings during on condition of Solar energy-based inverter circuit with different types of sensor

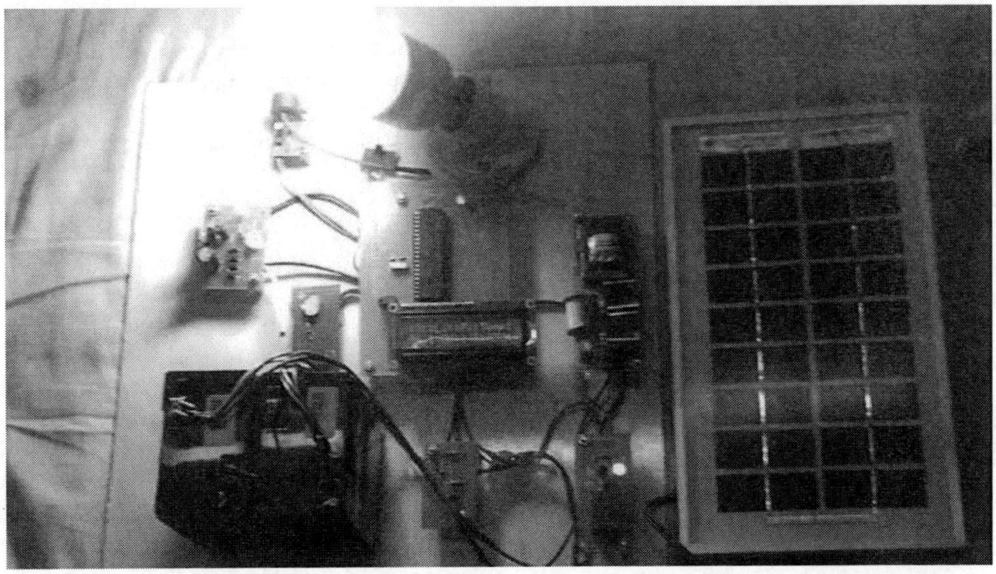

Fig 11. On condition of Solar energy-based inverter circuit with different types of sensor

4. Conclusion

1. In this project we constructed the solar LED light which resembles the ordinary solar street light but there are some potential advantages in these lights compared the ordinary street lights. The main advantage of the LED driver circuit, which guarantee the security of LED light, the driver circuit control the progression of current to the light and it resist the damage when the usual flow of current to the system and therefore the life time of the LED's will be high. Other advantage will be regulator circuit, by using their regulator circuit we can adjust the brightness of the LED in 6 ways its same like the motion sensor. The 6 ways adjustable brightness will be useful for the time we need low brightness and we need high brightness we can adjust the regulator for the maximum brightness. And other some advantages are low cost easily replaceable and low maintenance required for this LED light.

2. In this project we have constructed an inverter with different types of sensors including wireless sensor networks which can be used for data aggregation and multicast authenticated schemes.[41]. The solar panel produces a DC output and this DC is converted to AC by means of an inverter. This AC current is thus measured by CT coil which is connected to the inverter circuit. As the DC undergoes a transformation of AC by means of an inverter which is the basic principle behind it. One of the significant components of this project is the 7805-voltage regulator which manages the yield voltage at steady value. In addition to this we have fixed 2 sensors, one is LM35 that is the temperature sensor and a voltage sensor. These both sensors are fixed to measure the temperature and voltage at various intervals of time . As it varies based on climatic conditions as it reaches the peak value on sunny days and reaches lowest value on cloudy days and moderate when the intensity of sunlight is less.

5. Reference

1. D.Step Hani, 'prospects of SiC power devieces from the state of the art to future trends', proceedings of the 45th international power electronics conference, numberg, may 14-16, 2002,K-2
2. Jian-Song Chen, Sei-Hyung Ryu, Kornegay, K.T., IEEE journal of solid-state circuits, vol. 34, no.2, 1999,pp.192-202.
3. Chakrabarthi,S. (2000), Energy policy, Vol. 30, pp.33 42.
4. Denton, J. C. (1977) solar power systems, Energy Conversion, vol.16, pp.181-198.
5. Dracker, R. and P. De Laquil. (1996). Annual review of energy and environment, vol. 21, pp. 371 402.
6. Eduardo F Camacho, control of solar energy systems, London : Springer- verlag London Limited, 2012.
7. Puchenkin AV. Ivanovo : Ivanovo State Power University, 2011,-S.60-63.
8. Hohm DP, Progress in Photovoltaics : Research and Applications, vol. 11, no. 1, pp.47 -62, 2003.
9. Fermia N, Granozio D, Aerospace and Electronic Systems, IEEE Transactions on, vol. 43, no.3, July 2007.
10. Hamilton J. Careers in Solar Power.U.S. Bureau of Labor Statistics, Report 2 2011.
11. Jager W. Stimulating the diffusion of photovoltaic systems : a behavioural prespective. Energy Policy 2006; 34: 1935-43.
12. Nielsen TD, Solar Energy Materials and Solar cells 2010; 94 (10): 1553-1571.
13. Bruce T. Research Dissertation, ENG 4111/4112 Project 2011.
14. IEA- PVPS. Trends in Photovoltaic Applications a survey report of selected IEA Countries between 1992 and 2010. Report IEA-PVPS T1-20 2011.
15. Yilmaz S, Renewable and sustainable energy reviews 2015; 52: 1015-1024.
16. The data sheets of solar panel brands, 2016.
17. N. Sahraei S, J. Appl, Phys., Vol. 56, No. 8S2, p.08ME01, 2017.
18. BP Statistical Review of World Energy, << BP Energy Outlook,>> London, 2017.
19. M.Rosa- Clot, Solar Energy, vol.133, pp.3005-3014,2016.
20. J.A. Muaddi e M. A. Jamal, Renewable Energy, vol. 1,n. 1,pp. 31-352, 1991.
21. Y.Ueda, 23rd European Photovoltaic Solar Energy Conference and Exhibiton, Valentia, 2008.
22. R. Lanzafame, IEEE transactions on industrial electronics, vol.57,n. 7,pp. 2492-2498, 2010.
23. G. Tina, Energy 39 (1), 17-26,vol. 39, n. 1, pp.17-26, 2012.
24. M.Rosa -Clot, Solat Energy, vol.85, pp.2433-2442, 2011.
25. G. Tina, Journal of Solar Energy Engineering 139, vol. 139, p. 213, 2017.
26. H. Zondag, Renewable Sustainable Energy Review, vol.12, pp.891-959, 2008.
27. P.Dash eN.Gupta, Int.Journal of Engeneering Research and Applications, vol.5, pp.148-151,2015.
28. H.Khaing, Int. Journal of Electr Comput Electron Electron Commun Eng, vol. 8,p. 1328-32, 2014.
29. G.Notton, Renewable and Sustainable Energy Reviews, vol. 33, pp. 6070, 2014.
30. A. Aboushi e A.Raed, International Proceedings of Chemical, Biological and Environmental Engineering, vol. 87, p. 90, 2015.
31. S. Jimenez, TECCIENCIA DOI: http:/dx.doi.org/10.18180/tecciencia.2015.19.4, 2015.
32. Greitemann, J.; Systematishe Identifikation von Produktionstechnologien In: Zeitschrift fur wirtshaftlichen Fabrikbetrieb. P. 7-8, 2015.
33. Seidlitz, H.: Entwicklung von kraftflussgerechten Verbindungstechniken fur Mischbauweisen mit thermoplastischen Faserverbunden und Metallen. Dissertation. Technishe Universitat Chemnitz, 2013.
34. Gerstenberger, C. et al.: Materials Science Engineering Congress 2014 (MSE 2014). Darmstadt, 23rd-25th of September, 2014.

35. Weilin, H.: Beitrag zum Kerbspannungsverhalten anisotropy verstarkter Mehrschichtverbunde unter Berucksichtigung hygrothermisher Einflusse. Dissertation. Technische Universitat Chemnitz, VDI Verlag Nr, 342, 2016.
36. Cuntze, R.: Computer Science and Technology Bd. 64 (2004), pp. 487-516, 2004.
37. S. Surendran, V. SenthilKumar, E. Sivasenthil, Jour of Adv Research in Dynamical & Control System, Vol. 10 14-Special Issue, 2018.
38. Sivasenthil .E, SenthilKumar .V ,International Journal of Applied Research 2016; 2(8):1680-171.
39. E. Sivasenthil, V. SenthilKumar ,International Journal of Innovative Research in Science, Engineering and Technology Vol.5,Issu8,DOI:10.15680/IJIRSET.2016.0508076.
40. N. Preetha, S. Padmavathi, M. Shobana, B. Mahalakshmi, E. Sivasenthil, International Journal of Advanced Science and Technology, Vol. 29, No.7,, (2020), pp. 2637-2640.
41. Dr.S.Saira Banu, Dr. E. Siva Senthil, International Journal for Research in Applied Science & Engineering Technology, Vol 8. Issue X Oct 2020,www.ijraset.com

IOP Conf. Series: Materials Science and Engineering 1126 (2021) 012030 doi:10.1088/1757-899X/1126/1/012030

Application of biosolids (sewage sludge) in agricultural soils: a case study for corn seed LVN10

Nguyen Kieu Lan Phuong[1,3], **Lam Van Tan**[1,4,**], **Lam Van Giang**[2], **Van Doan Thi**[5], **Le Thi Anh Hong**[6], **Thanh Tran**[1,2,*]

[1]Institute of Environmental Sciences, Nguyen Tat Thanh University, Ho Chi Minh City, Vietnam.
[2]Faculty of Environment and Natural Resources, Ho Chi Minh City University of Technology (HCMUT), VNU-HCM, Ho Chi Minh City 700000, Vietnam
[3]Faculty of Environmental and Food Engineering, Nguyen Tat Thanh University, Ho Chi Minh City, 70000, Vietnam.
[4]Department of Science and Technology of Ben Tre Province, 86000, Vietnam
[5]Faculty of Biology and Environmental Science, The University of Education, Danang
[6]Institute of Tropical Biology, Vietnam Academy of Science and Technology, Ho Chi Minh City, 70000, Vietnam.

Corresponding mail:* thanhtran2710@gmail.com ;** lvtan@ntt.edu.vn

Abstract. Reuse of biosolids aiming to take advantage of its nutrients and micronutrients for arable crops is an economical approach, in which wastes are considered potential resources and bring a variety of benefits. However, heavy metal residuals in sewage sludge have been gaining attention due to accumulation risk and adverse effects on the environment. This study experimented with sludges from Binh Hung municipal wastewater treatment plants treated by anaerobic co-digestion process on corn seed LVN10 to assess corn growth and heavy metals accumulation contents in various components of the corn and the soil. Consequently, within 90 days of using biosolids as treatments 1% and 3%, i.e., the corresponding amount of biosolids over the soil weight, there was no inhibition but growth stimulation for corn growth surveyed. Accumulations of Cu, Cd, Pb, Zn, Ni in roots, stalk, leaves, and mature corn and cultivated soil complied with the national regulations. The effectiveness of biosolids application for corn production in this study will benefit the depreciation of sludge production from municipal wastewater treatment plants and other applications for agricultural plants in the future.

1. Introduction

Sewage sludge management obtained from centralized municipal wastewater treatment plants (WWTPs) has faced critical issues due to its growing amount over time, particularly in developing countries. Ho Chi Minh City (HCMC), Vietnam, for example, is a hub for commercial activities with around 8.5 million inhabitants; however, there is only one centralized municipal WWTP, namely Binh Hung, with a treatment capacity 141,000 m^3/day in Phase I. According to Decision No.24/QD-TTG dated January 06, 2010, of the Prime Minister approval of adjusting General Planning of Ho Chi Minh City Construction to 2025 [1], the average total volume of wastewater in the city to 2025 based on the average water supply per capita will reach from 3.15 to 3.2 million m^3/day, leading to the amount of sludge from the urban WWTP be up to 125 tons/day.

Sludge treatment is a costly process, accounting for 50-60% of the total wastewater treatment cost [2]. Therefore, treating sludge in HCMC has not been entirely resolved, although several measures are suggested, such as landfill disposal at Da Phuoc, assigning to subcontract for handling [3]. Unfortunately, Binh Hung WWTP's operation continuously gives rise to foul odors, which annoys the neighbors [4] [5]. It is urgent work to propose appropriate solutions to tackle the problem.

According to Girovich [6], sludge disposal affects the environment and wastes a valuable resource because the composition of the sludge contains a high nutrient content and micronutrients, which can be used as organic fertilizer for agriculture. Application of sludge or biological solids (biosolids) raises the concern among scholars due to its fertilizer source from cheap raw materials and its sources of natural organic matter considered as organic fertilizer with a high content of total organic carbon and humus as well as nutrients such as nitrogen and phosphorus that plants readily absorb. Some benefits of biosolid reuse for soil quality improvement can be greater water-holding capacity, soil erosion reduction, increases in cation exchange capacity, earthworm and microbial activities, and correction of micronutrient deficiencies [7].

Although the sludge from municipal WWTP often contains toxic pollutants such as heavy metals, pathogenic microorganisms with content less than that of other hazardous and industrial wastes (Spinosa et al., 2011), commercial applications require adequate safety assessment, hefty metal accumulation. Several studies showed that the absorption of phytonutrients in the soil, in addition to nutrient absorption, there are potential risks of absorbing heavy metals such as cadmium, copper and zinc and pathogens, nematodes in soils that fertilize for grain products [8]. Additionally, Fu, et al. [9] claimed the existence of a variety of carcinogenic metals such as Cd, As, Cr, Hg, Pb in rice grains, while heavy metal residues were found in vegetables when using biosolids as fertilizer [10, 11].

This study is the subsequent work after assessing the benefits from the sludge treatment with biogas recovery and successfully applied to reduce the number of chemical fertilizers used by nearly 50% of the experiment on wet rice plants [12, 13]. In the current study, we aim to examine the effectiveness of organic fertilizer from biosolids on corn growth and heavy metal accumulation in the soil and some corn plants' parts. This will be a scientific basis and a new direction in biosolids management applications in agriculture in efforts to reduce the number of chemical fertilizers used toward sustainable agriculture for Vietnam.

2. Materials and research methods

2.1. Experimental materials

Sewage sludge was obtained from Binh Hung municipal WWTP. Sludge samples were taken from a centrifugal water separator with pH = 7.54, C/N ratio = 16.67, and humidity = 78%. The sludge samples were mixed with smoked rice hull that was finely ground at the ratio of 2:1 [13] so that the C: N ratio can meet the desired conditions for the anaerobic reaction in 20-40 [14]. Once the weight of the substrate has been changed, their proportions will be accordingly changed. The moisture content of the mixture was kept at 65-70% [15] for optimal decomposition. The process of creating fertilizer studied is running under conditions of mesophilic temperatures (30-35°C) [14]. The fertilizer application at the rate of 1% and 3%, i.e., the corresponding amount of biosolids over the soil weight of the estimated 20cm deep surface. The chemical fertilizer used in the study is commercial fertilizer NPK (20:20:15) for soil dressing. The surveyed corn seed is LVN10 with an average yield of 65-75 quintals per ha.

2.2. Study area

The study conducts in a cornfield in Hoi Gia Hamlet, My Phong Commune, My Tho City, Tien Giang Province. The soil category is alluvial with moderate organic content (2.09%) and total protein (0.27%), insignificant Al^{3+} and Fe^{3+} contents. Generally, the soil is rich in clay and organic matter and is most favorable for agriculture.

2.3. Experimental design and methodology

The corn cultivation process and research contents are as follows:

Table 1. Growth stage and research procedure

Days after planting	Growth stage	Cultivation	Indicator(s)
0	Field preparation and planting	Soil preparation, making beds, and spraying herbicides to kill germs. Fertilizing the soil by biosolids and then sowing corn seed LVN10.	Heavy metal content in the soil.
7	Germination	Recording beds with more than 50% of plants have sprouts breaking the soil surface.	-
12	3-6 leaves	Soil dressing by commercial fertilizer NPK for beds. Recording beds with more than 50% of plants have 3-6 leaves	Growth rate
26	8-10 leaves	Recording beds with more than 50% of plants have 8-10 leaves	Growth rate
40	Silking	Recording beds with more than 50% of plants have 12 leaves	Growth rate, kernel count, and kernel weight
52	Blister	Dimension measurement of stalks. Monitoring diseases.	Kernel number and kernel weight
61	Milk	Plant height measurement. Ear height measurement.	Kernel number and kernel weight
90	Maturity	Harvest and measurements of mature corn ear number per plant, length of mature corn ear, the dimension of mature corn ear, kernel row number per ear, kernel number per row, and weight of 1000 kernels.	Corn yield and heavy metals

The indicators for evaluating the growth rate and development degree rely on QCVN01-56:2011/BNNPTNT - *National technical regulation on testing for Value of Cultivation and Use of Maize varieties*, including plant height (cm), ear height (cm), mature corn ear number per plant, length of ripe corn (cm), the dimension of mature corn ear(cm), kernel row number per ear, kernel number per row and weight of 1000 kernels (g). The allowable limits of heavy metals in the soils and food area based on QCVN 03-MT:2015/BTNMT - *National technical regulation on the permissible limits of heavy metals in the soil*, and QCVN 8-2:2011/BYT - *National technical regulation on the limits of heavy metals contamination in food*, respectively.

Sample treatment and heavy metal analysis referred to the procedures of TCVN 6649:2000 (ISO 11466:1995) *Soil quality - Insoluble elements in hydrothermal water*, EPA Method 3051A - *Microwave Assisted Acid Digestion of Sediments, Sludges, Soils, and Oil* (the United States Environmental Protection Agency, 2014), EPA Method 6020B - *Inductively Coupled Plasma—Mass Spectrometry*, and make some modifications to fit the samples

We employ Microsoft Excel 2013 software for descriptive statistics and IBM SPSS Statistics 20.0 software for assessing the differences with a significance level of 95%.

3. Result and discussion

3.1. Properties of biosolids after treatment

The concentration of N, P, K analyzed after incubation has the concentration close to TCVN 562: 2002/BNN & PTNT as follows:

Table 2. Nutrient content in biosolids

Specie	Biosolids after incubation treatment	TCVN 562
N	2.60 %	> 2.5 %
P	1.32 %	> 1.5 %
K	1.61 %	> 1.5 %

After incubation and treatment, the heavy metals contents decreased and were lower than the limit according to QCVN 03-MT:2015/BTNMT and EPA 503, as shown in Table 3.

Table 3. Heavy metals contents in the cultivation

Specie (mg/kg)	Sewage sludge	After the mixing process (2:0.5 ratio)	After incubation treatment (2:0.5 ratio)	Treatment 1%	Treatment 3%	QCVN 03	EPA 503
Pb	14.01	10.51	7.88	0.08	0.24	70	300
Cd	0.42	0.32	0.25	<1.00	<1.00	1.5	39
Cu	18.69	14.02	13.30	0.13	0.40	100	1500
Ni	6.98	5.24	6.07	0.06	0.18	-	420
Zn	176.17	132.13	101.36	1.01	3.04	200	2,800

3.2. Evaluation of the experimental corn quality and corn yield

Regarding the growth rate criteria, with increasing biosolids concentrations, the growth rate of corn plants is happening faster. The results showed that corn uses stored nutrients from seeds in the initial stage of growth (3-leaf corn). However, we can see clearly that the growth rate is significantly different at fertilizer concentrations of 0%, 1%, 3% when applying to the soil. By the stage of nutritional growth (corn plants reach 9-12 leaves), corn will assimilate nutrients from the soil and absorb CO_2 for photosynthesis.

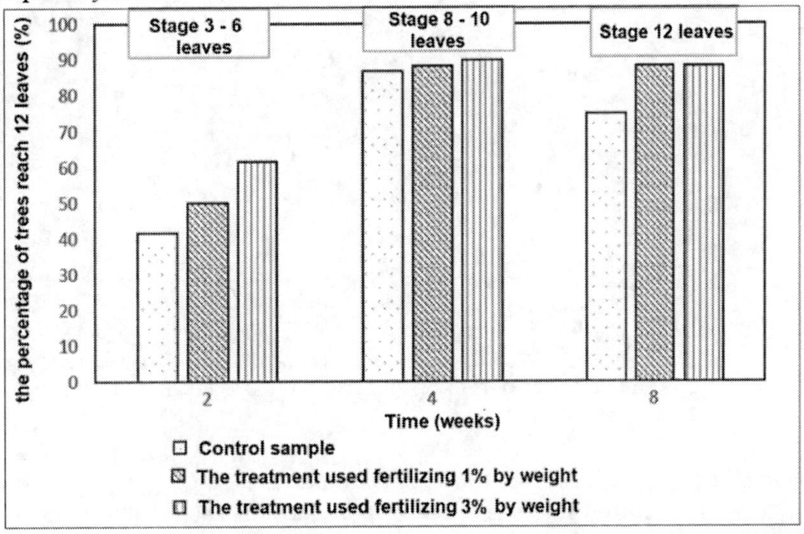

Figure 1. Effect of biosolids concentration on growth time of corn seed LVN10

Based on Figure 1, we can see that the growth rates of 1% and 3% concentrations have the metabolism but is still faster than the 0% concentration. This proves that the corn plant is growing well, and there is a difference between the concentrations of 0%, 1%, 3% when entering the real growth phase. At the concentration of 1% and 3% at the 8th week, more than 80% of the plants reached 12 leaves. As for the control sample, about 70% of plants achieved 12 leaves. Hence, the growth rate of 12 leaves in the control sample is slower than in the two samples, concentrations 1%, 3%.

IOP Conf. Series: Materials Science and Engineering 1126 (2021) 012030 doi:10.1088/1757-899X/1126/1/012030

Table 4. Differences among plant height of samples with additional biosolids concentrations

Sample	Plant height (cm)	p-value
Control (0%)	190.90	-
Treatment 1%	222.30	0.000(a)
Treatment 3%	226.28	0.000(b); 0.851(c)

(a)Difference in plant height between treatment 1% and control;
(b)Difference in plant height between treatment 3% and control;
(c)Difference in plant height between treatment 3% and treatment 1%

Among the plant height of three samples, treatment 3% had the highest value (226.28 cm), followed by treatment 1% (222.30) and control (190.90). This means that the experimental biosolids with concentrations of 1%, 3% are suitable to stimulate the growth of corn plants. In terms of morphological indicators, there was a statistical difference in plant height between treatment 1% and control and treatment 3% and control with $P = 0.000 < 0.05$. Similarly, the whole results of corn yield factors in Table 5 revealed the highest values of additional biosolids 3% compared with additional biosolids 1% and control sample. An example of mature corn ear at the different biosolids concentrations is shown in Figure 2.

Table 5. Effects of additional biosolids concentrations on corn yield factors

Indicator	Control sample	Treatment 1%	Treatment 3%
Mature corn ear number (per plant)	1.4	1.9	2
Length of mature corn ear (cm)	27.86	28.11	30.18
Dimension of mature corn ear (cm)	4.09	4.11	4.31
Kernel row number (per ear)	11.07	11.42	12.14
Kernel number (per row)	21.5	23.5	29
Weight of 1000 kernels (g)	329.3	330.34	334.78

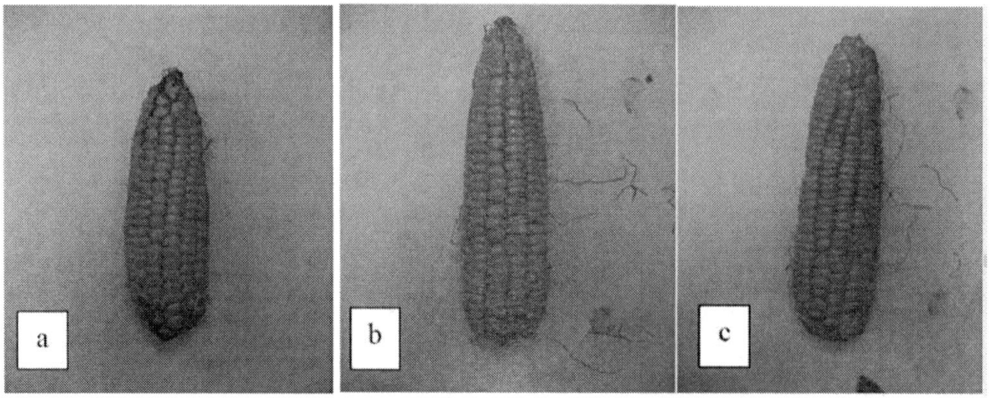

Figure 2. Mature corn ears after harvest with fertilized biosolids at: (a) 0%; (b) 1%; (c) 3%

3.3. Assessment of heavy metal residues and accumulation

In order to assess the safety of biosolids reuse, the study measured the content of heavy metal accumulated in soils and mature corn ears after harvest, see Tables 6 and 7.

Table 6. Heavy metal contents in soils before and after harvest (mg/kg)

Specie	Soil background concentration	Treatment 1%	Treatment 3%	Treatment 1% after harvest	Treatment 3% after harvest	QCVN 03
Cu	21.95	0.13	0.40	22.00	22.78	100

Zn	61.18	1.01	3.04	62.63	76.31	200
Ni	24.24	0.06	0.18	21.00	21.05	-
Pb	32.00	0.08	0.24	29.27	27.14	70
Cd	0.18	<0.1	<0.1	<1	<1	1.5

The results in Table 6 have demonstrated that heavy metal accumulation in soils was not significant, lower than the allowance standard. However, the study should keep tracking the stacks in the next seasons for a better assessment. For heavy metal residues in parts of the corn plant after harvest in Table 7, the results revealed Zn content accumulates most in roots, followed by in ear, in leaves, and the stalk. Zn supports the synthesis of growth substances and yeast systems necessary for plant growth [16]. Thus the highest accumulation of Zn in roots makes sense. The highest value was found in leaves for Cu accumulation, followed by roots, stalk, and ear. The accumulation of a high level of Cu in the leaves helps in chlorophyll formation in the plant. Other accumulations of Cd, Pb, Ni are not significant. Overall, the accumulations of heavy metals in corn plant parts are at the safety threshold.

Table 7. Average heavy metals in separate parts of corn plant after harvest (mg/kg)

Specie	Roots		Stalk		Leaves		Mature corn ear		QCVN 8
	1%	3%	1%	3%	1%	3%	1%	3%	
Cu	1.03	2.93	2.38	2.91	1.46	2.98	1.04	1.27	-
Cd	0.01	0.01	0.02	0.02	0.03	0.04	-	-	0.1
Pb	0.06	0.03	0.05	0.03	0.09	0.07	-	0.01	0.2
Zn	8.37	10.31	1.06	2.62	5.35	5.54	0.66	0.87	-
Ni	0.55	0.28	0.53	0.61	0.44	0.58	0.25	0.27	-

4. Conclusion

At concentrations of 3% and 1% of experimental biosolids used, there was no inhibition but growth stimulation for corn seed LVN10. The yield and quality of surveyed corn after harvest are higher than that of the sample control and the reference yield. Heavy metal contents in soils and biosolids after treatments are lower than the standard QCVN 03-MT:2015/BTNMT, although it appears an accumulation of Cu and Zn in soils with additional biosolids 1% and 3%. In corn plant parts, Zn accumulated most in roots, while the highest value of Cu was found in leaves. The accumulations of heavy metals in corn plants are at the safety threshold, but the study should keep tracking the accumulations in the next seasons for a better assessment. The findings in this study tackle the problem of odor annoyance of WWT system from sewage sludge partly as well as propose a new direction in applications of biosolids management in agriculture to minimize chemical fertilizers use toward sustainable agriculture for Vietnam. Although the research is still undergoing different stages, its benefits of economy and environment are promising.

References

[1] H. T. Hai 2010 *Approval of the adjustment plan on construction of Ho Chi Minh City to 2025* Ho Chi Minh City: People's Committee of Ho Chi Minh City.

[2] Y. Wei, R. T. Van Houten, A. R. Borger, D. H. Eikelboom, and Y. Fan 2003 *Water Resear,* vol. 37, pp. 4453-4467

[3] C. Hung 2010 *Moi Truong Online.* Available: http://bit.ly/2IhCQmW

[4] M. Khanh 2013 *News Onl.* Available: http://bit.ly/2JZH0Ns

[5] M. Hương 2016 *Social Onl* Available: http://bit.ly/2rqScvE

[6] M. J. Girovich 1996 *Taylor & Francis,* pp 105 - 150

[7] D. M. Sullivan 2015 Pacific Northwest pp 180 - 205

[8] D. Pritchard, N. Penney, M. McLaughli, H. Rigby, and K. Schwarz 2010 *Water Sci Tech,* vol. 62, pp. 48-57

[9] J. Fu, Q. Zhou, J. Liu, W. Liu, T. Wang, Q. Zhang, *et al 2008 Chemos.,* vol. 71, pp. 1269-1275

[10] M. L. A. Silveira, L. R. F. Alleoni, and L. R. G. Guilherme 2003 *Scien. Agri.,* vol. 60, pp. 793-806

[11] P. Zhuang, M. B. McBride, H. Xia, N. Li, and Z. Li 2009 *Sci. Total Environ.,* vol. 407, pp. 1551-1561

[12] T. Tran, H. N. Do, P. L. N. Nguyen, T. B. Linh, P. H. Diem, L. H. Quy, *et al.* 2016 *Adv. Tech. and Sustain. Develop.*, Industrial University of HCM city pp. 1065 - 1072.

[13] T. Tran, P. L. N. Nguyen, L. T. A. Hong, P. N. K. Lan, L. T. D. Thien, and G. L. Van 2016 *J BioTech.,* vol. 14, pp. 563 - 571

[14] N. T. Phong 2009 *HCMC National University Publisher*, pp 85 - 89

[15] H. Bouallagui, H. Lahdheb, E. B. Romdan, B. Rachdi, and M. Hamdi 2009 *J. Environ. Manage.* vol. 90, pp. 1844-1849

[16] B. J. Alloway 2004 *Belgium and Paris, France: International Zinc Association Brussels* p. 135.

IOP Conf. Series: Materials Science and Engineering 1126 (2021) 012031 doi:10.1088/1757-899X/1126/1/012031

Energy recovery of sewage sludge treatment by anaerobic co-digestion

Thanh Tran[1,2,3], Nguyen Kieu Lan Phuong[3], Van Doan Thi[4], Lam Van Giang[2], Lam Van Tan[1,5,*]

[1]Institute of Environmental Sciences, Nguyen Tat Thanh University, Ho Chi Minh City, Vietnam.
[2]Faculty of Environment and Natural Resources, Ho Chi Minh City University of Technology (HCMUT), VNU-HCM, Ho Chi Minh City 700000, Vietnam.
[3]Faculty of Environmental and Food Engineering, Nguyen Tat Thanh University, Ho Chi Minh City, 70000, Vietnam.
[4]Faculty of Biology and Environmental Science, The University of Education, Danang
[5]Department of Science and Technology of Ben Tre Province, 86000, Vietnam

*Corresponding mail: lvtan@ntt.edu.vn

Abstract. Sewage sludge produced from wastewater treatment is one of the challenges facing many countries in their development process; therefore, the sewage sludge needs treating suitably. Besides, recovering energy from waste to save operating costs in treatment plants is also being considered. The study was conducted by mixing urban sludge samples with substrates (agricultural residues) to create different C/N ratios from 16.6 to 32.5, then evaluating the quantity and quality of methane) the gas generated from the test pieces. This method not only solved the environmental problem but also had benefited from biogas produced from the digestion process. The initial results showed that the quantity of biogas from anaerobic co-digestion of sewage sludge and rice husk (ratio 4:1) are 160ml/g VS. Furthermore, the quality of methane can be up to 60% and near twice as much as mono-digestion. The study results estimate that the potential of recovering electricity from the sludge for the wastewater treatment plant of Binh Hung (Vietnam) is about 0.2 MW/year or higher.

1. Introduction

Nowadays, the management and treatment of sewage sludge produced from wastewater treatment plants is the most serious biological treatment challenge. Fresh and untreated sludge will have many pathogens and normally odorous, putrid. Sludge treatment accounts for half, even up 60% of the total wastewater treatment cost **[1]**. The challenge facing the sludge managers is finding a suitable solution, effective cost, and environmental and economical, and social benefits.

Currently, Ho Chi Minh City only has Binh Hung wastewater treatment plants, which treated 150.000m^3 every day, and the wastewater treatment process will generate approximately 30 tons of sludge **[2]**. According to the sewage system plan by 2025 **[3]**, 12 wastewater treatment plants will be completed with a total processing capacity of up to 2,912,000 m^3/day; therefore, the amount of sludge generated can be up to nearly 200 tons/day. Nowadays, the sludge of Binh Hung Plant is treated by stabilization, de-watering and drying process, this process was odorous, putrid and air pollution. The anaerobic digestion can be solved this problem - one of the challenges facing sludge treatment.

The sludge treatment process by anaerobic digestion was accounted for about 38% of sewage sludge, the second most popular treatment process in India **[4]**. Anaerobic digestion seems to be the most popular stabilization method in European countries, applying in 24 countries **[5]**. Sewage sludge-treatment scenario with anaerobic digestion was the most environmentally and economically suitable method to treat sewage sludge because of energy recovery **[6]**.

Simultaneously, a massive quantity of wastes from the agricultural process all over the world. These wastes account for a significant proportion of agricultural production. However, the treatment of agricultural wastes is disposed of by incineration or landfill. This process caused damage that affected the environment. Furthermore, agricultural wastes are valuable resources, which can be recovered and diverse economic, social and environmental purposes **[7]**. The increasing materials such as rice straw, rice husk, water hyacinth, and cassava peel can be paid more attention to researchers, especially recycling and waste to energy issues.

Finding a suitable solution for sewage sludge and mixing materials is one of the most urgent challenges facing many countries **[8]**. Both anaerobic and aerobic techniques can be used for sludge treatment. However, anaerobic digestion has more advantages, such as it consumes little or even no energy. The feedstock can be stabilized, and the last stage's product can be fertilizer to improve the soil **[9]**. Besides, it is one of the most suitable technology to produce renewable energy from wastes. This technique converts organic matter, which is in sludge and agricultural wastes, into biogas by microorganisms, and biogas is used for producing electricity by microorganisms.

2. Materials and research methods

2.1. Experimental materials

The sample of sewage sludge was taken from the wastewater treatment plant urban Binh Hung in Ho Chi Minh City (Vietnam) – urban wastewater treatment plant on the input material. All samples were mixed so that the C: N ratio could meet the desirable condition for anaerobic reactions is between 20 and 40 [10]. When the weight of substrate was changed, their rates would be adjusted respectively. The mixture's moisture was controlled at 65-70% for best decomposition [11] to produce methane gas optimally.

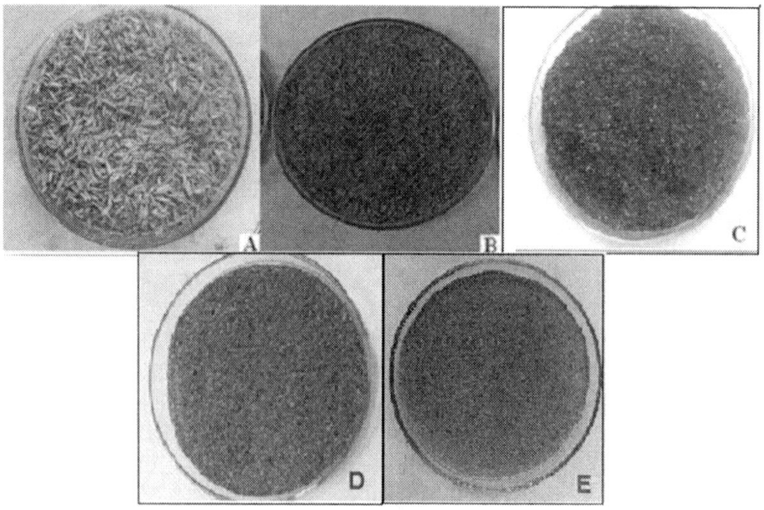

Figure 1. The mixing materials after pre-treatment (A: rice husk, B: rice straw, C: rice husk charcoal, D: water hyacinth, E: cassava peel)

The materials include rice husk, rice straw, water hyacinth, cassava peels were used in the research. Rice husk and rice husk were bought from the ornamental plant's store, and water hyacinth was collected from the river; cassava peels are waste from Hung Duy Company - Cassava Starch Plant in Tay Ninh Province, Vietnam.

These materials were pre-treated before the experiment in many steps. In the first step, these materials were drying by solar. After that, they were cut into small pieces, which were dimensions from 1-3cm. Finally, they were grind out fine by QE 500 and sieved to the size of approximately 0.154mm.

2.2. Experimental design

The mixing materials and sewage sludge were analyzed about total organic carbon (TOC), total nitrogen (TN), volatile solids (VS), total phosphorus (TP), humidity, drying coefficient (K). The results were shown in Table 1.

Table 1. Analysis structure used in this research about TOC, TN, TP, Moisture, K, VS

Structure	Humidity	K	TOC (%)	TN (%)	TP (%)	VS (%)
Sewage sludge	74	1.74	17	1	0.29	47.6
Rice husk	13	1.13	22	3	0.12	75.2
Rice husk charcoal	2	1.02	21	2.6	0.20	4.64
Rice straw	8	1.08	29	0.2	0.10	92.7
Water hyacinth	60	1.90	23	1.6	0.11	84.5
Cassava peel	12	1.12	21	5	0.34	81.6

Experiment 1: Analyzing nutrient ingredient of materials and chose ratio for mixing the sample

Experiment 2: Comparing the quantity of the biogas produced from the mono digestion and co-digestion with various materials.

Experiment 3: Comparing the quality of the biogas produced from the mono digestion and co-digestion with various materials.

2.3. Experimental conditions

After mix controllably in about 70% moisture, samples were kept in a sealed glass bottle 300ml (which are placed in Yamato incubator) with mesophilic condition 30°C in total time experiment. Each ingredient had ten glass bottles with the same conditions about the weight of sewage sludge and mixing materials, temperature.

The digestion ability of samples was shown through the volume of generated biogas per total volatile solids (TVS) (ml/g VS). Biogas was collected every two days. C/N ratio, biogas generation performance (ml/g VS), and time for biogas generation were criteria to evaluate the experimental design. Biogas is collected by equilibrating the ambient pressure when water in turned-over bottles is pushed out. CH_4 is analyzed immediately after collection.

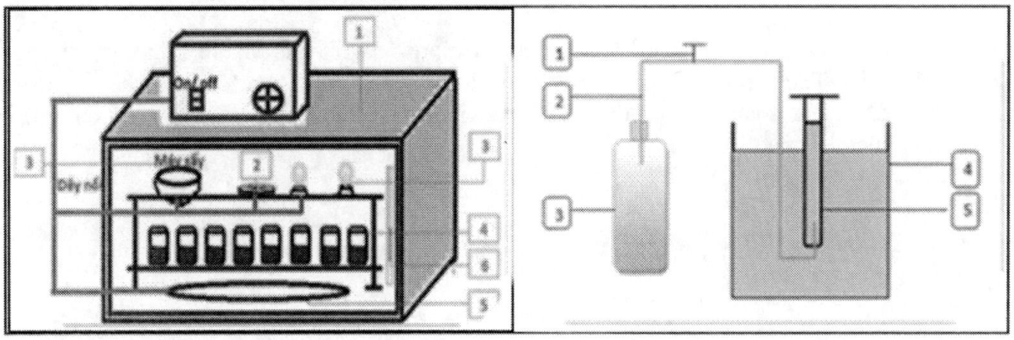

Figure 2. The experiment model about anaerobic digestion and collecting the biogas

Biogas was analyzed by IMR 2800P equipment – combustion gas analyzes equipment. IMR 2800P can be continuously and simultaneously measured up to 18 parameters such as temperature, CO_2, O_2, CO, H_2S, hydrocarbon, CH_4, NO_2, NO.

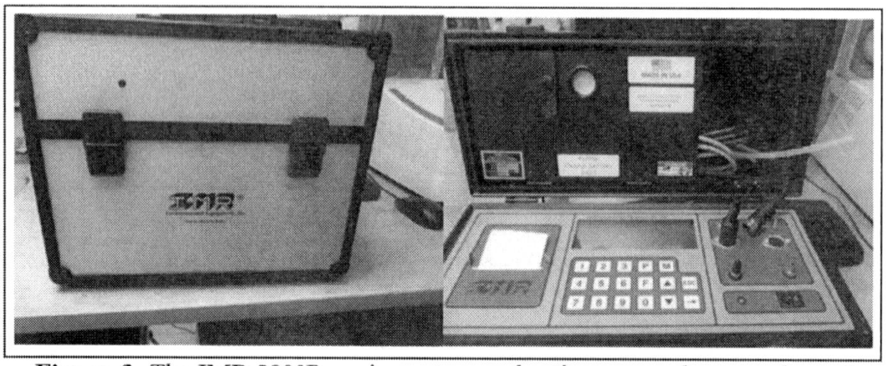

Figure 3. The IMR 2800P equipment – combustion gas analyzes equipment

3. Results and discussion

3.1. Nutrient ingredient analysis of samples

The sewage sludge and mixing material ratio was 4:1, which had a C/N rate suitable for anaerobic digestion. The result of the analysis was shown in Table 2.

Table 2. Different of the structure on the mixture in anaerobic co-digestion about TOC, TN, TP, Humidity, K, VS, and C/N ratio

Code Sample	Ingredient	Humidity	K	TOC%	TN%	TP (%)	VS (%)	C/N
M1	Sewage sludge	78%	1.74	20	1.2	0.26	37.6	16.67
M2	Sewage sludge and rice husk	61%	1.61	52	1.6	0.32	62.1	32.5
M3	Sewage sludge and rice husk charcoal	64%	1.61	55	2.1	0.58	48.9	26.19
M4	Sewage sludge and rice straw	68%	1.68	63	3	0.25	63.3	21
M5	Sewage sludge and water hyacinth	76%	1.76	69.7	2.7	0.34	51.4	23.23
M6	Sewage sludge and cassava peel	61%	1.61	76	3.8	0.53	65.2	20

3.2. Comparison of the quantity of the biogas producing from the mono digestion and co-digestion with various materials.

Based on initial results, mono-digestion generated a more little biogas volume than co-digestion with various materials. The detailed result was shown in Figure 1. Therefore, in the perspective of performance digestion, anaerobic co-digestion was more effective than mono-digestion and brought significant economic value.

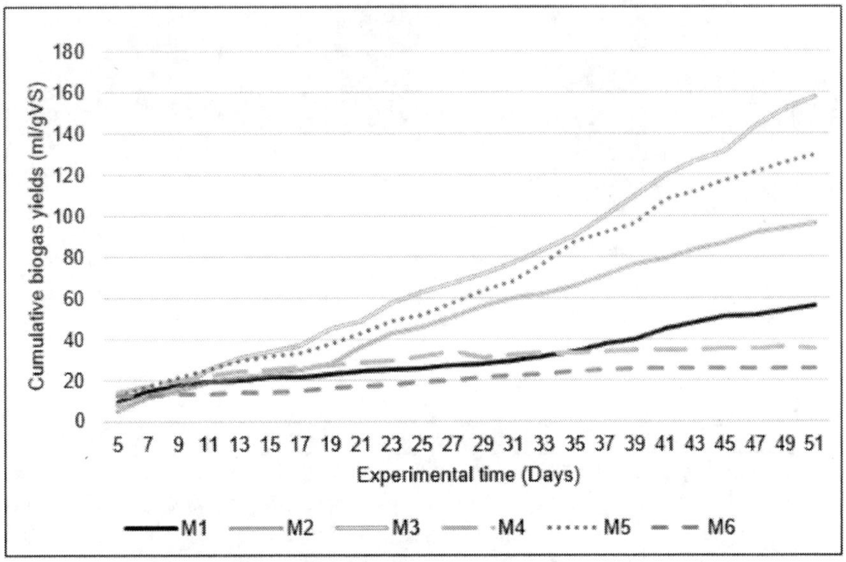

Figure 4. Diagram of the cumulative biogas yields of samples

Figure 4 presents the daily biogas production results from anaerobic co-digestion of sewage sludge and various material under mesophilic conditions until the end of the process. From the figure, the higher volume of the biogas yields was found to be M3 and M5, about 160ml/g VS and 130ml/g VS, higher than M1 (single digestion – only sewage sludge). From the figure, it is clear that the cumulative biogas yields from anaerobic co-digestion of sewage sludge and various material though 51 days are higher than that was obtained from single digestion. After the day of 51st, biogas speed generate is decelerated, so we finish this experiment.

However, the biogas produced from M4 and M6 was low, and the process that produced the biogas was ended soon, approximately ten days. These results were conformable with the C/N ratio of M4 and M6 because this ratio of about 21 and 20 was not suitable with the anaerobic digestion process, which is 20-40 **[11]**. M6 is the mixture of sewage sludge and cassava peels; furthermore, the cassava peels had cyanide. This cyanide can be inhibited the convert organic matter into biogas of a consortium of microorganisms. So that, cassava peels need to cyanide treated before mixing with sludge.

According to research abroad, the quantity of the biogas producing of the anaerobic digestion was 310ml/g VS after 30 days **[12]**. However, in Vietnam, the study about the anaerobic digestion of sewage sludge was not well-known. Thus, the quantitative review of our research in the country is promising, but the research is still undergoing different stages to increase biogas production.

3.3. Comparison of the quality of the biogas producing from the mono digestion and co-digestion with various materials.

Choosing the samples M3 and M5 which the biogas' quantity was the highest cumulative biogas yields, were analyzed and compared with single digestion.

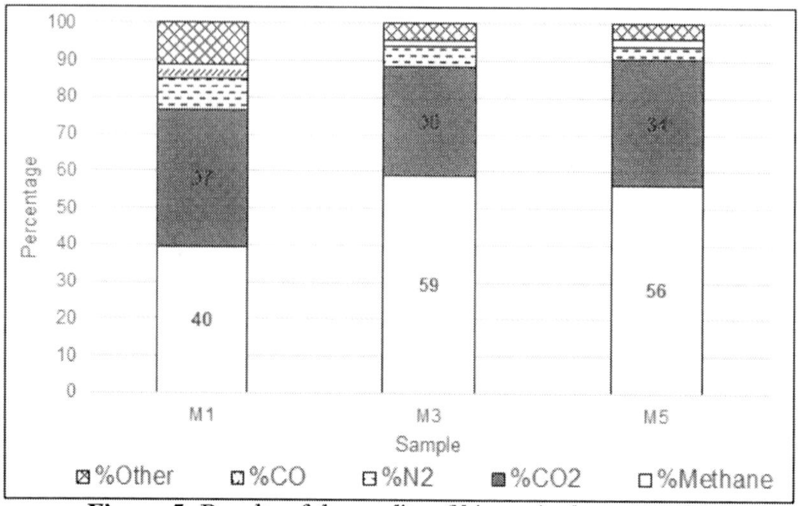

Figure 5. Results of the quality of biogas in three samples

The cumulative methane yields from the M3 sample were 58.7%, and the M5 sample was 56.2%, higher than M1 sample – 39.5%. The initial result showed the quality of methane from co-digestion was twice as much as mono-digestion. It can be seen that energy recovery of sewage sludge treatment by anaerobic co-digestion is potential.

3.4. Estimating the potential of electricity recovery

In this experiment, the highest volume of biogas producing was about 4800 ml biogas/200g sewage sludge and time digestion was 45 days; approximately 0.5 m^3 biogas/ton sewage sludge/day. Binh Hung wastewater treatment plants can be produced 120-ton sludge every day.

Estimate the quantity of biogas obtained in 1 year **[10],** the estimated parameters including calorific value and performance of electricity production of methane

120 (tons sludge/day) x 0.5 (m^3 biogas/ton sludge/day) x 45 (days for digestion) x 365 (days/year) x 75% (methane/biogas) = **739125** (m^3 methane/year) ≈ **740.000** (m^3methane/year).

The methane's calorific value is 37.78 MJ/$m^{3,}$ and the conversion factor from CH_4 to 1 kWh electricity is 30%. **[13]**

Estimate the amount of electricity that can be obtained from methane

740.000 (m^3 methane/year) x 37.78 (MJ/m^3 methane) x 30% = 8387160 (MJ/year) = **2330 (MWh/year) = 0.2MW**

So, the plant electrification project, which capacity is 0.2 MW, potential can be set up. 20% of electricity from this plant can be used to heat for the anaerobic digestion process and operate this plant. About 80% of electricity can be used to demand wastewater treatment system of Binh Hung Plant for reducing operating costs.

4. Conclusion

The analyses results of the C/N ratio of mono – digestion sample was not good enough to treat the sewage sludge by anaerobic digestion, so the cumulative biogas yields were medium volume (58 ml/g VS) and lower than the mixture of sewage sludge and rice husk charcoal and water hyacinth, is 160 ml/g VS and 130 ml/g VS respectively which the ratio of the sewage sludge and mixing material is 4:1. The cumulative methane yields producing from anaerobic co-digestion with rice husk charcoal and water hyacinth can be up to 56 – 59%, the potential electrification plant project can be set up, which a capacity is 0.2 MW/year. The next step of the study is to research the microorganism, which can be support for faster digestion process and higher quality of methane. Then, we will utilize sewage sludge after treatment for agriculture application. Although the research is still undergoing different stages, its benefits of economy and environment are promising.

References

[1] Wei Y, Houten R, Borger A, Eikelboom D and Fan Y 2003 *Water Res* 37(18) pp 4453-4467
[2] Steering Center of the Urban Flood Control Program (2013). *Report test results of running water separator centrifuge* (Ho Chi Minh City)
[3] Hoang Trung Hai 2010 *People's Committee of Ho Chi Minh City – The decision of the government 24/QĐ-TTG.*
[4] Ministry of Environmental Protection 2010 *Best Available Techniques Directive for Treatment and Disposal of Sludge from Wastewater Treatment Plant* (in Chinese).
[5] Alexandros K and Athanasios S. Stasinakis 2012 *Wast. Manage. J.* 32 pp. 1186-1195
[6] Changqing X., Wei C. and Jinglan H. 2014 *J. Clean. Pro.* 67 pp. 79-87
[7] FAOSTAT 2015 *Food and agricultural commodities production*, vol. 5
[8] Rizzardini C and Goi D 2014 Sustain. 6 pp 2424-2434
[9] Zhang W, Wei Q, Wub S, Qi D, Li W, Zou Z, et al (2014). *Appl. Ener.* 128 pp 175-183
[10] Nguyen Tan P. 2009 *HCMC National university Publisher*, Ho Chi Minh City.
[11] Bouallagui H., Lahdheb H., Romdan E., Rachdi B., Hamdi M. 2009 *J. Environ. Manage* 90 pp. 1844-1849.
[12] Xiaohu D, Xiaoshuai L, Dong Z, Yinguang C and Lingling D 2016 *Biores Tech* Vol 216 pp. 323-330.
[13] Richard A 2010 *Faculty of Mechanical and Agricultural Engineering, Kwame Nkrumah University of Science and Technology* pp. 113 -120

IOP Conf. Series: Materials Science and Engineering 1126 (2021) 012032 doi:10.1088/1757-899X/1126/1/012032

On Some Cases of Plastic Deformations Under Complex Loading

AbirovR.A.

Institute of mechanics and earthquake engineering, 100125 Tashkent, Uzbekistan
rustam_abirov@mail.ru

Abstract. The experimental results under complex loading in view of theory of processes are explained in this issue. The opportunity for reducing of external efforts at metal forming is shown here. These results and investigations can be expanded for anisotropic material and for spatial processes of material forming.

1. Introduction

At present the energy saving technologies and optimal control of deformation at metallurgy has a great importance. The use of new isotropic and polymer materials demands to investigate more widely their mechanical properties and ways for optimal reinforcing. The account of complex loading processes is vital here. It has been experimentally established that at some complex loading processes the force of plastic deformation is less than at simple loading (like tension or torque). The Il'yushin's theory of elasto-plastic processes can explain this effect [1]. So, at complex loading the energy at metal deformation can be reduced compared with simple loading. The possibility and mathematical explanation of this problem are described in this issue. The theory of elasto-plastic processes by Il'yushin is the base for this approach. For deformation trajectories with bend points the approach for decreasing of energy at metal processing are shown. Given effect can be used at development of new technology at plastic processing of metals and application it in practice. Thereby, taking into account the complex loading and investigated effect appearing here, aside from definition of reserves of bearing capability of structures, allows to minimize the work of external forces at technological processes by optimal controlling of strain path at metal forming processing.

Mathematical problems of plasticity under complex loading [2] cannot be solved by creating the universal constitutive relations. The models [3-7] which developed for complex loading processes can be applicable only for specific processes of loading and for different materials in each case separately.

2. Main part

Many boundary plasticity problems have not reliable solution [8,9]. Particularly, reason for it the initial errors in gradient plasticity positions. The Drucker's Postulate [10] and its gradient principle can't always at any loading adequately reflect the processes of strain-stress state in continua. For instance are structured-unstable materials, material with falling "strengthening" diagram, etc. At investigation of complex loading for isotropic metals also appear the breaches of the Drucker's Postulate. This is due to not only deformational anisotropy. One of them is a breach of the gradient principle. In generalized Hodge and Prager's view [11] for increments of full deformation are:

$$d\varepsilon_{ij} = H_{ijkl}d\sigma_{kl} + 0.5[P(J_2,J_3)s_{ij} + Q(J_2,J_3)\partial J_3/ds_{ij}]df$$

here first summand is elastic increments of strain, the second summand is plastic component of one, J_2 and J_3 are invariants of stress deviators, s_{ij} are components of stress deviators, H_{ijkl} is matrix of elastic components, df is yield surface, $d\varepsilon_{ij}$ and $d\sigma_{kl}$ are increments of strain tensors and stresses respectively.

As possible see when moving on yield surface (neutral loading) have to be absent the increments of plastic forming deformation. The yield surface in strain space gives the same results for neutral deformation. However, experiments and theoretical investigations [12-14] under neutral loading both in stress space and in strain space (neutral deformation) shows that this condition is not executed. Under neutral loading on yield surfaces lead to strain growing. Under loading under constant intensity deformation the situation is vice versa, fall of stress intensity. The typical process of stress falling is expressed on Figure 1 at different value of plastic deformation. These experimental data give the cause to consider influence of complex loading on value of work under plastic deformation can be greatly lower, under simple loading.

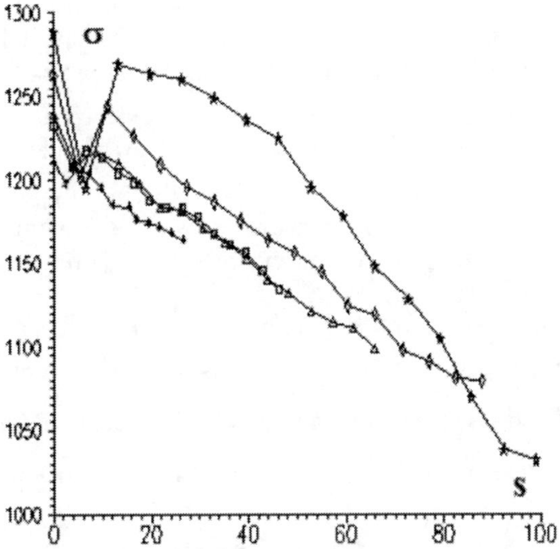

Figure 1. Intrinsic behavior of materials at neutral deformation.

Also at process of the metal processing by pressure was noticed that work of external power under different path of deformation has smaller importance, than under simple loading. Particularly is obviously at processes of loading or deformation with bend points. This effect can be explained on base of plastic theories of processes by Il'yushin [1] and principle of delay of vector and scalar characteristic of materials.

Lemma: Minimum of the work of external force for achievement of the given form under plastic deforming is reached only at realization of the strain processes with bend-point.

Proof: On the base of generalized Clapeyron theorem the work of external force on corresponding displacement is equal to accumulated potential energy of deformed rigid:

$$A=\int_V\int_0^s \sigma_{ij} d\varepsilon_{ij} dV = \int_V 3\int_0^s \sigma_0 d\varepsilon_0 dV + \int_V\int_0^s s_{ij} de_{ij} dV$$

here s is length of strain's path, de_{ij} are components of strain deviator's increments, σ_0 is a module of spherical part of stress tensor and $d\varepsilon_0$ is increment of the average relative elongation, s_{ij} - components of deviator of stress tensor.

Since work of forces on elastic deformation does not depend on strain path because field of the stress vector on deformation before start yielding is conservative. The work of hydrostatic pressure on three dimensional compressions is conservative also. It is follow assumption to fact that material is plastic incompressible.

Herein:

$$\int_0^s s_{ij} de_{ij} = \int_0^s \bar{\sigma} d\bar{Э} = \int_0^s \sigma \cos\vartheta ds = \int_0^{e_z} \sigma ds + \int_{e_z}^s \sigma \cos\vartheta ds$$

here ϑ is angle of deviation between stress vector $\bar{\sigma}$ and vector of strain's increment $d\bar{Э}$.

At active loading process the minimum of functional of the work is reached when angle of deviation is $\pm\pi/2$. It is possible only at normalcy of stress vector to Frenét's ort $\bar{p}_1$ (tangent to strain paths). Follow to principle of delay the angle of deflection cannot be constant on all deformation process. It is means that at loading before exhausting of trace of delay of vector characteristics $\bar{\lambda}$ of material is strain path must has break point.

Conclusion: At metal forming processing by pressure, stampings, etc. important is aside from the whole the possibility of optimum management for loading process. Thereby, follows it in condition should be put the optimum quantity of section of streak line for optimal solving of problem. Thereof possible to put define on which stage to produce the break for strain path. Here must know aside from classical physical parameter of material, as well as its trace of delay $\bar{\lambda}$.

Remark: At realization of the loading processes in stress space with bend-point appears the situation when scalar product of stress vector $\bar{\sigma}$ and strain's increments vector $\bar{p}_1$ can have a minimum also. However, difference between loading processes in strain and stress spaces consist in change vector characteristic material in the first event because here occurs dropping in stresses and it gives minimum work at strain process of loading.

Let's consider the multilink strain process of homogenous specimen and evaluate its work value. At plastic incompressibility of material the first invariants of stress tensor σ_{ij} and Lagrangian strain tensor ε_{ij} can be considered as external parameters. Then, plastic deformation occur cause forming only. Deviator of the stresses $\tilde{\sigma}_{ij}$ and strains $\tilde{\varepsilon}_{ij}$ are linear dependent. And only five of theirs are independent.

Let's consider the processes with break-point in Il'yushin vector space $Э^5$ [1] which generated by strain vector. The stress vector in this space is presented as accompanying to strain path.

The forming work can be expressed in the next view:

$$\int_0^s \bar{\sigma} d\bar{Э} = \int_0^s \sigma \cos\vartheta ds$$

Obviously, that under determined strain path the work at complex loading wills less than at simple one. However, impose the certain restrictions on strain path geometry are necessary. Follows previous lemma for minimization of forming work it is necessary to consider the processes with bend-point. At such choice in general case a quantity of section n, lengths of section $\Delta s^{(l)}$ and corner-angles $\beta^{(l)}$ are unknown. In real processes to reach of given points in strain or stress space are required.

Here considered the strain trajectory Figure 2 where all section of the deformation process and strain vector are coplanar. It is correspond to plane problems of plasticity.

Without restriction of generality, according to Isotropy postulate, let's consider processes in $Э_1$~$Э_2$ plane. In this case problem of the determination directing cosine of stress vector possible on the base of the Convergence Law. Enter the following values: $\alpha^{(l)}$ (l=1,n) are angles of inclination of strain path to axis O$Э_1$, $\beta^{(l)} = \alpha^{(l-1)} - \alpha^{(l)}$ (l=2,n) are corner-angles of strain path, $\gamma_l = [\bar{\sigma}\Delta\bar{Э}^{(l)}]$ (l=1,n) are angles of inclination of stress vector in bend-point to previous section of strain path, $\theta^{(l+1)} = \gamma_l + \beta^{(l+1)}$ (l=1,n-1) are angles of inclination of stress vector in bend-point to next section of strain path. The positive direction angles are defined if tumbling from previous to the following section goes clockwise.

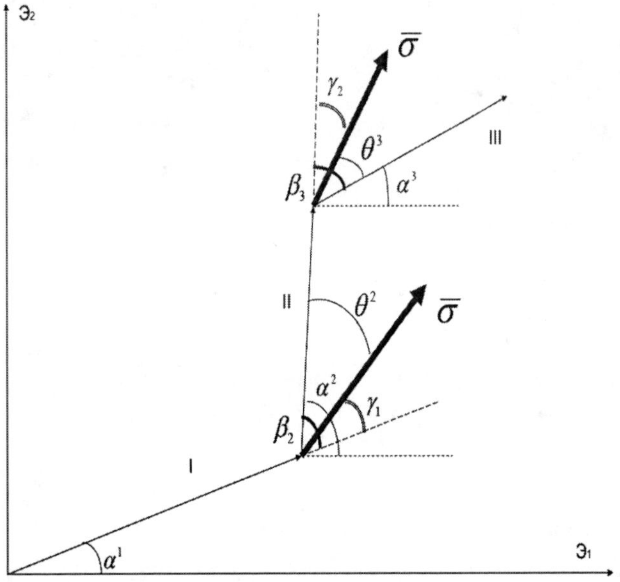

Figure 2. Strain path.

As was it earlier noted, in experiments was noticed versatility of ϑ function for each material under same break-angles in trajectory of deformation. For given reasons the empirical formulas from experiments as (1-3) for approximation of ϑ function were offered. On the base of above mentioned approximations the graph is built. Here the convergence process of stress vector and strain path is shown on Figure 3. The trace of delay accepted as: $\overline{\lambda} = 5\varepsilon_s$.

$$\vartheta = \theta / \exp(k(1 + \frac{1}{\overline{\lambda}})\Delta s), \tag{1}$$

here $k = \ln 16/(1+\overline{\lambda})$, θ - angle of bend

$$\vartheta = \exp\left(\frac{l}{\Delta s - \overline{\lambda} b}\right) - H, \tag{2}$$

here $H = 1$, $b = \frac{\ln(\theta/16+1)}{\ln(\theta/16+1) - \ln(\theta+1)}$, θ - angle of bend

$$\vartheta = C arcctg\left(\frac{r}{\overline{\lambda}}\Delta s\right), \tag{3}$$

here $C = \frac{2\theta}{\pi}$, $r = ctg\left(\frac{\pi}{32}\right)$, θ - angle of bend

There is possibility for reduction of the plastic forming work value is possible see from graph. The ratio as (1) better reflects the vector characteristic of the material before exhausting the trace of delay $\overline{\lambda}$, but quicker decreases after passing of it than in experiment. The approximations as (2) and (3) on the contrary quicker decrease, than experimental curves, but after exhausting of trace of delay $\overline{\lambda}$ approaches to experimental curve.

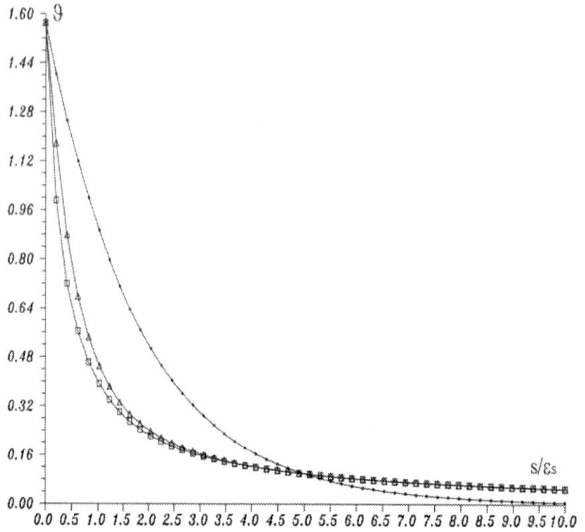

Figure 3. * - 6, □ - 7, Δ – 8, θ = π/2.

For ideal plastic material:

$\int_0^s \bar{\sigma} d\bar{Э} = \sigma_t Э[n]$ - at simple loading;

$\int_0^s \bar{\sigma} d\bar{Э} = \sigma_t \int_0^s \cos\vartheta ds$ - at complex loading.

If accept all section of multi-link trajectory equal to Δs_o all angles of deflection on each section is identical (i.e. angle of alteration are changes by one law), that:

$$\sigma_t \int_0^s \cos\vartheta ds = \sigma_t n \int_0^\Delta \cos\vartheta ds .$$

Let's show the opportunity to find the multi-link strain path with work of stress on increments of deformations will less than at same deformation under simple loading and condition when it can be reached. For simplification, consider the processes with equal angles in bend points on strain paths in the manner of zigzag ($\beta_i = \pm x$). In this case the value of strain vector is:

$|Э| = n\Delta\sqrt{\frac{1+\cos(x-\alpha_1)}{2}}$ at even quantity of sections.

$|Э| = \Delta\sqrt{\frac{n^2+1}{2} + \frac{n^2-1}{2}\cos(x-\alpha_1)}$ at odd quantity of sections.

Not limiting generalities are possible to accept α_1=0. Then for multi-link zigzag trajectory with angles of breaks equal to x the condition when work at complex loading at plastic deformation less than one under simple loading is for ideal plastic material:

$$\sigma_t n \int_0^\Delta \cos\vartheta ds \le \sigma_t n\Delta\sqrt{\frac{1+\cos(y)}{2}}$$

or $\sigma_t n \int_0^\Delta \cos\vartheta ds \le \sigma_t \Delta\sqrt{\frac{1+n^2}{2} + \cos(y)\frac{n^2-1}{2}}$

Simplifying get:

$$\int_0^{\Delta} \cos\vartheta ds \leq \Delta\sqrt{\frac{1+\cos(\gamma)}{2}} \quad (4)$$

Remark: Active processes of loading are considered here, i.e. at $\beta_i \leq \pi/2$.

3. Conclusion

This condition is sufficient, but not necessary one. It is possible to show as at case of strengthening material this condition is sufficient too. The condition as (4) even under "worst" variant is wholly workable. For it reason is enough to constraint of corner-angle with length of each link of strain path. Thereby, it is shown opportunity in spite of increase the strain path under complex loading compare with simple one to reduce external efforts for achieving the same form at plastic deformation.

References

[1] Il'yushin A A 2016 *Plasticity. Foundation of the general mathematical theory* (URSS Press, Moscow)
[2] Zubchaninov V G 2000 Problems of the mathematical theory of plasticity *Strength Mater.*
[3] Rauch E F, Gracio J J, Barlat F and Vincze G 2011 Modelling the plastic behaviour of metals under complex loading conditions *Model. Simul. Mater. Sci. Eng.*
[4] Vilks U K and Kregers A F 1967 Complex loading of a polymer material with nonlinear creep *Polym. Mech.*
[5] Bodmann E, Breuer H J, Raule G and Roedig M 1984 Material behavior under complex loading. *Nucl. Technol.*
[6] Loukidis D and Salgado R 2009 Modeling sand response using two-surface plasticity *Comput. Geotech.*
[7] Dafalias Y F and Popov E P 1975 A model of nonlinearly hardening materials for complex loading *Acta Mech.*
[8] Abirov R A 2008 On the physical reliability and taking complex loading into account in plasticity Mater. Sci.
[9] Abirov R A 2009 On the constitutive relations in plasticity at complex loading *ASME International Mechanical Engineering Congress and Exposition, Proceeding*
[10] Drucker D C 1958 Variational principles in the mathematical theory of plasticity
[11] Hodge P and Prager W 1948 A Variational Principle for Plastic Materials with Strain-Hardening J. Math. Phys.
[12] Babamuratov K S and Abirov R A 2001 On physical reliability in the theory of plasticity *Strength Mater.*
[13] Doshinski GA and Koreneva AM 1970 Plastic yielding at constant stress intensity *Mechanics of solids.*
[14] Zubchaninov V G and Okhlopkov N L 1997 Experimental investigation of the laws of plastic strain of metals along two-dimensional curvilinear trajectories *Int. Appl. Mech.*

Dynamic mechanical behavior of unfilled and graphite filled carbon epoxy composites

SudarshanRao K

Professor, Mechanical Engineering Department, ShriMadhwaVadiraja Institute of Technology and Management, Bantakal Karnataka, India

sudakapu@gmail.com

Abstract. In the present study, the viscoelastic behavior of graphite filled and unfilled carbon epoxy composite is investigated through dynamic mechanical analysis. The investigation has been conducted using a three-point bend test (flexure mode) in the range of temperature from 30 to 200° C at one Hz frequency at arate of heating of 2° C/min.The viscoelastic characteristics including loss-modulus, storage modulus, damping factor, and composite glass transition temperature have been determined. It was found that the dynamic mechanical characteristics of the carbon fabric reinforced epoxy are highly dependent on the presence of graphite particulate composites. The storage modulus, glass-transition temperature, and loss modulus of filled graphite composites were higher than the unfilled ones. However, combining a graphite filler with carbon epoxy composite limits polymer molecules movement that leads to a decrease in tan δ.

Keywords: *carbon, graphite, viscoelastic, tan delta, loss modulus, storage modulus.*

1. Introduction

Because of its high basic strength and special rigidity, composites have not grown only in the aerospace weight-sensitive field into attractive building materials, as well as in the railway, structural engineering sectors, marine, and automobile. A careful matrix selection and reinforcement process will produce a composite with overall strength and modulus equivalent or much stronger than those of traditional metallic materials. The mechanical, thermal, and physical properties could be further changed by applying a solid filler process to the matrix body during the fabrication of the composite, which would offer a synergism in terms of enhanced properties and efficiency and reduce material cost.

The in-plane tensile characteristics of the composites are generally described by fiber reinforced polymer, although the compression characteristics along the thickness direction are determined through the properties of the matrix resin. The most frequently utilized polymer matrix for advanced composite components is epoxy resin. Over the years, several experiments were formed to change epoxy by inserting either nano or micro fillers to enhance the matrix-dominated characteristics of the composites. Fiber is believed to be the most powerful strengthening stage in polymer materials and high mechanical performance of the fiber-reinforced composites. Particularly carbon fibers have excellent characteristics like greater strength and modulus, better electrical and thermal properties than other fibers. Thus, the most commonly used reinforced composites today are carbon fibers.

Dynamic mechanical measurements generally offer a better understanding of the behavior of the material under load than other tests, although other mechanical tests theoretically may convey the same information.These research methods were commonly used to analyze the structures and viscoelastic properties of polymeric materials to assess irrelevant hardness and damping properties for different applications. The rigidity depends on the mechanical characteristics of the material and is transformed into a modulus. Tan delta is represented in damping and is related to the quantity of energy that a material can store. The DMA is the most sensitive technique for regulating events like glass transitions because when relaxation behavior is detected, its mechanical properties change drastically (Nielsen [1]). When calculated on a variety of temperatures and frequents, the dynamic properties of polymer materials have significant functional importance.

The three essential parameters that must be achieved during DMA are: (1) storage modulus-that calculates the peak energy contained in a material during one oscillation period; (2) loss modulus-proportional to the quantity of energy which has been dissipated through the sample as heat; and (3) Mechanical tan delta damping term, which is the ratio of loss to the storage modulus, which relates to the degree of molecular mobility of the polymer material. Dynamic tests are adaptive to all types of resin matrix relaxation and transitions mechanisms as well as the composite morphology. Data and information produced can then be used to locate polymer systems of the glass transition temperature (Tg). Tg is the optimum transition in either the loss modulus or loss tangent curve (Nielsen [1]). Daniel and Yasmin [2] investigated the mechanical and thermal characteristics of graphite-epoxy platelet composites, suggesting that composite storage modulus and the glass transfer temperature were increased by increasing the concentration of graphite platelet. They have also found that higher graphite concentrations improve composite heat stability than pure epoxy. Huimin et al. [3] investigated the impact of nano SiO_2 on polyamide 66 composite morphology, dynamic mechanical characteristics, and crystallization. In comparison with the cleanly polyamide 66, dynamic mechanical properties showed substantial improvements in the storage and loss modulus. The tan δ peak showing the temperature of glass-transition to the maximum temperature of nanocomposites. Frederico et al. have studied the complex mechanical parameters of ramie fiber-enhanced epoxy composites [4]. In a flexural mode, studies have been considered in the temperature ranges from 20 to 200°C at 1Hz below the nitrogen level. Based on the experimental findings, they noted that the inclusion of ramie fiber appears to enhance the viscoelastic rigidity and epoxy matrix damping ability. Zhou et al. [5] conducted unfilled and multi-walled CNT (Carbon Nano Tube) filledepoxy and carbon composites mechanical and thermal characterization.The findings illustrate that the temperature of glass transition, the flexural strengths, and the temperature of decomposition have been increased by infusing CNT. Nirmal and Amar [6] have observed the viscoelastic activity of polyethylene fiber (PEF) and unidirectional fiber-glass reinforced polymethyl methacrylate composites for different relative PEF volume fractions with a mechanical dynamic analyzer. It had been reported that the relative volume of PEF increased and Tg enhanced to a higher region with a damping efficiency in the hybrid composite.

Mostly, the literature reports on the impact of nanofillers and nanofibers on the dynamic characteristics of polymers as well as polymer composites; the dynamic behavior of reinforced particulate fiber composites is being assessed with little effort. Therefore, in the current investigation, an attempt is made to research the effect of graphite filler on the viscoelastic behavior of carbon epoxy composite.

2. Experimental

2.1 Material

Unfilled and graphite filled carbon epoxy composites were fabricated by hand layup technique, followed by compression molding, were manufactured for the present study. The composite consists of a two-way plain-woven carbon ($200g/m^2$density) fabric comprising poly acryl nitrile (PAN) carbon fiber, strengthened with the matrix content of the resin epoxy (LAPOX -12), at room temperature cure polyamine hardener (K5). The filler material was used with graphite particulates 40-60 µm. The weight fraction of a graphite filler is used to produce four composite laminates. Table 1 shows the details of the samples composite manufactured with reinforcement, fillers, and a weight fraction of the matrix.

Table 1. Details of composite samples prepared.

Sample	Matrix (% Epoxy)	Reinforcement (%Carbon Fabric)	Filler (% Graphite)
Unfilled	40	60	0
2 % filled	38	60	2
4 % filled	36	60	4
6 % filled	34	60	6

2.2 Dynamic Mechanical Analysis

The mechanical characteristics of materials are measured according to temperature, frequency, and time by DMA (Dynamic Mechanical Analysis). The mechanical reaction of a sample subject to a particular temperature program is a thermal analytical process. In general, this technique is used to qualitatively characterize the viscoelastic properties of composites and polymers.

DMA measures stiffness & damping and are recorded as tan delta and modulus. The modulus represented as a phase component, a phase-out component, and storage modulus are the loss modulus since sinusoidal force is added. The viscoelastic polymers storage and loss modulus test the energy stored and lost during oscillation respectively. The first demonstrates the elastic section, and the second demonstrates the vicious section. The loss-to-storage modulus ratio is tan delta and is also referred to as damping. It is an indicator of the material's energy dissipation. For researchers, manufacturers, and consumers of polymers and composites, modulus knowledge extracted from dynamic mechanical testing is really useful for structural purposes.

The moduli for storage (E′) and loss (E″) are represented as:

$$E' = \frac{\sigma_o}{\varepsilon_o} \cos\delta \quad N/mm^2 \tag{1}$$

$$E'' = \frac{\sigma_o}{\varepsilon_o} \sin\delta \quad N/mm^2 \tag{2}$$

here strain ε_o. and dynamic stress σ_o

$$\tan\delta = \frac{E''}{E'} \tag{3}$$

A three-point bending test according to ASTM D5023[7] has been carried out in Tri-Tec 2000B (Triton Technology, UK). The parameters like storage modulus, damping factor (Tan δ), glass transition temperature, and loss modulus are measured from the DMA tests. Samples have been measured with a rate of heating of 2° C per minute at 1 Hz frequency in the range of temperature from 30-200° C.

3. Results and discussion

The design and the distribution of the matrix material, characteristics of fiber, the presence of additives like filler, the nature of the interface fiber matrix, and the mode of testing effects the dynamic mechanical behavior of polymer composites. The chemical and physical structure of the fiber in a particular matrix and the relative composition variations of each fiber and matrix can also lead to major changes in the overall dynamic mechanical characteristics of the composite (Kevin [8]).

Influence of graphite filler on dynamic properties of carbon epoxy composite

In the current research, graphite filler affects on Loss modulus (E"), glass transition temperature (T_g), storage modulus (E'), and tan δ of composites of carbon epoxy are assessed. Table 2 provides unfilled and graphite filled composites with the determined performance of the dynamic properties (Storage Modulus (E'), Loss Modulus (E"), tan δ, along with Tg).

Table 2. Dynamic properties of filled andunfilled graphite composites.

Composite Sample	Storage Modulus (E')(GPa)	Loss Modulus (E")(GPa)	Tan δ	Tg (°C)	
				By E"peal	By tan δ peak
Unfilled	16.11	3.962	0.3766	118.1	25.0
2 % filled	23.86	5.518	0.3424	118.3	25.1
4 % filled	24.72	4.210	0.2468	129.1	33.0
6 % filled	28.18	5.732	0.3703	124.2	30.6

3.1 Storage modulus (E')

The analysis of the storage modulus for a range of temperatures considered to be extremely in the evaluation of the polymer structure changes in performance characteristics. In dynamic load conditions, the storage modulus (E') indicates the inherent rigidity of the material. The storage modulus vs. temperature curve is used for determining the mechanical characteristics of the material on the molecular basis, because the molecular weight, the level of cross-connection, and the interfacial fiber-matrix bonding are very sensitive to structural changes (Pothan et al. [9]). The variance of carbon epoxy composites storage modulus with a frequency temperature of 1 Hz is shown in figure 1 for different weight percent of graphite filler.The effect of the inclusion of graphite particles in the carbon epoxy composite causes a variation of storage modulus.

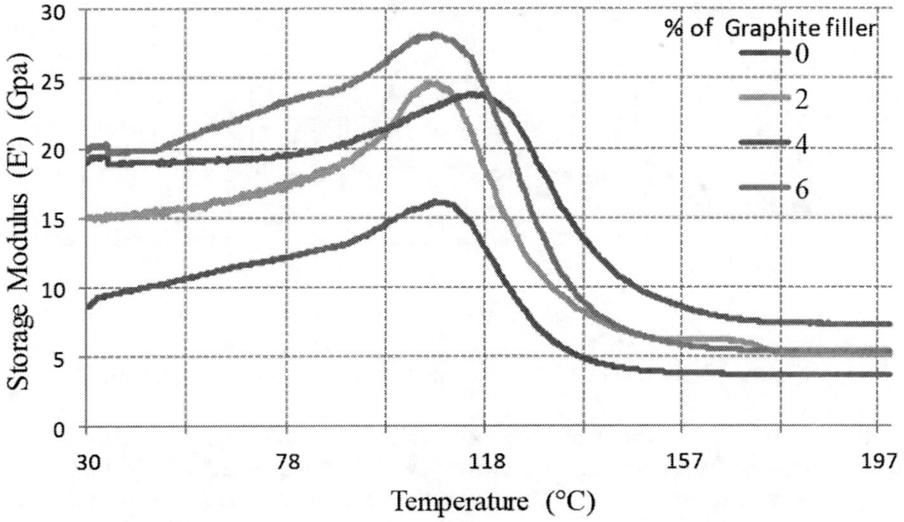

Figure 1 Variation "of E' with temperature for composite samples

The integration of the carbon epoxy composite graphite filler had a good impact on the storage modulus both at glass transition temperature (Tg) and beyond that, which confirms improved elastic characteristics of the composite by incorporating the graphite filler. This indicates that graphite incorporation increased the rigidity of the composite. The storage modulus rises to the limit in the glassy region (temperature <Tg), then falls inside the rubbery region (Tg<temperature). The decrease in modulus relates to the shift of materials from a glassy to a rubbery state.

In the glassy area E' relies on the loading of graphited that the gap between composite sample storage modulus and a different filler weight percentage in the glassy area more, but not important in the rubbery region and characteristic curves overlap. Composite filled with six percent graphite filler initially illustrated a maximum E' at room temperature and the temperature of glass transition (Tg), while for all composite samples E' declined in the rubbery region with a maximum E' of a graphite filler with 4 percent. But the graphite filled composite was higher E' than the non-filled in both the glassy and rubbery regions. The E' value was increased with the graphite particle percentage increase as shown in Table 2. The findings from Asma and Isaac [10] are close. They found that the storing modulus of pure epoxy was increased through the integration of graphite particles into epoxy nanocomposites; they also stated that the storage modulus increased consistently with the rise in the concentration of graphite particles.

3.2 Loss modulus (E")

The loss modulus is described as the energy that is dissipated or lost during material deformation as heat. It is the material's viscous response. The loss modulus is most susceptible to molecular movements. The loss factor for polymer composites varies by temperature and exceeds optimum levels at Tg since their inner damping arises from the polymer matrix's viscoelasticity. The increased damping capacity without compromising the storage module is achieved. The difference in carbon epoxy composite temperature loss modulus shows in Figure 2.

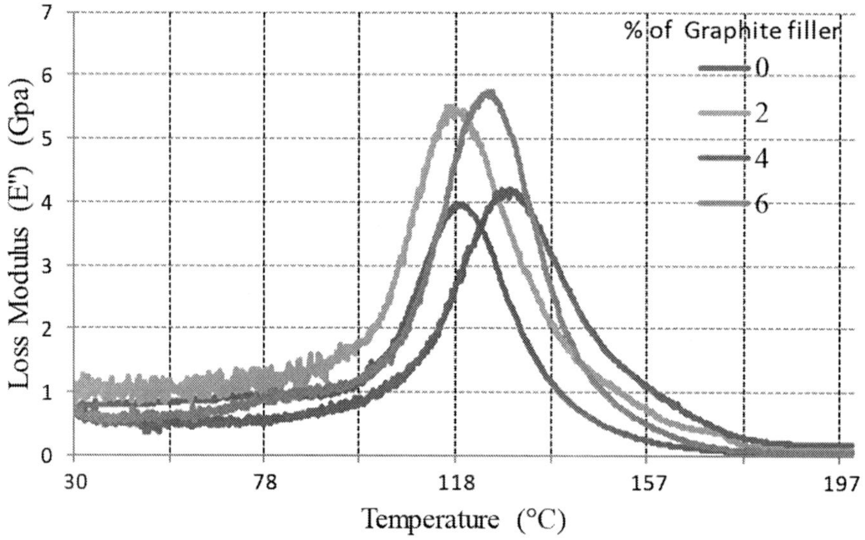

Figure 2 Variation of E"" with temperature for composite samples.

For the graphite filled composites, the loss module (E"), that represents the power for energy dissipation, was substantially higher than the unfilled one across the entire temperature range, with the highest increase at approximately Tg. This shows that energy dissipation also increases with the incorporation of graphite into the carbon epoxy composite. The graph represents that E" is minimum at room temperature and almost equal in all composite samples, so it increases as the temperature rises to Tg and then continues to decrease and it eventually reaches zero. The highest E" peak exhibited 6 percent graphite filled composite (Table 2). Huimin et al. [3] found a major improvement in storage and loss modulus by the incorporation of nano-SiO_2 when experimenting with the nano-SiO_2 filled polyamide 66 composites.

3.3 Loss tangent (Tan δ)

Damping characteristics are related to the molecular movements and phase transitions of the dynamic loss modulus (or viscous) to the dynamic storage modulus (or elastic). The damping peak happens in the region of the glass transition from a stiff state to a more elastic condition where the material is related to the movement of small molecular groups and chains within the polymer structure. Thus, Tan δ is susceptible to all molecular motions occurring in polymers. The integration of fibers into a composite system will impact damping. This is primarily due to the accumulation of shear stress at the ends of the fiber along with the further viscoelastic energy dissipation of the matrix material. The temperature function of Tan δ is illustrated in Figure 3 for all composite samples.

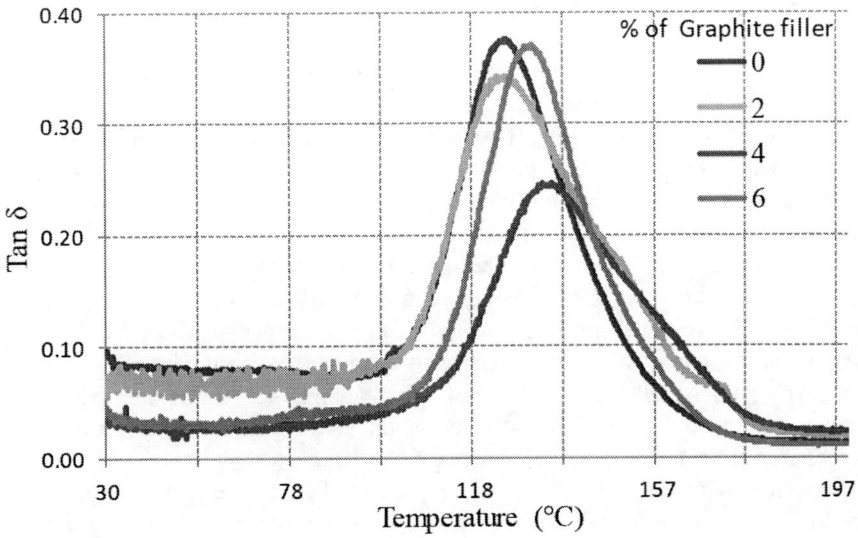

Figure 3 Variation of tan'δ"with temperature for composite samples

The thermogram indicates that tan'δ stays at the beginning nearly constant up to a certain temperature, then rises and reaches a maximum value and then reduces. This polymer behavior is typical throughout the transitional area. The integration of graphite filler in the carbon epoxy composite reduces the motion of polymer molecules which results in a lower tan δ value (Table 2). It is noted that the value of tan δ with an increased % filler content does not differ linearly. The maximum value of unfilled composite was recorded and the Tan δ'decrease was recorded at 2 and 4% and the value was further increased by the weight of graphite filling into composites by 6%. More dampness at the interface, less adhesion at the interface. Therefore, a strong interface for graphite-filled composites adhesive is reported. Basavaraj et al. [11] have stated, although substantiating the above findings, that tan δ decreases with an increase in polyamide 66/ carbon black composite molybdenum disulphide material.

3.4 Glass transition temperature (Tg)

Tg is usually known as the temperature at which the loss tangent (tan δ) peaks occur; the temperature at loss modulus peaks (E") versus the temperature curve is also often defined. Tg as a result of E" peak and not tan δ peak results in a way that is in line with the standard of stress safety of the matrix with fibers in composites. The temperature of Tg via E" peak shows even more specifically that stiffness has deteriorated significantly. Thus the, Tg by E" is more reliable and more acceptable than the one dependent on the tan δ peak for advanced composite materials (Akay [12]). Table 2 indicates the value of Tg defined by the peak E" and the peak tan δ. From the table, it can be noted that the tan δ peak temperature is above than peak of E". Tg difference calculated for these 2 maximum values is between 4 and 8° C. Such as tan δ, for Tg there is no linear trend with an increase in the proportion of graphite filler is observed, the maximum value of 4% is reported as a graphite filled composite.These findings comply with those of Akay [12] which recorded similar results for the single-direction carbon epoxy or carbon fiber laminates. Furthermore, he observed that glass transition temperatures filled with graphite composites are more than unfilled. Kumaresan et al. analyzed the dynamic mechanical behavior of the silane-treated SiC particulate filled carbon epoxy composites[13]. From the outcomes, it was identified that the SiC-filled composite glass transition (Tg) temperature was higher than the unfilled composite. Basavaraj et al. [14] stated that polycarbonate and carbon black composite glass transition temperature increased as molybdenum disulfide content increased [14].

4. Conclusions

The following conclusions are made from the dynamic mechanical tests of the carbon epoxy composite:

- The dynamic mechanical characteristics of the composites heavily dependent on the presence of graphite particles.
- The incorporation of the graphite filler has had a positive impact on the storage modulus above and below Tg, suggesting that composite elastic characteristics have been enhanced due to the inclusion of the graphite filler. The value of E' increased by an increase in graphite particulate percentage.
- The loss modulus that represents the potential of energy discharge for graphite filled composites, has been significantly higher than the unfilled over the entire temperature range examined and the rise was maximal around Tg. This shows that energy dissipation is also increased with the introduction of graphite into carbon epoxy composites.
- Addition of graphite filler to the carbon epoxy composite limits the motion of polymer molecules to reduce tan δ. Higher the damping at the interface, the poorer the adhesion interface. Therefore, the graphite filled composites have a strong interface connection.
- The glass transition temperature (Tg) is represented as the temperature of the tan δ peak reached E". The gap in Tg value determined by these two peak values is between 4 and 8°C. Further, Graphite filled composites have more Tg than unfilled ones.

References

1. Nielsen L.E. (1974). Mechanical Properties of Polymers and Composites, Marcel Dekker: New York.
2. Yasmin A. and Daniel I.M. (2004). Mechanical and thermal properties of graphite platelet/epoxy composites, *Polymer*, **45**: 8211–8219.
3. Huimin Lu, XiangminXu, Xiaohong Li, and Zhijun Zhang. (2006). Morphology crystallization and dynamic mechanical properties of PA66/nano-SiO_2 composites, *Bulletin of Material Science*,**29**(5): 485–490.
4. FredericoMuylaertMargem, Sergio NevesMonteiro, Jarbas Bravo Neto, Ruben Jesus Sanchéz Rodriguez, and Bluma Guenther Soares. (2010). The dynamic-mechanical behavior of epoxy matrix composites reinforced with ramie fibers, *RevistaMateria*,**15**(2): 164–171.
5. Yuanxin Zhou, FarhanaPervin, Lance Lewis andShaikJeelani. (2008). Fabrication and characterization of carbon/epoxy composites mixed with multi-walled carbon nanotubes, *Materials Science and Engineering A*,**475**: 157–165.
6. NirmalSaha and AmarNath Banerjee. (1996). Dynamic mechanical study on unidirectional polyethylene–glass fibers: PMMA hybrid composite laminates, *Journal of Applied Polymer Science*, **62**(8):1199–1204.
7. ASTM D5023.(2007). Standard Test Method for Plastics: Dynamic Mechanical Properties: In Flexure (Three-Point Bending), ASTM Annual Book of Standards, West Conshohocken, PA, Vol. 08.02.

8. Kevin P.Menard. (2008). Dynamic Mechanical Analysis: A Practical Introduction, Second Edition, CRC Press.
9. Pothan L.A., Oommen Z., and Thomas S. (2004). Dynamic mechanical analysis of banana fiber reinforced polyester composites, *Composite Science and Technology*,**63**: 283-293.
10. AsmaYasmin and Isaac M. Daniel. (2004). Mechanical and thermal properties of graphite platelet/epoxy composites, *Polymer*, **45**: 8211–8219.
11. E. Basavaraj, B. Ramaraj, Joong-Hee Lee, and Siddaramaiah. (2013). Microstructure, Thermal, Physico–Mechanical and Tribological Characteristics of Molybdenum Disulphide-Filled Polyamide 66/Carbon Black Composites, *Polymer Engineering, and Science*,**53**(8): 1676-1686.
12. M. Akay. (1993). Aspects of dynamic mechanical analysis in polymeric composites, Composites Science and Technology, 47(4): 419–423.
13. Kumaresan K., Chandramohan G.,Senthilkumar M., and Suresha B. (2012).Dynamic mechanical analysis and three-body wear of carbon–epoxy composite filled with SiC particles, *Journal of Reinforced Plastics and Composites*, **31**(21): 1435-1448.
14. E. Basavaraj, B. Ramaraj, and Siddaramaiah. (2012). Polycarbonate / Molybdenum Disulfide/Carbon Black Composites: Physicomechanical, Thermal, Wear, and Morphological Properties, *Polymer Composites*,**33**(4): 619-628.

The Effects of the Preservation Conditions on Physical Properties and Bio-active Substance of Garlic (*Allium sativum L.*)

T C Q Ngo [1,2,*], D P Nguyen[2], V K Tran[3,4], H D Le[5], V T Pham[5] and B L Tran[6]

[1]Faculty of Environmental and Food Engineering, Nguyen Tat Thanh University,Ho Chi Minh City, Vietnam
[2]Institute of Environmental Sciences, Nguyen Tat Thanh University, Ho Chi Minh City, Vietnam
[3]Nong Lam University, Ho Chi Minh City, Vietnam
[4]Department of Science and Technology of Ninh Thuan Province, Vietnam
[5]Faculty of Food Science and Technology, Ho Chi Minh City University of Food Industry, Vietnam
[6]College of Agriculture, Can Tho University, Can Tho City, Vietnam

*Corresponding author: ntcquyen@ntt.edu.vn

Abstract: Garlic (Allium sativum L.) is a plant species native to Central Asia and has been widely grown in other regions such as Mediterranean regions as well as Asia, Europe and Africa. Determining the optimal time of storage is very important in maintaining stable garlic agricultural supply chain. In this study, we aim to identify optimal storage conditions and packaging materials that allow maintaining highest nutritional values of garlic. Garlic was harvested and stored at different temperatures and storage time. A wide range of quality parameters of garlic were taken into account including total polyphenol content, total flavonoid content, antioxidant activity, free radical scavenging activity, moisture and weight loss. The results of the study show that, at 0 °C, garlic bulbs showed minimal mass weight loss (0.49%) and stabilized polyphenol content. The change of DPPH free radical activity and the total flavonoid content in garlic bulbs stored under different temperature conditions were minor. Lastly, net cloth packaging introduced favorable loss of weight of garlic bulbs during prolonged storage.

Keywords: *Garlic (Allium sativum L.), Storage conditions, Physical properties, Bio-active substance*

1. Introduction

Garlic (*Allium sativum L.*) is a plant species of the Onion family, meaning that is related to onions, onions, purple onions, leeks ... etc, originating in Central Asia with history. now more than 7,000 years in the field of food, medicine ... [1]. The latest study (2003) has published that eight distinct groups of garlic have evolved; 6 groups of hardnecked garlics (hard-necked garlic) are called Asiatic, Creole, Purple Stripe, Marbled Purple Stripe, Porcelain, Rocambole and 2 groups of softnecked garlics (soft-neck garlic) are Artichoke and Silverskin. A recent study found that 17 different types of garlic evolved from the eight groups above. In fact, around the world, there are hundreds of different types of garlic developed from the 17 basic species above. They have different properties because they are grown in different soil conditions such as: soil fertility, rainfall, temperature, altitude above sea level, time of year sowing and watering regime. ...

The most commonly used part of the whole garlic plant is the garlic bulb. Garlic bulb has many cloves. Each clove of garlic as well as the whole garlic bulb has a thin protective skin. Garlic grows well in hot and humid environments. Store in a cool, dry place. Garlic is believed to have antibiotic properties and increase the ability to prevent cancer, fight high blood pressure, and blood fat in humans. However, it is necessary to peel the garlic and leave it in the air for a while and then eat it raw, it will have a higher anti-cancer effect. Some peoples of the world believe that garlic helps them fight evil and evil.

In the word, 2006, Vázquez-Barrios studied and predicted quality changes of garlic (Allium Sativum L.) stored at different temperatures. The results of Vázquez-Barrios show that it is possible to estimate the shelf life of garlic using the internal germination index [2-3].

In 2014, Alexandra Zakarova and Ji Yeon Seo in Korea studied that sprouted garlic extract for 5 days had the highest antioxidant activity, while raw garlic extract had relatively low antioxidant activity. Furthermore, sprouting changed the metabolic profile of garlic: the metabolite profile of garlic sprouted in 5 - 6 days was different from the metabolic profile of germinated garlic in 0 - 4 days, edema. is consistent with the finding that garlic sprouted for 5 days had the highest antioxidant activity. Therefore, sprouting can be a useful way to improve garlic's antioxidant potential [4]. In 2015, Alena Tomšik also studied the problem of optimizing the extraction with ultrasonic assistance of bioactive compounds from garlic [5].

In the Vietnamese, in 2016 the research in VietNam was studied on changes in the physical and chemical properties of garlic bulbs during maturity and storage. In addition, other research group also studied the changes of some chemical components and bioactive compounds in garlic cloves (Allium sativum L.) during storage for industrial processing. future garlic products such as black garlic, pickled garlic [6-7].

Garlic is a miraculous natural healing herb. According to traditional Vietnamese and Chinese medicine, garlic has a spicy taste, temperament, a slightly poisonous taste, goes into two cans and tastes. Garlic has the effect of clearing heat, detoxifying, inducing, disinfecting, treating ice, vestibular, blood dysentery, disinfecting, clearing, killing boils and lymph nodes in the lungs, eliminating sputum, bloating, and having difficulty urinating, cholera, dysentery, ... Since ancient times, people have known that eating garlic will remove the fluke from the intestine, used as an antiseptic in all kinds of asthma, pneumonia, intestinal infections, skin infections. Garlic is very useful for people who have snake bites and hemorrhoids if they are cut with alcohol and then drink. With salt and garlic oil cure all skin rashes; mixed with honey to make garlic ointment cure psoriasis, scabies and red spots caused by weak liver. Oriental medicine advises not to use garlic in negative, internal heat, maternity, eye pain, hat .... [8]

Currently, in many countries around the world, scientists have researched and discovered that in garlic there are silen and microelements containing antibacterial allin that enhance the ability of the immune system, reduce blood. High pressure and many health-promoting diseases. Some results in research obtained by scientists have been documented [9] such as: reduce cholesterol to prevent cardiovascular disease, prevent cancer, reduce arthritis pain, in addition have beneficial effects on longevity, slow aging, ...

Therefore, given the importance of garlic bulbs, in this study we determine the appropriate harvesting time and storage conditions (temperature and packaging) in order to select high quality and limited sources of garlic raw materials. losses to the lowest level through a change in physical and chemical properties.

2. Material and method

2.1. Material

Solo garlic material was purchased at the Hoc Mon agricultural wholesale market in Ho Chi Minh City, Vietnam.

Figure 1. Solo garlic

2.2. Method

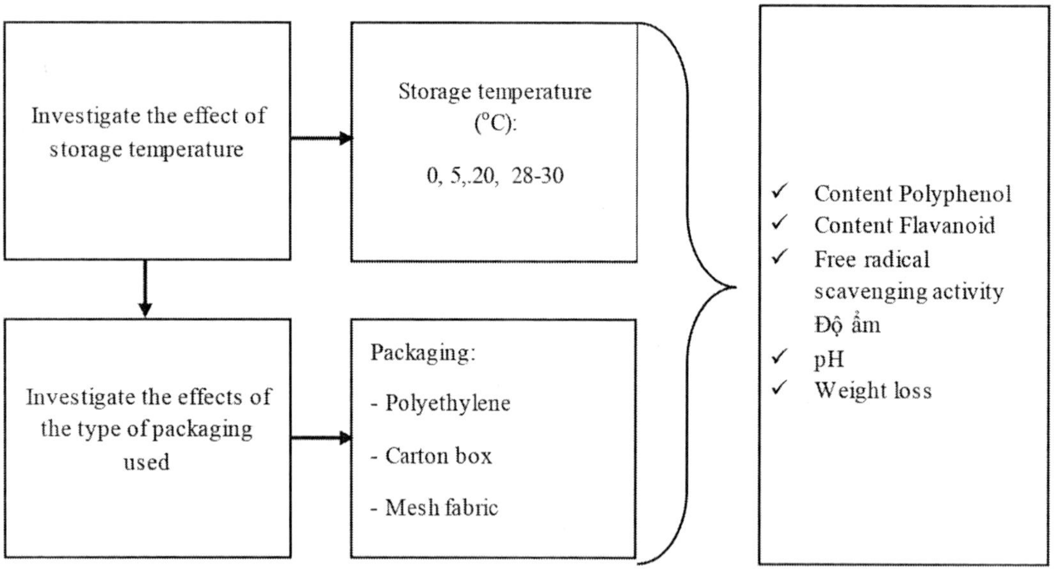

Figure 2. Research diagrams

2.3. Antioxidant activity determination (DPPH Test)

Principle: Based on the 2,2-diphenyl-1-picrylhydrazyl free radical capture capacity of the 2,2-diphenyl-1-picrylhydrazyl antioxidant which is a characteristic purple, stable free radical in solvent and have maximum absorption at 517 nm [10]. In this method, antioxidants yield hydrogen atoms to reduce the stable free radicals in 2,2-diphenyl-1-picrylhydrazyl. When antioxidants clear the free radical activity, the characteristic purple color fades away and turns yellow, the more yellow the color changes, the stronger the resistance to oxidation. The absorbance is measured at 517nm to determine the antioxidant capacity of 2,2-diphenyl-1-picrylhydrazyl. Vitamin C is used as the standard, the results are expressed in terms of vitamin C equivalent content.

2.4. Determination of polyphenol content

Principle: In the composition of the reagent Folin - Ciocalteu has a complex of phosphorus - wolfarm - phosphomolybdat. This complex will be reduced by polyphenol compounds to create a reaction product with blue color, maximum absorption at 765 nm [10]. The polyphenol content in the sample is directly proportional to the sample strength and is calculated as gallic acid.

2.5. Determination of total flavanoid content (TFC)

Principle: Based on the strong color properties of flavonoids with metals, Al3+ is often used for investigations because it is a non-toxic blue complexing metal. This complex has the strongest absorption at 510nm. The color intensity of the reaction mixture is proportional to the concentration of flavonoid compounds present in the trich and is measured by a colorimetric spectrophotometer. Quercetin is used as a standard plot to determine total flavonoid content in analytical sample with a provincial unit of mg QE/g dry matter used for extract collection.

2.6. Determination of moisture

According to TCVN 10788 - 2015

Principle: Heat is used to evaporate the water present in the sample. From the difference in mass of the sample before and after drying, calculate the moisture content of the sample. Dry the sample at 100 - 105oC to constant weight, then the free water in the sample will evaporate.

2.7. Loss of mass

Determine initial mass (md) of sample and weight after storage time (mc).

Loss of mass (%) is calculated by:

$$\omega = \frac{m_d - m_c}{m_d} \times 100$$

Inside:

- md: Initial weight (g)
- mc: Weight after storage time (g)

2.8. Methods of pH determination

Method of use: Use a benchtop pH meter - inoLab pH 7110

3. Results and discussion

3.1. Effect of storage temperature on changes in physical and chemical composition of garlic bulbs

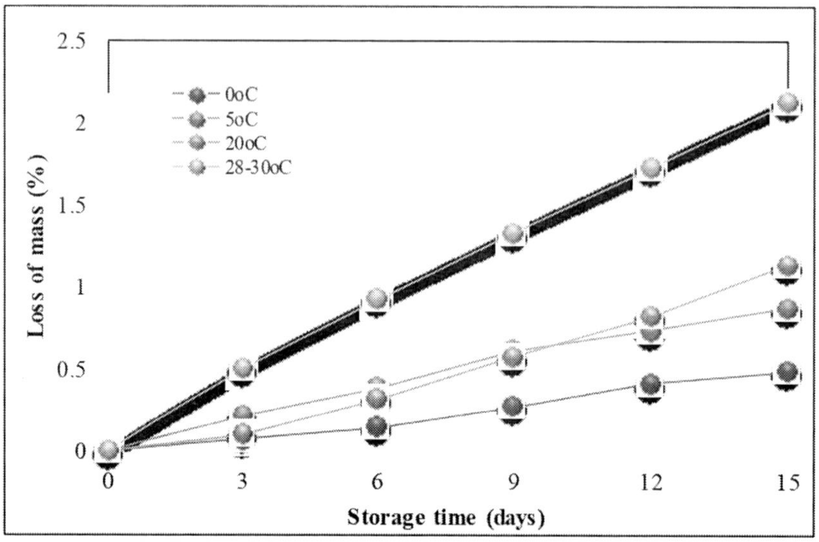

Figure 3. Weight loss (%) of garlic bulbs at storage temperatures

Through Figure 3, we can see that the weight loss of garlic bulb increases during storage at different temperature conditions. As we can see, the higher the temperature, the greater the respiration of the plant, the more the loss of matter will lead to the increase. So this result is completely consistent. 0oC (0.49%). Garlic is insensitive to freezing and temperatures of 0°C remain above garlic's freezing point (Cantwell et al., 2003) [11]. A higher storage temperature results in a fast rate of water evaporation. Although the RH under low storage conditions, due to the composition of the garlic bulb is covered with many layers of silk and thus minimizes dehydration. To explain this, the authors Iglesias-Enriquez and Fraga (1998) argued that: the physiological process of garlic bulbs interact with the environment and affect the transport of water vapor, germ and root growth changes. The shape of the garlic bulb, which causes tension and disruption of surface tissue, facilitates dehydration [12].
Figure 4 shows that, the total polyphenol content varies complexly at the storage temperatures and tends to decrease. The total polyphenol content was less variable at 0°C storage temperature than at remaining storage temperatures. In particular, when storing garlic bulbs at 20oC, the most variable total polyphenol content. This is probably due to the ongoing physiological sprouting in the garlic bulb, altering the polyphenol content (Zakarova et al., 2014) [13].

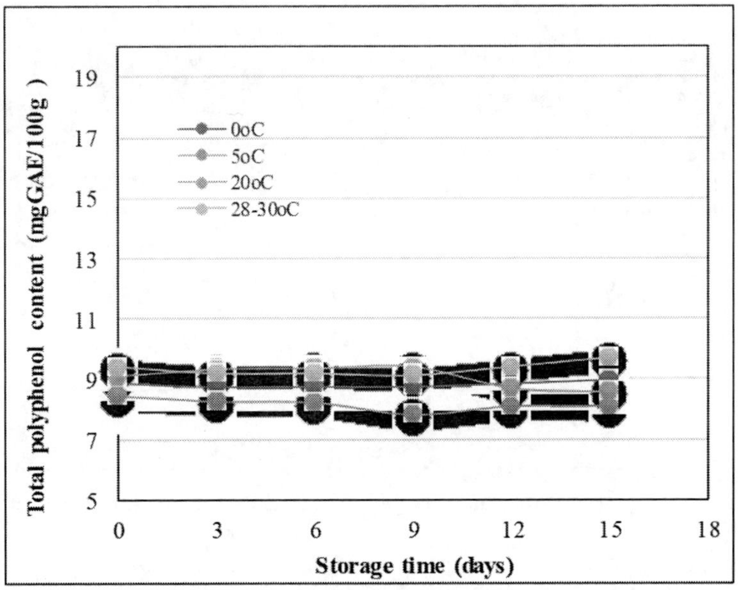

Figure 4. Changes in total polyphenol content in garlic bulbs stored under temperature conditions

Besides a decrease in the total polyphenol content, the antioxidant activity of the garlic bulb remained stable during storage. Analysis results showed that the free radical scavenging activity of garlic bulbs stored at temperatures of 0, 5, 20 and 28 - 30oC represents 7.07-7.5% respectively; 6.76-6.91%; 6.54-7.02%; 6.73-7.55% is shown in Figure 5. In addition to polyphenols with antioxidant activity, the organic sulfur compounds in garlic bulbs also function similarly S. Gorinstein, 2006 [14].

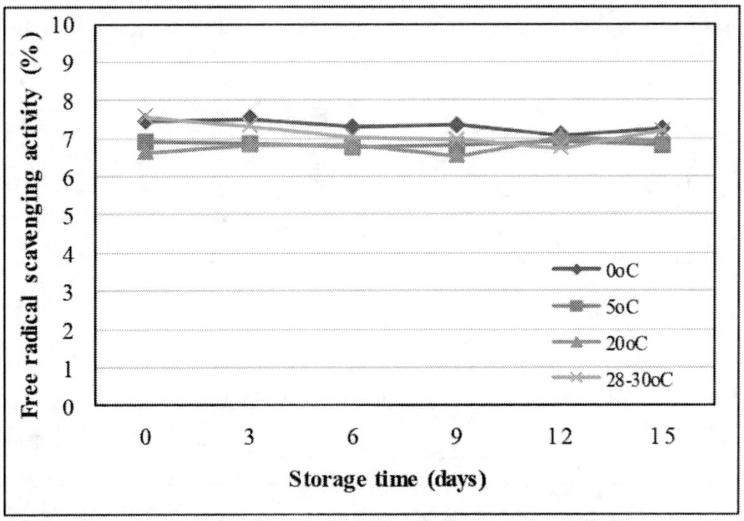

Figure 5. Changes in free radical scavenging activity in garlic bulbs stored under temperature conditions

Flavonoids measurement results are shown in Figure 6. In general, the flavanoids content of garlic bulbs was little changed during storage at 0, 5, 20 and 28-30oC. They were in the range 0.659-0.774 mg QE/g compared with studies with much smaller total flavanoids such as in Tomsik's study where flavanoids were 1.3 mg QE/g [15].

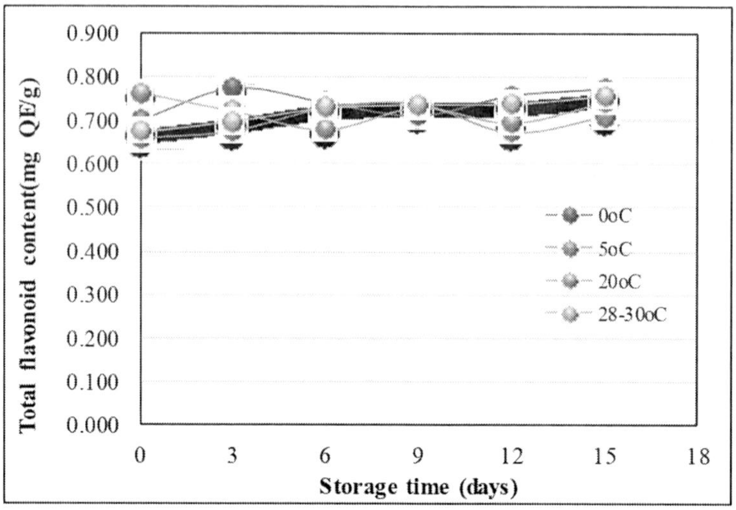

Figure 6. Total flavonoid content in garlic bulb stored under temperature conditions

3.2. The effects of the packaging types on the weight loss of the garlic bulbs during storage

Effect of packaging type on garlic bulb weight loss during storage is shown in Figure 7. Weight loss increased with time of storage in all 3 packaging types. Net cloth packaging recorded the highest loss of volume (1.01%), while carton packaging recorded the lowest loss of moisture (0.66%). However, the type of packaging did not have a significant effect on the loss of mass. On the other hand, mesh fabric packaging shows usability, cheapness and good ventilation compared to the other two types of packaging.

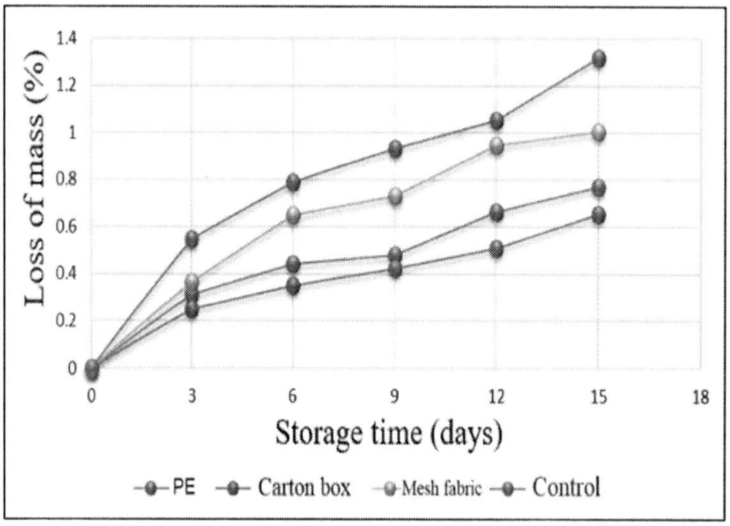

Figure 7. Weight loss of garlic bulbs stored in different packaging at 0oC

4. Conclusions

This study has evaluated the effect of the storage parameters on the change of the physical and bioactive properties of the garlic bulb. Regarding the effect of storage temperature on the changes in the physical and chemical composition of garlic bulbs, the storage of 0 °C seemed to result in lowest mass loss of garlic bulbs (0.49%). The change of DPPH free radical activity in garlic bulbs stored under temperature conditions and total flavonoid content in garlic bulbs stored under different temperature conditions had changed but not significantly. As for the change in total polyphenol content in garlic bulbs stored at temperature conditions, the polyphenol content was stable during storage at 0 °C. Regarding the effect of the packaging type on the weight loss of the garlic bulbs during storage, net cloth packaging offered good garlic shelf-life in terms of mass loss and and takes up less volume than cardboard cartons and PE during storage andtransport.

Acknowledgement: This research is funded by Nguyen Tat Thanh University, Ho Chi Minh City, Vietnam.

References

[1] Ankri S and Mirelman D 1999 Antimicrobial properties of allicin from garlic *Microbes and Infection* **1** 125–9
[2] Pham T N, Nguyen V T, Toan T Q, Cang M H, Bach L G and Van Muoi N 2020 Effects of Various Processing Parameters on Polyphenols, Flavonoids, and Antioxidant Activities of Codonopsis javanica Root Extract *Nat. Prod. Commun* 15 1934578X2095327
[3] Vázquez-Barrios M E, López-Echevarría G, Mercado-Silva E, Castaño-Tostado E and León-González F 2006 Study and prediction of quality changes in garlic cv. Perla (Allium sativum L.) stored at different temperatures *Scientia Horticulturae* **108** 127–32
[4] Ichikawa M, Ide N and Ono K 2006 Changes in Organosulfur Compounds in Garlic Cloves during Storage *J. Agric. Food Chem.* **54** 4849–54
[5] Zakarova A, Seo J Y, Kim H Y, Kim J H, Shin J-H, Cho K M, Lee C H and Kim J-S 2014 Garlic Sprouting Is Associated with Increased Antioxidant Activity and Concomitant Changes in the Metabolite Profile *J. Agric. Food Chem.* **62** 1875–80
[6] Pham T N, Nhan L T H, Cang M H, Lam T D, Thu D N A, Nguyen T Q and Danh V T 2020 Evaluation of polyphenol, anthocyanin content and antioxidant capacity of some Vietnamese fruits *IOP Conf. Ser.: Mater. Sci. Eng.* 736 062016
[7] Dao TP, Nguyen DC, Nguyen DT, et al. 2019 Extraction process of essential oil from Plectranthus amboinicus using microwaveassisted hydrodistillation and evaluation of it's antibacterial activity. *Asian J Chem.* 31 977-981
[8] Pham TN, Tran BP, Tran TH, et al. 2019 Response surface modeling and optimizing conditions for anthocyanins extraction from purple sweet potato (Ipomoea batatas (L.) Lam) grown in Lam Dong Province, Vietnam. *IOP Conf Ser Mater Sci Eng*. **479** 012012.
[9] Nguyen N Q et al 2020 Phytochemical screening and antioxidant potential of crude drug "Cao Khai" in Ninh Thuan Province, Vietnam *IOP Conf. Ser.: Mater. Sci. Eng.* **991** 012016
[10] Nyangena I O, Owino W O, Imathiu S and Ambuko J 2019 Effect of pretreatments prior to drying on antioxidant properties of dried mango slices *Scientific African* **6** e00148
[11] Cantwell M I, Kang J and Hong G 2003 Heat treatments control sprouting and rooting of garlic cloves *Postharvest Biology and Technology* **30** 57–65
[12] Iglesias-Enriquez I, Fraga R Envase y forma de almacenamiento adecuado para la conservacion poscosecha del ajo irradiado y sin irradiar 1998.
[13] Zakarova A, Seo J Y, Kim H Y, Kim J H, Shin J-H, Cho K M, Lee C H and Kim J-S 2014 Garlic Sprouting Is Associated with Increased Antioxidant Activity and Concomitant Changes in the Metabolite Profile *J. Agric. Food Chem.* **62** 1875–80
[14] Gorinstein S, Leontowicz H, Leontowicz M, Drzewiecki J, Najman K, Katrich E, Barasch D, Yamamoto K and Trakhtenberg S 2006 Raw and boiled garlic enhances plasma antioxidant activity and improves plasma lipid metabolism in cholesterol-fed rats *Life Sciences* **78** 655–63
[15] Tomšik A, Pavlić B, Vladić J, Ramić M, Brindza J and Vidović S 2016 Optimization of ultrasound-assisted extraction of bioactive compounds from wild garlic (Allium ursinum L.) *Ultrasonics Sonochemistry* **29** 502–11

Optimisation of Drilling Parameters of Metal Matrix Composites using Genetic Algorithm in the Taguchi Method

C V Subba Rao[1*], Y Sesha Rao[1*], P Marimuthu[12]*, R Jeyapaul[3], N S Kalyan Chakravarthy[4]* and P Murugesan[5]

[1]Department of Mechanical Engineering,
[3]Department of Production Engineering, National Institute of Technology, Tiruchirappalli, India
[4] Department of Data Science and Business Systems
*QIS College of Engineering and Technology, Vengamukkapalem, Ongole, Prakasam District, Andhrapradesh-523 272, India
[5] Department of Mechanical Engineering, College of Engineering, Al Imam Mohammed Ibn Saud Islamic University, P.O Box 5701, Riyadh 11432, Kingdom of Saudi Arabia
[2]Email: pmarimuthu69@gmail.com (Corresponding author)

Abstract. This work presents the machining parameters optimization on machining metal matrix composites using genetic algorithm in the Taguchi method with the objectives of minimization of surface roughness, cutting force, maximization of metal removal rate, cutting power, specific cutting energy, cutting power and machining time. Experiments were conducted on aluminium silicon carbide Metal Matrix composite using TiN coated HSS twist drills. Optimal setting of parameters have been identified using Taguchi Technique and Genetic algorithm. Genetic algorithm is developed in 'C' language. The presented work gives a better results and it can be used for any machining parameter optimization in solving multiple response problems.

1. Introduction

Optimization of the operating parameters for radial drilling machine is an important step in order to utilize the machine effectively and efficiently. Selection of machining parameters relies heavily on the operator's skills and his experience. Tables delivered by the machine builder cannot come across the operator requirements. So in order to optimize the machining parameters, it is required to frame the machining process and conditions as an optimization problem.

Metal matrix composites (MMC) are the materials, which are having a hard ceramic reinforcement. MMC a novel class of engineering materials which has sparked a great interest amidst the design engineers for its improved properties. Aluminium-matrix composites are generally used in automobile industries, aerospace industries and precision application etc. Due to the flaring range of application of MMC's the machining of these materials has become a very essential subject for research. Previous studies provided limited information on drilling parameters of optimization of AL/Sic Composites.

The optimum setting parameters such as Cutting speed, Feed, and Point angle are obtained by using Genetic Algorithm and Taguchi method. Multi responses such as minimization of surface roughness, allowable Cutting force, minimization of cutting power, specific cutting energy, torque, machining time and maximization of metal removal rate are considered in this work.

2. Literature survey

Advanced wear resistance with increased oxidation resistance of TiN coating acts as thermal barriers between the tool and the chip for dry drilling machining. Feed rate influence more followed by speed for HSS drills in Al/SiCp on burr height. Francicus and Juze Balic, (2004) & Kishalay Minra and Ravi Gopinath (2004) have optimized the cutting process by Genetic Algorithms. They stated that Genetic Algorithm approach can obtain near-optimal solution. It can be used for obtaining optimal setting machining parameters of complex machined parts which required many machining constraints.

Davim (2003) used Taguchi method for studying optimized parameters and cutting time on drilling MMC's. The hardness of the tool material influenced on the cutting edge wear and on the drilling torque, surface finish and thrust force. Marimuthu and Chandrasekaran (2011) used Taguchi and Neural network for optimal machining parameter study. Kaviarasan et al (2019) used Neural Network for drilling parameter optimization of Derling.

Saravanan et al. (2003) uses the genetic algorithm and simulated annealing algorithm for machining parameters of turning operation. The range of independent variables such as speed, feed and point angle was selected based on the machining data hand book (Alok Nayar 2002), past experience and the available resource. Ranganathan and Senthilvelan (2011) used Multi-response optimization of machining hot turning using grey analysis. Suresh et al. (2014), conducted experiments for finding optimum major intervening parameters in microEDM of Stainless Steel (SS) 316L. Tripathy and Tripathy (2017) conducted experiments on of H-11 die steel for Multi-response optimization using grey relational analysis and Topsis. Sanjay Krishna and Pavan Kalyan (2018) used Multi-Response Optimization of Wire-EDM Process Variables for Machining Process. So far very few have discussed the multi response optimization for machining parameter optimization. Genetic algorithm in the Taguchi method is not used by the previous researchers for drilling Metal Matrix Composites. Ram Prasad et al. (2019), conducted experimentation for solving Multi-Response Optimization of Machining Process Parameters for Wire Electrical Discharge Machining of Lead-Induced Ti-6Al-4V Alloy Using AHP–TOPSIS Method. Multi-response optimization of wire electric discharge machining parameters of Ti-3Al- 2.5V alloy using Taguchi integrated grey relational analysis" was used by Babu et al. (2020).

Taguchi method is the method for solving a single response problem. Multi response problems are solved using Taguchi method. More than one quality characteristic is to be solved in actual practice (Jeyapaul et al. (2006)).

3. Experimental setup (Noorul haq et al 2008)

Experiments are conducted using radial drilling machine. Aluminum silicon carbide metal matrix composites plates were used as wokpiece. Blackened HSS twist drill coated with TiN of 10 mm diameter were used throughout experiment. No cutting fluid during drilling. The surface finish of drilled hole was measured with the aid of a surf test III.

4. Plan of experiments

L_9 orthogonal array is used at three factors at three levels which is shown in the table 1. (Noorul haq et al. 2008).

Table 1 Parameters and level (Noorul haq et al. (2008)

PARAMETERS	LEVELS		
	1	2	3
Cutting speed (m/min)	35.18	56.54	87.96
Feed rate (mm/rev)	0.050	0.125	0.200
Point angle (deg)	90	115	140

5. Methodology

Step by step procedure involved in GA to solve multi response problems is detailed below.
Step 1: Calculate the S/N ratio and it is to be normalized using the formula (Noorul haq et al (2008)
Step 2
Genetic algorithm:
Process of Calculating fitness value using the normalized S/N ratio values, Initialization, Evaluation, Selection, Crossover, Mutation and Termination test are carried out as per Kishalay Minra and Ravi Gopinath (2004)
Step 3
Calculate weighted S/N ratio value using the following formula

$$WSN_i = W_1Z_{11} + W_2Z_{12} + ... + W_jZ_{ij}$$

Step 4
Find the optimal setting of parameters: To calculate the effect of parameter i, calculate the average of weighted S/N ratio values (WSN) for each level j, denoted as WSNij, then the effect, Ei, is defined as:

Ei=max (WSNij) – min (WSNij) (5.1)

If the factor i is controllable, the best level j*, is determined by

j*= max (WSNij) (5.2)

Step 5
Execute ANOVA to identify the significant parameters.

6. Implementation of Genetic Algorithm

Population of 20 chromosomes is generated to a feasibility condition and the fitness value is calculated using the following formula.

$$f(x) = \sum_{j=1}^{k}\sum_{i=1}^{n} W_jZ_{ij} \quad (6.1)$$

Where f(x) is the total weighted SN (WSN) ratio and it is to be maximized, Wj = weights, Zij = normalized SN ratio values, n=number of observations, and k = number of responses.
The individuals are sorted for ranking selection. Ps is allotted to the best individual and rank 1 to the worst individual
To find the optimal weights, a modified two-point crossover (MTPX) operation used in the algorithm.

6.1. Check for feasibility

To continue feasibility, the following algorithm is used. Randomly chosen two crossover points is indicated by the arrow.

6.1.1. Algorithm for obtaining feasible condition

$$\sum_{i=1}^{n} Wi = 1$$

Step1:W1+W2+W3+.......Wn= δ

Step 2: if δ = 1 then

Go to **mutation**

Step 3: if $\delta < 0$ then

m = 1 - δ and l = m/ (n-2)

Add l to Wi for i=2 to n-1

Step 5: if $\delta > 0$ then

m = δ -1 and l = m/ (n-2)

Subtract l from Wi for i=2 to n-1

Parent 1: feasible condition

↓ ↓

0.1	0.15	0.15		0.2	0.3	0.05	0.05	1.0

Parent 2 : feasible condition

↓ ↓

0.05	0.1	0.2	0.15	0.05	0.3	0.15	1.0

After interchanging the parent chromosomes, the following offspring are produced.

Offspring 1: feasible condition

0.05	0.15	0.25	0.2	0.3	0.05	0.15	1.05

Offspring 2: feasible condition

0.1	0.1	0.2	0.15	0.05	0.3	0.05	0.95

After crossover, the feasible condition should be verified as per Jeyapaul et al. (2006).

Offspring 1: feasible condition

0.05	0.14	0.14	0.19	0.29	0.04	0.15	1.0

Offspring 2: feasible condition

0.1	0.11	0.21	0.16	0.06	0.31	0.05	1.0

6.2. Mutation
Mutation process gives new offspring as per the following.

Before mutation: feasible condition

↓ ↓

0.05	0.14	0.14	0.19	0.29	0.04	0.15	1.0

After mutation : feasible condition

0.05	0.04	0.14	0.19	0.29	0.14	0.15	1.0

6.3. Stopping condition
In this case, the stopping condition is fixed at 10,000 number of generations.

6.4. Optimal setting of drilling parameters using Genetic Algorithm in the Taguchi Method

6.4.1. Identification of drilling parameters and their levels.

The identification of drilling parameters is important for the success of any industrial experiment. S/N ratio is used for measuring the functional performance. If the S/N ratio is great, then the product quality will be high (Ross, 1996). An L_9 orthogonal array is used for experimentation. The trials are conducted and the results are given in Table 2. The optimal weights obtained are:

[0.3, 0.0, 0.0, 0.2, 0.3, 0.05, 0.15]

Table 2 Experimental output data (Noorul haq et al. 2008)

Trial no	Ra (μm)	Fc (N)	B (N-m)	Ka (N/mm^2)	P (Watts)	MRR (mm^3/min)	Tm (min)
1	7.83	107.87	0.88	1408.0	103.21	4398.23	0.4464
2	4.01	254.96	2.06	1318.4	241.61	10995.57	0.1656
3	2.22	470.67	2.26	904.0	265.07	17592.92	0.0974
4	6.7	186.31	1.96	3136	269.45	7068.58	0.2577
5	5.8	539.33	1.28	898.2	241.27	17671.46	0.09698
6	6.09	1186.53	3.24	1296.0	610.73	28274.33	0.0694
7	6.01	274.57	0.69	1104.0	202.32	10995.57	0.1559
8	8.27	1078.66	2.55	1632.0	747.7	27488.94	0.0714
9	6.2	1274.78	2.16	864	633.35	43982.3	0.0414

Consequently from step 4 we have,

$WSN_{i1} = 0.3Z_{i1}+0.0xZ_{i2}+0.0xZ_{i3}+0.2Z_{i4}+0.3Z_{i5}$
$+0.05Z_{i6}+0.15Z_{i7}$

where Zi1, Zi2, Zi3, Zi4 Zi5,Zi6 and Zi7 represents the normalized S/N ratio for the responses Ra, Fc, B, Ka, P, MRR, Tm at i^{th} trail correspondingly. The WSN values are calculated and it is shown in last column of Table 3.

Table 3: Weighted S/N ratio (WSN) values

Trail no	Ra	Fc	B	Ka	P	MRR	Tm	WSN
1	0.9584	0	0.1572	0.4035	0	1.0000	0	0.4182
2	0.4496	0.3483	0.7072	0.3534	0.4295	0.5830	0.3979	0.5079
3	0.	0.5966	0.7671	0.0734	0.4763	0.3598	0.6021	0.5044
4	0.8399	0.2213	0.6750	1.0000	0.6440	0.76895	0.2061	0.6827
5	0.7302	0.6517	0.3995	0.	0.4288	0.3580	0.6460	0.3911
6	0.7673	0.9709	1.0000	0.3417	0.8978	0.2179	0.8081	0.7750
7	0.7573	0.3783	0.	0.2222	0.3399	0.5576	0.3979	0.3339
8	1.0000	0.9324	0.8452	0.5134	1.0000	0.2292	0.7959	0.7407
9	0.7809	1.0000	0.7378	0.0397	0.9162	0	1.0000	0.6625

Table 4 Main effects on WSN

Parameters	1	2	3	Max-Min
V	0.476	**0.616**	0.5790	0.14
F	0.478	0.546	**0.647**	0.169
PA	**0.644**	0.617	0.409	0.236

Table 4 shows the main effects on WSN. The larger value of WSN gives the better the quality. From table 4, we identified the optimal setting condition in the order $\mathbf{V_2\ F_3\ PA_1}$.

6.5. Analysis of variance (ANOVA)

Table 5 shows the results of pooled ANOVA. From the table 5, point angle contributes more that affect the multi response characteristics. Point angle contribute 45.21% followed by Feed with the contribution of 21.5 % and the cutting speed with the contribution of 15.4 %. The optimized values for drilling parameters of metal matrix composites using Genetic algorithm are Point angle 90°, Feed 0.20 mm/rev and Cutting speed 56.54 m/min.

Table 5 Results of the pooled ANOVA on WSN

Parameters	DOF	SS	MS	F - Test	% Contribution
V	2	0.031253	0.01563	0.86	15.4
F	2	0.0434	0.0217	1.2	21.5
PA	2	0.0913	0.04565	2.5242	45.21
Error	2	0.03617	0.0181	1	17.9
Total	8	0.202123	0.0253	-	100

7. Conclusions

Optimum setting of machining parameters for drilling metal matrix composites has been identified in this work. Taguchi method and Genetic algorithm used in this work were played major role for solving multi response problems. Optimal setting of machining parameters for drilling MMC is identified which are V_1 (56.54m/min) F_3 (0.2mm/rev), PA_1 (90°). This methodology can be used for solving any machining parameters optimization with the multi responses.

References

[1] R. Jeyapaul, P. Shahabudeen and K. Krishnaiah (2006), "Simultaneous optimization of multi response problems in the Taguchi method using genetic algorithm" Springer-Verlag London Limited, pp. 870-878.

[2] A. Noorul haq, P. Marimuthu and R. Jeyapaul (2008), "Multi response optimization of machining parameters of drilling Al/SiC metal matrix composite using grey relational analysis

in the Taguchi method", The International Journal of Advanced Manufacturing Technology, Vol. 37, pp. 250–255.

[3] S. Ranganathan and T. Senthilvelan (2011), "Multi-response optimization of machining parameters in hot turning using grey analysis", The International Journal of Advanced Manufacturing Technology, Vol. 56, issue 5, pp.455-462.

[4] P Suresh, R Venkatesan, T Sekar, N Elango and V Sathiyamoorthy (2014), "Optimization of intervening variables in microEDM of SS 316L using a genetic algorithm and response surface methodology", Strojniški vestnik-Journal of Mechanical Engineering, Vol. 60, Issue10, pp.656-664.

[5] S. Tripathy and D. K. Tripathy (2017) "Multi-response optimization of machining process parameters for powder mixed electro-discharge machining of H-11 die steel using grey relational analysis and topsis", Machining Science and Technology, An International Journal, Vol. 21, Issue 3, pp. 362-384.

[6] G. R. Sanjay Krishna and K. Pavan Kalyan (2018), "Multi-Response Optimization of Wire-EDM Process Variables for Machining Process", Journal of Advanced Research in Dynamical and Control Systems, Issue: 06-Special Issue, pp. 46-55.

[7] A.V.S. Ram Prasad, Koona Ramji, Murahari Kolli and G. Vamsi Krishna (2019), "Multi-Response Optimization of Machining Process Parameters for Wire Electrical Discharge Machining of Lead-Induced Ti-6Al-4V Alloy Using AHP–TOPSIS Method", Journal of Advanced Manufacturing Systems, Vol. 18, No. 02, pp. 213-236.

[8] N. Babu, S. M. Vinu kumar andA. Megalingam Murugan (2020) "Multi-response optimization of process parameters in wire electric discharge machining of Ti-3Al- 2.5V alloy using Taguchi integrated grey relational analysis", International Journal of Mechanical and Production Engineering Research and Development, Vol. 9, Special Issue 1, pp. 233-242.

[9] Francicus and Juze Balic (2004) "Optimization of cutting process by GA approach. Robotics and computer integrated manufacturing", Vol. 19, Issues 1–2, pp. 113-121.

[10] Kishalay Minra and Ravi Gopinath (2004) "Multiobjective Optimization of an industrial grinding operation using elitist non-dominated sorting Genetic algorithm", Chemical Engineering Science, Vol. 59, Issue 2, pp.385-396.

[11] Davim JP, (2003), "Study of drilling metal matrix composites, based on the Taguchi technique", Material Processing Technology. Vol.132, pp. 250-254.

[12] P Marimuthu and K Chandrasekaran, (2011), "Experimental study on stainless steel for optimal setting of machining parameters using Taguchi and neural network", ARPN Journal of Engineering and Applied Sciences, Vol.6, issue 10, pp.119-127.

[13] V Kaviarasan, R Venkatesan and Elango Natarajan (2019), "Prediction of surface quality and optimization of process parameters in drilling of Delrin using neural network", Journal of Progress in Rubber, Plastics and Recycling Technology, Vol.35, issue 3, pp. 149-169

[14] R. Saravanan, P. Asokan and K. Vijayakumar (2003) "Machining parameters optimization for turning cylindrical stock into a continuous finished profile using genetic algorithm and simulated annealing", International Journal of Advanced Manufacturing Technology. Vol.21, pp. 1-9.

[15] Phillip J Ross (1996)" Taguchi Techniques for quality Engineering.", New York: Mc Graw-Hill,

Design and Optimization of Hybrid Interface Joint in a Composite Fan Blade of Aircraft Engine

Prakash Jadhav[1], **Yella Gruha Lakshmi**[2]

[1,2] Mechanical Engineering Department, School of Engineering and Sciences, SRM University -AP, Andhra Pradesh – 522240, India
[1]prakash.j@srmap.edu.in, [2]gruhalakshmi_kandru@srmap.edu.in
[1] corresponding author

Abstract. Fan blades are one of the most important components of an aircraft engine. Bird strikes on fan blades have always been a cause of worry and it can cause slices of birds hitting other parts of the engine which may lead to greater damages. Bird strikes cannot be completely avoided. However, reduction of the impact of a bird on jet engines can be achieved by suitable design and manufacturing, through the simulation analysis. Although current composite blades can withstand the bird strike impact, some delamination failures are still observed on the trailing edge side of the blade, possibly due to vibration bending modes This paper talks about using two fibers in composite blade instead of the current single fiber one. For this to be feasible, two fiber joints at various locations on the blade have to be properly designed. The design criteria used here is the lowest inter-laminar shear strain level at critical joint locations. Suitable size coupon models in FEA with hybrid joints inside are used to simulate the blade bending with appropriate boundary conditions. The coupon models were developed by using the ANSYS APDL and linear static analysis was performed for different repetitive layups and varying layer by layer joint locations. The coupon joint designs with minimum inter-laminar shear strains are shortlisted to be recommended for use in sub-element and blade models.

Keywords: Hybrid Composite Joints; Bird Strike; Fan Blade; Interface

1. Introduction:

A jet engine is a component of an aircraft's propulsion system and is a machine that converts liquid fuel into a powerful thrust called pushing force. A plane is pushed forward by the thrust from one or more engines. The jet engine works by drawing in some of the air through which the aircraft is moving, compressing it, combining it with fuel, heating it, and finally ejecting the ensuing gas with such a force that the plane is propelled forward. The power produced by such engines is expressed in terms of thrust. In an aircraft engine, fan blades are one of the most important components that produce thrust. The fan is the frontside component of the engine. The large spinning fan sucks in large quantities of air. Mostly the blades of the fan are made of titanium. In an engine, the primary function of fan blades is to produce a huge amount of thrust. The 75 - 80 % of the thrust is produced from the fan through the bypass duct in a turbofan jet engine, and the remaining 25 - 20 % is produced as exhaust from the nozzle from the engine core.

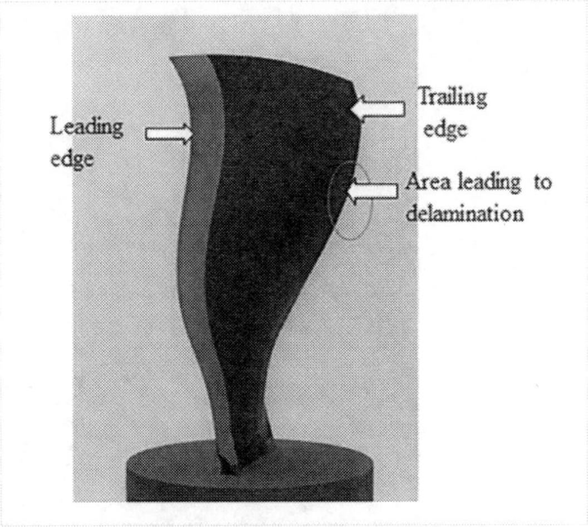

Figure 1: Fan blade [1]

The rate of air intake varies and is dictated by factors such as airfoil geometry, angle of attack, air density, and the speed at which the airfoil moves through the air. As shown in Figure 1, fan blades can be made of composite blade and consist of a blade body, a rounded leading edge (LE), a sharp trailing edge (TE), an integrated platform, and a dovetail or single tooth attachment. The swept area diameters and chord lengths of blades have increased in recent years, leading to an increase in the size and thrust of jet engines. Jet engine fan blades come in either a solid or a hollow blade design. The design criteria for the fan blade is the damage tolerance under bird strike at takeoff or landing situation. The successful application of carbon fiber reinforced polymer matrix composite (PMC) fan blades by GE has significantly reduced engine weight; relative to a structure using titanium blades. The GE90 fan blade's uniquely curved design makes it larger, lighter, and more aerodynamic than traditional titanium blades for reduced engine weight and less fuel burn [2]. The author has worked on various technologies which can reduce the weight of the composite fan blades or increase their aero efficiency [3-11]. The composite fan blades are made with IM7/8551-7 unidirectional carbon epoxy prepreg and manufactured in desired layer wise orientations to improve the impact resistance of the blade. The author of this paper has tried various technologies like investigating the effect of ply drop locations on the failure of composite blade and ways to come up with design allowable for this[3, 9]; the feasibility of using lightweight foam inserts inside the composite blade and their geometry or shape optimization to avoid delamination type of failure [4-5]; the feasibility of use of carbon composite material for the compressor blade in aircraft engine in order to further reduce the weight [6]; the feasibility of using passive morphing technology in the composite fan blade using unsymmetric layup which will result in increased aero efficiency[7, 10]; the delamination modeling to understand the failure of fan blade composite using cohesive zone methodology [8]; the feasibility of wavy trailing edge fan blades and coming up with the failure criteria for these blades using basic concepts along with the experimentation on coupon level [11]. Some of the technology ideas were very promising and were filed as a patent and also were implemented in the design of the real composite fan blade.

To avoid or reduce the impact of the bird on the fan blades and engine, it is proposed here to redesign the composite fan blade. A delamination type of failures was observed on the trailing edge of the fan blade in the bird strike tests of the real blades. Although some level of delamination is allowed under bird strike test as a damage tolerant design, it's always better to avoid the any damage initiation altogether. The softer material (higher strain capability) is proposed to be used in the trailing edge region of the fan blade. It means introducing softer fiber like glass inside the carbon fan blade by choosing the appropriate interface design of the laminated composite joint of the two fibers. This will lead to many design trials and multiple simulation cases.

No directly relevant research has been done in the past or available in published literature on the design improvement of composite fan blades using more than two fibers.

2. Methodology:

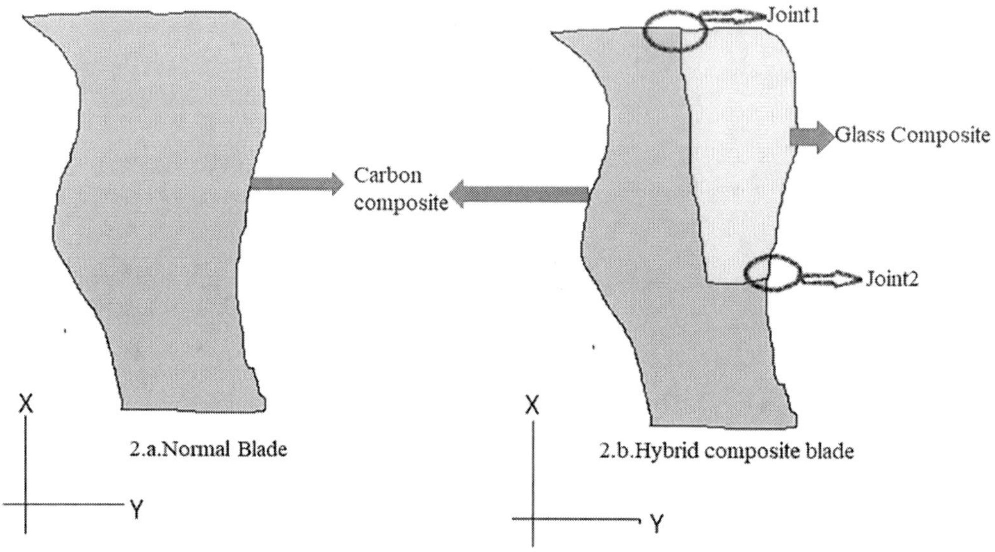

Figure 2: Proposed Solution

Figure 2(a) shows the sketch of the original fan blade made of carbon composite fan blade and 2(b) shows the sketch of the proposed solution to the delamination problem in hand. In this figure, the blue region represents the carbon fiber composite and the yellow region represents the glass fiber composite. The yellow region area size is not yet fixed; it will be decided based on the blade level simulations later. The interface between the carbon and glass composite regions in the blade seems like a straight line (plane section). In reality, this is not possible because it may lead to separation under loading. To avoid this, a better interface needs to be designed for the interface of carbon and glass regions. We will call the proposed interface as a laminated splice joint.

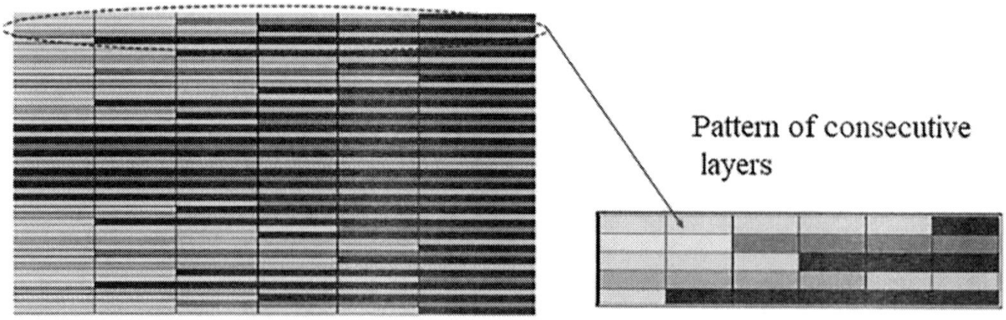

Figure 3: Sample sketch of splice joint (interface region) with 48 layers

A sample laminated hybrid composite sample spliced joint design is shown in Figure 3. It is assumed here that the blade is made of 48 layers at the interface region based on the real thickness of composite blade in the interface region. The shades in the color indicate the different orientations. It is proposed to divide this interface region into 5 lengthwise sections as shown in Figure 3. The length of glass and carbon fibers in the joint region varies in every layer as shown depending upon the joint location. It is assumed here that the location of joints will vary in 5 consecutive layers in a pattern as shown in Figure 3, which will be repeated over and over again throughout the thickness. This may lead to 100+ different possible combinations of joint designs. Table 1 shows that the properties of IM7/8551-7 carbon fiber/epoxy and S2-glass fiber/epoxy composites used here as input to FEA [12,13].

Table 1: Material properties

Material/ Property	**E_x (GPa)**	**E_Y (GPa)**	**E_Z (GPa)**	**PR_{XY}**	**PR_{XZ}**	**PR_{YZ}**	**G_{XY} (GPa)**	**G_{XZ} (GPa)**	**G_{YZ} (GPa)**
Carbon composite	150.99	9.17	8.27	0.35	0.32	0.37	4.34	4.82	3.03
Glass composite	43.29	11.58	12.13	0.28	0.28	0.36	3.37	4.68	3.24

The two different layups were considered initially in FEA simulation to find the most effective one out of the [0°/45°/0°/-45°] and [0°/90°/45°/-45°] quasi-isotropic layers. 3D coupon FEA models were built using Ansys parametric scripting (APDL) so that the joint location and orientation of fibers in the model can be changed parametrically. The size of the FEA coupon models used was 275 mm X 12.7 mm x 7 mm (designed as per the blade dimensions). It is a half symmetry, 3D model with rollers under four point bend boundary condition. In the mid thickness region of the coupon, the continuous carbon composite layer materials with different orientations are used throughout the blue and yellow regions to give sufficient strength to the coupon. It is seen in the video of the testing that the blade tends to bend both spanwise and chordwise under the bird strike loading. To simulate that kind of bending, the coupons are subjected to four-point bend kind of loading using the steel rollers with suitable contact definitions and half symmetry boundary condition at one end.

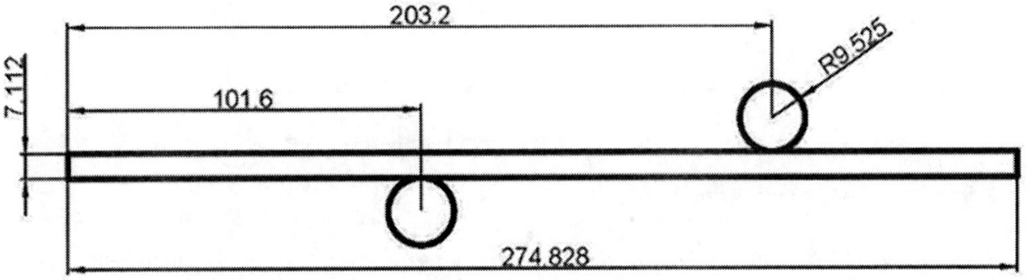

Figure 4: Dimensions of the coupon (All dimensions are in mm)

Figure 4 shows the half symmetry 2-D sketch of the coupon model with dimensions. 3D model coupons were created in Ansys by extruding the 2D elements. The element type Solid 185 was used to create the 3D coupon with the above dimensions as shown in Figure 5. IM7/8551-7 carbon fiber epoxy composite and S2-Glass epoxy composite properties were assigned for the corresponding material zone elements and epoxy-resin material property was assigned to the interface elements between two fibers in every layer. Each layer is one element thick (0.14mm) and the 600 elements were used along the length of the layer. Using a local-coordinate system, the planned fiber orientation was assigned in every layer. Linear static structural analysis was performed; and the longitudinal strains, transverse strains, shear strains, and inter-laminar shear strain levels were monitored at and near the critical joint locations in all layers for all cases. While measuring the strain values, all resin elements were unselected to see the strain distribution around the joint because we are interested to see the strain contour in the fiber composite elements. As for composites, the strain based failure criteria is most popular and widely used in the industry, hence here too strain levels are considered to be the measure of better composite structure designs.

Displacement loading of 16.51 mm was applied to all the coupons. This was based on a few laboratory level coupon tests done earlier using an interface joint of a typical configuration which failed at that loading level under four-point bend boundary condition. Due to space limitation that result is not discussed here. Another way to come up with the appropriate loading is, displacement loading level in such a level that it produces longitudinal direction (0) strain of about 1.1%, as this is the most common level of longitudinal failure strain for carbon composite. As the similar loading is applied in all the cases, it is easy to compare the results i.e strain levels in all the cases.

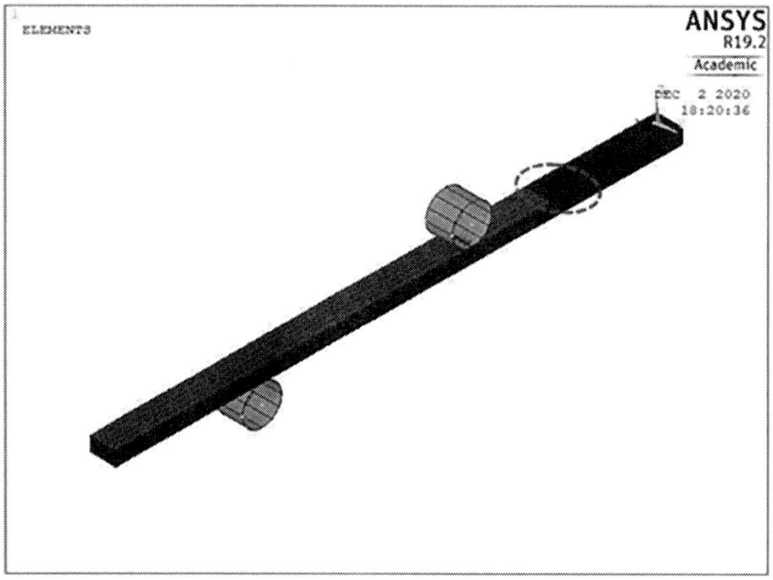

Figure 5: 3D model of the coupon

2.1 Different cases that were analyzed are:

a. Coupons with $[0°/45°/0°/-45°]_{6S}$: In this configuration, for providing the axial strength and stiffness to the coupon, the more no. of 0° orientation fibers are used. Glass and carbon regions are distinctly separated as yellow and blue color shown in the Figure 3.
b. Coupons with quasi-isotropic $[0°/90°/45°/-45°]_{6S}$: By using 90° orientation fibers in the coupon, we can provide for the better shear strength and shear stiffness. Glass and carbon regions are distinctly separated as yellow and blue color shown in the Figure 3.
c. Coupons with alternate glass and carbon layers in the yellow region with $[0°/45°/0°/-45°]_{6S}$ lay up: By using the alternate fiber layers, we can get better integration of two fiber zones.
d. Coupons with alternate glass and carbon in the yellow region with $[0°/90°/45°/-45°]_{6S}$ layup.

3. Results and Discussion:

3.1. Coupons with [0°/45°/0°/-45°]$_{6S}$ orientation:

The coupons were designed with [0°/45°/0°/-45°]$_{6s}$ orientation and load was applied using appropriate half symmetry four-point bend boundary conditions, At the max displacement load level, all the directional strains were monitored for 120 possible configurations at the critical joint locations. To avoid considering the wrongful strain concentration at the top and bottom layer due to the surface effect, the corresponding elements were removed while analyzing the results. The coupon joint designs with a lowest level of maximum inter-laminar strain at the critical joints are shortlisted (best 5) as shown in Table 2 to be recommended for use in sub-element and full blade models.

Table 2: Strain comparison for [0°/45°/0°/-45°] orientation

Joint model No	Displacement	Strain (%)						Location	
	Z(mm)	X	Y	Z	XY	YZ	XZ- ILS	orientation	Layer no
1002	12.27	0.76	0.21	0.66	0.16	0.14	1.03	0°	11
1006	12.27	0.76	0.21	0.66	0.16	0.14	1.03	0°	11
1009	12.27	0.76	0.21	0.66	0.16	0.14	1.03	0°	11
1011	12.27	0.75	0.21	0.66	0.16	0.14	1.03	0°	11
1015	12.27	0.74	0.21	0.66	0.16	0.14	1.03	0°	11

Table 2 represents the longitudinal strain, transverse strain, in-plane shear strain, and inter-laminar shear strain for the shortlisted cases of specific orientation set coupons. The last two columns also show the orientation of the layer where max inter-laminar strain level is occurring and that layer number from the top. All these five shortlisted cases show the consistency of occurrence of max ILS strain level inside the 11th layer with a 0-degree orientation.

As shown in Figure. 6, the result of one such shortlisted case no 1002 shows the maximum inter-laminar shear strains occur in the immediate next 45-degree layer i.e. 0-45 interface. All other cases in this category show the consistently similar location of maximum inter-laminar shear strain.

3.2. [0°/90°/45°/-45°]$_{6S}$ orientation:

The coupons were designed with [0°/90°/45°/-45°]$_{6S}$ orientation and the appropriate boundary conditions were used for simulation. Similar to the previous case, all 120 possible configurations were analyzed. The coupon joint designs with minimum level of max inter-laminar strain levels are shortlisted (best 5) as shown in Table 3 to be recommended for use in sub-element and full blade models.

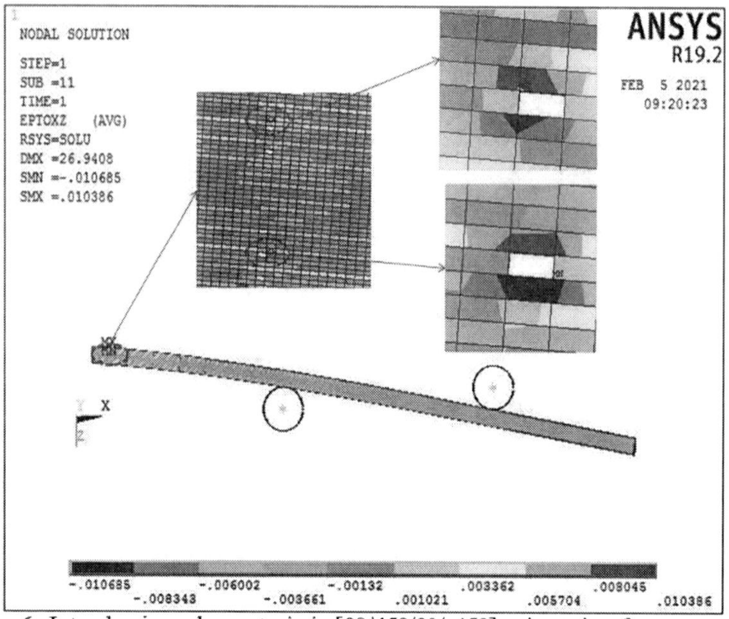

Figure 6: Inter-laminar shear strain in [0°/45°/0°/-45°] orientation for case no.1002

Table 3: Strain comparison for $[0°/90°/45°/-45°]_{6S}$ orientation

Joint model No	Displacement	Strain (%)						Location	
	Z(mm)	X	Y	Z	XY	YZ	XZ- ILS	Orientation	Layer no
1013	12.48	0.61	0.19	0.62	0.15	0.11	0.79	0°	9
1019	12.48	0.61	0.21	0.71	0.15	0.11	0.84	0°	5
2001	12.42	0.62	0.18	0.57	0.12	0.11	0.84	-45°	4
2002	12.43	0.62	0.18	0.63	0.12	0.11	0.86	0°	5
2008	12.41	0.61	0.16	0.71	0.11	0.11	0.86	-45°	4

Table 3 represents the longitudinal strain, transverse strain, in-plane shear strain, and inter-laminar shear strain levels for the shortlisted cases of the specific orientation coupon sets. Some of these five shortlisted cases show the consistency of the occurrence of max ILS strain level inside the 5th layer with 0-degree orientation, some of them show the consistency of max strain level inside the 4th layer with negative 45-degree orientation and one of the shortlisted case shows the max strain level inside the 9th layer.

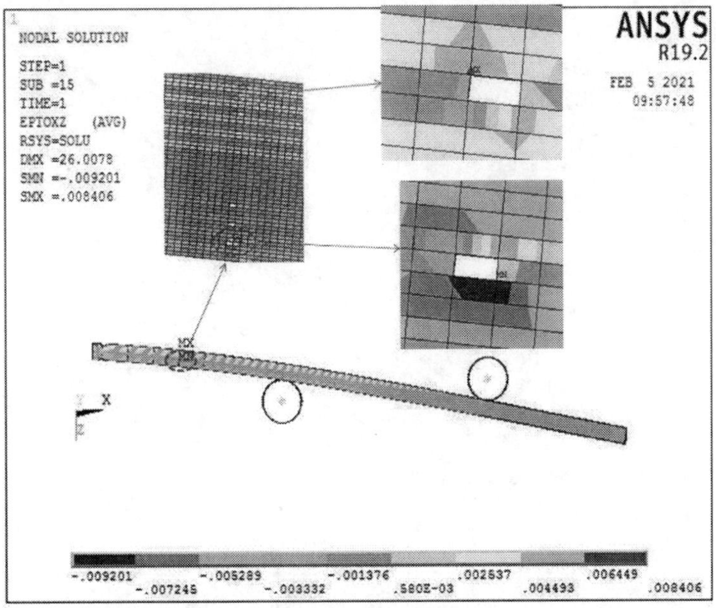

Figure 7: Inter-laminar shear strain in [0°/90°/45°/-45°] orientation for case no.1019

As shown in Figure. 7, the result of one of the shortlisted case no 1019 shows the maximum inter-laminar shear strains in the immediate next 90-degree layer i.e. 0-90 interface. In all the cases, the max strain is occurring either in the 0-90 layer interface or 0-45 layer interface.

3.3. Alternate glass and carbon with [0°/45°/0°/-45°] $_{6S}$:

Some coupons were designed with alternate layers of glass and carbon in the yellow region only while maintaining all carbon in the blue region to maintain the stiffness. The appropriate boundary conditions were used in simulation. Here too, all the 120 configurations were analyzed just like the previous case. Figure no 8 shows the stiffness (load vs displacement plots) variations study for the representative cases of use of alternate layers along with the earlier designs that we have discussed and all carbon coupons cases too for the comparison purpose. Alternate layers CGGC (carbon-glass-glass-carbon. repetitive layup in the yellow region) shows the closest stiffness to that of the original hybrid composite fan blade layup i.e., carbon-glass (0/45/0/-45). Although some of the alternate layer designs do show higher stiffness, it is to be noted here that lower compliance material is required to be used here in the yellow trailing edge region. Hence higher stiffness cases can not be used. CGGC case is recommended because its stiffness is closest to the earlier baseline design. Although quasi-isotropic layup shows lower stiffness than that of baseline design, it may not be sufficient enough for the fan blade application.

Table 4 represents the longitudinal strain, transverse strain, in-plane shear strain, and inter-laminar shear strain for the shortlisted cases of the specific orientation coupon sets. Some of these shortlisted cases show the consistency of the occurrence of max strain level with 0-degree orientation layers, while one of them shows the max strain level inside the 2nd layer with 45-degree orientation. As shown in Figure. 9, result of the case CGGC (most promising case as discussed) shows the maximum inter-laminar shear strains in the immediate next layer which is a 45-degree orientation or in the 0-45 interface.

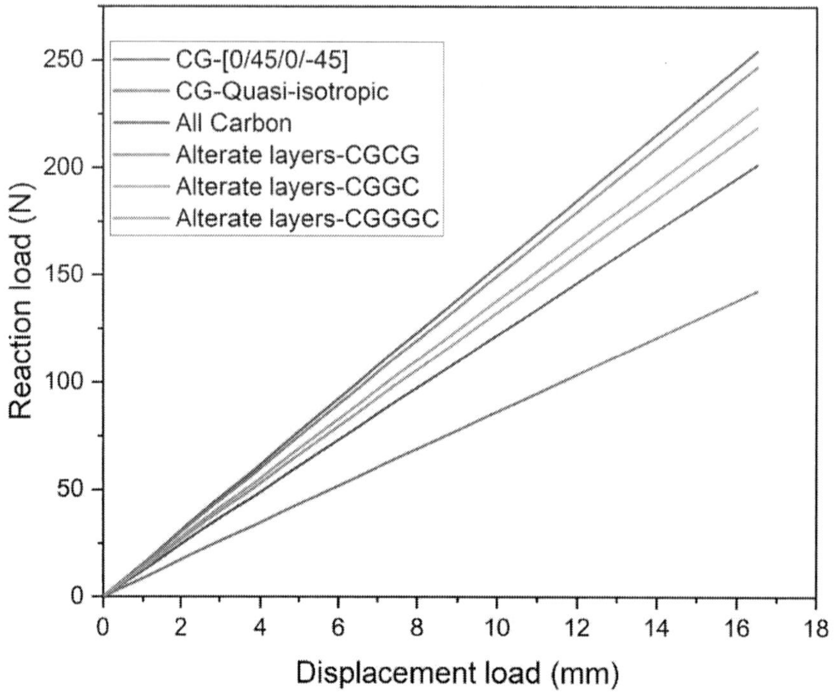

Figure 8: Alternate layers -Stiffness plots

Table 4: Strain comparison for alternate layers with [0°/45°/0°/-45°] $_{6S}$ orientation

Joint model No	Displacement	Strain (%)						Location	
	Z(mm)	X	Y	Z	XY	YZ	XZ-ILS	Orientation	Layer no
CGCG	12.43	0.31	0.12	0.22	0.03	0.05	0.43	45°	2
CGGC	12.33	0.41	0.16	0.56	0.08	0.08	1.04	0°	5
CGGGC	12.37	0.38	0.14	0.58	0.06	0.05	0.84	0°	45
CGGGGC	12.33	0.46	0.15	0.77	0.11	0.12	1.33	0°	41

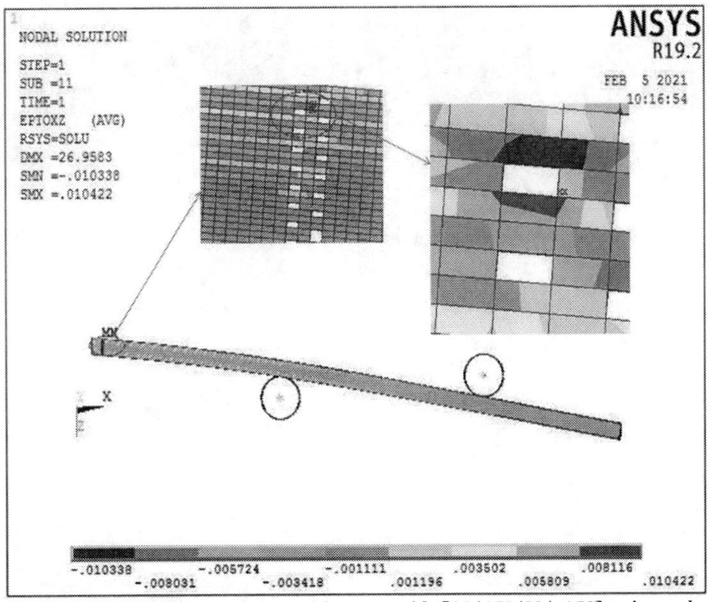

Figure 9: Inter-laminar shear strain in alternate layers with [0°/45°/0°/-45°] orientation for case CGGC

3.4. Alternate glass and carbon with [0°/90°/45°/-45°] $_{6S}$:

Table 5: Strain comparison for [0°/90°/45°/-45°] $_{6S}$ orientation

Joint model No	Displacement	Strain (%)						Location	
	Z(mm)	X	Y	Z	XY	YZ	XZ- ILS	Orientation	Layer no
CGCG	12.67	0.36	0.17	0.39	0.11	0.12	0.49	0°	1
CGGC	12.58	0.41	0.18	0.88	0.08	0.08	1.06	0°	5
CGGGC	12.46	0.51	0.12	0.37	0.05	0.06	0.57	-45°	4
CGGGGC	12.39	0.81	0.17	0.61	0.11	0.12	1.42	0°	39

Table 5 represents the longitudinal strain, transverse strain, in-plane shear strain, and inter-laminar shear strain for the shortlisted cases of the coupon sets of specific orientation. Some of these shortlisted cases show the consistency of the occurrence of max strain level within 0-degree orientation layer or 0-45 or 0-90 interface, one of them shows the consistency of occurrence of max strain level inside the 4th layer with negative 45-degree orientation.

As shown in Figure. 10, the result of the case CGGC shows the maximum inter-laminar shear strains in the immediate next layer which is 90-degree orientation. Some other cases in this category show the immediate next layer which is 0-degree orientation.

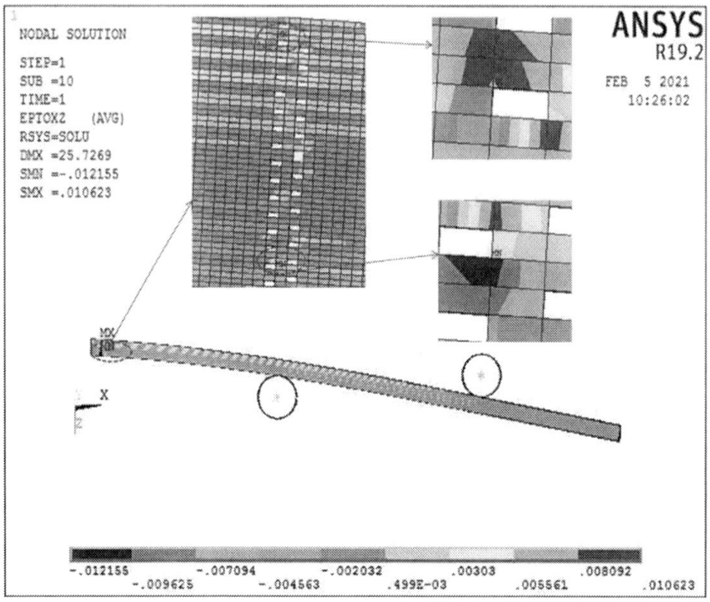

Figure 10: Inter-laminar shear strain in alternate layers with [0°/90°/45°/-45°] orientation for case CGGC

The linear static analysis performed here is used here as just the preliminary filter criteria for a large number of possible cases. This has been proved by the author [2-11] that the simulation and testing of composite coupons can be used for studying the feasibility of any improvement in technology idea by incorporating proposed changes in the coupons and try to prove the feasibility. After proving the feasibility on the coupon level, the same idea can be implemented on sub element and blade level models and tests, and the idea can be proved to work on the real components. Only precaution which need to be taken is, coupons have to be designed very carefully and proposed technology has to be implemented in the coupon carefully along with the most appropriate boundary condition/loading. For all the shortlisted cases, sub-element coupon and blade level analysis will be performed under bird strike loading and further shortlisting will be done based on the interlaminar strain levels. The bird strike analysis of sub-element coupons/plates and blades is underway and will be discussed in the follow-up paper. After that, the future work will be to test the sub element plates under bird strike and also do test the hybrid blades under bird strike.

4. Conclusion:

Two fiber hybrid interfaces (spliced joints) were designed using a parametric 3D coupon FEA model using an appropriate boundary condition and the optimization was conducted to find out the best possible design which can generate a minimum level of max interlaminar shear strains which will reduce the chances of failure through delamination. The S-glass composite was incorporated inside the carbon composite coupon to increase the compliance capability of certain regions through a proper design without affecting the stiffness and strength of the coupon. Coupon level analysis acts as a filter to find out the most promising cases and saves lot of time and cost by avoiding the blade level analysis and testing of multiple cases. Shortlisted designs will be used for further sub-element plate and blade model level analysis and testing to implement the hybrid interface joint design in the composite fan blade of an aircraft engine. This is supposed to eliminate the delamination issues on the trailing edge.

5. References:

1. GE Aircraft Engines, Jet Engine Fan Blade (model GE90-115B). 2001: MoMA. https://www.moma.org/collection/works/93637
2. T. General et al., "General Electric GE90," pp. 1–8, 2019.
3. Prakash Jadhav, Effect of ply drop on aerospace composite structures, Key Engineering Materials, vol 847, pp 46-51, June 2020.
4. Prakash Jadhav, Innovative designs of embedded foam inserts in aerospace composite structures, Materials Today Proceedings, vol 21, part 2, pp 1164-1168, 2020.
5. Spoonire Ross Ashely et.al, Hybrid turbine blade including multiple insert sections, US Patent US 9309772B2, April 12, 2016.
6. Kray Joseph Nicholas et al., Composite compressor blade and method of assembling, US Patent 10247014 B2, April 2, 2019.
7. Shim Dong Jin et. al, Fan Blade, US Patent US 2013/02875588A1, Oct 31, 201.
8. P. Jadhav, Zheng Li, S. Prasad, Cohesive zone modeling to model delamination in composite structures under static and dynamic loading, Proceedings of SAMPE Conference May 2016, California.
9. P. Jadhav, Zheng Li, S. Prasad, Evaluation of ply drop allowable in composites, Proceedings of COMTEST Conference April 2015, IMDEA, Spain.
10. Prakash Jadhav, Passive morphing in aerospace composite structures, Key Engineering Materials, 2021, in press
11. Prakash Jadhav, Failure criteria for composite blades with wavy edge in aerospace applications, Lecture notes in mechanical engineering, 2021, in press
12. Swapnil, A. Salgar, B. Sathe, Sandip, P. ChaudhariBapu, and S. Jagadale Vishal. "Experimental investigation of mechanical properties of glass fibre/epoxy composites with variable volume fraction." materials today: proceedings 4, no. 9 (2017): 9487-9490.
13. Colvin, G. E. "Mechanical characterization of IM 7/85517 carbon/epoxy under biaxial stress." Recent advances in the macro- and micro-mechanics of composite materials structures (1988): 235-241.

Production and characterization of wax and grease from waste plastic

P R Kumbar [1], **V S Patil** [1], ***S R Kumbar** [2], **R B Kumbar** [3] and **Dr. P Bhange** [2]

[1]Sanjay Ghodawat International School, Atigre- 416118, Maharashtra.

[2]Sanjay Ghodawat University, Atigre- 416118, Maharashtra.

[3]Sanjay Ghodawat Polytechnic, Atigre- 416118, Maharashtra.

* **Corresponding author** Email address: shobha.kumbhar@sanjayghodawatuniversity.ac.in

Abstract: World has opened up to start a plastic-free zone. Plastic reusing is the greatest test for us all. During reusing it is important to avoid the potential risk of environmental degradation and hazard. Consuming plastics makes it perilous to the general public. The major focus is to utilize waste grade 4 plastic and generate grease and wax so efficiently that there is no release of harmful oxides (CO_2, CO) in the environment and it is very economical. Utilizing waste plastic, the creation of grease and wax is completed. It is valuable to the general public without making the risk to condition. The portrayal of the grease and wax is featured right now. Results show that grease can be utilized for substantial vehicles and with certain added substances, it may be very well utilized for some applications.

Keywords: Production; Characterization; Plastic; Wax; Grease.

1. Introduction

The use of plastic nowadays is constantly decreasing but the plastic which already exists in the world is very difficult to manage and the industrial processes are not much economical and neither are eco-friendly. An alternative approach to processing plastic solid waste is chemical recycling, the success of which relies on the affordability of processes [1]. Many processes are followed to get the useful product out of waste like wax and grease out of waste plastics. Plastic derived diesel is suitable as a blend component for petroleum diesel fuel [2], oil [3, 4], wax and grease are the main. Production of grease and additives to be added must be realized for the purpose. Proper thickener in the grease will exhibit the service characteristics [5]. Grease can be more than a lubricant; it is often expected to perform as a seal, corrosion inhibitor, shock absorber and a noise suppressant [6]. Different characteristics like FTIR and SEM is performed to check the material produced.

Objectives & Motivation:

The major focus of the paper is to use grade 4 plastic (Low-Density Polyethylene) and generate useful products more economically and in an eco-friendly method where there is no release of harmful oxides during the procedure (Ex: CO_2, CO). The release of toxic oxides is cut due to the addition of camphor to the mixture while heating, which when reacts with the mixture, the oxides are consumed by the camphor, as camphor requires oxides to burn out. When the oxides are exhausted the remaining camphor remains in the mixture and becomes a part of the products. So, if the products (wax and grease) are heated in the future the camphor will also burn hence preventing toxic oxide production in the future. After the oxides are consumed there is a byproduct formed which is carbon settlement, this can be used in making inks, etc. When the mixture is heated at the end, we get a combination of solid particles and semi-liquid below. The solid particles are the wax and the semi-liquid turns into grease after cooling down. The process is very economical compared to the industrial process and can be completed with the materials available at home.

2. Methodology

2.1. Collection, Sorting, and Washing

Plastics are collected from surroundings. After plastics are collected, they are sorted to obtain the grade-4 plastic. Just like with clothes, fruits/vegetables, and many other things, plastics must be washed before they are further processed. The goal of this step is to remove impurities and everything that is not made from plastic.

2.2. Heating Process and Collection

Engine oil is taken and boiled at 150°C and grade-4 plastic is added into the oil. To reduce the environmental hazards camphor is added to the mixture. The end product is having two forms. The solid one is wax and semi-liquid after cooling is grease. The final step is the collection of wax and grease into glass jars. Cost analysis of the material is tabulated in the table.

Table 1: Material and cost of Project 1.

S. NO.	Material	Availability	Cost
1.	Waste Grade 4 plastic (Low-Density Plastic)	Waste plastic available for free in the environment.	0 Rs
2.	Used oil	The oil that is already used for cooking available in homes or used engine oils in garages.	0 Rs
3.	Camphor	In shops	15 Rs – 20 Rs per 100 g
4.	Gas or electricity for heating	In-home	Use of resources for heating, 20 min usage of Gas: 1.5 Rs (approx), 20 min usage of electricity: 2 Rs (approx)
5.	Total Cost		21.5 Rs – usage of gas as a heating method. 22 Rs – usage of electricity for a heating method

3. Results and Discussions

3.1 IR studies

IR spectra of the Grease and wax samples were recorded with FT/IR-4600, Jasco, Japan Spectrometer. FT- IR characterization of wax and grease are done and the graphs are as mentioned in Figure 1 and Figure 2 respectively.

3.1.1 Wax

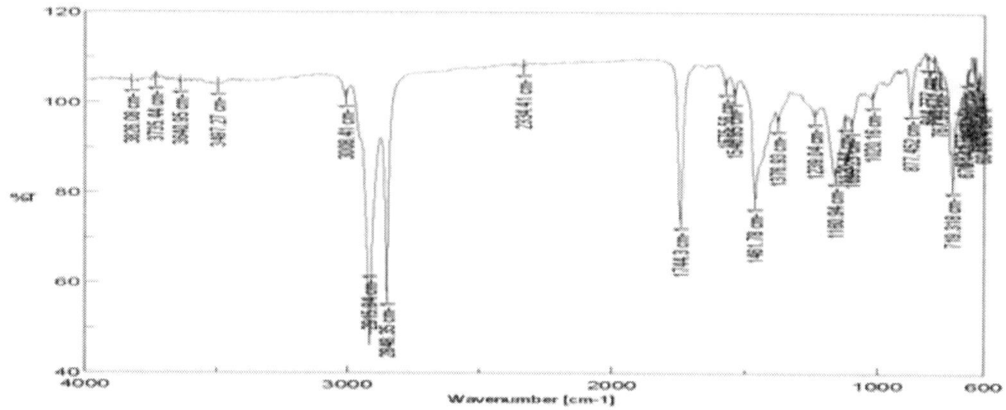

Fig.1. FTIR spectra of the wax sample

In Fig.1. It is been observed that wax which is produced by the Plastic is having wavelength 2916 and 2848, 1744, 1462, 1376, 1239, 1161, 719 cm^{-1} respectively and is compared with biofuel oil by cracking Rice Bran Wax at 600°C, in order to confirm the Py-GC-MS results. It was not liquid at room temperature. It was solid. When the temperature was elevated to 40°C, liquid material was obtained. which was analyzed using FTIR [7]. The strong peaks at 2916 cm^{-1} and 2848 cm^{-1} were attributed to the presence of methylene groups (CH_2) [8, 9]. The peak at 1744 cm^{-1} corresponding to CO was attributed to the presence of the aldehydes, ketones, ester groups or carboxylic acids [10]. Liu et al. carried out Py -GC-MS analysis and confirm that C=O mainly arise from the acid and ester compounds in the bio-fuel oil [11]. The peak located at 1462 cm^{-1} was attributed to the asymmetric vibrations of CH_2 and CH_3 groups. In addition, the weak peak at 1376 cm^{-1} arises due to the bending vibration of CH_3. The peak located at 1240 cm^{-1} was CO stretching vibration of acid or ester. Many papers thought the strong peak at 720 cm^{-1} was attributed to aromatic stretching vibration. Another probability was that the vibration at 720 cm^{-1} was attributed to long chain compounds [9]. The vibration located at 910 cm^{-1} was attributed to the di-substituted benzene. The vibration located at 1160 cm^{-1} was attributed to the stretching vibration of saturated ester [9].

Table 2: Comparison of wax produced and paraffin wax

Sr. No.	Wax produced	Paraffin Wax
1	Is more economical and efficient	Is comparatively less economical and efficient
2	Can be made for different uses by changing oil only	Has to follow entire procedure to be made for different purposes
3	It is less harmful gases when burnt.	It produces comparatively more harmful gases when burnt.
4	Wax does not melt easily and is strong comparatively	Melts easily and weak comparatively.

Pyrolysis process enlightens a well-organized, clean and operative means of not only removing plastic wreckage from the environment but also explore some innovative ways of turning waste into valuable products like Petrol, High-Speed Petrol, Diesel and Gas etc. [12]. From FTIR results this project, confirmed that the pyrolysis products of RBW were main alkane and alkene. Few oxygen containing compounds and aromatic compounds also was found. It needed other experiments to assist confirm the components of bio-fuel oil. Following are the features of the wax produced and compared with paraffin wax.

3.1.2 Grease

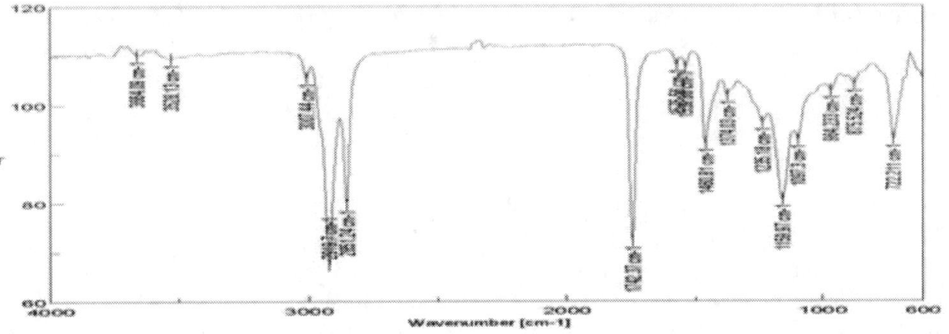

Fig.2. FTIR spectra of the grease sample

Fig. 2 shows FTIR spectra of a grease sample show the peaks at 2919 and 2851 cm^{-1} correspond to asymmetrical and symmetrical stretching vibration of CH_2 group. We have observed a new peaks at 1742 cm^{-1} which is a characteristic of a C=O bond arising from esters formed by oxidation of the sample. The peaks at 1576 and 1540 cm^{-1} were attributed to the asymmetrical and symmetrical stretching vibration of COO group which arises from the composition of the thickener. The peaks at 1460 cm^{-1} arises from the C–H asymmetrical vibration of CH_3 group which comes from Base oil, while bands at 1377, 1307, 722 cm^{-1} correspond to the bending, twisting, rocking vibrations of CH_2 groups [10]. Grease produced tested for lubrication at Siddhartha Grease and Lubricants PVT Ltd .It is found that with some additives can be used for heavy vehicles as a lubricant.

4. Conclusions

Grade 4 plastic is a great threat to the environment and its process is very complex mainly hazardous to the environment. To overcome such a problem, method mentioned of processing is the efficient and more effective than other processes. Wax and grease produced from waste engine oil and grade 4 plastic are characterized. Further, grease can be used with some additives for heavy vehicles.

Acknowledgement

We are grateful to Mr. Anil Rao, consultant and Dr. E. Sayanna, Chief Technological officer, Siddhartha Grease and Lubricants Pvt Ltd Guragaon for his continuous support for testing the grease. Also, special thanks to Mr. Pradeep Patil, HOD Physics, Mrs. Sasmita Mohanty, Director Principal Sanjay Ghodawat International School, Dr. H.M. Naveen Principal, SGIS, Mrs. Shobha Naveen, Dept of Chemistry and Dr. Lohar, Faculty of Chemistry department, Sanjay Ghodawat University for their encouragement.

Conflict of interest

The authors declare that they have no conflict of interest.

References

[1] Rahimi, A., García, J. *Chemical recycling of waste plastics for new materials production. Nat Rev Chem* **1,** 0046 (2017). https://doi.org/10.1038/s41570-017-0046.

[2] Brajendra K. Sharma, Bryan R. Moser, Karl E. Vermillion, Kenneth M. Doll, Nandakishore Rajagopalan,*Production, characterization and fuel properties of alternative diesel fuel from*

pyrolysis of waste plastic grocery bags,Fuel Processing Technology,Volume 122,2014,Pages 79-90,ISSN 0378-3820,https://doi.org/10.1016/j.fuproc.2014.01.019

[3] Williams PT,Williams EA, Recycling plastic waste by pyrolysis, Journal of the Institute of Energy, Vol.71, No.487, 81-93, 1998.

[4] Miller, Stephen J. Shah, Naresh, Huffman, Gerald P. Conversion of Waste Plastic to Lubricating Base Oil Energy Fuels 2005, 19, 4, 1580–1586.

[5] Yu.L. Ishchuk - Lubrication and lubricants New Age International, 2006.

[6] Gow G. (2010) Lubricating Grease. In: Mortier R., Fox M., Orszulik S. (eds) Chemistry and Technology of Lubricants. Springer, Dordrecht.https://doi.org/10.1023/b105569_14

[7] Qing Liua, Peng Liub, Zhi Xiang Xua, Zhi-Xia Hec, Qian Wanga, Bio-fuel oil characteristic of rice bran wax pyrolysis, Renewable Energy. December 2017 DOI:10.1016/j.renene.2017.12.012.M.L.S. Albuquerque, I. Guedes Jr., P. Alcantra, et al., Vib. Spectrosc. 33 (2003) 127.

[8] Xu Junming, Jiang Jianchun, Lu Yanju, Chen Jie,Liquid hydrocarbon fuels obtained by the pyrolysis of soybean oils, Bioresource Technology,Volume 100, Issue 20,2009,Pages 4867-4870,ISSN 0960-4,https://doi.org/10.1016/j.biortech.2009.04.055.

[9] Pan Jiabao, Cheng Yanhai and Yang Jinyong, Effect of heat treatment on the lubricating properties of lithium lubricating grease, RSC Advances, June 2015.

[10] B.S.S. Phanisankar, N. Vasudeva Rao, J.E. Manikanta, Conversion of waste plastic to fuel products, Materials Today: Proceedings, 2020.

[11] Trevor M. Letcher, Chapter 1 - Introduction to plastic waste and recycling, Editor(s):Trevor M. Letcher, Plastic Waste and Recycling, Academic Press, 2020, Pages 3-12,ISBN 9780128178805.

[12] Kundan Kumar Jha, T.T.M. Kannan, Recycling of plastic waste into fuel by pyrolysis- a review, Materials Today: Proceedings, 2020.

IOP Conf. Series: Materials Science and Engineering 1126 (2021) 012038 doi:10.1088/1757-899X/1126/1/012038

Structural, optical and wettability properties of zirconium oxide thin films deposited at various helium partial pressure

Nicky P. Patel[1] and Kamlesh V. Chauhan[1]

[1]CHAMOS Matrusanstha Department of Mechanical Engineering,
Chandubhai S. Patel Institute of Technology (CSPIT),
Charotar University of Science and Technology (CHARUSAT),
Changa-388421, Gujarat, India.

Corresponding author: kamleshchauhan.me@charusat.ac.in

Abstract. The aim of the present work is to deposit zirconium oxide thin films by reactive magnetron sputtering on glass substrates. The influence of helium partial pressure on structural, optical and wettability properties are investigated. The structural properties were investigated using X-Ray Diffraction, which shows that the films deposited has (111) peak of Zirconium Oxide. The intensity of the peak increased with the increase in helium partial pressure. UV-Vis-NIR spectrophotometer was used to study the optical properties of the films; the deposited films were transparent and had excellent transmittance of 95%. Surface roughness measured was in the range of 20.9nm to 21.8nm, the roughness of the deposited films decreased with increase in helium partial pressure. The wettability property measured using contact angle goniometer had a maximum contact angle of 94.1^0, which proves that the deposited film has hydrophobic property when measured using water.

KEY WORDS: Zirconium Oxide; Hydrophobicity; Sputtering; Transmittance; Roughness

1. Introduction

Metal oxides belong to the group of materials that have picked up consideration because to their extraordinary and fascinating optical properties in recent years. Thin films that are coloured and are resistant to corrosion and scratch are attracting more and more research work [1]. From an optical perspective, zirconium oxide has a high transparency, wide band gap and high refractive index. Due to these great optical properties, zirconium oxide has been generally utilized as a part of different optical applications, including broadband interference filters, high reflectivity mirrors and low-loss waveguides [2,3]. Optical, mechanical and structural properties of ZrO_2, is influenced by various parameters and deposition technique. Various structure of deposited films like amorphous, cubic, monoclinic, tetragonal and orthorombic were observed for various deposition techniques including Vacuum Arc Deposition (VAD), Sol–gel, Sputtering, Chemical Vapour Deposition (CVD) and Electron Beam Physical Vapour Deposition (EB-PVD) [4]. The most frequently utilized deposition technique is Magnetron sputtering, because of the prevalence in giving a uniform film on an enormous substrate [5].

2. Experimental Details

Reactive Magnetron Sputtering was used to sputter deposit Zirconium (Zr) metallic target (99.99% pure) on corning glass substrates. The distance between target and substrate was consistent (50 mm) during deposition. The cleaning of the substrates were done by rinsing the substrates in ultrasonic baths of acetone and methanol. For depositing thin films, helium was used as inert gas and oxygen was used as reactive gas and the flow of gas was controlled by Mass Flow Controller. The partial pressure of helium was varied from 45.50% to 57% and that of reactive gas was kept steady at 15sccm. The deposition time for all the coatings were 60minutes and the sputtering pressure was kept constant at 2.5Pa. The parameters for deposition is given in table 1 below:

Table 1: Parameters for depositing zirconium oxide thin films

Target	Zr (99.999%)
Substrate	Corning glass
Target-substrate distance	50 mm
Vacuum pressure	6 x 10^{-4} Pa
Working Pressure	2.5Pa
Gas Ratio (He:O_2 in sccm)	12.5:15, 15:15, 17.5:15, 20:15
Sample names	Z12.5, Z15, Z17.5, Z20
Rf Power	225W
Substrate temperature	550°C

Bruker D2 phaser, advance diffractometer with Cu-Kα radiation having wavelength 1.54Å was used to study the structural properties of the deposited thin films. UV–Vis–NIR spectrophotometer (Shimadzu 3600) was used to study optical transmittance and the wettability studies was carried out using Contact angle measurement system (Rame Hart 290).

3. Results and Discussion

The deposited zirconium oxide films were analysed by XRD to observe the structural properties obtained at different helium partial pressure. Figure 1. shows the X-Ray Diffraction patterns for zirconium oxide films deposited at various helium partial pressure.

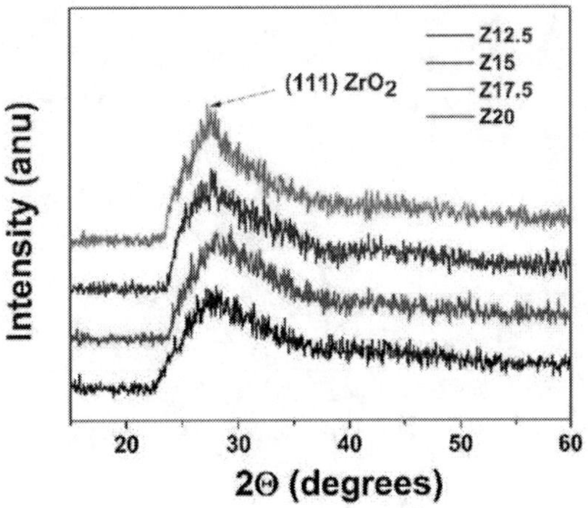

Figure 1. X-Ray Diffraction patterns for zirconium oxide thin films deposited at various helium partial pressure

The deposited films of zirconium oxide showed (111) peak of ZrO_2 phase (JCPDS card No. 89-9066). When the partial pressure of helium was 45.50% a low intensity peak of (111) ZrO_2 was observed. As the helium partial pressure increased from 50% to 57% the intensity of the peak also increased. Maximum intensity of the peak (111) for ZrO_2 was observed at 57% partial pressure. Goedicke *et al.* *(2000)* deposited thin films using reactive pulsed magnetron sputtering and studied the variation of argon flow rate on structural properties and concluded that sputtering pressure had a significant influence on the films coated. XRD results showed only single strong and relatively broad diffraction peak of (111) which is of monoclinic phase of ZrO_2. Also tried to get rough estimation of grain size of (111) reflex of monoclinic phase using full width and half maximum [6].

Kuhaili and Durrani observed that zirconium oxide films deposited were amorphous which attributed to high heat which formed ZrO_2, it obstructs the re-arrangement of oxygen–metal bonds at low temperatures and that results in formation of a crystalline phase. The predominance of the monoclinic phase can be perceived based on the thermodynamics of crystal growth. Mostly, the crystal plane that has the lowest Gibbs free energy will grow preferentially [6].

The crystalline size was assessed by FWHM (full width at half maximum) of the XRD peaks using Scherrer's method [2]. The results showed that as the intensity of the partial pressure of helium increased the average crystalline size also increased. The average crystalline size of the thin films changed from ~1 to ~2nm as intensity of the helium partial pressure increased.

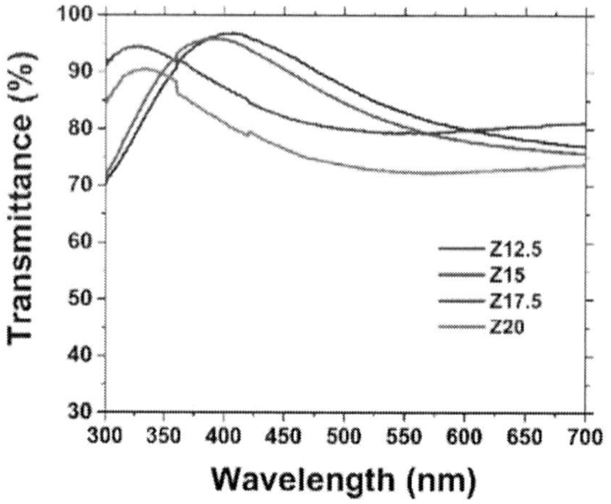

Figure 2. Spectral transmission curve of zirconium oxide thin films deposited at various helium partial pressure

The spectral transmission curve of zirconium oxide thin films is showed in Figure 2. The average transmittance of the deposited films are in the range of 85-95%, which means that the deposited films were transparent. It can be observed from the graph that the transmittance for films deposited at 45.50% helium partial pressure had a maximum transmittance of 95%, which reduced to 90% as the helium partial pressure increased to 57%.

Zhao *et al. (2014)* deposited zirconium oxide thin films and observed the four group of oxygen compositions using spectral transmission curves. It was noticed from the analysis that as the oxygen flow rate increased to 4.5sccm the transmittance of the films improved. At 4.5sccm flow rate average transmittance observed was 89.16% and the curves did not had any critical variation in the range of 340~400 nm. High OMR ratio was responsible for the full combination of zirconic and oxygenic atoms, prompting a reduced film layer with strengthening crystallization density [5]. The zirconium oxide films synthesized at higher oxygen ratio values displayed higher transmittance [7]. So the optical properties observed in this research are similar to as observed in literature.

The wettability property and roughness of the films synthesized is shown in Figure 3. It was noticed from the results that contact angle increased with decrease in helium partial pressure. The zirconium oxide films coated at 45.50% partial pressure had a maximum contact angle of 94.1° when measured with water, which showed that the films deposited as varying partial pressure had a hydrophobic property. It is seen from the graph that the lowest contact angle value of 67.5° was found at 57.0% partial pressure. Also from the analysis it was confirmed that the surface energies of the deposited samples decreased as the conatct angle increased.

Surface roughness measured is also aligned to the results of contact angle measurement. The roughness of the films deposited decreased from 21.8nm for 45.50% partial pressure to 20.90nm for

57.0% partial pressure. Patel *et al.* (2016) studied the effect of varying nitrogen gas flow on zirconium nitride films and found that the films deposited had a hydrophobic property when measured with water and the results showed that the surface energy increases with decrease in contact angle [8].

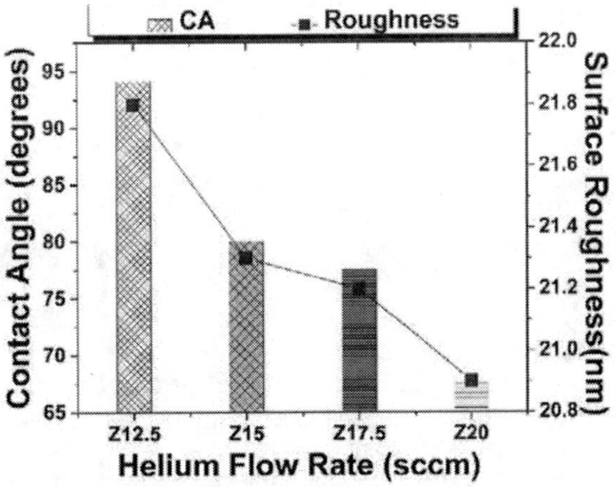

Figure 3. Contact angle and roughness of zirconium oxide thin films deposited at various helium partial pressure

Rawal *et al.* (2011) observed that structure surface is important to achieve the hydrophobic property. Besides the surface roughness, the structure formed on the surface is equally important to decide the wettability of the films [9]. When the films are hydrophilic, the surface energies are higher when compared to films, which are hydrophobic [10, 11, 12]. Hence, the results achieved by our research are in agreement with the literatures.

4. Conclusion

Zirconium oxide thin films were deposited using variation in helium partial pressure and the deposited films showed a high intensity (111) peak of ZrO_2 and as the partial pressure of helium decreased from 57.0% to 45.50% the intensity of the peak also decreased. The average crystalline size of the films were in the range ~1 to ~2nm and as the intensity of the helium partial pressure increased the crystalline size also increased. The maximum transmittance observed was 95% for the least partial pressure of helium and the average transmittance decreased to 90% as the partial pressure increased to 57.0%, which proves that the deposited films are transparent. The contact angle measurement for 45.50% helium partial pressure had a maximum contact angle of 94.1° when measured with water and the roughness was 21.8nm which reduced to 20.9nm which was validated with result of contact angle which also reduced to 67.0^0 for 57.0% helium partial pressure.

5. Acknowledgement
This research work is upheld by AICTE grant number 20/AICTE/RIFD/RPS (POLICY-III) 24/2012-13 approved under Research Promotion Scheme (RPS). We are grateful to Head, Dr. K. C. Patel Research and Development Centre (KRADLE) subsidiary to Charotar University of Science and Technology (CHARUSAT), India for conceding authorization to utilize different equipment's available in their characterization laboratory.

6. References

[1] S. K. Rawal, A. Kumar, V. Chawla, R. Jayaganthan, and R. Chandra, "Structural, optical and hydrophobic properties of sputter deposited zirconium oxynitride films," Mater. Sci. Eng. B, vol. 172, no. 3, pp. 259–266, 2010. doi:10.1016/j.mseb.2010.05.027.

[2] Al-Kuhaili and Durrani Effect of annealing on pulsed laser deposited zirconium oxide thin films. Journal of Alloys and Compounds, 509(39), 9536–9541, 2011. doi:10.1016/j.jallcom.2011.07.062.

[3] S. Zhao, F. Ma, K.W. Xu, H.F. Liang, J. Alloys Compd. 453 (2008) 453.

[4] O. Goldberg, E. Goldenberg, V. N. Zhitomirsky, S. R. Cohen, and R. L. Boxman, Surface & Coatings Technology Zirconium vacuum arc operation in a mixture of Ar and O 2 gases : Ar effect on the arcing characteristics, deposition rate and coating properties. Surf Coat Technol. 2012;206(21):4417-4424. doi:10.1016/j.surfcoat.2012.04.082.

[5] X. Zhao, J. Jin, J. Cheng, J. Lee, and K. Wu, "NU SC," Thin Solid Films, 2014. doi:10.1016/j.tsf.2014.05.060.

[6] K. Goedicke, U. Liebig, O. Zywitzki, and H. Sahm, "Influence of process parameters on the structure and the properties of ZrO 2 coatings deposited by reactive pulsed magnetron sputtering Ž PMS .," pp. 37–42, 2000.

[7] D. Pamu, K. Sudheendran, M. G. Krishna, K. C. J. Raju, and A. K. Bhatnagar, "Ambient temperature stabilization of crystalline zirconia thin fi lms deposited by direct current magnetron sputtering," Thin Solid Films, vol. 517, no. 5, pp. 1581–1585, 2009. doi:10.1016/j.tsf.2008.09.074.

[8] N. P. Patel, K. V. Chauhan, J. M. Kapopara, N. N. Jariwala & S. K. Rawal, Characterization of sputtered zirconium nitride films deposited at various argon : nitrogen ratio 10.1088/1757-899X/149/1/012015

[9] S. K. Rawal, A. K. Chawla, V. Chawla, R. Jayaganthan, and R. Chandra, "Characterization of bi-phased Zr2ON2-ZrO2coatings deposited by RF magnetron sputtering," *Thin Solid Films*, vol. 520, no. 5, pp. 1589–1596, 2011.

[10] S. K. Rawal, A. K. Chawla, R. Jayaganthan, and R. Chandra, "Structural, Wettability and Optical Investigation of Titanium Oxynitride Coatings: Effect of Various Sputtering Parameters," *J. Mater. Sci. Technol.*, vol. 28, no. 6, pp. 512–523, 2012.

[11] S. K. Rawal and R. Chandra, "Wettability and Optical Studies of Films Prepared from Power Variation of Co-sputtered Cr and Zr Targets by Sputtering," *Procedia Technol.*, vol. 14, pp. 304–311, 2014.

[12] H. V Patel, H. N. Patel, P. A. Soni, H. D. Parmar, P. Nicky, and K. V Chauhan, "Characterization of sputtered zirconium nitride thin films deposited at various RF power and sputtering pressure" vol. 020093, no. October, 2019. doi:10.1063/1.5130303

Enhancement of photo luminescent and magnetic properties of surface modified zinc sulphide nanoparticles by manganese doping

I.A. Shaikh[1*] , Adhish V. Raval[1] , D.V. Shah[1]

[1]Applied Physics Department, SVNIT, Surat, Gujrat-395007, India.
[*]e-mail address: isrars_74@yahoo.co.in

Abstract: In the current paper we have reported the structural, optical and magnetic properties of manganese doped (1% to 10%) zinc sulphide nanoparticles synthesised by simple chemical co-precipitation method at room temperature. The crystal structure of the synthesised nanoparticles has been confirmed by X-ray diffraction and the average crystallite size calculated by Scherrer formula was about 3.0 nm. A slight quenching in the crystallite size with the doping was also observed. The FESEM images showed spherical shape particles with uniform size distribution. The elemental composition of pure and manganese doped zinc sulphide was confirmed by EDAX attached to FESEM. The UV-Vis spectra has an absorption edge at 321 nm with sharp fall in the absorption confirming the particles of uniform size distribution with no shift in the absorption edge with the doping. The band gap widening was observed (4.4 eV) and the particle size (3.4 nm) calculated was by effective mass approximation. The photoluminescence spectra has wide emission spectra with the peaks at 397 nm, 414 nm and 487 nm with increase in the intensity with the doping concentration giving a peak at 1% of doping and decrease in emission intensity after that. The hysteresis curve shows the ferromagnetism and the saturation magnetisation increases with the doping concentration.

1. Introduction

Semiconductors are one of the most appreciated materials in the field science and technology owing to their use in the field of electronics, chemical and biomedical. In general the traditional semiconductors have band gap less than 2 eV, which put several restriction in their applications in the field power electronics. On other hand, the wide band gap semiconductors with higher band gap allow the development of smaller, faster, reliable and efficient devices when compared to normal semiconductors [1]. Amongst all the wide band semiconductors, zinc sulphide (ZnS) is a nontoxic and chemically [2] stable group II-VI material having found applications in the field of UV-light emitting diode, solar cells, dye degradation, bioimaging and sensors [3-4]. The bulk zinc sulphite shows the zinc-blende and wurzite crystalline form with direct band gap of 3.68 eV and 3.77 eV respectively.
It has been observed that the properties of zinc sulphide nanoparticles are dramatically different from its bulk counter part due to quantum confinement effect [5]. As the size of the nanoparticles is reduced to exciton Bohr radius it exhibits the quantum confinement effect resulting in broadening of the band gap with decrease in the size of the nanoparticles.

Doping of the zinc sulphide nanoparticles with transition metals are easy and introduce deep levels of the band gap and thereby influencing the physical properties – optical, electrical and magnetic. In 1994 Bhargava et al. reported the high luminescent efficiency and lifetime shortening of the manganese doped zinc sulphide nanoparticles which further encouraged the research in the field. The doping of the group II-VI semiconductors with rare earth elements is not preferred as their chemical properties are totally different than the host metals, like cadmium and zinc. But their chemical properties are similar to transition metals such as manganese, cobalt, copper etc. and therefore can be easily doped in zinc sulphide nanoparticles.

Emission in the zinc sulphide contributed by the sulphur ion interstitial and vacancies is in the blue region (around 450 nm). Whereas, the zinc ion interstitials and vacancies give emission around 490 nm [6]. Upon doping with the manganese impurity orange (around 600 nm) luminescence has been also documented by many groups [7]. Thus, one is compelled to conclude that the overall spectra of manganese doped zinc sulphide nanoparticles have minimum three peak contributed by each element present in the composite.

The pure zinc sulphide exhibit diamagnetism at room temperature but the introduction of transition metal impurity induces superparamagnetism showing weak ferromagnetism [8]. But at nano size the surface defects resulting due to higher surface to volume ratio may show some room temperature ferromagnetism [9]. The doping induced magnetism depends on the type impurity introduced and the saturation magnetisation increases with the increase in the concentration of the impurity.

Upon doping the zinc sulphide nanoparticles having particle size in the range of 2.2-2.7 nm with manganese with the doping concentration from 1.5 to 5%, three photoluminescence peaks were observed at 445 nm, 476 nm and 520 nm in with slight red shift in all the samples. Also the intensity of orange peak increased with the increase in the manganese concentration [10]. But some reports show the drastic changes in the properties of particle size of 20 nm to 50 nm and photoluminescence showing orange peak (587 nm) along with blue peak (463 nm) with shift to 449 nm when doped with the manganese [11]. The single orange peak (602 nm) has been reported in some work with either missing blue peak or negligible intensity [12] and showing a shift in the emission peak with the surface modification [13,14]. Induction of ferromagnetism in zinc sulphide nanoparticles when doped with transition metals has been reported. The doping (1, 3 and 10%) with the zinc sulphide nanoparticles with iron in the size range of 2-5 nm have shown the weak ferromagnetism and increase in the saturation magnetisation with doping percentage. Similarly, doping with the copper (1 to 10%) has shown the origin of ferromagnetism [15].

In view of these, in the current work, a comprehensive study of manganese doped zinc sulphide nanoparticles has been carried out for the different concentration of the impurity. In the current work doping induced changes in the structural, optical and magnetic properties have been investigated.

2. Experimental

2.1 Synthesis

The manganese doped zinc sulphide nanoparticles with different manganese concentration were synthesised by facile chemical co-precipitation method. The chemicals used for synthesis were of analytical grade, locally sourced and used without further purification [16]. For the synthesis of pure zinc sulphide nanoparticles, 0.1 M of Zinc sulphate ($ZnSO_4.H_2O$) solution was prepared in 50 ml deionised water was mixed with 1 gm of polyvinypyrrolidone (PVP) and stirred for proper mixing. Then 0.1 M sodium sulphide (Na_2S) solution prepared in 50 ml deionised water was added drop wise with continuous stirring. The addition of sodium sulphide results in white precipitate, the stirring at ambient condition was continued for 2 hours for the proper dispersion. The solution was collected, centrifuged, washed several times using the deionised water as the medium and the finally the centrifuged precipitate was dried overnight in a hot air oven to obtain the powder. For doping with manganese, manganese sulphate ($MnSO_4$) in calculated proportion was added along with the zinc sulphate in the beginning of the reaction keeping the rest of the procedure same.

2.2 Characterisation

The X-ray diffractometry (XRD) of all the samples was performed on Rikagu Ultima IV operating at 40 kV-40 mA (λ = 1.5406 Å) in the range of 20° to 80°. The scanning electron microscopy (FESEM-FEI Nova NanoSEM 450) was used to study the morphology and determine the particle size. The elemental analysis was performed by Bruker XFlash 6130 attached to FESEM. The optical absorption spectra were recorded by LABINDIA: UV-3000 in the range of 200 nm to 800 nm. The photo luminescence (PL) of the samples was recorded by spectroflourometer, Flouromax-4. The magnetisation was measured by M-H hysteresis curve for the samples at room temperature using physical property measurement system (PPMS) Quantum Design (USA).

2.3 Results and discussion

The fig. 1 shows the X-ray diffractogram of pure and manganese doped zinc sulphide nanoparticles. The diffraction pattern has broadened peaks at 2θ values of 28.66°, 47.98° and 56.51° corresponding to the diffraction planes (111), (220) and (311) respectively [17]. The observed data matches with the JCPDS Card No. 05-0566 confirming the formation of zinc sulphide wurtzite crystals. Other than this no other peak is observed confirming the formation of no other crystalline state upon the doping of manganese.

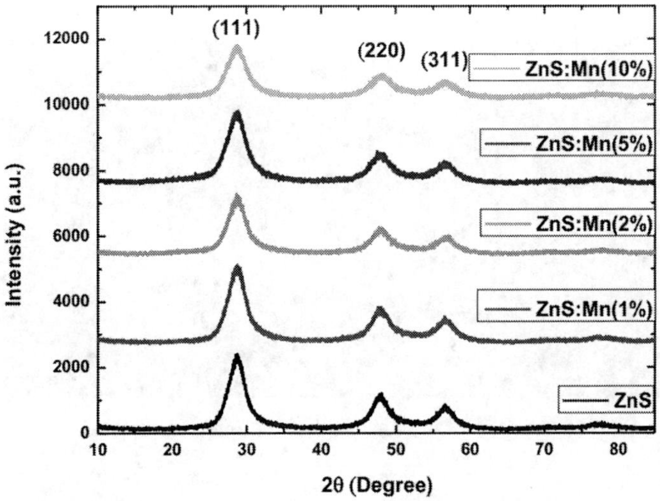

Figure 1. XRD of pure and manganese doped zinc sulphide nanoparticles

To calculate the diameter of the nanoparticles Scherrer formula [18] used is,

$$D = \frac{0.9\lambda}{\beta cos\theta} \quad (1)$$

Where D is the average particle size, λ is wavelength of X-ray and value is 1.5406Å, β is the full width at half maxima and θ is the Bragg's angle. The particle size calculated by the above formula is in range of 2.7 nm to 2.9 nm and shows decrease in the particle size with the increase in the doping.

The other crystal parameters are calculated by the formulae,

Interplanar distance,

$$d = \frac{\lambda}{2sin\theta} \quad (2)$$

Lattice parameter or constant,

$$d = \frac{a}{\sqrt{h^2+k^2+l^2}} \quad (3)$$

where h, k and l are Miller's indices.

Microstrain,

$$\varepsilon = \frac{\beta cos\theta}{4} \quad (4)$$

Dislocation density,

$$\delta = \frac{1}{D^2} \quad (5)$$

Stacking fault.

$$SF = \left[\frac{2\pi^2}{45(3tan\theta)]^{\frac{1}{2}}}\right]\beta \quad (6)$$

The values calculated by above formulae have been tabulated in Table 1. The lattice defect have decreasing values with the increase in the doping percentages, confirms the better crystallisation.

Table 1. Structural properties of the pure and doped zinc sulphide nanoparticles

Doping (%)	$2\theta^o$	FWHM (β^o)	Crystallite size (D) (nm)	Dislocation density (δ) $\times 10^{-1}$ nm^{-2}	Micro strain (ε) $\times 10^{-2}$	Interplanar distance $d_{(111)}$ (Å)	Lattice Constant (Å)	Stacking fault (SF) $\times 10^{-2}$
0	28.6587	2.7935	2.9350	1.1608	1.1810	3.1123	5.3906	2.4403
1	28.7567	2.8817	2.8460	1.2346	1.2177	3.1020	5.3728	2.5129
2	28.7558	2.8943	2.8430	1.2372	1.2230	3.1020	5.3728	2.5239
5	28.7949	2.8153	2.9130	1.1784	1.1895	3.0979	5.3657	2.4532
10	28.7448	3.0143	2.7210	1.3506	1.2738	3.1032	5.3749	2.6294

The FESEM images of pure and manganese doped zinc sulphide nanoparticles are shown in the fig. 2. The images show highly aggregated particles due to high surface energy. The nanoparticles are spherical in nature with narrow size distribution. The crystallinity of the nanoparticles shows improvement upon doping with the manganese [19].

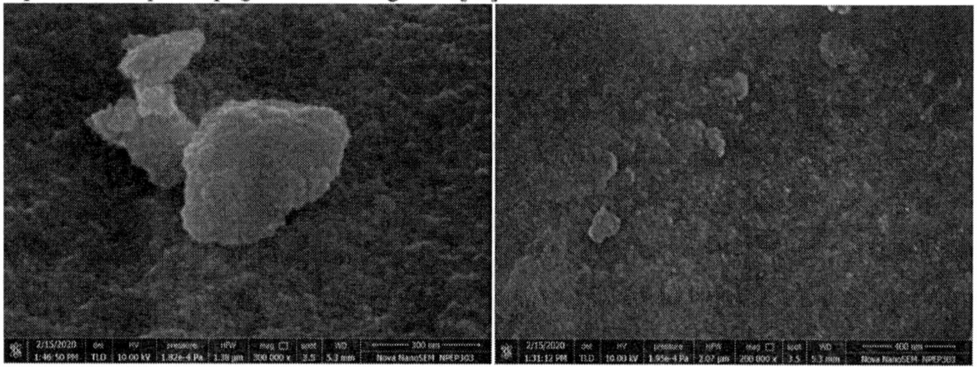

Figure 2. FESEM images of (i) Pure and (ii) doped (10%) zinc sulphide nanoparticles

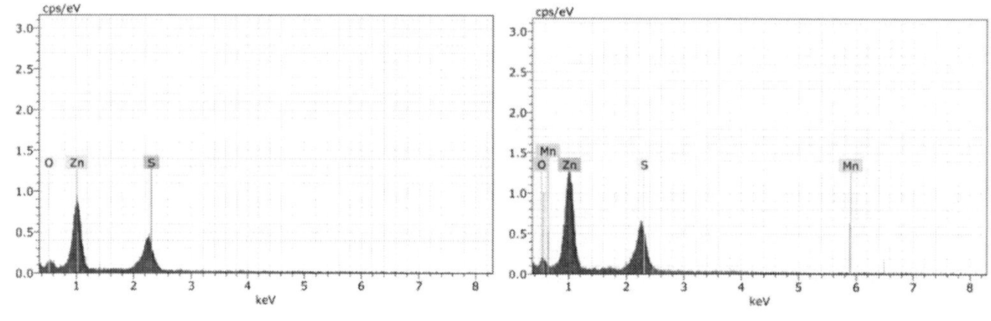

Figure 3. EDAX of (i) Pure and (ii) doped (10%) zinc sulphide nanoparticles

The EDAX spectra (fig. 3) of the samples reveal the presence of all individual elements with suitable composition [20] confirming the purity and the doping of the samples.

The fig. 4 shows the UV-Vis absorption spectra of pure and manganese doped zinc sulphide nanoparticles. The absorption edge of pure as well as doped nanoparticles is at 321 nm and there is no noticeable shift in the absorption edge upon doping or the doping concentration [21]. Tauc's plot is employed to estimate the band gap of the synthesised nanoparticles. The widening of band gap (4.4 eV) is observed which is result of quantum confinement effect.

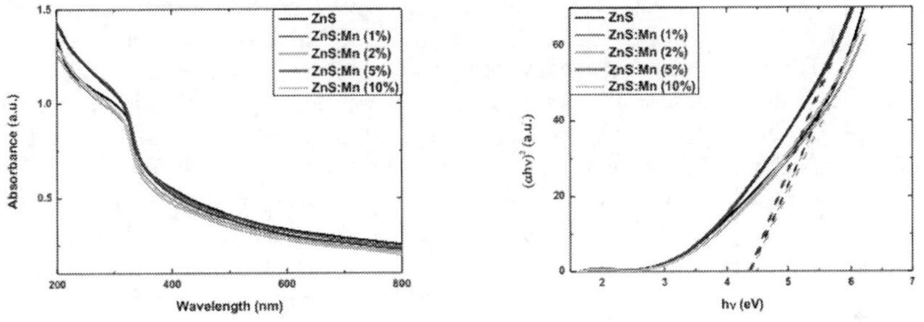

Figure 4. UV-visible spectra of (i) Pure and (ii) Tauc's plot of zinc sulphide nanoparticles

The particle size of synthesised nanoparticles were also calculated by effective mass approximation,

$$E(R) = E_{bulk} + \frac{h^2}{8R^2}\left[\frac{1}{m_e^*} + \frac{1}{m_h^*}\right] - \frac{18e^2}{\varepsilon R} \quad (7)$$

Where E(R) and E_{bulk} are the energy band gap of nanoparticles and bulk semiconductor, m_e^* and m_h^* are the effective mass of electron and hole respectively and the third term which is due to Coulomb's attraction and is negligible. The particle size calculated by above formula comes out to be 3.4 nm which is in good agreement with value obtained by the XRD data.

The photoluminescence emission spectra for 265 nm excitation of pure and manganese doped zinc sulphide nanoparticles are shown in fig. 5. The photoluminescence spectra are highly asymmetric with multiple peaks at 397 nm, 414 nm and 487 nm [22]. The 414nm peak has the maximam intensity while the 487 nm peak has the least intensity. The multiple peaks are due to involvement of different transition involving the energy levels of interstitial sulphur, zinc interstitials [23]. In the spectra the orange emission was not observed and it can be concluded the manganese impurity is just acting as the sensitising agent, improving the overall emission intensity rather than giving additional emission peak. The intensity of the luminescence reaches a peak value for 1% of the impurity and decreases thereafter confirming the quenching [24, 25].

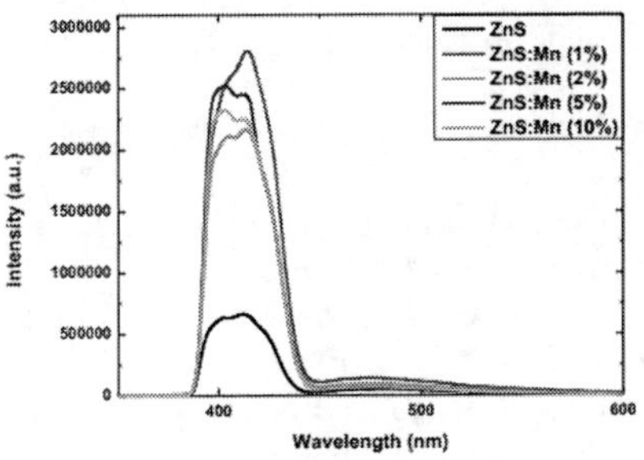

Figure 5. Photoluminescence of pure and zinc sulphide nanoparticles

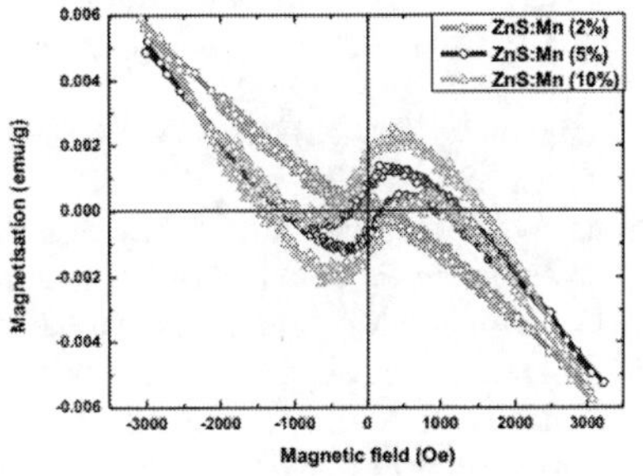

Figure 6. Magnetic hysteresis curve of doped zinc sulphide nanoparticles

The fig. 6 shows well defined magnetic hysteresis curve for manganese doped zinc sulphide nanoparticles at room temperature. The pure zinc sulphide nanoparticles do not show the ferromagnetism at room temperature owing to the absence of unpaired electrons in the 'd' orbital [26, 27] and is not shown in the figure. But the doped nanoparticles show the ferromagnetism which increases with manganese concentration. At relatively low doping concentrations the manganese atoms replace the host lattice cations and can be considered as an isolated atom, but at the higher concentrations clusters of manganese atoms are formed and the antiferromagnetic behaviour starts dominating [28]. The overall observation is that the saturation magnetisation increases with the doping concentration giving maximum at 10% of the doping.

Conclusion

The manganese doped zinc sulphide nanoparticles have been synthesised by simple one step chemical co-precipitation method. The synthesised nanoparticles have wurtzite crystal structure with average particle size of about 3 nm which was calculated by using X-Ray diffraction spectra and Tauc's plot. Upon doping with the manganese no extra peak was observed confirming formation of no other crystalline phase. With increase in the doping concentration slight quenching in the particle size was observed. The UV-Vis spectra showed no shift in the absorption edge and band gap with the doping and its amount. Widening in the band gap (4.4 eV) due to quantum confinement effect was also observed. The photoluminescence showed peaks at 397 nm, 414 nm and 487 nm and the intensity changed considerably with doping with no additional peaks. Initially, the intensity increased and decreased thereafter with the doping giving a maximum at 1% of doping. The magnetic hysteresis curve exhibited weak ferromagnetism and the saturation magnetisation increased with the doping concentration. The present study showed that the photo luminescence and ferromagnetism may be tuned by balancing the doping concentration indicating the possible applications in photoluminescent and spintronic devices.

References

[1] Parvaneh Iranmanesh, Samira Saeednia and Mohsen Nourzpoor 2015 *Chin. Phys. B* **24(4)** 046104-1

[2] Christian Kirchner, Tim Liedl, Stefan Kudera, Teresa Pellegrino, Almudena Munoz Javier, Hermann E. Gaub, Sonja Stolzle, N. Fertig, and Wolfgang J. Parak 2005 *Nano Lett.* **5(2)** 331

[3] M. Jothibas, C. Manoharan, S. Johnson Jeyakumar, P. Praveen, I. Kartharinal Punithavathy, J. Prince Richard *Sol. Energy* 2018 **159** 434

[4] Hamid Reza Rajabi and Mohammad Farsi *Journal of Molecular Catalysis A: Chemical* 2015 **(399)** 53

[5] He Hu and Weihua Zhang *Opt. Mater.* 2006 **28** 536

[6] S. Radhu and C. Vijayan *Mater. Chem. Phys.* 2011 **129** 1132

[7] W.Q. Peng, G.W. Cong, S.C. Qu and Z.G. Wang *Opt. Mat.*2006 **29** 313

[8] Sunil Kumar and N. K. Verma *J. Mater Sci: Mater. Electron* 2015 **26** 2754

[9] D. Saikia and J. P. Borah *J. Mater. Sci: Mater. Electron* 2017 **28** 8029

[10] A. K. Kole and P. Kumbhakar *Appl. Nanosci.* 2012 **2(15)** 23

[11] Kamal Bera, Satyajit Saha and Paresh Chandra Jana *Mater. Today: Proc.* 2018 **5** 6321

[12] Rahizana Mohd Ibrahim, Masturah Markom and Huda Abdullah *ECS J. of Solid State Sci and Technol.* 2015 **4 (2)** R31

[13] Hoon-Young Kong, Byung-Kwan Song, Jonghoe Byun, and Cheong-Soo Hwang *Bull. Korean Chem. Soc.* 2013 **34(4)** 1181

[14] Song Wei Lua, Burtrand I. Lee, Zhong Lin Wang, Wusheng Tong, Brent K. Wagnerc, Wounjhang Park and Christopher J. Summers *J. of Lumin.*2001 **92** 73

[15] N.G. Imam and Mohamed Bakr Mohamed *Superlattices and Microstructures* 2014 **73** 203

[16] N. V. Desai, I. A. Shaikh, K. G. Rawal, and D. V. Shah *AIP Conf. Proc.*2018 **1953** 030149

[17] B. Saraswathi Amma, K. Manzoor, K. Ramakrishna, Manjunatha Pattabi *Mater. Chem. and Phys.* 2008 **112** 789

[18] Khalid T. Al- Rasoul, Nada K.Abbas and Zainb J. Shanan *Int. J. Electrochem. Sci.* 2013 **8** 5594

[19] G. Murugadoss and M. Rajesh Kumar *Appl. Nanosci* 2014 **4** 67

[20] Gang-Juan Lee, Hui-Chuan Chen and Jerry J. Wu *Int. J. of Hydrogen Energ.* 2019 **44(1)** 110

[21] G. Murugadoss *J. of Lumin.* 2010 **130** 2207

[22] R. Sarkar, C.S.Tiwary, P.Kumbhakar, A.K.Mitra *Physica B* 2009 **404** 3855

[23] Pule V. Raleaooa, Andreas Roodt, Gugu G.Mhlongo, David E. Motaung and Odireleng M. Ntwaeaborwa *Optik - International Journal for Light and Electron Optics* 2018 **153** 31

[24] P. H. Borse, N. Deshmukh, R. F. Shinde, S. K. Date and S. K. Kulkarni *J. Mater. Sci.* 1999 **34** 6087

[25] Lata Wanjari, D. P. Bisen and Raunak Kumar Tamrakar *Advance Physics Letter* 2015 **2(3)** 12

[26] B. Sreenivasulu, S. Venkatramana Reddy and P. Venkateswara Reddy *J. Mater. Sci.: Mater. Electron.* 2018 **29(251)** 259
[27] B. Poornaprakash , D. Amaranatha Reddy , G. Murali, N. Madhusudhana Rao, R.P. Vijayalakshmi and B.K. Reddy *J. Alloys Compd.* 2013 **577** 79
[28] Tania Jahanbin, Meriem Gaceur, Helene Gros-Dagnac, Soraya Benderbous and Souad Ammar Merah *J. Nanopart. Res.* 2015 **17** 258

IOP Conf. Series: Materials Science and Engineering 1126 (2021) 012040 doi:10.1088/1757-899X/1126/1/012040

Conductivity tensor in cylindrical quantum wire with infinite potential for the case of electron-optical phonon scattering

Hoang Van Ngoc

Institute of Applied Technology, Thu Dau Mot University, No 6, Tran Van On Street, Thu Dau Mot city, Binh Duong province, Vietnam

Email: ngochv@tdmu.edu.vn

Abstract. Conductivity tensor characterizes the conductivity of the material, this work calculates the conductivity tensor in a cylindrical quantum wire with infinite potential in the presence of two external fields, the laser field and linearly polarized electromagnetic wave field. The problem is also only considered for the case of electron-optical phonon scattering. By using a quantum kinetic equation for electrons in a quantum wire in the presence of two external fields, the conductivity tensor is calculated. The expression for conductivity tensor shows its dependence on the frequency and amplitude of the external field as well as other parameters specific to the system. The figure of the dependence of the conductivity tensor to the external field will be investigated and plotted for the quantum wire GaAs / GaAsAl.

1. Introduction

With the continuous development of science and technology, nanotechnology is an important issue in the research of materials. The low-dimensional systems that are studied a lot are quantum wire [1-5], quantum well [6-11], quantum dot, superlattice [12-15]. When an external field acts on charged particles, many new effects appear [1,7,14]. The external field carries energy and momentum, which greatly changes the properties in the material [1,7,16-18].

In this study, we put the particle system into two external fields, the laser field, and the linearly polarized electromagnetic wave field. Under the action of two external fields, charged particles will interact with the phonon, using quantum kinetic equations for electrons in a cylindrical quantum wire with infinite potential, the conductivity tensor expression will be calculated. This work also considers only electron-optical phonon scattering.

The conductivity tensor expression indicates its dependence on the external field and system-specific parameters. Quantum wire is examined and plotted as GaAs / GaAsAl quantum wire. The dependence of the conductivity tensor to the external field at each different frequency range is different, this dependence is nonlinear. The dependence of the conductivity tensor on the laser field amplitude is significant, it shows that the impact of the laser field is largely due to the large energized laser field.

2. Calculating conductivity tensor

Consider a cylindrical quantum wire with radius R, the confining potential V(r) of the string has the form:

$$\begin{cases} V(\vec{r}) = 0 \Leftrightarrow r < R \\ V(\vec{r}) = \infty \Leftrightarrow r > R \end{cases} \tag{1}$$

The wave function of the electron has the form:

$$\begin{cases} \psi_{n,l,\vec{p}}(\vec{r},\phi,z) = 0 \Leftrightarrow r > R \\ \psi_{n,l,\vec{p}}(\vec{r},\phi,z) = \dfrac{1}{\sqrt{V_0}} e^{in\phi} e^{i\vec{p}z} \psi_{n,l}(\vec{r}) \Leftrightarrow r < R \end{cases} \tag{2}$$

here $\psi_{n,l}(\vec{r}) = \dfrac{1}{J_{n+1}(B_{n,l})} J_n\left(B_{n,l}\dfrac{r}{R}\right)$; $B_{n,l}$ is the solution of the Bassel function n level corresponding to the equation $J_n(B_{n,l}) = 0$; $V_0 = \pi R^2 L_z$ is the volume of the quantum wire; $\vec{p}$ is momentum of electron.

The energy spectrum of the electron takes the form: $E_{n,l}(\vec{p}) = \dfrac{p_z^2}{2m} + \dfrac{B_{n,l}^2}{2mR^2}$ (3)

Assume the system is placed in a laser field and a linearly polarized electromagnetic field:

$$\vec{F}(t) = \vec{F}\sin(\Omega t) \text{ and } \begin{cases} \vec{E}(t) = \vec{E}(e^{-i\omega t} + e^{i\omega t}) \\ \vec{H} = [\vec{n}, \vec{E}(t)] \end{cases} \tag{4}$$

here ω is the electromagnetic wave frequency; F is the laser field amplitude; Ω is the laser field frequency.

Quantum kinetic equations for the system ($\hbar = 1$) [1,16,17]:

$$\begin{aligned} &\frac{\partial f_{n,l,\vec{p}_z}(t)}{\partial t} + \left(e\vec{E}(t) + \omega_H\left[\vec{p}_z, \vec{h}(t)\right], \frac{\partial f_{n,l,\vec{p}_z}(t)}{\partial \vec{p}_z}\right) = \\ &= 2\pi \sum_{n,l,n',l',\vec{q}} M_{n,n',l,l'}(\vec{q}) \sum_{L=-\infty}^{+\infty} J_L^2(\vec{a},\vec{q})\left[f_{n',l',\vec{p}_z+\vec{q}}(t) - f_{n,l,\vec{p}_z}(t)\right]\delta\left(\varepsilon_{n',l',\vec{p}_z+\vec{q}} - \varepsilon_{n,l,\vec{p}_z} - L\Omega\right) \end{aligned} \tag{5}$$

$f_{n,l,\vec{p}_z}(t)$ is distribution function of electrons; $\vec{q}$ is momentum of phonon;

$f_{n,l}(\vec{p}_z,t) = f_0(\varepsilon_{n,l,\vec{p}_z}) + f_1(\vec{p}_z,t)$; $f_0(\varepsilon_{n,l,\vec{p}_z}) = \theta(\varepsilon_F - \varepsilon_{n,l,\vec{p}_z})$; $f_1(\vec{p}_z,t) = -\vec{p}_z\vec{\chi}(t)f_0'\left(\varepsilon_{n,l,\vec{p}_z}\right)$;

$f_0'\left(\varepsilon_{n,l,\vec{p}_z}\right) = \dfrac{\partial f_0(\varepsilon_{n,l,\vec{p}_z})}{\partial \varepsilon_{n,l,\vec{p}_z}}$; $f_1(\vec{p}_z,t) = f_1(\vec{p}_z)e^{-i\omega t} + f_1^*(\vec{p}_z)e^{i\omega t}$; $f_1(\vec{p}_z) = -\vec{p}_z\vec{\chi}f_0'\left(\varepsilon_{n,l,\vec{p}_z}\right)$;

$\vec{\chi}(t) = \vec{\chi}e^{-i\omega t} + \vec{\chi}^* e^{i\omega t}$; $\vec{\chi} = \dfrac{e}{m}\vec{E}\dfrac{\tau(\varepsilon)}{1 - i\omega\tau(\varepsilon)}$; $\vec{\chi}^* = \dfrac{e}{m}\vec{E}\dfrac{\tau(\varepsilon)}{1 + i\omega\tau(\varepsilon)}$;

$$\begin{aligned} \frac{\partial f_{n,l}(\vec{p}_z,t)}{\partial t} &= \frac{\partial}{\partial t}\left[f_0\left(\varepsilon_{n,l,\vec{p}_z}\right) + f_1(\vec{p}_z)e^{-i\omega t} + f_1^*(\vec{p}_z)e^{i\omega t}\right] = \\ &= -i\omega f_1(\vec{p}_z)e^{-i\omega t} + i\omega f_1^*(\vec{p}_z)e^{i\omega t} \end{aligned} \tag{6}$$

$$M_{n,l,n',l'}(\vec{q}) = \left|C_{\vec{q}}\right|^2 \left|I_{n,l,n',l'}\right|^2 N_{\vec{q}} \tag{7}$$

With optical phonon [7]:

$$\left|C_{\vec{q}}\right|^2 = \frac{2\pi e^2 \omega_q}{\varepsilon_0 V_0 q^2}\left(\frac{1}{\chi_\infty} - \frac{1}{\chi_0}\right); \varepsilon_0 = \frac{10^{-9}}{36\pi}; N_q = \frac{k_B T}{\omega_q} \quad (8)$$

Here χ is the dielectric constant; V_0 is the volume of the wire; ω_q is the frequency of optical phonon; k_B is Boltzmann contant.

$I_{n,l,n',l'}\left(\vec{q}_{h,j}\right) = \frac{2}{R^2}\int_0^R J_{|n-n'|}(q_{h,j}R)\psi_{n,l}(r)\psi^*_{n',l'}(r)rdr$ is form factor;

$\varepsilon_{n,l,\vec{p}_z+\vec{q}} = \frac{(\vec{p}_z+\vec{q})^2}{2m} + \frac{B_{n,l}^2}{2mR^2}$; $J_l^2(\vec{a},\vec{q})$ is the Bessel function of real argument; $\vec{a} = \frac{e\vec{F}}{m\Omega^2}$; $\omega_H = \frac{eH}{mc}$; $\vec{h}(t) = \frac{\vec{H}(t)}{H}$.

We consider only the simple case: $l = 0, \pm 1$: $J_0^2 \approx 1$; $J_\pm^2 \approx \frac{(\vec{a},\vec{q})^2}{4}$;

so [16]:

$$\begin{aligned}\vec{j}(t=0) &= \int(\vec{R}(\varepsilon)+\vec{R}^*(\varepsilon))d\varepsilon = \vec{j}_1 + \vec{j}_1^* \\ &= \frac{4e^2 n_0}{m}\frac{\tau(\varepsilon_F)}{1+\omega^2\tau^2(\varepsilon_F)}\left\{\left[\varepsilon_F - \frac{B_{n,l}^2}{2mR^2}\right]I + \lambda\frac{\tau(\Omega)\left[1-\omega^2\tau(\Omega)\tau(\varepsilon_F)\right]}{1+\omega^2\tau^2(\Omega)} + \right. \\ &\left. -A\frac{\tau(\varepsilon_F)\left[1-\omega^2\tau^2(\varepsilon_F)\right]}{1+\omega^2\tau^2(\varepsilon_F)}\right\}\vec{E}\end{aligned} \quad (9)$$

Here $\vec{R}$ is the expression of the partical current density:

$$\vec{R}(\varepsilon) = \frac{\tau(\varepsilon)}{1-i\omega\tau(\varepsilon)}\left(\vec{Q}+\vec{S}(\varepsilon)\right) \quad (10)$$

$$\vec{R}^*(\varepsilon) = \frac{\tau(\varepsilon)}{1+i\omega\tau(\varepsilon)}\left(\vec{Q}+\vec{S}(\varepsilon)\right) \quad (11)$$

$$\vec{Q} = \frac{e^2}{m}\sum_{n,l,\vec{p}_z}\left(\vec{E}, \frac{\partial f(\vec{p}_z,t)}{\partial \vec{p}_z}\right)\vec{p}_z\delta\left(\varepsilon - \varepsilon_{n,l,\vec{p}_z}\right) \quad (12)$$

$$\begin{aligned}\vec{S}(\varepsilon) = &-\frac{e}{m}2\pi\sum_{n,n',l,l',\vec{q}} M_{n,n',l,l'}(q)\frac{(\vec{a},\vec{q})^2}{4}\sum_{\vec{p}_z} f_1(\vec{p}_z)\times \\ &\times\left[\delta\left(\varepsilon_{n',l',\vec{p}_z+\vec{q}} - \varepsilon_{n,l,\vec{p}_z} - \Omega\right) + \delta\left(\varepsilon_{n',l',\vec{p}_z+\vec{q}} - \varepsilon_{n,l,\vec{p}_z} + \Omega\right)\right]\times \\ &\times\left[(\vec{p}_z+\vec{q})\delta\left(\varepsilon - \varepsilon_{n,l,\vec{p}_z+\vec{q}}\right) - \vec{p}_z\delta\left(\varepsilon - \varepsilon_{n,l,\vec{p}_z}\right)\right]\end{aligned} \quad (13)$$

$$A = \frac{e^2F^2}{2m\Omega^3}M_{n,n',l,l'}\left(\sqrt{2m\Omega}\right)\sqrt{2m\left(\varepsilon_F - \frac{B_{n,l}^2}{2mR^2}\right)}$$
$$= \frac{\pi e^4 k_B T F^2}{2m^2\varepsilon_0^2 V_0 \Omega^4}\left(\frac{1}{\chi_\infty} - \frac{1}{\chi_0}\right)\sqrt{2m\left(\varepsilon_F - \frac{B_{n,l}^2}{2mR^2}\right)} \quad (14)$$

$$\lambda = \frac{e^2F^2}{2m\Omega^3}M_{n,n',l,l'}\left(\sqrt{2m\Omega}\right)\left\{\sqrt{2m\left(\varepsilon_F - \frac{B_{n,l}^2}{2mR^2}\right)}\left[\sqrt{2m\left(\Omega - \frac{B_{n,l}^2}{2mR^2}\right)} - 1\right]\right\}$$
$$= \frac{\pi e^4 k_B T F^2}{2m^2\varepsilon_0^2 V_0 \Omega^4}\left(\frac{1}{\chi_\infty} - \frac{1}{\chi_0}\right)\left\{\sqrt{2m\left(\varepsilon_F - \frac{B_{n,l}^2}{2mR^2}\right)}\left[\sqrt{2m\left(\Omega - \frac{B_{n,l}^2}{2mR^2}\right)} - 1\right]\right\} \quad (15)$$

n_0 is particle density; ε_F is the Fermi energy;

with $\vec{j}(t=0) = 2\sigma\vec{E}$ [3]; so:

$$\sigma_{ik} = \frac{2e^2 n_0}{m}\frac{\tau(\varepsilon_F)}{1+\omega^2\tau^2(\varepsilon_F)}\left\{\left[\varepsilon_F - \frac{B_{n,l}^2}{2mR^2}\right]\delta_{ij} + \frac{\tau(\Omega)\left[1-\omega^2\tau(\Omega)\tau(\varepsilon_F)\right]}{1+\omega^2\tau^2(\Omega)}\lambda + \right.$$
$$\left. - \frac{\tau(\varepsilon_F)\left[1-\omega^2\tau^2(\varepsilon_F)\right]}{1+\omega^2\tau^2(\varepsilon_F)}A\right\} \quad (16)$$

This is the conductivity tensor expression to be calculated with δ_{ik} is the Kronecker symbol.

3. Numerical results and discussion

Values of the quantities used when plotting [1,2,5]: m = 0,0665m$_0$ (m$_0$ is the mass of free electron); ε_F = 50 meV; $\tau(\varepsilon_F) \sim 10^{-11}$ s^{-1}; $n_0 = 10^{23} m^{-3}$; $\rho = 5.3\times10^3 kg/m^3$; R = $100\overset{0}{A}$; $T = 300 \div 340K$; $\chi_\infty = 10.8$; $\chi_0 = 13.1$; $q = 2.10^5 m^{-1}$; $V_0 = 1$.

Figure 1 and figure 2 show the dependence of σ on Ω. Looking at the figures, they see that as the laser field frequency increases, so does the tensor. When frequency $\Omega > 3*10^7 s^{-1}$ then the conductivity tensor increases very small as frequency increases, when frequency increases to $\Omega > 3.3*10^{10} s^{-1}$ then tensor does not increase anymore, figure is a straight line. With three different values of temperature gives three separate plots. The higher the temperature, the larger the conductivity tensor for the same laser field frequency value.

Figures 3, 4, and 5 show the dependence of σ on ω. At different frequency ranges the dependence of the conductivity tensor is different. When the frequency is less than 10^4, the conductivity tensor does not depend on the electromagnetic wave frequency, which is suitable in theory because at a small frequency the energy of the electromagnetic wave is very small so it does not affect the conductivity of the material. When frequency increases $\omega > 10^4 s^{-1}$ conductivity tensor decreases, conductivity tensor decreases rapidly when frequency $\omega > 10^{12} s^{-1}$.

 doi:10.1088/1757-899X/1126/1/012040

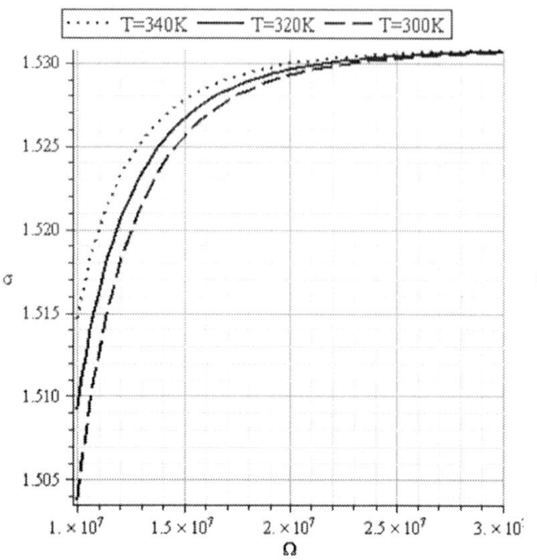

Figure 1. The dependence of σ on Ω when $\Omega < 3.10^7$

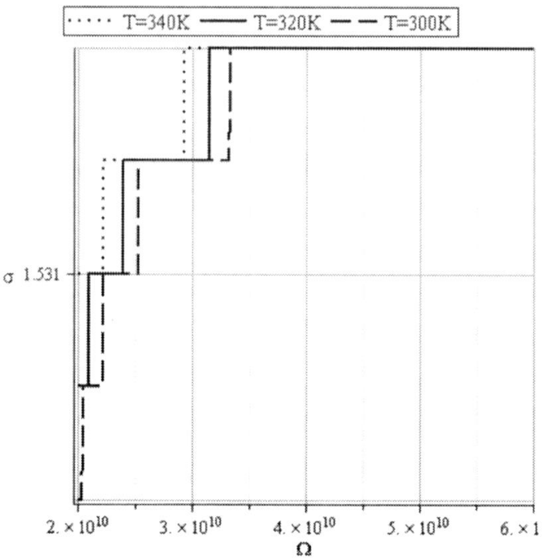

Figure 2. The dependence of σ on Ω when $\Omega > 2*10^{10}$

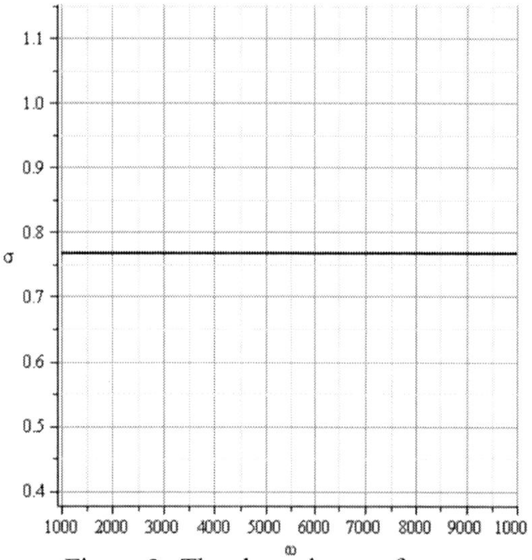

Figure 3. The dependence of σ on ω when $\omega < 10^4$

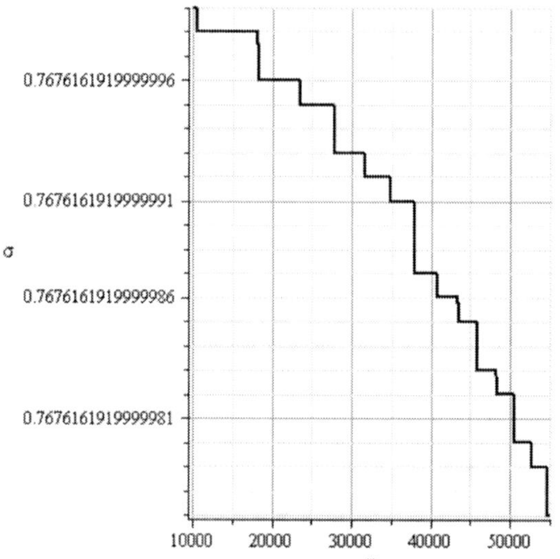

Figure 4. The dependence of σ on ω when $10^4 < \omega < 5*10^4$

Figures 6, 7, and 8 show the dependence of σ on F. When the laser field amplitude is less than 10^4, the plot is straight-line, the conductivity tensor does not depend on the amplitude. When the laser field amplitude increases $F > 2*10^4$ the conductivity tensor begins to decrease but the decrease is small. Corresponding to the three values of temperature gives three separate plots, the higher the temperature the larger the conductivity tensor value.

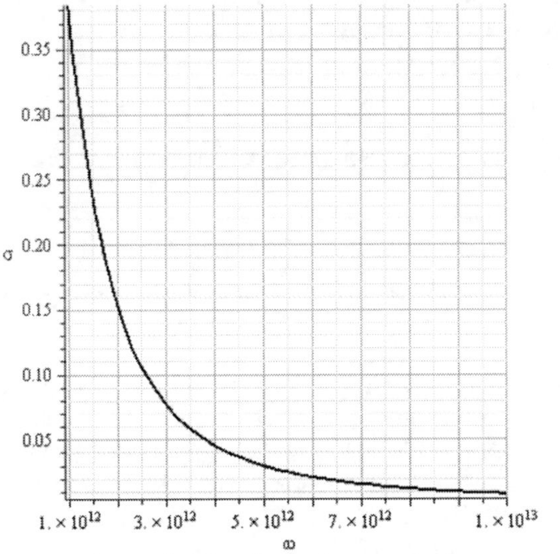

Figure 5. The dependence of σ on ω when $\omega > 10^{12}$

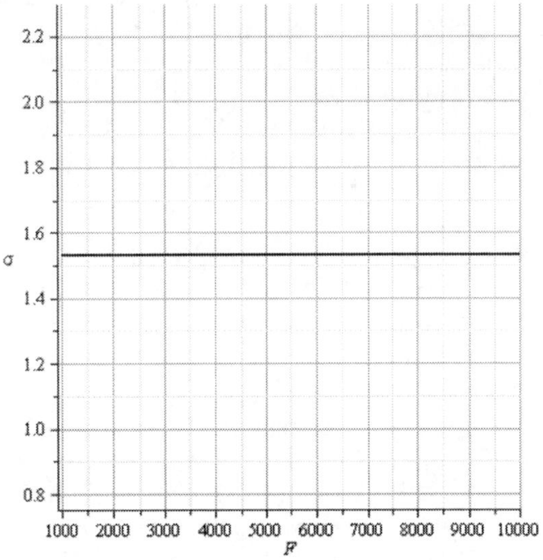

Figure 6. The dependence of σ on F when $F < 10^4$

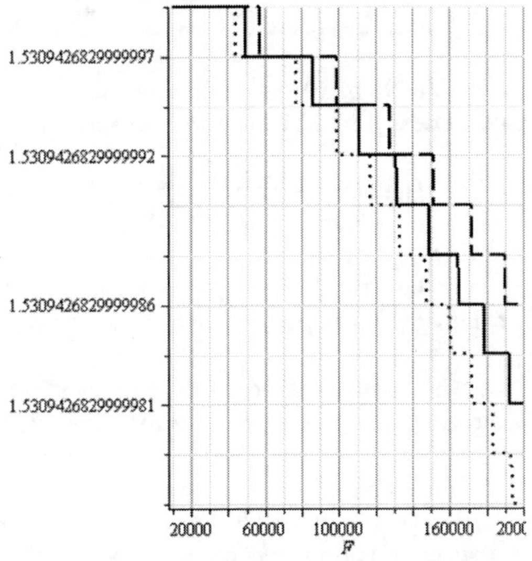

Figure 7. The dependence of σ on F when $10^4 < F < 2*10^5$

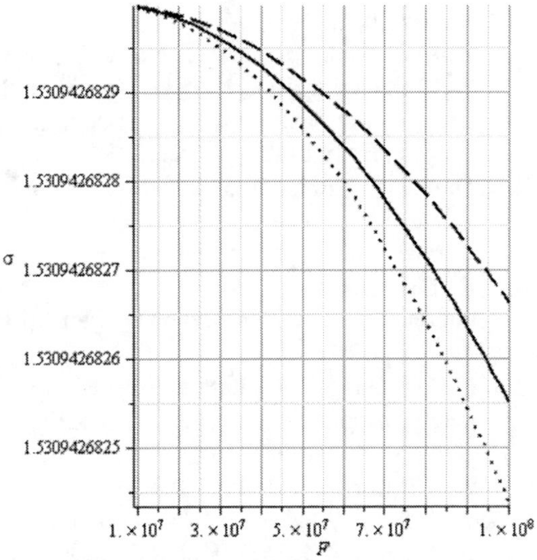

Figure 8. The dependence of σ on F when $F > 10^7$

4. Conclusions

Conductivity Tensor is an important concept in materials science, this paper studies conductivity tensors in cylindrical quantum wires with infinite potential in the presence of two external fields, laser fields and linearly electromagnetic wave fields, the problem is also only considered for the case of electron-optical phonon scattering. We have calculated the expression of conductivity tensor, plotted the dependence of the conductivity tensor on Ω, F, and ω corresponding to the GaAs/GaAsAl quantum wire. The conductivity tensor changes as the frequency and amplitude of the external field change. The conductivity tensor increases as the laser field frequency increases and eventually reaches a constant value. At small electromagnetic wave frequencies, the conductivity tensor is independent of ω. Temperature also affects the conductivity tensor of quantum wire, at a higher temperature the greater the tensor of conductivity. This is perfectly suited because the higher the temperature, the larger the molecules move, which changes the conductivity tensor of the material.

Acknowledgement

This research is funded by Thu Dau Mot University, Binh Duong Province, Vietnam.

References

[1] Hoang Van Ngoc, Nguyen Vu Nhan, Dinh Quoc Vương 2017 The photon – drag effect in retangular quantum wire with an infinite potential *VNU Journal of Science, Mathematics – Physics,* No. 33 (1) pp 53-59.

[2] Hoang Van Ngoc, Nguyen Vu Nhan, Dinh Quoc Vuong 2017 The photon – drag effect in cylindrical quantum wire with an infinite potential for the case of electrons – acoustic phonon scattering *VNU Journal of Science, Mathematics – Physics*, No. 33 (4) pp 80-86.

[3] M. Blencowe and A. Skik 1997 Acoustoconductivity of quantum wires *Superlattices and Microstructures,* No. 22 (2) p 209-212.

[4] H.D. Trien and N.V. Nhan 2011 The nonlinear absorption of a strong electromagnetic waves caused by confined electrons in cylindrical quantum wire *Progress In Electromagnetics Research Letters,* No. 20 p 87-96.

[5] Nguyen Quang Bau and Hoang Dinh Trien 2009 The Nonlinear Absorption Coefficient of Strong Electromagnetic Waves Caused by Electrons Confined in Quantum Wires *Journal-Korean Physical Society,* No. 56 (1) p 120-127.

[6] Nguyen Quang Bau, Nguyen Van Hieu, Nguyen Vu Nhan 2012 Calculations of acoustoelectric current in a quantum well by using a quantum kinetic equation *Journal of the Korean Physical society*, No. 61 (12) pp 2026-2031.

[7] Dao Thu Hang, Dao Thu Ha, Duong Thi Thu Thanh, and Nguyen Quang Bau 2016 The Ettingshausen coefficient in quantum wells under the influence of laser radiation in the case of electron-optical phonon interaction *Photonics leters of Poland,* No. 8 (3) pp 79-81.

[8] B. D. Hung, N. V. Nhan, L. V. Tung, and N. Q. Bau 2012 Photostimulated quantum effects in quantum wells with a parabolic potential *Proceeding national conference theoretical physics*, No. 37 pp 168-173.

[9] N. Q. Bau, D. M. Hung and N. B. Ngoc 2009 The nonlinear absorption coefficient of a strong electromagnetic wave caused by confined electrons in quantum wells *Journal- Korean Physical Society,* No 54 (2) p 765-773.

[10] A. Grinberg and Luryi 1988 Theory of the photon - drag effect in a two-dimensional electron gas *Physical review B,* No. 38 (1) pp 87-96.

[11] N. Q. Bau and B. D. Hoi 2012 Influence of a strong electromagnetic wave (Laser radiation) on the Hall effect in quantum well with a parabolic potential *Journal- Korean Physical Society,* No. 60 (1) pp 59-64.

[12] N. Q. Bau, D. M. Hung and L. T. Hung 2010 The influences of confined phonons on the nonlinear absorption coefficient of a strong electromagnetic wave by confined electrons in doping superlattices *PIER Letters,* No. 15 p 175-185.

[13] N. Q. Bau and D. M. Hung 2010 Calculating of the nonlinear absorption coefficient of a strong electromagnetic wave by confined electrons in doping superlattices *PIER B,* No. 15 p 175-185.

[14] S. V. Kryuchkov, E. I. Kukhar' and E. S. Sivashova 2008 Radioelectric effect in a superlattice under the action of an elliptically polarized electromagnetic wave *Physics of the Solid State,* No. 50 (6) p 1150.

[15] G. M. Shmelev, L. A. Chaikovskii and N. Q. Bau 1978 HF conduction in semiconductors superlattices *Soviet Physics and Technics of Semiconductors,* No. 12 (10) p 1932.

[16] Nguyen Quoc Anh, Nguyen Thi Tu Uyen 1995 Electric conductivity tensor of semiconductors in a variable magnetic field *VNU journal of science,* No. 01 p19-23.
[17] G. M. Shmelev, G. I. Tsurkan and É. M. Épshtein 1982 Photostumilated radioelectrical transverse effect in semiconductors *Physica Status Solidi (b),* No. 109 p53.
[18] V. L. Malevich Izv. 1977 High-frequency conductivity of semiconductors in a laser radiation field *Radiophysics and quantum electronics*, No. 20 (1) p 98-101.

IOP Conf. Series: Materials Science and Engineering 1126 (2021) 012041 doi:10.1088/1757-899X/1126/1/012041

Comparison between Conventional Fast Multipliers and Improved Fast Multipliers using PTL Logic

Malti Bansal*, Vishal Bharti, Vibhor Chander

Department of Electronics and Communication Engineering, Delhi Technological University (DTU), Delhi-110042, India
*maltibansal @gmail.com

Abstract. Multiplier circuit is an important element in majority of the arithmetic operations. In the domain of VLSI industry, obtaining low power consumption by the components and high speed has become a crucial concern for the IC designers. It plays a crucial part in the speed of the ALUs and DSP and dissipates an enormous amount of power. Consequently the functioning operation of the desired circuit can be enhanced by its optimization. By using shift and add operation the multiplication process is also implemented in hardware. So we improved the multiplier with the use of efficient adder circuit. In this paper, the circuits are simulated in Cadence Virtuoso using 90nm technology. The proposed array multiplier circuit and Wallace tree multiplier offers a reduction of approximately 56% and 91% in power consumption and approximately 48% and 90% in product of power and delay respectively.

1. Introduction

With the increase in integration scale, a lot of advanced and compact signal processing systems area unit are required to be actualized on VLSI chip. Processing involve in these applications consume high amount of power and need sensible computation capability. With performance as well as area, power dissipation has also become an additional concern issue for the designing of integrated circuits. There are basically two major factors which led to this building of low power systems. Firstly, increase in the integration which led to increase in processing capacity due to which huge amount of flow of currents takes place that leads to heating up of the chip. And secondly, in portable electronic devices the battery life is restricted and hence prolonged operation of those portable devices are often obtained by achieving low power design. It is known that in most of the signal processing algorithms, multiplication shows an elementary role to play. The performance of system is usually dependent on the multiplier's performance, as a result multiplier has greatest delay within the system. However, it also consumes most of the area of that system. Hence, enhancing its speed and optimizing its area could also be a serious design issue. All the multipliers use full adders and hence are often optimized using the modified full adders. Here in this paper, we have simulated a proposed and a compact PTL Multiplier structure to fulfil daily needs of high speed applications which consumes low power. Here, improved adder designed and implemented using pass transistor logic is used for the proposed multiplier having 90nm technology. The proposed adder consumes lesser power and area than that of conventional techniques as discussed in literature. The paper consist of following sections which are as follows: Section 2 and 3 discuss the conventional approaches. Section 4 presents the proposed Wallace Tree structures and Array circuit. The Working is discussed in 5th section and results are compiled in 6th section. The last section is conclusion.

2. Wallace Tree Multiplier

The major design consideration for any chip designer are delay, area and power consumption. Speed of the system is totally dependent on the delay of the multiplier. Therefore, to improve the speed of multiplier many research has been done. Wallace Tree has serious significance in high speed applications because of great speed and area efficient multiplier [2]. It multiplies integers using the column compression technique that implements simple and efficient hardware methodology for Wallace multiplier. As compare to array multiplier having linear dependency, Wallace tree offers a quick speed which results in the entire delay is directly proportional to the logarithm of length of word of the operand of multiplier.

Steps involve for the operation of Multiplier are mentioned below:

First: Formation of partial products,
Second: Grouping of those formed partial products,
Third: Addition using adders.

To improve the performance of multipliers numerous analysis and research has been done [3-9]. In Wallace multiplier, to reduce the area and latency booth encoding with compressor approach is employed [4]. In conventional multiplier, a full adder circuit is built using 4:1 multiplexer [7]. In [8], full adder circuit is built using 2:1 multiplexer which also results in reducing power. These approaches of implementing full adder has led to power reduction but results in the critical path delay which is more than that of [9] Wallace tree multiplier. From all the previous literature review of multipliers, [9] offers the finest performance that are based on ground of area, power and its speed. The Figure below shows the full adder employed in [9].

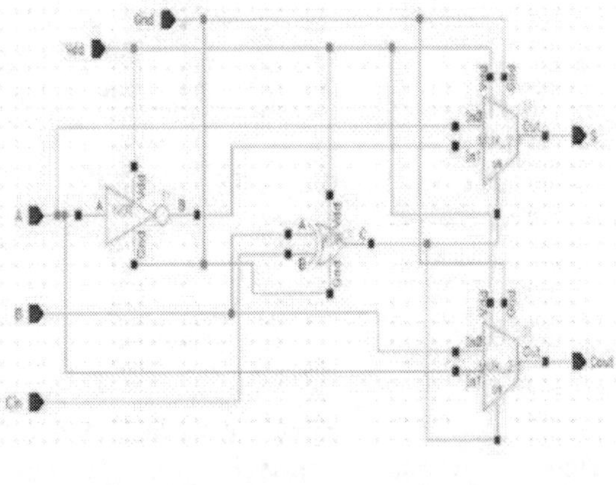

Fig. 1: Conventional Full Adder [9]

3. Array Multiplier

Array multiplier is easier to design because it is very regular in structure. "For multiplication of unsigned numbers Array multiplier is used. Half adders and full adders are connected in diagonally, horizontally and vertically to obtain the partial products sum. Cin will be taken as '0' if the implementation of partial product's first row is obtained using only full adders. By simple routing, partial products are properly aligned. Partial products are added to the sum by each row of adders, generating an order of carriers and a new partial sum" [1]. n half adders, n (n-2) full adders and n^2 AND gates are required in a nxn Array multiplier.

Total time consumed by the signals to propagate through full adders, half adders and by each AND gate is the effective delay of the Array multiplier. They are very large in size and this is the main disadvantage of an array multiplier. Due to increase in number of operands, the array arise linear in size at a rate equal to the square of the operand size. In design the third input is considered as 0 if the full adder have two inputs data only. The array multiplier with Carry save addition is shown below in Figure 2.

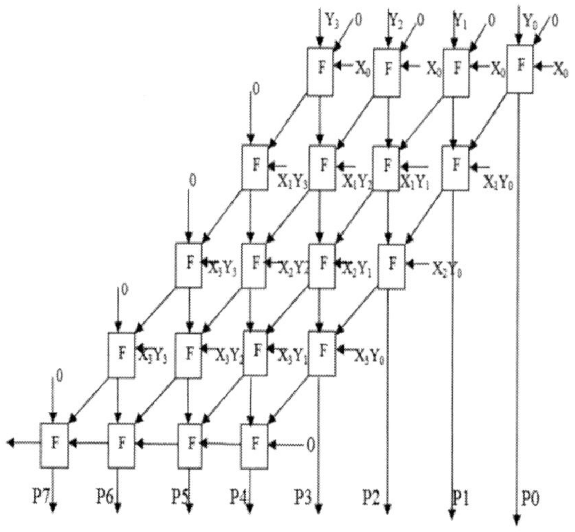

Fig. 2: Conventional Array Multiplier [1]

4. Methodology

As mentioned above Wallace Tree multiplier and Array multipliers offers best speed compared to other multipliers, therefore we tried to compare both Wallace and Array multiplier based on their existing and proposed structure and various other techniques so to enhance its structure that are proposed in several researches [3-9]. In this paper, a modified structure of Wallace tree and Array Multiplier has been proposed and simulated which results more better performance as compared to existing approaches. The proposed schematics are designed taking care of to reduce the complexity algorithm [10-11] and modified adder sub-circuit which process the intermediate addition of bits. The Figure 4 shows the schematic of improved adder that is designed using Pass Transistor Logic which is based 2:1 multiplexer. Fig.3 shows the Pass Transistor Logic based 2:1 multiplexer. Because of PTL, there is considerable decrease in number of transistors and area. . Apart from this it has the best advantage of least static leakage. The short circuit power is also least because there is very few V_{dd} to ground connections during the switching. The full adder circuit modified expressions are given as following:

$$\begin{aligned} SUM &= A\,\mathrm{XOR}\,B\,\mathrm{XOR}\,C \\ &= (A\,\mathrm{XOR}\,B)\,\mathrm{XOR}\,C \\ &=\left((AB)' + AB\right)C + \left(A'B + AB'\right)C \\ &=(AB)'C + ABC + A'BC' + A(BC)' \\ &=A\left(B'C + BC'\right) + A'\left(B'C + BC'\right) \\ &=\left((B\,\mathrm{XOR}\,C)'\right)A + (B\,\mathrm{XOR}\,C)A' \end{aligned} \quad (1)$$

$$\begin{aligned} CARRY &= AB + BC + CA \\ &= C(B+A) + AB \\ &= C(B+A)(B+B') + AB \\ &= B(C+A)(C+C') + ACB \\ &= BC + ABC' + ACB' \\ &= B\left((B\,XOR\,C)'\right) + A(B\,XOR\,C) \end{aligned} \quad (2)$$

(Expression for Sum and Carry)

The sum and carry expression of improved adder circuit are given by Eq. (1) and Eq. (2)

5. Working

Working of the circuit is explained as follows and is verified from the truth table as shown below.

If B = C = 0/1 then Sum = A and Carry = B.
If B! = C then Sum = A! and Carry = A.

In this proposed structure, modified adder is employed having low complexity algorithm for multipliers [10-11]. On comparable to typical conventional multipliers during which both full adders and half adders are used to process three and two bits respectively, it only uses full adders until and unless the amount of stages remain adequate to that of conventional Wallace algorithm and similar too with Array multiplier. The Functioning operation of multiplier is not suffering from eliminating half adders as they don't compress the amount of partial bits, two bits added gives two bits in output (Sum and Carry).

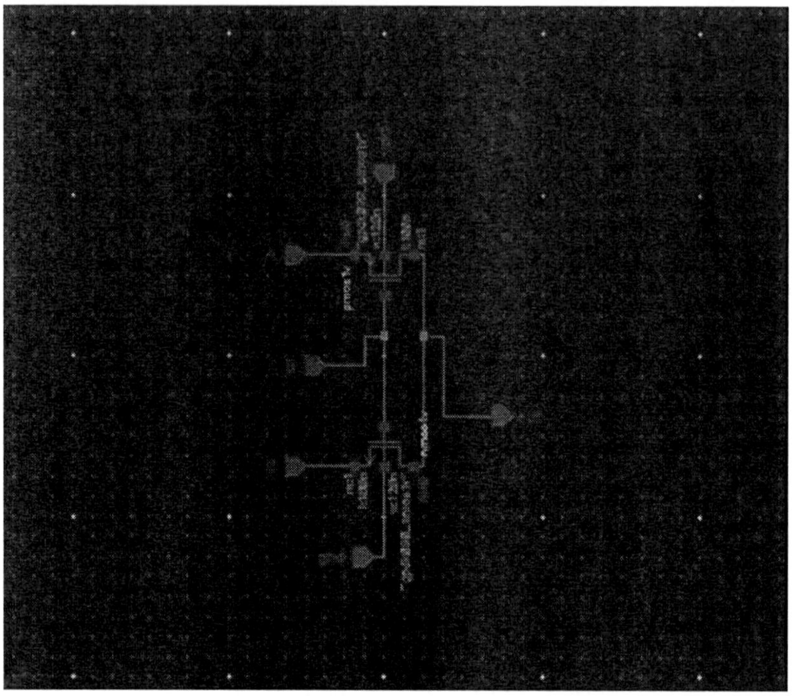

Fig. 3: Pass Transistor Logic Multiplexer

Table 1: Truth Table of Improved Adder

A	B	C	Sum	Carry
0	0	0	0(A)	0(B)
0	0	1	1(~A)	0(A)
0	1	0	1(~A)	0(A)
0	1	1	0(A)	1(B)
1	0	0	1(A)	0(B)
1	0	1	0(~A)	1(A)
1	1	0	0(~A)	1(A)
1	1	1	1(A)	1(B)

IOP Conf. Series: Materials Science and Engineering 1126 (2021) 012041 doi:10.1088/1757-899X/1126/1/012041

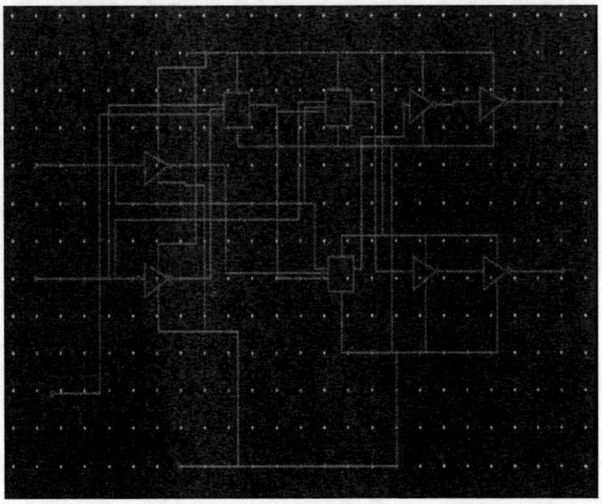

Fig. 4: Schematic of improved adder

Fig.5 shows 4x4 bit multiplication using low level complexity algorithm. It often seen that only S3 and C2 are processed as two its bits so that number of stages does not exceed the conventional approach of multiplication. The intermediate additions are performed by using the proposed adder.

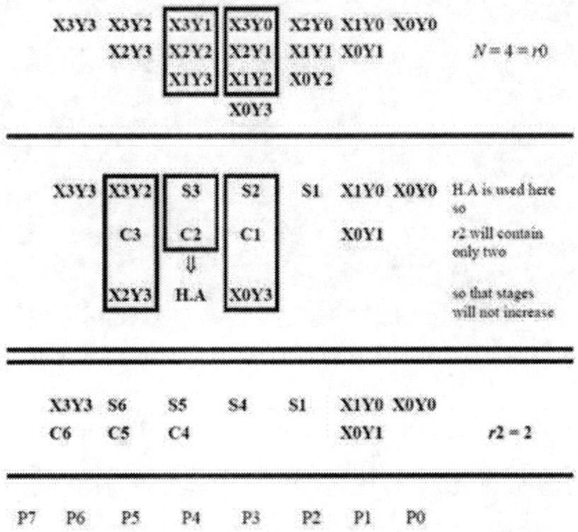

Fig. 5: 4x4 bit Multiplication using reduced complexity Wallace Algorithm

All partial product rows of proposed Array multiplier are implemented identical to the conventional array multiplier but instead of conventional adder the improved adder is used. The use of improved adder led to considerable reduce in power consumption in array multiplier. Figure (6) shows the suggested Array Multiplier.

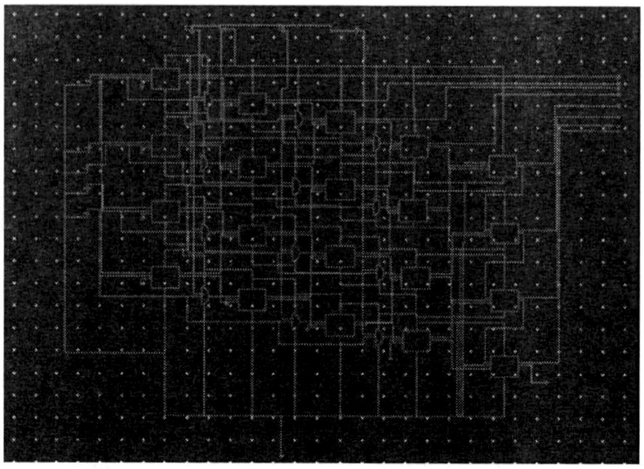

Fig. 6: 4x4 bit Proposed Array multiplier

6. Results

The simulated results of circuits are shown below. Cadence Virtuoso is used by us using 90nm technology. Here various aspects of proposed adder has been compared with the conventional existing adder in table (2). Also result of proposed adder shown in Fig 7.

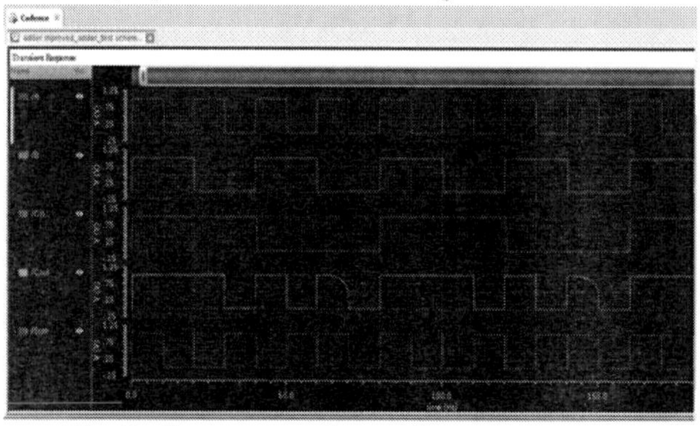

Fig. 7: Output waveform of proposed adder

Adder Circuit	Existing	Proposed
Logic used	CMOS XOR gate + Mux	Pass Transistor Logic
Number of transistor	56	18
Power	0.124 mW	0.043 mW
Delay	60.58 ps	48.48 ps
Power Delay Product	7.511	2.084

Table 2: Comparative Result of Adders

IOP Conf. Series: Materials Science and Engineering 1126 (2021) 012041 doi:10.1088/1757-899X/1126/1/012041

It is observed that the area i.e., number of transistors and therefore the delay of improved adder reduced considerably compared to the conventional adder. Simulated results of the proposed Array Multiplier and Wallace Tree Multiplier are shown below in Figure 8 and Figure 9 respectively. The Table.3 summarizes the obtained results of the Array Multiplier. It can be seen that number of transistors are considerably decreased in proposed Array Multiplier. Table.4 summarizes the obtained results of the Wallace Tree Multiplier. Due to less number of transistors the power dissipation in both proposed multiplier structures i.e. the power that is dissipated due to switching activity of the PT logic is less compared to power that is dissipated by the extra number of transistor in the conventional structures of the multipliers. Also the power delay product is a smaller in proposed multipliers as compared to the conventional multipliers.

Array Multiplier	**Existing**	**Proposed**
Power	0.389 mW	0.170 mW
Number of transistor	1192	432
Power delay product	0.326	0.169

Table 3: Analysis of Array Multiplier Using PTL.

Fig. 8: Waveform of Array Multiplier

Wallace Tree Multiplier	Existing	Proposed
Power	2.283 mW	0.192 mW
Number of Transistor	768	312
Power delay product	473.72	46.28

Table 4: Analysis of Wallace Tree Multiplier Using PTL

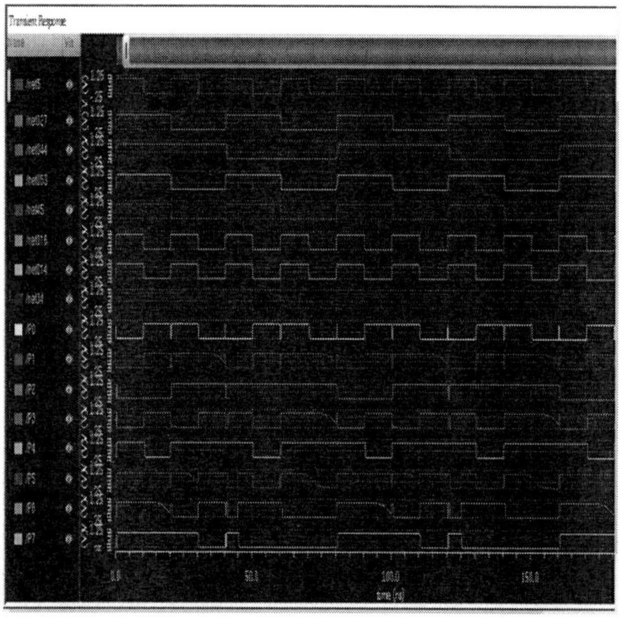

Fig. 9: Waveform of Wallace Tree Multiplier

According to our study, using the pass transistor logic in the 2:1 mux, improved adder of both multipliers, the power dissipation, area and product of power and delay has been minimized by a considerable magnitude. This will results in the increase in demand for greater speed with minimal battery usage and cover less area. This improved adder proves beneficial in all the applications that performs mathematical operations using multiplier which are used in ALU's and Digital Signal Processors.

7. Conclusion

In this paper, the improved output results of the conventional full adder were proposed along with conventional multipliers on the basis of number of transistor, power dissipation, product of the power and delay. Both multiplier has been studied and then modified multipliers circuit were proposed and simulated using 90nm technology in Cadence Virtuoso. Initially we simulated the inverters and the existing adder [9]. Then we simulated a new mentioned improved adder that uses the pass transistor logic based 2:1 multiplexer. Here, the number of transistors employed in adder is comparatively less than existing adder. This results in the area which gets minimized and offers a less delay. The proposed Array Multiplier offers an improvement of 56.29% in reduction of power, 48.15% of reduction in power delay product and 63.75% diminishment in the number of transistors used i.e. area. The proposed Wallace tree Multiplier offers an improvement of 91.5% in reduction of power, 90.23% of diminishment in power delay product and 59.37% diminishment in the number of transistors. The consumption of power, area, and product of power and delay of the proposed Multipliers has been minimized by a considerable magnitude. Our proposed structures will prove beneficial in all the applications that perform mathematical operations using multipliers. Some of the applications in which it can be widely used are ALU's and DSP structures [13-28].

References

[1] Corchado Rodríguez, Juan & Mitra, Sushmita & Thampi, Sabu & El-Alfy, El-Sayed. (2016). "Intelligent Systems Technologies and Applications".
[2] C.S. Wallace, "A Suggestion for A Fast Multiplier", IEEE Transaction on Electronic Computers, Vol. 13, No.1,pp.1417,1964.
[3] S.Rajaram and K.Vanithamani, "Improvement of Wallace Multipliers using Parallel Prefix Adders", Proceedings of IEEE International Conference on Signal Processing, Communication, Computing and Networking Technologies, pp. 781-784, 2011.
[4] M.J. Rao and S. Dubey, "A High Speed and Area Efficient Booth Recoded Wallace Tree Multiplier for Fast Arithmetic Circuits", Proceedings of Asia Pacific Conference on Postgraduate Research in Microelectronics and Electronics, pp. 220-223, 2012.
[5] S. Karthick, S. Karthika and S. Valannathy, "Design and Analysis of Low Power Compressors", International Journal of Advanced Research in Electrical, Electronics and Instrumentation Engineering, Vol. 1, No. 6, pp. 487-493, 2012.
[6] Saeeid Tahmasbi Oskuii, Per Gunnar Kjeldsberg and Oscar Gustafsson, "Power Optimized Partial Product Reduction Interconnect Ordering in Parallel Multipliers", Proceedings of Nordic Circuits and Systems Conference, pp. 1-6, 2007.
[7] S. Murugeswari and S.K. Mohideen, "Design of Area Efficient and Low Power Multipliers using Multiplexer based Full Adder", Proceedings of 2ndInternational Conference on Current Trends in Engineering and Technology, pp. 388-392, 2014.
[8] Yingtao Jiang, Abdulkarim Al-Sheraidah, Yuke Wang, Edwin Sha and Jin-Gyun Chung, "A Novel Multiplexer-based Low-Power Full Adder", IEEE Transactions on Circuits and Systems, Vol. 51, No.7, pp.345-348, 2004.
[9] Kokila Bharti Jaiswal,Nitish Kumar,Pavithra Seshadri and G. Laxminarayan, "Low Power Wallace Tree Multiplier using Modified Full Adder", Proceedings of 3rdInternational Conference on Signal Processing, Communication And Networking, pp. 1-4, 2015.
[10] R.S. Waters and E.E. Swartzlander, "A Reduced Complexity Wallace Multiplier Reduction", IEEE Transactions on Computers, Vol. 59, No. 8, pp. 1134-1137, 2010.
[11] Sandeep Kakde, Shahebaj Khan, Pravin Dakhole and Shailendra Badwaik, "Design of Area and Power Aware Reduced Complexity Wallace Tree Multiplier", Proceedings of International Conference on Pervasive Computing, pp. 1-6, 2015.
[12] K.K. Senthil Kumar, S. Yuvaraj, R. Seshasayanan, "High-Performance Wallace Tree Multiplier" International Journal of Computer Techniques — Volume 7, Issue 01, February, 2020.
[13] M. Bansal and J. Singh, "Qualitative Analysis of CMOS Logic Full Adder and GDI Logic Full Adder using 18 nm FinFET Technology," 2019 3rd International Conference on Recent Developments in Control, Automation & Power Engineering (RDCAPE), Noida, India, 2019, pp. 404-407, doi:

10.1109/RDCAPE47089.2019.8979087.

[14] N. Raghav and M. Bansal, "Analysis of Power Efficient 6-T SRAM Cell with Performance Measurements," 2017 International Conference on Innovations in Control, Communication and Information Systems (ICICCI), Greater Noida, India, 2017, pp. 1-4, doi: 10.1109/ICICCIS.2017.8660819.

[15] M. Rawat, M. Bansal, " Voltage Mode Two-Phase and Four-Phase Sinusoidal Oscillator using VDCC", International Journal of Advanced Research in Electronics and Communication Engineering, Vol. 4, Issue 6, pp. 1687-1691.

[16] D. Ruhela, M. Bansal, "Adiabatic Vedic Multipler Design Using Chinese Abacus Approach", International Journal of Advanced Research in Computer and Communication Engineering, Vol. 4, Issue 4, pp. 81-85, 2015.

[17] M. Bansal, D. Ruhela, "High Speed and Area Efficient Vedic Multiplier using Adiabatic Logic", Journal of Basic and Applied Engineering Research, Vol. 1, Issue 11, pp. 14-17, 2014.

[18] N. Raghav, M. Bansal, "Analytical Study of Full Adder Circuit using Modified Glitch Free Cascadable Adiabatic Logic", International Conference on Advanced Production and Industrial Engineering (ICAPIE-2017), 2017.

[19] M. Bansal and J. Singh, "Qualitative Analysis of 2-bit CMOS Magnitude Comparator and GDI Magnitude Comparator using FinFET Technology (18nm)," 2020 International Conference on Smart Electronics and Communication (ICOSEC), Trichy, India, 2020, pp. 1323-1327, doi: 10.1109/ICOSEC49089.2020.9215251.

[20] M. Bansal and J. Singh, "Comparative Analysis of 4-bit CMOS Vedic Multiplier and GDI Vedic Multiplier using 18nm FinFET Technology," 2020 International Conference on Smart Electronics and Communication (ICOSEC), Trichy, India, 2020, pp. 1328-1332, doi: 10.1109/ICOSEC49089.2020.9215317.

[21] M. Bansal, H. Saxena, "Power-Delay Behaviour of Digital Circuits Designed using MOSFET, CNTFET and FinFET, with the Scope of Miniaturization in these Transistors", International Journal of Advanced Research in Computer and Communication Engineering (IJARCCE), Vol. 9, Issue 2, pp. 64-71 , 2020.

[22] M. Bansal, H. Saxena, "Comparative Analysis and Behavior of Digital Applications simulated using MOSFET, CNTFET and FinFET Transistors in HSPICE", International Journal of Advanced Research in Computer and Communication Engineering (IJARCCE), Vol. 9, Issue 2, pp. 54-58, 2020.

[23] M. Bansal, H. Saxena, "High Speed and Low Power Consumption Multipliers using FinFET Technology", International Journal of Advanced Research in Computer and Communication Engineering (IJARCCE), Vol. 9, Issue 2, pp. 59-63, 2020.

[24] M. Bansal, J. Singh, "CMOS MOSFET, FinFET and CNTFET: Qualitative Analysis", National Conference on Emerging Trends in Electronics & Communication-2019, pp. 79-84, 2019.

[25] N. Raghav and M. Bansal, "Analytical study of high performance flip-flop circuits based on performance measurements," 2017 International Conference on Computing, Communication and Automation (ICCCA), Greater Noida, India, 2017, pp. 1543-1548, doi: 10.1109/CCAA.2017.8230047.

[26] D. Ruhela, M. Bansal, "Vedic Multiplier with Chinese Abacus Adder Design using Reversible

Logic Gates", 2nd International Conference on VLSI, Communication and Networks (VCAN-2015), pp. 9-12, 2015.

[27] M. Rawat, M. Bansal, "A Voltage mode biquad with lowpass, bandpass and notch outputs using Voltage Differencing Current Conveyor", International Journal of Advanced Research in Computer and Communication Engineering, Vol. 4, Issue 5, pp. 506-508, 2015.

[28] M. Bansal, D. Ruhela, "High Speed, Energy Efficient Vedic Multiplier using Adiabatic Logic Gates", International Conference on Advances in Electrical, Power Control, Electronics & Communication Engineering (AEPCECE-2014), 2014.

Investigation on Impact and Thermogravimetric Analysis of Aramid and Carbon based Nanocomposites

S.Charvani[1], Ch.Gopal Reddy[2],G.Narendar[3]

[1,2]Department of Physics, University College of Science, Osmania University, Hyderabad, Telangana, India

[3]Department of Mechanical Engineering, University College of Engineering, Osmania University, Hyderabad, Telangana,India

Email: ouprofessor2000@gmail.com

Abstract. The demand from industrial sector for light weight and high strength materials is ever increasing. Researchers are trying to fulfil the industrial requirements by fabricating tailor-made materials with required properties. The recent developments in the nanotechnology and nanomaterials have opened new doors in the field of nanocomposites. Nanoparticles are the fascinating materials due to their unique mechanical, thermal and structural properties paved the way for exploring its use in various industrial applications. Nanocomposite materials can be fabricated to modify and control the properties via innovative synthetic approaches. Polymeric nanocomposites combine the lightness and low cost manufacturing of polymers with alumina nanoparticles have high performance properties due to interfacial characteristics of individual constituents, being an attractive alternative to heavier metals or expensive ceramics and other materials.

Researchers are exploring for the enhancement of the properties such as mechanical properties viz., impact strength, tensile strength, flexural strength and thermal stability etc. Nanocomposites are fabricated imbedded with alumina (Al_2O_3) nanoparticles in to the binding resin matrix and reinforce materials as Aramid fibre, Carbon fibre, and hybrid of Carbon and Aramid fibres. Nanoparticles of Alumina of less than 30 nm size are dispersed in the matrix during preparation of the composites.

Mechanical and structural characterization of polymer nanocomposites with different proportions of nanoparticles, different fibres and also with plain epoxy polymer composite was evaluated. Structural characterisation confirmed the formation of nanocomposites and thermal stability. Mechanical impact test revealed the improved impact resistance of the nanocomposites. The tests established that these nanocomposites exhibited improved properties.

1. Introduction

Composite materials are prepared with minimum of two materials, one is binding material called matrix and the other one is called reinforcing material that bears the load, and gives strength to the composite material. The matrix holds together both the materials. This new material will now exhibit properties of both the materials i.e. reinforcement and matrix which holds each other and work together. The two materials are entirely different in their physical and chemical properties also they do not blend or dissolve in to each other.

The Composite materials are fabricated by combining two or more different materials to make use of their better properties and thereby minimize their deficiencies. The properties of composites can be tailor-made as per the requirement by using lighter reinforcement materials and fillers [1, 2]. Each of the constituent materials retains its physical, structural or chemical properties separate and distinct within the final product. The reinforcement or reinforcing fibres in the composites are strong and carries the load and the fibre is kept in place by the matrix material [3,4].

The main function of reinforcement in the composite is providing strength and stiffness. It also helps in supporting of structural load and stress [5, 6]. The matrix material works as a binder and surrounds the reinforcement. The matrix protects the reinforcement from the surrounding environment. It also keeps the position and orientation of the reinforcement to balance loads between the reinforcements. The matrix provides shape and form to the composite structure [7].

The Composite materials are attracted and being used in industries due to their light weight and accompanied with high strength and other useful properties. The aim of preparing composites is to produce materials that have high strength, light weight, creep resistance, toughness, and high flexural and high tensile strength. Composites are light in weight, very strong and stiff, so ratios of stiffness/weight and strength/weight are many times high compared with other materials. Their wear resistance is generally better and Toughness of the composites is also high [8, 9]. Composites have the advantage that they do not corrode like metals and possible to get required properties that cannot be attainable with metals, polymers and ceramics [10, 11].

To meet industrial requirements, several varieties of composite materials are being developed with tailor-made properties. New types of composites are developed to attain superior properties. Composites generally prepared by using artificial fibers like glass fiber, boron fibre, carbon fibers, Aramid (Kevlar) fiber, and natural fibers. The matrix materials generally used are vinyl ester, epoxy resin, polyamide and phenolic etc. The matrix material acts as a binding material. It also transfers the stresses and loads to the reinforcing fiber material. The mechanical, physical, structural and thermal properties of composites are improved with the presence fiber material in the composites. The matrix and reinforcement volume fractions are adjusted to get the desired properties [12, 13, 14].

When nanoparticles are imbedded in a composite, then they are called nanocomposites. In nanocomposites, the particles have at least one of the dimension less than 100 nm. Among the phases of composites, the matrix exhibits high toughness and the reinforcement exhibits low density, high stiffness and strength. The designer can select different fibers and matrices as per the particular requirement. Because of the ease in fabricating the composites and low cost, they have become so attractive in the present upcoming applications [15, 16].

1.1 Types of Fibers

The fiber which is the reinforcement that can be a continuous fiber, short fiber and particulate fiber or particulate matter used in composites. The continuous fiber can exist as a single fiber or as a bundle of fibers or a bundle of fibers weaved into a matter. The fibers can be further classified into unidirectional and bi-directional fibers. Unidirectional mat is fabricated in one direction. In bi-directional fibers, the mat is woven in both directions. Due to this, there is a uniform amount of distribution of stresses, so they have a very high strength. In by- directional fabric can be woven cross ply or angle ply. Fiber used in the composites carry the applied load that provides the required strength, stiffness, other structural properties and thermal stability to the composite material. It can also provide electrical conductivity or insulation depending upon the choice of fiber.

The fibres govern and control the basic mechanical properties. The fiber increases the surface interaction. The amount of fiber can vary from two percent up to even ninety percent. Ninety percent reinforced composites for special types of application are also being used in industrial sector. The orientation of the fiber also plays an important role. The fiber can be oriented in the required direction to get the best property out of orientation.

The properties governing or controlling factor in the composites are the surface interaction between the fiber and the resin and the fiber volume fraction i.e. amount of fiber present in the composite material and the orientation fibre in the composite. The characteristic features of a composite can be varied with respect to fiber content and matrix content. The strength properties can be keep changing, moving towards lower side or higher side by keep doing it depending upon the volume fraction. So, volume fraction in the composite material plays an important role on the characteristic features of the composite materials.

The influence of matrix with respect to tensile stress and strain: The matrix has lot amount of strain, but the strength or the stress is very low. For glass fiber, there is not much of strain, but the strength goes very high. Therefore, a composite strength is somewhere in between these two, a composite which has a property less than the tensile stress of the fiber glass and above the matrix.

1.1.1 Aramid Fibers
Aramid fibers known under their trademark name as Kevlar is the most important fiber used nowadays because it has the high strength and toughness properties. Aramid is a material formed by combining the paraphernalia in diamine and territory chlorate. It is a high strength material and it is a very light weight when compared to steel, and in strength it has a very high tensile strength. High modulus, high strength and highly crystalline are some of the properties. Aramid fibre is shown in fig. 1. Aramid fibres can be obtained as a thread, so that it can be woven into fabric etc.

Fig. 1 Aramid Fibre Woven Mat

Fig.2 Carbon Fibre Woven Mat

Fig.3 Hybrid of Aramid and Carbon Woven Mat

1.1.2 Carbon Fibers
Carbon fibers have very high modulus, very high strength, low coefficient of thermal expansion and offers a good wear resistance at very high temperatures. It gives excellent modulus and strength on top of it is light weight it finds lot of applications. Carbon fiber a good alternative material for bulletproof jackets, rocket applications, helmets which are used by defence personalities all these places carbon fiber plays a very important role.

1.1.3 Hybrid Fibers
Fibre reinforced composites fabricated with only single type of reinforcement fibre such as glass fibre, carbon fibre, aramid fibre etc., have some limitations related to the single fibre property. These limitations can be overcome by hybrid fibres in the same binding matrix. Due to hybridization required properties can be obtained, enhanced and can be tailor-made suitable for applications.

Hybrid of Carbon fibre and Aramid fibre woven mat is shown in Fig.3. The hybrid fibre is made from a hybrid of carbon and aramid. The weave is a very attractive appearance and better strength properties but a little heavier than carbon fibre. It can be used in products with high strength, toughness and durability. The hybrid of carbon and aramid fiber are used to compensate the weak points of aramid in compression resistance and carbon in impact resistance.

1.2 Matrix Materials
The most common type of matrix is thermosetting resin. Thermosetting resins chemically react under certain temperature and time conditions. The resin becomes fused into a solid after the reaction takes and will not melt on heating. Epoxy resins are widely used resins in fabrication of advanced composites. Other resins used as matrix are Polyester, phenolic, Vinyl ester, epoxy, polyamide, novolac, bismaleimide, etc. Each resin has its own practical application. The resins are selected as per the required performance in application and manufacturing method.

2. Materials and Methodology

In preparation of nanocomposite materials, binding resin matrix, reinforcing fibres, nanoparticles are required. In the present research, Epoxy resin (Araldite LY-556) and curing agent (hardener HY-951) are used for the matrix material. Alumina (Al_2O_3) nanoparticles are imbedded in to the binding resin matrix. The reinforcement fiber used in preparation of the nanocomposite material are (i) Carbon fibre woven mat (ii) Aramid fibre woven mat and (iii) hybrid of aramid fibre and carbon fibre woven mat.

The composite laminates are fabricated with alumina nanoparticle loading of 1, 3 and 5 percent by weight and composite laminate without nanoparticle loading are also fabricated for evaluating the properties. The description of nanocomposite samples is presented in Table 1. The sample is identified with the following designation: first digit is percentage of nanoparticles used in the sample, second digit is the fiber type used in the composite and the last digit represents the sample number. The percentages of nanoparticles used in the fabrication of the composites are: 0%, 1%, 3% and 5% and P, A, C and H in the sample represents the type of fiber used which are plain (without nanoparticles), Aramid fiber, Carbon fiber and Hybrid fiber respectively.

2.1 Preparation of Nanocomposites
In preparation of nanocomposites, the alumina nanoparticles added into the composites with 1, 3 and 5% by weight. The alumina nanoparticles were first dispersed into methanol by mixing with high speed stirrer for one hour. Then the mixture was sonicated using a probe-sonicator for another hour to break agglomerations of nanoparticles and was added to epoxy resin. The mixture is preheated. After mixing process, methanol was removed from the epoxy and alumina mixture by heating at above boiling point of methanol and drying process continued in the oven. The epoxy and alumina mixture is dried-up by placing it in an oven at 70°C for an hour. Lastly, the hardener was added and thoroughly mixed into the epoxy and alumina mixture in ten to one proportion by weight. The epoxy - alumina and hardener mixture were thoroughly stirred for uniform mixing and used in the production of nanocomposite laminates. The mould was allowed to cure at room temperature for one day. After then, the samples were taken out of the mould, and kept in hot air oven for post-curing at around 80°C. After post curing in the oven, the cured hybrid alumina nanocomposite laminates were cut into required sizes for testing and characterization.

Table 1. Description of nanocomposites with various proportions of nanoparticle loading

S.No.	Test Sample	Description of test sample	S.No.	Test Sample	Description of test sample
1	P0	Plain composite with 0% nanoparticle addition	9	C0	Carbon composite with 0% nanoparticle addition
2	P1	Plain composite with 1% nanoparticle addition	10	C1	Carbon composite with 1% nanoparticle addition
3	P3	Plain composite with 3% nanoparticle addition	11	C3	Carbon composite with 3% nanoparticle addition

4	P5	Plain composite with 5% nanoparticle addition	12	C5	Carbon composite with 5% nanoparticle addition
5	A0	Aramid composite with 0% nanoparticle addition	13	H0	Hybrid composite with 0% nanoparticle addition
6	A1	Aramid composite with 1% nanoparticle addition	14	H1	Hybrid composite with 1% nanoparticle addition
7	A3	Aramid composite with 3% nanoparticle addition	15	H3	Hybrid composite with 3% nanoparticle addition
8	A5	Aramid composite with 5% nanoparticle addition	16	H5	Hybrid composite with 5% nanoparticle addition

3. Characterisation of Composites

3.1 Thermo Gravimetric Analysis (TGA)

The Thermogravimetric Analysis is helpful in the study of thermal stability and decomposition of the composite materials at elevated temperatures. The investigation on the thermal behaviour of reinforcement is very important parameter for the usage of these composites. The thermogram of composites were recorded using Thermo gravimetric analyzer (Perk-Elmer TGA-7). The test was performed in the temperature range of surrounding temperature to 800°C at 10°C/min of heating rate. The thermograms of A0, A1, A3, A5, C0, C1, C3 and C5 samples were recorded.

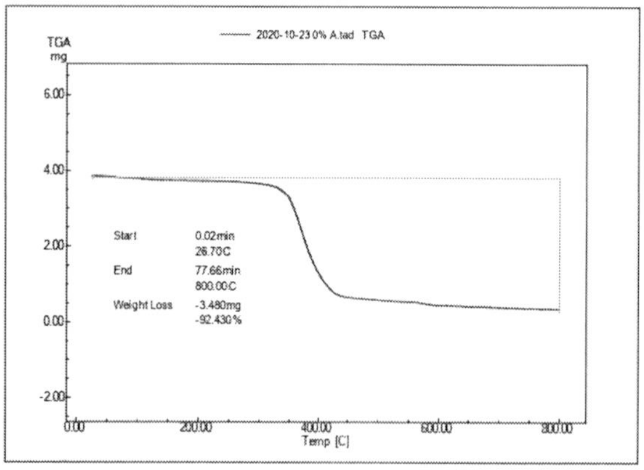

Fig.4a TGA curves of Aramid nanocomposite sample with 0% nanoparticles

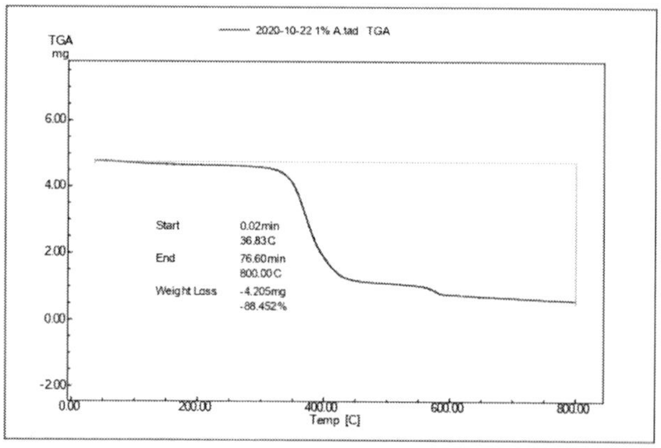

Fig.4b. TGA curves of Aramid nanocomposite sample with 1% nanoparticles

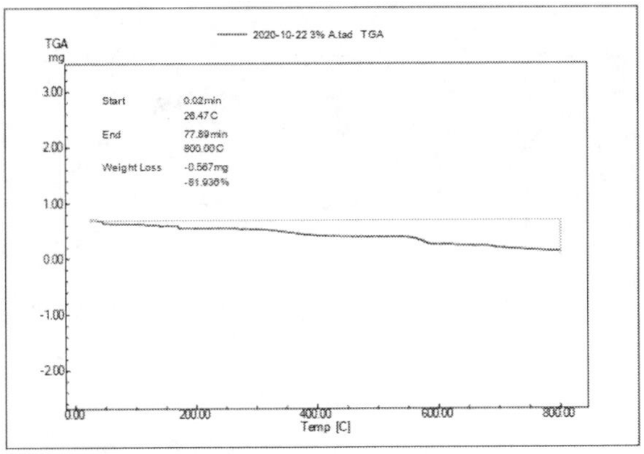

Fig.4c TGA curves of Aramid nanocomposite sample with 3% nanoparticles

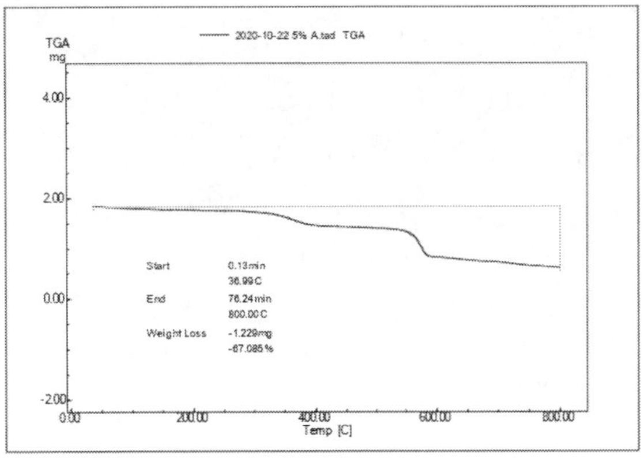

Fig.4d TGA curves of Aramid nanocomposite sample with 5% nanoparticles

Fig.4 TGA graphs of Aramid nanocomposites with different proportions of nanoparticles

The material sample exhibited a trend of three stage decomposition of weight loss. The thermogram of fig. 4 represents weight loss is initiated at around 30°C (surrounding temperature) this is due to loss of moisture present on the surface. This weight loss proportion is very much less and is around 5%. Later, there is a rapid degradation stage is found in the range 300°C to 400°C. The weight loss is more of around 85% is in this small range of temperature. Final degradation stage occurred in the temperature range of 400°C to 800°C that is the residual part of around 10% in the test sample. During the TGA investigation on the A0, A1, A3 and A5 samples, the total material losses are 92.43%, 88.45%, 81.93% and 67.08%. This reveals that the thermal stability increased with increase in nanoparticle loading.

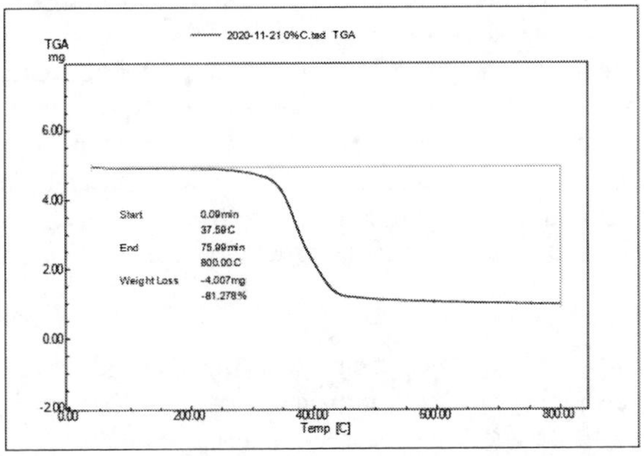

Fig.5a TGA curves of Carbon nanocomposite sample with 0% nanoparticles

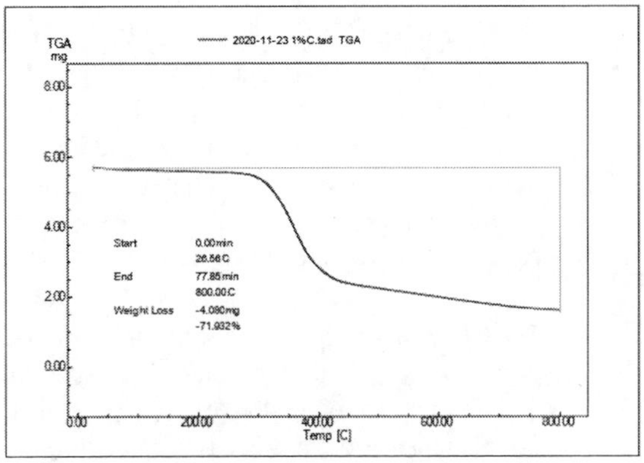

Fig.5b TGA curves of Carbon nanocomposite sample with 1% nanoparticles

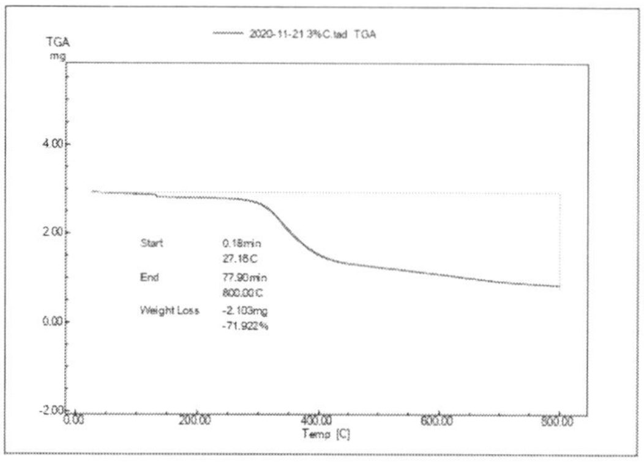

Fig.5c TGA curves of Carbon nanocomposite sample with 3% nanoparticles

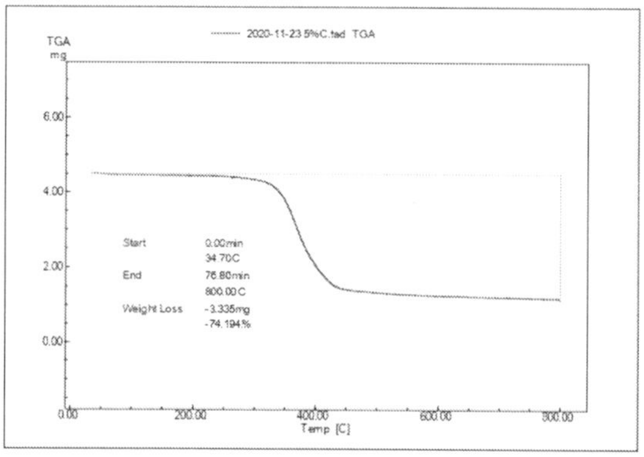

Fig.5d TGA curves of Carbon anocomposite sample with 5% nanoparticles

Fig.5 TGA graphs of Carbon nanocomposites with different proportions of nanoparticles

During the TGA test, the carbon composite material sample also exhibited a trend of three stage decomposition of weight loss. The thermogram in Fig.5 shows a weight loss initiated at around 30°C (surrounding temperature) this weight loss is due to absorbed moisture on the composite. This weight loss proportion is very less till 300°C and is around 5 percent. Later, degradation is increased from 300°C to 420°C temperature. The loss of weight proportion is more which is around 90% in this range of temperature. Final degradation stage occurred after 420°C to the final temperature of 800°C. This is the residual part of the composite. During the TGA investigation on the C0, C1, C3 and C5 samples, the total material losses are 81.27%, 71.93%, 71.92% and 74.19%. This reveals that nanoparticle loading and carbon fibre in the composite promoted thermal stability.

The weight loss percent during Thermo gravimetric Analysis test on the composite test samples reveals that, among all the samples, the loss of weight proportion is more for A0 in which nano particles were not added. The weight loss is least for A5 and next is C3 due to its more nanoparticle content and fibre nature. The range of weight loss is more in case of Aramid fibre nanocomposites of 25.35% where as for Carbon fibre nanocomposites the range of weight loss is 7.08%.

From the TGA investigation, it is found that the weight loss for the fiber material, matrix and composite started at around 300°C where the matrix, fiber and the composite degradation has started. It is found that the final degradation temperatures are increased with increase of nanoparticle percentage in the nanocomposite material. Increase in thermal stability is attributed to high thermal stability characteristics of Alumina nanoparticles and the integration of nanoparticles with the binding resin matrix and thermal stability was found to be increasing with nanoparticle loading.

3.2 Characterisation of Mechanical Properties

Application of engineering materials is evaluated by analyzing their mechanical properties. The mechanical tests are very useful in understanding the behaviour of new materials and developing new materials of required properties. The engineering materials should be strong enough to withstand the loads that it will undergo during its life time. Therefore, many experimental techniques have been developed for testing the materials subjected to various types of loads. Mechanical properties are analysed to check the strength of the composites under different forces, stresses, impact loading etc. The mechanical properties and performance of the nanocomposites depend on all the three constituents viz., fiber, matrix and filler material interface.

3.2.1 Impact Test

Impact test is used to find out the ability of a material to withstand suddenly applied impact load. Impact test is used to estimate the toughness of the composite material. The machine on which this impact tests will be performed is having a measuring rating scale and a hammer or pendulum for applying impact load. The height of the pendulum before hitting the specimen and after hitting the specimen can be taken to evaluate the energy absorbed.

The test specimen is placed in the middle of the machine vice. The pendulum is dropped from a predefined height. The striking position of the hammer after striking the specimen and the reading obtained is an indication of energy absorbed. When pendulum strikes on the specimen this will lose energy and this specimen will gain energy. So height of the pendulum will be changed according to absorbing energy by the specimen and this reading gives the toughness of the material.

The energy required to fracture/break the specimen is estimate using the expression.

Impact energy, $E = W*(h_1-h_2)$

Where, W = weight of pendulum

h_1, h_2 = height of the pendulum before and after striking the specimen

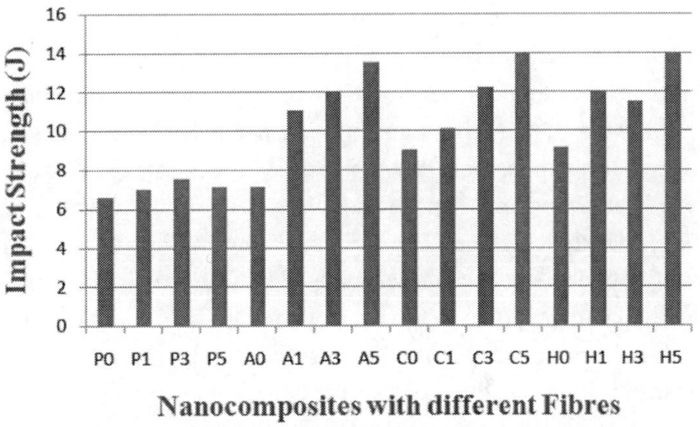

Fig.6 Impact Strength of Nanocomposites with different fibers

The impact strength increased with increase in nanoparticle loading. The gradual increase in strength of nanocomposites reveals that mechanical stresses are efficiently transferred via the interface in the composites. This improvement in impact strength is attributed to the increased inter-laminar fracture toughness of the fibres.

Because of bidirectional woven fibres there is a high internal damping, due to this impact strength is better in the composites. Hybrid composite with stiffened binding matrix laterally supports the slender fibres, strengthens the composites and increases the impact absorption. Due to better resistance to impact damage of aramid, carbon fiber, the hybrid composite exhibits better impact strength in high nanoparticle loading of composites.

Mechanical tests were carried out to find the impact strength properties of nannocomposites. It is observed that dispersion of nanoparticles into epoxy resin has great impact on the mechanical properties of nanocomposites. It is found that the impact strength is increased when Carbon fibre and aramid fibre is added in the nanocomposite. The plain epoxy shows the low impact strength value compared with hybrid nanocomposites which indicate that the fiber and nanoparticles plays a major role in impact strength. The highest strength is obtained for three percent and five percent nanoparticle addition. Good impact strength indicates a good interfacial interaction. This is due to the best interlayer and inter- laminar properties.

4. Conclusions

The tests are performed on the composites material samples of plain epoxy and with Alumina nanoparticle loading of 0, 1, 3, and 5% with fiber woven mats of Carbon, Aramid and Hybrid of Carbon and Aramid fibres. The results depicted that there is a strong bonding is observed between reinforcements and the binding matrix.

The Thermal stability of the nanocomposite is higher than the plain matrix material and fiber composite material and observed that thermal stability increased with nanoparticle loading. The results showed an improvement in properties. From the tests conducted on the composites, the following conclusions are drawn:

- The thermal stability of composites with plain epoxy compoiste and with nanoparticle proportion of 1, 3, and 5% is studied with woven fibre mats.
- The TGA graphs showed that, there is a right ward shift in the temperature values. This depicts the increase in thermal stability of the composites with increase in the nanofiller percentage.
- It is found that the impact strength is increased when Carbon fibre and Aramid fiber are added in the nanocomposite. The plain epoxy composite showed the low impact value compared to fiber reinforced hybrid composites which indicate that the fiber and nanoparticles played a major role in improving the impact strength.
- The highest strength is obtained for 3% and 5% addition of alumina nanoparticles. Thus it is found that the best interlayer and intralaminar adhesion is obtained for this composition.
- The impact strength increased with increase in nanoparticle loading. The gradual increase in strength of nanocomposites reveals that mechanical loads and stresses are efficiently transferred via the interface. This increased impact strength is attributed to the better inter-laminar fracture resistance of the fibers. Because of bidirectional woven fibers there is a high internal damping, due to this impact strength is better in the composites.
- The Aramid, carbon and hybrid composites with 3% and 5% nanofiller showed the maximum impact strength among all the composite samples.

Acknowledgements

The authors would like to express sincere thanks to TEQIP-III funding agency and Principal, College of Engineering, Osmania University, Hyderabad for encouraging and financially supporting through TEQIP-III to carry out the Research and Publish the Technical Paper.

References

[1] Arpitha G R, Sanjay M R, L Laxmana Naik, B Yogesha, Mechanical Properties of Epoxy Based Hybrid Composites Reinforced with Sisal/SIC/Glass Fibers, International Journal of Engineering Research and General Science Volume 2, Issue 5, 2014

[2] Manickavasagam. V M, Vijayaramnath. B, Elanchezhian. C, Mechanical Characteristics and Wear behavior of Hybrid Composites with Areca and Neem Fibers, International Journal of Mechanical and Production Engineering Research and Development, Vol. 10, Issue 1, Feb 2020, 375–388

[3] G. Musalaiah, Y Shireesha, M.V Babu Tanneru, U Pranavi Flexural Strength Enhancement of Glass Fibre Reinforeced Polymer Composite, International Journal of Mechanical Engineering and Technology, Volume 8, Issue 7, 240–246

[4] TP Sathishkumar, J Naveen and S Satheeshkumar, Hybrid fiber reinforced polymer composites – a review, Journal of Reinforced Plastics and Composites 2014, Vol. 33(5) 454–471
[5] R Yahaya, SM Sapuan, M Jawaid, Z Leman and ES Zainudin, Mechanical performance of woven kenaf-Kevlar hybrid composites, Journal of Reinforced Plastics and Composites, (2014) 1-13
[6] Pedro Henrique Cury Camargo, Kestur Gundappa Satyanarayana, Fernando Wypych , Nanocomposites: Synthesis, Structure, Properties and New Application Opportunities, Materials Research, Vol. 12, No. 1, 1-39, 2009
[7] R.V. Kurahatti, A.O. Surendranathan, S. A. Kori, Nirbhay Singh., A.V. Ramesh Kumar., Saurabh Srivastava, Defence Applications of Polymer Nanocomposites, Defence Science Journal, Vol. 60, No. 5, September 2010, pp. 551-563
[8] T. Seshaiah, K. Vijaya Kumar Reddy, Effect of Fiber Orientation on the Mechanical
[9] behavior of E-Glass Fibre Reinforced Epoxy Composite Materials, International Journal of Mechanical and Production Engineering Research and Development , Vol. 8, Issue 4, Aug 2018, 379-396
[10] K. P. Ashik, Ramesh S. Sharma, A Review on Mechanical Properties of Natural Fiber Reinforced Hybrid Polymer Composites, Journal of Minerals and Materials Characterization and Engineering, 2015, 3, 420-426
[11] Pradeep, P., Edwin Raja Dhas, J, Evaluation of Mechanical Property On Palm/Coir Based Polymer Matrix Composites, Advances in Materials Science and Engineering: An International Journal, Vol. 2, No. 3, September 2015
[12] Pedro Henrique Cury Camargo, Kestur Gundappa Satyanarayana, Fernando Wypych, Nanocomposites: Synthesis, Structure, Properties and New Application Opportunities, Materials Research, Vol. 12, No. 1, 1-39, 2009
[13] V. Fiore , G. Di Bella, A. Valenza, The effect of alkaline treatment on mechanical properties of kenaf fibers and their epoxy composites, Composites: Part B 68 (2015) 14–21
[14] NRR. Anbusagar a, P.K. Giridharan b, K. Palanikumar, Effect of nanomodified polyester resin on hybrid sandwich laminates, Materials and Design 54 (2014) 507–514
[15] J. Njuguna, K. Pielichowski, J. R. Alcock, Epoxy-Based Fibre Reinforced Nanocomposites: Current Status, Advanced Engineering Materials , vol 9, iss10 pp.835-847
[16] Shilpi Tiwari, Kavita Srivastava, Studies on Mechanical and Thermal Properties of Epoxy/Fly Ash/Nanofiller Nanocomposite: A Review, International Journal of Civil Engineering and Technology, Volume 11, Issue 2, February 2020, pp. 120-139

Optimization of Parameters for Nitronic-60 on Wire-Cut Electrical Discharge Machining using Zinc-Coated Wire

V. Srinivasa Reddy[1*], V. Gopinath[1*], M. Haritha[1*] P Marimuthu[12]*, K.Chandrasekaran[3] and N S Kalyan Chakravarthy[4]*

[1]Department of Mechanical Engineering,
[3]Department of Mechanical Engineering, MAM School of Engineering, Tiruchirappalli, India
[4]Department of Data Science and Business Systems
*QIS College of Engineering and Technology, Vengamukkapalem, Ongole, Prakasam
District, Andhrapradesh-523 272, India
[2]Email: pmarimuthu69@gmail.com (Corresponding author)

Abstract. The objectives of this work are to research the effective parameters of pulse ON time; pulse OFF time; voltage & wire feed whereas machined on Nitronic-60 exploitation wire-cut discharge machining exploitation zinc-coated wire. During this work, metal removal rate, surface roughness, and kerfwidth are taken as output parameters. The experimental results planned an optimum combination of parameters that provide the utmost material removal rate, minimum surface roughness, and kerf dimension. Finally, confirmation experiments were disbursed to spot the effectiveness of the planned technique. Multivariate analysis and ANOVA are performed exploitation Response Surface Linear Model.

1. Introduction

Machining eliminates certain components of work piece items to alter them to finished components. Machining are classified into ancient machining and Non-traditional machining. Ancient Machining, conjointly referred to as typical machining needs the tougher tool than the work piece. This tool ought to be pierced within the workpiece to a definite depth. Besides, a motion between tool and workpiece is to blame for creating or producing the desired form. The nonappearance of any of those parts in machining method like the absence of tool and workpiece contact or relative motion makes the method a non-traditional one.

2. Literature survey

Vinod Kumar et al (2015) have planned a mathematical model and experimented by taking six WEDM method parameters. They have used D2 alloy steel employing Zn coated Copper wire conductor. Output such as MRR, Surface Roughness, and Kerf were used for further analysis. Soundrarajan et al (2015) have mentioned and investigated the analysis of various parameters using RSM. ANOVA is employed to search out the share contribution of serious method parameters. WEDM was used for machining A413 alloy which are making tools.

Pujari Srinivasa et al (2016) experimented the work on Ti6AL4V alloy for analysing the result of voltage, material pressure, and pulse on time and pulse off-time. It is concluded that pulse on-time and pulse off time are important parameters that affect the spark gap of WEDM. Antar et al (2011) studied the various factors that affect the WEDM of Inconel-600 using RSM method. They have used four input method parameters of WEDM to review the method performance. Brajaesh lodhi and Sanjay Agarwal (2014) investigated the results of the responses by varying WEDM method parameters on the machining AISI 304 chrome steel and got the optimized results from the analysis.

Panchal Yogesh Ratilal et al (2018) used L_{18} OA for conduction of experiments for solving single response problem. H-21 die alloy steel was taken as work piece material and Zn coated brass wire was taken as tool material. Herbert et al (2012) have conducted experiments on Nickel based super alloy for the evaluation of surface integrity in hole making process.

A detailed review was given by Chennakeseva Reddy Alavala and Naresh Baki (2016) in the area of application of CNC EDT for Aerospace application. RSM is the powerful tool that can be used for optimization of WEDM process (Gopalakannan et al 2012) and (Balasubramanian and Senthilvelan 2014). Sharanya Nair and Nehal Joshi (2014) reviewed the various works carried out on composite materials using WEDM. Kamal Jangara (2012) conducted the experiments in WEDM for studying unmachined area in intricate machining after roughcut.

Several machining parameters optimization were carried out by the previous researchers on different materials using WEDM and RSM was used for the development of mathematical models. But there is a lacking in WEDM study of Nitronic 60 material which is used for making stack liners, ducts, dampers, scrubbers etc. So in this work experiments were conducted on Nitronic 60 material in WEDM and RSM method is used for developing mathematical model for the important responses such as machining time and surface roughness and kerfwidth.

3. Experimental details

Experiments were conducted on corrosion materials of Nitronic 60 using a wire electrical discharge machine (WEDM). Zinc wires is used as tool.

Table 1 shows the composition of Nitronic 60 and Table 2 shows its properties.

Table 1 Composition of Nitronic 60

MATERIALS	COMPOSITION (%)
Ni	8-9
Cr	16-18
N	0.08-0.18
Mn	7-9 Max.
C	0.1 Max.
Si	3.5-4.5 Max.

Table 2 Properties of Nitronic 60

Density @ 72° F	7700-7700 lb/in. ^{3}
Melting Temperature	1450° F to 1510° F
Specific heat	460-460 Btu/lb. ° F
Electrical Resistivity @ 75° F	0.55-0.55 μΩ.m

Zinc anodes are sensibly protected and can be concentrated also. They yield unexpected voltages in comparison to the copper cells. In a specific, magnesium or copper cells can create voltage as extensive as 1.6v in lemon cells. This voltage is bigger than zinc or copper cells. It is similar to that of standard family unit 1.5v which helps control gadgets. Zinc is a somewhat blue-white, radiant metal. It is weak at normal temperatures yet pliable at 373-545k. It is a reasonable transmitter of power and consumes noticeable all around at high red warmth with an assessment of white billows of oxides. The metal is utilized in various amalgams with different metals. Metal, nickel, silver, business bronze, and aluminium are probably the most significant amalgams. Enormous amounts of zinc are utilized to deliver bite the dust projecting, which are utilized widely via auto, electrical, and equipment businesses. So Zinc terminals are utilized for wire electrical release machining of Nitronic 60. The test results have appeared in Table 3. Reaction Surface Methodology is utilized for additional examination.

4. Result and discussions

The primary objectives of this work is to analyze various input parameters in WEDM for machining Nitronic 60. Ton, Toff, Voltage and Wire feed were taken as input parameters and Machining time, Surface roughness and Kerfwidth were taken as responses. RSM was used for further analysis. Table 4 and Figure 1 shows the regression analysis for machining time.

Table 3 Experimental Results

T on	T off	Voltage	Wire feed	Machining time	Surface roughness	Kerfwidth
4	2	40	2	477.38	2.18	0.247
4	4	45	4	557.07	2.03	0.291
4	6	50	6	671.96	1.8	0.289
4	8	55	8	796.43	1.64	0.294
6	2	50	4	465.75	1.69	0.297
6	4	55	6	562.08	1.85	0.32
6	6	40	8	500.24	2.03	0.289
6	8	45	2	556.47	1.99	0.29
8	2	55	6	463.32	1.86	0.292
8	4	50	8	472.12	1.83	0.293
8	6	45	2	488.27	1.84	0.355
8	8	40	4	208.37	2.17	0.344
10	2	45	8	445.53	2	0.332
10	4	40	6	459.38	2.26	0.357
10	6	55	4	568.47	1.69	0.368
10	8	50	2	559.75	1.79	0.367

IOP Conf. Series: Materials Science and Engineering 1126 (2021) 012043 doi:10.1088/1757-899X/1126/1/012043

Table 4 Regression analysis for machining time

Response	1	Machining tim			
ANOVA					
	Sum of		Mean	F	p-va
Source	Squares	Df	Square	Value	Pr
Model	[illegible]	10			[illegible]
A-T on	31950	1			0.01
B-T off		1			-
C-		1			[illegible]
D-		1			
AB		1			
AC		1			
AD		1			
BC		1			
BD		1			
CD		1			
		5			
		15			
			R-		
			-		
			-		

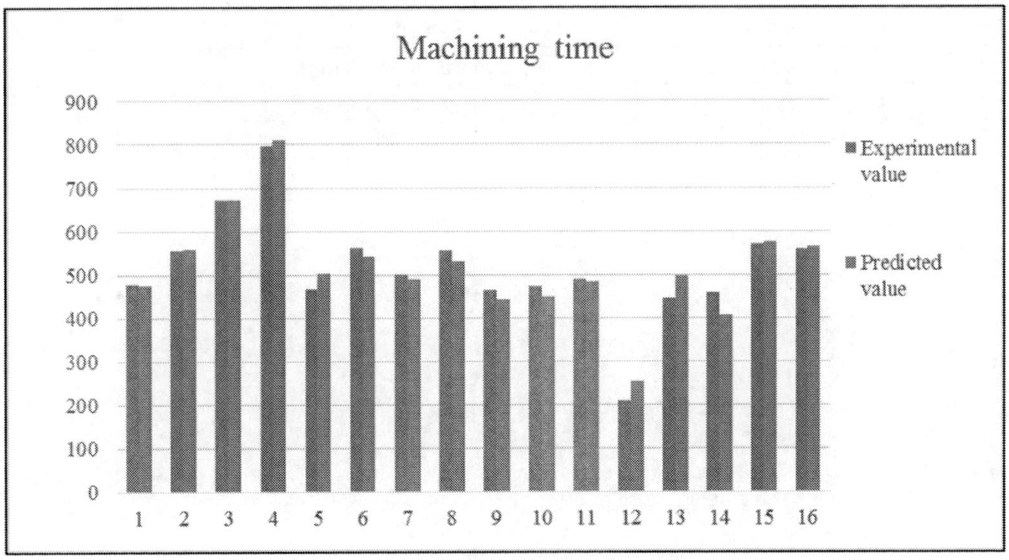

Figure 1 Machining time

From Table 4 and figure 1, the Pred R-Squared" is 0.2164 and Adj R-Squared is 0.8508 "Adeq Precision" measures the signal-to-noise ratio. A ratio greater than 4 is desirable. A ratio of

14.180 indicates satisfactory signal. The model developed for machining time from this study is given below

Machining time = +685.03319+22.19034 * T on-192.37313 * T off-13.26234 * Voltage+196.16216 * Wirefeed-9.66693 * T on * T off+0.76868 * T on * Voltage-8.40988 * T on * Wirefeed+5.90026 * T off * Voltage-4.90879 * T off * Wirefeed-2.58316 *Voltage * Wire feed

From Table 5 and Figure 2, it is observed that Model F-value is 9.30 that implies the model is significant. There is only a 0.15% chance that a "Model F-value" this large could occur due to noise.

The "Pred R-Squared" of 0.5652 is reasonable with the "Adj R-Squared" of 0.6889. A ratio of 8.519 indicates satisfactory signal. The model for surface roughness is shown below

Surface roughness=+3.25645+5.12500E-003 * T on-0.012348 * T off-0.027961 * Voltage+2.63587E-003 * Wire feed

Table 5 Regression analysis for surface roughness

Response	2	Surface roughness			
ANOVA					
	Sum of		Mean	F	p-value
Source	Squares	Df	Square	Value	Prob> F
Model	0.400905109	4	0.100226277	9.304597	0.0015
A-T on	0.00210125	1	0.00210125	0.195071	0.6673
B-T off	0.011689275	1	0.011689275	1.085184	0.3199
C-Voltage	0.3746174	1	0.3746174	34.77794	0.0001
D-Wirefeed	0.000511359	1	0.000511359	0.047472	0.8315
Residual	0.118488641	11	0.010771695		
Cor Total	0.51939375	15			
Std. Dev.	0.103786775		R-Squared	0.771871	
Mean	1.915625		Adj R-Squared	0.688915	
C.V. %	5.417906688		Pred R-Squared	0.565183	
PRESS	0.225841212		Adeq Precision	8.519386	

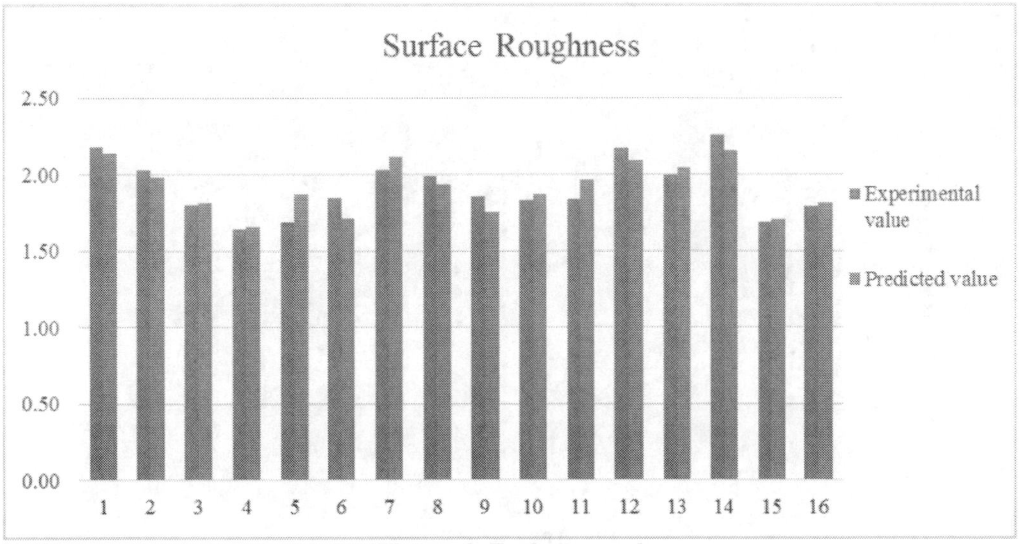

Figure 2 Surface roughness

Table 6 Regression analysis for Kerfwidth

Response	3	Kerfwidth			
ANOVA					
	Sum of		Mean	F	p-value
Source	Squares	Df	Square	Value	Prob> F
Model	0.014964	4	0.003741	10.26591	0.001
A-T on	0.012425	1	0.012425	34.09576	0.0001
B-T off	0.001851	1	0.001851	5.078177	0.0456
C-Voltage	0.000164	1	0.000164	0.449651	0.5163
D-Wirefeed	0.000225	1	0.000225	0.616601	0.4489
Residual	0.004009	11	0.000364		
Cor Total	0.018973	15			
Std. Dev.	0.01909		R-Squared	0.78872	
Mean	0.314063		Adj R-Squared	0.711891	
C.V. %	6.078325		Pred R-Squared	0.586968	
PRESS	0.007836		Adeq Precision	10.31731	

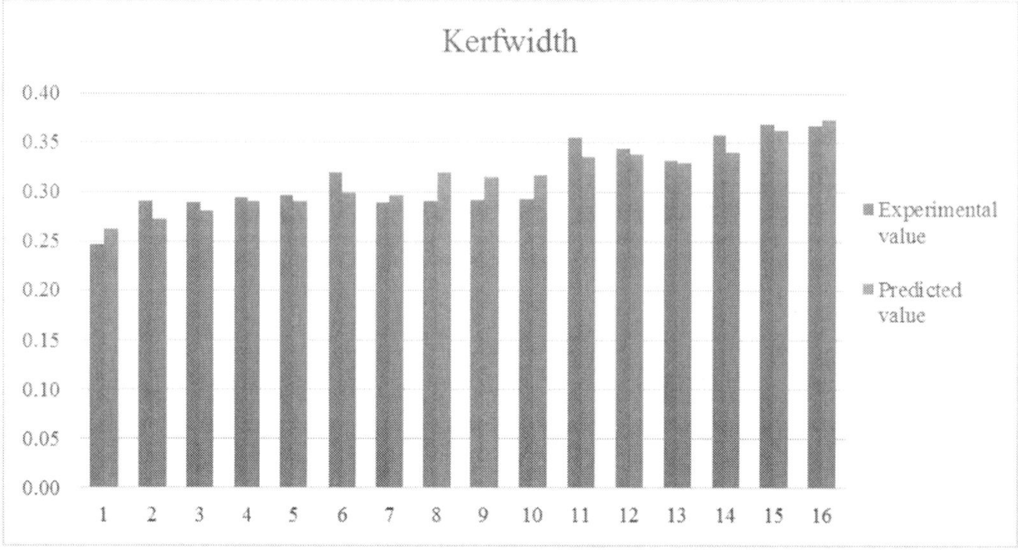

Figure 3 kerfwidth

Table 6 and Figure 3 shows the Regression analysis for Kerfwidth. From Table 6 and Figure 3, it is observed that The Model F-value is 10.27 that implies the model is significant. are many insignificant model terms (not counting those required to support hierarchy), model reduction may improve your model.

The Pred R-Squared" of 0.5870 is in reasonable with the Adj R-Squared" of 0.7119.

A ratio of 10.317 indicates an adequate signal. The developed model for Kerfwidth is shown below.

Kerfwidth=+0.18322+0.012463 * T on+4.91304E-003 * T off+5.84783E-004 * Voltage-1.74728E-003 *Wire feed

5. Conclusion

This experimental work uncovers the accompanying ends on the WEDM procedure on Nitronic 60 workpiece material. The main principle objective was to build up the experimental model utilizing RSM. The reaction surface philosophy is probably the best method to recognize the impacts of machining boundaries on the WEDM cycle. The voltage and heartbeat off time significantly affect machining time. The more elevated level of current produces lower machining time. The reaction surface models were created dependent on the plan of the analysis with current, voltage, and beat on schedule, and heartbeat off time as info, and machining time and surface harshness were the reactions. The reaction surface model has a more modest deviation from test information and affirms that the created model can be utilized to foresee the machining time and surface unpleasantness esteem adequately.

References

[1] Vinod Kumar, Vikas Kumar, and Kamal Kumar Jangra (2015), "An Experimental Analysis and Optimization Of Machining Rate And Surface Characteristics In Wedm Of Monel-400 Using RSM And Desirability Approach", Journal of Industrial Engineering International, Vol. 11, pp. 297-307.

[2] R. Soundrarajan, A. Ramesh, N. Mohanraj and N. Parthasarathi (2016), "An investigation of material removal rate and surface roughness of squeeze cast a413 alloy a WEDM by multi-response optimization using RSM", Journal Of Alloy And Compounds, Vol. 685, pp. 533-545.
[3] Pujari Srinivasa Rao, Koona Ramji, and Beela Satyanarayan (2016), "Effect Of Wire EDM Conditions On Generation Of Residual Stresses In Machining Of Aluminium 2014 T6 Alloy" Alexandria Engineering Journal, Vol. 55 issues 2, pp. 1077-1084.
[4] M. T. Antar, S. L. Soo, D. K. Aspinwall, D. Jones and R. Perez (2011), "Productivity And Workpiece Surface Integrity When WEDM Aerospace Alloys Using Coated Wires", Procedia Engineering, Vol. 19, pp. 3-8.
[5] Brajaesh lodhi and Sanjay Agarwal (2014), "Modeling Of Wire Electrical Discharge Machining Of AISI D3 Steel Using Response Surface Methodology" 5th International & 26th All India Manufacturing Technology, Design, and Research Conference (AIMTDR 2014), IIT Guwahati, Assam, India
[6] Panchal Yogesh Ratilal, Nilesh K.Patel and Kapil S.Banker (2018), "Review Of Research Work In Wire Cut Electrical Discharge Machining On Metal Matrix Composite Materials" Procedia Manufacturing, Vol. 20, pp. 253-258.
[7] C. R. J. Herbert, J. Kwong, D. A. Axinte, M. C. Hardy and P.J. Witters (2012), "Evaluation Of The Evaluation Of Work Piece Surface Integrity In Hole Making Operations For A Nickel-Based Superalloy" Journal Of Materials Processing Technology, Vol. 212, Issue 8, pp.1723-1730.
[8] Chennakesava Reddy Alavala and Naresh Baki (2016), "Application of CNC Electrical Discharge Turning for Aerospace Materials – A Review", International Journal of science and research, Vol.5, issue 11, pp.1647-1653
[9] T. Gopalakannan, S. Senthilvelan and S. Ranganathan (2012), "Modeling and Optimization of EDM Process Parameters on Machining of Al 7075-B4C MMC Using RSM", Procedia Engineering, Vol. 38, pp. 685-690.
[10] P.Balasubramanian and T.Senthilvelan (2014), "Optimization of Machining Parameters in EDM Process Using Cast and Sintered Copper Electrodes", Procedia Materials Science, Vol. 6, pp. 1292-1302.
[11] Sharanya Nair and Nehal Joshi (2014), "A review on wire electrical discharge machining (WEDM) of composite materials", International Journal of Engineering Trends and Technology
Vol. 17, Number 9, Pp. 415-420.
[12] Kamal Jangra (2012), "Study of the unmachined area in intricate machining after rough cut in WEDM", International Journal of Industrial Engineering Computations, 3 Issue 5 pp. 887-892.

IOP Conf. Series: Materials Science and Engineering 1126 (2021) 012044 doi:10.1088/1757-899X/1126/1/012044

Effect of annealing temperature and cutting angle of specimens on metallurgical and mechanical properties of Mg-7Li-1Zn alloy via Taguchi approach and response surface analysis

M Pahlavani[1], **J Marzbanrad**[1] **and K Reza Kashyzadeh**[2,3]
[1]School of Automotive Engineering, Iran University of Science and Technology, Tehran, Iran
[2]Department of Transport, Academy of Engineering, Peoples' Friendship University of Russia (RUDN University), 6 Miklukho-Maklaya Street, Moscow, 117198, Russian Federation
[3]Department of Mechanical and Instrumental Engineering, Academy of Engineering, Peoples' Friendship University of Russia (RUDN University), 6 Miklukho-Maklaya Street, Moscow, 117198, Russian Federation
*Corresponding author, email address: reza_kashi_zade_ka@rudn.ru & kazem.kashyzadeh@gmail.com

Abstract. The main aim of this research is to assess the effect of annealing temperature and cutting angle of specimens on the metallurgical and mechanical properties of superlight Mg-Li-Zn alloy thin sheets. To this end, as-rolled LZ71 (Mg-7%Li-1%Zn) alloy was annealed at different temperatures (0, 250, and 350°C) for 3 hours. After that, the test specimens were cut at different angles relative to the rolling direction (0, 45, and 90°). Next, several experiments were accomplished to study the microhardness, microstructure, grain size, and tensile behavior of the LZ71 alloy. Finally, different Design of Experiment (DOE) techniques were used to determine the individual and interaction effects of annealing temperature and cutting angle of specimens on the material characterization of LZ71 alloy. In this regard, annealing temperature and cutting angle were considered as input parameters in both Taguchi Approach (TA) and Response Surface Analysis (RSA). Also, key parameters of the tensile test including Ultimate Tensile Strength (UTS), Yield Stress (YS), Elastic Modulus (EM), and Strain at BreakPoint (SBP) were considered as outputs. It was concluded that annealing temperature influence on the static behavior of LZ71 alloy is several times greater than the effect of cutting angle (i.e., the effect of annealing temperature on the UTS, YS, EM, and SBP is 92%, 76%, 61%, and 66%, respectively). Moreover, the results of the Response Surface Optimization Algorithm (RSOA) showed that to achieve the best static properties in different directions, the annealing temperature should be selected to be 350°C.

Keywords: Magnesium-lithium alloy, Microstructure, Mechanical properties, Taguchi approach, Response surface analysis, Optimization.

1. Introduction

Magnesium alloy applications are experiencing a sharp rise in various industries such as automotive, aerospace, and electronics due to their mechanical behavior [1-3]. Moreover, these alloys are well known as the lightest structural alloys having a density of 1.8 gr/cm3 [4, 5]. Pure magnesium is one of the most abundant metals and has considerable properties namely: high durability, very high strength

to weight ratio, high recyclability, desirable heat loss, and electromagnetic shielding [6, 7]. Magnesium-Lithium composite is one of the most ideal designing metals because of its incredible properties like high formability and ultra-low density [2]. Amongst all magnesium alloys, Mg-Li compounds are considerably more interesting than others due to their elongation besides the obvious low density [8]. To reach higher yield stress, adding rare earth elements to the compound is an effective solution. However, Mg-Li with around 5 to 12 weight percentage of lithium has the best structural behavior because, with this proportion of lithium, the microstructure of the alloy will be dual-phase with α-Mg (Mg solid solution, hcp structure) and β-Li (Li solid solution, bcc structure) [9]. Studies on microstructures and mechanical properties of magnesium alloys have been done because of their unique behavior under different situations of casting, forming, and annealing [5, 10-12]. Tang et al. have investigated the influences of the annealing process on microstructural and mechanical properties of Mg-XLi-3Al-2Zn-0.2Y [11]. In this study, experimental results have been reported taking into account three different values for the X parameter (5, 8, and 11). In other words, the effect of the percentage of lithium element in the alloy has been analyzed on the materials' microstructure and mechanical properties. The results indicated that the ultimate and yield stresses increase and the elongation at breakpoint decreases by raising the weight percentage of lithium. Moreover, the surface microhardness reduces by increasing the lithium value. Also, they have shown that the surface microhardness variations have a direct relationship with the temperature if the amount of lithium in the alloy was high enough. Zou et al. have improved the mechanical behavior of superlight Mg-Li-Al alloys (LA36, LA96, and LA156) by duplex phases and fine precipitates [12]. The experimental results (tensile, compression, XRD, SEM, and TEM images) have been compared to show that the mechanical behavior of different Mg alloys is better than the pure magnesium. In the other words, the alloying with Li to improve ductility and the Al addition to enhance strength are proofed.
To improve the mechanical and microstructural characteristics of as-cast alloys, extrusion and rolling processes have been known as the most prevalent work hardening methods [13-20]. Meng et al. had a survey on mechanical properties of Mg–8Li–1Al–1Zn, Mg–8Li–1Zn, and Mg–8Li–1Al [9]. According to their study, the strength of Mg–8Li–1Al–1Zn and Mg–8Li–1Al alloys improved due to the forming process. Also, Yu et al. have studied the rolling process temperature effects on mechanical properties and microstructures of AZ910 alloy [21]. They reported that the cutting angle impact on the results of tensile tests of the fine-grained rolled AZT910 alloy sheets is practically insignificant during deformation at 300 °C. In other studies, the microstructural and textural evaluation of Mg-X%Li-1%Zn (X=6, 8, and 12) has been performed [22]. Recently, the study on the fatigue behavior and fracture mechanisms of magnesium alloys has been considered by many scholars [23-29]. Also, in the previous study, the authors reported the high-cycle fatigue curves of LZ71 magnesium alloy and compared it with fatigue properties of pure magnesium. Moreover, they studied the high-temperature tensile behavior of LZ71 magnesium alloy in the laboratory [30].
In the present paper, for the first time, the individual and interactive effects of annealing temperature and cutting angle of specimens on the material characterization of as-rolled LZ71 alloy sheets were investigated using TA and RSA. To this end, test specimens were provided in three different cutting angles relative to the rolling direction and various annealing temperatures. Afterward, Optical Microscopy (OM) observations, X-ray Diffraction (XRD) analysis, microhardness measurement, and tensile tests were carried out to study the metallurgical and mechanical properties of the material. Eventually, the most effective parameter was reported based on the TA results. And, the optimal values were extracted using RSOA.

2. Methodology

In the present study, the commercial pure Magnesium (Mg), Zinc (Zn), and Lithium (Li) were used to produce LZ71 alloy through casting at almost 770°C under a protective atmosphere of Argon gas. The chemical composition of the cast alloy was Mg-7(Wt%)Li-1(Wt%)Zn. The initial ingot was flat-rolled from an initial thickness of 10 mm to a final thickness of 2 mm (total thickness reduction of almost 80% including 8 passes by 10% reduction in each pass) at a rolling temperature of 300°C [30]. Then, the sheets were annealed at two different temperatures including 250 and 350°C for 3 hours. Next, tensile specimens were fabricated from the sheets of annealed and as-rolled materials with a thickness of 2 mm according to ASTM E8M [31] standard. Specimens were provided in different directions including 0, 45, and 90 degrees. The angle zero means that the longitudinal directions of specimens are parallel to the direction of the rolling process. Eventually, Figure 1 shows the working flowchart to clarify the research methodology in this study.

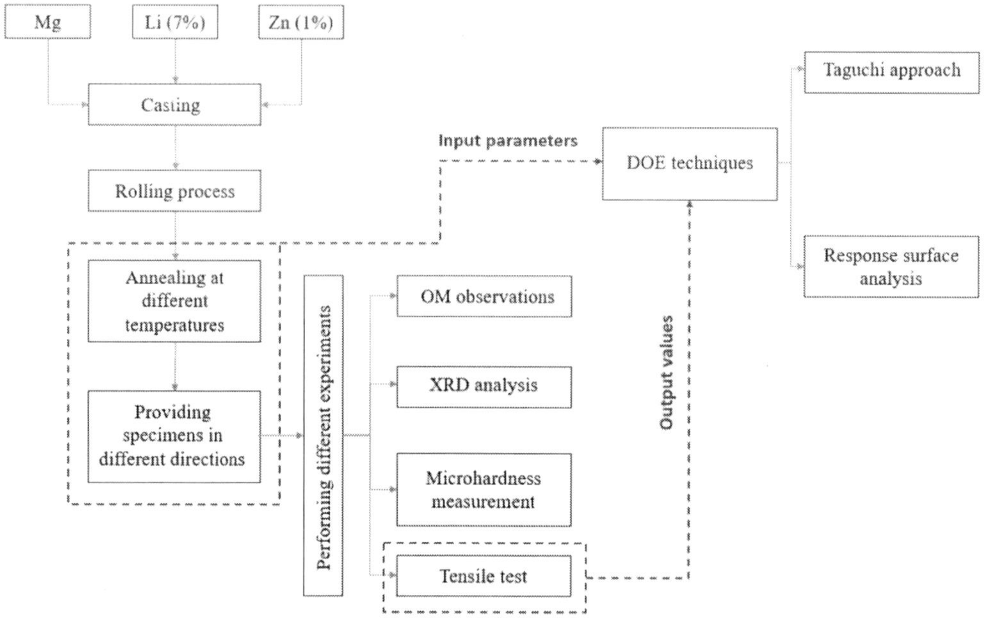

Figure 1. The working flowchart used in this study

3. Experimental procedure

3.1. OM observations
Microstructures of the specimens were observed utilizing the Optical Microscope (OM) before and after the annealing process at different temperatures. The specimens have been etched using a solution of CH3COOH (2 ml), HNO3 (2 ml), H2C2O4 (2 gr), and water (310 ml).

3.2. Microhardness measurement
Microhardness measurements were accomplished utilizing a SHAAB M5 tester. The applied load was set to 50 gf and the time duration was 15 seconds using Vickers indenter. To evaluate the trace of indenter accurately, surfaces of the specimens were prepared carefully, and four random indents have been considered for each specimen. Vickers microhardness of each indent was calculated based on HV=1.854P/d^2. In this equation, HV is Vickers microhardness value with the unit of Kgf/mm^2, P is the applied force with the unit of Kgf, and d is the average length of the diagonal remain by the indenter in millimeters.

3.3. XRD analysis
To obtain the grain size of the material before and after applying the annealing process at different temperatures, XRD measurements were carried out. X'Pert PRO MPD (PANalytical) X-ray diffractometer system was employed with Cu Kα radiation (λ=1.54060 Å). Also, Scherer's equation was employed to calculate the crystallite size [32-34].

3.4. Tensile test
The STM-250 SANTAM universal testing machine was used to carry out the tensile test. All tests were performed at the constant strain rate of 0.1 mm/mm/min by applying a short travel extensometer

with an accuracy of 0.001 mm at room temperature. All environmental conditions were controlled during the test following the ISO-17025 [35] standard.

4. Statistical analysis

Two different DOE techniques including TA and RSA were used to evaluate the data collected from experiments. Also, each of these methods was used for a specific purpose. However, the same conceptual structure was considered for both methods (Figure 2).

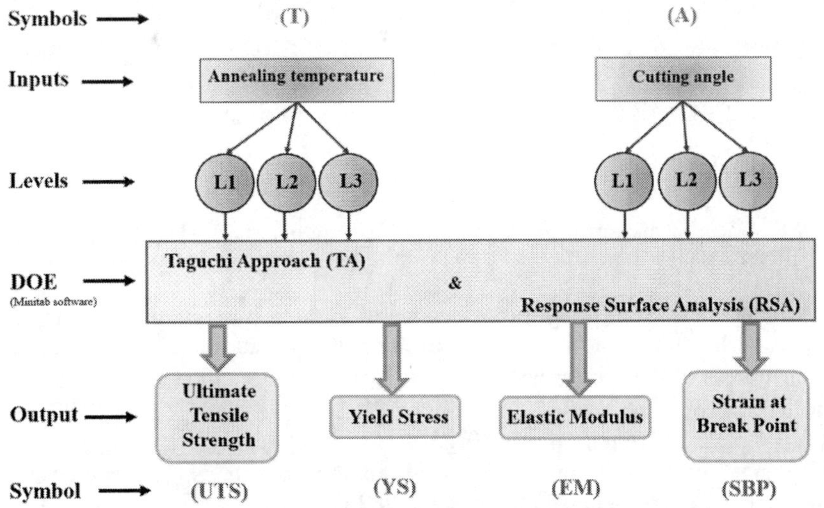

Figure 2. Conceptual structural of DOE techniques according to the considered input and output parameters

Taguchi sensitivity analysis was performed to find the most effective parameter (annealing temperature or cutting angle) on the static properties of the LZ71 superlight alloy sheet. Among all DOE techniques, the Taguchi algorithm uses the least number of samples for sensitivity analysis [36-39]. Considering the number of input parameters and their levels, the L9 orthogonal matrix was used to perform Taguchi analysis. Moreover, the current research seeks to improve the static behavior of the material (increasing the UTS, YS, EM, and SBP). Therefore, the viewpoint of larger is better was considered to calculate the mean of signal-to-noise ratio [40]. Next, response surface analysis was done to investigate the interaction effects of parameters on the static behavior of the material. Eventually, the RSM-based optimization algorithm [41] was employed to determine the optimal conditions to achieve the best static properties in different directions.

5. Results and discussion

5.1. Experimental results

0 presents the OM images of annealed specimens of the Mg-7Li-1Zn alloy. On the OM images of the sample before the annealing process, two different phases including α-Mg (bright regions) and β-Li (dark regions) are identified. Also, it could be noticed that the volume fraction of the alpha phase increase by raising the temperature for the annealing process. When the material is as-received (annealing temperature of zero or without annealing treatment), the microstructure is fiber shaped rolling formation. Then, the material can also find partial recrystallization by increasing the annealing temperature to 250 °C. Finally, consider the temperature of 350 °C for annealing treatment caused to create complete recrystallization, forming a fine recrystallized grains, and fibrous microstructure completely dis-appeared. According to the published researches, the alloy has completed the recrystallization, if continues to increasing annealing temperature, some grains will appear irregular

growth and into re-crystallization grain coarsening and growth stage [29]. In other words, the microstructure finds a more homogeneous distribution, because static recrystallizations occurred at the annealing temperature of 350° C. Thus, mechanical properties for annealed samples at 350 °C are expected to be the same in all directions.

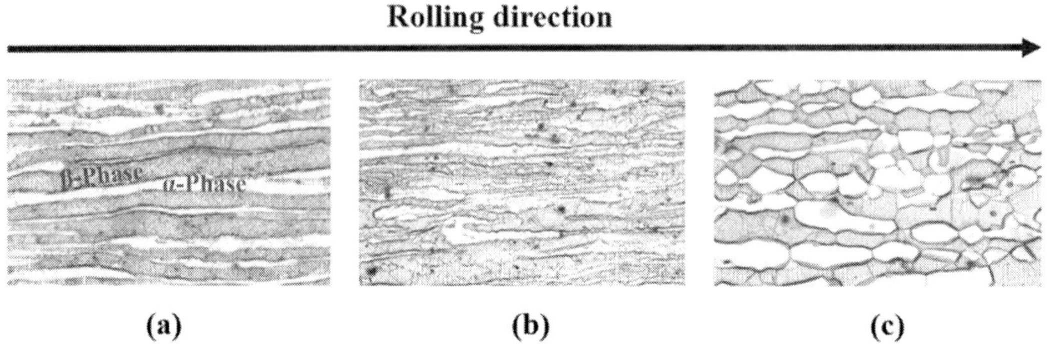

Figure 3. OM images of the different heat-treated specimens of LZ71: (a) Without annealing process, (b) Annealing temperature of 250 °C, and (c) Annealing temperature of 350 °C

After microstructural characterization, distributions of surface microhardness of the treated specimen were measured in 4 different points. It is clear from the results of hardness measurements that the surface microhardness of the specimens decreases by raising the temperature of the annealing process. Average surface microhardness of 65.51 ± 1.82, 62.12 ± 1.73, and 52.34 ± 3.77 have been obtained for annealed specimens at temperatures of 0, 250, and 350°C, respectively.

The crystallite size of the top surface layer of treated specimens calculated using XRD analysis (Figure 4) and Scherrer-Debye equation are presented in Table 1.

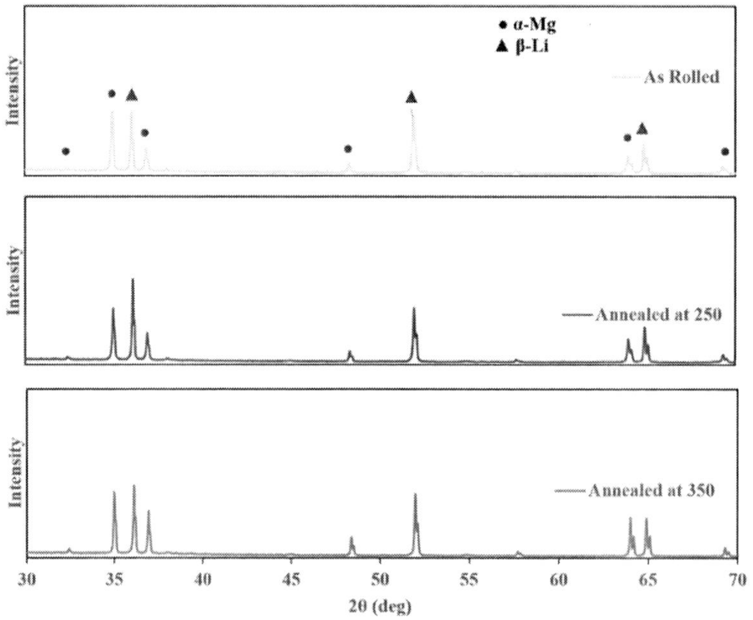

Figure 4. XRD results of the annealed specimens at different temperatures

Table 1. Crystallite size of the treated specimens

Sample No.	Treatment	Temperature (°C)	Crystallite size (nm)	
			hcp	bcc
1	No annealing process	0	70.13	89.94
2	Annealing process	250	79.07	103.37
3	Annealing process	350	94.32	104.77

The key parameters of stress-strain diagrams as a result of tensile tests including ultimate stress, yield stress, elastic modulus, and elongation at breakpoint are reported in Table 2. The yield stress was calculated based on the strain offset of 0.2%. All tensile tests were repeated 3 times and the average values of key parameters with standard deviation calculation are presented in this section.

Table 2. Key parameters of tensile testing results

Annealing temperature	Sheet angle relative to the rolling process	Ultimate stress (MPa)	Yield stress (MPa)	Elastic modulus (GPa)	Strain (%)
	0	196.64 ± 1.22	119.81 ± 0.78	12.44 ± 0.93	49.2 ± 1.09
As-rolled	45	192.51 ± 1.22	122.97 ± 0.84	14.44 ± 1.83	42.4 ± 0.94
	90	188.48 ± 1.19	110.71 ± 0.92	11.81 ± 1.98	33.46 ± 0.8
	0	170.39 ± 1.08	96.33 ± 0.97	12.22 ± 0.78	54.8 ± 0.91
250 °C	45	163.57 ± 1.09	110.11 ± 1.17	13.03 ± 0.84	48.9 ± 0.81
	90	168.99 ± 1.07	93.26 ± 1.42	10.93 ± 0.93	42.4 ± 0.7
	0	142.9 ± 0.9	87.7 ± 0.63	13.49 ± 0.47	61.4 ± 0.38
350 °C	45	140.68 ± 0.99	90.5 ± 0.69	14.88 ± 0.52	61.1 ± 0.42
	90	138.85 ± 0.89	92.18 ± 0.7	15.19 ± 0.53	62.7 ± 0.41

It is clear from the tensile testing results that the values of ultimate stress, yield stress, and elastic modulus are approximately the same at different sheet angles from each group and there is a significant change in their elongation at the breakpoint. The maximum and minimum values of elongation are for sheet angles of 0 and 90 relatives to the rolling process, respectively. Finally, the experimental results indicated that the static properties of the sheet become homogeneous in different directions by raising the annealing temperature. Therefore, it can be stated that the material is utterly isotropic at the annealing temperature of 350 °C.

5.2. Results of statistical analysis

Taguchi sensitivity analysis was performed for each of the key parameters of the stress-strain diagram. Figure 5 presents the influences of S/N ratios at every level. Also, the ranking of the influence of different parameters (annealing temperature and cutting angle) is depicted in Figure 6. The results showed that the annealing temperature influence on the static behavior of LZ71 alloy is several times greater than the effect of cutting angle. Moreover, this parameter also has the greatest effect on the UTS (92%). Also, the effect of cutting angle is more visible on EM and SBP.
The response surface 2D graphs of annealing temperature and sheet angle on the various responses are demonstrated in 0. Also, Figure 7-a and Figure 7-b display the stress reduction in ascending order through the severity of the factors individually. On the other hand, the minimum value of stress components (ultimate and yield) is related to the maximum temperature of the annealing process. However, the minimum value of elastic modulus is related to a particular range including annealing temperature of 125-200°C and 70-220°C for sheet angle close to 0 and 90, respectively. However, the maximum elastic modulus occurs at the maximum annealing temperature as shown in 0-c.

Furthermore, 0-d shows that the two parameters have a different effect on the failure strain of the material. In the lower annealing temperature (less than 150°C), the sheeting angle has the most effect. But, its effect significantly decreases by increasing the annealing temperature.

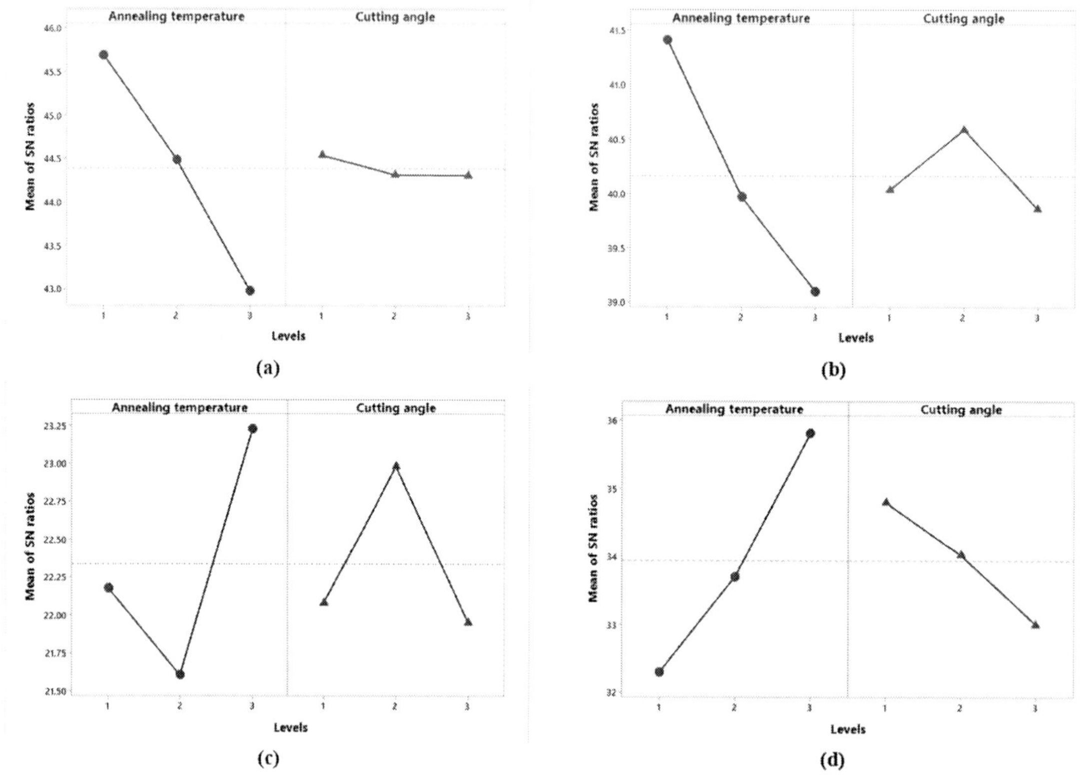

Figure 5. Influences of S/N ratios of all parameters related to a) UTS, b) YS, c) EM, and d) SBP

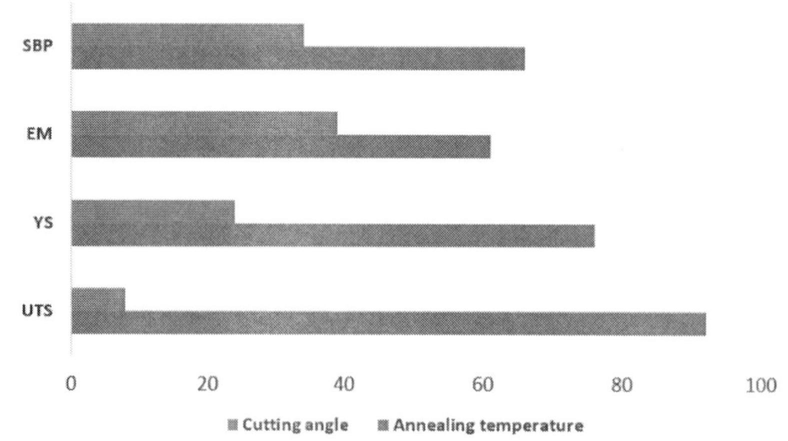

Figure 6. The ranking of the influence of different parameters on static properties of LZ71 alloy sheet

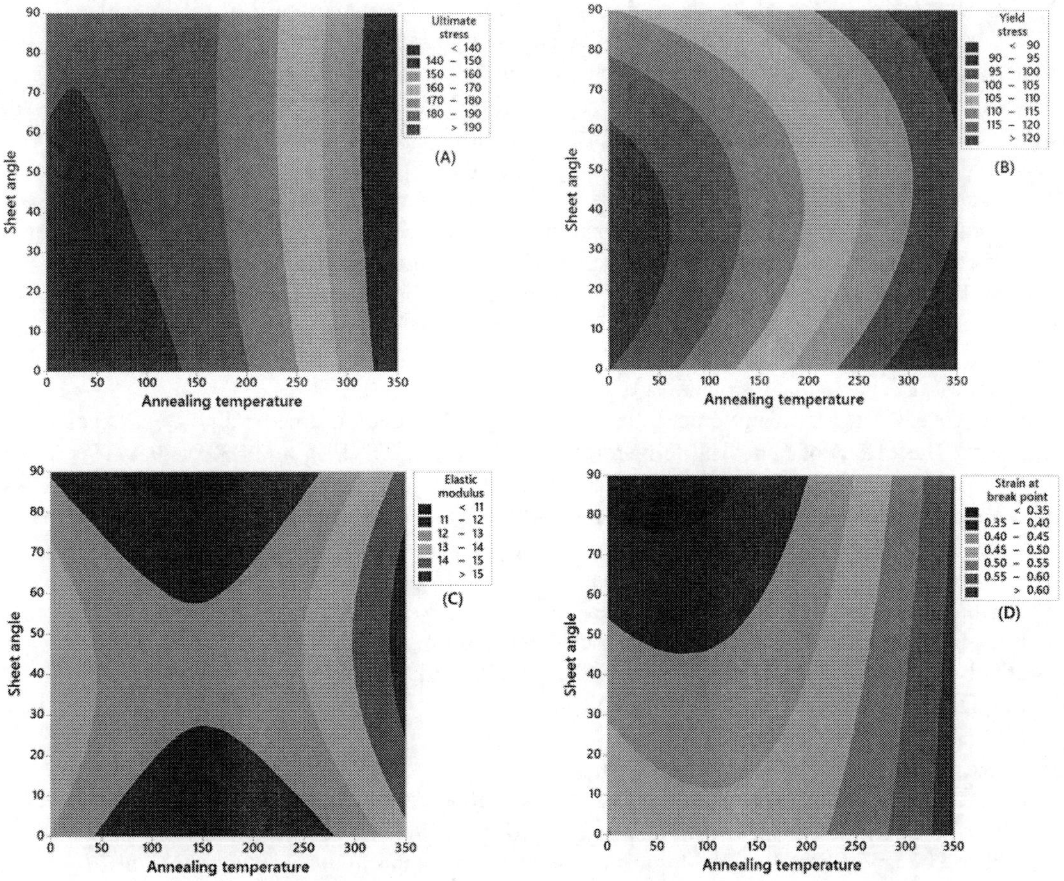

Figure 7. Response surface 2D graphs of annealing temperature and sheet angle on the various responses: a) ultimate stress, b) yield stress, c) elastic modulus, and d) strain at the breakpoint

It is obvious that if the target function is to increase the elongation, the sample should be used with a sheet angle of zero and an annealing temperature of 350°C which has a significant disadvantage. This process reduces ultimate and yield stresses. In other words, it causes to reduces the static strength of the material. Response optimization was used to distinguish the combinations of variable settings which optimize a single response or a bunch of responses. While assessing the influence of numerous factors on a response is wanted, response optimization is altogether helpful [41-42]. Therefore, the optimization was provided by an RSM-based optimization algorithm, the annealing temperature of 350°C, and sheet angle of 90° are to be optimized with a composite desirability value of 0.7308 by considering the maximizing all the key parameters of the tensile test results as optimization target. Due to the optimized values, ultimate and yield stresses are optimized with 194.6347 and 122.7793 MPa, respectively by overcoming the desirability value of 0.96. elastic modulus is optimized with 13.5568 GPa, and elongation reaches a level of 47.58%.

6. Conclusion

In the current research, cast Mg-7Li-1Zn alloy specimens were fabricated from different sheet angles relative to the rolling direction. Then, the specimens were annealed at different temperatures. The study was aimed at investigating the effect of annealing temperature and cutting angle on the microstructure, microhardness, grain size, and mechanical properties of the material. Various experimental approaches were employed to explore the microstructural and mechanical properties of the treated specimens. Next, the Taguchi approach and response surface method were used to analyze and optimize.

The volume fraction of the alpha phase increase by raising the temperature of the annealing process and the microstructure finds a more homogeneous distribution. Thus, mechanical properties for annealed samples at 350 °C are expected to be the same in all directions. The surface microhardness of the specimens decreases by raising the temperature of the annealing process. Moreover, the crystallite size increase by raising the annealing temperature in both hcp and bcc structures.

Tensile testing results showed that ultimate and yield stresses and elastic modulus are approximately the same at different sheet angles and there is a significant change in their elongation at the breakpoint. The maximum and minimum values of elongation are for sheet angles of 0 and 90 relatives to the rolling process, respectively. However, the value of this parameter converges in different directions by increasing the temperature of the annealing process. Also, ultimate stress decreases by increasing the temperature. Also, the static properties of the sheet become homogeneous in different directions by raising the annealing temperature. Therefore, it can be stated that the material is completely homogeneous at the annealing temperature of 350°C.

TA results reveal that annealing temperature is the most effective parameter on the static behavior of LZ71 alloy. The effect of annealing temperature on the UTS, YS, EM, and SBP is 92%, 76%, 61%, and 66%, respectively. Moreover, RMS results indicated that the minimum value of elastic modulus is related to the annealing temperature range of 125-200°C and 70-220°C for sheet angles close to 0 and 90, respectively. However, the maximum elastic modulus occurs at the maximum annealing temperature. Moreover, the sheeting angle has the most effect in the annealing temperature less than 150°C. But, its effect significantly decreases by increasing the annealing temperature. Due to the optimized factors, ultimate and yield stresses are optimized with 194.6347, and 122.7793 MPa, respectively by overcoming the desirability value of 0.96. Elastic modulus is optimized with 13.5568 GPa, and failure strain reaches a level of 47.58%.

References

[1] Kulekci M K 2008 Magnesium and its alloys applications in automotive industry *The International Journal of Advanced Manufacturing Technology*, 39 851-865

[2] Mordike B L and Ebert T 2001 Magnesium: Properties - applications - potential *Materials Science and Engineering: A*, 302 37-45

[3] Sahoo K and Poddar P 2017 Studies on Magnesium Alloys–Properties and Potential for Automotive and Aerospace Applications *Transaction of 65th Indian foundry Congress*, 151-156

[4] Tao W, Milin Z, Zhongyi N and Bin L 2006 Influence of Rare Earth Elements on Microstructure and Mechanical Properties of Mg-Li Alloys *Journal of Rare Earths*, 24 797-800

[5] Tang Y, Le Q, Misra R D K, Su G and Cui J 2018 Influence of extruding temperature and heat treatment process on microstructure and mechanical properties of three structures containing Mg-Li alloy bars *Materials Science and Engineering: A*, 712 266-280

[6] Lin M C, Lin S Q and Uan J Y 2014 Effect of annealing temperature on the microstructure and mechanical properties of an as-rolled Mg-9wt.% Li-3wt.% Al-1wt.% Zn alloy sheet *Frontiers of Materials Science*, 8 271-280

[7] Pandey R, Tekumalla S and Gupta M 2018 Effect of defects on electromagnetic interference shielding effectiveness of magnesium *Journal of Materials Science: Materials in Electronics*, 1-12

[8] Wang T, Zhang M and Wu R 2008 Microstructure and properties of Mg–8Li–1Al–1Ce alloy *Materials Letters*, 62 1846-1848

[9] Meng X, Wu R, Zhang M, Wu L and Cui C 2009 Microstructures and properties of superlight Mg–Li–Al–Zn wrought alloys *Journal of Alloys and Compounds*, 486 722-725

[10] Chang L L, Shi C C and Cui H W 2018 Enhancement of mechanical properties of duplex Mg-9Li-3Al alloy by Sn and Y addition *Transactions of Nonferrous Metals Society of China*, 28 30-35

[11] Tang Y, Le Q, Jia W, Liu X and Cui J 2018 Influences of warm rolling and annealing processes on microstructure and mechanical properties of three parent structures containing Mg-Li alloys *Materials Science and Engineering: A*, 711 1-11
[12] Zou Y, Zhang L, Li Y, Wang H, Liu J, Liaw P K, Bei H and Zhang Z 2018 Improvement of mechanical behaviors of a superlight Mg-Li base alloy by duplex phases and fine precipitates *Journal of Alloys and Compounds*, 735 2625-2633
[13] Zeng Z, Stanford N, Davies C H J, Nie J F and Birbilis N 2018 Magnesium extrusion alloys: a review of developments and prospects *International Materials Reviews*, 1-36
[14] Lee J U, Song S W, Kim Y, Kim S H, Kim Y J and Park S H 2018 Effects of {10–12} Twins on Dynamic Torsional Properties of Extruded AZ31 Magnesium Alloy *Metals and Materials International*, 24 283-289
[15] Liu Y, Li F, Li X W and Shi W Y 2018 Properties of Rolled AZ31 Magnesium Alloy Sheet Fabricated by Continuous Variable Cross-Section Direct Extrusion *Journal of Materials Engineering and Performance*, 27 1334-1342
[16] Wang W, Chen W, Zhang W, Cui G and Wang E 2018 Effect of deformation temperature on texture and mechanical properties of ZK60 magnesium alloy sheet rolled by multi-pass lowered-temperature rolling *Materials Science and Engineering: A*, 712 608-615
[17] Jiang J, Wu J, Ni S, Yan H and Song M 2018 Improving the mechanical properties of a ZM61 magnesium alloy by pre-rolling and high strain rate rolling *Materials Science and Engineering: A*, 712 478-484
[18] Guo L and Fujita F 2018 Modeling the microstructure evolution in AZ31 magnesium alloys during hot rolling *Journal of Materials Processing Technology*, 255 716-723
[19] Jahedi M, McWilliams B A, Kellogg F R, Beyerlein I J and Knezevic M 2018 Rate and temperature dependent deformation behavior of as-cast WE43 magnesium-rare earth alloy manufactured by direct-chill casting *Materials Science and Engineering: A*, 712 50-64
[20] Tang W, Huang S, Li D and Peng Y 2015 Mechanical anisotropy and deep drawing behaviors of AZ31 magnesium alloy sheets produced by unidirectional and cross rolling *Journal of Materials Processing Technology*, 215 320-326
[21] Yu Z P, Yan Y H, Yao J, Wang C, Zha M, Xu X Y, Liu Y, Wang H Y and Jiang Q C 2018 Effect of tensile direction on mechanical properties and microstructural evolutions of rolled Mg-Al-Zn-Sn magnesium alloy sheets at room and elevated temperatures *Journal of Alloys and Compounds*, 744 211-219
[22] Karami M and Mahmudi R 2013 The Microstructural, Textural, and Mechanical Properties of Extruded and Equal Channel Angularly Pressed Mg-Li-Zn Alloys *Metallurgical and Materials Transactions A*, 44 3934-3946
[23] Shih T S, Liu W S and Chen Y J 2002 Fatigue of as-extruded AZ61A magnesium alloy *Materials Science and Engineering: A*, 325 152-162
[24] Wu L, Jain A, Brown D W, Stoica G M, Agnew S R, Clausen B, Fielden D E and Liaw P K 2008 Twinning–detwinning behavior during the strain-controlled low-cycle fatigue testing of a wrought magnesium alloy *ZK60A, Acta Materialia*, 56 688-695
[25] Eisenmeier G, Holzwarth B, Höppel H W and Mughrabi H 2001 Cyclic deformation and fatigue behaviour of the magnesium alloy AZ91 *Materials Science and Engineering: A*, 319-321 578-582
[26] Ro Y, Agnew S R, Bray G H and Gangloff R P 2007 Environment-exposure-dependent fatigue crack growth kinetics for Al–Cu–Mg/Li *Materials Science and Engineering: A*, 468-470 88-97
[27] Potzies C and Kainer K U 2004 Fatigue of magnesium alloys *Advanced Engineering Materials*, 6 281-289
[28] Pahlavani M, Marzbanrad J, Rahmatabadi D, Hashemi R and Bayati A 2019 A comprehensive study on the effect of heat treatment on the fracture behaviors and structural properties of Mg-Li alloys using RSM *Materials Research Express*
[29] Rahmatabadi D, Pahlavani M, Bayati A, Hashemi R and Marzbanrad J 2019 Evaluation of fracture toughness and rupture energy absorption capacity of as-rolled LZ71 and LZ91 Mg alloy sheet *Materials Research Express*, 6 036517

[30] Kashyzadeh K R, Pahlavani M and Marzbanrad J 2020 High-temperature tensile behavior and high-cycle fatigue properties of Mg-7Li-1Zn alloy *Materials Today: Proceedings*
[31] ASTM E8 / E8M-16ae1 2016 Standard Test Methods for Tension Testing of Metallic Materials, *ASTM International*, West Conshohocken, PA, www.astm.org
[32] Maleki E, Unal O and Kashyzadeh K R 2018 Effects of conventional, severe, over, and re-shot peening processes on the fatigue behavior of mild carbon steel *Surface and Coatings Technology*, 344 62-74
[33] Maleki E, Unal O and Kashyzadeh K R 2018 Fatigue behavior prediction and analysis of shot peened mild carbon steels *International Journal of Fatigue*, 116 48-67
[34] Maleki E, Unal O and Kashyzadeh K R 2019 Surface layer nanocrystallization of carbon steels subjected to severe shot peening: Analysis and optimization *Materials Characterization*, 157 109877
[35] ISO/IEC 17025:2005 General requirements for the competence of testing and calibration laboratories, www.iso.org
[36] Farrahi G H, Reza Kashyzadeh K, Minaei M, Sharifpour A and Riazi S 2020 Analysis of resistance spot welding process parameters on the weld quality of three-steel sheets used in automotive industry: Experimental and finite element simulation *International Journal of Engineering, IJE TRANSACTION A: Basis*, 33(1) 148–157
[37] Reza Kashyzadeh K, Ghorbani S and Forouzanmehr M 2020 Effects of drying temperature and aggregate shape on the concrete compressive strength: Experiments and Data mining techniques *International Journal of Engineering (IJE), IJE TRANSACTIONS C: Aspects,* 33(9) 1780-1791
[38] Mousavi-Bafrouyia S S, Reza Kashyzadeh K and Khorsandijoua S M 2021 Effects of Road Roughness, Aerodynamics, and Weather Conditions on Automotive Wheel Force *International Journal of Engineering (IJE), IJE TRANSACTIONS B: Applications*, 34(2) 536-546
[39] Kashyzadeh K R and Ghorbani S 2020 Numerical study of free vibration behaviour of filled tool holder using epoxy-granite *In Journal of Physics: Conference Series*, 1687(1) 012025
[40] Maleki E, Unal O and Reza Kashyzadeh K 2019 Efficiency analysis of shot peening parameters on variations of hardness, grain size and residual stress via taguchi approach *Metals and Materials International*, 25(6) 1436-1447
[41] Kashyzadeh K R, Rahimian Koloor S S, Omidi Bidgoli M, Petrů M and Amiri Asfarjani A 2021 An Optimum Fatigue Design of Polymer Composite Compressed Natural Gas Tank Using Hybrid Finite Element-Response Surface Methods *Polymers*, 13(4) 483
[42] Ghorbani S, Ghorbani S and Reza Kashyzadeh K 2020 Taguchi approach and response surface analysis to design of a high-performance SWCNT Bundle Interconnects in a full Adder *International Journal of Engineering (IJE), IJE TRANSACTIONS B: Applications*, 33(8) 1598-1607

Characterization of fly ash based bricks, alternative to normal clay bricks in building materials and other construction

Amit Kumar Yadav and Krishna Rawat*

School of Environment and Sustainable Development
Central University of Gujarat, Gandhinagar-382030, India
Corresponding author E-mail*: krishnarwt2@gmail.com

Abstract. Fly ash is a byproduct that is released after coal burned in thermal power plants. The fly ash is capable of utilizing in value added product such as fly ash bricks used as an alternative to normal clay bricks. Fly ash has been used as a raw material for the manufacture of fly ash bricks in recent years, it seems like a logical solution that helps for the conservation of natural resources, reduce further pollution and avoid requirement of land for disposal of fly ash. In this study, the possibility of using fly ash based bricks in building materials and other construction was analyzed by studying the physico chemical parameters of the material. The fly ash bricks sample taken from the manufacturing unit to evaluate the value added capacity. For this purpose, mineralogical composition, chemical composition, structure, morphology and functional group analysis were carried out using XRD, SEM-EDS, XRF and FTIR. Results obtained confirmed better future for using Fly ash based bricks as alternative material to be used in building materials and other construction. Keywords: Fly ash bricks, characterization, building materials.

1. Introduction

With the rapid growth of industrialization, urbanization and development demand of electricity has been increase day by day in India. In India, most of the electricity is produced from the coal-fired power plant which produces a large amount of fly ash. Fly ash is a fine powder which is a by-product of the combustion of pulverized coal in thermal power plants which is mainly composed of ferro-aluminosilicate materials. All over the world, coal-fired power plants are the major sources of electricity that generate large amount of fly ash as a by- product during the combustion of coal. About 160 million tones of fly ash are produced annually in India, of which around 63% is utilized beneficially. The rate of production of fly ash exceeds with the rate of increasing growth rate of electricity production of its user. The disposal and management of fly ash is a major problem in coal-fired power plants. In India, most of the fly ash is disposed of in ash ponds and landfills (Bhattacharjee and Kandpal, 2002) and only 63% is utilized. Disposal of fly ash from coal-fired power station poses environmental problems. With growing concerns about pollution and rising landfill costs, the study on the utilization and application of fly ash has increased worldwide. The target of 95% use of fly ash likely brings in to existence in the upcoming next ten years. The generation of huge amount of fly ash has increased the demands for utilization of fly ash as a value-added product, such as bricks and in other construction materials (Yadav and Fulekar, 2018). In recent years, the use of fly ash has increased in different industries for many economic, technological and environmental reasons. In construction industries mainly clay bricks were used where top layer of soil was excavated for making bricks, due to which soil erosion, emission from burning coal or wood heating causes deforestation are the serious problems posed by brick industry. Due to the continued use of clay bricks in the construction industry, a great loss of the top fertile soil will occur (Bansode, 2012). The materials required for the fly ash bricks such as fly ash, gypsum and lime are easily available near each other in many areas. Gypsum and lime are commonly available from mineral sources or can be obtained from industrial wastes (Verma and Gadling, 2016). High fineness, low CaO content, rounded particle and spherical shape are in general, favorable properties for use in construction materials (Kumar, 2002). Fly ash has pozzolanic property that becomes cementitious when combined with water, however be considered for use in the construction materials (Wesche, 2005, Giuseppe and Eduardo, 2009, Rawat and Yadav, 2020). Fly ash is used as an essential additive in some sectors to improve the properties of final products, such as the special pozzolan additive in the concrete and cement industry (Ahmaruzzaman, 2010). Fly ash is also used as a raw material in the production of bricks. Therefore, a study was conducted to assess the characterization of fly ash bricks and assess their potential use in construction materials.

The aim of this study is to investigate the possibility of using fly ash bricks as an alternative to normal clay bricks. Analysis such as X-ray diffraction (XRD), Scanning electron microscopy-energy-dispersive spectroscopy (SEM-EDS), X-ray fluorescence spectroscopy (XRF) and Fourier transform infrared spectroscopy (FTIR) were carried out to characterize the fly ash bricks. Based on the results of the analysis, the morphology, functional group and the chemical composition of fly ash bricks was studied.

2. Experimental

A. Samples

The samples used in this study included fly ash bricks (FAB). For characterization studies, FAB samples were collected from a nearby manufacturing site, Gurukrupa Fly Ash Bricks and Acta Cem (Gandhinagar) is located at opposite Divine Child School, Sabarmati-Kalol Highway, Nr Narmada Canal, Adalaj, Ahmedabad. The samples were put in polyethylene bags and were transported to the laboratory.

B. Treatment of FAB samples and preparation

FAB samples were brought to laboratory and air dried. The FAB samples used for this investigation were dried at 110°C-160°C in hot air oven for 24 hours, homogenized and sieved through 2 mm sieve and stored in an air tight container in laboratory for later use. Fly ash bricks (FAB) sample were crushed and made fine powdered in a mortar.

C. Characterization of FAB samples

Dried representative samples were ground less than 100 μm and characterization of FAB samples were performed by using different equipments/techniques such as X-Ray diffraction (XRD), Scanning electron microscopy/Energy-dispersive spectroscopy (SEM-EDS), X-Ray fluorescence spectroscopy (XRF) and Fourier transform infrared spectroscopy (FTIR).

i) X-ray diffraction (XRD) analysis:
X-ray powder diffraction is the most widely used technique to study the mineralogical composition, different phases and characteristics of crystalline structure of materials. X-ray diffractometry is attractive for its speed and ease of performance, because it requires only small amounts of material and also because it is non-destructive. The kinematic theory (which considers the scattering from each atom to be independent of the scattering from all other atoms and that once scattered, the X-rays pass beyond without further scattering) of X-ray diffraction is used for fly ash materials in fine powder form (Cullity, 1978). The crystalline structure and qualitative mineralogical composition of the samples were determined by conducting High Resolution X-Ray Diffractometer (HR-XRD) system using a Rigaku Rint 2200 powder diffractometer equipped with a Cu Ka radiation source generated at 30 kV, 15 mA and a graphite monochromator.

ii) Scanning electron microscope and energy dispersive spectroscopy (SEM-EDS) analysis:
A high resolution scanning electron microscope (HR-SEM) with energy dispersive spectroscopy (EDS) was used to study the morphology, texture and elemental composition of FAB samples. SEM was used for the characterization of micro structural changes (porosity, cavities, and pits), determination of particle size and morphology of compacts sample. Scanning Electron Micro analyzer was used to take the micrograph of the sample. The sample was mounted on aluminum stubs with conductive glue, and then covered with a thin layer of carbon. Magnetic fraction samples were gold-coated for SEM observation with qualitative EDS analysis. The EDS was used to determine the elemental composition of manually chosen areas in the fly ash particles. Qualitative and quantitative analyses of the elemental spectra were performed using aWD-5 software program.

iii) X-ray fluorescence (XRF) analysis:
X-ray fluorescence technology has been used to determine the chemical composition of materials. This technology provides one of the simplest, most accurate and economical analytical methods for determining the chemical composition of many types of material. It is a fast, precise, cost-effective, multi-elements, non-destructive and reliable technique that requires no or very little sample preparation and is suitable for powdered, solid and liquid samples. XRF spectroscopy is widely used for quantitative and qualitative elemental and chemical analysis. A Horiba XGT-2700 analytical X-ray microscope equipped with an X-ray tube with an Rh target detector, a high purity silicon detector (XEROPHY) was used. A vacuum was used a means of analyses to avoid the interaction of X-rays with air particles. Samples were ground to <75 μm in a tungsten carbide grinding vessel. Then, a mixture of 1g sample and 6g $Li_2B_4O_7$ was melted on a glass slide. The analyses of measure elements were carried out on the fused bead.

iv) Fourier transforms infrared (FTIR) analysis:
The FTIR spectroscopic technique is used to understand the surface chemistry of fly ash bricks in thermally active state as well as different state of the mineral phases, H_2O and –OH group on silica and alumina. This method has been widely used for the qualitative determination of the major components of minerals and can be considered as a finger print for identification. The features of the structure, like family of the mineral, nature of isomorphic substitution, extent of structural regularity, distinction of molecular water form constitutional water, and the presence of impurities (crystalline and non-crystalline), etc., can be well documented by IR study (Farmer, 1979). In the present work, the Perkin Elmer FTIR spectrometer was used to record the FTIR spectra of the samples at room temperature. FTIR was used in identification of functional groups and chemical bonds of samples to calculate the transmission percentage of infrared.

3. Results

Fly ash bricks characterization

A. XRD analysis of FAB samples

X-ray diffraction analysis was carried out to obtain mineral composition present in the fly ash brick materials. XRD patterns of FAB shown in Fig.1. It shows that the bricks materials are mainly X-ray amorphous in character where the diffraction crystals were those of the original materials, mainly berlinite. An amorphous hump is observed in the diffraction diagram between 2θ values of approximately 20° to 30°, which could be due to the presence of amorphous vitreous/glassy materials. The minerals identified in FAB are shown in Table 1.

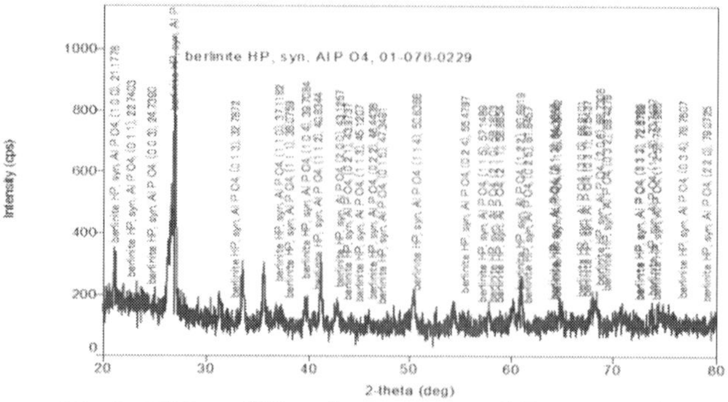

Fig.1: X-Ray diffraction patterns of FAB sample

Table 1: Mineral identified in FAB sample

Mineral Name	Chemical Formula
Berlinite	$AlPO_4$

B. Surface topography of FAB sample (SEM analysis)

The surface topography of the FAB materials was identified by scanning electron microscopic (SEM). The SEM micrographs are shown in Fig.2. SEM image of the FAB reveals the particles are spherical in shape and were homogeneous. The micro structure of FAB powder also shows a complex formation of molecules indicating the absence of porous formation. The brick-powder particles were fused well together and cavities were present in them.

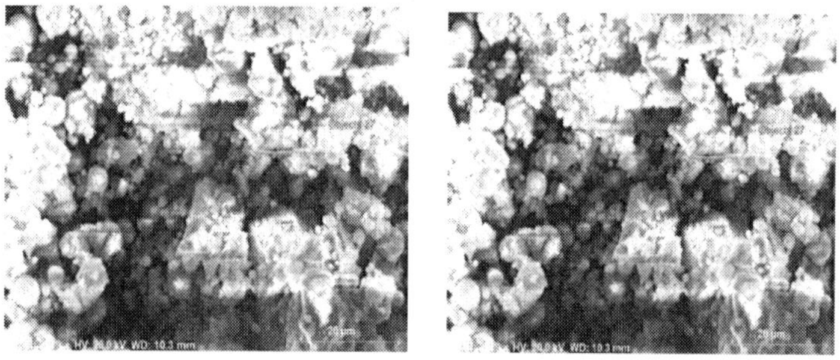

Fig. 2: SEM image of FAB sample

C. Elemental composition of FAB sample from SEM-EDS analysis

EDS spectra show the presence of C, O, Si, Al, Fe, Na, Ti, Mg, Na, K and Cl as the major elements in the FAB samples (Fig.3). The weight % and atom % of the FAB samples, as measured by EDX, are given in Table 2. In this case, O is the predominant element, but C, Si and Al vary widely from one type of FAB samples.

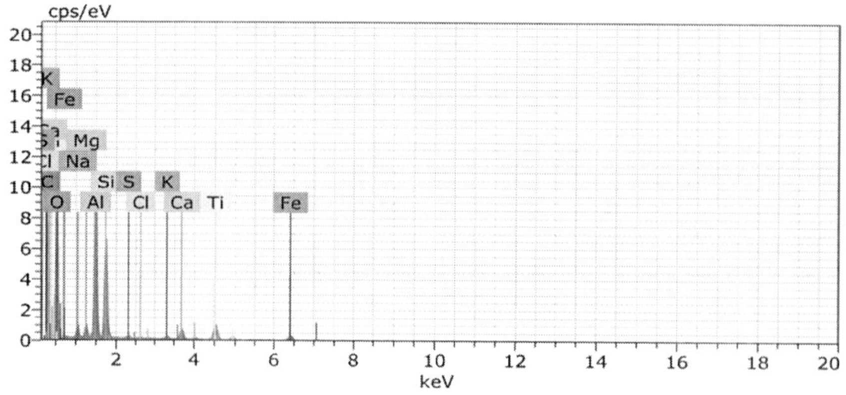

Fig. 3: EDS spectra of FAB sample

Table 2: Elemental composition of FAB sample

Element	Weight %	Atomic %
C	4.12	6.39
O	62.29	72.50
Si	8.29	5.49
Al	14.49	10.00
Fe	2.02	0.67
Na	2.01	1.63
Ti	3.52	1.37
Mg	1.36	1.04
Ca	1.53	0.71
K	0.19	0.09
Cl	0.11	0.06

The elemental predominance of the FAB materials is O > Al > C> Si > Na > Ti > Mg > Ca > Fe > K > Cl. Fly ash bricks is elementally composed of a high concentration of oxygen, aluminum, carbon and silicon with lighter amount of iron, sodium, magnesium, titanium, potassium and calcium.

D. XRF measurements

The compositions of the FAB samples obtained from XRF are given in Table 3. The relative distribution of the oxides in the FA Brick materials is also shown in Fig. 4. XRF spectra show the presence of Si, Al, Fe, Na, Ti, K, Ca, Mn, V, Zn, Sr, Cr, Co, Mg, Ni, Cu, As and Br as the bulk oxide of major elements in the FAB.

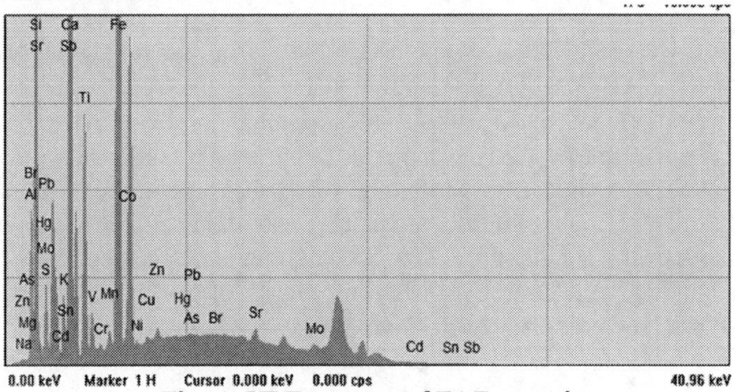

Fig. 4: XRF spectra of FAB sample

Table 3: Chemical composition of FAB sample

Compound	**Mass [%]**
SiO_2	54.86
Al_2O_3	19.79
Fe_2O_3	7.38
Na_2O	0.73
TiO_2	2.37
K_2O	0.76
CaO	11.61
MnO_2	0.081
V_2O_5	0.029
SO_3	1.29
ZnO	0.035

SrO	0.034
Cr_2O_3	0.036
CoO	0.025
MgO	0.88
NiO	0.012
CuO	0.029
As_2O_5	0.005
Br_2O	0.004
LOI	1.04
Cr_2O_3	0.036

From the results, the following conclusions may be made about the compositions of the FA Brick materials:

FAB samples have high loss on ignition (LOI) value, indicating that the sample still contained some amount of volatile materials (must be organic in nature). The chemical composition is dominated by silica which constitutes nearly 54% by wt of the total composition. Overall, the composition of the oxides in the FA Brick materials in the order of $SiO_2 > Al_2O_3 > CaO > Fe_2O_3 > TiO_2 > SO_3 > MgO > Na_2O > K_2O > MnO_2 > Cr_2O_3 > ZnO > SrO > CuO \sim V_2O_5 > CoO > NiO > As_2O_5 > Br_2O$. SiO_2, Al_2O_3, Cao and Fe_2O_3 are the four most important constituents of FAB materials and they together make up 93 % by wt of the FAB. The composition of the minor oxides, MgO, Na_2O, K_2O, MnO_2, Cr_2O_3, ZnO, SrO, CuO, V_2O_5, CoO, NiO, As_2O_5 and Br_2O show some variation in the FA Bricks, as seen above.

The composition of FAB sample is shown in Table 3. From the table it can be observed that the main constituents are silica such as SiO_2 about 54%, alumina as Al_2O_3 about 19%, calcium oxide like CaO about 11% and iron oxide as about 7% Fe_2O_3. The chemical composition of FAB depends mainly on the composition of fly ash, gypsum and calcium oxide.

E. FTIR results

An FTIR investigation was conducted to obtain information on the functional groups of FAB samples. The FTIR results of the FAB powders are shown in Fig.5. Table 4 lists the position of wave numbers for the molecular vibrations of the various bonds present in the FAB sample with their possible assignments. The IR spectrum of FAB sample shows transmission bands at 3420, 1484, 1084, 874, 855, 795 and 692cm^{-1}. The significant broad bands were observed at the 3420, 1484 and 1084 cm^{-1} for stretching vibrations of OH, asymmetric stretching vibration of SiO and asymmetric stretching vibration of AlO respectively. Peaks at the 874, 855 and 795 cm^{-1} respectively, attributed to the Si-OH stretching, Al-O-H bending, and stretching vibration quartz of Si–O–Si. In addition, the peak at 692 cm^{-1} was associated with the presence of Si-O vibration.

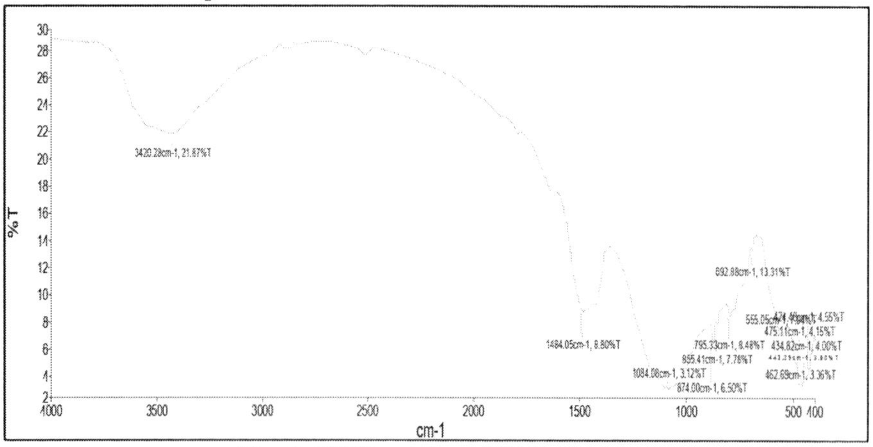

Fig. 5: FTIR spectra of FAB sample

Table 4: Band assignments for different minerals of FAB sample

Frequency(cm^{-1})	Tentative assignment
3420	stretching vibrations of O-H
1484	asymmetric stretching vibration of Si–O
1084	asymmetric stretching vibration of Al–O
874	Si-OH stretching
855	Al-O-H bending
795	stretching vibration quartz of Si–O–Si
692	Si-O vibration

4. Discussion

In present study the fly ash bricks materials were characterized by different advance instrumentation and their characterization briefly discussed.

X-ray diffraction analysis was performed to obtain crystalline nature and mineral composition present in the fly ash bricks materials. XRD patterns of FAB in Fig.1. The XRD patterns of FAB show that the brick materials are mainly X-ray amorphous in nature where the diffraction crystals were those of the original materials, mainly berlinite. An amorphous hump is observed in the diffraction pattern between 2θ values of approximately 20° to 30°, which could be due to the presence of amorphous vitreous/glassy materials.

The surface topography of the fly ash bricks materials was identified by scanning electron microscopic (SEM). The SEM micrographs of FAB are shown in Fig.2. SEM image of the FAB reveals the particles are spherical in shape and were homogeneous. The micro structure of the FAB powder also shows a complex formation of molecules indicating the absence of porous formation. The brick-powder particles were fused well together and cavities were present in them.

The elemental composition of the FAB materials was studied by EDS. EDS spectra showed the presence of O, Si, Al, and Ca as the major elements in the fly ash bricks samples (Fig.3). The weight % and atom % of the FAB samples, as measured by EDS, are given in Table 2. FAB samples are elementally composed of a high concentration of oxygen, aluminum, carbon and silicon with lighter amount of iron, sodium, magnesium, titanium, potassium and calcium (Fig.3). The elemental predominance of the FAB samples is O > Al > C> Si > Na > Ti > Mg > Ca > Fe > K > Cl. In this case, O is the predominant element, but C, Si and Al vary widely from one type of FAB samples.

The compositions of the fly ash bricks materials obtained from XRF are given in Table 3. The relative distribution of the oxides in the FAB materials is also shown in Fig.4. XRF spectra show the presence of Si, Al, Fe, Ti, K, Ca, Mn, Zn, Sr, Cr, Mg, Ni, Cu, As and Pb as the bulk oxide of major elements in the FAB materials. The chemical composition of fly ash bricks is dominated by silica which constitutes nearly 50% by wt of the total composition. Overall, the composition of the oxides in the fly ash bricks materials in the order of $SiO_2 > Al_2O_3 > CaO > Fe_2O_3 > TiO_2 > SO_3 > MgO > Na_2O > K_2O > MnO_2 > Cr_2O_3 > ZnO$. SiO_2, Al_2O_3, Cao and Fe_2O_3 are the four most important constituents of fly ash bricks materials and they together make up 80 to 90 % by wt of the FAB materials. The chemical composition of FAB depends mainly on the composition of fly ash, gypsum and calcium oxide.

FTIR investigation was conducted to obtain information on functional groups of FAB materials. The FTIR result of the FAB materials is shown in Fig.5 while Table 4 lists the position of the wave numbers for the molecular vibrations of the various bonds present in the FAB materials with their possible assignments. The assignment of bands to different structural characteristics is done according to Farmer, 1979. The IR spectrum of FAB shows the significant broad bands were observed at the 3420, 1484 and 1084 cm^{-1} for stretching vibrations of O-H, asymmetric stretching vibration of Si–O and asymmetric stretching vibration of Al–O respectively. Peaks at 874, 855 and 795 cm^{-1} respectively, attributed to the Si-OH stretching, Al-O-H bending and stretching vibration quartz of Si–O–Si (Lyon, 1967). In addition, the peak at 692 cm^{-1} was associated with the presence of Si-O vibration.

5. Conclusions

In this study, FAB can be used as an alternative to normal clay bricks in building materials and other construction. Results obtained in this investigation show that the nature of fly ash bricks was amorphous and the highest concentrations of the SiO_2 were present in the fly ash bricks. XRD analysis showed that the maximum percentage of the FAB sample is in the amorphous glass phase and that the main crystalline phases are berlinite. XRF analysis revealed that FAB particles mainly consist of silica and alumina with less percentage of Fe_2O_3, Cao and others. SEM images revealed the amorphous nature of the FAB and a clear difference in the morphology of the samples. The spherical morphology and small particle size of FAB can improve its applicability as a filler in polymers. The EDS spectra showed the percentage distribution of the three principal elements, O, Si and Al, in the FAB samples. FTIR measurements indicated that fly ash bricks sample showed shifts in frequencies of major bands and also modification of their intensities.

Acknowledgements

Authors would like to thank the University Grant Commission (UGC), Govt. of India, New Delhi for providing financial support in the form of JRF and SRF to carrying out this research work. Authors also thank the authorities of the CIF-IIT Delhi, CSIR-CECRI, CIF-CUG for providing XRD SEM-EDS, XRF and FTIR analysis results.

Conflict of interest

The authors declare that they have no conflict of interest.

References

1. Ahmaruzzaman, M. (2010). A review on the utilization of fly ash. Progress in energy and combustion science, 36(3), 327-363.
2. Bansode, S. S. R. (2012). Comparative Analysis between Properties of Steel Slag, Fly Ash, and Clay Bricks" Geo-Congress (ASCE) pp. 3816-3825.
3. Bhattacharjee, U., and Kandpal, T. C. (2002). Potential of fly ash utilisation in India. Energy, 27(2), 151-166.
4. Cullity, B. D. (1978). Elements of X-ray Diffraction.
5. Farmer, V. C. (1979). Infrared spectroscopy. Data Handbook for Clay Materials and other Non-metallic Minerals, 285-337.
6. Giuseppe, C., Sebastian, E. (2009). Fly ash addition in clayey materials to improve the quality of solid bricks" published in Journal of Construction and Building Material 23, pp.1178-1184.
7. Kumar, S. (2002). A perspective study on fly ash–lime–gypsum bricks and hollow blocks for low-cost housing development" Construction and Building Materials 16, pp. 519–525.
8. Lyon, R. J. P. (1967). Physical Methods in Determinative Mineralogy, edited by Zussman J., 371-403.
9. Rawat, K. and Yadav, A. K. (2020). Characterization of coal and fly ash (generated) at coal based thermal power plant. Materials Today: Proceedings, 26(2), 1406-1411.
10. Varma, M. B. and Gadling, P. P. (2016). Additive to Cement –A Pozzolanic Material-Fly Ash" International Journal of Engineering Research ISSN: 2319-6890(online), 2347-5013(print) Volume No.5 Issue: Special 3, 27-28 Feb. pp. 558-564.
11. Yadav, A. K. and Fulekar, M. H. (2018). Characterization of fly ash from coal thermal power station. International Journal of Technical Innovation in Modern Engineering & Science (ITIMES), 4(6), 775-783.
12. Wesche, K. (2005). Fly Ash in Concrete. Edition published in the Taylor & Francis e-Library, ISBN 0-203-62641-9 Master e-book ISBN, pp. 168.

Chemically synthesized ZnO and Cd-ZnO thick films as Ethanol sensor

Yogita Patil[1], R. B. Pedhekar[2], Swaraj Patil[3], Swapnil Kosalge[4], F. C. Raghuvanshi[5]

[1]Government Engineering College, Jalgaon. 425002
[2] M. J. Fule Commerce, Science and V. Raut Arts College, Bhatkuli, Dist. Amravati 444602
[3]COE, Pune 411005,
[4]Guru Nanak College of Arts, Science and Commerc, Mumbai-400037
[5]S. G. B. Amravati University, Amravati 444602

Abstract: In this paper, we report precipitation method for preparation of pure and Cd-ZnO nanorod using different doing concentration. Thick films are fabricated by screen printing technique. For structural, morphological, compositional and optical characteristics of fabricated thick films X-ray diffraction technique (XRD), field-emission scanning electron microscopy (FESEM), energy dispersive X-ray technique and UV–visible spectroscopy were used. To check gas sensing properties, hydrogen sulphide (H_2S), carbon dioxide (CO_2), liquid petroleum gas (LPG), ammonia vapour (NH_3), ethanol vapour and hydrogen (H_2) gases were used. The excellent gas response was obtained for 3.0 wt.% Cd- ZnO towards ethanol among all Cd-ZnO thick films at operating temperature 573 °K.
Keywords: Cd doped ZnO, nanorods, gas sensor

1. Introduction:

Ethanol is high odorous gas and is explosively used for scientific, industrial applications and for beverages. Ethanol is toxic in nature [1-5]. There is heavy risk of cancer upon disclosure and use of alcoholic beverages [6]. There is a risk of digestive track cancer and respiratory system for the workers working for the synthesis of ethanol. Commercially important area is accurate detection of ethanol vapour. One of the applications of ethanol sensor is to detect ethanol vapour level in human breathing (breathalyser) which is linked to the concentration of alcohol in blood.

In recent years, the study of gas sensors has given great attention in order to detect harmful gases because of their effect on medical diagnosis, personal security, detection of toxins and pollutants, industrial transport [7, 8]. It is known that gas sensing performance of gas sensors not only depends on their gas response, selectivity, response or recovery speed and stability but also its reproducibility. Hence, sensors with low cost, increasing sensitivity and selectivity are in demand for safety in an industrial areas and residential areas [9–15], semiconductor metal oxide gas sensors are in demand now a day. They are widely used for detection of inflammable and poisonous gases [16–22] because of their low cost. From literature review, pure metal oxide materials have lower gas sensing performance as compared with doped metal oxide materials [23-26]. Metal oxide semiconductors, SnO_2, ZnO, Fe_2O_3 etc. are commonly used as gas sensing materials from ten years back because of their enhanced sensitivity to the target gases and simple synthesis process. Researchers have showed that the crystallite size, surface morphology and chemical composition of metal oxide nanoparticles are related to their gas sensing properties [27-29].

Zinc oxide is associated with II–VI group semiconductor with larger band gap and larger exciton energy of 60 meV [30]. The properties of ZnO such as its chemical stability, piezoelectricity, and visible region optical transparency, biocompatibility and high voltage current non-linearity etc. make it's position unique among all the materials. The resistivity of ZnO can be varied in large range (10^{-3} Ωcm to 10^5 Ωcm) it is used in many industries in solar cells, transport electrodes, photodiodes, gas sensors, surface acoustic wave devices and transistors, etc [31-35]. The high chemical and physical stability, low cost, easy methods of preparation are the cause of attraction in the practical gas sensors [36]. Sensing mechanism of ZnO is surface control type of reaction. There is change in resistivity of metal oxide semiconductors with adsorption/desorption for gas molecules of target gas on material's surface. By decrease in concentration of adsorbed oxygen, resistance of zinc oxide sensor decreases and electron concentration increases when exposed to target gas. The surface states are correlated to grain size and surface morphology [37-39], both are important for gas sensing performances.

In the meantime, the different morphologies, porosity and impurity-doping are two effective ways to considerably increase the gas sensing response of ZnO [40-46]. In majority of the papers, ZnO is doped with noble metals that are platinum (Pt) and palladium (Pd) for improving gas performances of sensor [47-49]. In many papers, the properties of pure ZnO and doped ZnO have been reported but the results of Cadmium doping in zinc oxide is less reported.

The present work deals with synthesis of ZnO and Cd-ZnO by precipitation method, screen printing fabrication method for preparation of thick films and studies on their various properties such as structural, morphologic, optical and gas sensing for H_2S, CO_2, LPG, Ammonia, Ethanol vapour and H_2 gases.

2. Experimental

2.1 Synthesis process for ZnO and Cd- ZnO

Zinc nitrate hexahydrate [$Zn(NO_3)_2.6H_2O$] and cadmium nitrate tetrahydrate [$Cd(NO_3)_2.4H_2O$] used in the experiment were of AR grade. Precipitation method was used for preparing ZnO and Cd-ZnO nanorod. Pure ZnO and cadmium doped ZnO were synthesized by dissolving certain amount of $Zn(NO_3)_2.6H_2O$ and $Cd(NO_3)_2.4H_2O$ in deionized water. The doping concentration was varied as 1, 3, 6, 9 and 12 wt.%, these solutions were stirred magnetically for half an hour for homogeneous solution of zinc and cadmium precursors. Sodium hydroxide solution was prepared in deionised water with continuous stirring, heated at about 328 °K. At room temperature sodium hydroxide solution added in prepared precursor solution at high speed constant stirring. The precipitates were obtained in this reaction and washed with absolute ethanol and deionised water many times to remove the contaminations and then obtained powders were dried in oven at 343 °K. Then the prepared powders were annealed at 523 °K. and named as pure zinc oxide, 1wt.% Cd - ZnO, 3wt.% Cd - ZnO, 6wt.% Cd - ZnO, 9wt.% Cd - ZnO and 12 wt.% Cd -ZnO.

2.2. Fabrication and gas sensing measurements of thick films

Ethyl cellulose mixed with mixture organic solvents such as butyl carbitol acetate, butyl cellulose and turpineol for preparing thixotropic paste [50]. Then the prepared samples were mixed with this paste. The inorganic to organic part was maintained as 80:20 while formation of the paste. The material binding with class substrate was performed by this temporary binder. The screen printing technique was used for fabricating the thick films. The uniform pressure was applied while robbing the paste on glass substrate. . The made-up thick films were then fired at 773 °K for half hour.

In Gas sensing system silver electrodes were used for electrical contacts between thick films and gas sensing system. The experiment was performed for fixed gas concentration 600 ppm. The gas sensor response was noted in presence of air and the target gases and the sensor response was calculated as the ratio of resistance of the film in air (R_a) to its resistance in target gas (R_g) [51].

$$S = R_a/R_g \qquad (1)$$

3. Results and discussion

3.1. Characterization of thick film sensors

3.1.1. Structural analysis by XRD:

For examining crystalline structure of pure and Cd-ZnO, XRD analysis was performed.

Figure 1 shows the XRD patterns of (a) pure ZnO, (b) 1 wt.% Cd -ZnO, (c) 3 wt.% Cd -ZnO, (d) 6 wt.% Cd -ZnO ,(e) 9 wt.% Cd -ZnO and (f)12 wt.% Cd -ZnO.

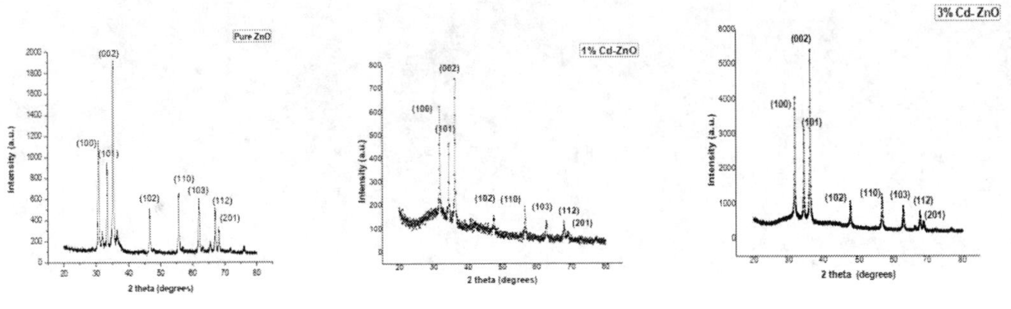

(a) (b) (c)

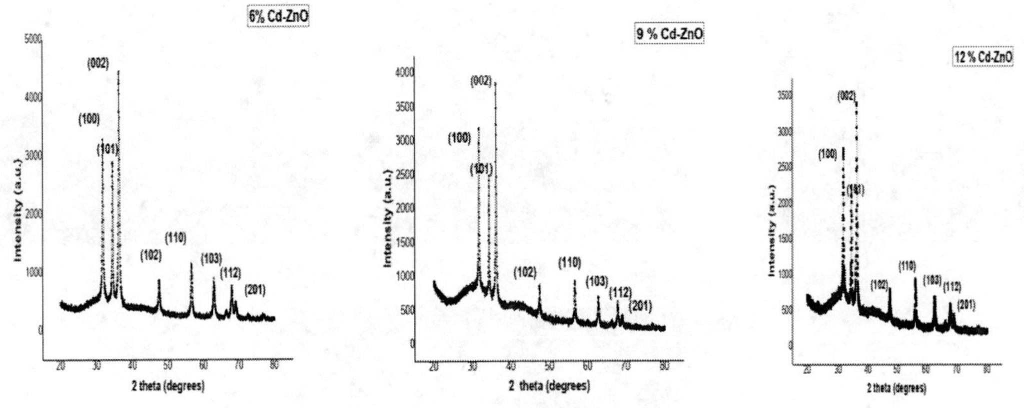

(d) (e) (f)

Fig. 1:XRD pattern of (a) ZnO, (b) 1 wt.% Cd-ZnO, (c) 3 wt.% Cd-ZnO, (d) 6 wt.% Cd-ZnO , (e) 9 wt.% Cd-ZnO and (f)12 wt.% Cd-ZnO.

From XRD spectras, polycrystalline nature of the thick films are having wurtzite crystal structure with a preferential oriented growth along (0 0 2) plane. No significant detection of Cd or CdO phase is observed in all films as Cd content is very small. Single ZnO phase is observed for all thick films. From XRD spectras, it is observed that intensities of peaks (002) decrease with slight broadening of full width half maxima. For 3% Cd doped ZnO maximum intensity of peak (002) is 5401a.u. which is greater than intensity of 002 peak of other Cd doped ZnO thick films which suggest that 3% Cd doped ZnO thick film is more crystalline as compared to other Cd doped ZnO thick films. In addition to this, increase of Cd content increases FWHM, lattice defects and grain boundary defects due to Cd doping [52, 53]. The crystallite size was calculated by using Debye Scherrer equation [54]:

$$\mathbf{D} = \frac{0.9\,\lambda}{\beta\ \mathrm{Cos}\ \theta} \qquad (2)$$

In above formula, λ= 1.5406 A°, β = Full width at half maxima, θ =Bragg's angle

The average crystallite size of fabricated films was of the order of 24 nm - 42nm. As the doping cadmium doping concentration in zinc oxide increases, the grain size is observed to be decreased.

3.1.2. Morphological analysis FE-SEM

Figure 2 shows FE-SEM images of (a)ZnO, (b)1wt. % Cd-ZnO, (c)3wt. % Cd-ZnO, (d)6wt. % Cd-ZnO, (e)9wt. % Cd-ZnO, (f)12 wt. % Cd-ZnO thick films.

Fig.2: FE-SEM pictures of (a) pure ZnO, (b) 1 wt.% Cd -ZnO, (c) 3 wt.% Cd -ZnO, (d) 6 wt.% Cd -ZnO, (e) 9 wt. % Cd -ZnO and (f)12 wt.% Cd-ZnO, (g) Size of 3 wt.% Cd –ZnO

The images show that there is formation of nano-rod like structures for both pure and Cd doped ZnO thick film samples and Cd-doped ZnOnanorods are obviously very small than pure ZnO nanorods. The nanorods are randomly distributed in samples and the diameter and length of the nanorods changes with change in concentration of Cadmium doping. The maximum length and diameter of the rod is 203 nm and 26.3 nm respectively (fig. g). As concentration of doping increases, length of the rods decreases. Creation of sub nanometre crystallites of various sizes are also observed in SEM images that shows agglomeration in Cd doped thick films.

3.1.3. Elemental analysis by EDAX

The elemental analysis of pure and Cd-ZnO thick films were investigated by using EDAX. Fig. 3 shows EDAX patterns of ZnO and Cd-ZnO thick films. From these graphs, it can be concluded that no other impurity elements are present except Zn, O and Cd.

IOP Conf. Series: Materials Science and Engineering 1126 (2021) 012046 doi:10.1088/1757-899X/1126/1/012046

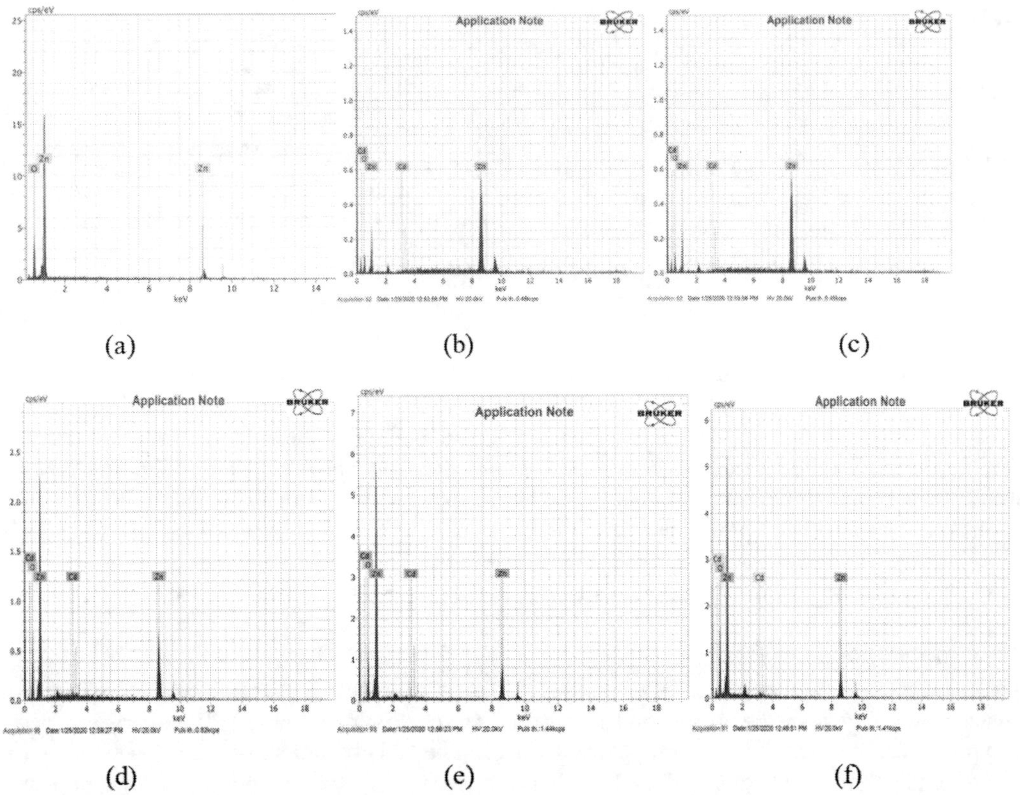

Fig. 3. EDAX of (a) ZnO, (b) 1 wt. % Cd-ZnO, (c) 3 wt.% Cd-ZnO, (d) 6 wt.% Cd-ZnO, (e) 9 wt.% Cd-ZnO, (f)12 wt.% Cd-ZnO.

From figure 3, it was observed that sharp signals were found for zinc and oxygen which confirms the formation of ZnO nanoparticles. For 1wt. %, 3wt. %, 6wt. %, 9wt. %, and 12wt. % along with zinc and oxygen, Cadmium signals are also found which confirms the Cd doping in ZnO. As the Cd doping percentage is increasing, the intensity of peaks in EDAX is increased.
The elemental composition of the ZnO and Cd-ZnO are shown in table 1.

Table 1: EDAX of ZnO and Cd-ZnO thick films

Element (mass %)	Cd doping in ZnO					
	0 wt. %	1 wt. %	3 wt. %	6 wt. %	9 wt. %	12 wt. %
Zn	68.88	76.46	76.87	66.04	68.18	60.17
O	31.12	22.63	19.59	27.96	23.55	28.68
Cd	00	0.91	3.54	6.00	8.27	11.15

3.1.4 Optical Properties by UV-vis. Spectroscopy:

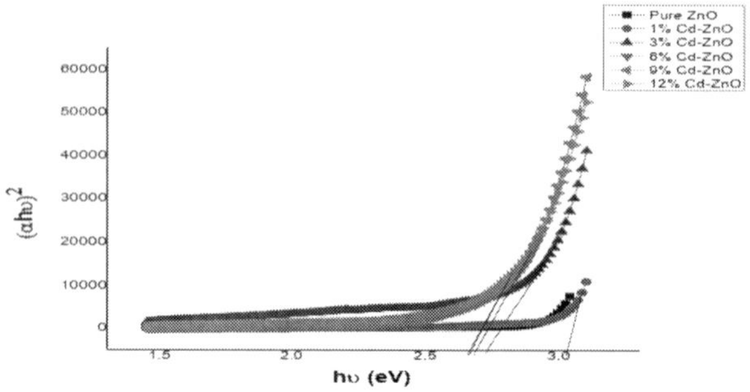

Fig. 4 .Absorbance spectras of $(\alpha h\nu)^2$ versus (hν) for ((a) ZnO, (b) 1 wt. % Cd-ZnO, (c) 3 wt.% Cd-ZnO, (d) 6 wt.% Cd-ZnO, (e) 9 wt.% Cd-ZnO and (f)12 wt.% Cd-ZnO.

Figure 4 depicts Absorbance spectres of pure and Cd -ZnO thick films. The pure ZnO and Cadmium doped ZnO , both are the materials with direct band gap [55]. The band gap energies calculated by extrapolating the linear portion of an $(\alpha h\nu)^2$verses (hν) graph. Thus from these results, it can be concluded that the band gap is decreased with increase in Cd concentration in ZnO thick films.

3.2 Study of gas response:

3.2.1 Current-voltage measurements:

Figure 5 depicts volt-ampere characteristics of pure and cadmium doped zinc oxide thick films. By increasing biased voltage the corresponding current was observed and noted after every 5 volts. The same procedure was carried out for negative voltage. The I-V characteristics graph was plotted for both positive and negative voltages and the I-V characteristics curves obtained in the graph for pure and cadmium doped ZnO are same in nature. Hence it can be concluded that the silver contacts made between probes of the static gas sensing system and the thick films having ohmic contact [56].

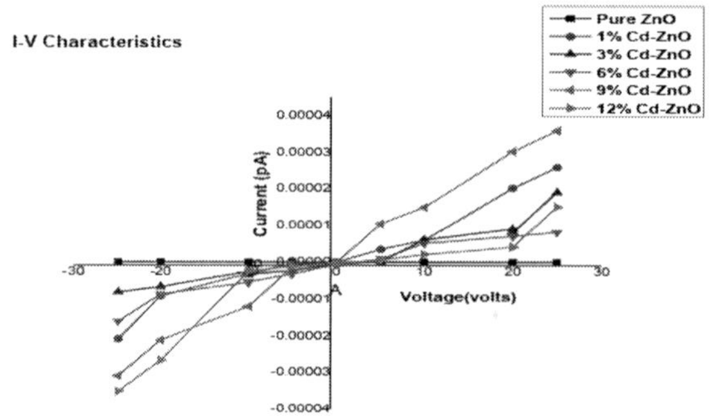

Fig.5: Current voltage graph of (a) ZnO, (b) 1 wt.% Cd-ZnO, (c) 3 wt.% Cd-ZnO, (d) 6 wt.% Cd-ZnO , (e) 9 wt.% Cd-ZnO, (f)12 wt.% Cd-ZnO thick films.

3.2.2Electrical conductivity:

Figure 6 indicates electrical conductivity of ZnO, 1, 3, 6, 9 and 12 wt.% Cd–ZnO thick films. It is observed that as operating temperature increase, electrical conductivity increase. The semiconducting nature of ZnO and Cd-ZnO thick films is confirmed from the linear increase in conductivity.

IOP Conf. Series: Materials Science and Engineering 1126 (2021) 012046 doi:10.1088/1757-899X/1126/1/012046

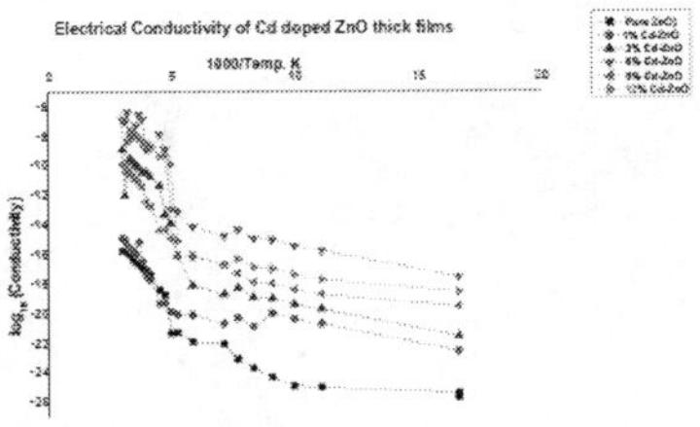

Fig. 6: Conductivity- temperature profile of (a)ZnO, (b) 1 wt.% Cd-ZnO, (c) 3 wt.% Cd-ZnO, (d) 6 wt.% Cd-ZnO ,(e) 9 wt.% Cd-ZnO and (f)12 wt.% Cd-ZnO films.

3.2.3 Gas Sensing:

Figure 6(a-f) gives the gas response of ZnO and Cd-ZnO thick films as a function of operating temperature upon exposure of 600ppm H_2S, CO_2, LPG, Ammonia, Ethanol and H_2 gases. From figure it was seen that Cd doped ZnO thick films are respond to Ethanol vapour than other gases H_2S, CO_2, LPG, Ammonia, and H_2 gases. Among all Cd - ZnO thick films, 3 wt.% Cd doped ZnO thick films showed maximum response to Ethanol vapour at 573 K. The higher gas response of 3 wt.% Cd doped ZnO may be due to larger effective surface area of 3 wt.% Cd doped ZnO thick film to react with gas molecules and optimum porosity of film. There may be optimum number of CdO misfits in 3 wt. % Cd doped ZnO thick film to adsorb oxygen.

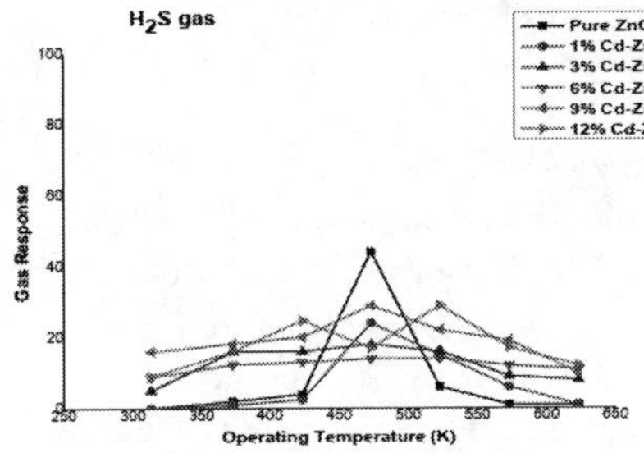

(a)

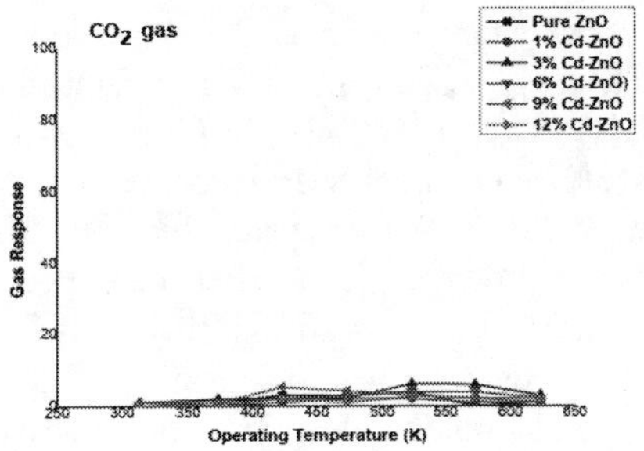

(b)

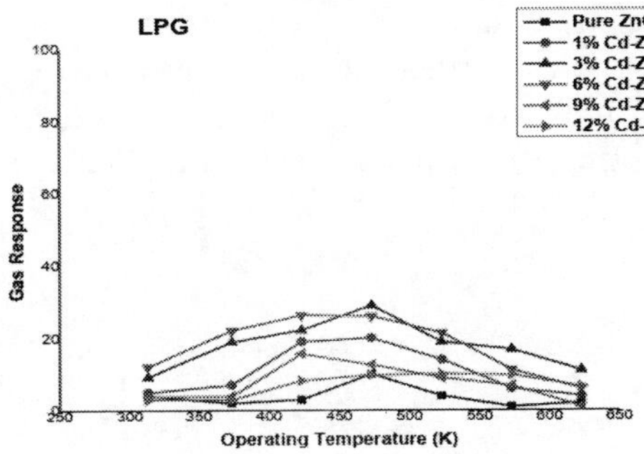

(c)

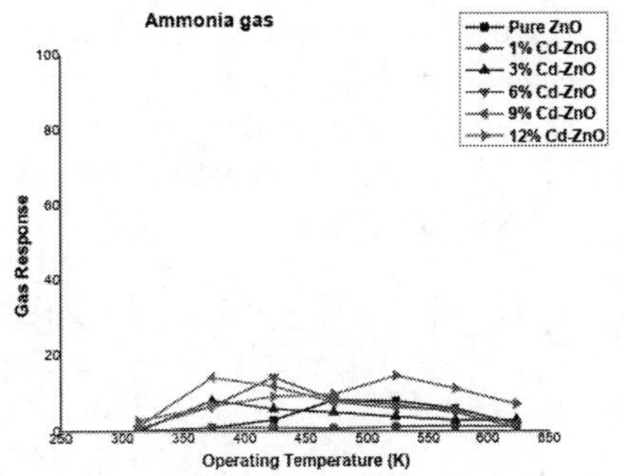

(d)

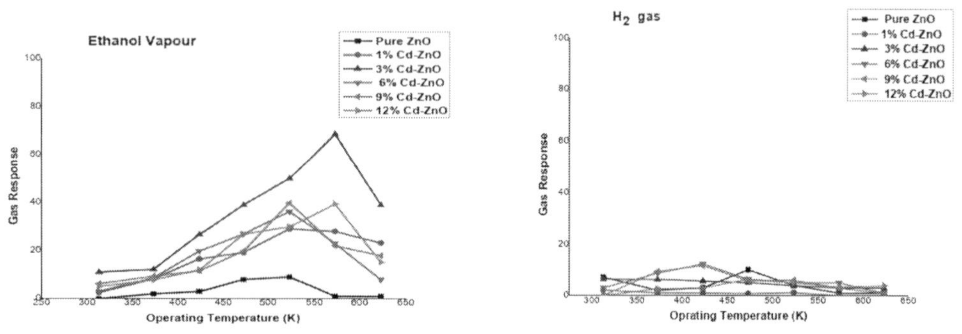

(e) **(f)**

Fig. 7 (a-f): Gas response of ZnO and Cd-ZnO thick films for 600 ppm H_2S, CO_2,LPG, NH_3, Ethanol, and H_2 gas.

3.2.4 Ethanol vapours sensing mechanism:

The gas sensing mechanism to detect ethanol vapours can be described by two different ways. The first way includes formation of CH_3CHO intermediate due to oxidation of ethanol by dehydrogenation and second way includes formation of C_2H_4 due to dehydration of ethanol. The acidbase properties of metal oxide surface helps to determine the selectivity of these two reactions. For basic oxide surfaces, dehydrogenation process is suitable while for acidic surfaces dehydration process is favoured. Dehydrogenation process take place at higher temperatures while dehydration processes takes place at lower temperatures. The possible reactions are as follows [57, 56]:

$$CH_3CH_2OH_{(gas)} + [O]_{(ads)} \rightarrow CH_3CHO(ads) + H_2O \quad (1)$$

$$CH_3CH_2OH_{(gas)} + [O]_{(ads)} \rightarrow C_2H_4 + H_2O \quad (2)$$

Where surface oxygen species are represented by $O_{(ads)}$. Ethanol dehydrogenates and the formation of acetaldehyde (CH3CHO) took place by dehydrogenation of ethanol at about 573 K and adsorbed hydrogen atom is produced [56]. Then oxidation of adsorbed hydrogen occurs at ZnO surface. The produced protons associate with the surface-adsorbed oxygen ions forming OH(ads).Elimination of a H_2O molecule occurs due to condensation of two adsorbed OH groups, leaving an adsorbed oxygen anion. In the conduction band of ZnO, one net electron is released during the above dehydrogenation process and thus reduction in resistance take place [57]. The adsorbed O on the sensor surface reacts with the ethanol and releasing electrons back to the conduction band of the material which contribute to the decrease in resistance of the sensor material [58, 59].

In air ambient, as temperature increases, activation energy decreases and hence it can be concluded that less energy is required to activate the material at higher temperature in air ambient. But in gaseous medium, as operating temperature increases activation energy also increases and hence upon exposure of ethanol, larger energy is acquired by electrons. The sufficient energy releases for trapped electrons due to decomposition of ethanol molecules to jump into the conduction band of 3% Cd-ZnO thick film.

$$C_2H_5OH + 6\,O^- \rightarrow 2\,CO_2 + 3\,H_2O + 6\,e^- \quad (3)$$

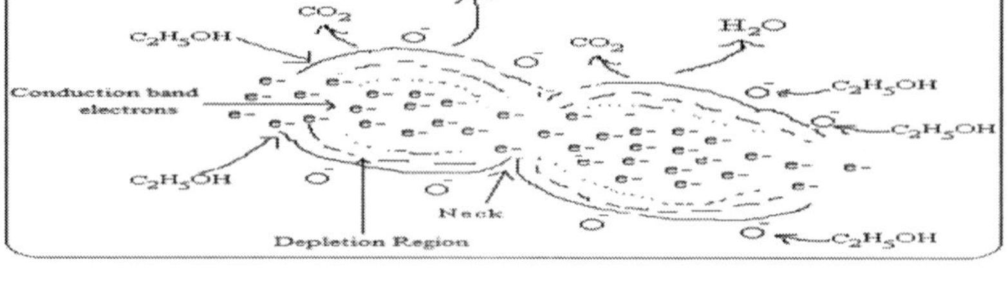

3.2.5 Ethanol vapour seelctivity compared to H_2S, CO_2, LPG, Ammonia, and H_2 gas:

From Figure 8, it is observed that 3 wt.% Cd-ZnO sensor provides maximum response to ethanol vapours at 573 K. The thick film sensors indicated highest selectivity for ethanol among all other targeted gases viz. H_2S, CO_2, liquid petroleum gas and H_2.

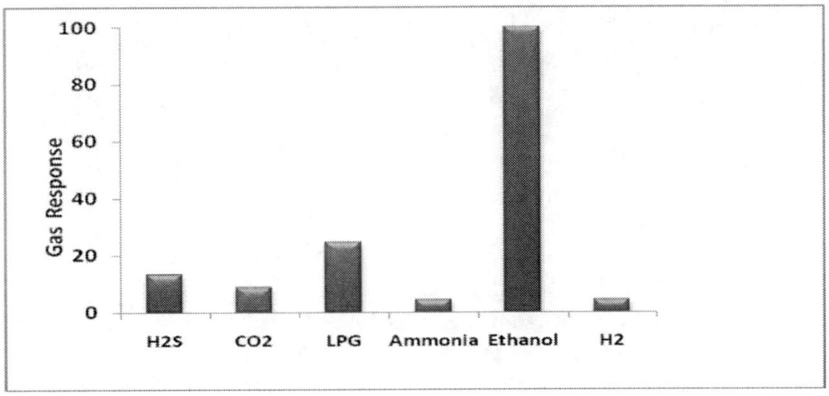

Fig. 8 : Gas response of 3 wt.% Cd-ZnO film towards H_2S, CO_2,LPG, Ammonia. Ethanol and H_2 at 573 °K.

3.2.6 Response time and Recovery time of thick film sensor

The response time of 3 wt. % Cd-ZnO film and recovery time 3 wt. % Cd-ZnO film are represented in Figure 9. The response was quick (~ 19 s) for 600 ppm concentration of ethanol while the recovery time was fast (~ 30 s). Because of the fast oxidation of gases, the response may be fast.

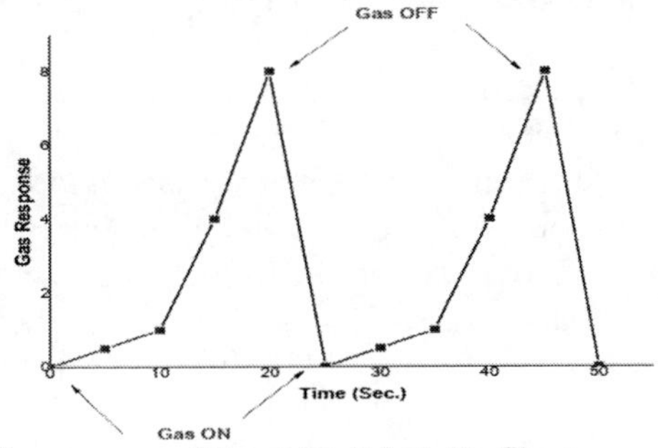

Fig. 9: Response and recovery of 3 wt. % Cd-ZnO film towards Ethanol at 573 °K.

4. Conclusion:

Following are the conclusions are drawn on the basis of experimental results

- Pure ZnO was less sensitive to ethanol vapours.
- Cd doping in ZnO is good to promote the ethanol sensing.
- 3 wt% Cd -ZnO was found to be most sensitive to ethanol vapour at 573 °K.
- 3 wt. % Cd-ZnO film respond ethanol highly at 573 °K temperature with quick response time and rapid recovery.
- 3 wt. % Cd-ZnO film was found to be selectivity for ethanol vapour at 573 °K against H_2S, CO_2, liquid petroleum gas and H_2.

References:

[1]. Solomons, Graham 2007, *John Wiley and Sons*, 497-498.
[2]. Jeong, Hae In, Seonyoung Park, Hae In Yang, and Woong Choi. 2019, *Materials Letters* 253 209-212.
[3]. D. R. Patil, L. A. Patil, D. P. Amalnerkar, 2007, Bull. Mater.Sci.-Springer link 30, 553-559.
[4]. R. B. Pedhekar, F. C. Raghuwanshi, 2017 , *International Journal of Engineering, Science Invention.* 6(8), 20-28
[5]. G. S. Sodhi, 2002, *Narosa Publishing House, New Delhi*, 1, 135.
[6]. R. A. Michaels, Environ. 1999, *Health Perspective*, 107, 617-627.
[7]. J J. Kong, N.R. Franklin, C. Zhou, M.G. Chapline, S. Peng, K. Cho, H. Dai, 2000, *Science* 287, 622–625.
[8]. A. Kolmakov, M. Moskovits, Annu. 2004, *Rev. Mater. Res*. 34, 151–180.
[9]. C. Hagleitner, A. Hierlemann, D. Lauge, A. Kimmer, N. Kerness, O. Brand, H. Baltes, 2001, *Nature* 414, 293–296.
[10]. Y. Cui, Q.Q. Wie, H.K. Park, C.M. Lieber, 2001, *Science* 293, 1289–1292.
[11]. Z.W. Pan, Z.R. Dai, Z.L. Wang, 2001, Science 291, 1947–1949.
[12]. R. B. Pedhekar, F. C. Raghuwanshi, 2015, *International journal of pure and applied research in engineering and technology*. 3(9), 30-38.
[13]. C. Li, D.H. Zhang, X.L. Liu, S. Han, T. Tang, J. Han, C.W. Zhou, 2003, *Appl. Phys. Lett.* 82, 1613–1615.
[14]. H.T. Wang, B.S. Kang, F. Ren, L.C. Tien, P.W. Sadik, D.P. Norton, S.J. Pearton, J. Lin, 2005, *Appl. Phys. Lett.* 86, 243503.
[15]. I. Heller, A.M. Janssens, J. Mannik, E.D. Minot, S.G. Lemay, C. Dekker, 2008, *Nano Lett.* 8, 591–595
[16]. K. Park, K.U. Jang 2005, *Journl of Alloys and Compounds* 391, 123–128.
[17]. K. Arshak, I. Gaidan, 2005, *Materials Science and Engineering B* 118, 44–49.
[18]. L. Satyanarayana, K.M. Reddy, S.V. Manorama, 2003, *Materials Chemistry and Physics* 82, 21–26.
[19]. S.I. Rembeza, E.S. Rambeza, T.V. Svistova, O.I. Borsiakova, 2000, *Physica Status solidi (a)* 179, 147.
[20]. J.G. Partridge, M.R. Field, J.L. Peng, A.Z. Sadek, K. Kalantar-zadeh, J. Du Plessis, D.G. McCulloch, 2008, *Nanotechnology* 19, 125504 5.
[21]. K. Mukherjee, S.B. Majumder, 2010, *Electrochemical and Solid State* 13, J25–J27.
[22]. S.K. Biswas, P. Pramanik, 2008, Sensors and Actuators B 133, 449–455.
[23]. R. B. Pedhekar, F. C. Raghuwanshi, G. N. Raut, 2014, *International journal of pure and applied research in engineering and technology.* 2(9), 127-135.
[24]. M. Zhao, X. Wang, L. Ning, J. Jia, X. Li, L. Cao, 2011, *Sensors and Actuators B* 156 , 588–592.
[25]. N.J. Dayan, S.R. Sainkar, R.N. Karekar, R.C. Aiyer, 1998, *Thin Solid Films* 325, 254–258.
[26]. N. Han, Y. Tian, X. Wu, Y. Chen, 2009, *Sensors and Actuators B* 138, 228–235.
[27]. R.B. Pedhekar, F.C. Raghuwanshi, V.D. Kapse, 2016, *Mater. Sci. Poland* 34 (3), 571–581.
[28]. W. Chung , C. Shim , S. Choi , D. Lee, 1994, *Sensor Actuat. B-Chem.* 20, 139.
[29]. V. D. Kapse, S. Ghosh., G. Chaudhari., F. C. Raghuvanshi, 2008, *Talanta*, 76, 610 .
[30]. M. Mihailovic, A.-L. Henneghien, S. Faure, P. Disseix, J. Leymarie, A. Vasson, D.A. Buell, F. Semond, C. Morhain, J. Zuniga Perez, 2009, *Opt. Mater*. 31, 532–536.
[31]. D.R. Sahu, Shin-Yuan Lin, Jow-Lay Huang, 2006, *Appl. Surf. Sci*. 252, 7509–7514.
[32]. S. Ilican, Y. Caglar, M. Caglar, M. Kundakci, A. Ates, 2009, *Int. J. Hydrogen Energy* 34, 5201–5207.
[33]. M.C. Carotta, A. Cervi, V. di Natale, S. Gherardi, A. Giberti, V. Guidi, D. Puzzovio, B. Vendemiati, G. Martinelli, M. Sacerdoti, D. Calestani, A. Zappettini, M. Zha, L. Zanotti, 2009, *Sensors Actuators B* 137, 164–169.
[34]. Alivov, Ya I., et al. 2005, *Superlattices, and Microstructures* 38.4-6 , 439-445.
[35]. D.-J. Yun, S.W. Rhee, 2009, *Thin Solid Films* 517, 4644–4649.

[36]. S. H. Lee, H. J. Lee, H. Shiku, T. Yao, T. Matsue, 2012, *Electron. Mater. Lett.* 8, 511–513.
[37]. M. H. Cao, Y. D. Wang, T. Chen, M. Antonietti, M. Niederberger, 2008, *Chem. Mater.* 20, 5781–5786.
[38]. Y. D. Wang, T. Chen, Q. Y. Mu, G. F. Wang, 2009, *Scr. Mater.* 61, 935–938.
[39]. Z. H. Jing, J. H. Zhan, 2008, *Adv. Mater.* 20, 4547–4551.
[40]. S. P. Huang, T. Wang, Q. Xiao, 2015, *J. Phys. Chem. Solids* 76, 51–58.
[41]. W. Wen, J. M. Wu, Y. D. Wang, 2012, *Appl. Phys. Lett.* 100, 262111.
[42]. W. Wen, J. M. Wu, Y. D. Wang, 2013, *RSC Adv.* 3, 12052–12055.
[43]. W. Wen, J. M. Wu, Y. D. Wang, 2013, *Sens. Actuat. B: Chem.* 184, 78–84.
[44]. A. Patil, C. Dighavkar, R. Borse, J. Optoelectron. 2011, *Adv. Mater.* 13, 1331–1337.
[45]. R. Dhahri, S. G. Leonardi, M. Hjiri,2017, *Sens. Actuat. B: Chem.* 239, 36–44.
[46]. R. Dhahri, M. Hjiri, L. El Mir, J. 2016, *Phys. D: Appl. Phys.* 49, 135502.
[47]. M. Shafiei, J. Yu, R., Arsat, K. Kalantar-zadeh, E. Comini,M. Ferroni, G. Sberveglieri, and W. Wlodarski, 2010, *Sensors and Actuators B: Chemical,* 146(2), 507-512.
[48]. Y. S., Huang, Y. Y Chen,., &T. T. Wu, 2010. *Nanotechnology,* 21(9), 095503.
[49]. L. C., Tien, P. W. Sadik, D. P. Norton, L. F. Voss, S. J. Pearton, H. T. Wang, B. S. Kang, F. Ren, J. Jun, and J. Lin. 2005, *Applied Physics Letters* 87, no. 22, 222106.
[50]. Yogita Patil, R.B. Pedhekar, Swaraj Patil, F.C. Raghuwanshi. 2020, *Materials Today: Proceedings,* 28, 1865 – 1871.
[51]. R. B. Pedhekar, F. C. Raghuwanshi, V. D. Kapse, 2015, *International Journal of Materials Science and Engineering.* 3, 219-230.
[52]. J. Cembrero, Elmanouni, Hartiti, M. Mollar,Marí, 2004, *Thin Solid Films* 451, 198.
[53]. B. Marí, M. Tortosa, M.Mollar, J.V. Boscà, H.N. Cui, 2010, *Optical Materials* 32, 1423–1426
[54]. M. Kashif, M.E.Ali, Syed M. Usman Ali and U. Hashim, 2013, *Ceramics International,* 39(6), 6461-6466
[55]. H. Derraz-Tabet, N. Benramdane, D. Nacer, A. Bouzidi, M. Medles, 2002, *Sol. Energy Mater. Sol. Cells* 73, 249
[56]. R. B. Pedhekar, F. C. Raghuwanshi, V. D. Kapse, 2015, *International Journal of Materials Science and Engineering.* 3, 219-230.
[57]. B. B. Rao, 2000, *Mater. Chem. Phys.* 64, 62-65.
[58]. H. W. Cheong, H. W.; Lee, M. J. J. 2006, *Ceram.Process. Res.* 7, 183-191.
[59]. A. V. Rajgure, L. K. Bagal, R. C. Pawar, I. S. Mulla, S. S. Suryavanshi, 2013, *Ceramics International,* 39(4), 4383-4390.

Modeling of sediment transport of Capo Negro bay

Ayad Hassan; Mouhid Mustapha; Mouakkir Laila;
El Bouni Brahim; Chagdali Mohamed

Polymer Physics and Critical Phenomena Laboratory, Faculty of Sciences Ben M'sik, P.O.BOX 7955, University Hassan II, Casablanca, Morocco.
ayadhassan01@gmail.com; mouhidmustapha@gmail.com

Abstract. The coastline is a curve / line representing the intersection between land and sea. The interaction between the atmosphere, the sea and the continent give the coastal fringe a multiphasic character that is very complex and difficult to manage. The concentration of human activity in coastal areas requires looking at problems that may endanger beach stability, coastal development, and consequently loss of human heritage. The most widespread problem on a global scale is erosion, especially with the current trend of rising sea levels. And given the complexity of this process and for better management of coastal areas, mathematical modeling is an indispensable tool.

In this work, we study the evolution of the Cabo Negro coastline and present a methodology based on two complementary approaches: first, an oceanographic, granulometric and sedimentological study of the site. Secondly, numerical modeling to analyze the evolution of the coastline and the prediction of its long-term position by using a mathematical model: the GENESIS calculation model.

.**Keywords:** Numerical modeling, sediment transport, GENESIS, coastline.

1. Introduction

The coast (or littoral) defines the strip of land between a maritime area and the mainland. Depending on the scales selected, this area can range from a few hundred to several meters. These limited, coveted and attractive spaces, favorable to the various flows (trade, travel, tourism, etc.) receive a large part of the world's population. This coactualization is an ancient process but has taken on an important and global dimension. According to Bird [1985] [1] more than 70% of the world's sandy or gravel beaches are eroded. These beaches play an important role in human development and may show a retreat of several meters per year. Thus, all the cities and activities located near these coasts are directly threatened.

The retreat of the coasts can be explained by the accumulation of different natural and anthropogenic effects [Paskoff, 1998a] [2]. Schematically, the position of the coastline can be seen as a balance between the maritime and land sedimentary contributions, on the one hand under the action of waves and tides and on the other hand transported by rivers and wind.

Many researchers work on the development of modeling of the coastal sediment and changes in beach morphology. Margvellashi et al. (2008) [3] developed one-dimensional vertical and three dimensional

fine-resolution numerical models of sediment transport. Nam et al. (2009) [4] proposed a two-dimensional numerical model of nearshore waves, currents and sediment transport.
The objective of this study, which focuses on the coastline near Cabo Negro (Martil Bay) about 15 km north of Tetouan in northern Morocco, is to correctly model and predict the evolution of the coasts in order to better Manage and protect them. To answer this question, our study concerns the observation of the natural bay and the modeling of these beaches by mathematical model by using the coastline evolution calculation code: GENESIS.

2. Theoretical formula

The bay of Martil extends about 9 km from the groin of Cabo Negro in the north to sidi Abdessalam in the south.

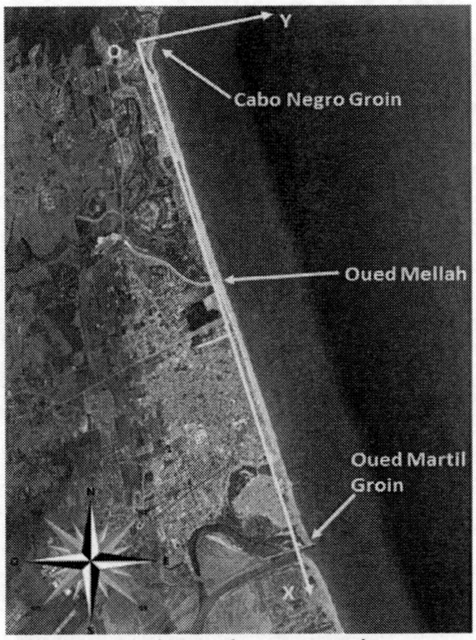

Figure 1. site study presentation

The hydrographic network of the bay is composed by the two wadis:

- The Wadi Martil, whose sea mouth position was moved northward in the late 1970s with the construction of a left-handed groin;
- Wadi Mellah which flows into the sea, the mouth of this wadi is not fixed.

Since the beginning of the 80s and even before, this bay has been under a severe sand minin (2 100 000 m³ since 1981) in various points: Sidi Abdessalam, Oued Martil mouth, oued Mellah mouth. In 1992, 107 000 m³ of sand was removed from the south of the bay.
After the construction of the waterfront protecting the residential area in 1992, the width of the beach at the right of developments has dropped sharply, a phenomenon that, if continued, may endanger the sustainability of the urbanized area.

2.1. Winds

The tide is of semi-diurnal type (period 12h30). The tide characteristics are given in the following table:

Table 1. The tide in the Cabo Negro area

	Low tide level (m NGM)	High tide level (m NGM)
MVE	-0.60	+0.47
MVEE	-0.88	+0.70

2.2. Swell

The swell regime was established on the basis of an analysis of vessel observations made between 1960 and 1980 in the maritime area at the site.

- The main points emerging from the analysis of these observations are:
- Calm accounts for 60% of observations (219 days / year),
- Swells greater than or equal to 3 m have a frequency of 2% (7 days / year)
- The eastern sector is predominant in frequency of occurrence and it is also from this sector that swells with higher heights come.

2.3. Granulometry of the Study Area

The beach and the seabed of this zone are constituted by sands. The spatial distribution of the particle sizes along 4 profiles between Cabo Negro and the Wadi Martil is given in the following figure:

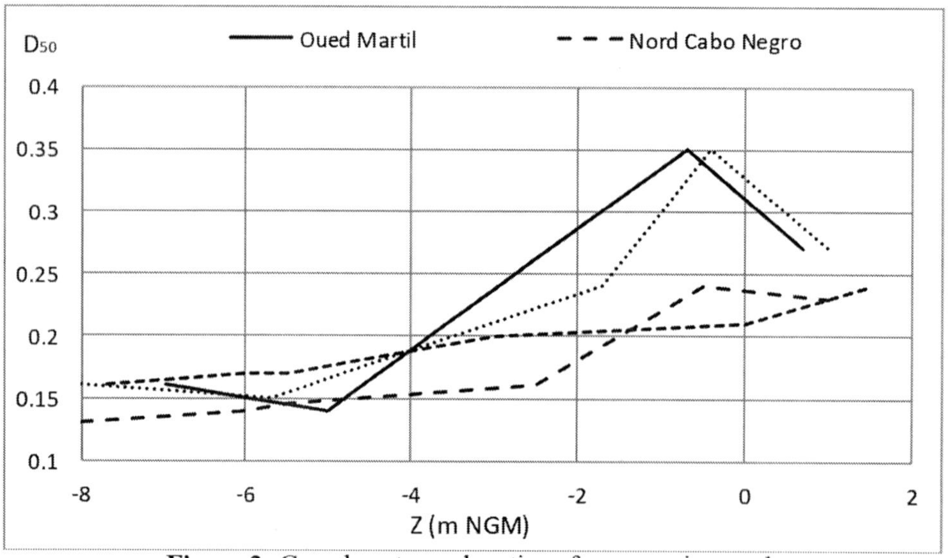

Figure 2. Granulometry and sorting of area marine sands

These curves show:
A refinement of D50 of the foreshore sands from south to north which indicates a resulting transit in this direction.
A granulometric sorting in the profile, the D50 passing from 0.25-0.35 on the foreshore to 0.15 mm in bottoms of -6 to -8 m, such sorting is characteristic of the action of the waves.

2.4. Sediment transit generated waves

Swell (waves) is the essential driver of sediment movements.
The movements in the profile (cross-shore) have a virtually zero result. These movements decrease rapidly with depth. It can be considered that from the bottoms of -5 m the sedimentary movements are very weak.
The longshore transit which is the transport generated by the waves when they reach the coast with a certain incidence has been evaluated on the basis of the swell regimes by the use of the GENESIS coastline evolution software.
The orders of magnitude of the coastal transit that emerge from this estimate are as follows:

Table 2. Coastal Transit in the Cabo Negro Region

Transit (m^3/y)	
Southward	40 100
Northward	(-) 55 800
Total	95 900
Resulting (from South to North)	(-) 15 700

Martil Bay receives sand inputs from the Martil Wadi which are less than 10 000 m^3 /year, before the construction of the two groins (Martil and Cabo Negro) and sand mining, the resulting transit from the south to the north was 15,000 to 25,000 m^3 / year.
Thus, in total, the bay was in dynamic equilibrium and the long-term evolutions had to be very weak. There could, however, be seasonal advances and retreats.
Human interventions have chronologically consisted of:

- The construction of Cabo Negro groin at the beginning of the sixties, which resulted in an advance of the coastline to the south of groin,

- The construction of the Martil groin in the late 1970s with the aim of fixing a new mouth of the wadi Martil,

Sand extraction the south of Wadi Martil (1,250,000 m^3 since 1981), at Wadi Mellah (900,000 m^3 between 1986 and 1992).
The study on a mathematical model of the coastline evolution in the area between the Wadi Martil groin and that of Cabo Negro will allow to specify the evolution if on the one hand no development is carried out And on the other hand for the construction of a new groin at the mouth of the Oued mellah.

3. Modeling of the coastline

The longshore sediment transport along Martil bay has been modeled considering the existent coastal structures with the use of GENESIS numerical model (Hanson and Craus, 1991) [5]. GENESIS model was developed by US Army Corps of Engineers and it is one of the most common used sediment models in the literature. It is applied for the sediment transport due to wave effects. Several numbers and combinations of groins, breakwaters, and seawalls can be examined by GENESIS model. The effects of combined T and Y type groins can be investigated and the sand erosion and deposition around groins are taken into account for different wave heights, wave periods and incident wave angles. But GENESIS model has some limitations such as neglecting wave reflection on the structures, tides.
The model covers a coastal zone of 7300 m extending from the Cabo Negro groin in the north to the Wadi Martil groin in the south. The model was built in local landmark Oxy.
The calibration of the model consisted of reproducing the evolution of the bay between 1981 and 1993. At the Wadi Mellah level, 793 000 m^3 of sand were extracted between 1986 and the end of 1991 and 107 000 m^3 between April and May 1992.
After performing the calibration. This model can be used to predict future developments for various management scenarios. To this end, two scenarios have been established, each one consisting of

reproducing the evolution of the coastline for 10 annual cycles (10 years), starting from the current state of the coastline. The two tests are carried out as follows:

- Scenario 1: Evolution in 10 years if no development is carried out and if there is no sand removal at the Wadi Mellah.
- Scenario 2: construction of a sand-stop groin to depths of -5m with a beach nourishment of 250 000 m³ to the north of the jetty.

4. Result and discussions

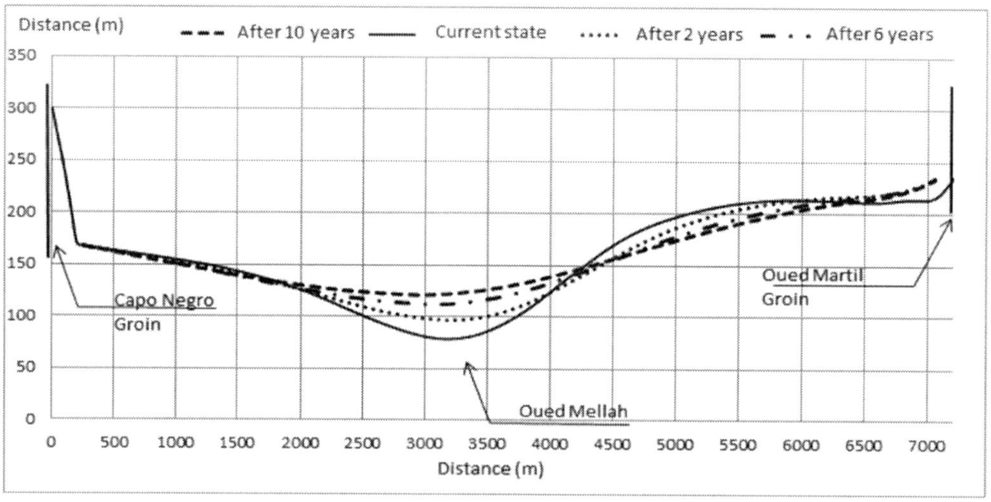

Fig 3. coastline evolution for Scenario 1

This test gives the expected evolution within 10 years if no other development is carried out (maintenance of the structures in their present situation) and if the sand excavation is totally stopped. The results of this trial show that the coastline will tend to realign with:

- Progress (advance) located on the 500 m north of the wadi Martil groin with a peak that exceeds 15 m to disappear 1000 m north of the cob.
- A coastline retreat of more than 20 m in a zone 700 m south of Wadi Mellah, this decline decreases in directions, north and south.
- Progress of the coastline in the zone ranging from 200 m to the north of the wadi Mellah up to 700 m in the south of the wadi Mellah. This advance can reach 45 m at the level of Wadi Mellah.
- Slight tendency to retreat in the immediate south of Cabo Negro ear, less than 5 m over 10 years.

Thus, if nothing is done (no development, no input and no excavation), the zone between the coordinates 2000 and 4500 would progress at the expense of the zone that is further south (figure 3).

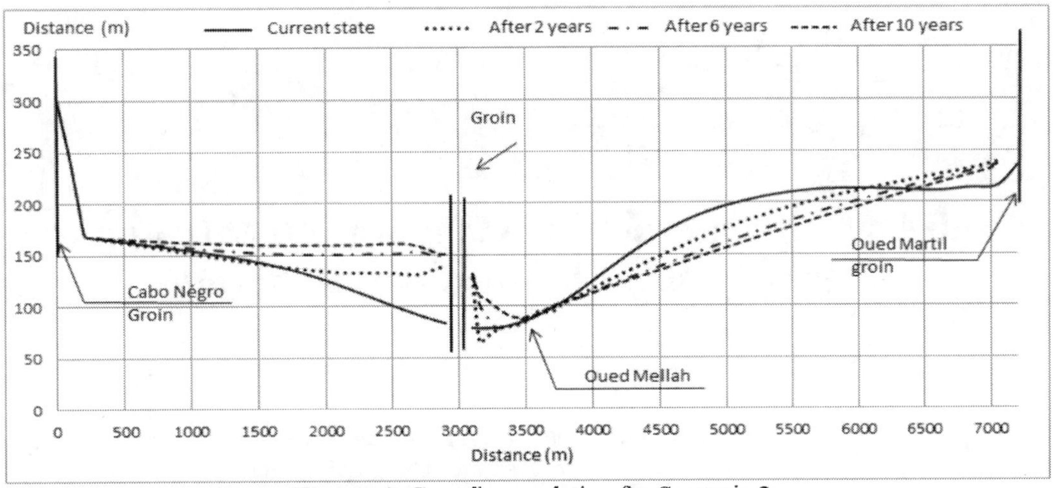

Figure 4. Coastline evolution for Scenario 2

For the scenario 2, we observe the following changes:

- A 40 m erosion south of Wadi Mellah, this retreat decreases on both sides of the peak point located at 2000 m south of the southern jetty.
- An advance to the south of the groin due to the effect of swells diffraction.
- Sand deposits allow advancing of the coastline more than 60 m at the north of the groin, this advance decrease going north to the Cabo Negro groin.

5. Conclusion

The evolution and morphology of beaches change under the effects of coastal structures, wind, current and waves. Coastal structures and human interventions significantly affect the natural balance of sediment transport in coastal areas. To study sediment transport and shoreline changes, it is necessary to fully understand the coastal hydrodynamics, waves and the current system in the region. The coastal circulations and long-term shoreline changes of Capo Negro Beach located at north of morocco, have been modeled by numerical models. The GENESIS model was applied after calibration to the coastal zone. Numerical simulations were able to provide reliable representations of the coastline trend. These representations show that if no remedial intervention (stopping sand mining, breakwaters construction, beach nourishment…) has been carried out, the Bay of Capo Negro risks undergoing severe changes in its morphology especially with the growth of its urbanization.

References

[1] E. C. F Bird. Coastline changes. John Wiley and Sons, New York, 1985.

[2] R. Paskoff. La crise des plages : pénurie de sédiment. MappeMonde, 52 :1115, 1998a.

[3] Margvelashvili, N., Saint- Cast, F., Condie, S., 2008, Numerical modeling of the suspended sediment transport in Torres Strait, Continental Shelf Research, 28 (16), 2241-2256.

[4] Nam, P.T., Larson, M., Hanson, H., Hoan, L.X., 2009., Numerical model of nearshore waves, currents, and sediment transport, Coastal Engineering, 56 (11-12), 1084-1096.

[5] Hanson, H. and Kraus, N.C., 1991. GENESIS: Generalized model for simulating shoreline change: Report 2, Technical reference, CERC-89- 19, Vicksburg.

IOP Conf. Series: Materials Science and Engineering 1126 (2021) 012048 doi:10.1088/1757-899X/1126/1/012048

Modeling wave interaction in long wave theory during its spread over a stratified porous bottom

Mouhid Mustapha; Ayad Hassan; Mouakkir Laila; Mordane Soumia El Bouni Brahim; Chagdali Mohamed

Polymer Physics and Critical Phenomena Laboratory, Faculty of Sciences Ben M'sik, P.O.BOX 7955, University Hassan II, Casablanca, Morocco.

mouhidmustapha@gmail.com

Abstract. In this work, we are interested in the study of the interaction between a gravity wave and a porous media made up of multi layers of different geometrical characteristics. The dispersion relation and the reflection coefficient are calculated and analysed according to the porosity of the porous media. A generalization of these relations is proposed in the case of long waves. In terms of application, we analyse the most adequate configuration for better behaviour of the porous media under the gravity wave. The wave is calculated within the potential theoretical framework and the Forchheimer model is selected in order to take into account the inertial aspect of the flow in the porous media.

Keywords: Interaction; Wave; Porous media; Forchheimer model; Analytical Formulation; Long Wave Theory

1. Introduction

When the swell propagates on seabed's, considered as porous media, which permeability induces internal flow and non-negligible flux through the porous medium interface - external medium. This type of problem has been the subject of several investigations in coastal engineering. The theory of SOLITT & CROSS (1972) [1] served as a theoretical basis for modeling diverse problems in this area. This approach has also been applied to long wave analysis by MADSEN (1983) [2]. LEE and LAN (1995) [3] have developed a generalized method for multi-layers and multi-region porous media. BROSSARD and al. (2004) [4] have proposed, by an analytical approach validated experimentally in a wave flume, the dispersion relation of the flow in a porous monodisperse and rigid media. In this work, the dispersion relation is generalized to a multi-layer laminated (stratified) porous media based on the hypothesis of the long wave propagation hypothesis.

Forchheimer model is used as a theoretical model. This model is adapted to unsteady flows and it allows to take into account the friction forces and the terms of inertia. We proceed by a first-order linear analytical approach by adopting the Lorentz principle (SOLITT & CROSS, 1972).

2. Theoretical formula

A porous medium consisting of several horizontal layers in the presence of the swell is shown in Figure 1. This porous medium consists of porous multilayers of different characteristics. The orthogonal axes Ox and Oz coincide with the free surface at rest. The depths of the fluid and the porous layers are respectively h0, h1, h2... hi.

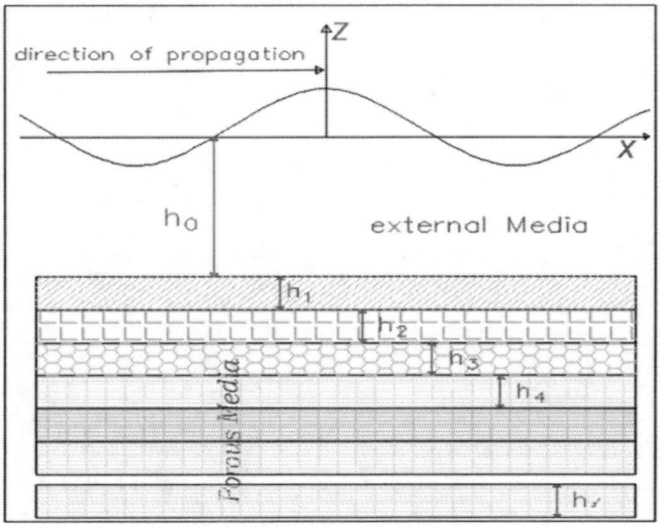

Figure 1. Schemes of study

We consider a monochromatic swell propagating in the presence of a porous bottom of different characteristics. The field of study consists of an external flow:

$-\infty < x < +\infty; -h_0 \leq z \leq 0$

The forced flow is defined in the domain defining the porous bottom: $-\infty < x < +\infty; -h_i \leq z \leq -h_{i-1}$; (i=1,2,3,4). The index used takes the values 1,2,3,4.

Note:

$Z(h_i) = s(h_i) - i f(h_i)$. The impedance in domain i (i = 1.2)

$Z(h_0) = s(h_0) - i f(h_0) = 1 - i\,0$.

i is the complex number; ε the porosity of the medium; φ the velocity potential; η the elevation of the free surface; ρ the density of the fluid; z the vertical rating; g gravity; $\vec{\nabla}$ designates the gradient operator; $\vec{n}_f$ the normal vector at the interface; s is the coefficient of inertia; f can be assimilated to a coefficient of friction or damping; ω represents the pulsation of the harmonic motion of period T; Z is a coefficient characterizing the porous layer in terms of added mass and coefficient of inertia.

Equations of the movement

$$\Delta\varphi = 0 \qquad -h_0 \leq z \leq 0 \qquad (1)$$

$$\vec{\nabla}(\varepsilon\vec{\nabla}\varphi) = 0 \qquad -h_i \leq z \leq h_{i-1} \qquad (2)$$

$$\frac{\partial\varphi}{\partial z}-\frac{\omega^2}{g}\phi=0 \qquad z=0 \tag{3}$$

Boundary condition

$$\frac{\partial\varphi}{\partial z}-\frac{\omega^2}{g}\varphi=0 \qquad z=0 \tag{4}$$

$$\eta(x,t)=-\frac{1}{g}\frac{\partial\varphi_0}{\partial t} \qquad z=0 \tag{5}$$

$$Z(-h_i^-)\,\phi(-h_i^-)=Z(-h_i^+)\,\phi(-h_i^+) \qquad z=-h_i \tag{6}$$

$$\vec{\nabla}\varphi(-h_i^-)\cdot\vec{n}_i=\vec{\nabla}\varphi(-h_i^+)\cdot\vec{n}_i \qquad z=-h_i \tag{7}$$

$$\vec{\nabla}\varphi\cdot\vec{n}_f=0 \qquad z=-h_2 \tag{8}$$

The evaluation of the coefficient f is based on the Lorentz hypothesis of equivalent work which consists in equalizing over a period, the work of the non-linear term with that of a linear term. The term f is used (Dingemans, 1996) [5]:

$$f\omega=\frac{\nu\varepsilon}{k}\left(1+\frac{C_f\,\varepsilon\sqrt{k}}{\nu}\tilde{f}\right) \tag{9}$$

$$\tilde{f}=\frac{\int_{\tau}\left[\int_0^T|\vec{q}|^3\,dt\right]d\tau}{\int_{\tau}\left[\int_0^T|\vec{q}|^2\,dt\right]d\tau}=\frac{8\sqrt{a+b}}{3\pi}\left\{E(m)-\frac{1}{4}\left(1-\frac{b}{a}\right)K(m)\right\} \tag{10}$$

Or: $a=\frac{1}{T}\int_0^T|q|^2\,dt \qquad b=Max(|\vec{q}|^2-a) \qquad m=\frac{2b}{a+b}$

$\vec{q}$ is the actual interstitial velocity or mean velocity of the fluid, $K(m)$ and $E(m)$ represent elliptic integrals of first and second species respectively. It is therefore sufficient to use an iterative method combining the two expressions of f and the Forchheimer equation to obtain the velocity field in the porous medium.

In the following, we note ( ) j the variable relative to each layer j, where the index j denotes the j multi porous layer ($j=1,2$).

3. Analytical resolution

The procedure of the analytical calculation (The analytical calculation process) developed in this work, consists of three steps:

- Subdivision of the field of study into several sub-domains. The swell propagates in the external domain. The other sub-domains represent the three layers of the porous media (fig.1).
- Analytical resolution in each subdomain. In the external domain, the form of the general solution is obtained by solving the problem in the first order with long wave approximation. In the porous media, the theory of SOLLITT & CROSS (1972) will be considered as a basis to model the form of the general solution.
- In the third step, the solutions are connected between all the subdomains using the continuity of flows and pressures.

By this procedure, we calculate the dispersion relation of the wave parameters considering a porous multilayer bottom in long wave theory as well as the kinematics of the flow generated by the swell in the porous media.

In particular, we are interested in the depth of closure of the swell's impact in the porous media. The analysis will be done according to two parameters: The porosity and the thickness of the laminated layers.

3.1. Dispersion relation and velocity potential

External media

- Dispersion relation

$$\frac{\omega^2}{K_0 g} = t_0 = th(K_0 h_0) \tag{11}$$

- Velocity potential

$$\phi_0(x,z,t) = \text{Re}\left\{ \frac{jag}{\omega Z_0} [\frac{\omega^2}{gK_0} sh(K_0 z) + ch(K_0 z)] e^{j(\omega t - Kx)} \right\} \tag{12}$$

Single layer

- Dispersion relation

$$\frac{\omega^2}{K_1 g} = \frac{t_0 + \frac{t_{10}}{Z_1}}{1 + t_0 \frac{t_{10}}{Z_1}} \tag{13}$$

- Velocity potential (external media)

$$\phi_{10}(x,z,t) = \text{Re}\left\{ \frac{jag}{\omega Z_0} [\frac{\omega^2}{gK_1} sh(K_1 z) + ch(K_1 z)] e^{j(\omega t - Kx)} \right\} = \psi_0(K_1) \tag{14}$$

- Velocity potential (Porous media)

$$\phi_{11}(x,z,t) = \text{Re}\left\{ \frac{jagc_0}{\omega Z_1} [Z_1(\frac{\omega^2}{gK_1} - t_0) shK_1(z + h_0) - (\frac{\omega^2}{gK_1} t_0 - 1) chK_1(z + h_0)] e^{j(\omega t - Kx)} \right\} \tag{15}$$

Two layers

- Dispersion relation

$$\frac{\omega^2}{K_2 g} = \frac{t_0 + \frac{t_{10}}{Z_1} + \frac{t_{21}}{Z_2} + \frac{Z_1}{Z_2} t_0 t_{10} t_{21}}{1 + t_0 \frac{t_{10}}{Z_1} + t_0 \frac{t_{21}}{Z_2} + \frac{Z_1}{Z_2} t_{10} t_{21}} \tag{16}$$

- Velocity potential (external media)

$$\varphi_{20}(x,z,t) = \text{Re}\left\{ \frac{jag}{\omega Z_0} \left[\frac{\omega^2}{gK_2} sh(K_2 z) + ch(K_2 z) \right] e^{j(\omega t - Kx)} \right\} \tag{17}$$

- Velocity potential (Mono-layer)

$$\varphi_{21}(x,z,t) = \text{Re}\left\{ \frac{jagc_0}{\omega Z_1} \left[Z_1 \left(\frac{\omega^2}{gK_2} - t_0 \right) sh\left[K_2(z + h_0)\right] - \left(\frac{\omega^2}{gK_2} t_0 - 1 \right) ch\left[K_2(z + h_0)\right] \right] e^{j(\omega t - Kx)} \right\} \tag{18}$$

- Velocity potential (two-layers)

$$\varphi_{22}(x,z,t) = \text{Re}\left\{ \begin{array}{l} \frac{jagc_0 c_{10}}{\omega Z_1} \left\{ \left[t_{10} \left(\frac{\omega^2}{gK_2} t_0 - 1 \right) + Z_1 \left(\frac{\omega^2}{gK_2} - t_0 \right) \right] sh\left[K_2(z + h_1)\right] - \right. \\ \left. \frac{1}{Z_2} \left[Z_1 \left(\frac{\omega^2}{gK_2} t_0 - 1 \right) + Z_1^2 t_{10} \left(\frac{\omega^2}{gK_2} - t_0 \right) \right] ch\left[K_2(z + h_1)\right] \right\} e^{j(\omega t - Kx)} \end{array} \right\} \tag{19}$$

three layers

- Dispersion relation

$$\frac{\omega^2}{K_3 g}=\frac{t_0+\frac{t_{10}}{Z_1}+\frac{t_{21}}{Z_2}+\frac{t_{32}}{Z_3}+\frac{Z_1}{Z_2}t_0t_{10}t_{21}+\frac{Z_1}{Z_3}t_0t_{10}t_{32}+\frac{Z_2}{Z_3}t_0t_{21}t_{32}+\frac{Z_2}{Z_1Z_3}t_{10}t_{21}t_{32}}{1+t_0\frac{t_{10}}{Z_1}+t_0\frac{t_{21}}{Z_2}+\frac{Z_1}{Z_2}t_{10}t_{21}+\frac{Z_1}{Z_2}t_{10}t_{21}+\frac{Z_1}{Z_3}t_{10}t_{32}+\frac{Z_2}{Z_3}t_{21}t_{32}+\frac{Z_2}{Z_1Z_3}t_0t_{10}t_{21}t_{32}} \quad (20)$$

- Velocity potential (external media)

$$\phi_{30}(x,z,t)=\mathrm{Re}\left\{\frac{jag}{\omega Z_0}[\frac{\omega^2}{gK_3}sh(K_3z)+ch(K_3z)]e^{j(\omega t-Kx)}\right\} \quad (21)$$

- Velocity potential (Mono-layer)

$$\phi_{31}(x,z,t)=\mathrm{Re}\left\{\frac{jagc_0}{\omega Z_1}[Z_1(\frac{\omega^2}{gK_3}-t_0)shK_3(z+h_0)-(\frac{\omega^2}{gK_3}t_0-1)chK_3(z+h_0)]e^{j(\omega t-Kx)}\right\} \quad (22)$$

- Velocity potential (Two-layers)

$$\phi_{32}(x,z,t)=\mathrm{Re}\left\{\begin{array}{l}\frac{jagc_0c_{10}}{\omega Z_1}([t_{10}(\frac{\omega^2}{gK_3}t_0-1)+Z_1(\frac{\omega^2}{gK_3}-t_0)]shK_3(z+h_1)-\\ \frac{1}{Z_2}[(\frac{\omega^2}{gK_3}t_0-1)+Z_1(\frac{\omega^2}{gK_3}-t_0)]chK_3(z+h_1)])e^{j(\omega t-Kx)}\end{array}\right\} \quad (23)$$

- Velocity potential (Three-layers)

$$\phi_{33}(x,z,t)=\mathrm{Re}\left\{\begin{array}{l}\frac{jagc_0c_{10}c_{21}}{\omega Z_1Z_2Z_3}(Z_3[(\frac{\omega^2}{gK_3}t_0-1)(Z_1t_{21}+Z_2t_{10})+Z_1(\frac{\omega^2}{gK_3}-t_0)(Z_1t_{10}t_{21}+Z_2)]shK_3(z+h_2)-\\ -Z_2[(\frac{\omega^2}{gK_3}t_0-1)(Z_1+Z_2t_{10}t_{21})+Z_1(\frac{\omega^2}{gK_3}-t_0)(Z_1t_{10}+Z_{21}t_{21})]chK_3(z+h_1)])e^{j(\omega t-Kx)}\end{array}\right\} \quad (24)$$

Four layers

- Dispersion relation

$$\frac{\omega^2}{K_3 g}=\frac{\begin{array}{l}t_0+\frac{t_{10}}{Z_1}+\frac{t_{21}}{Z_2}+\frac{t_{32}}{Z_3}+\frac{t_{43}}{Z_4}+\frac{Z_1}{Z_2}t_0t_{10}t_{21}+\frac{Z_1}{Z_3}t_0t_{10}t_{32}+\frac{Z_2}{Z_3}t_0t_{21}t_{32}+\frac{Z_2}{Z_1Z_3}t_{10}t_{21}t_{32}+\frac{Z_1}{Z_4}t_0t_{21}t_{32}t_{43}\\ +\frac{Z_2}{Z_4}t_0t_{21}t_{32}t_{43}+\frac{Z_3}{Z_4}t_0t_{21}t_{32}t_{43}\end{array}}{\begin{array}{l}1+t_0\frac{t_{10}}{Z_1}+t_0\frac{t_{21}}{Z_2}+\frac{Z_1}{Z_2}t_{10}t_{21}+\frac{Z_1}{Z_2}t_{10}t_{21}+\frac{Z_1}{Z_3}t_{10}t_{32}+\frac{Z_2}{Z_3}t_{21}t_{32}+\frac{Z_1}{Z_4}t_{10}t_{43}+\frac{Z_2}{Z_4}t_{21}t_{42}+\frac{Z_3}{Z_4}t_{32}t_{43}\\ +\frac{Z_2}{Z_1Z_3}t_0t_{10}t_{21}t_{32}+\frac{Z_3}{Z_1Z_4}t_0t_{10}t_{21}t_{32}t_{43}\end{array}} \quad (25)$$

- Velocity potential (external media)

$$\phi_{40}(x,z,t)=\mathrm{Re}\left\{\frac{jag}{\omega Z_0}[\frac{\omega^2}{gK_4}sh(K_4z)+ch(K_4z)]e^{j(\omega t-Kx)}\right\} \quad (26)$$

- Velocity potential (Mono-layer)

$$\phi_{41}(x,z,t)=\mathrm{Re}\left\{\frac{jagc_0}{\omega Z_1}[Z_1(\frac{\omega^2}{gK_4}-t_0)shK_4(z+h_0)-(\frac{\omega^2}{gK_4}t_0-1)chK_4(z+h_0)]e^{j(\omega t-Kx)}\right\} \quad (27)$$

- Velocity potential (Two-layers)

IOP Conf. Series: Materials Science and Engineering 1126 (2021) 012048 doi:10.1088/1757-899X/1126/1/012048

$$\phi_{42}(x,z,t)=\mathrm{Re}\left\{\begin{array}{l}\frac{jagc_0c_{10}}{\omega Z_1}([t_{10}(\frac{\omega^2}{gK_4}t_0-1)+Z_1(\frac{\omega^2}{gK_4}-t_0)]shK_4(z+h_1)-\\ \frac{1}{Z_2}[(\frac{\omega^2}{gK_4}t_0-1)+Z_1(\frac{\omega^2}{gK_4}-t_0)]chK_4(z+h_1)])e^{j(\omega t-Kx)}\end{array}\right\} \tag{28}$$

- <u>Velocity potential (Three-layers)</u>

$$\phi_{43}(x,z,t)=\mathrm{Re}\left\{\begin{array}{l}\frac{jagc_0c_{10}c_{21}}{\omega Z_1Z_2Z_3}(Z_3[(\frac{\omega^2}{gK_4}t_0-1)(Z_1t_{21}+Z_2t_{10})+Z_1(\frac{\omega^2}{gK_4}-t_0)(Z_1t_{10}t_{21}+Z_2)]shK_4(z+h_2)-\\ -Z_2[(\frac{\omega^2}{gK_4}t_0-1)(Z_1+Z_2t_{10}t_{21})+Z_1(\frac{\omega^2}{gK_4}-t_0)(Z_1t_{10}+Z_{21}t_{21})]chK_4(z+h_1)])e^{j(\omega t-Kx)}\end{array}\right\} \tag{29}$$

- <u>Velocity potential (Four-layers)</u>

$$\phi_{44}(x,z,t)=\mathrm{Re}\left\{\frac{jagc_0c_{10}c_{21}c_{32}}{\omega Z_1Z_2Z_3Z_4}[Z_4(Z_2t_{32}\xi+Z_3\zeta)]shK_4(z+h_3)+Z_3(Z_2\xi+Z_3t_{32}\zeta)chK_4(z+h_1)]e^{j(\omega t-Kx)}\right\}=\psi_4(K_4) \tag{30}$$

Or

$$\left\{\begin{array}{l}\xi\\ \zeta\end{array}\right\}=\left\{\begin{array}{l}(\frac{\omega^2}{gK_4}t_0-1)(Z_1+Z_2t_{10}t_{21})+Z_1(\frac{\omega^2}{gK_4}-t_0)(Z_1t_{10}+Z_2t_{21})\\ (\frac{\omega^2}{gK_4}t_0-1)(Z_1t_{21}+Z_2t_{10})+Z_1(\frac{\omega^2}{gK_4}-t_0)(Z_1t_{10}t_{21}+Z_2)\end{array}\right\}$$

$$Z_i=s_i-jf_i \qquad t_{ii-1}=thK_i(h_i-h_{i-1})$$

$$c_0=ch\left[Kh_0\right]\ ;\ c_{i\ i-1}=ch\left[K(h_i-h_{i-1})\right]\ ;\ t_0=ch\left[Kh_0\right]\ ;\ t_{i\ i-1}=th\left[K(h_i-h_{i-1})\right].$$

3.2. *Generalization of the dispersion relation and the velocity potential of a multi-layer medium*

A generalized dispersion relationship of a multi-layer can be obtained in the following form:

$$\frac{\omega^2}{Kg}=\frac{(\gamma_0\Gamma_0+\gamma_1(1+\Gamma_1))}{(\gamma_2\Gamma_0+\gamma_3(1+\Gamma_1))} \tag{31}$$

Or

$$\gamma_0=Z_1(1+Z_1t_0t_{10})\ ;\quad \gamma_1=Z_1t_0+t_{10}\ ;\quad \gamma_2=Z_1(Z_1t_{10}+t_0)\ ;\quad \gamma_3=Z_1+t_0t_{10}$$

$$\Gamma_0=\sum_{l=0}^{E(\frac{n-1}{2})}\sum_{k_1=1}^{n-(2l+1)}\sum_{k_2=k_1+1}^{n-(2l+1)+1}\cdots\cdots\sum_{k_p=k_{p-1}+1}^{n-(2l+1)+(p-1)}\cdots\cdots\sum_{k_{2l+1}=k_{2l}+1}^{n-1}\prod_{j=1}^{l}(Z_{k_{2j+1}})^2\prod_{j=1}^{2l+1}Y_{k_j+1}$$

$$\Gamma_1=\sum_{l=1}^{E(\frac{n-1}{2})}\sum_{k_1=1}^{n-2l}\sum_{k_2=k_1+1}^{n-2l+1}\cdots\cdots\sum_{k_p=k_{p-1}+1}^{n-2l+(p-1)}\cdots\cdots\sum_{k_{2l}=k_{2l-1}+1}^{n-1}\prod_{j=0}^{l-1}(Z_{k_{2j+1}+1})^2\prod_{j=1}^{2l}Y_{k_j+1}$$

$$Y_i=Y_{i,i-1}=\frac{t_{i,i-1}}{Z_i}$$

- <u>Velocity potential</u>

$$\phi_{n,l}(x,z,t)=\mathrm{Re}\left\{\frac{jagc_0}{\omega}\prod_{k=1}^{l-1}c_{k,k-1}[F_{n,l}sh(K_n(z+h_{l-1})+\frac{1}{Z_l}G_{n,l}ch(K_n(z+h_{l-1})]e^{j(\omega t-Kx)}\right\} \tag{32}$$

Or

$$l=0,1,\ldots\ldots\ldots n\ ;\qquad F_{n,l}=\sum_{k=l}^{n}\frac{t_{i,j-1}}{Z_i}\ ;\ G_{n,l}=P_{n,l}+Q_{n,l}\ ;\ P_{n,l}=1-t_0^2-(t_0+\sum_{k=1}^{l-1}Z_kt_{k,k-1})\sum_{j=1}^{n}\frac{t_{j,j-1}}{Z_j}$$

$$P_{n,l}=1-t_0^2-(t_0+\sum_{k=1}^{l-1}Z_kt_{k,k-1})\sum_{j=1}^{n}\frac{t_{j,j-1}}{Z_j}\ ;\ Q_{nl}=\sum_{k=1}^{l-2}\sum_{j=1}^{l-k-1}\frac{Z_{l-k}}{Z_j}t_{l-k,l-k-1}\,t_{j,j-1}$$

3.3. Long wave theory approximation for a multilayer media

In the case of long wavelength waves (kh<<1), (TEMPERVILLE, 1985) [6], using a limited development in the vicinity of the term Kh0 a neglecting order greater than one, the dispersion relation of a multilayer can be written in a homogeneous form:

$$\frac{\omega^2}{Kg} \approx KH \qquad or \qquad H = h_0 + \sum_{i=1}^{n} \frac{h_{n-1}}{Z_i} \tag{33}$$

A dispersion relationship is then obtained with an equivalent medium (external medium + porous) of equivalent height H in order to homogenize the multilayer into a single layer.

If the thicknesses of the porous layers are identical to a thickness e, the multilayer is equivalent to a monolayer whose dispersion relation is:

$$\frac{\omega^2}{Kg} Z_e \approx K(h_0 + ne) \approx KHZ_e \tag{34}$$

An equivalent thickness he and an equivalent impedance Ze (equivalent permeability) are defined as:

$$h_e = \frac{ne}{Z_e} \quad avec \quad H = h_0 + h_e \quad et \quad \frac{1}{Z_e} = \frac{1}{n}\sum_{i=1}^{n}\frac{1}{Z_i}$$

$$Z_e = s_e - if_e \quad or \quad s_e = \frac{1}{\varepsilon_e} + \frac{1-\varepsilon_e}{\varepsilon_e^2} C_{me} \quad ; \quad f_e = \frac{\nu}{\omega}\frac{1}{\sigma_e}$$

n is the number of layers and e is the index for "equivalent".

3.4. Graphical results

the profile of the free surface in the case of a single-layer bottom and a bi-layer and tri-layer bottom are presented respectively for the following data:

$T=0.5$; $a = 0.012$; $h_0=0.15$; $K=(16.287\text{-}6.95{*}10^{-3})$; $f_1=1$; $\varepsilon_1=0.30$; $C_{m1}=1$; $h_1=0.17$; $f_2=0.8$; $\varepsilon_2=0.40$; $C_{m2}=1$; $h_2=0.185$; $f_3=1$; $\varepsilon_3=0.50$; $C_{m3}=1$; $h_3=0.20$

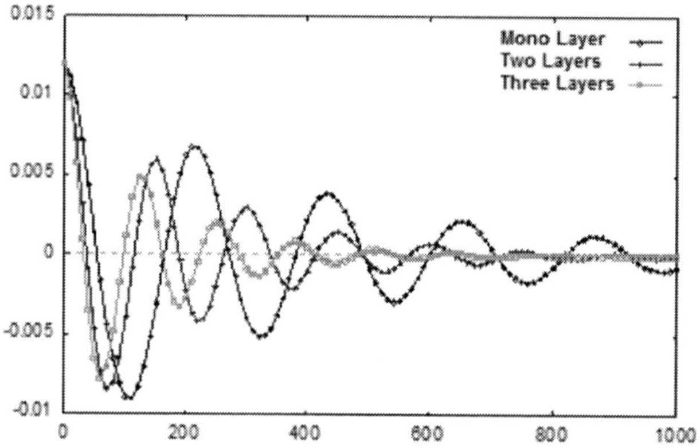

Figure 2. Free surface profile as a function of length (m)

The damping coefficient and the horizontal velocity profile are presented in different cases for the following data:

$T=1$; $a = 0.012$; $h_0=0.15$; $K= 5.691\text{-}6.7856{*}10^{-3})$; $f_1=1$; $\varepsilon_1=0.30$; $C_{m1}=0.86$; $h_1=0.20$;

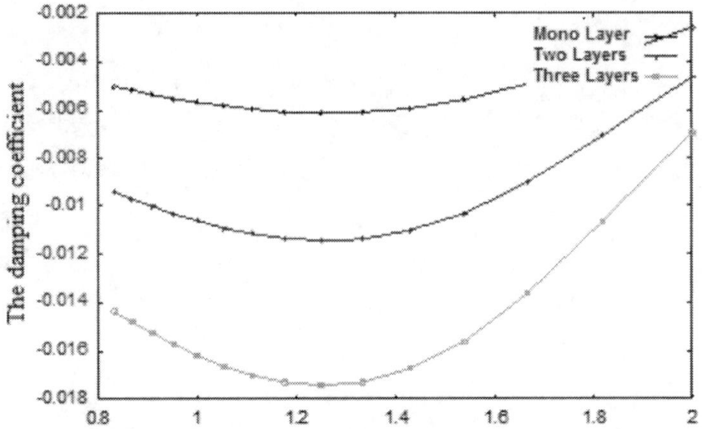

Figure 3. Damping Coefficient as a Function of Frequency (Hz)

the horizontal velocity profile is presented in different cases for the following data:
$T=1$; $a = 0.012$; $h_0=0.15$; $K= (5.64\text{-}1.56*10^{-3})$; $f_1=1$; $\varepsilon_1=0.30$; $C_{m1}=0.86$; $h_1=0.166$; $f_2=0.8$; $\varepsilon_2=0.40$; $C_{m2}=1$; $h_2=1.183$; $f_3=0.8$; $\varepsilon_3=0.50$; $C_{m3}=1$; $h_3=0.20$; $x=0.$

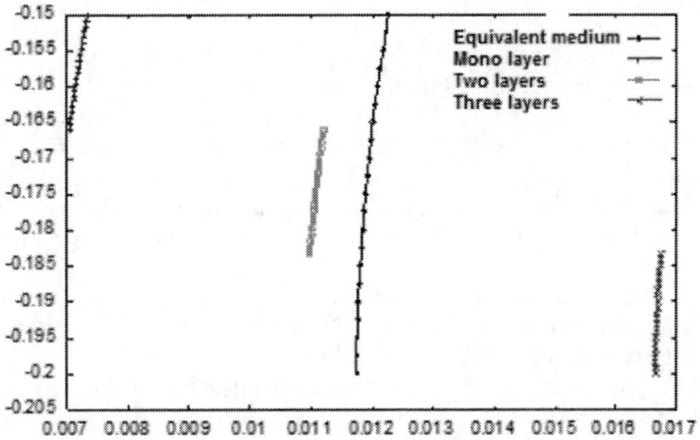

Figure 4. Horizontal velocity profile in a different medium and its equivalent medium

Concept of the depth of closure

The analytical calculation of the velocity potential makes it possible to define the closing depth, that is to say the depth limit of remobilization of sediments in the porous medium.
We present the evolution of the damping coefficient as a function of the frequency in the presence of a porous medium composed of monolayers, two layers, three layers and four layers of different characteristics.

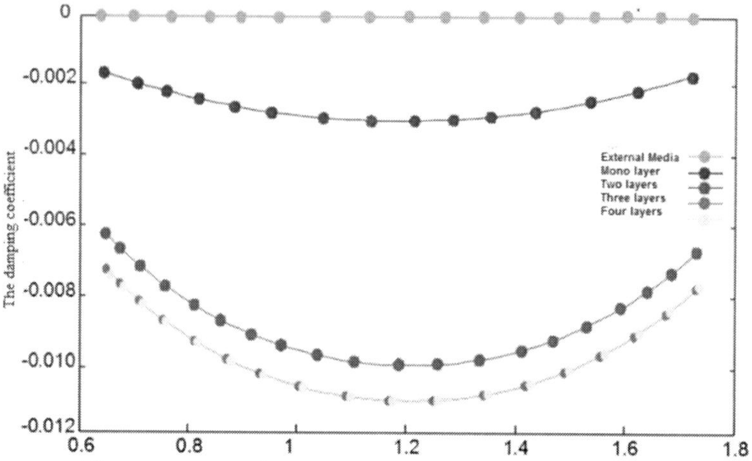

Figure 5. Damping Coefficient as a Function of Frequency (Hz)

4. Conclusion

We modelled the interaction of the linear wave with a multi-layer porous medium by adapting the Forchheimer model in the instantaneous mode based on the flat wave hypotheses and the long wave approximation, perfect flow, incompressible and isothermal, the porous medium is rigid and undeformable. We have calculated analytically by the potential theory the dispersion relation and the velocity potentials. We note that:

- The wavelength decreases exponentially as the frequency increases, increases with the height of the outer medium and little influenced by the thickness of the porous layer;
- The variation of the coefficient of depreciation as a function of the frequency is not monotonous;
- The evolution of the free surface shows that a multilayer porous structure attenuates and shifts the free surface more than a single layer;
- The definition of an equivalent medium and an equivalent impedance (equivalent permeability) can influence the flow behavior in the two layers and three layers, replaced by this equivalent medium in order to homogenize the multi to a mono;
- The horizontal velocity profile presents a discontinuity at the interfaces, we will question the model of a perfect fluid, we had to take into account tangential constraints, a viscous fluid and the Forchheimer model is a more realistic approach;
- From the fourth porous layer, there is no longer any effect on the damping coefficient and therefore on the attenuation of the swell

References

[1] SOLITT C. K., CROSS R. H. (1972). *Wave transmission through permeable breakwaters.* 13th. Coastal Engineering Conf., vol.3, ASCE. Vancouver, pp.1827. https://doi.org/10.1061/9780872620490.106

[2] MADSEN P. A. (1983). *Wave reflection from a vertical permeable wave absorber.* Coastal Engineering 7, 381-396. doi: 10.1016/0378-3839(83)90005-4

[3] LEE J. F. AND LAN Y. J., 1995, "A second – order solution of waves passing porous structures". Department of Hydraulics and Ocean Engineering, National Cheng Kung University, Tainan, Taiwan, R. O. C.

[4] BROSSARD.J., CHAGDALI.M., REY V. (2004). *Écoulement engendré par la houle dans un milieu poreux.* VIII èmes Journées Nationales Génie Côtier-Génie Civil, Compiègne 2004. doi :10.5150/jngcgc.2004.006-b

[5] Dingemans MW (1996). Water wave propagation over uneven bottoms. World Scientific Publishing Co, London, 700 p.

[6] TEMPERVILLE A. (1985). *Contribution à la théorie des ondes de gravité en eau peu profonde.* Thèse d'Etat, Université Scientifique et médicale et Institut National Polytechnique de

Grenoble.

Stability study for the rehabilitation of the protective dike of the port El Jadida (Morocco)

Mouakkir Laila; Mouhid Mustapha; Ayad Hassan; El Bouni Brahim; Chagdali Mohamed

Polymer Physics and Critical Phenomena Laboratory, Faculty of Sciences Ben M'sik, P.O.BOX 7955, University Hassan II, Casablanca, Morocco.
mouakkir_laila@yahoo.fr ; mouhidmustapha@gmail.com

Abstract. In this study, we are interested in the stability study and reinforcement of protective structures represented by the breakwater of the port of El Jadida. This study is based on the numerical modeling of the interaction of the swell with an embankment dam. The swell simulation parameters were simulated using the SWAN model [1]. The results of this modeling show that the most constraining swells likely to reach the project site are the swells of 14s and 18s of 50 years and 100 years return period and that the residual swell heights are between 1.63 m and 1.96 m maximum. The redefinition of the project swell based on this numerical modeling makes it possible to resize the components of the protective dike by the Hudson formula [2].
Keywords: Morocco, El Jadida Port, Island, Stability, Dike, Swell, Numerical modeling, SWAN model.

1. Introduction

The EL JADIDA port is located on the Atlantic coast 100 km southwest of Casablanca and 170 km northeast of Safi. The harbor, which is primarily dedicated to fishing, is located on a portion of the coastline that extends south-east/north-west. (Fig.1) [3]. The location of the port is 33°15'N-8°29'O. Its vocation is fishing and it is sheltered from the prevailing swells by the main breakwater of 560 m, rooted at the corner of the Portuguese citadel.

The port of El Jadida began to be built in 1922-1924, with the two converging dikes marking a 40 m wide passage. The excavation to make a rectangular dock dug at -1,5 m ZH (fishing basin). The current configuration of the harbor has not changed since 1949, the date of the last works.

In 1982, a project to reinforce the main dike was carried out. The elements of the main dike have been replaced by tetrapod's and a large number of rehabilitation actions.

Figure 1. Aerial view of the port of El Jadida

2. Expertise of the state of the protection dyke

The visit is based on the assessment of the state of degradation of the structures and their causes, the preparation of a visual diagnosis of the structures and the identification of the degradations and anomalies by type of structure.

Figure 2. degradation of the protective dike

The works concerned by the diagnosis are the main dam and the protection dam. These structures work with dissipation properties against swell and they also have a number of disadvantages:

- Internal erosion of the central core of the dike by water flow and departure of the fines thus opening a breach on the dike.
- Scouring at the bottom of the dike at the level of the abutment causing its ruin and consequently destabilization.

At the beginning of the degradation, the effect of the swell is manifested by the tearing away of the carapace blocks leaving the guard wall and the inner layers unprotected. In the second class of degradation, the erosion of the base of the slope or the absence of a protective riser causes the collapse and descent of the tetrapods on the slope and the creation of a void between the guard wall and the protective structures. The deterioration of the feet of the structure puts the elements of the structure directly on the bottom which leads to a strong erosion and consequently the instability of the embankments.

In order to better understand the problem of dike degradation, it seems important to us to make a return of the experiments on the dike degradation. For this type of structure, there are four types of unstable dikes (Fig.3):

- Instability of the carapace that manifests as settling, falling blocks and breaking blocks. These instabilities result in a disorder of the inner slope, the crossing, the run-up, displacement and deterioration of the guard wall as well as the cavitation of the core and its settlement.
- Instability of the core and under layer are manifested by a disorder of the armour, a disorder of the inner slope as well as the disorder, settling and cavitation of the nucleus. These instabilities lead to the displacement and deterioration of the guard wall.
- Instability of the crown wall is manifested by a deterioration of the scour protection.
- Toe instability manifests a disorder of the carapace. The consequences are the instabilities of the entire dam.

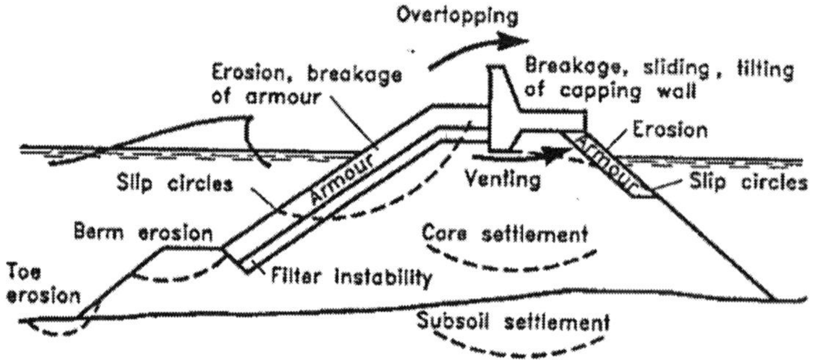

Figure 3. Standard rubble mound breakwater failure modes [4]

3. Significant wave height

The redefinition of the project swell by transposing the swell from the open sea to the structures of the ports concerned must be based on a credible and exhaustive database and on a mathematical wave propagation model taking into account:

- The refraction, the effect of small depths on an irregular bathymetry.
- Dissipation by breaking and rubbing at the bottom.
  Wave generation under the influence of the wind.
- Diagrams of joint occurrences, Hs/Tp, Hs/Dirp, et Tp/Dirp.
- Diagrams of joint occurrences by sector Hs/Tp.
- Databases of reduced parameters of sea states and winds, over a period of more than 20 years, broken down according to independent energy peaks.

Tr: Return period, or recurrence interval (year).
Dir: Mean direction of waves, usually to grid north [(rad or (°)].
Tp: Wave period (s).
Hs: Significant wave height, Hs = $H_{1/3}$ (Mean height of highest 1/3 fraction of waves).

4. SWAN Model

The Simulating WAves Nearshore (SWAN) model is a numerical wave model used to obtain realistic estimates of wave parameters in coastal areas, lakes, and estuaries from given wind, bottom, and current conditions. The model is based on the wave action balance equation (or energy balance in the absence of currents) with sources and sinks. SWAN is a third-generation wave model with first-, second-, and third-generation options.

<u>Equations of the physical problem</u>:

In SWAN the waves are described with the two-dimensional spectrum of the action density of the waves N [Manuel SWAN] [5]:

$$N(x, y, \sigma, \theta)= E(x, y, \sigma, \theta)/\square \quad (1)$$

N(r, h): Wave action density (= qg.E(q, h)/r[J.s/m^2]).
E(r, h): spectral variance.
where x and y are horizontal Cartesian coordinates.
θ : is the propagation direction of each wave component.
σ: relative angular frequency (relative to current) [Hz].
In cartesian coordinates, this equation is written in the following form (Hasselman et al. 1973) [6]

$$\frac{\partial N}{\partial t}+\frac{\partial}{\partial x}(C_x N)+\frac{\partial}{\partial y}(C_y N)+\frac{\partial}{\partial \sigma}(C_\sigma N)+\frac{\partial}{\partial \theta}(C_{y\theta} N)=\frac{S_{en}}{\sigma} \quad (2)$$

σ: relative angular frequency (relative to current) [Hz]
C_x, C_y, C_θ, C_ρ: propagation speeds according to x, y and θ and angular frequency
S_{en}: term of energy sources [J/m^2]
C_x, C_y, C_σ, and C_θ are the propagation velocity in x-space, y-space, σ-space, and θ-space respectively. Details of these processes can be found in the SWAN technical documentation manual.

5. Model simulation

5.1. Bathymetric data

The bathymetry databases are:

- GEBCO: (General Bathymetric Chart of the Oceans): site www.gebco.net
- Etopo2 – Site : http://www.ngdc.noaa.gov/mgg/image/2minrelief.html
- Carte marine: Oceano Atlantico Norto RADA Y PUORTO DE MAZAGN (EL JADIDA)

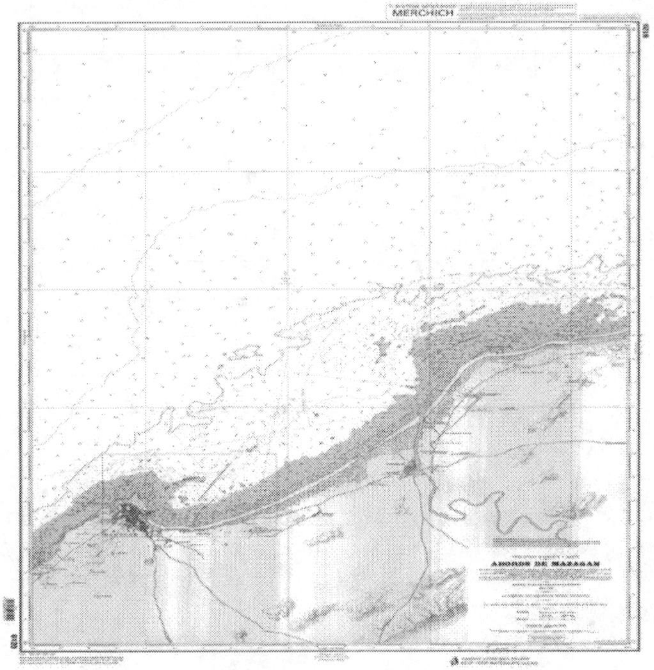

Figure 4. Nautical chart of EL JADIDA port [7]

Figure 5. Bathymetry of port EL JADIDA [8]

5.2. swell of the site

The various statistical studies on the nearshore of El Jadida and Safi lead to the conclusion that the orders of magnitude to be considered for the project swells are the following:

northwest sector

Table 1. Estimated of Residual Wave Heights

Return period	Hs offshore	Hs to -30m	Tp – Dir(°N)
1	6.5 m	6.5 m	14/20s – 310°N
10	8.75 m	9 m	15/18s - 310°N
50	10.25 m	10.5 m	16/18s -310°N
100	10.75 m	11 m	16/18s -310°N

southwest sector

Table 2. Estimated of Residual Wave Heights

Return period	Hs offshore	Hs to -30m	Tp – Dir(°N)
1	3.25 m	2.75 m	9s – 230°N
10	6 m	4.25 m	10s – 230°N
50	7 m	5 m	11s – 230°N
100	7.25 m	5.25m	11s – 230°N

5.3. Current of de site

General and tidal currents are generally low in the area. The average speeds are of the order of 0.15 to 0.20 m/s with moderate winds fairly regular without exceeding values of 0.5 m/s. It often happens that the currents are very low (0.10 m/s) or non-existent. Currents therefore do not constitute a constraint for a project in the site.

5.4. Simulation results

The Global right-of-way of numerical modelling is defined so that all geographical areas within which waves can be propagated to the coast are well taken into account. From the SHOM marine chart, the bathymetry was extracted and interpolated. It is considered to depths of more than 100m. The bathymetry is built in Lambert area 93-Morocco and water levels are referenced in NGM-Morocco

[9]. The mesh consists of T3 triangular elements. It is regular with a step of 60 meters up to the coast. SWAN wave propagation model was used to simulate effects between 20 meters deep and the coastal zone of EL Jadida. These effects include local creation of waves by wind, refraction, shoals, friction dissipation on the bottom and breaking. In this project, SWAN was used in stationary mode.

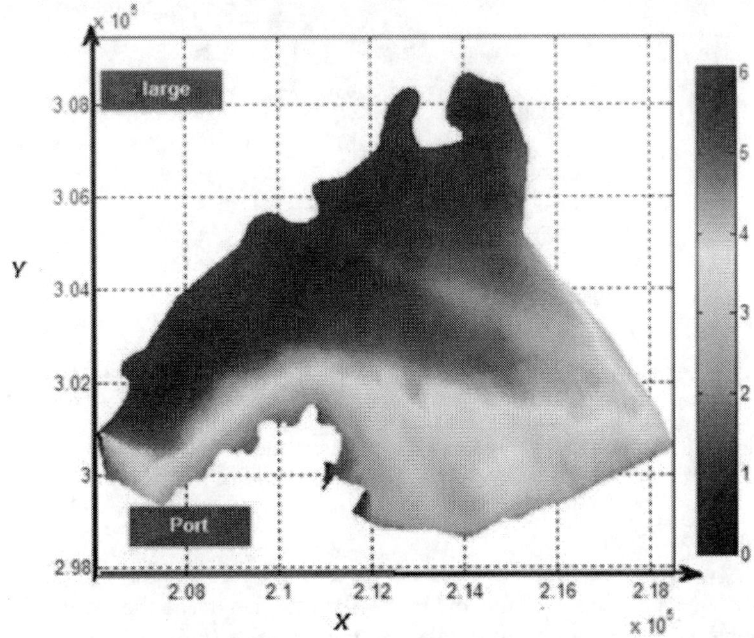

Fig.6. Directional representation of Hs in the coastal domain Dirp=NNW

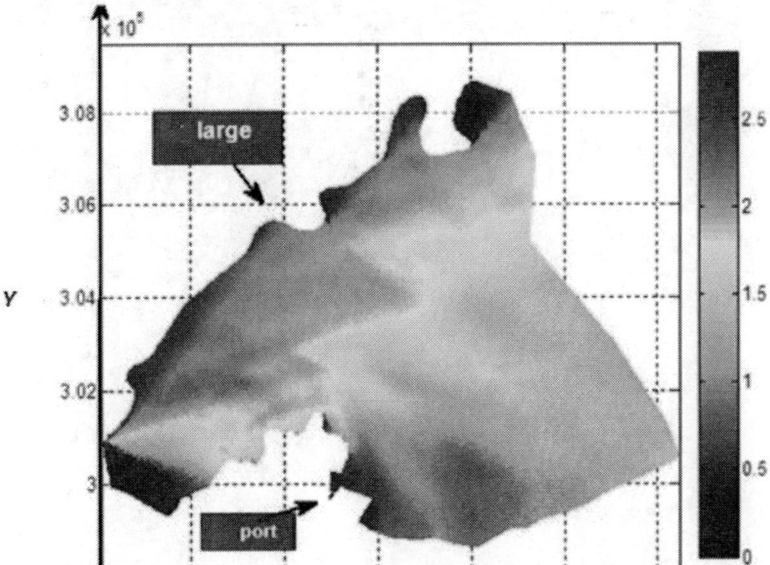

Fig.7. Directional representation of Hs in the coastal domain Dirp=West

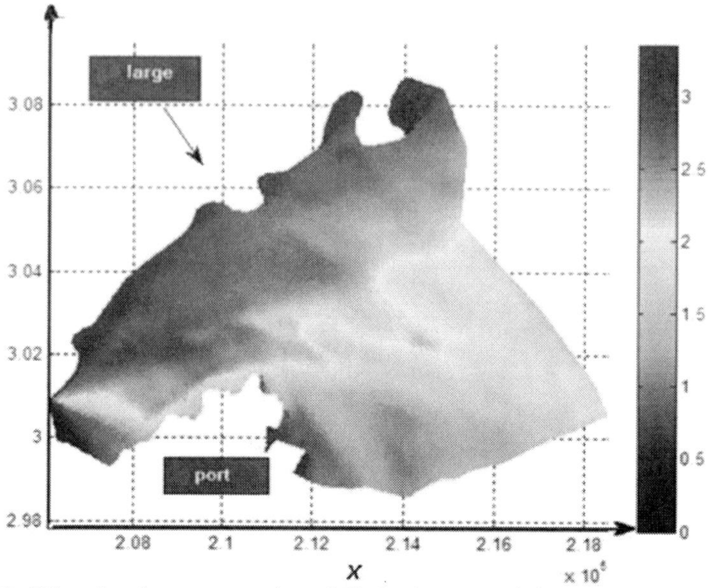

Figure 8. Directional representation of Hs in the coastal domain Dirp=WNW

5.5. *SWAN Simulation Points*

The NNW area incident swell characteristics off the El Jadida Port on the outskirts of the project area, offshore, the average inlet depth of the model (Z) and the height of the swell (Hs), calculated, were recorded in 9 points (Fig.9) and are reported in a summary table (table.3).

Figure 9. Simulation points for NNW direction

Table 3. Simulation points SWAN

Simulation points	Depth(m)	Hs(m)	Direction (deg)
JD2	3.3112	1.90158	328.338
JD3	6.6430	1.96076	321.765
JD4	7.3837	1.94945	325.108
JD5	7.9875	1.92774	327.270
JD6	8.5293	1.86711	327.349
JD7	8.4827	1.84535	327.349
JD8	8.6581	1.82231	327.349
JD9	8.6922	1.82543	327.349

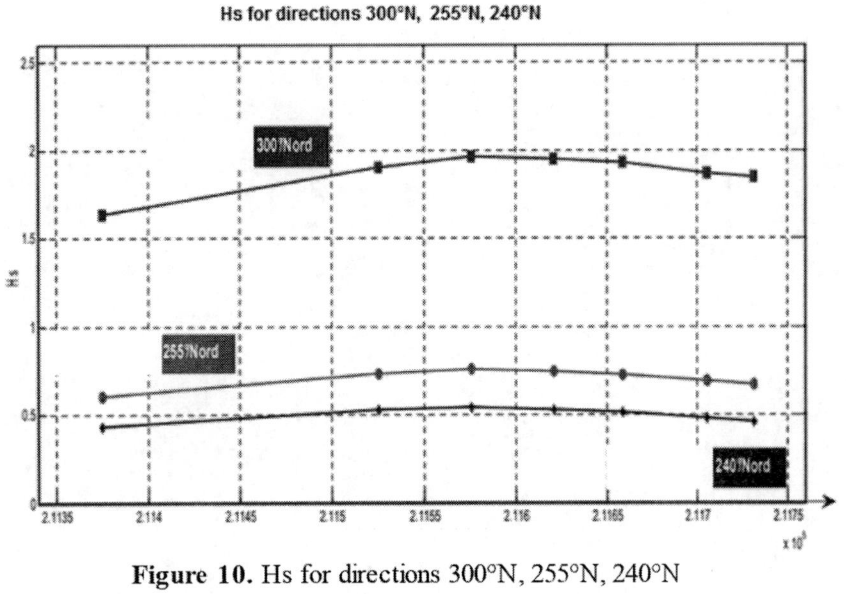

Figure 10. Hs for directions 300°N, 255°N, 240°N

6. Conclusion

The results of this modelling show that the most constraining swells likely to reach the project site are the swells of 14s and 18s of 50-year and 100-year return period and that the residual swell heights are between 1.63 m and 1.96 m maximum. The main dike is subject to repeated swell attacks, under its cuttings, some of the most exposed artificial blocks erode and the riprap forming the protective carapace moves, revealing breaches that can quickly enlarge under the action of the swell. these results are used to define a maintenance method that guarantees a significant time for repairs. The chosen solution consists of a reloading by riprap at dimensioning to be determined according to the empirical formulae (HUDSON), including a reprofiling of the slope to strengthen the foundation of the carapace blocks from the stop below sea level.

A regulation of the external slope is essential to maintain continuity and adapt to the profile of the dike on the whole of the dike.

References

[1] N. Booij, R.C. Ris, L.H. Holthuijsen, A third-generation wave model for coastal regions. 1, Model description and validation, J. Geophysics. Res. 104 (C4) (1999) 7649-7666.
[2] Hudson, R Y (1953,1959). "Wave forces on breakwaters". Trans Am Civ Engrs, vol 118, pp 653-674. "Laboratory investigations of rubble mound breakwaters". J Water & harbors Div, Am Civ Engrs, vol 85, no 2171, pp 93-121.
[3] Aerial image from Google earth.
[4] CIRIA, CUR, CETMEF (2007). The Rock Manual. The use of rock in hydraulic engineering. Ch 5 (2nd edition).https://www.kennisbank-waterbouw.nl/ SWAN SCIENTIFIC AND TECHNICAL DOCUMENTATION; Delft University of Technology, Faculty of Civil Engineering and Geosciences, Environmental Fluid Mechanics Section, The Netherlands, http://swanmodel.sourceforge.net/download/zip/swantech.pdf.
[5] Hasselmann, K., R.P. Barnett, E. Bouws, H. Carlsen, D.E. Cartwright, K.Enke, J.A.Ewing, H.Gienapp, D.E. Hasselmann, P. Kruseman, A. Meerburg, P. Mueller, D.J. Olbers, K. Richter, W. Sell and H. Walden, 1973, Measurements of wind-wave growth and swell decay during the Joint North Sea Wave Project (JONSWAP), Dtsch. Hydrogr. Z., A8, No. 12.
[6] SHOM, Le service hydrographique et océanographique de la Marine.www.shom.fr
[7] ANP, Agence Nationale des Ports de Maroc.
[8] Nivelement Générale Marocain.

Study of the importance of core cylinders in the monitoring of the compressive strength of the concrete

J De La Iglesia[1*], M Murillo[1], D Borrero Restrepo[1]

[1] Department of Civil and Environmental Affairs, Universidad de la Costa, Barranquilla, 080002, Colombia

Abstract. The fulfillment of the resistance to the compression as a fundamental property of the concrete, during the process of execution of a work, constitutes great work for all the factors that intervene in the constructive process. The same concrete cannot be suitable for all structures, many times when using the same mixture, failures can occur with relevant damage. This situation frequently occurs at construction sites and resources must be allocated to repair structures to guarantee their useful life. This research aims to focus on the importance of performing compressive strength tests and taking enough samples for reliable results. For this purpose, 36 concrete test samples were taken from different structural elements of work according to the requirements of the standard. Subsequently, the laboratory tests were carried out at 7, 14, 28, and 56 days with a control cylinder, to analyze the behavior of each sample according to the concrete specifications requested for the different types of structures. From the results obtained, we could observe the differences in the percentages achieved by each of the samples and the type of failure presented, in turn, the importance of the control cylinder, for the monitoring of the strength of the concrete.

1. Introduction

Concrete is the most used construction material in the world, manufactured by man, occupies the second place after the water in its use, more than six thousand tons are produced annually, for its manufacture raw materials are used as cement, sand, and aggregates, the aggregates are inert and rigid materials that form a granular skeleton, constitute approximately 65% to 75% of the total volume of concrete, besides, water, additives that represent the remaining volume. New materials have been incorporated into the concrete, among them minerals, chemicals, fiber; therefore, the concrete is a heterogeneous product and its mechanical behavior [1].

In concrete setting times, several influencing factors affect this process such as water-cement ratio, cement type, chemical additives, additive addition time, mixing, and temperature, which is one of the most influential in the development of the resistance of the mixture, the resistance increases at high temperatures and must be kept uniform in the setting process to avoid fractures by a thermal shock, because in curing times there is a release of heat energy that can generate losses of heat.
To meet the technical requirements related to the situation in which the concrete was used, it is necessary to define some properties related to its cool and hardened state and it is, therefore, necessary

that of the appropriate proportion of its components [1]. It is increasingly important to determine the resistance to the impact load and in turn to determine the strength properties of cement-based materials at an experimental level before they can be considered for use in construction [2].

A lot of times specialized machinery and tools are needed when performing the emptying and a skilled workforce for the sampling of the cylinders, which is very involved in the result, making it necessary to use a control cylinder which will serve to corroborate the strength in case the required strength is not achieved after testing at 7, 14 and 28 days is not able to obtain the required strength.

Cylindrical cores for compression tests at 56 days are extracted with a probe equipped with diamond drills when the concrete has acquired sufficient strength so that during the cut the adhesion between the aggregate and the paste is not lost. In all cases, the concrete must have at least 14 days of placement.

The extraction must be done perpendicular to the surface, taking care that in the area there are no joints, nor are they near the edge. The compressive strength obtained when testing a tell-tale shall be corrected concerning the last test performed, and account shall be taken of the differences between the tell-tales and test-pieces in aspects such as age, dimensions, the state of saturation at the time of the test, the presence of reinforcing steel inside and the fact that the controls are subjected to the aggressive extraction process. Differences associated with material properties, such as exudation, should also be corrected. [3] by carrying out a correct sampling and paying attention to the mixtures of the concrete and the cycles through which it passes.

Various causes can motivate the evaluation of the concrete in the structure. In works under construction, it may be due to low results of resistance to compression obtained in the test pieces at the age of 28 days; in finished works, it may be associated with the presence of pathologies in the structure, to the change of use (an increase of the load), among others. [3], resistance is often influential on surface wettability.

Concrete surfaces are one of the substrates on which the mortar coating can be applied, which is widely used in the building. Frequently, when there are incompatibilities between the substrate and the mortar, the durability of the coating system can be compromised, causing pathological manifestations such as detachments and displacements [4].

Compressive strength can be defined as the maximum measured strength of a concrete specimen under axial compression load and is expressed as force per unit cross-sectional area [5], a more in-depth study of the influence of the addition of minerals is necessary to find the conditions where the auto compaction and the compressive strength of a mixture are simultaneously improved by looking for correlations between these [6].

The requirements to perform the test for resistance to simple compression of the test pieces are The height/diameter ratio from 2.5 to 3.0, Diameter greater than 10 times the maximum grain size of the rock (Diameter not less than 50 mm -Tolerance of the test piece faces). Such requirements are important because in some cases they are very difficult to achieve or make the trial very costly [7].
The most widely used aggregates in the world come from natural deposits formed in riverbeds or floodplains and are relatively inexpensive as they generally do not require any industrial process. However, some regions lack rivers, where source rocks are extracted which give rise to aggregates directly from the Earth's crust, using explosives and/or excavating machines; these rocks are subsequently fragmented by mechanized crushing. Such aggregates are called shredded aggregates and require a screening and grading process for particulates [8]. For more than 70 years in American practice, the most widely used test for concrete has been the standard cylinder compression test. The

test method is relatively easy to perform in terms of sampling, specimen preparation, and force determination, when performed correctly, this test has a low variation of resistance within the laboratory, and therefore easily lends itself to be used as a standard. The compressive force thus obtained is modified by specified factors and used to verify the nominal forces of structural members. This force value is therefore an essential parameter in design codes.

The test is mainly used as a basis for quality control to ensure that project requirement are secured. It is not intended to determine the strength on the ground of the concrete, since none of the effects of placement, compaction, or curing are considered. Traditionally, a certain measure of the strength of the concrete in the structure has been obtained using in situ cured cylinders. These supposedly heal on-site under the same conditions as the concrete in the structure. However, the measured strength of the cylinders cured in situ is often significantly different from the strength on the ground because it is difficult, and often impossible, to have an identical indentation, compaction, and have the same curing conditions for concrete in cylinders as for concrete in structures. In-situ cured specimens are also prone to errors due to incorrect address or inadequate storage, which can generate inaccurate data for critical operations.

The purpose of on-site testing is to estimate concrete characteristics in the structure. The desired characteristic is very often compressive force. To estimate strength, it is necessary to have a known link between the in-situ test result and the strength of the concrete. For a running construction, this loop is generally established empirically in the laboratory [9], Axial compression strength is the greatest of the strengths of concrete and its most appreciated mechanical attribute; however, stress resistance is also important as it is required to predict cracks in this material [10].

The materials as aggregates are made granulometry, to determine that they are suitable in the preparation of the mixture; another topic is the composition, in percentage, of the different sizes of aggregate in a sample, this proportion is usually indicated from larger to a smaller size, by a figure representing, by weight, the partial percentage of each size that passed or was retained in the different sieves that are compulsorily used for such measurement.

The granulometry and maximum size of the aggregate affect the relative proportions of the same in the designs of mixtures, as well as the requirements of water and cement, impacting the resistance, workability, pumping capacity, economy, porosity, shrinkage, and durability of the concrete. It is known that all classical methods of mixing design such as Fuller, Bolomey, O'Reilly, etc., are based on the granulometry of aggregates [11]. The addition of fine materials improves concrete strength providing the workability must be controlled. [12].

The objective of this study is to demonstrate the methodological process to determine the resistance of a mixture, through tests of resistance to compression, which will be carried out according to what is established by the ASTM C39 Standard, considering all the factors involved in the process from sampling to the test where the resistance reaches at least 100%. which are analyzed at different times, in the case of non-compliance, a control cylinder is tested at 56 days of age to corroborate the strength of the structure, essay of great importance, and that gives a part of tranquility in the work that is being carried out.

2. Experimental procedure

The concrete mixture was initially prepared, in a mixing truck with a drum from 70 to 100 rpm and at a speed from 6 to 18 rpm. This mixture was designed with a resistance of 4000 psi, it was sent to the discharge, which was made 30 minutes after the preparation of the mixture, 6 samples are taken from each structure, 2 samples during the discharge process, one at the start and the other sample after 15 minutes, and then tested in the laboratory to test their resistance at the age of 7, 14, 28 and 56 days. Two samples were tested per age and the average obtained from the two samples was the result of resistance.

In the case of two of the samples where the required strength was not achieved, samples of hardened concrete were taken and extracted directly from the structure, this was done 56 days after the first sampling, to corroborate compliance.

For the compression test of the concrete, the mixture was prepared with a resistance of 4000 psi, then cylindrical test specimens of 6" X 12" were molded, the curing process is performed according to the ASTM C31 standard, finally, they were stored in the field until the concrete reached a solid-state, at a temperature of 24 C, and transported to the laboratory for the next 48 hours; other properties of fresh concrete were tested such as temperature, settlement, density, and air content, the settling test was conducted following ASTMC 139, in the laboratory the cylinders were subjected to the compression strength test, analyzing their percentage of advance according to the age of the concrete, and the type of fractures or faults that they presented at the time of testing, This was performed per the ASTM C39 standard test for compressive strength of cylindrical specimens of concrete. After obtaining the results released by the certified laboratory, we were able to observe that several of the samples did not meet their required resistance at 28 days, therefore, it is necessary to analyze the control cylinder at 56 days; The results obtained show several types of failures, among which we find the transverse failure one of the most common in this type of tests.

2.1 **Materials and Methods**

Various types of materials were used for the samples taken fine sand, coarse sand of different sizes, Structural Cement, water, and additive required for the type of concrete of 4000 psi that was produced. Among the tests carried out, the settlement was considered, which was carried out using the Abrams cone method.
For the materials used as fine and coarse sand are performed the granulometry and absorption density tests, with which we obtain that they are suitable materials, to achieve the required strength compliance 4000 psi. See Table 1 granulometry and 2 characterizations for fine agglomeration, see Table 3 coarse aggregate granulometry and Table 4 characterization for coarse agglomeration.

Table 1. Fine sand granulometry used.

Sieve Inches	Mass Withheld (g)	Percentage Withheld (%)	Percentage Ret Accumulated (%)	Percentage Pass (%)	Limits NTC 174. 2018 Min	Max
1/2"		0.0%	0.0%	100.0%	100%	100%
3/8"	3.9	0.3%	0.3%	99.7%	100%	100%
N° 4	63.9	5.20%	5.6%	94.4%	95%	100%
N° 8	92.5	7.6%	13.2%	86.8%	80%	100%
N° 16	84.0	6.9%	20.1%	79.9%	50%	85%
N° 30	158.4	13.0%	33.1%	66.9%	25%	60%
N° 50	404.9	33.4%	66.5%	33.5%	5%	30%
N° 100	309.8	25.5%	92.0%	8.0%	0%	10%
N° 200	57.6	4.7%	96.8%	3.2%	0%	3%
Mondo Masa	0.1	0.0%	96.8%	3.2%		
Total	1174.8	96.8%				

Table 2. Density and absorption of the fine aggregate.

MASS OF THE SAMPLE SSS, Mass (g)	503.4	503.4
MASS OF THE DRY SAMPLE, MS (g)	495.8	495.8
PYCNOMETER MASS WITH SAMPLE AND WATER (g)	974.5	974.5
PYCNOMETER MASS FILLED WITH WATER (g)	663.4	663.4
NOMINAL DENSITY (g/cm3),23 C	2.68	2.68
BULK DENSITY (g/cm3),23 C	2.57	2.57
DENSITY SSS (g/cm3),23 C	2.61	2.61
PERCENTAGE OF ABSORPTION (%)	1.5	1.5

Table 3. Granulometry of coarse sand used.

Sieve	Mass Withheld	Percentage	Percentage	Percentage	Limits NTC 174. 2018	
		Withheld	Ret Accumulated		Min	Max
Inches	(g)	(%)	(%)	Pass (%)		
1 1/2"			0.0%	100%	100%	100%
1"	89.4	1.6%	1.6%	98.4%	100%	100%
3/4"	741.9	13.1%	14.7%	85.3%	90%	100%
1/2"	2381.7	42.0%	56.7%	43.3%	20%	55%
3/8"	1131.9	20.0%	76.6%	23.4%	0%	15%
N°4	1443.0	20.2%	96.8%	3.2%	0%	5%
N°8	18.1	0.3%	97.1%	2.9%	0%	0%
N°16	3.0	0.1%	97.2%	2.8%	0%	0%
N°30	3.8	0.1%	97.2%	2.8%	0%	0%
N°50	5.2	0.1%	97.3%	2.7%	0%	0%
N°100	7.2	0.1%	97.5%	2.5%	0%	0%
N°200	9.3	0.2%	97.5%	2.4%	0%	0%
Mondo	134.5	2.4%	100.0%	0.0%		
Masa Total	5669.0	100.2%				

Table 4. Density and absorption of coarse aggregate

MASS OF THE DRY SAMPLE, MS (g)	3371.5
MASS OF THE SATURATED AND SUPERFICIALLY DRY SAMPLE (g)	3496.5
SAMPLE MASS IMMERSED IN WATER (g)	2113.5
NOMINAL DENSITY (g/cm3),23 C	2.67
BULK DENSITY (g/cm3),23 C	2.43
SATURATED AND SUPERFICIALLY DRY DENSITY (g/cm3) 23C	2.52
PERCENTAGE OF ABSORPTION (%)	3.7

The fine and coarse aggregate is always baked to constant mass at a temperature of 110 +- 5C before being immersed in water for 24 hours. After this, calculations were made to determine the unit mass values of the aggregates. The results are presented in Table 5.

Table 5. Determination of the unit mass of the aggregates used.

COMPACT UNIT MASS BY TAMPING	
	SAND
SAMPLE MASS + MOULD (g)	6550
MOLD MASS (g)	1800
NET MASS OF SAMPLE (g)	4750
MOLD VOLUME (cm3)	2807
UNIT LOOSE MASS (kg/m3)	1690
UNIT LOOSE MASS	
SAMPLE MASS + MOULD (g)	6100
MOLD MASS (g)	1800
NET MASS OF SAMPLE (g)	4300
MOLD VOLUME (cm3)	2807
UNIT LOOSE MASS (kg/m3)	1530

2.2 Methodology

2.2.1. Settlement test.

To the newly mixed concrete on site, the respective samples were taken to verify the fulfillment of their design parameters. In this way, to verify its consistency condition, the settling test was carried out in a fresh state. To analyze the manageability of the concrete it is necessary to be governed under the ASTM C31, this is fundamental at the time of the acceptance or rejection of the mixture. The results for this trial were favorable, meeting a settlement value of 5 ± 1.

Figure 2 illustrates the procedure carried out for conducting the settling test for the concrete prepared for this investigation.

Figure2. Illustration of the settlement test procedure.

2.2.2. Compressive strength tests:
The samples were carried out in a certified laboratory, with the respective tests of compressive strength. The test machine, properly calibrated and certified by the laboratory, was used and the tests were carried out according to the ASTM C39 standard

3. Analysis of Results

After submitting the cylinders to the tests of resistance to compression, it is evident in sample 1 of Table 6 that after 14 days their resistance reaches 102 %, this resistance is average of the % obtained, in the two cylinders tested. No further tests are performed for these samples because the required strength is achieved.

In sample 2 of Table 7, we can show that trials are performed until the age of 28 days when the mixture reaches 113% of the resistance. In samples 3 and 4 of Table 8 and 9 respectively, the control cylinder was tested, at 7, 14, and 28 days, yielding resistances below 100%, so we used the control cylinder and analyzed at the age of 56 days, and check the resistance.

After analyzing the control, it is observed that the resistance is achieved with 128% for sample 3 and 103% for sample 4, demonstrating the importance of the control cylinder test, and in turn giving complete reliability to the structure.

Tabl3 6. Resistance 14 days

SAMPLE	1		
Dimension	6"x12"		
Age	psi	%	%
7	2.79	69	70
7	2.81	71	
14	3.6	100	102
14	4.17	105	

Table 7. Resistance 28 days

SAMPLE	2		
Dimension	6"x12"		
Age	psi	%	%
7	2.81	71	73
7	2.86	74	
14	3.12	78	84
14	3.55	89	
28	4.6	115	113
28	4.37	110	

Table 8. Resistance for control cylinder

SAMPLE	3		
Dimension	6"x12"		
Age	psi	%	%
7	1.5	40	41
7	1.9	42	
14	1.9	42	45
14	2.2	47	
28	3.55	89	88
28	3.44	86	
56	4.68	125	128
56	5.41	131	

Table 9. Resistance for control cylinder

SAMPLE	4		
Dimension	6"x12"		
Age	psi	%	%
7	2.2	47	45
7	1.9	42	
14	3.12	78	84
14	3.55	89	
28	3.6	90	91
28	3.62	91	
56	4.47	107	103
56	3.93	98	

4. Conclusions and Recommendations

From the results of this research, we can conclude the importance of compressive strength. In each of the samples taken, we show the evolution of resistance for the days. Two of the four samples failed to reach their maximum percentage at 28 days, having the need to make use of a witness cylinder of hardened concrete, to have the follow-up of the behavior, and checking that, when 56 days, achieved a positive result, achieving the required resistance. Laboratory tests represent a cost subject to regulatory procedures that ensure their full application and compliance with specifications.

Concrete is a product developed technically and under strict quality controls. If we carry out the tests with poor sample management, the results will give rise to doubts that will not benefit the structure in terms of quality and durability; and it is precisely this that leads us to carry out tests with witnesses, which are of great help in decision-making, with their analysis and review give continuity to the constructive process.

Acknowledgments

The authors thank the company AE INGENIEROS CIVILES SAS, for facilitating the on-site follow-up in the sampling and testing of cylinders, to carry out the development of this research.

References

[1] J. F. Moretti, C. R. Minussi, J. L. Akasaki, C. F. Fioriti, J. L. P. Melges, and M. M. Tashima, "Previsão do módulo de elasticidade e da resistência à compressão de corpos de prova de concreto por meio de redes neurais artificiais," *Acta Sci. - Technol.*, vol. 38, no. 1, pp. 65–70, 2016, doi: 10.4025/actascitechnol.v38i1.27194.

[2] S. K. Sateshkumar, P. O. Awoyera, T. Kandasamy, S. Nagaraj, P. Murugesan, and B. Ponnusamy, "Impact resistance of high strength chopped basalt fibre-reinforced concrete," *Rev. la Constr.*, vol. 17, no. 2, pp. 240–249, 2018, doi: 10.7764/RDLC.17.2.240.

[3] A. F. Domínguez and J. J. H. Albear, "Factores de corrección a la resistencia a compresión de testigos de hormigón. Análisis crítico de las normas cubanas e internacionales," *Inf. la Constr.*, vol. 69, no. 547, pp. 1–9, 2017, doi: 10.3989/id53723.

[4] E. Cappelari, C. Giodani, S. T. H. Dapper, and A. B. Masuero, "Influence of concretes compressive strength on superficial wettability," *Rev. Mater.*, vol. 25, no. 1, 2020, doi: 10.1590/s1517-707620200001.0907.

[5] R. Chaupis and M. Angel, "UNIVERSIDAD SAN PEDRO FACULTAD DE INGENIERIA Resistencia a compresión de un concreto f ´ c = 210 kg / cm2 al sustituir al cemento en 4 %, 6 % y 8 % por cascara de huevo," 2019.

[6] J. A. Correa-Yepes, N. Rojas-Reyes, and J. I. Tobón, "Effect of fly ash and silica fume on rheology, compressive strength and self-compacting in cement mixtures," *DYNA*, vol. 85, no. 206, pp. 59–68, 2018, doi: 10.15446/dyna.v85n206.68960.

[7] M. Galván-Ceballos and I. A. Restrepo, "Resistance correlation with compression uniaxial effective humidity and porous rocks," *Dyna*, vol. 83, no. 198, pp. 9–15, 2016, doi: 10.15446/dyna.v83n198.53151.

[8] Abazid, Mohammad, Hüseyin Gökçekuş, and Tahir Celik. "Implementation of TQM and the Integration of BIM in the Construction Management Sector in Saudi Arabia Validated with Hybridized Emerging Harris Hawks Optimization (HHO)." (2021).

[9] M. Antonio, "RESISTENCIA A LA COMPRENSIÓN ULTRASONIDO Marco Antonio Céspedes García," p. 112, 2003.

[10] E. I. Moreno and J. Varela-rivera, "Resistencia a Tensión Del Concreto Tensile Strength of Concrete With High Absorption Limestone Aggregate," vol. 8, 2016, [Online]. Available: http://www.scielo.org.mx/pdf/ccid/v8n1/2007-3011-ccid-8-01-00035.pdf.

[11] J. Toirac Corral, "Caracterización granulométrica de las plantas productoras de arena en la República Dominicana, su impacto en la calidad y costo del hormigón," *Cienc. Soc.*, vol. 37, no. 3, pp. 293–334, 2012, doi: 10.22206/cys.2012.v37i3.pp293-334.

[12] A. Alfeehan, M. Mohammed, M. Jasim, U. Fadehl, and F. Habeeb, "Utilizing industrial metal wastes in one-way ribbed reinforced concrete panels," *Rev. Ing. Constr.*, vol. 35, no. 3, pp. 246–256, 2020, doi: 10.4067/S0718-50732020000300246.

Nano crystalline $Sr_6Sb_4NbO_{18}$ -Energy Material in Solid Oxide Fuel Cells

Fergy John[a], Shemim S S[b], Rinu Sam S[a]

[a] Department of Physics, St.Gregorios College, Kottarakara, Kollam, Kerala.
[b] Department of Physics, TKM college of engineering, Kollam, Kerala.
Corresponding author: fergyphd@gmail.com

Abstract. Because of the technological applications, Niobium based ceramic is one of the most research interest. Thus, there is numerous research works in recent years on their application as electronic and optical devices. Sintered oxide ceramics, having the required shape, micro structured size can be used for different electronic applications. In this paper, nano crystalline $Sr_6Sb_4NbO_{18}$ ceramic was prepared by combustion method. The sample was analysed by the X-ray diffraction, TEM, radio frequency and Impedance spectroscopic studies. The XRD pattern of as prepared $Sr_6Sb_4NbO_{18}$ sample reveals, the material has rhombohedral perovskite structure. All the peaks were broad when compared with the bulk, which indicates the reduction in the crystallite size. This is due to the micro strain and nano size in the crystallites. The crystallite size has been calculated from the XRD data, using Scherrer formula and it was obtained as 21 nm. The surface morphology of as prepared sample was studied using TEM and the particle size was found to be 19 nm. The sample was sintered at 1050 ^{0}C, obtained the 98 % of theoretical density. An impedance spectroscopic study of the sample carried out with a temperature range of , 200 °C to 500 °C shows that the ions are the main source for the conduction. The reason for the conductivity of the material might be because of the grain and grain boundary effects which can be clearly seen in the impedance plot. The impedance study confirms that the material is a possible candidate to fabricate solid oxide fuel cell.
Keywords: Nanomaterial, Dielectric studies, Impedance spectroscopy, Solid Oxide Fuel Cell

1. Introduction

In the past few decades, the electronic market has undergone great changes mainly in the communication systems with the development of new technologies. To produce materials with superior properties, the mixing of metal oxides is used as an attractive strategy. The materials with $A_nB_{n-1}O_{3n}$ (n = 6) structure have gained more attraction in the recent years due to their technological application resulting from their dielectric and electrical properties. Numerous niobium based compounds have been reported with this structure, like $Sr_6Nb_5O_{18}$ [1], $Sr_6Nb_4TiO_{18}$ [2], $Ba_6Nb_4TiO_{18}$ [3], $Ba_5SrTa_4ZrO_{18}$ [4], $Ba_6Ti_{1-x}Sn_xNb_4O_{18}$ [5], and $Ba_6TiNb_4O_{18}$ [6]. Duivenboden et al. has reported the octahedral structure of $Ba_6TiNb_4O_{18}$ due to the formation of stacked hexagonal and cubic symmetries of BaO_3 layers [7]. Dielectric properties of various materials in the $A_6B_5O_{18}$ family in the microwave frequency region were reported by Zhang et al., Fang et al. and Drews et al. [8-10]. In this paper we have reported the sample preparation, electrical and dielectric properties of nano sized $Sr_6Sb_4NbO_{18}$ by auto ignition combustion method for the first time.

2. Experimental

In the present investigation auto ignition combustion method [11] was used to prepare the sample $Sr_6Sb_4NbO_{18}$ (SSN). The metal nitrate act as oxidizing agent and suitable fuel act as reducing agent in this method [12]. The solution containing ions of niobium, strontium, and antimony was prepared from high-purity Niobium penta chloride ($NbCl_5$, 99.9%) and nitrates of strontium, and antimony (99.99 %). To get a precursor complex, urea (NH_2CONH_2) was used as reducing agent and Citric acid as complexing agent. The stoichiometric amounts of the reagents were dissolved in a minimum volume of deionized water for obtaining the transparent aqueous solution. The solution was heated at 250°C within a combustion chamber, and it has undergone dehydration with the release of huge quantity of gases and finally got auto ignited. To obtain the pure and well crystalline powders of SSN the powder obtained after the combustion was calcined at 600 °C for 30 minutes

3. Characterization

The structure of the synthesized sample SSN, was determined by analysing the X-ray diffraction pattern (Model: Philips PW1710 diffractometer).The crystallite size of SSN was also calculated from the XRD data using Debye- Scherrer formula. Using TEM (Model: TM-300, Philips), the particle size of the sample SSN was calculated. The sample in the form of cylindrical pellet was sintered at 1050 °C for 2 hours. The Complex impedance, dielectric constant (ε_r) and dielectric loss (tan δ) at radio frequency region were measured in sintered pellet by using LCR meter (HIOKI, model 3532, Japan).

4. Results and discussion

The XRD pattern of the sample SSN after the combustion process is shown in figure 1. Each diffraction peaks are indexed as rhombohedral perovskite structure on comparing with the data of ICDD file 32-0095, for $Ba_6Nb_4ZrO_{18}$ ceramic. All the diffraction peaks are indexed and no other secondary phases are present in it which implies that phase pure SSN nanoparticles are formed in the single-step process. From the XRD pattern it is also clear that, the formation of SSN is complete in the combustion process without any high temperature treatment. The crystallite size of the sample SSN is calculated as 21 nm using the Debye-Scherrer formula.

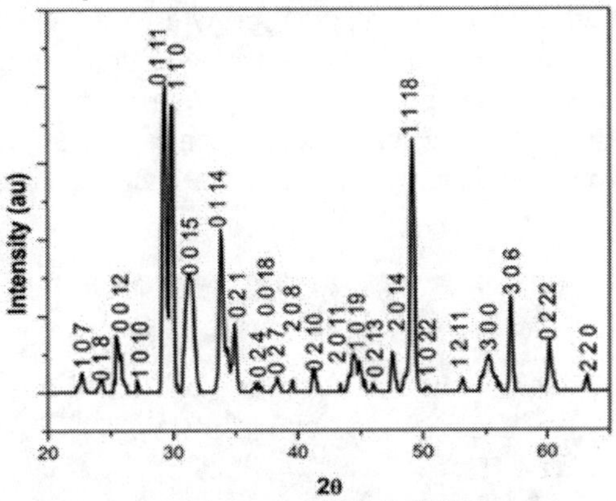

Figure 1 XRD of SSN

The particle size confirmation of SSN is done using TEM analysis. The TEM of as synthesized nano crystalline SSN is shown in the figure 2. In the TEM analysis, it is observed that almost-polycrystalline nano powder has been formed. It has been found that most of the nanoparticles have homogeneous morphology. The agglomeration is absent in this TEM image which clearly shows the crystallinity of the sample. Figure 3 shows the particle size distribution histogram of the nano sample SSN. The particles seem to be in 14 nm to 27 nm ranges, and the average particle size is about 19 nm.

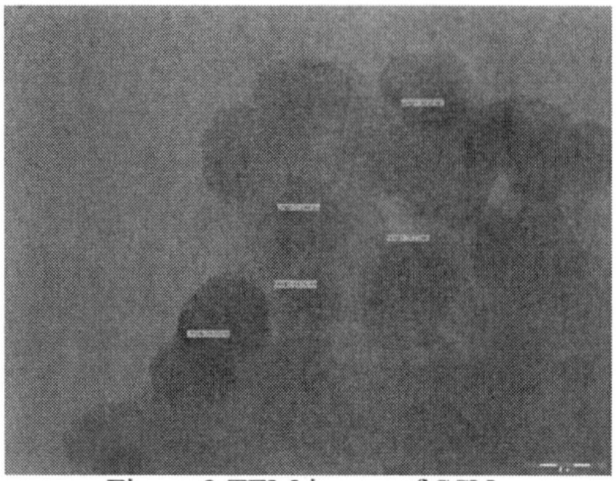

Figure 2 TEM image of SSN

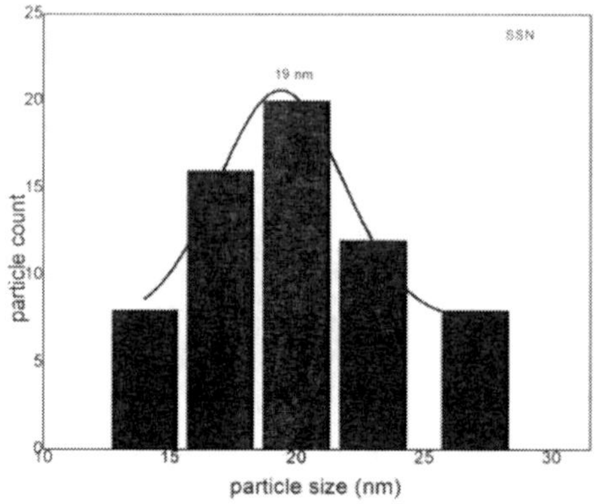

Figure 3 Particle size distribution of SSN

For the dielectric and impedance studies, highly dense compact form of the powder sample without any pores and air column has high significance. It has huge surface area due to its nano-size and it leads to excellent sintering behaviour. Nano sized SSN sintered at 1050 °C and obtained about 98 % of the theoretical density within the rhombohedral perovskite structure. Figure 4 shows the SEM images of SSN sample. It is clear that the sample is well sintered with no cracks on the surface due to densification.

When the temperature has increases the grain boundaries will merge and grain growth will also takes place. In the sintering, the rate of grain growth of the grains is directly proportional to its radius of curvature. The particle size is increased from nano to micro meter range during sintering as evidenced from the TEM and SEM images. The average grain size of the sintered samples is less than 2μm. Due to the auto combustion preparation method some sintering benefits are raised such as low sintering temperature, smooth and fast densification, absence of sintering aids, and low expensive sintering equipment.

Figure 4. SEM Image of SSN

The variation in the dielectric constant (ε_r), ac conductance (G) and loss factor (tan δ) in the radio frequency range is studied for the sample SSN. At high frequency region the dielectric constant is almost stable at 35 and it increases with the decrease of dielectric constant as shown in figure 5. This is due to interfacial polarization reported by Maxwell and Wagner [13]. Within the sample, the delay in polarization has occurred and it results the decrease in ε_r with the application of the electric field [14].

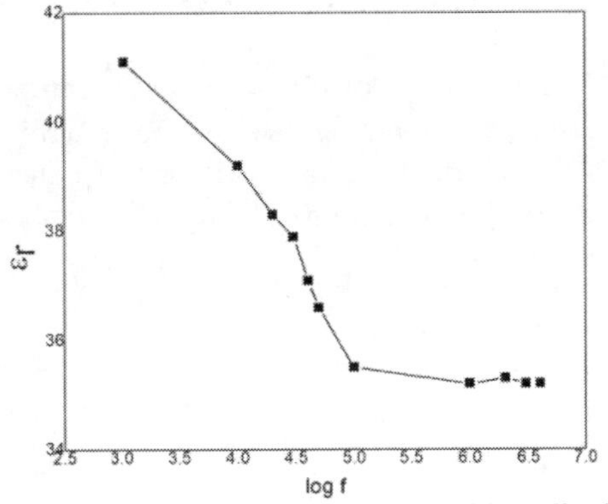

Figure 5. Plot of dielectric constant with radio frequency

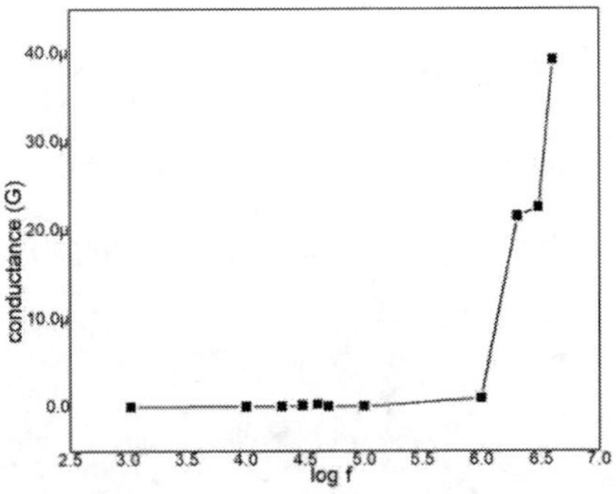

Figure 6. Plot of Conductance with radio frequency

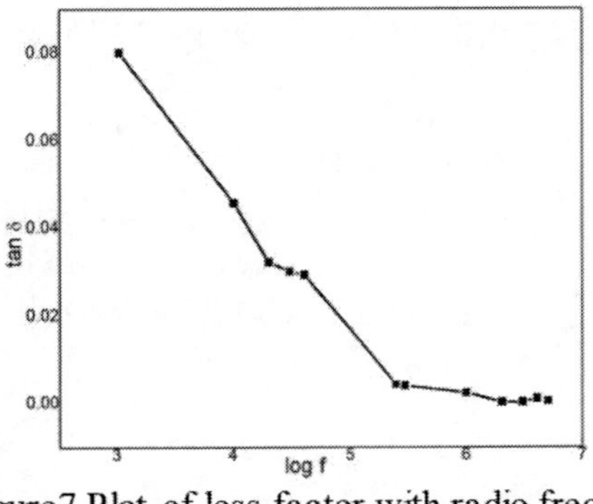

Figure7.Plot of loss factor with radio frequency

From the figure 6 it is clear that, at high frequency, ac conductance of the sample increases with the increase of frequency and will remain almost constant at low frequencies upto 1MHz .Therefore in communication systems, this material are useful for the capacitive applications. The variation of tan δ in radio frequency region is shown in figure 7. At higher frequencies, Dielectric loss remains constant with value as 2.60 x 10^{-3} and increases with decrease in frequencies. This behaviour has reported as Koops phenomenological model [15]. Due to the low loss in the radio frequency range, the dissipation of electrical energy of the sample is low. The studies of the variation of dielectric parameters in radio frequency region substantiate the feasibility of this material in dielectric resonator and substrate applications

Figure 8 shows the plot of the real part of impedance (Z′) versus logarithm of frequency at different temperatures 200-500 ^{0}C of the sample SSN. At low frequency region, the variation of Z′ is inversely proportional to the frequency and is in agreement with the negative temperature coefficient of resistance behaviour. This behaviour indicates that the conductivity of the oxygen ions increases with the frequency. All the curves at different temperatures coincide at the high frequency region. The value of Z' reaches a constant value which can be an indication of dc conduction.

Figure 9 shows the plot of Z'' as a function of frequency at the temperatures 200-500 ^{0}C. Asymmetric peaks are obtained at different temperatures with the peak maximum corresponding to the frequency f_z called as peak frequency. As the temperature rises, the peak frequency f_z, is shifted towards the high frequency region. This leads to the non-Debye type of relaxation behavior in the system. The broadening of the peak is attributed to the spread of relaxation time.

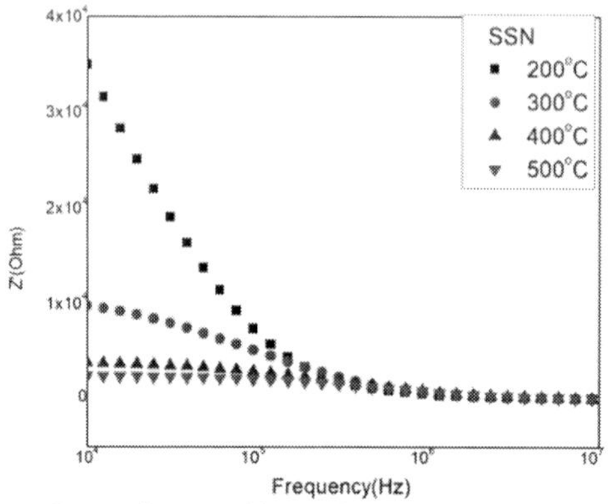

Figure 8. The real part of impedance with radio frequencies

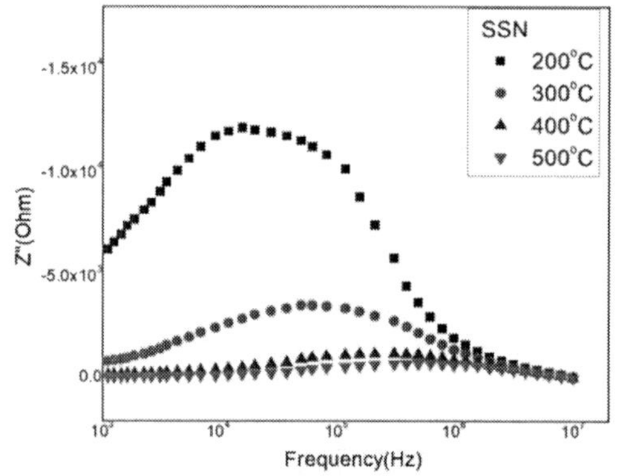

Figure 9. The imaginary part of impedance with radio frequencies

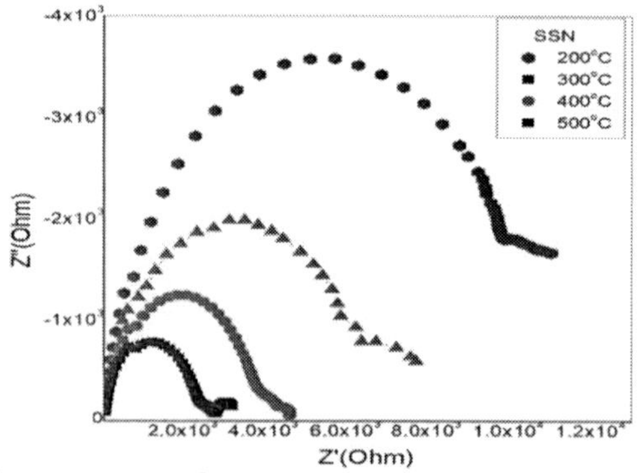

Figure 10.Complex impedance plot (cole- cole plot)

Figure 10 shows the impedance plot known as Cole-Cole plot of the sample SSN at the temperatures 200-500 ^{0}C. The centres of the semicircle arcs in the Cole- Cole Plot at different temperatures are found to be lying below the real axis. The first semi-circle is the contribution of grain and grain boundary effect and the plot shows the weak trace of the second semicircle is the contribution of the electrode effect. As the temperature increases, the peaks shift towards the low frequency side, which indicates a falloff in the grain resistance. This might be due to the prominent ionic conduction at high temperatures [16].

5. Conclusions

Nano-crystalline $Sr_6Sb_4NbO_{18}$ sample is synthesized through auto ignition combustion process. The XRD analysis confirms that the sample crystallises in rhombohedral perovskite structure. The average crystallite size calculated from FWHM is 21 nm. By analysing both XRD and TEM, The formation of nano sized $Sr_6Sb_4NbO_{18}$ is confirmed. The dielectric constant (ε_r) and loss factor (tan δ) of the sintered pellet at 1 MHz are about 35 and 2.60 x 10^{-3}, respectively. The behaviour of ac conductivity, dielectric loss confirms that sample can use as dielectric resonator in radio frequency communication systems. Also, the impedance spectroscopic studies shows that the sample is suitable to be used as electrolyte in SOFC.

References

[1] Lichtenberg F, Herrnberger A and Wiedenmann K 2008 Synthesis, structural, magnetic and transport properties of layered perovskite-related titanates, niobates and tantalates of the type $A_nB_nO_{3n+2}$, $A'A_{k-1}B_kO_{3k+1}$ and $A_mB_{m-1}O_{3m}$ *Progress in Solid State Chemi.*, **36**, 253–387.

[2] Drews AR, Wong-Ng W, Vanderah TA and Roth RS 1997 Preparation and crystal structure of $Sr_6TiNb_4O_{18}$ *J. of Alloys and Compounds*, **255**, 243-247.

[3] Duivenboden HCV, Zandbergen HW and Ijdo DJW 1986 Hexabarium titanium(IV) tetraniobate(V); a Rietveld refinement of neutron powder diffraction data *Acta Crystallographica Section C,* **42** 266-268.

[4] Abakumov AM, Shpanchenko RV, Antipov EV, Lebedev OI, Tendeloo G V and Amelinckx S 1998 Synthesis and structural study of hexagonal perovskites in the $Ba_5Ta_4O_{15}$-$MZrO_3$ (M = Ba, Sr) system *J. Soli. State Chemi.*, **141** 492-499

[5] Zhang H, Fang L, Su H, Yu Q and Wu B 2009 Improved microwave dielectric properties of $Ba_6Ti_{1-x}Sn_xNb_4O_{18}$ ceramics *J Am. Ceram. Soc.*, **92**, 238–240.

[6] Trolliard G, Teneze N, Boullay P, Manier M and Mercurio D 2003 HRTEM study of cation-deficient perovskite-related $A_nB_{n-\delta}O_{3n}$ ($n \geqslant 4\delta$) microphases in the $Ba_5Nb_4O_{15}$–$BaTiO_3$ system *J Solid State Chemi.*, **173** , 91-100.

[7] D ui venboden HCV, Zandbergen HW and Ijdo DJW 1986 Hexabarium titanium(IV) tetraniobate(V); a Rietveld refinement of neutron powder diffraction data *Acta Crystallographica Section C* **42** , 266-268

[8] Zhang H, Fang L, Su H, Yu Q and Wu B 2009 Improved microwave dielectric properties of $Ba_6Ti_{1-x}Sn_xNb_4O_{18}$ ceramics *J Am. Ceram. Soc.*, **92**, 238–240.

[9] Fang L, Zhang H, Yuan RZ and Liu HX 2005 $Ba_3La_3Ti_4NbO_{18}$: a new microwave dielectric of $A_6B_5O_{18}$-type cation-deficient perovskites *Materials Letters,* **59**, 535-538.

[10] Drews AR, Wong-Ng W, Vanderah T A and Roth R S 1997 Preparation and crystal structure of $Sr_6TiNb_4O_{18}$ *J Alloys.Comp.* **255**, 243-247.

[11] K.C. Patil and S.T. Aruna, Tanu Mimani 2002 Combustion synthesis: an update, *Curr. Opin. Solid. St. M.* **6** , 507.

[12] A.K Tyagi, S.V Chavan and R. D Purohit 2006 Visit to the fascinating world of nano-ceramic powders via solution combustion *Indian. J. Pure. Ap. Phy.* **44** 113

[13] J.C. Maxwell 1954 A treatise on Electricity and Magnetism Oxford University Press: Oxford U.K. **2** 328

[14] Raghavender A T and Jadhav K M 2009 Dielectric properties of Al-substituted Co ferrite nanoparticles *Bull. Materi. Science*, **32**, 575-578 .

[15] C.G. Koops 1951 On the Dispersion of Resistivity and Dielectric Constant of Some Semiconductors at Audiofrequencies, *Phys. Rev.* **83** 121.

[16] Rao K S, Prasad D M, Krishna P M, Tilak B and Varadarajulu K C 2006 Impedance and

modulus spectroscopy studies on $Ba_{0.1}Sr_{0.81}La_{0.06}Bi_2Nb_2O_9$ ceramic, *Materi. Science .Engi.: B,* **133**, 141-150.

Surfactant free synthesis and characterization of α- MoO_3 nanoplates: A feasibility study to remove methylene blue from aqueous medium

V M Jain[1,2], D V Shah[2], K K Patel[1], Y Doshi[2]

[1]Department of Physics, Uka Tarsadia University, Bardoli-394350, Dist-Surat, Gujarat, India
[2]Department of Physics, Sardar Vallabhbhai National Institute of Technology (SVNIT), Ichchanath, Surat-395007, Gujarat.India
Corresponding author's e-mail: **jvishu1990@gmail.com**

Abstract. Orthorhombic α-MoO_3 nanoplates have been synthesized by surfactant free chemical method. The prepared samples were analysed by various characterization techniques viz. X-ray diffraction (XRD), Scanning electron microscopy (SEM), Energy dispersive analysis of X-ray (EDAX), Fourier transform infrared spectroscopy (FTIR), Raman spectroscopy and Diffused reflectance spectroscopy (DRS). The XRD profile revealed the orthorhombic phase of the prepared material and obtained structural parameter confirms the formation of α- MoO_3. EDAX depicts the prepared material's compositional proportion and has no traces of the foreign elements as well. The topographic images of SEM indicate formation of nanoplate shaped MoO_3 having a thickness in range of 200 to 300 nm. FTIR and Raman, on the basis of vibrational behaviour of chemical bonds present in the material, confirm the formation of orthorhombic MoO_3. DRS study shows the band gap is about 2.95 eV. Further, the study of adsorption of methylene blue (MB) by synthesized MoO_3 shows its potential application in the waste water treatment.

1. Introduction

Molybdenum trioxide (MoO_3) occurs in varied crystal phases like monoclinic (β- MoO_3), hexagonal (h- MoO_3) and orthorhombic (α- MoO_3) [1]. Orthorhombic (α- MoO_3) is thermodynamically more stable in comparison to other phases and also possess unique layered structure. Each layer of it has octahedra MoO_6 as two sub-layers concorded with each other in (101) direction [2, 3]. Owing to this characteristic, it shows superior electrochemical, chemical, catalytic and electronic properties. As a result, α-MoO_3 had gained a significant importance in material world and have an various potential applications in the field of photo-catalysis [4], lithium ion batteries [5], electro-chromism [6], light emitting diodes [7], paints technology [8], as an antimicrobial agent [9] etc.

During the exploration of various synthesis routes for MoO_3, majority had preferred to employ non-vacuum or chemical methods to cut down the synthesis cost. While the complex methodology having consisting risky acids and sulphates pretended to a barricade towards the further deducing cost and high yield recipes as well [10-16]. Therefore, during the last couple of years researchers focused high throughput methodology by using simple techniques and harmless chemicals. In 2017, Haiyu Qin et al.[17] sysntheisied α- MoO_3 by solid state chemical reaction using polyethylene glycol and oxalic acid. While Neeraj Kumar et al. in 2017, synthesized hexagonal phase MoO_3 by co-precipitation method using nitric acid in the synthesis process [18]. By thermo-chemical method involving nitric acid, nanorod morphology of MoO_3 was reported by Azad Kumar et al. (2017) [19]. Nano needles morphology was synthesized by Rathnasamy et al. using sodium dodecyl sulfate in the precursor solution by the simple co-precipitation method [20]. Ja In Gu et al. (2018) prepared MoO_3 by hydrothermal method using nitric acid based precursor [21]. Guru Prakash et al. (2018) synthesized alpha phase MoO_3 by solution combustion method using urea and acetic acid [22]. Saihuan He et al. (2019) [23] used nitric acid as a solvent for the synthesis of orthorhombic MoO_3 by hydrothermal technique. Xin Guan et al. (2020) used hydro chloric acid (HCL) in synthesizing MoO_3 by hydrothermal method to study adsorption enhanced visible light photocatalytic activity [24]. Looking towards these recent studies; the different forms of acids were used in aqueous or non-aqueous medium to prepare pure MoO_3. We prepared orthorhombic α- MoO_3 nanoplates by surfactant free chemical method consisting only Ammonium heptamolybdate tetrahydrate (AHMT) and distilled water. Prepared material analyzed by various characterization techniques. In addition to that we had studied its adsorption potential to remove methylene blue (MB) from the aqueous medium, as part of feasibility test for the waste water treatment application. For instance, no one report such type of the method and its study.

2. Material and Methods

2.1. *Synthesis method*

The Analytical Reagent grade AHMT was purchased from Merck and was used further without any purification. 4% W/V concentration solution was prepared by dissolving AHMT in distilled water with continuously stirring the solution at 60 °C on 700 RPM for 30 minutes, so that AHMT decompose into MoO_3, NH_3 and H_2O [12].

$$(NH_4)_6Mo_7O_{24} \cdot 4H_2O + H_2O = 7MoO_3 + 6NH_3 + 8H_2O \quad (1)$$

Then after the solution was heated at 120 °C in order to evaporate the water from the solution and it results in a light green color powder left in the beaker. The left over powder was annealed at 400 °C in air. Annealed powder was then characterized by various sophisticated tools such as X-ray diffraction (XRD), Scanning electron microscopy (SEM), Energy dispersive analysis of X-ray (EDAX), Fourier transform infrared spectroscopy (FTIR), Raman spectroscopy and Diffused reflectance spectroscopy (DRS), to ensure the prepared material verification.

2.2. Adsorption measurement

The adsorption experiment was carried out in the dark environment at room temperature. Any form of external energy like heating, UV light exposure or stirring is not applied during the experiment, as these are the parameters may leads the increase adsorption rate. Annealed powder (i.e. MoO_3) was taken about 30 mg in 100 ml of solution of MB having concentration of 10 mg/L. Then after every 10 minute, 3 ml of solution was taken out for UV- Visible measurement. The 0.22 μm syringe filters was used for filtering MoO_3 particles during each UV-VIS measurement. Removal percentage of MB was calculated by following expression [25].

$$\text{Removal (\%)} = \frac{C_i - C_f}{C_i} \text{ x } 100\ (\%) \quad (2)$$

Where, C_i is initial and C_f is final concentration of MB.

3. Result and Discussion

3.1. Structure analysis

The structural properties of the prepared powder was investigated using Bruker make X-ray diffraction (XRD) having Cu Kα radiation of wavelength 1.54 Å. Figure 1 shows the XRD profile of the synthesized sample having major sharp peaks at 2θ values of 23.27 °, 25.65 ° and 27.25 °. The corresponding calculated d-spacing (3.80 Å, 3.47 Å, and 3.260 Å respectively) identified as (110), (040) and (021) planes using standard JCPDS card no. 05-0508 [8, 28]. That confirms, prepared material has the orthorhombic crystalline structure of MoO_3. Other minor peaks are also well matched with the standard JCPDS card. In addition to that, not a single peak was identified for other phases of MoO_3 viz. hexagonal and monoclinic [26-27].

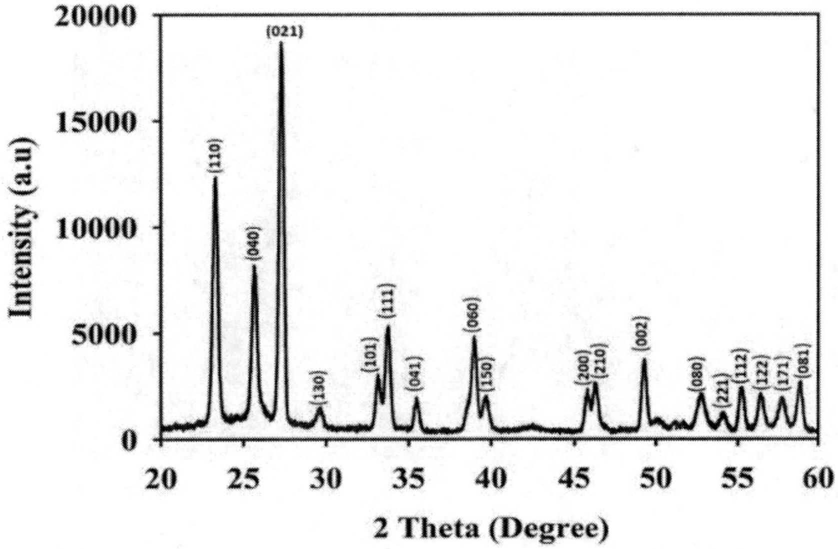

Figure 1 XRD pattern of α- MoO_3 sample annealed at 400 °C.

The d-spacing for prepared sample was evaluated using the Bragg's equation (3) [13],

$$n\lambda = 2d\sin\theta \tag{3}$$

Where, n represents the diffraction order, λ represent X-ray wavelength, θ represent angle of diffraction and d represents the interplanar distance of the parallel planes. Lattice parameters were evaluated using the following equation (4) [13].

$$\frac{1}{d^2} = \frac{h^2}{a^2} + \frac{K^2}{b^2} + \frac{l^2}{c^2} \tag{4}$$

The evaluated lattice parameters are a, b and c is 0.3959 nm, 1.3852 nm and 0.3694 nm respectively, which are well matched with JCPDS card no. 05-0508. The average crystallize size (i.e. 23 nm) was calculated for (021) plane at 2θ= 27.25° by using Debye- Scherer relation shown in equation (5) [13].

$$D = \frac{0.9\lambda}{\beta \mathrm{Cos}\theta} \tag{5}$$

Where, λ is the wavelength of X-ray used, θ is the position of major intense peak in degrees and β is the full width half maximum (FWHM) in radians.

The dislocation density and the micro strain were evaluated using below equation (6) & (7) [13] and were found 1.87 x10^{15} per m^2 and 0.0067 respectively.

$$\delta = \frac{1}{D^2} \quad (6)$$

$$\epsilon = \frac{\beta \cos\theta}{4} \quad (7)$$

3.2. Compositional Analysis

EDAX technique was utilized to measure the proportion of the elements present in the prepared sample. Molybdenum and oxygen contain in the sample was 25.5 % and 74.5 %, that match well with atomic percentage of its chemical formula and having the elemental ratio nearly equal to 3. Moreover, no other peak of any element was observed in the spectrum which represents the pure state of prepared MoO_3 sample.

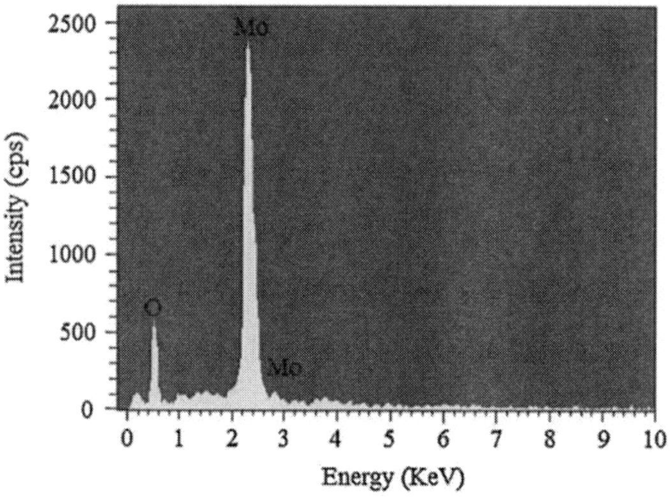

Figure 2 Elemental composition spectra of α-MoO_3 sample.

3.3. Morphological Analysis

Scanning electron microscopy (SEM), make JEOL Japan was used to study the surface morphology and elemental composition of the prepared MoO_3 sample. Figure 3 shows the SEM images of prepared MoO_3 having the resolution of 2 μm and 0.5 μm. It clearly reveals the uniform plate like nanostructure of the prepared sample. The length of the sample ranges from 1 μm to 5 μm and the thickness ranges from 200 nm to 300 nm. The layered type structure of MoO_3 is been reported in the literature [8, 17, 26]. In which one layer is been overlaid on the other layer. Such type of plate like structure may have more slender sub plate intercalated inside. The elongation in one direction could be due to the thermal annealing effect at 400 °C. As seen from SEM images, these elongated stacked layers seem to be a loosely bounded to each other. If separated, that can be used in any nano-sized optoelectronic devices.

Figure 3 Suraface morphology micrograph of α- MoO_3 sample.

3.4. FTIR Analysis

FTIR (Bruker make Alpha- T, FTIR) was used to assure the presence of different types chemical bonds and its basic behavior too. Figure 4 shows the FTIR spectrum of MoO_3 sample measured in the range of 4000-600 cm^{-1}. There were only two peaks in the spectrum, one at 993.3 cm^{-1} and other at 862 cm^{-1}. A peak at higher frequency (i.e. at 993.3 cm^{-1}), mostly reported for the stretched bond of Mo=O, and it is the significance of layered orthorhombic nature of α-MoO_3 [27, 28]. At the other end, lower frequency (i.e. 862 cm^{-1}) is generally assigned for Mo–O–Mo bond in which molybdenum is present in the +6 oxidation state [29]. Traces of N-H bond and water (H_2O) were not observed in the spectrum of sample.

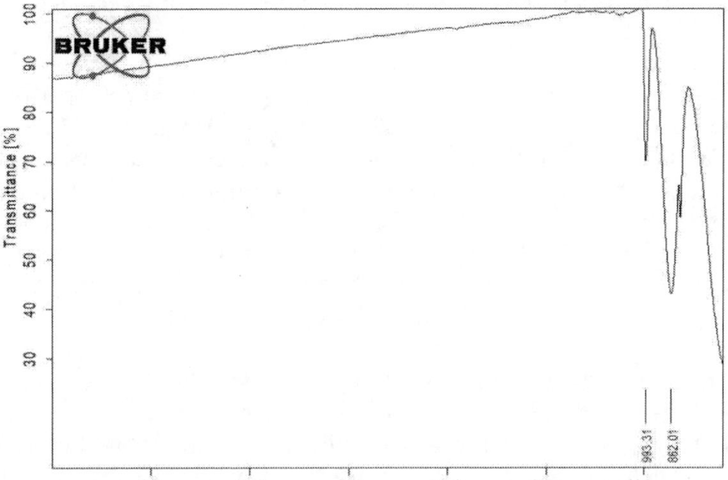

Figure 4 Fourier Transform-Infrared spectra of α- MoO_3 sample.

3.5. Raman Analysis

Raman analysis was accomplished by Renishaw make Invia Raman microscope. In Raman spectra, there are three main regions indicating the significant structural modes of vibration such as, lattice modes (at below 200 cm^{-1}), deformation mode (200 - 400 cm^{-1}) and stretching mode (600 - 1000 cm^{-1})

[30]. In the present study prepared sample shows (Figure 5) the intense peaks at 157.71 cm^{-1}, 287.21 cm^{-1}, 665.79 cm^{-1}, 818.25 cm^{-1} and 995.35 cm^{-1}. In the stretching mode we observed a three peaks; 995.35 cm^{-1}, 818.25 cm^{-1} and 665.79 cm^{-1}. These peaks correspond to stretching of the bonds Mo=O, Mo_2- O and Mo_3-O respectively [21, 30]. Hence, the peaks present in the stretching mode shows the sign of orthorhombic phase of MoO_3. The remaining peaks in the spectra i.e. 157.71 cm^{-1} and 287.21 cm^{-1}, are normally attribute to wagging modes of terminal oxygen [30].

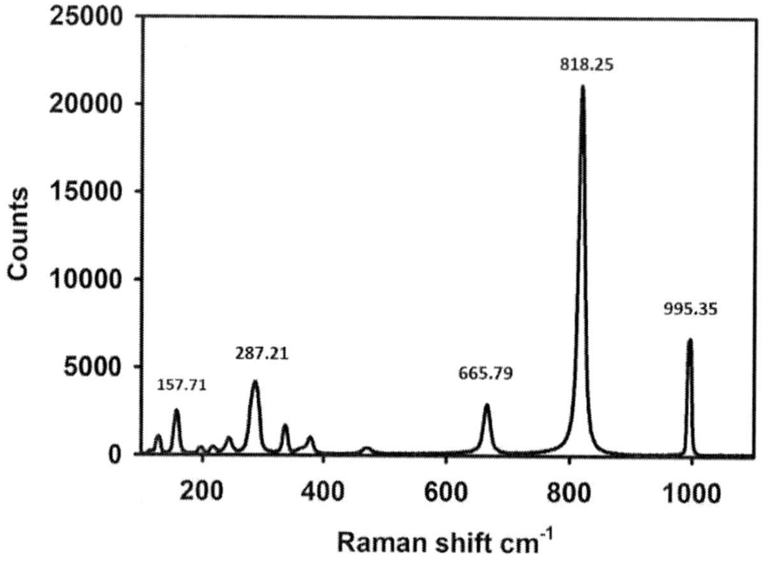

Figure 5 Raman spectra, vibrational mode identification of α- MoO_3 sample.

3.6. *Optical Properties*

In the study of optical properties, diffuse reflectance measurement was performed by UV-Visible spectrophotometer (Shimadzu make UV3800, Japan) of synthesized MoO_3 sample at room temperature. Figure 6(a) shows the diffuse reflectance spectrum in the wavelength range from 300 to 800 nm. The average reflectance intensity is about 55 % after 450 nm. In addition to this, a broad hump is clearly observed between 650 nm to 700 nm range, that may be due to inter-band transition in the structure arises because of defect bands. Due to the powder form of the synthesized sample, the Kubulka –Munk (K-M) function (equation 8) was used to calculate the band gap.

$$\frac{(1-R_\infty)^2}{2R_\infty} = F(R) = \frac{K}{S} \quad (8)$$

Where, F(R) represents the K-M function, k represents the absorption coefficient, S represents the scattering and R_∞ represents the reflectance at infinite thickness. The band gap of the prepared material is observed at 2.95 eV (Fig. 6(b)) that is well accordance with the reported values [26]. The band gap (2.95 eV) is also match with our visual color inspection of the powder i.e. light green color. The reduction of MoO_3 leads to creates defect states, and that contributes to the light green color, also well explained in the literature [31, 32].

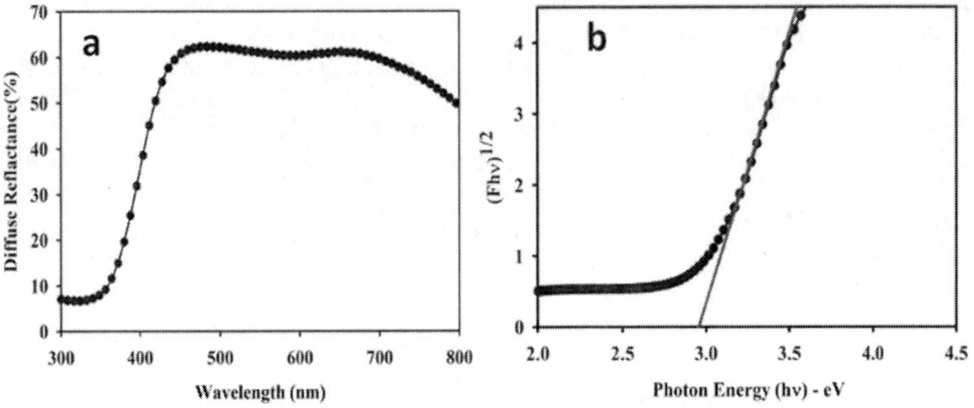

Figure 6 (a) Diffuse reflectance spectra of α- MoO_3 and (b) Tauc plot of α- MoO_3 sample.

3.7. Adsorption Activity

The adsorption activity of synthesized α- MoO_3 nanoplates has been studied under dark condition. Consider the 100 ml solution of MB containing 30 mg MoO_3 is a dye solution. This dye solution undergoes for UV-VIS spectroscopy measurement to check the adsorption (i.e. removal) of MB. In the Figure 7 (a), MB curve (pure MB solution) shows the highest absorption of visible light. Then after every 10 minute of interval, dye solution undergoes for UV-VIS measurement. As the UV-VIS spectra shows, there is a continuous decrease of the absorption values as the time increase. After around 130 minutes the absorption is seems to be zero. That indicates the MB is completely adsorbed by the MoO_3. In other words, we can say that MoO_3 is capable to remove the MB of aqueous solution. Figure 7(b) shows the removal of MB (in percentage) from dye solution. It clearly shows that up to around 90 minutes the removal rate is increases about linearly. Then after is slow down and saturate around 130 minutes.

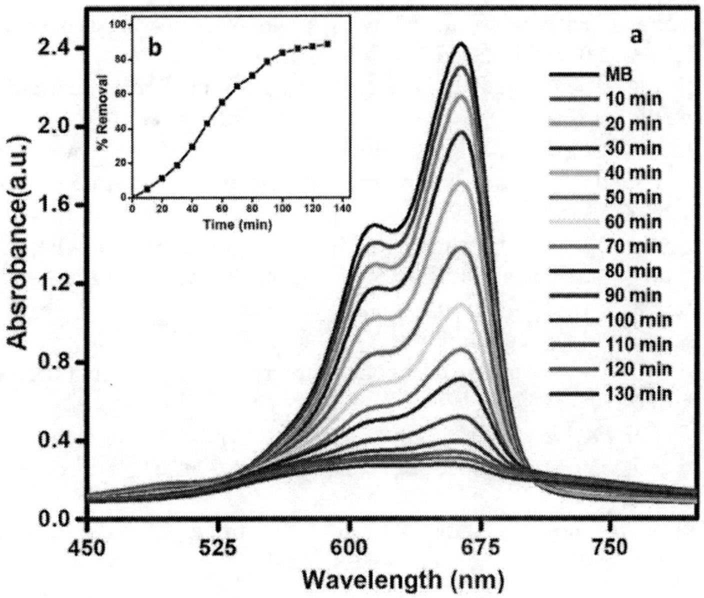

Figure 7 (a) Absorbance Vs. wavelength at every 10 minutes interval and (b) % Removal Vs. Time for removal of MB from dye solution.

4. Conclusion

Simplicity and easiness in synthesizing MoO_3 are the key features of this method. That leads to ecofriendly, economical and scalable synthesis of α- MoO_3 nanoplates without using any surfactant. The XRD confirms orthorhombic crystalline structure of α- MoO_3 with crystallize size of 23 nm. The stoichiometric composition of the MoO_3 was confirmed by EDAX. The SEM images reveal the formation of layered MoO_3 nanoplates. The thickness of the nanoplates ranges from 200 nm to 300 nm. FTIR peaks at 993 cm^{-1} shows Mo=O stretching and 862 cm^{-1} shows Mo–O–Mo vibration. Peaks at 665.79, 818.25 and 995.35 cm^{-1} in the Raman spectrum confirms the orthorhombic phase of MoO_3. The sample shows the direct band gap of 2.95 eV. The synthesized α-MoO_3 nanoplates have removed almost 90% of methylene blue from the aqueous solution, and might be utilize in the waste water treatment purpose.

5. Acknowledgment

The work was supported by Uka Tarsadia University, Bardoli, through B.U.Patel Research Promotion Scheme [UTU/RPS/1261/2018]. We would like to appreciate Mr. Harsh support for material characterization, USIC Delhi.

References

[1] H. Lunk, H. Hartl, M. Hartl, M. Fait, I. Shenderovich, M. Feist, A. Gurinov. "Hexagonal molybdenum trioxide" - Known for 100 years and still a fount of new discoveries. 2010 *Inorg Chem.* **49(20)** 9400–9408.

[2] L. Noerochim, J. Wang, D. Wexler, Z. Chao, H. Liu. Rapid synthesis of free-standing MoO_3/Graphene films by the microwave hydrothermal method as cathode for bendable lithium batteries. 2013 *J Power Sources.* **228** 198–205.

[3] J. Chen, Y. Cheah, S. Madhavi, X. Lou. Fast Synthesis of r-MoO_3 Nanorods with Controlled Aspect Ratios and Their Enhanced Lithium Storage Capabilities. 2010 *J. Phys. Chem. C* 8675–8678.

[4] A. Chithambararaj, S. Sanjini, A. Bose, S. Velmathi. Flower-like hierarchical h-MoO_3: New findings of efficient visible light driven nano photocatalyst for methylene blue degradation. 2013 *Catal Sci Technol.* **3(5)** 1405–1414.

[5] U. Sen, S. Mitra. Electrochemical activity of α-MoO_3 nano-belts as lithium-ion battery cathode. 2012 *RSC Adv.* **2(29)** 11123–11131.

[6] J. Yao, Y. Yang, B. Loo. Enhancement of photochromism and electrochromism in MoO_3/Au and MoO_3/Pt thin films. 1998 *J Phy Chem .* **102(11)** 1856–1860.

[7] Q. Fu, J. Chen, C. Shi, D. Ma. Room-temperature sol-gel derived molybdenum oxide thin films for efficient and stable solution-processed organic light-emitting diodes. 2013 *ACS Appl Mater Inter.* **5(13)** 6024–6029.

[8] K. Krishnamoorthy, M. Premanathan, M. Veerapandian, S. Jae Kim. Nanostructured molybdenum oxide-based antibacterial paint: Effective growth inhibition of various pathogenic bacteria. 2014 *Nanotechnology.* **25(31)**.

[9] K. Krishnamoorthy, M. Veerapandian, K. Yun, S. Kim. New function of molybdenum trioxide nanoplates: Toxicity towards pathogenic bacteria through membrane stress. 2013 *Colloids Surfaces B.* **112** 521–524.

[10] X. Chen, W. Lei, D. Liu, J. Hao, Q. Cui, G. Zou. Synthesis and characterization of hexagonal and truncated hexagonal shaped MoO_3 nanoplates. 2009 *J. Phys. Chem. C.* **113(52)** 21582–21585.

[11] Y. Zhao, J. Liu, Y. Zhou, Z. Zhang, Y. Xu, H. Naramoto, S. Yamamoto. Preparation of MoO_3 nanostructures and their optical properties. 2003 *J. Phys. Condens Matter.* **15(35)**

[12] S. Murase, Y. Yang, Solution processed MoO_3 interfacial layer for organic photovoltaics prepared by a facile synthesis method. 2012 *Adv Mater.* **24(18)** 2459–2462.

[13] S. Deshmukh, V. Kheraj. Comprehensive review on synthesis and characterizations of Cu_3BiS_3 thin films for solar photovoltaics. 2017 *Nanotechnol. Environ. Eng.* **2(15).**

[14] C. Subba Reddy, E. Walker, C. Wen, S. Mho. Hydrothermal synthesis of MoO_3 nanobelts utilizing poly(ethylene glycol). 2008 *J Power Sources*. **183(1)** 330–333.

[15] L. Fang, Y. Shu, A. Wang, T. Zhang. Template-free synthesis of molybdenum oxide-based hierarchical microstructures at low temperatures. 2008 *J Cryst Growth*. **310(21)** 4593–4600.

[16] X. Lou, H. Zeng. Hydrothermal synthesis of α-MoO_3 nanorods via acidification of ammonium heptamolybdate tetrahydrate. 2002 *Chem Mater*. **14(11)** 4781–4789.

[17] H. Qin, Y. Cao, J. Xie, H. Xu, D. Jia. Solid-state chemical synthesis and xylene-sensing properties of A-MoO_3 arrays assembled by nanoplates. 2017 *Sensor Actuat B: Chem.* **242** 769–776.

[18] N. Kumar, B. George, H. Abrahamse, V. Parashar, J. Ngila. Sustainable one-step synthesis of hierarchical microspheres of PEGylated MoS_2 nanosheets and MoO_3 nanorods: Their cytotoxicity towards lung and breast cancer cells. 2017 *Appl Surf Sci.* **396** 8–18.

[19] A. Kumar, G. Pandey. Synthesis, Characterization, Effect of Temperature on Band Gap Energy of Molybdenum Oxide Nano Rods and Their Antibacterial Activity. 2017 *American Journal of Applied and Industrial Chemistry*. **3(3)** 38-42.

[20] R. Rathnasamy, V. Alagan. A facile synthesis and characterization of α-MoO_3 nanoneedles and nanoplates for visible-light photocatalytic application. 2018 *Physica E.* **102** 146–152.

[21] J. Gu, J. Lee, C. Rhee, Y. Sohn. Enhanced electrochemical hydrogen evolution over defect-induced hybrid $MoO_3/Mo_3O_9 \cdot H_2O$ microrods. 2019 *Appl Surf Sci.***469** 348–356.

[22] N. Prakash, M. Dhananjaya, A. Narayana, D. Shaik, P. Rosaiah, O. Hussain. High Performance One Dimensional α-MoO_3 Nanorods for Supercapacitor Applications. 2018 *Ceram Int.*; **44(8)** 9967–9975.

[23] S.He, W. Li, L. Feng, W. Yang. Rational interaction between the aimed gas and oxide surfaces enabling high-performance sensor: The case of acidic α-MoO_3 nanorods for selective detection of triethylamine. 2019 *J Alloy Compd*. **783** 574–582.

[24] X. Guan, Y. Ren, S. Chen, J. Yan, G. Wang, H. Zhao, C. Liu. Charge separation and strong adsorption-enhanced MoO_3 visible light photocatalytic performance. 2020 *J Mate Sci.*; **55(14)** 5808–5822.

[25] S. Rakass, H. Hassani, M. Abboudi, F. Kooli, A. Mohmoud, A. Aljuhani, F. Wadaani. Molybdenum trioxide: Efficient nanosorbent for removal of methylene blue dye from aqueous solutions. 2018 *Molecules*. **23(9)** 16-21.

[26] A. Chithambararaj, A. Bose. Hydrothermal synthesis of hexagonal and orthorhombic MoO_3 nanoparticles. 2011 *J Alloy Compd.*; **509(31)** 8105–8110.

[27] JCPDS Card: 47-1081

[28] G. Nagabhushana, D. Samrat, G. Chandrappa. α-MoO_3 nanoparticles: Solution combustion synthesis, photocatalytic and electrochemical properties. 2014 *RSC Adv.***4(100)** 56784–56790.

[29] P. Wongkrua, T. Thongtem, S. Thongtem. Synthesis of h- and α -MoO_3 by refluxing and calcination combination: Phase and morphology transformation, photocatalysis, and photosensitization. 2013 *J Nanomater*. **2013** p-8.

[30] F. Emanuela, B. Francesca, T. Marco, R. Maria, C. Daniela, T. Antonio. Characterization of hierarchical α-MoO_3 plates toward resistive heating synthesis: electrochemical activity of α-MoO_3/Pt modified electrode toward methanol oxidation at neutral pH. 2017 *Nanotechnology*. **28(21)** 215601.

[31] A. Bararaj, D. Mathi, N. Yogamalar, A. Bose. Structural evolution and phase transition of $[NH_4]_6Mo_7O_{24}.4H_2O$ to 2D layered MoO_{3-x}. 2015 *Mater Res Express*. **2(5)** 1-10.

[32] M. Dieterle, G. Weinberg, G. Mestl. Raman spectroscopy of molybdenum oxides - Part I. Structural characterization of oxygen defects in MoO_{3-x} by DR UV/VIS, Raman spectroscopy and X-ray diffraction. 2020 *Phys Chem Chem Phys*. **4(5)** 812–821.

Influence of Cavity Materials and Selective Surface Coatings on the Performance of SCC

Dr. Lakshmipathy. B#[1], Dr. Senthil kumar. M[2], Dr. Sivakumar. K[3], Dr. Kajavali. A[4]

[1]Corresponding Author - Assistant Professor of Mechanical Engineering Department, Faculty of Engineering and Technology, Annamalai University, Annamalainagar, Pin - 608 002 - India, Deputed to Government polytechnic college, Kadathur, Dharmapuri District.
[2]Assistant Professor of Mechanical Engineering Department, Annamalai University. Deputed to Government polytechnic college, Dharmapuri
[3]Associate Professor-Department of Mechanical Engg- Annamalai University. At present working in Government polytechnic college, Nagercoil
[4] Associate Professor- Department of Mechanical Engg- Annamalai University.At present working in Government Engineering college, Bargur
#1. Email: blakshmipathy10@gmail.com, Contact: 9944627787/9080139880

Abstract

It is evident that the future energy demand can be compensated by using solar energy. The technological improvements are needed to trap the sunlight and have to be utilized in an efficient way. It mainly depends on the type of material which is to be used for this purpose. Solar energy can be effectively captured and utilized to heat the heat transfer fluids. Therefore the usage of solar energy is a right choice as it has the ample availability in countries like India.Solar cavity collector [SCC] is a kind of improvised, modified and redesigned construction of a solar Flat Plate Collector (FPC). In short, SCC is an upgraded and advanced version of FPC. A Copper pipe has been positioned concentrically fitted to a radius (outer) of 16 mm and placed into a metal box like structure having a size of 50 mm. In a similar way, five numbers of such cavities are formed. It has a provision for inlet and outlet water pipes. It has been designed, fabricated and experimented for its optimized level of performance. This paper explains about the usage of materials and solar selective surface coatings that are applicable to Solar Cavity Collector (SCC). The following experimental modifications are carried out in the SCC which are 1.Reciever materials of the SCC have been changed as aluminum and copper. 2. Cavity materials as Galvanized Iron (G.I) material and Mild Steel (M.S) material, both are experimented. 3. Solar selective coatings (SSC) of Black nickel have been used in SCC for the heat transfer improvement.
Keywords: Solar Energy, Cavity materials,Solar Cavity,Solar Selective coatings, cavity configuration.

Introduction

The Traditional usages of combustion of fossil fuels like wood, coal, Petroleum products, etc. have caused major air pollution in many areasglobally and associated with lot of problems and diseases.The reason for this is because of the localized release of large amounts of harmful gases including carbon-di-oxide into the atmosphere. And finally it results in the phenomenon of global warming which is now a focused matter of a greater concern.Among the renewable energies, solar energy has been a right candidate for its clean, abundant available, inexhaustible, non-polluting, efficient and more effective renewable energy source.

Solar energy incidental radiation can be easily transformed into Heat or thermal energy for different kind of applications. Mainly it is used for heating and production of hot water by using solar gadgets.With the list of researchers, the cavity collecting methods are mostly used only the following type of collectors such as concentrating and Fresnel lens type collectors. Obviously, a flat plate collector has to be improved in order to works more efficiently. A typical experimentation is needed to ensure the performance whether the cavity like structure can be utilized or not. After experimentations it came to know that, for optimizing the heat transfer characteristics cavity collectors has more impact than the other methods. One more advantage of SCC is that, inevitable kind of intermittent nature of solar radiations during at cloudy days also it works better when compared to Flat plate.The purpose of SCC which is implemented towards the modification on FPC leads to give better results even if it receives on the earth's surface is in the form of intermittent type of solar radiation because of climatical changes may occurs on time of the day.

Melchior et al. (1) creates and explains a cubical absorbing surface which is fixed to a cavity model receiver of absorber surface. This model has come with some modified solar reactor and it has been developed and validated experimentally. A tubular Ceramic absorber is used inside the solar chemical reactor that utilizes thermo chemical process. Because of this, the ability to produce higher temperature in a shorter duration of time. This implies to produce a generation of H_2 and other power sources. Performance analysis of cavity model with a cone structure fixed to a model has been described by Hahm et al (2). A cone is created and fixed within the cavity receiver and the model has been generated with some boundary conditions. The following observations are listed down. A conical surface having a small aperture at exit state suffers more Consequently high return/ rejected solar radiations from top. Therefore, lesser the losses from the surface. Similarly, if the surface has a larger aperture then the thermal losses are potentially more. Demichelis and Russo (3) have studied the determination of multi reflector effect on the slope surface macro cavity,cylinder in shape. This paper focuses on the design part of fundamentals of optical losses in cavity and also it explains about the cavity effect. Accordingly, a setoff observation has been expelled and usages of macro cavity were determined for the usage of concentrating collectors. Reyes et al (4) was presented the optical performance of solar collectors with Non-isothermal of flat plate type collector specifically. The feasibility studies of the model and in such a manner the creation of thermo-economic analysis of the proposed model have been reported. Also the determination of annual cost for the operational procedure of solar air heater has been calculated by the means of dimensionless and non- dimensional parameters. Also it includes empirical correlations, mass flow number, thermo economic feasible correlations, etc.

Singh et al (5) has studied and enumerates the convection mode of heat transfer mechanism happenings in a cavity. Also this can be a complex function to control all the parameters such as aspect ratio, shapes of the cavities, conditions of the boundary walls if it incorporates in regular shaped cavities. Natural convection at the end of walls also affects the shape of cavities like square, rectangular and cylindrical shaped surfaces. They analyzed and tested in the above aspects and the results shows while reducing the convective heat transfer significantly, performance of CPC (Compound Parabolic Collector) collectors can be improved. Also, they reported with different correlations which include mass number, Nusselt number and flow patterns used for their research. Bairi (6) form a parallelogrammic 2 Dim cavity numerically with certain wall conditions. The study includes the heat transfer and optimal operating parameters for the designed cavity system. He applied the heat exchanging possibilities between active and passive walls and the same have been compared with various Nusselt number correlations and several inclination angles (α).The thermal, optical losses and how it affects the system of the dish type collection system has been reported by Prakash et al (7). They conduct a performance test for the inlet fluid temperatures range from 50°C to 75°C.They made another more test with various receiver inclinations of 0°, 30°, 45°, 60° and 90°. The results were compiled by using the Fluent CFD software. They reported that that the Convective loss increases with the mean receiver temperature and decreases with increase in receiver inclination angles respectively. Kribus et al (8) procured an experimental setup which uses two heating stages namely, lower temperature stages and high interactive temperature receiver stages which is termed as directly irradiated receiver. That is, a partial ring is introduced in the intermediate temperature zones and further it divides a cavity tubular receiver into two stages.These two stages has been applied and implemented for the improvement of generation of high temperature. Their end results comprise the reduced convective heat losses and also the partitioning losses were minimized.

Lakshmipathy and Sivaraman (9) has been investigated the various influencing and inducing parameters of SCC. Furthermore, they investigate among the different working parameters which one guides more heat transfer than a flat plate SCC. Experimentations have been made with various lengths to diameter ratios (L/D) based on the commercially available pipe diameters. They concluded that the temperature of the water has an increase at exit if the ratio of L/D increases. While cavity area increases with the increase in diameter of the receiver decreases with the increase in L/D ratio and so on. Cruz et al (10) reported about the low cost passive water heater which can comprise the regulative and rejuvenate design, construction, and testing procedural activities. In their present study, they tested the model with a constant tilt angle of 45° especially designed for the climate in Portugal. Their conclusion explains about the energy saving largely depends onthermal stratification and model design aspects within the storage cavity. The experimentations with different inclination angle have been conducted by Lakshmipathy and Sivaraman (11) for the optimum performance of SCC. The results show that a maximum efficiency of 59% has been obtained with 11° inclination angle. At the same timethe 20° inclination angle yields an efficiency of 56%. Also, they conclude that the preferable/ optimum level of inclination tilt angle for SCC is 11° with the test results. Flores et al. (12) had made the analysis of heat transfer with in the cavity body and surfaces of the collector. They strongly agree with the fact that theconvective heat transfer does not plays a vital role when compared to the radiative heat transfer that in a cubic cavity. The analysis is made compared with the developed mathematical model and carried out a parametric study with specific conditions for various solar control coating (SCC) absorptances. This study is to enlightening the heat transfer characteristics of the model with coatings.Garcia et al. (13) has studied about the thermal analysis of solar cavities provided and fitted with baffle spacing. They deliberately explain and reported the thermal energy distortion in baffle gaps. A cavitywith baffles separated with some spacing provided has been fabricated and tested.Their results ensure that the specific dimensions of baffles spacing gives a better improvement on reduction of heat transfer considerably. Also the convection mode of heat transfer loss is much reduced in particularly.

Section 1.01 Aim of the Proposed present Work

Some of the aim/ objectives and salient features of this work are explained and listed below

- The performance test on SCC has to be conducted and to find out the instantaneous efficiencies for each modification.
- SCC has to be tested by varying such important working parameters which includes, changing the cavity materials, changing the receiver materials and different watermass flow rates.
- Cavity collector has been experimented for Solar Selective (absorber) receiver coatings.
- Comparisons have to be made with the modifications done and getting better appropriate solutions and finally finding out the optimum performance parameter.
- Design changes if any have been made to do such that, the reflection an immense effect on the performance of SCC.
- Also the parametric analysis of SCC has to make to get the better results for which parameter will influences much more between them.

Solar Cavity Collection - An Introduction

This section explains about the cavity fundamentals and its operations.

The cavity structure or simply called as a cavity of the collector. Usually, it enhances the availability of heat present inside in SCC. Furthermore, the flat plate collector's heat holding capacity increased considerably with the usage of cavity model of configuration. The net available heat energy inside the SCC decides the efficient capture of solar radiation and thus the increased level of total heat availability have been effectively transferred to the any kind of liquid often called as working fluid (water) that circulates in the receiver tubes. The available heat energy holds on cavity surfaces both inside and outside ultimately increases the Thermal and overall efficiency of the system as well. Existence of cavity inside the SCC compels the collector more effective during the time of late afternoon hours. It occurs because of the radiation multiple reflection effect of the light energy enters and gets reflected within the cavity. The main limitations of the conventional FPC are that they don't have the temperature stability throughout the time of the day. The fluctuations in outlet water temperature on part cloudy days might happen and whenever there is intermittent type of radiation is received. The heat fluctuations are compensated by the cavity structure itself to meet out the desired water outlet temperature. Therefore it ensures there is no sudden drop in the exit temperature of water by its virtue of operation. Cavity type configuration most often used in Fresnel lens collector and Concentrating type collectors. For experimentation purpose it has been tested for the flat plate configuration improvements.

Structure of a Specific view of a cavity (single) look like has been showed in Fig.1. It consists of three main parts which includes the receiver, cavity support structure and aperture. Receiver is a part that can transform the available solar radiation into heat energy. Cavity structure is used to withstand and holds the cavity. The aperture view on top is a small opening which allows the solar radiation rays inside to the cavity. Also it explains the location of the cavity and receiver tubes as well with the help of the figure. Fig.2 describes the solar radiation enters into the SCC and how the multi reflection is functioning inside the cavity. Based on the illustration, the heat is transferrred to the receiver tubes through the cavity can be understood. Solar radiation admitted through the aperture, then gets deflected and reflected by the cavity. Again, and again it re radiates back to the receiver tube. The reflection has been continued and prolonged inside the cavity. Some part of it may escaped from the cavity and cavity top through the aperture opening. The receiver absorbed the reflected light rays which comes in and receives the heat energy by virtue of its position and thus it heats up the operating substance like water.

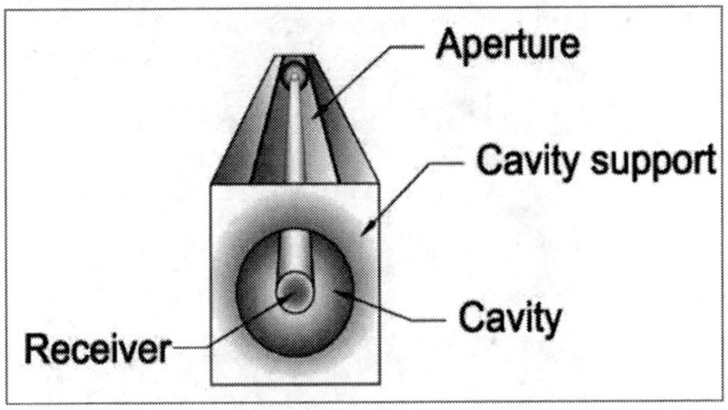

Fig.1. specific view of a cavity (single)

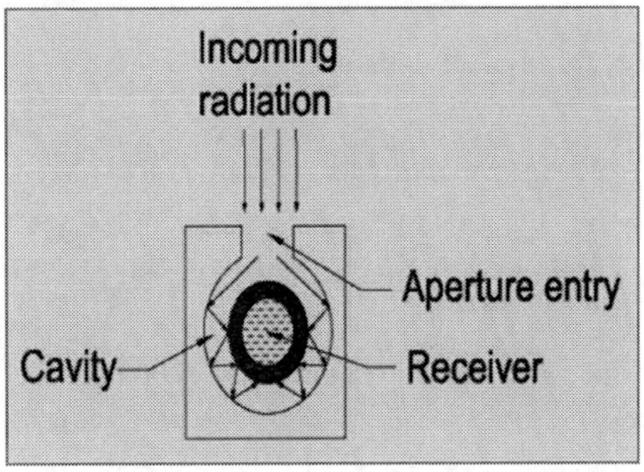

Fig.2 solar radiation multi reflection process in SCC

Section 1.02 Details of Experimental Set up And Testing Method

The SCC experimental setup consists of cylinder made up structure termed as a cavity.It is made up of the Aluminium and copper material. It has a 16 mm of radius with black painting coated and insulation made with glass wool material on bottom and sides of the collector setup. This comprises a single cavity and similarly the same procedure is followed for constructing five numbers of cavities. It has been put down in a square metal enclosure with identical distances. A tubular receiver cum absorber of radius 6.35mm is located in concentric all manners. The same is coated with the black paint as shown in the Fig.3. It manifests the photographic view of the cavity collector. Fig.4 illustrates the experimental arrangement of SCC with all dimensions. In order to set out as safeguarding shield for radiation, a single glass cover is fixed at the top surface. Also it reduces the top loss of heat from the collector to the surroundings. The top outlet tube is connected with a pipe having provisions for collecting the water and the bottom end is connected to the fresh water tank. The entire experimental setup is tilted to an inclination angle of 11° with respect to horizontal.

The instrument called Pyranometer is used to measure the global radiation at the location of Annamalai nagar and its Latitude is 11°N.Ambient temperature is recorded with a mercury thermometer has a precision of 0.1°C. Temperatures at different locations of cavity collector such as all cavities, glass plate, receivers, etc have been measured by thermocouples which are connected to a digital temperature indicator. Temperatures at different locations on SCC such as all cavities, glass plate, inlet and outlet pipes have been fixed with thermocouples. Thermocouples of Copper – constantan type are used for this purpose. The collector has been properly insulated at besides and bottom sides to reduce the heat losses escapes from the collector. SCC is kept in open yard facing south and exposed to sun with maximum possible solar radiations has to fall. Experiments are conducted on the time of the days from 9.30 AM to 5 PM. Observations have been made with a time interval of 10 minutes on different days with different mass flow rates of water. Performance investigations on the SCC have been made in order to optimizing the operating parameters. The parameters which have been carried out such as receiver materials of the SCC have been changed as aluminum and copper. 2. Cavity materials of both Galvanized iron (G.I) material and mild steel (M.S) plates are experimented to get better results. 3. Solar selective coatings (SSC) of black nickel have been used in SCC for the heat transfer improvement. The constructional details of SCC have been shown in Table 1. It enumerates the inner details of SCC in detail. Similarly, Fig.5 illustrates the Schematic sketch of a cavity collector.

Table 1 - Constructional details of SCC

Area of each cavity	: 0.101 m^2
Collector size	: 1.25 × 0.85 × 0.05 m
Diameter of the tube	: 0.0127m
Number of cavities	: 5
Receiver material	: copper
Receiver coatings	: Industrial mat black paint
Thickness of glass plate	: 0.004m
Collector insulation	: Glass wool

Fig.3 Photographic view of SCC

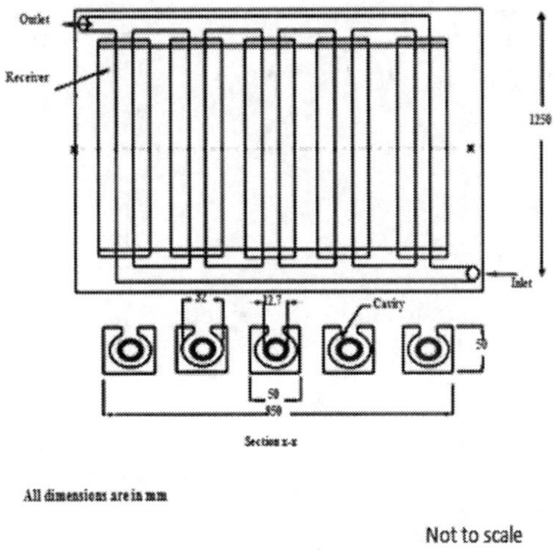

Fig.4. Experimental arrangement of cavity collector

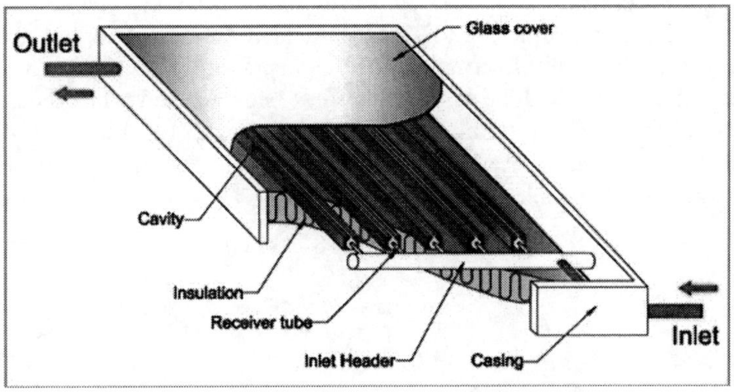

Fig.5. Schematic sketch of a cavity collector

Results and Discussion

An attempt to a new design and development towards Flat plate collector, SCC has been established and conducted experiments for different kind of working parameters with their end results are listed down.

For highly concentrated solar applications like solar power plant operation, conversion of water into steam, etc, normally can employ the cavity type arrangements and often equipped with any one type of concentrating collectors. The cavity receiver has enormous advantage over FPC as it has the radiative energy gets magnified using multiple reflection of inner surface of the cavity certainly. Especially a proper design of the cavity relatively increases effective capture of sun light enters in it. The air filled inside the cavity can absorbs the radiative heat; consequently it withstands the temperature inside and dispenses the heat to the surrounding operating fluid. Fig. 6 illustrates mass flow versus Water outlet temperature for both receiver materials copper and Aluminium. Also it is inferred that the usage of receiver materials especially copper works more efficiently than the aluminum. Maximum water outlet temperature of 68^0C has been obtained when copper material used.

It should be noted if copper being used as a receiver material, obviously there was no sudden decrease in temperature due to its capability of retains heat.

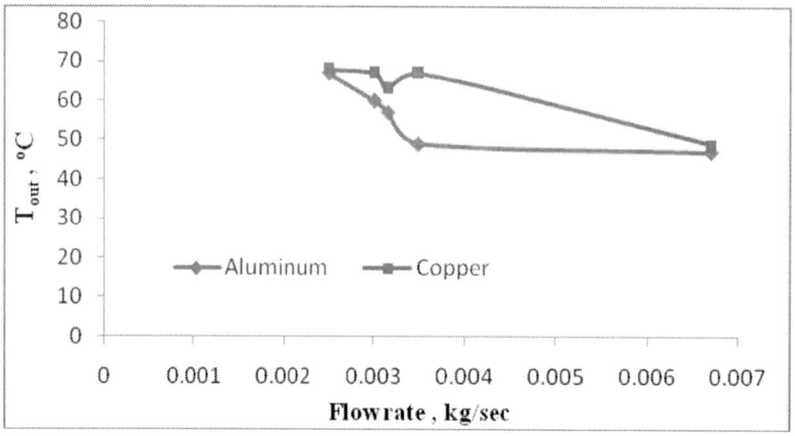

Fig.6. Differentiation of receiver materials being used

It is important to notify that the cavity case cover materials have a grater foot print on the performance of SCC. Performance curves for both G.I and M.S has been tested in SCC are shown in Fig. 7. Throughout the examination, a fixed water flow rate of 0.0025 kg/s has been kept constant for both the cases. Also from figure, the gradual efficiency increase can be observed by using G.I material with respect to time. But radical variations are appears in efficiency curve of M.S during the investigation. Also efficiency of 43 % of maximum has been derived while the other has a efficiency of 47%. It has been observed that with the use of G.I, up to 1:30 PM there is a constant increase in efficiency and after that it starts gradually decreasing trend. Therefore from the experimentation it came to know the better cavity case cover material is G.I.

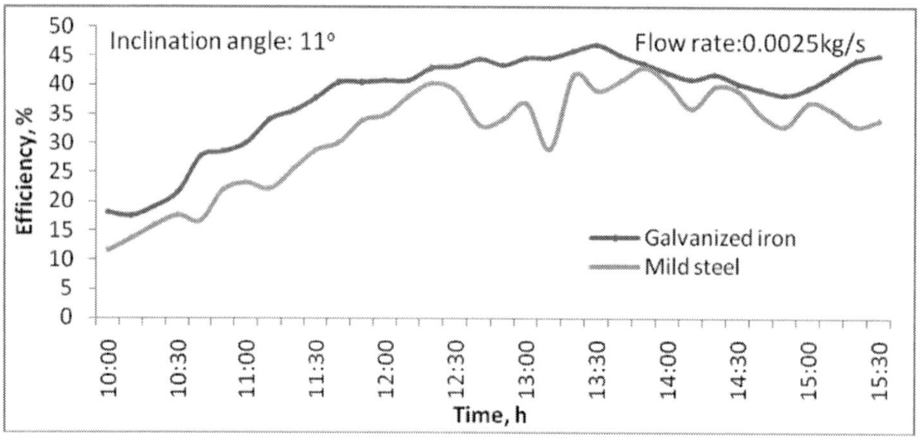

Fig.7. Comparison of efficiency with different cavity case covers materials

Fig.8. shows the outcome result of SSC on the water outlet temperature of SCC. It enumerates a typical curve of mass flow rate versus water outlet temperature for both with and without SSC coatings.For both the cases a constant decreasing trend of the curve, followed with respect to the mass flow rate of water. Also it is inferred from Fig.8 that the observation of maximum temperature of 74°C has been obtained with the usage of SSC coatings for water mass flow at 0.002 Kg/s. While on the other hand, 70°C of maximum temperature has been recorded without coatings.

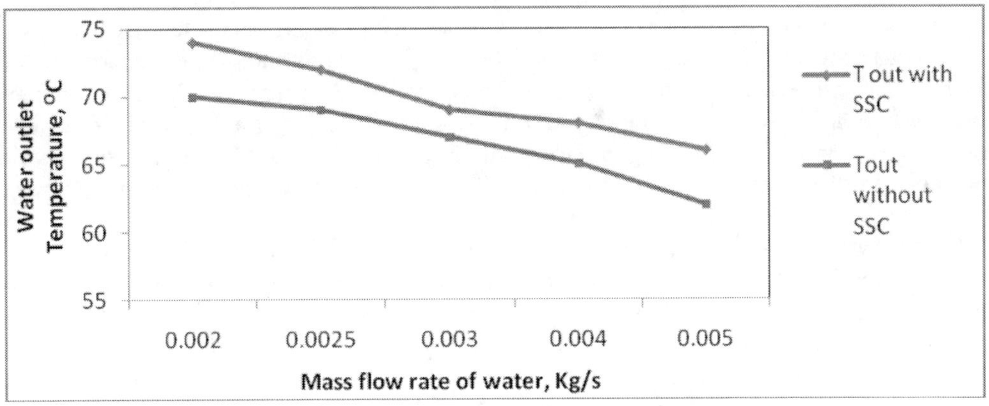

Fig.8 Outcome result of SSC on the water outlet temperature of SCC

The variation of efficiencies with respect to the SSC coatings has been illustrated in the Fig. 9. The efficiency curve goes on increasing trend for with SSC when compared to without SSC. That is, the differences in efficiency curve for both cases are visible with some constant values. Average efficiency of 59.21% has been achieved by the collector for the case of SCC with SSC coatings made. Moreover the trends followed by the SCC with SSC coatings are far better than that of without SSC coatings. Even though at the same time, all five numbers of cavities are exposed to sun, there are some variations in the difference in temperatures of both curves. the variations is not constant because of different parameters such as, slight pressure variation of the operating fluid and the restriction of flow inside on the receivers, flow pattern of air and water in and outside of the collector also influences the variation in temperatures. Moreover it mainly depends on the shadow effect of the cavities inside the SCC.

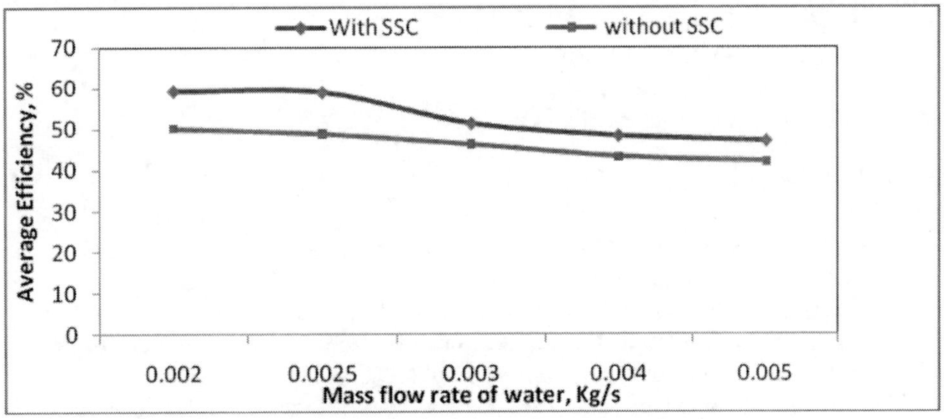

Fig.9. Variation of efficiencies with respect to the SSC

Summary and Conclusion

The main objective is, to enlarge a novel, more efficient, better conversion of radiation energy into heat and higher heat grasping capacity like configuration. Apparently one solution is a Cavity type collector or receiver methods used to trap the sunlight effectively. Moreover, conventional flat plate collector performs lesser efficient than the concentrating type collectors. But the problems associated with concentrating collectors are more costly and needs complicated tracking mechanism too. Therefore the solution made from both type of collectors is a cavity collector which has a numerous advantages of both flat plate and concentrating type collectors. Cavities like configurations are often used in concentrating type collectors as many researchers done their experimentations. But it has been strongly recommended for the development of FPC is a cavity collector. There is a significant improvement on the performance parameters on both efficiency and water outlet temperature. The need in changing an FPC into SCC, a least alteration is required and the conversion cost also too low. Investigations work has a center of attention to increase efficiency and should be on more effective collection methods. SCC works more efficiently during a certain period of time if there is no sunlight. Also it is best suited for cloudy and part cloudy days which in turn during the cloud interruptions are inevitable. Furthermore its efficiency has not dropped suddenly during afternoon hours as in the case of FPC.

Obviously the copper receiver gives better performance when compared to the aluminum material. Also with the heat transfer aspects it looks good enough. Experimentation end results clearly shows that the G.I has more potential than that of M.S as a cavity cover case material. Since the G.I sheets won't allow the heat from the cavity structure as it has the capability to retains the heat energy more when compared to Mild steel sheets. The usage of SSC coatings especially black nickel has a consequential effect on the experimentation of SCC. It increases not only the water outlet temperature but also improve efficiency of the whole system with a reasonable increase in cost.

Nomenclature

A_c	Area of the Collector, in m^2
Dim	Dimensional
I_t	Intensity of solar radiation, in W/m^2
η	Efficiency, in %
$\dot{m}$	Mass flow rate of water, in kg/s
H	Height of the collector, in m
α	Collector Inclination angle with respect to the horizontal axis ,in °
L	Length of the collector, in m
h	Height of the single cavity, in m
SCC	Solar Cavity Collector
T_{out}	Water outlet temperature, in °C
B	Breadth of the collector, in m
SSC	Solar Selective Coatings
l	Length of the single cavity, in m

List of References

[1] **Melchior T, Perkins C, Weimer A W and Steinfeld A,** 2008 A cavity-receiver containing a tubular absorber for high-temperature thermo chemical processing using concentrated solar energy. *International Journal of Thermal Sciences*, vol. 47, pp. 1496–1500.

[2] **Hahm T, Schmidt–traub H and Lebmann B,** 1999, A cone concentrator for high-temperature solar cavity receivers. *Solar Energy* vol.65 (1) pp.33–41.

[3] **Demichelis F and Russo G,**1979 Cavity-type surfaces for solar collectors, *Applied Physics* vol.18, pp. 307-309.

[4] **Reyes Torres, Salazar Ibarra and Cervantes De-G** ,2001 Thermo economic analysis at optimal performance of non-isothermal flat-plate solar collectors, *International journal ofApplied Thermodynamics*, vol.4(2), pp. 103-109.

[5] **Singh H and Eames P C,** 2011, A review of natural convective heat transfer correlations in rectangular cross-section cavities and their potential applications to compound parabolic concentrating (CPC) solar collector cavities, *Applied Thermal Engineering* ,vol.31, pp. 2186-2196.

[6] **Bairi A,** 2008, On the Nusselt number definition adapted to natural convection in parallelogrammic cavities. *Applied Thermal Engineering,* vol. 28, pp. 1267-1271, 2008.

[7] **Prakash M, Kedare S B, and Nayak J K**2009, Investigations on heat losses from a solar

cavity receiver. *Solar Energy,* vol.83, pp. 157–170.

[8] **Kribus A, Doron P, Rubin, Karni, Reuven, Duchan and Taragan.** 2000, A multistage solar receiver: the route to high temperature. *Solar Energy,* vol.67(1-3), pp. 3–11.

[9] **Lakshmipathy B and Sivaraman B (2015)**, Performance analysis of a solar cavity collector, *International Journal of Applied Engineering Research*, volume 10, number 8, pp.6070-6073.

[10] **Cruz J M S, Hammond G P, Reis A J P S ,** 2002, Thermal performance of a trapezoidal-shaped solar collector/energy store, *Applied Energy,* vol.73, pp. 195–212.

[11] **Lakshmipathy B and Sivaraman B,(2016b)**Performance analysis on working parameters of a flat plate solar cavity collector,*InternationalEnergyJournal*, Issue16, pp. 1-10.

[12] **Flores J J, Alvarez G and Xaman J P**2008, Thermal performance of a cubic cavity with a solar control coating deposited to a vertical semitransparent wall. *Solar Energy* vol.82, pp. 588–601.

[13] **Garcia R P, Bollini L E, Padilha A, and Scalon V L,** (2013), Thermal analysis of convection losses by changing baffle spacing in a cavity, In *22nd International Congress of Mechanical Engineering (COBEM).*

[14] **Tan Yuting, Li Zhao, Junjiang Bao, Qing Liu** 2014, Experimental investigation on heat loss of semi-spherical cavity receiver, *Energy Conversion and Management,* vol. 87, pp. 576–583.

[15] **Steinfeld A and Schubnell M**, 1993, Optimum aperture size and operating temperature of a solar cavity-receiver, *Solar Energy,* Vol. 50, No. 1, pp. 19-25.

[16] **Diver R B,** 1987, Receiver/Reactor concepts for thermo chemical transport of solar energy, *Solar Energy Engineering*, vol. 109, pp. 199-204.

[17] **Hunt A J, Ayer J, Hull P, Miller F, Noting J E, and Worth D**, 1986, Solar radiant heating of gas-particle mixtures, report no.LBL-22743, Lawrence Berkeley laboratory, University of California, Berkeley, CA 94720.

[18] **Steinfeld A and Fletcher E A,** 1990, A spherical receiver-reactor with specularly reflecting inner surfaces for high temperature solar furnace applications, Proc. 9th. International *Heat Transfer Conference,* August 19-24, 19-EN-03, Jerusalem, pp.127-132.

[19] **Siegel R and Howell JR**, 1981, Thermal radiation heat transfer, Hemisphere Publishing Company, Washington, DC, Ch. 2, 1981.

[20] **Harris J A and Lenz T G,** 1985 Thermal performance of solar concentrator/cavity receiver systems, *Solar Energy*, vol.34,

Effect of dispersing CTBN into Diglycidyl Ether of Bisphenol-A on Mixed mode Loading

J.V. Sai Prasanna Kumar[1], S. Baskar[2]

[1]*Department of Aeronautical Engineering*
[2]Department of Electrical Electronics Engineering
Veltech Rangarajan Dr. Sagunthala
R & D Institute of Science and Technology
Avadi, Chennai-600062, India

e-mail:
saiprasannajv1202@gmail.com

Abstract

Delamination of polymer composite is one of the major failure characteristic, in this context the fracture behavior of the tailored composites in combination of opening mode and shearing mode is assessed and its fracture robustness of the glass fiber reinforced laminate are determined experimentally, in this present work by developing a modified Arcan fixture using a single edge notched specimen for characterizing in various loading conditions, The fixture was made of industrial grade Aluminum ingot machined to an outer diameter of 150 mm, having a thickness of 10 mm. The fixture had provision to load the sample at different oblique angles up to 90 degrees in steps of 15 degrees. The samples were attached to the fixture by means of fasteners and were tested in tensile mode using a 40 KN Instron tensile tester at constant displacement and ambient conditions. Further, the resin was strengthened with Carboxyl Terminated Butadiene Arenonitrile (CTBN) and its effect on fracture parameters was studied by varying the loading angles. Results indicate that the fracture parameters vary with loading and at angles greater than 45 degrees, the GFRP sample's behavior changes from opening mode to shearing mode and the converse is also true. This pattern is similar, with the specimens even after modifying the resin with CTBN however, it was found that stress intensity factor increased corresponding to the increase of CTBN infusion. CTBN was dispersed into the Bisphenol base A (LY566 – A general purpose thermoset epoxy resin) in three different percent: 1%, 3% and 5% by weight. The layup sequence used in this work was cross ply. The effect of varying the loading angles on the stress intensity factors were studied.

Keywords: mixed, mode, modified arcan fixture, CTBN, loading angles

Introduction

Presently the polymeric composite has gained a steady entry in various field from sports industry to space technology. The researchers are always on their toes to seek a better material and contain its growth against crack growth. The failure can occur without warning and can begin with matrix cracking, fiber pull out and debonding of plies. The failure can be sudden and without a fore warning. [1-5]. One simple method to enhance the strength is to add some additives to the thermoset polymer otherwise prone to brittleness and weak in resisting the loads. These fillers are dispersed into the resin and with the addition of these institute deposition filler materials, enhances the strength of the polymer and thereby increases the fracture toughness, particularly the rubbery materials to the thermoset polymer. With the controlled polymerization reaction between the epoxy and CTBN, a phase separation ensues leading to the improved fracture characteristics also bettering thermal and mechanical characteristics. A number of studies [6-10] involved in the modification of the reactive rubber with epoxy resin particularly CTBN. This elastomer when dispersed into the resin, enhances the toughness of the neat resin specimens., while minimal modification led to thermal and mechanical characteristic. [10-12]

The main objective of this paper is: 1. To develop a fixture which can be utilized to determine the mixed mode charteristics. 2. Quantify the fracture parameter with the same sample under various loading angles and hence study the effect the of loading angles on the fracture behavior by strengthening the thermoset polymer by CTBN.

Materials and Methods

The diglycidyl ether of bisphenol-A (DGEBA)-based epoxy resin (viscosity: 1000–1500 mPa s at 27 °C) used in all experiments was LY 5566 with an epoxide equivalent weight of 192 g/equiv., as determined by titration. The curing agent was an aliphatic compound TETA, commercially available from M/S Huntsman, Chennai by the trade name HY951.The elastomer employed for this work was carboxyl terminated butadiene acrylonitrile (CTBN) was obtained from M/S Zibo chemicals, Shandon, China, and it had a viscosity of viscosity: 625,000 mPa s at 27 °C, Figure 1., with a molecular weight of 3500 containing acrylonitrile content 27% and carboxyl content 32%. A 450 GSM stitched woven unidirectional E-Glass mat was supplied by M/S S. T. Composites, Chennai

Fig. 1. CTBN from Zibo Chemicals

Fixture development

One of the important problems in material testing is to design a fixture and specimens to produce a plane state of stress in all materials and more so for the composite materials, since the composites are often employed as laminates. Perhaps, if a single specimen is available, for off-axis loading, and preferably, if the fibers are oriented at an angle with respect loading angle then plane stress condition ensues. Originally in 1978, Arcan developed this fixture, where the sample had a notch and the fixture can be rotated, [Arcan, 1978]. If the notch of the sample is 90 degrees then the sample is undergoing Mode-I type of loading and if the fibers are parallel to the loading angles, the samples suffer mode-II loading. All fixtures are the successors of the fixture developed by Arcan, Figure 2(a-d). The configuration in figure 2d is the fixture developed for the mixed mode testing of the GFRP samples strengthened with CTBN. An Aluminum block was obtained from M/S Alli Metals, Guindy, Chennai. The block was machined to dimensions having diameter 150 mm with a wall thickness of 10 mm. The machined fixture shown in figure 2d.

a. Earliest Fixture for adhesive joint

b. Fixture developed for mixed mode bolted joints

c. Fixture to test rock samples

d. Modified fixture

Fig. 2 Development of the Fixture

A state of pure shear can be obtained by varying the loading angle and a combined stress state is achieved in the test section. Jurf and Pipes (1982) used this Arcan fixture to produce fracture mechanics data for adhesively bonded composite joints but the samples were bonded to the fixture figure. But in this present work the earlier fixture is modified to hold a single edge notched specimen by means of fasteners. The modified fixture had two pairs of 6.4 mm thick aluminum parts. Each pair was equal to one half of the original fixture and was mounted from a circular disk. A 3.2 mm deep trapezoidal cut was machined in each part to host the specimen. As a result, a specimen of up to 6.4 mm thickness could be tested using a fixture. Three holes were drilled at the cut-out section in each part for tightening two parts together with bolts. In addition, the bolts could prevent the specimen from slipping and provide a smooth load transfer. In each part, several 6.4 mm diameter holes were drilled near the edge along the circumference of the plate. These holes were designed for load application through a pin yoke system and for tightening each pair of fixture plates together. They were located at angles starting from 0 degrees to 90 degrees and the angles were varied in steps of 22.5 degrees., figure 2d

Sample Fabrication

A known quantity of DGEBA was taken in a borosilicate graduated beaker and was thermally heated to a temperature of 100° C for an hour, also it was constantly agitated, in order to entrap the air bubbles. To his thermoset resin CTBN was added gently by weight in proportion to the resin. The various percent of CTBN was added to the Resin was: 1, 3 and 5 %. The mixture was cooled to the room temperature and then the polymerizing agent TETA was added again the ratio of 100: 100: 10;; resin : fiber: hardener. A laminate was demolded from an aluminum mould. Then these samples were post cured at 200 ° C for 2 hours After that the samples were machined into the required configuration as shown in t figure 4. The woven roving mat was utilized in casting the laminate of 9 layers each constituting of cross ply, and the laminate had a thickness of 6.5 mm with a fiber volume fraction of 0.45. Similar procedure was adopted to fabricate the samples after treating the DEGBA with CTBN by varying its percentage of about 1 %, 3 % and 5 % by weight of resin. Figure 3, shows the sample geometry without CTBMN modification

Fig. 3. Sample for mixed mode testing

Mathematical formulation and Modeling

Consider a portion of the fixture and the specimen attachment. F is the load appliued through the yoke, which transfers the load on the specimen geometry smoothly with th AB as the crack width and the farfield angle, the loads on the crack front can have two components., figure 4.

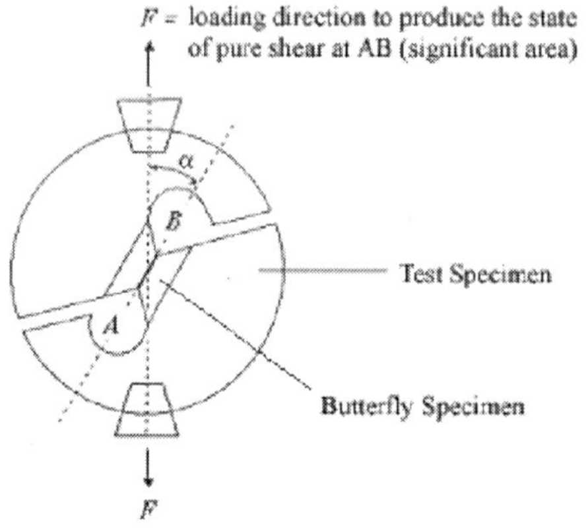

Fig.4. Specimen loading geometry

$\sigma_x = \sigma_A \text{ Sin } \alpha$ (1)

$\tau x = \sigma_A \text{ Cos } \alpha$ (2)

Where the stress is interms of load and area of the section where the starter crack is considered with a finite width. Basing on the normal and shear stress components the stress intensity factors for opening mode and the shearing mode can be obtaines as :

$$K_I = (P \sin \alpha \;/\; w.t)\; f_I\,(a/w) \quad (3)$$

$$K_{II} = (P \cos\alpha \;/\; w.t)\; f_{II}\,(a/w) \quad (4)$$

Where f_I and f_{II} are the calibration factors and they are obtained by numerical analysis. Numerical analysis was carried out by using the finite element software using Abaqus, the analysis is performed at constant load. The fixture and the specimen were modelled using a eight noded quadrilateral element generating a mesh, figure 5 and refining it at the crack tip, figure 6, with an elemental size of 0.02 mm.

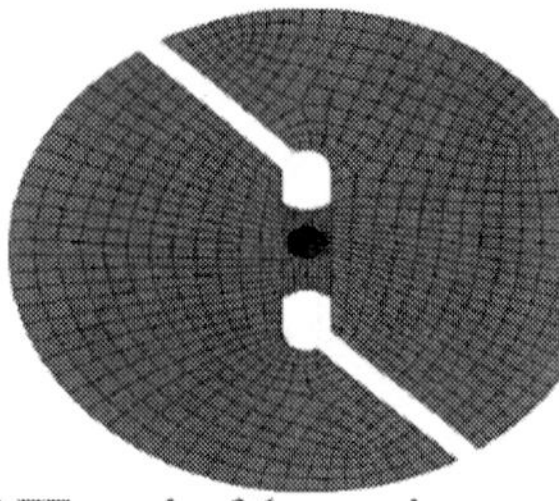

Fig. 5 FE mesh of the specimen and fixture

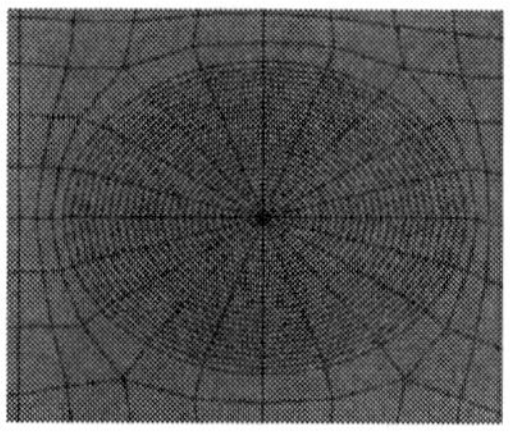

Fig.6. Mesh at the crack tip

A finite element analysis under plain strain condition was used to determine root square stress field singularity, and to obtain singularity term of the crack tip stress field, the elements around the crack tip and the mid side nodes were moved to a quarter point of each element side. And thus, the calibration factors were obtained as

$$f_I(a/w) = 1.12 - 0.231(a/w) + 10.35\,(a/w)^2 - 21.27\,(a/w)^3 + 30.39\,(a/w)^4$$

and

$$f_{II}(a/w) = (1.122 - 0.56\,(a/w) + 0.085\,(a/w)^2 + 0.18\,(a/w)^3)\,/\,(1 - (a/w))^{1/2}$$

From the expression 3 and 4 we can obser4ve that the ratio a/w can be varied to study the its influence on the corresponding stress intensity factor.Also noting that a is the initial length of the starter crack, w being the width of the geometry. Iandicating that when a=w, the sample has failed completely. While the mixed mode can be boytained by obtaining the ratio of K_I to K_{II}.

Test procedures

Prior to the sample preparation, reactivity test and thermal analysis were conducted the reactivity was performed to assess the gel and cure time, gel and cure temperature because the addition of CTBN into the thermoset resin will alter performance of the virgin resin. A known quantity of DGEBA and CTBN were mixed in a glass beaker and put in a water bath with a temperature of 70° C. A thermocouple was used to measure the temperature and the other parameters. Perkin Elmer differential scanning calorimeter was employed to determine the T_g of both untreated and treated samples. Samples of about 8–10 mg were heated at a rate 10°C min^{-1} in a nitrogen atmosphere over the temperature range from 30°C–100°C.

In fracture mechanics testing all the fracture modes are related to the loading either opening, shear and out of plane load conditions. A safe assumption the flaw propagates at the pre-existing crack plane, hence initially the loads at which the failure occurs were determined using the tensile tests and were noted for each layup sequence, these tests were repeated thrice and average values were used in determining the stress intensity factors. All test was done with open loop servo hydraulic universal testing machine with a constant cross head speed of 0.25 mm/min. From the test data the stress intensity factors were determined for pure opening mode, sliding mode and combination of both the modes.

Results and Discussions

The following properties like gel time, cure time, gel temperature and cure temperature were the object of the tests. It was found that the gel time and cure time increased with the CTBN content in the thermoset polymer, can perhaps be attributed due to the fact that movement of the reactive molecules. While the gel temperature and cure temperature were higher for all values of CTBN content, which was relatively higher than those of the untreated resin content, figure 7.

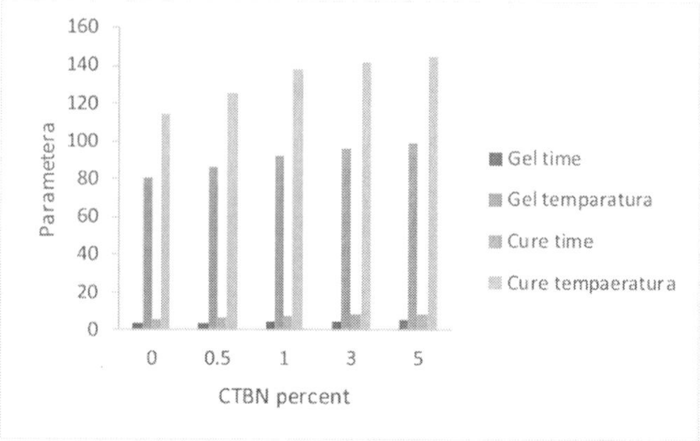

Fig 7. Effect of CTBN percent on gel time and temperature, cure time &temperature

The T_g values for all of the CTBN-modified epoxy samples are lower than those of the unmodified epoxy are. This decrease in T_g values can be attributed to the fact that a chemical interaction occurred between the CTBN and the polymer phase. This implies that a larger amount of CTBN led to better interaction with epoxy and hence batter crosslinking, Figure 8. In other words, the amine reactions formation of hydrogen bonds stabilizes the transition state while the reaction rates are accelerated. Larger addition of CTBN might have accelerated the curing rate. The carboxyl content present in the co-polymer has influenced the curing rate.

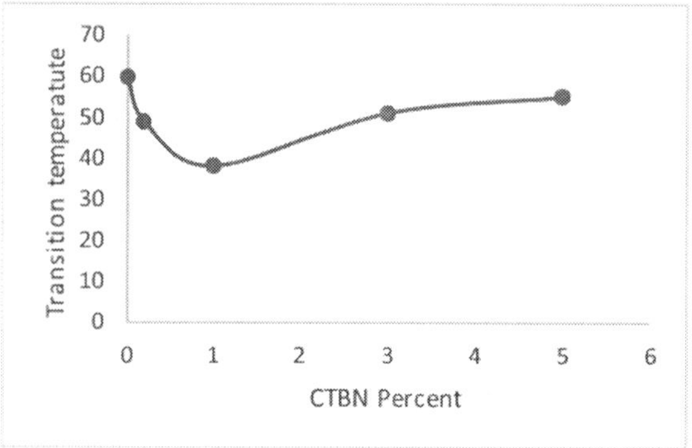

Fig. 8 Effect of CTBN percent on transition temperature

The relationship between the non-dimensional stress intensity factors and the loading angle is shown in figure 9 for a/w ratio of 0.8 It can be seen that for loading angles less than 60 degree, the mode-I fracture is dominant and as the mode-II loading contribution increases, the mode-I stress intensity factor decreases and the mode-II stress intensity factor increases. For loading angles greater than 60-degree mode-II fracture becomes dominant. This pattern is true for the laminates enhanced with CTBN

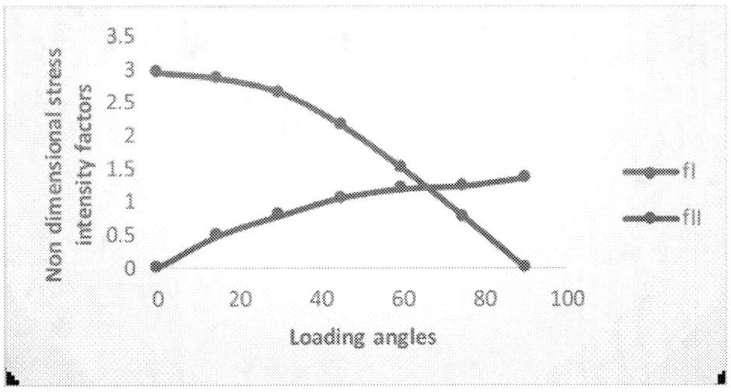

Figure 9 Effect of the calibration parameters on the loading angle

All the sample for different stacking sequence were mounted on to the fixture and were tested on a open loop servo Universal test machine, with constant displacement mode. Gradually the fixture was rotated the samples were tested for different loading angles. Peak loads were obtained in static mode and were used to determine the stress intensity factors using equations 3 and 4. The mixed mode results are obtained by considering the natural logarithms of the ratio of KI/ KII. The stacking sequences were considered along with various CTBN percent. The result is presented for mode -I, mode -II and mixed mode for various CTBN infusion and for each stacking mode. Fracture tests were carried out on UTM in tensile mode, the samples were fastened to the fixture and tested under constant displacement at the rate of 0.5 mm/min. The tests were repeated for consistency. The average values were utilized for determining the stress intensity factors. The laminate was loaded under pure mode-I loading, in mode-II. The mixed mode behavior was exhibited by the samples when the loading angles were between 55 and 65 degrees. All the graphs were plotted for the stacking sequence of 0/90 with 9 layers and at a CTBN dispersion percent of 1, 3 and 5 by weight. It can be noticed that the stress intensity behavior was same irrespective of the percent CTBN diffusion. The calibration factors helped in understanding the fracture behavior. Then the mode-I stress intensity factors were obtained by fitting a polynomial inclusive of the critical loading. The mode-II stress intensity factors increased.

IOP Conf. Series: Materials Science and Engineering 1126 (2021) 012054 doi:10.1088/1757-899X/1126/1/012054

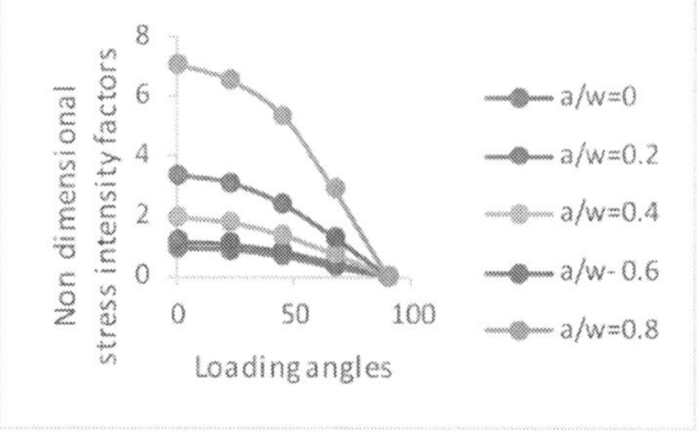

Fig. 10. Unmodified stress intensity factors for Mode-I

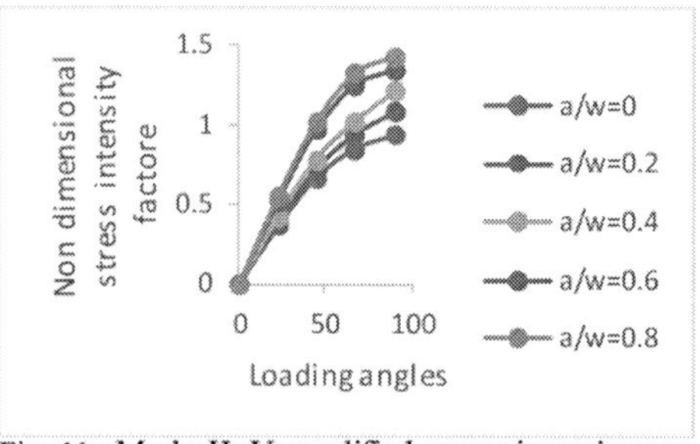

Fig. 11. Mode II- Unmodified stress intensity factors

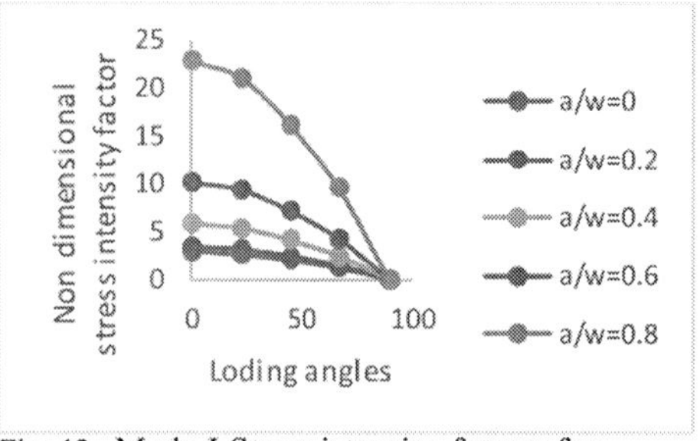

Fig. 12. Mode-I Stress intensity factors for 1% CTBN dispersion

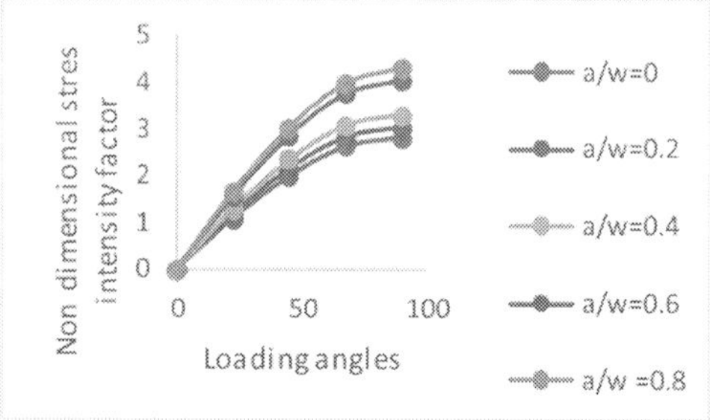

Fig. 13. Mode -II stress intensity factors for 1 % CTBN dispersion

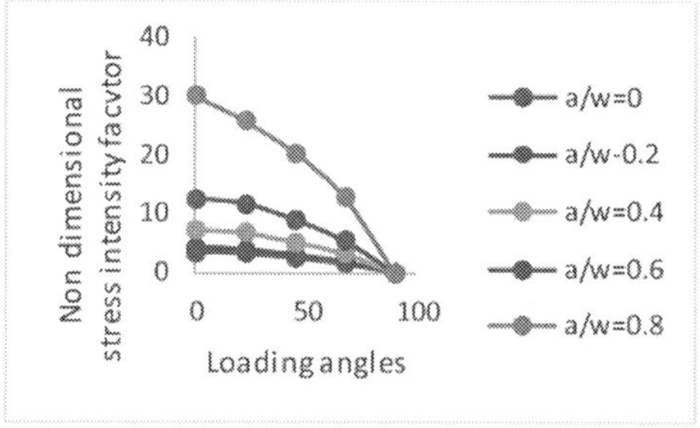

Fig. 14. Stress intensity factors for 3 % CTBN dispersion for Mode-I

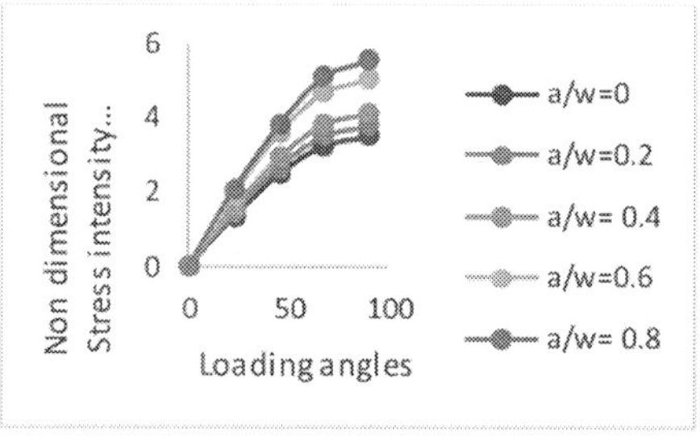

Fig.15. Mode-II stress intensity factors for 3 % CTBN dispersion.

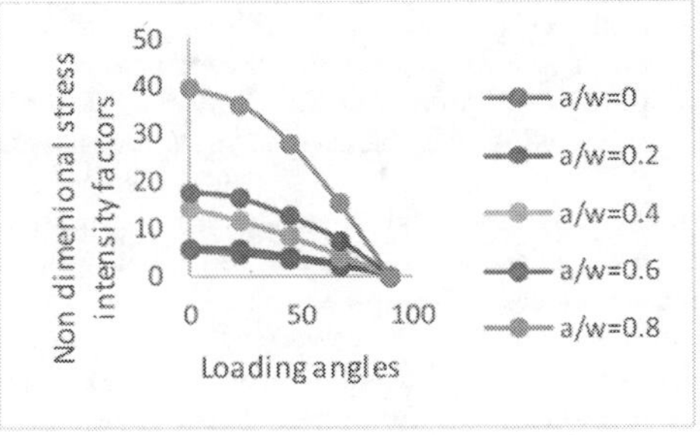

Fig. 16. Mode-I Stress intensity factors 5for 5 % CTBN

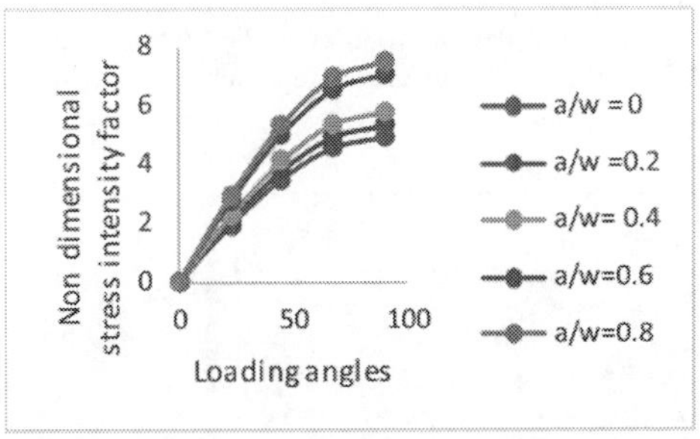

Fig. 17. Mode-II stress intensity factors for 5 % CTBN

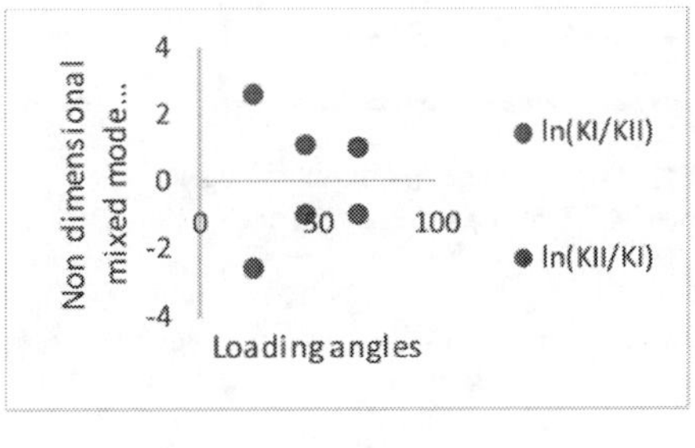

Fig. 18. Mixed mode stress intensity factors

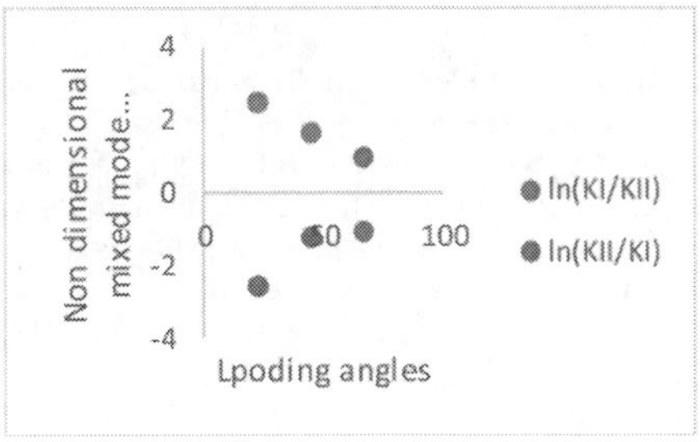

Fig. 19 Mixed mode factors for 1% CTBN

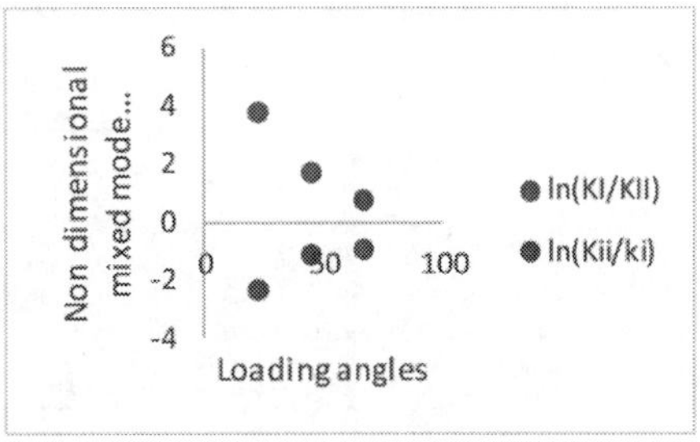

Fig.20. Mixed mode factors for 3% CTBN

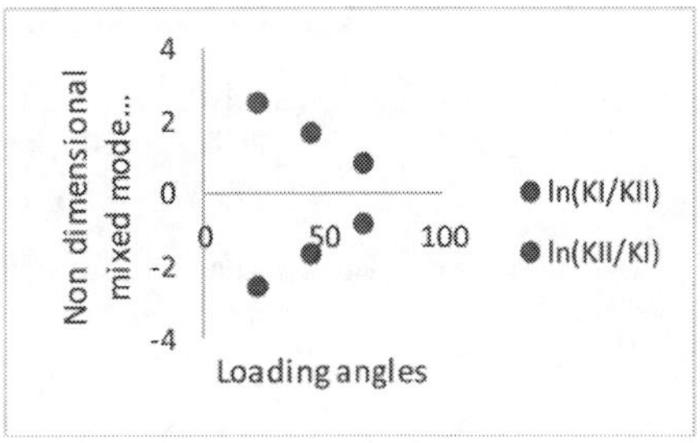

Fig. 21 Mixed mode factors for 5 % CTBN

The data presented in figures 9-10 show the variation of the non-dimensional stress intensity factors against the loading angles by rotating the fixture and the effect of varying a/w. Series of curves are obtained. As the loading varies the opening mode values decrease, while the shearing mode values increase. Figures 11, 13 and 15 represent the variation of the non-dimensional stress intensity factors for mode-I for different CTBN infusion into the resin system, comparing the figures the stress intensity factors increase against the unmodified thermoset resins. While the figure 12, 14 and 16 represent the sliding mode stress intensity factors for different CTBN dispersion at 1 %, 3 % and 5% respectively. Again, as expected the fracture parameters improved significantly. The mixed mode parameters are presented in figure 17, indicating that beyond the angle 65 degrees the mode-I behavior changes to Mode-I, the figures 18-20 represent the non-dimensional mixed mode stress intensity factors. Curing of the resins is one major contribution to forming crosslinking with the reactive agents, and mechanical properties are dependent on the level of curing. And lowering the Tg which is an established fact. Improving the mechanical properties can be achieved by the addition of the fillers to the resin systems, the size, topology of the fillers varies in size and shape, another requirement is that the filler must dissolve in the resin and participate in the curing process. Carbaryl terminated butadiene acrylonitrile satisfies the both requirements. The addition of CTBN reduces the Tg, figure, stiffness, increases CTE and plasticizes the resin system, and increases the fracture toughness. These particles yield, tear, stretch, bifurcate, crack bridging, debonding, crack path deflection and pinning, thus crack tip bluntness ensues and leading to increase in the fracture toughness.[12], all these phenomena are represented schematically in figure 21.

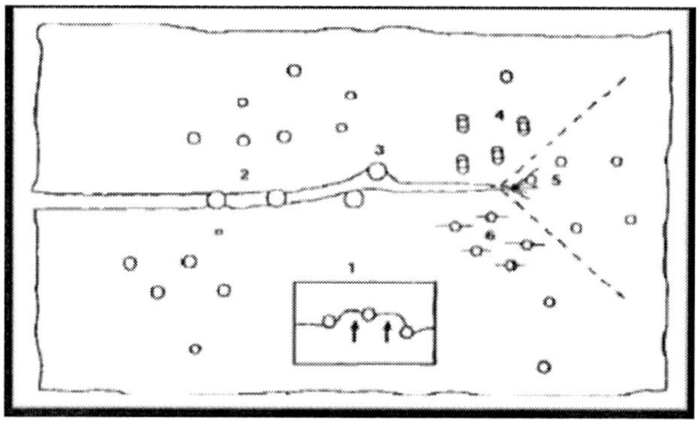

Fig. 22 Crack propagation model:(1) Crack pinning, (2) Particle bridging, (3) Crack-path-deflection, (4) Debonding, (5) Bifurcation, and (6) Microcracking [12]

The unmodified values and the modified stress intensity factors in opening and sliding mode is presented in table 1.

Table 1 Mixed mode stress intensity factors

Loading Angles	KI $MPa\,m^{0.5}$				KII $MPa\,m^{0.5}$			
	UM	1%	3%	5%	UM	1%	3%	5%
0	7.09	22.79	28.911	39.86	0	0	0	0
22.5	6.589	21.055	21.316	38.29	0.5500	1.650	2.0633	2.887
45	5.3686	16.114	20.301	28.286	1.0100	3.01	3.789	5.311
67.5	2.89	9.634	12.01	15.288	1.321	3.974	5.113	6.952
90	0	0	0	0	1.428	4.30	5.556	7.499

Conclusions

The mixed mode fracture mechanics parameters were determined for CTBN enhanced GFRP samples using the modified Arcan test specimen. Various percent of CTBN was infused into the brittle epoxy thus toughening it. Finite element analysis is used to evaluate the effect of crack length on fracture criterion. The geometric calibration factors were obtained for both the modes of loading. The same calibration parameters were used for both modified and unmodified coupons. CTBN enhanced the fracture parameters but did decrease the Tg, and the particles provided a meandering path for the crack to propagate, ensuing the crack tip bluntness and hence the increase in fracture toughness.

Acknowledgements

The author thanks the management, dean and the head, aeronautical engineering department for allowing him to utilize the facilities in conducting the research

References

[1] S. Zhao, M. Gadke and R. Prinz. Mixed-mode delamination behavior of carbon/epoxy composites. *J Reinf Plast Compos* 1995; 14: 804-826. https://doi.org/10.1177/073168449501400803

[2] Y. J. Yum and H. You. Pure mode I, II and mixed mode interlaminar fracture of graphite/epoxy composite materials. *J Reinf Plast Compos* 2001; 20: 794-808. https://doi.org/10.1177/073168401772678571

[3] H.Y. Yeh and C. H. Kim. The mixed mode fracture analysis of unidirectional composites. *J Reinf Plast Compos* 1994; 13: 498-508. https://doi.org/10.1177/073168449401300602

[4] D. B. Lee, T. Ikeda, N. Miyazaki et al. Damage zone around crack tip and fracture toughness of rubber-modified epoxy resin under mixed-mode conditions. *Eng Fract Mech* 2002; 69: 1363-1375. https://doi.org/10.1016/S0013-7944(01)00151-5

[5] R. A. Pearson and A. F. Yee. Influence of particle size and particle size distribution on toughening mechanisms in rubber-modified epoxies. *J Mater Sci* 1991; 26: 3828-3844. https://doi.org/10.1007/BF01184979

[6] R. Bagheri and R. A. Pearson. Role of particle cavitation in rubber-toughened epoxies, part I: micro void toughening. *Polymer* 1996; 37: 4529-4538. https://doi.org/10.1016/0032-3861(96)00295-9
[7] G. D.Tracy, P. Feraboli and K. T. Kedward. A new mixed mode test for carbon/epoxy composite systems. *Compos A* 2003; 34: 1125-1131. https://doi.org/10.1016/S1359-835X(03)00205-7
[8] R. Marat-Mendes and M. Freitas. Failure criteria for mixed mode delamination in glass fiber epoxy composites. *Compos Struct* 2010; 92: 2292-2298. https://doi.org/10.1016/j.compstruct.2009.07.017
[9] R. Marat-Mendes and M. Freitas. DCB, ENF and ECT tests for interlaminar fracture of glass/epoxy unidirectional laminates. In: Proceedings of 10th Portuguese conference on fracture, Guimaraes, Portugal, 2006, pp. 22-24. Guimara˜ es, Portugal: University of Minho.
[10] R. Marat-Mendes and M. Freitas. Mode I, mode II, mode III and mixed mode interlaminar fracture of glass/epoxy unidirectional laminates. In: Proceedings of EUROMECH 473 conference, Porto, Portugal, October 2005, pp. 27-29.
[11] J. G. Williams. End correction for orthotropic DCB specimens. *Compos Sci Technol* 1989; 35: 367-376. https://doi.org/10.1016/0266-3538(89)90058-4
[12] Y. Wang and J. G. William. Corrections for mode II fracture-toughness specimens of composites materials. *Compos Sci Technol* 1992; 43: 251-256. https://doi.org/10.1016/0266-3538(92)90096-L

Names of referees

1. Dr. V. Arumugam, Professor, Anna University, India, arumugam.mitaero@gmail.com
2. Dr. S. Venkatachalam, Karunya University, India, aerovenkat74@gmail.com
3. Dr. Carlos Santali, School of Science and Technology, Universita degli Studi di Camerino, Italy, calo.santulli@unicam.it
4. Dr. R. Velmurgan, IIT(M), India, rvel@iitm.ac.in
5. Dr, S. Vishnuvardhan, FFL, CSIR-SERC, Tiramani, India, svvardhan@serc.res.in

Design and analysis of hybrid composites using adaptive neuro fuzzy inference system

[1]C. kavitha,[2]T.S. Frank gladson

[1]Department of Mathematics, [2]Department of Mechanical Engineering
[1]Sathyabama Institute of Science and Technology, [2]Velammal Engineering College
ceekavi@gmail.com, frankgladson@velammal.edu.in

Abstract. This article shows an adaptive neuro-fuzzy inference system (ANFIS) model that can predict the cyclic response under a mixture of different parameters. The ANFIS model used for prediction can accurately predict the cyclic reaction of blends with different parameters. It is used for evaluation to predict and analyze the relationship between independent machining variables cutting speed, feed rate, depth of cut and machinability cutting force, cutting power, specific cutting pressure standards. In this evaluation of PEEK CF 30 processed with TiN-coated cutting tools, experiments were conducted and the results obtained were used to apply the ANFIS method to prediction. The primary motive on this research is the affect regarding it factors of cutting force, cutting power and specific cutting pressure.
Keywords: ANFIS, FIS, cutting force, cutting power, specific cutting pressure, cutting speed, feed rate, depth of cut

1. Introduction

Surface roughness plays an essential role in the manufacturing industry, since that perform have an effect on the mechanical properties, cost and exorcism of the product. There are numerous statistical methods that can be used to find the participating parameters, optimize the parameters and estimate the cause in the research work in accordance with the surface roughness by transferring the hybrid contract on the material, without doing a lot about experimental data.

Adel T. Abbas et.al tested then in contrast the experimental data over surface roughness concerning one of a kind cutting parameters among CNC turning operations including regression analysis and ANFIS. Kanhu Charan Nayak et.al proposed that the ANFIS model together with Gaussian number is a predictive tool for milling forces. Djordje cica et.al used ANN, ANFIS models in imitation of check the cutting force, Feed force, and the prediction over the pushed force. The estimated worth do nearly predict the force between the turning operation. Mahdi S. Alajmi et.al proposed a approach because predicting the surface roughness values concerning AISI 304 stainless steel dry then low-temperature the usage of ANFIS-QPSO machine learning method, who perform insure a excessive degree over truth in predicting surface roughness. To determine the degree of association between cutting parameters, cutting force and surface roughness, Ibrahem Maher et.al used the ANFIS model. An ANFIS system has been shown by Ishwer Shivakoti et.al that can use processing parameters such as feed, revolution per minute and cutting depth to predict process response such as material removal rate, surface roughness and cutting force. Vineet Jain et.al revolve around an commentary provision so do makes use of the adaptive neuro-fuzzy inference system (ANFIS) to perceive cutting force based about cutting parameters certain namely spindle speed, feed and depth of cut. C.Kavitha et.al used constants and linear types to take a look at the ANFIS models including special membership function types, and used subtractive clustering according to generate a fuzzy inference system to predict the surface roughness on the hallosyte nanotubes regarding end-milling concerning aluminium reinforced epoxy hybrid composites. C.kavitha studied the low prediction oblivion of the gbell membership function and other membership functions within ANFIS.

This article considers ternary machining parameters, namely, cutting speed, feed rate and depth of cut. The ranges of these parameters are selected based on preliminary investigations. The research intention establishes a mathematical model of the machining process in order according to uses an ANFIS according to predict machining standards.

2. Methodology

ANFIS was once proposed by means of Jang in 1993, but then Geoffrey Boothroyd et.al proposed that ANFIS is a hybrid of dual technologies, such as adaptive neural network (ANN) and Takagi-Sugeno fuzzy inference system. ANFIS is a technology that focuses on neuro-fuzzy input / output data. The mapping of input parameters below output parameters relating to relevant parameters in a neural network depend on fuzzy logic. The input/output parameter is modified with the aid of the fuzzy technique within the training data. It relies on human information as a fuzzy inference system then generates a single output. Fuzzy technique follows IF–THEN governance to determine the superiority of the membership function. Through back-propagation, gradient descent, least squares method by reducing the sum of square errors, the neural network is used to predict the output, so as to train and then test the input and then output data.

Ishwer Shivakoti et.al regards ANFIS as like comprising on five network layers which are used to perform fuzzy inference systems. Each layer has unique capabilities, for example, Fuzzy layer, Product layer, Normalized layer, Defuzzify layer then Output layer. The layer incorporates couple distinctive nodes, such as square and circle. The square node is an adaptive node that can change the factor, the circle is a fixed factor, and its parameters will change during the training process, and the fixed node is a fixed node. Input nodes are being used by Muhammad Rizal et.al to talk about training values, output nodes talk about expected values and some nodes serve as membership functions and rules in the hidden layer. The present layer's input is obtained from the nodes in the previous layer. The Sugeno fuzzy inference system's IF–THEN rules are used as the ANFIS system's rule base and these rules are fixed as

Rule 1: IF x is A_1 AND y is B_1, THEN f is $f_1(x, y)$
Rule 2: IF x is A_2 AND y is B_2, THEN f is $f_2(x, y)$,

where the inputs of ANFIS are x and y. Fuzzy sets are A_i, B_i and the output of the first order Sugeno fuzzy inference system is $f_i(x, y) = p_i x + q_i y + r_i$, p_i , q_i, r_i are design parameters that are determined during the training process, f_i is a first order polynomial.

Jang J is used to the characteristics of each layer in ANFIS, as described as below:

Layer I: The first layer is a layer of fuzzification with in ANFIS network. It is a pre processing step for information and each node in this layer is adaptive. Converts the numeric input values into fuzzy output values between 0 and 1 with membership functions such as $O_{1,i} = \mu_{A_i}(x_1)$, for node i= 1, 2 and $O_{1,i} = \mu_{B_{i-2}}(x_2)$ for node i= 3, 4. Here x_1 and x_2 are the input to node i, A_i and B_{i-2} are fuzzy sets which contain linguistic terms such as small, large etc. $\mu_{A_i}(x_1)$ and $\mu_{B_{i-2}}(x_2)$ are the membership functions. Various types of membership function have been used such as trapezoidal, triangular, Gaussian etc. Parameters are called introduce parameter.

Layer 2: Each node of product layer is a fixed node, which is marked by a circle marked as Π, In addition, it can also perform T-norm operation on the input, but it can also calculates the rule firing strength w_i . The output on each rule node is a multiplication concerning membership degrees of the previous layer.

$O_{2,i} = w_i = \mu_{A_i}(x_1)\mu_{B_i}(x_2)$ for i=1,2

Layer 3: Each node on normalized layer is fixed nodes, marked by means of a circle is labeled hence N. Each ith node on it layer calculates the ratio about i^{th} node firing strength according to the sum of all rules firing strengths. Output concerning the ith node of this layer is normalized firing strength yet such do stay represented as $O_{3,i} = \overline{w} = \frac{w_i}{w_1 + w_2}$ $for\ i = 1,2$

Layer 4: Every node in that layer has adaptive nodes which are called as like defuzzify layer marked through a square yet labeled by way of D. This layer computes the product about normalized firing strength beside previous layer then first order polynomial [sugeno model]. Takagi_sugeno type output of this layer can be represented as $O_{4,i} = \overline{w}_i f_i = \overline{w}_i(p_i x_1 + q_i x_2 + r_i)$where the evaluation of right hand side polynomials perform consequent parameter set as $\{p_i, q_i, r_i\}$, $\overline{w}_i$ is the normalized weighting factor of the ith rule, f_i is the output of the ith rule of this node.

Layer 5: Each node among this output layer has a fixed node, as is separated by using a circle and named afterward it. This wide variety represents the general output as much the sum regarding the membership function and a firing strength yet obtains the genuine estimation about the ANFIS framework.

$$O_{5,i} = \overline{w}_i f_i = \frac{\sum_i w_i f_i}{\sum_i w_i}$$

Hybrid learning algorithm use least square technique in conformity with identify the consequent parameters then premise parameters are updated by using gradient descent. If the premise parameters are fixed then the general output can stand expressed as much a linear combination given by the following equation

$$O_{5,i} = \frac{w_1}{w_1 + w_2} f_1 + \frac{w_2}{w_1 + w_2} f_2$$

$$i.e., O_{5,i} = \overline{w}_1(p_1 x_1 + q_1 x_2 + r_1) + \overline{w}_2(p_2 x_1 + q_2 x_2 + r_2)$$

3. Results then Discussion - Prediction concerning the machining of variables through ANFIS access:

In order to prophesy the cutting force, cutting power and specific cutting pressure of PEEK CF 30 processed with TiN coated cutting tools, the technique used is ANFIS. As well as this study, the data were extracted from Issam Hanafi et.al and shown in Table 1. The input parameters are machining parameters certain so cutting speed v (m / min), depth of cut d (mm), feed rate f (mm/rev) and the outputs are the Cutting force Fm (N), Cutting power Pc (W), Specific cutting pressure Ks (N/mm ^{2}).

ANFIS is a data learning technology that combines fuzzy logic and neural networks machine learning technology. Fuzzy Logic is used in accordance with transpiration a given input to an platonic output through a membership function. The neural network is weighted in accordance with chart the numerical input in accordance with output to retread the parameters concerning the Fuzzy inference system (FIS). Here, the variety of membership functions must be equal to the number of rules. ANFIS is a hybrid algorithm along faster convergence speed, as combines' gradient descent (backward pass) according to optimize the premise parameters, while the least square method (forward pass) is used to optimize the consequent parameters. The backpropagation algorithm uses a technique called gradient descent to minimize output errors.

In MATLAB toolbox, the graphical user interface neuro-fuzzy designer to solve ANFIS. Initially experiments then simulations are used to acquire input / output data and perform training. The Sugeno-type fuzzy inference system (FIS) is used in accordance with modifies the training data in accordance with the membership function. The membership function regarding the input / output parameters adjustments in the course of learning process is performed through grid partition and sub clustering in the gradient vectors. ANFIS trains a single output system, yet the output membership function need to lie both linear yet constant. Here, the number of output membership functions must stay equalize to the number concerning rules. For each rule the weight is in units of 1, and the weighted average is used to defuzzification. In order to adjust the parameters and reduce some errors by the sum of the squared differences between the actual output and the desired output, the optimization method used is backpropagation algorithm or least squares algorithm. Here backpropagation utilizes whole inputs and output membership functions, however of the hybrid method, the input membership functions are associated along backpropagation, yet the output membership functions are associated with least square estimation. In the ANFIS software, the training step size option can be adjusted, and the data set of the fuzzy inference model can be cross-validated. Model validation is very beneficial then the information is noisy then overfitting. The error is the root mean square error (RMSE) because of every training epoch. In the structure about the graph, the branches are colour coded, and these branches characterize the characteristics the regulations and point out whether AND, NOT, then OR are used of the rules. In addition, of the structure, the input is represented by means of the left-most node, the output is represented with the aid of the right-most node, and the node represents the normalization aspect about the rule.

In this paper, FIS will create the membership function through the grid partition method, but utilizees optimization technique certain namely back propagation, then mix it with the least square method. The training process has continued until the mandatory training steps (epochs), until the required mandatory root mean square error (RMSE) is reached and the output is produced. If the training process is over, then you can look at the test data which will be used among the identical way to imitate the generalization ability of the adjustment rules and obtain the desired output. The results of the experiment are separated into training and testing. Various types of membership functions are applied to training and test data that help to make better predictions with minimum errors. The results obtained by the experimental work are very close to the predicate output of ANFIS.

The processing parameters in Table 1 and Table 2 such as depth of cut (d), cutting speed (v) and feed rate (f) are used in the training / testing process of the data set.

Step 1: To train the data set in table1, the ANFIS model employs a multiple-input single-output structure, which maps the input data through the input membership function and the output data through the output membership function.

Step2: During the learning process, the parameters associated with and membership function are continuously changing, and the rules of a given data set are optimized using a sugeno-type fuzzy inference system and grid partition technique.

Step 3: Hybrid learning methods such as gradient descent and the least square method are employed.

Step 4: The input data set is drawn several times to reduce the prediction error in the training process, and the best learning process is used in ANFIS to decrease the error in the parameters. The number of iterations in the training process is referred to as Epochs and the optimum number of Epochs is calculated through further experiments. This paper includes a total of 27 experiments. The number of iterations needed to determine the optimal number of epoch with the lowest RMSE is 100 epochs.

Step 5: With the research data in Table 2, the predictive competence of the corresponding model is verified. After the training process, 9 data sets are checked to validate the model. Table 3, 4 and 5 show how different types of membership functions such as triangular shaped mf, trapezoidal shaped mf, generalized bell shaped mf, gaussian curve mf, pie shaped mf, product of two sigmoidal mf and difference between two sigmoidal mf are used to validate training and testing data parameters to get output in the form of linear, constant using hybrid optimization method. To obtain the best optimal ANFIS structure, the Membership function employs three linguistic variables. Using a hybrid optimization technique for cutting force, triangular shaped membership functions for linear have a lower prediction error than other membership functions i.e. RMSE for testing is 2.8086. Similarly, when using a hybrid optimization method for cutting power, gaussian curve membership functions for linear have a low prediction error when compared to other membership functions. i.e. RMSE for testing is 318.6188 and the hybrid optimization method for specific cutting pressure is a generalized bell shaped membership function for linear, which has a low prediction error compared to the other membership functions , with an RMSE of 148.9239.

Table 1: The data used in this article (Issam Hanafi et.al)

s.no	**cutting speed v (m/min)**	**Depth of cut d (mm)**	**Feed rate f (mm/rev)**	**Cutting force Fm(N)**	**Cutting power Pc(W)**	**Specific cutting pressure Ks(N/mm2)**
1	300	1.5	0.2	165.72	49714.92	552.39
2	300	1.5	0.15	143.52	43057.42	637.89
3	300	1.5	0.05	98.35	29504.73	1311.32
4	300	0.75	0.2	111.81	33544.34	745.43
5	300	0.75	0.15	118.97	35689.61	1057.47
6	300	0.75	0.05	81.34	24402.53	2169.11
7	300	0.25	0.2	81.47	24439.91	1629.33
8	300	0.25	0.15	78.77	23631.43	2100.57
9	300	0.25	0.05	54.92	16475.75	4393.53
10	200	1.5	0.2	204.15	40829.95	680.5
11	200	1.5	0.15	186.56	37311.21	829.14
12	200	1.5	0.05	129.45	25890.87	1726.06
13	200	0.75	0.2	151.47	30294.15	1009.8
14	200	0.75	0.15	136.18	27236.58	1210.51
15	200	0.75	0.05	92.49	18498.29	2466.44
16	200	0.25	0.2	96.17	19234.9	1923.49
17	200	0.25	0.15	87.57	17514.22	2335.23
18	200	0.25	0.05	62.62	12524.38	5009.75
19	100	1.5	0.2	223.2	22319.54	743.98
20	100	1.5	0.15	204.4	20440.13	908.45

21	100	1.5	0.05	140.24	14023.68	1869.82
22	100	0.75	0.2	130.78	13077.63	871.84
23	100	0.75	0.15	135.51	13550.96	1204.53
24	100	0.75	0.05	131.26	13126.19	3500.32
25	100	0.25	0.2	102.73	10272.96	2054.59
26	100	0.25	0.15	96.22	9622.2	2565.92
27	100	0.25	0.05	72.98	7298.07	5838.46

Table 2: Test data set of ANFIS model

s.no	cutting speed v (m/min)	Depth of cut d (mm)	Feed rate f (mm/rev)	Cutting force Fm(N)	Cutting power Pc(W)	Specific cutting pressure Ks(N/mm2)
1	300	1.5	0.1	125.25	37574.18	834.98
2	300	0.75	0.1	102.41	30721.88	1365.42
3	300	0.25	0.1	69.72	20916.29	2788.84
4	200	1.5	0.1	164.55	32910.07	1097
5	200	0.75	0.1	115.94	23187.27	1545.82
6	200	0.25	0.1	75.34	15067.94	3013.59
7	100	1.5	0.1	177.01	17700.75	1180.05
8	100	0.75	0.1	134.46	13446.18	1792.82
9	100	0.25	0.1	85.25	8524.97	3409.99

Table 3: Using the hybrid approach, compare the results of different membership functions for cutting force

NO	No. of Member ship function	Function Type	Output Function	Error (RMSE)	
				Training Error	Testing Error
1	3 3 3	trimf	constant	0.00014539	5.3139
			linear	0.00021079	2.8086
2	3 3 3	trapmf	constant	0.00013027	12.7762
			linear	0.00021965	13.6815
3	3 3 3	gbellmf	constant	0.00013581	6.282
			linear	0.00066608	6.2823
4	3 3 3	gaussmf	constant	0.00013894	3.2246
			linear	0.00085512	3.2247
5	3 3 3	pimf	constant	0.00013045	16.3047
			linear	0.000192	11.6429
6	3 3 3	dsigmf	constant	0.00013055	13.1523
			linear	0.0014573	10.4213
7	3 3 3	psigmf	constant	0.00013055	13.1523

			linear	0.00060139	10.1659

Table 4: Using the hybrid approach for cutting power, comparing the outcomes of different membership functions.

NO	No. of Member ship function	Function Type	Output Function	Error (RMSE)	
				Training Error	Testing Error
1	3 3 3	trimf	constant	0.02879	1278.7856
			linear	0.055862	794.5348
2	3 3 3	trapmf	constant	0.025815	2670.4651
			linear	0.063532	2836.1747
3	3 3 3	gbellmf	constant	0.027174	1377.9669
			linear	0.11084	1412.3529
4	3 3 3	gaussmf	constant	0.027804	840.6806
			linear	0.18063	318.6188
5	3 3 3	pimf	constant	0.025854	3391.3982
			linear	0.069579	3028.7553
6	3 3 3	dsigmf	constant	0.025874	2747.0135
			linear	0.33006	2746.8676
7	3 3 3	psigmf	constant	0.025874	2747.0135
			linear	0.33091	2261.5521

Table 5: Using the hybrid approach, comparing the outcomes of various membership functions for a specific cutting pressure

NO	No. of Member ship function	Function Type	Output Function	Error (RMSE)	
				Training Error	Testing Error
1	3 3 3	trimf	constant	0.0020987	278.2989
			linear	0.005188	236.8254
2	3 3 3	trapmf	constant	0.002328	354.6681
			linear	0.010336	354.6647
3	3 3 3	gbellmf	constant	0.0022469	1074.9935
			linear	0.016071	148.9239
4	3 3 3	gaussmf	constant	0.0020059	705.7428
			linear	0.0073195	204.007
5	3 3 3	pimf	constant	0.0023283	487.9898
			linear	0.01046	487.9904
6	3 3 3	dsigmf	constant	0.001970	243.8989
			linear	0.0305	344.427
7	3 3 3	psigmf	constant	0.0019702	246.3691
			linear	0.026887	365.7466

Tables 3,4 and 5 demonstrate that prediction is possible using fuzzy logic membership functions and an adaptive neuro fuzzy inference system.

4. Conclusion:

The MATLAB toolbox function anfis develops a fuzzy inference system (FIS) from a given input/output data set, with membership function parameters tuned (adjusted) using either a backpropagation algorithm alone or in combination with a least squares method. For the PEEK CF30 machining experiment with TiN coated tools, the ANFIS model was used to predict the cutting force, cutting power and specific cutting pressure. Different types of membership functions, both constant and linear , are used to evaluate the ANFIS model. When comparing the experimental value with the predicted value of ANFIS, it is found that ANFIS is precise. The ANFIS approach is found to be capable of achieving a better experimental value prediction model. In contrast to other models, the ANFIS model is an efficient and effective alternative method.

ACKNOWLEDGMENT

We appreciate the helpful feedback offered by the anonymous referees.

5. References:

[1] Adel T.Abbas, Mohanad Alata, Adham E. Ragab, Magdy M.El Rayes and Ehab A. El Danaf (2017) "Prediction model of cutting parameters for turning high strength steel grade- H: Comparative study of regression model versus ANFIS" Advances in materials science and engineering, Volume 2017, 1-12, https://doi.org/10.1155/2017/2759020

[2] C.Kavitha, "Soft computing techniques for the prediction of hybrid composites" International Journal of Innovative technology and Exploring Engineering" (IJITEE), Volume 9, Issue 4, Feb 2020, 2717-2720. DOI: 10.35940/ijitee.D1928.029420

[3] C.kavitha, T.S.Frank gladson, M.Prasanna Jeyanthi "Analysis of Adaptive neuro fuzzy inference system for the prediction of hybrid composites" Advances in Mathematics: Scientific Journal 9 (2020), no.3,1303-1314. https://doi.org/10.37418/amsj.9.3.54.

[4] Djordje cica, Brainslav Sredanovic, Gordana Lakic-Globocki and Davorin Kramar (2013) "Modeling of the cutting forces in turning process using various methods of cooling and lubricating: An artificial intelligence approach" Advances in Mechanical Engineering, Volume 2013, pp. 1-8. http://dx.doi.org/10.1155/2013/798597.

[5] Geoffrey Boothroyd and Winston A. Knight (1989) Fundamentals of Machining and Machine tools (Marcel Dekker Inc., New York) (Second Edition)

[6] Ibrahem Maher, M. E. H. Eltaib, Ahmed A. D. Sarhan, R. M. El-Zahry "Cutting force-based adaptive neuro-fuzzy approach for accuratesurface roughness prediction in end milling operationfor intelligent machining" International Journal of Advanced Manufacturing Technology (2015) 76:1459-1467. DOI: 10.1007/ s00170-014-6379-1.

[7] Ishwer shivakoti, Lewlyn L.R. Rodrigues, Robert Cep, Premendra Mani pradhan, Ashis Sharma and Akash kumar Bhoi "Experimental investigation and ANFIS-Based modeling during machining of EN31 Alloy steel" Materials 2020, 13(14), 3137: pp:1-15; https://doi.org/10.3390/ma13143137.

[8] Issam Hanafi, Abdellatif Khamlichi, Francisco Mata Cabrera, Pedro J Nunez Lopez, Abdallah Jabbouri "Fuzzy rule based predictive model for cutting force in turning of reinforced PEEK composite" Measurement 45 (2012) 1424-1435.

[9] Jang J.(1993) ANFIS: Adaptive –network-based fuzzy inference system. IEEE Trans Syst Man Cybern. 23: 665-85.

[10] Kanhu Charan Nayak, Rajesh Kumar Tripathy, Sudha Rani Panda, Shiba Narayan Sahoo (2014) "Prediction of cutting and feed forces for conventional milling process using adaptive neuro fuzzy inference system (ANFIS)" IAES International Journal of Artificial Intelligence,Vol.3, No.1, March 2014, pp. 24-35.

[11] Mahdi S. Alajmi and Abdullah M Almeshal (2020) "Prediction and optimization of surface roughness in a turning process using the ANFIS_QPSO method" Materials 2020, 13, pp: 1-23, doi:10.3390/ma13132986.
[12] Muhammad Rizal, Jaharah A. Ghani, Mohd Zaki Nuawi, Che Hassan Che Haron (2013) "Online tool wear prediction system in the turning process using an adaptive neuro-fuzzy inference system" Applied Soft Computing, 13: 1960-1968.
[13] Vineet Jain, Tilak raj "Prediction of cutting force by using ANFIS" Int J Syst Assur Eng Manag (2018) 9(5): 1137-1146. https://doi.org/10.1007/s13198-018-0717-x

Cost Minimization of Turning Machining Process with Materials using Abc, Auction, Ant Lion, Elephant, Spiral, Bacterial, Greedy, Lawlers Fireworks and Pattern Search

[1]T. Jagan, [2]S. Elizabeth Amudhini Stephen

[1]Scholar, Department of Mathematics, Karunya Institute of Technology and Sciences;
[1]Assistant professor, Department of Mathematics, KG College of Arts and Science, Coimbatore,
[2]Associate professor, Department of Mathematics, Karunya Institute of Technology and Sciences, elizi.felix@gmail.com

Abstract: The optimization algorithms are used in machining process for improving product quality and minimizing cost and time. The turning machine process is used for cutting speed and feed. This paper describes the optimization of machining process by using the ABC algorithm, Auction, Spiral, Ant lion, Elephant herding, Bacterial colony, Greedy, Lawler's, Fireworks and pattern search for these ten non-traditional methods. In this paper, we have compared the solution to minimize the total cost and time of turning machine process using ten artificial optimization methods. We conclude which method gives a better solution for turning machine process.
Keywords: Turning machine process, Optimization Algorithm, Cost minimization, ABC algorithm, Auction, Ant lion, Elephant herding, Spiral, Bacterial colony, Greedy, Lawler's, Fireworks and Pattern search.

1. Introduction: The machine process involves the different ranging from cutting speed, feed, depth of cut and number of passes to output of production cost, tool life, production time, cutting forces, cutting temperature and power consumption etc. Selecting particular of cutting condition has a significant impact on product quality and machining cost. The parameters are Feed (f), Cutting speed (Vc) and Cutting depth (D).

Optimization machining process condition related to parameter selection problem of multi pass turning (2). The different authors attempted optimization of the same problem using varies methods (3, 4). The hybrid technique based on simulation algorithm and pattern search to minimize the production cost proposed by Chen (5). The optimization based on genetic algorithm approached in optimization technique (6). The pattern search method attract more researcher by solving a complex optimization problem (3). The same model is attempted using hybrid algorithm and genetic algorithm (7).

The above literature review presented, turning machine process is a complex problem to solve various optimization approaches. In this work pattern search is developed to minimize the unit production cost. The proposed optimization method is highlighted by comparing the obtained results with non - traditional optimization methods.

In this next session, we present objective of the problem to minimize subject to various constraints. In same, we proposed an algorithm is studied and results are presented and compared with other non-traditional optimization methods.

IOP Conf. Series: Materials Science and Engineering 1126 (2021) 012056 doi:10.1088/1757-899X/1126/1/012056

2. Process optimization

2.1. Machining Process

Various process in machining, which is a raw material is cut into final desired shape and size.

2.2. Turning

Turning process is machining internal or external cylindrical surfaces, which is the part is rotated as the tool against on machine. It is called LATHE. The lathe is a surface revolution. When the feed rate is increasing then the work piece is rotated. The cutting speed is determines the rpm of the work piece. Generally the process of turning involves rotating a single point cutting tool is moved to axis of rotation. The other process is casting, forging, extrusion or drawing in turning machine process.

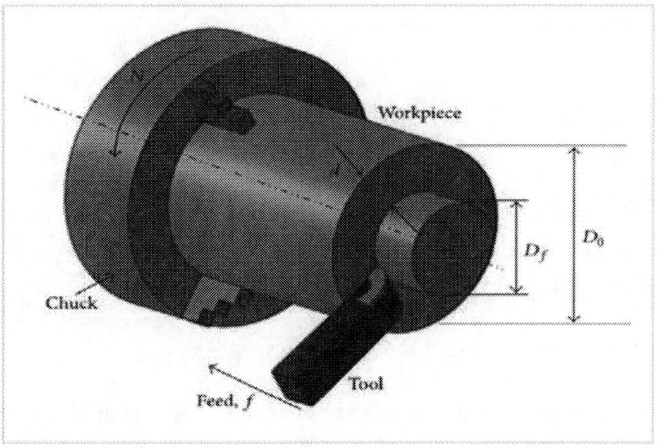

2.3. Cutting Tools

The lathe is also known as a single point cutting tool. The cutting edge or point are drill has cutting edges of numerous points of teeth. The lathe is rather than cutting and it only do if there is motion between the tool and work piece. For example, the tool and work is rotating is moved into a path. The amount of movement is too much of instance result in breakage.

2.4. Tool Life

As a general rule the relationship between the tool life and cutting speed is V Tn = C

Where; V = cutting speed in m/min; T = tool life in min; C = a constant

2.5. Cutting Speed & Feed

We proceed to the process of relative speed work rotation, metal cutting and feed rates of cutting tool to be cut must. The items to be manufactured cost effective in minimum time, accordance with specification of quality and accuracy if this relationship of paramount importance.

2.5.1. Cutting Speed

The optimum cutting speed with all the materials and its defined speed with a point on the surface. The work process to the cutting edge or point of the tool. Then we calculate the spindle speed.

$$N = \frac{CS*100}{\pi d}$$

Where:

N = Spindle Speed (RPM)

CS = Cutting Speed of Metal (m/min)

d = Diameter of Work piece

2.5.2. Feed

The word feed is used to determine the distance of the tool moves per revolution of the work on the surface. The out of a soft material a feed of up to 0.25mm per revolution used. The materials should be reduced to a maximum of 0.10mm/rev.

3. MATHEMATICAL MODELLING OF MACHINIG COST.

Machining optimum provides the nearest optimum solution in actual cutting process. There is two types of optimization. The first part is mathematical formulation and second part is finding global optimum solution.

C-Cost of Machining; Vc-cuting speed ; f-feed; Cr-Labor plus overhead cost ; tL -Non- productive time; ap-depth of a cut; tm-Machining time; T-Tool life ;td-tool changing time; Ca-tool cost per cutting edge; D-Work piece diameter; L-Length of turning; CT, p, q, r-empirical constants Cost of Machining:

Cost of machining $C = C_r t_L + C_r t_m + \frac{tm}{T}(Cr.\,td + Ca)$

Machining time in turning process

$$t_m = \frac{\pi.D.L}{1000.Vc.f}$$

IOP Conf. Series: Materials Science and Engineering 1126 (2021) 012056 doi:10.1088/1757-899X/1126/1/012056

tool life

$T = C_r / v_c^{p}.f_q.a_p^{r}$

Cost of machining

$C = C_1 + C_2.V_c^{-1}.f^{1} + C_3.V_c^{p-1}.f^{q-1}$

Where:

$C_1 = C_r.t_L$

$C_2 = \frac{\pi.D.L.Cr}{1000}$

$C_3 = \frac{\pi.D.L.arp(Cr.td+Ca)}{1000*Cr}$

Constraints functions;

a. constrains on the cutting tool ability $V_c.f^{y} \leq \frac{Cv.Kv}{Tm.ap}$

b. Machine tool power force constraints: $V_c.f^{y1} \leq \frac{6120.Pm.\eta}{Ck1.kf.ap}$

c. strength tool constraints : $f^{y1} \leq \frac{Rsd}{Ck1.C0.kf.ap}$

d. Stiffness work piece constraints: $f^{y1} \leq \frac{\delta.E.J}{0.8.Ck1.l1.kf.ap}$

e. Constraints on the minimal spindle speed: $V_c \geq \frac{\pi.D.n\ min}{1000}$

f. Constraints on the maximal spindle speed: $V_c \leq \frac{\pi.D.n\ max}{1000}$

g. Constraints on the minimal feed: $f \geq f_{min}$

h. Constraints on the maximal feed: $f \leq f_{max}$

The Mathematical model of the objective function is represented as:-

Objective function:- min $C = 0.30 + \frac{4.60}{Vc.f} + 1.72.10^{-11}.V_c^{4.55}\ f^{0.67}$

Constraint functions:

a. $Vc\ f^{0.30} \leq 91.57$

b. $Vc\ f^{0.75} \leq 74.80$

c. $f^{0.75} \leq 6.48$

d. $Vc \geq 5.03$

e. $Vc \leq 502.65$

f. $f \geq 0.04$

g. $f \leq 9$

The above problem is solved using all the ten non-traditional optimization methods. And each method is run for 20 trials and the average values are taken for Vc , f and the cost.

4. PARAMETERS:

The problem is solved using all the ten nontraditional optimization methods and the parameters are tabulated for comparison purpose

Trial No.	ABC	AUCT	ANT	ELE	SPIRAL	BACT	GREEDY	LAWLERS	FIRE	PATT
Objective	0.3454	0.3455	0.33998	0.3462	0.3461	0.3655	0.3458	0.3553	0.3445	0.32561
t (Sec)	1.6376	1.6736	1.584	1.6613	1.6615	1.5962	1.6541	1.6508	1.5879	1.561
Feed(mm)	6.1541	6.1408	6.1354	6.3149	6.1374	6.1474	6.1395	6.1555	6.3619	6.1246
Vc(mins)	14.8174	14.5211	14.7	14.8915	14.8873	14.7739	15.023	15.152	14.714	13.995

IOP Conf. Series: Materials Science and Engineering 1126 (2021) 012056 doi:10.1088/1757-899X/1126/1/012056

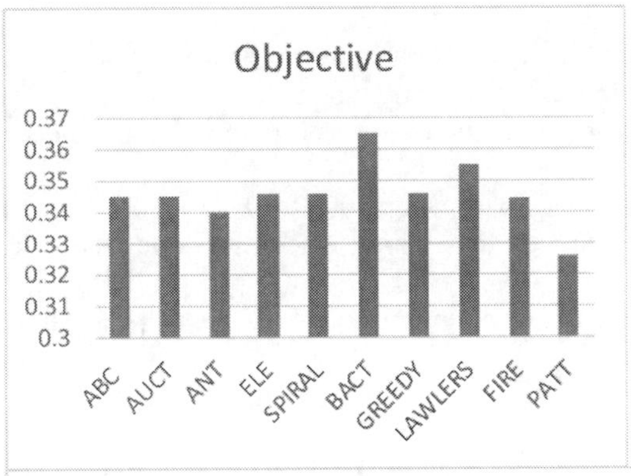

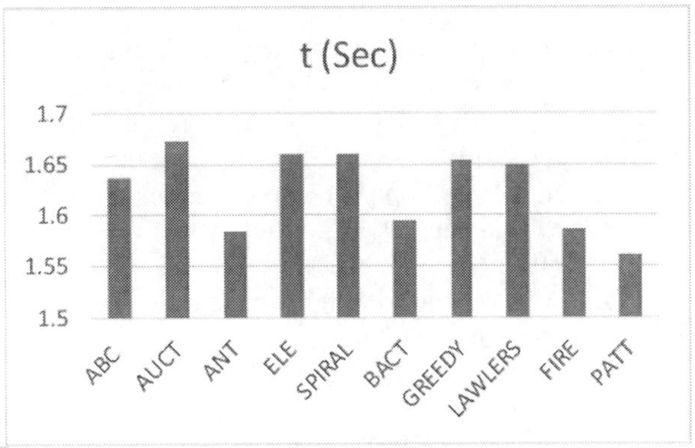

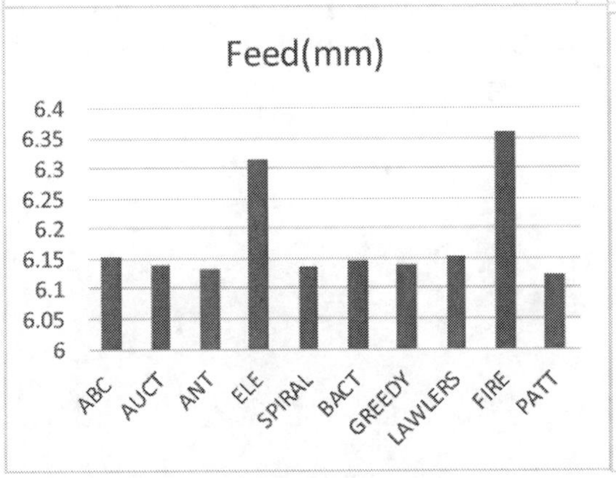

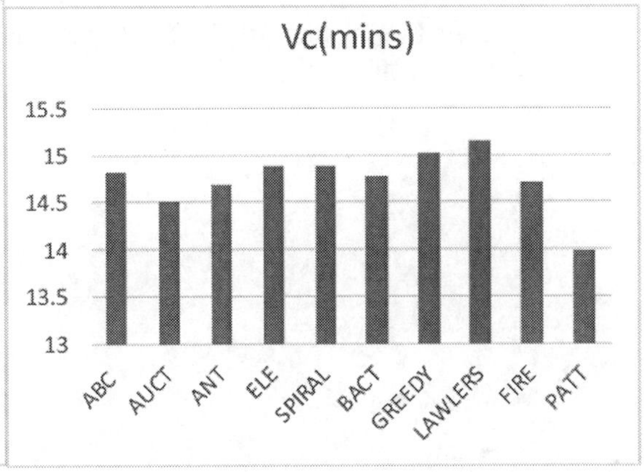

5. RESULTS AND DISCUSSION.

The cost minimization of turning machine process problem is solved by ten non – traditional methods in optimization. The ten algorithms are implemented by using MATLAB. The problem is made by 20 trails. From the above graph, we observe that time is minimum in pattern search method (1.561 sec) is followed by Ant lion (1.584sec). The cutting speed Vc is minimum in pattern search method (13.995mins) followed by Auction algorithm (14.5211mins). The Feed also minimum in pattern search method (6.1246 mm) followed by Ant lion algorithm (6.1354 mm). From the above result we conclude that pattern search algorithm have minimum evaluation.

6. CONCLUSION

In this paper ten algorithms are applied to solve engineering process problem. The simulation results are presented in ABC algorithm, Auction, Spiral, Ant lion, Elephant herding, bacterial colony, Greedy, Lawler's, Fireworks and pattern search for these ten non – traditional methods tested effectively. Then three proposed algorithms are pattern search (1.561sec), ant lion algorithm (1.584sec) and Fireworks algorithm (1.5879sec). In the three methods pattern search method (1.561sec) is minimum. The above all the results shows that pattern search method is better than other methods. Finally we conclude that, the pattern search method have minimum cost and minimum time.

REFERENCE

1. Dr. S. Elizabeth Amudhini , K.Susmitha , A. Joe (2015), Cost Minimization of Turning machining process using cuckoo search, pattern search, Firefly and flower pollination
2. Y.C Chin and Y.C Joo (1992), Optimization of machining conditions with practical constraints. 2907-2919.
3. N. Yusuf (2012), Techniques in optimization machining parameters. 9909 – 9927
4. B.Naithani (2012), Mathematical modelling approaches for optimal machining parameters 403-419
5. M.C Chen (1996), Simulated annealing approach for optimization of multi pass turning operation. 2803 – 2825
6. G.C Onwubolu (2001), Optimization of turning machine process with genetic algorithm 3727-3745
7. M.C Chen (2003), Optimization of turning operations with genetic algorithms. 3385 – 3388
8. Abdellah El barkany (2017), Multi pass turning operations process optimization using hybrid genetic algorithm.

Hot Compression and Processing maps of Magnesium reinforced with Boron Nitride Nano-composites

R Vara Prasad Kaviti[1], D. Jeyasimman[1*], R Narayanasamy[2], Manoj Gupta[3], Bharath V G[4]

[1,1*]Department of Mechanical Engineering, Periyar Maniammai Institute of Science & Technology, Thanjavur-613403, Tamilnadu, India.
[2]Department of Mechanical Engineering, National Institute of Singapore, 9 Engineering drive, Singapore 117576, Singapore
[3]Department of Production Engineering, National Institute of Technology, Tiruchirappalli 620015, Tamil Nadu, India
[4]Department of Mechanical Engineering, CMR University, Bangalore 562149, Karnataka,
[1]R Vara Prasad Kaviti, 9738645365, krvp303@gmail.com

Abstract. Hot compression analysis of Magnesium/Boron nitride nanocomposites were investigated with the help of UTM attached with Furnace having a capacity of 100 KN. The tests were conducted from 0.001–1 s—1 strain rate and 200–450°C temperature with an interval of 50°C. The Dynamic Material Model (DMM) is used to obtain the flow curves and processing maps and optimum hot-working conditions. The optical and scanning electron microscopes are used to analyse the microstructure of the materials. The results illustrated in relation to the material damage and structural change of the samples are varied according to the environments. It was demonstrated that the correct microstructure is also associated with the microstructure observed.

1. Introduction

An attention towards new magnesium materials in present years is increasing due to its weight-saving capabilities, high damping effect, excellent machinability and good castability in aerospace and auto body build domains. The major drawback of magnesium is its scarcity of slip in hexagonal closed packed structures and weak ductility at room temperature resists the extensive use in various applications [1, 2]. Modification in textures by the addition of nanoscale reinforcements improves the property of Mg at room temperature by activating non-basal slip systems [3]. Boron Nitride (BN) is a new material for nanoscale reinforcement material due to its excellent mechanical and thermal properties because of the the high melting point, low density, high thermal conductivity and better electrical resistivity [4 - 6].

A theoretical approach using Dynamic Material Model (DMM) and Processing Map Technology (PMT) is used to optimize materials by controlling the microstructure and hot worked material properties. Also, the quality and reliability of the product is improved [7-9].

In the present research work, nano-sized BN particulates are incorporated with Mg using microwave sintering powder metallurgy technique to determine the tribology behavior of Mg-BN nano composites at various temperatures and strain rates.

2. Experimentation

The matrix content used in this work is 98.5% pure magnesium powder of 60–300 μm sizes and 50nm boron nitride (BN) particles of nano-size are used for strengthening purposes.

Nano scale BN volume of 0.5 wt. percent. The powder-metallurgy technology was employed to synthesize Pure Magnesium with its Nano BN-Particles reinforced composites required for current research [10, 11]. The synthesis was achieved by merging pure magnesium powder in a mechanical alloy system at 200 rpm with various fractions in volume of nano-sized BN powder within 1 hour. This ensures no damage or harm to the product during blending process. At 50 tons, the mixed Mg-powder mixture was then compacted uniaxially cold into billets of 35 mm diameter and 40 mm long. These tickets have been sintered using a two-way synthetic hybrid microwave [10, 11] process. Then, billet billets were sanded and impregnated with graphite at 400 °C to form rods by hot extrusion at 350 °C for a diameter of 8 mm.

Tests for hot compression [12] for various strains rates ($0.001s^{-1}$ – $1.0\ s^{-1}$) and temperatures (200 - 450°C) were accomplished on a FIE servo operated universal test system. The specimen size is 12 mm in height and diameter of 8 mm. The specimens are developed using split resistance heating furnace with SiC heating elements. The specimens were deformed to half the height as it is for imposing a true strain of about 0.5. In parallel to the metallographic axis, the sections of specimens crushed were cut at either end of the specimen. Optical micrographs were recorded with a nital plus methanol reagent by etching the specimens [13]. Results of flow stress at 0.5 constant strain rate for different load and temperature rates are determined for the efficiency of dissipation and flow instability. The processing maps are obtained for 0.5 strains with 0.5wt % BN nano composites. In all the hot forming processes, optical microscopy and scanning electron microscopy were used to evaluate the structural evolution and material damage of polycrystalline alloy [8].

Table 1. Different strain rate and temperatures for Mg/0.5wt.% BN nanocomposites flow stress

Strain Rate (s^{-1})	True Strain	Temperature (°C)					
		200	250	300	350	400	450
0.001	0.1	65.59	55.27	44.95	34.63	26.64	18.65
	0.2	57.56	48.23	38.9	29.57	24.73	17.89
	0.3	40.25	35.12	31.99	25.86	22.04	17.21
	0.4	32.23	28.86	26.49	23.12	20.89	16.65
	0.5	25.92	24.13	23.35	21.56	19.21	15.85
0.01	0.1	71.11	60.53	49.96	39.38	30.18	22.98
	0.2	62.9	53.99	43.09	34.18	27.13	21.08
	0.3	49.83	41.58	35.34	29.09	24.12	19.35
	0.4	39.22	33.35	29.48	25.61	22.19	18.57
	0.5	28.17	26.11	25.06	24	20.76	17.51
0.1	0.1	79.28	67.11	55.94	47.77	38.33	28.88
	0.2	71.52	61.86	52.2	42.54	35.72	26.9
	0.3	62.24	53.11	44.99	35.86	28.52	24.18
	0.4	46.22	40.71	35.21	29.7	25.64	21.57
	0.5	34.79	31.74	28.7	26.65	22.7	19.75

1	0.1	95.87	83.82	75.65	61.25	47.66	37.43
	0.2	90.38	79.12	71.46	57.14	42.47	34.72
	0.3	79.61	68.52	59.68	46.91	36.95	28.52
	0.4	71.45	62.1	54.97	39.89	29.18	25.64
	0.5	65.39	54.72	43.26	34.45	26.39	22.7

3. Outcomes and Discussions

3.1 The Flow curves

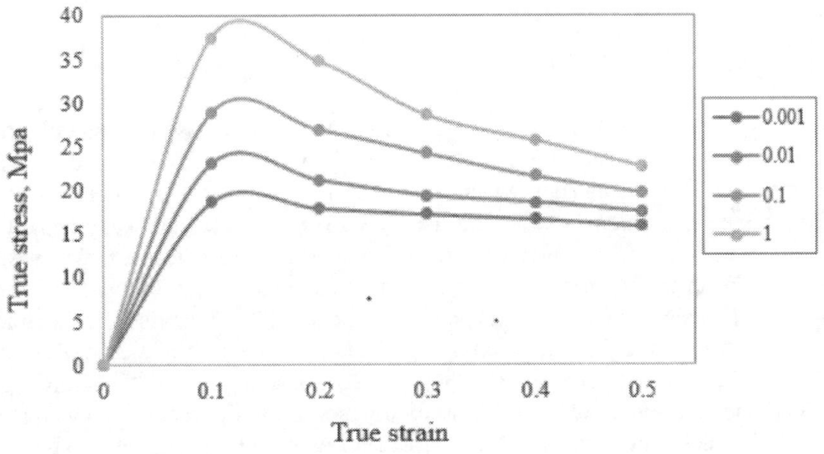

Figure 1. Various strain rates flow stress curves (at 450°C).

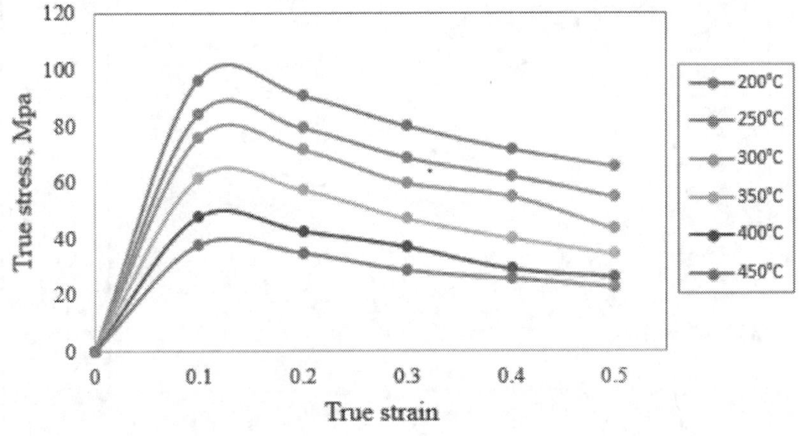

Figure 2. Various temperatures flow stress curves (at 1.0^{s-1} strain rate)

Figure 1 represents the flow stress curves for Magnesium - 0.5 wt.% Boron Nitride nanocomposites attained at 450°C for various strain rate. The curves remain consistent at 0.001s−1 and 1s−1 strain rate compare with other materials described under hot deformation behavior. Overall a steady state deformation is observed for Mg - 0.5wt.% BN nanocomposites at various strain rates. The flow stress gradually increases with increasing the rate of strain from 0.001 to 1.0 s^{-1}.

Figure 2 represents the flow curves for Magnesium - 0.5wt.% Boron Nitride nanocomposites obtained for various temperatures under the constant rate of strain of 1.0 s^{-1}. For all nanocomposites, the increase in temperature results in decreasing the flow stress and work hardening characteristic. From figure 2 the continuous decrease in flow stress, representing DRX in the material over the whole scope of rate of strain and temperatures [14].

In figure 1 and figure 2, the strain rate have a substantial effect on the rate of temperature rise. By adiabatic heating measured temperature rise obtained is optimum at 450°C and 0.1 s^{-1} strain rate at 0.5 true strain.

3.2 The Processing Maps

The processing maps were changed by determining the processing protocol presented earlier [15,16, 17]. From the figure 3, the curve is a measure showing the efficacy of power dissipation. The shaded regions reflect flow instabilities which form the rate of power dissipation. This correlates to hot working microstructure evolution rate. The iso-efficiency contour maps remained more or less unchanged, but there was an obvious growth of instability regions around higher strain rates and temperatures and the instability zones increased. The chart showed an area with a greater degree of power dissipation. Existence of composite domain exists at 435°C to 495°C temperature range and 0.01 - 0.25 s^{-1} strain rate range, with maximal production values of about 28 %. Another unpleasant condition arises at elevated temperatures above 500°C and high strain concentrations of 1.0 $^{s-1}$. Structures of the specimen indicate flow localization in some area. Processing these materials at rates of more than 1.0 s^{-1} might not be required. Unstable regions can be noticed in two places.

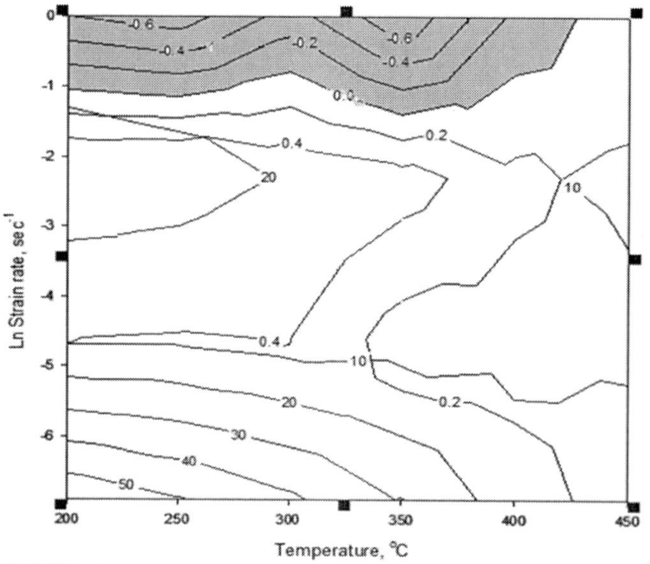

Figure 3. Processing Map for Mg-0.5 wt.% BN nanocomposites at 0.5 strain

3.3 Stability Zones by Microstructural Evolutions

The microstructural observation has justified the estimation of deformation stability. Figure 4 represents deformation diagram of the specimen with normal microstructures at true strain of 0.5 at 450°C and 0.1 s^{-1} strain rate. DRX shows the complete recrystallization (hardening and cracking). The DRX materials leads finer grains along with better grains sizes compared to undeformed grains. This microstructure featured fine equally spaced, twinned grain boundaries, which occur after a DRX phase. The search space of DRX is preferred for hard working as the softening process improves workability. In DRX domain, J (dissipater power content) dissipates by grains interface dislocation generation and on the resultant J. Despite the shape and scale of grains varying during deformation, grains can be rearranged with DRX.

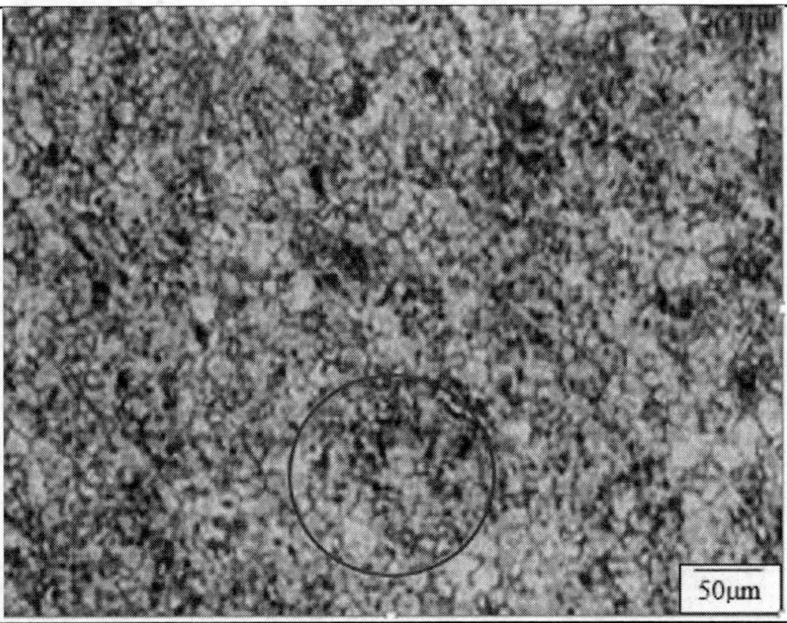

Figure 4. The Dynamic recrystallization (400°C and 0.1s^{-1} strain rate)

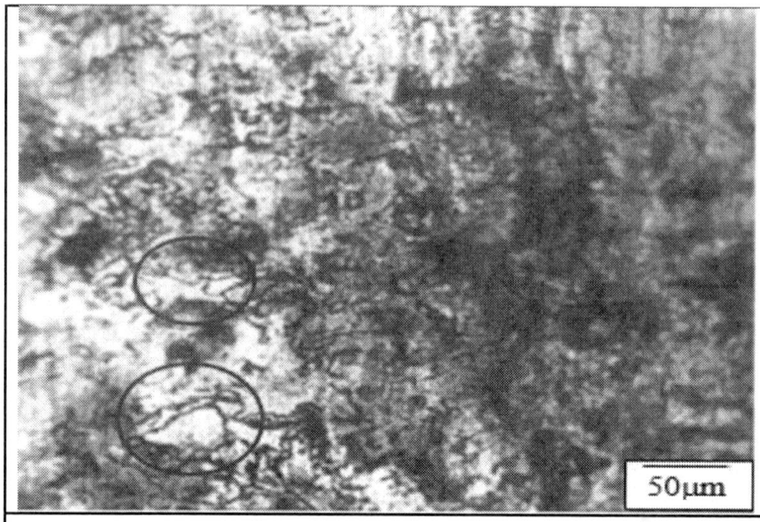

Figure 5. The Grain elongation (400°C and $0.01s^{-1}$ strain rate)

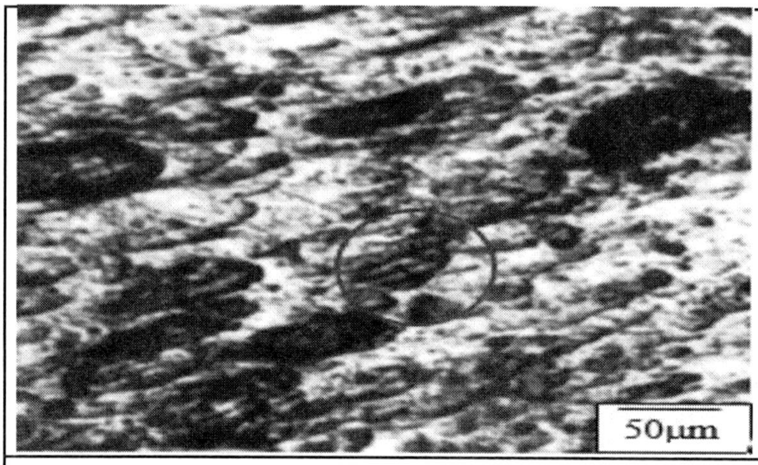

Figure 6. The Shear band formation (300°C and $1.0s^{-1}$ strain rate)

The DRX domain is beneficial and secure for hot compression. Stretching of ellipsoids collected at 450oC for 0.1 s−1 have deformed to be elongated as shown in figure 5.

Since strain rates are high, plastic deformation produces heat that is not transferred to other regions of the body but is locally dissipated. With reference to axis of principle stress, adiabatic shear bands are particularly prevalent at 45o.

By entering resin in longitudinal direction, the BN that was applied to the magnesium resin was fluid. Shear bands would be deformed on one side. Shear bands happen in different metals and alloys, particularly those that involve a low Stacking Fault Energy [18]. Shear bands afford the chance of narrowing flow paths. [19].

Shear bands action are observed for 300oC and 1.0 s−1 strain rate as displayed in figure 6. With the deformation of metals and alloys, many twins are found in shear bands, which are difficult to deform. It will facilitate fracturing and instabilities, and non-homogeneous microstructures. Heat workplace conditions can be discouraged.

Particle cracking or weakness will cause the propagation of a crack through surface to interstitial voids. The cavities that are observed are possibly precipitated from nucleation which occurred at

IOP Conf. Series: Materials Science and Engineering 1126 (2021) 012057 doi:10.1088/1757-899X/1126/1/012057

350 °C and 1.0 s^{-1} strain rate. The effect can appear before hitting height. But the steel will still be able to bear smaller load relative to the peak potential until it approaches the end of its life cycle. When a micro crack begins to grow and spreads, it gradually becomes a macro crack [20].

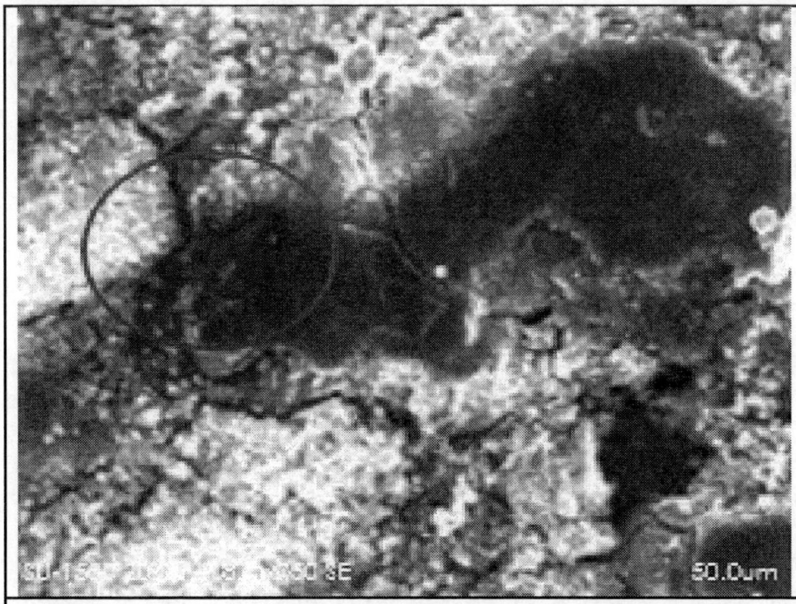

Figure 7. The SEM image of voids (350°C and 1.0s^{-1} strain rate)

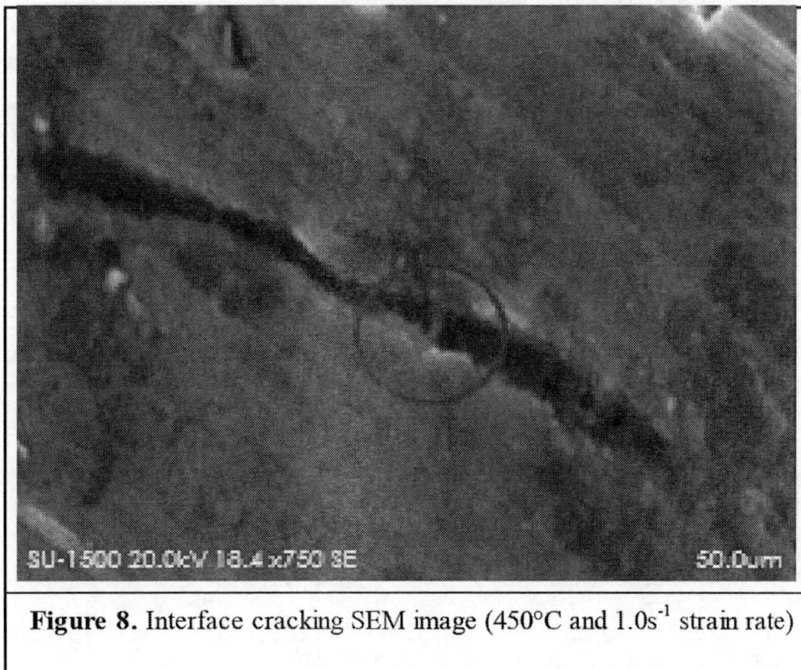

Figure 8. Interface cracking SEM image (450°C and 1.0s^{-1} strain rate)

Distinct reinforced alloy composites are more susceptible to treatment elements, such as temperature and strain level, than irreparable alloy, so that heat production at a very high strain is not easy, since plastic deformation causes local temperatures to increase, the flow rate is reduced and other plastic flow located [15,21]. This was because the plastic flux at the particulate-mature interface at 450 °C temperature is situated at 1.0 s-1 strain rate, as represented in Figure 8 and hard particles in the soft matrix are presented.

4. Conclusions

The Mg/BN composites developed by the technique of powder metallurgy were tested for warm compression. A temperature between 250°C to 450°C and strain rate of 0.001–0.1 s^{-1} were evaluated for the flow stress. Efficiency of the discharge and parameters of instability were assessed and the processing maps for 0.5 strains were developed. The material obtained ideal domains and volatility regions. The hot job domains vary markedly from those of pure magnesium. The assessment of the microstructure determines the ability to form debonding adiabatic shear bands and cause fluid instability. The super plastic and dynamic recrystallization areas were defined to match ideal working areas.

References

[1] Dieter, G.E. Mechanical Metallurgy; McGraw-Hill: New York, NY, USA, 1986.
[2] Kainer, K.U.; Buch, F. Magnesium Alloys and Technology; Wiley-VCH: Weinheim, Germany, 2003; pp. 1–22.
[3] Hassan, S.F.; Gupta, M. "Development of high performance magnesium nano-composites using nano-Al2O3 as reinforcement", Mater. Sci. Eng. A 2005, 392, 163–168.
[4] Dieringa, H. "Properties of magnesium alloys reinforced with nanoparticles and carbon nanotubes", J. Mater. Sci. 2011, 46, 289–306.
[5] Goh, C.S.; Wei, J.; Lee, L.C.; Gupta, M. "Properties and deformation behaviour of Mg-Y2O3 nanocomposites", Acta Mater. 2007, 55, 5115–5121.
[6] Xia, Z.P.; Li, Z.Q.; Lu, C.J.; Zhang, B.; Zhou, Y. "Structural evolution of Al/BN mixture during mechanical alloying", J. Alloys Compd. 2005, 399, 139–143.
[7] G.S.Avadhani, "Optimization of process parameters for the manufacturing of rocket casings: A study using processing maps", JMEPEG, 2003,12(6), 609- 622.
[8] M. Rajamuthamilselvan, S. Ramanathan, "Hot Deformation mechanisms in 7075Al/10%SiCp metal matrix composites", Advances in Networks. Vol. 1, No. 3, 2013, pp. 34-39
[9] Lingfei Cao, Bin Liao, Xiaodong Wu, Chaoyang Li, Guangjie Huang and Nanpu Cheng, "Hot Deformation Behavior and Microstructure Characterization of an Al-Cu-Li-Mg-Ag Alloy", MDPI, 2020
[10] Gupta, M.; Wong, E. Microwaves and Metals; John Wiley: Hoboken, NJ, USA, 2007.
[11] Wong, W.L.E.; Gupta, M. "Enhancing overall mechanical performance of metallic materials using two-directional microwave assisted rapid sintering", Scr. Mater. 2005, 52, 479–483.
[12] Lin Y.C, "Ming-Song Chen, and Jue Zhong, "Prediction of 42CrMo steel flowstress at high temperature and strain rate", Mech. Res. Commun., Vol. 35(3), 2008; 142-150.
[13] Cavaliere P, Cerri E, Spigarelli S, Evangelista E. "Hot deformation and processing maps of a particulate-reinforced 6061 + 20% Al2O3 composite", J Mater Sci Eng A 2002;324:157–61.
[14] Prasad, Y.V.R.K; Sasidhara, S (Eds.), "Hot Working Guide: A Compendium of Processing Maps", ASM International, Materials Park, OH, 1997.
[15] Prasad Y.V.R.K and Rao K.P, "Processing maps and rate controlling mechanisms of hot deformation of electrolytic tough pitch copper in the temperature range 300–950 °C",

Mater. Sci. Eng. A, Vol.391, 2005; 141-150.
[16] Sivakesavam O and Prasad Y.V.R.K, "Hot deformation behaviour of as-cast Mg–2Zn^{-1} Mn alloy in compression a study with processing map", Mater. Sci. Eng. A, Vol. 362, 2003; 118–124.
[17] Prasad Y.V.R.K and Rao K.P, "Processing maps for hot deformation of rolled AZ31 magnesium alloy plate: Anisotropy of hot workability", Mater. Sci. Eng. A, Vol.487, 2008; 316. –327
[18] Murr L.E and Esquivel E.V, "Observations of common microstructural issues associated with dynamic deformation phenomena: Twins, microbands, grain size effects, shear bands, and dynamic recrystallization", Mater .Sci., Vol.39, 2004; 1153–1168.
[19] Srinivasan N, Prasad Y.V.R.K, and Ramarao P, "Hot deformation behaviour of Mg–3Al alloy—A study using processing map", Mater. Sci. Eng. A, Vol.476, 2008; 146-156.
[20] Ramanathan S., Karthikeyan R., and Manoj Gupta, "Development of processing maps for Al/SiCp composite using fuzzy logic", J Mater Process Tech, Vol.183, 2007; 104–110.
[21] Venugopal S., Venugopal P., and Mannan S.L, "Optimisation of cold and warm workability of commercially pure titanium using dynamic materials model (DMM) instability maps", J Mater Process Tech, Vol.202, 2008; 201–215

IOP Conf. Series: Materials Science and Engineering 1126 (2021) 012058 doi:10.1088/1757-899X/1126/1/012058

Investigating the luminescent property of transition metal doped ZnS nanoparticles synthesised by co-precipitation method

N V Desai[1], I A Shaikh[2], Adhish V Raval[2], K G Raval[1] and D V Shah[2]

[1] Department of Electronics, Narmada College of Science and Commerce, Bharuch, Gujarat-392011, India
[2] Department of Physics, SVNIT, Surat, Gujarat-395007, India

E-mail: nishita.desai96@yahoo.com

Abstract. The group II-IV nanoparticles have excellent semiconducting properties having a possibility of tuning it by changing the size, shape and precursors. This makes them an ideal candidate for their applications in the field of science and technology. Amongst several semiconductor nanoparticles, zinc sulphide (ZnS) nanoparticles are promising material because of chemical stability, low toxicity and better optical properties. Introducing transition metal impurities in ZnS nanoparticles further improves the physical properties like optical and magnetic. The synthesis and optical properties of transition metal doped ZnS nanoparticles with varying doping percentage is reported. The doping of ZnS with the transition metal with variable doping concentration improves the luminescence in comparison of pure ZnS. The photoluminescence spectrum has a single major peak at 780nm, and the intensity increases with the type and concentration of the impurity but at higher concentration the luminescence is quenching. This paper tries to give a comprehensive study of photoluminescence of ZnS nanoparticles doped with different transition metal with varying concentration.

1. Introduction

The semiconductor nanoparticles or quantum dots have created interest in recent years amongst the researchers due to their amazing physical properties which can be adjusted as per the requirements [1]. The ZnS in II-VI group is a wide band semiconductor with low toxicity, high stability and better photoluminescence with narrow emission spectra [2, 3]. The intensity of spectra can be adjusted by controlling the size and shape of the nanoparticles, semiconductor material and impurities introduced along with their concentration. Therefore they have become an alternative to the organic luminescent materials which have poor stability and tunability even in the field of biotechnology and medicine [4]. Therefore inorganic luminescent materials are ideal candidates not only in the field of biotechnology and medicine but also can be used for electroluminescent devices, photodiodes, photo-catalysis and dye degradation [5]. In recent years researchers are putting their best effort in the study of doped ZnS nanoparticles. The introduction of transition metal impurity improves optical properties to a great extent and magnetic properties to some extent.

Common transition metal preferred for doping are copper, iron, nickel, manganese and cobalt. The copper (Cu) doping in ZnS nanoparticles gives a photoluminescence peaks at 420nm and 510nm whose intensity almost got doubled upon the introduction of the impurity [6]. With the increase in the impurity percentage the luminescence shows quenching of the luminescence which has a maximum at 0.4% at wavelength of 445nm. Also the luminescence intensity and stability are improved by the addition of the surfactants [7]. With the doping of iron (Fe) ions photoluminescence intensity has a peak at 400nm with additional peak at 620nm. The photoluminescence peak has red shift if the doping percentage is increased. Also, the peak intensity increases with increase of Fe concentration which peaks at 3%. [8]. But some of the reports have mentioned only the quenching of the photoluminescence [9]. Doping with nickel (Ni) enhances the photoluminescence intensity having a peak at 445nm for undoped, 472nm and 498nm for doped. With the increase in doping concentration the intensity increases which peaks at 0.5% and quenching thereafter [10]. In some of the studies, multiple peaks in the range of 400nm to 550nm have been reported [11]. The introduction of manganese (Mn) ion in the ZnS nanoparticle lattice enhances the photoluminescence intensity with red shift with a peak at 600nm [12]. In Mn doping, also, quenching was reported [13]. Another promising dopant from the transition group is cobalt (Co) which enhances the photoluminescence intensity which has a peak at 450nm with quenching of intensity [14].

It is an overall observation that the photoluminescence increases with the doping of the transition metals to some percentage and then shows quenching. All the above papers have reported blue-green emission for pure and doped ZnS nanoparticles. In some cases doping of the transition metal changes the emission wavelength, like, blue-green for copper and orange for Mn doping but in most of the cases the dopants only serve as the sensitising agent and do not change the emission wavelength.

In the above studies the samples were prepared by different chemical methods with varying precursors and physical conditions. A variety of surfactants in different concentration were used to avoid the agglomeration, controlling the crystal growth and improving the photoluminescence. This resulted in variety of particle size with different surface conditions, thus the studies undertaken were on different platforms and was not giving any information of actual change in the luminescent intensity. Also, we got a sharp peak for red emission for all the impurities and their different percentages. The present study is under taken to select the best impurity and its concentration for the best photoluminescence.

2. Experimental

2.1. Synthesis

Pure and doped ZnS nanoparticles were synthesized by chemical co-precipitation method in ambient condition. All chemicals used were of analytical grade with no further purification and locally sourced. The pure ZnS nanoparticles were synthesized by the precipitation of 0.1M of zinc (0.8972gm of $ZnSO_4.H_2O$ in 50ml of deionized water) and 0.1M of Sulphur (0.6604 gm of Na_2S in 50ml of deionized water). The zinc sulfate solution was taken in a beaker stirred continuously at 300rpm while the sodium sulfide solution was added drop wise with the help of a burette. The solutions were prepared in deionized water and the reaction was carried out at room temperature in ambient environment. The capping agent 1gm PVP was introduced in the reaction along with zinc sulfate to control the crystal growth and avoid agglomeration. A white precipitate appeared at the end of the reaction. The solution was continuously stirred at constant rpm for 2 hours. The precipitate formed in the reaction was collected and centrifuged at 5000rpm for 20 minutes. The ZnS nanoparticles got settled at the bottom of the centrifugation tube which was redispersed in deionized water with the help of sonication. For removing the impurities and unreacted chemicals, the process is repeated several times. Finally, the precipitate is collected and kept for drying in hot air oven for overnight. The dried sample was collected and crushed to fine powder in mortar-pestle.

The chemical reaction is as given below,

$$ZnSO_4 + Na_2S \longrightarrow ZnS + Na_2SO_4$$

For doping, the salts of transition metals (1%, 2%, 5% and 10% of the molar value of $MnSO_4$ for Mn, $NiCl_2$ for Ni, $FeCl_3$ for Fe, $CuSO_4$ for Cu, $CoCl_2$ for Co) is added along with the zinc sulfate. Other steps of the synthesis remain unchanged.

2.2. Characterizations

Analysis of nanoparticles is done by using analytical techniques such as optical absorption, photoluminescence, XRD, EDAX and TEM.

The optical absorption spectra of nanoparticles were recorded by Labindia (UV-3000) UV–VIS spectrometer in the range of 200nm to 800nm. The Agilent (Cary Eclipse) was used to record the photoluminescence spectra. The XRD analysis was done by Rigaku (Ultima IV) operating at 40 KV-40 mA using CuKα radiation (λ = 1.5406 Å). The morphology of the synthesized nanoparticles was examined by using transmission electron microscope (TEM) Technai (G^2 20U Twin). The elemental analyses (EDAX) was carried out to confirm the formation and purity of the ZnS nanoparticles. For EDAX Bruker Quantax EDS attached with Sigma Zeiss HV field effect scanning electron microscope (FESEM) was used.

3. Results and discussion

3.1. Optical absorption studies

Figure 1 shows the absorption spectra for pure and transition metal doped (2%) ZnS nanoparticles. The spectra showed a sharp absorption edge with a shoulder at 320nm. The doping of the transition metals showed a minor blue shift in the absorption shoulder. But there is no noticeable shift due to the type of doping [15]. The band gap was calculated by using Tauc plot and is found to be 4.05eV for pure and 4.20 to 4.51eV for doped ZnS nanoparticles. The calculated band gap was higher than the bulk ZnS (3.68eV) which is by virtue of quantum confinement [16].

The radius was calculated by the equation.

$$E(r) = E_{bulk} + \frac{h^2}{8r^2}\left[\frac{1}{m_e^*} + \frac{1}{m_h^*}\right] - \frac{1.8e^2}{4\pi\varepsilon r} \quad (1)$$

Where, E(r) is the band gap nanoparticles, E_{bulk} is the band gap of the bulk, m_e^* and m_h^* are the effective mass of electron and hole respectively. The third term is the Coulomb's attractive force between electrons and holes which is negligible. The diameter of the nanoparticles calculated from above equation is approximately 3nm. A small decrease in the size of nanoparticles with the doping is observed which is due to quenching in the growth of nanoparticles [8].

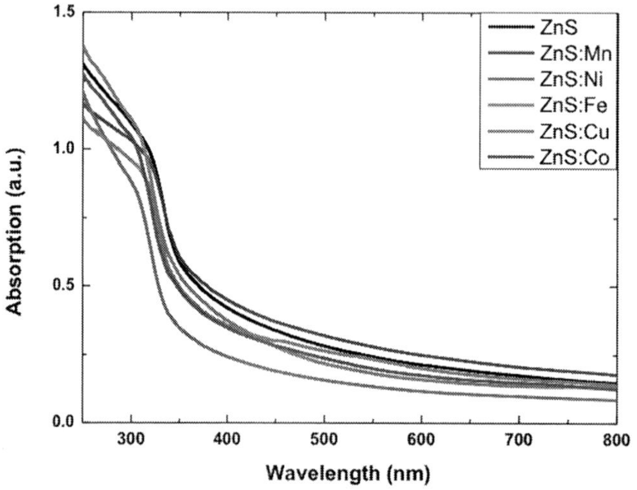

Figure 1. UV-VIS spectra of pure and doped (2%) ZnS nanoparticles

3.2. Photoluminescence

In pure ZnS nanoparticle the photoluminescence is affected by the quantum confinement effect, thus the emission spectra of nanoparticles are size dependent, shifting to the blue with decrease in particle size. However, transition metal doped ZnS nanoparticles exihibited red-shift in the photoluminescent emission spectrum [17]. The photoluminescence measurement was performed at room temperature on pure and doped ZnS nanoparticles at an exciting wavelength of 390 nm which yields two peaks, one at 440nm and another centered at 780nm. The observed wavelengths are much higher than the reported values in most of the papers [18, 19] which may be due to the result of oxygen atoms trapped in the nanostructures. The oxygen atoms may have been trapped either due to atmospheric oxygen as the reaction was carried out in ambient condition or the capping agent PVP. The presence of oxygen atoms was confirmed by the energy dispersive spectroscopy (EDS). These oxygen atoms form a trap level in the band gap of the ZnS nanoparticles, resulting new transition levels [20]. The figure 2 shows possible transition in ZnS nanoparticles. The emission at 440nm is due to the transition from the conduction band to the energy levels of interstitial sulphur (S_S). The sharp red emission is due the transition between the energy level of trap oxygen (O_T) to the energy levels of interstitial sulphur (S_S).

With the increase of doping percentage the luminescence increases up to 1% doping and then decreases showing the quenching of the luminescence. Upon doping with transition metals it is observed that the peak wavelength value remains the same only the magnitude of the intensity changes. The emission peaks at different doping percentage for different doping materials. The quenching of luminescence has shown different behaviour for different doping percentage, the fall in the luminescence is slow in case of Mn doping and steep in the Cu doping. The higher doping concentration the dopant transition metal acts as electron trapping centres. This increases the non-radiative recombination reducing the radiative transitions [21-24]. The figure 3 shows the photoluminescence and relative photoluminescence of doped and undoped ZnS nanoparticles.

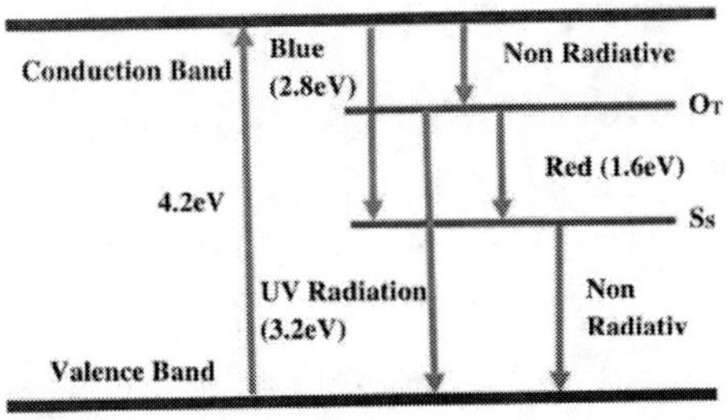

Figure 2. Possible transitions in ZnS nanoparticles

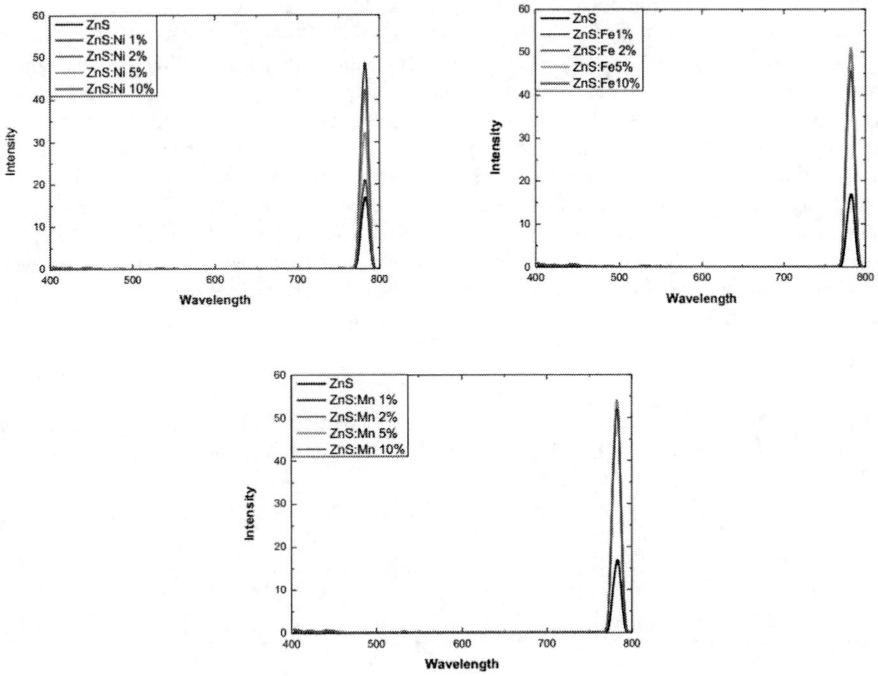

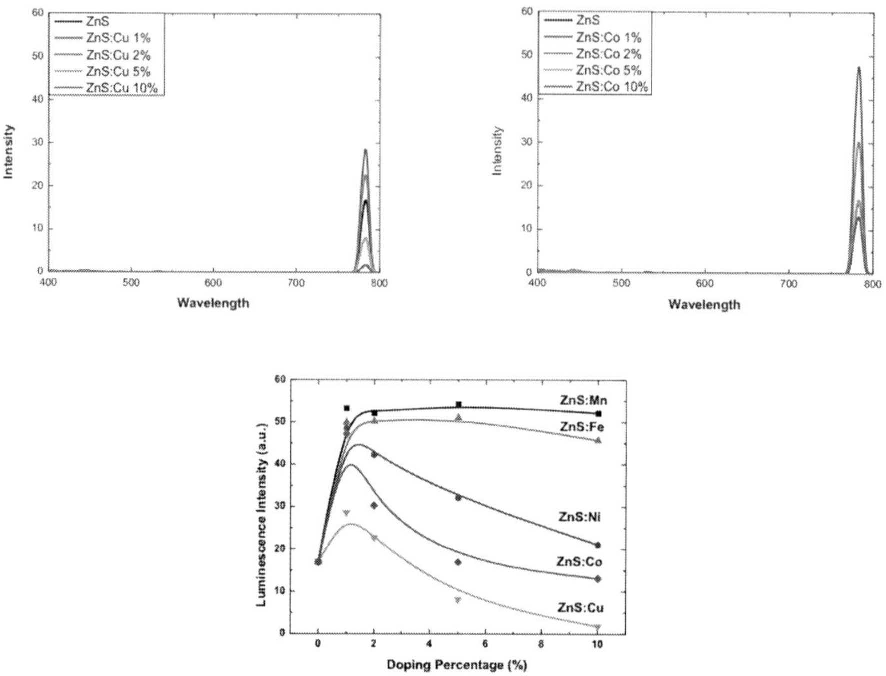

Figure 3. Photoluminescence of pure and doped ZnS nanoparticles

3.3. Structural and morphological studies

The XRD patterns of pure and doped ZnS nanoparticles have been shown in figure 4. The diffraction pattern has three well marked peaks for all samples at 2θ = 28.6°, 47.8°, and 56.6° for the miller planes (111), (220), and (311) respectively [25]. The XRD pattern matched with the cubic zinc blende structure, confirming the formation of the ZnS crystals. The addition of the impurity does not introduce any additional peaks confirming that the impurity atom is placed in the lattice structure of the ZnS crystal. The broadening of XRD peak is also observed which is due to nanosized particles [26]. Doping of the transition metal impurities resulted in a small decrease in the particles size of ZnS nanoparticles. The nanoparticle size quenching may be because of different ionic radii among ions of different elements. The ionic radii of are 0.74, 0.83, 0.69, 0.645, 0.73 and 0.65Å for Zn^{2+}, Mn^{2+}, Ni^{2+}, Fe^{2+}, Cu^{2+} and Co^{2+} ions, respectively. As the ionic radii of Zn^{2+} are different than the ionic radii of the dopants this may be causing the nanoparticles to shrink in their size. The crystallite size was calculated by Debye Scherrer's formula [27, 28],

$$d = \frac{k\lambda}{\beta cos\theta} \quad (2)$$

Where d is the diameter of nanocrystal, k is the constant (= 0.9), λ is wavelength of the X-ray (= 1.54056 Å), β is the full width half maxima (FWHM), and θ is the Bragg angle. The average crystallite size was calculated to be in the range of 3nm.

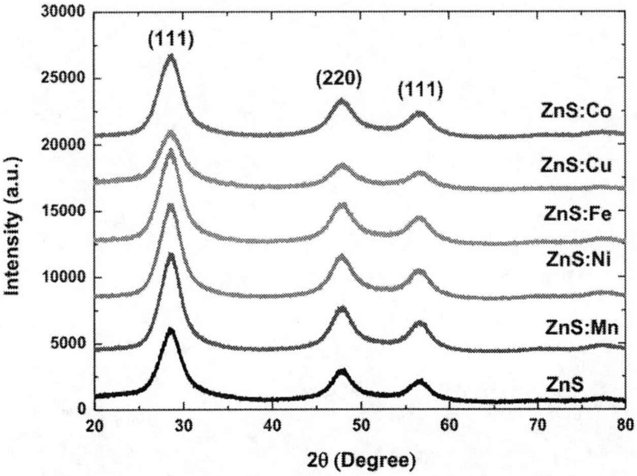

Figure 4. XRD of pure and transition metal doped (2%) ZnS nanoparticles

The purity and the doping of the synthesised nanoparticles were confirmed by energy dispersive spectroscopy (EDS) spectra and the corresponding micrographs are shown in figure 5. The EDS spectrum of pure ZnS has the peaks corresponding to Zn and S which confirms the purity of the sample prepared. Whereas the transition metal doped ZnS nanoparticles have peaks corresponding to Zn, S and the doped transition metal confirming the doping of the synthesised nanoparticles [29, 30].

As per the calculations, the size of the synthesized nanoparticles are in the range of 3nm so FESEM has resolution limit and cannot be used for the morphological study. So, for the morphological studies transmission electron microscope (TEM) was used. The TEM images of pure ZnS nanoparticles, as shown in figure 6. It show that the nanoparticles are spherical in shape and monodispersed. The average size of the nanoparticles is well in agreement with the calculated value [31].

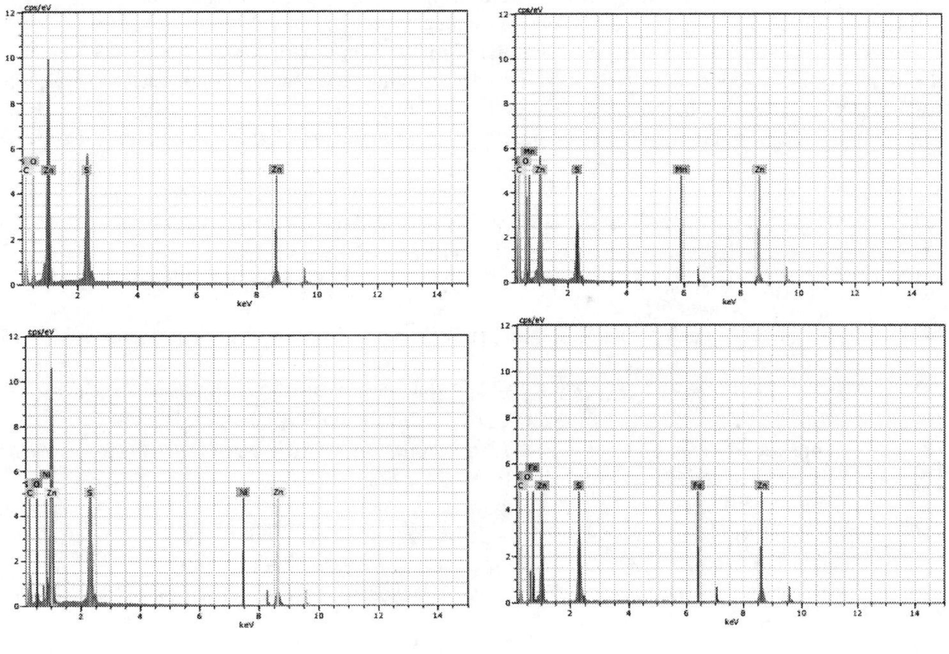

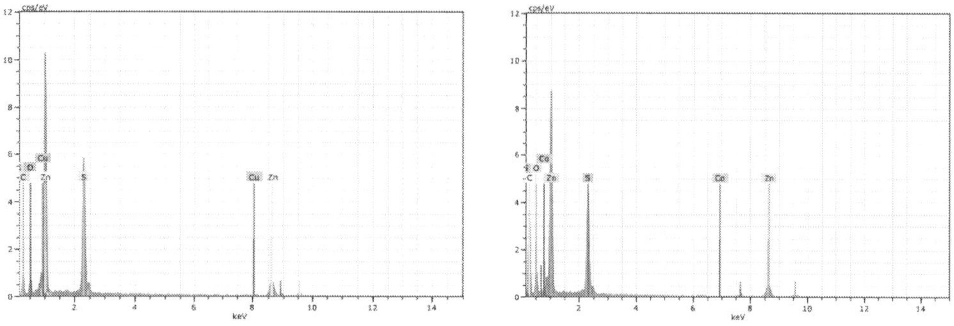

Figure 5. EDS of pure and doped ZnS nanoparticles

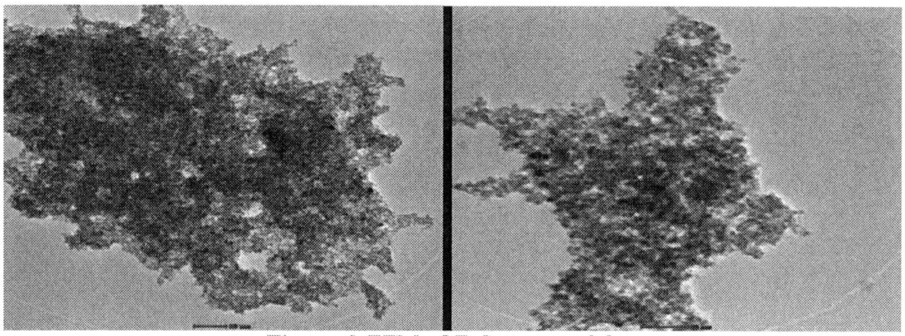

Figure 6. TEM of ZnS nanoparticles

4. Conclusion

The pure and doped ZnS nanoparticles with an average size of 3nm were synthesized by chemical co-precipitation method at room temperature by using deionised water as solvent. The UV–visible absorption spectra showed a strong quantum confinement effect confirming the formation of nanoparticles. XRD showed cubic phase of all the synthesized samples without any changes in the crystal structure. The room temperature photoluminescence spectra of all the samples had peak positions at 440 and 780nm. The emission peak intensity increased with doping concentration initially and decreased beyond a critical value. The maximum intensity and the critical doping percentage were different for different doping materials. The highest luminescence was obtained for Mn (5%), which is sufficiently high as compared to other doping impurities. The sharp and single peak of photoluminescence makes it an ideal candidate for the imaging and tagging in biomedical application.

References

[1] Al- Rasoul K, Abbas N and Shanan Z 2013 *Int. J. Electrochem. Sci.* **8** 5594
[2] Kole A and Kumbhakar P 2012 *Appl. Nanosci.* **2** 15
[3] Pons T, Pic E, Lequeux N, Cassette E, Bezdetnaya L, Guillemin F, Marchal F and Dubertret B 2010 *ACS nano* **4(5)** 2531
[4] Kirchner C, Liedl T, Kudera S, Pellegrino T, Javier A, Gaub H, Stolzle S, Fertig N and Parak W 2005 *Nano Lett.* **5(2)** 331
[5] Amirian F, Molaei M, Karimipour M and Bahador A 2018 *J. of Luminescence* **196** 174
[6] Wanjari L, Bisen D and Tamrakar R 2015 *Advance Physics Letter* **2(3)** 2349
[7] Kuppayee M, Vanathi Nachiyar G K and Ramasamy V 2011 *Appl. Surface Sci.* **257** 6779
[8] Saikia D, Raland R and Borah J 2016 *Physica E* **83** 56
[9] Kumar S and Verma N 2015 *J Mater. Sci: Mater Electron* **26** 2754
[10] Murugadoss G and Rajesh Kumar M 2014 *Appl. Nanosci.* **4** 67
[11] Saikia D and Borah J 2017 *J. Mater. Sci: Mater Electron* **28(11)** 8029
[12] Wei Lu S, Lee B, Wang Z, Tong W, Wagner B, Park W and Summers C 2001 *J. of Luminescence* **92** 73

[13] Jahanbin T, Gaceur M, Gros-Dagnac H, Benderbous S and Merah S 2015 *J. Nanopart. Res.* **17** 258
[14] Yang P, Lu M, Xu D, Yuan D, Song C and Zhou G 2001 *J. of Phys. and Chem. of Solids* **62** 1181
[15] Desai N, Shaikh I, Rawal K and Shah D **2018** *AIP Conf. Proc.* **1953** 030149-1
[16] Borse P, Deshmukh N, Shinde R, Date S and Kulkarni S 1999 *J. Mater. Sci.* **34** 6087
[17] Hu H and Zhang W 2006 *Optical Materials* **28** 536
[18] Murugadoss G 2010 *J. of Luminescence* **130** 2207
[19] Kong H, Song B, Byun J and Hwang C 2013 *Bull. Korean Chem.* **34(4)** 1181
[20] Radhu S and Vijayan C 2011 Mater. Chem. and Phys. **129** 1132
[21] Jothibasa M, Manoharan C, Johnson Jeyakumar S, Praveen P, Punithavathy I and Richard J 2018 *Solar Energy* **159** 434
[22] Eryong N, Donglai L, Yunsen Z, Xue B, Liang Y, Yong J, Zhifeng J and Xiaosong S 2011 Appl. Surface Sci. **257** 8762
[23] Peng W, Cong G, Qu S and Wang Z 2006 *Optical Materials* **29** 313
[24] Poornaprakash B, Reddy D, Murali G, Rao N, Vijayalakshmi R and Reddy B 2013 *J. of Alloys and Compounds* **577** 79
[25] Priyadharsini N, Elango M, Vairam S and Thamilselvan M 2016 *Mater. Sci. in Semiconductor Processing* **49** 68
[26] Ibrahim R, Markom M and Abdullah H 2015 *ECS J. of Solid State Sci. and Tech.* **4(2)** R31
[27] Rajabi H and Farsi M 2015 *J. of Molecular Catalysis A: Chemical* **399** 53
[28] Raval A, Shaikh I, Jain V, Shastri N, Patel P, Saini L and Shah D 2020 *J. Nano- Electron. Phys.* **12** 02010
[29] Sreenivasulu B, Reddy S and Reddy V 2018 *J. of Mater. Sci.: Mater. in Electron.* **29** 251
[30] Kumar S and Verma N **26(5)** 2754
[31] Soltani N, Saion E, Yunus W, Erfani M, Navasery M, Bahmanrokh G and Rezaee K 2014 *Appl. Surface Sci.* **290** 440

Evaluation of Mechanical Properties of High Strength (M40) Fibre Reinforced Concrete using Admixtures

[1]Gopu Ganesh Naidu, [2]Maila Nagaraju, [3]K.Ashok Kumar

[1] Associate Professor and Head of the department, Civil Engineering, Pace institute of technology & sciences(A), Valluru, Prakasam, Andhra Pradesh, India.
[2] PG Student ,Structural Engineering, Pace institute of technology & sciences(A), Valluru, Prakasam, Andhra Pradesh, India. Email: mailanagaraju@gmail.com
[3] Assistant professor, Department of Civil Engineering, Pace institute of technology & sciences(A), Valluru, Prakasam, Andhra Pradesh, India.

Abstract. This research concentrates on the evaluation of mechanical properties of high strength concrete using fibers and admixtures. Now-a-days High strength concrete is used in the construction to get the more strength and durability. High strength concrete mix of M40 grade is designed by adding fibers and admixtures. Admixtures like Super plasticizer & master glenim are used to replace the cement content in the mix. Chemical admixtures are used to decrease the water content in the concrete such that its workability and compressive strength increased. Along with that as the concrete is weak in tension, to overcome that, fibers of polypropylene were used. Fibers can increase the tensile behaviour of the concrete beam. The specimens were prepared by adding the fibers and mineral admixtures to the high strength (M40) concrete and conventional concrete and tested for compressive strength, Slump cone test, Modulus of elasticity and flexural strength at 7,14, 28 & 56 days. Based on the experimental values, it is clearly identified that all the results were obtained target mean strength. Experimental data shows that using fiber content from 0.5% to 1% can increase all mechanical properties of the concrete but using more than above mentioned content will leads to decrement of strength parameters of the concrete.
Keywords: Fibers, High strength concrete, super plasticizers, Target mean strength, Green house gasses, poly propylene, Modulus of elasticity.

1. Introduction

Now-a-days concrete plays a vital role in construction & widely used in various constructions. Cement produces high amount of Green house gasses like CO_2 while its production, which causes harmful effects to the environment [1]. The materials used in high strength concrete are as same as normal concrete but additional to it there are some admixtures were added. To reduce the usage of concrete with same compressive and flexural strengths we use different kinds of admixtures in the mixing. By adding the mineral admixtures like, GGBS, Fly ash, Silica fume, Master Glenim etc, we can enhance the stability, workability and compressive strength of the concrete [2,3]. Some of these admixtures have very fine particles than cement. So we can easily replace cement with these admixtures and can lead to the increase of durability and reduce the porosity [5]. Fibers are used in this concrete to get the concrete strong in tension which results well in fire resistance and high compressibility [6]. Fiber reinforced concrete is a mixture of tiny discrete particles which are distributed uniformly in concrete mix and oriented transversely. The ingredients of fiber reinforced concrete are concrete, steel, cellulose, polypropylene. The amount of fibers added to the total volume may be of 0.1% to 0.3% [7].

2. Materials and Properties

2.1. Cement

Cement is one of the major binding materials which is used in the construction. In this research we used the Ordinary Portland Cement (OPC) of 53-grade confirming to IS: 12269-2013 [11]. Several experimental tests were conducted on the taken cement to know the quality of cement as per the codal provisions of IS: 8112-1989 and get the following results.

Table: 1 Properties of Cement

Property	Test Result
Compressive Strength(M pa)	47.35
Fineness (%)	6.1
Specific Gravity	3.02

2.2. Ground Granulated Blast furnace Slag (GGBS)

GGBS is used in the preparation of concrete as an admixture. Now-a-days it is popularly used as admixture rather than using as slag cement. Major constituents of blast furnace slag are Silica oxide, magnesium oxide, calcium oxide [4]. By using the GGBS in concrete it can increase the slump value, reduce the heat hydration etc.

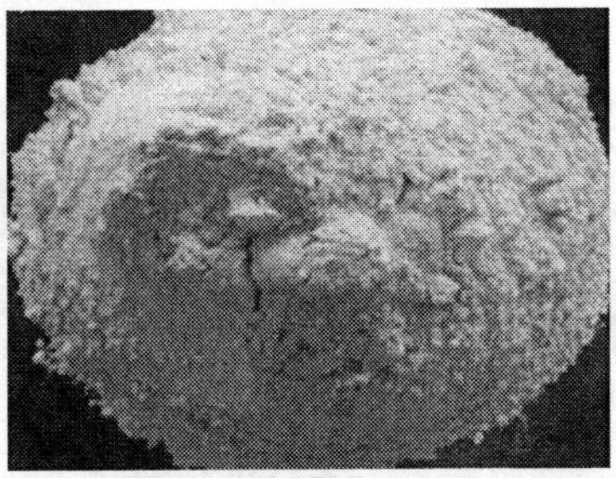

Fig: 1 GGBS

Table: 2 Chemical compositions of GGBS and cement

Property	GGBS	Cement
Cao	46	60.1
MgO	1	2.1
Al2O3	10	15
N2O	0.25	0.45
Specific Gravity	2.77	3.02

2.3. Poly propylene

The poly propylene fibers are used in the concrete to get increases the tensile strength of concrete [8]. Surface bleeding and aggregate settlement in case of fresh concrete are the basic reasons for the formation of cracks. To reduce these cracks we use poly propylene fiber in concrete. The poly propylene concrete is cheap in cost [10]. The fibers are available in three different sizes i.e., 6mm, 12mm and 24mm.

Fig: 2 Poly propylene

Table: 3 Properties of Poly propylene

Property	Poly propylene
Unit weight(gm /cm3)	0.9-0.91
Tensile strength (MPa)	300-400
Thermal conductivity(W/m/K)	0.12
Reaction with water	Hydrophobic
Elongation at break (%)	100-600

3. Concrete Mix Proportion

Batching of concrete is done at room temperature [14]. Different percentages of fibers were mixed in the concrete like 0.5%, 1% and 1.5% and a reference mix conventional concrete is designed as per the codal provision of IS: 10262-2009 and obtained the following mix proportions.

Mix Ratio (C: FA: CA) : 1:1.2:2.3
Water Cement Ratio : 0.32

Fig: 3 Placing and Compacting

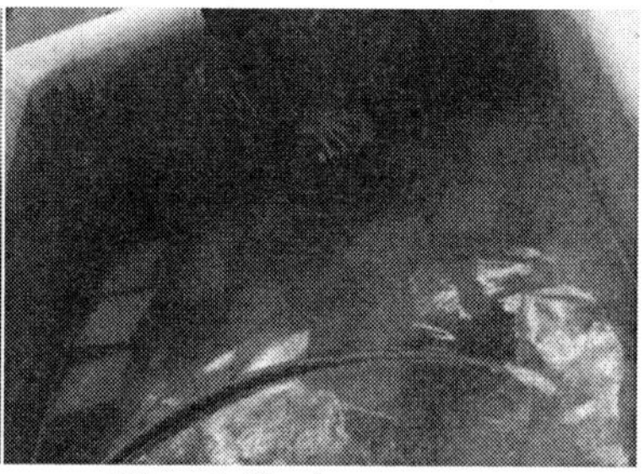

Fig: 4 Curing

4. Methodology

4.1. Workability Test

4.1.1. Slump cone Test

The slump cone test was conducted for fresh fiber reinforced concrete with mineral admixtures and for conventional concrete. Obtained results were compared and analyzed .By adding the mineral admixtures the slump values are giving good [11]. The workability of the fiber reinforced concrete with mineral admixtures increased by adding the admixtures.

Fig: 5 Slump cone test

4.2. *Compressive Strength*

The compressive strength of fiber reinforced concrete with mineral admixtures was finding out by conducting the compressive strength test using Compressive Testing Machine (CTM) for the different compositions of fibers and admixtures. For conducting the tests the specimens (cubes) were prepared of standard dimension 150*150*150mm [13]. And the test results were compared.

Fig: 6 Compressive strength test on specimen

4.3. *Modulus of Elasticity Test*

Modulus of Elasticity of concrete is defined as the ratio of stress applied on the concrete to the respective strain caused. To get the modulus of elasticity of concrete the compression test should be done on a cylindrical concrete specimen by using a compressometer [15]. From the stress-strain (load-deflection) curve slope and 45% compressive strength value the modulus of elasticity value is obtained [9]. The cylindrical moulds were prepared and tested for 28 days as per IS: 516-1959 codal provisions.

4.4. *Flexural strength test*

For the prepared specimens the flexural strength test was conducted for both the conventional concrete and fibre reinforced concrete with mineral admixtures of different proportions. And the test results were compared [12]. The flexural behaviour of the concrete can be determined by using this test.

5. Results and Discussion

5.1. *Workability Test*

5.1.1. *Slump cone Test*

By adding mineral admixtures to the concrete we can increase the durability property of the concrete and fibers can increase the strength of the concrete. The water content of the concrete can be decreased by adding the chemical admixtures to the concrete. The heat of hydration also decreases by adding the admixtures and fibers to the concrete.

Table: 4 Slump values for different % of fiber

% Fiber	**Slump values in mm**
0	60
0.5	64
1	71
1.5	78

5.2. Compressive Strength

Based on the compressive strength test results, it is clearly observed that the Fiber reinforced mineral admixture concrete gives the better results compared to the nominal concrete. Especially at 1% of fiber and admixtures replaced by the cement in the concrete gives more strength values compared to the remaining proportions. Experimentally proved that by adding fibers and admixtures we can improve the strength and durability of the concrete [16]. Based on the test results it is not advised to add more than 1% of admixtures and fibers to the concrete. If one add more amount of admixtures to the concrete which leads to the more amount of swelling and shrinkage.

Table 5: Variation of Compressive strength values with various percentages of fibers

% Fiber	**7days**	**14days**	**28days**	**56days**
0	30	37	44	49
0.5	28	34	40	43
1	31	38.5	45	49.6
1.5	26	33	41	42

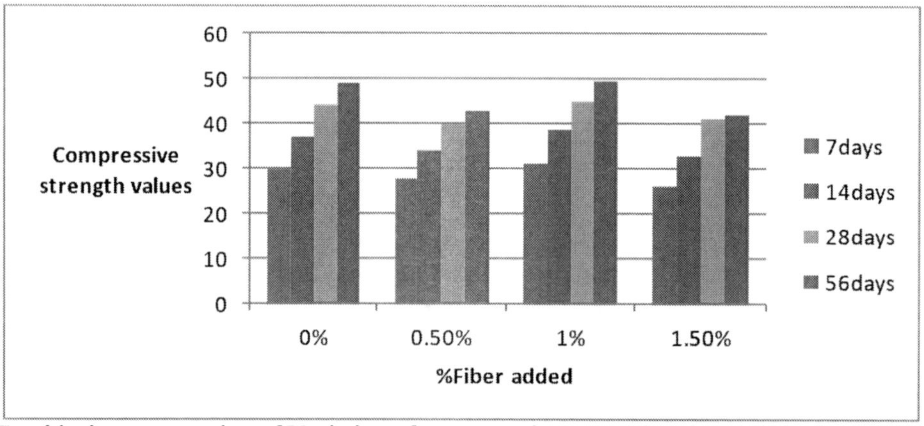

Fig 8: Graphical representation of Variation of Compressive strength values with various percentages of fibers

5.3. *Flexural strength test*

After conducting the flexural strength test is it observed that the conventional concrete gives less flexural strength values compared to the fiber reinforced mineral admixtured concrete. Flexural strength test also gives good results at 1% adding of the fiber and admixtures to the concrete. The concrete get more strength when fibers are added to it compared to the normal concrete. If fibers and admixtures content increases more than 1% in the concrete which gives less strength values, because the chemical admixtures content increases in the concrete which leads to the reactions in the concrete mix which may cause decrease the strength of the concrete.

Table 6: Variation of Flexural strength values with various percentages of fibers

% Fiber	7days	14days	28days	56days
0	3.4	4.1	4.4	4.9
0.5	3.1	3.7	4.1	4.3
1	3.5	4.15	4.5	5.1
1.5	3.1	3.8	4.15	4.26

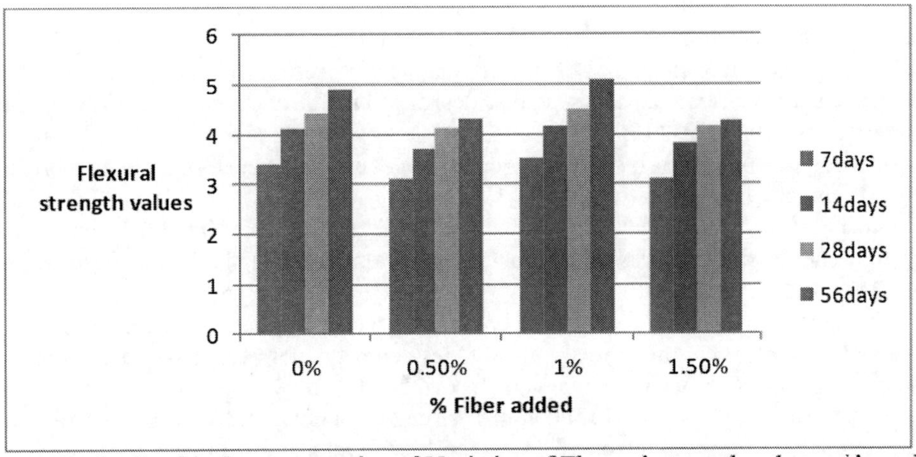

Fig 9: Graphical representation of Variation of Flexural strength values with various percentages of fibers

6. Conclusions

From the experimental results the following conclusions were drawn.

- Plasticizers have the capacity to increase the workability by reducing the w/c ratio and which helps in gaining the high strength concrete.
- It is observed that the target mean strengths are achieved when fibers and admixtures were added to the concrete.
- High target mean strengths are achieved when 1% fiber content is added to the concrete mix.
- If further increase in fiber content which leads to the decrease in strength parameters.
- By using the mineral admixtures in the concrete the slump values are more than compared to the conventional concrete. And which can decrease the heat of hydration.
- The compressive strength, flexural strength and slump values of the fiber reinforced concrete with mineral admixtures gives best results up to 1% of fiber content beyond that gives less results.
- Based on the above discussion it is advised that up to 1% adding fibers to the mineral admixture concrete is preferable, more than that is not preferable.

IOP Conf. Series: Materials Science and Engineering 1126 (2021) 012059 doi:10.1088/1757-899X/1126/1/012059

References

[1] Tara Rahmani, Bahnam Kiani, Farzanesh sami, "Durability of Glass, Polypropylene and Steel Fiber Reinforced" Concrete, ACI Journal, Vol.83,374-382

[2] A.A.Ramezanianpourab, M.Esmaeilic, S.A.GhahariahM.H. Najafic laboratory Study on the effect of polypropylene fibre on durability, physical and mechanical characteristics of concrete for application in sleepers, Volume 44,July 2013

[3] O.KayalIM.NHaqueBZhu, Study on cement and concrete composites some characteristics of high strength fiber reinforced lightweight aggregate,, concrete mati .Volume 25,issue2, February,P.

[4] XinchengPu and Chong Wang ,Contribution of silica fume to the strength and flow ability of 150MPa superhigh strength flowable concrete,, 2000, journal no.2:845-52.

[5] SaeidKakooei The corrosion investigation of rebar embedded in the fibers reinforced concrete"., Construction and Building materials 35 (2012) ,pp:564-570.

[6] Gopu Ganesh Naidu, Kannella Edukondalu, "Mechanical Properties of Concrete by Replacing Cement with Eggshell powder and Fly ash". *International Journal of Innovative Technology and Exploring Engineering (IJITEE)* ISSN: 2278-3075, Volume-9 Issue-4, February 2020, DOI: 10.35940/ijitee.D1715.029420

[7] Dinakar " Effects of metakaolin content on the properties of high strength concrete",, international journal of concrete structures and materials,207(3):215-23.

[8] patil "Strength and durability property of high performance concrete incorporating high resistivity metakaolin", ,international journal of modern engineering research 2012,2(3):1099-104..

[9] Permul and soundararajan Analysis on mix design of high strength concrete (M100), IJET, Volume:05 Issue:06 June-2018.

[10] Dr.Aignesberger "Micro silica by-product to get the better performance than ordinary Portland cement when compared",.

[11] KolliRamujee A study on strength properties of polypropylene fiber reinforced concrete Polypropylene fibers, compressive strength, , split tensile strength. Vol.2, Issue 8, Appears on pages:3409-3412, August 2013.S

[12] ZiadBayasi and jack Zeng workability A study on air entrainment; compression; fiber reinforced concrete; flexural strength; impact; permeability; polypropylene fiber;; Materials Research.Volume:90, issue:6, Appears on pages:605-610, 1993.

[13] UmeshSharma "Use of M sand in high strength concrete and high performance concrete"., Indian Journal of science and technology, 28, ipl1063, October 2013.

[14] IS:456-2000.Code of practice for plain and reinforced concrete , Bureau of Indian Standards, New Delhi.

[15] IS:10262-2009.Code of practice for concrete mix design proportioning, Bureau of Indian Standard.

[16] M.S.Shetty, Concrete Technology, 3rd Edition, S.Chand & Company Limited, Delhi, 1992

Strength Characteristics of Concrete by Partial Replacement of Coarse Aggregate with Coconut Shells & Cement with Glass Powder

[1]Gopu Ganesh Naidu, [2] Nettikoppula Ramesh Babu, [3] Pappula Ravikumar

[1] Associate Professor and Head of the department, Civil Engineering, Pace institute of technology & sciences (A), Valluru, Prakasam, Andhra Pradesh, India.
[2] PG Student, Structural Engineering, Pace institute of technology & sciences(A), Valluru, Prakasam, Andhra Pradesh, India. Email: rummy3399@gmail.com
[3] Assistant professor, Department of Civil Engineering, Pace institute of technology & sciences(A), Valluru, Prakasam, Andhra Pradesh, India.

Abstract. In this study coarse aggregates and cement are partially replaced in the concrete with similar materials like coconut shells and glass powder. While production of the cement there is a lot of pollution to the environment and which causes health issues to the society. In the same way while production of the course aggregates which releases lot of dust into the atmosphere which causes damage to human lives. We can obtain some advantages by partially replacing the coarse aggregates and cement in the concrete. As we all know that the Cement and the coarse aggregates are the main ingredients in the concrete. The cost of these materials is getting increased day by day. There are some of the materials proposed by many researchers to lesser the cost of the concrete and to improve the strength characteristics of the concrete. For coarse aggregates, the coconut shell is the most preferable and it has more potential to the partial replacement of the coarse aggregates in concrete. As well as there is a similar material which used to partially replacement of the cement is waste glass powder. As we observe at many places there is a lot of wastage of glass will produce throughout the year. Mainly glass is the composition of silica. When the glass is crushed into very small size as powder it experienced to the pozzolanic reactions with hydrates in the cement, and calcium silicate hydrate will be formed. The main aim of this research is utilization of the waste material in the concrete as well as decreasing the cost of the construction by using these waste materials in the concrete, by performing all these things we can easily reduce the pollution of the environment and clean the environment by using these waste materials in the concrete..
Keywords: Coconut shells, Crushed Glass powder, Calcium Silicate Hydrate, Replacement, Splitting tensile strength, Pollution, Green house gasses.

1. Introduction

In the construction work concrete is the main element, in the concrete the coarse aggregates and cement are the major ingredients [1]. By using the high amount of cement in the concrete it will produce CO_2 and some other green house gasses into the atmosphere which leads to the damage of the human lives and in the same way the cost of the cement production is also getting expensive now-a-days [2]. Coconut shells are the waste products which are going very high growth on day by day in the daily use. The wastage of these coconut shells are occupying lot of space on the earth and which produces toxic gases into the atmosphere which is very dangerous to the all biological lives [3,4]. To mitigate those problems, to reduce these waste on the earth, in this research we suggest to use a substitute material in the concrete to partially replacing the cement with glass powder and Course aggregates with coconut shells [5]. By crushing the waste glass into very fine powder it gains pozzolonic actions, which gives strength to the concrete. And it can replace the cement and contributes development in the strength characteristics. Both cement and glass powder has nearly equal chemical composition and the fineness of the glass powder is nearly equal to the cement that is the major reason to use glass powder in the concrete as a partial replacement of cement [9]. The partial replacement of these two materials at a time in the concrete gives good results. The crushed coconut shell gives nearly same properties like coarse aggregates. Both conventional concrete and the concrete with the partial replacement of the material were prepared and tested at room temperature which gives better results than conventional concrete at some % replacement of these materials [12].

2. Materials and Properties

2.1. Coconut shells

Coconut shells are the waste material which is available very easy in the atmosphere; in this research we used coconut shells as a partial replacement of the coarse aggregate. The cost of one unit of the coarse aggregates is more than compared to the preparation of the same amount of coconut shells [6]. By adding the coconut shells to the concrete it will gain more workability due to the smoothen surface of the one side of the shell. Comparing with the conventional concrete the coconut shell concrete has more impact resistance. Water absorption capacity of the coconut shell is higher than the nominal aggregates [16].

Figure: 1 Coconut shells Grinding

Table: 1 Properties of coconut shell

Property	Value
Lignin	28.6
Ash	0.56
Cellulose	26.9
Moisture	8.2
Solvents & Anhydrides	7.5

2.2. Course Aggregates

The shape and texture of coarse aggregate affects the properties of fresh concrete more than hardened concrete. A smooth surface can improve workability, yet a rougher surface generates a stronger bond between the paste and the aggregate creating a higher strength. In coarse aggregates there is a possibility to have both smooth and rough surfaces by that it can increase both workability and bonding of the concrete [8]. Coarse aggregates can increases the dimensional stability and strength of the concrete. The volume and weight of the concrete can be increased by the coarse aggregates.

Figure: 2 Coarse Aggregates

2.3. Cement

The 53-grade (OPC) ordinary Portland cement was used in the designed concrete mix. As per **IS: 12269-2013** the quality of the cement is determined by the test which are shown below table. If the calcium content is present more in the cement it will give high strength properties to the concrete [10]. As we all know that the cement has a very fine particle in size which can increase the binding property in the concrete. There are different grades of the cement available among them in this study we used 53-grade cement.

Table: 2 Cement properties

Property	Test Result
Fineness (%)	6.78
Specific Gravity	2.56
Compressive Strength(M pa)	46.36

2.4. Glass powder

Glass powder has more pozzolonic properties when it has milled into very fine powder. By using the waste glass powder in the concrete mix the strength characteristics, binding properties will be increased and heat of hydration will reduces. Glass powder has nearly equal composition like cement that is the major reason that we can comfortably use this glass powder as a partial replacement of cement in the concrete. This glass powder is the waste material which is freely available at the glass design shops [11]. By using this crushed glass powder as a partial replacement of cement in the concrete we can reduce the cost of the preparation of concrete and can minimize the amount of pollution while production of the cement.

Figure: 3. Glass Powder

Table: 3 Properties of glass powder

Compound	**Glass powder**
silicon dioxide	68.1
Aluminum oxide	0.9
Calcium oxide	14.5
Magnesium oxide	1.8
Potassium oxide	0.6
Sodium oxide	0.8
Sulfur trioxide	12.2

3. Concrete Mix Proportion

At the room temperature the concrete batching will be done. Different percentages of coconut shells and glass powder were mixed in the concrete like 10%, 20%, 30% and 40% and a standard mix of nominal concrete is mixed and the following mix proportions are designed as per the codal provisions.

Mix Ratio: 1:1.72:2.69
W.C Ratio : 0.42

Figure: 4 Placing, Compacting and Curing

4. Tests and Results - Comparison

4.1. Workability Test

4.1.1. Slump cone Test

By the test results obtained from the workability test, the workability is increasing with respect to the increasing percentage amount of the partial replacement of the coarse aggregates with coconut shells and cement with crushed glass powder [7]. The coconut shell has smooth surface and smaller in size compared to the nominal coarse aggregates, that is the major reason that replaced material concrete gives more workability than standard nominal mix [13].

Figure: 5 Workability Test

Table: 4 Test results

%Replacement of CA &cement	**Slump cone values in mm**
0	3.6
10	4.3
20	6.2
30	9.2
40	10

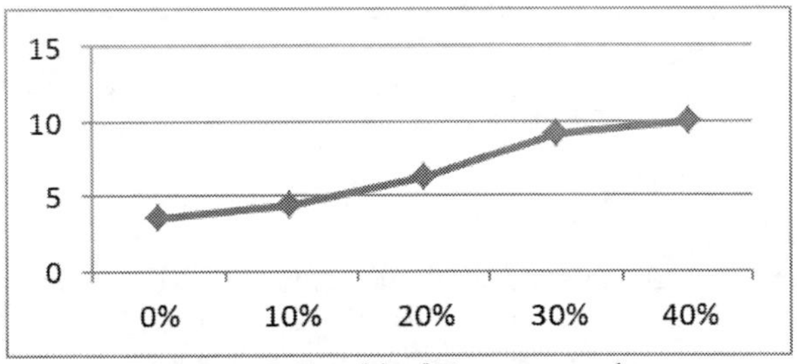

Figure: 6 Graphical Representations

4.2. Compressive Strength

The resistance to the external loading of the material can be determined by using the compressive strength test [14]. In this study, the 20% replacement of the material mix got a better compressive strength than normal concrete. The test values allow that the coconut shell and glass powder concrete may used at high strength applications by replacing around 20%, if we replace more than 20% which leads to the decrease of strength.

Figure: 7 Samples

Table: 5 Test results

%Replacement of CA &cement	**1week**	**2 weeks**	**4weeks**	**8weeks**
0	29	38	43	48.6
10	28	33.1	39.8	42.4
20	30	37.3	44	49.8
30	27	34	40.6	41.4
40	25	33	39	40

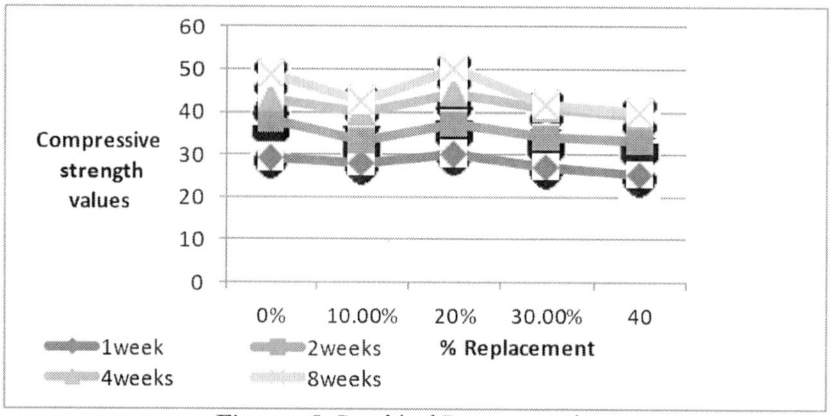

Figure: 8 Graphical Representations

4.3. Splitting Tensile strength test

The tensile behaviour of the concrete under external loading can be determined by the Splitting tensile strength [15]. The result may allow that the coconut shell and glass powder concrete mostly used in high strength applications by replacement around 20%. The 20% replacement of material mix got a better splitting tensile strength than normal concrete, if we replace more than 20% which leads to the decrease of strength.

Figure: 9 Samples

Table: 6 Test results

%Replacement of CA &cement	1week	2weeks	4weeks	8weeks
0	2.67	3.7	4.12	4.76
10	2.47	3.42	3.8	4.23
20	2.73	3.78	4.2	5.1
30	2.48	3.44	3.85	4.26
40	2.36	3.25	3.62	3.92

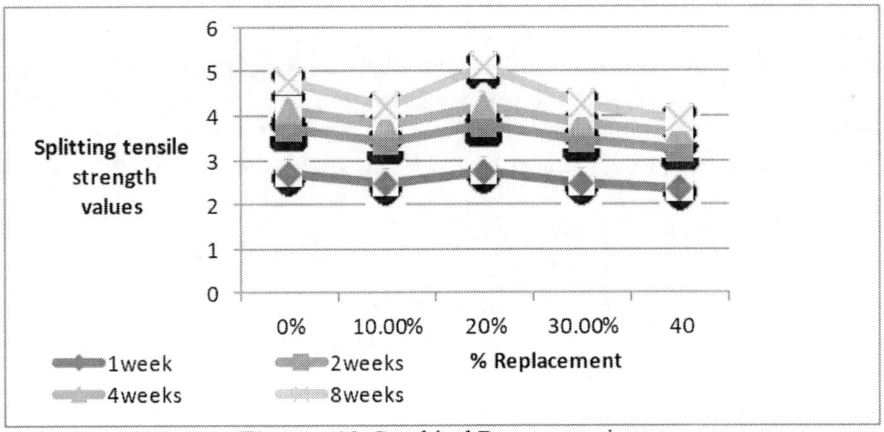

Figure: 10 Graphical Representations

4.4. Flexural strength test

The bending nature of the concrete can be determined by performing this Flexural strength test. The concrete is mixed as per the recommended proportions and tested, Comparison of the results obtained by the flexural test on the prepared specimen with different mixes of concrete by partially replacing the coconut shells and glass powder in the place of coarse aggregates and cement i.e., 0%, 10%, 20%, 30%, 40%.. The test results give good results at 20% replacement.

Table: 7 Test results

%Replacement of CA &cement	1week	2weeks	4weeks	8weeks
0	2.77	3.8	4.2	4.7
10	2.53	3.52	3.8	4.2
20	2.63	3.68	4.3	4.9
30	2.46	3.34	3.9	4.18
40	2.23	3.35	3.7	3.9

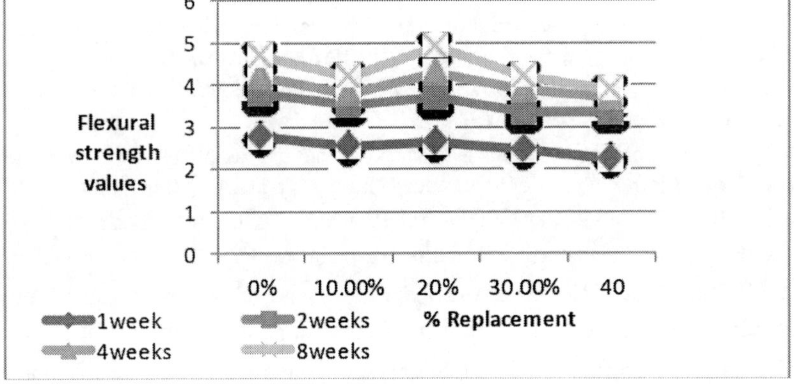

Figure: 11. Graphical Representation

5. Conclusion

From the experimental results the following conclusions were drawn.

- In the concrete mix by adding 20% of coconut shells and glass powder as the partial replacement of coarse aggregates and cement, we can get good strength characteristics to the concrete than the nominal mix.
- The results concluded that the usage of the materials like coconut shells and glass powder is acceptable in the concrete.
- The recycling of the material called coconut shells and waste glass are done.
- If more than 25% of coconut shells and glass powder are partially replaced in the concrete, then the strength characteristics are decreasing gradually.
- To reduce the cost of the materials used in the concrete like cement and coarse aggregates, the coconut shells and glass powder are to be used.
- The workability and the strength characteristics of the concrete is getting increased up to 20-30% replacement of these materials, if it is greater than 25% the performance of the concrete is getting decreased.
- To get good strength characteristics to the concrete, 20% to 30% of coconut shells and glass powder are proposed to replace by this study.
- By using the glass powder and coconut shells in construction materials, it leads to reduce the environmental pollution issues.

References

[1] ACI 311-05. "Specifications for Structural Concrete". American Concrete Institute, Farmington Hills.

[2] ACI 211.2-98. "Standard Practice for Selecting Proportions for Structural Lightweight Concrete". American Concrete Institute, Detroit, Michigan.

[3] ACI 213R-87. "Guide for Structural Lightweight Aggregate Concrete". American Concrete Institute, Detroit, Michigan.

[4] IS: 383 (1970). "Indian Standard Specification for Coarse and Fine Aggregates from Natural Sources for Concrete (Second Revision)." Bureau of Indian standards, ManakBhavan, 9 Bahadur Shah Zafar Marg, New Delhi 110002.

[5] Ghambhir, M. L. (2003). Concrete Technology (2nd edition), Huang, B., Li, G., Pang, S., and Eggers, J., (2004). "Investigation into Waste Tire Rubber- Filled Concrete. Journal of Materials in Civil Engineering, 16(3), 187-194.

[6] Gopu Ganesh Naidu, Kannella Edukondalu,. Mechanical Properties of Concrete by Replacing Cement with Eggshell powder and Fly ash. *International Journal of Innovative Technology and Exploring Engineering (IJITEE)* ISSN: 2278-3075, Volume-9 Issue-4, February 2020, DOI: 10.35940/ijitee.D1715.029420

[7] IS: 10262 (1982) (Reaffirmed 2004). "Recommended Guidelines for Concrete Mix Bureau of Indian standards, ManakBhavan, 9 Bahadur Shah Zafar Marg, New Delhi 110002.

[8] IS: 456 (2000). "Indian Standard Plain and Reinforced Concrete Code of Practice." Bureau of Indian standards, ManakBhavan, 9 Bahadur Shah Zafar Marg, New Delhi 110002.

[9] IS: 4031 (1996). "Indian Standard method of physical tests for hydraulic cement." Bureau of Indian standards, ManakBhavan, 9 Bahadur Shah Zafar Marg, New Delhi 110002.

[10] "Advantages of Structural Lightweight Aggregate Concrete". Expanded Clay, Shale and Slate Institute,

[11] ASTM C496-11. "Standard Test Method for Splitting Tensile Strength of Cylindrical Concrete Specimens".

[12] ASTM C39-16b. "Standard Test Method for Compressive Strength of Cylindrical Concrete Specimens".
[13] Ganesh Naidu, G., Sri Durga Vara Prasad, M. and Venkata Sai Pavani, A. (2019) 'Impact of chloride attack on basalt fibre reinforced concrete', International Journal of Innovative Technology and Exploring Engineering, 8(12). doi: 10.35940/ijitee.L3502.1081219.
[14] Sofi, A. and Naidu Gopu, G. (2019) 'Influence of steel fibre, electrical waste copper wire fibre and electrical waste glass fibre on mechanical properties of concrete', in IOP Conference Series: Materials Science and Engineering. doi: 10.1088/1757-899X/513/1/012023.
[15] ASTM C143-15a. "Standard Test Method for Slump of Hydraulic-Cement Concrete".
[16] IS: 516 (1959). "Indian Standard methods of tests for strength of concrete." Bureau of Indian standards, ManakBhavan, 9 Bahadur Shah Zafar Marg, New Delhi 110002.

Applications of EPS Geo-foam and Geo-membranes in Construction Industry – Bridges, Embankments

[1] Gopu Ganesh Naidu, [2]Kota Sai Manohar

[1] Associate Professor and Head of the department, Civil Engineering, Pace institute of technology & sciences(A), Valluru, Prakasam, Andhra Pradesh, India.
[2] PG Student, Structural Engineering, Pace institute of technology & sciences(A), Valluru, Prakasam, Andhra Pradesh, India. Email: kotasaimanohar@gmail.com

Abstract. This paper mainly concentrates on the uses and applications of EPS geo-foams and geo-membranes in construction industry like Bridges and Embankments over weak soil (compressible soils). If any external load is acting on weak soil, it will undergo settlement and the soil will be compressed. To overcome these settlements and compression of the weak soils, in this research we suggest EPS geo-foams. The EPS means "Expanded Poly Styrene", it is a light weight material which is 1% weight of soil and it is less than 10% weight of other light weight materials and it is available in different densities. It has high compressive resistance and it reduces settlement of weak soil. It is generally used as a soil fill replacement material. If the soil contains any moisture content, the geo membrane acts like a water barrier for the geo-foam material. Geo-membrane is a thin continues polymeric sheets and it is a geo synthetic material, it is generally used in construction industry and geotechnical applications. If we place different types of densities of EPS geo-foam blocks and geo-membranes over weak soil in bridges and embankments, the soil behaviour is changing and attaining stability. Finally based upon the behaviour of weak soil we conclude that, at which density of the EPS blocks the soil attain more strength and how other strength parameters has been related to density by conducting the different tests.
Keywords: EPS Geo-foam, Geo-membrane, Weak soil, compressive resistance, settlement, soil fill replacement.

1. Introduction

Now a days all countries are developing more rapidly in the construction side. To construct any structure on weak soil, if the soil is not suitable for construction, then there is a need to go for alternative materials [3]. Especially while the construction of the structures like Brides and Embankments, it is mandatory that the soil should have enough strength (Bearing capacity) to with stand the external loads and when we required to construct heavy structures like bridges and embankments on weak soils one need to increase the strength and stability of the soil. Stabilizing of soils is not economical and it requires more time [7, 8]. In this research we suggest the best alternative material is EPS geo-foam means "Expanded Poly Styrene" generally called as "Thermocol" is used in the weak soils to get more stability [10]. It contains 95% of air and the strength of EPS is related to the density of EPS geo-foam. In foreign countries they were used these EPS blocks in the construction of Bridges and embankments to increase the strength and stability of the weak soils and to decrease the settlements of the soils under loading. The usage of these EPS geo-foams in Bridge abutments is best economical alternative than a piled embankment inside the abutment. We can use different qualities of EPS geo-foams based on the requirement in the soils .Generally by placing different densities over weak soil and comparing the behavior of weak soil based on the results obtained [11].

2. Materials and Properties

2.1. EPS Geo-foam

EPS Geo-foam blocks means Expanded Poly-Styrene Geo-foam blocks, which is a geo-synthetic material generally used for packaging and insulation purposes over compressible soils, but now a days it is also used in the construction field [6]. The EPS geo-foams can be used in the bridge abutments and the construction of embankments to increase the strength and stability of the weak soils. The use of EPS geo-foam over weak soils was best suggestion because it was found to be the most economical geotechnical solution for the project [5]. EPS geo-foam blocks were therefore used to make up the filling inside the abutments and the embankments while the extension of highways. The EPS geo-foam weight is very less, it can carry easily without any special equipment and can be available in different shapes and densities.

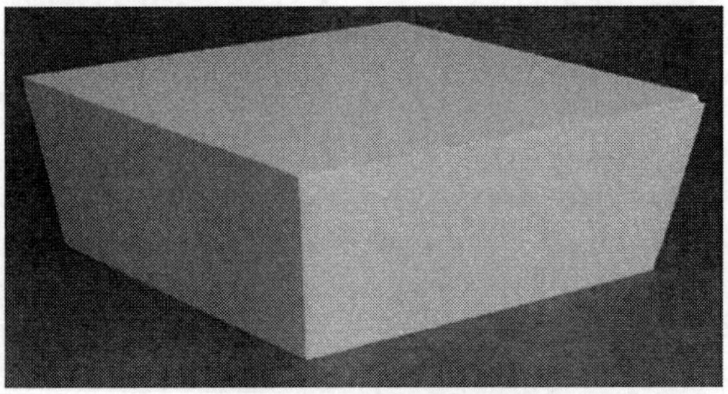

Fig: 1 EPS Geo-foam

Table: 1 Properties of EPS Geo-foam

Property	Value
Density (kg/m^3)	15.8
Compressive strength (K Pa)	27
Flexural strength (K Pa)	168
Water Absorption (%)	3.84
Density (kg/m^3)	15.8

2.2. Geo-membrane

A geo-membrane is a very less permeable membrane, it will not allow water through it and act like a barrier for other material which is placed over geo membrane [4]. It is majorly used over weak soil and it is act like a barrier between soil and EPS geo-foam. The main purpose of these geo membranes is to hold the EPS geo-foam blocks on the soil and to minimize the passage of the water through it. And by placing these geo membranes on the side fills we can mitigate the settlement and sliding of the soil.

Fig: 2 Geo-membranes

Table: 2 Properties of Geo membrane

Property	Value
Thickness	12-100 miles
Specific gravity	0.9 -1.6
Tear resistance	4-32 pounds
Impact resistance	0.5-13 ft-lb
Puncture resistance	10-100 pounds

2.3. *Compressible soils*

Weak soils generally called as "compressible soils" such as clay or silty clay soil having low bearing capacity. When loads act on this type of soils it undergoes easily settlement [2]. To increase the properties of weak soil instead of stabilizing, if we place geo-foam members over weak soil it will give best results more than stabilizing techniques and the cost of the usage of these EPS blocks in the soils is very less compared to the stabilization of the soil by using any stabilizing materials [9]. The below figure shows the properties of weak soil.

Fig: 3 Compressible soil

Table: 3 Properties of soil

Property	Value
Liquid limit	62%
Plastic limit	41%
Plasticity index	22%
Specific gravity	2.34

3. Test Results and Comparison

3.1. Density

EPS geo-foam densities are taken as 10, 12, 14, 16, 18, 20kg/m^{3}and comparing the stability of soil at various densities. The stability of soil depends upon the compressive strength and shear strength values. The test results are shown in the following graph. The below graph shows the variation between different density values of EPS geo-foam to the other strength parameters of EPS geo-foam after placing over weak.

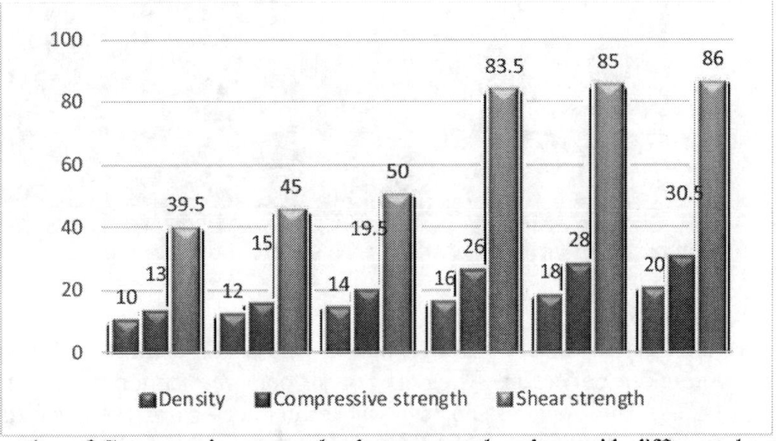

Fig 4: Comparing of Compressive strength, shear strength values with different density values of EPS geo-foam.

3.2. California Bearing Ratio Test

The CBR test is generally used for evaluating the sub grade strength of the roads and pavements. But in case of construction of structures over weak soil by placing EPS geo-foam, we need to perform CBR ratio test on geo-foam to find the penetration strength [1]. CBR ratio test has been performed on different densities of geo-foam as 10, 15, 20, 25 kg/m^3. By placing EPS geo-foam in cylindrical sections of CBR mould. The CBR test is penetrated into EPS geo-foam with a standard rate of 1.28 mm/min as per ASTM D1883 [12]. This test is performed on different densities and obtained different CBR ratio values. Finally graph is drawn between density and CBR ratio values as shown as below.

Fig: 5 CBR test specimen

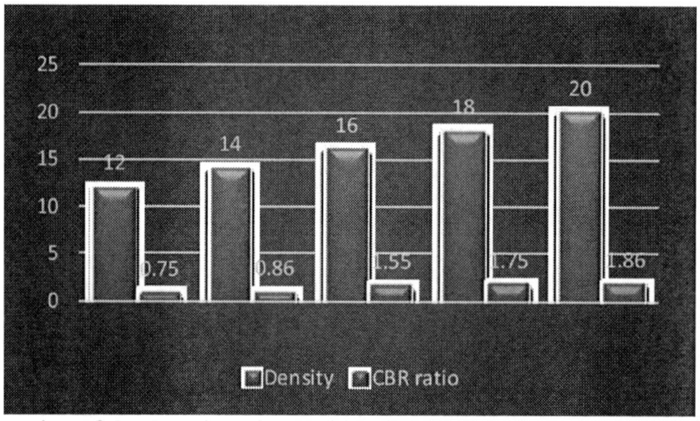

Fig 6: Comparing of CBR ratio values with different density values of EPS geo-foam

3.3. *Compression Test*

Compressive strength is the test which is performed on geo-foam to find the ability, to carry the loads without any crack or deflection. A material under compression tends to reduce the size, while in tension, size elongates. The compression test will be conducted for the calculation of compressive strength of the soil by placing densities of EPS geo-foam as 12, 14,16,18, 20kg/m^3 over the weak soil. EPS geo-foam blocks with compressive strength of 25 kN/m^2 at 4% deformation constituted large part of the filling. Higher strength blocks were also used at selected sections and these had a strength of 28 kN/m^2 at 5% deformation. The results are shown in the below graph.

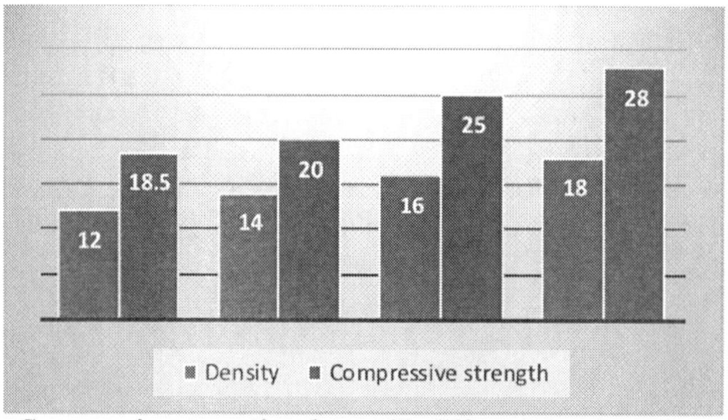

Fig 7: Comparing Compressive strength values with different density values of EPS geo-foam.

3.4. *Shear Strength*

Shear strength is the maximum internal resistance to shearing stresses and a consequent tendency for shear deformation. It has been tested over different densities as 14, 16,18,20 kg/m^3 of EPS geo-foam under tri axial testing. EPS geo-foam blocks with shear strength of 45 kpa at 5% deformation constituted large part of the filling. Higher strength blocks were also used at selected sections and these had a strength of 62 kpa at 5% deformation. Below figure shows the test result values.

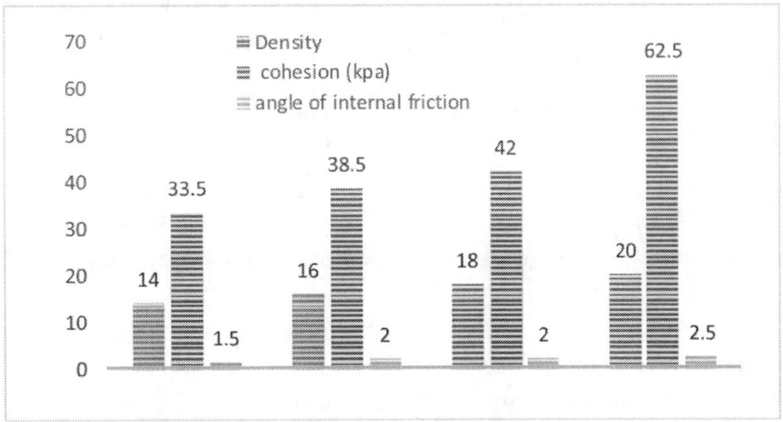

Fig 8: Comparing of strength parameters with different density values of EPS geo-foam.

4. Conclusion

The following are the list of the conclusions drawn from the above investigation,

- Based upon above results, we can conclude that the weak soil can get strength by using these EPS Geo-foams and Geo membranes.
- This technique is more economical compared to the stabilization of soil by using stabilized materials, because the cost of Thermocol is very less compared to the stabilizing materials.
- Behavior of weak soil depends upon the different density values of geo-foam used in the construction.
- It is clear that from the above test results, density of EPS Geo-foam should be not less than 15.
- If the density of Geo-foam is less than 15 then there is no effect of these EPS Geo-foams in the strength properties of the soil.
- A full pledge inspection of the embankment during construction plays vital role in the quality checking. Overall, the lightweight fill in bridge abutments and embankments are safe with regard to settlement and stability concerns.
- Construction of bridge abutments and highway embankments over soft soil using expanded polystyrene (EPS) block (geo-foam block) to decrease consolidation settlement is a recent technology and has been implemented by various countries around the world.
- By using EPS blocks, bridge abutments, infrastructural embankments and extension of highways can now be done cheaply and quickly without unnecessary embankment slopes. This was implemented in the southeast of the Netherlands.
- If the density value is less than 15 the strength will not increase more, if the value is more than 20the compressive strength, shear strength and CBR ratio values increases and it is best for any type of construction.

References

[1] Mohammad Rafiullah Md. H. Mansoori, Diwan Usama MDHaris."Geofoam Construction and Application to Industry" (IJREAT).ISSN(online) :2347-2812,volume-4,issue-3,2016.

[2] Dharmendra Kumar Srivastava, Amit Srivastava, Anil Kumar Misra and Vaishali Sahu"Sustainabilityassessment ofEPS-geofoam in road construction: a casestudy" International journal of sustainable engineering.https://doi.org/10.1080/19397038.2018.1508319.
[3] Y. Z. Beju, J. N. Mandal"Expanded polystyrene (EPS) geofoam: preliminary characteristic evaluation"
Precedia engineering (2017)239-246 :<www.sciencedirect.com>.
[4] Y. Zou, J.C. Small and C.J. Leo ,"Behavior of EPS Geo-foam as Flexible Pavement Sub-grade Material In Model Tests"Geosynthetic International,2000 volume-7.
[5] Milan Duskov and Johan de Jongh"Dutch A4all Tramway EPS Embankment with Vertical Sides".
5th international conference on geofoam applications in civil engineering.
[6] Jan Vaslestad, JornEinarLiveroed, TsedayDamtew, StaaleSingstad and Dag Loevstad "Geofoam EPS Used in Bridge Abutment, E18 Farris Bridge".
[7] Abdullah TolgaOzer and EmreAkmay "First Geofoam Roadway Embankment Application in Turkey"5th international conference on eps geofoam in civil engineering applications, 2019,pp 71-80.
[8] H. Hazarika, "Stress-strain modeling of EPS geofoam for large strain applications, Geotextile, Geomembrane". 24 (2006) 79-90.
[9] A.H. Padade, J.N. Mandal, Behavior of EPS geofoam under triaxial loading condition, Elec. J. Geotech. Eng. 17(2012) 2543-2553.
[10] "Deformability and tensile strength of expanded polystyrene (EPS) under short term loading, Polymer Testing",I.Y. Gnip, S. Vejelis, V. Kersulis, S. Vaitkus, 26 (2007) 886-895.
[11] "40 years of experience with the use of EPS geofoam blocks in road construction".Aaboe R (2011) Paper presented at the 4th international conference on geofoam blocks in construction applications,Lillestrom, Norway, 6–8 June 2011.
[12] "Use of expanded polystyrene in road embankments—design, construction and quality assurance".Norwegian Road Research Laboratory (NRRL) (1992) ,Public Roads Administration, Oslo, Norway.

Application of Multiple Linear Regression Models and Adaptive Neuro-Fuzzy Inference System Models to estimate the Compressive Strength of Concrete

Mohammed Hussain[1], Y Kamala Raju[2] and V Kamakshi Prasad[3]

[1,2]Department of Civil Engineering,Gokaraju Rangaraju Institute of Engineering and Technology(GRIET) Hyderabad, India
[3]Department of Computer Science and Engineering, Jawaharlal Nehru Technological University Hyderabad(JNTUH),Hyderabad, India
[1] mohd_hussain@griet.ac.in; [2]kamalaraju@griet.ac.in; [3]kamakshiprasad@jntuh.ac.in

Abstract . The most widely used composite construction material is concrete . Compressive strength is an important monitoring parameter for quality assurance at construction site. In this paper,two different models of Multiple Linear Regression (MLR) and Adaptive Neuro-Fuzzy Inference System (ANFIS) are developed to predict the 28 day compressive strength of concrete using the experimental datasets available . These two models are compared . Sensitivity Analysis (SA) is carried out for two different sets of parameters.ANFIS models predict more effectively as their coefficients of multiple determination(R Square) are higher than those of MLR models.This shows that the nonlinear correlation among input variables is better represented in ANFIS models than in MLR models .Goal 12 of United Nations Sustainable Development deals with the sustainable consumption and production patterns and it is taken into account as environmentally degrading materials(flyash and blast furnace slag) are used.

1. Introduction

Quality control at construction sites can only be possible by constantly monitoring the compressive strength of concrete before placement. In this study , two methods of prediction of compressive strength of concrete by Multiple Linear Regression (MLR) in Microsoft Excel and Adaptive Neuro Fuzzy Inference System (ANFIS) in MATLAB are discussed .The application of ANFIS models for the prediction of Compressive strength of concrete is discussed [1],[2], [3], [4],[5],[6] [8] and [12]. A detailed note on ANFIS gives the detailed structure and components of ANFIS [7]. To ensure sustainable consumption and production patterns is Goal 12 of Sustainable Development of United Nations [11].This paper is concerned with this goal as environmentally degrading materials (flyash and blast furnace slag) are used.

1.1 *Multiple Linear Regression Models*

Models fitted to data in a statistical process are Regression models. Multiple Linear Regression involves more than one input variable. The general form of a multiple linear regression statistical model is given below:

$$\hat{Y} = a_0 + \sum_{j=1}^{m} a_j X_j,$$

$\hat{Y}$ is the dependent output variable. In the present study, it is 28 day Compressive strength of concrete.

X_j (X_1 , X_2 ,..........$X_{m)}$are independent input variables.

X_1: Fine Aggregate in Kg/m^3

X_2: Cement in Kg/m^3

X_3: Coarse Aggregate in Kg/m^3

X_4: Fly Ash in Kg/m^3

X_5: Water in Kg/m^3

X_6: Blast Furnace Slag in Kg/m^3

X_7: Super Plasticizer in Kg/m^3

$\hat{Y}$: Concrete 28 day Compressive Strength in M Pa (Mega Pascals)

a_0,a_1,a_2,a_3,..........a_m are partial regression coefficients.

The multiple linear regression model brings out correlation between 28 day compressive strength and seven parameters of concrete.

1.2 Adaptive Neuro Fuzzy Inference System (ANFIS)

A brief introduction to ANFIS is given in Wikipedia[9].It is a kind of artificial neural network that is based on Takagi-Sugeno fuzzy inference system. It integrates the benefits of both fuzzy logic and neural networks. A set of fuzzy IF-THEN rules is defined in this inference system. Nonlinear functions can be approximated by learning capability of ANFIS.

Supervised learning is used in ANFIS. ANFIS consists of a learning algorithm, multilayer and feed-forward nerworks that include input/ouput variables and a fuzzy rule[4]

ANFIS architecture consists of five layers. The first layer is commonly called Fuzzification layer which takes the input values and determines the membership functions belonging to them. The second layer, known as rule layer, generates the firing strengths for rules.The third layer normalizes the computed firing strengths.The fourth layer, which is known as the adaptive layer, takes as input the normalized values.The fifth layer is called the de-fuzzification layer[9]. Figure 1 shows the ANFIS structure.

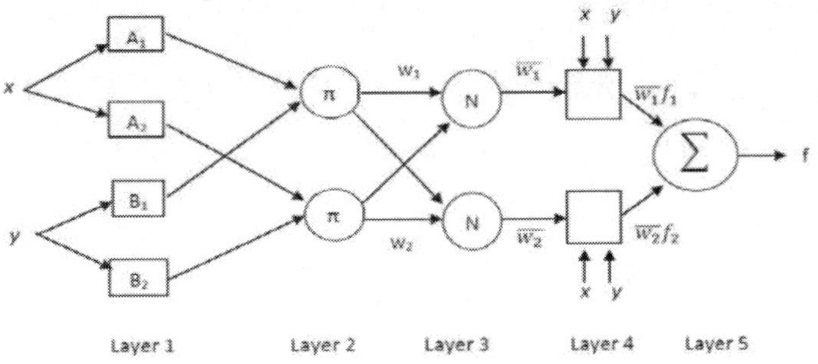

Figure 1. Structure of ANFIS based on Takagi-Sugeno

2.Materials and Data Collection

The 28 day compressive strength of concrete values of 350 in number are taken from the experimental datasets of Machine Learning Laboratory of University of California , Irvine [10].The observed values of seven constituents of Concrete (Cement, Fine aggregate , Superplasticizer , Blast Furnace Slag, Fly ash, ,Coarse aggregate and Water) and 28 day compressive strength of Concrete are taken into consideration.Table 1 and Table 2 show the characteristics of the dataset.

Table 1. Variation of Maximum and Minimum Values of Eight Parameters of Concrete of 350 Datasets.

Number	Parameters	Unit	Maximum	Minimum
1	Cement	kg/m^3	540	102
2	Blast Furnace Slag	kg/m^3	359.4	0
3	Fly Ash	kg/m^3	200.1	0
4	Water	kg/m^3	247	121.8
5	Superplasticizer	kg/m^3	32.2	0
6	Coarse Aggregate	kg/m^3	1130	801
7	Fine Aggregate	kg/m^3	992.6	594
8	Compressive strength of Concrete	MPa(Mega Pascals)	81.75	8.54

Table 2 .Variation of Standard Deviation and Mean Values of Eight Parameters of Concrete of 350 Datasets.

Number	Parameters	Standard deviation	Mean
1	Cement	106.08	267.3
2	Blast Furnace Slag	88.74	86.5
3	Fly Ash	64.56	59.1
4	Water	19.13	182.9
5	Superplasticizer	5.48	6.9
6	Coarse Aggregate	83.54	954.8
7	Fine Aggregate	70.62	767
8	Compressive strength of Concrete	14.4671	37.32

3. Methodology adopted

The methodology adopted in this study is as shown in the following diagram.

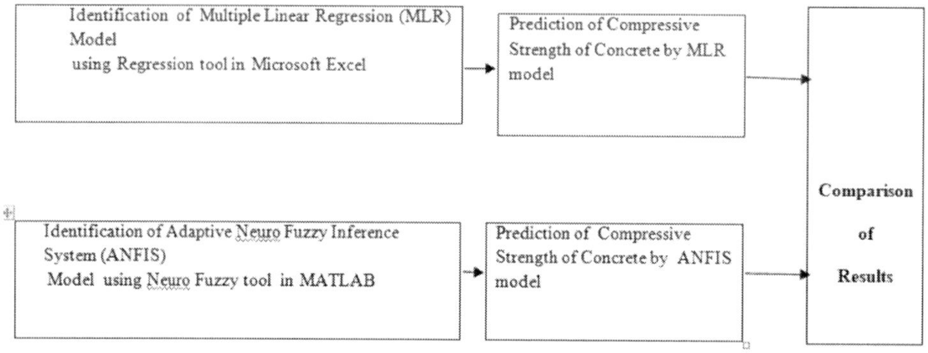

Figure 2. Schematic diagram showing the methodology adopted

As it is clear from the above diagram, Multiple Linear Regression model using Microsoft Excel is fitted for 350 values of 28 day compressive strength with the associated seven other parameters. The datasets are taken from the experimental datasets of Machine Learning Laboratory of University of California , Irvine [10].

ANFIS model is fitted using Neuro Fuzzy tool in MATLAB. The two model building methods in MATLAB ANFIS toolbox are Grid Partition and Subtractive Clustering. In grid partition method, as the number of inputs grows, membership function shows exponential growth that may stop the calculation system .This exponential explosion can be prevented and easy input –output rule model can be generated by using subtractive clustering method[1]. Subtractive clustering method is used in this study with the following four parameters. Out of total number of 350 values, 280 values (80% of data) are for training data , 35 values (10% of data) are for checking data and another 35 values (10% of data) are for testing the data in ANFIS model. Hybrid Optimization is considered.

Reject Ratio: 0.15
Accept Ratio : 0.5
Squash Factor : 1.25
Range of Influence : 0.5

4. Results and Discussion

The Multiple Linear Regression fitted is shown in Table 4 with the Coefficient of Determination (R square) as 0.769 and Standard Error of 6.95.Figure 2 shows the variation of Observed Concrete Compressive Strength and Predicted Compressive Strength in Mega Pascals

Table 4 . Results of Multiple Linear Regression to predict 28 day Compressive Strength.

Multiple R : 0.877
R Square : 0.769
Standard Error : 6.95
Observations :350
Multiple Linear Regression Equation Fitted :

$$X_8 = -121.84 + 0.174X_1 + 0.147X_2 + 0.112X_3 - 0.015X_4 + 0.153X_5 + 0.049X_6 + 0.063X_7$$

X_1: Cement in Kg/m^3

X_2: Blast Furnace Slag in Kg/m^3

X_3: Fly Ash in Kg/m^3

X_4: Water in Kg/m^3

X_5: Super Plasticizer in Kg/m^3

X_6: Coarse Aggregate in Kg/m^3

X_7: Fine Aggregate in Kg/m^3

X_8 : 28 day Compressive Strength of Concrete

Figure 2 shows the methodology adopted in this study.
Figure 3 shows the variation of Observed Concrete Compressive Strength and Predicted Compressive Strength in Mega Pascals in the fitted MLR model with R^2=0.7698.
Figure 4 shows the variation of Predicted and Observed Concrete Compressive strengths with the number of Observations in the fitted Multiple Linear Regression model
Figure 5 shows the ANFIS structure used in the present study.all the five layers are indicated.
Figure 6 shows the variation of Fuzzy Inference System (FIS) output parameters and the target parameters for Training Data in ANFIS model of MATLAB
Figure 7 shows the Variation of Fuzzy Inference System (FIS) output parameters and the target parameters for Validation Data in ANFIS model
Figure 8 shows the Variation of Fuzzy Inference System (FIS) output parameters and the target parameters for Testing Data in ANFIS model
Figure 9 shows the variation of Predicted Compressive strength using ANFIS model and Observed Compressive Strength with R^2=0.9623.

The confidence level adopted is 95%. The Coefficient of Multiple determination (R Square) is the ratio of explained variance to total variance. In MLR model , R Square is 0.77 which indicates that 23% of the total variance only is unexplained. In ANFIS model, R Square is 0.96 which indicates that 4% of the total variance is unexplained.

The Coefficient of Multiple Correlation Coefficient (R) is the positive square root of the coefficient of multiple determination. Its value is 0.88 in MLR model. The standard error of estimate is the square root of sum of the square of the errors divided by the degrees of freedom. Its value is 6.95 in MLR model.

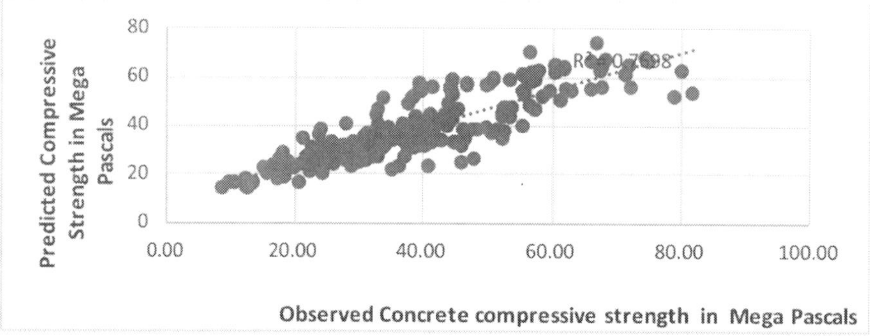

Figure 3 . Variation of Observed Concrete Compressive Strength and Predicted Compressive Strength in Mega Pascals

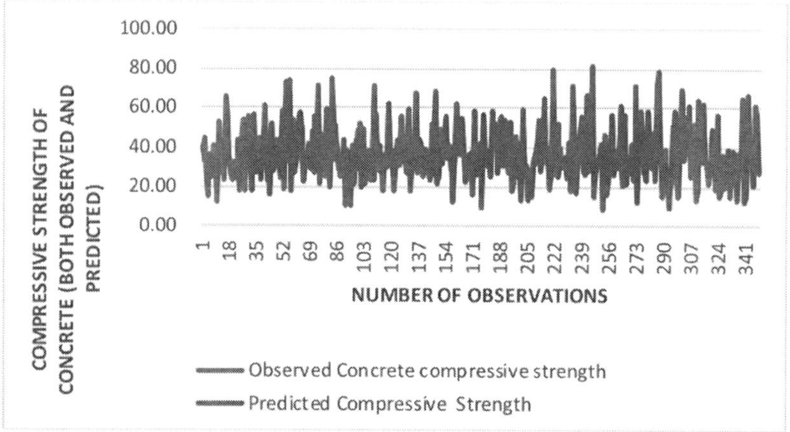

Figure 4 . Variation of Predicted and Observed Concrete Compressive strengths with the number of Observations

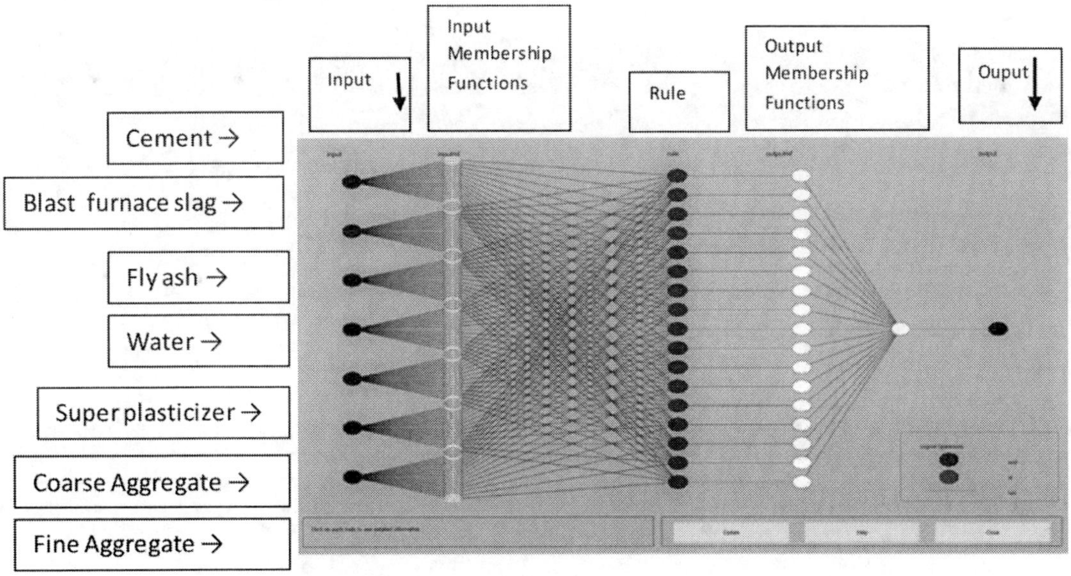

Figure 5. Structure of the ANFIS Model in the present study

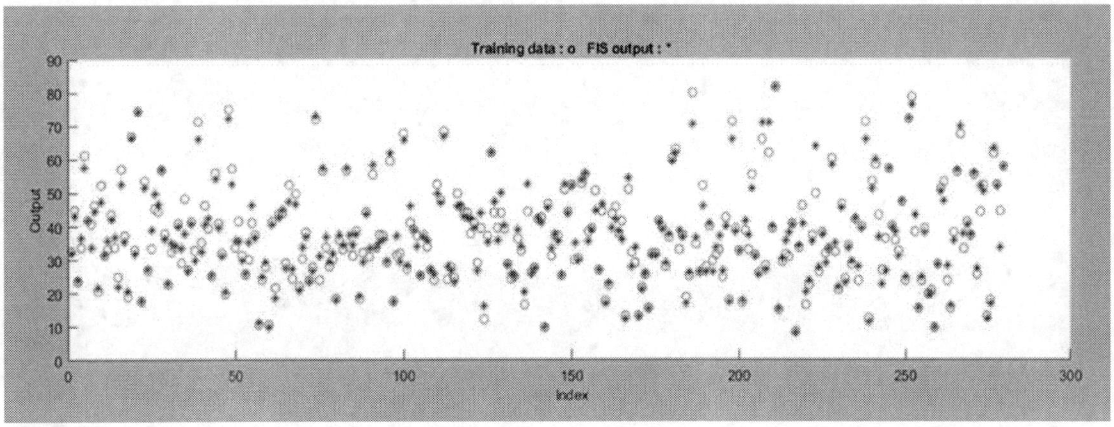

Figure 6 . Variation of Fuzzy Inference System (FIS) output parameters and the target parameters for Training Data in ANFIS model

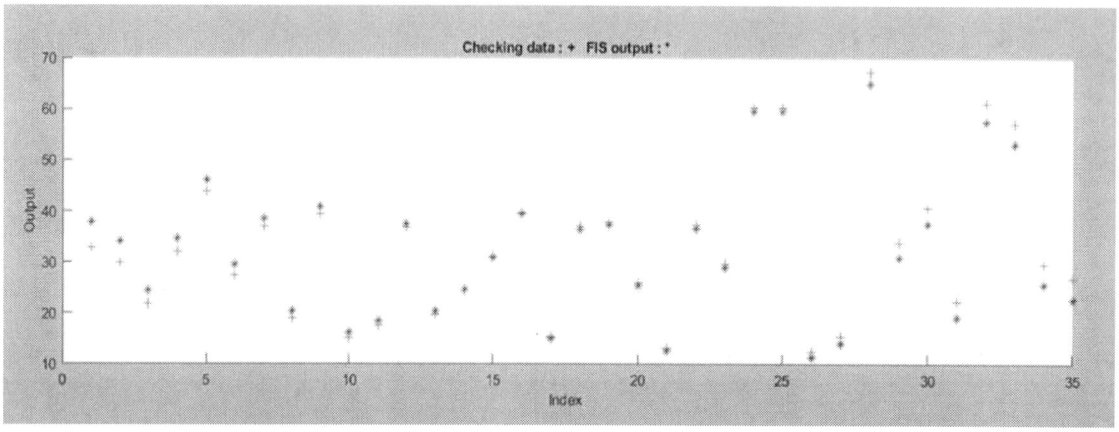

Figure 7. Variation of Fuzzy Inference System (FIS) output parameters and the target parameters for Validation Data in ANFIS model

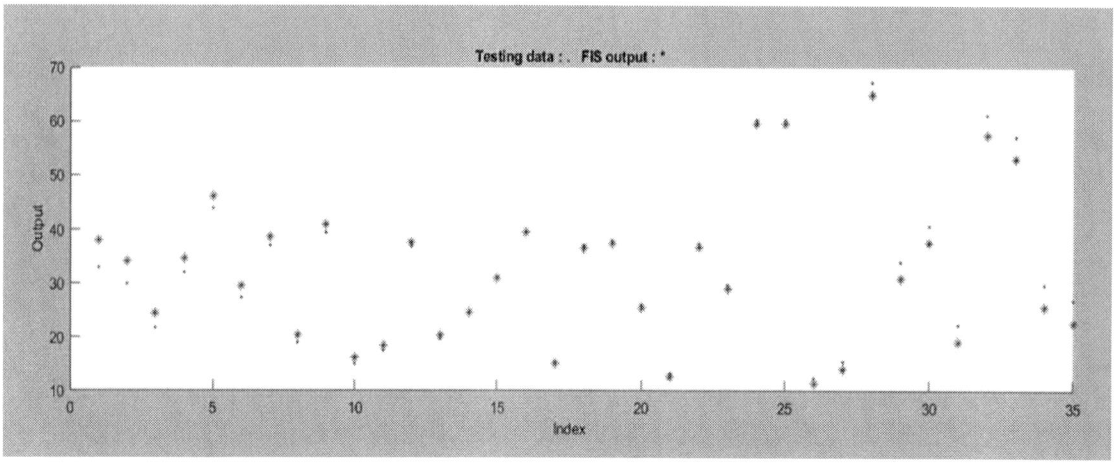

Figure 8. Variation of Fuzzy Inference System (FIS) output parameters and the target parameters for Testing Data in ANFIS model

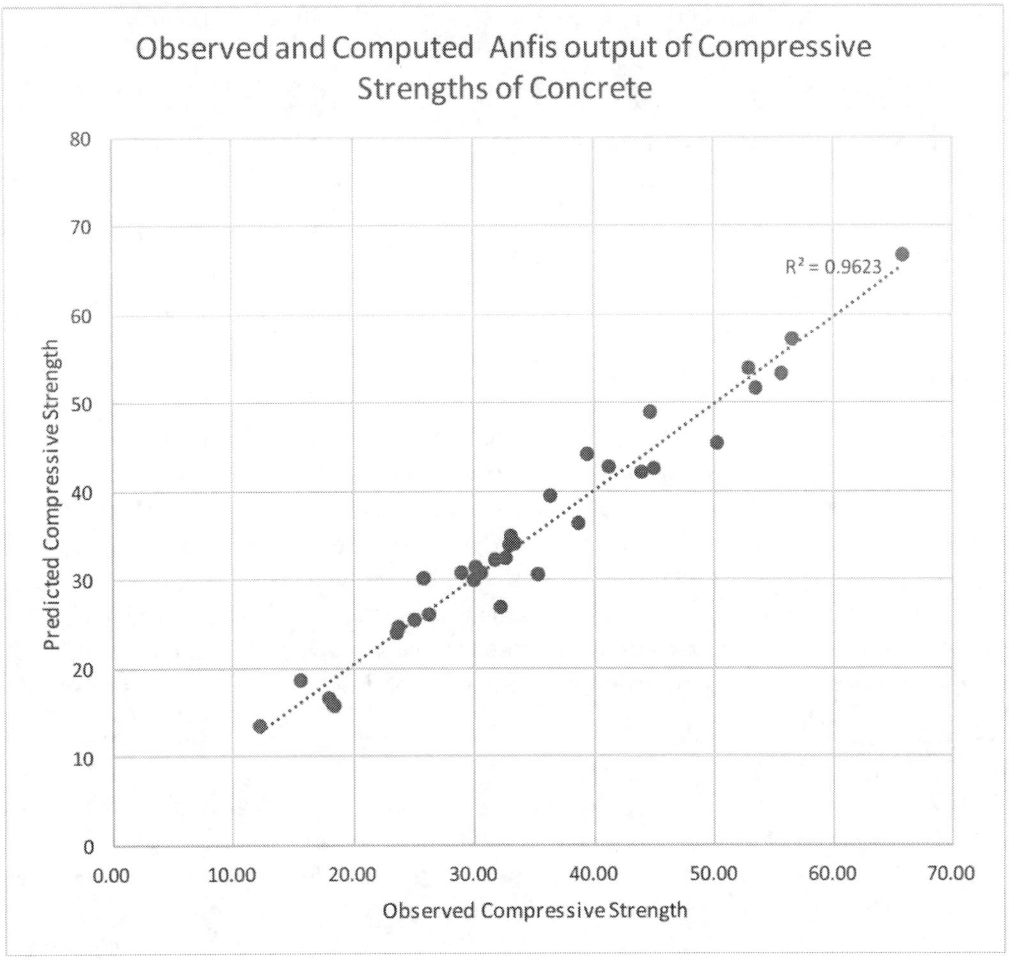

Figure 9. Variation of Predicted Compressive strength using ANFIS model and Observed Compressive Strength of concrete

5.Sensitivity Analysis

Sensitivity Analysis (SA) is an analysis of changes in model response with changes in the model input parameters[8].Two sets of parameters are considered to do the sensitivity analysis .First set excludes the flyash parameter (with zero values in the original experimental dataset) and second set excludes Blast Furnace Slag parameter(with zero values in the original experimental dataset).

Table3: Sensitivity Analysis on MLR and ANFIS Models

Model Name	Coefficient of Determination (R Square) for Set 1 (exclusive of Blast Furnace Slag)	Coefficient of Determination (R Square) for Set 2 (exclusive of Flyash)
MLR	0.836	0.775
ANFIS	0.952	0.954

In this paper, it is clear that ANFIS models with their higher Coefficient of Determination (R Square) are more effective in prediction when compared to MLR models.

6. Conclusions

In this study, multiple linear regression and ANFIS models are developed to predict the 28 day Compressive strength of Concrete of 350 concrete samples. The 28 day compressive strength is dependent on the contents of Cement, Blast Furnace Slag,Fly ash,Water, Superplasticizer,Coarse Aggregate and Fine Aggregate. It is observed that ANFIS models predict the 28 day compressive strength of concrete more effectively than MLR models as is evident from higher values of RSquare for ANFIS models. The nonlinear correlation among input variables is better represented in ANFIS models than in MLR models. This avoids costly field methods . The sensitivity analysis on two different sets of input variables indicates the importance of ANFIS models for their more reliability in prediction.

4. Acknowledgements

The authors thank Professor I-Cheng Yeh for his Datasets of Compressive strength provided at Machine Learning Repository of University of California , Irvine .The authors also thank concerned authorities of Jawaharlal Technological University Hyderabad for providing the latest version of MATLAB 2020b

References

[1]Zhe Yuan, Lin-Na Wang and Xu Ji 2014 Prediction of concrete compressive strength: Research on hybrid models genetic based algorithms and ANFIS *Advances in Engineering Software* **67** 156-163

[2] Z. Keshavarz and H. Torkian 2018 Application of ANN and ANFIS Models in Determining Compressive Strength of Concrete *Journal of Soft Computing in Civil Engineerin,* **2-1** 62-70

[3] Faezehossadat Khademi and Sayed Mohammadmehdi Jamal 2017 Estimating the compressive strength of concrete using multiple linear regression and adaptive neuro-fuzzy inference system *International Journal of Structural Engineering* **8** 20-31

[4] Loan T. Q. Ngo, Yu-Ren Wang and Yi-Ming Chen 2018 Applying Adaptive Neural Fuzzy Inference System to ImproveConcrete Strength Estimation in Ultrasonic Pulse Velocity Tests *Advances in Civil Engineering* 1-11

[5] Hai-Bang Ly, Binh Thai Pham et al.2019 Improvement of ANFIS Model for Prediction of Compressive Strength of Manufactured Sand Concrete *Applied Sciences MDPI* **9** 1-16

[6] Behnam Vakhshouri, Shami Nejadi 2018 Prediction of compressive strength of self-compacting concrete by ANFIS models *Neurocomputing* **280** 13–22.

[7] Cristobal Fresno and Elmer A. Fernandez 2012 *ANFIS vignette*, Catholic University of Cordoba, Argentina **1** -2
[8] Faezehossadat khademi, Mahmoud akbari, Sayed Mohammadmehdi jamal, Mehdi nikoo 2017 Multiple linear regression artificial neural network and fuzzy logic prediction of 28 days compressive strength of Concrete, *Frontiers of Structural and Civil Engineering* **11** 90–99
[9] Adaptive Neuro Fuzzy Inference System, ***Wikipedia***
[10] Rafiei, Mohammad Hossein, Waleed H. Khushefati, Ramazan Demirboga, and Hojjat Adeli. "Supervised Deep Restricted Boltzmann Machine for Estimation of Concrete." *ACI Materials Journal* 114, no. 2 (2017).
[11] un.org/sustainable development/sustainable –consumption-production/
[12] Khademi F, Akbari M and Jamal S M 2015 Prediction of Compressive Strength of Concrete by data driven models *I-Manager's Journal on Civil Engineering* **5** 16-23

Improve the efficiency of Battery Cooling System in E Vehicle by Nano-fluids with CFD Analysis

Vinoth Kumar H[1], Saravanakumar S [2],Sivakumar A[3], Gokul M [4], Jagadheeswaran S[5], Manikandan M[6]

[1] Assistant Professor, Department of Mechanical Engineering, M.Kumarasamy College of Engineering, Karur.

[2] Assistant Professor, Department of Mechanical Engineering, M.Kumarasamy College of Engineering, Karur.

[3] Associate Professor, Department of Mechanical Engineering, Kongu Engineering College, Erode.

[4] Student, Department of Mechanical Engineering, M.Kumarasamy College of Engineering, Karur.

[5] Student, Department of Mechanical Engineering, M.Kumarasamy College of Engineering, Karur.

[6] Student, Department of Mechanical Engineering, M.Kumarasamy College of Engineering, Karur.

[1]h.vinothkumar@gmail.com

[2]saravananmeksr@gmail.com

[3]askmech@kongu.ac.in

[4]mvgokul648@gmail.com

[5]jagadheeswaran02@gmail.com

[6]manikandan99mkce@gmail.com

Abstract. To run the electric vehicle, battery system is used. In battery system internal heat is generated due to various reason such as close arrangement of cells, periodic recharge and discharge of batteries. This heat can cause reduction of driving range, System safety, Life time of battery and Efficiency. To eliminate the heat by introduction of Battery thermal management system. In Battery thermal management system various cooling arrangement configurations (1s,2s,Parallel & Parallel with smooth tubes) are done to maintain the optimum temperature level of battery. This work is done by a Computational fluid analysis to characterize the temperature distribution of the battery under various configuration using Ethylene glycol and also Nano particles and Phase change material are introduced in the Lithium ion battery system. Inlet velocity of fluid is given as 0.1m/s and the temperature is 300K. Heat generated by the cell is 4 watts per cell and the heat flux is 1.66*e^5 w/m^2[5].

Keywords—Battery Colling, Lithium Ion Battery, Ethylene Glycol, Nano Particles, Phase Change Material.

I. INTRODUCTION

In day-to-day life E-vehicles are increased in numbers to avoid pollution. Electric vehicles use large amount of batteries to store energy. Battery system is one of the major part in electric vehicles, on behalf of that this project has been explored[1]. Lithium ion battery system is used in the most of the electric vehicle. Lithium ion battery contains low self discharge rate, light weight and high energy density. Despite their auspicious features, though, there are certain battery devices hurdles such as temperature related problems. Existence, efficiency and protection can be minimized by temperature[2]. In a battery pack, temperature variations between the modules and cells must be managed otherwise thermal run away may occur. In order of excess temperature in the battery to vanish, an appropriate battery thermal management device is also required. In addition, heating is needed in low temperature cases to ensure the maximum efficiency[3]. The purpose of battery thermal management system is to increase the life time of Lithium ion cells and there by control the degree and distribution of temperature. The Battery efficiency of vehicle is operated in very high or low ambient temperature battery efficiency will decrease[5]. A BTMS is especially necessary when the cells are susceptible to high rates of charging and discharging. A geing is affected not only by the normal or mean temperature in a cell, but also by the temperature gradient in the cell[6]. The premature aging of a single cell degrades the module's efficiency[5]. The heat is produced by the heating law of joule. The high temperature of the cell contributes to an increase in the cell's internal resistence, which decreases the output power[7]. More heat energy would be generated in them as the battery pack increases in size and charge / discharge rate. If this heat is not sufficiently dissipated it will stay in the battery pack. It can be achieved to establish the distribution of convective heat transfer temperature[8].The goal of this project to examine and compare the pipe structure and the efficiency of the various cooling arrangements used in the module of the lithium ion battery. The relation is achieved by simulating the modules's output using tools for computational fluid dynamics [9]. This program relicates the cell and flow propagation properties. To analyze result such as maximum temperature, temperature difference and the temperature distribution within the cell are compared to each other[10].

II. METHODOLOGY

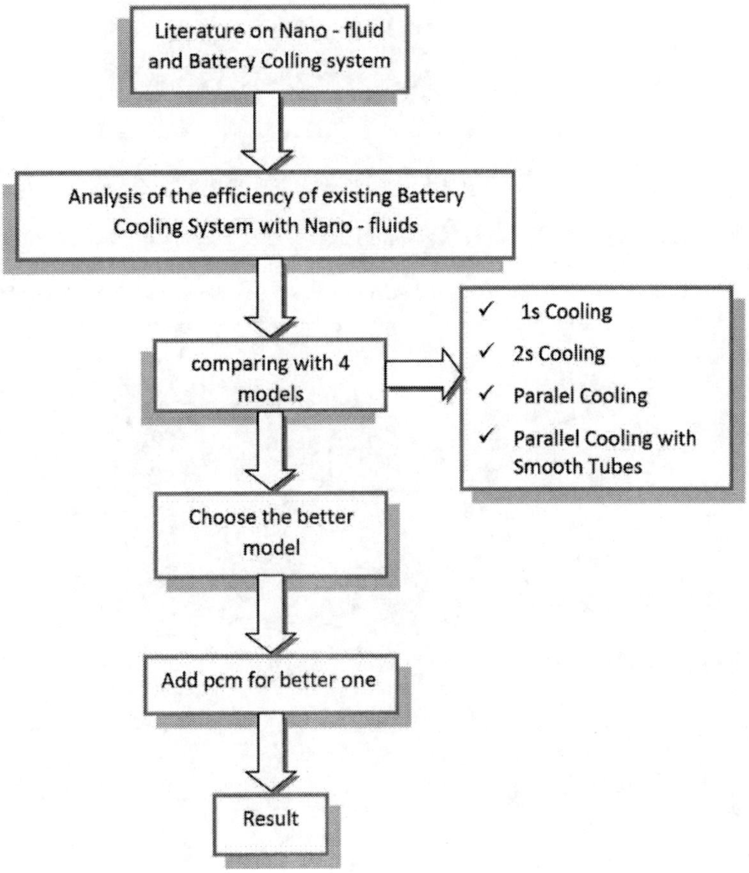

The main objective of this is to maintain the energy dissipation of each cell in the module in order to control the temperature level of batteries. The dissipation rate of the cells used in the module plays a major role in the energy storage. Due to this reason the usage of flow of coolant through various channels is to control the temperature of the battery moduleand to improve the life of the battery. Since the battery is less affected by the temperature change there will be a huge difference in the life time working of the battery. In this use the constant heat flux generated in the cells and to find the optimal solution. The fluids in the channel can observe the heat of the cell. So, optimize the fluid channel is necessary. The fluids are flow in various channel, as shown in the Fig. 1-4.

2.1 1-s Cooling Tube (1-Inlet & 1-Outlet)

This channel consist of one inlet and one outlet and it is a primitive type. A single channel can flow a entire cell, as shown in Fig.1. A first cell has a minimum temperature and last cell has a maximum temperature. It has a high pressure pump to flow a fluid at a edge.

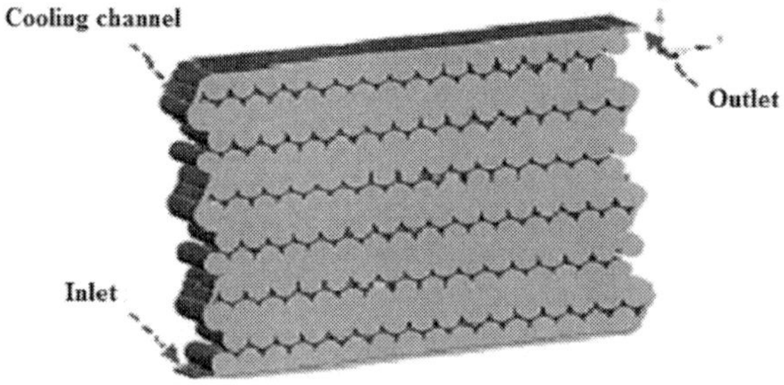

Fig. 1 1-s Cooling Tube (1-Inlet & 1-Outlet)

2.2 2-s Cooling Tube (2-Inlet & 2-Outlet)

This channel consist of two inlet and two outlet. A single channel can flow a half of the module, as shown in Fig.2. It has a less pressure pump than 1S configuration to flow a fluid at a edge.

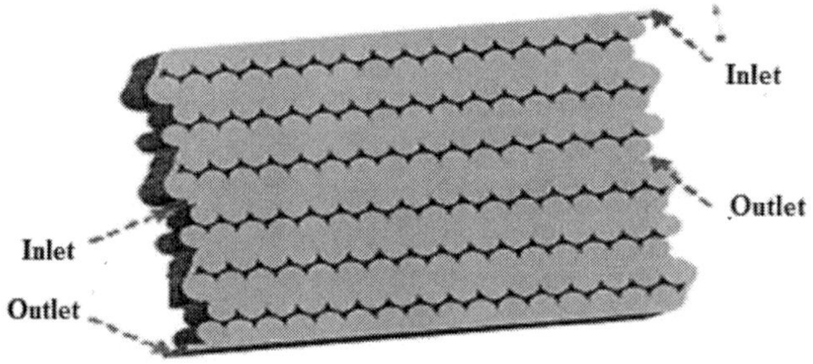

Fig. 2 2-s Cooling Tube (2-Inlet & 2-Outlet)

2.3 *Parallel Cooling Tube*

In this type of channel has also one inlet and one outlet. The fluid will not fulfilled to travel in the pipe. The pipe is fitted in entire row in the module, as shown in Fig.3.

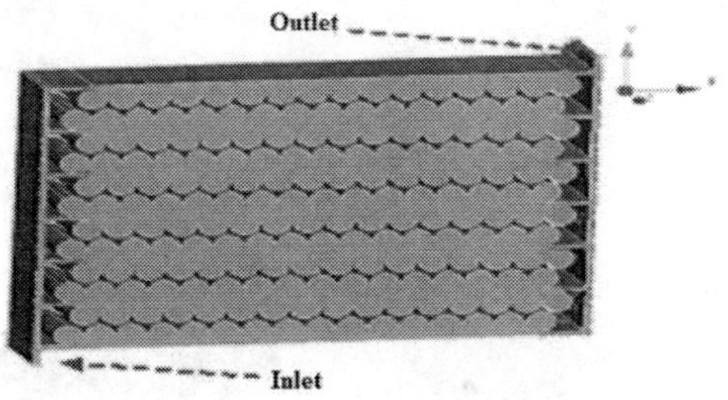

Fig. 3 Parallel cooling tube

2.4 *Parallel Cooling with Smooth Tubes*

In this configuration has also one inlet and one outlet, as shown in Fig. 4. The fluid is fully developed flow in the channel. The fluid flow in the smooth way in the tube. So, it is called as Parallel Cooling with Smooth Tubes.

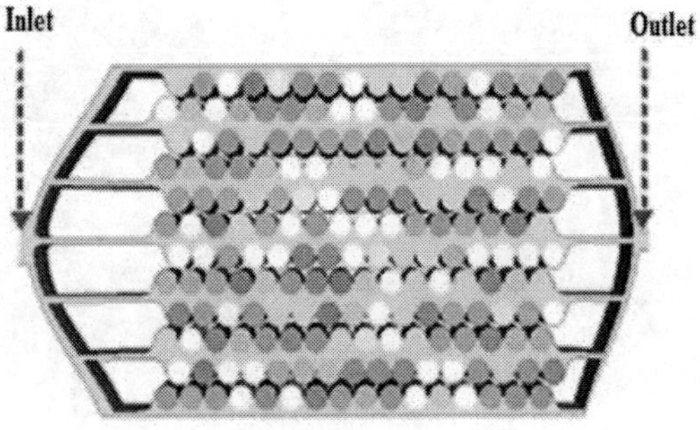

Fig. 4 Parallel Cooling with Smooth Tubes

III. ANALYSIS

3.1 *Surface Meshig:*

After cleaning up the geometry surface mesh is generated in ANSA tool. All the surfaces are discredited using triangular surface element .As the geometry has some complicated and skewed surfaces triangular dominant surface elements are used to capture the geometry. The following figure shows the surface meshes and volume mesh in a cut plane. The meshing of various configuration as shown in Fig. 5-8. The surface mesh count of the entire flow ,maximum surface skewness, the volume count and maximum volume skewness are shown in Table 1.

IOP Conf. Series: Materials Science and Engineering 1126 (2021) 012063 doi:10.1088/1757-899X/1126/1/012063

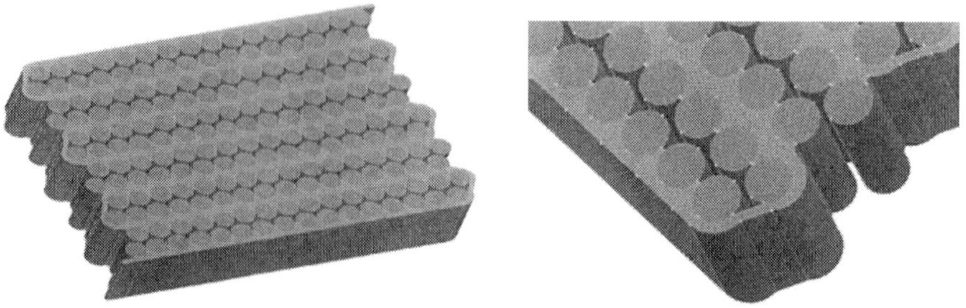

Fig. 5 Surface Mesh for 1-s Cooling Tube (1-inlet & 1-outlet)

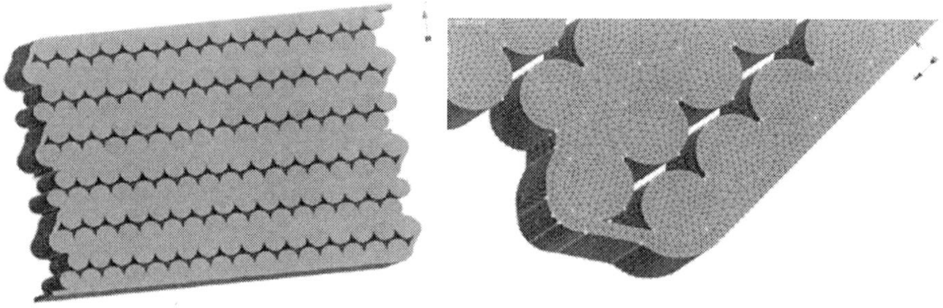

Fig. 6 Surface Mesh for 2-s Cooling Tube (2-inlet & 2-outlet)

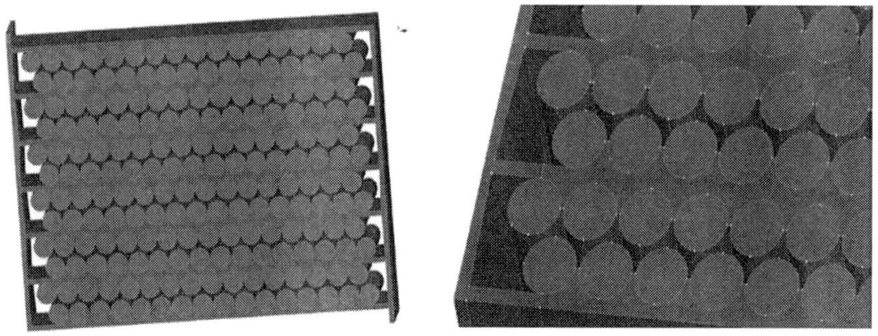

Fig.7 Surface Mesh for Parallel Cooling Tube

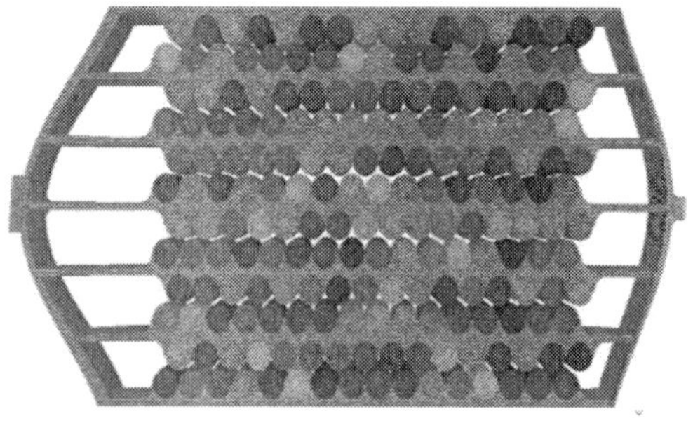

Fig.8 Surface Mesh for Parallel Cooling with Smooth Tubes

3.2 MESH-DETAILS

The meshing of these configuration is very complicated because it has two domain solid and fluid, as shown in Fig.5-8. By using the ANSA software surface mesh is generated. The meshing details are shown in Table 1.

Parameter	1S Configuration	2S Configuration	Parallel Cooling Tube	Smooth Parallel Cooling Tube
Surface Mesh Count	5,10,709	5,31,691	4,84,276	6,08,489
Surface Mesh Quality	0.7	0.7	0.7	0.7
Volume Mesh Count	11,98,690	11,92,709	12,86,761	2,26,997
Volume Mesh Quality	0.9	0.91	. 0.91	0.91

TABLE 1 : Mesh Details

3.3 SOLVER-SETUP

Coolant Fluid is assumed to be Ethylene-Glycol, Flow is assumed to be 3D,Turbulent and compressible, Flow is assumed to be velocity inlet with 0.1 m/s, Air temperature is assumed to be at STP, Cell volumes are assumed to be at constant heat flux of 1.66*e^5 W/m2.

3.4. MATERIAL PROPERTIES

To choose four different material and has an different fluid properties as shown in Table 2. In PCM material has both fluid and solid properties and the Nano particle is combined with Ethylene-Glycol to become Nano fluid. The Nano fluid has special fluid properties based on the concentration.

Properties/Material	Symbols	Ethylene-Glycol (Fluid)	Lithium ion Battery (Solid)	Nano fluid	PCM
Density(Kg/m3)	ρ	1097	2560	1169.25	857
Specific Heat (j/Kg-K)	Cp	2200	1000	3500	680
Thermal Conuctivity (W/m-K)	k	0.254	25	0.326	30
Viscosity (Kg/m-S)	μ	0.001003		0.001095	0.00553
Latent Heat (j/Kg)					223000
Liquidus Temperature (K)					317
Solidus Temperature (K)					317.1

TABLE 2: PROPERTIES OF MATERIAL

II. RESULTS AND FINDINGS

4.1 Temperature contours for different geometrical configuration

1) 1-S Cooling Tube (1-Inlet & 1-Outlet): For this configuration there is a considerable variation of temperature some cells with high temperature (315K) near outlet and low temperature near the inlet. This is because of the reason that the heat is continuously added to the cooling liquid. While the high temperature cells produce unfavorable condition, too much temperature variation cause thermal stresses in the Battery system. A single channel can flow a entire cell, as shown in Fig. 1. A first cell has a minimum temperature and last cell has a maximum temperature. Maximum temperature is 315K. Minimum temperature is 300K. Temperature difference is 15K, as shown in Fig. 9.

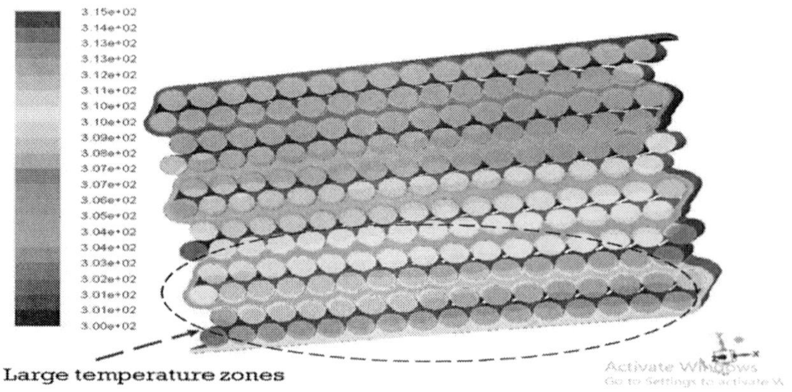

Fig. 9Temperature Contour of 1-s Cooling Tube

4.2 2-S Cooling Tube (2-Inlet & 2-Outlet): In this configuration temperature has reduced by 3 K from 1S configuration.Still the high temperature cells produce unfavourable condition. This too much temperature variation cause thermal stresses in the Battery system in this configuration too. Maximum temperature is 312K. Minimum temperature is 300K. Temperature difference is 12K ,as shown in Fig. 10.

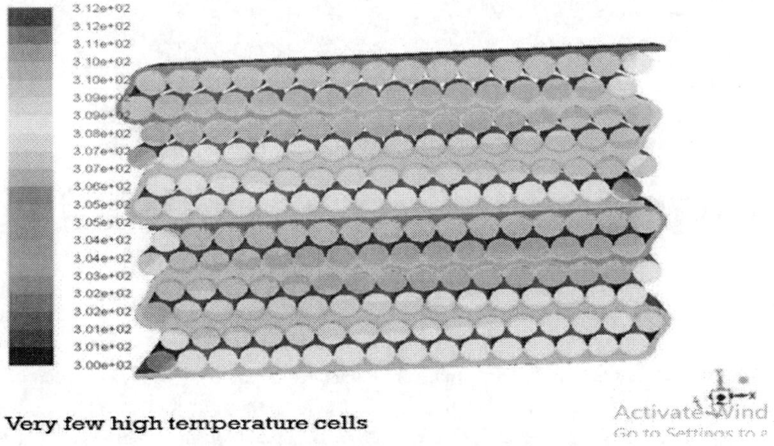

Fig. 10 Temperature Contour of 2-s Cooling Tube

4.3 Parallel Cooling Tube: In parallel configuration most of the cells have minimum temperature of 307 K. At the same time few cells near the outlet region have very high temperature. Maximum temperature is 337K.Minimum temperature is 300K.Temperature difference is 37K, as shown in Fig. 11.

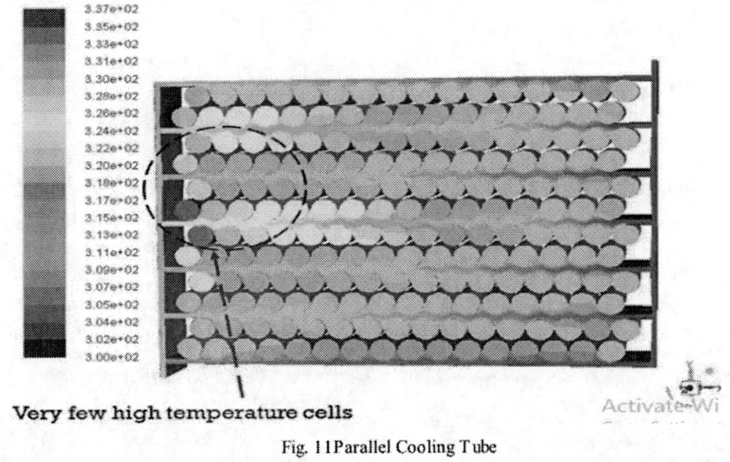

Fig. 11Parallel Cooling Tube

4.4 Optimized Parallel Cooling with Smooth Tube: This parallel configuration with smooth coolant flow channels eliminates high temperature cells very efficiently.Maximum temperature is 308K. Minimum temperature is 300K. Temperature difference is 8K, as shown in Fig. 12.

IOP Conf. Series: Materials Science and Engineering 1126 (2021) 012063 doi:10.1088/1757-899X/1126/1/012063

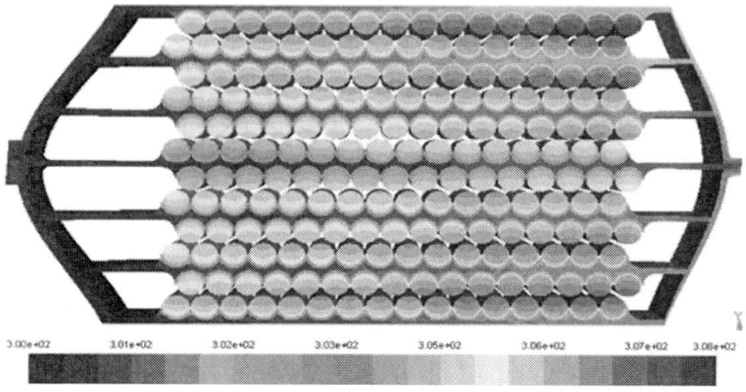

Fig. 12 Parallel Cooling with Smooth Tubes

4.5 *Findings from the analysis*

In the base 1S configuration, Heat transfer characteristics are comparably poor due to "Continuously decreasing thermal gradient between Cold fluid and Hot cells". Thus the Cells near the outlet have higher temperatures and in-turn less fatigue life. Where as in 2S configuration, the thermal gradient is high in both loops that result in better heat removal. Few cells registered high temperatures due to improper design that can be still improved. The average temperature distribution is very low comparing to other configurations and it would be the optimum model structure for better heat removal. In parallel configuration the smooth configuration provides the optimum temperature distribution. Thus this geometrical configuration is taken for the further studies.

4.6 *Finding the Optimum Nano Particle Concentration*

With parallel configuration with smooth coolant path is found to be optimum where the heat transfer is maximum, a study is executed with Nano particles are added in to the base fluid. The volume concentration of Nano particles are varied from 0.1 to 0.4, as shown in Fig. 13-16 and CFD results are taken and compared.

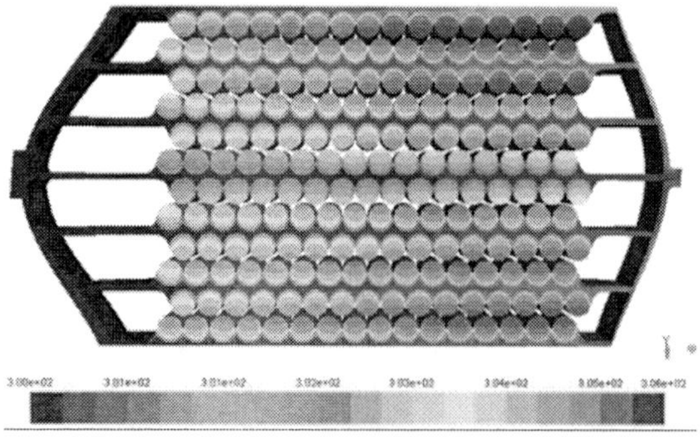

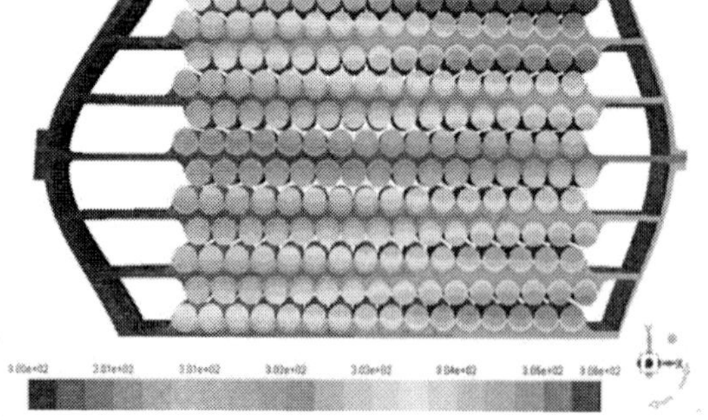

Fig.13 0.1% Nano particle concentrationFig.14 0.2% Nano particle concentration

IOP Conf. Series: Materials Science and Engineering 1126 (2021) 012063 doi:10.1088/1757-899X/1126/1/012063

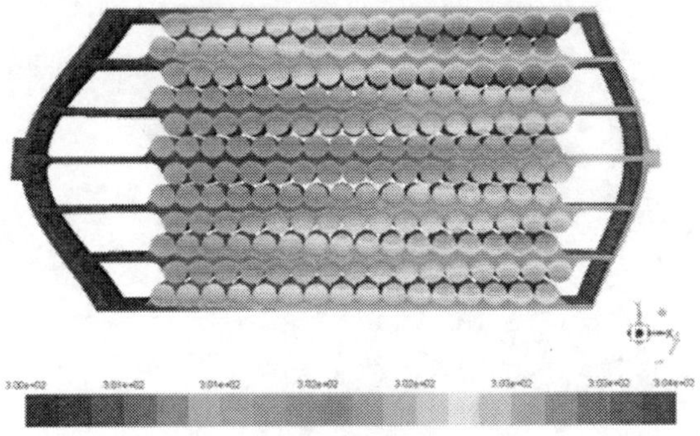

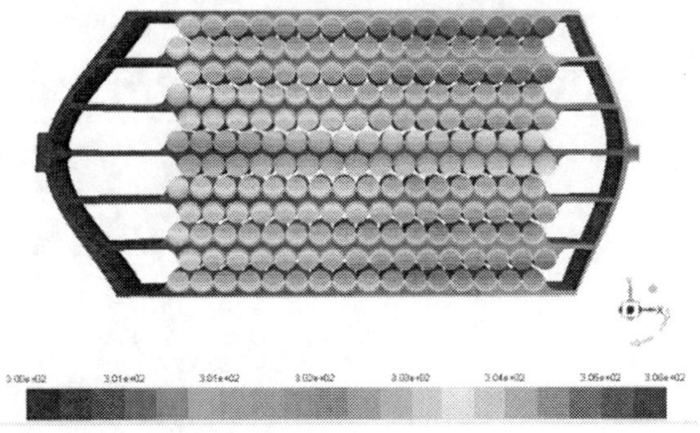

Fig.15 0.3% Nano particle concentrationFig.16 0.4% Nano particle concentration

4.7 *Findings from Nano Analysis*

With the concentration of Nano particles increase heat transfer rate increases and after 0.3 again heat transfer rate decreases. This is because of the reason the specific heat capacity (Cp) decreases after Nano particles concentration of 0.3 drastically. Thus 0.3 would be optimum concentration of Nano particles as shown in Fig. 15.

4.8 *Model with PCM*

The PCM used for thermal management have a melting point in the optimum performing range of lithium cells. The PCM has high latent heat and acts as a heat sink during battery discharge. When the cells are on standby, the PCM releases heat to the cells an environment. The PCM consist of paraffin mixed with graphite flakes. The PCM is introduced in parallel cooling with smooth tube configuration and also Nano fluid is introduced in the channel, as shown in Fig.17. The fluid properties are in Table 2.It act as a hybrid thermal management system.

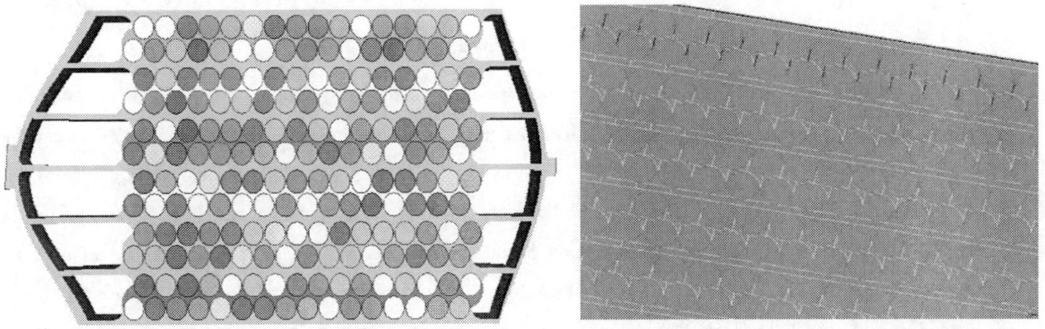

Fig. 17 PCM Cells

4.9 Domain Discretization with PCM

The meshing of parallel cooling with smooth tubes and PCM material are shown in Fig. 18. With the inclusion of PCM temperature distribution on the cells has found to be uniform and minimum at 304K, as shown in the Fig.19. But due to the space available very little PCM can be introduced for the given arrangement of cells.

Fig.18 Mesh on PCM

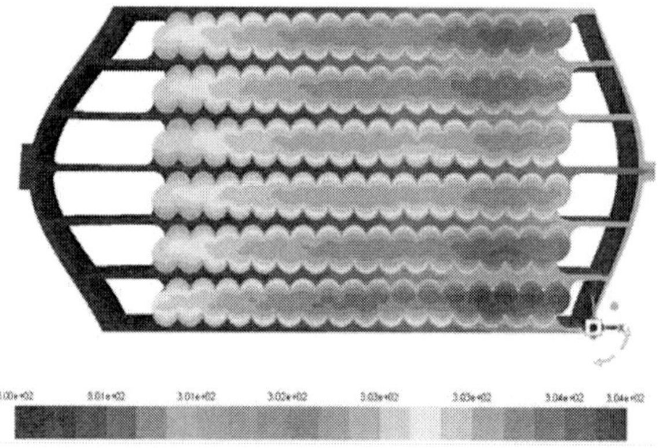

Fig.19 Temperature Contour of PCM with 0.3% Concentration of Nano fluid

V. CONCLUSION

Thus an exclusive and extensive analysis is executed using computational fluid dynamics (CFD) for the understanding the impact of enhancing the rate of heat transfer by varying geometrical configuration, using nana particle with the base fluid and adding PCM in the available space. From a detailed analysis of EV using CFD for various geometrical configuration, different concentration of Nano particles and PCM it was found that Parallel configuration with smooth coolant path found to better. A Nano particle concentration of 0.3% was found to be optimum for higher heat transfer rate.PCM concentration shows good heat transfer characterises still PCM materials are costlier.Thus Parallel configuration with smooth coolant path allied by 0.3 concentration of Nano particles found to be the optimum model.

VI. REFERENCES

[1] Alaoui,*C., 2013. Solid-state thermal management for lithium-ion ev batteries.IEEE Trans. Veh. Technol..*
[2] Alaoui,C., Salameh, Z.M., 2003. Modeling and simulation of a thermal man-agement system for electric vehicles. In: IECON Proceedings (IndustrialElectronics Conference), vol. 1.
[3] Bandhauer,T.M., Garimella, S., Fuller, T.F., 2011. A critical review of thermalissuesinlithium-ionbatteries.J.Electrochem.Soc.158(3),R1.
[4] Ellabban,O., Abu-Rub, H., Blaabjerg, F., 2014. Renewable energy resources:Current status, future prospects, and their enabling technology. Renew.Sustain. EnergyRev..
[5] Esfahanian, V., et al., 2013. Design and Simulation of Air Cooled Battery ThermalManagement System using Thermoelectric for a Hybrid Electric Bus, vol. 191.In: LNEE, vol. 3.
[6] Gadsden,S.A., Al-Shabi, M., Habibi, S.R., 2011. Estimation Strategies for theCondition Monitoring of a Battery System in a Hybrid Electric Vehicle. ISRNSignal Process. 2011.
[7] Kim, S.Y., Lee, K., Park, S., Kim, J., 2014. Thermal design analysis and perfor- mance test of a 1kW thermoelectric battery cooler. In: Thermomechanical Phenomena in Electronic Systems-Proceedings of the Intersociety Conference.
[8] Li, X., et al., 2019. Experimental investigation on a thermoelectric cooler for thermal management of a lithium-ion battery module. Int. J. Photoenergy 1–10.http://dx.doi.org/10.1155/2019/3725364.
[9] Liu,X., Deng, Y.D., Chen, S., Wang, W.S., Xu, Y., Su, C.Q., 2014a. A case studyon compatibility of automotive exhaust thermoelectric generation system, catalytic converter and muffler. Case Stud.Therm.Eng..
[10] Liu, Y., Yang, S., Guo, B., Deng, C., 2014b. Numerical analysis and design of ther-mal management system for lithium ion battery pack using thermoelectriccoolers. Adv. Mech. Eng..
[11] Maral, Y., Aktas, M., Erol, O., Ongun, R., Polat, F., 2017. An examinaion aboutthermal capacities of thermoelectric coolers in battery cooling systems. J.Eng.Res.Appl.Sci.6,703–710.
[12] Pesaran, A.A., 2002. Battery thermal models for hybrid vehicle simulations. J. Power Sources 110 (2), 377–382.
[13] Rao, Z., Wang, S., 2011. A review of power battery thermal energy management.Renew. Sustain. Energy Rev..
[14] Shukla, A.K., Sudhakar, K., Baredar, P., 2016. Exergetic analysis of building integrated semitransparent photovoltaic module in clear sky condition at Bhopal India. Case Stud. Therm. Eng..

IOP Conf. Series: Materials Science and Engineering 1126 (2021) 012064 doi:10.1088/1757-899X/1126/1/012064

Investigation on the Mechanical Property and Crack Arresting Mechanism of Natural Jute Fiber used as Reinforcement for Lightweight Concrete Masonry Prism

Vijayalakshmi R*

* Associate Professor, Sri Sivasubramaniya Nadar College of Engineering, Department of civil Engineering, Chennai, Tamilnadu, India
Email: vijayalakshmir@ssn.edu.in

Abstract. Concrete being the widely used material in construction work is strong in compression and brittle in nature. Addition of fibres to the cementitious mix helps to improve the tensile property of the material, particularly after reaching the yield load. In addition, fiber reinforced cement composites exhibit better toughness, crack resistance, flexural capacity, and ductility than non-reinforced cement materials. In developing countries natural fibers are used in the cement composites to produce a low-cost building construction. Previous research works, carried out using sugarcane, hemp, jute, banana and coconut were found to improve the tensile and compressive strength of concrete composite elements. Main aim of this work is to study the performance of jute fiber reinforcement in lightweight foam concrete masonry structure. To study the load deformation patter, jute fiber reinforced masonry prism were subjected to axial compression load. From the stress-strain behaviour it can be concluded that jute fiber reinforcement in foam concrete help to improve the pre cracking behaviour and prevents the formation of major crack plane in masonry unit. Masonry unit with different dosages of fiber were tested and results were plotted. From the study it can be concluded that light weight foam concrete with fiber reinforcement can be used as a better alternative to clay bricks and aerated concrete blocks.
Key words: lightweight concrete, CLC prism, Jute fiber, CLC mortar, Micro fiber

1. Introduction

Cellular Light weight Concrete (CLC) are formed by mixing, cement and fly ash with appropriate amount of water and foaming agent. Addition of foaming agent does not produce any hazardous chemical reaction; it only involves in the generation of tiny air bubbles which will not make the concrete to expand. The concrete product after hardening does not expand and maintain its density. Nowadays CLC blocks have replaced conventional clay bricks in building construction. The energy consumed in the production of lightweight blocks is only a fraction of amount when compared to the red bricks and it does not emit pollutants or any toxic products or by-products [1]. CLC blocks has less dead weight, good thermal insulation and also good mechanical strength. In seismic region, application of lightweight concrete without reinforcement have only less ability to dissipate energy, which is really a concern. A large number of non- reinforced masonry buildings are constructed around the world, such unreinforced building has limited ability to withstand seismic load and undergoes sudden failure. The mode of failure is brittle in nature and sudden, which leads to loss of life and property. In order to minimise such failures and prevent sudden collapse of buildings it is really important to focus on low-cost lightweight masonry structure with better tensile and shear resistance.

In developing countries like India, natural fibre reinforced construction products are much effective, for the development of low-cost building construction [2]. Naturally available fibres such as Roselle, sugar-cane bagasse, sisal, coconut, hemp, and jute are reported to yield improved compressive and tensile strength of the cement-based composites [3-5]. During early days, the naturally available fibers were used for improving the tensile and shear strength of construction. When compare to synthetic fibers, natural fibers are low cost, easily available, highly flexible and produces less health hazards.

On the other hand, synthetic fibers have high strength, absorbs less amount of water and available in low density. From the literature it can be concluded that a considerable amount of work has been carried out in synthetic fiber reinforced light weight concrete, but only a handful amount of work has been done on natural fiber reinforced light weight foam concrete. Therefore, the main aim of this work is to develop a low cost fiber reinforced masonry unit, suitable for construction in seismic region.

2. Literature study

Oraimi and Seibi [6] conducted experiments using nonwoven coir mesh matting in mortar and concluded that coir mesh mortar showed improved performance in composite post-cracking flexural stress, toughness, ductility, and toughness index, compared to plain mortar. Merta and Tschegg [7] carried out work on natural fiber reinforced composites and concluded that the fracture energy was enhanced by the addition of natural fibers. Wang et al [8] studied the tensile property of concrete composited reinforced with acrylic, nylon, and aramid fibers and concluded that the tensile properties of concrete composites were enhanced due to the addition of synthetic fibers. Joshi et al. [9] compared the performance of natural fiber reinforced composites with glass fiber reinforced composites and concluded that natural fiber are much better than synthetic fiber. Fang et al [10] used synthetic fiber to improve the mechanical strength of fiber reinforced concrete and reduced the crack formation in canal lining. Jute fibers when added to concrete, arrest cracks, shares major amount of tensile load and absorbs large amount of fracture energy in concrete composites. [11-14]. Ramaswamy et al. [15] studied the tensile property and breaking strength of jute fiber which was around 2260 kg/cm^2. It had been concluded from many researches that adding jute fibers contributes in the increase of toughness, tensile and flexural strengths of concrete composites ([16,7&14]. The author conducted research work using polypropylene and banana fiber in light weight concrete prism [17&18] and reported positive response in the crack resistance mechanism of CLC prism.

Chandar and Balaji [19] reported that there was around 44% and 12% increase in flexural strength and tensile strength of concrete due to the addition of jute fibers. Rasheed and Prakash [20&21] used poly olefin micro and macro fibers as reinforcement for cellular lightweight concrete. Lightweight concrete prism with different dosages of macro fibers and combination of micro and macro fibers were tested. The density of concrete specimen was around 800-900 Kg/m^3. The micro fibers increases the ductile nature of masonry in the post peak region while the micro fibers arrest the crack at micro level and improves the pre cracking behaviour. Addition of fibers increases the performance of concrete by reducing the permeability, increasing the structural integrity and thereby increasing the durability and service life of concrete structure. Therefore, synthetic and natural fibers increases the ductility and post peak residual strength of composite specimens. From the literature it can be concluded that only limited information is available on the mechanical property of jute fiber reinforced concrete specimen of density around 800-900 Kg/m^3. Therefore, a thorough knowledge is required to study the failure pattern and compression behaviour of Jute Fibre Reinforced Cellular Light Weight Concrete (JFRCLC) masonry prism. The detailed objective of this study is to plot the stress strain behaviour, and finally develop a high strength, low cost fiber reinforced lightweight concrete masonry unit.

3. Test programme

Experimental investigation was first carried out in light weight foam concrete block, cylinders and prism without any fibers, those specimens were treated as control specimen. In the next phase of study, jute fibers were added to light weight foam concrete in four different dosages. Those fiber reinforced CLC blocks and prism were subjected to compression load and the stress strain plot was obtained. The effectiveness of jute fiber as reinforcement in CLC blocks and the energy dissipation capacity of the same was studied. In order to study the property of mortars used in bed joints, mortar cylinder was cast in the proportion of 1:6 and the stress strain plot were obtained. Thereafter, reinforced CLC prisms were cast using jute fiber and the influence of different jute fibre (JF) dosage on the CLC prism were studied. The added jute fiber increases the tensile strength and shear resistance capacity of the CLC prism. The added jute fiber does not play a major role in the compressive strength but it helps to improve the performance of the prism in the post peak region.

3.1 Material properties

Ordinary Portland cement, class F-fly ash, water and foaming agent were mixed in appropriate proportion to prepare the fresh foam concrete mix. The mix proportion of cement: Fly ash: water is 1:3:1. The amount of foaming agent was 1.4 kg/m^3. Water binder ratio is 0.4. Class F fly ash with less lime content was used to prepare CLC. Synthetic foaming agent was diluted in the ratio of 1:40 and then the mixture was fed into the foam generator to produce an aerated mix of density 75 kg/m^3. The maximum dosage of fibers in the mix is around 5 Kg/m^3 of CLC mix, which corresponds to 0.55 % of volume fraction. The different fiber dosages is calculated based on the ratio of volume of jute fiber to the total volume of CLC mix and jute fiber. The jute fiber were cut to a length of 6 - 8 mm (Figure 1) and mixed with CLC. The properties of the fiber are listed in Table 1. Specimens with fiber dosages of 0% (control), 0.22%, 0.33%, 0.44%, 0.55% of jute micro-fibers were cast in different t batch and cured.

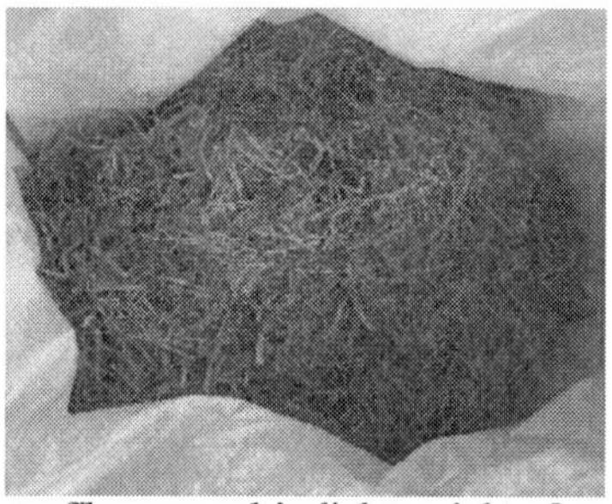

Figure 1 Jute fibers used in lightweight foam concrete

Table1. Properties of Jute Fibers

Property	Raw material	Type	Length of fiber	Width	Water adsorption	Tensile strength	Stiffness
Range	Jute fiber	Single filament	9 mm	0.08 mm	84 %	400-800 Mpa	10-30 KN/mm^2

3.2 Specimen details

Test specimen includes, plain CLC block without fibers, CLC blocks with fibers of different volume fraction, CLC cylinder with and without fibers, CLC prism without fibers, CLC prism with fibers and finally mortar cylinders. Figure 2 shows the dimension details of CLC block prism. The dimension of each block is 600 x 200 X 150. Test specimen details, specimen ID and volume fraction of fiber are listed in Table 2. Series I specimen consist of control Specimen without fiber dosage. Control specimens are CLC block, CLC cylinder, Mortar Cylinder (1:6) and CLC prism. Series II consist of CLC cylinders with different volume fraction of jute fibers. Series III consist of CLC prism built using different volume fraction of jute fiber reinforced CLC blocks. Specimen name JF-mi-0.22% stands for CLC prism built with Jute micro-fiber-with 0.22 % volume fraction CLC blocks.

3.3 Preparation of specimen

Dry raw materials, were added to the mixer and thoroughly mixed for 5 minutes. After thorough mixing, water was added to the mixture and a wet mix was obtained. Then the diluted and aerated synthetic foaming agent was added to the wet mix. The foaming agent was then introduced into the mixture, in increments at the rate of 35 g/s for a duration of 40 sec. For fiber reinforced lightweight mix, jute fibers of length approximately 9 mm were added to the wet mix and mixed for another five minutes. After thorough mixing, the CLC slurry was poured into the assembled mould of required dimension and curing was done. The specimens were demoulded after 24 hours and cured for 28 days. The detailed procedure of manufacturing of lightweight foam concrete blocks in CLC plant is shown in Figure 3. Light weight fiber does not affect the density of CLC blocks. Wee et al [22] has stated that at an optimum air content of around 40%, the strength to weight ratio for foam concrete is achieved. A total void ratio of 0.35 is adopted in this study to arrive at a concrete density of 900 kg/m^3. The water absorption tests revealed that CLC block masonry absorbs more or less same amount of water as clay brick masonry. Four CLC block were joined using 10 mm thick cement mortar to form the prism. The finished CLC prism with and without fibers are shown in Figure 4.

3.4 Test methods

According to ASTM, testing of masonry prism assembly is sufficient to predict the strength of masonry and It is also more economical than full-scale testing. The top surface of the specimen was levelled and uniform smooth surface was formed before applying load. The stress-strain plot was obtained using the reading recorded in Data Acquisition (DAQ) System. Testing of specimen was done in compression testing machine of capacity 2000KN. Fiber board were used on the top of the specimen to ensure uniform distribution of load. Initially Testing of the specimens were done by applying the load at very small increment, after reaching about 75% of the peak load the rate of loading was 0.001mm/sec. The axial deformation was measured using LVDTs along the direction of loading. The test setup is shown in Figure 5.

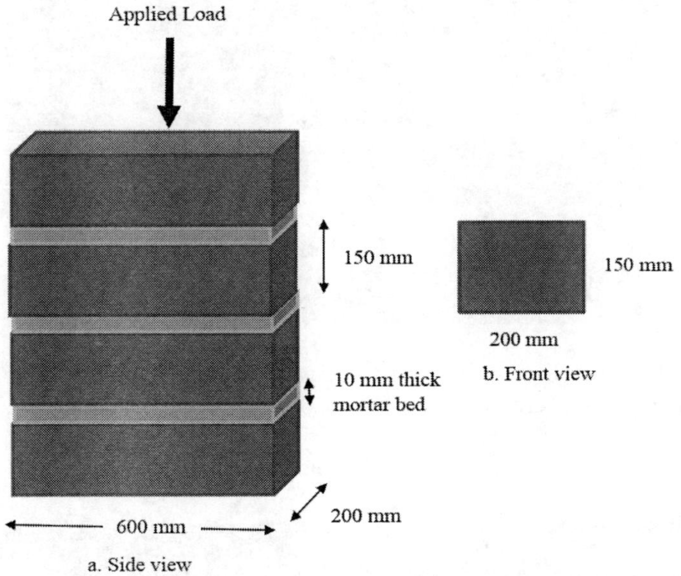

Figure 2. Dimension of CLC blocks prism subjected to compression

Table 2 Details of test specimen and fiber dosage

Specimen type	Series	Specimen ID	Quantity	Volume Fraction (%)
Mortar cylinder	I	Control	3	0
CLC Cylinder		Control	3	0
CLC Blocks		Control	3	0
CLC Prism		Control	3	0
CLC cylinder with jute fibers	II	Cl-0.22%	3	0.22
		Cl-0.33%	3	0.33

		Cl-0.44%	3	0.44
		Cl-0.55%	3	0.55
CLC Prism with jute fibers	III	JF-mi-0.22	3	0.22
		JF-mi-0.33	3	0.33
		JF-mi-0.44	3	0.44
		JF-mi-0.55	3	0.55

(a. synthetic foaming agent; b. adding foam to cement and fly ash; c. CLC mixture; d. CLC mould preparation; e. placing of mixture in mould; f. CLC blocks ready for curing; g. cured CLC blocks; h. CLC cylinders)

Figure 3. Process of manufacture of CLC blocks

Figure 4. Fabricated jute fiber reinforced CLC prism

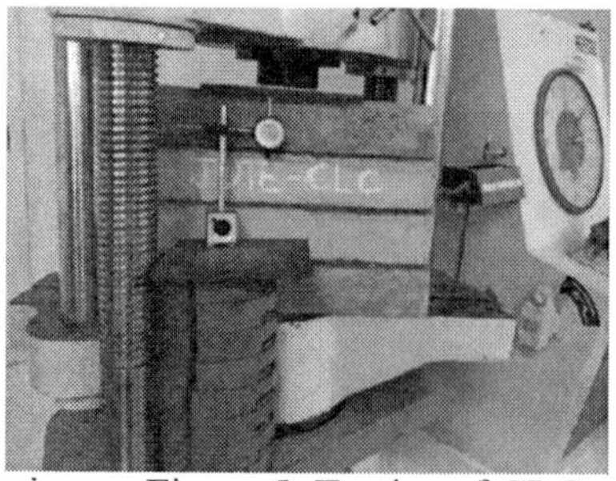

Figure 5. Testing of CLC prism

4. Result and Discussion

Lightweight foam concrete specimen was casted and tested in three series. Series I consist of control specimens without any fibers. CLC cylinder with four different fiber dosages form the series II specimens. Series III had jute fiber reinforcement CLC Prism with four different fiber dosages (0.22%,0.33%,0.44%,0.55%). Three specimens were tested for each case and the average reading was taken. Test results of control specimen, CLC block, CLC cylinder and CLC prisms in compression are presented in Table 3.

4.1 Compression behaviour of Control CLC block and prism

The stress-strain curve for control specimens is shown in Figure 6. Stress-strain plot of Mortar cylinder, CLC block and CLC prism were compared. The elastic modulus of mortar cylinder was much higher when compared to CLC prism, on reaching the maximum load CLC prism showed minimum resistance to the applied compression load and a sudden failure occurred. The stress strain plot of CLC prism showed linear behaviour till it reaches 30% of peak load. After which the crack formation takes place and failure of specimen occurred. The strength of CLC prism (3.7 MPa) is much closer to that of CLC block. The elastic modulus of mortar cylinder was around 12200Mpa while that of CLC block and CLC prism was around 4100 Mpa and 2100 Mpa respectively. From the elastic modulus values, it is clear that mortar cylinders are much stiffer than CLC block and CLC prism. In CLC prism, the blocks are subjected to tri-axial compression and mortar joints are subjected to uniaxial compression and bi-axial tension. The failure started from the mortar joint followed by crack formation which is transmitted to the CLC blocks and results in the failure of prism.

Table 3 Compressive strength of CLC cylinder and CLC prism with and without fibers

Specimen type	Series	Specimen ID	Compressive Strength (Peak) (Mpa)			Average Strength (Mpa)
			1	2	3	
Mortar cylinder	I	Control	4.63	4.52	4.91	4.71
CLC Cylinder		Control	7.3	7.2	6.8	7.1
CLC Prism		Control	3.42	3.52	3.32	3.52
CLC Blocks		Control	3.5	3.8	3.2	3.5
Jute Fiber Reinforced CLC Cylinder	II	Cl-0.22%	7.51	7.43	7.81	7.54
		Cl-0.33%	8.01	7.91	7.54	7.79
		Cl-0.44%	8.40	8.50	8.48	8.46
		Cl-0.55%	8.9	8.91	8.75	8.62
Jute Fiber Reinforced Prism	III	JF-mi-0.22%	3.56	3.77	3.54	3.46
		JF-mi-0.33%	3.61	3.52	3.63	3.61
		JF-mi-0.44%	3.87	3.54	4.1	3.74
		JF-mi-0.55%	4.02	4.03	4.01	4.03

IOP Conf. Series: Materials Science and Engineering 1126 (2021) 012064 doi:10.1088/1757-899X/1126/1/012064

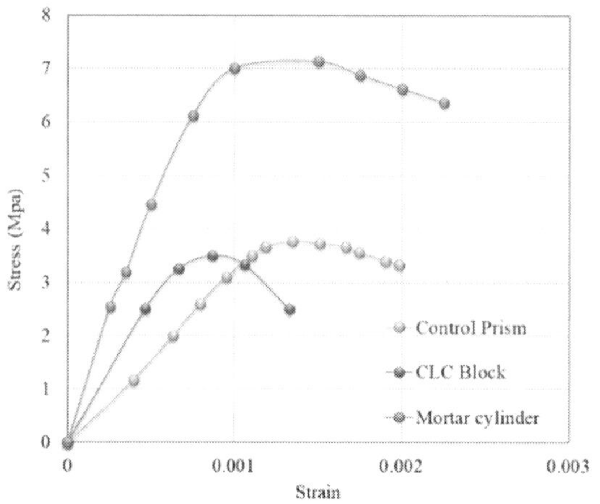

Figure 6 Stress strain plot of CLC block, control prism and mortar cylinder

4.2 Compression behaviour of CLC cylinders

On comparing the compression behaviour of JFRCLC cylinder with control cylinder, the peak strength of jute fiber reinforced cylinder is increased by 12-15% with the addition of fibers. The stress-strain plot of CLC cylinders is shown in Figure 7. In unreinforced CLC cylinders (CLC cylinders without fibers) when subjected to compression load, the stress strain behaviour was linear up to 30 % of the peak load after that the curve become nonlinear due to lack of fiber reinforcement. Resistance offered by the prism to the applied load decreases with the increasing axial load particularly after the peak load, and collapsed suddenly. The pre-cracking behaviour was same for both control CLC cylinders and JFRCLC cylinder. A mild increase in peak load carrying capacity and increase in elastic stiffness was observed with the increase in fibre dosage. As the load increases, after reaching the peak value, the load remains constant which indicates a less amount of degradation or increased stiffness after post peak region (Vijayalakshmi and Ramanagopal 2020 (a)). The peak load in JFRCLC cylinder increased up to about 20% with respect to that of cylinders without fibres. The CLC cylinder reached upto to a maximum of 4-8Mpa strength.

4.3 CLC prism with different jute fibre dosage

The stress-strain curves for prisms reinforced with different dosage of jute fibres are compared in Figure 8. Jute fiber reinforcement, increase the elastic modulus of CLC prism and reduces the stiffness degradation in the post peak region and provides a softening behaviour to the constructed prism. The values of elastic modulus of fiber reinforced CLC prism lies between CLC block and mortar cylinder. As the fiber dosage increases, the peak strength of CLC prism also increases, this increase in peak strength is mainly due arresting of cracks at the micro level and results in the increased performance after post peak region. The residual load carrying capacity and toughness is increases due to the addition of fibers in lightweight foam concrete. Therefore, fiber reinforced light weight foam concrete can be used as a better alternative for clay bricks.

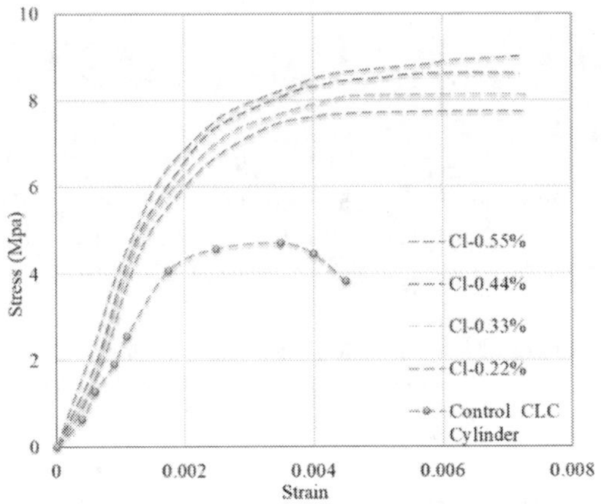

Figure 7 Stress Strain plot for JFR CLC cylinders

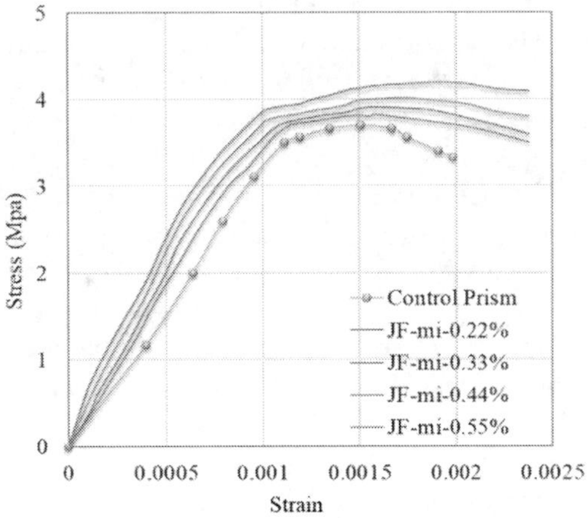

Figure 8 Stress Strain plot for JFRCLC prism

4.4 Crack formation and failure pattern

Autoclaved Aerated Blocks (AAB) are subjected to high temperatures during the manufacturing process, so addition of fibers is not feasible in AAB. The fiber added to AAB will melt at such high temperature, therefore light weight foam concrete block with fibers is a better alternative. In the control prism the crack formation was a single explicit crack (Figure 9). Stress concentrations in particular region led to the formation of wider cracks which cannot be distributed to the entire section of the prism due to the absence of fibers. In control prism the mortar joints form the weak section and crack formation starts from the joint. The crack are then transmitted to the CLC blocks and sudden failure of prism occurs. In case of fiber reinforced CLC, the jute fibers involves in the crack arresting mechanism and distributes the cracks at micro level (Figure 10). once cracked formed in the matric of CLC, jute micro fibres get engaged in the crack arresting process, as a result, large number of micro cracks emerges. The formation of major cracks is arrested by the fibres in the CLC blocks which forms a closed network. Fiber reinforcement also improves the post peak strength of CLC. Therefore, it can be concluded that fiber helps to improve the strength and ductility of CLC matrix under compression load, shear and tension load. From this study it can be suggested that fiber reinforced CLC can be used as reinforced masonry unit in seismic zone.

Figure 9. Failure mode of Control CLC prism

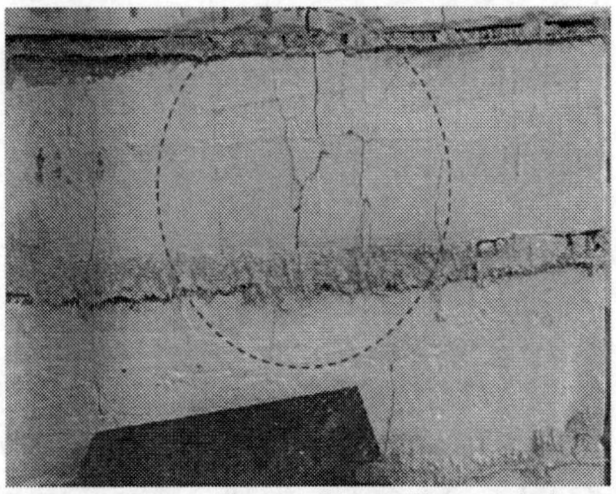

Figure 10. Failure of jute fiber reinforced CLC prism

5. Conclusion

The effect of adding natural jute fibres to lightweight foam concrete was tested by adding different volume fraction of fiber to CLC block and testing the CLC prism under compression load. The following conclusion was derived from the study:

- The behaviour of CLC prism was more or less same as that of the CLC cylinder when tested under compression load.
- The peak strength of CLC prism with different fiber dosage increase with the increase in the volume fraction of fibers. A maximum of 15% increase in peak strength was observed for 0.55% VF when compared to control prism.
- Failure in CLC prism without fiber was sudden and a single explicit crack was formed across the section of prism. But in case of jute fiber reinforced CLC prism a large number of micro cracks was formed and formation of major weak plane was prevented. Jute fibers embedded in the matrix of CLC, distributes the crack and increases the peak strength of prism. It also increases the stiffness of prism in the post peak region.
- Jute fiber reinforcement converts the CLC block to exhibit a composite behaviour. Addition of fibers improves the stiffness and strength of CLC block which is more or less similar to mortar cylinder.
- Addition of fibers increases only 10% of the total manufacturing cost, which in turn helps to improve the overall life and durability of the structure.

REFERENCES

1. Yang KH and Lee KH (2015) Tests on high-performance aerated concrete with a lower density. Constr. Build. Mater. 57(4): 109–117.
2. Ali M (2016) Use of coconut fibre reinforced concrete and coconut-fibre ropes for seismic-resistant construction. Mater. Constr 66(32): e073.
3. Cook DJ, Pama RP and Weerasingle HL (1978) Coir fiber reinforced cement as a low-cost roofing material. Build. Environ.13(3):193–198.
4. Ramakrishna G, and Sundararajan T (2005) Impact strength of a few natural fiber reinforced cement mortar slabs: a comparative study. Cem. Concr. Compos 27(5): 547–553.
5. Ali M (2014) Seismic performance of coconut-fibre-reinforced-concrete columns with different reinforcement configurations of coconut-fibre ropes. Constr. Build. Mater. 70: 226–230.
6. Oraimi SK, Seibi AC (1995) Mechanical characterisation and impact behaviour of concrete reinforced with natural fibers. Compos. Struct. 32 (4):165– 171.
7. Merta EK and Tschegg (2013) Fracture energy of natural fiber reinforced concrete. Constr. Build. Mater 40: 991–997.
8. Wang Y,Backer S and Li VC (1987) An experimental study of synthetic fiber reinforced cementitious composites. J. Mater. Sci. 22(12): 4281–4291.
9. Joshi SV, Drzal LT, Mohanty Ak and Arora S (2004) Are natural fiber composites environmentally superior to glass fiber reinforced composites. Composites Part A. 35(3): 371–376
10. Fang HG, Liu, PH, and Zhang T (2011) Experimental investigation on the polypropylene fiber concrete performance of yellow river canal lining in the middle line of south-to-north water transfer project. Appl. Mech. Mater 52:1987– 1991.
11. Gupta ND, Paramasivam P and Lee SL (1978). Mechanical properties of coir reinforced cement paste composites. Hous. Sci. 2(6): 391–406.
12. Mansur MA, and Aziz MA (1982) A study of jute fiber reinforced cement composites. Int. J. Cem. Compos. Lightweight Concrete. 4(2): 75–82.
13. Zhou X , Ghaffar SH, Dong W, Oladiran O and Fan M (2013) Fracture and impact properties of short discrete jute fiber-reinforced cementitious composites. Mater. Des. 49: 35–47.
14. Zakaria M, Ahmed M, Hoque MM and Islam S (2017) Scope of using jute fiber for the reinforcement of concrete material. Text. Clothing Sustain. 2 (11).
15. Ramaswamy HS, Ahuja BM and Krishnamoorthy S (1983) Behaviour of concrete reinforced with jute, coir and bamboo fibers. Int. J. Cem. Compos. Lightweight Concrete 5(1): 3–13.

16. Liu B, Zhang LZ, Liu QX and Ji T (2013) Study on behaviors of jute fiber reinforced cement-based materials. Appl. Mech. Mater. 25(3): 508–511.
17. Vijayalakshmi R and Ramanagopal S (a) (2020) Compression behaviour of polypropylene fibre reinforced cellular light weight concrete masonry prism. Civil and Environmental Engineering Reports 30(1): 145-160.
18. Vijayalakshmi R and Ramanagopal S (b) (2020). Experimental Investigation on Banana fiber reinforced lightweight Concrete Masonry Prism sandwitched with GFRP sheet. Civil and Environmental Engineering Reports 30(2): 15-31.
19. Chandar SP and Balaji CJ (2015) Experimental study on the mechanical properties of concrete mixed with jute fiber and steel fiber. Int. Res. J. Eng. Technol. 1(2):77–82.
20. Rasheed MA and Prakash SS (2015) Mechanical behaviour of sustainable hybrid- synthetic fiber reinforced cellular light weight concrete for structural applications of masonry. Construction & Building Materials 98: 631–640.
21. Rasheed MA and Prakash SS (2018) Behavior of Hybrid-Synthetic Fiber Reinforced Cellular Lightweight Concrete under Uni-axial Tension - Experimental and Analytical Studies. Construction and Building Materials 162: 857-870.
22. Wee TH, Babu DS and Tamilselvan TLH (2006) Air-void systems of foamed concrete and its effect on mechanical properties. ACI Materials Journal 103(1): 245–52.

Density, Molar Volume and IR Studies of Lithium Doped Bismuth Lead Borate Glasses

Sangeeta B. Kolavekar[a*] and N. H. Ayachit[b]

[a]Department of Physics, KLE Technological University, Hubli-580031.Karnataka, India.
[b] Center for Material Science, KLE Technological University, Hubli-580031. Karnataka, India.
Corresponding author: *sangeeta_k@kletech.ac.in

Abstract. Structural analysis of $15Bi_2O_3$-(50-x)PbO-$35B_2O_3$-xLi_2CO_3, x = 0,5,10,15 and 20) of lithium doped bismuth lead borate glass system (BPBL) using density, molar volume, and IR absorption spectra has been carried out for These glasses were prepared by the most reliable conventional melt quenching technique. Densities and molar volumes have been measured as a function of Li_2CO_3 content. Almost linear variation in the density and the molar volume has been observed in the present glasses and is attributed to the role of Li^+ in these glasses. IR spectra of these glasses exhibit characteristic vibrations due to the borate network. The band at 400 cm^{-1} - 600 cm^{-1} due to absorption refers to metallic and cationic vibrations of metallic cations of Bi^{3+} and Li^+.

1. Introduction

Glasses made using ionic conducting materials like Li+, Bi^{3+}, etc., have gained broad technological importance [1]. The addition of different alkaline metals, transition metals, and alkaline earth leads to the formation of suitable structures in the borate glasses, making these glass structures promising materials for various technological applications [2, 3]. Heavy metal oxides doped glasses containing lead, bismuth, etc., are of possible attention due to their exclusive optical properties like high RI and excellent non-linear properties. Also, studies of alkali borate glasses cover numerous industrial and technical uses [4]. Further, it has been reported that lead borate glasses have many applications comprising radiation shields, optical and thermal properties [5]. Similarly, bismuth-containing oxide glasses have advantages such as host materials for lasers, long IR cut-off, glass ceramics, memory, and switching devices [6]. A lot of work on the physical and optical properties of borate and phosphate glasses has been reported by Sangeeta et al. [5-11]. Like borate glasses, bismuth glasses also show interesting structural features. The present paper is based on the structural study of lead borate glasses containing 15 Bi2O3-(50-x) PbO-35B2O3 -x Li2CO3 (x = 0, 5, 10, 15, and 20.) to study the role of Li2CO3 in structure modifications. Already the structural part of Bi_2O_3 in $15Li_2CO_3$-(50-x) PbO-35B2O3 -xBi2O3 (x=0, 5, 10, 15, and 20.) has been reported in IOP conference Proceedings [5].

2. Experimental approach

2.1 Sample Synthesis

The glass system of $15Bi_2O_3$-(50-x) PbO-$35B_2O_3$-xLi_2CO_3, x = 0, 5, 10, 15, and 20 was selected and samples were synthesized using most reliable conventional melt-quench technique. Starting materials like lead monoxide (PbO), boric acid (H_3BO_3), lithium carbonate Li_2CO_3 of AR grade, and bismuth oxide (Bi_2O_3) have been used to prepare the samples. Through mixing and grinding of the chemicals is for about ½ an hour using agate mortar. The batch was moved into a porcelain crucible kept in a muffle furnace (electric) at 900°C for about ½ an hour, depending on composition. Details are referenced already [9-10]. Highly viscous melt was then poured into brass molds. Light yellowish-colored glasses were formed.

2.2 Density and Molar Volume

At room temperature, using toluene as the immersion liquid, density was measured for all glass samples. Applied Archimedes principle to measure the density of the present glasses [5, 8].

$$\rho = \frac{W_A}{W_A - W_L}\rho_b \quad (1)$$

Can calculate Molar volume of the glass samples from the following expression,

$$V_m = \frac{M}{\rho} \quad (2)$$

Here, V_m is molar volume, ρ is the density of the sample and M is the molecular weight of the sample.

3. Results and Discussion

3.1 Density and Molar Volume

Figure.1 shows the Variation of density (ρ) and molar volume (V_m) with mole% of Li_2CO_3 for all BPBL glass samples. For the present samples, the density is found to decrease with an increase in the Li_2CO_3 concentration. Molar volume is found to increase with increasing concentration of Li_2CO_3 compared to LPBB glasses reported earlier [5]. Values are listed in TABLE.1.This trend indicates there is a direct bearing on the glass network. Initially, a decrease in density is due to inter substitution Li on PbO and Bi_2O_3 (heavier element by lighter), which modifies the network [5] and increases the number of non-bridging oxygen's (NBOs) [8]. However, at a high concentration of Li2CO3, there is a clear indication of the modification of the network (increase in molar volume), which can be associated with the increasing number of bridging oxygen (BO) groups [11].

TABLE 1. Composition (mole%) with sample code, density and molar volume values of $15Bi_2O_3$-(50-x)PbO-$35B_2O_3$-xLi_2CO_3, x = 0, 5, 10, 15, and 20) glass system

Sr. No.	Composition Li_2CO_3 mole%	Molar Mass (M) g	Density (ρ) g/cm^3	Molar Volume (V_m)
1	0 (1BPBL)	144.52	7.06	20.47
2	5 (2BPBL)	156.54	6.66	23.50
3	10 (3BPBL)	168.75	6.52	25.88
4	15 (4BPBL)	180.87	6.33	28.57
5	20 (5BPBL)	192.99	6.05	31.89

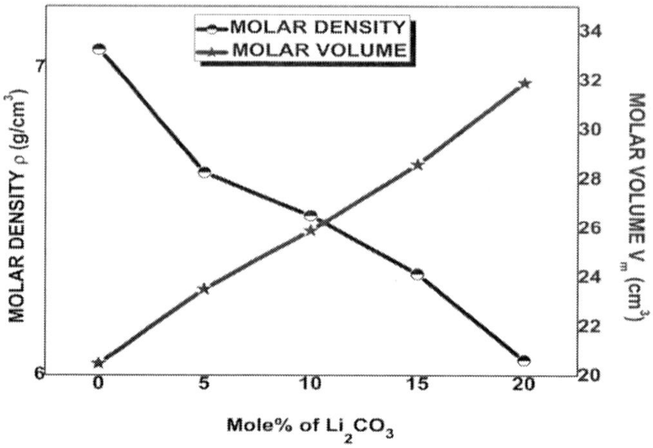

Figure1. Molar density and molar volume of present BPBL glasses as a function of Li_2CO_3 content.

3.2 IR Spectra

Structural analysis of the prepared samples was done using the IR Spectra noted at 300K using the KBr pellet method in the region of wave number 400 cm-1–2000 cm-1, as shown in Figure.2. The region 698-804 cm-1 represents the bending vibrations of B-O-B, 1100 cm-1-1200 cm-1 relates to vibrations of [BO3] triangles, and the region 1000 cm-1-1100 cm-1 corresponds to vibration in [BO]4 units [5, 12]. The vigilant observation indicates that the band intensities steadily increase with the increase in Li_2CO_3 content, which indicates an increase in the population of bridging oxygen (BOs). Further, the bands' width increases, leading to enhanced randomness in the network with an increase in Li_2CO_3 concentration. The spectra also show a shifting of bands with increasing concentration of Li_2CO_3 due to vibrations by mixed bodings such as Li-O-B, Pb-O-B, etc. The band at 400- 600 cm-1 due to absorption refers to metallic and cationic vibrations of metallic cations of Bi^{3+} and Li^{2+} [13]. The bands give B-O-B bending vibrations of BO3 structural units of the network at 600-800 cm^{-1}[5, 14]. All the investigated glass samples show prominent peaks corresponding to the borate network. Different modes of vibrations of Bi-O are due to hexahedral (BiO6) units [15]. The band at 804 cm-1 is because of the bonds of stretching vibrations Bi-O in Bi_2O_3. The presence of BiO3 and BiO6 units of Bi2O3 represents the role of bismuth as network former and modifier [13]. The exchange of BO4 units into BO4 structural units leads to structural changes described from the IR band area enclosed due to BO3 and BO4 structural units decreasing the density, causing an increase in NBOs. Bands around 1266 cm^{-1} prove the existence of Bi-O-Bi groups of vibrations of [BiO6] octahedral units [5]. Peaks at 400 cm^{-1} are due to coop vibrations of Pb^{2+}, Bi^{3+}.and Li^{2+} [5, 13].

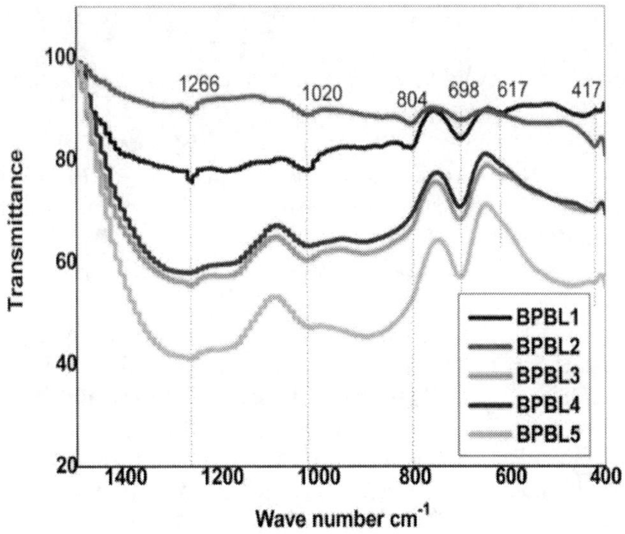

Figure 2. I.R spectra of investigated BPBL glass system.

4. CONCLUSIONS

Prepared glasses are light yellowish. The density of the glasses varies inversely with mole-concentration of lithium carbonate in the present BPBL glasses in comparison to LPBB glasses reported earlier [5]. While their molar volume values increase linearly with Li_2CO_3 concentration in present BPBL glass samples in contrast to LPBB glasses said more before [5]. The FTIR study of ($15Bi_2O_3$-(50-x) PbO-$35B_2O_3$-xLi_2CO_3, x = 0, 5, 10, 15, and 20) glasses show that the network structure of the present samples is primarily based on the BO_3 and BO_4 units and PbO_4 units. In the lithium bismuth borate glass samples (BPBL glasses), the band at 400- 600 cm^{-1} due to absorption refers to metallic and cationic vibrations of metallic cations of Bi^{3+} and Li^{2+}. The exchange of BO_4 units into BO_4 structural units leads to structural changes described from the IR band area enclosed due to BO_3 and BO_4 structural units decreasing the density, causing an increase in NBOs. Peaks at 400 cm^{-1} are due to coop vibrations of Pb^{2+}, Bi^{3+} and Li^{2+}. The IR bands 400 cm^{-1} in BPBL glasses indicate the existence of coop vibrations of Pb^{2+} groups. The addition of PbO and Bi_2O_3 leads to the conversion of [BO_3] units to [BO_4] units in borate glasses; this indicates PbO and Bi_2O_3 acts as a glass modifier, tending to decrease the density and increase in molar volume, causing an increase in BOs.

ACKNOWLEDGMENTS

Authors thank KLE Society, Principal, BVBCET, Hubballi, and Dean R&D, KLE Technological University, and Hubballi for providing financial sustenance through Capacity Building Fund Projects TEQIP grants. Thanks to Physics department, KLE Technological University, Hubballi for support.

References

[1] R . Komiya, A. Hayashi, H. Mrimoto, M. Tatsumisago, T. Mianmi, Solid State Ionics, 140(2001) 83.

[2] I. D. Pye, V. D. Frechette, N. J. Kreidl, Borate glasses: structure, Properties, Applications, Springer Science and Buisiness Media, (2012) 617-622.

[3] S. Cetinkaya Colak, I. Akyuz, F. Atay, on the dual role of ZnO in zinc-borate glasses, journal of non-crystalline solids, 432,(2016) 406-412.
[4] Rajyasree Ch, Krishna Rao D (2011). J. Non Cryst., Solids 357, 836–41.
[5] S. B. Kolavekar and N. H. Ayachit, IOP Conf. Series: Journal of Physics: Conf. Series 1455 (2020) 012018
[6] S. B. Kolavekar, N.H.Ayachit, Journal of materials science: Materials in Electronics 30(1), 432-449.
[7] S. B. Kolavekar, RV Anavekar, N.H Ayachit, bulletin of material science, 40(3), 523-526.
[8] S. B.Kolavekar and N. H. Ayachit, Journal of materiomics, 5 (3), (2019) 455-462.
[9] S. B. Kolavekar, N.H.Ayachit, G. Jagannath, K. Naga Krishnakanth and S. Venugopal Rao, Optical Materials, 83 (2018) 34–42
[10] S. B. Kolavekar, N.H.Ayachit, R. Rajaramakrishnac, Pramod N G and J. Kaewkhao, Journal of Luminescence, 226 (2020) 117498
[11] S. B. Kolavekar and N.H.Ayachit, Materials Chemistry and Physics 257, (2020)123796.
[12] Motke SG, Yawale SP, Yawale SS, Bull Mater, (2002), Sci 25, 75–8.
[13] A. Yadav, S. Khasa, M. S. Dahiya, S. Dalal, A. Hooda and A. Agarwal. Phys. Chem. Glasses: Eur. J. Glass Sci. Technol. B 57(3), (2016), 146-152.
[14] C. Bocker , Christian Russel journal of the European ceramic society 29, (2009), 1221-1225.
[15] P. Narwal, M. S. Dahiya, A. Yadav, A. Hooda, A. Agarwal, S. Khasa, Ceram. Int. 43, (2017),11132-11141

Stress evaluation of CSZ and SiAlON under Thermo-Mechanical loading condition

M. Vigneshwaran[1], S. Srisanthosh[2], J. Nandhakumar[3], R. Madhubalan[4], Vaisak R Chandran[5]

[1]Assistant Professor, Department of Mechanical Engineering, Sri Krishna College of Engineering and Technology, Coimbatore, India, vigneshwaranm@skcet.ac.in
[2, 4, 5]PG student, Sri Krishna College of Engineering and Technology, Coimbatore, India 19eped004@skcet.ac.in, [4]19eped003@skcet.ac.in, [5]19eped005@skcet.ac.in
[3]UG student, Sri Krishna College of Engineering and Technology, Coimbatore. 18eumc519@skcet.ac.in

Abstract— This work deals with the Thermo- mechanical analysis of Thermal barrier coating over Aluminum 2026 alloy through Finite Element Method in ANSYS 15.0R software. The coating material is Nano composite (metal matrix). It consists of double ceramic layer (TC1, TC2) with a layer of bond coat (BC) between the substrate (SUB) and the ceramic layers. The top Ceramic layer is of a new kind of refractory material with better oxidation resistant and corrosion resistant above 1000oC. The inside ceramic layer is well known for thermal resistant. In order to give a better bonding strength, the grain size of the coating powder is maintained at Nano scale. A Finite Element model has been developed to identify the intermittent temperature of the coatings under thermal loading conditions. From this FEA result the model was subjected to thermo-mechanical stress analysis to find out the stress distribution over the coatings and the substrate. We analyzed with three different temperatures in the same model. The result shows that the double ceramic layer will reduce the Substrate temperature effectively. The interfacing area and the edges of the coating experience more stresses when compared to other area. The paper reveals that the coated material's stress concentration is very less when compared to the uncoated material.
Keywords—Thermal barrier coating, Finite Element Method, Nano composites, Double ceramic layer, Thermo-Mechanical loadings.

I. INTRODUCTION

The resources like fossil fuels are depleting faster rate recently. This demands improving the efficiency of high temperature components which is used in the application of automobiles, aerospace and power industries. When a material is subjected to a challenging thermal environment it must have good material properties. The increasing temperature in these environments will increase the efficiency and at the same time the life time of the material is reduce [6]. The applications such as automobile engines, Gas turbines are operating in a high temperature environment facing such problems [1]. In order to sustain the materials in such conditions Thermal Barrier coatings is provided to those materials. Though ceramic materials having good thermal resistant properties we are going that material for coating [2]. Thermal barrier coating applied to the combustion chamber reduces the fuel consumptions as well as pollution and also improve the fatigue life of the component. [9] The factors influencing the life time of the coating are formation of oxides, corrosion, and thermal mismatches etc. [1]. One single coating cannot able to solve such problems, so we are going for a Double layer.

Thermal barrier coatings are normally made of a top coat with zirconium oxide and a metallic bond coat. In this paper the top coat is of two layers. The first layer (TC1) is made of synthesized SiAlON a new kind of refractory material, the second layer (TC2) is of CSZ (ZrO_2 7-8 Wt. % CeO_2 and that of bond coat is NiCrAlY. Double ceramic layer will effectively protect the substrate material. Research shows that the total of 450nm thickness of coating including bond coat gives a better results [5]. Coating can be applied by mainly two process. They are diffusion process and overlay process. However the plasma spraying process is comes under the overlay process which gives a better results in ceramic coatings [1]. During thermal spray deposition processes a high amount of heat is transferred to the substrate material and the coatings. This can lead to a distribution of residual stress in these area [11]. The coating sintering temperature also plays a role in the property of the coating material, it shows that coating sintering temperature makes more contribution to the thermal conductivity of the Nano structured coating material. [12].

II. Experemental Procedure

In this model first we investigated the temperature distribution in the coating with three different temperatures. From this result we switch over to thermo-mechanical stress analysis and the inputs are from the previous thermal analysis. Doing this we calculated the stresses present in each layer. A finite element numerical model was created and analyzed with one side is maintained at room temperature and the other side is exposed to three different temperature.

III. Simulation Parameters

There are so many ways to analyze the temperature distribution of thermal barrier coatings by numerical method. For the simplification of our problem consider the 3D model as a 2D-plane model. In this the Heat transfer is occurred along the thickness direction of the coating. For simulation the element type used in ANSYS 15.0 is, for thermal analysis SOLID Q 4 node 55 and for structural analysis SOLID Q 4 node 182 are used.

a) *Physical Model*

The Fig.1. Shows the physical model of the TBC coated Aluminum 2026 substrate. In this heat conduction and heat convection mechanisms are considered. From the surrounding to the first layer (TC1) convection takes place and heat conduction takes place at the intermittent layers.

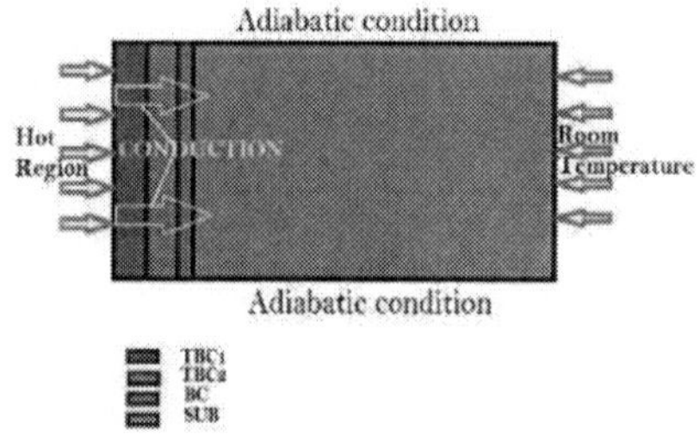

Fig.1

b) *Finite Element Model And Boundary Conditions*

The finite element model was shown in Fig.2. it consist of four layer, Top ceramic layer(TC1), inside ceramic layer (TC2), bond coat layer(BC) and a substrate layer(SUB). The thickness of the coatings is taken from previous literature. In this model the heat transfer is occur in the thickness direction. For analyzing this model in ANSYS the following assumptions are made.

i. All layers are Homogenous and isotropic.
ii. The top and bottom of this model is considered as adiabatic condition.

IOP Conf. Series: Materials Science and Engineering 1126 (2021) 012066 doi:10.1088/1757-899X/1126/1/012066

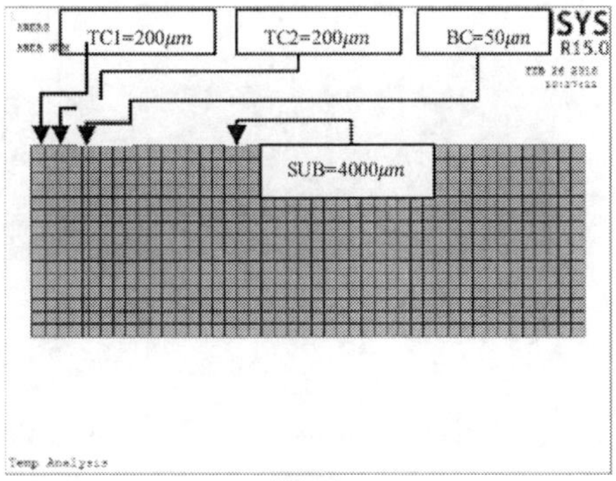

Fig.2

Three different temperature of500°C, 700°C, 1000°C are taken into consideration. The convective heat transfer Co-efficient in the hot side is 3000 W/m^2 k and the other side is maintained at 30°C and the respective heat transfer co-efficient is 2500 W/m^{2}k.

c) Material Properties

The thermal conductivity of the top layers TC1 and TC2 are calculated by using the formula

$$\frac{K}{K0} = 1 - \frac{4}{3} \in$$

which is presented in the literature [5].

K is the effective thermal conductivity; K0 is the thermal conductivity of dense body and ∈ is the porosity. The remaining are taken from others work [6].

Table I. Material Properties

Properties	TC1	TC2	BC	SUB
Conductivity w/mk	11	0.2	32.1	190
Density gm/m^3	3.24	6.2	8.21	2.77
Expansion co- efficient x 10^{-6}/°c	3	10	13.8	22.9
Elastic modulus GPa	250	200	190	72.4
Poisson ratio	0.28	0.27	0.3	0.33

IV. Result And Discussion.

a) Thermo mechanical analysis at 500oC

Left side of the model is subjected to 500°C and the other side is at room temperature. There is no heat transfer takes place at the upper and the lower side of the model. The Fig.3. Shows the temperature distribution and their respective stress present in the model at 500°C.

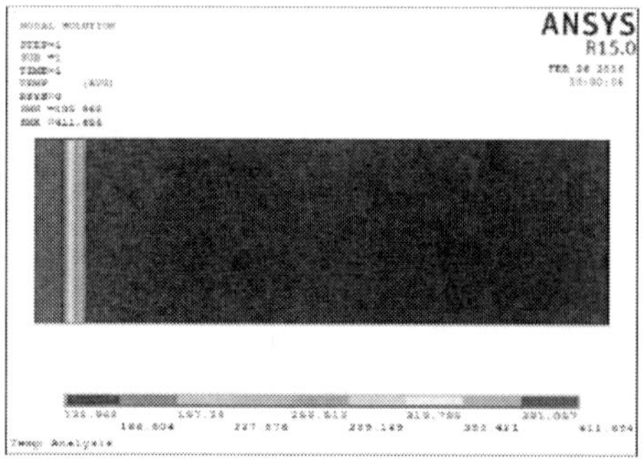

Fig.3(a). Temperature Distribution at $500^{O}C$

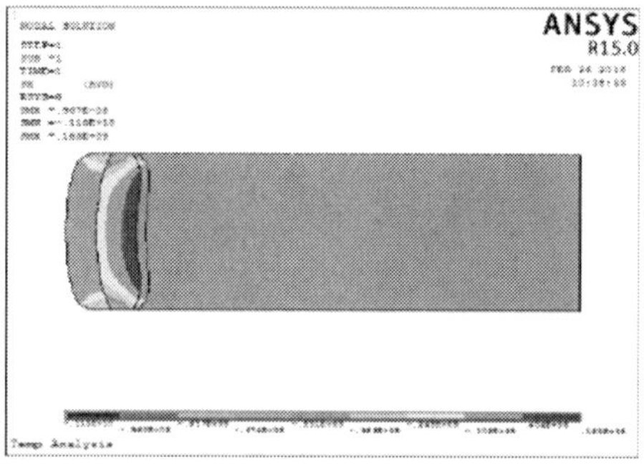

Fig.3(b) stress in axial

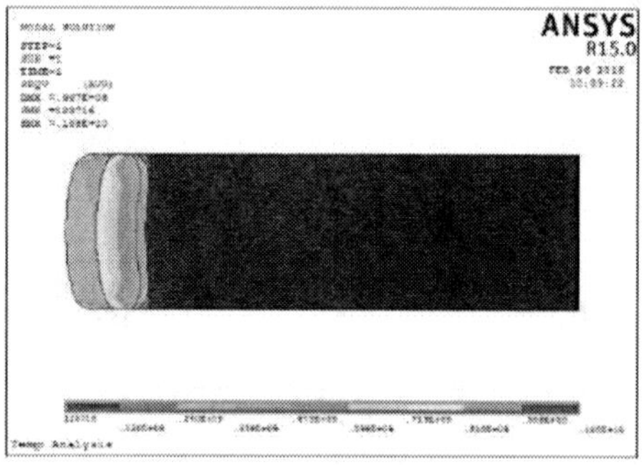

Fig.3(c) vonmisses stress

There is no micro cracks formed in this stress concentration. Almost the base body's temperature is room temperature only. Presence of coating at this temperature effectively reduces the stress intensity factor in the substrate. The stress formed in the interface is mainly due to the result of different in the value of co-efficient of thermal expansion of the materials.

IOP Conf. Series: Materials Science and Engineering 1126 (2021) 012066 doi:10.1088/1757-899X/1126/1/012066

Table II. Temperature Distribution

Variable	TC1	TC2	BC	SUB
Temp oC	381	289	227	166

b) Thermo-mechanical analysis at 700oC

Fig.4. shows the temperature distribution and their respective stresses present at the temperature of 700 ^{O}C.

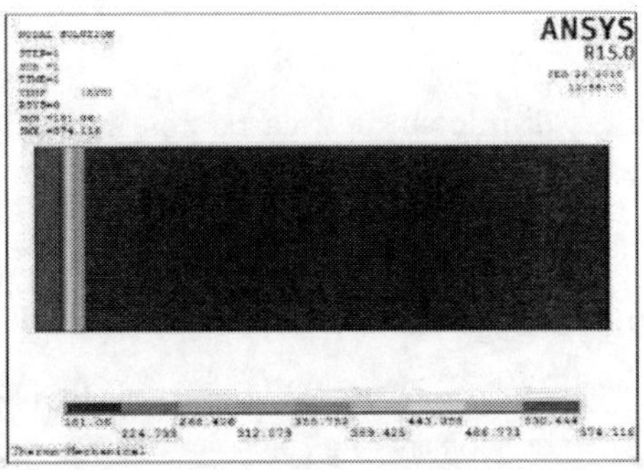

Fig.4(a). Temperature Distribution at 700°C

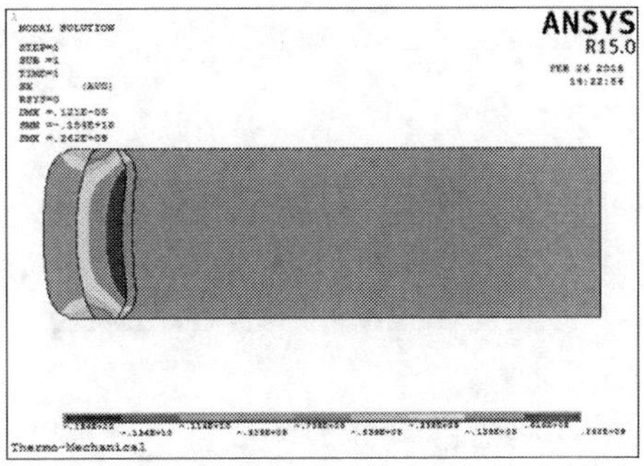

Fig.4(b) stress in axial

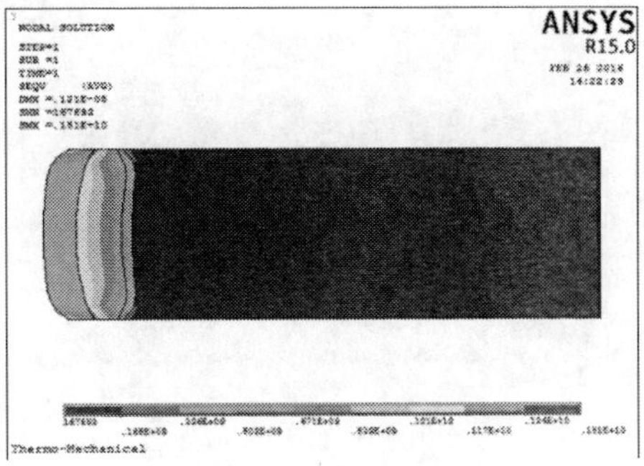

Fig.4(c) vonmisses stress

Table III. Temperature Distribution

Variable	TC1	TC2	BC	SUB
Temp oC	530	399	312	224

The interface temperature between the BC/SUB and BC/TC has random in nature. The idealized result of the single simulation may not give the real behavior of the coating. It is observed that

the failure will comes in the interfaces between the SUB and the BC during high temperatures. The interfaces stress between the SUB and the BC experiences less amount. This range will not affect the base material to fail.

c) Thermo mechanical analysis at 1200oC

A transient coupled temperature-displacement is carried out in this model. A finer mesh is applied in the intermittent layer where the stress is maximum. For this temperature range considering the absence of the crack propagation because results shows that above 1200°C only crack will propagate in coating. The influence of the thermal distribution cycle of the thermal barrier coating also not important so we are considering a single load cycle. In the first stage of the temperature analysis is the actual working condition of the IC engine outlet port, because the outlet temperature is below 500°C. Temperature at 1000°C we are considering only temperature and stress distribution without considering oxidation, up to this range no oxides forms in the coating. Fig.5. shows that the temperature distribution and their respective stresses present at the temperature of 1000 ^{o}C.

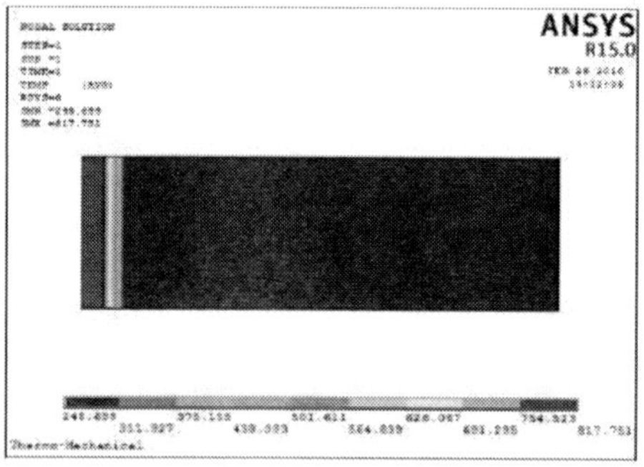

Fig.5(a). Temperature Distribution at 1000°C

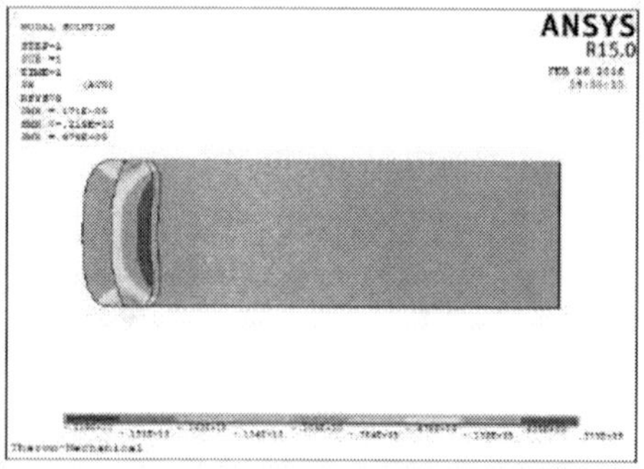

Fig.5(b) stress in axial

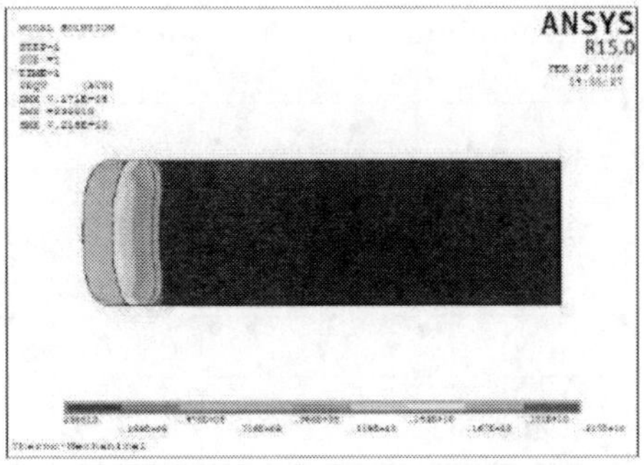

Fig.5(c) vonmisses stress

Table IV. Temperature Distribution

Variable	TC1	TC2	BC	SUB
Temp oC	754	564	438	311

To ensure the model is working properly, a comparison between with coating and without coating also analyzed under the same thermo- mechanical loading condition. In the first stage we identified the critical stress area. In the second stage the size of mesh in the interface area is increased.

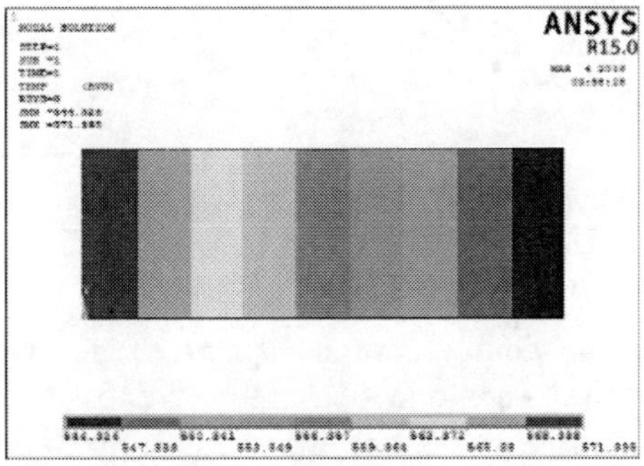

Fig.6(a) Temperature distribution in SUB

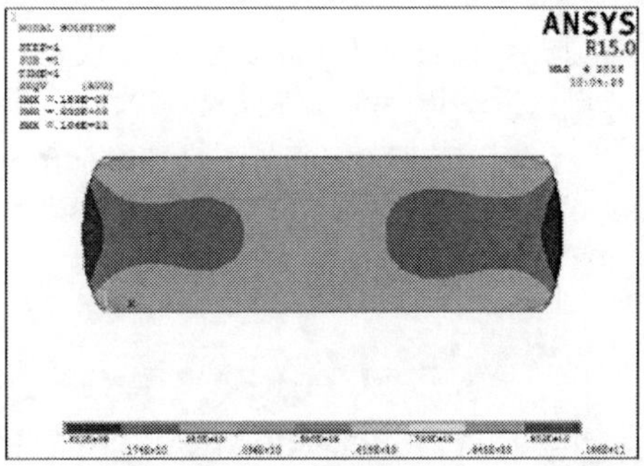

Fig.6(b) Von misses stress in SUB

Table V. Maximum and Minimum Vonmisses stress

Temp ^{0}C	Max Stress(GPa)	Min Stress(GPa)
500	1.08	0.000123716
700	1.51	0.000167693
1000	2.15	0.000235813
1000 (Only SUB)	10.6	0.632000000

Fig.7. shows the Temperature vs Stress graph at 500^{0}C, 700^{0}C, 1000^{0}C. It is noted that the temperature range increases the stress experience by the material also increases.

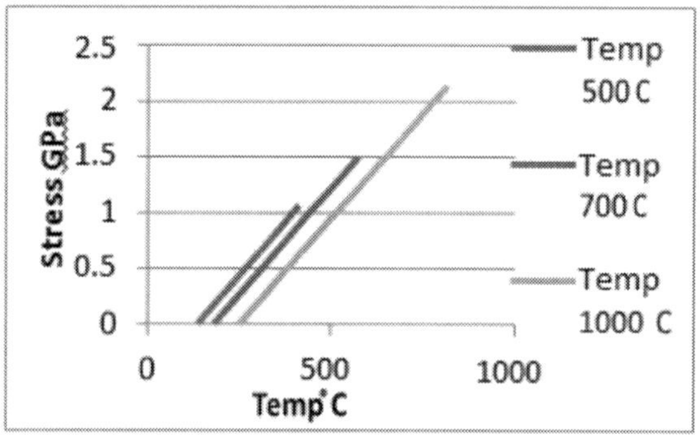

Fig.7. Temp vs Stress

V. CONCLUSION

In this study we investigated the double ceramic layer coating applied to the aluminum substrate. Thermo-Mechanical stress analysis in x-direction and y-direction are taken into account. It is found that stress in x-direction is a tensile stress and the stress in y-direction is a compressive stress. The compressive stress role is very less so it is neglected. From the above figure it is identified that the stress in the Bond coat is more when compared to the Top coat layer. From the results we came to know that CSZ having better thermal resistant and it gives a better result even in 1000^{0}C also. The first layer SiAlON effectively reduce the stress due to the effect of this temperature. We conclude that the Double ceramic layer of CSZ and SiAlON will give it optimum performance above 1000^{0}C also. However the maximum stress obtained by this model is less and these stresses will not affect the working condition of the base material.

VI. FUTURE WORK

The same model presented in this paper is fabricate, and this will be analyzed to find the thermal resistant, oxidation resistant and corrosion resistant of the coatings experimentally.

APPENDIX

TBC-Thermal Barrier Coating.
TC- Top Coat
BC- Bond Coat SUB- Substrate
CSZ- Ceria Stabilized Zirconia.

ACKNOWLEDGMENT

"Author thanks the management of Sri Krishna College of Engineering and Technology Coimbatore for their valuable support for this project."

REFERENCES

[1] R.Rajendran (2012) "Gas Turbine Coatings-An overview" in *Engineering Failure Analysis*, ,pp.355- 369.

[2] S.C.Sharma, N.M.Gokhale, Rajiv Dayal, Ramji Lal (2002) "Synthesis microstructure and mechanical properties of Ceria stabilized tetragonal Zirconia prepared by spray drying technique" in *Material Science* vol 25,No.1, pp. 15-20.

[3] Meng Han, Guodong Zhou, Jihua Huang, ShuHai Chen, (2013), " A Parametric study of double-ceramic layer thermal barrier coatings", in *Surface and Coating Technologies*, 236, pp.500-509.

[4] C.Zhang, K.Komeya, J.Tatami, T.Meguro, Y.B.Cheng (2000), " Synthesis of Mg-α SiAlON powders from talc and halolysite clay minerals", in *Journal of European ceramic society 20*, pp.1809-1814.

[5] L.Wang, X.H.Zhong, Y.X.Zhao, S.Y.Tao, W.Zhang, Y.Wang, X.G.Sun, (2014), "Design and optimization of coating structure of the thermal barrier coating fabricated by atmospheric plasma spraying via Finite element method" in *Journal of Asian Ceramic Society, vol 2*, pp.102-116.

[6] M.Rezvani Rad, G.H.Ferrahi, M.Azhadi, M.Ghodrati, (2015), "Stress analysis of thermal barrier coating system subjected to out of phase of thermo-mechanical loadings considering roughness and porosity effects", in *Surface and coating technology* 262, pp. 77-86.

[7] A.K.Saini, (2013),"Simulated thermal analysis of 8Y- PSZ/NiCrAlY coating over ALSI alloy", in International journal of Advanced research *in Engineering and Applied Sciences*, vol.2.No.8.

[8] Natalja Zilinska, Ilmars Zalite, JanisKrastins, (2012),"Investigation of production of fine grained SiAlON ceramic from nano powders", in *Material science vol.18*.No.3.

[9] O.Kovaric, P.Hausild, J.Siejl, Z.Pala, J.Matejicek, V.Davidov, (2014), " The influence of plasma sprayed multilayer of Cr2O3 and Ni10Wt.%Al on fatigue resistant", in *Surface and coating technology* 251,pp.143-150.

[10] Irina Bannikova, Sergey uvarov, Marina Davidova, Oleg Naimark,(2014),"Study of ceramic tube fragmentation under shock wave loading", in Procedia Material Science.vol.3.pp.592-597.

[11] Ming Song, Yue Ma, Sheng Kai Gong,(2011),"Analysis of residual stress distribution along interface asperity of thermal barrier coating system on macro curved surface", *in Progress in natural sciences: Materials international*.vol 21.PP.262-267.

[12] Wang Kai, Peng Hui, GUO Hongbo, Gong Shengkai.(2012)," Effect of sintering on thermal conductivity and thermal barrier effects of thermal barrier coatings", *in Chinese journal of aeronautics* 25. Pp.811-816.

[13] Hasan Koruk a, JasonT.Dreyer, RajendraSingh a (2013) on "modal analysis of thin cylindrical shells with cardboard liners and estimation of los factor", mechanical systems and signal processing

[14] T. Subramani, Athulya Sugathan (2012) on "Finite Element Analysis of Thin Walled- Shell Structures by ANSYS and LS- DYNA", Volume Issue.4, July-Aug. 2012 Pages 1576-1587

[15] Matin Latifi, Fatemeh Farhatnia, Mahmoud Kadkhodaei (2013) on "Buckling analysis of rectangular functionally graded plates under various edge conditions using Fourier series expansion", Volume 41, September–October 2013, Pages 16–27

A Composite Material an alternative for manufacturing of Automotive Disc Brake: A Review

Milind K Wasekar[1,2] **& Mohan P Khond**[1,]

[1]College of Engineering Pune,India
[2]wasekarmk14.mech@coep.ac.in.

Abstract. The brake disc is an indispensable component of an automobile especially in this trending era of electric mobility, where gasoline engines are superseded by electric vehicles in which prominent criteria is to lower down the weight to enhance the performance of the vehicle. Aluminum Metal matrix composites (MMC) play an important role because of researchers as these are not only used in automotive industries but aircraft and locomotives also. In most vehicles, these brakes are made up of Cast Iron which has well anti-wear properties and cheap too. But in certain other cases such as high-performance vehicles, these brake discs are not up to their standards of high performance because the cast iron brake discs are heavy and so reduces the vehicle's performance to a particular extent. This paper reviews on fundamental function of Disc brake and the crucial challenges found in this review are, the problem of evolution of hazardous wear particles in an atmosphere, thermoelectric friction problem during braking, friction-induced vibration which causes squeal, the behavior of abrasive particles, and dynamic stability analysis, etc. an alternative to cease these problems possibly is to adopt novel material

1. Introduction

Disc brakes: Shoes or pads contract on the outer surface of a spinning disc, having compressive frictional force. The brake disc is a rounded metal disc on which the pads are mounted. It typically consists of material made from cast iron. Depending on the application, amount of exposure, thermal properties of the material and the amount of heat dissipation desired when brakes are applied, and the total mass to be stopped, the configuration of disc brakes varies. Based on its workings functions of brakes can be summered as:

1.1 Function of brakes: The basic functions of the brake system must be described predictably. Circumstances, at fair cost and lifetime of brake wear, thus providing directional stability and reasonable use of tire-road friction. All relevant safety requirements must comply with the braking system. During Safety requirements are considered minimum performance in most situations and Specifications. Brakes must operate safely in all fairly imaginable operating conditions, including slick, wet, and dry roads; with a vehicle that is light or fully loaded, straight or turning brakes, new or worn brakes, beginner or professional drivers, on smooth or rough roads. Mainly brake functions can be divided as follows:

1.1.1 Slowing and/or Stopping: Large values of thermal conductivity, specific heat, and density for brake rotors yield lower swept

1.1.2 Maintaining Speed on a Downgrade: The potential energy is transformed into thermal energy by the brakes during constant-speed downgrade operation, resulting in increased brake temperature. As long as the operating conditions are such that the steady-state brake temperature reached is less than the brake-specific critical temperature vehicle will be able to safely descend the downgrade.

1.1.3 Holding Stationary on a Downgrade: Keeping a vehicle stationary primarily depends on the transmission of force between the application lever and brake tires, or mechanical advantage. Security in Safety In general, requirements requires a defined capacity for hill-holding. However, a parking brake can be used when the service brake has failed in an emergency.

In the present era, metals are extensively used, and there are many ways of producing a product using only pure molten metal or some other phase of the metal. When it comes to production processes, the following are the most common ones used in large industries Previously, these brake discs were made by melting the molten metals, according to the application and feasibility. the main drawback of this method was brittleness and high density. To get rid of this challenge, researchers observed that by adding smart contents into the molten metals these difficulties may be eliminated. Hence following materials are proposed for the manufacturing of the brake discs.

- Metal Matrix Composites (MMC)
- Carbon-Carbon Composites
- Aluminum Fly ash (ALFA) composites
- Titanium Alloys.

After the brake disc shape is achieved by heating and cooling the mixture of the parent materials, the addition of other contents into the matrix is carried out to harden the brake disc, resulting in a new material known as composite Material. While manufacturing of brake disc various properties that are required to be considered are as follows:

1.2 Properties for brake disc rotor:

The temperatures induced by friction heating are directly related to the properties of the material. Apart from sliding speed, brake disc material properties need to be addressed. In addition to these, materials should possess the following specifications

a. Corrosion-resistant

b. Ability to withstand high temperature

c. Lightweight,

d. Long life,

e. Low noise,

f. Stable-friction

g. Low wear

h. Fair cost.

i. Imperviousness for atmospheric conditions.

A comprehensive account of the literature review is given in the report. To classify the previous research activities and directions relevant to our focal field. The review-related topic is discussed in chronological order so that the underlying pattern of the development of thoughts and ideas in that area is also suggested. Care was taken to preserve the original terminology used by the authors, to the Freepost address, to retain the originality of the opinions. The crucial challenges found in this review were the problem of evolution of hazardous wear particles in an atmosphere, thermoelectric friction problem during braking, friction-induced vibration which causes squeal, the behavior of abrasive particles and dynamic stability analysis, etc. The analysis is summarized below.

2. Researcher's Experimental approaches

Mikael Eriksson et al., [2001] observed that the contact region between a brake pad specimen and a glass disc generated great insights into the dynamics and mechanics of the contact situation. The softer and more porous areas around the contact plateaus are worn out by three-body abrasion. A significant number of small particles (wear debris) are transported through the narrow labyrinth between the surfaces while the disc is slipping against the plateaus, milling down the weaker constituents.

K.Shin et al., [2002] the purpose of the present research were to use the basic model to analyze the effect of damping on a disc-brake system's stick-slip vibration. When the natural frequencies of a pad and disc are nearby, a brake is more likely to be noisy. The amount and distribution of damping in the system is a key factor. To know the mechanisms of brake noise two Degree of freedom model was taken into consideration. It was observed that the damping of the disc is as significant as that of the pad. To illustrate different limit cycles in step space, non-linear analysis is often performed. Further, the machine can become more unstable and therefore noisy by adding damping to either the disc or the pad alone.

N.M. Kinkaid et al., [2003] in the automotive field, disc brake squeal remains an elusive issue. Many researchers have investigated the issue with experimental, computational, and computer techniques since the early 20th century, but there is still no way to effectively suppress disc brake squeal. It is well recognized that a non-linear model is expected to address the cause of instability by the in-depth study.

Ji-Hoon Choi, In Lee, [2004] the study aims for thermoelastic characteristics of carbon-carbon composite disk brakes. It was noted that, due to uniform and mild pressure distributions, the orthotropic disk brakes can provide better brake output than the isotropic ones. This outcome can be used as a design tool for improving braking performance.

M. Kermca, et al., [2005] In this study new Material has been introduced into a braking system in which performance of temperature and coefficient of friction were analyzed It was observed that The temperatures produced by ceramic-carbon composite C/C SiC and metal-matrix composite (MMC) contacts are very high; they reached more than 800 ^{0} C under the test conditions, which is more than two times higher than traditional contacts with grey cast iron. Compared to grey cast iron, the coefficient of friction in contacts with ceramic-carbon composite C/C SiC was two times higher and more uniform wear.

P. Liu, et al. [2007] demonstrated the new functionality of ABAQUS/Standard, which allows for a nonlinear analysis to identify the stability of brake systems. The study suggests that to investigate the effects on the disc squeal of device various parameters, such as the hydraulic strain, the rotational velocity of the disc, the coefficient of friction of the contact interactions between the pads and the disc, the stiffness of the disc, and the stiffness of the pads backplates are responsible. The simulation results suggest that the disc brake squeal could be responsible for major pad bending vibration. By decreasing the coefficient of friction, improving the stiffness of the disc, using damping material on the backplates of the pads, and changing the shape of the brake pads, the squeal can be decreased.

Zmago Stadler, et al.[2007] In this paper, Sintered metallic brake wear and friction linings were studied on a composite brake disc C/C-SiC. The addition of SiC increases by more than 50 %. The hardness of the base metallic lining, while its effect on the mean COF is less important. Despite the higher hardness, the wear rate of this lining material is 30 % greater than the base metallic lining. In comparison, the addition of graphite significantly decreases the hardness of the padding (by 40 %) Since graphite is a lubricant, the COF of the lining increases by 10%, possibly because of the lower hardness. Also, it shows a lower wear rate.

M. Nouby, And K. Srinivasan [2009] FEM is commonly used for simulation and prediction of disc brake squeal using complex Eigenvalue analysis where positive part indicates corresponding Eigenmode which was found to be unstable. So to reduce squeal various modifications have been made such as changing the young's modulus of the backplate, applying chamfers to the pad. It has been discovered that squeal can be reduced by, decreasing the stiffness of the backplate of the pad and by providing different chamfers at the pad side.

Anders Soderberg and Soren Andersson [2009] this paper explore how wear can be predicted using general-purpose finite element analysis tools for the pad-to-rotor interface. To measure the pressure distribution in the pad-to-rotor touch, a three-dimensional finite element model of the brake pad and the rotor is developed. A wear simulation method is used to simulate the wear of the brake pad under steady-state drag conditions based on a simplified form of Archard's wear law and explicit Euler integration. Wear at the pad side is been calculated.it has resulted that deepest Wear occurs with high pressure in the field, although almost there is wear. No, wear around the edge at all. After approximately 1600 revolutions, the method a steady state seems to have reached where all points on the wear rate of surfaces are the same without any change in the distribution of pressure or the surface profile. The distribution of steady-state pressure indicates a small negative gradient. In the radial direction, correction for sliding variance Spacing over the interface.

Jens Wahlströma, et al.[2010] conducted research to identify airborne particle effects due to wear in the atmosphere. As a piece of test equipment, a pin-on-disc tribometer fitted with particle counting instruments was used. The findings suggested that the low metallic pads cause more wear than the NAO (Non-Asbestos Organics) pads to the rotor material, resulting in higher airborne wear particle concentrations. Similar size distributions have been obtained, although there are variations in the measured particle concentrations. Independent of pad content.

M.A. Maleque, et al. [2010] in this study the material assessment tools for the design and application of automotive brake disc were developed. For the initial screening of the candidate materials using Ashby's selection chart of materials, the functional properties of the brake discs or rotors were considered. The digital logic method showed the highest performance index for AMC 2 material and was listed as the optimum material among the candidate brake disc materials. From a technological and economic point of view, higher friction coefficient and lower density are advantageous.

Ali Belhocine Mostefa Bouchetara [2012] The study was carried out for computational simulation of the thermal behavior of a full and ventilated disc in the transient state employing the computer code ANSYS three variants of materials were taken into consideration i.e. cast iron (AL FG 25, FG 20, FG 15) and obtained thermal behavior for a determined braking mode. It was derived that several parameters are responsible for the quality of outcomes such as several elements, Braking mode, and materials.

Faiz Ahmad et al., [2013] it has been demonstrated that aluminum alloy reinforced with alumina particles (AMC) were tested for various conditions of wear at a different speed. Weight loss was found to be 52% less in the case of AMC compared to Cast Iron due to reinforcement of hard alumina particles which increases stiffness, strength, and fatigue resistance of AMCs.

Dattatray Galhe and Vishnu Wakchaure [2013] in this study it was concluded that the brake squeal is caused by various parameters. Such as friction-induced vibrations, braking pressure, temperature, and roughness, which are directly proportional to the vibrations produced during braking. Checked here are grey iron grade 250, high carbon grey iron, and lightweight graphite iron used in the brake disc rotor. The braking operation requires low braking pressure from these lightweight graphite iron discs. This material is also proposed for the brake disc rotor.

Zhiqiang Li, et al. [2014] there are various fatigue cracks on the surface of the brake disc, including crackles, radial cracks, and circumferential cracks. Conduct initiation and dissemination under various braking conditions. radial cracks arise and spread on the friction surface of 300 km/h brake discs. Only crackles occur at 200 km/h after repeated emergency braking and no crack propagation during normal braking has been observed. Breaking fatigue crack tests showed that oxides were covering the broken surface of fatigue cracks. The form of the surfaces of the fractures was elliptic.

A.A. Yevtushenko [2014] the research was carried out to compare the transient temperature fields of the pad and the disc, the braking time, and the braking distance at a constant and temperature-dependent friction coefficient. It was concluded that it did not take into account the temperature dependence of the friction coefficient for the studied materials, the component dimensions of the brake, and the operating parameters. Braking distance and wear, although the maximum temperature remains almost unchanged. Due to its low thermal conductivity value for the FC-16L pad material, the high temperature is observed in the vicinity of the friction surface. Cast Iron material was used for the Disc surface.

Ali Belhocine et al. [2015] concluded that the best for the current application is the ventilated style disc brake. Based on the parameters of strength and rigidity, the brake disc design is therefore stable. Compared to the trailing side, interaction pressure is expected to be greater on the leading side and its value increases significantly. Compared to the trailing side, interaction pressure is estimated to be better on the leading side and its value increases significantly with an increase in the speeds of disc rotation. In terms of the variance of the coefficient of friction, there is no important improvement in disc-pad deformation. When the device is in the thermomechanical coupling, the total stresses, and deformations in the disc increase in a notable way.

R.K.Pohane and S.C.Kongare [2016] in this study, the material selection methods for the design and application of automotive brake disc were developed. Based on the highest material performance index listed among the Alternative Brake Disc, Ti-64 Material was found to be an optimal material. Furthermore, the PROMETHEE Rainbow showed that Ti-64 is superior in good compressive strength, higher friction coefficient, density, lightweight, good thermal ability, economically viable, and lacking optimum hardness, the Performance Aggregated Score for Ti-64 was found to be highest.

Lijesh K.P. et al. [2017] observed that a desirable alternative to replacing the traditional disc brake is the Magneto-Rheological (MR) brake. The performance of the experiment was assessed by calculating the weight loss of the blocks before and after the tests. Different hardness discs were suggested to obtain a reliable braking performance of the MR brake, as the brake-disc hardness affects the MR brake performance. MR fluid particles deform the softer disc surface and deteriorate the output of the brakes. To reduce the wear of the disc, an increase in disc-hardness is required. The disc of 63 HRC has lesser impingement of iron particle, wear, and surface cracks compared to the 15HRC disc. Higher hardness of disc leads to enhance of the braking effect

N.K.Kharate and S.S.Chaudhari [2018] The goal of this research was to establish the FE model of each part of the disc brake system to understand the impact of material properties on the squeal of the disc brake and to validate it with experimental modal analysis. The purpose of the current analysis was to predict the squeal using a more practical model at the previous stage of invention. For four distinct materials such as Gray C.I-1, Gray C.I-2, carbon-ceramic, and steel, modal brake rotor analysis has been performed as it is one of the critical components of the brake system that contributes more to the Noise Generation. For the rotor, the first ten modes were extracted to predict the natural frequencies The FEM and EMA results suggested that the various material properties of the components of the disc brake components play a sustainable role in the development of the brake squeal. It was found that, Concerning the number of modes and disc stiffness, the ratio between Young's module and density increases the normal vibration frequencies of the disc.

Satish Pujari and S. Srikiran [2019] To make the brake pads, the use of asbestos content is avoided as it is dangerous and poisonous. It also contributes to numerous health conditions, such as asbestosis, mesothelioma, and lung cancers. Natural fibers such as Palm Kernel are proposed to replace these brake pads. Natural fibers such as Palm Kernel (0-50%), Nile Roses (0-15%), and Wheat (0-10%) can be substituted with additives such as aluminum oxide (5-20 %) and graphite powder (10-35%)and 35% phenolic resin is used as a binder. by using these contents composite material have been formulated and tested for its various properties like physical, mechanical tribological. It has been observed that to increase the friction coefficient, particulate Nile roses are used and wheat powder is used to decrease the wear rate. In nature, aluminum oxide and graphite are abrasive. This helps to generate high friction coefficient brake pads with low noise emissions and a lower wear rate.

M.Vijaya and K.Srinivas [2019] Stated that the Liquid Metallurgy route with distinct weight percent of reinforcements effectively-prepared AA 6351/SiC/Gr hybrid composite and AA6351/SiC metal matrix material. The density of reinforced Al/SiC & Al/SiC/Gr hybrid composites with different weight fractions has been measured and the density increases with SiC and decreases with SiC/Gr hybrid particles so that this Al/SiC/Gr hybrid composite can be regarded as a suitable lightweight material for all applications. Also, it has been concluded from the results of various specimens that, 6 wt.% SiC and Graphite reinforced AA6351 to be the best of the tested materials for an experiment.

Vikas Verma, et al. [2019] study involves the processing through the stir casting route of Al MMC reinforced with fly ash and Al_2O_3 particles in various proportions. Their wear activity was further investigated at varying loads (15-25 N), time (3-5 minutes), and velocity (700-900 rpm). Electron micrograph (SEM) scanning of treated composites showed uniform dispersion of coarser particles and agglomeration. It was found that an increase in the hardness of 48 %, the strength of 63%, and bending strength of 58%, and substantial reduction in weight loss calculated after wearing 5 % fly ash and 12 % Al_2O_3 processed Al composite. SEM pure Al alloy micrograph worn surface showed crack and fracture presence at high load, longer time, and at high speed. In composite, the presence of the adhered oxide layer is observed with 5% fly ash and 12% fly ash Al_2O_3, which limited the composite to more wear.

A.A.Agbeleye et al.,[2020] research aim to see how different weight fractions of aluminosilicate clay affect the mechanical, friction, and wear properties of Al-clay composites used in brake disc rotors. The UTS of the composite samples increases as the addition of clay increases to a peak value of 133.98 MPa at 15 wt.% addition of clay particles over the conventional AA6063 of 104 MPa. The addition of clay particles was increased to 30 wt.% clay, resulting in a decrease in the UTS to 122.29 MPa.The Al-clay composite's Vickers hardness values were calculated. An increment in the percentage of clay particles added was accompanied by an increase in hardness values. At 15 wt.%, it reached a high of 76.7 HV, after which the hardness value dropped to 67.3 at 30 wt. % Strain values typically decrease as clay particle addition improves, with a minimum value of 5.7 percent at 20 wt. percent.

Matheus Henrique Pires Miranda et al., (2020) during the interaction of braking, the temperature plays an important role because it influences the coefficient of friction between the disc and the pads on direct proportionality. Both velocity and pressure are inversely proportional to the coefficient of friction, so a decrease in either one causes an increase in the coefficient of friction, rendering the device vulnerable to squealing.

3. Conclusion

It can be noted here that the quality of the mechanical product, such as the disc brake, depends greatly on the attribute of the material it is made of. While several studies have been undertaken to enhance disc brake efficiency, the deprivation of standardized material presents countless challenges for the automotive industry. There is a good chance that the disc brakes currently used will be replaced by composite material, which has many advantages over them. Such as good thermal capability, good compressive strength, stiffness, and lightweight, etc. Therefore, it is evident from the literature review that a comprehensive analysis to assess the reliability of disc brakes made up of various types of composite material must be undertaken.it can be summarized that Aluminium Material Matrix composite with the addition of fly ash gives better strength, heat stress stability. Apart from these merits addition of fly ash and SiC contents reduces greenhouse gases which are evolved during the manufacturing of Aluminium alloy.

4. References

[1] A.A.Agbeleye, D.E.Esezobor,, S.Abalogun, J.O.Agunsoye, J. Solis and. Nevill *Journal of King Saud University – Science* **32** 21–28.,2020
[2] A.A.Yevtushenko, P.Grzes *International Communications in Heat and Mass Transfer* **57** 341–346. 2014
[3] Ali Belhocine, Abd Rahim Abu Bakar & Mostefa Bouchetara,.*Australian Journal of Mechanical Engineering*, 2015 Vol.14, No.1, 26–38
[4] Anders Soderberg and Soren Andersson, *Wear* **267** 2243–2251,2009
[5] Dattatray Galhe and Vishnu Wakchaure, *International Journal of Science, Engineering and Technology Research*, vol. II issue III 2013
[6] Faiz Ahmad, S.H. Jason Lob, Muhammad Aslama, Ahmad Haziqa, *Procedia Engineering* **68** 674 – 680,2013
[7] Jens Wahlstroma, D.Gventsadze, L.Olander, E.Kutelia & Ulf Olofssonaet, *Wear* **268** 763–769,2010
[8] Ji-Hoon Choi & In Lee, *Wear* **257** 47–58,2004
[9] K. Shin K., J. E. Oh, M. J. Brennan & C. J. Harris, *Journal of Sound and Vibration* **254**, 837-848.2002
[10] Lijesh K.P Deepak Kumar and Harish Hirani, *Engineering Failure Analysis* **74** 228–238.2017
[11] M. Kermca, M. Kalina & J. Vi˘zintina, *Wear* **259** 1079–1087,2005
[12] M. Nouby and K. Srinivasan, *Proceedings of International Conference on Recent and Emerging Advanced Technologies in Engineering* (ICREATE-2009). Kuala Lumpur, Malaysia
[13] M.A. Maleque, S.Dyuti & M.M. Rahman, *Proceedings of the World Congress on Engineering* Vol III, London, U.K.,2010
[14] M.Vijaya and K.Srinivas, *International Journal of Recent Technology and Engineering* vol. 8 issue 1, 2019.

[15] Matheus Henrique Pires Miranda, Rômulo do Nascimento Rodrigues,Roberto de Araújo Bezerra Pierre Maurice Christophe Lamary & Raimundo Arrais de Oliveira Neto, *Journal of the Brazilian Society of Mechanical Sciences and Engineering* 42:239,2020
[16] Mikael Eriksson, John Lord & Staffan Jacobson , Wear **249** (2001) 272–278.
[17] N.K.Kharate & S.S.Chaudhari, *Materials Today: Proceedings* **5** 4986–4994.2018
[18] N.M. Kinkaid, O.M. O'Reilly & P. Papadopoulos, *Journal of Sound and Vibration* **267** 105–166.2003
[19] P. Liu, H. Zheng C. Cai, Y.Y. Wang, C. Lu, K.H. Ang & G.R. Liu, *Applied Acoustics* 68 603–615.2007
[20] R.K.Pohane and S.C.Kongare, *International Journal of Advanced Research and Innovative Ideas in Education*, Vol 2 Issue 3.2016
[21] Satish Pujari and S.Srikiran, *Defence Technology* **15** 295-299.2019
[22] Vikas Verma, P. C. Tewari, Roshan Zameer Ahmed,& Syed Youseef Ahmed, *Procedia Structural Integrity* **14** 68-77. 2109
[23] Zhiqiang Li, Jianmin Han, Zhiyong Yang & Like Pan, *Engineering Failure Analysis* **44** 272–284. 2014
[24] Zmago Stadler, Kristoffer Krnel, Toma z & Kosmac , *Wear* **265** (2008) 278–285

IOP Conf. Series: Materials Science and Engineering 1126 (2021) 012068 doi:10.1088/1757-899X/1126/1/012068

Latent heat energy storage system using phase change materials and techniques for their performance improvement: A Review

S S Mundra[1], **S S Pardeshi**[2]

[1]Research Scholar, Mechanical Engineering, College of engineering Pune, MS, India
[2]Associate Professor, Mechanical Engineering, College of engineering Pune, MS, India

ssm.mech@coep.ac.in

Abstract. Entire world is struggling to explore newer methods to overcome the energy crisis. With limited and exhausting energy resources researchers around the world are finding efficient ways to store thermal energy. One of the efficient ways is to store thermal energy in the form of latent heat energy using phase change materials (PCMs). Latent heat storage (LHS) units have been widely adopted owing to superior energy storage density and constant operating temperature. But the thermal performance of these systems is limited due to low thermal conductivity of the PCMs. Researchers across the globe working on approaches to improve the thermal performance of these systems. The present review paper is concerned with a review on PCMs, methods to improve performance of LHS system using PCMs with special emphasis on use of various additives such as nano particles and porous materials to increase the thermal performance of PCMs and types of PCMs for specific applications. The factual data presented in this paper would be helpful in selection of appropriate PCM for specific application, and to adopt suitable performance enhancement technique to obtain most optimal LHS system.

Key words: Solar energy, Energy storage and recovery, Porous material, PCM-Nano composite, thermo physical properties, Thermal Conductivity

1. Introduction

Owing to ever increasing demand of energy and increasing levels of green house emissions, energy conservation and storage become more imperative. Many researchers are working towards effective harnessing of renewable energy for replenishing the depleting levels of energy. Solar energy is considered an important and abundant source of energy but efficient harnessing is still a challenge. Another way of meeting energy demands is energy storage in various forms in a way that it can be reused. For example storing and reusing different energies such as solar thermal energy, waste heat energy from processing plants, energy from electronic devices. One effective way is thermal energy storage (TES) by use of PCMs. Amongst available TES methods, LHS is most widely adopted methods due to salient benefits such as good energy storage density, excellent repeatability and little variation of temperature associated with the process. LHS uses PCM as thermal energy storage medium. PCM undergoes phase change from solid to liquid and vise versa, and alternatively absorbs and releases energy.

Successful implementation of LHS is highly dependent on configuration of energy storage device and properties of PCMs such as thermal conductivity, specific volume, heat of fusion etc. Challenges associated with system and PCMs need to be addressed systematically to get high performing LHS systems. Available TES systems, PCMs for specific applications have been exhaustively investigated and reviewed by Sharma et. al. [1]. PCMs were classified by Rathod et. Al [2] into inorganic, organic and eutectics. Thermal stability of PCMs was reviewed by them and also carried out repeated thermal cycling tests to create holistic database. Experimental investigation of five grades of paraffin wax was conducted by Ukrainczyk et [3] to determine thermo-physical properties. Thermal properties such as specific heat capacities, melting temperature and latent heat were measured by DSC tests. The issue of density change in liquid and solid phase, corresponding buoyancy effect was also addressed in detail.

PCMs typically are characterized by low thermal conductivity that drastically slows down the rate of charging and discharging. Lot of research work has been undertaken in recent years towards improvement in performance of LHS system by different adopting techniques. A detailed review has been carried by Jegadheeswaran and Pohekar [4] about various performance enhancement techniques reported in literature. A comparative review on PCMs, their characteristics and improvement techniques was done by Y.B. Tao et al. [5]. Certain research gaps were also identified by them. The review had classified enhancement techniques into certain categories and suitable suggestions were made for suitable selection of enhancement technique for particular applications.

2. Phase Change Materials (PCMs) for Energy Storage

PCMs are essential component of LHS as they store latent heat very efficiently. However they should possess certain desirable thermo physical, chemical and kinetic properties to be classified as energy storage materials[6][7]. PCMs can be organic, inorganic and eutectics as shown in Figure1. Organic PCMs can undergo large number of melting and solidification cycles without degradation of thermal properties. Their thermal behavior do not change even after its repeated use. Fatty acids and paraffins fall in this category are mostly used.

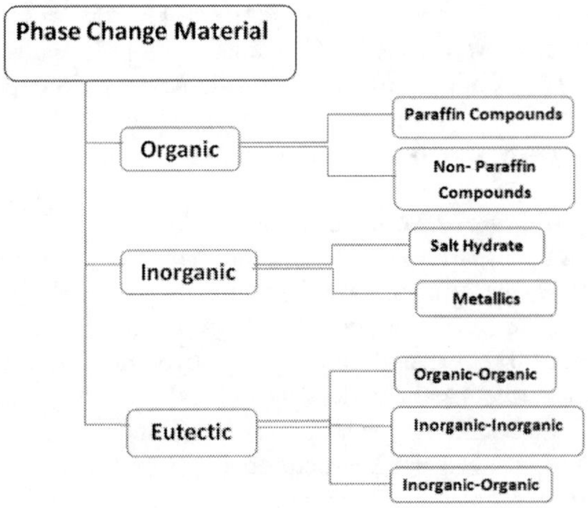

Figure 1: Classification of PCMs

Figure 1 shows broad classification of PCMs. Inorganic PCMs are characterized by good heat of fusion per unit volume and also possess good thermal conductivity compared to organic materials. But they bring adverse corrosive effect on metals that are in contact. Eutectic is a mixture of two or more elements. They melt and freez cohesively without separating from each other in the process of crystallization [8].

3. Performance enhancement techniques for LHS systems

PCMS typically possess low thermal conductivity and hence the heat transfer is slow during charging and discharging. LHS systems can be made more efficient by implementing techniques like modifications in geometry and design, enhance thermal conductivity of PCM and optimize process parameters. The performance evaluation of LHS system can be carried on the basis of energy analysis as well as by exergy analysis. Energy analysis is a well developed method that is widely used in thermodynamic analysis of thermal systems including LHS systems. Ezra et a. [9] have undertaken numerical investigation and optimization of multiple PCMs based LHS system. A mathematical model was developed and effect of operating parameters such as temperature and flow rate of HTF, number and arrangement of PCMs was studied. Tao and Carey [10] utilized orthogonal experimental technique to evaluate effect of PCM thermal properties on LHS system. Range and Regression analysis, optimization was carried out that helped them validate the results and establish selection criteria for PCMs for shell and tube LHS unit.

Review on exergy based evaluation of LHS systems was done by Jegadheeswaran et al. [11]. The effect of design and process parameters on exergy accumulated/retrieved and optimization was mainly addressed in the paper. The economics involved with the exergy based evaluation of LHS systems was also presented. Kousksou et al.[12] proposed a conceptual model for investigation and optimization of a PCM based solar system. Attempt was made to understand the system based on energy and exergy analyses. Heat transfer during a phase change process primarily is either conduction controlled or convection controlled or conduction and convection controlled modes. Heat transfer occurs by conduction during solidification and by convection during a melting process. Buoyancy effect also places a significant role during melting process [13].

From detailed literature studies, it has been found that researchers have focused on various geometrical configurations of LHS units, improvement strategies and applications. LHS have been developed using cylindrical, cuboidal and spherical containers. Prominent improvement methods in use are incorporation of fins, more number of tubes, collective PCMs, inclusion of high thermal conductivity particles.

Akgun et al. [14] studied the melting and solidification processes by altering the geometrical configuration. The process parameters considered are HTF flow rates and Stefan number. Thermal properties of PCM was measured using DSC. Kadri et al. [15] analyzed the progress of PCM melt front during the process. Yang et al. [16] worked on shell type LHS experimentally and confirmed that melting time in composite PCM is lesser compared to plain paraffin wax. Figure 2 shows pictorial representations of prevailing heat transfer improvement techniques.

3.1. Enhancement by addition of Nano material

PCM can be added with nano particles with high thermal conductivity to accelerate the charging and discharging process. This helps in improving the performance of the LHS system. Li-Wu Fan et al. [17] have prepared composite PCMs dispersed with graphene nanoplatlets (GNP) upto 3% by weight. Measurement of properties like thermal conductivity, dynamic viscosity and heat of fusion was for composite PCMs. It is observed that 3% loading of GNP increased the thermal conductivity by two times. Effect of temperature boundary conditions was also studied.

Fazel Yavari et al. [18] studied nano structured PCM graphene and 1-octadecanol composite for different fraction of graphene content. There is an increase in thermal conductivity of nano composite by 140% when 4% (by weight) graphene is added. Reduction in latent heat was 15%. Min Li [19] prepared Nano-graphite/paraffin composites to improve heat transfer rate. Microstructure studies were undertaken to study the dispersion pattern of nano material. A review about experimental studies is undertaken by Kibria et al. [20]. Effect of variation in thermophysical properties on thermal performance is reviewed by them. Additives of carbon based nano materials can increase thermal conductivity of PCM , the homogeneity of dispersion being a significant factor.

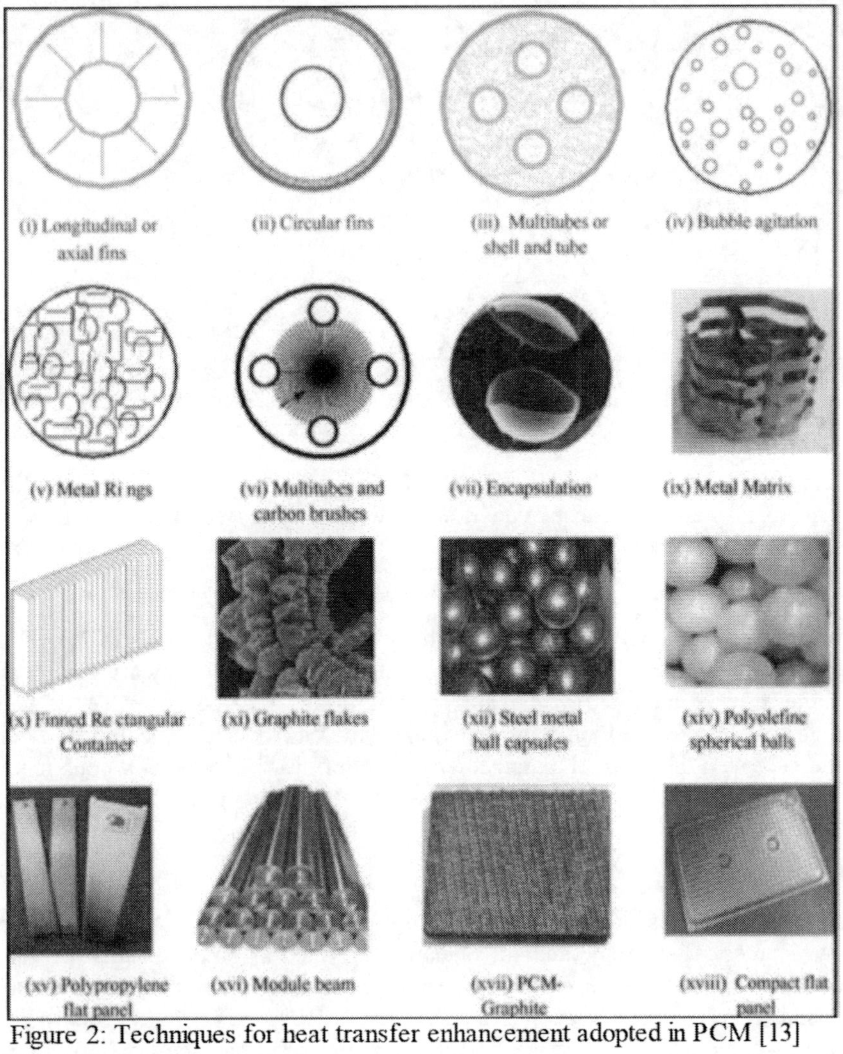

Figure 2: Techniques for heat transfer enhancement adopted in PCM [13]

Thermal conductivity of various composites made with addition of carbon allotropes like multi-walled carbon nanotubes, graphene etc is studied using transient hot wire method and results interpreted in terms of thermal performance enhancement [21]. Nourani et al. [22] analyzed sodium stearoyl lactylate as surfactant added to PCM. It helps better dispersion of alumina nano particles in paraffin. Thermal behavior of paraffin wax embedded with copper oxide particles was interpreted in terms of melting and solidification profiles. Results indicate that thermal conductivity as well as dynamic viscosity increase with increase in loading fraction of CuO particles. Addition of CuO nano particles augments conduction as well as natural convection [23].

3.2. PCM with Microencapsulation

In this technique, the area exposed to heat transfer in LHS is extended. The PCM is surrounded with thin polymer layer in the form of capsule. This creates a seal tight enclosure and hence restricts the PCM leakage during phase transformation. PCM microcapsule techniques, manufacturing methods, microstructure characterization methods were studied by Xia et. al [24]. In this study, encapsulation of paraffin as PCM was done using inorganic polymer. Experimental studies on paraffin reveal that performance of PCM depends on encapsulated PCM mass and energy storage during phase change [25].

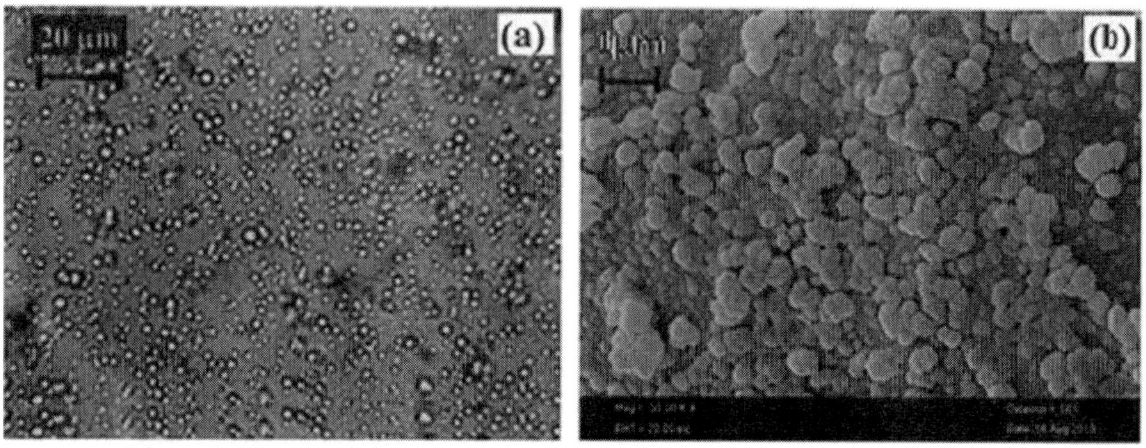

Figure 3: Microscopic image of Polystyrene/*n*-heptadecane microcapsules [26]

High temperature encapsulated NaNO3 as PCM was developed for high temperature applications for solar energy systems. Encapsulation is done with stainless steel. The studies were carried out experimentally and numerically [27] [28].

3.3. Use of porous material

Porous material having high thermal conductivity such as metal foams is excellent candidate for PCMs. Some commonly used are Cu foam, Al foam and Ni foam. They are used with organic PCMs such as paraffin wax having low melting point [29][30]. Composite PCM of paraffin wax with aluminium foam is used in Li-ion battery. Aluminium foam has significantly acclerated the melting process and improved the thermal management of battery [31]. Although porous material as additives enhance the thermal performance of PCM, its porosity and pore density have significant effect on performance.

3.4. Use of multiple PCMs

Incorporating multi PCMs in LHS ensures uniform heat transfer in the direction of transfer fluid (HTF) flow. Numerical and experimental validation for multi PCM based solar power units suggested that use of multiple PCMs produce better performing results [32]. Performance of such LHS systems greatly depends upon melting temperatures and layout of PCMs. The PCM proportions and their phase change temperatures largely affect the performance of shell and tube type LHS unit [33]. The two solid-liquid boundary movement is tracked and liquid fractions were analyzed for a three stage LHS unit and suggestions were given about the optimum lengths of stages [34].

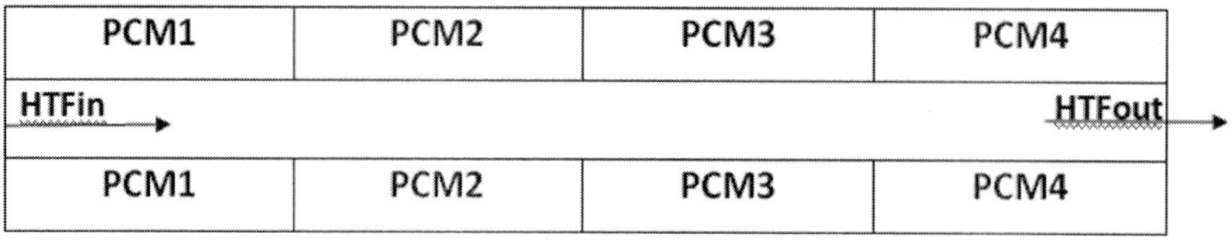

Figure 4: Schematic for using multiple PCMs in LHS system

4. Application of PCMs

Model creation and analysis of Shape stabilized PCM was undertaken for solar building. Effect of parameters like melting temperature, layer thickness, latent heat conducting performance of PCM was explored [35]. Analysis of solar heater with and without PCM was done based on type of storage and collector [36]. Thermoelectric cooling was studied using matrix based enthalpy formulation in 3D perspective. Aluminum containers were used as the heat storage unit; Water flows in inner tube and PCM in a porous aluminum matrix is contained in the outer one. The charging and discharging process are analyzed. A direct application of the studied system is vaccine conservation in solar powered thermoelectric cooling systems [37].

5. Conclusion

This review paper presents the overview on PCMs used for energy storage systems, methods to enhance thermal performance of LHS units with special emphasis to improvement in thermal conductivity by use of additive materials. Based on requirement by a particular LHS unit, suitable enhancement techniques can be adopted. Additives accelerate the heat transfer rates their by making the melting and freezing processes faster, in turn improve the efficiency of the LHS systems The additives vary for low and high temperature applications. Carbon nano materials are suitable for low as well as high temperature range.

References

[1] Sharma A, Tyagi V V, C Chen, Buddhi D 2009 Review on thermal energy storage with phase change materials and applications Renewable &Sustainable Energy Rev 13 318–45
[2] Rathod M K Banerjee 2013 Thermal stability of phase change materials used in latent heat energy storage systems: A review, Renewable and Sustainable Energy Rev 18 246–58
[3] Ukrainczyk , Kurajica S, and Šipušiæ,J 2010 Thermo-physical Comparison of Five Commercial Paraffin Waxes as Latent Heat Storage Materials, Chem. Biochem. Eng. Q. 24 (2) 129–37
[4] Jegadheeswaran S, Pohekar S D 2009 Performance enhancement in latent heat thermal storage system: A review, Renewable and Sustainable Energy Rev 13, 2225–44
[5] Y.B. Tao, Ya-Ling He 2018 A review of phase change material and performance enhancement method for latent heat storage system, Renewable and Sustainable Energy Rev 93 245–59
[6] Abhat A. 1981 Low temperature latent heat thermal energy storage: heat storage materials Solar Energy 30(4) 313–32
[7] Buddhi D, Sawhney R L 1994 Proceedings on thermal energy storage and energy conversion
[8] George A. 1989 Hand book of thermal design. In: Guyer C, editor. Phase change thermal storage materials. McGraw Hill Book Co. chapter 1
[9] Ezra M, Kozak M, Ziskind G 2016 Analysis and optimization of melting temperature span for a multiple-PCM latent heat thermal energy storage unit, Applied Thermal Engineering 93, 315-29
[10] Y.B Tao, V.P Carey, 2016 Effects of PCM thermophysical properties on thermal storage performance of a shell-and-tube latent heat storage unit Applied energy 179 203-10
[11] Jegadheeswaran S, Pohekar S D, Kousksou T 2010 Exergy based performance evaluation of latent heat thermal storage system: A review, Renewable and Sustainable Energy Reviews, 14, 2580–95
[12] T.Kousksou, F.Strub, J.Castaing Lasvignottes, A.Jamil, J.P.Bédécarrats, 2007 Second law analysis of latent thermal storage for solar system, Solar Energy Materials & Solar Cells, 91:14 1275-81
[13] Agyenim F, Eames P, Smyth M 2010 Heat transfer enhancement in medium temperature thermal energy storage system using a multi tube heat transfer array Renewable Energy 35 198–207

[14] Akgun M, Aydın O, Kaygusuz K, 2008 Thermal energy storage performance of paraffin in a novel tube-in-shell system Applied Thermal Engineering 28 405–13
[15] Kadri S, Dhifaoui,B, Dutil Y, Jabrallah S B, Rousse D R 2015 Large-Scale Experimental Study of a Phase Change Material: Shape Identification for the Solid–Liquid Interface Int J Thermophys
[16] Yang J, Yang L, Xu C, Du X 2016 Experimental study on enhancement of thermal energy storage with phase-change material Applied Energy 169 164–76
[17] Li-Wu Fan, Zi Qin Zhu, Yi Zeng, Qian Lu, Zi-Tao Yu 2014 Heat transfer during melting of graphene-based composite phase change materials heated from below, International Journal of Heat and Mass Transfe 79 94-104
[18] Fazel Y, Hafez R Fard, Kamyar P, Mohammad A R, Amir Z, Zhongzhen Y, Rahmi O, Theodorian B, Nikhil K 2011 Enhanced Thermal Conductivity in a Nanostructured Phase Change Composite due to Low Concentration Graphene Additives, The journal of Physical Chemistry 115 8753-58
[19] Min Li 2013 A nano-graphite/paraffin phase change material with high thermal conductivity, Applied Energy 106 25-30
[20] Kibria M A, Anisur M R, Mahfuz M H, Saidur R, Metselaar I 2015 A review on thermo-physical properties of nanoparticle dispersed phase change materials Energy Conversion and Management 95:69–89
[21] Choi D H, Lee J, Hong H, Kang YT 2014 Thermal conductivity and heat transfer performance enhancement of Phase Change Materials (PCM) containing carbon additives for heat storage application, International Journal of Refrigeration
[22] Nourani M, Hamdami N, Keramat J, Moheb A, Shahedi M 2016 Thermal behavior of paraffin-nano-Al_2O_3 stabilized by sodium stearoyl lactylate as a stable phase change material with high thermal conductivity Renewable Energy 88 474-82
[23] Jesumathy S, Udayakumar M, Suresh S 2012 Experimental study of enhanced heat transfer by addition of CuO nanoparticle, Heat Mass Transfer 48: 965–78
[24] Xia D, Xiaodong S 2006 Research on microcapsules of phase change materials Rare Metals 25 393-99
[25] Alkan C 2006 Enthalpy of melting and solidification of sulfonated paraffins as phase change materials for thermal energy storage, Thermochimica Acta 451 126–30
[26] Ahmet S, Cemil A, Derya K, Alper B 2014 Micro/nano-encapsulated n-heptadecane with polystyrene shell for latent heat thermal energy storage Solar Energy Materials & Solar Cells 126 42-50
[27] Zheng Y, Zhao W, Sabol J C,Tuzla K, Neti S, Oztekin A , Chen J C 2013 Encapsulated phase change materials for energy storage- Characterization by calorimetry Solar energy 87:117-26
[28] Zheng Y, Barton Ji, Tuzla K, Chen JC, Neti S, Oztekin A, Misiolek WZ, 2015 Experimental and computational study of thermal energy storage with encapsulated $NaNO_3$ for high temperature applicatrions solar energy 115:180-94
[29] Wang C, Lin T, Li N, Zheng H 2016 Heat transfer enhancement of phase change composite material: copper foam/paraffin, Journal of Renewable energy, 96:960-65
[30] Zhao Y, Zhao C Y, Xu Z G, Xu H J 2016 2016 Modeling metal foam enhanced phase change heat transfer in thermal energy storage by using phase field method International Journal of Heat & Mass transfer 99:170-81
[31] Wang Z, Zang Z, Jia I, Yang I 2015 Paraffin and paraffin aluminium foam composite phase change material heat storage experimental study based on thermal management of Li-ion battery Applied thermal enginenering 78:428-36
[32] Michels H, Pitz Paal R 2007 Cascaded latent heat storage for parabolic trough solar power plants Solar energy 81:829-37
[33] Fang M, Chen G, Effects of different multiple PCMs on the performance of a latent thermal energy storage system, Applied Thermal Engineering 2007, 27:994-1000

[34] Li YQ, He YI, Song HJ, Xu C, Wang WW, Numerical analysis and parameters optimization of shell and tube heat storage unit three phase change materials. Renewable energy 2013, 59:92-99
[35] Xu X, Zhang Y, Lin K, Di H, Yang,R 2005 Modeling and simulation on the thermal performance of shape-stabilized phase change material floor used in passive solar buildings Energy and Buildings 37 1084–91
[36] Shukla A, Buddhi D, Sawhney R L 2009 Solar water heaters with phase change material thermal energy storage medium: A review Renewable and Sustainable Energy Reviews 13 2119–25
[37] Trelles JP, Dufly J 2003 Numerical simulation of porous Latent Heat Storage for thermoelectric cooling Applied Thermal Engineering 23 1647–64

Molecular Dynamics Simulation of the Tensile Properties of an Al6063 – SiO_2 Composite under Varying Temperature Conditions

N Idusuyi, A H Adekoya and T T Olugasa

Mechanical Engineering Department, University of Ibadan, Ibadan, Nigeria
nosaidus@gmail.com

Abstract. The tensile stress behaviour of Aluminium Metal Matrix Composites (AMCs) reinforced with SiO_2 was evaluated using Molecular Dynamics (MD). A cubic model was used for the simulation, while the Modified Embedded Atom Method (MEAM) was implemented to describe the atomic interactions for the MD simulation. From the MD simulation the presence of the SiO_2 was found to substantially disrupt the FCC crystallography of the Al and reduce the ductility of the Al while substantially increasing the yield strength of the composite. A Maximum Elastic modulus of 43.3GPa was obtained at a temperature of 350K and 10wt% SiO_2.

1. Introduction

The development of Aluminium Metal Matrix Composites (AMCs) is one of the most significant achievements in the history of materials research and processing [1]. AMCs which exhibit high strength, high elastic modulus and good wear resistance have been developed to meet the requirements of materials for aerospace and automobile applications [2,3]. However, the high costs associated with producing AMCs limits its applications [4,5]. The properties of AMCs are influenced by type of reinforcement (Synthetic ceramic particles, Industrial wastes and agro-wastes) as well as other properties like shape (particulate, volume fraction, bonding type and size [6,7]. Silicon Oxide (SiO_2) is the dominant compound in Rice Husk Ash and other agricultural waste products which have found usefulness as a less expensive reinforcement alternative for the production of MMCs [8–11]. In order to rationalize the experimental observations, researchers have resorted to a theoretical analysis of MMCs by way of atomic simulation[12]. In a study of Al/SiC interfaces by Li et.al [13], it was reported that the bond strength between Al and Al were generally lower than SiC and Al, with the later been up to 1.5 times stronger. Jena [14] however, used a self-consistent quantum mechanical method to determine the binding energies and bond lengths of Al–Al, Al-Si, Al–C and Si–C dimers. Their study discovered that the bonding of Al–C dimers was stronger than others studied, thereby leading to the formation of Al_4C_3 at the interface. Similar observations were made by Lu et al.[15] based on an extended Hückel theory, that the Al–C interaction was stronger and the Al–SiC interaction was confined to a narrow region at the interface. Molecular Dynamics is a powerful technique that utilizes Newton's laws to model the interatomic interactions between the particles on a Nano scale [16]. MD simulations are based on interatomic potentials which can be used to study systems involving millions of atoms and to provide macroscopically predictive results. Some of the more popular MD softwares are Large Scale Atomic/Molecular Massively Parallel Simulator (LAMMPS), Tinker, Materials Studio and Chemistry at Harvard Macromolecular Mechanics (CHARMM), etc.[12,17,18]. At present, there are numerous models of these interatomic potentials. These potentials are generally divided into two broad categories: (1) Pair Potentials, the most common example of which is the Lennard-Jones Potential, and (2) Many body potentials which include Tersoff, Embedded Atomic Potential (EAM) and Modified Embedded Atomic Potential (MEAM). The potentials are developed individually to describe the interactions between specific elements/atoms [19–24]. However, the true interatomic interactions as described by the Schrödinger equation or Dirac equation for all electrons and nuclei. Ward et.al [25] used MD simulation to study the mechanisms of deformation and failure in Al-Si Nano composites. It was reported that a clear difference in the fracture mechanisms was observed in the Al-Si composites as compared to single-phase Al. It was deduced that the Al/Si interfaces controlled the mechanical behaviour in the Nano composites, reducing the role of bulk metal deformation modes [25]. Other researchers have corroborated the importance of the Matrix/reinforcement interface to the properties of the composite[26–29]. Sedigh et.al [30] used MD simulation to study the influence of nanofiller geometry, temperature and defects on the mechanical properties of Aluminium reinforced with Boron-Nitride Nano sheets and nanotubes respectively. It was concluded that Boron-Nitride Nano sheets was a better reinforcement with higher fracture toughness. From the study, the authors concluded that Boron Nitride/Al Nano composites could be useful in producing lightweight materials critical to the aerospace and automobile industry. In another study, the mechanical properties of light nonporous Carbon nanotubes (CNT)-Aluminium (Al) composites were investigated using atomistic tensile simulations. The reinforcements (CNT) were embedded in Al matrix composites to reduce the weight of Al by 23%, thus large volume fractions of large size CNT were used. Pure aluminium structure was simulated using a FCC lattice structure with 4.046 Å as the lattice constant. A pore with a diameter of 2.58nm was drilled at the centre of the pure aluminium to prepare nonporous Al. The nonporous CNT-Al was prepared by inserting CNT into pure aluminium. The aluminium atoms which were inside the CNT radius were removed by considering van der Waals radius carbon atoms and aluminium atoms. The final core hole sizes of the NP Al and NP CNT-Al were configured with a diameter of 2.58 nm. LAMMPS was the software of choice for the simulations, while Open Visualisation Tool (OVITO) was used for visualising and analysing the LAMMPS output. It was observed that the lightweight nanoporous CNT-Al composite had 105.8, 246.9 and 243.7% improvements in tensile strength, fracture toughness, and elastic modulus, respectively [31]. Literature is sparse on the analysis of SiO_2 reinforced AMCs using Molecular dynamics. In this work, the effects of temperature on the tensile stress behaviour of Aluminium Metal Matrix Composites reinforced with SiO_2 were studied using Molecular Dynamics (MD).

2.0 MD Simulation Steps

2.1 Molecular dynamics simulation

The MEAM potentials used in this study for Al and O were obtained from Angelo and Baskes [22], while the Al–O interaction was obtained from Baskes [23]. The elastic constants calculated by the MEAM potential for Al were C11 =111.09 GPa and C12 = 61.68 GPa. Which correspond almost exactly with the values reported by Dandekar et.al [32].

2.2 The model for the study

The structure of pure aluminium used in this study was a Face Centred Cubic (FCC) structure with a lattice parameter of 4.05Å [33]. SiO_2 has a Diamond structure with a lattice constant of 5.43Â.The model was taken as a cuboid of aluminium atoms with spherical inclusions of silica placed at the centre of the cube as depicted in Figure 1(a –c). The dimensions of the simulation box were fixed at 50Â by 100Â by 50Â, while the spherical infusions of silica had a diameter of 5.4Â. The molecular model containing 15,625 atoms was set up using LAMMPS (Large Scale Atomic/Molecular Massively Parallel Simulator). The interaction between the particles were modelled using MEAM potentials for the Al-Si particles, as well as Tersoff potentials for the SiO_2 particles. The Hybrid method was utilized for the pair style command in LAMMPS to combine the effects of the two potential methods, thus providing a comprehensive model. Subsequently, after setting up the system, the atoms were relaxed to find global minimum energy of the system by doing a systematic conformational search via the Polak-Ribiere version of the conjugate gradient (CG) algorithm. At each iteration the force gradient is combined with the previous iteration information and used to compute a new search direction which is perpendicular (conjugate) to the previous search direction. The atoms were assigned velocities following a Maxwell-Boltzmann distribution around the desired temperature. An MD simulation was performed on the model at a time step of 0.001 ps (Picosecond), under the isothermal canonical ensemble NVT (constant volume - constant temperature) using a Nosė-Hoover thermostat to control the temperature and keep it constant at 300K. The simulations were performed for models with 0wt%, 2.5wt%, 5wt%, 7.5wt% and 10wt% reinforcements respectively. The Open Visualization Tool (OVITO) was used to visualise the outputs of the simulations. The Elastic modulus of the tensile plots was determined by differentiating a second-degree polynomial equation which was fitted to the Stress-Strain graph.

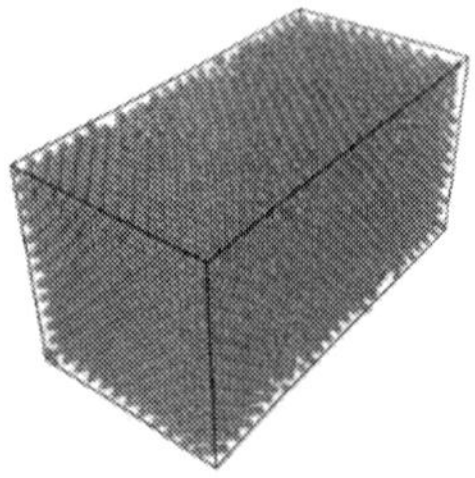

(a)

(b)

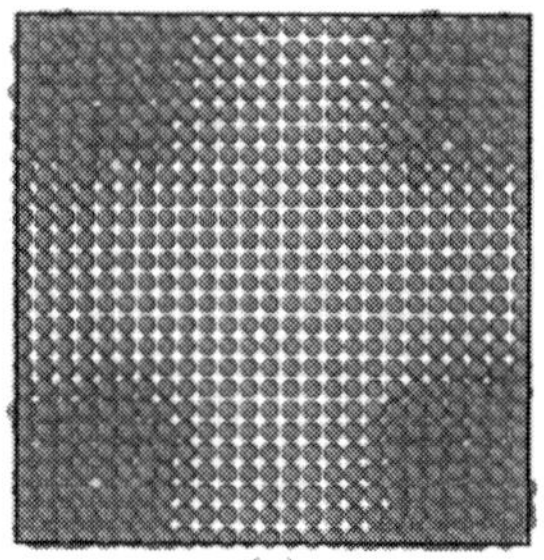

(c).

Figure 1 (a-c) : Aluminium Model for MD Simulation – (a) Cubic Model, (b) Single Unit Cell of model, (c) Agglomeration of Al in Composite Model

Table 1 shows the assumed reinforcement volumes used for the study.

Table 1: Volume Fraction of Composites

Composite	wt.% Al6063	Al6063 (Vol. fraction)	wt.% SiO_2	SiO_2 (Vol. fraction)
Al	100.0	1.0000	0.0	0.0000
Al-2.5wt%SiO_2	97.5	0.9657	2.5	0.0325
Al-5wt% SiO_2	95.0	0.9350	5.0	0.0650
Al-7.5wt% SiO_2	92.5	0.9025	7.5	0.0975
Al-10wt% SiO_2	90.0	0.8700	10.0	0.1300

2.3 *Tensile test*

The models were stretched at a constant engineering strain of 0.01/femto second under the NVT ensemble. A very low strain rate is desired to ensure the stability of the system. With higher stain rates

IOP Conf. Series: Materials Science and Engineering 1126 (2021) 012069 doi:10.1088/1757-899X/1126/1/012069

(for instance 0.02/femto second), it was observed that the system was unstable under the chosen conditions resulting in either rapid fluctuations in the temperature of the system as the atoms are not given sufficient time to equilibrate before the next step of the iteration. In the other case, the physical stability of the system completely failed giving a "lost atoms" error. The problems with the stability of the system were solved by reducing the time step and reducing the strain rate. A suitable compromise was sought between the larger computational costs required for low time steps and the instability associated with larger time steps. The tests were performed at three different temperatures; 300K, 325K, 350K and kept at constant temperature during the process. The system was then analysed to determine the stress present in the model.

3. Results and Discussion

As opposed to the stress-strain diagrams derived from the conventional tensile testing, which presents a smooth curve with a linear slope in the elastic region; the stress-strain plot of the MD simulation is characterized by a series of steps of rise and fall. This behaviour can be attributed to strain hardening and subsequent softening of the material due to dislocation build up and rapid motion of the dislocations that follows [34]. A plot of the stress-strain curves for the composites at various temperatures are presented in Figure 2 (a-c). It can be observed that there is no clear distinction between the linear elastic range, the plastic range or the yield point. This lack of clear delineation creates difficulties in obtaining the values of the elastic moduli with appreciable degree of certainty. In this case, the modulus was obtained by fitting a second-degree polynomial to portions of the stress–strain diagram. All three graphs display similar trends. It is observed that the presence of the clusters of SiO_2 increased the yield strength of the base material (pure Al) which is subsequently reduced with the addition of more SiO_2. Further increasing the percentage of SiO_2 in the Al matrix uniformly reduces the ductility of the model.

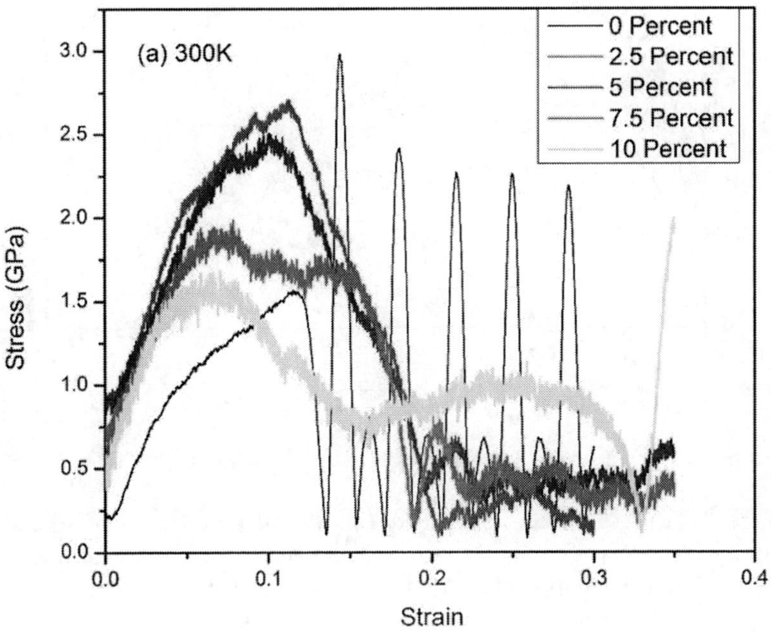

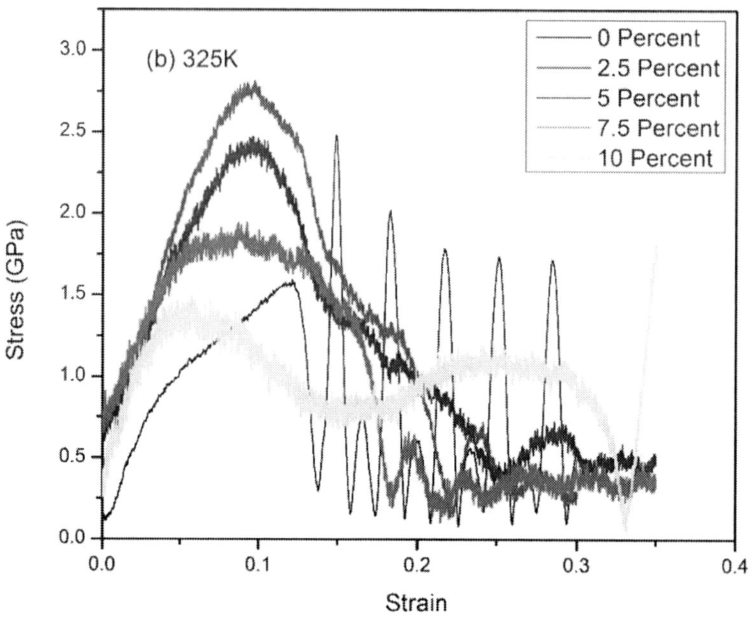

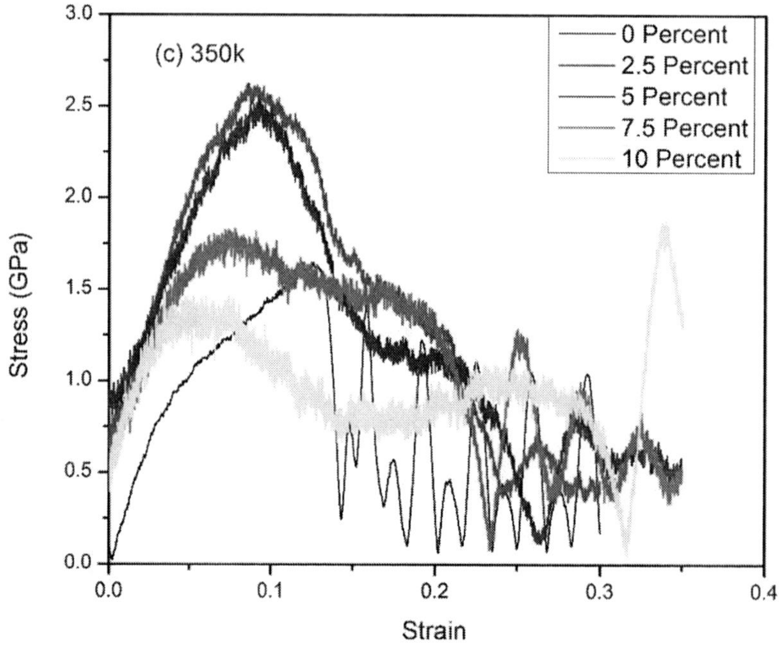

Figure 2: Tensile stress – Strain Plots at a) 300K. b) 325K. c) 350K

As shown in Figure 2(a-c), the clusters of SiO_2 disrupt the lattice of the pure Al and result in stress concentrations at the sites of infusion. As a result, an increase in the volume of SiO_2 used as reinforcement and an increase in temperature tends to reduce the tensile strength of the composites. To further understand this phenomenon, the Centrosymmetry tool in the OVITO software was used to visualise the change in the crystal structure of the composite by the addition of the SiO_2 clusters. The rendering of the OVITO analysis is presented in *Figure 3*. Clearly, the effects of the reinforcing phase on the crystallography of the base material can be observed. For instance, the sample containing 0wt%

SiO_2 had as high as 95% of the particles positioned in the standard FCC orientation which is expected of pure Aluminium. The sample with 2.5wt% SiO_2 had about 77.3% in the FCC configuration, the sample containing 5wt% SiO_2 had 62%, the sample containing 7.5wt %SiO_2 had 47.7% while the sample containing 10wt% SiO_2 had as low as 24.8% of the particles positioned in the FCC orientation. Quite clearly, with every increase in the percentage of SiO_2 in the composite, the crystal arrangement of the Al atoms was further disturbed. These disruptions occurred far beyond the local area/sites of infusion of SiO_2. However, because of the scale of samples studied using molecular dynamics, it is quite possible that the effective disruption to the crystallography of the base material on a much larger scale might be relatively less substantial than what was observed in this study. Animation of the tensile tests using OVITO showed that the atoms were substantially reorganized followed by rapid movement likely due to the generation of dislocations.

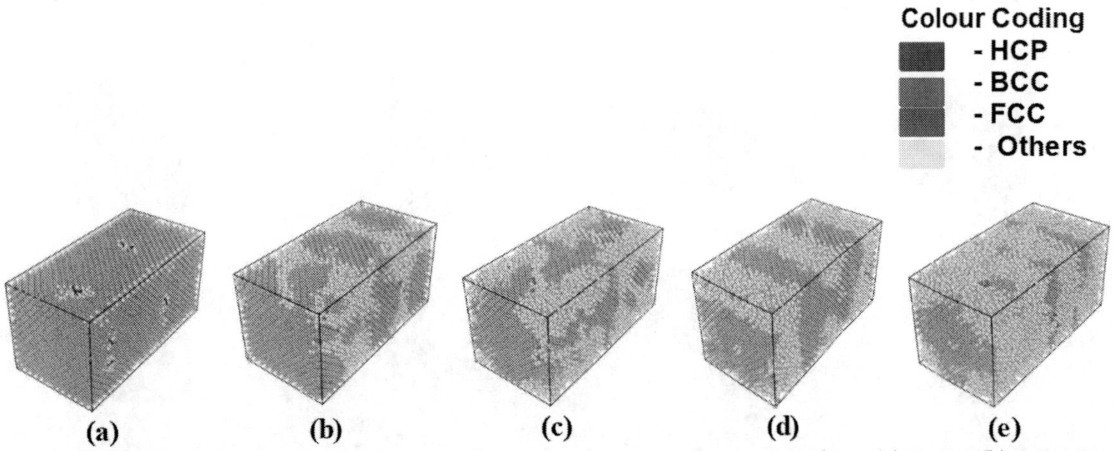

Figure 3: Distribution of Crystal Structures in composites at various wt% SiO_2; (a) 0wt% (b) 2.5wt%, (c) 5wt%, (d) 7.5wt%, (e) 10wt%

A snapshot of the process at different engineering strain values is displayed below in Figure 4 (a-e) for the different samples at 325K. Observation of the OVITO snapshots at similar strain values reveals that the pure aluminium (0 wt% SiO_2) fractures at lower strain value than the others. Furthermore, increasing the volume of SiO_2 in the composite results in lower strain values (i.e, increased brittleness) before the process of separation/fracture is observed.

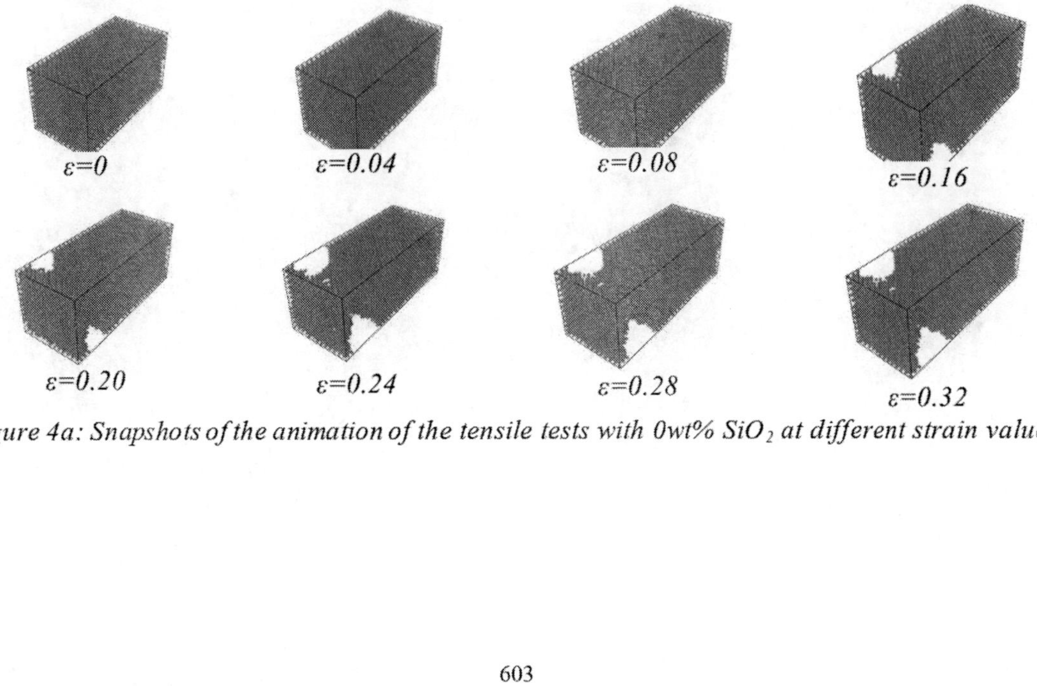

Figure 4a: Snapshots of the animation of the tensile tests with 0wt% SiO_2 at different strain values.

IOP Conf. Series: Materials Science and Engineering 1126 (2021) 012069 doi:10.1088/1757-899X/1126/1/012069

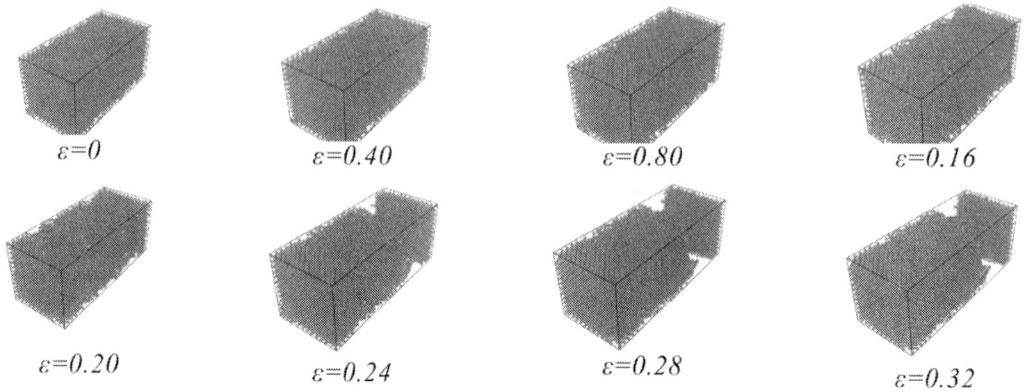

Figure 4b: Snapshots of the animation of the tensile tests with 2.5wt% SiO_2 at different strain values.

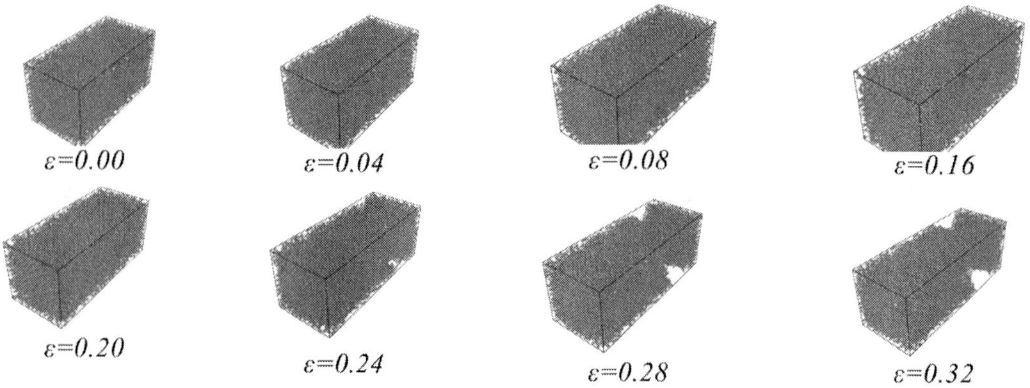

Figure 4c: Snapshots of the animation of the tensile tests with 5wt % SiO_2 *at different strain values.*

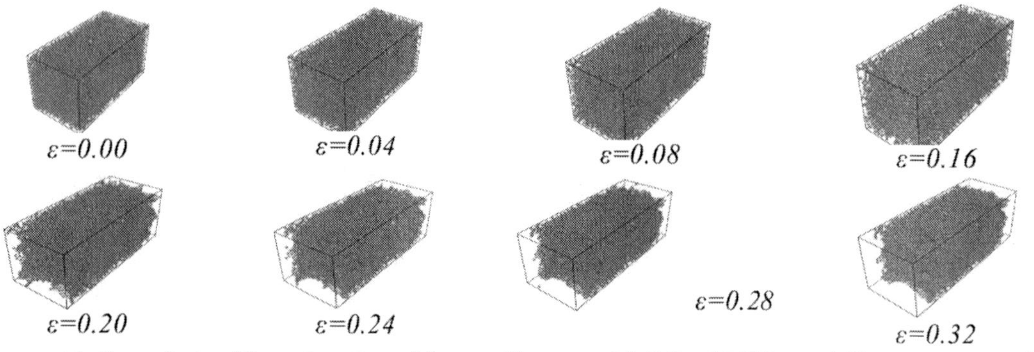

Figure 4d: Snapshots of the animation of the tensile tests with 7.5wt % SiO_2 *at different strain values.*

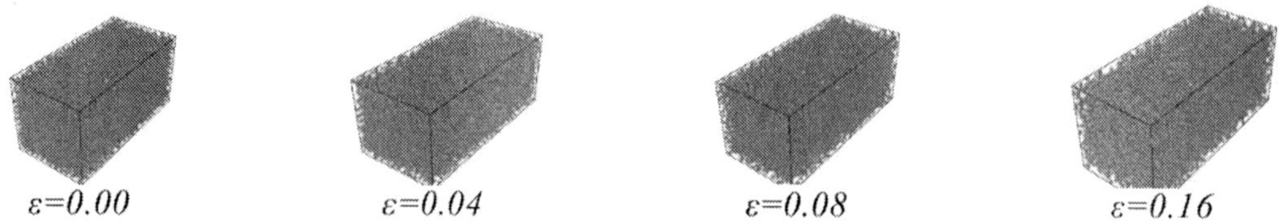

ε=0.20 ε=0.24 ε=0.28 ε=0.32

Figure 4e: Snapshots of the animation of the tensile tests with 10wt% SiO_2 *at different strain values.*

As mentioned earlier, there is substantial agglomeration around regions of SiO_2 with the reinforcing particles causing significant disruption to the otherwise perfect lattice of Al. As the sample is further stressed, the stress concentrations on those sites result in local stresses much higher than the average stress of the body, resulting in the initiation of cracks, propagation of the cracks and eventually resulting in the fracture/separation of the sample. The entire process can be easily monitored using the various tools available in OVITO. Figure 5 shows the Maximum stress distribution obtained at different temperatures. The results show that the maximum stress obtained for the composites decreased as temperature and wt% of SiO_2 increased. A maximum stress of 2.8141 GPa was obtained at 2.5wt% SiO_2 and a temperature of 325K. The lowest value of stress was 1.4946 GPa obtained at 325K and 10wt% SiO_2 reinforcement. At a temperature of 300K, the percentage change in stress were 73%, 60%, 27% and 7% at 2.5wt%SiO_2, 5wt%SiO_2, 7.5wt%SiO_2 and 10 wt%SiO_2 respectively.

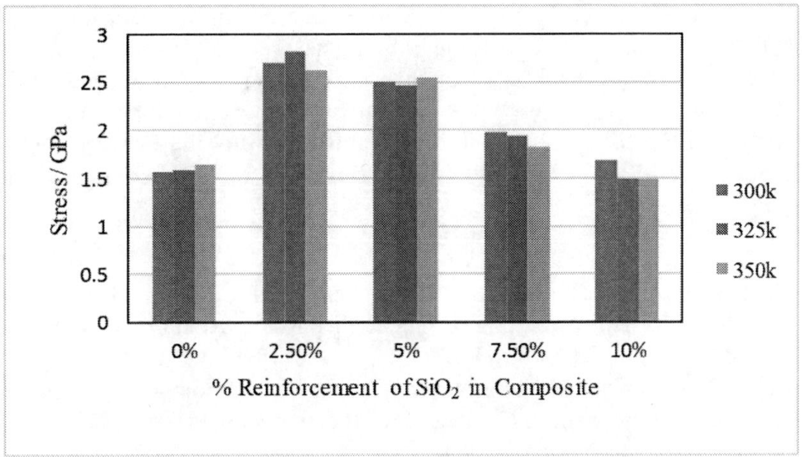

Figure 5: Maximum stress at 300K, 325K and 350K

From Figure 6, the Elastic modulus was observed to increase with increase in temperature, but decreased slightly with an increase in wt% of SiO_2. At 300K, 325k and 350k the Maximum elastic modulus of 25.30 GPa, 28.95 GPa and 43.23 GPa were obtained at 7.5wt% SiO_2, 2.5wt% SiO_2 and 10wt% SiO_2 respectively.

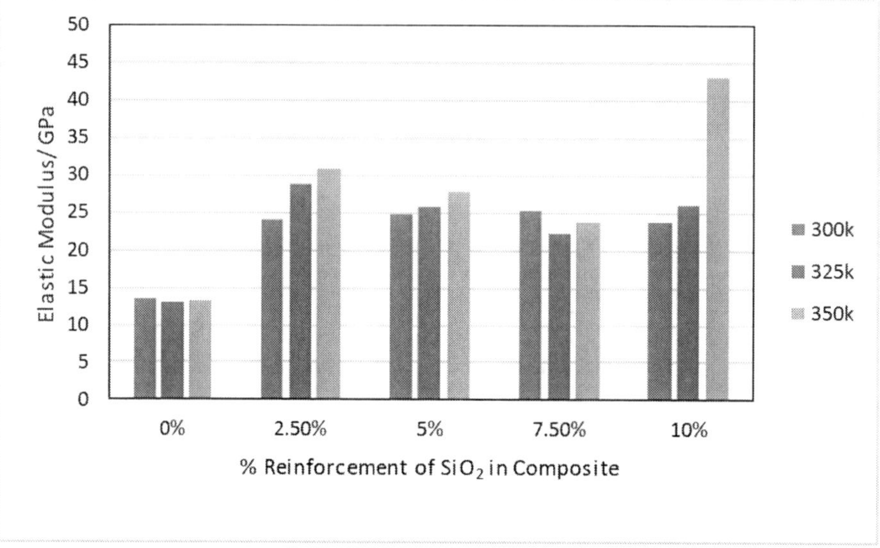

Figure 6: Elastic modulus at 300K, 325K and 350K

4. Conclusions

In this work, the elastic response of an Al- SiO_2 composite was studied. Molecular dynamics was used to characterise the behaviour of the composite. The results generally indicate that SiO_2 is suitable as a reinforcement for Al, and can result in substantial improvement in the mechanical properties of the base metal. A Maximum Elastic modulus of 43.3GPa was obtained at temperature of 350K and 10wt% SiO_2. The presence of the SiO_2 was found to substantially disrupt the FCC crystallography of the Al and reduce the ductility of the Al while substantially increasing the yield strength of the composite.

References

[1] R. Malik, S. Bhandari, A. Pant, A. Saxena, N. Kumar, N. Chotrani, D. Gunwant, P.L. Sah, Fabrication and Mechanical Testing of Egg Shell Particles Reinforced Al-Si Composites, Int. J. Math. Eng. Manag. Sci. 2 (2017) 53–62.

[2] S. Banerjee, B. V. Sankar, Mechanical properties of hybrid composites using finite element method based micromechanics, Compos. Part B Eng. 58 (2014) 318–327. doi:10.1016/j.compositesb.2013.10.065.

[3] Mruthyunjagouda, G. Manavendra, Experimental Investigation of Thermal behaviour on Rice Husk Ash-Alumina Reinforced with Al6063 Alloy Matrix, Int. J. Innov. Res. Sci. Eng. Technol. 6 (2017) 5454–5460. doi:10.15680/IJIRSET.2017.0604049.

[4] S.B. Hassan, V.S. Aigbodion, Effects of eggshell on the microstructures and properties of Al – Cu – Mg / eggshell particulate composites, J. King Saud Univ. - Eng. Sci. 27 (2015) 49–56. doi:10.1016/j.jksues.2013.03.001.

[5] J.A.K. Gladston, N.M. Sheriff, I. Dinaharan, J.D.R. Selvam, Production and characterization of rich husk ash particulate reinforced AA6061 aluminum alloy composites by compocasting, Trans. Nonferrous Met. Soc. China. 25 (2015) 683–691. doi:10.1016/S1003-6326(15)63653-6.

[6] S.S. Mukadam, A Review on Mechanical Behavior of Aluminium Metal with Different Reinforcing Materials, (2017) 3–6.

[7] R. Umunakwe, D.J. Olaleye, A. Oyetunji, O.C. Okoye, I.J. Umunakwe, Assessment of some Mechanical Properties and Microstructure of Particulate Periwinkle Shell-Aluminum 6063 Metal Matrix Composite (PPS-ALMMC) Produced by Two-Step Casting, Niger. J. Technol. 36 (2017) 421–427. doi:10.4314/njt.v36i2.14.

[8] K. Sinulingga, M. Sirait, A.M. Siregar, Addition of Nano Particles Effect of Rice Husk Ash as a Mixture On The Bricks Strength, IOP Conf. Ser. J. Phys. Conf. Ser. 1120 (2018) 1–9.

doi:10.1088/1742-6596/1120/1/012090.
[9] K. Kanayo, P. Apata, Original article Corrosion and wear behaviour of rice husk ash — Alumina reinforced Al – Mg – Si alloy matrix hybrid composites, J. Mater. Res. Technol. 2 (2013) 188–194. doi:10.1016/j.jmrt.2013.02.005.
[10] K.K. Alaneme, M.O. Bodunrin, A.A. Awe, Microstructure, mechanical and fracture properties of groundnut shell ash and silicon carbide dispersion strengthened aluminium matrix composites, J. King Saud Univ. - Eng. Sci. (2016). doi:10.1016/j.jksues.2016.01.001.
[11] K.K. Alaneme, O.J. Ajayi, Microstructure and mechanical behavior of stir-cast Zn–27Al based composites reinforced with rice husk ash, silicon carbide, and graphite, J. King Saud Univ. - Eng. Sci. (2015) 6–11. doi:10.1016/j.jksues.2015.06.004.
[12] H. Gu, X.-L. Gao, X.C. Li, Molecular Dynamics Study on Mechanical Properties and Interfacial Morphology of an Aluminum Matrix Nanocomposite Reinforced by -Silicon Carbide Nanoparticles, J. Comput. Theor. Nanosci. 6 (2009) 61–72. doi:https://doi.org/10.1166/jctn.2009.1007.
[13] S. Li, R.J. Arsenault, P. Jena, Quantum chemical study of adhesion at the SiC/Al interface, J. Appl. Phys. 64 (1988) 6246–6253. doi:10.1063/1.342082.
[14] B.K. Rao, P. Jena, Molecular view of the interfacial adhesion in aluminum-silicon carbide metal-matrix composites, Appl. Phys. Lett. 57 (1990) 2308–2310. doi:10.1063/1.103878.
[15] W. Lu, K. Zhang, X. Xie, Adsorption of a monolayer of iron on β-SiC(100) surfaces, Phys. Rev. B. 48 (1993) 18159–18163. doi:10.1103/PhysRevB.48.18159.
[16] S. Sharma, S.K. Tiwari, S. Shakya, Mechanical properties and thermal conductivity of pristine and functionalized carbon nanotube reinforced metallic glass composites: A molecular dynamics approach, Def. Technol. (2020). doi:https://doi.org/10.1016/j.dt.2020.04.004.
[17] H. He, Y. Rong, L. Zhang, Molecular dynamics studies on the sintering and mechanical behaviors of graphene nanoplatelet reinforced aluminum matrix composites, Model. Simul. Mater. Sci. Eng. 27 (2019) 18. doi:10.1088/1361-651X/ab2095.
[18] A. Stukowski, Visualization and analysis of atomistic simulation data with OVITO – the Open Visualization Tool, Model. Simul. Mater. Sci. Eng. 18 (2010) 1–8. doi:10.1088/0965-0393/18/1/015012.
[19] J.E. Lennard-Jones, On the determination of molecular fields.—I. From the variation of the viscosity of a gas with temperature, Proc. R. Soc. London. 106 (1924) 441–462. doi:https://doi.org/10.1098/rspa.1924.0081.
[20] J.E. Lennard-Jones, On the Determination of Molecular Fields. — II. From the Equation of State of a Gas., Proc. R. Soc. London A. 106 (1924) 463–477. doi:https://doi.org/10.1098/rspa.1924.0082.
[21] M.S. Daw, M.I. Baskes, Embedded-atom method: Derivation and application to impurities, surfaces, and other defects in metals, Phys. Rev. B. 29 (1984) 6443–6453. doi:10.1103/PhysRevB.29.6443.
[22] M. Baskes, Modified embedded atom method calculations of interfaces., Sandia. 96 (1996).
[23] J.E. Angelo, M.I. Baskes, Interfacial Studies Using the EAM and MEAM, Interface Sci. 63 (1996) 47–63.
[24] K. Gall, M.F. Horstemeyer, M. Van Schilfgaarde, M.I. Baskes, Atomistic simulations on the tensile debonding of an aluminum – silicon interface, J. Mech. Phys. Solids. 48 (2000) 2183–2212.
[25] D.K. Ward, W.A. Curtin, Y. Qi, Mechanical behavior of aluminum – silicon nanocomposites : A molecular dynamics study, Acta Mater. 54 (2006) 4441–4451. doi:10.1016/j.actamat.2006.05.022.
[26] J.-K. Kim, S.C. Tjong, Y.-W. Mai, Effect of Interface Strength on Metal Matrix Composites Properties, Elsevier Ltd., 2018. doi:10.1016/B978-0-12-803581-8.09982-3.
[27] N. Yedla, R. Nalla, S. Pal, P. Gupta, M. Meraj, Molecular Dynamics Studies on the Prediction of Interface Strength of Cu(metal)-CuZr (metallic glass) Metal Matrix Composites, (n.d.).
[28] P. Gupta, N. Yedla, High Temperature Mechanical Behavior of Aluminum- Cu50Zr50 Metallic Glass Interface, IOP Conf. Ser. Mater. Sci. Eng. 115 (2016). doi:10.1088/1757-

899X/115/1/012024.
[29] S. Schmauder, S. Hocker, P. Kumar, Molecular dynamics simulations of tensile tests and interfacial fracture in, ResearchGate. (2010) 1–9.
[30] P. Sedigh, A. Zare, A. Montazeri, Evolution in aluminum applications by numerically-designed high strength boron-nitride / Al nanocomposites, Comput. Mater. Sci. 171 (2020) 109227. doi:10.1016/j.commatsci.2019.109227.
[31] M.E. Suk, Enhanced tensile properties of weight-reduced nanoporous carbon nanotube-aluminum composites, Mater. Express. 9 (2019) 801–807. doi:10.1166/mex.2019.1561.
[32] C.R. Dandekar, Y.C. Shin, Effect of porosity on the interface behavior of an Al 2 O 3 – aluminum composite : A molecular dynamics study, Compos. Sci. Technol. 71 (2011) 350–356. doi:10.1016/j.compscitech.2010.11.029.
[33] C.R. Dandekar, Y.C. Shin, Effect of porosity on the interface behavior of an Al 2 O 3 – aluminum composite : A molecular dynamics study, Compos. Sci. Technol. 71 (2011) 350–356. doi:10.1016/j.compscitech.2010.11.029.
[34] R. Komanduri, N. Chandrasekaran, L.M. Raff, Molecular dynamics (MD) simulation of uniaxial tension of some single-crystal cubic metals at nanolevel, Int. J. Mech. Sci. 43 (2001) 2237–2260.

Strengthening of Soil by adding Lime and Glass Fiber as Stabilizing Materials for the Construction of High Rise Buildings

[1]Gopu Ganesh Naidu, [2]Ravulapalli Ramakrishna, [3]G Ezri Babu

[1]Associate Professor and Head of the department, Civil Engineering, Pace institute of technology & sciences (A), Valluru, Prakasam, Andhra Pradesh, India.
[2] PG Student ,Structural Engineering, Pace institute of technology & sciences (A), Valluru, Prakasam, Andhra Pradesh, India.
[3] Assistant professor, Department of Civil Engineering, Pace institute of technology & sciences (A), Valluru, Prakasam, Andhra Pradesh, India.

Abstract. The main aim of this research is to increase the strength properties of the weak soil by using stabilizing materials for the construction of high raised buildings on weak soils. As we all know that the most important thing in construction industry is soil. In this research we are adding stabilizing materials like lime and glass fibers to the weak soil to increase the strength parameters of the soil. The addition of stabilizing materials and admixtures to the soil leads to the increase in the strength parameters of the soil. There are various Infrastructural projects which are constructed on the soil, in many of these cases the black cotton soil is not suitable, because the Black cotton soil is weak in strength and which has low engineering properties. Black cotton soil has more swelling property and high shrinkage nature. To overcome these problems and to improve the engineering properties of the black cotton soil, we are adding Lime and Glass fibers as the stabilizing materials in the weak black cotton soil. The engineering properties like strength, compressibility and bearing capacity of the soil were increased by stabilizing the weak soil these stabilizing materials. This work mainly concentrates on the feasibility of the usage of lime and glass fibers as the stabilizing materials for the black cotton soil. The percentage of Lime and glass Fibers used in black cotton soil varies in different proportions like 2%, 4%, 6%, 8% and 10%. All the necessary tests were conducted to get the results. The properties like Liquid limit, Plastic limit, Compaction characteristics values and Differential free swell index of soil have been studied in this research.
Keywords: stabilization, admixtures, lime, glass fibers, strengthening, black cotton soil, montmorillonite, swelling, expansive.

1. Introduction

For the development of infrastructure, the footings should be strong enough to bear the large loads. For footings to be strong the soil below and around the footings should be strong enough to bear the large amount of loads [1]. The mixture of organic matter, minerals, water, air and some organisms form soils. We have different types of soils available in INDIA e.g. black cotton soils, mountain soils, alluvial soils; laterite soils among all the soils available in India, black cotton soils are very poor in bearing capacity and in engineering properties [2]. The soil in nature is very useful for agricultural and engineering point of view. From an engineering perspective, soil plays a different role in buildings, roads, railways, airports and in other construction works. The strength of the structure depends upon the type of soil, if the soil has low strength for a particular structure on site, stabilization techniques are needed to increase the strength of the soil [3]. Therefore, before construction of any structure, an engineer must know about the characteristics of the soil below and around the footings, properties as well as the factors affecting their behaviours. Soil stabilization is crucial to ensure soil properties [10]. In this research, lime and glass fibers are used as the strengthening materials for the black cotton soil, because the black cotton soil is not preferable for the construction of high rise buildings due to the presence of montmorillonite in black cotton soils which possesses large swelling and shrinkage when contact with water, as lime and glass fibers provide stabilization of the soil to strengthen the cracking [7]. Adding lime and glass fibers to the soil increases the elasticity modulus and toughness of the soil, which can also change the failure mode of stabilized soil completely from brittle to ductile in nature [9].

2. Materials and Properties

2.1. Black cotton soil

The soils which become sticky when water is added and contract when it is dry condition are known as black cotton soils [11]. Generally these soils contain montmorillonite, which may cause sudden swelling and shrinkage property when it is contacted with water [5]. These swelling and shrinkage properties of the black cotton soil are leads to the decrease of the strength and engineering properties of the soil. The black cotton soils are most preferable for agricultural purposes because of the properties it has like sudden swelling when contact with water but not preferable for the construction of high raised buildings, to overcome that we are stabilizing with the stabilizing materials to improve the strength and engineering properties.

Figure: 1 Black cotton soil

Table: 1 Black cotton soil properties

Property	Black cotton soil
Max dry density	15.23
Liquid limit (%)	63
Optimum moisture content (%)	24.31
Plastic limit (%)	29
Specific gravity	2.72
Shrinkage limit (%)	13

2.2. Lime

Lime is a multifunctional mineral that can be available in different forms and find application in environmental, metallurgical, architectural and chemical industries [7]. Due to the presence of the high Calcium content in the lime which gives more strength to the soil. The fastest growing lime issued in environmental applications where lime is used in compliance with air, drinking water, wastewater and solid waste regulations [12]. During the addition of lime in the soil, the clay particles convert natural clay particles as per the requirement such as interlocking metal line structure. Individual clay soil becomes drier and less vulnerable to water content.

Figure: 2 Lime

Table: 2 Lime Properties

Property	Poly propylene
Al_2O_3	<1.5
CaO	>82.4
MgO	<0.5
Specific gravity	2.2
So_3	<0.5

2.3. Glass fibers

Glass fibres are also called as fibre glass. Glass fibers consist of numerous fine fibers of glass. Glass fibers are the non-crystalline material with a short range network structure [12]. Generally these glass fibers are collected from the waste glass products from the elevation or interior design shops which work on the glass material [4]. The glass fibers act as reinforcement when it is added to the soil as a stabilizing material and it is very strong in tension. Waste glass pieces collected from them and used as a stabilizing material. And it is more economical.

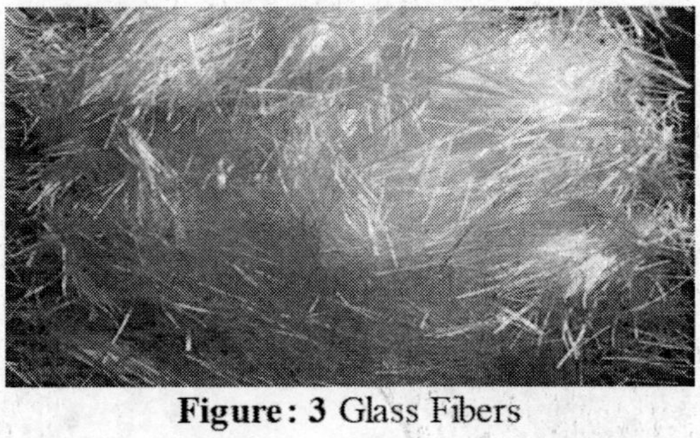

Figure: 3 Glass Fibers

Table: 3 Glass fibers Properties

Property	Glass fiber
SiO_2	53.8
Aluminum Oxide	27.3
CaO	12.62
Specific Gravity	2.31

IOP Conf. Series: Materials Science and Engineering 1126 (2021) 012070 doi:10.1088/1757-899X/1126/1/012070

3. Results and Discussion

3.1. Liquid Limit (LL) of Soil Test

This test is carried out to find out the soil liquid limit in accordance with IS: 2720 (Part 5) - 1985. This test is used to find the liquid limit of soil in which the soil behaves like liquid with smaller shear strength [6]. The number of blows utilized to contact the soil sample is noted. The graph is prepared by considering a number of blows towards abscissa and on the ordinates; water content has been taken [8]. The liquid limit is expressed by W_c.

Table: 4 Liquid limit values with different percentages of Glass fibers and Lime

%Lime	%Glass fiber	Liquid Limit%
0	0	58
2	1	57
4	2	56
6	3	55
8	4	56

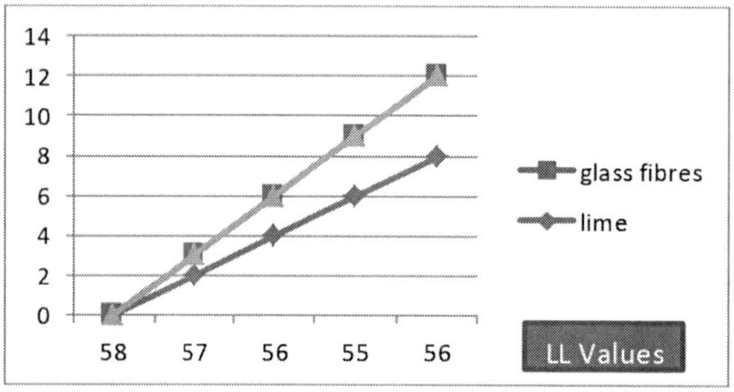

Figure: 4 Graphical Representation

3.2. Plastic Limit (PL) of Soil Test

This test is made to study the plastic constraints of the soil accordance with IS: 2720 (Part 5) - 1985. It is the value below which the soil behaves like plastic and when rolled into 3 mm diameter yarns begin to collapse [10]. The sample is prepared by passing through 425μ sieve and water is added to it after that which is rolled with hand on the glass plate at certain diameter it tends to get cracks and deformed after that we can determine its water content with the help of the oven.

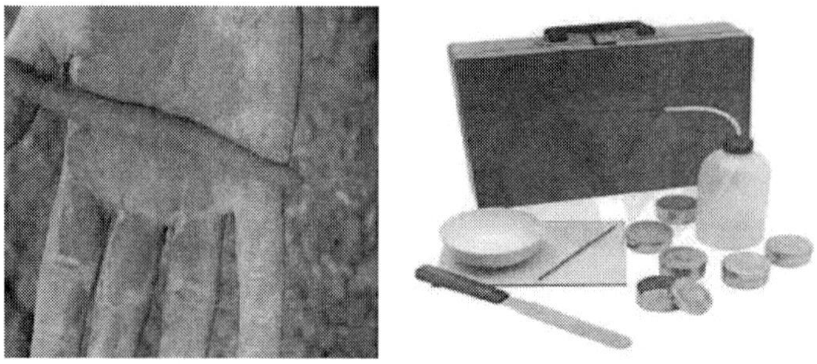

Figure: 5 Plastic limit test

Table 5: Variation of Compressive strength values with various percentages of fibers

%Lime	%Glass fiber	Liquid Limit%
0	0	53
2	1	52
4	2	50
6	3	48
8	4	51

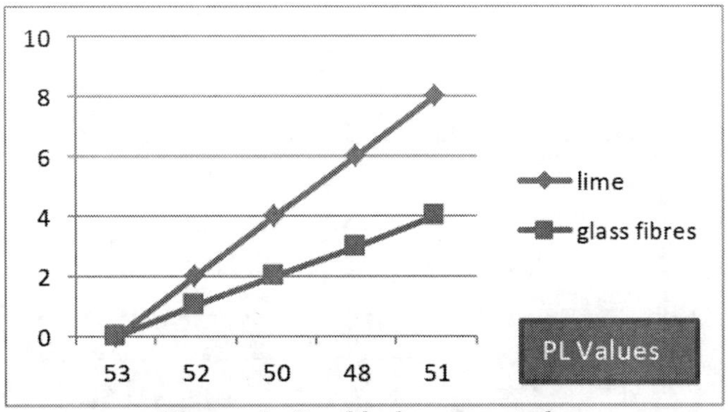

Figure 6: Graphical representation

3.3. Compaction Test

This test was conducted on version soil of 3 Kg sample. Maximum Dry Density (MDD) of the soil can be determined by conducting this test [4]. A sample of 3kg was prepared as per the codal provisions and certain content of water added to it and mixed thoroughly later on the prepared sample is filled in to the mould as layer by layer and each layer tampered with 25 blows by using a tampered rod [6]. The Maximum Dry Density can be achieved by plotting graph between the water content and dry density as abscissa and ordinate respectively. The graph of both the values with lime and glass fibers mixture soil and normal soil are shown below.

Fig: 7 Compaction Test

Table: 6 Compaction test values

Lime content	MDD
2	1.60
4	1.62
6	1.64
8	1.65
10	1.66
12	1.68

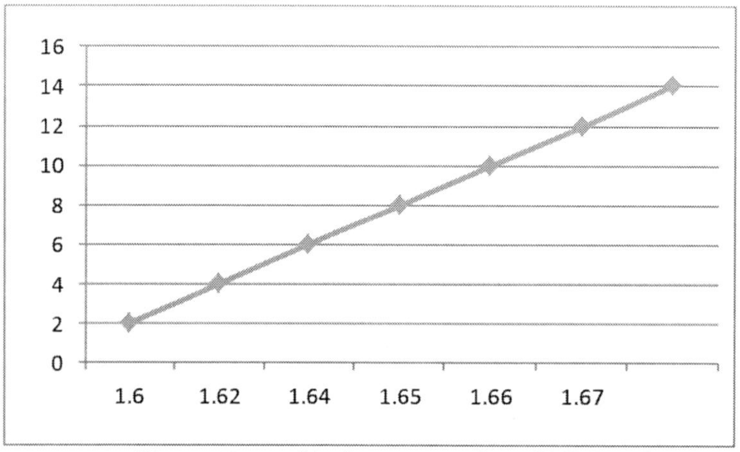

Figure 8: graphical representation of MDD values

After conducting the test, the results of the Maximum Dry Density were measured with respect to lime content (%) in the soil. From the above graphical representation, it is determined that Maximum Dry Density increases with increase of lime content in the soil due to the presence of the high amount of calcium in the lime. The highest relative increase in Maximum Dry Density is obtained from 4 % to 6 % of lime content as per the test results. Therefore we also perform a test on 5 % and obtained Maximum Dry Density of about 1.64. Therefore we consider 5 % lime is preferable as a stabilizing material in the soil in order to make structure economical.

Table: 7 Compaction test values

Glass fibers	MDD
0.2	1.70
0.4	1.71
0.6	1.72
0.8	1.73
1.0	1.71
1.2	1.70

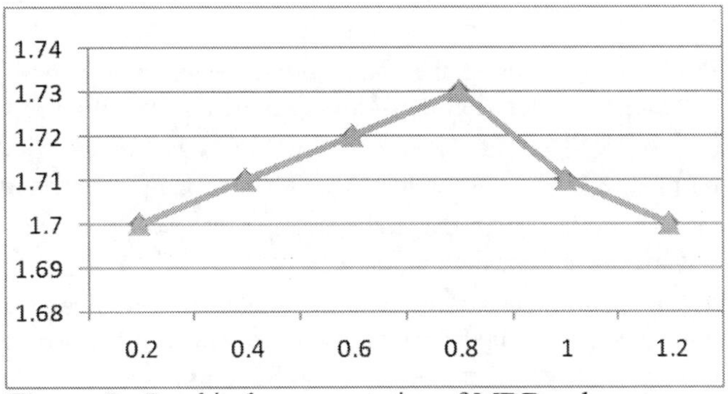

Figure 9: Graphical representation of MDD values

After conducting the test, the results of the Maximum Dry Density measured with respect to glass Fiber (%) content in the soil. Also the Maximum Dry Density has been analyzed by mixing 5 % lime with fiber content in different ratio (0.2 %, 0.4%, 0.6%, 0.8%, 1% and 1.2 %). From the experimental investigation, it is clear that Maximum Dry Density increases up to 0.8 % of polypropylene fiber content mixed in the soil after that it starts decreasing due to the presence of the tensile nature of the fibers. Higher is the Maximum Dry Density higher is the strength of the soil.

3.4. Differential Free swell Test

It is used to measure the enhanced volume of soil. It is determined using the equation written below [2]. The test was conducted for both normal soil and mixture of Lime and glass fibers in different percentages in the soil, as per IS code: 2720 part 40. The test sample was prepared as per the codal provisions, the soil which passes through the 425μ sieve and taken in to two jars one contains polar liquid and the other contains non-polar liquid.

Differential Free swell index (%) $= (V_d - V_k) / V_k * 100$

V_d = Vol. of the soil in distilled water.

V_k = Vol. of the soil in kerosene

Table: 8 Differential free swell index test values

Determination number	Parent Soil	Mixed soil
Mass of dry soil passing 425 μ sieve	10	10
Volume in water after 24 hrs V_d (cc)	14	13
Volume in kerosene after 24 hrs V_k (cc)	10	10
Average Free Swell Index	40%	30%

4. Conclusion

From the experimental investigation the below conclusions were drawn.

- It is clear that the glass fibers and lime can be used as stabilizing materials to the stabilization of black cotton soils for the construction of high raised buildings.
- The test results for the normal soil give less than compared to the stabilized soil by the glass fibers and lime powder up to certain % of the adding of these materials.

- Soil Stabilization using Glass fibers is better way to disposal of Glass instead of dumping glass in land. This gives the better strength properties than the normal soil.
- The engineering properties of the black cotton soil get improved up to certain % of glass fibers added as stabilizing material in the soil due to the tensile nature of the glass fibers.
- It is clearly observed that the lime and the waste glass powder are the best stabilizing materials for the soil to increase the strength and the engineering properties of the soil.
- The strength parameters of the soil get improved by adding the lime in the soil due to the presence of the calcium content in the lime.
- By performing experiment at varying percentage of Glass fibers and Lime Powder in the soil, we may concluded that Maximum strength properties of black cotton soil is achieved at 8% of Glass fibers and 10% of Lime Powder in black cotton soil.
- If we add more amount of lime and glass fibers to the soil which leads to the un-economical.

References

[1] Latifi, N., Eisazadeh, A., Marto, A., & Meehan, C. L. (2017). Tropical residual soil stabilization: A powder form material for increasing soil strength. *Construction and Building Materials*, 147, 827-836.

[2] IS: 2720 (Part 5,40) – 1985,Methods of test for soil, Determination of Liquid limit and Plastic limit of soil, Determination ofdifferential free swell index.

[3] Reddy, V. R., & Krishna, G. R. (2018). Improvement of Soil Characteristics Using Waste (plastic, steel and wood). *Journal of Geotechnical Engineering*, 4(3), 7-10.

[4] Razvi, S. S., Bhalke, S. B., Wadhave, M. D., Waghe, A. P., & Rathod, D. C. (2018). Soil Stabilization by using Lime. *International Journal of Engineering and Management Research (IJEMR)*, 8(2), 79-86.

[5] Khattab, S. A., Al-Mukhtar, M., & Fleureau, J. M. (2007). Long-term stability characteristics of a lime-treated plastic soil. *Journal of materials in civil engineering*, 19(4), 358-366

[6] Loehr, J. E., Bowders, J. J., Owen, J., Sommers, L., & Liew, W. (2000). Stabilization of slopes using recycled plastic pins. *Journal of the Transportation Research Board*, 1714, 1-8.

[7] Etim, R. K., Eberemu, A. O., & Osinubi, K. J. (2017). Stabilization of black cotton soil with lime and iron ore tailings admixture. *Transportation Geotechnics*, 10, 85-95.

[8] Harichane, K., Ghrici, M., & Kenai, S. (2018). Stabilization of Algerian clayey soils with natural Pozzolana and lime. *Periodica Polytechnica Civil Engineering*, 62(1), 1-10

[9] Wroth, C. P., & Wood, D. M. (1978). The correlation of index properties with some basic engineering properties of soils. *Canadian Geotechnical Journal*, 15(2), 137-145.

[10] Sofi, A. and Naidu Gopu, G. (2019) 'Influence of steel fibre, electrical waste copper wire fibre and electrical waste glass fibre on mechanical properties of concrete', in *IOP Conference Series: Materials Science and Engineering*. doi: 10.1088/1757-899X/513/1/012023.

[11] Satyam Tiwari & Nisheet Tiwari (2016). Soil stabilization using waste fiber materials. *International Journal Of Innovative Technology And Research* ,4(3), 2927 – 2930.

[12] Elias, T., George, S., Zachariah, A., Sulfi, S., & PS, S. (2016). Comparative Study of Soil Stabilization Using Human Hair and Lime. *International Journal of Scientific & Engineering Research*, 7(2).

Thermal analysis of fin materials for engine cylinder heat transfer enhancement

Shankar Durgam*, **Aniket Kale, Nikhil Kene, Akshay Khedkar, Sanket Palve, and Nikita M. Gawai**

Department of Mechanical Engineering, College of Engineering Pune - 411005, India

E-mail: *sod.mech@coep.ac.in

Abstract. In automobiles the internal combustion engine cylinder is one of the major component which is subjected to high thermal stresses and temperature variations. Engine cylinders come up with fins or extended surfaces. The heat transfer enhancement from engine cylinder is mainly depends on fin material and shape. We can use different fin materials and fin geometries in order to increase heat transfer rate. In this work heat transfer rate through fins with different materials was evaluated. The fin geometries such as the rectangular fins attached to the cylinder block. In this analysis different physical and thermal properties along with cost associated with the fin materials are considered. The 3D model of geometries are created using CATIA - V5, and thermal properties are analyzed using commercial software ANSYS FLUENT 2020 R1. From the results of steady state thermal analysis of engine cylinder it is found that the value of heat flux increased through cylinder block to the surroundings. The weighted point method applied to find performance of fins based on material that is best suited for manufacturing of fins.

Keywords: Thermal conductivity, specific heat capacity, heat transfer enhancement, fins, ANSYS.

1. Introduction

The engine cylinder is to be cooled in order to prevent it from the ill effects of the high temperature and thermal stresses. Internal combustion engine cooling uses either air or liquid cooling to remove the heat from IC engine cylinder block. For increasing the rate of heat transfer we require either large surface area or high value of convective heat transfer coefficient, but increasing heat transfer area is easy and effective method. This is accompanied by using extended surfaces or Fins. Kummitha and Reddy [1] analyzed fins made up of different materials like aluminium alloys, magnesium alloys and gray cast iron and compared them on parameters like strength, weight, and concluded that aluminium A 380 is better among selected materials. Kanna [2] have numerically studied thermal performance of straight fin and wavy fin of different pitch for motorcycle engine. They found that 8 mm pitch straight fin is a better choice over wavy fins in heat transfer point of view. Thornhill et al. [3] conducted experiments for heat transfer from aluminum alloy engine cylinders with fins in to a free air. They found heat transfer enhancement with air velocity in the range 2 - 20 m/s with separated and shortened fins. Dubey et al. [4] studied thermal analysis of engine cylinder with thick tip fin having varying slot sizes and material. They found that 75 mm slotted fins have maximum heat transfer and Aluminium alloy 2014 having 75 mm slotted fin engine with thick tip fin have maximum heat transfer rate.

karthik et al. [5] analyzed the fin performance by taking into consideration four different materials and found that Aluminum alloys is best among them. Several authors [6, 7, 8, 9, 10, 11, 12] have studied heat transfer from different fin geometries and fin parameters for engine cylinder cooling applications. The above literature shows many studies are available on heat transfer enhancement from fins for engine convective air cooling applications. Though the studies on heat transfer studies from different fin materials are limited. Therefore the present study is focused on the heat transfer from the fins made up of different materials. In this work the properties of materials viz., density, thermal conductivity, specific heat capacity and strength are considered to show their effect on fin heat transfer enhancement. The studies have been performed on the thermal analysis using ANSYS software for finned surfaces having different materials, and considering cost as the main influencer on selection of the material.

2. Extended surfaces or Fins

Finned surfaces are commonly used in practice to enhance heat transfer, and they often increase the rate of heat transfer from a surface several fold. Fins are having different applications in heat exchanging devices such as economizers, heat exchangers etc. Applications of fins:

(i) Engine cylinder cooling, radiators in cars
(ii) Computer CPU heat sinks.
(iii) Fins are most important where convective heat transfer coefficient (h) have lower value.
(iv) Fins are commonly used as heat management in electrical applications.
(v) They are also used in newer technology such as hydrogen fuel cells and heat exchangers in power plants.

3. Methodology

The methodology followed in the present work is given in flow chart in Fig. 1

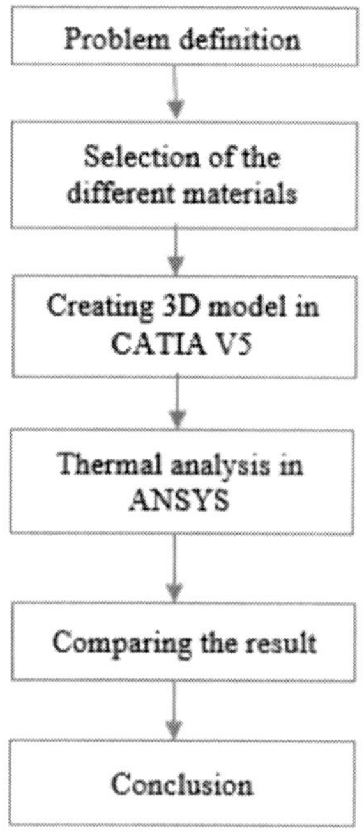

Figure 1. Flow chart showing methodology used in the study

IOP Conf. Series: Materials Science and Engineering 1126 (2021) 012071 doi:10.1088/1757-899X/1126/1/012071

4. Selection of different materials

The fins should be made up of such materials which have following properties:

- It should have high thermal conductivity
- t should have low specific heat capacity
- Relatively cheap and easily machined
- Must be having low density

We have chosen four different materials which satisfies most of the desired properties which mentioned above.

(i) Gray cast iron: Gray cast iron is best material which satisfies most of the desirable properties. It is generally used for making engine block because of its ability of dimensional stability under heating or thermal stress.

(ii) Aluminum (Al) alloy 6061 and Aluminum alloy 356: They are most widely used non-ferrous metals. They offer following advantages:
- Low specific gravity.
- Ease of fabrication.
- Corrosion resistance.
- High thermal conductivity.

(iii) Copper alloys: Copper possesses excellent electrical and thermal conductivity. It can be easily cast, machined and brazed. It has good corrosion resistance; but has poor strength. So it can be alloyed with zinc in various proportions for making substance according to the need of application.

Table 1. Thermophysical properties of various materials

Properties	Materials			
	Gray Cast Iron	Al alloy - 6061	Al alloy - 356	Brass C37700
Density, kg/m^3	7200	2713	2685	8267
UTS* Mpa	240	313.1	246.7	502
YTS** Mpa	230	259.2	172.8	367.4
Young's modulus, GPa	110	69.1	72.1	99.95
Sp. heat capacity, J/kg K	447	914.71	894.21	377.1
Thermal conductivity k, W/m K	52	155.12	158.75	107.66

UTS* Ultimste tensile strength, YTS** Yield tensile strength

5. Creation of 3D model in CATIA V5

IC Engine cylinder block along with fins was modelled in Catia V5. First, Cylinder block was modelled and then outer surface which is having fins were modelled. Dimensions of the model are as follow: Bore of cylinder = 50 mm
Stroke length = 64 mm
Length of fins = 13mm
Thickness of fins = 3 mm
Fir. 2 shows the 3D model of assembly of cylinder block and fins.

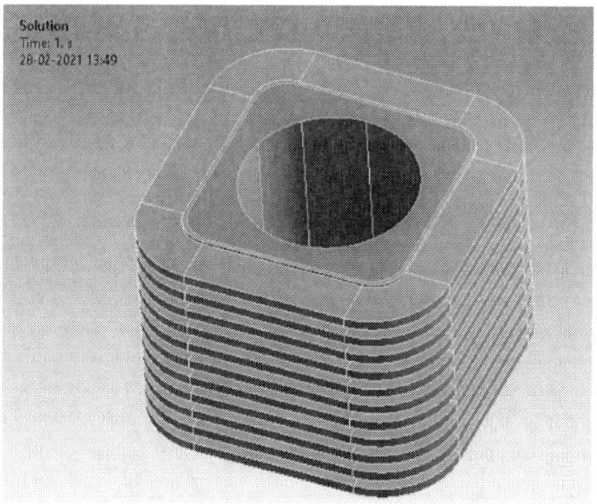

Figure 2. 3D model of cylinder block along with fins

6. CFD Analysis

The 3D geometry of model is then taken to the ANSYS FLUENT and the Tetrahedron meshing of the model is generated. Then the naming of sections is given in the model for computational purpose. The steady state thermal analysis of fin surface have been performed. The interior wall of engine cylinder block is chosen as inlet and outer surface along with fin surface as outlet. The inlet portion has given temperature of 300 $°C$. Outer surface is surrounded by stagnant air and has temperature of 27 $°C$. The four different materials considered are cast iron, aluminum alloy 6061, aluminum alloy 356 and brass C37700. In this analysis we have taken cylinder block of Gray cast iron and by changing the fin material for each subsequent analysis we found out simulation results. In the first case we applied fins of Gray cast iron over the cylinder block. Steady state thermal analysis using CFD software gives the results of temperature and heat flux distribution of cylinder block assembly attached with fins made of cast iron. Fig. 3 shows temperature distribution and heat flux distribution for grey cast tron fin material. From the Fig. 3(a) temperature distribution plot gives the maximum temperature of 300 $°C$ and minimum temperature of 291.2 $°C$. The heat flux distribution plot shown in Fig. 3(b) shows the minimum heat flux value of 1069 W/m^2 K and the maximum value of 24838 W/m^2 K. It is observed that the maximum values of both the temperature and heat flux occurs in the inside of the cylinder wall.

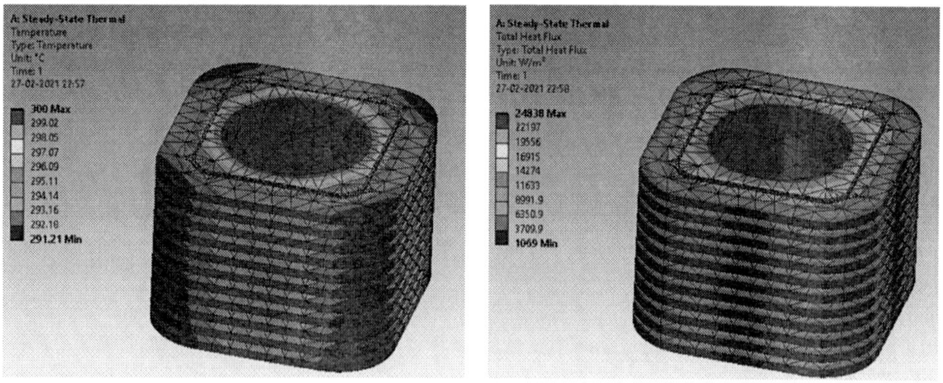

(a) Temperature distribution (b) Heat flux distribution

Figure 3. Temperature and heat flux distribution for grey cast iron fins

In second case fins of Aluminium alloy 6061 over the cylinder block are considered. Fig. 4 shows temperature distribution and heat flux distribution for aluminum alloy 6061 as fin material.

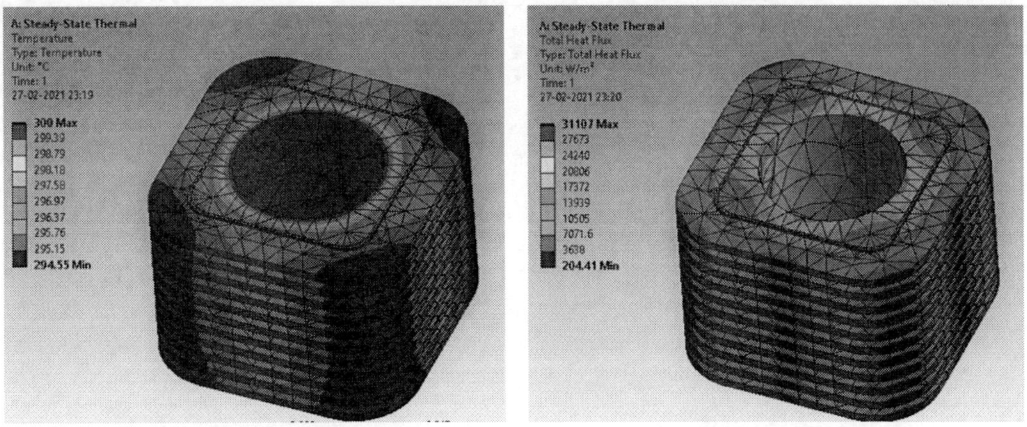

(a) Temperature distribution (b) Heat flux distribution

Figure 4. Temperature and heat flux distribution for aluminum alloy 6061 as fin material

Fig. 5 shows temperature distribution and heat flux distribution for aluminum alloy 356 as fin material.

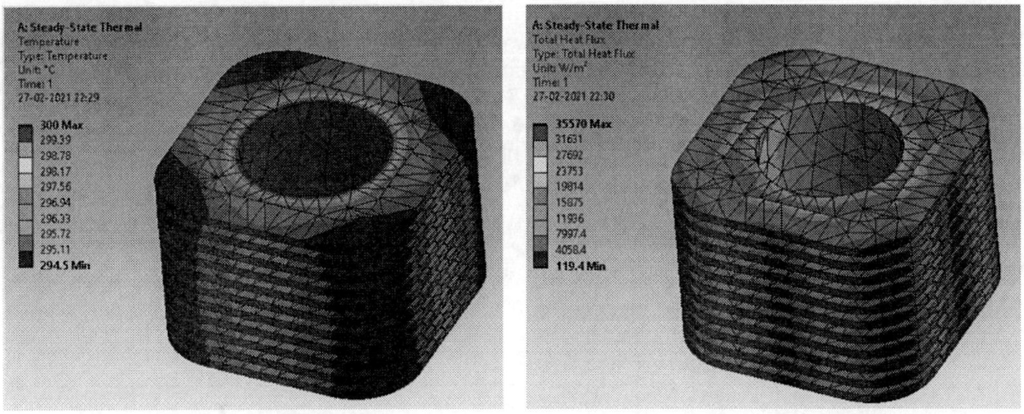

(a) Temperature distribution (b) Heat flux distribution

Figure 5. Temperature and heat flux distribution for aluminum alloy 356 as fin material

Fig. 6 temperature distribution and heat flux distribution for Brass C37700 as fin material

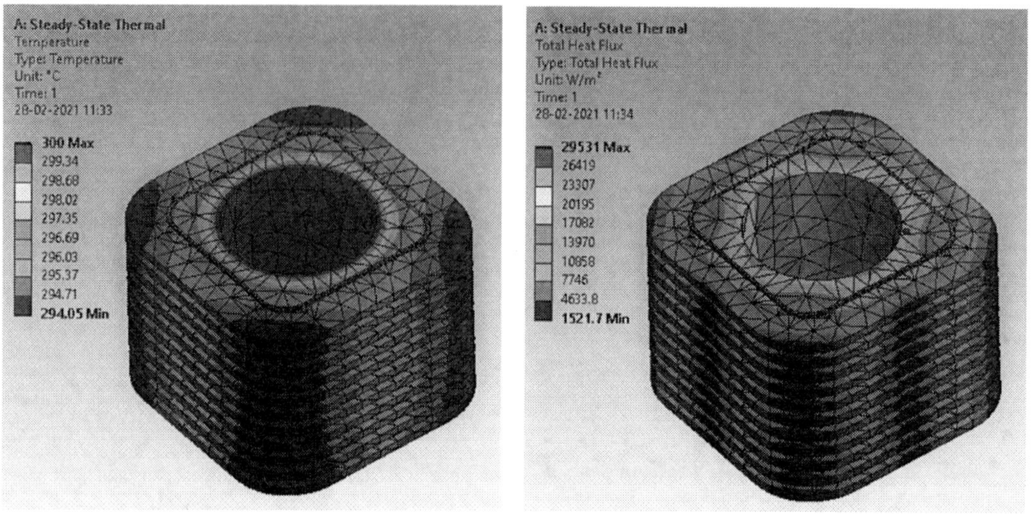

(a) Temperature distribution (b) Heat flux distribution

Figure 6. Temperature and heat flux distribution for Brass C37700 as fin material

7. CFD Results

Table 2 gives data of results from ANSYS simulation. The maximum and minimum temperature in each case is mentioned. The minimum temperature is clearly a function of thermal conductivity (k) and thermal diffusivity (α). The materials which have higher value of minimum temperature is a clear indicator of better material in terms of its thermal properties. So from the data aluminum alloy 356 is best material among the four materials which we have chosen. The average heat flux value also give idea about, which material is to be chosen; by this criterion also aluminum alloy 356 is better among all. The weight which is the function of density of material is also an important parameter which influence the selection of the material. The aluminum alloy 356 is having lowest density among the four materials considered in this study. So from above analysis aluminum alloy 356 seems to be a better material among the four materials to be used as material for manufacturing of fins. The simulation software results are compared with the weighted point method to find the better material.

Table 2. Temperature and heat flux distribution along with weight of fins for various materials from ANSYS report

Materials	Temperature $^{\circ}C$		Average heat flux, h, W/m^2	Weight of fins, kg
	Maximum	Minimum		
Grey cast iron	300	291.2	12352	0.99755
Al alloy 6061	300	294.6	132785	0.3758
Al alloy 356	300	294.6	14072	0.372
Brass C37700	300	294.1	13824	1.1454

8. Weighted point method

The weighted point method is a recent technique developed for selection of material. The method consist of different steps; first step consist of study of application and preparing list of material for that application. So we have chosen four different materials for comparison. The desirable properties then assigned some values according to their importance. We have selected five properties as influencers and their value is mentioned in respective bracket.

(i) Thermal conductivity, k (5)

(ii) Thermal diffusivity, α (4)

(iii) Cost (3)

(iv) Density, ρ (2)

(v) Ultimate tensile strength (UTS) (1)

The thermal conductivity and thermal diffusivity have given more weighted (value) that is 5 and 4 respectively because fins should be made up of materials having better thermal properties. Cost is also a main influencer in material selection; so it has given value 3. The fins should be of lighter weight so we have given some importance to density also. Similarly we have calculated the value of each material and the materials which have higher weight (value) is to be selected and other materials should be given priorities according to their total value. Table 3 gives data of total points for each material by using weighted point method.

Table 3. Total points for each material by using weighted point method

Material	Material properties contribution as per weighted point method					Total
	k	α	Cost	Density, ρ	UTS	
Cast iron	0.54907	0.36041	1.54639	0.31153	0.18436	2.95176
Al alloy 6061	1.63791	1.39434	0.41237	0.82677	0.24051	4.51190
Al alloy 356	1.67624	1.47491	0.78298	0.83539	0.18951	4.95903
Brass C37700	1.13678	0.77034	0.25773	0.26631	0.38562	2.57678
Sum	5	4	3	2	1	15

9. Conclusions

The studies on thermal analysis of engine cylinder having fins of different materials for heat transfer enhancement, and to find out the best material among the selected materials following conclusions were made. From the five main parameters which influence the selection of the material and by comparing this parameters we came to the final conclusion. From the thermal analysis which is done by using ANSYS it is clear that Aluminum alloys are better compared to the Gray cast iron and copper alloys. We also used weighted point method to select the best material among them. By doing the calculations according to weighted point method we made the final table which clearly depicts that Aluminum alloy 356 is best among the 4 materials. This method also gives us the selection preferences of the materials to be chosen. The results obtained from ANSYS analysis and weighted point method are complementary to each other and both shows us that Aluminum alloy 356 has better combination of required properties to be chosen as fin material.

References

[1] O. R. Kummitha, B. Reddy, Thermal analysis of cylinder block with fins for different materials using ansys, Materials Today: Proceedings 4 (8) (2017) 8142–8148.

[2] I. Vinoth Kanna, Modelling and thermal analysis of air-cooling system with fin pitch in ic engines, International Journal of Ambient Energy 41 (11) (2020) 1252–1260.

[3] D. Thornhill, A. Graham, G. Cunnigham, P. Troxler, R. Meyer, Experimental investigation into the free air-cooling of air-cooled cylinders, SAE transactions (2003) 2046–2057.

[4] D. Dubey, D. Singh, H. Thakur, et al., Thermal analysis of engine cylinder having thick tip fin with varying slot sizes and material, Materials Today: Proceedings 4 (8) (2017) 7636–7642.

[5] S. Karthik, K. Muralidharan, B. Anbarasan, material selection for fin based on thermal analysis using ansys and ann, International Journal of Mechanical Engineering and Technology (IJMET) 9 (11) (2018) 560–567.

[6] J. Pandit, M. Thompson, S. V. Ekkad, S. T. Huxtable, Effect of pin fin to channel height ratio and pin fin geometry on heat transfer performance for flow in rectangular channels, International Journal of heat and mass transfer 77 (2014) 359–368.

[7] P. S. Chaitanya, B. S. Rani, K. V. Kumar, Thermal analysis of engine cylinder fin by varying its geometry and material, IOSR Journal of Mechanical and Civil Engineering (IOSR-JMCE) 11 (6) (2014) 37–44.

[8] Z. K. Memon, T. Sundararajan, V. Lakshminarasimhan, Y. Babu, V. Harne, Parametric study on fin heat transfer for air cooled motorcycle engine, Tech. rep., SAE Technical Paper (2005).
[9] P. Sagar, P. Teotia, A. D. Sahlot, H. Thakur, Heat transfer analysis and optimization of engine fins of varying surface roughness, Materials Today: Proceedings 4 (8) (2017) 8565–8570.
[10] A. A. Al-Abidi, S. Mat, K. Sopian, M. Sulaiman, A. T. Mohammad, Internal and external fin heat transfer enhancement technique for latent heat thermal energy storage in triplex tube heat exchangers, Applied thermal engineering 53 (1) (2013) 147–156.
[11] G. Babu, M. Lavakumar, Heat transfer analysis and optimization of engine cylinder fins of varying geometry and material, IOSR Journal of Mechanical and Civil Engineering (IOSR-JMCE) e-ISSN (2013) 2278–1684.
[12] M. M. Alam, M. D. Sushmitha, Design and analysis of engine cylinder fins of varying geometry and materials, International Journal of Computer Engineering In Research Trends, ISSN (O) (2016) 2349–7084.

Materials selection for hybrid and electric vehicle battery pack thermal management: A review

Shankar Durgam*, Priyanka Datir, Onkar Tawase, Dipak Savant, Ganesh Tapkir, Ravi M. Warkhedkar, Nikita M. Gawai

Department of Mechanical Engineering, College of Engineering Pune - 411005, India

E-mail: *sod.mech@coep.ac.in

Abstract. This paper reviews materials for hybrid and electric vehicles battery pack thermal management required for efficient working of batteries in any climate conditions. Lithium-ion (Li-ion) battery cells are being used for electric or hybrid vehicles because they having high density of energy and long life cycle. Battery lifetime and performance of Li-ion batteries depends on operating temperature. The higher operating temperature increases capacity and decreases life of Battery. Thermal management strategies which are based on air, water, phase change material (PCM) and heat pipe are discussed and compared. After comparison, an efficient system of temperature management for battery pack is discussed. Phenomena of creating heat and temperature issues of Li-ion batteries are discussed in this paper. Then different strategies are reviewed and classified based on thermal cycle options. Improving the existing temperature control system and testing new technologies to reduce thermal issues is necessary, and efforts should be made to prioritize the thermal management system to improve temperature uniformity throughout battery pack, life of battery, and increase protection to battery packs.

Keywords: Lithium-ion battery, thermal characteristic, thermal management system, control strategies, Heat generation, phase change material.

1. Introduction

In recent times, electric vehicles are becoming more attractive and popular due to their aspects like reduce the energy loss, reduce the pollution, environment friendly. There are different types of batteries such as lithium-ion (Li-ion), lead–acid, nickel metal hydride (Ni-MH) are used for electric vehicles applications because, their high specific power and specific energy density. Compared to other batteries, Li-ion batteries mostly used in electric vehicles as well as in hybrid electric vehicles, because they have high power, light weight, does not required to charged continuously. When a lithium-ion battery operates at low temperature the lithium dendrites build up in battery anode and result is State of health (SoH) of the battery decreases significantly. [1, 2, 3, 4, 5]. When the battery temperature exceeds the fixed point, the most powerful reaction occurs out of control [6]. In addition, when a battery reaches a thermal escape, 12% of total heat released by battery is sufficient cause hot explosion in nearby batteries [7]. This is especially dangerous when using lithium-ion batteries. battery thermal management system (BTMS) is a mandatory for lithium-ion battery systems, at high temperatures [8, 9, 10, 11, 12]. There are two main purpose of battery thermal management system (BTMS): 1) keeping batteries running under optimal conditions and improving electrical performance and life of battery 2) preventing heat dissipation from occurring and improving safety. Most battery-powered heat transfer systems, like air conditioning, liquid cooling, and phase change (PCM). Therefore the main objective of this work is to explore the different materials for battery thermal management technologies available from the previous literature applicable for recent hybrid and electric vehicles.

2. Li-ion batteries for electric vehicles(EV) and Hybrid electric vehicle (HEV)

The most trusted battery for electric and hybrid vehicles, their general considerations, mechanism and configuration, market survey etc. are discussed in this section.

- General Considerations: The efficiency of electric vehicles (90%) is high compared to electric-powered vehicles (30%). And electric vehicles operate at high torque with low revolution speed. It has a fast torque feedback and converting kinetic energy into electricity from brake torque (Regenerative braking system to HEVs) is also attractive. The problem of heat loss in electric cars is less than that of an engine-driven car. Compared to liquefied petroleum or gasoline, the battery power of batteries is very low, which means that an electric car needs more battery cells to obtain the same performance given by an engine-powered car. Li-ion batteries widely used in electric vehicles because of their high energy content (140 Wh/kg and 500 Wh L respectively) compared to nickel-metal hydride (NiMH) batteries (65Wh/kg and 150 Wh/L), owns the HEV market.
- Mechanism and configuration: Lithium-ion battery method is given in Fig. 1 During charging, Lithium+ ions moves from negative electrode (cathode) to positive electrode (anode) through separator diaphragm to form discharge cycle, and vice versa. The cathode is make from graphite. An anode is made of lithium containing a substance such as: coated oxide (viz., lithium cobalt oxide - $LiCoO_2$), poly-anion (viz., lithium iron phosphate - $LiFePO_4$) or spinel (eg Lithium manganese oxide - $LiMn_2O_4$). Lithium salt solution which is not an aqueous solution such as ethylene carbonate or diethyl carbonate is used as an electrolyte. Current collector of negative and positive electrode is made from copper (Cu) and aluminum (Al) respectively. The mechanism of a lithium-ion battery with the charging and discharging mechanism is shown in Fig. 1.
- Market Survey: Future growth of the EV could predicted by introducing a great deal of work done by Special niche manufactures. It shows that NiCd and NiMH are the two most sought after EV batteries in 2000. The inconsistent intention set by manufacturers says that undeveloped battery automation, like limited proof of public electric vehicle weight production could be observed. In current years, lithium-ion batteries are well-known among EV manufacturers for their durability, high power output, high range of development, and fast charging time.

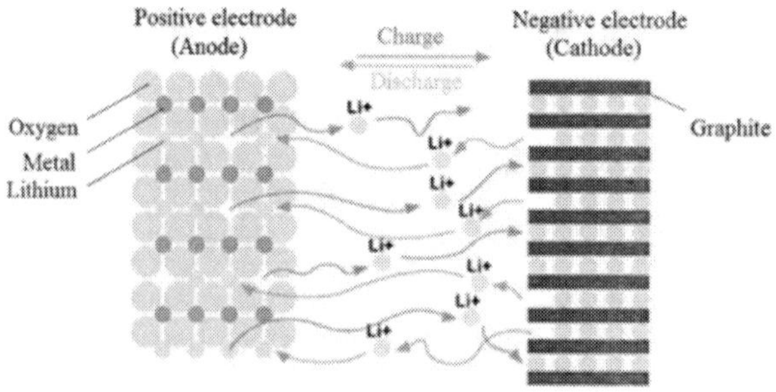

Figure 1. Lithium-ion battery mechanism (A123Systems. (2008))

IOP Conf. Series: Materials Science and Engineering 1126 (2021) 012072 doi:10.1088/1757-899X/1126/1/012072

3. Temperature problems in Li-ion battery

- The performance of this battery cells depending upon operating temperature and operating capacity. Li-Ion cells gives good result when it working within sufficient electrical, thermal energy. On the other hand, the cells will be damaged and will not be replaced.
- Charging power increases due to over-voltage over stable power supply, leading to current flow, which creates problems.
- With an anode, the current copper collector deteriorates. It causes an increasing battery discharging capacity also power of battery, however, ions of copper reduced as an irreversible copper. The condition is very dangerous and results in short circulation between the anode and the cathode.
- Chemical reaction rates are related to line and temperature. As operating temperature decreases the reaction rate and the current carrying capacity during charge or discharge decreases. We says that, power of battery reduced.
- Operating temperature exceeds the acceptable limit.
- Battery life is reduced due to low temperature.

4. Operating temperature

For-increasing battery performance parameters, it is important to avoid thermal issues like heat losses, overheating by controlling temperature within limits. Battery power and battery cycle life are affected by the temperature, as shown in Fig. 2, and Fig. 3. At the same time, temperatures divides same to ensure battery life, lifespan. That is the reason why a battery temperature control is required for a battery system. When the temperature rises between 20 $°C$ - 40 $°C$, power of battery reaches its peak, as shown in Fig. 2

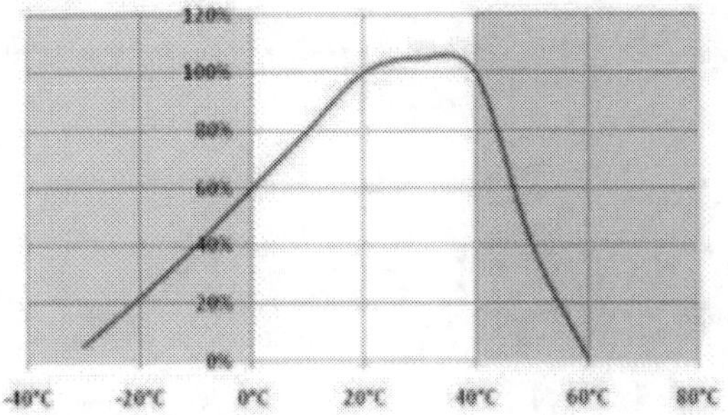

Figure 2. Grapth between power of and temperature [13]

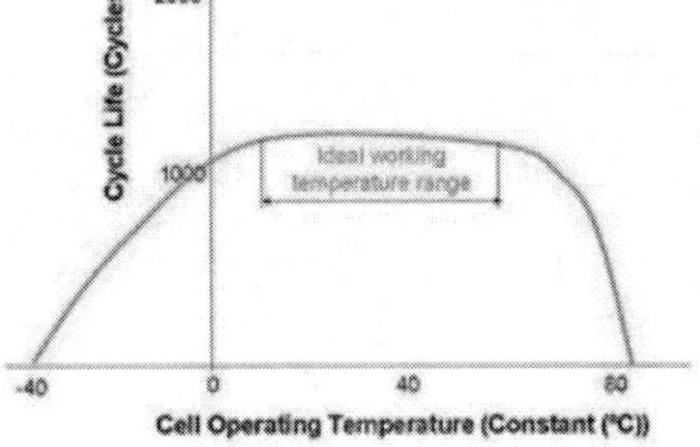

Figure 3. Graph between Battery Cycle life and temperature (Electropaedia, 2014)

Due to anode plating Lifecycle of battery decreases slightly below 10 $°C$ and decreases rapidly above 60 $°C$ because of damage to material of electrode, as shows in Fig. 4. Temperature can control between 20 $°C$ and 40 $°C$ protects function, health of the cycle.

5. Battery thermal management strategies (BTMS)

The performance of automobile (EV and HEV) is mainly depends upon the battery thermal management. There are various strategies available for battery thermal management for obtaining better performance and life of battery. These strategies includes: air cooling and heating, liquid cooling and heating, cooling and heating of refrigerant, phase change materials (PCM) are mainly discussed here.

5.1. Air cooling and heating

This program uses the spirit as a community. The air comes from the atmosphere (static air cooling/heating) and the modified air comes behind the heater or evaporator air conditioner (active air cooling/heating). The system is limited to certain hundreds of watts/cooling power watts while in the operating system, power restricted to 1 kW, (Valeo, 2010). In both of cases power is limited their supply supplied because of blower. There is no need to add additional air intake, as the air system provides heating / cooling and ventilation function. But waste air not connected again to this cabin. In some application, air-conditioning is installed back of battery to return heat to exhaust air. This prevents the combination of exhaust air, also provide more energy saving at the same time.

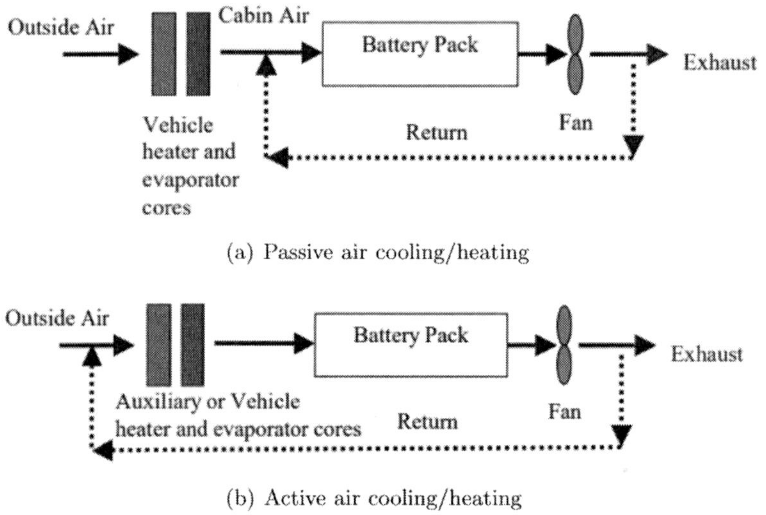

Figure 4. Passive and active air cooling/Heating [14].

5.2. Liquid cooling and heating

In this system the liquid is used which transfer heat. We have two types of beverages that can be used for hot management systems. One is dielectric liquid and the other is a liquid. Dielectric liquid interacts directly with battery cells for example mineral oil. The running liquid can only interact with battery cells indirectly, for example mixture of ethylene glycol and water.

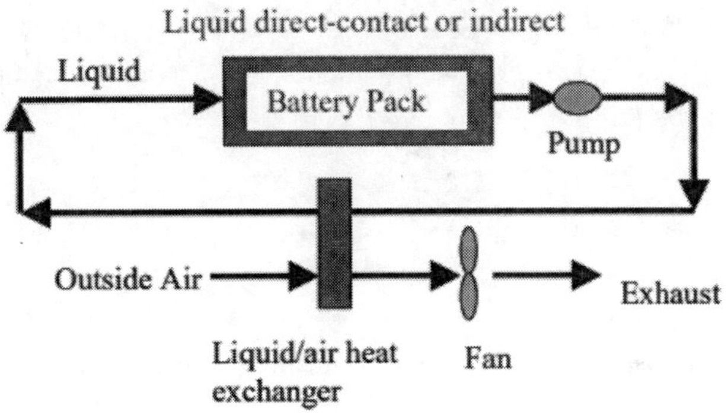

Figure 5. Passive liquid cooling/Heating [14].

Radiator is used as a cooling heat sink in the manufacturing process. Fig. 6 shows that the water system is inactive. In the case of a closed pump it is used to move the fluid. Battery release heat, it absorbed by fluid, it emits heat through radiator. Cooling capacity depends upon temperature difference between atmospheric air and battery. To improve performance of cooling fans are placed behind radiator. Liquid cooling system does not work if the dry air is above the battery temperature.

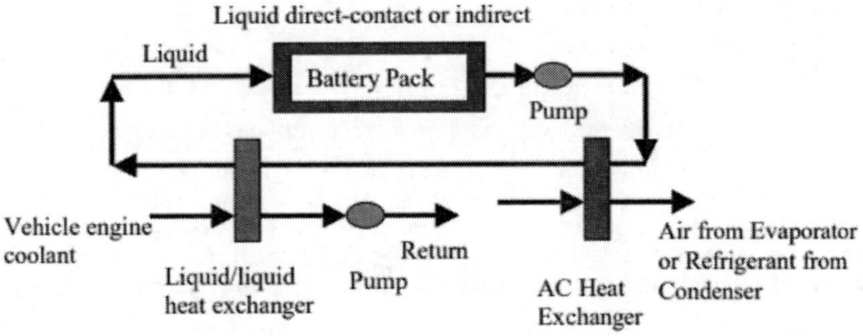

Figure 6. Active liquid cooling/Heating [14].

There are two pitfalls in this figure. The top loop called main loop. Bottom called second loop. Main loop similar to loop in an empty water system, in which fluid is transmitted through a pump. The second loop is air loop. High temperature switch acts as evaporator used to cooling operation, which used to connecting two loops together. In heating operation, four-way valve can replaced, high-temperature machine acts as condenser and the low-temperature machine acts as an evaporator. The heat function of the loop is known as heat pump loop.

5.3. Cooling and heating of refrigerant

The direct refrigeration system has A/C loop which similar to an active liquid system but in the refrigerator the system uses it directly as a battery-powered heat transfer fluid. The formal arrangement of ooling and heating of refrigerant is shown in Fig. 7.

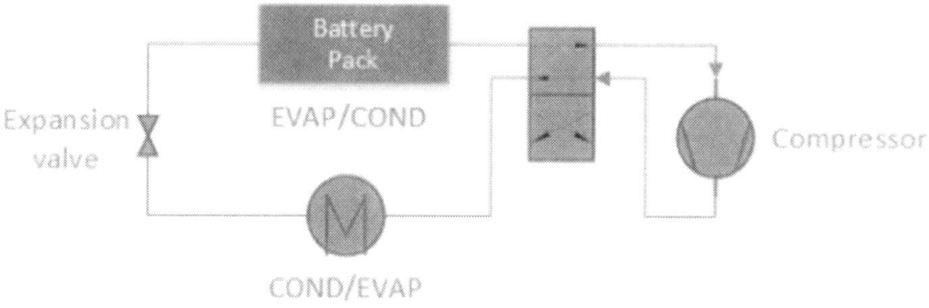

Figure 7. Direct refrigerant cooling system [14].

5.4. Phase change material (PCM)

Phase-change material (PCM) is a method of temperature management of battery packs. It has a high hidden temperature and has a desirable melting point due to its ability to store and produce high energy. Heat will transmitted from battery, It creates heat, then deliver to PCM. It eliminates need for effective cooling / heating during most of the work because it takes more time to increase in temperature when minimum temperature is low, during hot days. It keeps the battery below the proper temperature, decreases heat slope in battery, it resolves the difference between high heat retention efficiency, low thermal conductivity (0.26 W/m K), there are many ways to make composite PCMs that include:

(i) Installed metal matrix in PCM;
(ii) overheating of stress equipment
(iii) addition of thermal conductivity to paraffin and
(iv) Improving thermal energy storage systems with unfinished exhausted structures.

The working of PCM mechanism is shown in Fig. 8 for performance of battery cells. In melting condition, PCM absorbs heat and stored it as ambient temperature up to higher volume. Temperature maintained in melting point for specific duration also increase in temperature takes time. PCM is use as conductor, buffer in hot battery handling unit. Single-phase PCM used as combination of an air conditioning system or liquid cooling system for battery temperature Management.

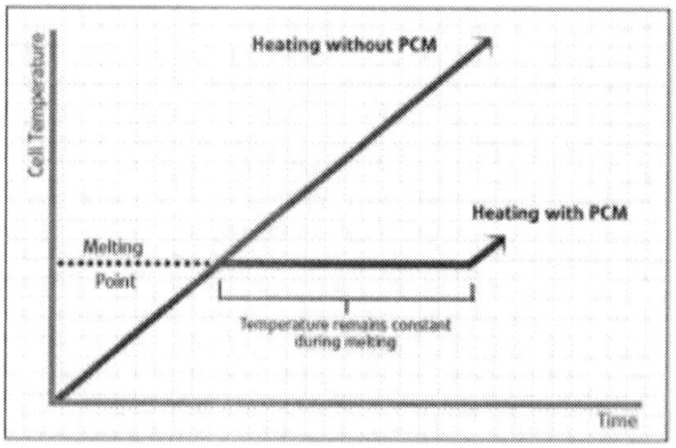

Figure 8. The working mechanism of PCM on battery cells (Charged, 2014).

5.5. Thermo-electric module

There are two ways to increase the cooling/heating capacity of a passive air system. Another is to use thermo-electric modules. Other use of heat pipe. Thermo-electric module is able to convert electrical energy into temperature differences and vice versa. The scheme diagram is shown in Fig. 9. To increase heat transfer by forced convection two fans are included. By combining air system and thermo-electric unit, system has capacity to chill battery which less than that of temperature of air, but power restricted 100 W, less than 1 kW. (Valeo, 2010). It is simpler to switch between cooling and heating temperature by reversing the size of the electrodes.

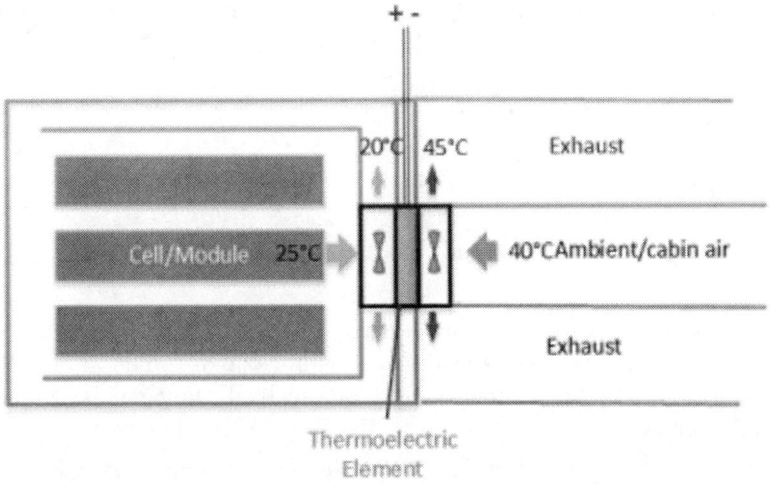

Figure 9. Thermoelectric cooling/heating system [14].

5.6. Heat pipe (HP)

Heating pipe is another way to improve the performance of synthetic air unit. Schematic of heating pipe is shown in Fig. 10. Flat copper part of heat pipe is under incomplete vacuum. To make capillary structure using sintered copper powder. Water is used as the active liquid in the heating pipe. Water on evaporator side absorb heat and evaporate below 100 ° C because of inside pressure is low. water inside condenser transfer heat to that area and returns to liquid state again. This process happens repetitively. Given Schematic indicates formation of cooling pipe which is inside cooling system. Battery is located at the bottom of the heating pipe on the evaporative side. The cooling wings used for heat absorbs which is located inside heat pipe on side of twist. According to experimental outcome (Tran, 2014), heat pipe reduces thermal resistance by 32% in natural convection compared to a system no use of heat pipe. Temperature losses reduce by using air velocity convection method are about 20%. Fig. 11 show the heat pipe cooling system.

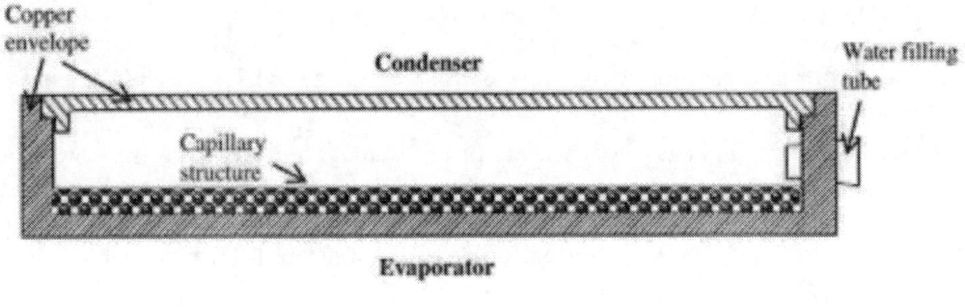

Figure 10. Structure of the heat pipe [15].

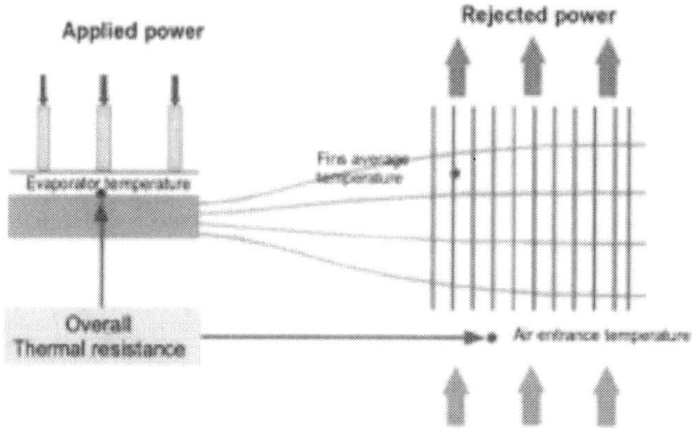

Figure 11. Scheme of heat pipe cooling system [16].

6. Observations from different technologies

(i) Forced-Air Unit: Forced-air cooling unit has simple, highly reliable, efficient and easy-to-use structure, but has some problems such as poor heat management (in addition to active temperature), unusual uniformity of heat distribution and hot escape.

(ii) Liquid System: The ambient temperature affects the idle cooling system, because the temperature of the atmosphere depends upon radiator, radiator diffuse heat by temperature difference between liquid and surrounding temperature. Under normal condition it works good, but at high temperatures not sufficiently work. Active cooling systems work best with heat. It can control temperature of battery within operating range and maintain distribution of temperature between cell parallels due to the high cohesive cooling temperature. Because of the many interlocking and operating parts, whole model is complex and difficult to maintain. As a result there are potential for leaks. Direct refrigeration cooling systems work better than an active cooling system because they can use the refrigerator directly for cooling purpose without using the refrigerator to cool down first, then coolant is used for cools system. Limitations of cooling system in the refrigerator are complex configuration, heavy storage, potential leaks and more.

(iii) PCM System: PCM cooling systems are very capable of heat management. Under atmospheric temperatures between 44 $°C$ - 52 $°C$, inside temperature of cell package remains below 56 $°C$ due to high temperatures and subtle temperatures. A great feature of PCM is it best suited for cold working conditions or surrounding. As temperature of battery pack increases above melting point of PCM, the temperature is maintained as ambient temperature. When atmospheric temperature maintained below melting point of the PCM it will be released to the battery module. Forming a PCM without new elements is easy, has less weight and requires less space. But it is hot and driven by electricity and volume changes due to the PCM-graphite feature.

(iv) Thermo-electrics: Thermo-electrics with minimal structure and light weight, it turns heating element into effective cooling agent with converting coating. It has no moving parts so it reduces wear, is reliable, durable and requires minimal care. And it is easy to replace it if it fails. Moreover, the performance is quiet and non-vibrating. Fig. 12 shows a combined liquid cooling system.

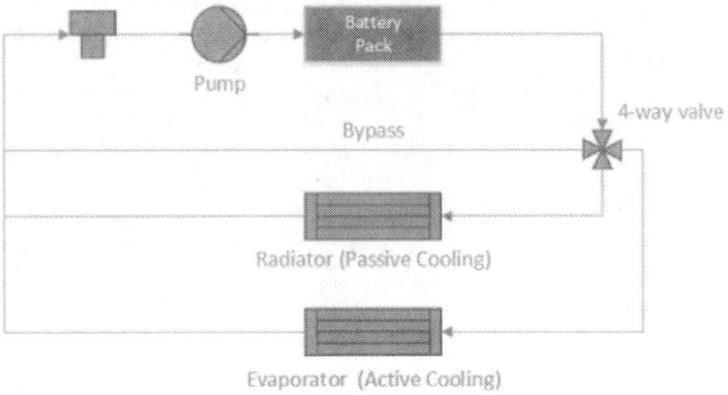

Figure 12. The combined liquid system [14].

7. Recommended cooling system

7.1. HP +PCM

Thermal battery management using a combination of heat pipe and PCM introduced. Temperature of battery of electric is kept below 49°C. When using a heat pipe only at temperatures of 22 W, lower the ground temperature from 45 ° C to 37 ° C, and at 50 W decrease from 79.2 ° C to 50 ° C. Compare the combination of heat pipe with phase switching by 20 W, lowering the maximum temperature from 45 ° C to 35.2 ° C and 80 ° C to 48 ° C -50 W. combination of heat pipe and beeswax PCM used as an effective application of battery cooling application..

7.2. PCM Model (CLS+PCM)

The integration of PCM material with CLS is another method that can be used for hot battery management. PCM layers situated in battery pack. Advantage of PCM is that it has good thermal response, and CLS compensates for moderate temperature of PCM, Fig. 13 The combined liquid system with PCM system

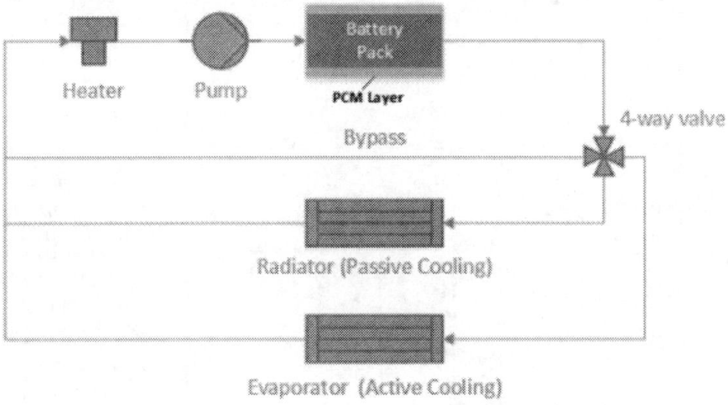

Figure 13. The combined liquid system with PCM system [17].

The critical analysis of battery thermal management strategies for Li-ion battery types are given in Table 1. The comparison between different batteries is given in Table 2.

Table 1. Battery thermal management strategies (BTMS) for Li-ion battery

BTMS	Battery type	Design Parameters	Experimental conditions	Obesrvations	Author /References
Air Cooling	Li-ion battery (72 prismatic cells)	capacity = 5.19 Ah	Ti = 40 °C Battery load - 4.81 W	T_{max} = 58.2°C	Park and Jung [18]
	Li-ion battery (24 pouch cell)	capacity = 2.2 Ah	Ti = 27 °C SOC(%) = 95 - 5 Discharge rate = 5 C	T_{max} = 58.2°C	Hong et al. [19]
	Li-ion battery (8 cylindrical cells)	capacity = 3.6 Ah	Ti = 20°C SOC (%) = 70 -100 Discharge rate = 7 C	T_{max} = 29.6 °C	Mahmud et al. [20]
Liquid cooling	Li-ion battery (3 cartridge heater)	Water(Oblique Channel plate)	Battery load = 220 W No. of channel = 01	T_{max} <50 °C	Jin et al. [21]
	Li-ion battery (rectangular cell)	capacity = 7 Ah Water(straight channel plate)	Discharge rate = 5 C No.of channel = 2.6 Ti = 25 C	T_{max} = 58.4 °C	Huo et al. [22]
PCM	12 Ah prismatic cells	PCM: Paraffin cell spacing = 1.5 mm; k_{pgs} = 800 W/m K	Ti = 25 °C Discharge rate = 5C	Expanded graphite mass fraction between 15 - 20% is recommended	Wu et al. [23]
	6 Ah LCO cell based on 4.4 Ah Hitachi battery	Paraffin + Graphene nano-membranes; Nanofillers' t = 10 nm	Ti = 25 °C Discharge rate = 0.2 C, 1 C and 3 C; initial SOC = 98 %	Graphite + Paraffin gives better performance than TMS designed using graphene nanoparticle	Mortazavi et al. [24]
Heat pipe	16 LFP 18 AH prismatic cells joined in series	flat micro heat pipe array t = 3mm; tilt angle = 5° Heat pipe: 200 x 60 mm Fin: 20x24x1mm	Ti = 27.5°C Discharge rate = 1C	Intracellular gradient smaller than 5 ° is noted	Ye et al. [25]
PCM + HP	Aluminum cylinders with heating rods	PCM: Paraffin/EG; EG wt fraction = 16% Sintered-copper HP, dia = 6mm	Ti = 20±0.5 °C Discharge rate = 5 C	Because of PCM addition to the module, maximum temperature reduces by 33.6 %	Zhao et al. [26]
PCM + CLS	5*6 1.1 Ah cylindrical cells collected in parallel	paraffin EG matrix with embedded flat plate pipes; k =10000 W/m K; cell spacing: 34 mm (x-dir) and 25 mm (z-dir) coolant: ethyl alcohol	Ti = 35 °C Discharge rates of 1 C, 2 C, and 3 C, 5 cycles	Liquid cooling + PCM take longer time to reach the maxi.setting temp of 44 °C in comparison to PCM + air cooled pipes	Huang et al. [27]

C = Coulomb's; means a fully charged batte[illegible] at 1Ah provides 1 Ampere for one hour.

IOP Conf. Series: Materials Science and Engineering 1126 (2021) 012072 doi:10.1088/1757-899X/1126/1/012072

Table 2. Comparison between different types of battery

Battery Type	NiCad	NiMH	Lead Acid	Li-ion
Sp. Energy (Wh/kg)	40-55	65	20-35	140
Energy Density (Wh/kg)	70-90	150	54-95	250-620
Sp. Power (W/kg)	125	200	250	300-1500
Internal resistance ($m\Omega$)	100-200 6V pack	200-300 6V pack	<100 12V pack	150-250 7.2V pack
Cycle life	2000	2000	4500	3500
Charging time	1h	2-4h	8-16h	2-4h
Self discharging rate	20%	30%	5%	10%
Cell Voltage (nominal)	1.25V	1.25V	2V	3.6V
Operating temperature	-40-60 $°C$	-20-60 $°C$	-20-60 $°C$	-20-60 $°C$
Maintenance	30-60 days	60-90 days	3-6 months	Not required
Typical battery (cost $/kWh)	280	500-1000	269	300-800
Advantages	High sp. power	30-40% higher capacity over a std. Ni-Cd	Low cost	High energy density
	Long life cycle	High energy density	availability of raw materials (lead, sulfur)	Self discharge
	Low self discharge	Simple storage and transportation	Easy for maintenance	Low maintenance
	Robust	—	good electro-mechanical characteristics	No requirement of priming
	Fast charge	—	—	Long life cycle
Disadvantages	Low cell voltage (about 1.2V)	Limited service life	Low energy density	Expensive
	High cost	high self discharge	Poor temperature characteristics	Protection required
	Cadmium is environmentally harmful	required more maintenance	Corrosive in nature	Transportation problem
Application	Replacement for flashlight battery	HEV's replacement for flashlight battery	Car battery, forklift, golf cart, backup power	Consumer electronics

Sp. - Specific; std. - standard; V - Volts; h - hour; $- US Dollar

8. Conclusions

From the study of materials selection for hybrid and electric vehicle battery pack thermal management and comparison between different battery packing heat management techniques, it is concluded that combination of PCM and heat pipe is most effective method of thermal packing of battery packs. The main conclusions can be made as follows:

(i) Combination of heat pipe and PCM is more effective and efficient, it was used in both EVs and HEVs. By using PCM and heat pipe, a significant decrease in temperature noticeable as compared to no cooling strategy especially during high emission levels. Also, the heat pipe + PCM module has longer operating time to reach set temperature of 50 $^{\circ}C$ as compared to PCM module only.

(ii) The maximum temperature controlled below 50 0C as the forced air assembly is used to improve the heat transfer rate.

(iii) In the case of cycle conditions, heat pipe + PCM module is reached at stable phase with the same temperature profile after the first cycle, while the only PCM module without HP continues to increase.

References

[1] S. A. Khateeb, M. M. Farid, J. R. Selman, S. Al-Hallaj, Design and simulation of a lithium-ion battery with a phase change material thermal management system for an electric scooter, Journal of Power Sources 128 (2) (2004) 292–307.
[2] A. Cooper, Development of a lead-acid battery for a hybrid electric vehicle, Journal of Power Sources 133 (1) (2004) 116–125.
[3] B. H. Lee, S. W. Kim, et al., Development of battery management system for nickel–metal hydride batteries in electric vehicle applications, Journal of Power Sources 109 (1) (2002) 1–10.
[4] T. Kojima, T. Ishizu, T. Horiba, M. Yoshikawa, Development of lithium-ion battery for fuel cell hybrid electric vehicle application, Journal of Power Sources 189 (1) (2009) 859–863.
[5] A. Friesen, F. Horsthemke, X. Mönnighoff, G. Brunklaus, R. Krafft, M. Börner, T. Risthaus, M. Winter, F. M. Schappacher, Impact of cycling at low temperatures on the safety behavior of 18650-type lithium ion cells: Combined study of mechanical and thermal abuse testing accompanied by post-mortem analysis, Journal of Power Sources 334 (2016) 1–11.
[6] Q. Wang, P. Ping, X. Zhao, G. Chu, J. Sun, C. Chen, Thermal runaway caused fire and explosion of lithium ion battery, Journal of power sources 208 (2012) 210–224.
[7] X. Feng, J. Sun, M. Ouyang, F. Wang, X. He, L. Lu, H. Peng, Characterization of penetration induced thermal runaway propagation process within a large format lithium ion battery module, Journal of Power Sources 275 (2015) 261–273.
[8] P. Amiribavandpour, W. Shen, D. Mu, A. Kapoor, An improved theoretical electrochemical-thermal modelling of lithium-ion battery packs in electric vehicles, Journal of Power Sources 284 (2015) 328–338.
[9] Z. Rao, S. Wang, A review of power battery thermal energy management, Renewable and Sustainable Energy Reviews 15 (9) (2011) 4554–4571.
[10] S. A. Khateeb, S. Amiruddin, M. Farid, J. R. Selman, S. Al-Hallaj, Thermal management of li-ion battery with phase change material for electric scooters: experimental validation, Journal of Power Sources 142 (1-2) (2005) 345–353.
[11] R. Kizilel, R. Sabbah, J. R. Selman, S. Al-Hallaj, An alternative cooling system to enhance the safety of li-ion battery packs, Journal of Power Sources 194 (2) (2009) 1105–1112.
[12] J. Li, Z. Zhu, Battery thermal management systems of electric vehicles, Master's thesis (2014).
[13] R. Matthe, L. Turner, H. Mettlach, Voltec battery system for electric vehicle with extended range, SAE International Journal of Engines 4 (1) (2011) 1944–1962.
[14] A. A. Pesaran, Battery thermal management in ev and hevs: issues and solutions, Battery Man 43 (5) (2001) 34–49.
[15] T.-H. Tran, S. Harmand, B. Sahut, Experimental investigation on heat pipe cooling for hybrid electric vehicle and electric vehicle lithium-ion battery, Journal of Power Sources 265 (2014) 262 – 272.
[16] G. Burban, V. Ayel, A. Alexandre, P. Lagonotte, Y. Bertin, C. Romestant, Experimental investigation of a pulsating heat pipe for hybrid vehicle applications, Applied Thermal Engineering 50 (1) (2013) 94–103.
[17] Battery thermal models for hybrid vehicle simulations, Journal of Power Sources 110 (2) (2002) 377 – 382.

[18] S. Park, D. Jung, Battery cell arrangement and heat transfer fluid effects on the parasitic power consumption and the cell temperature distribution in a hybrid electric vehicle, Journal of Power Sources 227 (2013) 191–198.
[19] S. Hong, X. Zhang, K. Chen, S. Wang, Design of flow configuration for parallel air-cooled battery thermal management system with secondary vent, International Journal of Heat and Mass Transfer 116 (2018) 1204–1212.
[20] R. Mahamud, C. Park, Reciprocating air flow for li-ion battery thermal management to improve temperature uniformity, Journal of Power Sources 196 (13) (2011) 5685–5696.
[21] L. Jin, P. Lee, X. Kong, Y. Fan, S. Chou, Ultra-thin minichannel lcp for ev battery thermal management, Applied energy 113 (2014) 1786–1794.
[22] Y. Huo, Z. Rao, X. Liu, J. Zhao, Investigation of power battery thermal management by using mini-channel cold plate, Energy Conversion and Management 89 (2015) 387–395.
[23] M.-S. Wu, K. Liu, Y.-Y. Wang, C.-C. Wan, Heat dissipation design for lithium-ion batteries, Journal of power sources 109 (1) (2002) 160–166.
[24] B. Mortazavi, H. Yang, F. Mohebbi, G. Cuniberti, T. Rabczuk, Graphene or h-bn paraffin composite structures for the thermal management of li-ion batteries: a multiscale investigation, Applied energy 202 (2017) 323–334.
[25] Y. Ye, Y. Shi, L. H. Saw, A. A. Tay, Performance assessment and optimization of a heat pipe thermal management system for fast charging lithium ion battery packs, International Journal of Heat and Mass Transfer 92 (2016) 893–903.
[26] Y. Zhao, B. Zou, C. Li, Y. Ding, Active cooling based battery thermal management using composite phase change materials, Energy Procedia 158 (2019) 4933–4940.
[27] C.-K. Huang, J. Sakamoto, J. Wolfenstine, S. Surampudi, The limits of low-temperature performance of li-ion cells, Journal of the Electrochemical Society 147 (8) (2000) 2893.

IOP Conf. Series: Materials Science and Engineering 1126 (2021) 012073 doi:10.1088/1757-899X/1126/1/012073

Experimental and Numerical Studies on Materials for Electric Vehicle Chassis

Shankar Durgam*, **Vikrant Mahesh Deshmukh**, **Akshay Prakash Jagtap**, **Mahendra J. Sable** and **Nikita M. Gawai**

Department of Mechanical Engineering, College of Engineering Pune - 411005, India

E-mail: *sod.mech@coep.ac.in

Abstract. This paper presents experimental and numerical studies on materials for electric vehicle chassis design, testing, and analysis. The chassis for a car should have good rigidity and should be light in weight which are very crucial parameters. To create any car frame is the basic and main structure. It connects rear and front suspension rigidly. Along with this it provides support to the various components of the car. This paper describes the simulation of electric vehicle (EV) Chassis frame with square beam for finding stress, strain and deformation properties in structural body by using finite element analysis (FEA) tools and strain gauge. In this case of chassis structure, we are considering engine weight, alternator weight, motor weight as well as battery pack weight. The total weight on chassis structure is approximately around 4 ton. We are working on finding maximum stress, maximum strain and deformation in structural body by using FEA tools as well as strain gauge.

Keywords: Electric Vehicle Chassis, 3D Modelling, Analysis in Ansys, Strain Gauge.

1. Introduction

Chassis of vehicle is nothing but structure of vehicle without including prime mover. Chassis of vehicle includes frame, power train, brakes, steering etc. subsystems attached to it. Function of chassis is to provide attachment for all components of vehicle such as suspension, steering, brakes and power train. Chassis of the vehicle must be strong enough in order to withstand loads acting on it in all possible cases of loading conditions. Chassis provides support to the control systems such as gearbox, steering, suspension and driver. Static loading conditions mostly involve weight of individual components acting over the frame and dynamic loads like braking, acceleration, cornering. Various loads acting on chassis are vertical bending, horizontal bending, horizontal lozenging, and torsion etc. There are various types of frame constructions used for chassis of vehicle.

(i) Space frame construction

(ii) Ladder frame construction

(iii) Monocoque Chassis

There is some loading condition like normal loading, lateral loading, bending loading torsional loading, combined bending and torsional loading has to be consider during designing new chassis structure. In this case of chassis structure, we are considering engine weight, alternator weight, motor weight as well as battery pack weight. The total weight on chassis structure is approximately around 3.5 ton. Riley and George [1] presented experimental and finite

IOP Conf. Series: Materials Science and Engineering 1126 (2021) 012073 doi:10.1088/1757-899X/1126/1/012073

element-based simulation study on formula (Society of Auomotive Engineers) SAE car chassis frame design and explained the torsion test method. Luque et al. [2] proposed algorithm for optimal chassis design of the electric vehicle. They found 20% reduction in battery weight with this method. Mat and Ghani [3] presented FEA base simulations for development of light weight chassis to withstand engine and drivers' weight and that protect the driver at the time of crash. Chinnaraj et al. [4] conducted experiments on dynamic stress strain response of components of heavy truck chassis. They compared experimental results with FEA based simulations and found this study is useful analysis of stresses in frame of truck rail. Kamble et al. [5] numerically studied design technique for Formula 3 sports car monocoque chassis. They optimized for minimum chassis weight, and high torsional rigidity. Pinheiro et al. [6] presented a skate concept to an electric car for low production cost, low weight, and security. They used sheet of different thickness for chassis parts and found reduction in chassis weight. Several authors [7, 8, 9, 10, 11, 12] studied structural and analysis of stress in ladder chassis frame design with various area of cross sections.
Above literature show that the stress strain analysis for the chassis is very limited. Therefore, this paper includes finding maximum stress, maximum strain and deformation in structural body by using FEA tools as well as strain gauge. The design consideration of chassis is weight. It should be as low as possible and having good strength. This paper also describes the design of Chassis with square beam with the material of Mild Steel Indian Standard (MS IS) 1161/2062 and its static and dynamic analysis. Table 1 shows mechanical properties of MS IS 1661/2062.

Table 1. Properties of MS IS 1661/2062

Youngs Modulus	2.1×10^5
Poisons Ratio	0.303
Bulk Modulus	1.7766×10^5 Mpa
Shear Modulus	80583 Mpa
Ultimate tensile strength	440 Mpa
Yield Strength	370 Mpa
Density	7.86×10^-6

2. Methodology

This section describes the design of an EV Chassis frame and simulation of design using Ansys software in order to obtain stress, strain and deformation values of the chassis frame at different load conditions.

2.1. Computer Aided Design (CAD) model

Fig. 1 shows the CAD model with magnitude of loads, moments, loading position, moment nature and fixed supports. The beams A, B, C and D are fixed supports. The 12260 N loads acting at point E, F, G and H. The load of 610 N is acting at point I and J, respectively.

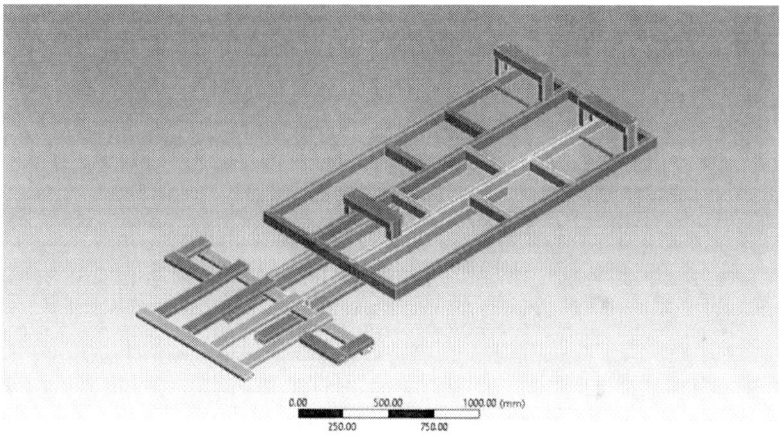

(a) CAD model

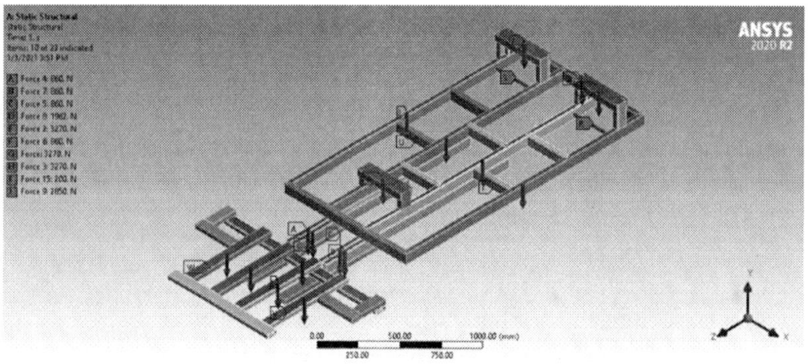

(b) Loading at different points

Figure 1. CAD model with loading at different points

2.2. Meshing

Fig. 2 shows mesh used in the present study with the expanded view of the mesh. Preprocessor generated hexa and quad elements were preferred for structural analysis. Areas with holes prone to stress concentration where mesh density was increased as shown in Fig. 2(a). FEA is helpful to do various calculations at limited number of points only. Further, these results are used for the entire structure. As continuous object has infinite number of degrees of freedom, it is impossible to solve the problem in such format. FEA reduces the degree of freedom of a structure from infinite to finite by using meshing. We have worked on meshing with 311932 nodes and 157330 elements. The convergence time for each case of loading condition is about one and half hour. This mesh is selected because it save time and cost of simulation. Fig. 2(b) shows the expanded view of the mesh used in the present study.

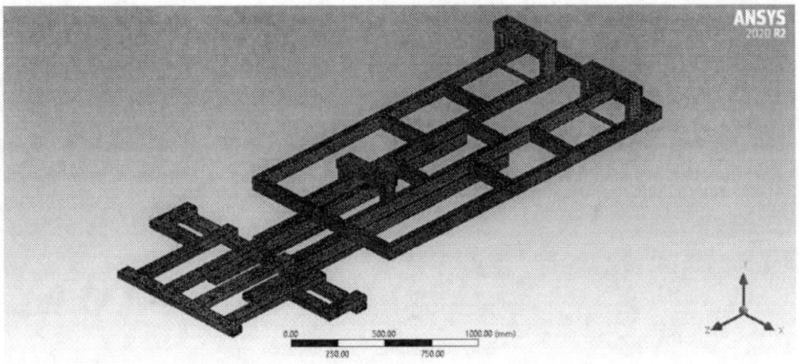

(a) Mesh used in the study

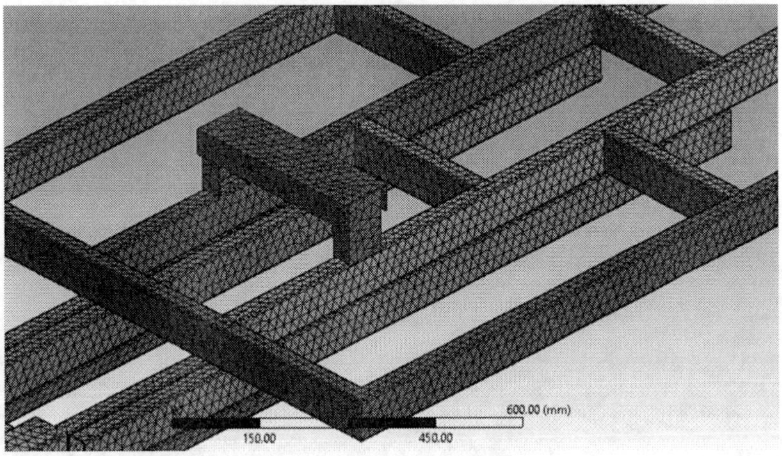

(b) Expanded view of the mesh

Figure 2. Mesh used in the study and the expanded view

3. Analysis in ANSYS

The analysis of chassis structure for studying normal stress, normal strain and total deformation in body at full load condition of 4000 kg is described in this section.

3.1. Normal Stress

Fig. 3 shows Normal stress distribution in x direction along chassis for static structural model. This analysis done at 4000kg load and we get maximum tensile stress value is 121.87 MPa and maximum compressive stress value is 124.44 MPa Fig. 3 shows the 3D model of assembly of cylinder block and fins.

IOP Conf. Series: Materials Science and Engineering 1126 (2021) 012073 doi:10.1088/1757-899X/1126/1/012073

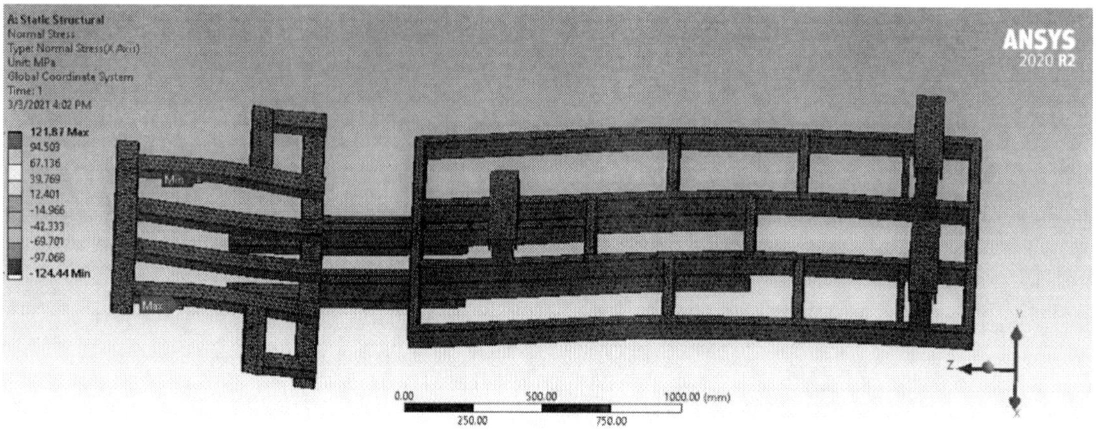

Figure 3. Normal stress for loading Condition of 4000Kg

3.2. Normal strain

Fig. shows Normal strain distribution along chassis for static structural model. This analysis done at 4000kg load and we get the maximum value of tensile strain value 0.00045324 and maximum value compressive strain value 0.00032846

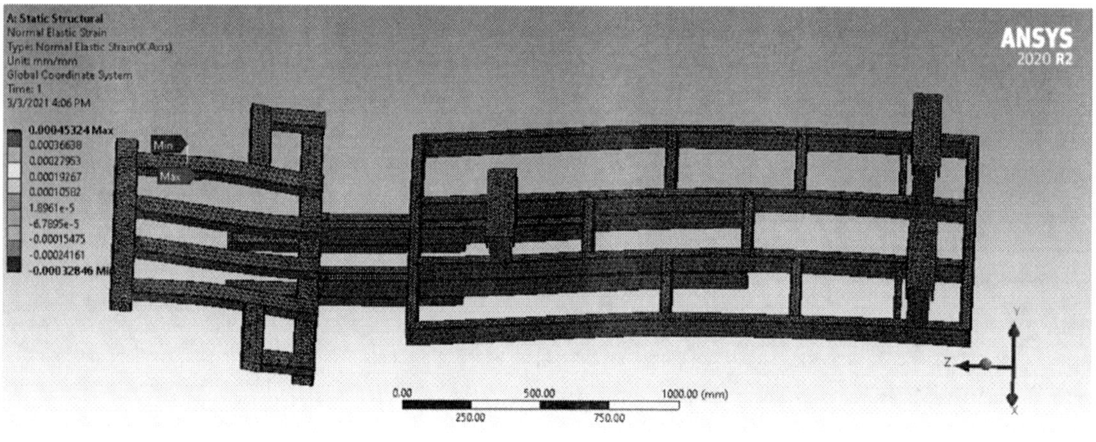

Figure 4. Normal strain for loading condition of 4000Kg

3.3. Total Deformation

Fig. 5 shows total diformation in static structural model of chassis. This analysis done at 4000kg load and we get maximum Total deformation value is 0.18702 mm.

IOP Conf. Series: Materials Science and Engineering 1126 (2021) 012073 doi:10.1088/1757-899X/1126/1/012073

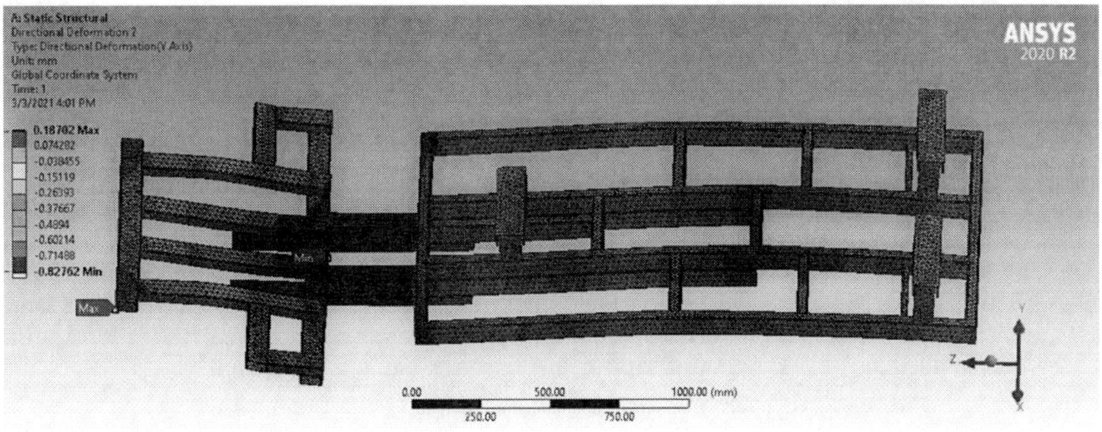

Figure 5. Total deformation for loading condition of 4000Kg

4. Experimental analysis using Strain Gauge

Strain gauge is working on the principle of measurement of change in electric resistance due to deformation of structure with applied load in order to get the value of rate of deformation. Strain gauge is useful for both tensile and compressive load condition. Strain gauge should be strong and light in weight for industry use. These Strain gauges are used in industries for pressure and weight measurement as a transducer. Strain gauges are widely used in study of strength of material such as vibration, residual lifetime, load condition, analysis of failures, etc. Strain gauge gives the values of stresses at those point only where strain gauge is located. Due to this selection of strain gauge location is important for analysis purpose. In order to validate the Ansys simulation result values strain gauge analysis is used. In this we calculate the values of strain induced at 20 different locations on Chassis structure as shown in Fig. 6.

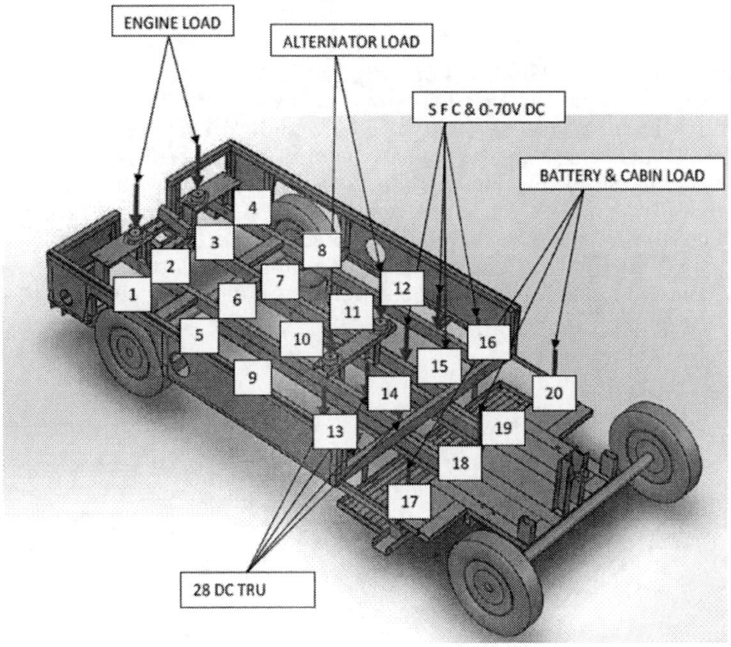

Figure 6. Total deformation for loading condition of 4000 kg

In order to validate the simulated results with the experimental results of strain gauge we have considered the result values at 10 locations. The experimental results using strain gauges in the form of strain which are further converted into stress values by using Young's modulus value.

5. Results and discussion

The Results of the above analysis chassis frame is discussed in this section.

(i) Based on Ansys Simulation The simulation values of strain are varying from maximum value of tensile strain 0.00045324 to the maximum value of compressive strain 0.00032846 with a load of 4000 kg. The simulation values of stress are varying from maximum value of tensile stress 121.87 MPa to the maximum value of compressive stress 124.44 MPa with a load of 4000 kg. The maximum total deformation value is 0.18702 mm.

(ii) Based on Strain Gauge (Experimental Values) The result of values of normal stress and normal strain in chassis structure is analyzed at 4000kg load condition by using strain gauge at different locations on chassis frame under study. The strain gauge values are varying from maximum value of tensile strain 0.00059828 to the maximum value of compressive strain 0.000404146 with a load of 4000 kg. The strain gauge values of stress are varying from maximum value of tensile stress 121.87 to the maximum value of compressive stress 124.44 with a load of 4000 kg.

Table 2 shows ANSYS based simulation results.

Table 2. ANSYS based simulation results

Item	Value
Maximum tensile stress	121.87 Mpa
Maximum compressive stress	124.44 Mpa
Maximum tensile strain	0.00045324
Maximum compressive strain	0.00032846
Total deformation value	0.18702

5.1. Validation of Stress Strain Values at 5000kg Load Condition

The validation of Ansys simulation-based strain values with the strain values obtained by using strain gauge at different locations at 4000 kg load condition are shown in Fig. 7. In this, a curve of Ansys simulation-based strain values varying from maximum to minimum values at different positions on chassis frame is shown. These are further comparing with experimental strain values obtained by using strain gauge varying from maximum strain value to minimum strain values. The comparison of Ansys based stress values and strain gauge based experimental stress values are shown in Fig. 8. The experimental and simulated stress strain values are in good agreements. Table 3 show the experimental and simulated strain values.

Table 3. The experimental and simulated strain values

Location	Simulated strain	Experimental strain
1	0.00045324	0.00059828
2	0.00036638	0.000390356
3	0.00027953	0.000278232
4	0.00019267	0.00019078
5	0.00010582	0.000135079
6	0.000018961	9.5494×10¯6
7	0.000067895	8.8295×10¯6
8	-0.00015475	-0.00015689
9	-0.00024161	-0.000235665
10	-0.00032846	-0.000404146

Table 4 showa the experimental and simulated stress values. It is found that the experimental and simulated values are in goob match.

Table 4. The experimental and simulated stress values

Location	Simulated stress	Experimental stress
1	212.87	125.64
2	94.503	84.87
3	67.136	58.43
4	39.769	32.91
5	12.401	18.54
6	-14.966	-18.36
7	-42.333	-40.06
8	-69.701	-54.88
9	-97.068	-87.18
10	-124.44	-118

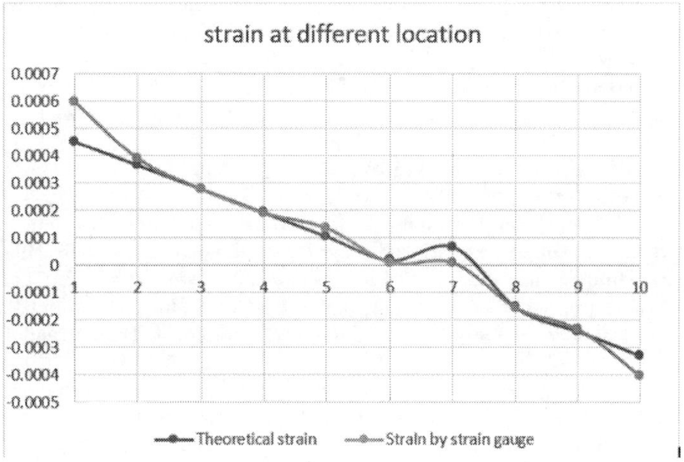

Figure 7. Strain validation at 4000 kg load

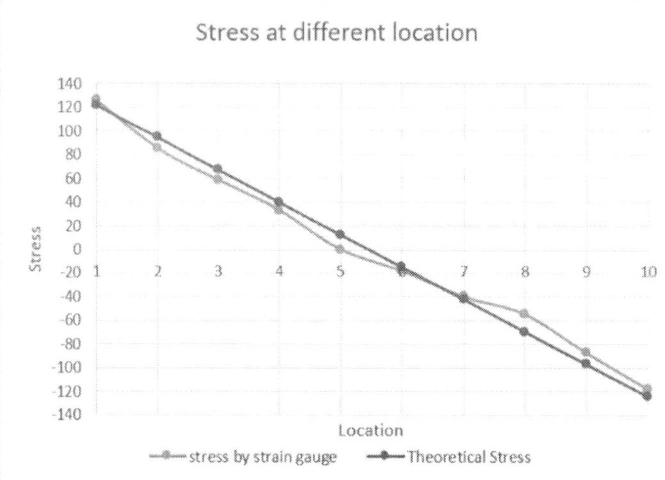

Figure 8. Stress validation at 4000 kg load

6. Conclusions

The bending moment that the beam can sustain is good for square beam. The frame having minimum deformation of beam for 4000 kg loading condition. Hence frame is safe to use up to that loading condition. To improve strength of frame, we can use cross members as well as diagonal support members. To optimize design, we can use skateboard type chassis structure and advance automotive material like high speed steel (HSS), dual phase (DP) Steel, ultra high speed steel (UHSS), transformation induced plasticity (TRIP) steel, Twin Steel where we can achieve more strength, crashworthiness.

References

[1] W. B. Riley, A. R. George, Design, analysis and testing of a formula sae car chassis, Tech. rep., SAE Technical Paper (2002).
[2] P. Luque, D. A. Mántaras, Á. Maradona, J. Roces, L. Sánchez, L. Castejón, H. Malón, Multi-objective evolutionary design of an electric vehicle chassis, Sensors 20 (13) (2020) 3633.
[3] M. H. Mat, A. R. A. Ghani, Design and analysis of 'eco'car chassis, Procedia Engineering 41 (2012) 1756–1760.
[4] K. Chinnaraj, M. Sathya Prasad, C. Lakshmana Rao, Experimental analysis and quasi-static numerical idealization of dynamic stresses on a heavy truck chassis frame assembly, in: Applied Mechanics and Materials, Vol. 13, Trans Tech Publ, 2008, pp. 271–280.
[5] M. Kamble, T. Shakfeh, R. Moheimani, H. Dalir, Optimization of a composite monocoque chassis for structural performance: a comprehensive approach, Journal of Failure Analysis and Prevention 19 (5) (2019) 1252–1263.
[6] L. Pinheiro, F. Passarinho, A. Teixeira, H. Silva, B. Ribeiro, New skate chassis concept for electric vehicles, Tech. rep., SAE Technical Paper (2012).
[7] N. A. Mohammed, N. Nandu, A. Krishnan, A. R. Nair, P. Sreedharan, Design, analysis, fabrication and testing of a formula car chassis, Materials Today: Proceedings 5 (11) (2018) 24944–24953.
[8] H. Radhwan, Z. Shayfull, S. Nasir, A. Irfan, et al., Design and analysis of solar car chassis, in: IOP Conference Series: Materials Science and Engineering, Vol. 864, IOP Publishing, 2020, p. 012144.
[9] V. V. Patel, R. Patel, Structural analysis of a ladder chassis frame, World Journal of science and Technology 2 (4) (2012) 05–08.
[10] K. Y. Patil, E. R. Deore, Stress analysis of ladder chassis with various cross sections, IOSR Journal of Mechanical and Civil Engineering (IOSR-JMCE) 12 (4) (2015) 111–116.
[11] W. G. Ferreira, F. Martins, S. Kameoka, A. S. Salloum, J. T. Kaeya, Structural optimization of automotive components applied to durability problems, Tech. rep., SAE Technical Paper (2003).
[12] L. L. Thompson, J. K. Lampert, E. H. Law, Design of a twist fixture to measure the torsional stiffness of a winston cup chassis, Tech. rep., SAE Technical Paper (1998).

An Interpretation of Performance and Emission Outcomes of the Compression Ignition Engine using Palm Oil Biodiesel-Diesel Blends

Nitin Dattatreya Kamitkar[1], Satishkumar[2*], Basavaraju A N[3], Shashikant Kushnoore[4], Deepa A B[5*]Marimuthu P[6*] and N S Kalyan Chakravarthy[7*]

[1] Research Scholar, Department of Automobile Engineering, Malnad College of Engineering, Hassan. Karnataka, India. And Assistant Professor, Department of Mechanical Engineering, KoneruLakshmaiah Education Foundation, Vaddeswaram (A.P.) India.
[2] Research Scholar, Department of Automobile Engineering, Malnad College of Engineering, Hassan. Karnataka, India
[3]Professor, Department of Automobile Engineering, Malnad College of Engineering, Hassan. Karnataka, India
[4]Assistant Professor, Department of Mechanical Engineering, KoneruLakshmaiah Education Foundation, Vaddeswaram (A.P.) India.
[5]Assistant Professor, Department of Electrical and Electronics Engineering,
[6]Professor, Department of Mechanical Engineering and Director Product Development
[7]Director, Department of Data Scienceand Business Systems
*QIS College of Engineering and Technology, Vengamukkapalem, Ongole, Prakasam District, Andhrapradesh-523 272, India
[6]Email: pmarimuthu69@gmail.com (Corresponding author)

Abstract. Research on alternative sources to replace fossil fuels has been started decades ago and still an uninterrupted process. Partial replacement rather than complete is justified with real-time applications. In this research, engine outcomes based on palm oil biodiesel blends are elaborated. The transesterification processed biodiesel from raw palm oil is used to run the engine with a blend proportion of P10D90 and P20D80. From the results obtained, the performance of an engine was degraded with an increase in biodiesel proportion and consumes higher fuel. At lower load diesel and P10D90 have almost equal thermal efficiency whereas at peak load it is found that 39.1% and 37.6% respectively. Apart from CO emissions, there is an increment in CO_2, HC, and NO_X with the load. NO_x emissions are in the margin with diesel for P10D90 at part load. P10D90 shown slightly lower in-cylinder pressure at full load compared to diesel.However, it is seen that P20D80 has given depreciated performance readings, lowered cylinder pressure 55.8 bar, and higher emissions at rated torque. From the experiment, it is noticed that above 10% of the biodiesel blend engine required suitable modifications and accessories to overcome demerits associated with blend characteristics.
Keywords: Biodiesel, Performance, Emission, P10D90, P20D80.

1. Introduction

Immense renewable sources can be supportive to eliminate full or partial conventional sources, biodiesel is an important kind among the same. Blends of 10% biodiesel with diesel when used in the engine it is observed that there is a reduction in brake power and an increase in fuel consumption with inferior combustion characteristics compares to diesel [1]. Neat biodiesel with methane gas as dual fuel mode increased thermal efficiency [2]. Blends of different biodiesel have no impact on engine performance and given improved combustion plots [3]. Increasing biodiesel and ethanol with diesel has given improved performance than biodiesel alone [4]. Castor seed biodiesel up to 20% blends can be used directly in a diesel engine without any modification in the injection system [5]. Exhaust gas circular with biodiesel up to certain limits can be an advantage to control the emissions but it affects engine brake thermal efficiency[6,7]. Carbon monoxide and hydrocarbons were reduced when 10-20% of pentanol is blended with biodiesel [8]. Higher thermal efficiency and NO_x, lower CO, CO_2, and HC emissions were recorded with clustered apple seed biodiesel [9]. The addition of Nan particles with biodiesel has reduced NO_x and hydrocarbon emissions with improved engine performance [10]. Corn seeds biodiesel of 20% blend with the diesel shown improved performance and combustion process at full load [11]. As Lord increases irreversibility process decreases and increases the unavailability of energy. At full load, there is a loss of 3. 68% of exergy with biodiesel compared to diesel [12].

Emission of NO_x depends on blends ration, B2O of soybean methyl ester emitted lower NO_x when tested on conventional engine [13]. Blend of 10% cashew nutshell and waste cooling oil with diesel at peak load has almost equal thermal efficiency as diesel and decreases NO_x emission whereas HC and CO are slightly increased [14]. Biodiesel from fish waste has similar performance and combustion characteristics as diesel. Apart from carbon monoxide, all other emissions are recorded with lower reading [15]. Jatropha biodiesel emission is decreased compared to diesel whereas NO_x quantity increased [16]. Cooled exhaust gas recirculation at a flow rate of 15% ensured a drastic reduction in NO_x but influenced the oxidization process that increases CO_2 [17]. Fuel consumption is increased by 4% when castor seed biodiesel is used, and a reduction of 2.2% thermal efficiency than diesel is observed [18]. Test conducted on the engine with neem biodiesel has given improved performance and reduced emission than that of diesel [19]. Higher density and viscosity of biodiesel influence the narrowing of the aerodynamic spray cone angel [20]. Blends of palm and animal fat biodiesel of 20% blend have almost equal performance attributes as of diesel but an increase in blends ration increase CO_2 and NO_x [21]. The combined effect of EGR and increased compression ratio have given improved performance characteristics and reduced NO_x compared to diesel [22].

Direct addition of waste fried oil without converting to biodiesel given lowered engine performance and increased emission[23]. Increasing injection pressure and the addition of nanoparticles in biodiesel-diesel blends have shown improved efficiency with reduced fuel consumption [24]. Engine tested with biodiesel at a higher compression ratio shown a reduction in HC and a slight increment in NO_x with improved thermal efficiency [25] Jamun seed powder and jackfruit seed powder mixed diesel has improved performance over plain diesel and there is a slight reduction in NO_x is recorded [26]. Higher injection pressure on the CRDI engine revealed that biodiesel-diesel blends can achieve higher thermal efficiency and lower CO, HC emissions as injection pressure increases [27]

From the research history, it is justified that increased blend ratio harm engine performance as well as on emissions. Hence the trial has made with a limited proportion up to 20% of palm oil biodiesel. In-cylinder pressure variation which directly proportional to power output is discussed.

2. Test fuels, Engine setup, and experimental methodology

2.1 Preparation of test fuels and Characterization

Biodiesel from raw palm oil is prepared by the transesterification process;a line diagram of the process is shown in figure 1. Prepared biodiesel is characterized according to ASTM standards to ensure the suitability of usage to run an engine without modification. Two proportions of blends are prepared namely P10D90 which is 10% of palm biodiesel and 90% of diesel similarly, P20D80 which is 20% of palm biodiesel, and 80% diesel.

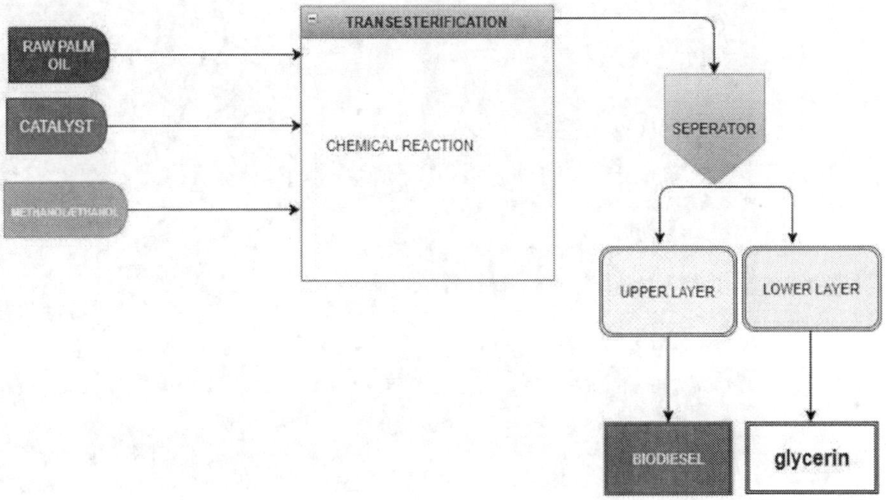

Fig.1 Line diagram of the biodiesel preparation process

Properties of diesel and test fuels are tabulated in table 1, characterized outcomes indicated that the addition of biodiesel to the diesel will reduce the calorific value, increases the flash and fire point of the fuel. Hence the addition of biodiesel is limited to 20% of total volume.

Table.1: Properties of fuels

Properties	DIESEL	P100	P10D90	P20D80
DENSITY (kg/m^3)	831	853	835	837
Kinematic Viscosity(cSt)	3.1	8.1	3.3	3.6
Flash Point (°C)	61	209	67	71
Calorific value(kJ/kg)	42000	39648	41778	41200

2.2 Test setup

The test engine shown in figure 2 is an integrated computerized setup which is connected to an eddy current dynamometer for loading purpose. Four-stroke natural air intake system with a water-cooled single-cylinder of capacity 35 kW, Kirloskar made prime mover is considered. The line diagram of the setup is shown in figure 3 with notations in table 2, and engine instruments with specifications are tabulated in table 3.

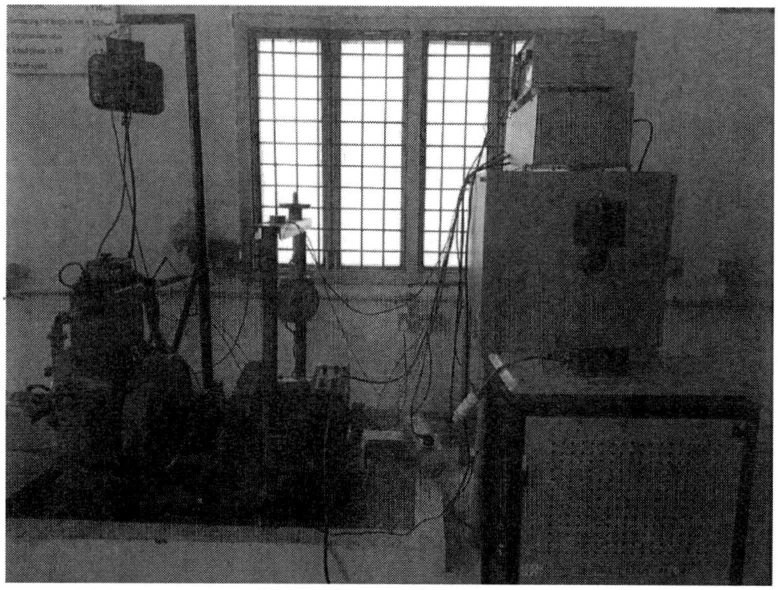

Fig.2 Test engine setup

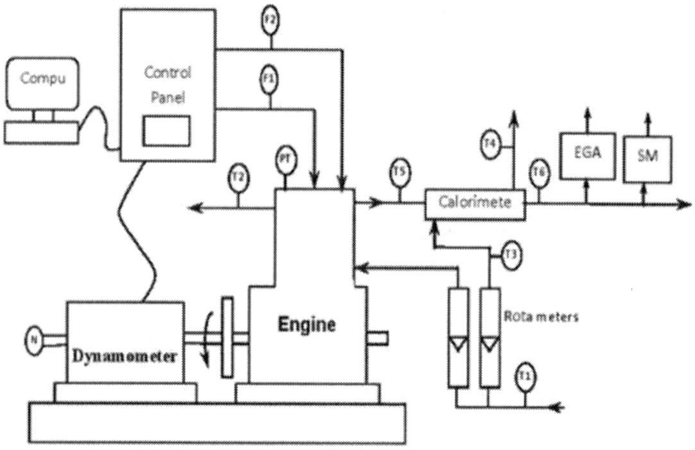

Fig.3 Line diagram of engine setup

Table 2: Engine notations

PT	Pressure transducer
N	Rotary encoder
F1, F2, F3, F4	Fuel flow, Airflow, Jacket water flow, Calorimeter water flow
T1&T2	Jacket water inlet and jacket water outlet temperature
T3&T4	Calorimeter water inlet and calorimeter water outlet temperature
T5&T6	Exhaust gas in & Exhaust gas out calorimeter temperature

Table 3: Test engine specifications

Manufacturer	Kirloskar oil engines Ltd. India
Model	TV-SR, naturally aspirated
Engine	Single cylinder, DI
Bore/stroke	87.5mm/110mm
C.R.	16.5:1
speed	1500 RPM, constant
Rated power	3.5 kW
Working cycle	Four-stroke
Injection pressure	200bar/23 degree before TDC
Type of sensor	Piezoelectric
Response time	4 micro-seconds

2.3 Test method and uncertainty analysis

The engine is tested with diesel, P10D90, and P20D80 fuels at a constant speed of 1500 rpm and load from zero to fullof an incremental order 25%. The lubrication oil level is ensured as per standards and a water flow rate of 42 cc/sec is maintained to avoid engine abnormalities.

Accuracy of an instrument used to measure the readings plays an important role to ensure validation of the experiment. Holman method is used to find uncertainty associated with pressure, speed, flow, and emission measuring instruments together which is about $\pm$1.9.

3.Results and discussion

Natural aspirated without modification of an engine is tested with three fuelsdiesel, P10D90, and P20D80. Results are discussed based on performance and emission categories. The pressure and crank angle plot at full load are summarized to justify the suitability and characteristics of test fuels.

3.1 Performance outcomes

3.1.1 Brake Thermal Efficiency

Characterization of fuels revealed that the calorific value of test fuels is lower than diesel which impacts brake thermal efficiency. From figure 4 it is observed that at 25% of load P10D90 and diesel have very close values 17.4% and 17.2% respectively. As load increases, this difference increased, and at full load diesel, P10D90 and P20D80 shown 39.1%, 37.26%, and 36.38% of brake thermal efficiency, respectively.

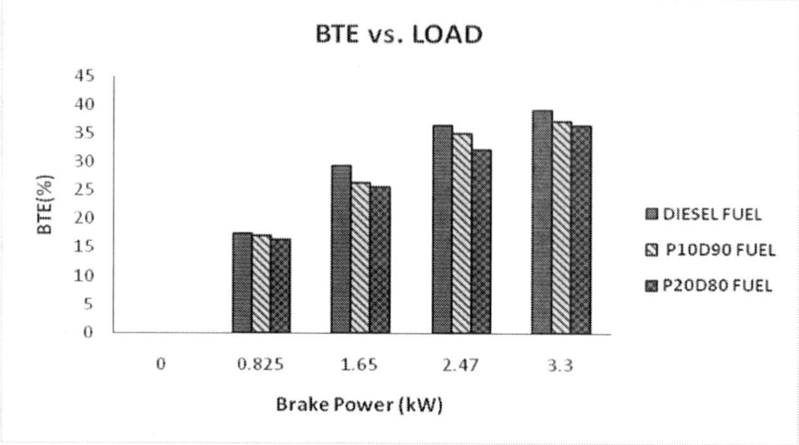

Fig.4 Brake thermal efficiency at various loads

3.1.2 Specific Fuel Consumption

Another important attribute which measured based on fuel burnt on that specific load influences rating of the fuel. Figure 5 shows the fuel expenditure of diesel, P10D90, and P20D80 at various loads. It is seen from the column chart, at lower load engine requires higher fuel quantity to run with rated speed. This gradually decreases with an increase in load. At peak load diesel and P10D90 have almost equal consumption of 0.220 and 0.226 kg/kW hr. As the calorific value decreases engine needs a higher chemical reaction rate to produce power which depends on the mass fraction of fuel burnt. Dilution of fuel quality by higher blend ratio increase SFC which is witnessed by P20D80 at peak load by consuming 0.236 kg/kW hr of fuel.

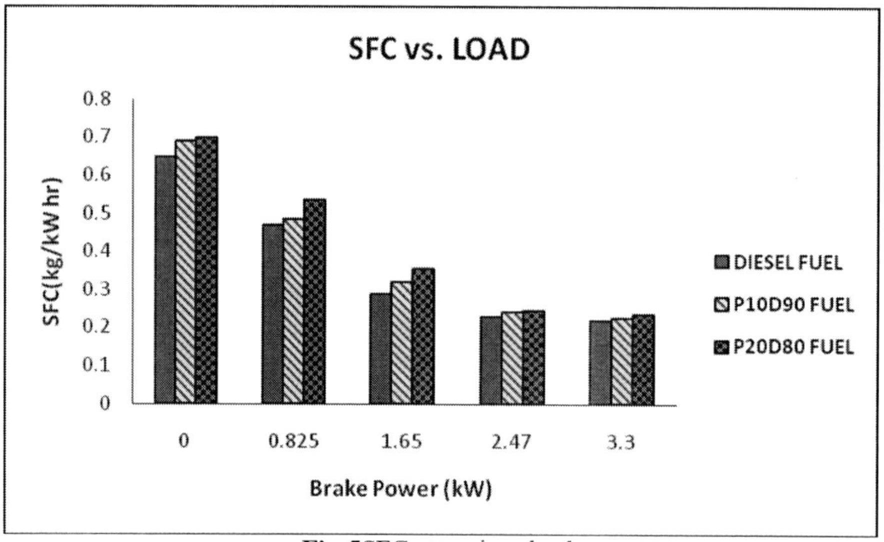

Fig.5SFC at various load

3.1.3 In-cylinder Pressure

A graphical representation of in-cylinder pressure versus crank angle at peak load is shown in figure 6. From the plot, it is noticed that. Diesel and P10D10 liberated the same pressure after a 14-degree crank angle in the expansion stroke. P10D90 has a higher flash and fire point with the moderated calorific value which influenced the delay period and proper mixing of charge. P20D80 was given a low-pressure rise due to higher density, flash point, viscosity, and lower heating value.

From the graph, it is noted that pressure rise for diesel, P10D90, and P20D80 at peak load is 62.08, 61.58, and 55.84 bar, respectively.

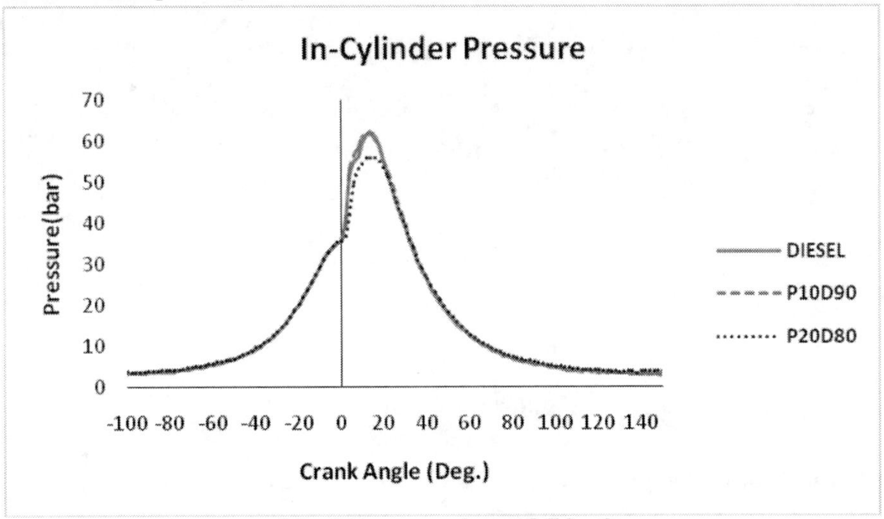

Fig.6 Pressure rise at full load

3.2 Emission Interpretation

3.2.1. Carbon Monoxide and Carbon Dioxide

Figure 7 and 8 show the variation of CO and CO_2 with load respectively, from the chart carbon monoxide emitted by test fuels, are less compared to diesel at all load conditions. This is because of the availability of oxygenated biodiesel that enhanced the combustion process. Atpeak load, P10D90 and P20D80 have the same quantity 0.04% volume of CO emission whereas diesel has a 3.5% higher contribution. At no load, diesel has 4.5% higher CO emissions than test fuels.

Carbon dioxide contribution is more with test fuels. As blends have an environment suitable for oxidization that triggers CO to CO_2. From the chart, it is seen that the increased blend ratio has higher CO_2. Diesel, P10D90, and P20D80 have contributed 2.35,5.63, and 5.73% volume of carbon dioxide at full load.

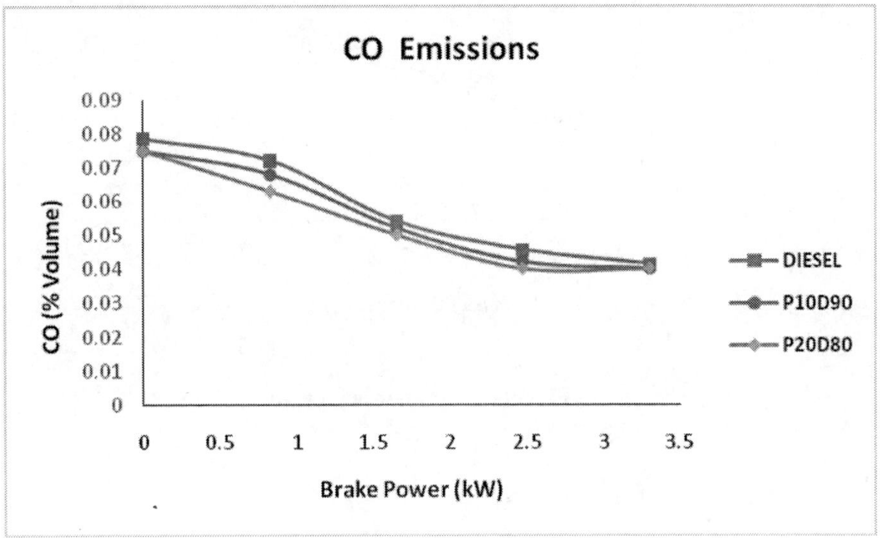

Fig.7CO variation with load

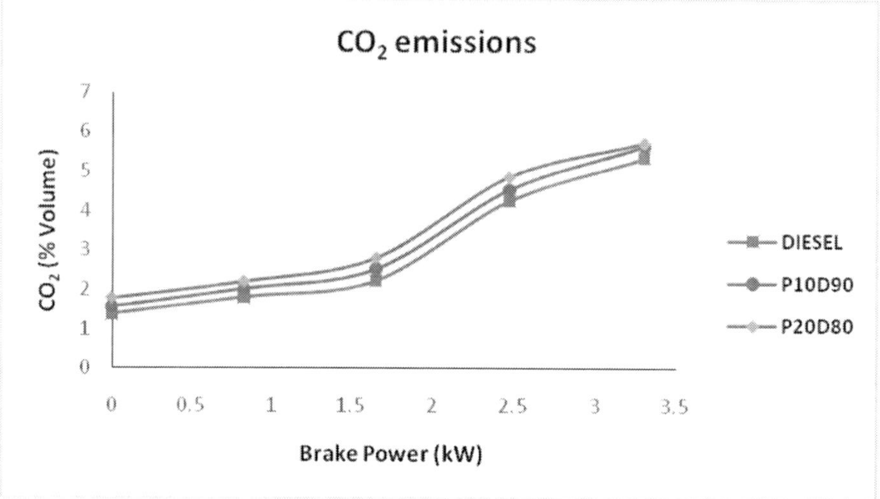

Fig.8 CO_2 variation with load

3.2.2 Un-burnt Hydrocarbon and NOx

Figure9 and 18 show the variation of HC and NOx emission with the load respectively. Improper combustion and constraints of combustion chamber design influence the fuel to produce more hydrocarbon. From the figure, it is observed that at peak load diesel, P10D90 and P20D80 have given 30, 31.2, and 34 ppm of hydrocarbons, respectively.

NOx emission is associated with many factors including fuel property, flame temperature, and time of stay. From the figure, it is revealed that higher NO_x is emitted with test fuels in all load conditions. At full load diesel, P10D90 and P20D80 emitted 1043, 1100, and 1160 ppm of NOx which is 5.4% and 11.2% higher than of diesel, respectively.

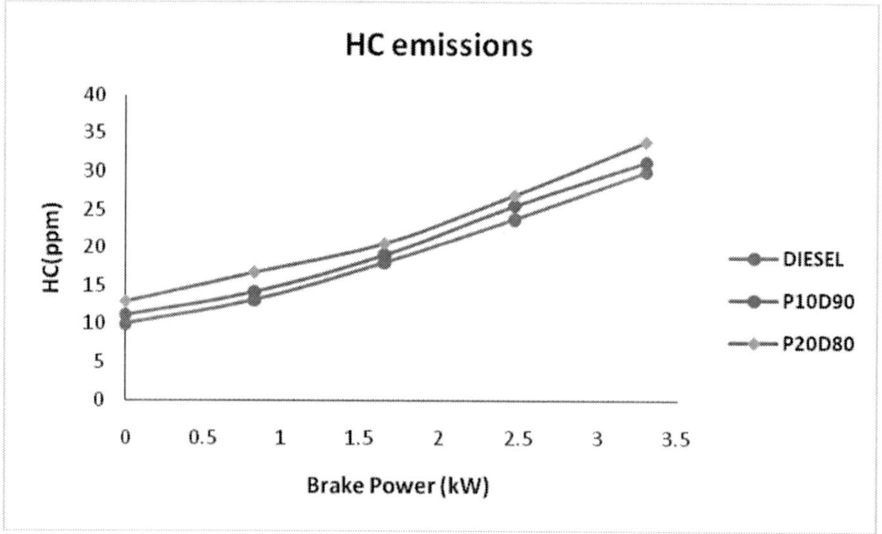

Fig.9 HC variation with load

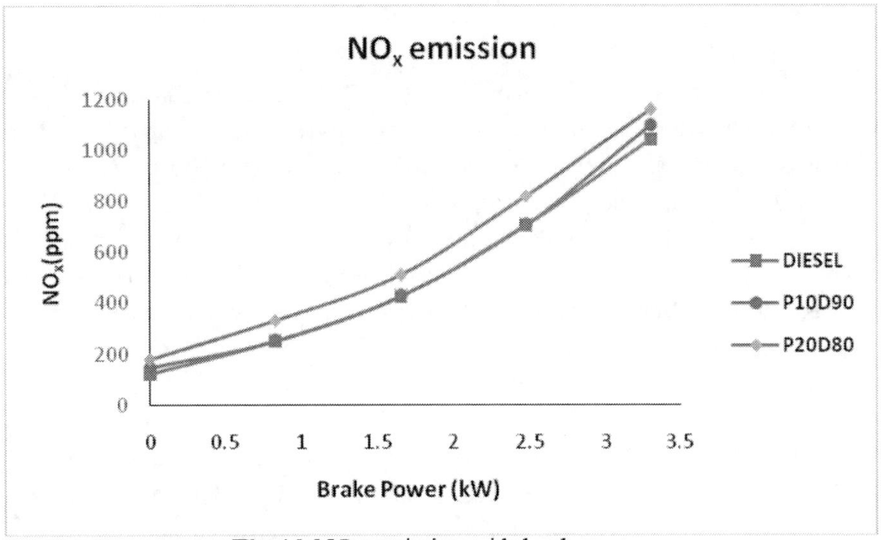

Fig.10 NO_x variation with load

4. Conclusion

From the research, it is summarized that biodiesel can be used to replace fossil fuel partially but not completely without engine modification. A blend ratio increases the performance of an engine degraded with increased pollutants. With a 10% blend, there is not much or negligible deviation in performance, and emission is recorded but with a 20% blend, rapid effect on engine outcome recorded. In this work, brake thermal efficiency at full load for P10D90 is reduced by 4.7 % and fuel consumption is increased by 2.7% than diesel. There is a slight deviation in pressure rise at peak load for P10D90 which affects mean effective pressure that interlinked to power output. Apart from CO all other pollutants CO_2, HC and NO_x are higher for blends than diesel.

From the experiment, it is concluded that above 10% of palm biodiesel-diesel blend test fuels cannot be used directly in the conventional engine without suitable modification and by providing necessary accessories.

DECLARATION

FUNDING-NOT APPLICABLE

COMFLICT OF INTEREST- All authors of an article declare that they have no conflict of interest.

Availability of data and material - Data available on request from the authors.

Code availability-Not applicable.

5.References

1. Naik, N. S., & Balakrishna, B. (2017). A comparative study of B10 biodiesel blends and its performance and combustion characteristics. *International Journal of Ambient Energy*, *39*(3), 257-263. https://doi.org/10.1080/01430750.2017.1303629
2. Justin Abraham Baby, S., Suresh Babu, S., & Devarajan, Y. (2018). Performance study of neat biodiesel-gas fuelled diesel engine. *International Journal of Ambient Energy*, 1-5. https://doi.org/10.1080/01430750.2018.1542625
3. Nabi, M., Rasul, M., Anwar, M., & Mullins, B. (2019). Energy, exergy, performance, emission and combustion characteristics of diesel engine using new series of non-edible biodiesels. *Renewable Energy*, *140*, 647-657. https://doi.org/10.1016/j.renene.2019.03.066

4. Srikanth, H. V., G. Sharanappa, Bhaskar Manne, and S. Bharath Kumar. "Niger seed oil biodiesel as an emulsifier in diesel-ethanol blends for compression ignition engine." *Renewable Energy* 163 (2021): 1467-1478.
5. Deep, A., Sandhu, S. S., &Chander, S. (2017). Experimental investigations on the influence of fuel injection timing and pressure on single cylinder C.I. engine fueled with 20% blend of castor biodiesel in diesel. *Fuel, 210*, 15-22. https://doi.org/10.1016/j.fuel.2017.08.023
6. Y V Hanumantha Rao, G Jamuna Rani, B Sri Krishna Kalyan,VAvinash, Ch Narendra and I Ravi Kiran (2018) Experimental Investigations on Performance and Emission Characteristics of a Direct Injection Diesel Engine Using Canola Oil Biodiesel with Exhaust Gas Recirculation (EGR), International Journal of mechanical Engineering and Technology, 9(4), 2018, pp. 192 200.
7. Satish Kumar, Nitin D. Kamitkar, Basavaraju A. N (2019). Effect of exhaust gas recirculation on performance and emissions of LHR diesel engine using jatropha biodiesel. *International Journal of Mechanical and Production Engineering Research and Development*,*9*(SI),118-125
8. Devaraj, A., Vinoth Kanna, I., Tamil Selvam, N., & Prabhu, A. (2020). Emission analysis of cashew nut biodiesel-pentanol blends in a diesel engine. *International Journal of Ambient Energy*, 1-5. https://doi.org/10.1080/01430750.2020.1725633
9. Sriram, V., Jeevahan, J., &Poovannan, A. (2017). Comparative study of performance of the DI diesel engine using Custard Apple biodiesel. *International Journal of Ambient Energy*, *40*(1), 54-56. https://doi.org/10.1080/01430750.2017.1360204
10. K. Venkatesan J. Rahul Rao (2018). Diesel engine performance and emission analysis using MosambiPeelpyro oil with Nano additive particles. *International Journal of Mechanical and Production Engineering Research and Development, Vol. 8, Issue 6, Dec 2018, 311-316*https://doi.org/10.24247/ijmperddec201835
11. Rama Krishna Reddy, E., Y. Subbalakshmi, V. Dhana Raju, K. Appa Rao, M. Harun Kumar, S. Rami Reddy, and P. Tharun Sai. "Assessment of performance, combustion and emission characteristics of the diesel engine powered with corn biodiesel blends." *International Journal of Ambient Energy* (2019): 1-9.
12. Yesilyurt, M. K. (2020). The examination of a compression-ignition engine powered by peanut oil biodiesel and diesel fuel in terms of energetic and exergetic performance parameters. *Fuel, 278*, 118319. https://doi.org/10.1016/j.fuel.2020.118319
13. Vieira da Silva, M. A., Lagnier Gil Ferreira, B., Da Costa Marques, L. G., Lamare Soares Murta, A., & Vasconcelos de Freitas, M. A. (2017). Comparative study of Nox emissions of biodiesel-diesel blends from soybean, palm and waste frying oils using methyl and Ethyl transesterification routes. *Fuel, 194*, 144-156. https://doi.org/10.1016/j.fuel.2016.12.084
14. Venkatesan Kuppusamy, SathyarajSankerlal (2018). Performance and emission characteristics of cashew nutshell pyrolysed oil - Waste cooking oil with diesel fuel in a four stoke Di diesel engine. *International Journal of Mechanical and Production Engineering Research and Development, 8*(1), 181-188. https://doi.org/10.24247/ijmperdfeb201821, Vol. 8, Issue 1, Feb 2018, 181-188.
15. Ramesh Kumar, R., Sathyaseelan, P., Alphonse, M., & Saleem, M. (2018). Performance analysis of biodiesel derived from fish waste. *International Journal of Ambient Energy*, *41*(2), 161-168. https://doi.org/10.1080/01430750.2018.1443501
16. Prabhu, A., Venkata Ramanan, M., &Jayaprabakar, J. (2018). Production, properties and engine characteristics of Jatropha biodiesel – a review. *International Journal of Ambient Energy*, 1-5. https://doi.org/10.1080/01430750.2018.1557548

17. K. V. Narasimha Rao Shaik NagulMeeravali, Nitin DattatreyaKamitkar. (2019). Effect of cooled EGR technique on performance and emissions of CRDI engine operated with biodiesel blends. *International Journal of Mechanical and Production Engineering Research and Development, Vol. 9, Issue 4, Aug 2019, 241-252*https://doi.org/10.24247/ijmperdaug201925.
18. Arunkumar, M., Kannan, M., & Murali, G. (2019). Experimental studies on engine performance and emission characteristics using castor biodiesel as fuel in CI engine. *Renewable Energy*, *131*, 737-744. https://doi.org/10.1016/j.renene.2018.07.096
19. Arunkumar, M. (2019). Research on performance and emission on compression ignition engine fuelled with blends of neem biodiesel.S*International Journal of Engineering and Advanced Technology*, *8*(6), 3732-3735. https://doi.org/10.35940/ijeat.f9382.088619
20. Yu, S., Yin, B., Jia, H., Wen, S., Li, X., & Yu, J. (2017). Theoretical and experimental comparison of internal flow and spray characteristics between diesel and biodiesel. *Fuel*, *208*, 20-29. https://doi.org/10.1016/j.fuel.2017.06.136
21. Shaik Khadar Basha, Satishkumar, Nitin D Kamitkar (2019). The performance evaluation and emission study of compression ignition engine operating with blends of animal fat and palm Oil based biodiesels, *International Journal of Mechanical and Production Engineering Research and Development*,9(4),199-206.
22. Ahamad Shaik, A., Rami Reddy, S., Dhana Raju, V., & Govindarajan, M. (2020). Combined influence of compression ratio and EGR on diverse characteristics of a research diesel engine fueled with waste mango seed biodiesel blend. *Energy Sources, Part A: Recovery, Utilization, and Environmental Effects*, 1-24. https://doi.org/10.1080/15567036.2020.1811809
23. Sanli, H. (2018). An experimental investigation on the usage of waste frying oil-diesel fuel blends with low viscosity in a common rail DI-diesel engine. *Fuel*, *222*, 434-443. https://doi.org/10.1016/j.fuel.2018.02.194
24. G. Jamuna Rani, Y.V. HanumanthaRao,B. Balakrishna (2018). Investigation of Fuel Injection Pressure Impact on Ci Engine Performance and Emissions using Biodiesel Blend with Alumina Nano Additives, International Journal of Mechanical Engineering and Technology, 9(6), 2018, pp. 922–928
25. Karikalan, L., Chandrasekaran, M., Venugopal, S., Jacob, S., & Baskar, S. (2019). Investigations on diesel engine characteristics with Pongamia biodiesel at dissimilar compression ratios. *International Journal of Ambient Energy*, 1-4. https://doi.org/10.1080/01430750.2019.1583131
26. Nitin. D. Kamitkar et al., N. D. (2018). Performance investigation and exhaust analysis of C.I engine fuelled by diesel mixed with Jamun seed powder and Jack fruit seed powder. *International Journal of Mechanical and Production Engineering Research and Development*, *8*(3), 339-350. https://doi.org/10.24247/ijmperdjun201838
27. Indrareddy, N., Venkateswarlu, K., &Konijeti, R. (2020). Experimental investigation of algae biofuel–diesel blends on performance of a CRDI diesel engine. *International Journal of Ambient Energy*, 1-8. https://doi.org/10.1080/01430750.2020.1725630

IOP Conf. Series: Materials Science and Engineering 1126 (2021) 012075 doi:10.1088/1757-899X/1126/1/012075

Effect of Nitrogen Dopant on Electrical Resistance of Hot Zone of SiC Heating Elements

S D Kavitha[1], N Suneel Kumar Kulakarni[2], R Suresh[3], and D M Shivanna[4]

[1] Mechanical Engineering, BTL Institute of Technology, Bangalore, India
[2] Mechanical Engineering, Cambridge institute of Technology, Bangalore, India
[3] Mechanical Engineering, MSR University of Advance Studies, Bangalore, India
[4] Mechanical Engineering, CMR institute of Technology, Bangalore, India

Abstract. In the present work gaseous nitrogen is used as dopant in silicon carbide heating element to study the effect of electrical resistance in heat generation portion of heating elements. The silicon carbide heating elements are used mainly in non-ferrous alloy melting and holding furnaces in automotive and aerospace industrial applications. The gaseous nitrogen plays a major role as dopant in achieving the electrical resistance of SiC heating elements. The green silicon carbide is used to prepare the heating element by powder metallurgy process because of its high mechanical strength, resistance to corrosion, and high thermal conductivity at elevated temperature. However, the nitrogen acts as dopant in silicon carbide to achieve the electrical resistivity in the heating element to generate uniform heat in hot zone portion of SiC heating elements while sintering. The hot zone of SiC heating element is connected with cold zone at both the terminals of heating element to pass the current to hot zone to generate the heat by its higher resistivity. A range of nitrogen gas flow rate ranges from76 lt/hr to 84 lt/hr with 99.99% purity of industrial grade passed to the sintering furnace, in turn it acts as dopant and creates inert atmosphere in the sintering furnace, then the SiC heating element tubes were made and analysed. The results indicate that the quantity of gaseous nitrogen dopant influenced the electrical resistivity of the heat generating portion of SiC heating element.

1. Introduction

Silicon carbide has numerous industrial applications such as heating elements, automobile brake pads, semiconductors, etc.. The electrical and mechanical properties of SiC are known to vary over large ranges depending on methods of manufacture, dopants and their mass fractions, and post treatment (such as siliconizing). Dopants can be solids such as Aluminium, Boron, Iron, etc., or gases such as Nitrogen. Methods of manufacture usually involve mixing of core ingredients, pressing, drying, sintering, and post treatment.

Methods of manufacture and resultant material properties of SiC are under continuing international study. Industrially produced SiC is referred to as Carborundum and crystolon [1]-[3]. β to α transition occurs rapidly at around 2300°C. High nitrogen pressures favour the reverse process [1]-[5]. Normal transformation of SiC is from cubic (3C) to hexagonal (6H) [6]. The atmospheric pressure difference between concave and convex surfaces of SiC particles can be of the order of 10^6 to 10^5 atmospheres [7]-[9]. Grain size was found to be dependent on isostatic pressing temperatures [10].

An improved production with low energy consumption is reported [11]. Use of boron as dopant with improved properties is reported [12]. Phase transitions between 3C-6H have been studied [1-12]. Effects of aluminium and zinc have been studied [13]. A method of tuning the electrical properties is reported [14]. A CVD formed SiC product [15]. Explosive compaction is investigated [16]. Impregnation with acid and coating with oxides, enamels, or glass to protect against oxidation has been investigated [17]. Effect of boron has been investigated. Efficient methods of production are investigated [18]. Doping with nitrogen (N2) has been investigated [19]. a method for imparting high insulation property and corrosion resistance. An improved production method is reported in. Additional variations are reported in [20][21] investigates doping by aluminium, additional carbon, oxygen, and nitrogen. A method to prepare wear and corrosion resistant SiC is reported, the preparation of SiC conjugate [22],[23].

2. Methodology

The following steps are followed for preparing samples;

1. Selection of SiC raw material grit size.
2. Preparing Ingredients and Blending: Mixing of SiC powder weight ratio along with binder and water to make lumps for pressing.
3. Extrusion: Prepared lumps pressed through horizontal cold extrusion to make samples.
4. Drying: Extruded samples are dried in the oven to remove moisture in the green samples.
5. Sintering: Dried samples passing through continuous carbon tube furnace to sinter the samples along with gaseous nitrogen inert atmosphere.
6. Testing: Sintered samples are tested for its electrical resistance.
7. Results and discussion: experimental results are discussed.

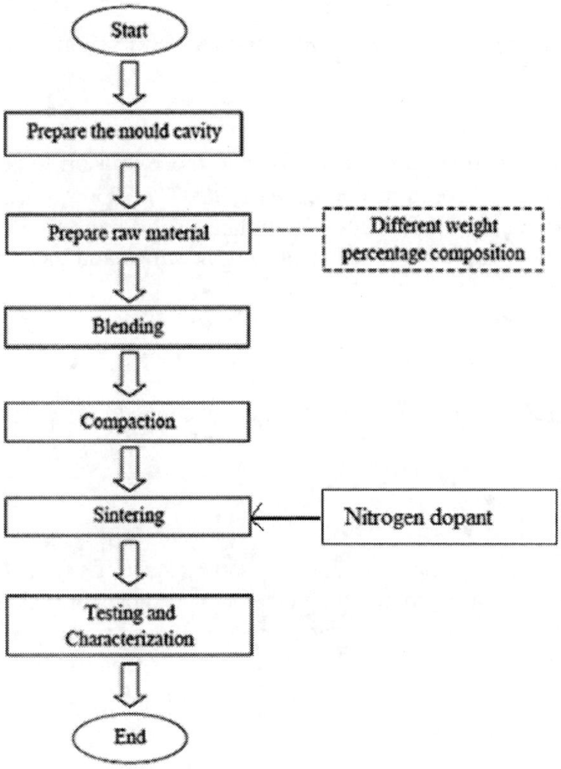

Figure 1. Flow chart showing heat generating zone of SiC heating elements.

3. Experimentation

3.1. Selection of SiC powder and blending

SiC powder with 99% purity in the form of green silicon carbide. The proportion of green SiC grit size of 24 # to #36 granularity from 5% to 10% of coarse grains, The grit size 80 # to 90 # of medium grains 45% to 50%, and the grit size in the form of fine grains 1000# to 1200 # of weight 30% to 40%. The methylcellulose as binder, range of viscosity 3500-5600 mPa. s (at 20 ° C, 2%), methoxy content 27% to 30%. The ingredients are mixed in a muller mixing machine for about 45 to 55 minutes, upon lumps formation after adding required quantity of water. Figure 2 shows the schematic process flow of SiC heating element tubes.

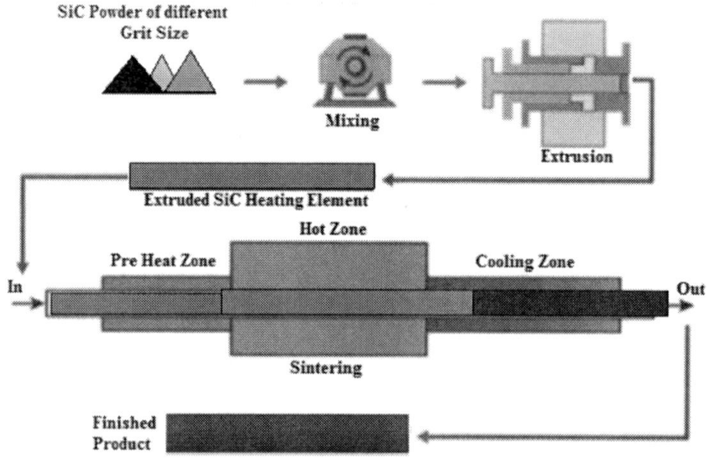

Figure 2. Process of hot zone portion of SiC heating elements tubes.

3.2. *Extrusion*

Lumps formed in the mixing machine is pressed in the extrusion to form tube of required diameter. Weigh 24 # 1500 grams of green silicon carbide, 1200 # 3000 grams, 300 # 1000 grams, 80 # 4500 grams, 420 #1750 grams, mixed for 15 minutes then added methyl cellulose of 2% with water of 8% to 9%. Mixing was continued upto 45 minutes to form lumps and then pressed at 85 Mpa to 95 Mpa to form tube of Ø25mm.

3.3. Drying

Formed tubes were dried at 20 ~ 40 °C for 24 hours at ambient temperature, then raised temperature to 180 ~ 200 °C for 3 hours.

3.4. *Sintering*

The dried SiC tubes were pushed through continuous tube furnace for sintering. Gaseous nitrogen dopants are introduced in the furnace to make inert atmosphere. A schematic arrangement of the sintering furnace is shown figure 3. The sintering furnace consists of pre-heat zone temperature 350° C-800° C, sintering zone temperature of 2450° C - 2550 ° C, carbon tube length 2000 mm, and cooling zone, advanced by constant pusher ramp rate, under variable gaseous nitrogen flow of 99.99% purity (industrial grade) from 76 lt/hr to 84lt/hr.

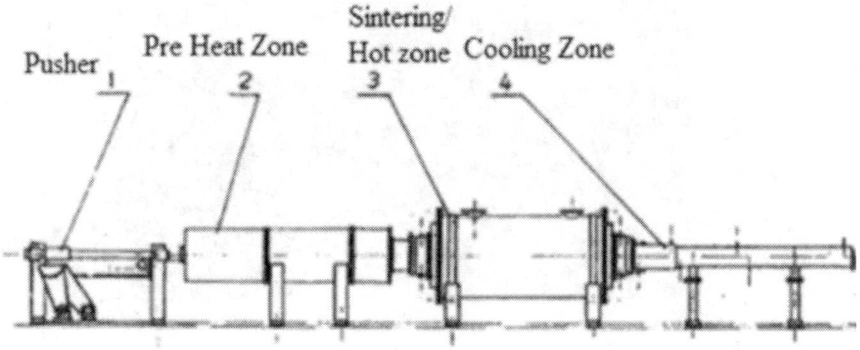

Figure 3. A schematic of the arrangement of sintering process.

Time spent in each of the zones is determined by the rate at which the product is being moved through the system (pusher rate). This is one of the parameters which can be chosen to help achieve the desired property for the finished product.

3.5. Testing

Sintered SiC heating elements tubes are tested for its electrical resistance by maintaining constant voltage. Samples of Ø25mm, length 500mm SiC tube has been produced at constant push rate and temperature process parameters with variable gaseous nitrogen flow from 76 lt/hr to 84lt/hr.

Figure 4 shows the uniform heat generation of hot zone of SiC heating elements.

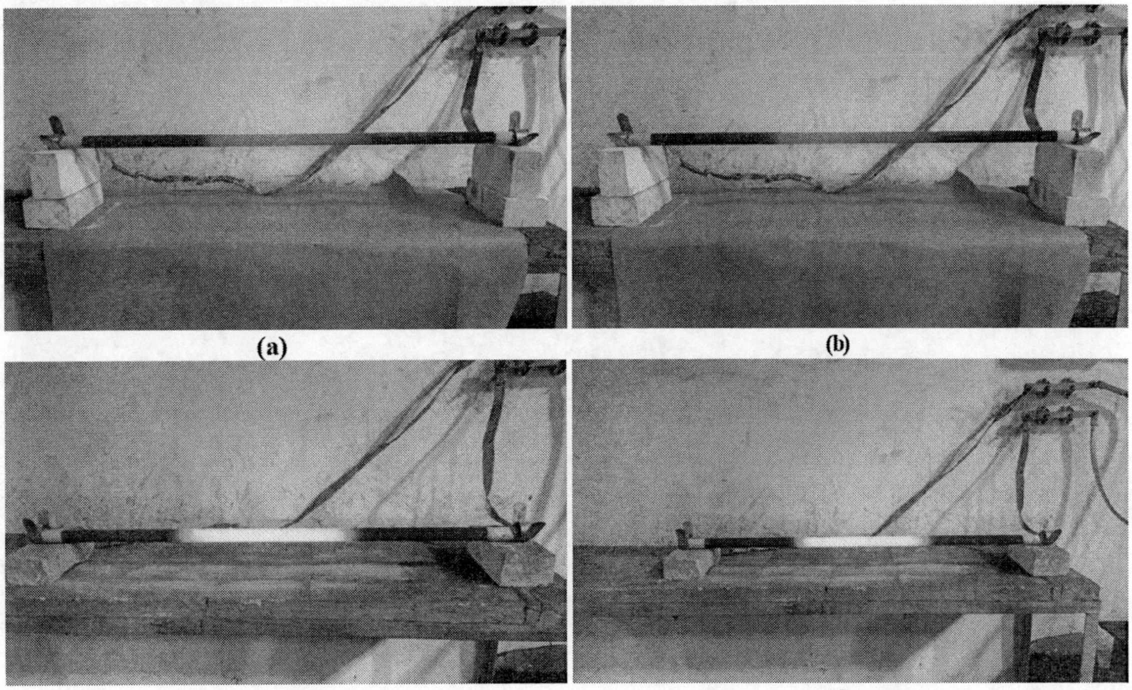

(a) (b)

(c) (d)

(e)

Figure 4 (a), (b), (c), (d), (e). Testing of hot zone of SiC heating element tubes.

After sintering, SiC heating tubes resistance has been measured and uniform glow of hot zone portion of heating elements has been observed as tabulated below in table 1 to 5. For measuring each SiC heating element tube resistance, the constant voltage maintained to note the current readings in amperes, the current and voltage indicates the electrical resistivity of sintered SiC heating element tubes. The resistance of the SiC tubes based on current flow and its porosity, sintering of SiC grains, nitrogen dopant.

Table 1. Hot Zone Ø25mm, length 500mm, constant pusher speed 20mm/min at Nitrogen gas flow rate 76 lt/hr & furnace temperature 2400°C.

Sample	Voltage (V)	Current (A)	Resistance (Ω)
1	55	56.04	0.981
2	55	52.01	1.057
3	55	50.06	1.098
4	55	53.00	1.037
5	55	49.11	1.119
6	55	52.18	1.054
7	55	48.10	1.143
8	55	48.55	1.132

Table 2. Hot Zone Ø25mm, length 500mm, constant pusher speed 20mm/min at Nitrogen gas flow rate 78 lt/hr & furnace temperature 2400°C.

Sample	Voltage (V)	Current (A)	Resistance (Ω)
1	55	65.12	0.844
2	55	69.3	0.793
3	55	68.43	0.803
4	55	65.53	0.839
5	55	66.8	0.823
6	55	65.93	0.834
7	55	64.95	0.846
8	55	67.15	0.819

Table 3. Ø25mm Hot Zone length 500mm, constant pusher speed 20mm/min at Nitrogen gas flow rate 80 lt/hr & furnace temperature 2400°C.

Sample	Voltage (V)	Current (A)	Resistance (Ω)
1	55	77.04	0.713
2	55	79.36	0.693
3	55	76.42	0.719
4	55	78.65	0.699

Table 4. Hot Zone Ø25mm, length 500mm, constant pusher speed 20mm/min at Nitrogen gas flow rate 82 lt/hr & furnace temperature 2400°C.

Sample	Voltag e (V)	Current (A)	Resistance (Ω)
1	55	83.23	0.660
2	55	83.96	0.655
3	55	84.3	0.652
4	55	83.51	0.658

5	55	79.53	0.691
6	55	80.22	0.685
7	55	79.94	0.688
8	55	80.54	0.682

5	55	82.86	0.663
6	55	80.24	0.685
7	55	83.71	0.657
8	55	83.31	0.660

Table 5. Hot Zone Ø25mm, length 500mm, constant pusher speed 20mm/min at Nitrogen gas flow rate 84 lt/hr & furnace temperature 2400°C.

Sample	Voltage (V)	Current (A)	Resistance (Ω)
1	55	85.23	0.645
2	55	86.96	0.632
3	55	87.38	0.629
4	55	85.51	0.643
5	55	88.16	0.623
6	55	86.21	0.637
7	55	85.73	0.641
8	55	86.75	0.634

4. Results and discussion

In order to investigate the effects of gaseous nitrogen dopant on hot zone portion of SiC heating element tube, a continuous carbon tube furnace was used. By varying the nitrogen gas flow rate through the furnace, the sintering time kept constant by maintaining the push ramp rate. Higher the nitrogen gas flow rate, the lesser the resistivity in the sintered SiC heating elements and uniform red hot eveness in the element.

Table 6. Dia 25mm Hot Zone, length 500mm, variable nitrogen gas flow rate 76 lt/hr to 84 lt/hr at furnace temperature 2400° C, Resistance in Ohm (Ω).cm.

Temp. in °C	Nitrogen Flow Rate				
	76 lt/hr	78 lt/hr	80 lt/hr	82 lt/hr	84 lt/hr
	Resistance in Ohm (Ω).cm				
30	0.0291	0.0232	0.0194	0.0165	0.0159
200	0.0265	0.0213	0.0175	0.0162	0.0153
400	0.0253	0.0192	0.0165	0.0146	0.0142
600	0.0241	0.0176	0.0143	0.0138	0.0137
800	0.022	0.0165	0.0137	0.0132	0.0126
900	0.02	0.0159	0.0128	0.0129	0.0118

From the table 1 to 6, the resistivity of SiC heating elements tubes for a overall length of 500mm heat generation portion of various test samples by varying nitrogen flow rate at temperature 900 ° C as noted in table 6. Temperature in ° C and resistivity Ohm per cm as tabulated in table 6 of sintered SiC heating elements passed through the continuous carbon tube furnace by varied nitrogen flow rate as explained in the results and discussion. The electrical resistivity of the SiC hot zone material sintered at varied nitrogen flow rate ranges from 76 lt/hr to 84 lt/hr at a furnace temperature of 2400°C with +/- 10°C with push ramp rate of 20mm/min.

Figure 5 shows graph of resistivity v/s temperature for heat generating portion of SiC heating element tubes extruded from SiC of constant grain size formed through a continuous carbon tube furnace at different nitrogen flow rate.

The electrical resistivity of the hot zone green silicon carbide material sintered at different nitrogen flow rate from 76 lt/hr to 84 lt/hr at a constant sintering furnace temperature range of 2400-2450°C. A nitrogen flow rate of 76 lt/hr possesses maximum resistivity. Figure 5 shows a graph of resistivity of hot zone of heating portion versus temperature when sintered at various nitrogen flow rate. The reduced electrical resistivity achieved by increasing the nitrogen gas flow rate from 82 lt/hr to 84 lt/hr is nominal when the gas flow rate is reduced from 78 lt/hr to 76 lt/hr. Although the nitrogen gas flow rate of 84 lt/hr showed the highest reduction in electrical resistivity, such a high nitrogen gas flow rate increases the cost of production. With the particular furnace used, the nitrogen gas flow rate of 82 lt/hr was considered optimum.

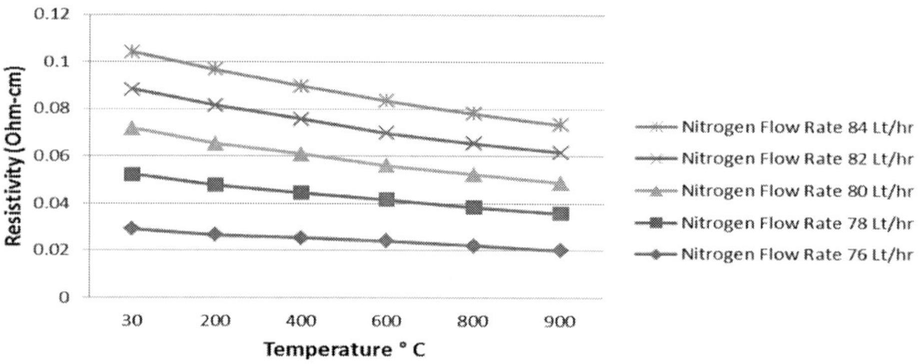

Figure 5. Resistivity versus temperature at different nitrogen flow rates for hot zone.

Conclusions

In this work, the processing of heat generating portion of SiC heating elements can be optimized by controlling the gaseous nitrogen flow in the sintering furnace to maintain inert atmosphere and as dopant in the sintered SiC heat generation portion, it has the advantage in stability of high resistivity value, performance, and improved yield of sintered hot zone of SiC heating element.

However, it has been found that the influence of nitrogen gas makes inert atmosphere in the sintering furnace and act as dopant in the sintered SiC hot zone of heating element reduces the resistivity of the hot zone of SiC heating element.

SiC Sintered temperature at 2400°C in optimum nitrogen gas flow rate (84 lt/hr) prominently reduced the resistivity (0.0126 Ω.cm at 800°C) when compared with nitrogen gas flow rate of 76 lt/hr. From this observed that nitrogen is doped in the SiC lattice while sintering in nitrogen inert atmosphere in the continuous carbon tube furnace, which transformed the β-Sic to α-SiC.

References

[1] Baumann H 1952 *J. Elect .Chem. Soc.* 99 109-144.

[2] Alliegro R Coffin L and Tinklepaugh J 1956 *J. Amer Ceram. Soc.*39 386-389.

[3] Hannam A L and Shaffer P T B 1969 *J. Appl. Cryst.***2** 45.

[4] Cobs L 1971 Springer Verlag New York Wein.

[5] Forrest C W 1972 *Special Ceramics 5 Proceedings of the fifth symposium on Special Ceramics held by The British Ceramic Research Association*(Ed: Popper) p 99-124.

[6] Nadeau J S 1973 *Amer. Ceram. Soc. Bull.* 52 [2] 170-76.

[7] Bind J N Biggers J V 1975 *J. Am. Ceram. Soc.* 58 304-306.

[8] Bind J 1977 Penn State University, University Park, Pennsylvania *USA JCPS Grant in Aid Report.*

[9] N A Likhomanova V F Kutukov 1979 *Eastern Scientific-Research and Design Institute of the Refractory Industry. Translated from Poroshkovaya Metallurgiya, Plenum Publishing Corporation* **2** (194) p 56-59.

[10] Heuer A H Fryburg G A Ogbuji L U And Mitchell T E 1978 *J. Am. Ceram. Soc.* **61** [9] 406-412.

[11] Jepps N W Page T F 1981 *J. Amer. Cer. Soc.* **64** (12) Cl 77-178.

[12] Sunil Duita 1984 *J. Mat. Sci.* **19** 1307-1313.

[13] Kriegesmann 1986 J. 18, (5), 341-343.

[14] Sunil Duita 1985 J. Am. Ceram. Soc., 68 [10] C-269 - C-270.

[15] Thibault N W 1944 Am. Mineral 29 249-278.

[16] Jihong S Jingkun G & Dongl Iang 1993 J Ceram. Int. 19 p 347-351.

[17] Watanabe Masakazu 1987 US patent no. US 4668452

[18] Gao Jiqiang 2007 CN Patent No: CN 1293020

[19] Shin Sugiyama 2001 J. Am. Ceram. Soc 3013–16.

[20] Henry N Baumann Jr 1951 research and developmenl division, the Carborundum company, Niagara falls, New York, Washington meeting, april 8 to 12.

[21] Martin Sandvik Materials Technology Uk.

[22] Beatson 2007 European patent no EP1645168.

[23] R J Carlson 1968 US patent no 3404202

IOP Conf. Series: Materials Science and Engineering 1126 (2021) 012076 doi:10.1088/1757-899X/1126/1/012076

Eco Bricks from Industrial Wastes such as Tannery Sludge and Sugarcane Bagasse Ash

Rose Enid Teresa A[1], Uma Maguesvari M[2], Yugasini S[3] and Muthaiyan P[4]

[1]Professor, Department of Civil Engineering, Rajalakshmi Engineering College, Chennai, India.
[2]Associate Professor, Department of Civil Engineering, Rajalakshmi Engineering College, Chennai, India.
[3,4]Assistant Professor, Department of Civil Engineering, Rajalakshmi Engineering College, Chennai, India.

Abstract. Clay bricks are one of the prominent construction materials. Due to the usage of Bricks in abundance in construction sector and manufacturing of quality clay bricks from fertile raw materials, admixtures etc., the world is in need of alternatives to prevent the depletion of clay from agricultural lands.. The tannery sludge which is disposed as a landfilling material, is a potential substitute for brick making. In addition, bagasse ash with high silica content, which is a waste disposal from sugarcane industry wastes, might be incorporated with sludge in brick manufacturing. In this paper, an attempt has been made to manufacture Eco Brick made from tannery sludge and bagasse ash with clay. Tests were carried out on Eco Bricks manufactured from various composition of Sludge and bagasse ash and it has been proven to be an effective alternative material in construction industry.

1. Introduction

Clay bricks play a vital role in the Construction Sector. In India, more than 20000 million bricks are used by Construction Industry, about 27% of natural energy is consumed for manufacturing the same. To make it cost effective and to reduce the energy consumption which affects the economy of the Nation as a whole, alternative efficient and cost-effective building materials are in high demand which would enable the every citizen to own their residence. Depending on the raw materials and the admixtures used, Bricks are of different categories and quality bricks are manufactured from clay available in agricultural fields. Over usage of these clays will lead to the exploitation of agricultural fields. In the near past, sludges from various industries [6] are used as an alternative material for manufacturing Clays. Aigbodion et.al [1] studied on the utilization of industrial and agricultural waste products (Bagasse ash) in the industry for economic, environmental, and technical reasons. Chihpin et.al. [4] investigated the use of sludge as partial substitute for clay in brick manufacturing. The produced brick properties were superior to those available in the Egyptian market. Divya et al [5] studied on the production of Eco bricks using tannery sludge. In this study, an attempt is made add tannery sludge with clay in various proportions to make porotherm brick block. Usage of tannery sludge [3] in an effective way in construction industry will prevent soil pollution and landfilling problems as well. The tannery sludge in powdered form was blended with clay and casted using mould manually. The addition of tannery sludge in powdered form does not affect the environment as various tests are done to know its properties. The authors concluded that the optimum percentage of tannery sludge with clay was found to be 20% Juel et al. [8], conducted a study on the Sustainable use of tannery sludge in brick manufacturing in Bangladesh, and the study indicated that tannery sludge can be sustainably stabilized in clay bricks and large-scale application of this technique can be envisaged in the context of Bangladesh where brick remains a dominant building material. Amsayazhi & Saravana Raja Mohan [9] carried out a study to utilize those sludge waste (SW) produced from the tannery effluent plant as a brick. The sludge brick (SB) mixtures were incorporated with many ratios of SW. The authors concluded that the brick with 20% utilization of SW is acceptable to produce good quality of brick. Dawit et al. [10] prepared bricks by incorporating textile sludge at different proportions (10–40%) and temperature (600, 900 and 1200 °C). Clay and sludge samples were collected from the Addis Ababa brick factory PLC and Hawassa Industrial Park. The authors concluded that, 10and 20% sludge bricks satisfied criteria of class "A" bricks as per Ethiopia standards, with the compressive strength of 30.43 and 29.10 Mpa, respectively, at 1200 °C. The authors suggested that rapid utilization of huge amount of partially dried sludge resources for low-cost and efficient large-scale brick production. Jurgita et al. [11] analysed the properties of recycled tannery sludge (TS) and the possibilities for using it in cement mortar mixture. The tests revealed that replacement of 6% of cement with tannery sludge in the mix increased flexural and compressive strength and UPV values, whereas water absorption decreased. SEM and XRD analysis revealed that specimens with tannery sludge contained lower amounts of ettringite and higher amounts of portlandite. Mangesh et al. [12] developed Sugaracne Bagasse Ash (SBA), quarry dust (QD), lime (L) bricks and tested for physicomechanical properties, functional (thermal conductivity, kk), durability (chloride, sulfate, and carbonation), and environmental [toxicity characteristics leaching protocol (TCLP)] properties as per recommendations. The developed (SBA-QD-L) bricks were further analyzed for technical feasibility with commercially available and accepted masonry products like clay and fly ash bricks. The experimental results showed that the SBA-QD-L combination brick is lighter in weight, durable, nonhazardous, energy efficient, has lower kk value, and meets the necessary physicomechanical properties of the standards.

From the Literature review, the use of sludge in construction industry is considered to be the most economic and environmentally sound option. In this paper, eco bricks are manufactured from sludge disposal from Pallavaram Tanners Industrial Effluent Treatment Co. Ltd (PTIET) and the efficiency of the eco brick was compared with the conventional bricks

2. Eco Brick using Tannery Sludge and Bagasse Ash as Replacement Materials

Globally, one of the most polluting industry which causes unfavourable impact on the environment is identified to be Leather industry. Tanneries in India are mainly located in states namely Punjab, Tamil Nadu, Uttar Pradesh and West Bengal. In Tamilnadu, around 900 tanneries are present. The Pallavaram Tanners Industrial Effluent Treatment Co. Ltd (PTIET) located in Pallavaram, Chennai treats around 3 MLD of waste water from the tanneries. The resulting sludge weighs about 1.5 tonnes and disposed by land filling. However, for highly organized cities, sludge disposal by land filling might not be appropriate due to land limitation as well as soil pollution. In this paper, an experimental investigation was conducted on Eco bricks manufactured using Tannery Sludge and Bagasse ash [2] as replacement materials. The work has led to evaluate the suitability of sludge and bagasse ash in partial replacement of clay in manufacture of Eco bricks. Test procedures are done to evaluate the properties of tannery sludge. The Sugarcane Bagasse ash and Tannery sludge are shown in Figures 1 & 2.

Figure 1. Bagasse Ash

Figure 2. Tannery Sludge

The chemical properties of Bagasse ash and characteristics of treated sludge are shown in Tables 1 & 2.

Table 1. Chemical Properties of Bagasse ash

SiO_2	87.40
Al_2O_3	3.6
Fe_2O_3	4.95
CaO	2.56
MgO	0.69
Na_2O	0.15
K_2O	0.47
S_3O	0.11
LOI	8.25

Table 2. Characteristics of treated sludge

Test Items	Units	Amounts	Limits
P^H	-	7	6 to 8
COD	mg/l	54	250 max
Oil and grease	mg/l	5	10 max
Suspended solids	mg/l	34	100 max
Kjeldahl's nitrogen	mg/l	34	100 max
Sulphide	mg/l	NIL	2 max
BOD	mg/l	18	30 max
Total dissolved solids	mg/l	111	2100 max
Ammonical nitrogen	mg/l	8	50 max

3. Materials used and Test Setup

The materials used for Eco Bricks are Bagasse Ash, Tannery Sludge in various proportions, clay, water sand. To know the quality of bricks following 8 tests (IS: 3495 –PART-1:1992) [7] were performed.

Few tests were performed in laboratory and the rest of it the field.

a. Size and Shape Test 20 bricks are randomly collected and staked along lengthwise, width wise, and height wise and then those are measured to know the variation of sizes as per standard. Bricks are closely viewed to check if its edges are sharp and straight and uniform in shape.
b. Colour Test: The bricks were checked for bright and uniform colour throughout.
c. Soundness Test: When the bricks are struck with each other, a give clear metallic ringing sound and non-breakage will exhibit good quality of bricks.

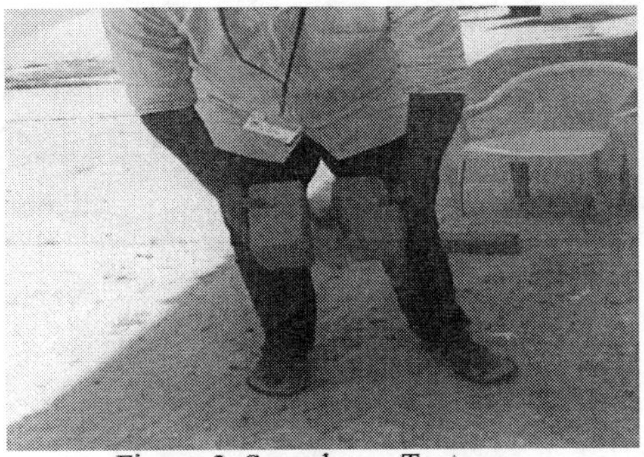

Figure 3. Soundness Test

Figure 4. Water Absorption Test

d. Water Absorption Test: Good quality brick doesn't absorb more than 20% water of its own weight.

e. Hardness Test: When scratched for abrasion, a good quality brick does not show any impression.

f. Compressive strength Test

Figure 5. Hardness Test

Figure 6. Compressive strength Test

4. Results and Discussion

4.1 Shape and Size

The Eco Bricks made from various sludge proportions such as 5%, 10%, 15% and bagasse ash proportions of 15%, 10% and 5% were tested for its shape and size and is shown in Table 1. All sizes of the bricks are as same as standard values of any proportions of tannery sludge and bagasse ash.

Table 1. Shape and Size

Sludge Proportion	Bagasse Ash Proportion	Length (mm)	Width (mm)	Height (mm)
5%	15%	190	90	90
10%	10%	190	90	90
15%	5%	190	90	90
STANDARDS		190	90	90

4.2 Water Absorption Results

The results are tabulated in Table 2 and it is found that, as percentage of bagasse ash increases and percentage of tannery sludge decreases, water absorption of bricks increases. Up to 15% of bagasse ash is adequate to replace.

Table 2. Water Absorption Results

Proportion	Weight of Dry Brick (W1 Kg)	Weight of Wet Brick (W2 Kg)	Weight of Water Absorbed (W2-W1)Kg	% Water Absorbed (Kg)	Average %
Standard	3.056	3.31	0.254	8.312	8.126
	3.048	3.28	0.232	7.612	
	3.288	3.566	0.278	8.455	
15% Sludge - 5% Ash	3.1	3.490	0.39	12.58	12.72
	3.20	3.590	0.39	12.18	
	3.28	3.72	0.44	13.4	
10% Sludge -10% Ash	2.660	3.001	0.341	12.8	14.29
	2.65	3.110	0.46	17.3	
	2.70	3.045	0.345	12.77	
5% Sludge - 15% Ash	2.62	3.050	0.43	15.925	16.228
	2.67	3.100	0.43	16.10	
	2.7	3.15	0.45	16.66	

4.3 Test for Compressive Strength

The test results are shown in Table 3. It is concluded that, for [15%ash-5%sludge] and [5%ash-15%sludge] Proportions, the compressive strength is as per requirement. For [10%ash -10%sludge] proportion, the strength reduces. It is desirable to use the [15%ash-5%sludge] and [5%ash-15%sludge] Proportions.

Table 3. Compressive Strength – Test Results

Proportion	Size of Brick (mm^3)	Surface Area (mm^2)	Load At Failure (P) in kN	Compressive Strength (P/A) in N/mm^2	Average
Conventional	190×90×90	17100	140	8.18	8.02
			135	7.89	
			137	8.01	
5%sludge-15%ash	190×90×90	17100	188	10.99	11.22
			196	11.46	
			192	11.22	
10%sludge-10%ash	190×90×90	17100	74	4.33	4.62
			85	4.97	
			78	4.56	
15%sludge-5%ash	190×90×90	17100	145	8.47	8.68
			153	8.88	
			149	8.71	

4.4 Colour Test : For all proportions of the tannery sludge and bagasse ash, bricks have good red uniform colour.

4.5 Soundness Test for Bricks : When two bricks are struck with one another, [15% ash-5% sludge] and [5%ash-15%sludge] proportions gave metallic sound. But [10%ash-10%sludge] proportion gave dull sound.

4.6 Hardness Test: For [15%ash-5%sludge] and [5%ash-15%sludge] Proportions, bricks were hard. When scratches were made, it doesn't leave any marks. But for the [10%ash-10%sludge] proportion, there was scratch marks left on brick surface.

5. Conclusion

Use of bagasse ash and industrial sludge in brick can solve the disposal problem; reduce cost and produce a 'greener' Eco friendly bricks for construction. A better measure by an innovative Construction Material is formed through this research. In this paper, different proportions of the industrial wastes and bagasse ash were added as a partial replacement for clay in the production of Eco bricks and the results were compared with the conventional clay bricks. An addition of bagasse ash and industrial sludge in the proportion of [15%ash-5%sludge] causes more water absorption, increases the compressive strength and provide hardness when under burnt. The hardness property and the soundness property of the bricks were also achieved. It is recommended to use partial replacement of clay by 15% of bagasse ash and 5% of industrial sludge for good quality Eco Bricks. This study helps in converting the non-valuable bagasse ash and industrial sludge into bricks and makes it valuable.

6. References

[1] V. S. Aigbodion, S. B. Hassan, T. Ause and G.B. Nyior, Potential Utilization of Solid Waste (Bagasse Ash), *Journal of Minerals and Materials Characterization and Engineering*, Vol. 9, No.1, pp.67-77, 2010.

[2] Apurva Kulkarni, Samruddha Raje , Mamta Rajgor. "Bagasse Ash As An Effective Replacement In Fly Ash Bricks". *International Journal of Engineering Trends and Technology (IJETT)*. V4(10):4484-4489 Oct 2013.

[3] Badr El-Din E. Hegazy, Hanan A. Fouad and Ahmed M. Hassanain, Brick Manufacturing From Water Treatment Sludge And Rice Husk Ash, *Australian Journal of Basic and Applied Sciences*, 6(3): 453-461, 2012 ISSN 1991-8178.

[4] Chihpin, H., P.J. Ruhsing and L. Yaorey, 2005. "Mixing Water Treatment Residual with Excavation Waste Soil in Brick and Artificial Aggregate Making", *Journal of Environmental Engineering*, 131(2): 272-277.

[5] G. Divya, Gokul Vasan M E, Sathya Narayanan G, Eco Bricks Using Tannery Sludge, *1st International Conference on Recent Innovation in Civil Engineering and Management (ICRICEM '18)*

[6] Maria P. Durante Ingunza; Anaxsandra C. L. Duarte; and Rubens M. Nascimento, Use of Sewage Sludge as Raw Material in the Manufacture of Soft-Mud Bricks, *Journal of Materials in Civil Engineering*, Volume 23, Issue 6.

[7] *IS: 3495 –PART-1:1992*: Methods of tests of burnt clay building bricks.

[8] Juel, M.A.I., et al. Sustainable use of tannery sludge in brick manufacturing in Bangladesh. *Waste Management* (2017), http://dx.doi.org/10.1016/j.wasman.2016.12.041.

[9] Amsayazhi P, Saravana Raja Mohan K, Use of Sludge Waste as Ingredient in Making of Brick, *International Journal of Engineering &Technology*, 7 (3.12) (2018) 419-422.

[10] Dawit Alemu Beshah, Girum Ayalneh Tiruye & Yedilfana Setarge Mekonnen, Characterization and recycling of textile sludge for energy-efficient, https://doi.org/10.1007/s11356-020-11878-7, Published online.

[11] Jurgita Malaiškiene, Olga Kizinievi c and Viktor Kizinievi, A Study on Tannery Sludge as a Raw Material for Cement Mortar, *Materials 2019, 12, 1562; doi:10.3390/ma12091562*

[12] Mangesh V Madurwar; Sachin A Mandavgane, and Rahul V Ralegaonkar, Development and Feasibility Analysis of Bagasse Ash Bricks, *Journal of Energy Engineering/Volume 141, Issue 3 – September 2015.*

Experimental investigation on Hydrogen embrittlement of EN47 Spring steel

H K Basukumar [1], **K V Arun** [2]

[1] Department of Mechanical engineering, KLE Institute of Technology, Hubballi 580027, Karnataka, India
[2] Department of Mechanical Engineering, Government Engineering college, Haveri 581110,
Karnataka,India
Basu2511@gmail.com

Abstract. Carbon steel is one of the significant engineering materials used in industrial and automotive sector. Most of the engineering metals are susceptible to hydrogen embrittlement (HE). It is an inherent phenomenon, which will occur during the course of service life and leads to pre mature failure of the component. In view of this, an experimental study was conducted to investigate the failure behavior of EN47 in a hydrogen simulated environment. The immersion type hydrogen charging method was adopted. The effect of HE was explored through conventional tensile testing. Experimentations were carried out according to ASTM E8 Standards. The tensile behavior of spring steel and effect of hydrogen interaction with materials was studied. The treated test material experience brittle to ductile transition exhibiting mixed mode fracture morphology on fracture surfaces, often considered being confirmation of failure through HE. Hydrogen charging causes the samples to become brittle, and reduced the plastic deformability. From the test results it can be interfered that the hydrogen charged specimen exhibits loss of ductility and also reduced the ultimate strength.

1. Introduction

Steel is an alloy of carbon, iron and other metals and non-metals. Because of their versatile achievable properties, availability and a relatively low production cost, they are most widely used in engineering and automotive industries viz., transportation, construction, Energy, Industry and appliances. They are especially important as engineering construction materials. They are widely used in boilers, pressure vessels, heat exchangers, piping, and other moderate-temperature service systems in which good strength and ductility are desired [1-3]. They are used due to their availability, relative inexpensiveness and they may be tailored to have a wide range of mechanical and physical properties. There is wide range of steels used as spring steel in automotive to aerospace industries for their resilience and elastic properties with high yield strength. Majority of the spring material requirement are met with the carbon steel, alloy steel and Corrosion resistant steels. Some of the desired mechanical properties are yield strength, toughness and fatigue strength to withstand the high stress, pulsating load, shock or impact loading. In these applications, the long-term static and dynamic behavior plays a prominent role. In the automotive industry, it is recognized that effective life of most components may be affected by yielding, not fracture.

Absorption and release of energy during the lifetime of the usage of a spring is a cyclic loading and unloading process. Therefore, spring failures generally occur by a fatigue mechanism although the root cause may be related to material and processing factors [4,5]. Damage of metals due to the influence of hydrogen is quite frequent and leads to dangerous failures as well as to loss of property and large compensational payments by insurance companies. High stresses combined with a corrosive environment can cause critical components inside power plants and other systems to crack and fail, sometimes with little warning. Hydrogen embrittlement is a delayed catastrophic failure that occur at ambient temperatures,in many steel and other metal alloys when when atomic hydrogen diffuses into the metal structure and interacts within the metal to form a brittle material. It is characterized by mechanical properties degradation due to hydrogen ingress, such as loss of ductility and, consequently, a loss of strength. Most of the engineering metals are susceptible to HE during manufacturing, processing and service environment. HE is considered as one of the primary cause of the premature failure in engineering structure and must be considered in determining the life expectancy of an engineering structure [6,7]. It can be a very complicated process and reliability life prediction is often difficult. This kind of intrusion of hydrogen will produce a disastrous reduction in the ductility and toughness of steel. For evaluation of HE resistance, hydrogen must be charged to specimens. The immersion type hydrogen charging method was adopted [8]. In the present study the effect of hydrogen on the mechanical behaviour of EN47 spring steel has been investigated by means of slow strain rate tensile test of pre-stressed and hydrogen charged specimens. The results are discussed concerning their tensile behavior to HE.

2. Material and Methods

EN 47 is selected as the testing material, as it is used extensively in the aviation, aerospace and automobile industries, among others, due to its good extrudability and weldability and excellent corrosion resistance. Spring steel is medium low alloy, medium carbon steel (EN47/AISI 6150/SUP 10) which is suitable for over-all engineering purposes with the higher strength than the mild steel. As procured material has been chemically tested for the compositions are represented in table 1. HE is the degradation of structural properties of metals due to interaction with hydrogen. It results in loss of ductility and reduced load carrying capacity which usually due to the presence of dissolved hydrogen. The interaction of hydrogen with microstructural features can strongly influence fracture resistance and sub-critical cracking of a material. As HE occurs in most metals, investigation of mechanisms failure under the tensile load is studied and mechanical properties are evaluated. Ammonium thiocyanate (NH4SCN) is one of the widely used reagent for charging steel materials with hydrogen because of its effect on promoting hydrogen absorption and its ease of use [7]. To charge hydrogen, the specimens were immersed in a 35% aqueous solution of ammonium thiocyanate at an ambient temperature for a period of 120 hours. The ammonium thyocynate solution for the immersion test was prepared by diluting with distilled water to concentration of 35%w/vol at room temperature. A typical test setup was shown in figure.1. Laboratory immersion tests were conducted in accordance with ASTM standard practice. The hydrogen charging carried out with a conventional laboratory method by immersing the pre-stressed specimens.

Table 1: Chemical composition of test specimen material EN47/SAE6510

Elements	Carbon	Silicon	Manganese	Phosphorus	Sulphur	Chromium	Vanadium
%	0.451	0.247	0.660	0.018	0.010	1.0	0.155

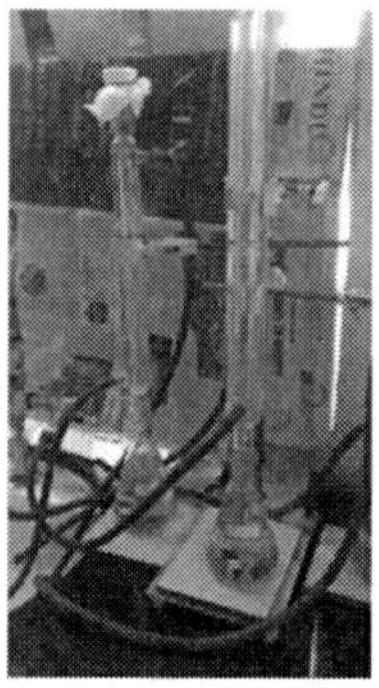
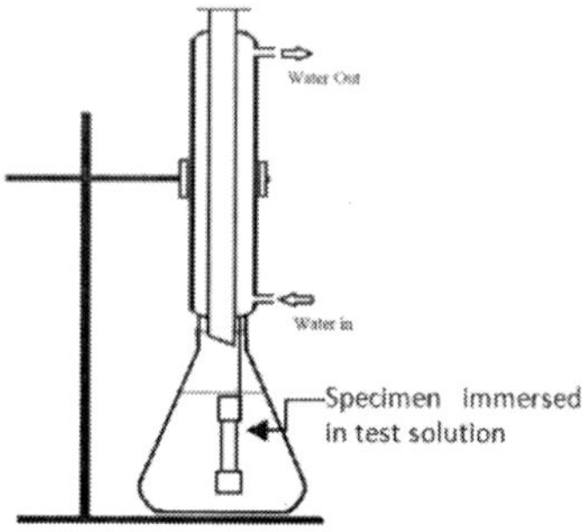

Figure 1 Typical experimental setup for immersion test

2.1 Test Specimens and Experimentation

The tensile tests were performed according to the ASTM E8 specification. The configuration and the dimensions of the smooth specimens used in tensile testing are shown in Figure 2. The test cylindrical bar samples of Alloy EN47, gage length '5d' with a diameter'd' of 6 mm were polished and cleaned to obtain the required metallographic finish before test. For the test a fully integrated universal testing machine 50 KN machine was used. The deformation rate was 0.3 mm/min at room temperature. . The tensile properties: yield stress, ultimate stress, percentage elongation were evaluated. The toughness is calculated from the area under the true stress–true strain curve. The tensile tests are performed on cylindrical specimens before end after exposed to a hydrogen charging environment. The decrease in mechanical properties between treated and untreated samples at the same strain rate is an indication of the hydrogen embrittlement susceptibility. In laboratory investigations, the specimens were pre-stressed beyond their yield stress to experience a plastic deformation (70% of σ_{ult}). These pre-stressed specimens were subjected to hydrogen charging in a simulated corrosion media and then subjected to complete failure by tensile testing. Failures, when they occurred, were in regions of stress concentrations resulting from the combined action of stress and the general attack of the corrodent. A typical untreated and hydrogen charged specimen is presented in figure 3.

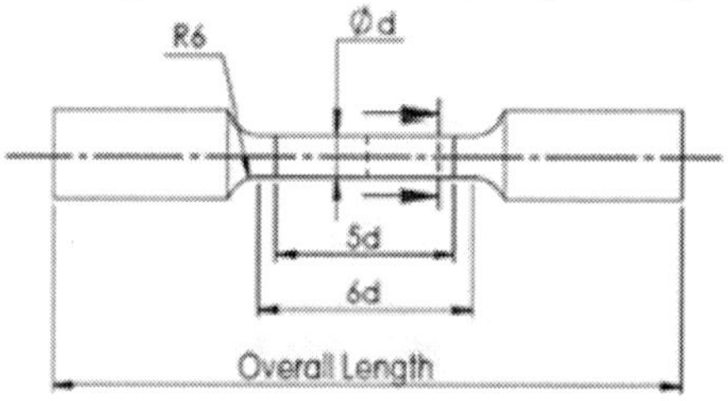

Figure 2 Typical Tensile Specimen

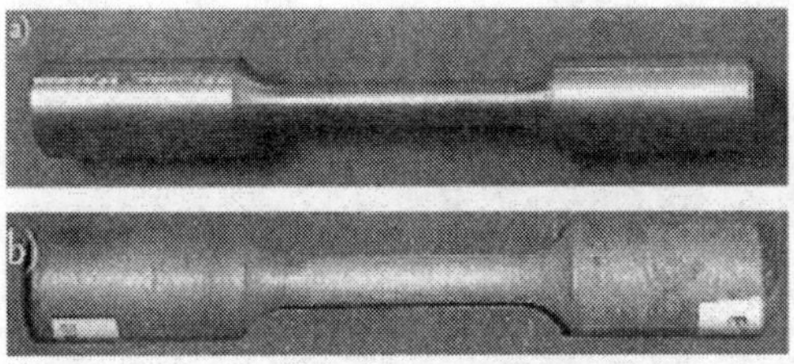

Figure 3. Standard Tensile specimens a) Untreated
b) Hydrogen embrittled

3. Results and Discussion

The research was based on the material's response to an externally applied load in conjunction with the laboratory simulated conditions. Under the applied load and service conditions, the macro/microstructural mechanical behaviour, the interaction between structure and mechanical properties, material instabilities, and fracture of the material were investigated. The tests were performed on spring steel specimens. The experimental study involves hydrogen charging the specimen and mechanical testing and characterization of the charged specimens and SEM observations of the deformation processes of materials at the micro- and macro-level.

3.1 Stress/strain behavior of treated and untreated material

The images in figure 4 (a) show the fracture surface of a steel with multiple cracks triggering portion (Radial Zone) that failed in a brittle manner with little deformation. It is roughly perpendicular to the applied tensile stress direction. The fan shaped ridges like pattern on the fracture surface indicates an unstable crack growth without increase in the applied stress. At the centre, a fibrous zone depicts overload fracture morphology. The figure 4(b) shows a fractured surface of pre stressed hydrogen charged specimen which depicts delayed fracture of brittle to ductile transition; arrow marks indicate the crack initiation site and shear lip formation and overload fracture. There is a mixed mode of fracture forming and flat surface with a shear lip at failure zone. Figure 4(c) illustrates the engineering stress/strain curves are obtained after hydrogen charging and untreated tensile test specimens. A stress/strain curve results in a linear relationship, as shown in Figure 4(c), of treated and untreated steel. It can notice that a variation in the yield strength, ultimate strength and final strain (ductility) exists. It is seen from Figure 4(c) that the tensile strength and the ductility decreased with reference to the untreated specimen. This is in accordance with the observations made in the test results. The slope of the initial linear portion of the stress strain curve shows that the ability to resist deformation in the treated specimen is increased indicating an increase in yield strength, but significant reduction in maximum load carrying capacity. The slope of the treated specimen curve indicates increased stiffness with greater slope confirms that material is embritttled. The figure 5(c), stress strain curve illustrates area under the curve, ability to absorb energy in the plastic range and is increased by approximately 10 percent. As the treated specimen is solution treated for 120 hours for hydrogen charging, might have led to precipitation of impurities which impede the movement of dislocation and to harden the material [9-12]. This might be the reason for increase in the yield strength which needs further investigation.

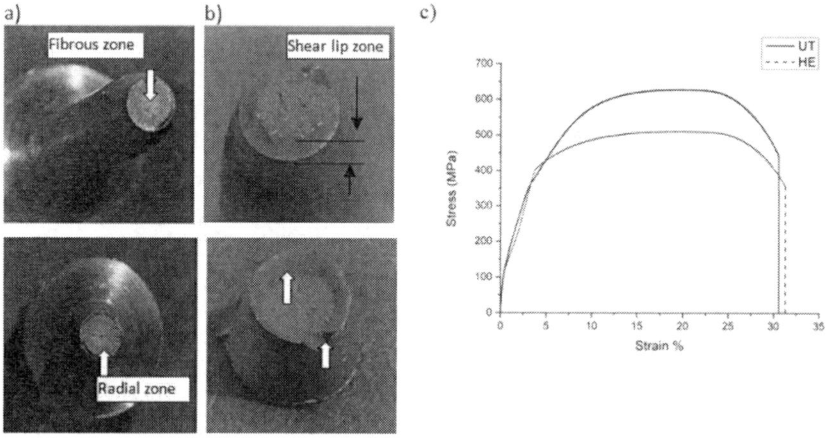

Figure 4. Facture surfaces: a) Untreated; b) treated; c) Stress Strain curve

3.2 Influence of HE on Volumetric changes and toughness

Displayed in figure 5, are the apparent values of the percentage elongation and area reduction; engineering yield stress and ultimate stresses; area under the curve as toughness of the untreated and pre-stressed hydrogen charged specimens. The hydrogen charging does affect the ductility of EN47 steel as depicted from figure 5(a) with variation in percentage change in elongation and reduction in area. The test result shows that the percentage of change in elongation and percentage reduction in area were 17 and 28 respectively. The effect embrittlement is greater upon the values of reduction of area than elongation.

The yield stress is given as 0.2% strain. The yield strength of the hydrogen charged specimen increased but reduced in ultimate strength pointing towards the inception of plastic deformation. From figure 4(c) and figure 5(c) has indicated Hydrogen charging causes the samples to become brittle, and reduced the plastic deformability. It is observed that a sharp drop in the ultimate strength of charged specimen, pointing towards embrittled failure. The interaction of hydrogen with the dislocations in the plastic zone of pre-cracks in specimen degraded the ultimate tensile strength (UTS).

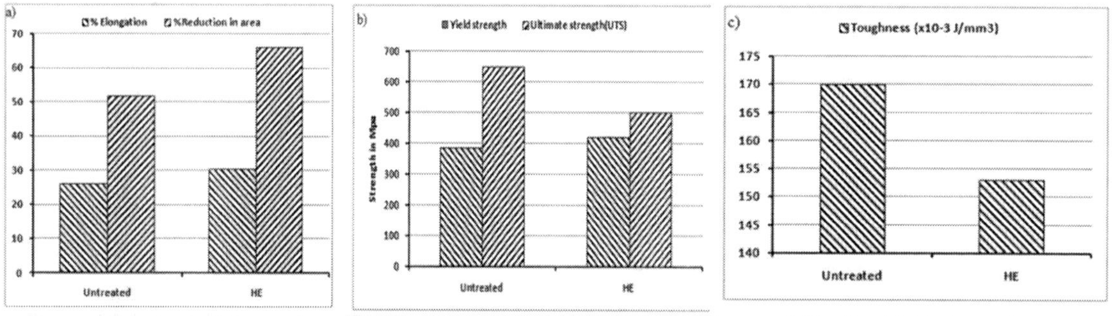

Figure 5 Mechanical Properties of tensile test untreated and treated specimens of EN 47; (a) % elongation and reduction in area, (b) Yield strength and Ultimate strength(Mpa) (c) Toughness (10^{-3} J/mm^3)

3.3 Micro-mechanism failure analysis

One of the characteristics that have convinced investigators that hydrogen embrittlement is related to the presence of internal molecular hydrogen evolved in a mono-atomic form as part of the cathode reaction of almost any electrochemical reaction, which are almost damaging cause of post-processing failure through hydrogen embrittlement[11].The introduction of excess hydrogen can result in crack nucleation and growth and in some cases failure under load, in both brittle and ductile materials.It is assumed that dissolution is concentrated at the crack tip because of the effects of yielding at that point, and that the cathodic reaction occurs only on the surface of the specimen. In general the stress at the tip of a crack exceeds the yield stress.

The susceptibility was attributed to segregation or precipitation at grain or sub-grain boundaries [12]. As discussed in the previous section, the macrograph provides substantial evidence of failure initiation site. The SEM micrographs analysis (Figure 6) shows the fracture morphologies at the failure area of untreated specimens. In an untreated material there won't be any preexisting cracks. By the virtue of the applied load, the material is pulled to certain extent with the slant surface because of induced crack. Here in this region there won't be any striation or any triggering sites. In this region the material act as aggregate till the crack initiates towards the core. From this point onwards the material will fail by the virtue of material constituents. At the mid region, where there is more resistance, the material will scoop. This is can be witnessed in the micrograph figure 6b, here it exhibits almost a flat surface beyond which normal shear cracks develops. The fracture occurred due to overload. The SEM micrographs analysis (Figure 7) shows the fracture morphologies at the failure area and pre existing cracks induced due the interaction of pre-stressed condition and hydrogen charging of specimens. Fracture surfaces of treated samples revealed pre-exist crack induced due the chemical treatment give raise to slant surface at the early stage itself. The cracks were induced because of embrittlement process Fracture surfaces of treated samples revealed distinctive mixed mode of fracture indicating a brittle–ductile fracture transition as in case of precipitate hardening. There are indication of cleavage facets and ductile tearing on the fracture surface. Both equiaxed and elongated quasi-cleavage facets were observed. The microvoid coalescence during ductile failure can be witnessed. The fracture surface of the treated specimen suffered from deep and uniformly distributed dimples, which indicates the ductile nature of fracture at the mid region [13,14].

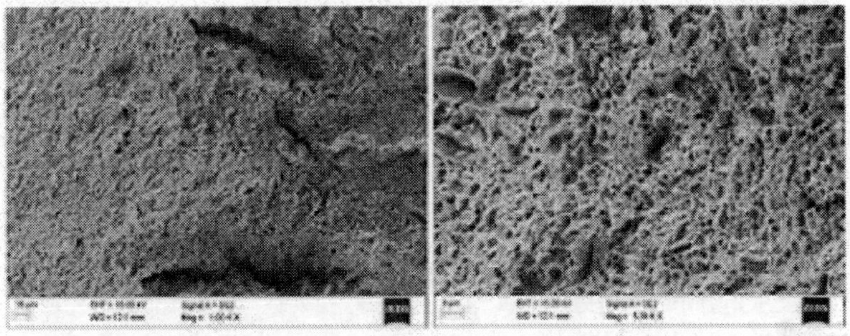

Figure 6. SEM micrograph of untreated fracture surface;;(a)Initiation of Multiple cracks due tensile loading (b) surface showing coalescence of void and fibrous fracture overload zone

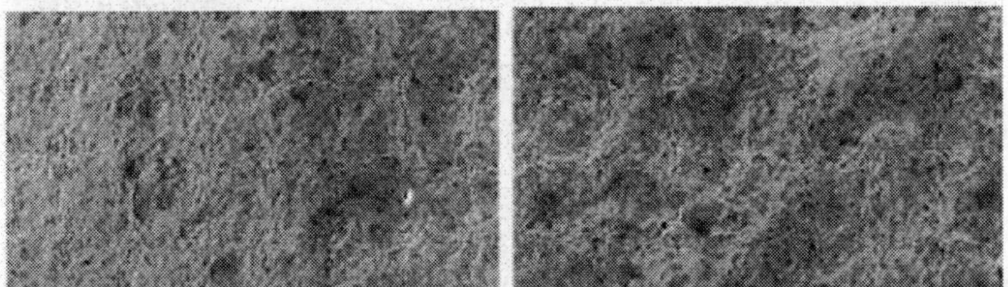

Figure 7. a) SEM micrograph treated fracture surface; (a) Showing dimples of varying size and a noticeable number of dimples. (b) distribution of voids and void coalescence on the overload fracture surface.

4. Conclusions

HE can be defined as the hydrogen-caused deterioration of the mechanical properties of most metallic materials and alloys. It leads loss of ductility, and tensile strength from the presence of dissolved hydrogen. Hardening by hydrogen is more general than softening for commercially pure iron, low-alloyed and stainless steels tested for bulky specimens at room temperature. Based on the tests, the following conclusions can be drawn:

- The effect of hydrogen charging is noticed in-terms of variation in yield strength, ultimate strength and ductility of the untreated and treated tensile test specimens. This could be due to Hardening by hydrogen.
- The failure is due to hydrogen embrittlement. It is observed that a sharp drop in the ultimate strength of charged specimen, pointing towards embrittled failure.
- The treated test material experience brittle to ductile transition exhibiting mixed mode fracture morphology on fracture surfaces.

References

[1] Basukumar H.K., Arun K V, 2016, *Int. J. Sci. Tech. and Sci,,* Experimental Approach to Study the Effect of Hydrogen Embrittlement on Structural Steel under the Influence of Impact Loading and Wear Characterization, 3(04), 108–113.
[2] Arikatti K C, Arun K V, Basavakumar H K., 2016, *Int. J. Eng. Sci. and Comput.,* Effect of Hydrogen Charging on the Wear Behavior of Spring Steel, 6(06), 7015-7020, ISSN (Online) 2456-1290
[3] Hardie, D., 1996, *Corrosion Science,* The Effect of Stress Concentration on Hydrogen Embrittlement of a Low Alloy Steel 38(5), 721–733, ISSN 0010-938X, https://doi.org/10.1016/0010-938X(96)00161-8.
[4] Canale, Lauralice & Penha, Renata & Totten, George & Canale, Antonio & Gasparini, Maria. 2007. *Int. J. Microstruct. Mater. Prop.,* 2(3/4), 262 – 309, 10.1504/IJMMP.2007.015310
[5] Spring Steels – IspatGuru; https://www.ispatguru.com/spring-steels (Accessed on June 6, 2020)
[6] The Primary Reasons Why Springs Fail - Yost Superior; https://www.yostsuperior.com/primary-reasons-springs-fail (Accessed on June 6, 2020)
[7] Milos B. Djukic, Gordana M. Bakic, Vera Sijacki Zeravcic, Aleksandar Sedmak, Bratislav Rajicic, 2016, *Corrosion,* 72(7), 943-961.
[8] Takagi, Shusaku & Toji, Yuki., 2012, *ISIJ International.* 52, 329-331, 10.2355/isijinternational.52.329.
[9] Louthan, M.R., 2008, *J Fail. Anal. and Preven.* 8, 289–307, https://doi.org/10.1007/s11668-008-9133-x
[10] T Suzuki, M Kubota, S Kozawa, H Miyamoto, 2019, technical report ,No. 122
[11] T. J. Carter and L. A. Cornish, 2001, *Eng. Fail. Anal.,* Hydrogen in metals, vol. 8, pp. 113–121.
[12] J. G. Hines, 1961, Corrosion Science, pp. 21-48
[13] Michihiko Nagumo, 2016, Fundamentals of Hydrogen Embrittlement, DOI 10.1007/978-981-10-0161-1
[14] Fu, Lei & Fang, Hongyuan.,2018, *MATEC Web of Conferences,* Interaction between plastic deformation and hydrogen damage behavior of 30CrMnSiNi2A steel, vol. 221,p. 01008.

CBR Characteristics of Cementitious Graded Gravel Soils by Curing Methods

Chandra Sekar Ganja[1*], Prof.P.V.V.Satyanarayana[2], P.Ganapati Naidu[3]

Assistant Professor,Department of Civil Engineering-Rgukt –Nuzvid and Ph.D Scholar at A.U ,chandu.g@rguktn.ac.in
Professor, pasalapudisat@yahoo.co.in, 3.Ph.D Scholar ,pvgnaidu@gmail.com
Department of Civil Engineering, Andhra University College of Engineering-Vishakhapatnam,Andhra Pradesh-India.

Abstract: The service life and performance of pavements mainly depends on the quality of materials used in the construction, Majority of local soils in their natural form lack in desired strength and durability required for the intended constructions, they are to be improved through the standard process of stabilization to make them suitable as one of the construction material, The cement treated soils are widely used in granular base and sub-base courses of pavements as construction material due to batter strength and durability. The current study there are three different locally available gravel soils has been selected and tested for their physical properties, compaction and strength characteristics, as per the test results the soils were classified as GM-GC, GC-1 and GC-2 and CBR% values were 28, 26 and 25 respectively, as per MORTH specifications CBR%>35 to use as base course material, hence the selected soils could not be used as base course material due to its low CBR values, so that cement stabilization has been used to improve the CBR% values. The different percentages of cement 0%, 1 %, 2% .....15% were added to GM-GC, GC-1 and GC-2 soils with gradation of G-50%(26.5 mm to 4.75mm IS-Sieves) and G - 50% (4.75mm to 0.002mm IS-Sieves) and tests were conducted to assess the compaction and strength characteristics by varying curing methods as 7days moisture, 7days water and 7days partial curing (3days maisture+4days water). As per the experimental results , it observed that among selected soils GM-GC soil has achieved CBR% of 85-101 with addition of 6-7% of cement under the partial curing (3M-4W) for 7days, it is more than appropriate to use this soil as base course material in road construction as per the MORTH (CBR %> 35)specifications.
Keywords: California Bearing Ratio (CBR) Value, Ministry of Road Transport and Highway (MORTH), Gradation, Stabilization, M-moisture is curing and W-water curing.

1. Introduction

Soil is a naturally occurring material above the ground surface, while physical verification of these soils it is observed that there are verity of soils are available in nature and these soils are classified into different groups based on its nature. It is very much needed to determine the index and engineering properties of locally available soils to construct the civil engineering structures like buildings, roads, dams etc. majority of local soils in their natural form lack in desired strength and durability required for the intended constructions, they are to be improved through the standard process of stabilization to make them suitable as one of the construction material. Among all available stabilization techniques cement stabilization is one of the best options as suits to all types of soils. Cement treated soils are widely used in granular base and sub-base courses of pavements as construction material due to batter strength and durability. Soils are classified into two types based on the grain size the grain size of the particles from 80mm to 4.75mm as course grained soils and the grain size of the particles from 4.75 mm to0.002mm as fine grained soils.

Compressive strengths in the samples under standard test increased with the increase of cement content up to some percentage of cement. But when cement content is increased above, the compressive strength increased at a slower rate, the percentage at which there is no more increment in strength, it is considered as optimal dosage of cement content [13]

The soil stabilized mechanism mainly depends on type of stabilizer applied. It's very from formation of new compounds, binding the finer particles of the soils, particle surface coating with stabilizer to limit the moisture sensitivity. In the presence of water the overall stabilization process can be divided into four deferent procedures cation exchange, flocculation and amalgamation, pozzolanic reaction and cementitious hydration. Cement hydration is the most important contributor to the improvement of the engineering properties of soil. Cement hydration is relatively fast and causes an immediate gain in the strength of the soils [17]

The soil gradation and curing method is very much influencing the compaction and strength characteristics of mix.

2. Materials and Experimental Study

2.1. Gravel Soils

The gravel soils are naturally available in huge quantity across the world, approximately 30 to 40% of gravel soils are available Eastern &Western Ghats and central plateau. Three locally available gravel soils are taken for the current investigation from Vishakhapatnam City-Andhra Pradesh at 17°43'47.3880"N latitude and 83°19'17.3820"E longitude. The physical, compaction and strength properties of gravel soils were determined from the experimental study, the data reported in Table-[1],figure no[1],[2] and references are [1], [2], [3] & [4] respectively.

Table-1: Physical Properties of Cement

S.no	Property of Gravel Soil	Sample-1	Sample-2	Sample-3
1.	Grain Size Distribution			
	a. Gravel Size% b. Sand Size% c. Fines%	50 35 15	50 30 20	50 25 25
2.	a. Liquid Limit% b. Plastic Limit % c. Plasticity Index%	24 18 06	27 19 08	30 20 10
3.	Soil Classification	GM-GC	GC-1	GC-2
4.	Compaction Characteristics a. OMC% b. MDD g/cc	7.20 2.08	8.60 2.06	9.00 2.05
5.	CBR %	28	26	25

Figure-1: Grain Size Distribution of Soils

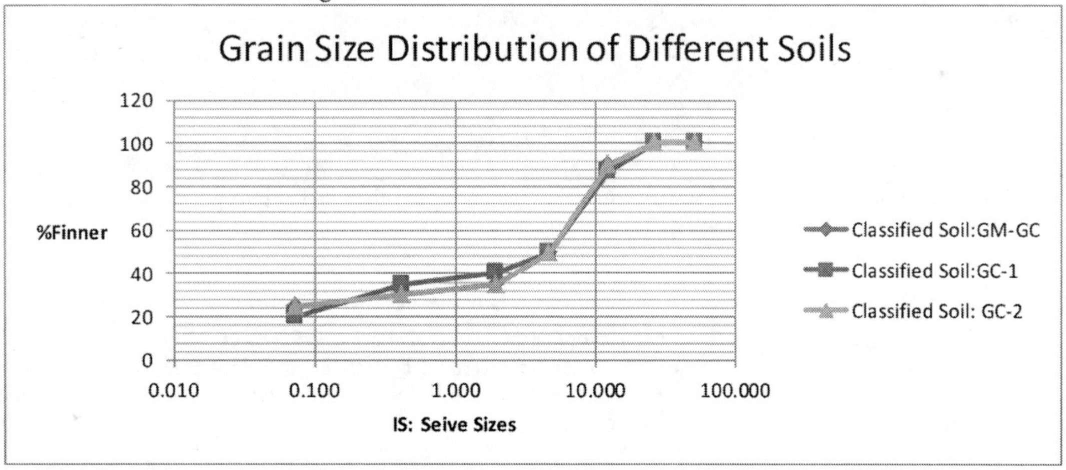

Figure-2: Plasticity Index (IP) Vs CBR% of Soils

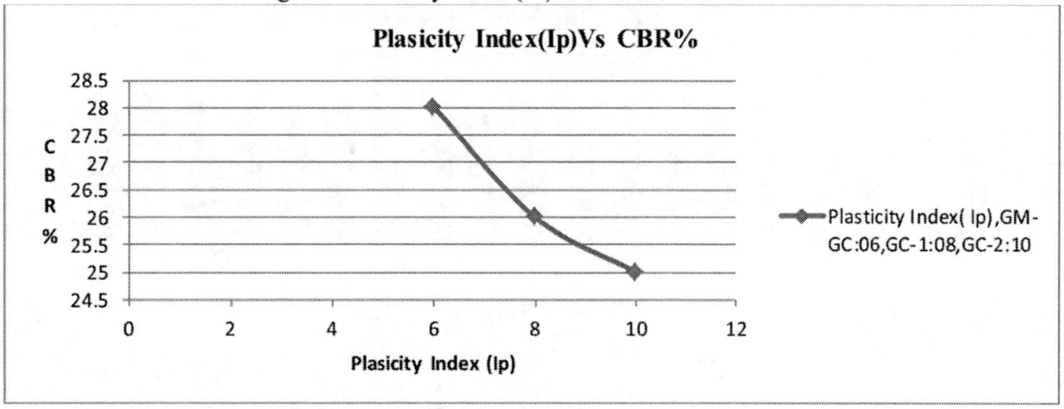

2.2. Physical Properties of Cement

The ordinary Portland cement of Ultra-Tech 53 grade cement is used in this study. The properties of cement determined by the laboratory tests are reported in the Table-[2] and references are [5] respectively.

Table-2: Physical Properties of Cement

S.No	Cement Property	Value
1.	Specific Gravity	**3.15**
2.	Initial Setting Time(Minutes)	**150**
3.	Final Setting Time(Minutes)	**350**
4.	Consistency %	**31**

5.	Compressive Strength(Mpa)	
	a. 7Days	**34.46**
	b. 28Days	**45.24**

2.3. Gradation of Gravel Soils
The gravel soils which are chosen for the investigation is graded into two categories for batter compaction characteristics and strength; one is G-50%the grain size of the gravel soil vary from 26.5 mm to 4.75mm as per IS sieve and second one is G-50% the grain size of gravel soil vary from 4.75 to 0.002 mm as per IS-sieves.

2.4. Compaction Characteristics of Cement Treated Graded Gravel Soils
The modified proctor test was carried out as per IS-4332 Part-3-1995 on cementitious graded gravel mix with 0%, 1%,2%.......up to 15% of cement content by weight of the total mix, to determine the optimum moisture content and maximum dry density of the mix. The test results are reported in the figure [3], [4] and references are [3] respectively.

Figure-3: Optimum Moisture Content of Soils

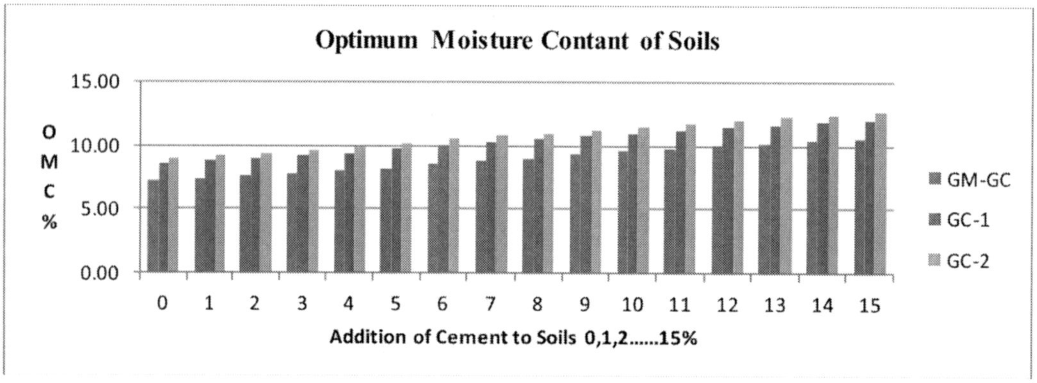

Figure-4: Maximum Dry Density of Soils

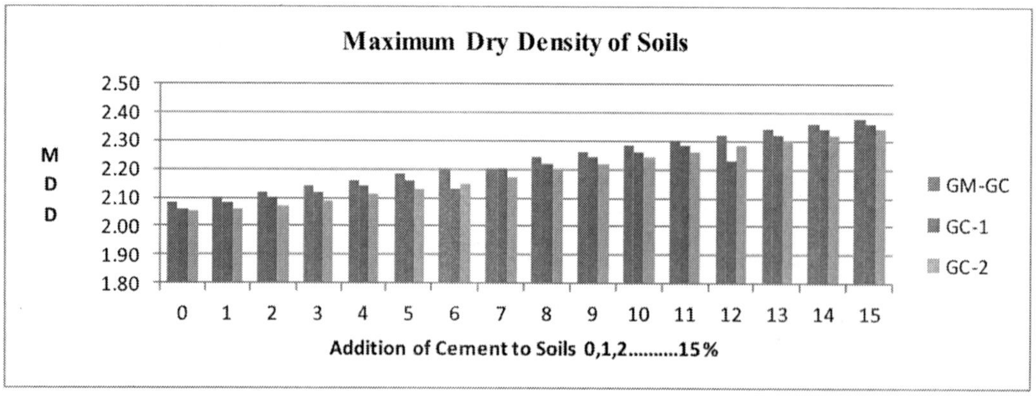

As per the rest results reported on figure no:[3], It is observed that with increase of cement content 0%, 1%, 2%.......15% to the graded gravel soils, the optimum moisture content (OMC) also increases respectively. In case of GM-GC soil has lowest OMC % 10.60 and GC-1:12.00&GC-2:12.60 respectively.

As per the rest results reported on figure no:[4], It is observed that with increase of cement content 0%, 1%, 2%.......15% to the graded gravel soils, the maximum dry density (MDD) also increases respectively. In case of GM-GC soil has more MDD value 2.38 g/cc and GC-1:2.36&GC-2:2.34respectively.

2.5. CBR %Values of Cement Treated Graded Gravel Soils with Different Curing Methods

The selected OPC-53 grade ultra-tech cement is added to the graded gravel soils from 0%, 1%, 2%, 3%.........15% in weight of the total mix. To determine the optimum dosage of cement with respective compaction characteristics, CBR values, its improvement ratio and different curing methods for 7days (Partial Curing Method (3M+4W), Moisture Curing Method and Water Curing Method)

$$\text{Improvement Ratio} = CBR_{Treated} / CBR_{Untreated}$$

$$\text{Percentage increase} = CBR_{T} - CBR_{UT} / CBR_{UT}$$

Figure-5:7Days Partial Curing (3M+4W)

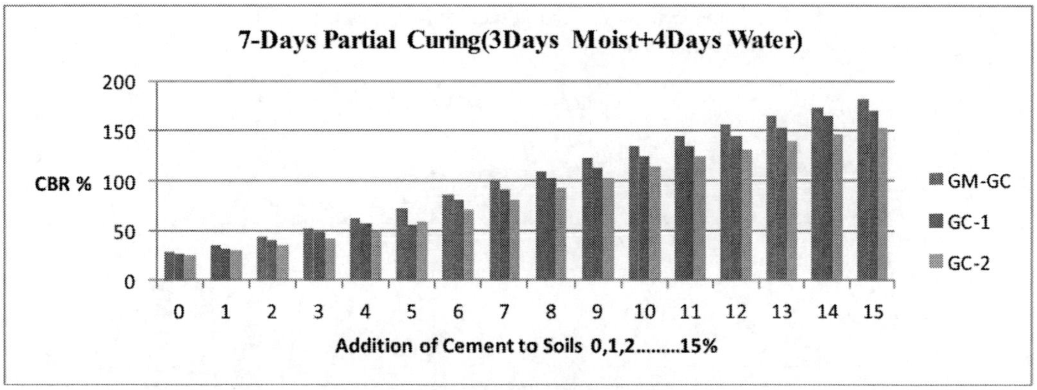

In partial curing method (3M+4W) the water content in the gravel soil due to presence of fine particles (slit and clay) and their corresponding plasticity characteristics slowdown and development of strong bonding between soil and cement particles in moist curing and it is continued in water curing also. Therefore, high CBR values are achieved at high percentages of cement. This behaviour is more pronounced further at high cement dosages also. In case of selected cement treated graded gravel soils the CBR% values are GM-MC: 182, GC-1:170, GC-2:153 respectively.

Figure-6:7Days Moisture Curing

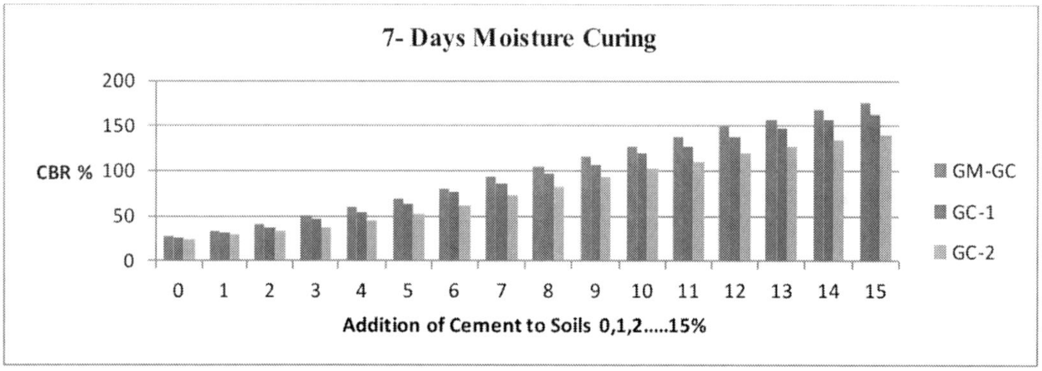

In case of moist curing (7M) development of strong bonding between soil particles and cement particles has been taken place during its initial curing period and not continued that effectively due to form requirement of some more water contain to form cementitious compounds under hydration process therefore relatively less and marginal reduction CBR values are reported w.r.t partially curing method. In case of selected cement treated graded gravel soils the CBR% values are GM-MC: 176, GC-1:163, GC-2:140 respectively.

Figure-7:7Days Water Curing

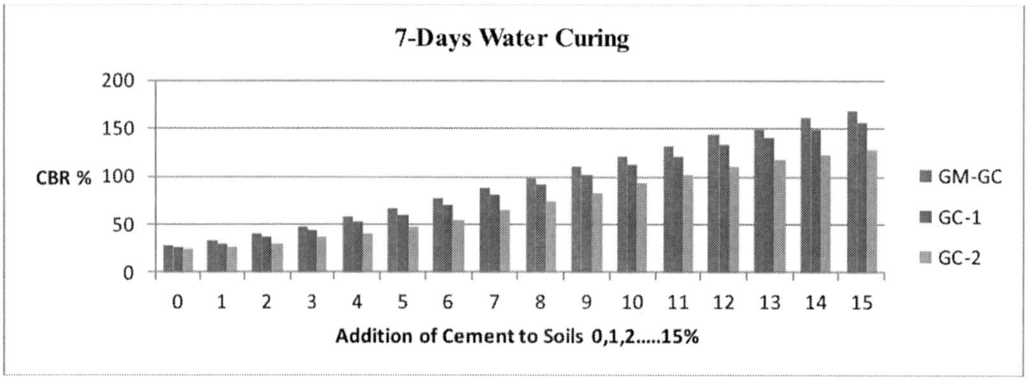

In case of water curing method (7W) the interaction between soil solids and cement particles is taking place at high water contents. Development of effective bonding is less at its initial curing period and continued with times it helps in formation of work bonds. Therefore, less CBR values are obtained compared is moist curing and partial curing methods this phenomenon is well observed at low dosages of cement and at high dosages of cement 10-15% strong bonds are to be formed due to availability of more percentage of cement and its hydration process due to high percentage of fine particle. In case of selected cement treated graded gravel soils the CBR% values are GM-MC: 168, GC-1:156, GC-2:128 respectively.

3. Conclusions

- By conducting laboratory tests on physical properties of selected soil samples, as per the test results, the selected soils are classified as GM-MC, GC-1and GC-2.

- Individually the selected soils have attained CBR% values GM-MC: 28, GC-1:26, GC-2:25 with their plasticity index (Ip) values 06, 08, 10 respectively, As per MORTH specifications CBR %> 35 to use as base course material. These soils could not be used as base course due to its low CBR% values, so that cement stabilization has used to improve the CBR% values.
- The different percentages of cement 0%, 1 %, 2% .....15% were added to GM-GC, GC-1 and GC-2 soils with gradation of G-50%(26.5 mm to 4.75mm IS-Sieves) and G - 50% (4.75mm to 0.002mm IS-Sieves) ,with reference of modified proctor test results it is observe that with increase of cement content in graded gravel soils, the optimum moisture content (OMC) also increases respectively. In case of GM-GC soil has lowest increment in the OMC than remaining GC-1&GC-2 soils.
- It is observed that with increase of cement content 0%, 1%, 2%.......15% to the graded gravel soils, the maximum dry density (MDD) also increases respectively. In case of GM-GC soil has more increment in the MDD than remaining GC-1&GC-2 soils.
- Cement treated graded gravel soil GC-1 has achieved CBR% 93-102 with addition of 10-11% of cement under the 7days water curing. In case of Moisture curing for 7days the CBR% is 93-103 with addition of 9-10% of cement. But in case of partial curing (3M-4W) for 7days the CBR% is 93-103 with addition of 8-9% of cement only.
- Cement treated graded gravel soil GC-2 has achieved CBR% 92-102 with addition of 8-9% of cement under the 7days water curing. In case of Moisture curing for 7days the CBR% is 97-107 with addition of 8-9% of cement. But in case of partial curing (3M-4W) for 7days the CBR% is 91-102 with addition of 7-8% of cement only.
- Cement treated graded gravel soil GM-GC has achieved CBR% 98-110 with addition of 8-9% of cement under the 7days water curing. In case of Moisture curing for 7days the CBR% is 93-104 with addition of 7-8% of cement. But in case of partial curing (3M-4W) for 7days the CBR% is 85-101 with addition of 6-7% of cement only.
- As per the experimental results, it is observed that among selected soils GM-GC soil has achieved CBR% of 85-101 with addition of 6-7% of cement under the partial curing (3M-4W) for 7days.it is more than appropriate to use this soil as base course material in road construction as per the MORTH (CBR %> 35)specifications.

4. References

[1] IS: 2720 (Part-IV) 1985, First Revision Methods of test for soils, part-4 and determination of grain size analysis, Bureau of Indian standards, New Delhi.
[2] IS 2720-5 (1985): Methods of test for soils, Part 5 Determination of liquid and plastic limit, Bureau of Indian standards, New Delhi.
[3] Determination optimum moisture content and Maximum Dry Density by Modified Proctor Test as per IS 2720(Part-08)-1987.
[4] Determination of California Bearing Ratio Value by CBR Test as per IS 2720(Part-16)-1987.
[5] Determine the chemical and physical requirements of 53 grade ordinary Portland cement as per IS: 269:2015 (Part-2, 3, 4, 5 & 6).
[6] IRC-37(2001), the general guidelines for the design of flexible pavements, Indian road congress, New Delhi.
[7] IRC-37(2012), the general guidelines for the design of flexible pavements, Third Revision, Indian road congress, New Delhi-2012.
[8]IRC: SP: 89 -2018 (Part-II) "Guidelines for the design of stabilized pavements". Published by Indian Roads Congress, New Delhi, India.
[9] MORTH - Specifications for Road & Bridge Works (5th Revision)
[10] Hamed Ahmadi Chenarboni and Seyad Hamid Lajevardi" The Effect of Ziolite and Cement Stabilization on the Mechanical Behavior of the Expansive Soils" Construction and Building Materials, Elsevier, Novermber-2020.

[11] Omid Amini and Mojtaba Ghasemi " Laboratory study of the effects of using magnesium slag on the geotechnical properties of cement stabilized soils" Construction and Building Materials, Elsevier,July-2019.

[12] Basha and Abd al-redhghani "Study of Cement Treated Base Aggregate Properties for Pavement Structure" International Journal of Information Research and Review, January 2018.

[13]Md Ali Ashraf, S. M. Shazeebur Rahman, Md Omar Faruk, Md Abul Bashar. "Determination of Optimum Cement Content for Treatment of Soft Soil and Durability Analysis of Soil Treated with Cement" *American Journal of Civil Engineering*. Vol. 6, No. 1, 2018.

[14] Ekwulo E.O and Eme D.B" Effect of aggregate gradation on compressive strength and elastic modulus of cement treated aggregate base material for highway pavement" IOSR Journal of Engineering Vol-07-Issue 10-2017.

[15] Yuanjie Xiao and Liuxin Chen" Laboratory validation of a gradation design concept for sustainable applications of unbound granular materials in pavement construction" Construction and Building Materials, Elsevier,October-2016.

[16] A.A Amadi and A.S.Osu " Effect of curing time and strength development in black cotton soil-Quarry fines composite stabilized with cement kiln dust(CKD),Journal of King Saud University-Engineering Sciences-April 2016.

[17] Murad Abu-Farsakha,and Sanjay Dhakalb, Qiming Chen, "Laboratory characterization of cementitiously treated poor sub grade soil under cyclic loadin", Published in Soil and Foundation,Elsevier-2014.

[18] Manasseh Joel and Isaac O. Agbede. "Mechanical-Cement Treatment of Laterite for Use as Flexible Pavement Materials" *Journal of Materials in Civil Engineering*, Vol. 23, No. 2, February 1, 2011. ©ASCE, ISSN 0899-1561/2011/2-146–152/$25.00.

Studies on factors influencing Low Velocity Impact of Composite Materials – A Review

Sunith Babu L [1,*], Pradeep Kumar K V [2] and JayaChristiyan K G [3]

[1,2,3] Department of Mechanical Engineering, Ramaiah Institute of Technology, Bengaluru 560054, Karnataka, India

[*] sunithbabu@cadprofessor.in

Abstract. Composite Materials has been the de-facto material of choice in the aerospace and automotive industry mainly due to its high strength to weight ratio. Composite materials fabricated by different methods have been used in the above fields, some have found to be resistant to failure while some have failed to live-up to the expected results. Therefore, it is evident that, although composite materials provide excellent directional properties, failure in composite materials is a key area that many researchers have been studying for many decades and may still continue with development of newer materials. One such failure that can occur in composite materials is due to impact loading and the focus of this paper is to review the factors associated with Low Velocity Impact. In general, damage in composite material is a serious issue and factors that contribute needs to be studied in depth, since this may lead to catastrophic failure and some failures that is generated, may not be visible to the naked eye. Failure arising from LVI is not limited to fibre matrix cracking, de-bonding, delamination, fibre pull out and possible ways to overcome these effects may be of significant importance particularly in the aerospace industry and possible ways to improve them is discussed in this paper.

Keywords: Low Velocity Impact, Damage Mechanism, Impactor, Stacking Sequence

1. Introduction

Composite Materials are more promising materials in the aerospace, automotive, consumer, wind power applications. One of the renowned benefits of composite materials is high strength to weight ratio and far superior mechanical characteristics as compared to metallic materials. Another promising features of composite materials is the net zero cost in the area of maintenance due to the fact that it has excellent corrosion resistance [1] enabling engineers to use such a material as an alternative to steel structures. Even though composites are extensively used in the above areas of application, it is essential to understand that composite materials are susceptible to damage and may lead to plastic deformation resulting in matrix cracking, delamination, fibre pull out. The damage mechanism may be caused due to impact and the type of impact can be categorised into three types (i) Low Velocity (ii) Intermediate / High Velocity (iii) Hyper Velocity Impact [2]. In one case study with reference to Boeing 747, it was found from a sample of 71 aircraft with over 600+ repair parameters, with a little over 13% of the repairs was found to be with reference Low Velocity Impact from foreign body impact (damage) [3]. Possible causes of Low Velocity Impact (LVI) may be from tool drop, maintenance and potential bird strike. Table 1 shows the classification of types of impact velocities.

Table 1. Impact Velocity Parameter

Velocity Type	Velocity (ms^{-1})
Low Velocity	0-6 m/s
Intermediate / High Velocity	> 6m/s
Ballistic Velocity	> 500 m/s
Hyper Velocity	> 1700 m/s

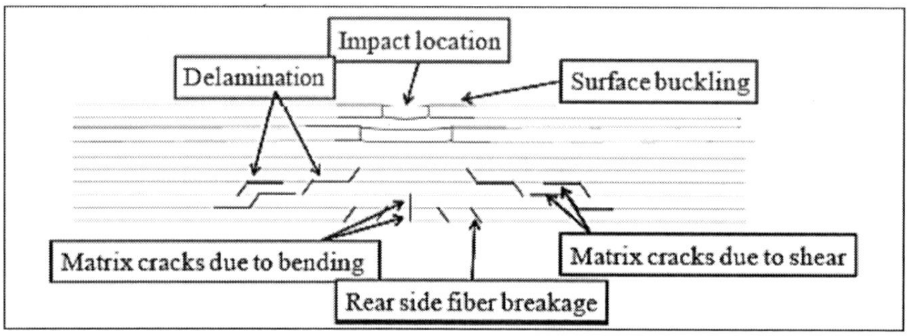

Figure 1. Damage due to LVI and its internal structure. [4]

As indicated above, a LVI can result in several types of damages and lead to catastrophic failure. With such a failure, the damage mechanism may not be visible at the surface and may result in exponential degradation of the composite panel and in turn reduce the overall performance of the composite panels [5]. In any impact process a list of prospective damage mechanism can be recorded and some of them are listed in Table 2.

Table 2. Prospective Damage Types due to Low Velocity Impact [6-7]

Sl. No.	Types of Damage
01	Delamination
02	Fibre Pull Out
03	Matrix Cracking
04	Damage / Dent at the location of Impact
05	Surface Buckling
06	Fibre Matrix Cracking
07	Inter-laminar Damage
08	Surface Buckling

When it comes to space scenario, low velocity impact is of significant importance, since there are several satellites that are orbiting around the earth and a chance of impact is very much possible, this is also not limited to re-entry scenario, this was the case when the space shuttle Columbia during its re-entry an impact with the foam on the wing lead to failure of the space shuttle [8]. For more than three decades the fabrication industry related to composites have been using thermosetting resin system. This however has some issues related to irrevocable exothermic reaction during the curing process [9]. It is worth noting that there are new systems such as thermoplastic resins that can improve the overall performance of composite materials in areas such as processing, recyclability etc. Refer to the Table 3 for comparison between TS (Thermosetting) and TP (Thermoplastic) resin

Table 3. TS and TP resin system [10]

Parameter	TS	TP
Toughness	Average	Excellent
Fabrication Temp	Poor	Excellent
Recycle Status	Average	Good
Service Temp	Excellent	Average

Researchers have done extensive work related to comparison between QSI and LVI related damage in PMC, it is found that composite laminates damage initiation due to the above two cases has led to similar results [11-14]. There are several factors that influence the impact properties of composite laminates which is shown in Figure 2.

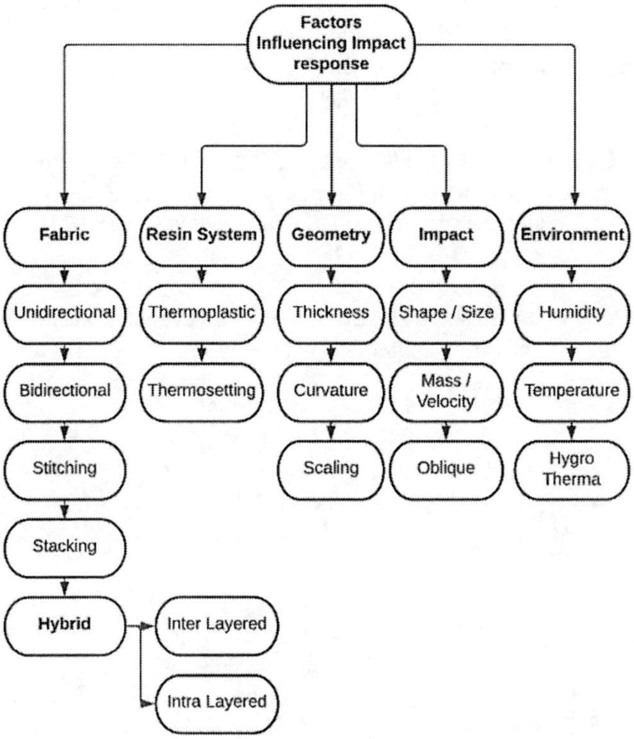

Figure 2. Factors Influencing Impact Response [15]

It is evident that above factors are not a dataset that influence the impact response of the composite laminates, there are several other factors which influence the impact damage performance of a composite materials. To review this, LVI process is always followed by CAI (Compression after Impact) test, which is essential since CAI assess the overall performance of the composite materials as this helps researchers in finding the residual strength left in the composite material after impact. It is always advised to carry out CBI (Compression before Impact) to enable researchers know the strength of the composite laminates before impact there by assessing its residual strength of the panel of interest. Further, impact response of the composite panels is also affected by chemical and mechanical characteristics which can lead to composite damage and resulting in different failure modes. Failure may also result due to laminate stacking sequence such as inter / intra ply.

Based on the several factors, this review paper will facilitate the user to understand the factors that can probably influence the impact damage of composite laminates due to Low Velocity Impact with varying parameters.

2. Parameters Affecting LVI Process

2.1. Process Parameter

In any low velocity impact test there are several process involved before the test can be performed. Process parameter usually involves fabrication related information, such as, type of resin involved as indicated in table 3, type of fabrication involved namely – Vacuum Bagging Process, VARTM – Vacuum Assisted Resin Transfer Molding, Compression Molding etc. The extent of damage is significantly noticed.

2.2. Material Parameter

The material parameter plays a significant role in determining the LVI process, parameters such as material fiber orientation - unidirectional, bi-directional. Material type - Glass, Carbon, Kevlar or Natural fibers. Stacking sequence – [0/90], [0/45/90/45/0] etc. and Inter / Intra Layered stacking sequence can significantly influence the impact properties of composite laminates.

2.3. Test Parameter

The test parameters associated with LVI process includes height of fall, types of impactors, mass of impactor. These parameters can significantly affect the LVI process based on the geometry of the composite laminate such as flat or curved, thickness of the panel and orientation of the fiber.

2.4. Standard Plot related to LVI

Any composite material when subjected to Low Velocity Impact undergoes various types of deformation and it experiences damage at different stages such as Contact, Impact, Deformation, Rebound, Perforation and finally Penetration. A typical LVI plot is shown in Figure 3, which illustrates the parameters as listed above.

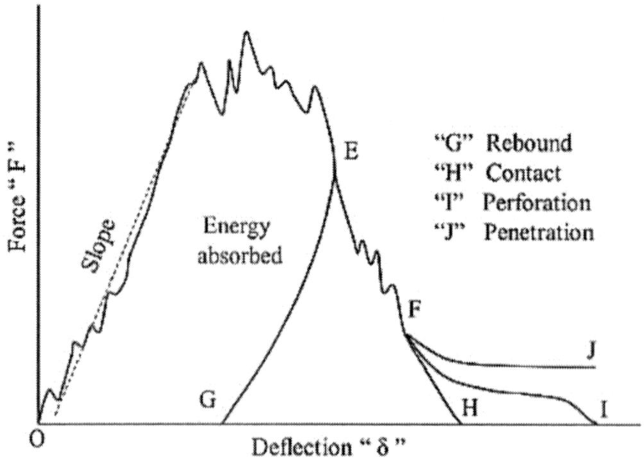

Figure 3. Typical LVI – Force vs Deflection Graph [15]

3. Basic Parameters Affecting LVI Process
During the service life of the composite material, it can be subjected to several internal or external damages. Some of them may not be limited to matrix cracking, fibre – matrix de-bonding, fibre breakage etc. Therefore, it is necessary to improve the overall resistance to damage of the composite material, in order to improve the LVI properties. There were several methods proposed to improve the properties related to composite materials such as matrix toughening, thickening of laminate, enhancing the woven fabric architecture via hybrid fibres using different material and via different stacking sequence as reported by kim [16].

3.1. Fibre Orientation on LVI process
As illustrated earlier, fiber orientation plays a significant role in enhancing the LVI properties. Composites provide better properties across in plane as compared to transvers directional properties and in turn undergo damage or deformation [17]. The delamination in a composite laminate is a serious scenario and needs better methods to overcome this issue and in turn enhance the residual strength of the composite laminate [18]. To address this issue, usage of UD, Woven composite laminates are on the rise and in turn better composite panels can be prepared to over the issues due to fiber fracture, fiber pull out and fiber matrix cracking when subjected to LVI process. Possible ways to over-come the effect of fiber failure could be to increase the fiber size of the cloth under consideration there by overcoming the potential effects of fiber failure due to localized and general damage of the composite panels. Another possible way to reduce the failure is to adopt z direction stitching of composites along with woven cloth there by enabling the overall impact damage resistance. All these is possible when a user can adopt 3D fabric along with warp and weft, transvers directional properties can be enhanced by adopting z directional reinforcements. It can be found that these reinforcements can enhance the structural integrity and delamination resistance along with greater energy absorption. Many researchers have found that 3D based composites have enhanced the overall performance of the composites as compared to 3D composites [19]. By switching over from 2D to 3D composite panels, the overall performance of composite laminates due to LVI has significantly increased [20]. Further research work carried out with reference to 2D and 3D composites showed a 50% less damage area, better structural performance and better impact properties [21] refer to the figure 4 below for better weaving process to enhance the damage resistance due to fiber architecture. In a nutshell the 3D composites provide better solution for all issues related to LVI.

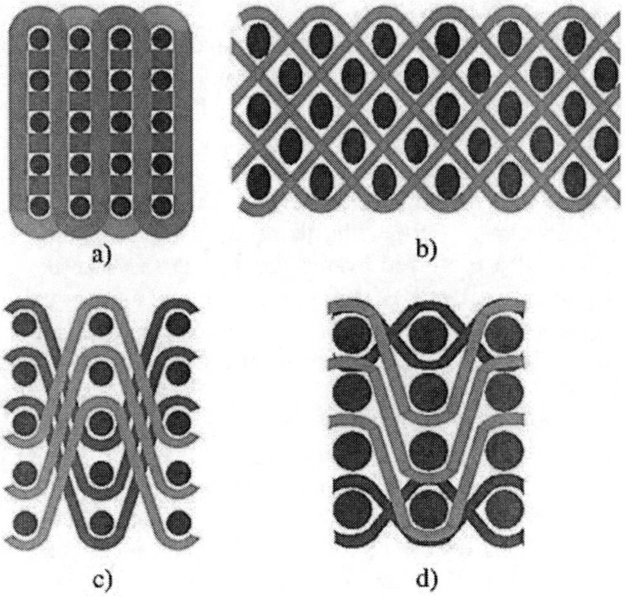

Figure 4. a) Orthogonal b) Interlock c) layered Interlock d) Modified Layered Interlock [22]

3.2. Resin and its role in LVI process

The resin is another important entity of the composite materials and also plays a significant. Resin related properties as listed in table 3 gives a clear understanding that TP type resins will provide better and superior impact damage resistance and is highly recommended for composite fabrication process. The one main advantage of TP type resins is that it ensures the fiber patter (orientation) in place and enables reduced fiber damage properties [23]. Another reason in adopting TP type resins in aerospace application is that it has superior damage resistance and exhibits toughness about ten times as compared to TS resin. Further the fatigue properties are also superior as compared to TS resin. But only one area where creep is considered the TS fares well [24]. It further noted that in terms of specific energy absorption properties of TS and TP, TP favors better roughly around 57.3% better than TS resins of the same configuration of composite panels when they were investigated in-situ using a digital microscopy and later analyzed via SEM [25]. Many researchers have concluded that TP will have better properties as compared to TS resin [26-28].

4. Other Parameters Affecting LVI Process

4.1 Fracture and its role in LVI process

As indicated earlier, TP resins enables better performance characteristics of LVI process, therefore toughness properties enhance [25]. With reference to temperature, the values increase as reported by Kim and Ye [29]. Further it can be noted that parameters such as fiber orientation and resin toughness can enhance the overall toughness of the composite materials. In addition to these factors, other factors that can potentially affect the fracture toughness are temperature, moisture, humidity.

4.2 Thickness and its role in LVI process

It is well understood that thickness plays a significant role in determining the effect of LVI process. It further understood that any thickness beyond 8 mm cannot be used as laminate structure and rather focus on sandwich panel type is preferred. But appropriately selecting the orientation of plies and selecting a suitable thickness enhances the overall performance due to LVI. But gradually increasing the thickness of composite laminates increases the failure threshold of the composites. When comparing with thin laminates thick ones favored better impact resistance. In a certain type of composites, the Hertzian value increased with increase in thickness [30].

4.3 Impactor and its role in LVI process

Impactors play a significant role in performing LVI studies. Several researchers have carried out studies on different impactors and the most commonly used impactor is hemispherical type and also it has a higher factor of influence on the level of damage. It is also clear that the tests performed on a controlled environment is different from a real time scenario, where the type of impact and the nature of impact can vary. To be more precise the impactor of size 12.7mm of hemispherical type is the most general configuration used in the impact studies as this provides a greater damage and higher penetration through the specimen. Further the damage tolerance increases with increased diameter. Impactors such as flat is also preferred by researchers as this provides a study related to surface damage mechanics and the effect of LVI process. Impactors such as truncated, spherical, ogival are also used for studying the impact damage assessment of composite structures, these are used based on the type of application. It is evident that higher impact results in deeper understanding of the energy absorption process [31-35].

4.4 Stacking Sequence and its role in LVI process

Stacking sequence plays a significant role. Composite materials offer unique advantage when they are tailored to meet specific design requirements by means of stacking process. In case of stacking of composite panels, adopting glass fiber cloth on top of the stacking enables better impact resistance as compared to any other fabric. By focusing on a changing stacking sequence delamination and deformation can be significantly reduced. In stacking sequence lay-up process has a dominant role as it alters the LVI process. Temperature and Stacking sequence does not have any role in enhancing the overall of impact resistance. Stacking process can be adopted to increase the stiffness of the composite panel, however, the damage due to impact may not have a significant role. By altering the ply orientation marginal increase in impact damage can be increased. By increasing the number of interface, damage tolerance can be marginally increased [36]. Adopting optimized stacking sequence for TS and TP layers and in turn inserting alternate TP layers among the TS and TP resulted in increased failure resistance by 15 times [37].

4.5 Temperature and its role in LVI process

The effect of impact properties and damage mechanics is greatly affected by temperature variation, humidity and hygro-thermal variations. TP and TS systems play a vital role in damage mechanism, TP composites at low temperature provides better performance of composites since they offer increased matrix ductility. Hygro-thermal ageing is another parameter that affects the LVI properties, at higher temperature and the residual strength reduced by 60-70%. Many researchers have carried out temperature related works on the Low Velocity Impact and the range of temperature were found to be between -80°C to +180°C and majority of the researchers have been concluded

- Reduction in Residual Strength
- Decrease in Energy Absorption
- At low Temperature Reduced Delamination
- Increase in BVID (Barely Visible Impact Damage)

5. Conclusion

There are several parameters that affect the performance of composite structures and LVI process is one such component that drastically reduces the service life of composites. In order to overcome these pitfalls, parameters that govern the LVI is essential to understand and adopt suitable methods to overcome them. Although in some cases, there are solutions that can improve, while in others an ideal parameter is yet in the developmental stage. Among all the listed parameters mentioned above, some of the key takeaway are listed below.

In case of process parameter related to preparation of composite panel, it is essential to prepare a laminate under controlled environment and in turn curing them using an autoclave in order to develop a high performance composite material. There by enhancing the performance of the composite material. Also by adopting a suitable fabrication process such as VARTM (Vacuum Assisted Resin Transfer Moulding), VB (Vacuum Bagging) process based on the type of application of composite, the overall shelf life of the panels can be enhanced.

The performance can be enhanced in case of composite laminates by choosing an appropriate material for a particular application. In case of general purpose applications, glass fibre could be adopted and in areas which require higher toughness, combination of Glass and Kevlar could be adopted or even combination of Glass Carbon and Kevlar could be used. Material selection could also be via a hybrid stacking sequence and also via 2D and 3D hybrid composites. This can enhance the service life of the composites and in turn increase the damage tolerance of composites.

All the factors that affect the damage tolerance, can be suitably addressed by selecting appropriate parameters that is relevant to the type of application. By choosing ideal resin such as TP over TS the overall performance increases including fracture toughness, by adopting a suitable stacking sequence the performance of composite panels can increase. Further by adopting a suitable lay-up with optimum thickness, the strength to weight ratio can be maintained. In case of impact region, by adopting inserts concepts, the extent of damage at the susceptible region can be reduced and in turn enhance the service life. In areas related to temperature, by focusing on temperature resistant material, the damage extent can be reduced due to LVI process. In areas related to residual strength, the extent of damage on the specimen may vary compared to the extent of damage on the structure and limited knowledge is available, this in turn needs better studies related to LVI process of structures and panels.

Acknowledgement
The authors would like to acknowledge the present review work carried out is part of an ongoing funded project from AICTE (RPS), New Delhi.

References

[1] K B Katnam, L F M Da Silva and T M Young 2013 Bonded repair of composite aircraft structures: a review of scientific challenges and opportunities, *Progressive Aerospace Sci.* **61** 26–42. DOI: https://doi.org/10.1016/j.paerosci.2013.03.003

[2] S. Abrate 1998 *Impact on Composite Structures* (United Kingdom: Cambridge University Press, Cambridge) DOI: https://doi.org/10.1017/CBO9780511574504

[3] Vogelesang L B and Vlot 2000 A Development of fiber metal laminates for advanced aerospace structures. *J. Mater. Process Technol.* **103** 1–5. DOI: https://doi.org/10.1016/S0924-0136(00)00411-8

[4] J. Jefferson Andrew, Sivakumar M Srinivasan, A Arockiarajan and Hom Nath Dhakal 2019 Parameters influencing the impact response of fiber-reinforced polymer matrix composite materials: A critical review *Composite Structures* **224** 111007. DOI: https://doi.org/10.1016/j.compstruct.2019.111007

[5] Richardson M and Wisheart M 1996 Review of low-velocity impact properties of composite materials *Compos. A. Appl. Sci. Manuf.* **27** 1123–31. DOI: https://doi.org/10.1016/1359-835X(96)00074-7

[6] Greenhalgh E and Hiley M 2003 The assessment of novel materials and processes for the impact tolerant design of stiffened composite aerospace structures. *Composites Part A* **34**(2) 151–61. DOI: https://doi.org/10.1016/S1359-835X(02)00188-4

[7] Andrew J Jefferson, Srinivasan Sivakumar M and Arockiarajan A 2018 Influence of patch lay-up configuration and hybridization on low velocity impact and post-impact tensile response of repaired glass fiber reinforced plastic composites *J. Compos. Mater.* **53**(1) 3-17. DOI: https://doi.org/10.1177/0021998318779430

[8] López-Puente J, Varas D, Loya J A and Zaera R 2009 Analytical modeling of high velocity impacts of cylindrical projectiles on carbon/epoxy laminates *Composites Part A*;**40**(8):1223–30. DOI: https://doi.org/10.1016/j.compositesa.2009.05.008

[9] Liu X, Yu W and Pan N 2011 Evaluation of high performance fabric under light irradiation *J. Appl. Polym. Sci.* **120** 552–6. DOI: https://doi.org/10.1002/app.33200

[10] Vishwas Mahesh, Sharnappa Joladarashi and Satyabodh M Kulkarni 2021 A comprehensive review on material selection for polymer matrix composites subjected to impact load *Defence Technology* **17**(1) 257-77. DOI: https://doi.org/10.1016/j.dt.2020.04.002

[11] Aoki Y, Suemasu H and Ishikawa T 2007 Damage propagation in CFRP laminates subjected to low velocity impact and static indentation *Adv. Compos. Mater. Off. J. Japan. Soc. Compos. Mater.* **16** 45–61. DOI: https://doi.org/10.1163/156855107779755318

[12] Lee SM and Zahuta P 1991 Instrumented Impact and Static Indentation of Composites *J. Compos. Mater.* **25** 204–22. DOI: https://doi.org/10.1177/002199839102500205

[13] Azwan S M S, Yazid Y M, Amran A and Abdi B 2014 Quasi-Static Indentation Behaviour of Glass Fibre Reinforced Polymer *Adv. Mater. Res.* **970** 317–9. DOI: https://doi.org/10.4028/www.scientific.net/AMR.970.317

[14] Kaczmarek H and Maison S 1994 Comparative ultrasonic analysis of damage in CFRP under static indentation and low-velocity impact *Compos. Sci. Technol.* **51** 11–26. DOI: https://doi.org/10.1016/0266-3538(94)90152-X

[15] S Z H Shah, S Karuppanan, P S M Megat-Yusoff and Z Sajid 2019 Impact resistance and damage tolerance of fiber reinforced composites: A review *Comp. Structures* **217** 100–21. DOI: https://doi.org/10.1016/j.compstruct.2019.03.021

[16] Kim, Jang Kyo. "Methods for improving impact damage resistance of CFRPs." In *Key Engineering Materials*, vol. 141, pp. 149-168. Trans Tech Publications Ltd, 1998.

[17] Cantwell W and Morton J 1991 The impact resistance of composite materials-a review *Compos.*

Mater. **22** 347–62. DOI: https://doi.org/10.1016/0010-4361(91)90549-V

[18] Wisnom M R 2012 The role of delamination in failure of fibre-reinforced composites *Phil. Trans. R. Soc. A* **370** 1850–70. DOI: https://doi.org/10.1098/rsta.2011.0441

[19] Alam, Md Ashraful, and Khalid Al Riyami. "Shear strengthening of reinforced concrete beam using natural fibre reinforced polymer laminates." *Construction and Building Materials* **162** (2018): 683-696.

[20] Bandaru A K, Sachan Y, Ahmad S, Alagirusamy R and Bhatnagar N 2017 On the mechanical response of 2D plain woven and 3D angle-interlock fabrics *Compos. B. Eng.* **118** 135–48. DOI: 10.1016/j.compositesb.2017.03.011

[21] Shah, S. Z. H., P. S. M. Megat-Yusoff, S. Karuppanan, R. S. Choudhry, F. Ahmad, Z. Sajid, P. Gerard, and K. Sharp. "Performance comparison of resin-infused thermoplastic and thermoset 3D fabric composites under impact loading." *International Journal of Mechanical Sciences* **189** (2021): 105984. DOI: 10.1016/j.ijmecsci.2020.105984

[22] P. Potluri, P Hogg, M. Arshad, Jetavat and P Jamshidi 2012 Influence of Fibre Architecture on Impact Damage Tolerance in 3D Woven Composites *Appl. Compos. Mater.* **19** 799–812. (https://link.springer.com/article/10.1007/s10443-012-9256-9)

[23] Bhatnagar A 2016 Lightweight ballistic composites: military and law-enforcement applications. (Woodhead Publishing)

[24] Sylvie Beland 1990 *High performance thermoplastic resins and their composites* (United States: Imprint: William Andrew).

[25] Tan W and Falzon B G 2016 Modelling the crush behaviour of thermoplastic composites. *Compos. Sci. Technol.* **134** 57–71. DOI: https://doi.org/10.1016/j.compscitech.2016.07.015

[26] Hart K R, Chia P X L, Sheridan L E, Wetzel E D, Sottos N R and White S R 2017 Comparison of compression-after-impact and flexure-after-impact protocols for 2D and 3D woven fiber-reinforced composites *Compos. A Appl. Sci. Manuf.* **101** 471–9. DOI: https://doi.org/10.1016/j.compositesa.2017.07.005

[27] Umer, R., H. Alhussein, J. Zhou, and W. J. Cantwell. "The mechanical properties of 3D woven composites." *Journal of Composite Materials* 51, no. **12** (2017): 1703-1716.

[28] Wang M, Cao M, Wang H, Siddique A, Gu B and Sun B 2017 Drop-weight impact behaviors of 3-D angle interlock woven composites after thermal oxidative aging *Compos. Struct.* **166** 239–55. DOI: https://doi.org/10.1016/j.compstruct.2017.01.046

[29] Kim K-Y and Ye L 2004 Interlaminar fracture toughness of CF/PEI composites at elevated temperatures: roles of matrix toughness and fibre/matrix adhesion. *Compos. A Appl. Sci. Manuf.* **35** 477–87. DOI: https://doi.org/10.1016/j.compositesa.2003.10.005

[30] Wang Wenjie, Chouw Nawawi and Jayaraman Krishnan 2016 Effect of thickness on the impact resistance of flax fibre-reinforced polymer J. of Reinf. Plastic Composites **35**(17) 1277–89. DOI: https://doi.org/10.1177/0731684416648780

[31] Nassir Nassier A, Guan Z W, Birch R S and Cantwell W J 2018 Damage initiation in composite materials under off-centre impact loading *Polym. Test.* **69** 456–461. DOI: 10.1016/j.polymertesting.2018.06.006

[32] Özen M 2017 Influence of stacking sequence on the impact and postimpact bending behavior of hybrid sandwich composites *Mech. Compos. Mater.* **52**(6) 759–66. DOI: 10.1007/s11029-017-9626-3

[33] Atas C and Sayman O 2008 An overall view on impact response of woven fabric composite plates *Compos. Struct.* **82**(3) 336–45. DOI: https://doi.org/10.1016/j.compstruct.2007.01.014

[34] Patel Shivdayal and Guedes Soares C 2018 Reliability assessment of glass epoxy composite plates due to low velocity impact. *Compos. Struct.* **200** 659–68. DOI: 10.1016/j.compstruct.2018.05.131

[35] Sutherland L S 2018 A review of impact testing on marine composite materials: Part III – Damage tolerance and durability *Compos. Struct.* **188** 512–18. DOI: https://doi.org/10.1016/j.compstruct.2018.01.042

[36] Reis L and De Freitas M 1997 Damage growth analysis of low velocity impacted composite panels *Compos. Struct.* **38** 509–15. DOI: https://doi.org/10.1016/S0263-8223(97)00086-X

[37] Sonnenfeld C, Mendil-Jakani H, Agogué R, Nunez P and Beauchêne P 2017 Thermoplastic/ thermoset multilayer composites: a way to improve the impact damage tolerance of thermosetting resin matrix composites *Compos. Struct.* **171** 298–305. DOI: https://doi.org/10.1016/j.compstruct.2017.03.044

Performance evaluation of AE-pulse of wire EDM process on Ti-10V-2Fe-3Al alloy by Taguchi GRA technique

Parameshwar K[1], Srinivasa Rao Nandam[2] and Dinesh G Thakur[3]
[1]MS (By Research) Scholar, DIAT, Pune, and Tech. Officer, DMRL, DRDO, India
[2]Scientist, DMRL, DRDO, Hyderabad, India
[3]Professor, Department of Mechanical Engineering, DIAT, Pune, India
Email:[1]paramesh.dmrl@gmail.com,[2]srinivas_nandam@dmrl.drdo.in
[3]thakur@diat.ac.in

Abstract. Titanium alloy, Ti-10V-2Fe-3Al is one of the recently developed materials having special characteristics of deep hardenability, ductility, fracture toughness including high strength to weight ratio. But, titanium alloys have poor machinability due to the formation of built-up edges and tool wear in conventional machining processes. Wire electrical discharge machining wire EDM is found to be the most preferred machining process among other non-traditional processes for cutting of the difficult-to-cut materials. The surface integrity aspects of the machined component play a crucial role in the functional performance of the component. The performance of an advanced AE-pulse (Anti Electrolysis Equi-energy) over E-pulse on surface integrity aspects in the wire EDM process is studied in detail. Experiments were carried with various levels of pulse-on-time, pulse-off-time, peak current setting, and servo voltage through a Taguchi L_9 orthogonal array. The responses such as kerf width, surface roughness, material removal rate, and white layer thickness were studied through gray relational analysis (GRA) technique. The optimized results were validated through experimental studies. It is observed that the optimized process parameter values were T_{on}: 12 µs T_{off}:40µs IP No.:10 SV:10V, which gave 28 % of the significant reduction of white layer thickness, 16% improvement of MRR, 5% reduction of surface roughness, and 4% of reduction of kerf width.
Keywords: WEDM, Ti-10V-2Fe-3Al alloy, Taguchi method, Grey relational analysis and Minitab

1. Introduction

Titanium alloys have excellent corrosion resistance, fatigue resistance, high strength to weight ratio. Therefore, these alloys are mainly used in automobiles and aerospace industries for high-temperature applications [1].Ti-6Al-4V, an alpha-beta alloy extensively used for both aero engine and airframe applications. In recent times, beta titanium alloy has been inducted in airframe applications because of their high strength and damage tolerance properties. Ti-10V-2Fe-3Al is a high strength near beta alloy being used in airframe and other applications like landing gears, wings, fuselage, doors, wing support structures and cargo handling structures [2].However, titanium and its alloys are difficult to cut materials due to low thermal conductivity, low modulus of elasticity and also very high chemical reactive with cutting tool materials [3]. Wire electrical discharge machining wire EDM is found to be the most preferred machining process among other non-traditional processes for cutting of the difficult-to-cut materials. It is ease in control of machining process and accuracy. Where electrical energy is used to generate electrical sparks and material removal mainly occurs due to the thermal energy of the sparks [4].The most important performance measures in WEDM are material removal rate (MRR), surface roughness, and kerf width. The pulse-on-time (T_{on}), pulse-off-time (T_{off}), setting peak current (Ip) and applied servo voltage(SV), speed of the wire, the feed rate of the wire, diameter of the wire, wire tension, and condition of the flushing dielectric are the machining variables that affect the performance measures. Several researchers have carried out various investigations on a standard pulse i.e., E-pulse to improve the WEDM performance and obtain better machining outputs like higher MRR, lower kerf, and good surface roughness on titanium and its alloy. Many research studies are in progress to improve the cutting time and surface roughness of the titanium alloys by the WEDM process.

Ghodsiyeh et al. [5] have used Taguchi's method and response surface methodology (RSM) to optimize the WLT (white layer thickness), HAZ (heat-affected zone), surface crack depth, and MRR on Ti-6Al-4Valloy by WEDM process. They observed that pulse-on-time is the most significant parameter on HAZ and peak current is the most influencing parameter on crack density. Sharma et al.[6]explained about control factors effective on the process of Ti-6Al-4Vby WEDM. L_9 OA (orthogonal array) was used for experimenting. The cutting speed and surface roughness was optimized. They observed that Toff is the most significant factor affecting the multi–characteristics. Jaskarn Singh et al.[7]have studied on review the effects of process parameters in the WEDM process. It was verified T_{on}, T_{off}, Ip, SV, wire feed rate, dielectric flow rate, wire tension as input factors and surface roughness and MRR were output responses. Wire-feed, wire-tension, and water flush have the least cause on the process. The higher level of current, the power of the spark is increased and results were elevated in MRR. The surface finish was acquired better by reducing T_{on} and T_{off} and current level. Feroze et al. [8] investigated the recast layer on Titanium alloy machined by the WEDM process. Taguchi L_9OA was used for experimentation. It was identified that table speed has a significant effect on the thickness of the recast layer and pulse off time has the least effect on recast layer thickness. Shivaprakasham et al. [9] worked on Ti-6Al-4V alloy considering MRR, Kerf width, and surface roughness (SR) as responses and voltage, capacitance, and feed rate as input parameters on micro-WEDM. Researchers used RSM and CCD to multi-objective optimizes the process parameters. Influencing factors were being found in the process by ANOVA (Analysis of variance). Interaction of voltage and capacitance has significant influenced on MRR. Prasad et al. [10] experimented with process variables to optimize the performance characteristics including surface finish and MRR on Ti-64Al-4V alloy. The effect of variables was setting peak current, pulse-on-time, pulse-off-time, and servo voltage setting. The experimental Taguchi's design was four factors and three levels. The output responses are MRR and surface roughness. ANOVA was indicated that peak current followed by pulse-on-time is the main contributing parameter. Mhatre et al.[11] studied EDM characteristics of Ti-6Al-4V alloy based on grey relational theory parameter optimization. The multiple responses optimized are MRR, electrode wear rate (EWR), and SR. It was found that pulse current is the most significant factor affecting MRR, dimensional accuracy, and surface integrity. Kumar et al. [12] investigated on surface integrity of pure titanium by Wire EDM based on Taguchi L_{27} OA. It was observed that pulse-on-time, pulse-off-time, and peak current setting no. declined the integrity of the machined surface. The micro-crack thickness was improved due to fast cooling and heating in the spark area. Wire rupturing and wire wear was observed due to a higher amount of discharge peak current and pulse-on-time duration. Goswamiet al. [13] investigated surface integrity, material removal rate, and wire wear ratio based on Taguchi's design of methodology. The observations are pulse-on-time period leads to a thicker recast deposit. At a lower level of pulse-on-time and high level of pulse-off-time, the machined surface had deposits of wire material. Kumar et al. [14] studied EDM characteristics of pure titanium based on Box-Behnken design RSM theory parameter optimization. The multiple responses optimized are cutting rate, dimensional deviation, surface roughness, and wire wear ratio. It was found that pulse-on-time, pulse-off-time, and peak current are the major factors affecting MRR, WW (wire wear) ratio, and dimensional deviation. It is observed that many researchers have worked on titanium alloy with E pulses under the WEDM process. They have found a considerable amount of surface integrity aspects such as HAZ, white layer thickness, and surface roughness on the machined surface. These surface integrity aspects play a crucial role in the functional performance of the machined component. Now, machine tool manufacturers have come up with an advanced pulse, anti-electrolysis equi-energy (AE) pulse. It is introduced for minimizing the interaction of stray current and contaminants on the workpiece surface in the WEDM process. Therefore, a systematic research study has been conducted to study the effect of pulse-on-time, pulse-off-time, peak current setting, and servo voltage while cutting of Ti-10V-2Fe-

IOP Conf. Series: Materials Science and Engineering 1126 (2021) 012080 doi:10.1088/1757-899X/1126/1/012080

3Al by AE pulse. The output responses considered for the study are MRR, SR, WLT, and Kerf width. Taguchi DOE technique was used for preparation experimental matrix and GRA technique was used for optimization.

2. Experimental details

2.1 Work material

Forged and rolled rod of Ti-10V-2Fe-3Al alloy of 1m long and 60 mm diameter was cut into 20 mm thick slices by WEDM process along the length with E-pulse. The rectangular workpiece of size 58 mm x 30mm x 20 mm was extracted from the round slice by the WEDM process. The chemical composition of Ti-10V-2Fe-3Al is shown in Table.1 and the physical properties of the work material are shown in Table.2.

Table 1. Chemical composition

Element	V	Fe	Al	O	N	C	H	Y	Ti
Weight%	9.0-11.0	1.6-2.2	2.6 -3.4	0.13	0.05	0.05	0.015	0.005	Balance

Table 2. Physical properties

Properties	Density	UTS	Thermal conductivity	Specific heat	Melting range	Electrical Resistivity
Units	4.61 g/cm^3	1240 MPa	7.8 W/mk	525 J/kg K	1649°±15° C	1.5 micro-ohm-m

2.2 Process parameters

The range of the identified process parameters such as pulse-on-time, pulse-off-time, peak current setting, and servo voltage were selected based on the feasible machining conditions through a set of trial experiments. The parameters and their levels for the experimental study are shown in Table 3.

Table 3. Experimental factors and levels.

Sl. No.	Machining parameters	Units	Coding	Levels		
				1	2	3
1	T_{on}(pulse-on-time)	µs	A	6	9	12
2	T_{off} (pulse-off-time)	µs	B	40	50	60
3	IP (peak current setting)	No.	C	10	11	12
4	SV (servo voltage)	V	D	10	15	20

Taguchi DOE technique was used for the preparation of the experimental matrix. The orthogonal Array (OA) and run order of L_9 is shown in Table 4.

Table 4. Orthogonal Array

Run Order	A	B	C	D
1	1	1	1	1
2	1	2	2	2
3	1	3	1	3
4	2	1	2	3
5	2	2	2	1
6	2	3	1	2
7	3	1	3	2
8	3	2	1	3
9	3	3	2	1

2.3 Experimental setup

A CNC Wire EDM machine (M/s Electronica, Pune, Model: Ultra Cut S_2) was used for experimental studies. The schematic representation of the WEDM process is shown in Fig 1. The workpiece was held on the WEDM table with clamps. The workpiece and the brass wire are connected to the anode and cathode respectively. The parameters used as fixed parameters are shown in Table 5.

Table 5. Fixed parameters

Sl.No	parameter	Unit	Condition
1	VP	V	1
2	WP	cm^2/kg	106
3	SF	mm/min	25
4	Brass wire	mm	Ø 0.25
5	Wire feed	m/min	10
6	Wire tension	grams	10

Experiments were conducted as per the run order along the width of the workpiece for 10 mm cutting length with an offset of 5 mm along the longitudinal distance as shown in Fig 2. by using E and AE-pulse on both sides of the workpiece. The experimented samples were extracted from the workpiece for further studies. The extracted cut samples are shown in Fig 3.

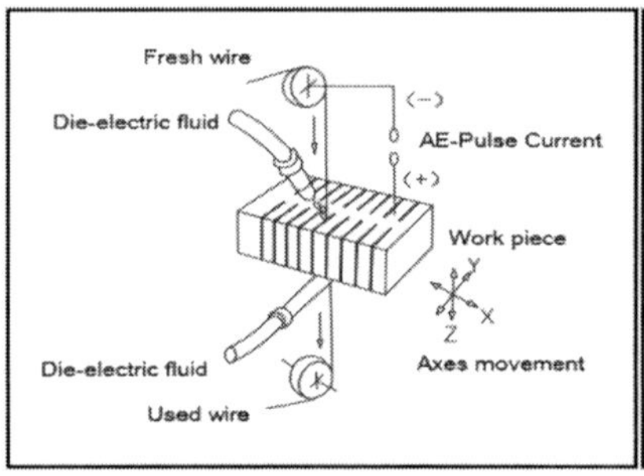

Fig 1. Schematic diagram of the Wire EDM Process

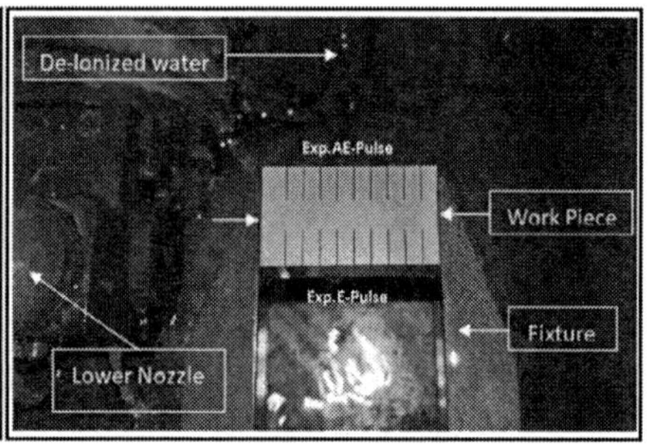

Fig 2. Experimental Setup

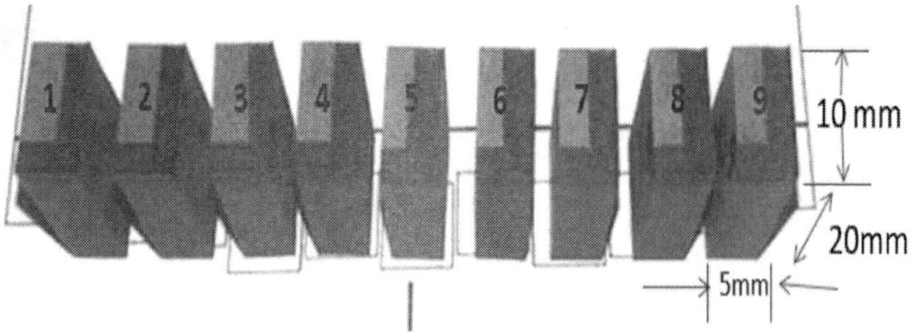

Fig 3. WEDM cut samples by AE-pulses

2.4 *Characterization Techniques:*

The MRR was calculated by using the below formula

MRR = CS x KW x HW ------------------------------- (1)

Where: CS: Cutting speed in mm/min, W: Kerf width in mm and H: Height of the workpiece in mm. The values of cutting speed were obtained from the machine controller while experimentation. This data was used for the calculation of MRR for each run order. Surface roughness was measured on the machined surface by using a stylus type surface roughness profile meter (M/s Taylor Hobson, UK, Model: Intra-II) with sample length: 6mm, cut-off length: 0.8mm and ISO Gaussian filter. Kerf width is the distance between the two surfaces of the wire pass. It was created by WEDM process. It was measured by using the OPP (M/s Baty, UK, Model: SM20).WLT was measured by digital optical microscope (M/s: Olympus Japan, Model: DSX-510) from the cross-section of cut samples.

3. **Results and discussion**

3.1 *Experimental design based on the Taguchi method*

The results of kerf, roughness, MRR and WLT from the experiments are indicated in Table 6.

Table 6. Experimental results.

Runs	Orthogonal Array				Kerf mm	Roughness µm	MRR mm^3/min	WLT µm
	A	B	C	D				
1	1	1	1	1	0.267	1.8278	0.961	5.7484
2	1	2	2	2	0.3 4	2.8094	10.554	10.631
3	1	3	1	3	0.336	2.9883	10.483	7.605
4	2	1	2	3	0.312	2.5377	7.238	9.8932
5	2	2	2	1	0.315	2.8419	8.064	6.3484
6	2	3	1	2	0.267	1.5902	0.748	6.2624
7	3	1	3	2	0.328	2.7832	10.562	7.0716
8	3	2	1	3	0.269	1.5083	0.807	4.5544
9	3	3	2	1	0.319	2.7633	8.804	5.7528

The low values of kerf width, surface roughness, white layer thickness, and high value of material removal rate are in wire EDM process are required to improve the part quality of the component. This multi-parametric optimization can be achieved through analysis the data by GRG.

3.2 *Grey relational analysis*

The normalized result values of kerf width, surface finish, white layer thickness and MRR are shown in Table 7.The deviation of sequences is shown in Table 8.

Table 7. Normalized results.

Runs	Kerf (mm)	Ra (μm)	MRR (mm^3/min)	WLT (μm)
1	1.0000	0.7841	0.9782	0.8035
2	0.0290	0.1209	0.9993	0.0000
3	0.0000	0.0000	0.9920	0.4980
4	0.3478	0.3045	0.6614	0.1214
5	0.3043	0.0990	0.7455	0.7048
6	1.0000	0.9447	0.0000	0.7189
7	0.1159	0.1386	1.0000	0.5858
8	0.9710	1.0000	0.0061	1.0000
9	0.2464	0.1520	0.8209	0.8028

Table 8. Deviation sequence.

Runs	Kerf (mm)	Ra (μm)	MRR (mm^3/min)	WLT (μm)
1	0.0000	0.2159	0.0218	0.1965
2	0.9710	0.8791	0.0007	1.0000
3	1.0000	1.0000	0.0080	0.5020
4	0.6522	0.6955	0.3386	0.8786
5	0.6957	0.9010	0.2545	0.2952
6	0.0000	0.0553	1.0000	0.2811
7	0.8841	0.8614	0.0000	0.4142
8	0.0290	0.0000	0.9939	0.0000
9	0.7536	0.8480	0.1791	0.1972

Table.9 shows the GRG and rank for each run order. The highest GRG is run order 1. This was indicated in text bold.

Table 9. Grey relational coefficients and GRG.

Runs	Kerf (mm)	Ra (μm)	MRR (mm^3/min)	WLT (μm)	GRG	R
1	**1.0000**	**0.6985**	**0.9583**	**0.7179**	**0.8437**	**1**
2	0.3399	0.3626	0.9985	0.3333	0.5086	8
3	0.3333	0.3333	0.9843	0.4990	0.5375	6
4	0.4340	0.4182	0.5962	0.3627	0.4528	9
5	0.4182	0.3569	0.6627	0.6287	0.5166	7
6	1.0000	0.9004	0.3333	0.6401	0.7185	3
7	0.3613	0.3673	1.0000	0.5469	0.5689	4
8	0.9452	1.0000	0.3347	1.0000	0.8200	2
9	0.3988	0.3709	0.7363	0.7171	0.5558	5

IOP Conf. Series: Materials Science and Engineering 1126 (2021) 012080 doi:10.1088/1757-899X/1126/1/012080

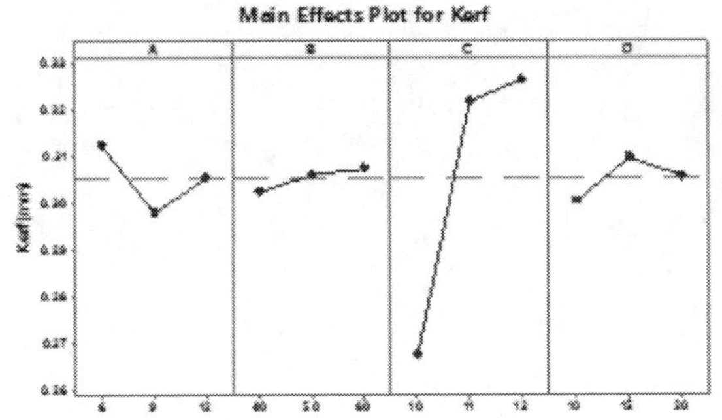

(a)

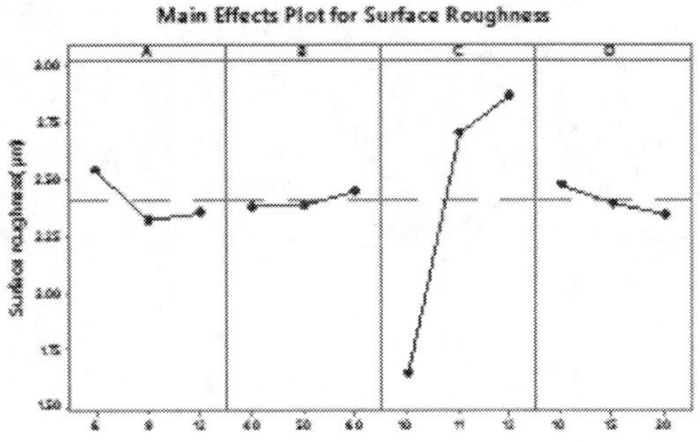

(b)

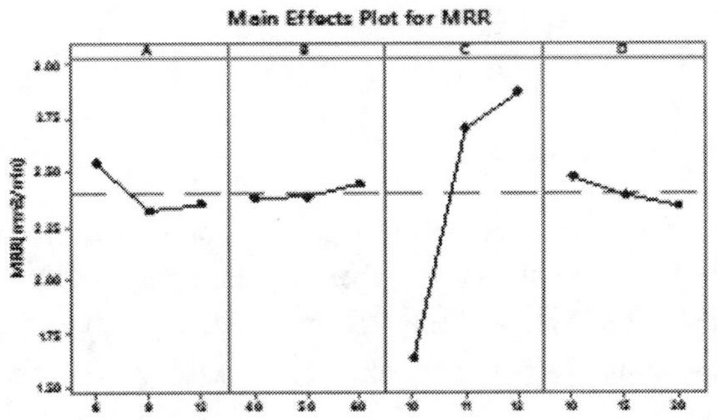

(b)

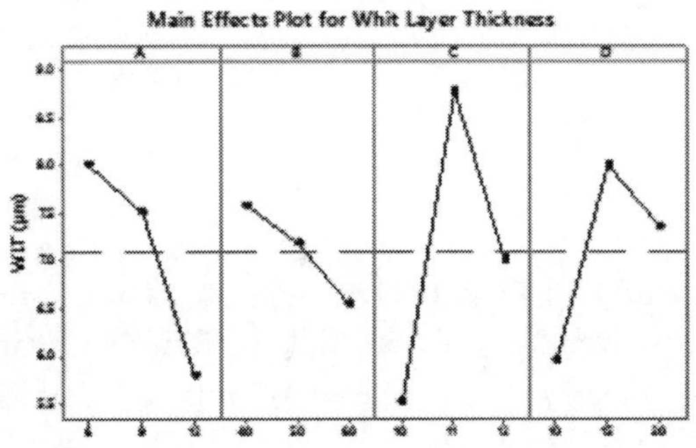

(d)

Fig 4.Main effect plot for (a)Kerf (b) Surface roughness (c) MRR (d) WLT on factors of Wire EDM

Fig 4. (a), (b), and (c) illustrate the main effects on individual performance parameter, where Kerf, Surface roughness, and MRR increases with peak current[4].(d) Shows that WLT increases with increasing peak current and servo voltage. Further when IP and SV increase from a specific limit (i.e 11 and 15V). The increase in peak current and servo voltage increases the discharge energy that promotes the melting and vaporization of the work material decrease the WLT[7].

Table 10. Mean Response table for a GRG.

Parameters	Level-1	Level-2	Level-3	Delta (Max-Min)	Rank
A	0.6299	0.5626	**0.6482**	0.0856	2
B	**0.6218**	0.6151	0.6039	0.0178	4
C	**0.7940**	0.5057	0.5410	0.2883	1
D	**0.6387**	0.5986	0.6034	0.0401	3
The total GRG mean value (γ_t)=0.6136					

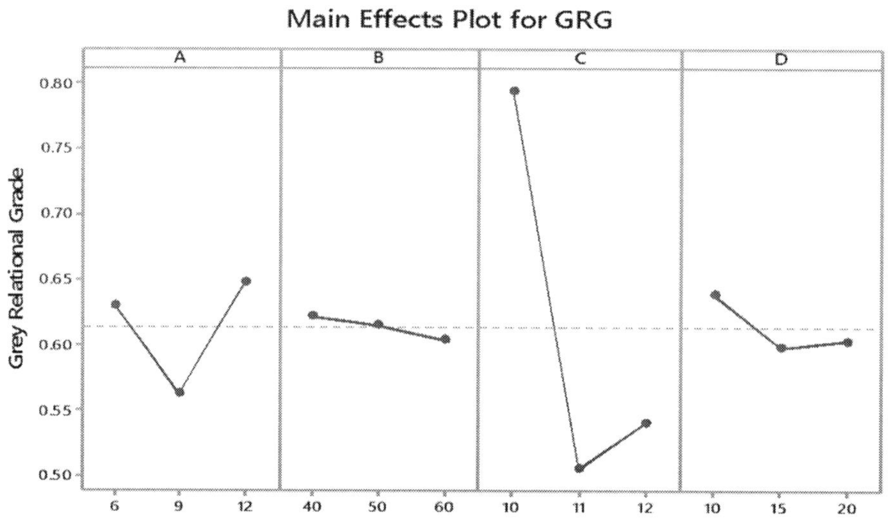

Fig 5. Grey relational grade graph

The means of the GRG for each level of variables considered from Tables 8 and 9 were detailed in Table 10. The higher GRG values (text bold) are shown in Table 10. The optimal values for improvement of the multiple performance characteristics were A3B1C1D1in as per Fig 5. i.e., pulse-on-time was 12μs (Level-3), pulse-off-time was 40μs (Level-1), peak current setting No. was 10 (Level- 1), and servo voltage was 10V (Level-1).

3.3 Analysis of variance for grey relational grade

Using Minitab software, the ANOVA was performed to determine which parameter significantly affects the performance characteristics.The results of ANOVA for GRG values in Table 11.

Table 11. Analysis of variance for a grey relational grade.

Source	DF	Seq SS	Adj MS	Contribution
A	2	0.012188	0.006094	7%
B	2	0.000487	0.000244	1%
C	2	0.148398	0.074199	90%
D	2	0.002873	0.001437	2%
Total	8	0.163947		100%

The results indicate that the percentage contribution of the pulse-on-time, pulse-off-time, peak current setting no., and servo voltage is 7%, 1%, 90%, and 2%, correspondingly. These two factors (peak current and pulse-on-time) are considerably influenced the GRG, and the peak current is the most important factor for multiple performance characteristics. This shows that the pulse-off-time and servo voltage has does not have a statistically significant result on multiple performance characteristics.

3.4 *Predicting the optimal GRG and verification test*

The best GRG is predicted at the selected controllable parameters. The most main variable with the optimal level was selected for AE-pulse as A3 (pulse-on-time), B1 (pulse-off-time), C1 (peak current setting no.) followed by D1 (servo voltage) the predicted GRG can be calculated as:

$$\hat{\gamma} = \gamma_t + \sum_{i=1}^{p}(\gamma_i - \gamma_t) \text{--} (2)$$

Where γ_t is the total mean of GRG, γ_i is the mean of the GRG at the optimal level. The A3B1C1D1is an optimal variable combination of the WEDM process obtained employing GRA and was well-thought-out as a verification test. Based on eq (2) the predicted GRG is presented in Table 12.

Table 12. Predicting GRG optimal value and verification test.

		Setting level	Kerf (mm)	Ra (μm)	MRR (mm^3)	WLT (μm)	GRG Value	Progress in GRG value
Initial control parameters		A1B1C1D1	0.267	1.8278	0.961	5.7484	0.8437	
Optimal control parameters	Prediction	A3B1C1D1					0.8864	
	Experiment	A3B1C1D1	0.254	1.7267	1.123	4.1236	0.9526	0.1089
Percentage	(%)		4.87	5.53	16.86	28.27	12.91	

4. Conclusion

Experiments were conducted on Ti-10V-2Fe-3Al alloy with AE and E-pulses by using various levels of pulse-on-time, pulse-off-time, peak current and servo voltage of Wire EDM process by Taguchi OA. The output responses of the kerf,surface roughness,MRR, and WLT are optimized by using the GRA. The following conclusions are made from the experimental studies.

1. AE-pulse gave better surface integrity results when compared with E-pulses due to the supply of controlled and localized discharge energy.
2. The percentage contributions of the machining variables were IP: 90%, T_{on}: 7%, SV: 2% and T_{off}: 1%.Peak current is identified as a highly contributing variable among others by the ANOVA technique.
3. The kerf, surface roughness, and MRR were highly influenced by IP and pulse-on-time.
4. The significant factors for the white layer thickness were IP and pulse-on-time followed by SV and pulse-off-time.
5. The best combination of AE-pulse variables to reduce the kerf, surface roughness and white layer thickness and to increase the MRR from optimum GRA mean grades were A3B1C1D1.
6. The predicted GRG value is verified by a confirmation test.
7. The white layer thickness was reduced by 28 %, the MRR was improved by 16%, the surface roughness was reduced by 5% and the kerf width was reduced by 4%.
8. The GRG of the experimental value was 12.91% higher than the initial value.
9. This method can be effectively used for the evaluation of optimal values from the multi-objective criteria.

IOP Conf. Series: Materials Science and Engineering 1126 (2021) 012080 doi:10.1088/1757-899X/1126/1/012080

Acknowledgments
The authors are sincerely thankful to Dr. G. Madhusudhan Reddy, Director, DMRL for his encouragement and kind permission to publish this work. The authors are grateful to Shri D Madhusudhan, TO-B and Shri M.N.Malleshwara Rao, TO-A, and other officers and staff of MEG for carrying out the experiments and evaluation of results. The authors are thankful to the officers and staff of TAG, SFAG, and EMG for their technical support. The authors are also thankful to Mr. Mahesh, Ph.D. scholar, DIAT for his technical inputs.

References
[1] Peters, Manfred, et al. "Titanium alloys for aerospace applications." *Advanced engineering materials* 5.6 (2003): 419-427.
[2] Boyer, Rodney R. "Aerospace applications of beta titanium alloys." *JOM* 46.7 (1994): 20-23.
[3] Niknam, Seyed Ali, Raid Khettabi, and Victor Songmene. "Machinability and machining of titanium alloys: a review." *Machining of titanium alloys* (2014): 1-30.
[4] Muthuramalingam, T., and B. Mohan. "A review on influence of electrical process parameters in EDM process." *Archives of civil and mechanical engineering* 15.1 (2015): 87-94.
[5] Ghodsiyeh, Danial, et al. "Experimental investigation of surface integrity after wire electro-Dischargemachining of Ti–6Al–4V." Sādhanā 44.9 (2019): 196.
[6] Sharma, Neeraj, et al. "Multi-quality characteristics optimization on WEDM for Ti-6Al-4V UsingTaguchi-grey relational theory." International Journal of Machining and Machinability of Materials 21.1-2 (2019): 66-81.
[7] Singh, Jaskarn, et al. "Review on effects of process parameters in wire-cut EDM and wire electrode development." International Journal of Innovative Research in Science, Engineering, and Technology 2 (2016): 701-706.
[8] Feroze, Waseem, V. R. Srinivasan, and C. P. S. Prakash. "Investigation of Recast Layer Formed on Titanium Alloy Machined by Wire Electric Discharge Machining."IJIRST,ISSN:2349-6010
[9] Sivaprakasam, P., P. Hariharan, and S. Gowri. "Modeling and analysis of the micro-WEDM process of titanium alloy (Ti–6Al–4V) using response surface approach." Engineering Science and Technology, an International Journal 17.4 (2014): 227-235.
[10] Prasad, AVS Ram, KoonaRamji, and G. L. Datta. "An experimental study of wire EDM on Ti-6Al-4V Alloy." *Procedia Materials Science* 5 (2014): 2567-2576.
[11] Mhatre, Mitali S., Sagar U. Sapkal, and Raju S. Pawade. "Electro discharge machining characteristics of Ti-6Al-4V alloy: A grey relational optimization." Procedia Materials Science 5(2014).
[12] Kumar, Anish, Vinod Kumar, and Jatinder Kumar. "Surface integrity and material transfer investigation of pure titanium for the rough-cut surface after wire electro-discharge machining." Proceedings of the Institution of Mechanical Engineers, Part B: Journal of Engineering Manufacture 228.8 (2014): 880-901.
[13] Goswami, Amitesh, and Jatinder Kumar. "Investigation of surface integrity, material removal rate and wire wear ratio for WEDM of Nimonic 80A alloy using GRA and Taguchi method." Engineering Science and Technology, an International Journal 17.4 (2014): 173-184.
[14] Kumar, Anish, Vinod Kumar, and Jatinder Kumar. "Multi-response optimization of process parameters based on response surface methodology for pure titanium using WEDM

process." The International Journal of Advanced Manufacturing Technology 68.9-12 (2013): 2645-2668.

[15] Nourbakhsh, Farnaz, et al. "Wire electro-discharge machining of titanium alloy." Procedia Cirp 5 (2013): 13-18.

[16] Alias, Aniza, Bulan Abdullah, and NorlianaMohd Abbas. "WEDM: Influence of machine feed rate in machining titanium Ti-6Al-4 V using brass wire and constant current (4A)." Procedia engineering 41(2012): 1812-1817.

[17] Ho, K. H., and S. T. Newman. "State of the art electrical discharge machining (EDM)." International Journal of Machine Tools and Manufacture 43.13 (2003): 1287-1300.

[18] Kao J. Y. et al. "Optimization of the EDM parameters on machining Ti–6Al–4V with multiple quality characteristics." *The International Journal of Advanced Manufacturing Technology* 47.1-4 (2010): 395-402.

[19] AhmetHascalik, and Caydas U. "Electrical discharge machining of titanium alloy (Ti–6Al–4V)." Applied surface science 253.22 (2007): 9007-9016.

[20] Kuriakose, Shajan, and M. S. Shunmugam. "Characteristics of wire-electro discharge machined Ti6Al4V surface." *Materials Letters* 58.17-18 (2004): 2231-2237.

Correlation of CFRP and AA7075 Aircraft Wing box structure using FEM

Mr.M.Vigneshwaran[1], Kumaran M[2], Praveen kumar K[3], Revanth K[4], Sathya prakash T[5]

[1]Assistant Professor, Department of Mechanical Engineering, Sri Krishna College of Engineering and Technology, Coimbatore.
[2,3,4,5]UG student Sri Krishna College of Engineering and Technology, Coimbatore.
[1]vigneshwaranm@skcet.ac.in

Abstract. Composite materials are becoming more important in the construction of Mechanical structures. The primary objective of this paper is to analyse the Carbon Fibre Reinforced Plastic (CFRP) and Aluminium alloy (AA 7075) aircraft wing box using finite element method. Both results has been correlated. Both CFRP and AA7075 structures modelling and numerical simulations were executed at various loading conditions. The deformed configuration of the CFRP and AA7075 aircraft wing box along with stress plot have been computed using analysis. Based on these studies, the CFRP and AA7075 central wing box, representing a weight saving of up to one and a half tonnes compared to the most advanced aluminium alloys. The results also expose the increasing strength with reduced weight.
Keywords. Aircraft wing box, Fabric Composite, Aluminium alloy (AA 7075), carbon/epoxy, FEM, ANSYS 15.0.

1. INTRODUCTION

Structural weight is the most important factor in aircraft manufacturing domain. From the total weight of aircraft wing box, 20% of weight had considered as payload, 40% of weight had considered as empty weight (Flight), remaining 40% of weight is considered as fuel weight. The main aim of an aircraft design engineer is to reduce the weight of the structures [1]. Different type of materials are used in aircraft manufacturing. The requirements of material for aircraft building are small weight, high specific strength, heat resistance, fatigue load resistance, crack resistance, corrosion resistance. Aluminium and its alloy are widely used in conventional subsonic aircraft. Currently a greater amount of composite is incorporated in to aircraft structures.

2. AIRCRAFT STRUCTURES

Wood was the original material of choice in the construction of early aircraft but it is no longer used in modern aircraft. The metallic and non- metallic were used, which have higher strength-to- weight ratios for lightweight and corrosion resistant structures. The metals used in the aircraft manufacturing industry include steel, aluminium, titanium and their alloys. Aluminium alloys were characterised by having lower density values compared to steel alloys (around one third), with good corrosion resistance properties.

In addition to metals, composite materials are currently used within the aircraft industry due to their strength [2], relatively low weight and corrosion resistance. Composites are created by the combination of different materials, which have been selected on the basis of their structural properties.

3. AIRCRAFT WING STRUCTURE

The wings of a modern aircraft can be designed as a combination of different types of materials, depending on their specific structural function. The structure of the wing (Shown in Fig. 3.1.) of an aircraft is comprised of several different elements, namely spars, skin and ribs, as well as control surfaces, such as ailerons and flaps [2]. Each of these components needs to support different loads and, thus, the right material needs to be selected. Steel and aluminium alloys can be used in the manufacture of ribs, whilst composite materials can be used in the design of the wing skin and the control surfaces.

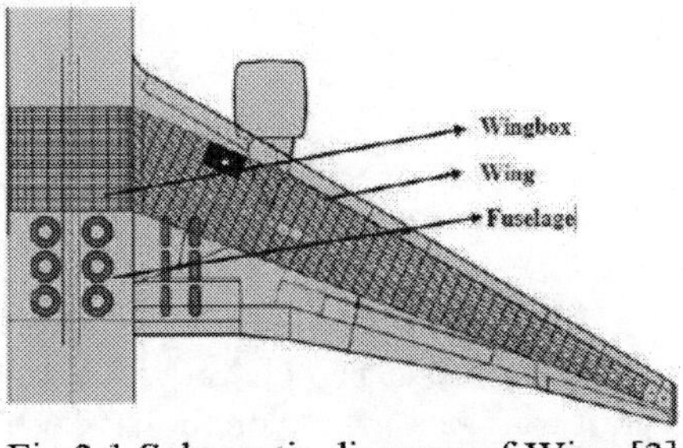

Fig 3.1 Schematic diagram of Wing [3]

4. MODELLING OF WINGBOX

Aircraft centre wing box is one of the essential part to join the both structures in aviation domain [2,3]. The centre wing box consists of connecting the main wings to fuselage, as well as the main landing gear door and the wing to body fairing.

The schematic sketch of the wing and wing box are represented in Fig 4.1. The wings are connected with centre wing box which is placed in the fuselage structure. The maximum stress value act on the wing box under bending load.

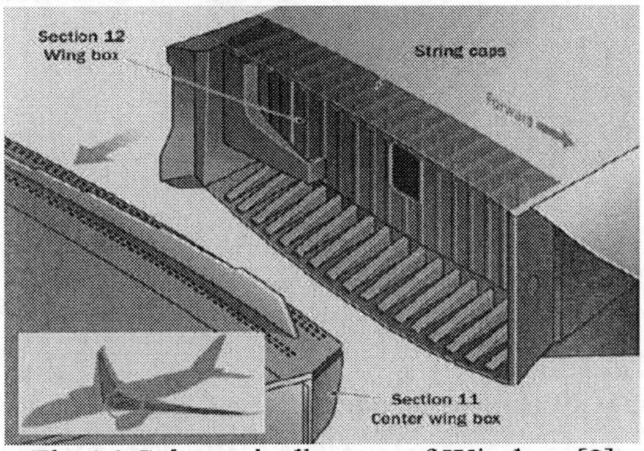

Fig 4.1 Schematic diagram of Wingbox [2]

(a) Solid model of wing box

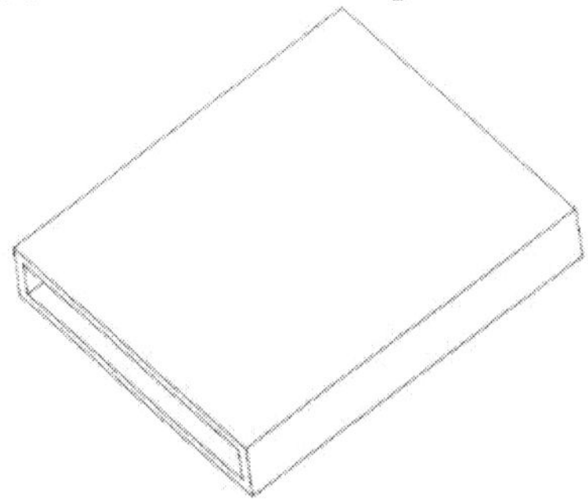

(b) Wireframe model of wing box

Fig 4.2 Modelling of Aircraft wing box

The solid modelling of the aircraft wing box [4] structures have been done using ProE® modelling software (Fig 4.2).

The structural configuration of the CFRP aircraft wing box are presented in Table 4.1.

Table 4.1 Structural Configuration of Aircraft wing box

SI.NO	Attribute	Specification
1	Aviation	
	Span length	60 m
	Length	57 m
	Height	170 m
2	Wingbox	
	Length	12m
	Width	5m
	Height	1m

5. FINITE ELEMENT ANALYSIS

The analysis and results evaluation of aircraft wingbox (CFRP & AA7075) has been analysed in the commercial available software ANSYS®15.0 [5,6]. The first step to start the analysis with the ANSYS is to select the type of analysis. The static structural analysis was selected for carrying out detail analysis of aircraft wingbox. The geometry of aircraft wingbox was created in ProE® software. After creating geometry material properties were applied to aircraft wingbox. The material properties of CFRP material [6] and Aluminium alloy 7075 were presented in Table 5.1. After creating geometry and applying material properties next step in ANSYS is to generate a meshing. The meshed FE model of aircraft wingbox has shown in Fig 5.1(a). After meshing the model, the boundary condition such as loads and constrains were imposed. The important factors to get accurate results is to apply the loads and boundary conditions on nods, on edges, on surface or elements [7,8].

Table 5.1 Material properties for FEA

SI.NO	Parameters	Values
1	Carbon/Epoxy	
	Density, ρ [g/cm3]	1.60
	E1 [GPa]	77
	E2 [GPa]	75
	E3 [GPa]	13.8
	υ12	0.06
	υ23	0.37
	υ13	0.50
	G12 [GPa]	6.5
	G23 [GPa]	4.1
	G13 [GPa]	5.1
2	Aluminium alloy (AA 7075)	
	E1 [GPa]	72
	Density, ρ [g/cm3]	2.850
	υ12	0.3

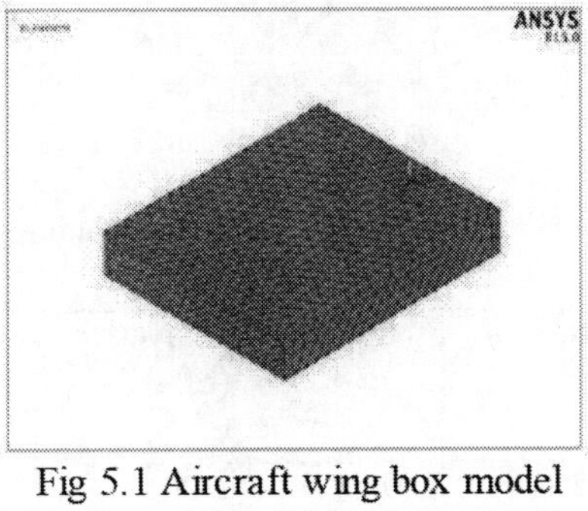

Fig 5.1 Aircraft wing box model

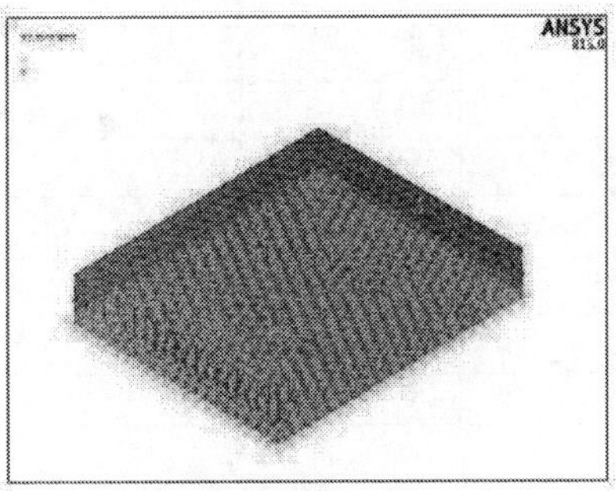

Fig 5.2 Loading and Boundary conditions

The boundary conditions and loads were shown in Fig 5.2. Final stage of analysis was running a solver to get the desired results.

A load were applied at both end of the wing box for actual wing loading conditions with an incremental load of 50 KN. The basement of the wing box has been arrested at all degree of freedom..

6. RESULTS AND DISCUSSION

In this study of finite element model the force has been applied at edges of the top surface and the bottom surface is kept fixed. The force applied at the aircraft wingbox from 100N to 500N. The displacement results of both CFRP and AA7075 were calculated for each different force.

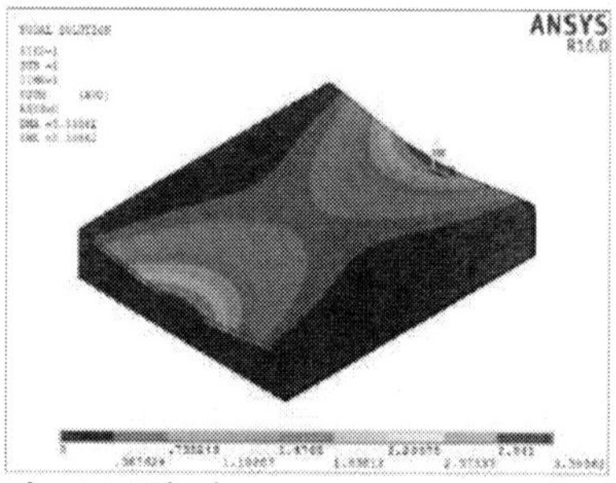

Fig 6.1 Displacement at 150 N load

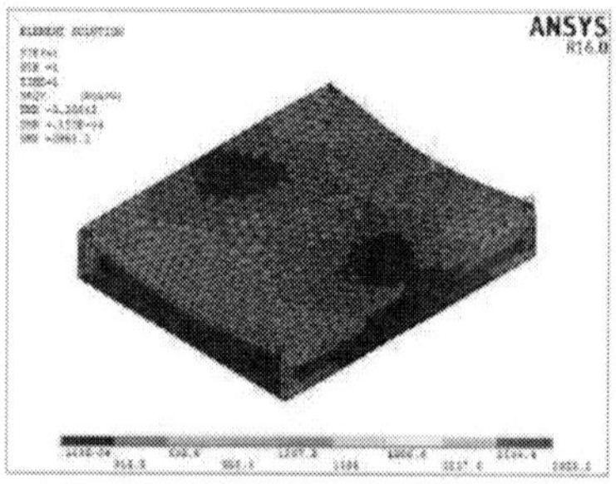

Fig 6.2 Stress at 150 N load

Table 6.1 FEA results of Aluminium alloy

SI.NO	Load(N)	Displacement (mm)
1	0	0
2	100	2.205
3	150	3.308
4	200	4.226
5	250	5.514
6	300	6.617
7	350	7.720
8	400	8.822
9	450	9.925
10	500	11.037

The displacement plot of Aluminium alloy (AA 7075) aircraft wingbox as shown in Fig 6.1 and Fig 6.2. The boundary conditions and loads were shown in Fig 5.2. Final stage of analysis was running a solver to get the desired results.

A load were applied at both end of the wing box for actual wing loading conditions with an incremental load of 50 KN. The basement of the wing box has been arrested at all degree of freedom. The displacement configuration of Aluminium alloy (AA 7075) results at various loading conditions have been mentioned in Table 6.1.

The displacement plot of CFRP aircraft wingbox as shown in Fig 6.3 and Fig 6.4.

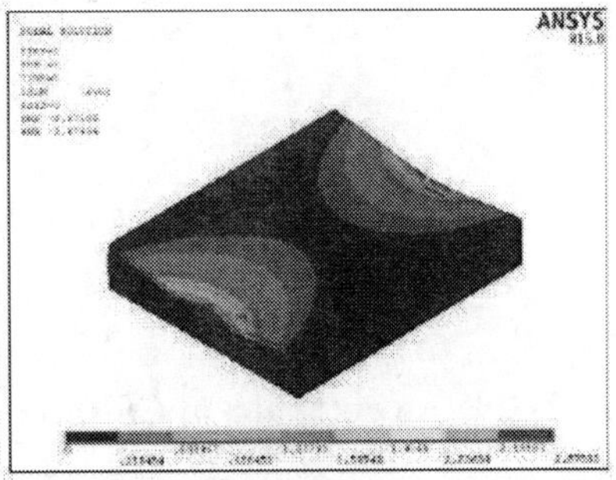

Fig 6.3 Displacement at 150 N load

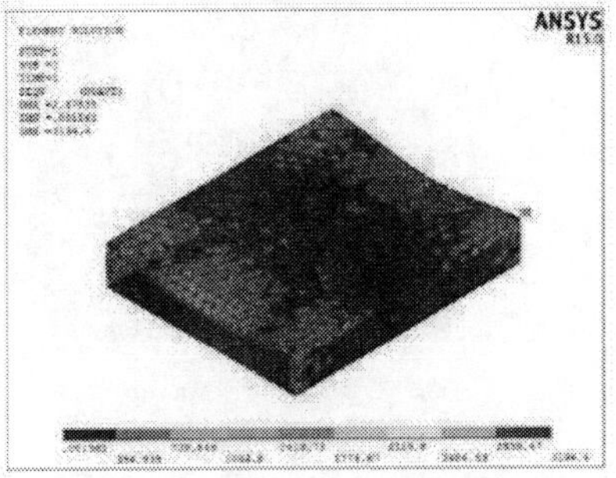

Fig 6.4 Stress at 150 N load

The displacement configuration of CFRP results at various loading conditions have been mentioned in Table 6.2.

Table 6.2 FEA results of CFRP

SI.NO	Load(N)	Displacement (mm)
1	0	0
2	100	1.916
3	150	2.870
4	200	3.833
5	250	4.800
6	300	5.750
7	350	6.700
8	400	7.660
9	450	8.626
10	500	9.163

The displacement results of CFRP and AA7075 aircraft wingbox were obtained from numerical analysis. Both numerical results were compared and plotted in the form of graph (Shown in Fig 6.5) using table 6.1 and 6.2.

IOP Conf. Series: Materials Science and Engineering 1126 (2021) 012081 doi:10.1088/1757-899X/1126/1/012081

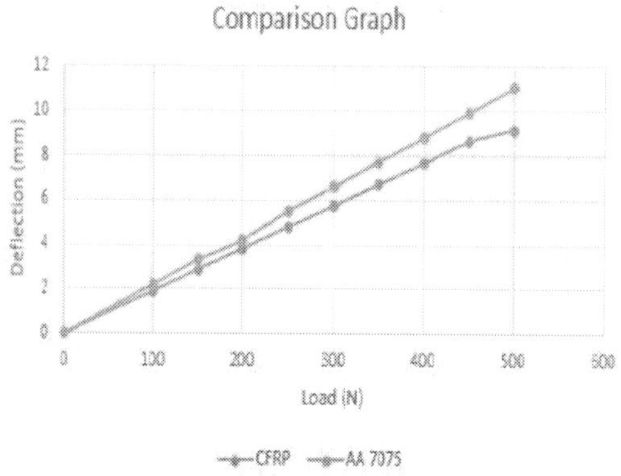

Fig 6.5 Displacement graph of CFRP & AA7075

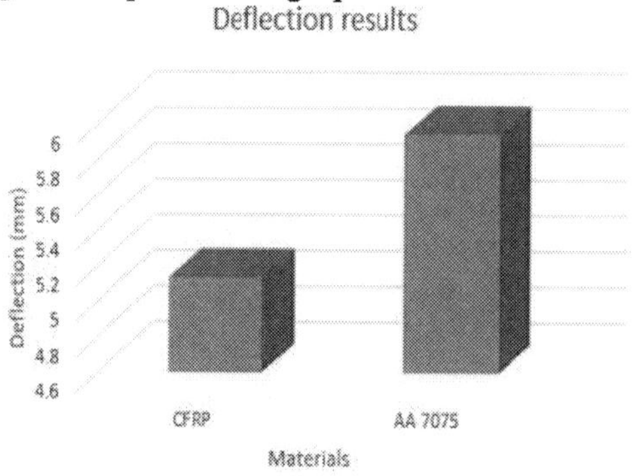

Fig 6.6 Comparison graph of CFRP & AA7075

The displacement comparison of CFRP and AA7075 results and plotted the maximum and minimum displacement results. The plotted chart is shown in Fig 6.6.

7. CONCLUSION

The static response of CFRP and AA7075 aircraft wingbox have been evaluated using ANSYS®15.0. The CFRP structure displacement results are less than the A7075 displacement results.

Upon comparison of both CFRP and AA7075 numerical results, it is seen that the Carbon/Epoxy structures exhibit more strength. The weight of the carbon fibre is less than the Aluminium alloy (AA 7075). So the strength to weight ratio of Carbon/Epoxy structure is more efficient and it is suitable for commercial aircraft structures.

APPENDIX

CFRP - Carbon Fibre Reinforced Plastic
AA7075 - Aluminium alloy (AA 7075)
ρ - Density [g/cm3]
E1 - Longitudinal modulus [GPa]
E2 – Transverse in plane modulus [GPa]
E3 – Transverse out of plane modulus [GPa]
G12 – In plane shear modulus [GPa]

G23 – Out of plane shear modulus [GPa]
G13 – Out of plane shear modulus [GPa]
υ12 – Major in plane Poisson's ratio
υ23 – Out of plane Poisson's ratio
υ13 – Out of plane Poisson's ratio

ACKNOWLEDGMENT

The author would like to thank the Department of Mechanical Engineering at Sri Krishna college of engineering and technology Coimbatore Tamilnadu, India for carrying out this research work.

References

[1] Maria Mrazova, "Advanced composite materials of the future in aerospace industry",INCAS Bulletin, Volume 5, (2013).
[2] W. Kang, I.S. Hwang, "Commercial transport aircraft composite wing box trade study".
[3] Immanuvel, Arulselvan, Maniiarasan, Senthilkumar, "Stress Analysis and Weight Optimization of a Wing Box Structure Subjected To Flight Loads", The International Journal Of Engineering And Science (IJES), 2014 33-40.
[4] G. Gruber, S. Wartzack, "Three-point bending analyses of short fiber reinforced thermoplastics: a comparison between simulation and test results", Volume 12, Issue 1, (2013).
[5] G.Rathnakar,H.K.Shivanan,"Experimental evaluation of strength and stiffness of fibre reinforced composites under flexural loading"(2013),Vol 2.
[6] Vasanthanathan Arunachalam, Dr. P. Nagaraj, "Correlation Study of IR TNDT Analysis with Structural Failure Modes of Carbon-Fabric-Reinforced Epoxy Composites", (2013).
[7] F. Hurlimann, R. Kelmb, M. Dugasb, G. Kressa, "Investigation of local load introduction methods in aircraft pre-design", Aerospace Science and Technology 21 (2012) 31–40.
[8] Chia-Chin Chiang, Vu Van Thuyet, Shih-Han Wang,Liren Tsai, "Analytical and Experimental Study of Residual Stresses in CFRP", Journal of Nanomaterials Volume 2013, Article ID 671469.
[9] R.Rajendran (2012) "Gas Turbine Coatings-An overview" in Engineering Failure Analysis, ,pp.355- 369.
[10] S.C.Sharma, N.M.Gokhale, Rajiv Dayal, Ramji Lal (2002) "Synthesis microstructure and mechanical properties of Ceria stabilized tetragonal Zirconia prepared by spray drying technique" in Material Science vol 25,No.1, pp. 15-20.
[11] Meng Han, Guodong Zhou, Jihua Huang, ShuHai Chen, (2013), " A Parametric study of double-ceramic layer thermal barrier coatings", in Surface and Coating Technologies, 236, pp.500-509.
[12] C.Zhang, K.Komeya, J.Tatami, T.Meguro, Y.B.Cheng (2000), " Synthesis of Mg-α SiAlON powders from talc and halolysite clay minerals", in Journal of European ceramic society 20, pp.1809-1814

Assessment of strength and durability parameters for concrete with partial replacement of coarse aggregates by iron slag and glass powder as an additive

Akula Prakash[1], A.N. Swaminathen[2], Akoju Ramu[3], Bypaneni Krishna Chaitanya[4]

[1]Assistant Professor, Civil Engineering Department, GRIET, Hyderabad, India. Corresponding author : akulprakash93@gmail.com
[2]Professor, Civil Engineering Department, Alamuri Ratnamala Institute of Engineering and Technology, Thane, Mumbai.
[3]Assistant Professor, Civil Engineering Department, Malla Reddy Engineering College, Hyderabad, India.
[4]Assistant Professor, Civil Engineering Department, RVR & JC College of Engineering, Guntur, Andhra Pradesh, India.

Abstract. Global warming and environmental destruction have become the major issue in recent years. Use of more and more environmentally friendly materials in any industry in general and construction industry, is of more importance. Preventing the depletion of natural resources and enhancing the usage of waste materials has become a challenge to scientists and engineers. Waste glass in one material when grounded to a very fine powder shows pozzolanic properties which can be used as additives, Iron slag rendered as scrap is used as partial replacement of aggregate in concrete. Iron slag is an industrial waste by-product of steel industry. The present investigation deals with the usage of Iron slag as a partial replacement of coarse aggregate and addition of glass powder to the concrete. The optimum percentage of iron slag to be replaced is found out by conducting 28 days compressive test on cubes of size 150mm*150mm*150mm with 5%, 10%, 15% replacement of coarse aggregates with iron slag. The glass powder is added with percentage of 5%, 10%. With the overall percentage of iron slag and glass powder the workability of concrete is checked and cubes are casted and tested for its compressive strength and durability parameters like saturated moisture content, effective porosity, sulphate attack and carbonation depth

1. Introduction

In civil engineering, concrete structures such as raft foundation, retaining walls and massive dam which are almost built by this material are very popular structural type. Such structures usually require large volumes of concrete. In the volume of concrete, the share of aggregates is typical in 70% and 80% which forms a major part of nearly 3 tons of concrete delivered each year for every human on the plant, making it the most manmade product in the world. The aggregates greatly influence different properties of concrete for example workability, dimensional, stability, strength, and durability.

It is the usual practice to use river and sand as fine aggregate and crushed granite, gravel and limestone of different sizes and shapes as coarse aggregate in the preparation of concrete. These materials are quarried from natural sources. Excessive utilization of such materials will result in depletion of natural resources as well as environmental degradation. Hence it is all more essential to search for alternatives to these materials. Towards this goal it is better to consider the wastes like iron slag, bottom ash, copper slag etc., generated by human activities. Use of such wastes as aggregates will to some extent will solve the problem of shortage aggregates in concrete making at several construction sites, and minimize the problem associated with aggregate mining and waste disposal. Even though some progress has been made in this direction, still, deeper research is required to evaluate the waste materials as aggregate in concrete. With this view in mind one of the wastes, namely iron slag scrap, was utilized as coarse aggregate in preparation of concrete.

The recycling and reuse of the waste has become the best other options and their removal issues of waste. The reuse of such waste will diminish the nature impact and is more economical. The energy required to reuse the recyclable material is not as much as that of virgin materials. The utilization of these waste products in construction industry is best option due to large numbers of construction site all over the world. Ordinarily glass does not hurt nature in any capacity since it doesn't emit contaminations, yet it can hurt just as creatures, if not managed cautiously and it is less benevolent to condition since it is non-biodegradable. Thus, the improvement of new advances has been required. The term glass contains several chemical diversities including soda- lime silicates glass, alkali-silicate glass, and borosilicate glass. Until this point, these kinds of glasses, glass powder have been generally utilized in cement and aggregate mixture as pozzolana for civil works. Presently the waste glass in and around the small shops is packed as a waste and arranged as landfill. Squander glass contain high silica (SiO_2)i.e., 72% waste glass when ground to fine powder reacts with alkalis in cement and cementitious product that help contribute to the strength development.

Khalid Raza et.al [1] in his study investigates that the compressive strength and flexural strength of M40 grade concrete with 0.45 water/cement ratio were investigated with coarse aggregate partially substituted with iron slag aggregate (ISA) at different proportions of 0 %, 10 %, 20 %, 30 %, 40 % and 50 % to evaluate and verify the compressive strength, flexural strength, and split tensile strength. J. Premalatha [8] et.al made the investigation on concrete properties with various % of glass powder varying from 0%, 5%, 10%, 15% and 20% were done on fresh and hardened concrete. Standard tests were used to examine the maximum stress-strain behaviour, mechanical strength, and toughness properties of concrete with partial replacement of natural river sand by glass powder. The optimum glass powder content was obtained by determining specimens produced with various concrete grades M20, M30, M40, M50, and M60.

2. Research Significance

The primary objective of this section is to assess the outcomes of distinct kinds and percentage of Iron Slag and Glass Powder proportioning on the mechanical properties of concrete such as Compressive and Flexural strength of reinforced concrete cube samples with M20 grade mix. According the Compressive strength of concrete of M20 Grade with water-binder ratio as 0.45 and mix proportion of 1:1.5:3. This study enhances to reduce the usage of natural resources like Coarse Aggregate by replacing it with alternately available materials or waste materials like Iron Slag. and to reduce the expense of construction and the material utilized for making concrete. It is also used for improving the the compressive strength of concrete by replacing Coarse Aggregate with Iron Slag and Glass Powder as an added substance to Concrete.

The main objectives of this present study to decide the ideal estimation of replacement of Iron Slag waste in ordinary concrete instead of aggregate to get the estimated workability and to decide the compressive strength of concrete at ideal replacement of waste Iron Slag and Extra Addition of Glass Powder as an added substance in ordinary concrete. Comparative investigation of workability and compressive strength of conventional concrete with concrete replaced with Iron Slag and Glass powder as an added substance.

3. Materials and Mix Proportioning

Materials utilized in this undertaking were OPC 53 Grade cement affirming to IS 8112, fine aggregate and coarse aggregates affirming to IS 383-1970. Designed concrete mix of M-20 grade having mix proportion 1:1.5:3 is taken with water cement ratio 0.5. Considering the above proportion the coarse aggregate is partially replaced by Iron slag and Glass powder is added to the concrete at various proportions (state Iron slag, Glass powder) 0%-0%, 5%-5%, 10%-5%, 15%-5%, 5%-10%,

10%-10%, 15%-10% were utilized in concrete. The concrete predominantly comprises of cement, coarse aggregate, and fine aggregate. In any case, as indicated by our project coarse aggregate is partially replaced by Iron slag and Glass powder is added to the concrete mix.

This study was conducted out using the Materials and Proportioning of M20 Mix with Portland Cement , widely available in India for concrete production, siliceous river sand with a fine aggregate having fineness modules of 2.6 and a proportional maximum size of 20 mm. In accordance with the Indian Standard IS 2386-1963, the coarse aggregate ingested cooperates with the Indian Standard IS 383-1970 code and is measured for its physical characteristics, its special gravity 2.75 as just an angular coarse aggregate. The free-state bulk density is 1554 kg/m^3 and the compressed state is 1594 kg/m^3. Potable water have been used in the investigations for mixing of specimens.

Calculation of Materials Proportioning of M20 Mix design

Volume of the concrete = 1 m3

Volume of cement = (Mass of the cement/specific gravity of the cement) x (1/1000)
= 320/(3.12)×1/1000 = 0.102m3

Volume of water = (Mass of water/specific gravity of water) x (1/1000)
= 197.16/1000 = 0.197m^3

Volume of all aggregates = 1 - [0.102 + 0.197] = 0.701m3

Mass of coarse aggregate = Abs. Volume x Vol. of CA x Sp. gravity of CA x 1000
= 0.701 x 0.56 x 2.65 x 1000 = 1040.28 kg

Mass of fine aggregate = Total Volume x Volume of FA x Sp.gravity of FA x 1000
= 0.701 x 0.44 x 2.65 x 1000 = 817.36 kg

For M20 grade of Concrete = 1:1.5:3

Volume of cement = 1/(1+1.5+3)*0.0243 = 4.418 x 10-3 m^3

Volume of fine aggregate (FA) = 1.5/(1+1.5+3)*0.0243 = 6.63 x 10-3 m^3

Volume of coarse aggregate (CA) = 3/(1+1.5+3)*0.0243 = 0.013 m^3

Mass of cement = density of cement x volume = 1440 x 4.418 x 10-3 = 6.36 kg

Mass of FA = density of fine aggregate x volume = 1680 x 6.63 x 10-3 = 11.53 kg

Mass of CA = density of coarse aggregate x volume = 1750 x 0.013 = 22.75 kg

Similarly Quantities for proportion when the coarse aggregate is partially replaced by Iron slag and Glass powder is added to the concrete at various proportions (state Iron slag, Glass powder) 0%-0%, 5%-5%, 10%-5%, 15%-5%, 5%-10%, 10%-10%, 15%-10% .

Table 1 : Quantities of Materials required for various Replacement Proportions

Proportions (Iron slag, Glass powder)	Wt. of Cement (kg)	Wt. of FA (kg)	Wt. of CA (kg)	Wt. of Iron Slag (kg)	Wt. of Glass Powder (kg)	Volume of Water (Litres)
Conventional	6.36	11.53	22.75	0	0	3.82
5%,5%	6.36	11.53	21.61	1.13	0.58	3.82
10%,5%	6.36	11.53	20.47	2.27	0.58	3.82
15%,5%	6.36	11.53	19.33	3.41	0.58	3.82
5%,10%	6.36	11.53	21.61	1.13	1.15	3.82
10%,10%	6.36	11.53	20.47	2.27	1.15	3.82
15%,10%	6.36	11.53	19.33	3.41	1.15	3.82

According to weights of the cement, fine aggregate and coarse aggregate are gauged. These materials are regularly blended in with the assistance of trowel in a tray, the mixture is moved to the pan mixer and permitted to blend for few moments. As per the w/c ratio the determined volume of water is added to the mixture and permitted to blend for few minutes to accomplish the consistency. At that point, the concrete mix is taken out in a tray and equitably blended in with the assistance of trowel.

Before casting the mould, oil is applied to the moulds, then the moulds are filled with concrete in 3 layers by giving 25 blows for each layer. Then the moulds are placed on the tabular vibrator for 5-10 minutes to keep away from voids or air bubbles. Then the concrete moulds are dried at room temperature 299K for 24hrs. After 24hrs moulds are demoulded and the concrete cubes are kept in restoring tank for 7days, 14days, 28days. For testing the cubes, the cubes are taken out from the restoring tank and left out for sun dry. At the point when the cubes are totally dried the cubes are prepared for testing.

Table 2 : Compressive strength for various Replacement Proportions

Proportions for Iron Slag and Glass powder	Compressive Strength (N/mm^2)									Target strength (N/mm^2)
	7 Days			*14 Days*			*28 Days*			
	S1	*S2*	*Average*	*S1*	*S2*	*Average*	*S1*	*S2*	*Average*	
Conventional	16.67	16.44	16.56	22.89	23.78	23.34	30.22	35.56	32.89	26.6
5%,5%	15.33	12.22	13.78	24.76	22.78	23.77	32.14	31.45	31.79	
10%,5%	10.44	17.55	14.00	23.78	23.12	23.45	34.56	34.24	34.31	
15%,5%	13.33	11.44	12.39	22.67	22.14	22.41	28.89	32.09	30.45	
5%,10%	15.76	11.67	13.72	23.33	21.45	22.39	26.11	31.54	28.85	
10%,10%	16.67	14.44	15.56	20.56	20.45	20.51	25.33	28.39	26.86	
15%,10%	10.44	12.67	11.56	21.22	22.12	21.67	25.56	26.78	26.17	

The outcomes got from the tests completed on 7, 14 and 28 days of curing were compared with the conventional concrete to decide the compressive strength. The strength characteristics of concrete mixtures have been figured by taking different percentages of Iron Slag as partial substitution of coarse aggregate and Glass powder as addition, and the partial replacement include 5%,10%,15% of coarse aggregate by Iron Slag and addition of Glass powder by 5%,10% respectively. Hence the following proportions are taken they are (5%, 5%), (10%, 5%), (15%, 5%), (5%, 10%), (10%, 10%), (15%, 10%).

The compressive strengths of various proportions have been compared with conventional concrete. From Table 3, At (5%, 5%) proportion, there is 26.93% increase after 14days, at (10%, 5%) proportion there is 18.41% increase, at (15%, 5%) proportion there is 10.62% increase, at (5%, 10%) proportion there is 13.43% increase, and at (10%, 10%) proportion there is 6.36% increase. By the test results the compressive strength tends to increase for all proportions. As the percentage of replacement of Iron slag and addition of Glass powder is increased the compressive strength is increased as shown in table 2.

Distinct mechanical properties, such as compressive strength and split tensile strength, must be investigated experimentally in different specimens. Within each mechanical characteristics, the size of the specimens has to be determined and analysed. The specimen descriptions and work scheme are shown in Table 2. The structural properties of the different mineral admixture replacements of specimens, such as 10%, 20%, 30%, 40%, and 50%, were investigated in this study. For 7 days, 14 days and 28 days, the respective samples were cast and examined.

4. Mechanical Behaviour of Alccofine based Self Compacting Concrete

Saturated moisture content, porosity, sulphate attack, carbonation depth, was moisture-curing with each combination after 14 and 28 days have shown in table 3. The Indian Standard mixtures were prepared (by weight) is used in mixes of traditional 1 : 1.5 : 3 after many trials for durability studies. For weight, the water/cement ratio were 0.45.

Table 3 : Durability parameters for Iron Slag and Glass powder replacements

Proportions for Iron Slag and Glass powder	Water Absorption (%)		Effective Porosity (%)		Penetration Depth (mm)		Average Loss in weight by Sulphate Attack (%)	
	14 days	28 days	14 days	28 days	14 days	28 days	14 days	28 days
Conventional	2.63	2.45	5.70	3.94	3.54	4.43	58	53

5%,5%	3.57	3.25	5.66	3.86	2.93	4.31	46	38
10%,5%	2.24	3.12	5.26	3.38	2.43	4.00	41	32
15%,5%	2.82	2.65	5.00	2.68	1.93	3.81	39	28
5%,10%	2.15	2.45	5.20	1.98	1.76	3.00	35	24
10%,10%	1.94	1.87	3.30	1.51	2.26	2.95	32	22
15%,10%	1.87	1.77	3.15	1.49	2.54	2.67	29	24

5. Conclusions

Based on the experimental the following conclusion is drawn within the limitation of test result. For different replacement, the fresh and hardened properties observed were good as compared to 5%,5%, 10%,5%, 15%,5%, 5%,10%, 10%,10% and 15%,10% Iron slag and glass powder ratios replacement. In view of results acquired, following conclusions can be drawn:

Based on test examination, it is discovered that Iron slag can be utilized as a partial replacement of coarse aggregate. Water required creating a similar workability stays same with the increase in the percentage of Iron slag and Glass powder. Usage of Iron slag and Glass powder in conventional concrete diminishes the expense of concrete because these are the waste materials from the industries, and they additionally reduce the issues of disposal and end up being environment friendly.

The greater compressive strength is increased up to 27% for 10% of Iron slag and 5% of Glass powder when compared to Conventional concrete. Meanwhile, it is noticed that the compressive strength is increased for all proportions from conventional concrete and Iron slag and glass powder ratios varying from 5%,5%, 10%,5%, 15%,5% 5%,10%, 10%,10% and 15%,10% The target strength is obtained at 14 days for all the proportions for some of the mixes which indicate the compressive strength have enhances by using this Iron slag and glass powder ratios in M20 grade of concrete. Thus, the development of concrete with iron slag and glass powder has given a new dimension in the disposal of waste by a cleaner technology. Also, using such waste reduces the strain in the extraction of natural aggregate from the earth.

Tests which conducted for the saturated moisture content, effective porosity, resistance to sulphate attack, carbonation depth, was moisture-curing with each combination after 14 and 28 days have been done. Due to these variation over the mixing proportioning, it is noticed that water absorption for 15% of Iron slag and 10 % of Glass powder lesser when compared to other mixes. Also, while finding out the acid resistance over the concrete, rate of loss in weight may increased in conventional concrete mix over other mixes. Effective porosity and carbonation depth tests for the concrete mixes have been increased for 14 days and decreases for 28 days in all mixes, which makes slight variation in the durability parameters.

6. References

[1] P. Vignesh Kumar, R. Ranjith Kumar "An Experimental Study on Partial Replacement of Coarse Aggregate by Iron slag with polypropylene fibre" *International Journal of Science and Research (IJSR)* ISSN (online):2319-7064, volume 5 issue 3, March 2016 paper ID:1031603

[2] Khalid Raza, Apoorv Singh, R.D. Patel "Strength Analysis of Concrete by Using Iron slag as a Partial Replacement of normal aggregate (coarse) in concrete" *International Journal of Science and Research (IJSR)* ISSN (online):2319-7064, volume 3 issue 10, October 2014,.

[3] IS 10262 : 2009 Concrete mix proportioning - guidelines, *Bureau of Indian Standards*, New Delhi, India

[4] T.S. Thandavamoorthy "Feasibility of making concrete with Iron slag scrap as coarse aggregate" *Construction Engineering* volume 4, 2016, doi:10.14355/ce.2016.04.001.

[5] Krati Gahoi, R. Kamal "Effect of waste glass powder on properties of concrete" *International Journal of Science and Research (IJSR)* ISSN (online):2319-7064, volume 5 issue 8, August 2016 paper ID: ART20161160.

[6] Ankur Meena, Randheer Singh "Comparative Study of waste glass powder as pozzalonic material in concrete" B.Tech thesis, National Institute of Technology, Rourkela, 2012.
[7] Veena V.Bhat, N.Bhavanishankar Rao "Influence of glass powder on the properties of concrete" *International Journal of Engineering Trends and Technology (IJEIT)* - vol 16 (5), October 2014.
[8] J. Premalatha, R.Srinivasan "Properties of concrete with waste glass powder as fine aggregate replacement" *International Journal of Recent Technology and Engineering (*IJRTE) ISSN: 2277-3878, Volume-8, Issue-2511, 2019.
[9] C Vivek Kumar, Patam Manisha, Pooja Sadula, Behavior of Monotonic Loading for Glass Fibre based High Performance Concrete in External Beam-Column Joint using ANSYS. Analysis, *E3S Web of Conferences*, 184 01088 (2020), DOI: 10.1051/e3sconf/202018401088, (2020).
[10] Keerthi Kumar B, Vignesh R, Srinithin S, Selva Kumar G, Ramesh Kumar A "Experimental study on partial replacement of coarse aggregate by Iron slag in cement concrete (M25)" *International Journal of ChemTech* Research ISSN:0974-4290, ISSN(online):2455-9555, vol.10 No-8, 2017.
[11] Anil Singh, Arjun Kumar, Sulekha and Harsimran Singh "Study of partial replacement of fine aggregate by Iron slag" *International Journal of Advanced Research*, ISSN: 2320-5407, volume 4, issue 7, 2016.
[12] IS 2386-1963 Part III, "Methods of Test for Aggregates for Concrete" *Bureau of Indian Standards* New Delhi.
[13] IS 516-1959, "Methods of Tests for Strength of Concrete", *Bureau of Indian Standards,* New Delhi.
[14] C Vivek Kumar, Mohammed Aiman Parvez, T Srinivas, Strength Behaviour of Blended Self Cured Concrete made with Cement by partial replacement of Supplementary Cementitious Materials (SCM's), *E3S Web of Conferences*, 184 01100 (2020), DOI:10.1051/e3sconf/ 202018401100, 2020.

Performance Evaluation of Durability and Flexural behaviour of Self Compacting Concrete blended with Alccofine

Vivek Kumar C[1], Palanisamy M[2], Sohang Debnath[3], Mahesh Munagala[4]

[1]Assistant Professor, Civil Engineering Department, GRIET, Hyderabad, India.
Corresponding author: vivekkumarcr@gmail.com
[2]Professor, Department of Civil Engineering, Balaji Institute of Technology and Science, Warangal, India.
[3]Assistant Professor, Civil Engineering Department, Malla Reddy Engineering College, Hyderabad, India.
[4]Assistant Professor, Civil Engineering Department, St. Peter's Engineering College, Hyderabad, India.

Abstract. Due to its superior performance over traditional concrete mixes, Self Compacting concrete (SCC) has increased in popularity in the construction industry in recent years. The major downside of using self-compacting concrete is that there is no universal design mix norm. The aim of this research is to determine the best mix proportioning and strength and durability parameters for M40 Alccofine based high strength self-compacted concrete. It often provides an exclusive application for self-compacting concrete since it can flow into any section of a heavily reinforced region without creating any vibration. Consequently, if using self-compacting concrete in beam-column joints does not adversely affect the frame's seismic efficiency, it could be used rather than traditional concrete. During earthquake excitation, the beam-column joints of reinforced concrete building frames play a very important role. As those are the highly congested areas of reinforced concrete framed buildings, effective concrete placement and compaction in these regions is extremely difficult. This study involves determining the impact of Alccofine with replacement levels from 10 to 50 percentages with cement is substituted for M40 grades with Viscosity Modifying Agent, with both the purpose of increasing compressive strength, lowering gas emissions, and making concrete more cost efficient. In specific, increasing the CaO volume of the Alccofine 1203 increases the binding capacity and compressive strength.

1. Introduction

Self-Compacting Concrete (SCC) is a revolutionary concrete that involves no placing and compaction while vibrating. Also, in the existence of congested reinforcement, it is free to flow through its own weight, completely filling the shapes and ensuring complete compaction. Self-compacting concrete, with faster construction times and easy to flow over reinforcement, requires a great rate of placing concrete. The incorporation to the concrete structure of Supplementary Cementitious Materials (SCM's) also including Flyash, Ground Granulated Furnace Slag, Alccofine and Silica Fume enhances its resilience, corrosion resistance, and lateral load carrying ability, thus minimising crack enhancement and dissemination at the same time. The key need for the development of self-compacting concrete is to develop a strong concrete structure that prevents from honey-combing and concrete surface irregularities. In its potential applications, such as nuclear industries where there are heavy dense reinforcements where traditional concrete does not fulfil its specifications, and a need for self-compacting concrete is very well recognised.

For self-compacting concrete, mineral admixtures have a benefit, such as improved workability with minimal cement content. As cement is the most costly part of concrete, an affordable alternative is to reduce the cement content. In addition, these pozzolanic materials can however strengthen the packaging of particles and decreases concrete permeability.

The adjunction of various supplementary cementitious materials to the concrete with cement emerging as the most appropriate binder is demonstrated by SCC dependent blended concrete. In contemporary time concrete practise, fly ash from power plants and metakaolin both have been critical. Increasing the level of SCMs such as Fly ash, Silica fume, Alccofine , Rice Husk Ash and Metakaolin in cement manufacturing and its consumption, guides to the conception of blended cements and concretes.

Viscosity Modifying Agent (VMA) makes the concrete more accommodating for changes in the mixture's water content and enhances the plastic viscosity and prevents segregation. Surface quality application can be achieved by viscosity modifying agent and the placement level could also be increased and also decreases segregation and bleeding rates. The main difference in the composition during manufacturing at the plant occurs in a well-proportioned SCC due to normal production at the plant changes in the moisture content of the sand and the coarse aggregates.

Variability of 1.5% moisture content were very common for the aggregate, leading to a change throughout the concrete mix of 10 to 15 liters /m3 of free water. This will result in significant differences in the concrete's flowing and cohesion properties being identified the same day or from one day to another from one batch to the next. By reorientating the mix design parameters for superior structural performance of concrete with thrust on controlling OPC content but increasing the whole cementitious material and controlling water content, consuming chemical admixture, for amended workability, for offsetting deliberate hydration with SCM's. Concrete incorporating SCM's will imply a new set of drift in the construction in the new millennium, due to the intensified use of performance built concrete.

2. Research Significance

The primary objective of this section is to assess the outcomes of distinct kinds and percentage of Alccofine 1203 on the mechanical properties of concrete such as Compressive and Flexural strength of reinforced concrete beam with M40 grade mix. According the Compressive strength of concrete of M30 Grade with water-binder ratio as 0.45 and mix proportion of 1 : 2.12 : 3.75 with addition of Glenium Stream 22 .

Studies have shown that the use of up to 10 percent of small percentages of superplasticisers in Self Curing Concrete including the use of appropriate amounts of high-range water to minimise admixture reduces the paste's viscosity, decreasing water demand and bleeding danger. The use of pozzolanic materials increases SCC's rheological properties and decreases concrete cracking due to decreased hydration heat and therefore can be used. In the development of conventional concrete, the addition of concrete additives and admixtures was the exception, although it is appropriate for SCC and their percentage at SCC is significantly greater.

The factors for improved SCC attribute efficiency to improved microstructure and homogeneity. Several tests, conducted by accurate microscopes including, seem to have an advanced microstructure of SCC compared to traditional concrete. Throughout the interfacial transition region, the void ratio of SCC between cement paste and aggregate was observed to be decreased and the pores were dispersed even more uniformly. For SCC mix design, there is no standard technique and several research organizations, admixture, ready-mixed, precast and contractual firms already established their own methods of mix proportioning. However, a higher proportion of ultrafine materials and also the incorporation of chemical admixtures were necessary to obtain the appropriate properties of fresh concrete throughout the SCC. An attempt has been made to performance characteristics of strength parameters of casting control samples with Glenium 33 incorporation as 1.0 percent in Blended Self Compacting concrete and M40 Grade concrete mix controlled Self Compacting concrete. Ordinary Portland Cement, Fine and Coarse Aggregate, Potable Water, Alccofine 1203 were the materials used in the current conclusions and identify the durability characteristics of SCM based Self Compacting Concrete.

3. Materials and Mix Proportioning

This study was conducted out using the Materials and Proportioning of M40 Blended Mix with Portland Cement and Alccofine, widely available in India for concrete production, siliceous river sand with a fine aggregate having fineness modules of 2.73 and a proportional maximum size of 20 mm. In accordance with the Indian Standard IS 2386-1963, the coarse aggregate ingested cooperates with the Indian Standard IS 383-1970 code and is measured for its physical characteristics, its special gravity 2.6 as just an angular coarse aggregate. The free-state bulk density is 1455 kg/m3 and the compressed state is 1529 kg/m3. Potable water have been used in the investigations for both mixing of SCM specimens.

Alccofine is more importantly resistant to the ingress of ions found in chloride and within concrete away with each other in the concrete mixtures, associated with decreased permeability having Silica content of 35.30%, Alumina as 21.4 % and Calcium Oxide content as 32.2 % was found and physical properties such as Specific Gravity as 2.70 and Bulk density as 680 kg/m^3

Consequently, the silica fume used has a 92.5 % SiO_2 content and was in fine and powdered form. In the present study, Glenium Stream 2 is a dominant liquid, organic, viscosity-modifying admixture (VMA) ready-to-use, specifically formulated for the development of concrete with strengthened viscosity and rheological regulated properties. The Glenium Stream 2 admixture comprising concrete possesses superior stability and regulated bleeding characteristics, thereby increasing segregation resistant and facilitating placement.

To validate the properties of the initial mix composition regarding the specified characteristics and groups, laboratory studies will be used Now that all the criteria have been met, the mixture should be checked on a full scale in the concrete plant and on site to check both fresh and hardened properties. By great selection and proportioning of the cement and additions, by limiting the Water/Powder ratio and then by applying a super plasticizer and viscosity modifying admixture, the fluidity and viscosity of the paste is modified and balanced. The method of achieving good filling capacity, passing ability and resistance to segregation is the regulation of SCC, its usability and interaction.

The paste is the mechanism for the transfer of the aggregate; thereby, the volume of the paste must be larger than the amount of the paste in the aggregate, such that the paste layer is sufficiently covered and lubricated by all the individual aggregate particles. This increases fluidity and decreases friction in aggregates. The coarse to fine aggregate ratio throughout the mix is minimized such that a layer of mortar is completely surrounded by individual coarse aggregate particles.

As the concrete enters through narrow openings or gaps throughout reinforcement and improves the SCC's passing capacity, this decreases aggregate interlock and structure. Although the specimen samples were cast of M40 grade blended concrete beams produced with optimal Alccofine replacement quantities. Tests for workability was obtained by Slump cone test and defined in Table 2.

Table 1. Mix Proportioning of Self Compacting Concrete for M40 Concrete

Materials	Cement	Fine Aggregate	Alccofine	Coarse Aggregate	Glenium Stream 2	Water	Water-Cement Ratio
Quantity	350	725.3	175	1125.5	3.5	190	0.4
Ratio	**1**	**2.07**	**0.5**	**3.21**	** bwp - by weight of powder		

Table 2. Fresh Properties of SCC Mixes

MIX	**Slump Flow (mm)**	**T50 (sec)**	**V- Funnel (Seconds)**	**U-Box (h2-h1) mm**	**L-Box (h2/h1) mm**
M CONTROL	550	5.5	12.05	30	0.82
Alccofine -10%	640	4.35	10.85	28	0.87
Alccofine -20%	655	3.75	9.25	25	0.92
Alccofine -30%	670	3.62	8.55	23	0.93
Alccofine -40%	655	3.55	8.34	24	0.93
Alccofine -50%	645	3.89	8.29	26	0.92

The Marsh cone test is a basic method for evaluating the rheological activity of cement pastes. To identify maximum super-plasticizer saturation point, whereas these dosage beyond which the flow time does not decline appreciably, this has already been used during cement-based materials design mix. Taking into the account the experiment used for the monitoring of post-tension grouts with just a 10 mm nozzle and a viscosity below 200 mPa, it can be shown that the flow time may not always be intimately proportional to the viscosity that use the past findings.

Table 3. Super Plasticizer vs Flow time

Ordinary Portland Cement (OPC 53 Grade)	
Super plasticizer (SP) dosage (% of cement mass)	**Flow time (in seconds)**
0.1	56.65
0.2	49.89
0.4	36.45
0.8	26.99
0.87	21.33
0.9	22.45
1	24.17

Distinct mechanical properties, such as compressive strength and split tensile strength, must be investigated experimentally in different specimens. Within each mechanical characteristics, the size of the specimens has to be determined and analysed. The SCC specimen descriptions and work scheme are shown in Table . The structural properties of the different mineral admixture replacements of specimens, such as 10%, 20%, 30%, 40%, and 50%, were investigated in this study. For 7 days, 28 days and 56 days, the respective samples were cast and examined. The functional necessity of a joint, which is also the beams and columns conjunction region, is to allow the adjacent members to build their ultimate ability and retain it. The demand on this finite-size factor is always strong, particularly during earthquakes. The joints must be solid and stiff enough to withstand the internal forces exerted by a framing members. Saturated moisture content, porosity, acid resistant, sulphate attack, carbonation depth, alkalinity assessment was moisture-curing with each combination after 28, 56 and 90 days. The Indian Standard mixtures were prepared (by weight) is used in mixes of traditional 1 : 2.07 : 0.5 : 2.82 after many trials for durability studies. For weight, the water/cement ratio were 0.45.

The aim of this durability study was to examine into drying and shrinkage, degradation, and water absorption for Alccofine -based self-compacting concrete in compliance with codal requirements. Saturated water absorption experiments were performed on 100 mm cube specimens at 28, 56, and 90 days after curing according to ASTM C 642. Figure displays the overall dimensions as well as the specifics of the beam-column joint reinforcement. The column were strengthened with four deformed high yield strength (HYSD) bars of 10 mm diameter and four HYSD bars of 10 mm diameter at the top and bottom were supplied to the beam. Transverse links in the columns and stirrups throughout the beams were made of 8 mm HYSD bars.

Table 4. Strength Parameters of Alccofine blended with SCC

MIX	Slump (mm)	Compressive Strength (MPa)			Split Tensile Strength (MPa)		
		7 days	28 days	56 days	7 days	28 days	56 days
Control Mix	550	18.48	48.21	54.21	1.99	3.4	5.41
Alccofine -10%	640	19.04	47.12	57.34	2.08	4.21	5.12
Alccofine -20%	655	20.79	51.16	53.16	1.97	4.18	4.98
Alccofine -30%	670	18.25	46.85	51.65	1.88	4.13	4.76
Alccofine -40%	655	17.28	45.21	49.21	1.76	4.03	4.58
Alccofine -50%	645	15.49	43.21	47.21	1.44	3.4	4.33

4. Analysis of Beam – Column Joint

Finite Element Methods are highly versatile and powerful, enabling designers to learn more about behaviour of complex structures under virtually any load. The obtained results must also be closely analysed before which can be used, considering the negligible advancements that have already been made in designing finite element packages. Four different replacements of ALCCOFINE as 20 % , 30 % , 40 % and 50 % were used. At 10 % and above 50 percentage of replacement of admixture, concrete mix became less workable, and hence, only the above mentioned mixes were considered in this study.

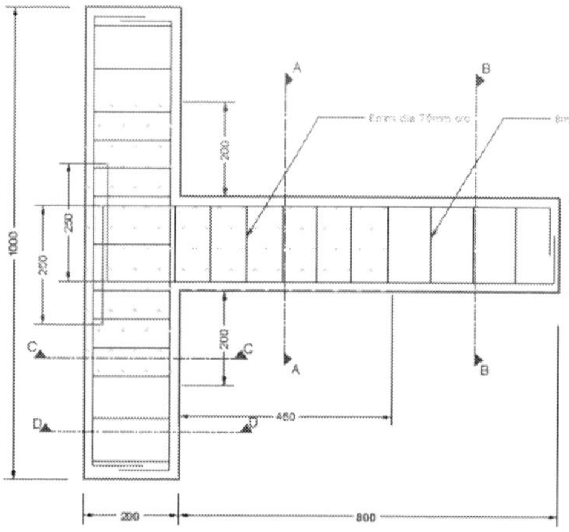

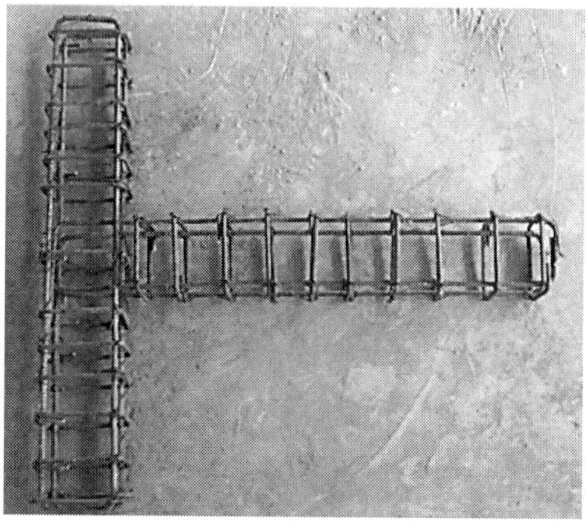

A total of five cast beam columns, one is a conventionally reinforced concrete beam column under the loading monotonic loading frame of 1000kN capacity, all beam columns were examined. With simply supported conditions under two-point loading, these beam-columns were checked. With LVDTs, deflections were assessed under the loading point and at the mid span. In reinforced concrete framed structures, beam-column joints are essential sub-assemblages that ensure structural contribution and move forces from one part to another. The specimens of the beam-column joint were subjected to monotonic loading at the beam tip. The aim of the loading pattern is to generate forces that replicate high levels of inelastic deformation during a serious earthquake.The testing results showed that a higher percentage of relatively fine powdered materials in the combination resulted in significantly greater value of slump (as seen in Fig) and self-compaction. The test results also confirmed that blends with a higher powder content and a lower coarse aggregate achieved a faster slump of 500 mm in less time in T 50 Slump test. The flow time was lowest for the mix containing less coarse aggregate and more fine powder materials, according to V-funnel test results. According to the findings of the L-box apparatus, the blocking was small for a mix containing minimal coarse aggregate, and the ratio was equal to 1.

5. Mechanical Behaviour of Alccofine based Self Compacting Concrete

The mechanical strength have been increased by 23 % when the control mix with 0% Alccofine has been compared to 10% cement which holds the greater strength parameters substituted by Alccofine after 28 days of curing among all the mixes , according to the experimental results. The results show which only a 10% substitution of cement with Alccofine material improves the strength. If the quantity of alccofine in the concrete is elevated above that amount, it functions as a filler and gives the concrete good workability. Concrete prices would be higher since the components are more costly than cement, but this can be changed during the prosecution of structures. Later, performing durability tests on the Alccofine content, it is suggested that it be used with cement.

Table 5a. Water Absorption, Porosity and Carbonation depth

ALCCOFINE (%)	Water Absorption (%)		Effective Porosity (%)		Penetration Depth (mm)	
	28 days	56 days	28days	56days	28days	56days
0%	3.63	3.45	6.70	6.43	58	51
10%	3.57	3.25	6.66	6.31	47	40
20%	3.24	3.12	6.26	6.00	42	34
30%	2.92	2.70	6.00	5.81	40	32
40%	2.45	2.41	5.20	5.00	37	30
50%	1.94	1.86	4.30	4.00	35	29

Table 5b. Alkalinity, Acid and Sulphate Resistance

ALCCOFINE (%)	Alkalinity P_H		Average Loss in weight by Acid Attack (%)		Average Loss in weight by Sulphate Attack (%)	
	28days	56days	28days	56days	28days	56days
0%	11.8	11.61	3.94	4.54	3.94	4.54
10%	11.0	10.57	3.86	3.93	3.86	3.93
20%	10.8	10.13	3.38	3.43	3.38	3.43
30%	10.4	9.94	2.68	2.93	2.68	2.93
40%	9.93	9.60	1.98	2.76	1.98	2.76
50%	9.79	9.25	1.51	2.26	1.51	2.26

Table 6 Load-Deflection reading for Control Mix.

	Deflection Control Mix		Deflection Alccofine 10%		Deflection Alccofine 20%		Deflection Alccofine 30%		Deflection Alccofine 40%		Deflection Alccofine 40%	
	End	Joint	End	End	End	Joint	Joint	Joint	End	Joint	End	Joint
0	0	0	0	0	0	0	0	0	0	0	0	0
5	0.86	0.15	0.60	0.42	0.56	0.10	0.07	0.10	0.56	0.10	0.68	0.12
10	1.23	0.21	1.39	1.06	1.02	0.17	0.18	0.24	1.02	0.17	1.42	0.24
15	2.08	0.35	2.30	2.02	2.00	0.34	0.34	0.40	2.00	0.34	2.26	0.38
20	3.48	0.60	2.69	2.82	2.42	0.41	0.92	0.45	2.42	0.41	2.36	0.72
25	5.17	0.88	3.10	3.00	3.00	0.51	1.09	0.53	3.00	0.51	3.38	0.87
30	6.93	1.18	4.90	4.00	3.88	0.66	1.28	0.84	3.88	0.66	4.36	0.94
35	7.61	1.30	6.53	7.12	5.42	0.92	1.41	1.11	5.42	0.92	5.18	1.19

The figure shows a typical load-deflection diagram through joining the peak points, the envelope curve is formed. On graphs, the loads and associated deflections for control mix, Alccofine 10%, Alccofine 20%, Alccofine 30%, Alccofine 40%, And Alccofine 50% specimens were mapped. These results were obtained by the load test performed on the corresponding samples.

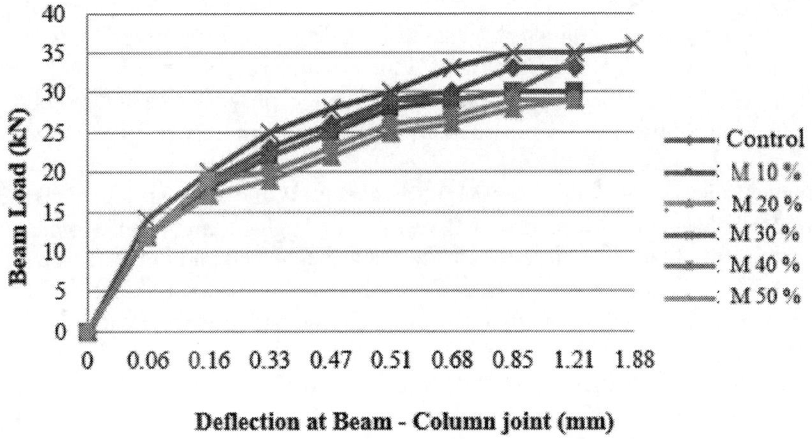

6. Conclusions

Based on the experimental the following conclusion is drawn within the limitation of test result. For 30% Alccofine replacement, the fresh properties observed were good as compared to 10%, 20%, 40% and 50% ALCCOFINE replacement. Hence if we increase the Alccofine replacement we can have a better workable concrete. The dosage of VMA should be properly designed as it may change the basic criterion of SCC. In other words, the flow ability may fall below 500 mm slump if the dosage of VMA is more than desired.

The locally available Viscosity Modifying Admixture (VMA) has a substantial influence on the fresh properties of SCC. A small change in VMA dose makes a substantial change in SCC properties i.e., flowing ability, passing ability, stability and segregation resistance. A sequential method of adjusting the mix proportions by increasing the fine aggregate content and decreasing the coarse aggregate content helps to obtain SCC mix without any segregation. Addition of Alccofine results in obtaining more viscous mix and also for 30% Alccofine replacement, the compressive strength observed higher as compared to 10% , 20% ,40% and 50% Alccofine replacement. The SCC mixes shows higher value of ultimate capacity of about 1.33 times than the control mix.

The SCC beam-column failed due to crushing of concrete with an explosive sound. The load carrying capacity of the beam-column increases up to 30% replacement of Alccofine mix, then decreases, from all the mixes, maximum ultimate load was obtained for Alccofine -30%. Compared to control beam-column joint, increase in first crack load was observed for beam-column joint with 20% and 30 % Alccofine respectively. Increasing the replacement rates of mineral admixtures have decreased compressive strength and increase both split tensile strength.

7. Discussions

[1] Darshan H R, M.Rame Gowda (2014) ‘Development and study of behavior of self - compacting concrete using GGBS’, International Journal of Advanced Technology in Engineering and Science, Volume No.02, Issue No. 07, July 2014.
[2] S.P.Sangeetha, P.S Joanna (2014) ‘Flexural behaviour of reinforced concrete beams with partial replacement of GGBS’, American Journal of Engineering Research (AJER) Volume - 03, Issue-01, Pages 119-127.
[3] M.Adams Joe, A.Maria Rajesh (2015) ‘Study on the effect of GGBS & M sand in self-compacting concrete’, The International Journal Of Engineering And Science (IJES), Volume 4, Issue 8, Pages 12-18.
[4] IS 383 : 1970 Specification for coarse and fine aggregates from natural sources for concrete, Bureau of Indian Standards
[5] EFNARC (European Federation of national trade associations representing producers and applicators of specialist building products)
[6] IS 10262 : 2009 Concrete mix proportioning - guidelines, Bureau of Indian Standards, New Delhi, India
[7] IS 13920:1993, “Indian Standard Ductile Detailing of Reinforced Concrete Structures subjected to Seismic forces”, Bureau of Indian Standards, New Delhi, India
[8] Mucteba Uysal, Mansur Sumer, (2011). “Performance of self-compacting concrete containing different mineral admixtures” Construction and Building materials Vol. 25 No. 11 pp. 4112-4120.
[9] Mallikarjuna Reddy V, Seshagiri Rao M V, Srilakshmi P and Sateesh Kumar B, “Effect of W/C Ratio on Workability and Me-chanical Properties of High Strength Self Compacting Con-crete (M70 Grade)” International Journal of Engineering Re-search and Development.
[10] Aravinth S. N – “Development of High Strength Self Compacting Concrete using Mineral and Chemical Admixture” – Journal of Civil Engineering and Environmental Technology, Vol. 1, Number 4, pp. 8 – 13.
[11] V Srinivasa Reddy, R Nirmala (2014) "Development of quaternary blended high performance concrete made with high reactivity metakaolin", International Journal of Engineering & Technology, ISSN: 2227-524X, 7 (2.1) (2018) pp. 79-83.

[12] B. H. Venkataram Pai, Maitreyee Nandy, A. Krishnamoorthy, Pradip Kumar Sarkar,C. Pramukh Ganapathy, Philip George, "Development of self compacting concrete with various mineral admixtures", American Journal of Civil Engineering, Vol. 2, No. 3, May 30, 2014.

IOP Conf. Series: Materials Science and Engineering 1126 (2021) 012084 doi:10.1088/1757-899X/1126/1/012084

SLRA and MLRA Based Correlations of Geotechnical Parameters of Fine Grained Soils

C Lavanya[1] **and Nandyala Darga Kumar**[2]

[1] Professor, Department of Civil Engineering, GRIET, Hyderabad-500090
[2] Assistant Professor & Head Department of Civil Engineering, JNTUHCE Manthani, Peddpally, Telangana

Abstract. Due to huge expenditure is involved in testing of soil by employing heavy machinery and costly equipment; it is necessitated to develop statistical correlations between various geotechnical parameters which can be used mostly in design of foundations and pavement constructions. Many techniques are developed for carrying out regression analysis to establish the relationships between response variable and explanatory variables as per the practice are concerned. Single linear regression analysis (SLRA) and multiple linear regression analysis (MLRA) are the modeling techniques commonly adopted to develop correlations between the required variables. The MLRA model creates a relationship in the form of linear line that best approximates all the individual data points. The MLRA is a statistical technique that uses several explanatory variables to predict the outcome of a response variable. This paper presents the relationships established between geotechnical parameters of fine grained soils (% of soil passing 0.075mm BIS sieve > 50%) using SLRA and MLRA technique. The geotechnical parameters of fine grained soils such as California Bearing Ratio (CBR), Liquid limit (LL), Plastic limit (PL), Plasticity Index (PI), Optimum Moisture Content (OMC), Maximum Dry Density (MDD), Free swell index (FSI) and % fine fraction (FF) are considered from the soil testing carried out on natural soil samples collected at different stretches of a road project from Khandi to Shadnagar in Telangana state. The SLRA and MLRA showed good fits between the geotechnical parameters of fine grained soils with regression coefficients showing above 0.80. The MLRA models developed can be utilized to estimate the CBR of fine grained soils whose % fines more than 50%.

Keywords: Fine grained soil, CBR, PI, OMC, MLRA.

1 Introduction

The suitability of soil is usually evaluated before start of construction activity on it. Geographical variability in soil conditions from one location to another makes it carry out investigation for its suitability to the intended construction [1]. In geotechnical engineering design, the most essential aspect that requires to be considered involves the soil moisture content, cohesion, friction angle, unit weight, saturation degree, porosity, plasticity index and grain size distribution. These properties are important elements especially in determination of CBR, bearing capacity and slope stability in soil. Soil investigation is laborious, expensive and invasive. Therefore, an alternate quick, non-destructive and environment friendly assessment of soil properties is very crucial. Interpretations of geotechnical data will involve a degree of uncertainty. It is more difficult to acquire a reliable set of representative and undisturbed samples of the various soil strata. Geotechnical investigations are expensive and time consuming activities. Soaked CBR value of a soil sample takes about a week, making CBR test expensive, time consuming and laborious. All these problems may result in delay in the progress of the project and lead to escalation of the project cost. To overcome these difficulties, it is imperative to predict CBR value of subgrade soil with index properties.

In order to cover the site investigation of larger area in a reduced time and cost, both geophysical and geotechnical methods appropriately conducted and the results obtained would be verified for the entire subsurface characterization. There are correlations available with high regression coefficient between P-wave values and engineering parameters of a ground such as the SPT-N values, rock quality, friction angle, velocity index, and density and penetration strength [2, 3]. The design and performance of a flexible pavement depend on the strength of sub-grade material such as California Bearing Ratio [4]. The use of the seismic piezocone test in geotechnical site investigation offers field assessment on stratigraphy and soil behavior and it is more convenient compared to traditional investigation methods [5]. Strength characteristics of the clay were correlated with various index properties [6]. Good agreement was obtained between samples and the CPT data using the new normalized charts [7]. One can use the relationships which were established between normalized shear wave velocity, normalized tip resistance, and mean grain size [8]. Multivariate regression model can be used to predict soil expansion index [9]. Based on the laboratory and field studies by employing CPT, SPT and seismic up-hole and down-hole tests, improved correlations were established for all types of penetrable soils [10]. Soil investigation is made easy by correlating the resistivity values with soil properties [11].

Empirical correlations play a key role in geotechnical engineering designs and analysis. By using appropriate empirical correlations, it is possible to derive many design parameters, thus limiting our reliance on these soil tests [12]. The relationship between CBR and water content is commonly presented in analyses [13]. Correlations were well agreed between California bearing ratio and the unconfined compressive strengths of sand- cement samples [14]. Based on the limited tests conducted, the correlations were established between CBR and soil index properties [15].

From the discussion, the correlations available between soil parameters are based on the limited testing in the laboratory and no correlations are established with respect to the % of fines (soil portions passing through 75micron sieve). In this paper, the correlations are developed between % fines and soil index properties and with compaction characteristics. The variations of CBR with PI, OMC and MDD are presented. Further the multi linear regression analysis is presented for CBR from other index properties of soil. A combination of index properties of soil obtained alongside the highway stretch of 55.94 km is used to develop statistical correlations.

2 Experimental Investigation

In order to have enough and reliable data for the target analysis, laboratory tests were conducted on soil samples obtained from different localities of Kandi-Shadnagar Highway project site in Telangana State, India. A total of three hundred disturbed samples were collected within reasonable sampling intervals (approximately 250m). Soils were collected from the respective localities at a depth of about 0.5 to 1m from the ground level after removing all the vegetation matter and were subjected to air dry as well pulverization. Afterwards the soil samples were placed in airtight bags and stored in the laboratory. From these samples again 132 samples whose % fines more than 50% were taken to establish correlations.

2.1 Index Property Tests

In nature soil occurs in a large variety. Engineers are continually searching for simplified tests that will increase the in knowledge of soils by employing a simple and rapid soil tests. These simplified tests which are indicative of the engineering properties of soils are called index properties. Index properties of cohesive soils are used to characterize the physical and mechanical behavior of soils by making use of parameters such as moisture content, particle size distribution, Atterberg limits and moisture density relationships. Such parameters are useful to classify cohesive soil sand provide correlations with engineering soil properties.

2.2 Tests Conducted

The soil samples which were pulverized using smooth wooden hammers were stored in airtight bags under the controlled temperature in the laboratory. The laboratory tests were conducted on 132 soil samples whose % fines more than 50. All the tests mentioned in the Table 1 were conducted as per the standard test procedures given in the respective code of practice of concerned test.

Table 1. List of Tests Conducted

Name of the test	Indian Standard Test Code of Practice
Liquid Limit and Plastic Limit Test	IS: 2720 (Part V) – 1985
Heavy Compaction Test	IS: 2720 (Part VIII) – 1983
Shrinkage limit Test	IS: 2720 (Part VI) – 1972
California Bearing Ratio Test	IS: 2720 (Part XVI) – 1987
Free Swell Index Test	IS: 2720 (Part XL) – 1977
Sieve Analysis	IS: 2720 (Part IV) - 1985

3 Results and Discussion

The obtained laboratory results pertinent to index properties (LL, PI and FSI), CBR and compaction characteristics such as OMC and MDD were analyzed using statistical analysis such as single linear regression analysis (SLRA) and multi linear regression analysis (MLRA). The discussion of results and correlations are presented in the following sections.

3.1 Single Linear Regression Analysis (SLRA)

Fig.1 presents the variation of free swell index (FSI) with the % fines. From this figure, it is noticed that as the % fines increases the FSI is increasing and for the soil samples of % fines from 80 to 100, the range of FSI noticed is 60 to 80%. The correlation between % fines and FSI is presented in Eqn.1. It has regression coefficient, $R^2 = 0.821$.

$$FSI = 0.72(FF) + 1.822 \quad (1)$$

FSI and FF are in %.

Fig.2 presents the variation of LL with the % fines. From this figure, it is noticed that as the % fines increases the LL is increasing and for the soil samples of % fines from 80 to 100, the range of liquid limit is 60 to 80%. The correlation between % fines and LL is presented in Eqn.2. It has regression coefficient, $R^2 = 0.822$.

$$LL= 0.465(FF) + 20.12 \quad (2)$$

LL and FF are in %.

Fig.3 presents the variation of PI with the % fines. From this figure, it is noticed that as the % fines increases the PI is increasing and for the soil samples of % fines from 80 to 100, the range of PI is 25 to 30%. This indicates that the soil samples whose % fines more than 80% are having PI more than 25 and these can be considered as high plastic clays. The correlation between % fines and PI is presented in Eqn.3. It has regression coefficient, $R^2 = 0.892$.

$$PI = 0.216(FF) + 8.108 \quad (3)$$

PI and FF are in %.

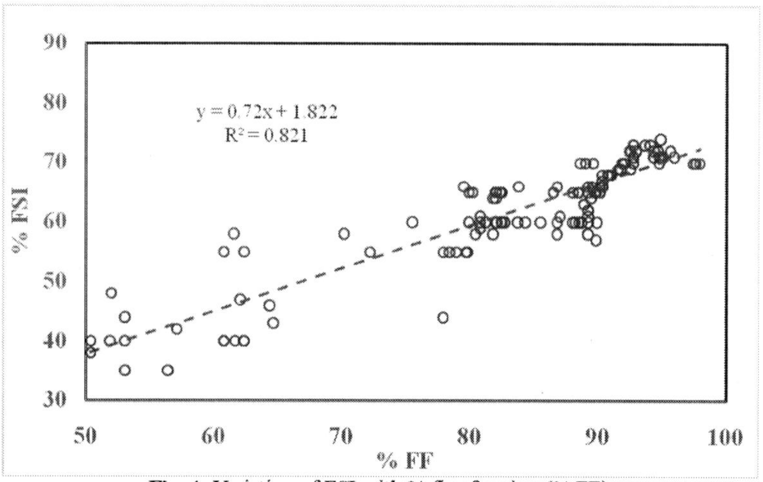

Fig. 1. Variation of FSI with % fine fraction (% FF)

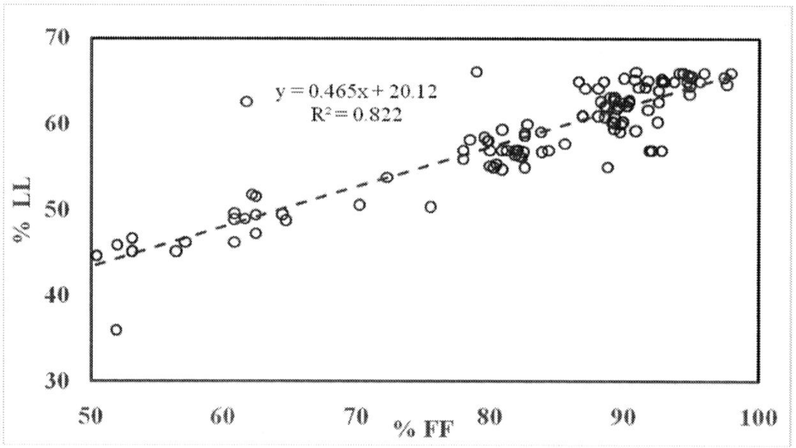

Fig. 2. Variation of LL with % fine fraction (% FF)

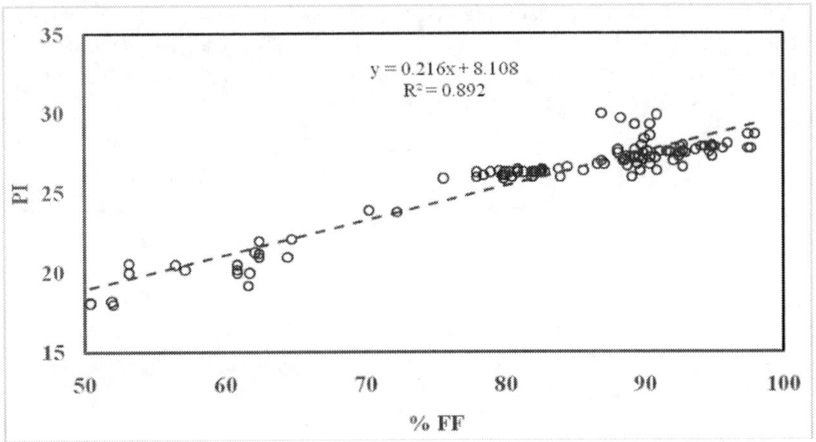

Fig. 3. Variation of PI with % fine fraction (% FF)

Fig.4 presents the variation of MDD with the % fines. From this figure, it is noticed that as the % fines increases the MDD is decreasing and for the soil samples of % fines from 80 to 100, the range of MDD is 15 to 12 kN/m^3. The correlation between % fines and MDD is presented in Eqn.4. It has regression coefficient, R^2 = 0.802.

$$MDD = -\ 0.18(FF) + 29.77 \tag{4}$$

MDD in kN/m^3 and FF in %.

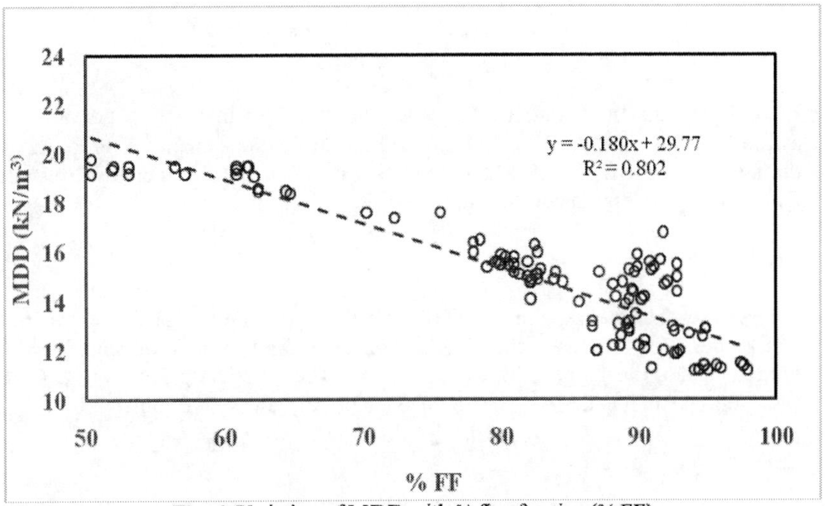

Fig. 4. Variation of MDD with % fine fraction (% FF)

Fig.5 presents the variation of OMC with the % fines. From this figure, it is noticed that as the % fines increases the OMC is increasing and for the soil samples of % fines from 80 to 100, the range of OMC is 20 to 26%. The correlation between % fines and OMC is presented in Eqn.5. It has regression coefficient, $R^2 = 0.846$.

$$OMC = 0.262(FF) - 1.620 \quad (5)$$

OMC and FF are in %.

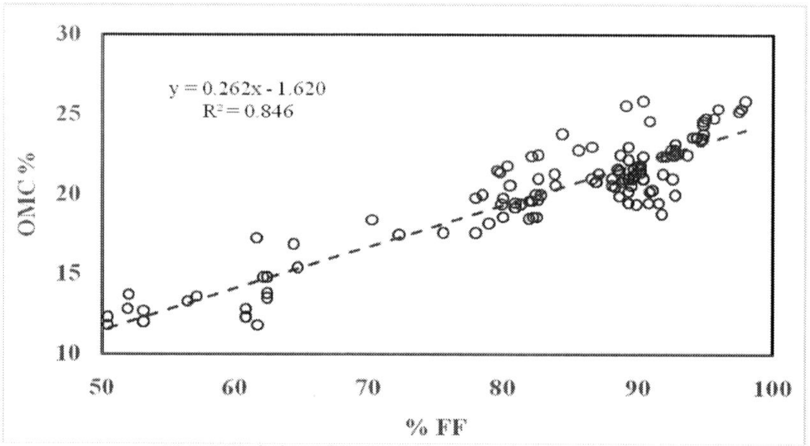

Fig. 5. Variation of OMC with % fine fraction (% FF)

Fig.6 presents the variation of CBR with the % fines. From this figure, it is noticed that as the % fines increases the CBR is decreasing and for the soil samples of % fines from 80 to 100, the range of CBR is 3 to 1%. The correlation between % fines and CBR is presented in Eqn.6. It has regression coefficient, $R^2 = 0.838$.

$$CBR = - 0.141(FF) + 14.94 \quad (6)$$

CBR and FF are in %.

Fig.7 presents the variation of CBR with the PI. From this figure, it is noticed that as the PI increases the CBR is decreasing and for the soil samples of PI from 26 to 30, the range of CBR is 3 to 1%. The correlation between % CBR and PI is presented in Eqn.7. It has regression coefficient, $R^2 = 0.735$.

$$CBR = - 0.578(PI) + 18.28 \quad (7)$$

CBR and PI are in %.

Fig.8 presents the variation of CBR with the OMC. From this figure, it is noticed that as the OMC increases the CBR is decreasing and for the soil samples of OMC from 20 to 26, the range of CBR is 3 to 1%. The correlation between % CBR and OMC is presented in Eqn.8. It has regression coefficient, $R^2 = 0.736$.

$$CBR = - 0.465(OMC) + 12.57 \quad (8)$$

CBR and OMC are in %.

Fig.9 presents the variation of CBR with the MDD. From this figure, it is noticed that as the MDD increases the CBR is increasing and for the soil samples of MDD from 11 to 14 kN/m^3, the range of CBR is 1 to 3%. The correlation between % CBR and MDD is presented in Eqn.9. It has regression coefficient, $R^2 = 0.806$.

$$CBR = 0.688(MDD) - 6.985 \quad (9)$$

CBR in % and MDD in kN/m^3.

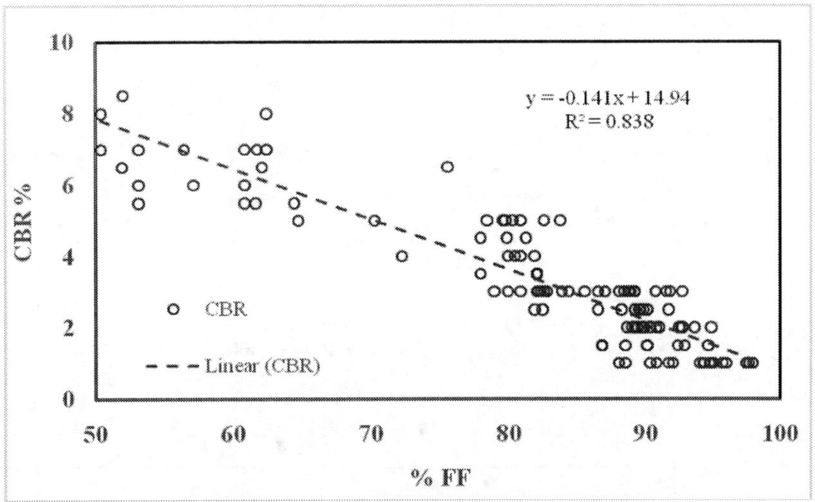

Fig. 6. Variation of CBR with % fine fraction (% FF)

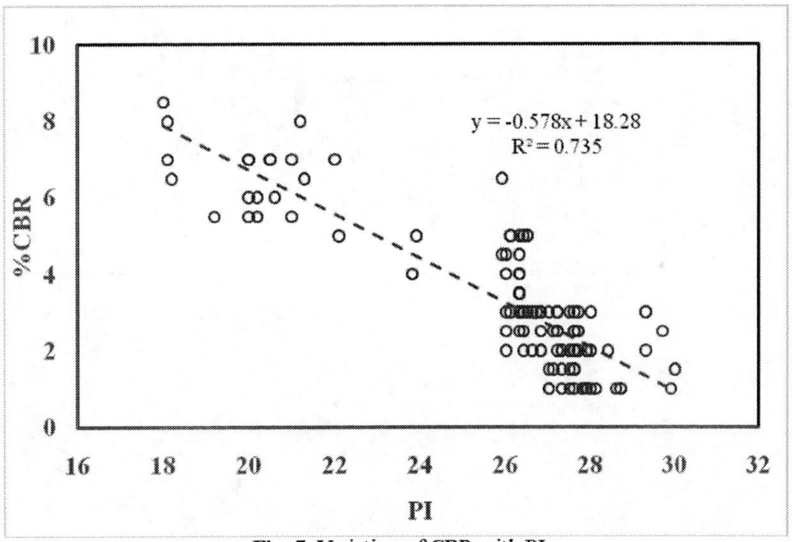

Fig. 7. Variation of CBR with PI

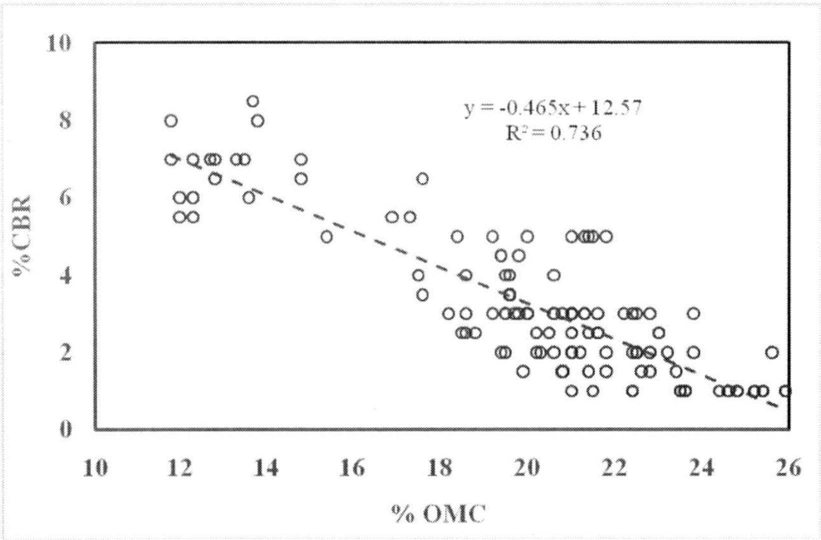

Fig. 8. Variation of CBR with OMC

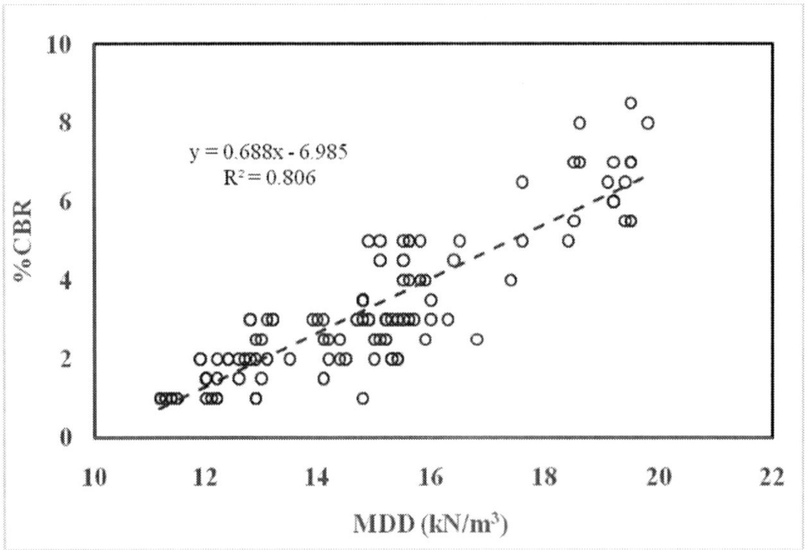

Fig. 9. Variation of CBR with MDD

IOP Conf. Series: Materials Science and Engineering 1126 (2021) 012084 doi:10.1088/1757-899X/1126/1/012084

3.2 Multiple Linear Regression Analysis (MLRA) Models for CBR

The goal of multiple linear regression (MLR) is to model the linear relationship between the explanatory (independent) variables and response (dependent) variable. Multiple regression is the extension of ordinary least-squares (OLS) regression that involves more than one explanatory variable. Multiple linear regression attempts to model the relationship between two or more explanatory variables and a response variable by fitting a linear equation to observed data. In words, the model is expressed as:

DATA = FIT + Residual
where the "FIT" term represents the expression $_0 + {}_1x_1 + {}_2x_2 + \ldots {}_p. x_p$.

3.2.1 MLRA Model for CBR with OMC, PI and FF

Eqn.10 presents the MLRA model for CBR with OMC, PI and FF. CBR is the response variable and OMC, FF and PI are the explanatory variables. From this equation it is noticed that the CBR is more sensitive with the % fines (FF) than the other two response variables PI and OMC. The corresponding R^2 as attained was 0.84.

$$CBR = -\ 0.055\ OMC + 0.041\ PI - 0.136\ FF + 14.52 \quad (10)$$

3.2.2 MLRA Model for CBR with LL, PL, PI, MDD and OMC

Eqn.11 presents the MLRA model for CBR with LL, PL, PI, MDD and OMC. CBR is the response variable and LL, PL, PI, OMC and MDD are the explanatory variables. From this equation it is noticed that the CBR is more sensitive with the MDD and not at all responsive with PI. The corresponding R^2 as attained was 0.84.

$$CBR = -\ 0.172\ LL + 0.128\ PL + (0)\ PI + 0.363\ MDD - 0.080\ OMC + 5.404 \quad (11)$$

3.2.3 MLRA Model for CBR with LL, PL and PI

Eqn.12 presents the MLRA model for CBR with LL, PL and PI. CBR is the response variable and LL, PL and PI are the explanatory variables. From this equation it is noticed that the CBR is more sensitive with the PL and with PI no response. The corresponding R^2 as attained was 0.77.

$$CBR = -\ 0.473LL + 0.358PL + (0)\ PI + 19.27 \quad (12)$$

3.2.4 MLRA Model for CBR with PI and FF

Eqn.13 presents the MLRA model for CBR with PI and FF. CBR is the response variable and PI and FF are the explanatory variables. From this equation it is noticed that the CBR is more sensitive with the FF. The corresponding R^2 as attained was 0.84.

$$CBR = 0.045\ PI - 0.151\ FF + 14.58 \quad (13)$$

3.2.5 MLRA Model for CBR with MDD, OMC and PI

Eqn.14 presents the MLRA model for CBR with MDD, OMC and PI. CBR is the response variable and MDD, OMC and PI are the explanatory variables. From this equation it is noticed that the CBR is more sensitive with the MDD. The corresponding R^2 as attained was 0.83.

$$CBR = 0.404\ MDD - 0.089\ OMC - 0.172\ PI + 3.51 \quad (14)$$

4 Conclusions

The SLR and MLR analysis carried out between the index properties, compaction characteristics and CBR showed good correlations. The regression coefficient, R^2 values for most of the cases is found to be above 0.8. The % fines (FF) obtained from the grain size analysis can be well utilized to ascertain the FSI, LL, PI and also MDD and OMC of fine grained soils, whose % fines more than 50%. From the MLR analysis, the equations established showed that the CBR is sensitive with the MDD and FF. The statistical correlations as established above from the MLR analysis can be utilized to obtain CBR for fine grained soil sub grades.

References

1. Charman, J H.: Laterite in road pavements. Construction Industry Research and Information Association, London (1988).
2. Bery, A.A. & Saad, R.: Correlation of Seismic P-wave Velocities with Engineering Parameters (N Value and Rock Quality) for Tropical Environmental Study, International Journal of Geosciences, 3, pp. 749- 757 (2012).
3. Akintorinwa, O.J and Adesoji, J.I: Application of Geophysical and geotechnical investigations in Engineering site Evaluation. International Journal of physical sciences, Vol. 4 (8). pp 443-454 (2009).
4. Rakaraddi, P. G., and Gomarsi, V.: Establishing Relationship between CBR with different Soil Properties. International Journal of Research in Engineering and Technology, 4 (2): 182 – 188 (2015).
5. Songyu Liu, Anand J. Puppala: Evaluation of geotechnical parameters of a lagoonal clay deposit in Jiangsu Lixia River area of China by seismic piezocone tests, KSCE Journal of Civil Engineering, July 2016, Volume 20, Issue 5, pp 1769–1782 (2016).
6. Springman, S. S., Trausch, J. G., Heil, H. M., and Heim, R.: Strength of soft Swiss lacustrine clay: Cone penetration and triaxial test data. Transportation Research Record, Vol. 1675, No. 1675, pp. 1–9 (1999).
7. P. K. Robertson: Soil classification using the cone penetration test, Canadian Geotechnical Journal, 1990, 27(1): 151-158 (1990).
8. Mourad Karray, Guy Lefebvre, Yannic Ethier, Annick Bigras: Influence of particle size on the correlation between shear wave velocity and cone tip resistance, Canadian Geotechnical Journal, 2011, 48(4): 599-615 (2011).
9. Fekerte Arega Yitagesu, Freek van der Meer, Harald van der Werff and Hadush Seged: Evaluation of Soil Expansion Index from Routinely Determined Geotechnical Parameters,

Soil Science Society of America Journal Abstract - Soil Physics, Vol.75 No. 5, p. 1640-1651 (2011).
10. A.Anagnostopoulos, George Koukis, N. Sabatakakis and George Tsiambaos : Empirical correlations of soil parameters based on Cone Penetration Tests (CPT) for Greek soils. Geotechnical and Geological Engineering, 21(4):377-387 (2003).
11. His yam Jusoh and Syed Baharom Syed Osman: The Correlation between Resistivity and Soil Properties as an Alternative to Soil Investigation. Indian Journal of Science and Technology, Vol 10(6), DOI: 10.17485/ijst/2017/v10i6/111205, (2017).
12. Jay Ameratunga, Nagaratnam Sivakugan and Braja M. Das: Correlations of Soil and Rock Properties in Geotechnical Engineering, A Text Book January 2016, DOI 10.1007/978-81-322-2629-1, ISSN 2364-5156, Springer Publisher (2015).
13. Yusep Muslih Purwana , Hamid Nikraz: The Correlation between the CBR and Shear Strength in Unsaturated Soil Conditions , International Journal of Transportation Engineering, Vol.1/ No.3/ Winter 2014, pp.211-222 (2014).
14. D.B.Eme, T.C Nwofor and S. Sule: Correlation between the California Bearing Ratio (CBR) and Unconfined Compressive Strength (UCS) of Stabilized Sand-Cement of the Niger Delta, SSRG International Journal of Civil Engineering (SSRG - IJCE), V3(2),7-13 March 2016. ISSN:2348 – 8352 (2016).
15. Okon Bassey Bassey, Imoh Christopher Attah, Edidiong Eseme Ambrose, Roland Kufre Etim: Correlation between CBR Values and Index Properties of Soils: A Case Study of Ibiono, Oron and Onna in Akwa Ibom State. Resources and Environment p-ISSN: 2163-2618, e-ISSN: 2163-2634, 7(4): 94-102 (2017).

Strength properties of double blend and triple blend self-compacting concrete subjected to different curing methods

Tadivaka Venkateswara Rao[1], Meduri V Seshagiri Rao[2], Pasumarthi Jagannadha Rao[3]

[1] Research Scholar, JNTUH CEH Hyderabad, India.
[2] Professor of Civil Engineering department, CVR College of Engineering, Professor (Retd.) JNTUH CEH Hyderabad, Telangana, India.
[3] Advisor, ACE Engineering College, Hyderabad, Telangana, India.

Abstract: The behaviour of Self Compacting Concrete is complex with the addition of both chemical and mineral admixtures. The use of supplementary cementing materials such as Metakaolin, FlyAsh, GGBS and Silica fume is required to improve the flow properties of fresh concrete, to decrease the cement contents, to increase the strength as well as durability of cement and to reduce cement consumption for environmental and economical concern. The hydration mechanism, strength, and setting characteristics of SCC are all influenced by the curing process. As a result, the findings of an investigation into the effects of curing methods on the compressive strength of SCC using Metakaolin as the primary supplementary cementing material are described in this paper. FlyAsh, GGBS and Silica fume are used individually for triple blend binders with OPC and Metakaolin. The Grade of concretes made were Standard concretes of M50. The concretes' compressive strengths were assessed at 7 and 28 days. The traditional water pond approach was used as a reference, and wax-based liquid membrane and polymer-based liquid membrane curing compounds were used for testing. To reduce the water cement ratio and increase the flow and filling capacity of SCC, a superplasticiser based on a polycarboxylic ether polymer with long lateral chains is used. The performance of the Polymer based membrane curing compound is better than Wax based membrane curing compound. In addition to this, the polymer based membrane allows plastering and painting over the concrete surface. The early gain in strength is observed in all the membrane cured cubes at 7 days, but the rate of gaining strength is reduced after 7 days when compared with the Water pond cured cubes.

Key words- Compressive strength, SCC, Metakaolin; Triple Blends, Membrane curing

1. Introduction

Concrete proportioning is just as critical as concrete curing. When day-to-day monitoring and inspection of curing is not feasible, water-curing concrete becomes more expensive and time-consuming. The foundations of towers and their copings for chimneys in scattered places, remote areas, deserts, wetlands, hilly terrains, and across rivers, concrete structures with vertical, sloped, and curved surfaces, tall chimneys, cooling towers in thermal power plants, and pillars of viaducts in cities are all suitable and cost-effective. Three different curing methods were used in this study: traditional water pond curing, wax-based liquid membrane forming compound, and polymer-based membrane curing compound. The use of self-compacting concrete (SCC) was favoured in this study due to its rapid development. Because of faster building, reduced site manpower, better surface finishes, easier placement in heavily reinforced buildings, increased reliability, greater design freedom, thinner concrete parts, lower noise levels, and absence of vibration, SCC has been dubbed the most innovative breakthrough in the concrete construction industry. For the experiments, 100mm concrete cubes made of M50 standard SCC with OPC and Metakaolin as supplementary cementing materials were prepared. For triple blend binders with OPC and Metakaolin, FlyAsh, GGBS, and Silica fume are used separately. Metakaolin is a pozzolanic mineral admixture that improves the efficiency of cement-based mortars, concretes, and related products in a variety of ways. Metakaolin is made from distilled kaolin clay, a white, amorphous alumino-silicate that reacts vigorously with calcium hydroxide, a by-product of normal cement hydration, to form cementitious compounds. In this plant, metakaolin was used as the primary supplementary cementing material.

2. Materials

The cement used was ordinary Portland Cement of Grade 53, which complies with IS12269-1987. Fine Aggregate is river sand that meets IS 383-1970 zone II requirements. Coarse Aggregate is 12mm HBG graded and meets IS 383-1970 requirements. Metakaolin was found to be white, highly pozzolanic, and highly reactive. The table 1 shows the physical and chemical properties. The liquid membrane forming curing compounds are wax-based compounds that, when applied to concrete, form a smooth film that helps to prevent water from evaporating through the concrete's capillaries. ASTM C 309 is met by this compound. This wax-based white pigmented compound lasts up to 28 days. The dose varies from 4 to 6m2/litre. Another liquid membrane forming curing compound is a high efficiency polymer based film forming curing agent conforming to BS 7542-1992. The recommended coverage rate is between 3.5 and 5.0 m^2 / litre. Superplasticisers based on polycarboxylic ether polymer with long lateral chains were used, which meet ASTM C 494-92 Type F and IS 9103 –2007 requirements.

Table 1. Physical and Chemical properties of Metakaolin

Physical properties						Chemical Properties		
1.	Brightness	85 ± 1	5.	Residue on 400μ	Nil	1.	SiO_2	52.8%
2.	Oil absorption (%)	60 ± 5	6.	Specific gravity	2 ± 0.1	2.	Al2O3	36.3%
3.	Moisture (%)	0.5	7.	Ph of 10%	6 ± 1	3.	Fe2O3	4.21%
4.	Particle size: Average	1 ± 1				4.	MgO	0.81%
	<10 Microns (%)	95 ± 2				5.	CaO	<0.10%
	< 2 Microns (%)	80 ± 1				6.	K_2O	1.41%
						7.	LOI	3.53%

3. Preparation Of SCC and Tests

"To proportion, blend, and test the fresh properties of SCC concrete, the European guidelines for self-compacting concrete - May 2005" were used. To compare the strength properties of differently cured cubes of 100mm size prepared with 53 grade OPC and Metakaolin, four different SCC mixes were chosen. There were a total of 96 cubes cast. Laboratory trials were used to check the properties of the initial mix composition in relation to the listed characteristics and classes specified in European guide lines and EFNARC. Adjustments to the mix composition were made to determine the demand for water and superplasticizer, sand, and the dose of Metakaolin to achieve the desired robustness, as well as to boost the paste's flow and stability. When the performance of the mixes was not sufficient, they were revamped.

After the tests on fresh concrete were finished, cubes were cast, and the top surfaces of 50% of the cubes were painted with liquid membrane forming curing compounds when the concrete was touch dry. Some of the cubes were submerged in water the next day, while the others were painted with liquid membrane forming curing compounds @ 4Sq.mt/Litre the next day. The compressive strengths of both water cured and membrane cured cubes of M50 are tabulated in Tables 3, 4, 5 & 6. The tests on hardened concrete were conducted at 7 days and 28 days, and the test results on fresh and hardened concrete are tabulated in Table 2. The comparison among strengths is shown in Bar Chart 1.

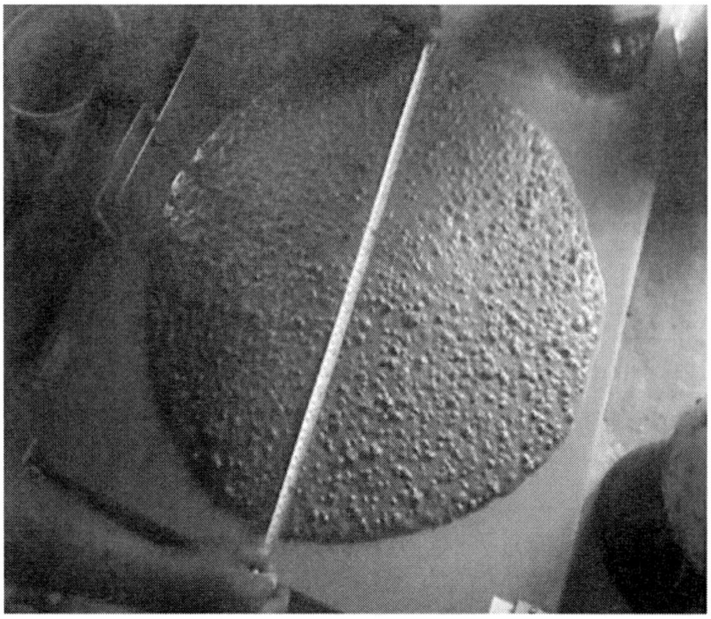

Table 2. SCC Mix proportions and compressive strengths

Slump-Flow classes Class	Slump-flow in mm	*Viscosity classes Class*	T_{500}, s	V-funnel time in s
SF1	550 to 650	VS1/ VF1	≤ 2	≤ 8
SF2	660 to 750	VS2/ VF2	> 2	9 to 25
SF3	760 to 850			
Passing ability classes (L-box) Class	Passing ability			

PA1	≥ 0,80 with 2 rebars			
PA2	≥ 0,80 with 3 rebars			

SCC with OPC(C), Metakaolin(MK), FlyAsh (FA), GGBS(GG) & Silica Fume(SF) Mix Proportions & Compressive Strengths						
			Mix-1 (C+MK)	Mix-1 (C+MK+Fa)	Mix-3 (C+MK+GG)	Mix-4 (C+MK+SF)
S.No.	Description	Unit	Quantity	Quantity	Quantity	Quantity
1	Fine aggregate	kg	949	949	949	949
2	Coarse aggregate	kg	779	779	779	779
3	Total Binder	kg	574	592	592	592
4	Cement	kg	267	239	239	239
5	Metakaolin	kg	307	148	148	148
6	FlyAsh	kg	0	205	0	0
7	GGBS	kg	0	0	205	0
8	Silica Fume	kg	0	0	0	205
9	Water	kg	228	213	206	276
10	water/binder	ratio	0.4	0.36	0.35	0.47
11	Super plasticizer	kg	4.4	6.9	4.7	10.72
	% SP1/Total binder=	%	0.77	1.17	0.8	1.81
	Ref. European Guide Lines(2005)		SF2, VS1/VF1, PA2 - Good	SF2, VS1/VF1, PA2 - Good	SF2, VS1/VF1, PA2 - Good	SF2, VS1/VF1, PA2 - Good
	CLASS(Ref: Appendix-A)					
12	Compressive Strength-7 days					
	Water Pond cured cubes	MPa	42.62(70%)	38.66 (70%)	64.34 (82%)	30.72 (62%)
	Wax Membrane cured cubes	MPa	44.26(81%)	40.11 (85%)	58.53 (90%)	29.01 (72%)
	Polymer Membrane cured cubes	MPa	46.79(83%)	48.76 (92%)	59.10 (83%)	29.73 (70%)
13	Compressive Strength-28 days					
	Water Pond cured cubes	MPa	60.76	55.76	78.17	49.96
	Wax Membrane cured cubes	MPa	54.81	47.07	64.88	40.16
	Polymer Membrane cured cubes	MPa	56.43	53.01	71.32	42.66

	Mix-1 (Cement+Metakaolin) M50				
Age of Concrete	Water Pond Cured Cubes (MPa)	Wax Membrane Cured Cubes (MPa)	Change in Strength	Polymer Membrane Cured Cubes (MPa)	Change in Strength
7 Days	42.62	44.26	+ 3.85 %	46.79	+ 7.44 %
28 Days	60.76	54.81	- 9.79 %	56.43	- 7.13 %

Table 3. Comparison of Compressive Strengths of SCC Mixes

	Mix-2 (Cement+Metakaolin+FlyAsh) M50				
Age of Concrete	Water Pond Cured Cubes (MPa)	Wax Membrane Cured Cubes (MPa)	Change in Strength	Polymer Membrane Cured Cubes (MPa)	Change in Strength
7 Days	38.66	40.11	+ 3.75 %	48.76	+ 26.13 %
28 Days	55.76	47.07	- 15.58 %	53.01	- 4.93 %

Table 4. Comparison of Compressive Strengths of SCC Mixes

	Mix-3 (Cement+Metakaolin+GGBS) M50				
Age of Concrete	Water Pond Cured Cubes (MPa)	Wax Membrane Cured Cubes (MPa)	Change in Strength	Polymer Membrane Cured Cubes (MPa)	Change in Strength
7 Days	64.34	58.53	- 9.03 %	59.10	- 8.14 %
28 Days	78.17	64.88	- 17.00 %	71.32	- 8.76 %

Table 5. Comparison of Compressive Strengths of SCC Mixes

	Mix-4 (Cement+Metakaolin+Silica Fume) M50				
Age of Concrete	Water Pond Cured Cubes (MPa)	Wax Membrane Cured Cubes (MPa)	Change in Strength	Polymer Membrane Cured Cubes (MPa)	Change in Strength
7 Days	30.72	29.01	- 5.56 %	29.73	- 3.22 %
28 Days	49.96	40.16	- 19.62 %	42.66	- 14.61 %

Table 6. Comparison of Compressive Strengths of SCC Mixes

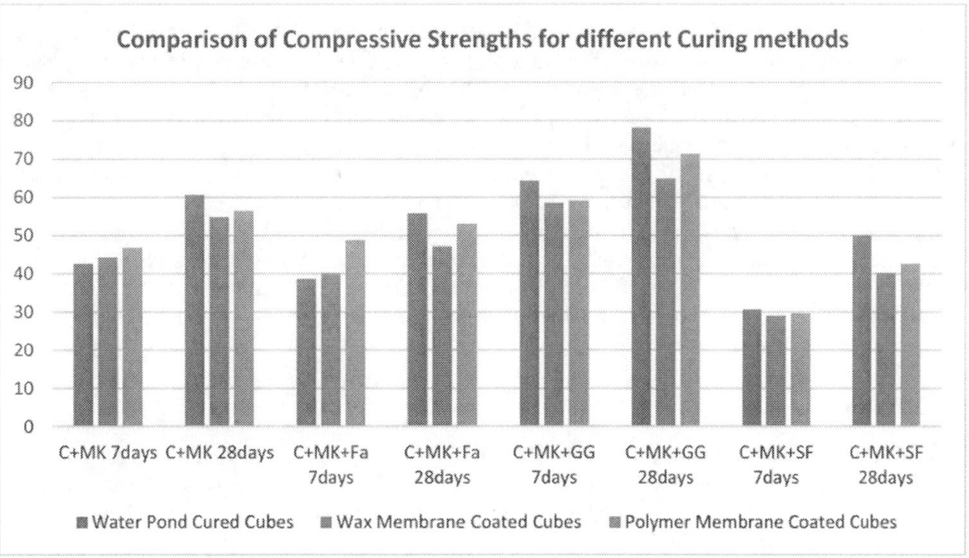

4. Analysis On Results

1. At 7 days, the compressive strength difference between water and wax based liquid membrane cured cubes of Mix-1 is +3.85 percent, but at 28 days, the difference is -9.79 percent. At 7 days, the difference in compressive strengths of water and polymer-based liquid membrane cured Mix-1 cubes is +7.44 percent, but at 28 days, the difference is -7.13 percent.
2. At 7 days, the difference in compressive strengths of water and wax based liquid membrane cured cubes of Mix-2 is +3.75%, but at 28 days, the difference is -15.58 percent. At 7 days, the difference in compressive strengths of water and polymer-based liquid membrane cured Mix-2 cubes is +26.13 percent, but at 28 days, the difference is -4.93 percent.
3. At 7 days, the difference in compressive strengths of water and wax based liquid membrane cured Mix-3 cubes is -9.03 percent, but at 28 days, the difference is -17.00 percent. At 7 days, the difference in compressive strengths of water and polymer-based liquid membrane cured Mix-3 cubes is -8.14 percent, but at 28 days, the difference is -8.76 percent.
4. At 7 days, the compressive strength difference between water and wax based liquid membrane cured cubes of Mix-4 is -5.56 percent, but at 28 days, the difference is -19.62 percent. At 7 days, the difference in compressive strengths of water and polymer-based liquid membrane cured Mix-4 cubes is -3.22 percent, but at 28 days, the difference is -14.61 percent.
5. Membrane cured cubes of Mix-1 and Mix-2 have higher 7-day compressive strengths than water pond cured cubes, while membrane cured cubes of Mix-3 and Mix-4 have lower 7-day compressive strengths than water pond cured cubes.
6. . An abnormal early compressive strength of +26.13% over the water cured cubes is observed for the polymer based liquid membrane cured cubes of Mix-2. The loss in compressive strengths are about -19.62% and -14.64% for wax and polymer membrane cured cubes, respectively for Mix-4 when compared to water pond cured cubes at 28 days.
7. The 7 days' strength of Water cured cubes of Mix-1 is 70% of the 28 days' strength, but it is 81% for Wax membrane cured cubes and 83% for Polymer membrane cured cubes. The 7 days' strength of Water cured cubes of Mix-2 is 69% of the 28 days' strength, but it is 85% for Wax membrane cured cubes and 92% for Polymer membrane cured cubes. The 7 days' strength of Water cured cubes of Mix-3 is 82% of the 28 days' strength, but it is 90% for Wax membrane cured cubes and 83% for Polymer membrane cured cubes. The 7 days' strength of water cured cubes of Mix-4 is 62% of the 28 days' strength, but it is 72% for Wax membrane cured cubes and 70% for Polymer membrane cured cubes.

8. Conclusions

1. The major finding out of the above results is "the performance of the Polymer based membrane curing compound is better than Wax based membrane curing compound". In addition to this, the polymer based membrane allows plastering and painting over the concrete surface.
2. The early gain in strength is observed in all the Membrane cured cubes at 7 days, but the rate of gaining strength is reduced at 28 days when compared with the Water pond cured cubes.
3. The early gain in strength of Membrane cured cubes of triple blends Mix-2(OPC+MK+FA) & Mix-3 (OPC+MK+GGBS) is more than double blend Mix-1 (OPC+MK).
4. The reasons for the early gain in strength for the membrane cured cubes are Duration of Hydration process, Heat of Hydration, Fineness of SCMs and Capillary water. Water pond cured cubes receive water through capillaries, hence the hydration process continues slowly. The Membrane cured cubes develop more heat of hydration which promotes early gain in strength and the capillary water, which helps in hydration process, evaporates and no further supply of water externally.

9. References

[1] Application of Statistical Analysis for Mixture Design of High-Strength Self-Consolidating Concrete Containing Metakaolin by Ahmed A.Abouhussien, M.A.Sc. and Assem A.A.Hassan, Ph.D., M.A.Sc., M.Sc. Faculty of Engineering and Applied Science, Memorial University of Newfoundland, St. John's, Newfoundland, Canada, ASCE Journal

[2] L.Garca1; M.Valcuende, Ph.D. S.Balasch, Ph.D. and J.Fernández-LLebrez-, Department of Architectural Constructions, Universidad Politecnica de Valencia, Spain, Study of Robustness of Self-Compacting Concretes Made with Low Fines Content, ASCE Journal, April 2013.

[3] Selcuk Turkel and Ali Kandemir, Department of Civil Engineering, Dokuz Eylul University, Izmir, Turkey, published an article in the ASCE Journal in October 2010 titled "New and Hardened Properties of SCC Made with Different Aggregate and Mineral Admixtures."

[4] Properties of Self-Consolidating Concrete Rendered with High Volumes of Supplementary Cementitious Materials by Hassan El-Chabib and Adnan Syed, Department of Civil Engineering and Construction, Bradley University, Peoria, ILLINOIS, published in the ASCE Journal in November 2013.

[5] Ahmed Ibrahim, Ph.D., A.M.ASCE; Hassan El-Chabib, Ph.D., and Ahmed Eisa, Ph.D., Department of Civil Engineering and Construction, Bradley Univ., Peoria, ILLINOIS, November 2013 - Ultra strength Flowable Concrete Made with High Volumes of Supplementary Cementitious Materials by Ahmed Ibrahim, Ph.D., A.M.ASCE; Hassan El-Chabib, Ph.D. Hydration mechanism and power of OPC and blended OPC with Fly ash in the presence of Metak in IJRET May-2015 Anamika Singh, Sukirti Gupta, Jahnavi Singh and N.P.Singh, Department of Chemistry, UPA college, Varanasi.

[6] M.V.Krishna Rao (CBIT), P.Ratish Kumar (NIT,Warangal), and Azhar M. Khan, FACTA UNIVERSITATIS-2010, A report on the effect of curing on the strength of a standard grade concrete mix (NIT,Warangal).

[7] M V Jagannadha Kumar, K Jagannadha Rao, B Dean Kumar and V Srinivasa Reddy, Development of Self-Curing Concrete Using Polyethylene Glycol as Internal Curing Agent, International Journal of Civil Engineering and Technology, 9(7), 2018, pp. 1133–1141.http://www.iaeme.com/ijciet/issues.asp?JType=IJCIET&VType=9&IType=7

[8] K Satya Sai Trimurty Naidu, M V Seshagiri Rao and V Srinivasa Reddy, Analytical Model for Predicting Stress-Strain Behaviour of Bacterial Concrete, International Journal of Civil Engineering and Technology, 9(11), 2018, pp. 2383–2393

[9] C Chandana Priya, M V Seshagiri Rao and V Srinivasa Reddy, Studies on Durability Properties of High Strength Self Review, International Journal of Civil Engineering and Technology (IJCIET) 2018, pp. 2218–2225.

[10] Srinivasa Reddy, V. Seshagiri Rao, M.V. Shrihari, S, Strength conversion factors for concrete based on specimen geometry, aggregate size and direction of loading, International Journal of Recent Technology and Engineering, 2019, 8(2), pp.2125-2130

[11] Naidu, K.S.S.T. Rao, M.V.S. Reddy, V.S. ,Microstructural characterization of calcite mineral precipitation in bacteria incorporated concrete, International Journal of Innovative Technology and Exploring Engineering, 2019, 8(9),Special Issue 2, pp. 641-642

[12] Supriya, Y. Srinivasa Reddy, V. Seshagiri Rao, M.V. Shrihari, S., Strength appraisal of light weight green concrete made with cold bonded fly ash coarse aggregate, International Journal of Recent Technology and Engineering, 2019, 8(3), pp.5381-5385
[13] Jyothi Kumari Ganta, M.V. Seshagiri Rao, Seyed Sina Mousavi, V. Srinivasa Reddy, Chandrasekhar Bhojaraju,
[14] Ganta, J.K. Seshagiri Rao, M.V. Mousavi, S.S. Srinivasa Reddy, V. Bhojaraju, C., Hybrid steel/glass fiber-reinforced self-consolidating concrete considering packing factor: Mechanical and durability characteristics, Structures, 2020, 28, pp. 956-972, ISSN 2352-0124, https://doi.org/10.1016/j.istruc.2020.09.042.

IOP Conf. Series: Materials Science and Engineering 1126 (2021) 012086 doi:10.1088/1757-899X/1126/1/012086

Experimental and Numerical Investigation of Residual Stresses in low pulsed Laser Shock Peening on AA7075 T651

M. ABEENS*[1]. E. RAJASEKARAN[1]. M. PREMNATH[1]. D. SARUKASAN[1]. R. MURUGANANDHAN[1]. K. THIRUMAVALAVAN[1].

[1] Department of Mechanical Engineering, Anna University, Chennai, Tamil Nadu, India.

Abstract

Laser shock peening (LSP) is the emerging technology among several severe surface treatment processes for improving mechanical characteristics of metallic materials by producing compressive residual stresses. Though several LSP experiments have been done on compressive residual stresses, scare amount of studies concentrated on numerical investigation of residual stresses in LSP. The objective of this article is investigation of the experimental and numerical study of compressive residual stresses on AA 7075 t651. In the experiment, residual stresses analysis is performed. The $sin^2\Psi$ method was utilized for measuring residual stresses. The three-dimensional (3-D) finite element method (FEM) analysis was applied to simulate the LSP on AA 7075 t651. The simulated and measured results showed good agreements. The numerical study was utilized to determine the compressive residual stresses.

1. Introduction

Laser shock peening (LSP) is a severe surface treatment process used to enhance the surface properties of the materials such as fatigue, wear and corrosion as result of induction in compressive residual stresses on the surface [1,2]. The objective of LSP is to acquaint compressive residual stresses in dangerous zones of fatigue. Though, the disadvantage of LSP is numerous process parameters which pay the resulting residual stress field. Therefore, the experimentation based on the optimization is a time-consuming process.

The numerical simulation on LSP such as FEA can speed up the optimization process and support to identify the mechanisms which are difficult to measure. Braisted and Brockman are reported on FEA analysis on LSP and found prediction of the residual stresses [3]. Dig and Ye studied the laser focus size and the influence of overlap and reported reasonable agreement to the experimental data [4]. Sathiyajith et. al investigated LSP on AA 6061, and reported high pulse laser shock peening process provides reliability issue due to its thermal effect [2]. Prabhakaran et. al reported low pulsed LSP process improved the mechanical properties of steel-based alloy [5].

From the literature survey, the low pulse energy on LSP is not done on non-ferrous materials and it provides better mechanical and surface properties so the study on low pulse LSP with non-ferrous material is important at this time and the simulation also necessary to predict the compressive residual stress with low pulse energy.

This study concentrates the Experimental and Numerical Investigation of Residual Stresses in low pulsed Laser Shock Peening on AA7075 T651.

2. Experimental techniques

2.1 Material

AA 7075 T651 material is thermo-mechanical treated so it possesses good mechanical property than other therefore it is selected for the experimental work. The sample was purchased from PMC Corporation,

Bangalore. The sample dimensions are 10mm×10mm×8 mm. before LSP, It was ensured that the surface of the sample is smooth and original. Table (a). and Table (b) shows the chemical composition and mechanical property of the sample.

2.2 Laser shock peening

The Nd: YAG laser with 1064 mm wave length was utilized in the LSP process. The beam quality factor (M^2) value is 2 and 08 mrad is beam divergence. Top hat profile was selected for shape of beam and a TEM00. Therefore the laser beam brightness is 345.04 mW $cm^2Sr^{-1}\mu m$; this methodology used form literature [6,-8]. The black paint coating with 1 mm thick is used as an ablative layer and the thin 1- 2 mm thick water layer was used confinement medium, which is also used for eliminate the ablated material constantly. The laser beam is passed on the material surface. The bi convex lens with 300 mm focal length and dichromatic mirror with kept 45^0 was utilized to deliver the laser beam on the target. The blower is placed nearer to the lens due to protect the lens from the water drops during LSP process. The sample holder was placed on XY translation stage, which is controlled by computer [9-12].

2.3 Test of Residual Stress

The magnitude compressive residual stress is calculated by X-ray Diffraction sin2Ψ method. X-pert Pro-system is used for computing the X-ray radiations of 4 mm^2 at the diffractive plane of (422). It is worked 40 mA and 45 kV voltage current utilizing Cu-Kα radiation (λ= 1.54 A°).

3. Modeling and Finite Element Analysis (FEA)

3.1 Simulation Parameters

The vital parameters involved in the FEA analysis on LSP process includes i) the laser spot shape and size, ii) the pressure pulse shape and duration, iii) geometric modelling and meshing, and iv) the material model. Except the mesh convergence study, the other parameters are selected on the basis of the experiments conducted in this research.

3..1.1. Laser Spot Shape & Size

If the spot diameter is in the range of 0.5 -1 mm it will create a high shock pressure, whereas the larger spot diameter will resulted in a lesser shock pressure. Hence, the laser spot size of 0.8 mm diameter (from Experiment) is used in this study and the shape of the laser spot used is the circle.

3.1.2. Pressure Pulse Magnitudes and Duration

A high amplitude and short- duration pulse are the key parameters with respected to the LSP processing. The magnitude of the shock pressure for the FEA is calculated as per the equation 2.1 and presented in Table 1.

Figure 1 depicts the pressure-time history. It is obtained from the experimental pulse duration and pressure formation. The simulation procedure does not change the explicit model of the physical process. Typically, the pressure pulse of the laser spot is relatively uniform over the complete surface, regardless of whether the laser spot is square or circular. It is owing to the narrow duration of the pressure pulse induced by LSP. The pressure--time history is demonstrated as a triangular ramp Gaussian temporal shape FWHM in the FEA simulation. The pressure increases linearly to a maximum value over the time of the pressure pulse FWHM and it then decays to zero.

Table 1. Magnitude of Shock pressure

Pulse Engergy (mJ)	**Magnitude of Shock Pressure (GPa)**
200	2.033
300	2.45
400	2.835

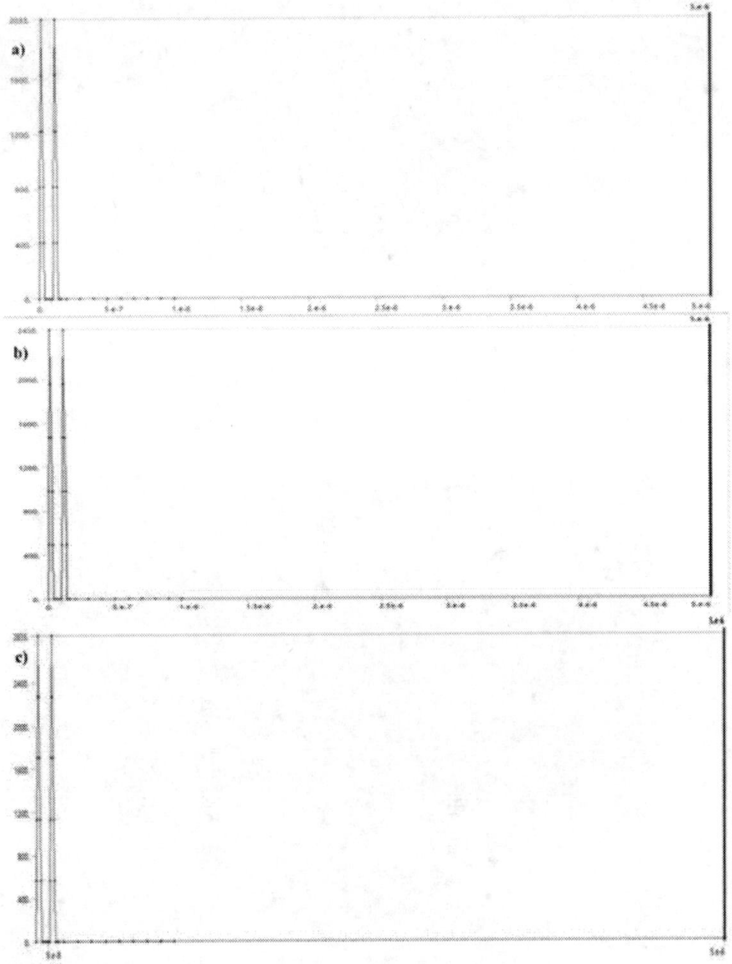

Figure 1. Pressure (y axis) vs time history (x axis) of pulse energies a) 200 mJ b) 300 mJ c) 400 mJ

3.1.3. Geometric Modeling and Meshing

The modelling is done using ANSYS workbench19.2, with a dimension 10 x 10 x 6 mm and laser spot diameter of 0.8 mm as shown in figure 2.

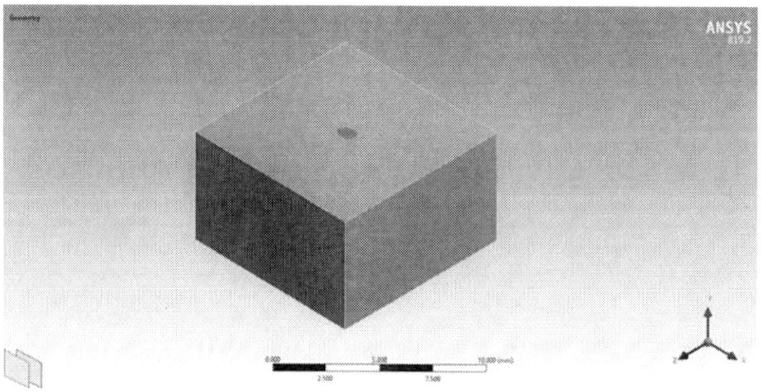

Figure 2. Geometric model of AA 7075 T651

Figure 3. Meshed model of AA 7075 T651

The model is then meshed using ANSYS mechanical with a tetrahedral shape, containing about 1 lakh elements. A finer mesh of face size 0.2 mm used in the region where the laser beam pressure is applied. Whereas the remaining regions are meshed using a biased condition as shown in Figure 3.

3.1.4 Material Modeling

LSP makes strain-rates exceeding 106 s^{-1} within the material target. The material model contributes a critical role in accurately FEA simulating the process with such a high strain-rate. In this research, two material models are explored for use in LSP simulation. The first model uses perfectly elastic-plastic material properties. In this model, Young's modulus and dynamic yield strength are used to define material properties. The dynamic yield strength depends upon the Hugoniot Elastic Limit of the material. While the second model uses a bi-linear isotropic model analysis.

3.1.5. Loading and boundary conditions

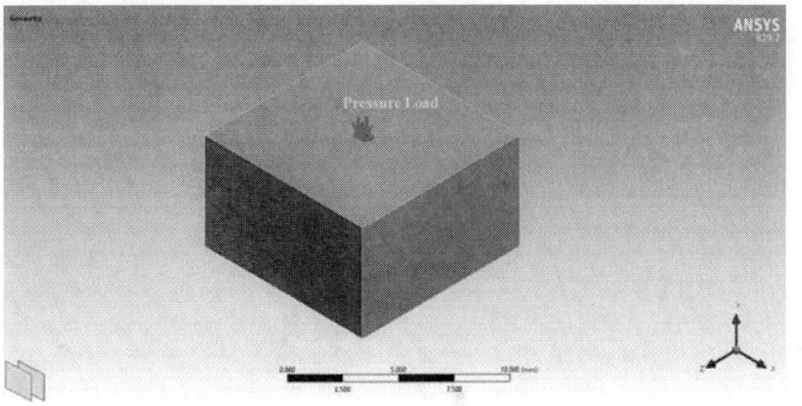

Figure 4. Load applied region of AA 7075 T651 model

A triangulated pulsed load is applied to the model on the region where it interacts with laser beam for this analysis with a varying peak pressure and 75 % overlapping. A triangulated pressure load of 2.033 GPa, 2.45 GPa and 2.835 GPa for a pulse energy of 200 mJ, 300mJ and 400 mJ respectively. The boundary conditions are fixed at bottom side of the model.

4. Results and Discussion

4.1 Residual Stress measurement

The maximum compressive residual stresses of unpeened and peened specimens with pulse energies 200 mJ, 300 mJ and 400 mJ are conferred in the Table (b)

Pulse energies (mJ)	Compressive residual stress(MPa)
Unpeened (0)	50
200	168.78
300	262.35
400	317.38

Table (b). Maximum residual values for diverse pulse energies

The T651 condition induces a compressive residual stress of around -50 MPa in the unpeened sample [4]. The compressive residual stress in the LSP phase, on the other hand, is much higher than in the unpeened sample. The degree of the compressive residual stress is directly proportional to the pulse energy, and at the same time, increase in depth for evaluating compressive residual stress with decreasing magnitude for the same, according to the trend result.

4.2 FEA Simulation of Residual Stress induced by LSP

FEA simulation of the compressive residual stress distribution in AA 7075 T651 peened with pulse energies of 200mJ, 300mJ and 400mJ are shown in Figure 5, 6 & 8 respectively. It can be observed that the maximum compressive residual stress value increases with the laser pulse energy, due to the increase in the energy imparted and profound compressive residual stress induced during the LSP.

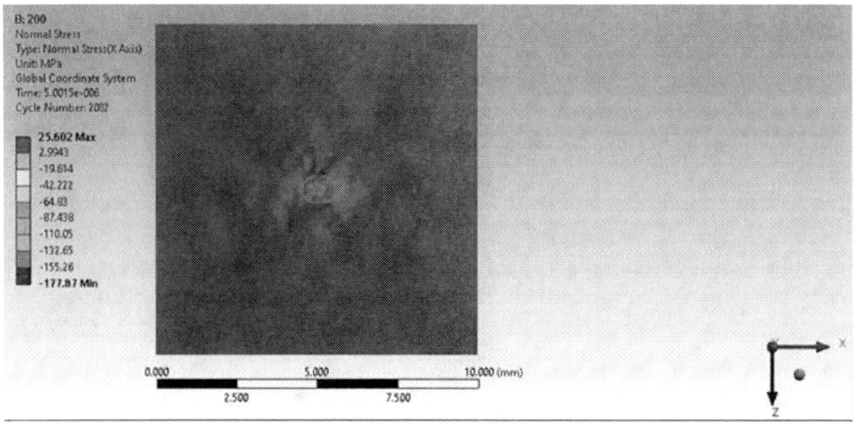

Figure 5. FEA effect of compressive residual stress at pulse energy 200 mJ

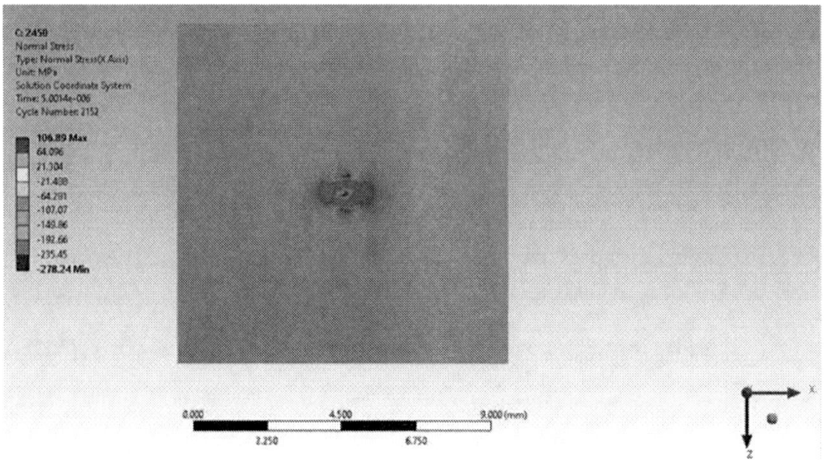

Figure 6. FEA effect of compressive residual stress at pulse energy 300 mJ

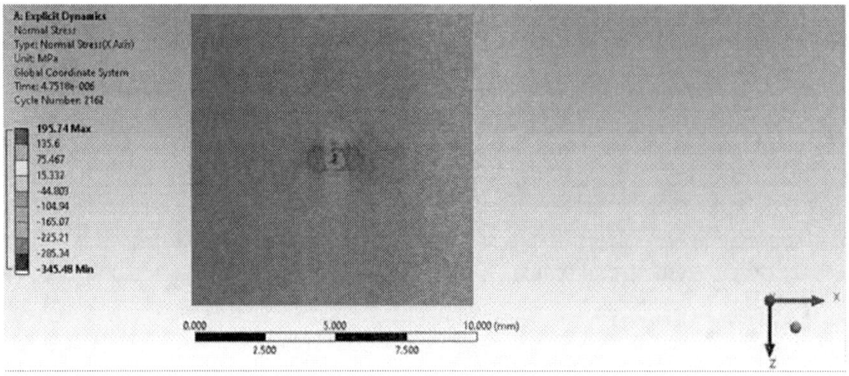

Figure 7. FEA effect of compressive residual stress at pulse energy 400 mJ

4.2.1 Comparison of Experimental and Simulation Results

The comparison of the magnitude of compressive residual stress in the FEA and the experimental values of the peened specimen with pulse energies of 200 mJ, 300 mJ and 400 mJ are shown the Figure 8. The percentage error of the stress is 8.8 %, 6% and 5.3 % respectively; therefore the magnitude of compressive residual stress is attained using FEA model is in good compatibility with the experimental value for LSP process.

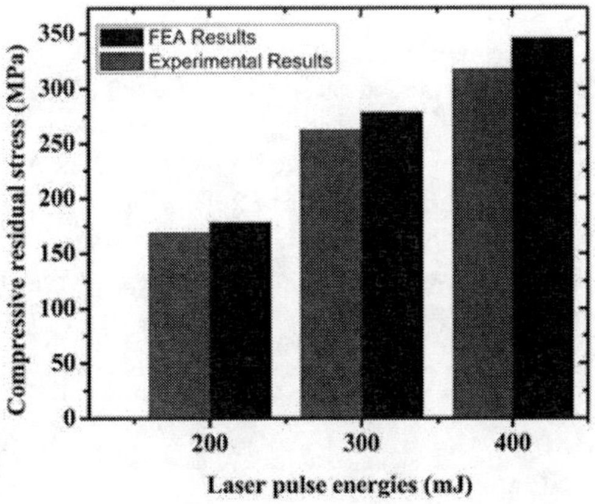

Figure 8. Comparison of compressive residual stress of FEA and Experiment results.

Conclusion

- The maximum residual stress induced in the 400 mJ used sample during LSP is 317 MPa which is 5.3 times that of compressive residual stress present in the as-received specimen.
- The prediction in maximum compressive residual stress in the 400 mJ used sample during LSP is -345 MPa and the variation from the experimental and predicted values obtained between 5 – 10 %.
- In FEA model analysis the magnitude of compressive residual stress is attained in good agreement with the experimental value for the LSP process

References

[1] Shukla, Pratik, Jonathan Lawrence, and Yu Zhang. "Understanding laser beam brightness: A review and new prospective in material processing." *Optics & Laser Technology* 75 (2015): 40-51.
[2] Sathyajith, S., S. Kalainathan, and S. Swaroop. "Laser peening without coating on aluminum alloy Al-6061-T6 using low energy Nd: YAG laser." Optics & Laser Technology 45 (2013): 389-394.
[3] Braisted, William, and Robert Brockman. "Finite element simulation of laser shock peening." International Journal of Fatigue 21.7 (1999): 719-724.
[4] Benedetti, M., et al. "Reverse bending fatigue of shot peened 7075-T651 aluminium alloy: The role of residual stress relaxation." *International Journal of Fatigue* 31.8-9 (2009): 1225-1236.

[5] Abeens, M., et al. "Surface modification of AA7075 T651 by laser shock peening to improve the wear characteristics." *Materials Research Express* 6.6 (2019): 066519.
[6] Thirumavalavan, K., et al. "Study on the influence of process parameters of severe surface mechanical treatment process on the surface properties of AA7075 T651 using TOPSIS and Taguchi analysis." *Materials Research Express* 6.11 (2019): 1165i1.
[7] Abeens, M., R. Muruganandhan, and K. Thirumavalavan. "Comparative analysis of different surface modification processes on AA 7075 T651." *IOP Conference Series: Materials Science and Engineering*. Vol. 574. No. 1. IOP Publishing, 2019.
[8] Sastry, C. Chandrasekhara, et al. "Microstructural analysis, radiography, tool wear characterization, induced residual stress and corrosion behavior of conventional and cryogenic trepanning of DSS 2507." *Journal of Mechanical Science and Technology* 34 (2020): 2535-2547.
[9] Abeens, M., R. Muruganatham, and S. Arulvel. "Friction-wear behavior of shot peened aluminium 7075-T651 alloy." (2019).
[10] Abeens, M., et al. "Experimental analysis of convective heat transfer on tubes using twisted tape inserts, louvered strip inserts and surface treated tube." *International Journal of Ambient Energy* 41.5 (2020): 540-546.
[11] Shendur, G. Shyam, et al. "Investigation of nano ceramics added bismuth antimony telluride for energy harvesting applications." *Materials Today: Proceedings* 22 (2020): 879-883.
[12] Abeens, M., R. Muruganandhan, and K. Thirumavalavan. "Effect of Low energy laser shock peening on plastic deformation, wettability and corrosion resistance of aluminum alloy 7075 T651." *Optik* 219 (2020): 165045.

AUTHOR INDEX